PHYLOGENY OF THE LIVING WORLD–*BACTERIA*

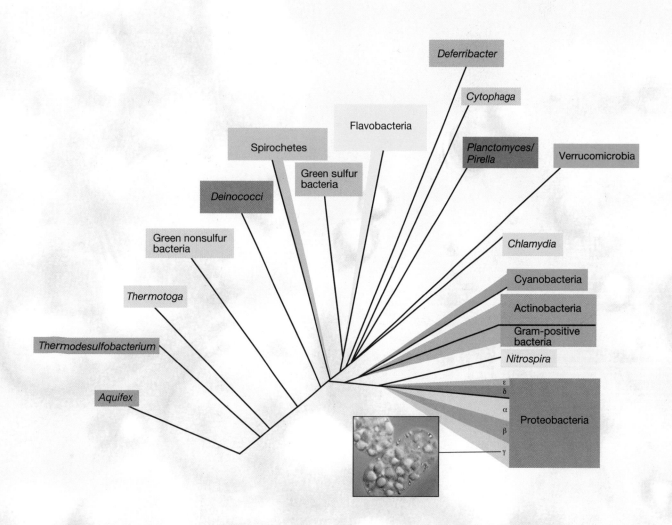

PHYLOGENETIC TREE OF *BACTERIA*. This tree is derived from 16S ribosomal RNA sequences. At least 17 major groups of *Bacteria* can be defined as indicated. See Sections 11.5–11.9 for further information on ribosomal RNA-based phylogenies. *Data for the tree obtained from the Ribosomal Database project* **http://rdp.cme.msu.edu**

CHINHYUN LEE (ERICA).

BROCK

BIOLOGY OF MICROORGANISMS

MICHAEL T. MADIGAN dedicates this book to the two old friends pictured with him below: Willie (left) and Plum (right). For the past 11 years these wonderful animals have given me the comfort and companionship that only a dog lover could understand. Willie is, and always has been, the paragon of a sweet dog. Plum (deceased April 16, 2004), by contrast, was your basic junkyard dog. But in reality, she had a heart of gold and loved people, and our paths never crossed that she wasn't thrilled to see me. Rest in peace amigo.

JOHN M. MARTINKO dedicates this book to his students, past and present. The best students continually present problems from new perspectives, inspiring me to not only teach, but to expand my knowledge and enhance my understanding. To all who I have had the pleasure to teach, thank you for teaching me!

MICHAEL T. MADIGAN received a bachelor's degree in biology and education from Wisconsin State University at Stevens Point in 1971 and M.S. and Ph.D. degrees in 1974 and 1976, respectively, from the University of Wisconsin, Madison, Department of Bacteriology. His graduate work involved the study of hot spring phototrophic bacteria under the direction of Thomas D. Brock. Following three years of postdoctoral training in the Department of Microbiology, Indiana University, where he worked on phototrophic bacteria with Howard Gest, he moved to Southern Illinois University Carbondale, where he is a Professor of Microbiology. He has been a coauthor of *Biology of Microorganisms* since the fourth edition (1984) and teaches courses in introductory microbiology, bacterial diversity, and diagnostic and applied microbiology. In 1988 he was selected as the outstanding teacher in the College of Science, and in 1993 its outstanding researcher. In 2001 he received the university's Outstanding Scholar Award. In 2003 he received the Carski Award for Distinguished Teaching of Undergraduates from the American Society for Microbiology. His research has dealt almost exclusively with anoxygenic phototrophic bacteria, especially those species that inhabit extreme environments. He has published over 100 research papers, has coedited a major treatise on phototrophic bacteria, and has served as editor and chief editor of the journal *Archives of Microbiology*. His nonscientific interests include reading, hiking, tree planting, and caring for his dogs and horses. He lives beside a quiet lake about five miles from the SIU campus with his wife, Nancy, two dogs, Willie and Pupagano, and Springer and Feivel (horses).

JOHN M. MARTINKO received his B.S. in Biology from The Cleveland State University. As an undergraduate student he participated in a cooperative education program, gaining experience in several microbiology and immunology laboratories. He then worked for two years at Case Western Reserve University as a laboratory manager, conducting research on the structure, serology and epidemiology of *Streptococcus pyogenes*. He did his graduate work at the State University of New York at Buffalo, investigating antibody specificity and antibody idiotypes for his M.A. and Ph.D. in Microbiology. As a postdoctoral fellow, he worked at Albert Einstein College of Medicine in New York on the structure of major histocompatibility complex proteins. Since 1981, he has been in the Department of Microbiology at Southern Illinois University Carbondale where he is an Associate Professor and Chair. His research interests include the effects of growth hormone in the immune response, the development of immunodiagnostic tests for soybean brown stem rot disease, and the investigation of structural mutations that alter function in peptide-major histocompatibility protein complexes. His teaching interests include undergraduate and graduate courses in immunology. He also teaches immunology, host defense, and infectious disease topics in a general microbiology course. In 2004, he was selected the outstanding teacher in the College of Science. He is also an avid golfer and cyclist. He lives in Carbondale with his wife, Judy, a high school science teacher.

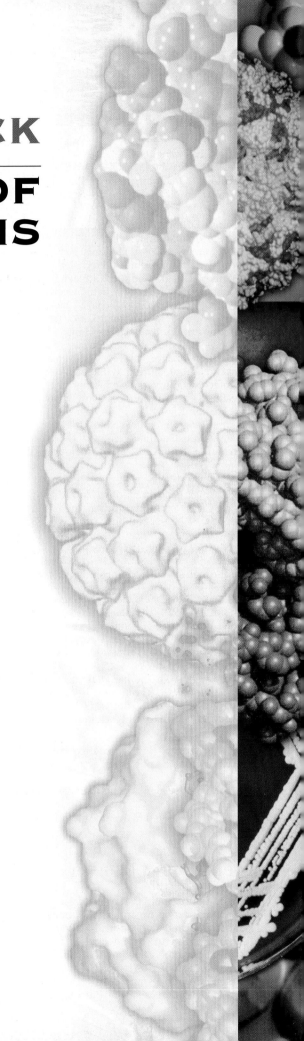

Eleventh Edition

BROCK
BIOLOGY OF
MICROORGANISMS

Michael T. Madigan

John M. Martinko

Southern Illinois University Carbondale

PEARSON

Prentice
Hall

Upper Saddle River, NJ 07458

Library of Congress Cataloging-in-Publication Data

MADIGAN, MICHAEL T.

Brock biology of microorganisms / Michael T. Madigan, John M. Martinko.—11th ed.
p. ; cm.
Includes index.
ISBN 0-13-144329-1
1. Microbiology.
[DNLM: 1. Microbiology. QW 4 M182b 2006] I. Title: Biology of microorganisms. II. Martinko, John M. III. Brock, Thomas D.
Biology of microorganisms. IV. Title.
QR41.2.M336 2006   579—dc22   2004026995

Executive Editor: Gary Carlson
Editor-in-Chief: John Challice
Project Manager: Crissy Dudonis
Production Editor: Debra A. Wechsler
Executive Managing Editor: Kathleen Schiaparelli
Assistant Managing Editor: Beth Sweeten
Managing Editor, Media: Nicole M. Jackson
Editor-in-Chief, Development: Carol Trueheart
Development Editor: Jonathan Haber
Senior Media Editor: Patrick Shriner
Marketing Manager: Andrew Gilfillan
Assistant Maunfacturing Manager: Michael Bell
Manufacturing Buyer: Alan Fischer
Director of Creative Services: Paul Belfanti
Art Director: Kenny Beck
Interior Design: Wanda España
Cover Design: Geoffrey Cassar
Managing Editor, Audio and Visual Assets: Patricia Burns

AV Production Manager: Ronda Whitson
AV Production Editor: Denise Keller/Sean Hogan
Art Studio: J/B Woolsey Associates
Manager, Composition: Allyson Graesser
Desktop Administration: Clara Bartunek
Electronic Page Makeup: Clara Bartunek, Julie Nazario, Jude Wilkens
Director, Image Resource Center: Melinda Reo
Manager, Rights and Permissions: Zina Arabia
Interior Image Specialist: Beth Brenzel
Cover Image Specialist: Karen Sanatar
Image Permission Coordinator: Debbie Latronica
Editorial Assistant: Jennifer Hart
Cover Image: Cells of a filamentous cyanobacterium (yellow) from a soda lake biofilm (see back cover text). Photo courtesy of Gernot Arp and Christian Boeker, Carl Zeiss, Jena, Germany

© 2006, 2003, 2000, 1997, 1994, 1991, 1988, 1974, 1970 by Pearson Education, Inc.
Pearson Prentice Hall
Pearson Education, Inc.
Upper Saddle River, NJ 07458

© 1984, 1979 by Thomas D. Brock

First through seventh editions titled *Biology of Microorganisms*

Editions of Biology of Microorganisms

First Edition, 1970, Thomas D. Brock
Second Edition, 1974, Thomas D. Brock
Third Edition, 1979, Thomas D. Brock
Fourth Edition, 1984, Thomas D. Brock, David W. Smith, and Michael T. Madigan
Fifth Edition, 1988, Thomas D. Brock and Michael T. Madigan
Sixth Edition, 1991, Thomas D. Brock and Michael T. Madigan
Seventh Edition, 1994, Thomas D. Brock, Michael T. Madigan, John M. Martinko, and Jack Parker

Eighth Edition, 1997, Michael T. Madigan, John M. Martinko, and Jack Parker
Ninth Edition, 2000, Michael T. Madigan, John M. Martinko, and Jack Parker
Tenth Edition, 2003, Michael T. Madigan, John M. Martinko, and Jack Parker
Eleventh Edition, 2006, Michael T. Madigan and John M. Martinko

Pearson Prentice Hall™ is a trademark of Pearson Education, Inc.

Printed in the United States of America

10 9 8 7 6 5 4 3 2

ISBN 0-13-144329-1

Pearson Education LTD., *London*
Pearson Education Australia PTY, Limited, *Sydney*
Pearson Education Singapore, Pte. Ltd.
Pearson Education North Asia Ltd., *Hong Kong*
Pearson Education Canada, Ltd., *Toronto*
Pearson Educatión de Mexico, S.A. de C.V.
Pearson Education—Japan, *Tokyo*
Pearson Education Malaysia, Pte. Ltd.

BRIEF CONTENTS

ELEVENTH EDITION OVERVIEW

Each edition creates an opportunity to clarify concepts through the graphics, and *BBOM* has the most exceptional **Illustration Program** in microbiology. Students rely on graphics more each year, and *BBOM* remains dedicated to providing figures, photos and tables that are not only visually engaging, but focused on helping students to better understand the material. ▼

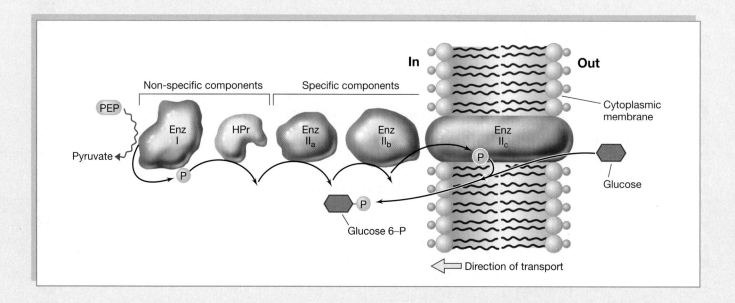

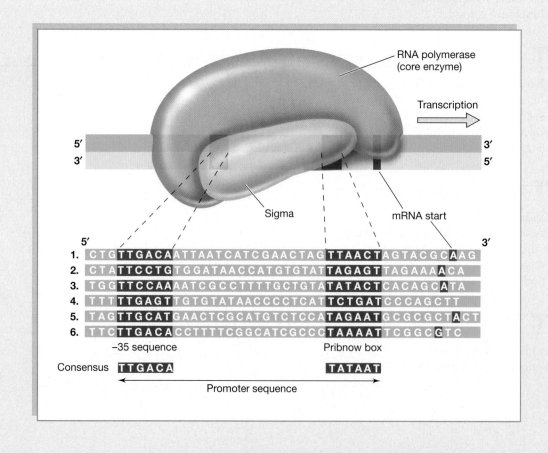

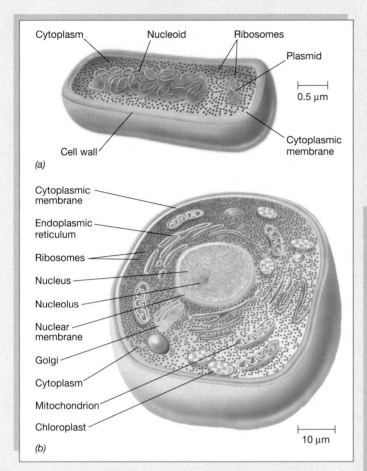

(a)

Cytoplasm   Nucleoid   Ribosomes

Plasmid

0.5 µm

Cell wall

Cytoplasmic membrane

Cytoplasmic membrane

Endoplasmic reticulum

Ribosomes

Nucleus

Nucleolus

Nuclear membrane

Golgi

Cytoplasm

Mitochondrion

Chloroplast

10 µm

(b)

▲ Geared toward today's visual learners, hundreds of graphics in the 11th edition have been completely redesigned and improved upon to provide greater depth and realism.

Wherever possible, techniques and processes include photomicrographs that link abstract representations to biological reality. ▶

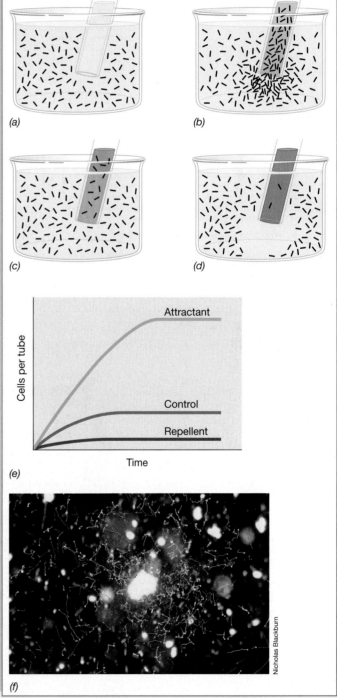

(a)

(b)

(c)

(d)

Attractant

Cells per tube

Control

Repellent

Time

(e)

(f)

Nicholas Blackburn

COMPREHENSION

The **Working Glossary** is the students' guide to the language of microbiology. By beginning each chapter with the glossary, students can more easily master terminology and increase their understanding of key concepts. ▶

WORKING GLOSSARY

Bacteriophage a virus that infects prokaryotic cells

Early protein a protein synthesized soon after virus infection

Late protein a protein synthesized toward the end of virus infection

Lysogen a bacterium containing a prophage

Lysogenic pathway a series of steps that, after virus infection, lead to a state (lysogeny) where the viral genome is replicated as a prophage along with that of the host

Lytic pathway a series of steps after virus infection that leads to virus replication and the destruction (lysis) of the host cell

Minus (negative)-strand virus a virus with an RNA genome in which the RNA strand has the opposite sense of (is complementary to) the mRNA of the virus

Nucleocapsid the complex of nucleic acid and proteins of a virus

Oncogene a gene whose expression causes formation of a tumor

Plaque a zone of lysis or cell inhibition caused by virus infection of a lawn of sensitive cells

Plus (positive)-strand virus a virus with an RNA or DNA genome in which the genome has the same complementarity as the mRNA of the virus

Prion an infectious protein whose extracellular form contains no nucleic acid

Provirus (prophage) the genome of a temperate virus when it is replicating with, and usually integrated into, the host chromosome

Retrovirus a virus whose RNA genome has a DNA intermediate as part of its replication cycle

Reverse transcription the process of copying information found in RNA into DNA

by the enzyme reverse transcriptase

Temperate virus a virus whose genome is able to replicate along with that of its host and not cause cell death in a state called lysogeny

Transformation in eukaryotes, a process by which a normal cell becomes a cancer cell (but see alternative usage in Chapter 10)

Virion the complete virus particle; the nucleic acid surrounded by a protein coat and in some cases other material

Virulent virus a virus that lyses or kills the host cell after infection; a nontemperate virus

Virus a genetic element containing either RNA or DNA that replicates in cells but is characterized by having an extracellular state

Viroid small, circular, single-stranded RNA that causes various plant diseases

---

1.7 Concept Check

Beijerinck and Winogradsky studied bacteria in soil and water and developed the enrichment culture technique for the isolation of representatives of various physiological groups. Major new concepts in microbiology emerged during this period, including enrichment cultures, chemolithotrophy, chemoautotrophy, and nitrogen fixation.

◆ What is the enrichment culture technique?

◆ In examining Figure 1.16, describe why sulfur oxidation and nitrification are considered chemolithotrophic processes while nitrogen fixation is not. (*Hint:* Look at the reactions involving ATP in each case.)

◀ **Concept Checks** encourage students to stop and assess their understanding of key concepts before moving on. A new "stop sign" icon indicates when a concept check appears in the narrative.

Microbial Sidebar ◆ RNA Editing

Microbial Sidebars replace the boxed inserts. These illustrated vignettes were designed and written to be interesting, timely and related to each chapter's central theme. ▶

In Chapters 7 and 14 we saw that certain genes have coding regions that are split by noncoding regions called *introns*. Typically, introns are removed after transcription to form a mature mRNA, a process called *splicing* (Section 14.8). Interestingly, there is a phenomenon found in the genomes of organelles that is almost the opposite of splicing: *RNA editing*.

RNA editing involves either the insertion or deletion of nucleotides *into the final mRNA* that were not present in the DNA transcribed. Editing can also involve the chemical *modification* of a base in the mRNA that changes it from one base to another. In either case, RNA editing can alter codons in such a way that one or more different amino acids are inserted in a polypeptide than those encoded by its gene.

In the mitochondria of trypanosomes and related protozoa (Section 14.10) some mitochondrial transcripts are edited such that large numbers (hundreds in some cases) of uridylates are added or, more rarely, deleted. An example of this type of RNA editing is shown in Figure 1 ●. RNA editing is precisely controlled by short sequences present in the mRNA that "guide" the enzymes involved in their specific edits. Obviously this process must be very precisely controlled. Inserting

too many or too few bases would yield a frameshift product that would likely be nonfunctional.

The other type of RNA editing, the changing of one base into another, is common in the mitochondria and chloroplasts of higher plants. At specific sites in some mRNAs, a C will be converted to a U by oxidative deamination (the opposite modification is more rare). There are at least 25 sites of C to U conversion in the maize chloroplast. Although mostly found in organellar genomes, an example of the programmed conversion of a C to a U is also known for a mammalian nuclear gene. Depending on the location of the edit, a new codon may be formed, leading to formation of a protein sequence not predictable from the gene that encodes it.

RNA editing, although a curious phenomenon, was not a significant obstacle in analyzing *organellar* genomes. This is because the number of proteins they encode is small and the proteins highly conserved. By contrast, had RNA editing been a widespread phenomenon in *cells*, genomic annotations and the identification of orthologous genes in different organisms could have been an even more formidable challenge than it has been to date.

The function and origin of RNA editing is unknown. But some scientists have pointed out that this process may be yet another remnant, along with ribozymes (Section 14.8) and other catalytic RNAs, of the RNA World (Section 11.2). ■

Protein	...Leu Cys Phe Trp Phe Arg Phe Phe Cys...
mRNA	...uuG uGu UUU UGG uuu AGG uuu uuu uGu...
DNA	... G G TTT TCC AGG G ...
	... C C AAA AGG TCC C ...

Figure 1 RNA editing. *The upper part of the figure shows a portion of the amino acid sequence of subunit III of the enzyme cytochrome oxidase from the protozoan Trypanosoma brucei (Section 14.10). This protein is encoded by mitochondria. Beneath the amino acid sequence, the sequence of the messenger RNA (mRNA) for this region is shown. The bases in uppercase letters are those transcribed from the gene, which is shown below. The bases in the mRNA in lowercase have been inserted into the transcript by RNA editing. Although the DNA has many informational gaps, there are no actual gaps in the molecule itself. The spaces between the base pairs are simply to aid in visualization.*

NAVIGATION

Microorganisms are microscopic and independently living cells that, like humans, live in communities.

1

◀ **Section Numbers** keyed to page references provide an easy way to assign readings as well as an organized outline for students to review and take notes.

---

rier that separates the inside from the outside called the **cytoplasmic membrane** (Figure 2.1●). It is through the cytoplasmic membrane that nutrients and other substances needed by the cell enter, and waste materials and other cell products exit. Within a cell, and bounded by the cytoplasmic membrane, is a complex mixture of sub-

◀ **Figure Reference Locators** occur next to key figure references in the narrative. Requested by students, these red dots help them study more effectively by allowing them to quickly return to the narrative after viewing a figure.

Concept Links, signaled with a blue chain link (∞), alert students that a concept is related to material from other areas of the text. Each link refers students to a section number for quickly reviewing the related material, helping them to continually make connections between concepts. ▶

In a typical Winogradsky column a mixture of organisms develops. Algae and cyanobacteria appear quickly in the upper portions of the water column; by producing O_2 these organisms help to keep this zone oxic. Decomposition processes in the mud lead to the production of organic acids, alcohols, and H_2, suitable substrates for sulfate-reducing bacteria (∞ Sections 12.18, 13.7, 17.15, and 19.13). Sulfide from the sulfate reducers triggers development of purple and green sulfur bacteria (anoxygenic phototrophs, ∞ Sections 12.2 and 12.32) that use sulfide as a photosynthetic electron donor. These organisms typically appear in

INSTRUCTOR RESOURCE CENTER CD/DVD

Instructor Resource Center on CD/DVD (0-13-144340-2)

The multi-disc IRC on CD/DVD package offers everything an instructor needs to prepare and present lectures. Easy-to-use menus and exporting functions enable you to locate resources and build presentations efficiently. IRC on CD/DVD resources include:

- JPEG files of all drawings, photos, and tables from the text optimized for projection purposes

- JPEG files of all text figures and tables in comprehensive PowerPoint® presentations for each chapter

- A second set of PowerPoint® presentations consisting of lecture notes and key figures for each chapter

- ALL NEW Instructor Animations adding depth and clarity to key topics, dynamic processes, and techniques described in the 11th Edition

- A comprehensive set of CRS/In-Class questions

- Word files of the Instrutor's Manual and the assessment materials including the Test Bank

- Printable PDF files of a Media Integration Guide describing each chapters media offerings

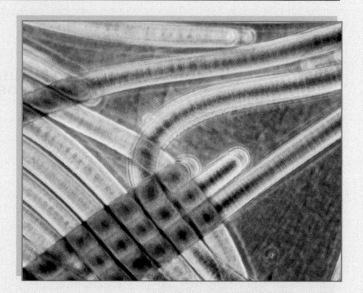

COMPANION WEBSITE

> Home > Macromolecules > Online Study Guide > Review Question
>
> **Chapter 3: Online Study Guide**
> **Review Question 2**
>
> Hydrophobic interactions are important in maintaining [Hint]
>
> ○ the structure of the cytoplasmic membrane.
> ○ interactions between proteins in multisubunit enzymes.
> ○ protein structure.
> ○ all of the above.
>
> Submit Answer for Grading

Companion Website

www.prenhall.com/madigan

Valuable study tools to aid student comprehension of the 11th Edition are offered on the Companion Website. Features include:

- Online Study Guide provides a focused section-by-section review of topic coverage featuring summaries, key illustrations, and review questions

- "Track Your Progress" tool helps students get a better sense of where to focus study time shows them how successful their efforts have been

- **ALL NEW** Web Tutorials help students visualize key topics, processes, and techniques

ANIMATION RESOURCES

Instructor Animation and Student Web Tutorial topics:

Pasteur's Experiment
Koch's Postulates
The Gram Stain
The Prokaryotic Flagellum
Aseptic Transfer and the Streak Plate Method
Direct Microscopic Counting Procedure (Petroff-Hausser Chamber)
DNA Replication
The Polymerase Chain Reaction (PCR)
Transcription
Translation
Enzyme Regulation
Negative Control of Transcription and the *lac* Operon
Attenuation and the Tryptophan Operon
A Temperate Bacteriophage
The Molecular Basis for Mutations
Replica Plating
The Molecular Basis for Mutations
Generating Phylogenetic Trees from RNA Sequences
Cell Division in Conventional, Budding, and Stalked Bacteria
Bacteriorhodopsin and Light-Mediated ATP Synthesis
Life Cycle and Mating Type Switching in a Typical Yeast
DNA Chips
Replication of Poliovirus
Electron Transport Processes: Aerobic and Anaerobic Conditions
Enrichment Cultures
Serial Dilutions and a Most-Probable Number Analysis

Root Nodule Bacteria and Symbiosis with Legumes
Antibiotic Modes of Action
Diphtheria and Cholera Toxins
Antigen Presentation
Producing Monoclonal Antibodies
The ELISA Test
HIV Replication
Life Cycle of the Malaria Parasite
Isolation and Screening of Antibiotic Producers
Production of Recombinant Vaccinia Virus

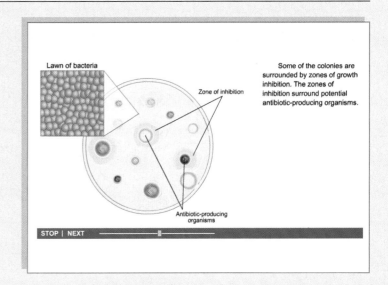

Lawn of bacteria

Zone of inhibition

Some of the colonies are surrounded by zones of growth inhibition. The zones of inhibition surround potential antibiotic-producing organisms.

Antibiotic-producing organisms

STOP | NEXT

OTHER RESOURCES

Instructor's Resource Manual with Tests (0-13-144342-9)
This manual and test bank contains over 2000 questions an instructor can use to prepare exams. This resource also provides chapter summaries as well as the answers to the end of chapter review and application questions.

Test Gen EQ Computerized Testing Software (0-13-144337-2)
In addition to the printed volume, the test questions are available as part of the Test Gen EQ Testing Software, a text specific testing program that is networkable for administering tests. It also allows instructors to view and edit questions, export the questions as tests, and print them out in a variety of formats.

Transparencies (0-13-144341-0)
400 figures from the text are included in this transparency package. The font sizes of the labels have been increased for easy viewing from the back of the classroom.

Research Navigator www.researchnavigator.com
Prentice Hall's Research Navigator™ gives your students access to the most current information available for a wide array of subjects via EBSCO's Content Select™ Academic Journal Database, The New York Times Search by Subject Archive,

"Best of the Web" Link Library, and information on the latest news and current events. This valuable tool helps students find the most useful articles and journals, cite sources, and write effective papers for research assignments.

SafariX Textbooks Online is an exciting new choice for students looking to save money. As an alternative to purchasing the print textbook, students can subscribe to the same content online and  save up to 50% off the suggested list price of the print text. With a SafariX WebBook, students can search the text, make notes online, print out reading assignments that incorporate lecture notes, and bookmark important passages for later review. For more information, or to subscribe to the SafariX WebBook, visit www.safarix.com.

Course Management
Prentice Hall offers all of the instructor and student media for the 11th Edition through your preferred course management platform. For details visit http://cms.prenhall.com.

CONTENTS

PREFACE

We are living in the age of microbiology. Almost daily come reports of new discoveries in this exciting science—new emerging infections, new organisms, and new tools to facilitate discovery. Such is the pace of the science of microbiology today. And here to bring you the up-to-the-minute picture of microbiology today is the eleventh edition of *Brock Biology of Microorganisms (BBOM)*

This textbook has roots going back over 35 years. Since the publication of the first edition of *Biology of Microorganisms* by Thomas D. Brock (Prentice Hall, 1970), this book has had a single mission: to present the principles of microbiology within the framework of modern science. *BBOM 11/e* maintains this tradition, and speaks with the same accuracy and authority of the previous ten editions.

Microbiology today places unusual demands on students and instructors alike. The amount of information is enormous, the background required in supporting sciences is significant, and introductory classes in microbiology are bursting at the seams. The authors of *BBOM 11/e* are keenly aware of these problems and have worked hard to craft a textbook of microbiology where the principles stand up and shout, the details are complementary, and the supporting concepts are well integrated. We hope you will agree.

What's New in the 11th Edition?

Those who have taught from *BBOM* in the past will find the new edition the same old friend they knew before. However, instructors will find *BBOM 11/e* more teachable, and students will find it a more invaluable learning resource, than ever before.

BBOM 11/e is pedagogically geared towards today's visual learner. The design of *BBOM 11/e* starts where the tenth edition left off, but charts new ground in terms of presentation, use of color, and art. The chapters are organized around the logical and helpful numbered outline system, present in this book from the first edition. But we have now organized each chapter into several blocks of related information, integrating the concepts into more digestible bites. As usual, a working glossary—the student's dictionary of essential terms—opens each chapter, to bring the language of microbiology where it needs to be, up front and center. Concept checks remain and are now signaled by a bright red "stop sign!" Concept checks signal the student to stop, review, and assess, before proceeding to the next concept. As usual, challenging review and application study questions are present at the end of each chapter. A comprehensive glossary and index in the back of the book, wrap up the package.

Traditional "boxed" material is presented in *BBOM 11/e* in our new *Microbial Sidebars*. These richly illustrated vignettes were designed and written to be "fun reads" of enrichment material related to a chapter's central theme. Tables have been completely redesigned in *BBOM 11/e* to make the information in them easier to follow and better organized. Since a science like microbiology relies heavily on tabular resources, the new table design should be a winner with both students and instructors alike. Many other pedagogically useful features will make themselves obvious to the reader as s/he proceeds through this book. These include more distinctive heads, an eye-catching red dot icon that leads the reader's eye from text to figures and back again, and review questions keyed to section number. The latter will make it easier for students to refresh their memories before answering each question.

Pervading the entire book is a spectacular art program, with every piece of art redone by a new art studio. The result is brighter, more distinctive art, which is also more colorful, appealing, impeccably consistent, and instructive than ever before. Moreover, the use for the first time of a high quality, glossy paper in the eleventh edition, has brought out the best in the outstanding photomicrographs and other photos that have been a tradition in this book since the first edition. Users will quickly recognize new pedagogical aids, such as our "energy arrows," built right into the art. Cellular reactions that produce or consume ATP are often key ones. Energy arrows—bright red wavy arrows—signal these reactions and bring them to a student's attention.

Although *BBOM 11/e* is actually shorter than the previous edition, it contains substantial new content. Indeed, new material can be found in every chapter and we give only a taste of what's in store here: Toxic Forms of Oxygen (Chapter 6); Diversity of Sigma Factors, Consensus Sequences, and Other RNA Polymerases (Chapter 7); The Stringent Response (Chapter 8); RNA Regulation and Riboswitches (Chapter 8); Sub-Viral Particles (Chapter 9); The Carbon and Energy Metabolism of Primitive Life Forms (Chapter 11); The Biology of *Nanoarchaeum* (Chapter 13); RNA Processing and Ribozymes (Chapter 14); Replication of Linear DNA (Chapter 14); Annotating the Genome (Chapter 15); Bioinformatic Analyses and Gene Distribution in Prokaryotes (Chapter 15); Microarrays and the Transcriptome (Chapter 15); Viruses of *Archaea* (Chapter 16); Environmental Genomics (Chapter 18); Host Risk Factors for Infection (Chapter 21); Inflammation, Fever, and Septic Shock (Chapter 22); Natural Immunity (Chapter 22); Receptors and Immunity (Chapter 23); West Nile Virus (Chapter 27); Microbial Sampling and Food Poisoning (Chapter 29); Severe Acute Respiratory Syndrome (SARS) (Chapter 25); Anthrax as a Biological Weapon (Chapter 25); and Fermented Foods (Chapter 29).

Several supplements accompany *BBOM 11/e*. These include a website (*www.prenhall.com/madigan*) containing online media resources (flagged by an icon in the text), practice exam questions, and additional content resources. For instructors, a CD is available that contains *all* of the tables and figures in the book arranged in PowerPoint format for ease in organizing classroom learning activities. Overhead transparencies are also available for those who use this format in the classroom. Indeed, the *BBOM 11/e* instructor package offers every necessary tool for developing clear, compelling, and stimulating presentations.

Acknowledgments

The final product you see before you is the collective effort of many people. These include a number of folks at Prentice Hall/Pearson Publishing, but especially Executive Editor Gary Carlson and his assistants Susan Zeigler and Jennifer Hart, and our outstanding production editor, Debra Wechsler. Gary was the guiding light for this edition, while Susan/Jennifer and Debra maintained the pace of the project and were the "glue" in editorial and production, respectively. Ed Thomas (production) is also acknowledged for production assistance early in the project. The authors give a hearty thanks to the input of our design and art editors, Kenny Beck and Jay McElroy, and our superb media editors, Patrick Shriner and Crissy Dudonis. Excellent copy-editing of the entire manuscript was done by Jane Loftus (Clackamas, OR).

The authors especially wish to thank the thorough and very helpful input early on in the project of the development editor, Jon Haber (New York, NY), and the expert assistance with electronic photographs provided by Deborah O. Jung (Carbondale, IL). The excellent input of Elizabeth McPherson (University of Tennessee) and David Crowley (University of California-Riverside) as accuracy checkers during the production of *BBOM 11/e* is heartily acknowledged. Special thanks are also extended to Gernot Arp and Christian Boeker, Universität Göttingen, Germany and Carl Zeiss, Inc., Jena, Germany for the gorgeous confocal micrograph that graces the covers of this book.

Both authors also wish to thank their graduate students, colleagues, and Department of Microbiology staff for their assistance and patience during the busy time of preparing a book of this scope. The seemingly endless patience of our wives, Nancy and Judy, for the countless hours spent away from them during the development and production of *BBOM 11/e*, is also acknowledged. Their love, understanding, and support allowed the authors to devote the time necessary for such a massive project.

Finally, we are extremely grateful for the kind help of so many individuals that provided reviewer input into the draft manuscript or who supplied new photos for the 11th edition. They are listed below.

Laurie Achenbach, *Southern Illinois University*
Richard Adler, *University of Michigan-Dearborn*
Karen Aguirre, *Clarkson University*
Stephen Aley, *University of Texas-El Paso*
Mary Allen, *Hartwick College*
Ricardo Amils, *University Autonoma, Madrid, Spain*
Robert Andrews, *Iowa State University*
Michael Benedik, *Texas A&M University*
David Boone, *Portland State University*
Matt Boulton, *University of Michigan*
Cheryl Broadie, *Southern Illinois University*
Jean Cardinale, *Alfred University*
Jannice Carr, *Centers for Disease Control and Prevention-Atlanta*
David Clark, *Southern Illinois University*
Rhonda Clark, *University of Calgary*
Morris Cooper, *Southern Illinois University School of Medicine*
David Crowley, *University of California-Riverside*
Mark Davis, *University of Evansville*
Michael Davis, *Central Connecticut State University*
Dennis Dean, *Virginia Tech University*
Arvind Dhople, *Florida Institute of Technology*
Biao Ding, *Ohio State University*
Rodney Donlan, *Centers for Disease Control and Prevention-Atlanta*
Paul Dunlap, *University of Michigan*
Paul Edmonds, *Georgia Institute of Technology*
Elizabeth Ehrenfeld, *Southern Maine Community College*
Bruce Farnham, *Metropolitan State University-Denver*
Rebecca Ferrell, *Metropolitan State University-Denver*
Doug Fix, *Southern Illinois University*
Niels-Ulrik Frigaard, *Pennsylvania State University*
George Garrity, *Michigan State University*
Claire Geslin, *Université de Bretagne Occidentale, France*
Eric Grafman, *Centers for Disease Control and Prevention-Atlanta*
Bonita Glatz, *Iowa State University*
Ricardo Guerrero, *University of Barcelona, Spain*
John Haddock, *Southern Illinois University*
Martin Hanczyc, *Harvard University*
Ernest Hanning, *University of Texas-Dallas*
Pamela Hathorn, *Midwestern University*
John Hayes, *Woods Hole Oceanographic Institution*
Lee Hughs, *University of North Texas*
Michael Ibba, *Ohio State University*
Johannes Imhoff, *Universität Kiel, Germany*
Mary Johnson, *Indiana University*
Deborah Jung, *Southern Illinois University*
Judy Kandel, *California State University-Fullerton*
Patrick Keeling, *University of British Columbia*
Joan Kiely, *SUNY-Stonybrook*
Arthur Koch, *Indiana University*
Allan Konopka, *Purdue University*
Vikki Kourkouliotis, *National Renewable Energy Laboratory*
Susan F. Koval, *University of Western Ontario*
Harry Kurtz, *Clemson University*
Sharon Long, *University of Massachusetts*

Bonnie Lustigman, *Montclair State University*
Mark Martin, *Occidental College*
William McCleary, *Brigham Young University*
Elizabeth McPherson, *University of Tennessee*
Ohad Medalia, *Max Planck Institut für Biochemie, Germany*
Jianghang Meng, *University of Maryland*
Eric Miller, *North Carolina State University*
Abraham Minisky, *Weizmann Institute of Science, Israel*
Ivan Oresnik, *University of Manitoba*
Jörg Overmann, *University of Munich, Germany*
Norm Pace, *University of Colorado*
Jack Parker, *Southern Illinois University*
Laurence Pelletier, *Max Planck Institut für Cell Biologie und Genetiks, Germany*
Michael Pfaller, *University of Iowa*
Reinhard Rachel, *Universität Regensburg, Germany*
Michael Rappe, *Oregon State University*
Chris Rensing, *University of Arizona*
Frank Roberto, *Idaho Environmental and Engineering Labs*
Craig Rouskey, *Southern Illinois University*
Jill Zeilstra Ryalls, *Oakland University*
Herb Schelhorn, *McMaster University*
Bernhard Schink, *Universität Konstanz, Germany*
Heide Schulz, *University of California-Davis*
Kate Scow, *University of California-Davis*
James Shapleigh, *Cornell University*
Jolynn Smith, *Southern Illinois University*

Jerry Sipe, *Anderson University*
Nancy Spear, *Murphysboro, Illinois*
Julia Thompson, *American Society for Microbiology*
Sonia Tiquia, *University of Michigan*
David Tison, *Multicare Health Systems*
Paul Tomasek, *California State University-Northridge*
Amy Treonis, *Creighton University*
Michael Wagner, *University of Vienna, Austria*
David Ward, *Montana State University*
Joy Watts, *University of Maryland*
Mary Watwood, *Northern Arizona University*
Susan Wells, *Affymetrix, Santa Clara, California*
Carl Woese, *University of Illinois*
Gordon Wolfe, *California State University-Chico*
Alexander Worden, *University of Miami*
Mark Young, *Montana State University*
Stephen Zinder, *Cornell University*

Any errors in this book, either ones of commission or omission, are solely the responsibility of the authors. In past editions, users have been so kind as to contact us when they came across an error. Although we hope that *BBOM 11/e* is error-free, no book ever is. So users should feel free to contact either author about errors or concerns. We hope you enjoy *BBOM 11/e*.

Michael T. Madigan (*madigan@micro.siu.edu*)
John M. Martinko (*martinko@micro.siu.edu*)

BROCK

BIOLOGY OF
MICROORGANISMS

1

MICROORGANISMS AND MICROBIOLOGY

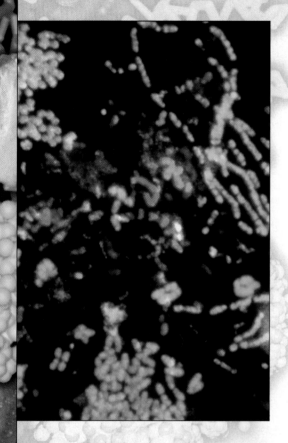

Microorganisms are microscopic and independently living cells that, like humans, live in communities.

WORKING GLOSSARY

Cell the fundamental unit of living matter
Cytoplasm the fluid portion of a cell, bounded by the cell membrane but excluding the nucleus (if present)
DNA deoxyribonucleic acid, the hereditary material of cells and some viruses
Ecology the study of organisms in their natural environments
Ecosystem organisms plus their nonliving environment
Enrichment culture a method for isolating microorganisms from nature using specific culture media and incubation conditions

Enzyme a protein catalyst that functions to speed up chemical reactions
Habitat the location in an environment where a microbial population resides
Koch's Postulates a set of criteria for proving that a given microorganism causes a given disease
Metabolism all biochemical reactions in a cell
Microorganism a microscopic organism consisting of a single cell or cell cluster, including the viruses

Pathogen a disease-causing microorganism
Pure culture a culture containing a single kind of microorganism
RNA ribonucleic acid, involved in protein synthesis as messenger RNA, transfer RNA, and ribosomal RNA
Spontaneous generation the hypothesis that living organisms can originate from nonliving matter
Sterile absence of all living organisms and viruses

Welcome to microbiology—the study of microorganisms. Microorganisms include a large and diverse group of microscopic organisms that exist as single cells or cell clusters, and the viruses, which are microscopic but not cellular.

What is microbiology all about? Microbiology is about cells and how they work, especially the bacteria, a large group of cells of enormous basic and practical significance (Figure 1.1●). Microbiology is about microbial diversity and evolution, about how different kinds of microorganisms arose and why. It is about what microorganisms do in the world at large, in human society, in the human body, and in animals and plants. One way or the other, microorganisms affect all other life forms on Earth (Figure 1.1b), and thus the science of microbiology is of enormous importance.

Microorganisms are distinct from the cells of animals and plants—macroorganisms. Cells of plants and animals are unable to live alone in nature and exist only as parts of multicellular structures, such as the organ systems of animals or the structural components of plants. By contrast, most microorganisms can carry out their life processes of growth, energy generation, and reproduction independently of other cells, either of the same kind or of a different kind.

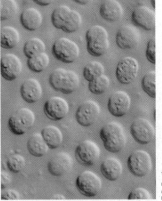

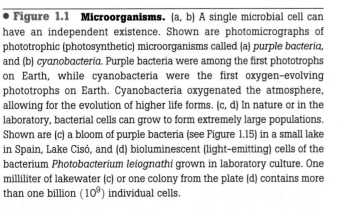

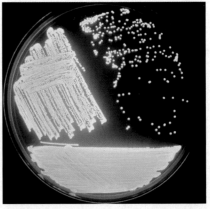

● **Figure 1.1 Microorganisms.** (a, b) A single microbial cell can have an independent existence. Shown are photomicrographs of phototrophic (photosynthetic) microorganisms called (a) *purple bacteria*, and (b) *cyanobacteria*. Purple bacteria were among the first phototrophs on Earth, while cyanobacteria were the first oxygen-evolving phototrophs on Earth. Cyanobacteria oxygenated the atmosphere, allowing for the evolution of higher life forms. (c, d) In nature or in the laboratory, bacterial cells can grow to form extremely large populations. Shown are (c) a bloom of purple bacteria (see Figure 1.15) in a small lake in Spain, Lake Cisó, and (d) bioluminescent (light-emitting) cells of the bacterium *Photobacterium leiognathi* grown in laboratory culture. One milliliter of lakewater (c) or one colony from the plate (d) contains more than one billion (10^9) individual cells.

This chapter begins our journey to the microbial world. We discover what microorganisms are and their impact on life. We set the stage for a consideration of the structure and evolution of microorganisms that will unfold in the next chapter. We also place microbiology in historical perspective, as a process of scientific discovery. Thanks to the landmark contributions of both early microbiologists and scientists practicing today, we can see the ramifications of microbiology for medicine, agriculture, and the environment.

I INTRODUCTION TO MICROBIOLOGY

In the first four sections of this chapter we introduce the subject of microbiology, look at microorganisms as cells, examine where and how microorganisms live in nature, and review the impact that microorganisms have had and continue to have on human affairs.

1.1 Microbiology

Microbiology revolves around two basic themes: (1) the basic science of understanding life, and (2) the applications of science to human needs.

As a *basic biological science*, microbiology provides tools for probing the processes of life. Our most sophisticated understanding of the chemical and physical basis of life has arisen from studies of microorganisms. Microbial cells share many biochemical properties with cells of multicellular organisms; indeed, *all* cells have much in common. Moreover, microbial cells can grow to extremely high densities in laboratory culture (Figure 1.1*d*) and are readily amenable to biochemical and genetic study. All of these features make microorganisms excellent models for understanding cell function in higher organisms, including humans.

As an *applied biological science*, microbiology deals with many important practical problems in medicine, agriculture, and industry. Some of the most important diseases of humans, other animals, and plants are caused by microorganisms. Microorganisms also play major roles in soil fertility and domestic animal production. In addition, many large-scale industrial processes, such as the production of antibiotics or human proteins, are microbially based.

The Importance of Microorganisms

As this book unfolds, we will see the central role that microorganisms play in both human activities and the web of life on Earth. In the absence of microorganisms, for instance, higher life forms would never have arisen and could not now be sustained. Indeed, the very oxygen we breathe is the result of past microbial activity (Figure 1.1*b*). Moreover, we will see how humans, plants, and animals are intimately tied to microbial activities for the recycling of key nutrients and for degrading organic matter. No other life forms are as important as microorganisms for the support and maintenance of life on Earth.

Microorganisms existed on Earth for billions of years before plants and animals appeared. Thus, their evolutionary diversity has far outpaced that of higher organisms. This huge diversity accounts for some of the spectacular properties of microorganisms. For example, we will see how microorganisms can live in places unsuitable for higher organisms and how their diverse physiological capacities rank them as Earth's greatest chemists. We will also see how microorganisms have established relationships with higher organisms that can be either beneficial or harmful. In humans, for instance, the very health we enjoy and many of the diseases we suffer are due to microorganisms.

Microorganisms are clearly central to the very functioning of the biosphere. Hence, the science of microbiology is the foundation of all the biological sciences. We begin our study of this important science with a consideration of microorganisms as cellular entities.

1.2 Microorganisms as Cells

The **cell** is the fundamental unit of life. A single cell is an entity, isolated from other cells by a membrane (and perhaps a cell wall) and contains within it a variety of chemicals and subcellular structures (Figure 1.2●). Compartmentalization is a prerequisite for life. This is because the chemical components of life must remain at concentrations sufficient to allow necessary chemical reactions to occur. But the cell is not a closed system. Instead, the cell is an open, dynamic entity. Cells communicate and exchange materials with their environments, and they are constantly undergoing change. We will study cell structure and function in detail in Chapters 2, 4, and 14.

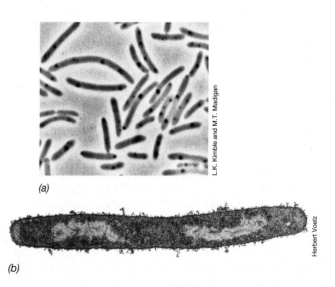

(a)

L.K. Kimble and M.T. Madigan

(b)

Herbert Voelz

● **Figure 1.2   Cells.** (a) Photomicrograph of rod-shaped bacterial cells as seen in the light microscope; a single cell is about $1\,\mu m$ in diameter. (b) Longitudinal section through a bacterial cell as viewed with an electron microscope. The two lighter areas are the nucleoid, regions in the cell containing aggregated DNA.

Cell Chemistry and Key Structures

Cells are highly organized structures consisting primarily of four chemical components: **proteins, nucleic acids, lipids**, and **polysaccharides**. Collectively, these large molecules are called **macromolecules**. It is the chemistry and arrangement of macromolecules in different cells that make organisms distinct from one another. Thus, chemically speaking, all cells share much in common, whether they are parts of a plant or animal or whether they are microorganisms.

There are a number of key structures in a cell. The **cytoplasmic membrane** is the barrier that separates the inside of the cell from the outside. Inside the cell membrane are various structures and chemicals suspended or dissolved in the **cytoplasm**. The latter is where the *machinery* for cell growth and function is present. Key structures here include the **nucleus** or **nucleoid**, where the cells' *genetic information*—deoxyribonucleic acid (**DNA**)—is stored, and **ribosomes**, the structures upon which new proteins are made in the cell.

Characteristics of Living Systems

What are the essential characteristics of life? What differentiates cells from inanimate objects? Our concept of what is alive is constrained by what we observe on Earth today or can deduce from the fossil record. But from our knowledge of biology thus far, we can identify several characteristics shared by most living systems. These are summarized in Figure 1.3●.

All cellular organisms show some form of **metabolism**. That is, cells take up nutrients from their environment and transform them—conserving some of the energy present in these substances in a form the cell can use—and then eliminate waste products. All cells show **reproduction**. That is, a cell can direct a series of biochemical events that results in growth and division to form two cells. Many cells undergo **differentiation**, a process by which new substances or structures are formed. Cell differentiation is often part of a cellular life cycle in which cells form special structures, such as spores, involved in reproduction, dispersal, or survival.

Cells respond to chemical signals in their environment, including those produced by other cells. Cells can thus undergo **communication**. Cells can even assess their own numbers by way of small diffusible molecules passed between neighboring cells, a process call *quorum sensing*. Although not universal, living organisms are often capable of **movement** by self-propulsion; in the microbial world we will see several different mechanisms of motility. Finally, unlike nonliving structures, cells can *evolve*. Through the process of **evolution**, cells can change their characteristics and transmit these new properties to their offspring.

Cells as Machines and as Coding Devices

Cells can be viewed in two ways. On one hand, cells can be considered *machines* that carry out chemical

1. Metabolism
Uptake of nutrients from the environment, their transformation within the cell, and elimination of wastes into the environment. The cell is thus an *open* system.

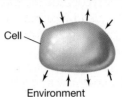

Cell

Environment

2. Reproduction (growth)
Chemicals from the environment are turned into new cells under the direction of preexisting cells.

3. Differentiation
Formation of a new cell structure such as a spore, usually as part of a cellular *life cycle*.

Spore

4. Communication
Cells *communicate* or *interact* primarily by means of chemicals that are released or taken up.

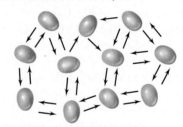

5. Movement
Living organisms are often capable of self-propulsion.

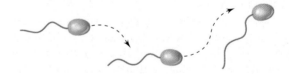

6. Evolution
Cells contain genes and *evolve* to display new biological properties. Phylogenetic trees show the evolutionary relationships between cells.

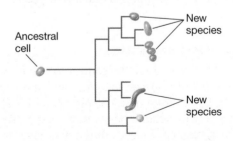

New species

Ancestral cell

New species

● **Figure 1.3 The hallmarks of cellular life.** Differentiation and motility are not properties of all microbial cells.

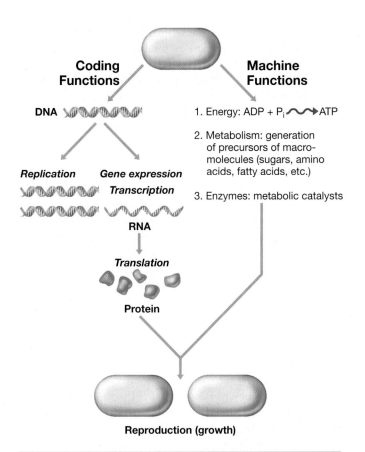

Coding Functions

Machine Functions

DNA

1. Energy: ADP + P$_i$ ⟿ ATP

2. Metabolism: generation of precursors of macromolecules (sugars, amino acids, fatty acids, etc.)

Replication   *Gene expression*

Transcription

3. Enzymes: metabolic catalysts

RNA

Translation

Protein

Reproduction (growth)

● **Figure 1.4 The machine and coding functions of the cell.** In order for a cell to reproduce itself there must be (1) an adequate supply of energy and precursors for the synthesis of new macromolecules, (2) the genetic instructions replicated such that upon division each cell receives a copy, and (3) gene expression (the processes of transcription and translation) to form the proper amounts of necessary proteins and other macromolecules that will make up the new cell.

transformations. The catalysts of this chemical machine are **enzymes**, proteins capable of greatly accelerating the rate of chemical reactions (Figure 1.4●). On the other hand, cells can be considered *coding devices*, analogous to computers, which store and process genetic information (DNA) that is eventually passed on to offspring during reproduction (Figure 1.4). Replication and processing of the stored genetic information will be considered in Chapter 7, where we cover the core cellular functions of *DNA replication, transcription*—the production of **RNA**—and *translation*—the production of **protein**.

In reality, cells are both chemical machines *and* coding devices. The link between these two cellular attributes is *growth*. Under proper conditions, a cell will grow larger and then divide to form two cells (Figure 1.4). In the orderly process that results, the amount of all the constituents of the cell must *double*. This requires the chemical machinery of the cell to supply energy and precursors for biosynthesis of macromolecules. Also, when a cell divides, each of the two resulting cells must contain all of the genetic information necessary for the formation of more cells. Thus, during the growth process

there must also be a replication of DNA (Figure 1.4). The machine and coding functions of the cell must therefore be highly coordinated in order for the cell to reproduce itself faithfully. Indeed, we will see later that in addition to *coordination*, the various machine and coding functions of the cell are subject to *regulation*. This ensures that the cell stays optimally attuned to its environment.

The First Cells

Where did the first cells come from? Because all cells are constructed in similar ways, it is hypothesized that all cells have descended from a common ancestor, the *universal ancestor* of all life (∞ Chapter 11). In some way the first cell must have come from a noncell, something before the cell, a *procellular* structure. Evolution of the first cell on Earth—over 3.8 billion years ago—may have taken several hundred million years to occur. Yet once the first cells arose, their growth and division formed populations. Evolution could then select for improvements and diversification of these early life forms. Through billions of years of evolution, a tremendous diversity of cell types appeared. We will get a taste of this microbial diversity in Chapter 2, and then consider the topic in detail in Chapters 12–16.

1.2 Concept Check

The cell has a barrier, the cytoplasmic membrane, that separates the cytoplasm from the environment. Other cell features include the nucleus or nucleoid, and cytoplasm. Metabolism and reproduction are associated with the living state, and cells can be considered both chemical machines and coding devices. Life has been present on Earth for nearly 4 billion years.

◆ List the four classes of cellular macromolecules.

◆ List six features associated with living organisms. Why might each feature be important to the survival of a cell?

◆ Compare the machine and coding functions of a microbial cell. Why is neither of value to a cell without the other?

1.3 Microorganisms and Their Natural Environments

In nature, cells live in association with other cells in assemblages called **populations**. Populations are composed of groups of cells derived from successive cell divisions from a single parent cell. The location in an environment where a microbial population lives is called the **habitat**. In microbial habitats, a population of cells rarely lives alone. Rather, a population of cells lives and interacts with other populations in **microbial communities** (Figure 1.5●). The components and cell numbers in a microbial community are governed by the resources and conditions that exist in the habitat. The study of microorganisms in their natural habitats is called **microbial ecology**.

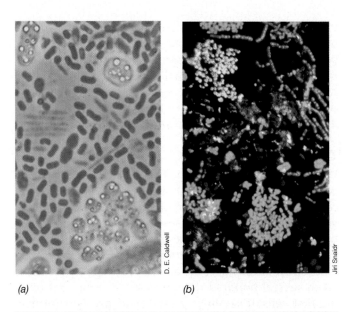

(a)   (b)

D. E. Caldwell

Jiri Snaidr

● **Figure 1.5** **Examples of microbial communities.** (a) Photomicrograph of a bacterial community that developed in the depths of a small lake (Wintergreen Lake, Michigan), showing cells of various sizes. (b) A bacterial community in a sewage sludge sample. The sample was stained with a series of dyes, each of which stained a different bacterial group (∞Section 18.4 and Figure 18.11b for details of the staining process). From R. Amann, J. Snaidr, M. Wagner, W. Ludwig, and K.-H. Schleifer, 1996. *Journal of Bacteriology* 178: 3496–3500, Fig. 2b. © 1996 American Society for Microbiology.

The Effect of Organisms on Each Other and on Their Habitats

Populations in microbial communities interact in various ways, both harmful and beneficial. In many cases microbial populations interact and cooperate. For example, the waste products of the metabolic activities of some cells can be nutrients for others. Organisms in a habitat also interact with their physical and chemical environment. Habitats differ markedly in their characteristics, and a habitat that is favorable for the growth of one organism may actually be harmful for another organism. Collectively, we speak of the living organisms together with the physical and chemical constituents of their environment as an **ecosystem**. Major microbial ecosystems include aquatic (oceans, ponds, lakes, streams, ice, hot springs), terrestrial (soil, deep subsurface), and higher organisms, both plant and animal.

Ecosystems are controlled to a significant extent by microbial activities. Organisms carrying out metabolic processes remove nutrients from the environment and use them to build new cells. At the same time, organisms excrete waste products of their metabolism into the environment. Thus, over time, a microbial ecosystem will gradually change, both chemically and physically, through the cycling of nutrients by microorganisms. The environmental changes may allow for other microorganisms to grow. For example, molecular oxygen (O_2) is a vital nutrient for some microorganisms, but a poison to others. However, the oxygen-consuming activities of one group of organisms (aerobes) can make an oxic habitat

anoxic (O_2-free) and suitable for growth of anaerobic organisms that were previously unable to grow.

In later chapters, after we have learned some of the basic features of microbial structure and function, genetics, evolution, and diversity, we will return to a discussion of the ways in which microorganisms affect animals, plants, and the whole global ecosystem.

The Extent of Microbial Life

Microorganisms are small, but their biomass on Earth is huge, even when compared with the biomass of higher organisms. Examination of natural material such as soil or water always reveals microbial cells. Although such tiny cells may seem inconsequential, single cells are capable of multiplying rapidly and producing large populations that may have a major effect on the habitat (see Figure 1.1c). Microorganisms are thus extremely important and quantitatively significant parts of virtually every ecosystem.

Estimates of total microbial cell numbers on Earth and specifically, the numbers of **prokaryotes** (**bacteria**, ∞Chapter 2) show this number to be on the order of 5×10^{30} cells. The total amount of carbon present in this very large number of very small cells equals that of all plants on Earth (and plant carbon far surpasses animal carbon). In addition, the collective contents of nitrogen and phosphorous in prokaryotic cells is over 10 times that in all plant biomass.

Thus prokaryotic cells, small as they are, constitute the *major portion of biomass on Earth*, and are key reservoirs of essential nutrients for life. An equally startling revelation is the realization that most prokaryotic cells do not reside on Earth's surface, but instead *lie underground* in the oceanic and terrestrial subsurfaces. Because these habitats are relatively unexplored, there is much left for microbiologists to discover and understand about the life forms that dominate Earth.

 1.3 Concept Check

Microorganisms exist in nature in populations that interact with other populations in microbial communities. The activities of microbial communities can greatly affect the chemical and physical properties of their habitats. Most of the biomass on Earth is microbial.

◆ What is a *microbial habitat?*

◆ How do microorganisms change the chemical and physical properties of their habitats?

◆ Where are most prokaryotic cells located on Earth?

1.4 The Impact of Microorganisms on Humans

One goal of the microbiologist is to understand how microorganisms work and devise ways in which the benefits of microorganisms can be increased and their harmful

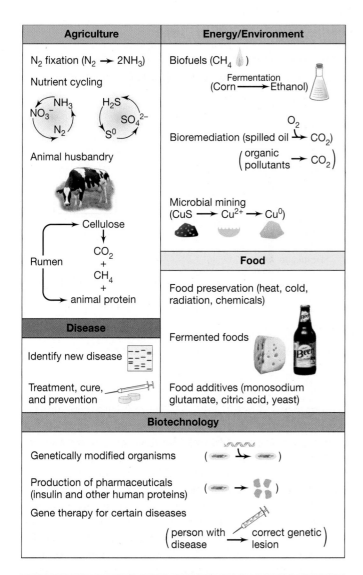

● Figure 1.6   The impact of microorganisms on human affairs. Although many people think of microorganisms in the context of infectious diseases, few microorganisms actually cause disease. Microorganisms affect many aspects of our lives in addition to playing a role as disease agents.

effects curtailed. Microbiologists have been highly successful in achieving these goals, and microbiology has played a major role in the advancement of human health and welfare. An overview of the impact of microorganisms on human affairs is shown in Figure 1.6●.

Microorganisms as Disease Agents

One measure of the microbiologist's success in controlling microorganisms is shown by the statistics in Figure 1.7●. These data compare the present causes of death in the United States to those of 100 years ago. At the beginning of the twentieth century, the major causes of death were infectious diseases (Figure 1.7). Infectious diseases are caused by **pathogens**. Children and the aged in particular succumbed in large numbers to microbial diseases. Today, infectious diseases

are of much less importance in terms of mortality rates in developed countries. Control of infectious disease has come as a result of our comprehensive understanding of disease processes, as well as improved sanitary practices and the discovery and use of antimicrobial agents. As we will see later in this chapter, microbiology as a science had its roots in the study of infectious disease.

Although now many infectious diseases can be controlled, microorganisms can still be a major threat to survival. Consider here the individual dying slowly of a microbial infection as a consequence of acquired immune deficiency syndrome (AIDS) or the individual infected with a multiple-drug-resistant pathogen. Furthermore, microbial diseases are still the major causes of death in many developing countries of the world. While eradication of smallpox from the world was a stunning triumph for medical science, millions still die yearly from such pervasive microbial diseases as malaria, tuberculosis, cholera, African sleeping sickness, and severe diarrheal syndromes.

Clearly, microorganisms are still serious threats to human health. However, microbiology has shown that most microorganisms are *not* harmful to humans. In fact, most microorganisms cause no harm to higher organisms and instead are beneficial, carrying out processes that are of immense value to human society. We consider some of these processes now.

Microorganisms and Agriculture

Our whole system of *agriculture* depends in many important ways on microbial activities (Figure 1.6). For example, a number of major crops are members of a plant group called the **legumes**, which live in close association with bacteria that form structures called **nodules** on their roots. In the root nodules, these bacteria convert atmospheric nitrogen (N_2) into fixed nitrogen (NH_3) that the plants use for growth. In this way, the activities of the bacteria reduce the need for costly and polluting plant fertilizer.

Also of major agricultural importance are the microorganisms that are essential for the digestive process in ruminant animals such as cattle and sheep. These important farm animals have a special digestive vessel called the **rumen** in which microorganisms carry out the digestion of cellulose, the major component of plants. Without these microorganisms, cattle and sheep could not thrive on cellulose-rich but otherwise nutrient-poor substances like grass and hay.

Microorganisms also play key roles in the cycling of important nutrients in plant nutrition, particularly those of *carbon, nitrogen*, and *sulfur*. Microbial activities in soil and water convert these elements to forms that are readily accessible to plants. In addition to benefitting agriculture, microorganisms can also have harmful effects. Microbial diseases of plants and animals have major economic impacts on the agricultural industry. For example,

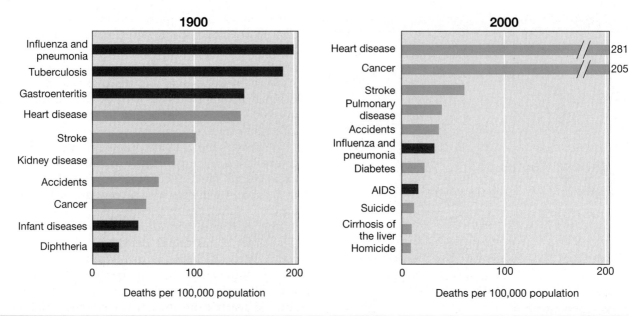

● **Figure 1.7** **Death rates for the 10 leading causes of death in the United States: 1900 and 2000.** Infectious diseases were the leading causes of death in 1900, whereas today they are much less significant. Microbial diseases are shown in red, nonmicrobial causes of death in green. Data from the United States National Center for Health Statistics.

the mad cow disease (∞Section 29.10) incident in the United States in 2003 shut down exports of beef to several major foreign markets, causing a temporary yet major impact to the U.S. beef industry.

Microorganisms and Food

Once food crops and animals are produced, they must be delivered in wholesome form to consumers. Microorganisms play important roles in the food industry (Figure 1.6). *Food spoilage* alone results in huge economic losses each year. Indeed, the canning, frozen-food, and dried-food industries exist to prepare foods in such ways that they will not undergo microbial spoilage. *Foodborne disease* is also a consideration. Because food fit for human consumption can support the growth of many microorganisms, foods must be properly prepared and monitored to avoid transmission of disease.

However, not all microorganisms in foods have harmful effects on food products or those who eat them. For example, dairy products that benefit from microbial activity include cheese, yogurt, and buttermilk, all products of major economic value. Similarly, sauerkraut, pickles, and some sausages also owe their existence to microbial fermentations. Moreover, baked goods and alcoholic beverages are based on the fermentative activities of yeast. Many of these topics are covered in Chapter 30 of this book.

Microorganisms, Energy, and the Environment

Microorganisms play major roles in energy production (Figure 1.6). Natural gas (**methane**) is a product of bacterial activity, arising from the metabolism of *methanogenic* mi-

croorganisms. Phototrophic microorganisms can harvest light energy for the production of **biomass**, energy stored in living organisms. Microbial biomass and existing waste materials such as domestic refuse, surplus grain, and animal wastes, can be converted to "biofuels," such as methane and ethanol, by the activities of microorganisms.

Microorganisms can also be used to help clean up pollution created by human activities, a process called **bioremediation** (Figure 1.6). Various microorganisms can be used to consume spilled oil, solvents, pesticides, and other environmentally toxic pollutants. The great diversity of microorganisms on Earth contains vast genetic resources to mediate solutions for cleaning up the environment, and much research in this area is taking place at present.

Microorganisms and the Future

Biotechnology entails the use of microorganisms in industrial biosyntheses, typically by microorganisms that have been genetically modified to synthesize products of high commercial value (∞Chapter 31). Biotechnology is highly dependent on **genetic engineering**—the artificial manipulation of genes and their products (Figure 1.6). Genes from any source can be manipulated and modified using microorganisms and their enzymes as molecular tools. For instance, human insulin, a hormone found in abnormally low amounts in people with the disease diabetes, can be produced microbiologically from a human insulin gene engineered into a microorganism. And now, using genomic techniques (∞Chapter 15), one can search hundreds of genetic blueprints for genes encoding proteins of interest, clone the genes into a suitable host, and then produce the proteins commercially.

The overwhelming influence of microorganisms in human society is clear. Indeed, we have many reasons to be aware of microorganisms and their activities (Figure 1.6). As the eminent French scientist Louis Pasteur, one of the founders of microbiology, expressed it: "The role of the infinitely small in nature is infinitely large." Before we begin our study of microbiology in earnest, let us briefly consider the contributions that Pasteur and other early microbiologists made to the development of the science of microbiology.

 ### 1.4 *Concept Check*

Microorganisms can be both beneficial and harmful to humans. Although we tend to emphasize harmful microorganisms (infectious disease agents), many more microorganisms in nature are beneficial than harmful.

◆ In what ways are microorganisms important in the food and agricultural industries?

◆ What fuels can be made by microorganisms?

◆ What is *biotechnology* and how might it improve the lives of humans?

 # II PATHWAYS OF DISCOVERY IN MICROBIOLOGY

Like any science, modern microbiology owes much to its past. Although claiming very early roots, the science of microbiology didn't really develop until the nineteenth century. Since that time, the field has exploded and spawned several new but related fields. We retrace these pathways of discovery now.

1.5 The Historical Roots of Microbiology: Hooke, van Leeuwenhoek, and Cohn

Although the existence of creatures too small to be seen with the naked eye had long been suspected, their discovery was linked to the invention of the microscope. Robert Hooke described the fruiting structures of molds in 1665 (Figure 1.8●), and by doing so was the first person to describe microorganisms. The first person to see bacteria was the Dutch draper and amateur microscope builder Antoni van Leeuwenhoek. In 1684, van Leeuwenhoek, who was aware of the work of Hooke, used extremely simple microscopes of his own construction to examine a variety of natural substances for their microbial content (Figure 1.9●).

Van Leeuwenhoek's microscopes were crude by today's standards, but by careful manipulation and focusing he was able to see organisms as small as bacteria. His discovery of bacteria occurred in 1676 while

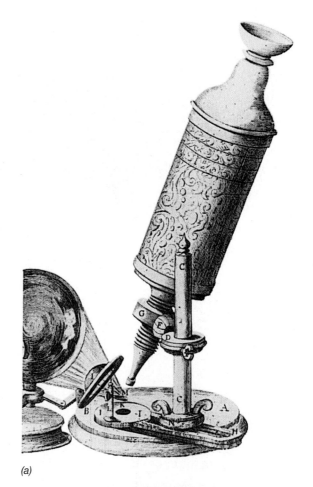

(a)

(b)

● **Figure 1.8 Early microscopy.** (a) A drawing of the microscope used by Robert Hooke in 1664. The objective lens was fitted at the end of an adjustable bellows (G), with illumination focused on the specimen by a single lens (1). (b) A drawing by Robert Hooke. This drawing, published in *Micrographia* in 1655, is the first description of a microorganism. The organism is a bluish-colored mold growing on the surface of leather. The round structures (sporangia) contain spores of the mold.

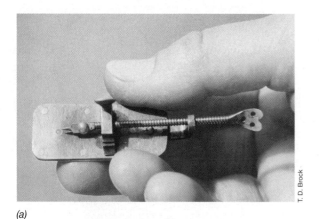

(a)

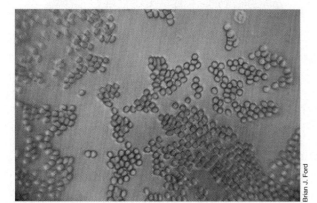

(b)

(c)

● **Figure 1.9   The van Leeuwenhoek microscope.** (a) Photograph of a replica of van Leeuwenhoek's microscope. The lens is mounted in the brass plate adjacent to the tip of the adjustable focusing screw. (b) van Leeuwenhoek's drawings of bacteria, published in 1684. Even from these relatively crude drawings we can recognize several morphological types of common bacteria. A, C, F, and G, rod-shaped; E, spherical or coccus-shaped; H, cocci packets (∞ Figure 4.11). (c) Photomicrograph of a human blood smear taken through a van Leeuwenhoek microscope. Red blood cells are clearly apparent.

studying pepper-water infusions. van Leeuwenhoek reported his observations in a series of letters to the Royal Society of London, which published them in 1684 in English translation. Drawings of some of van Leeuwenhoek's "wee animalcules," as he referred to them, are shown in Figure 1.9*b*.

As years went by, van Leeuwenhoek's observations were confirmed by others, but progress in understanding the nature and importance of these tiny organisms remained slow for nearly the next 150 years. Only in the nineteenth century did improved microscopes become widely distributed, and about this time the extent and nature of microbial life forms became more apparent.

In the mid- to late nineteenth century major advances in the new science of microbiology occurred, primarily because of a focus on two major questions that pervaded biology and medicine at the time: *spontaneous generation* and *the nature of infectious disease*. Answers to these seminal questions were provided by two giants in the fledgling field of microbiology: the French chemist *Louis Pasteur* and the German physician *Robert Koch*. But before we explore their work, let us briefly consider the work of a German botanist, *Ferdinand Cohn*, a contemporary of Pasteur and Koch, and the founder of the field we now call *bacteriology*.

Ferdinand Cohn and the Science of Bacteriology

Ferdinand Cohn (1828–1898) was trained as a botanist, and his interests in microscopy naturally led him toward the study of unicellular plants—the algae—and later to photosynthetic bacteria. Cohn believed that all bacteria, even those lacking photosynthetic pigments, were members of the plant kingdom, and his microscopic studies gradually drifted away from plants and algae to a variety of different bacteria, including the large sulfur bacterium *Beggiatoa* (Figure 1.10●).

Cohn became particularly interested in heat-resistant forms of bacteria, which led him to discover the genus *Bacillus* and the process of endospore formation. We now know that bacterial endospores are extremely heat-resistant. Cohn described the entire life cycle of *Bacillus* (vegetative cell → endospore → vegetative cell; ∞ Section 4.15) and discovered that vegetative cells but not endospores are killed by boiling. Indeed, Cohn's findings helped explain why earlier scientists, such as John Tyndall, had found boiling to be an unreliable means of sterilization.

Cohn continued to work with bacteria until his retirement. During this time he laid the groundwork for a scheme of bacterial classification and founded a major scientific journal. Cohn was also a strong advocate of the techniques and research of the first medical microbiologist, Robert Koch. Cohn also is credited with helping devise simple but very effective methods for preventing the contamination of sterile culture media, such as the use of cotton for closing flasks and tubes. These methods

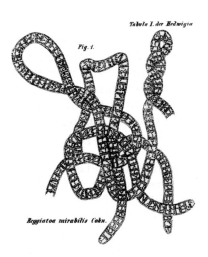

● **Figure 1.10** **Drawing by Ferdinand Cohn made in 1866 of the filamentous sulfur-oxidizing bacterium *Beggiatoa mirabilis*.** The small granules inside the cell consist of elemental sulfur, produced from the oxidation of hydrogen sulfide (H_2S). Cohn was the first to identify the granules as sulfur.

were later used by Koch and allowed him to make rapid progress in the isolation and characterization of several disease-causing bacteria (see Section 1.6).

 1.5 Concept Check

Robert Hooke was the first to describe microorganisms and Antoni van Leeuwenhoek was the first to describe bacteria. Ferdinand Cohn founded the field of bacteriology and discovered bacterial endospores.

◆ About what period was van Leeuwenhoek discovering bacteria? When was Cohn working on endospores?

◆ What prevented the science of microbiology from developing before the era of van Leeuwenhoek?

1.6 Pasteur, Koch, and Pure Cultures

The mid- to late nineteenth century saw the science of microbiology blossom. The idea of spontaneous generation was crushed, and pure culture microbiology came to the forefront.

Pasteur and the Downfall of Spontaneous Generation

The concept of **spontaneous generation** existed since biblical times. The basic idea of spontaneous generation can easily be understood. For example, if food is allowed to stand for some time, it putrefies. When the putrefied material is examined microscopically, it is found to be teeming with bacteria, and perhaps even higher organisms like maggots and worms. Where do

these organisms come from, since they are not apparent in the fresh food? Some people said they developed from seeds or germs that had entered the food from the air, whereas others said they arose spontaneously from nonliving materials. Keen insight was needed to solve this problem.

Louis Pasteur (1822–1895) was a powerful opponent of spontaneous generation. Pasteur first showed that microorganisms were present in air that closely resembled those seen in putrefying materials. Pasteur concluded that the organisms found in putrefying materials originated from microorganisms present in the air. He postulated that these cells are constantly being deposited on all objects and grow when conditions are favorable. Pasteur reasoned that if food were treated in such a way as to destroy all living organisms contaminating it, that is, rendered **sterile**, and then protected from further contamination, it should not putrefy.

Since it had already been established that heat effectively kills microorganisms, Pasteur used heat to eliminate contaminants. In fact, other workers had shown that when a nutrient solution was sealed in a glass flask and heated to boiling, it did not support microbial growth. Killing all the bacteria or other microorganisms in or on objects is a process we now call **sterilization**.

Proponents of spontaneous generation criticized such experiments by declaring that fresh air was necessary for spontaneous generation. Boiling, so they claimed, in some way affected the air in the sealed flask so that it could no longer support spontaneous generation. In 1864 Pasteur side-stepped this objection simply and brilliantly by constructing a swan-necked flask, now called a *Pasteur flask* (Figure 1.11●). In such a flask nutrient solutions could be heated to boiling. However, after the flask was cooled, air could reenter, but bends in the neck (the "swan neck" design) prevented particulate matter (containing microbial contaminants) from entering the main body of the flask and growing.

Broth sterilized in a Pasteur flask did not putrefy, and microorganisms never appeared in the flask as long as the neck did not contact the sterile liquid. If, however, the flask was tipped to allow the sterile liquid to contact the contaminated neck of the flask (Figure 1.11*c*), putrefaction occurred and the liquid soon teemed with microorganisms. This simple experiment effectively settled the controversy surrounding the theory of spontaneous generation, and the science of microbiology was able to move ahead on firm footing. Not incidentally, Pasteur's work also led to the development of effective sterilization procedures, which were eventually refined and carried over into both basic and applied research. Food science also owes a debt to Pasteur, as his principles are applied today in the canning and preservation of many foods (*pasteurization*).

Pasteur went on to many other triumphs in microbiology and medicine. Chief among these was his development of vaccines for the diseases anthrax, fowl

cultures. Since not all organisms grow on potato slices, however, Koch devised more uniform and reproducible nutrient solutions solidified with gelatin and later, agar (see the Microbial Sidebar).

A Test of Koch's Postulates: Tuberculosis

Koch's greatest accomplishment in medical bacteriology was his discovery of the causative agent of tuberculosis. At the time Koch began this work (1881), one-seventh of all reported human deaths were caused by tuberculosis (Figure 1.7). There was strong evidence that tuberculosis was a contagious disease, but the suspected causal organism had never been seen, either in diseased tissues or in culture. Koch was determined to demonstrate the causal agent of tuberculosis, and to this end he brought together all of the methods he had so carefully developed in his previous studies: microscopy, staining of tissues, pure culture isolation, and an animal model system.

As is now well known, the bacterium *Mycobacterium tuberculosis* is very difficult to stain because of large amounts of waxy lipid present in its cell wall. But Koch devised a staining procedure for *M. tuberculosis* in tissue samples using alkaline methylene blue in conjunction with a second stain (bismarck brown) that stained only the tissue (Koch's method was the forerunner of the acid-fast stain used today for staining bacteria such as *M. tuberculosis*; ⬡⬡Section 12.23). Using this method, Koch observed bright-blue, rod-shaped cells of *M. tuberculosis* in tuberculous tissues, the tissue itself staining a light brown (Figure 1.13●). However, from his previous work on anthrax, Koch realized that simply *identifying* an organism associated with tuberculosis was not enough. He knew he must *culture* the organism in order to prove that it was the specific cause of tuberculosis.

Producing cultures of *M. tuberculosis* was not easy, but eventually Koch was successful in growing colonies of this organism on a medium containing coagulated blood serum. Later he used agar, which had just been introduced as a solidifying agent (see the Microbial Sidebar). Under the best of conditions, *M. tuberculosis* grows slowly in culture, but Koch's persistence and patience eventually led to pure cultures of this organism from a variety of human and animal sources.

From here it was relatively easy to use his postulates to obtain definitive proof that the organism was the cause of the disease tuberculosis. Guinea pigs can be readily infected with *M. tuberculosis* and eventually succumb to systemic tuberculosis. Koch showed that diseased guinea pigs contained masses of *M. tuberculosis* cells in their tissues and that pure cultures obtained from such animals transmitted the disease to uninfected animals. Thus, Koch successfully satisfied

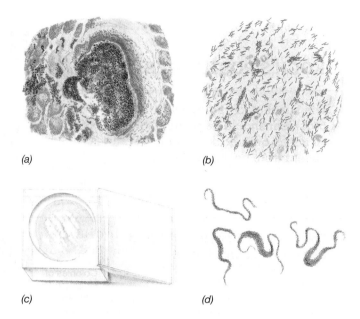

(a)   (b)

(c)   (d)

● **Figure 1.13 Robert Koch's drawings of cells of *Mycobacterium tuberculosis* in tissues and in laboratory culture.** Robert Koch was the first to isolate *M. tuberculosis* and show it to cause tuberculosis. (a) Section through a tubercle from lung tissue. Cells of *M. tuberculosis* stain blue, whereas the lung tissue stains brown. (b) Cells of *M. tuberculosis* in a sputum sample of a tuberculous patient. (c, d) Growth of *M. tuberculosis* in pure culture. (c) Growth of *M. tuberculosis* on a glass plate of coagulated blood serum inside a glass box (with lid open). (d) A colony of *M. tuberculosis* cells taken from the plate in (c) and observed microscopically at 700×; cells appear as long "cordlike" forms (compare with Figure 12.70*b*). Original drawings appeared in Koch, R. 1884. "Die Aetiologie der Tuberkulose." *Mittheilungen aus dem Kaiserlichen Gesundheitsamte* 2:1–88.

all four of his postulates (Figure 1.12), and the cause of tuberculosis was understood. For his contributions on tuberculosis, Robert Koch was awarded the 1905 Nobel Prize for physiology or medicine.

 1.6 Concept Check

Louis Pasteur's work on spontaneous generation led to the development of methods for controlling the growth of microorganisms. Robert Koch developed criteria for the study of infectious microorganisms and developed the first methods for growth of pure cultures of microorganisms.

◆ How did Pasteur's famous experiment defeat the theory of spontaneous generation?

◆ How can Koch's postulates prove cause and effect in a disease?

◆ What advantages do solid media offer for the culture of microorganisms?

Microbial Sidebar ◆ Solid Media, the Petri Plate, and Pure Cultures

Robert Koch was the first to grow bacteria on solid culture media. Koch's early use of potato slices as solid media was fraught with problems. Besides being rather selective in terms of which bacteria would grow on the slices, the slices would frequently get overgrown with fungi. Koch thus needed a more reliable and reproducible means of growing bacteria on solid media, and he found the answer in agar.

Koch initially employed gelatin as a solidifying agent for the various nutrient fluids he used to culture pathogenic bacteria and developed a method for preparing horizontal slabs of solid media that were kept free of contamination by covering them with a bell jar or glass box (see Figure 1.13c). Nutrient gelatin was a good culture medium for the isolation and study of various bacteria, but it had several drawbacks, the most important being that it did not remain solid at 37°C, the optimum temperature for growth of most human pathogens. Thus, a more versatile solidifying agent was needed, and this turned out to be agar.

Agar is a polysaccharide derived from red algae. It was used widely in the nineteenth century as a gelling agent. Walter Hesse first used agar as a solidifying agent for bacteriological culture media. The actual suggestion that agar be used instead of gelatin was made by Hesse's wife, Fannie. Fannie Hesse had used agar in the preparation of fruit jellies, and when it was tried as a solidifying agent in nutrient media, its superior qualities were immediately evident. Hesse wrote to Koch about this discovery, and Koch quickly adapted agar to his own studies, including his classic studies on the isolation of the bacterium *Mycobacterium tuberculosis*, the cause

of the disease tuberculosis (see text and Figure 1.13).

Agar had many other properties that made it desirable as a gelling agent for microbial culture media. In particular, agar remained solid at body temperature, and after melting during the sterilization process, remained liquid to about 45°C, at which time it could be poured into sterile vessels. In addition, unlike gelatin, which many bacteria can hydrolyze causing the medium to liquify, agar is not degraded by the vast majority of bacteria. Hence, agar found its place early in the annals of microbiology and is still used today for obtaining and maintaining pure cultures of bacteria.

In 1887 Richard Petri published a brief paper describing a modification of Koch's flat plate technique. Petri's enhancement, which turned out to be amazingly useful, was the development of the double-sided dishes that bear his name. The advantages of Petri dishes were immediately apparent. They could easily be stacked and sterilized separately from the medium, and, following the addition of molten medium to the smaller of the two double dishes, the larger dish could be used as a cover to prevent contamination. Colonies that formed on the surface of the agar in the Petri dish remained fully exposed to air and could easily be manipulated for further study. The original idea of Petri has not been improved on to this day, as the Petri dish, made either of reusable glass and sterilized by dry heat or of disposable plastic and sterilized by ethylene oxide (a gaseous sterilant), is a mainstay of the microbiology laboratory.

Finally, it should also be noted that Koch was keenly aware of the implications his pure culture methods had for the study of microbial

systematics. Koch observed that different colonies (differing in color, morphology, size, and the like, see Fig. 1) developed on solid media exposed to a contaminated object, and that these colonial forms bred true and could be distinguished from one another by their colony characteristics. Cells from different colonies also differed microscopically and often in their temperature or nutrient requirements as well. Koch realized that these differences among microorganisms met all the requirements that taxonomists had established for the classification of larger organisms, such as plant and animal species. In Koch's own words (translated from the German): "*All bacteria which maintain the characteristics which differentiate one from another when they are cultured on the same medium and under the same conditions, should be designated as species, varieties, forms, or other suitable designation.*" Koch also realized from the study of pure cultures that one could show that specific organisms have specific effects, not only in causing disease, but in other capacities as well. Such insightful thinking was significant in the relatively rapid acceptance of microbiology as an independent biological science.

Koch's discovery of solid culture media and his emphasis on pure culture microbiology reached far beyond the realm of medical bacteriology. His discoveries supplied critically needed tools for development of the fields of bacterial taxonomy, genetics, and several other subdisciplines. Indeed, the entire field of microbiology owes much to Robert Koch and his associates for the intuition they displayed in grasping the significance of pure cultures and developing some of the most basic methods in microbiology. ■

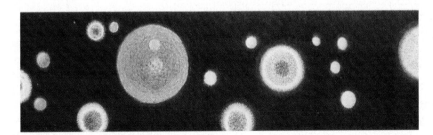

Figure 1 *A hand-colored photograph of colonies formed on agar taken by Walter Hesse, an associate of Robert Koch. The colonies include those of fungi (molds) and bacteria and were obtained during studies Hesse initiated on the microbiological content of air in Berlin, Germany, in 1882.* From Hesse, W. 1884. "Ueber quantitative Bestimmung der in der Luft enthaltenen Mikroorganismen," in Struck (ed.), *Mittheilungen aus dem Kaiserlichen Gesundheitsamte.* August Hirschwald.

1.7 Microbial Diversity and the Rise of General Microbiology

As microbiology advanced from the nineteenth to the twentieth century, our understanding of microbial diversity improved significantly. During this period several subdisciplines arose in microbiology, leading to the present era of "molecular microbiology."

Two giants in the field led this transition, the Dutchman Martinus Beijerinck and the Russian Sergei Winogradsky. Both of these early microbiologists were interested in bacteria that inhabit soil and water, and both are remembered primarily for their contributions to bacterial diversity. This period and the years thereafter until the era of molecular biology was ushered in were the heyday of *general microbiology*. The latter deals primarily with the nonmedical aspects of microbiology, with an emphasis on the diversity and physiology of microorganisms.

Martinus Beijerinck and the Enrichment Culture Technique

Martinus Beijerinck (1851–1931) was a professor at the Delft Polytechnic School in Holland but was originally trained in botany. He thus began his career in microbiology studying the microbiology of plants. Perhaps Beijerinck's greatest contribution to the field of microbiology was his clear formulation of the **enrichment culture**. In the enrichment culture, microorganisms are isolated from natural samples in a *selective* fashion, paying careful attention to nutrient and incubation requirements (∞Section 18.1).

Using his enrichment culture technique, Beijerinck isolated the first pure cultures of many soil and aquatic microorganisms, including aerobic nitrogen-fixing bacteria (Figure 1.14●), sulfate-reducing and sulfur-oxidizing bacteria, nitrogen-fixing root nodule bacteria, *Lactobacillus* species, green algae, and many other microorganisms. Using selective filter techniques in his studies of tobacco mosaic disease, Beijerinck showed that the infectious agent (a virus) was not bacterial, but somehow became incorporated into the cells of the host plant and required the living plant to reproduce. In this insightful work, Beijerinck described not only the first virus, but also the basic tenets of virology.

Sergei Winogradsky and the Concept of Chemolithotrophy

Sergei Winogradsky (1856–1953) had scientific interests similar to Beijerinck's and was also successful in isolating several key bacteria for the first time. Winogradsky was particularly interested in bacteria involved in the cycling of nitrogen and sulfur compounds (Figures 1.15 and 1.16●). He showed in this work that bacteria can be important biogeochemical agents. Moreover, Winogradsky's keen insight revealed the *metabolic significance* of biogeochemical processes. For example, from his studies of the sulfur-oxidizing bacteria, Winogradsky proposed the concept of **chemolithotrophy**, the oxidation of *inorganic* compounds linked to energy conservation (Figure 1.16a). And from his studies of the chemolithotrophic process of *nitrification* (the oxidation of ammonia to nitrate), Winogradsky concluded that the organisms responsible—the nitrifying bacteria—

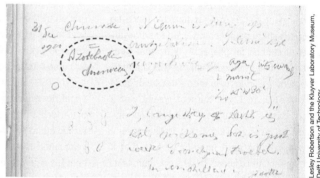

(a)

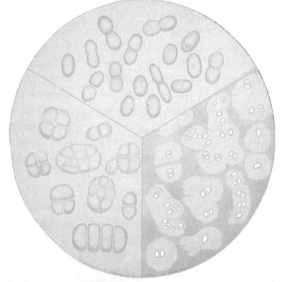

● Figure 1.14 Martinus Beijerinck and *Azotobacter.* (a) Portion of a page from the laboratory notebook of M. Beijerinck dated December 31, 1900, that describes his observations on the aerobic nitrogen-fixing bacterium *Azotobacter chroococcum* (name circled in red). It is on this page that Beijerinck uses this name for the first time. Compare Beijerinck's drawings of pairs of *A. chroococcum* cells with a photomicrograph of cells of *Azotobacter* shown in Figure 12.19a. **(b)** A painting by M. Beijerinck's sister, Henriëtte Beijerinck, showing cells of *Azotobacter chroococcum*. Beijerinck used such paintings to illustrate his lectures, because this was long before the days of the overhead, slide, and computer projectors used in lectures today.

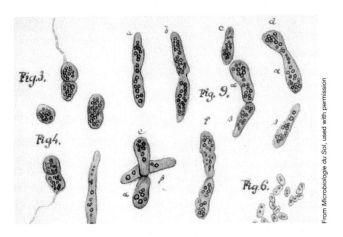

• **Figure 1.15 Hand-colored drawings of cells of purple sulfur phototrophic bacteria.** The original drawings were made by Sergei Winogradsky about 1887 and then copied and hand-colored by his wife Hélène. These drawings included cells of the genus *Chromatium*, such as *C. okenii* (Figs. 3 and 4). This species is still recognized today. Compare with a photomicrograph of cells of *C. okenii* shown in Figure 12.4*a*.

 From Microbiologie du Sol, used with permission

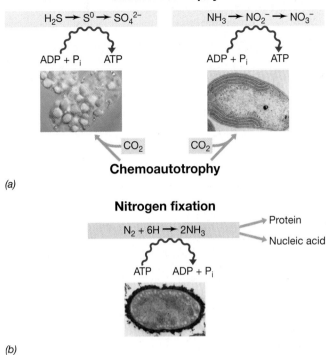

(a)

(b)

• **Figure 1.16 Major concepts developed by Sergei Winogradsky.** (a) Chemolithotrophy and chemoautotrophy. Oxidation of the sulfur or nitrogen compounds yield energy (ATP), while cell carbon is obtained from CO_2. Photos, left, *Achromatium*; right, *Nitrobacter*. (b) Nitrogen fixation. This process consumes ATP but allows the cell to use gaseous nitrogen for all of its nitrogen needs. Photo, *Azotobacter*, an aerobic nitrogen-fixer (see also Figure 1.14).

obtained their carbon from CO_2. Winogradsky thus proposed that these organisms were autotrophs, now called *chemoautotrophs*, to distinguish them from autotrophic organisms that are phototrophic (Figure 1.16*a*).

Using an enrichment method, Winogradsky also isolated the first nitrogen-fixing bacterium (the anaerobe *Clostridium pasteurianum*) and formulated the concept of bacterial N_2 fixation (Figure 1.16*b*). Winogradsky lived to be almost 100, publishing many scientific papers, along with a major monograph, *Microbiologie du Sol* (*Soil Microbiology*). The latter work, a milestone in microbiology, contained original drawings of many of the organisms he had isolated or otherwise studied in enrichment culture or natural material during his career (Figure 1.15).

Table 1.1 summarizes some of the important discoveries in the field of microbiology, from van Leeuwenhoek to the present.

1.7 Concept Check

Beijerinck and Winogradsky studied bacteria in soil and water and developed the enrichment culture technique for the isolation of representatives of various physiological groups. Major new concepts in microbiology emerged during this period, including enrichment cultures, chemolithotrophy, chemoautotrophy, and nitrogen fixation.

◆ What is the enrichment culture technique?

◆ In examining Figure 1.16, describe why sulfur oxidation and nitrification are considered chemolithotrophic processes while nitrogen fixation is not. (*Hint:* Look at the reactions involving ATP in each case.)

1.8 The Modern Era of Microbiology

In the twentieth century, the field of microbiology developed rapidly in two distinct directions—applied microbiology and basic microbiology.

Development of the Major Subdisciplines of Applied Microbiology

The practical advances of Koch led to extensive developments in applied microbiology. For example, the fields of *medical microbiology* and *immunology* emerged. Along with these disciplines came the discovery of many new bacterial pathogens and elucidation of the mechanisms by which these pathogens infect the body and are resisted by the body's defenses. Other early practical advances, bolstered by the discoveries of Beijerinck and Winogradsky, were in the field of *agricultural microbiology*, which led to an understanding of microbial processes in the soil, such as nitrogen fixation, that benefit plant growth. Later in the

Table 1.1 Three hundred years of microbiology: Some key papers in microbiology, 1684–2000[a]

Year	Investigator(s)	Discovery
1684	Antoni van Leeuwenhoek	Discovery of bacteria
1798	Edward Jenner	Smallpox vaccination
1857	Louis Pasteur	Microbiology of lactic acid fermentation
1860	Louis Pasteur	Role of yeast in alcoholic fermentation
1864	Louis Pasteur	Settled spontaneous generation controversy
1867	Robert Lister	Antiseptic principles in surgery
1876	Ferdinand Cohn	Discovery of endospores
1881	Robert Koch	Methods for study of bacteria in pure culture
1882	Robert Koch	Discovery of cause of tuberculosis
1882	Élie Metchnikoff	Phagocytosis
1884	Robert Koch	Koch's postulates
1884	Christian Gram	Gram-staining method
1885	Louis Pasteur	Rabies vaccine
1889	Sergei Winogradsky	Concept of chemolithotrophy
1889	Martinus Beijerinck	Concept of a virus
1890	Emil von Behring and Shibasaburo Kitasato	Diphtheria antitoxin
1890	Sergei Winogradsky	Autotrophic growth of chemolithotrophs
1901	Martinus Beijerinck	Enrichment culture method
1901	Karl Landsteiner	Human blood groups
1908	Paul Ehrlich	Chemotherapeutic agents
1911	Francis Rous	First cancer virus
1915 / 1917	Frederick Twort / Felix d'Hérelle	Discovery of bacterial viruses (bacteriophage)
1928	Frederick Griffith	Discovery of pneumococcus transformation
1929	Alexander Fleming	Discovery of penicillin
1931	Cornelius van Niel	H_2S (sulfide) as electron donor in anoxygenic photosynthesis
1935	Gerhard Domagk	Sulfa drugs
1935	Wendall Stanley	Crystallization of tobacco mosaic virus
1941	George Beadle and Edward Tatum	One gene–one enzyme hypothesis
1943	Max Delbruck and Salvador Luria	Inheritance of genetic characters in bacteria
1944	Oswald Avery, Colin Macleod, Maclyn McCarty	Explanation of Griffith's work—DNA is genetic material
1944	Selman Waksman and Albert Schatz	Discovery of streptomycin
1946	Edward Tatum and Joshua Lederberg	Bacterial conjugation
1951	Barbara McClintock	Discovery of transposable elements
1952	Joshua Lederberg and Norton Zinder	Bacterial transduction
1953	James Watson, Francis Crick, Rosalind Franklin	Structure of DNA
1959	Arthur Pardee, François Jacob, Jacques Monod	Gene regulation by a repressor protein
1959	Rodney Porter	Immunoglobulin structure
1959	F. Macfarlane Burnet	Clonal selection theory
1960	François Jacob, David Perrin, Carmon Sanchez, Jacques Monod	Concept of an operon
1960	Rosalyn Yalow and Solomon Bernson	Development of radioimmunoassay (RIA)
1961	Sydney Brenner, François Jacob, and Matthew Meselson	Messenger RNA and ribosomes as the site of protein synthesis
1966	Marshall Nirenberg and H. Gobind Khorana	Discovery of the genetic code
1967	Thomas Brock	Discovery of bacteria growing in boiling hot springs
1969	Howard Temin, David Baltimore, Renato Dulbecco	Discovery of retroviruses/reverse transcriptase
1969	Thomas Brock and Hudson Freeze	Isolation of *Thermus aquaticus*, source of *Taq* DNA polymerase
1970	Hamilton Smith	Specificity of action of restriction enzymes
1973	Stanley Cohen, Annie Chang, Robert Helling, and Herbert Boyer	Recombinant DNA
1975	Georges Kohler, Cesar Milstein	Monoclonal antibodies
1976	Susumu Tonegawa	Rearrangement of immunoglobulin genes
1977	Carl Woese and George Fox	Discovery of the *Archaea*
1977	Fred Sanger, Steven Niklen, Alan Coulson	Methods for sequencing DNA
1981	Stanley Prusiner	Characterization of prions
1982	Karl Stetter	Isolation of first prokaryote with temperature optimum >100°C
1983	Luc Montagnier	Discovery of HIV, the cause of AIDS
1985	Kary Mullis	Invention of the polymerase chain reaction (PCR)
1986	Norman Pace	Molecular microbial ecology
1992	Jed Fuhrman and Edward DeLong	Discovery of marine *Archaea*
1995	Craig Venter and Hamilton Smith	Complete sequence of a bacterial genome
1999	The Institute for Genomic Research (TIGR), and others	Over 100 microbial genomes sequenced or in progress
2000	Edward DeLong	Discovery of proteorhodopsin
2004	Craig Venter and others	First large scale environmental genome: the Sargasso Sea

[a] Major reference sources here include Brock, T. D. (1961), *Milestones in Microbiology*, Prentice Hall, Englewood Cliffs, NJ; Brock, T. D. (1990). *The Emergence of Bacterial Genetics*, Cold Spring Harbor Press, Cold Spring Harbor, NY. *Year* refers to the year in which the discovery was published.

twentieth century, studies of soil microorganisms led to the discovery of antibiotics and other important chemicals. This led to the field of *industrial microbiology*, the large-scale growth of microorganisms for the production of commercial products.

Advances in soil microbiology also provided the foundation for studies on microorganisms in lakes, rivers, and the oceans—the fields of *aquatic microbiology* and *marine microbiology*. One branch of aquatic microbiology deals with the important processes of treating sewage and providing safe water for humans. As interest in the biodiversity and activities of microorganisms in their natural environments grew, the field of *microbial ecology* emerged in the 1960s and 1970s. Microbial ecology is now enjoying a second "golden era" with the influx of molecular biology into the discipline (Figure 1.17●).

Basic Science Subdisciplines in Microbiology

In addition to advances in *applied* areas of microbiology, the twentieth century saw many new *basic science* areas develop, particularly those employing the concepts or tools of molecular biology. Some of the landmarks in the past 65 years are summarized in Figure 1.17.

Since World War II many new microorganisms have been discovered and classified, resulting in considerable refinement of *microbial systematics* and construction of the phylogenetic tree of life (Figure 1.17). Study of the nutrients that microorganisms require and the products that they make has advanced the field of *microbial physiology*. Enhanced understanding of the physical and chemical structure of microorganisms (*cytology*) and the discovery of microbial enzymes and the chemical reactions they carry out (*microbial biochemistry*), have also influenced how microbiology is practiced today.

A key area of basic research that moved forward rapidly in the mid-twentieth century was the study of heredity and variation in bacteria, the discipline of *bacterial genetics*. Although some aspects of bacterial genetics were known early in the twentieth century, it was not until the discovery of genetic exchange in bacteria around 1950 that bacterial genetics became a major field of study. Bacterial genetics, biochemistry, and physiology developed mainly during the 1950s. By the early 1960s, these fields had provided an advanced understanding of DNA, RNA, and protein synthesis. The field of *molecular biology* arose to a great extent from these bacterial studies (Figure 1.17).

The study of viruses also blossomed in the twentieth century. Although Beijerinck discovered the first virus over 100 years ago, it was not until the middle of the twentieth century that the true nature of viruses was understood. Much of this work involved the study of viruses that infect bacteria, called *bacteriophages*. Scientists realized that virus infection was analogous to genetic transfer, and the relationship between viruses and other genetic elements was worked out primarily from research on bacteriophages.

The Era of Molecular Microbiology

By the 1970s, our knowledge of bacterial physiology, biochemistry, and genetics had advanced to such an extent that it was possible to experimentally manipulate the genetic material of cells. With the discovery of restriction enzymes, it also became possible to introduce DNA from foreign sources into bacteria and control its replication. This led to development of the field of *biotechnology*. At about this same time, nucleic acid sequencing was developed, and its ramifications were felt in all areas of biology. In microbiology, sequencing technology helped to reveal phylogenetic (evolutionary) relationships among prokaryotes, which led to revolutionary new concepts in the field of biological classification. Sequencing also gave birth to the field of **genomics**—the comparative analysis of the genes of different organisms.

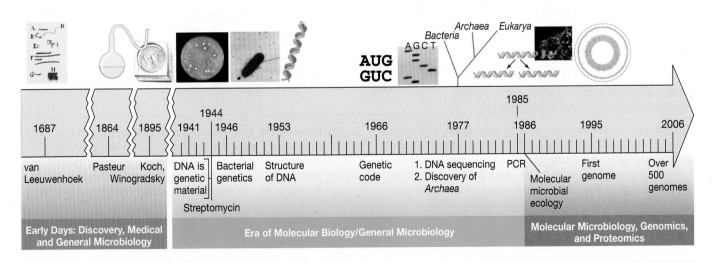

● **Figure 1.17   Some landmarks in microbiology in the past 65 years.** The icons or photos above the dates are symbolic of the discoveries; each has been covered here or will be revisited in later chapters.

The huge amounts of genomic information now in hand are fueling major advances in medicine, microbial ecology, industrial microbiology, and many other areas of biology. Indeed, the genomics era is upon us in full force (Figure 1.17) and has already given birth to a new subdiscipline, **proteomics**, the study of protein expression in cells. Both genomics and proteomics are considered in more detail in Chapter 15.

New Frontiers

Not only has nearly 350 years of microbiology brought us startling insight into the biology of microorganisms, it has brought new challenges, both good and bad. On the one hand, new emerging diseases, such as SARS, and before that AIDS, seem to appear without warning and have challenged even our most sophisticated understanding of microbial diseases. On the other hand, new research discoveries have put us on the doorstep of understanding how a cell works at the most fundamental of levels, and newly discovered bacteria stretch our already overwhelming picture of microbial diversity. With genomics poised to reveal the minimalist genome—the minimum complement of genes necessary for life—we may soon be able to precisely define the prerequisites for life. Can the laboratory creation of a cell be that far off?

As the late evolutionary biologist Stephen Jay Gould put it, this is the "age of bacteria." What an exciting time to be learning the science! Welcome to microbiology!

 1.8 Concept Check

In the middle to latter part of the twentieth century basic and applied microbiology worked hand-in-hand to usher in the current era of molecular microbiology.

◆ List the subdisciplines of microbiology whose focus is the following: metabolism; enzymology; nucleic acid and protein synthesis; microorganisms and their natural environments.

REVIEW QUESTIONS

1. List six key properties associated with the living state. Which of these are properties of *all* cells? Which are properties of only *some types* of cells (Sections 1.1 and 1.2)?

2. Cells can be thought of as both machines and coding devices. Explain how these two attributes of a cell differ (Section 1.2).

3. What is needed for translation to occur in a cell? What is the product of the translational process (Section 1.2)?

4. What is an ecosystem? Do microorganisms live in pure cultures in an ecosystem? What effects can microorganisms have on their ecosystems (Section 1.3)?

5. How would you convince a friend that microorganisms are much more than just agents of disease (Section 1.4)?

6. For what contributions are Hooke and van Leeuwenhoek remembered in microbiology? How did Ferdinand Cohn contribute to bacteriology (Section 1.5)?

7. What is a pure culture and how can one be obtained? Why was knowledge of how to obtain a pure culture important for development of the science of microbiology (Section 1.6)?

8. Explain the principle behind the use of the Pasteur flask in studies on spontaneous generation (Section 1.6).

9. Explain why the invention of solid culture media was of great importance to the development of microbiology as a science (Section 1.6).

10. What are Koch's postulates and how did they influence the development of microbiology? Why are they still relevant today (Section 1.6)?

11. Describe a major contribution to microbiology of the early microbiologist Martinus Beijerinck (Section 1.7).

12. What major concepts do we owe to Sergei Winogradsky (Section 1.7)?

13. What major advances in microbiology have occurred since World War II (Section 1.8)?

APPLICATION QUESTIONS

1. Pasteur's experiments on spontaneous generation were of enormous importance for the advance of microbiology, having an impact on the methodology of microbiology, ideas on the origin of life, and the preservation of food, to name just a few. Explain briefly how the impact of his experiments was felt on each of these topics.

2. Describe the various lines of proof Robert Koch used to definitively associate the bacterium *Mycobacterium tuberculosis* with the disease tuberculosis. How would his proof have been flawed if any of the tools he developed for studying bacterial diseases had not been available for his study of tuberculosis?

2

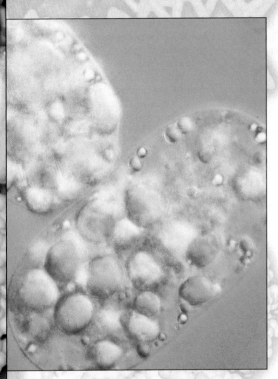

AN OVERVIEW OF MICROBIAL LIFE

The scope of microbial diversity is enormous, and microorganisms have exploited every means of making a living consistent with the laws of chemistry and physics.

WORKING GLOSSARY

Archaea phylogenetically related prokaryotes distinct from *Bacteria*

Bacteria phylogenetically related prokaryotes distinct from *Archaea*

Chemolithotroph an organism obtaining its energy from the oxidation of inorganic compounds

Chemoorganotroph an organism obtaining its energy from the oxidation of organic compounds

Chromosome a genetic element containing genes essential to cell function

Cytoplasm cellular contents inside the cytoplasmic membrane, excluding the nucleus (if present)

Domain the highest level of biological classification

Endosymbiosis the process by which mitochondria and chloroplasts originated from descendants of *Bacteria*

Eukarya the domain of life that includes all eukaryotic cells

Eukaryote a cell having a membrane-bound nucleus and usually other organelles

Evolution change in a line of descent over time leading to new species or varieties within a species

Extremophile an organism that grows optimally under one or more environmental extremes

Genome the complement of genes in an organism

Morphology cell shape

Nucleoid the aggregated mass of DNA that constitutes the chromosome of cells of *Bacteria* and *Archaea*

Nucleus a membrane-enclosed structure that contains the chromosomes in eukaryotic cells

Organelle a unit membrane-enclosed structure such as a mitochondrion or chloroplast present in the cytoplasm of eukaryotic cells

Phototroph an organism that obtains its energy from light

Phylogeny the evolutionary relationships between organisms

Plasmid an extrachromosomal genetic element nonessential for growth

Prokaryote a cell that lacks a membrane-enclosed nucleus and other organelles

Proteobacteria a large phylum of *Bacteria* that includes many of the common gram-negative bacteria, such as *Escherichia coli*

Ribosome a cytoplasmic particle that functions in protein synthesis

CELL STRUCTURE AND EVOLUTIONARY HISTORY

This chapter introduces key concepts of cell structure and function and microbial diversity that will carry through the remainder of this book. Here we compare the internal architecture of microbial cells, differentiate cells from viruses, explore the evolutionary tree of life, and consider some of the major groups of microorganisms that affect our lives and our planet.

Much of what we know about cell structure has come from microscopy, including studies with both light and electron microscopes. So in this chapter we present several light and electron *micrographs* (photographs taken under the microscope). However, we reserve formal discussion of the microscopes themselves until Chapter 4, using it as a prelude to a more detailed consideration of cell structure.

2.1 Elements of Cell and Viral Structure

All cells have much in common and contain many of the same or functionally same elements. All cells have a barrier that separates the inside from the outside called the **cytoplasmic membrane** (Figure 2.1●). It is through the cytoplasmic membrane that nutrients and other substances needed by the cell enter, and waste materials and other cell products exit. Within a cell, and bounded by the cytoplasmic membrane, is a complex mixture of substances and structures called the **cytoplasm**. These materials and structures, either dissolved or suspended in water, carry out the functions of the cell.

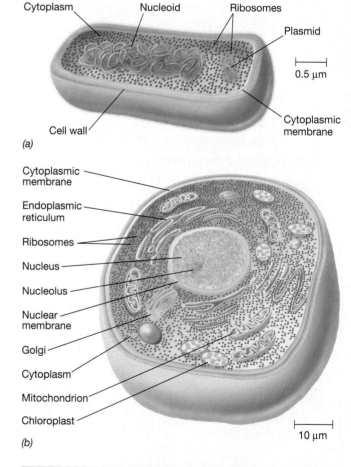

● **Figure 2.1 Internal structure of microbial cells.** (a) Diagram of a prokaryotic cell. (b) Diagram of a eukaryotic cell. Note differences in scale and internal structure.

The major components dissolved in the cytoplasm include **macromolecules** (two especially important classes of which are the *proteins* and *nucleic acids*), small organic molecules (mainly precursors of macromolecules), and various inorganic ions. **Ribosomes**—the cell's protein-synthesizing factories—are particulate structures composed of ribonucleic acid (RNA) and various proteins and are suspended in the cytoplasm. Ribosomes interact with several cytoplasmic proteins and messenger and transfer RNAs in the key process of *protein synthesis* (translation) (∞ Figure 1.4).

The **cell wall** gives structural strength to a cell. The cell wall is relatively permeable and located outside the membrane (Figure 2.1a); it is a much stronger layer than the membrane itself. Plant cells and most microorganisms have cell walls, while animal cells for the most part do not. Animal cells are instead reinforced by scaffolding within the cytoplasm called the *cytoskeleton*.

Eukaryotic Cells

Examination of the internal structure of cells reveals two structural types: the **prokaryote** and the **eukaryote** (Figure 2.1). Eukaryotic cells are generally larger and structurally more complex than prokaryotic cells. Eukaryotic microorganisms include **algae**, **fungi**, and **protozoa** (see Figures 2.23 and 2.24). All multicellular plants and animals are constructed of eukaryotic cells. We consider eukaryotic cells in more detail in Chapter 14.

A major feature of eukaryotic cells, absent from prokaryotic cells, is the presence of membrane-enclosed structures called **organelles.** These include, first and foremost, the *nucleus*, but also *mitochondria*, and *chloroplasts* (the latter in photosynthetic cells only) (Figures 2.1b, 2.2c●). The nucleus is the repository of the cells' genetic information (DNA, "the genome"), and is also the site of transcription in eukaryotic cells. Mitochondria and chloroplasts play specific roles in energy generation by carrying out respiration and photosynthesis, respectively.

Prokaryotic Cells

In contrast to eukaryotic cells, prokaryotic cells have a simpler internal structure, lacking membrane-enclosed organelles (Figures 2.1a and 2.2a, b). Prokaryotes consist of the **Bacteria** and the **Archaea**. Although species of *Bacteria* and *Archaea* share a prokaryotic cell structure, they differ dramatically in their evolutionary history. In this book, the term *bacteria*, written with a lower case "b," is synonymous with the term *prokaryote*. By contrast, the term *Bacteria* (written with a capital "B" and set in italics) refers to the phylogenetically related group of prokaryotes distinct from the *Archaea* (see Section 2.3).

In general, microbial cells are very small, particularly prokaryotic cells. For example, a rod-shaped prokaryote is typically about 1–5 micrometers (μm) long and about 1 μm wide (a micrometer is 10^{-6} of a meter) and thus is invisible to the naked eye. To conceive of how small a bacterium is, consider that 500 bacteria each 1 μm long could be placed end-to-end across the period at the end of this sentence. Eukaryotic cells are typically much larger than prokaryotic cells, but the range of sizes in eukaryotic cells can vary dramatically, from as small as three to several hundred micrometers in diameter. We revisit the subject of cell size in more detail in Chapter 4.

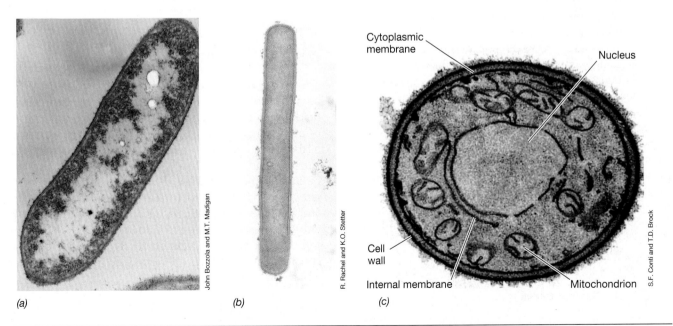

(a)   *(b)*   *(c)*

● **Figure 2.2** **Electron micrographs of sectioned cells from each of the domains of living organisms.** (a) *Heliobacterium modesticaldum* (*Bacteria*); the cell measures 1 × 3 μm. (b) *Methanopyrus kandleri* (*Archaea*); the cell measures 0.5 × 4 μm. [Reinhard Rachel and Karl O. Stetter, 1981. *Archives of Microbiology* 128:288–293. © Springer-Verlag Gmb H & Co. KG]. (c) *Saccharomyces cerevisiae* (*Eukarya*); the cell measures 8 μm in diameter.

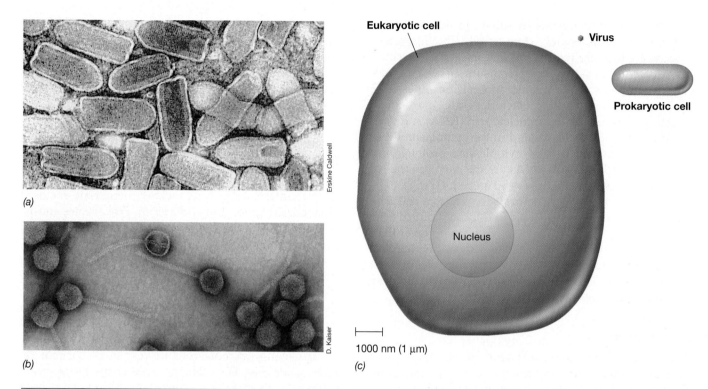

(a)

(b)

(c)

⊢————⊣
1000 nm (1 μm)

● **Figure 2.3 Virus structure and size comparisons of viruses and cells.** (a) Rhabdovirus (a virus of eukaryotes) particles. A single virus particle is about 65 nm (0.065 μm) in diameter. (b) Bacterial virus (bacteriophage) lambda. The head of each particle is about 65 nm in diameter. (c) The size of the viruses shown in (a) and (b) in comparison to a bacterial and eukaryotic cell.

Viruses

Viruses are a major class of microorganisms, but they are not cells (Figure 2.3●). Viruses lack many of the attributes of cells, the most important of which is that they are not dynamic open systems, taking in nutrients and expelling wastes. Instead, a virus particle is a static structure, quite stable and unable to change or replace its parts. Only when it infects a cell does a virus acquire the key attribute of a living system—reproduction. Unlike cells, viruses have no metabolic abilities of their own. And, although they contain their own genomes, viruses lack ribosomes and therefore depend totally on the cell's biosynthetic machinery for protein synthesis.

Viruses are known to infect all cells, including microbial cells. Many viruses cause disease in the organisms they infect. However, viral infection can have many profound effects on cells, including genetic alterations that can actually improve the capabilities of the cell. Viruses are also much smaller than cells, even much smaller than prokaryotic cells (Figure 2.3), with the smallest known viruses only some 10 nanometers (0.010 μm) in diameter.

⬡ 2.1 Concept Check

All microbial cells share certain basic structures in common such as a cytoplasmic membrane, ribosomes, and (usually) a cell wall. Two structural types of cells are recognized: the prokaryote and the eukaryote. Viruses are not cells but depend on cells for their replication.

◆ By looking inside a cell how could you tell if it was *prokaryotic* or *eukaryotic*?

◆ What important function do *ribosomes* play in cells?

◆ How long is a typical rod-shaped bacterial cell? How much larger are you than this single cell?

2.2 Arrangement of DNA in Microbial Cells

The living processes of all cells are governed by their complement of genes (the **genome**). In cells, a gene can be defined as a segment of DNA that encodes a protein (via messenger RNA) or another RNA molecule, such as a ribosomal RNA or transfer RNA. In Chapter 15 we will consider the rapid advances that have been made in sequencing and analyzing the genomes of living organisms, from bacteria to humans. These advances have yielded detailed genetic blueprints of hundreds of different organisms, and have allowed for extensive and rather revealing comparisons to be made. Here we just consider how genomes are organized in prokaryotic and eukaryotic cells.

Nucleus vs. Nucleoid

The genomes of prokaryotic and eukaryotic cells are organized differently. In prokaryotic cells, DNA is present in a large double-stranded molecule called the *bacterial chromosome*. The chromosome aggregates to form a visible

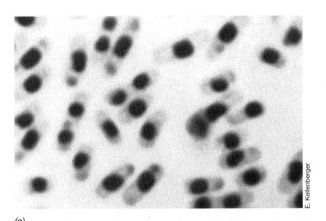

(a)

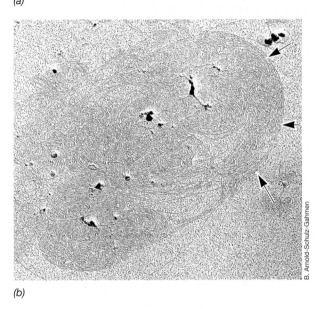

(b)

● **Figure 2.4 The nucleoid.** (a) Photomicrograph of cells of *Escherichia coli* treated in such a way as to make the nucleoid visible. A single cell is about 3 μm in length. (b) Electron micrograph of an isolated nucleoid released from a cell of *E. coli*. The cell was gently lysed to allow the highly compacted nucleoid to emerge intact. Arrows point to the edge of DNA strands.

mass called the **nucleoid** (Figure 2.4●). We will see in Chapter 7 that DNA is *circular* in most prokaryotes. Most prokaryotes have only a *single* chromosome. Because of this, they typically contain only a *single copy* of each gene and are therefore genetically *haploid*. Many prokaryotes also contain small amounts of circular extrachromosomal DNA called **plasmids**. Plasmids typically contain genes that confer special properties (such as unique metabolic properties) on a cell. This is in contrast to essential ("housekeeping") genes, which are needed for basic survival and are located on the chromosome.

In eukaryotes, DNA is present in linear molecules within the nucleus, packaged and organized in **chromosomes**. Chromosome number varies with the organism. For example, the baker's yeast *Saccharomyces cerevisiae* contains 16 chromosomes arranged in 8 pairs, while human cells contain 46 (23 pairs). Chromosomes in eukaryotes contain more than just DNA, however. They

also include proteins that assist in folding and packing the DNA and other proteins that are required for gene expression. A key genetic difference between prokaryotes and eukaryotes is that eukaryotes typically contain *two copies* of each gene and are thus genetically *diploid*. During cell division in eukaryotic cells the nucleus divides (following a doubling of chromosome number) in the process called **mitosis** (Figure 2.5●). Two identical daughter cells result, and each daughter cell receives a nucleus with a full complement of genes.

The diploid genome of eukaryotic cells is halved in the process of **meiosis** to form haploid gametes for sexual reproduction. Fusion of two gametes during zygote formation restores the cell to the diploid state. We discuss these processes in more detail in Chapter 14.

Genes, Genomes, and Proteins

How many genes and proteins does a cell have? The genome of *Escherichia coli*, a typical bacterium, is a single circular chromosome of 4.68 million base pairs of DNA. Because the *E. coli* genome has been completely sequenced, we also know that it contains about 4,300 genes. The genomes of some bacterial cells have nearly three times this number of genes while the genomes of others have fewer than one-eighth this number. Eukaryotic cells typically have much larger genomes than prokaryotes. A human cell, for example, contains over 1,000 times as much DNA as a cell of *E. coli* and about 7 times as many genes (we will see later that much of the DNA in eukaryotic cells is noncoding DNA).

A single cell of *E. coli* contains about 1,900 *different kinds* of proteins and about 2.4 million *total* protein molecules (⊙⊙ Table 3.2). Some proteins in *E. coli* are very abundant, others are only moderately abundant, and some are present in only one or a very few copies. Thus,

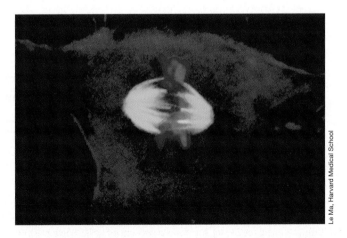

● **Figure 2.5 Mitosis in stained kangaroo rat cells.** The cell was photographed while in the *metaphase* stage of mitotic division. The green color stains a protein called *tubulin*, important in pulling chromosomes apart (⊙⊙ Section 14.5). The blue color is from a DNA-binding dye and shows the chromosomes. Although an integral part of the cell cycle of eukaryotic cells, mitosis does not occur in prokaryotic cells.

E. coli has mechanisms for *controlling the expression* of its genes so that not all genes are expressed (transcribed and translated, ⟳ Figure 1.4) to the same extent or at the same time. This is a common observation in all cells—prokaryotic and eukaryotic—and we focus on the major mechanisms of gene expression in Chapter 8.

 2.2 Concept Check

Genes govern the properties of cells, and a cell's complement of genes is called its genome. DNA is arranged in cells to form chromosomes. In prokaryotes there is usually a single circular chromosome, while in eukaryotes, several linear chromosomes exist.

◆ Differentiate between the *nucleus* and the *nucleoid*.

◆ How do *plasmids* differ from *chromosomes*?

◆ Why does it make sense that a human cell would have more genes than a bacterial cell?

2.3 The Tree of Life

Evolution is the change in a line of descent over time leading to new species or varieties. Is cell structure an evolutionary determinant? The answer to this question is both "yes" and "no." The evolutionary relationships between life forms are the subject of the science of **phylogeny**. On the one hand, we can say that all known prokaryotic cells are phylogenetically distinct from eukaryotic cells. But on the other hand, not all prokaryotic cells are closely related in an evolutionary sense; *Bacteria* and *Archaea* are themselves phylogenetically distinct.

Phylogenetic relationships can be deduced by comparing sequences of certain macromolecules. For reasons to be discussed in Chapter 11, macromolecules that form the ribosome, in particular *ribosomal RNAs*, are excellent tools for determining evolutionary relationships. And because all cells contain ribosomes (and thus ribosomal RNA), this molecule can and has been used to construct a phylogenetic tree of all life forms, prokaryotic and eukaryotic (see Figure 2.7). Recognition of ribosomal RNA as a tool for constructing phylogenetic relationships was first made by Carl Woese, an American microbiologist.

The steps involved in generating an RNA-based phylogenetic tree are outlined in Figure 2.6●. In brief, genes encoding ribosomal RNA from two or more organisms are sequenced and the sequences aligned and inspected, base by base, in a computer. The greater the difference in ribosomal RNA gene sequence between two or more organisms, the greater their evolutionary distance. These distances are then expressed in the form of a phylogenetic tree (Figure 2.6).

The Three Domains of Life

From comparative ribosomal RNA sequencing, three phylogenetically distinct lineages of cells have been identified. Two of these lineages contain only prokaryotic cells, while the third is composed of eukaryotes. The lineages, called **domains**, are the *Bacteria*, the *Archaea*, and the *Eukarya* (eukaryotes) (Figure 2.7●). The domains are thought to have diverged from a common ancestral organism or community of organisms early in the history of life on Earth (⟳ Section 11.7).

Besides clearly showing that all prokaryotes are *not* phylogenetically closely related, the tree of life reveals

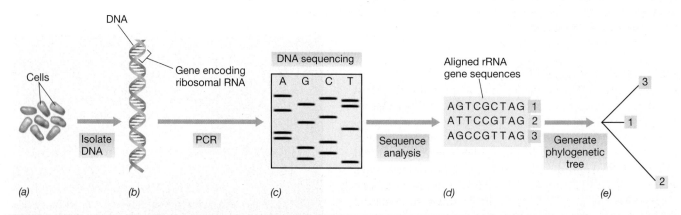

● **Figure 2.6 Ribosomal RNA (rRNA) gene sequencing and phylogeny.** (a) Cells, either from a pure culture or from an environmental sample, are broken open; (b) the gene-encoding rRNA is isolated, and many identical copies are made by the technique called the polymerase chain reaction (PCR); ⟳ Section 7.9. (c) The gene is sequenced (⟳ Section 10.13), and (d) the sequences obtained are aligned by computer. An algorithm makes pairwise comparisons and generates a tree (e) that depicts the differences in rRNA sequence between the organisms analyzed. In the example shown, the sequence differences are as follows: organism 1 versus organism 2, three changes; 1 versus 3, two changes; 2 versus 3, four changes. Thus organisms 1 and 3 are closer relatives than are 2 and 3 or 1 and 2. If an environmental sample is used, the isolated rRNA genes from the different microorganisms in the sample must first be sorted out (cloned) before copies are made and sequencing is performed. For further discussion of these methods, see Sections 11.5 and 18.5.

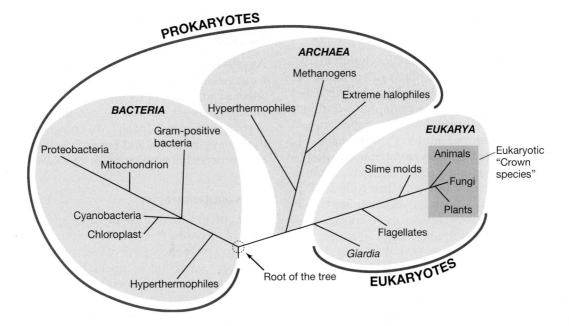

● **Figure 2.7 The phylogenetic tree of life as defined by comparative rRNA gene sequencing.** The tree consists of three domains of organisms: the *Bacteria* and the *Archaea*, cells of which are prokaryotic, and the *Eukarya* (eukaryotes). Only a few of the groups of organisms within each domain are shown. See more detailed trees of each domain in Figures 2.9, 2.18, and 2.22, and the phylogenetic trees in Chapters 11–14. Hyperthermophiles are prokaryotes that grow best at temperatures of 80°C or higher. The groups shaded in red are *macro*organisms. All other organisms on the tree of life are *micro*organisms.

another important evolutionary fact: *Archaea* are more closely related to eukaryotes than are species of *Bacteria* (Figure 2.7). Thus, evolutionary diversification from the universal ancestor initially went in two directions: *Bacteria* and a second main branch. The latter eventually diverged to yield the *Archaea* and the *Eukarya*.

Eukarya

Because the cells of higher animals and plants are all eukaryotic, it follows that eukaryotic microorganisms were the ancestors of multicellular organisms. The tree of life clearly bears this out. As expected, microbial eukaryotes branch off early on the eukaryotic lineage while plants and animals branch near the crown (Figure 2.7). However, molecular sequencing, as well as several other lines of evidence, have shown that eukaryotic cells contain genes from cells of two domains. In addition to the genome in the chromosomes of the nucleus, mitochondria and chloroplasts of eukaryotes contain their own genomes (DNA arranged in circular fashion, as in *Bacteria*) and ribosomes. Using ribosomal RNA sequencing technology (Figure 2.6), these organelles have been shown to be highly derived ancestors of specific lineages of *Bacteria* (Figure 2.7). Mitochondria and chloroplasts were thus once free-living cells that, for protection or other reasons, established stable residency in cells of *Eukarya* eons ago. The process by which this stable arrangement developed is known as *endosymbiosis* and is discussed in later chapters (∞ Sections 11.3 and 14.4).

Contributions of Molecular Sequencing to Microbiology

Molecular phylogenies have revealed the evolutionary connections between all cells. Equally important, however, the technology has created an evolutionary framework for the prokaryotes, something that the science of microbiology had been without since its founding. In addition, RNA-based phylogenies have spawned new tools that have impacted many subdisciplines of microbiology. These include, in particular, microbial classification, microbial ecology, and clinical diagnostics. Within these areas molecular phylogeny has helped shape our concept of a bacterial species and has allowed microbial ecologists and clinical microbiologists to identify organisms without the need to culture them. We will consider these developments in later chapters (∞ Chapters 11, 18, and 24).

 2.3 Concept Check

Comparative ribosomal RNA sequencing has defined the three domains of life: *Bacteria*, *Archaea*, and *Eukarya*. Molecular sequencing has also shown that the major organelles of *Eukarya* have evolutionary roots in the *Bacteria* and has yielded new tools for microbial ecology and clinical microbiology.

◆ How can *Bacteria* and *Archaea* be differentiated? In what ways are they similar?

◆ What molecular evidence supports the theory of endosymbiosis?

II MICROBIAL DIVERSITY

Evolution has molded all life on Earth. The diversity we see in microbial cells today is the result of nearly 4 billion years of evolutionary change. Microbial diversity can be seen in many ways. For example, variations occur in cell size and **cell morphology (shape)**, metabolic strategies (physiology), motility, mechanisms of cell division, pathogenicity, developmental biology, adaptation to environmental extremes, and many other aspects of cell biology. Indeed, taken collectively, the diversity of microorganisms is overwhelming. And each year new research reveals ever more spectacular examples of microbial ingenuity, many of which we will cover in this book.

In the following sections we paint a picture of microbial diversity with a broad brush. We will return to the theme of microbial diversity in more detail in Chapters 12–15. We preface our discussion of *microbial* diversity in this chapter with a brief consideration of *metabolic* diversity. The two are closely linked. Microorganisms have exploited every conceivable means of "making a living" consistent with the laws of chemistry and physics. This great metabolic capacity made available new habitats for microbial colonization, and this in turn helped fuel evolutionary diversification. Metabolic diversity will be covered in more detail in Chapters 5, 6, and 17.

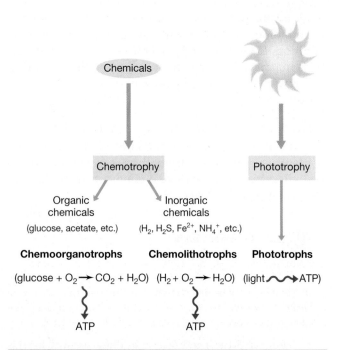

● **Figure 2.8 Metabolic options for obtaining energy.** The organic and inorganic chemicals listed here are just a few of the many different chemicals used by various chemotrophic organisms. Oxidation of the organic or inorganic chemicals yields ATP in chemotrophic organisms while conversion of solar energy to chemical energy (again, in the form of ATP) occurs in phototrophic organisms.

2.4 Physiological Diversity of Microorganisms

All cells require energy and a means to trap it. All cells also require genetic mechanisms to allow for replication and to adapt to variations in their environments. These processes are highly energy demanding, and thus energy sources are of prime importance for all cells. Energy can be obtained in three ways in nature, and Figure 2.8● summarizes the options: *organic* chemicals, *inorganic* chemicals, or *light*.

Chemoorganotrophs

Many of the thousands of different organic chemicals present on Earth can be used by one microorganism or another to obtain energy. All natural and even most synthetic organic compounds can be broken down by one or more microorganisms. Energy is obtained by *oxidizing* (removing electrons from) the compound and is conserved in the cell as the energy-rich compound, **adenosine triphosphate** (ATP) (Figure 2.8). Some microorganisms can extract energy from the compound only in the presence of oxygen; these organisms are called **aerobes**. Others can extract energy only in the absence of oxygen (**anaerobes**). Still others can break down organic compounds in either the presence or absence of oxygen. Organisms that obtain energy from *organic* compounds are called **chemoorganotrophs** (Figure 2.8). Most microorganisms that have been cultured are chemoorganotrophs.

Chemolithotrophs

A number of prokaryotes can tap the energy available in *inorganic* compounds. This is a form of metabolism called *chemolithotrophy* (discovered by Winogradsky, ∞ Section 1.7) and is carried out by organisms called **chemolithotrophs** (Figure 2.8). This form of energy-yielding metabolism is found only in prokaryotes and is widely distributed among *Bacteria* and *Archaea*. The spectrum of different inorganic compounds used is quite broad, but as a rule, a particular prokaryote specializes in the use of one or a related group of inorganic compounds.

It should be obvious why the capacity to extract energy from inorganic chemicals might be advantageous: Competition with chemoorganotrophs is not an issue. In addition, many of the inorganic compounds oxidized, for example H_2 (Figure 2.8) and H_2S, are actually the *waste products* of chemoorganotrophs. Thus, chemolithotrophs have evolved strategies for exploiting resources that most organisms are unable to use.

Phototrophs

Phototrophic microorganisms contain pigments that allow them to use light as an energy source, and thus their cells are colored (see Figure 2.10a). Unlike chemotrophic organisms, **phototrophs** do not require chemicals as a source of energy; ATP is made from the energy of sunlight. This is obviously a significant advantage, since competition for energy with chemotrophs is not a problem and light is available in a wide variety of microbial habitats.

Two major forms of phototrophy are known in prokaryotes. In one form, called *oxygenic photosynthesis*, O$_2$ is evolved. Oxygenic photosynthesis is characteristic of cyanobacteria and their phylogenetic relatives. The other form, *anoxygenic photosynthesis*, occurs in the purple and green bacteria, and does not result in O$_2$ evolution. Both groups of phototrophs use light to make ATP, however, and the similarities in their mechanisms of ATP synthesis are quite striking. Indeed, the evolutionary roots of oxygenic photosynthesis lie in the much simpler anoxygenic process. We consider photosynthesis in more detail in Chapter 17.

Heterotrophs and Autotrophs

All cells require *carbon* as a major nutrient. Microbial cells are either **heterotrophs**, requiring one or more organic compounds as their carbon source, or **autotrophs**, where CO$_2$ is the carbon source. Chemoorganotrophs are also clearly heterotrophs. By contrast, many chemolithotrophs and virtually all phototrophs are autotrophs. Autotrophs are sometimes called *primary producers* because they synthesize organic matter from CO$_2$ for both their own benefit and that of chemoorganotrophs. The latter either feed directly on the primary producers or live off products they excrete. All organic matter on Earth has been synthesized by primary producers, in particular by phototrophic organisms.

Habitats and Extreme Environments

Microorganisms are present everywhere on Earth that will support life. These include common habitats we are all familiar with—soil, water, animals, and plants—as well as on and in virtually any structures made by humans. Indeed, sterility (the absence of life forms) in a natural sample is a very rare occurrence.

Some microbial environments are ones that we as humans would find too extreme for life. Although these environments pose particular challenges to microorganisms, in many cases extreme environments are awash in microbial life. Organisms inhabiting extreme environments are called **extremophiles**, a remarkable group of primarily prokaryotes that, collectively, define the physiochemical limits of life.

Extremophilic prokaryotes abound in such harsh environments as boiling hot springs, on or within the ice covering lakes, glaciers, or the polar seas, in extremely salty bodies of water, and in soils and waters having a pH lower than 0 or as high as 12. Interestingly, these prokaryotes are not just *tolerant* of these extremes, but actually *require* the environmental extreme in order to grow. That is why they are called *extremophiles* (the suffix "phile" means "loving"). Table 2.1 summarizes the current "record holders" for extremophilic prokaryotes and lists the types of habitats in which they reside. We will revisit many of these species again in later chapters (Chapters 6, 12, and 13).

⬣ **2.4 Concept Check**

Carbon and energy sources are needed by all cells. The terms chemoorganotroph, chemolithotroph, and phototroph refer to cells that use organic chemicals, inorganic chemicals, or light, respectively, as their source of energy. Autotrophic organisms use CO$_2$ as their carbon source, while heterotrophs use organic carbon. Extremophiles thrive under environmental conditions that higher organisms cannot.

Table 2.1   Classes and examples of extremophiles[a]

Extreme	Descriptive term	Genus/species	Domain	Habitat	Minimum	Optimum	Maximum
Temperature							
High	Hyperthermophile	*Pyrolobus fumarii*	*Archaea*	Hot, undersea hydrothermal vents	90°C	**106°C**	113°C[b]
Low	Psychrophile	*Polaromonas vacuolata*	*Bacteria*	Sea ice	0°C	**4°C**	12°C
pH							
Low	Acidophile	*Picrophilus oshimae*	*Archaea*	Acidic hot springs	−0.06	**0.7[c]**	4
High	Alkaliphile	*Natronobacterium gregoryi*	*Archaea*	Soda lakes	8.5	**10[d]**	12
Pressure	Barophile	*Moritella yayanosii[e]*	*Bacteria*	Deep ocean sediments	500 atm	**700 atm**	>1000 atm
Salt (NaCl)	Halophile	*Halobacterium salinarum*	*Archaea*	Salterns	15%	**25%**	32% (saturation)

[a] In each category the organism listed is the current "record holder" for requiring a particular extreme condition for growth.

[b] A newly isolated archaeon can apparently grow up to 121°C.

[c] *P. oshimae* is also a thermophile, growing optimally at 60°C.

[d] *N. gregoryi* is also an extreme halophile, growing optimally at 20% NaCl.

[e] *Moritella yayanosii* is also a psychrophile, growing optimally at about 4°C.

◆ How might you distinguish a *phototrophic* microorganism from a *chemotrophic* one by simply looking at it under a microscope?

◆ What are *extremophiles*?

2.5 Prokaryotic Diversity

As we have seen, prokaryotes cluster into two phylogenetically distinct domains, the *Archaea* and the *Bacteria* (Figure 2.7). In this section we will explore the phylogenetic tree and briefly consider some major organisms found there. Most of the prokaryotes that are familiar to the beginning student of microbiology are found in the domain *Bacteria*, and so we begin here.

Bacteria

The domain *Bacteria* contains an enormous variety of prokaryotes. All known disease-causing (pathogenic) prokaryotes are *Bacteria*, as well as thousands of nonpathogenic species. Moreover, a large variety of morphologies and physiologies occur in this domain. The **Proteobacteria** is the largest division (called a *phylum*) of *Bacteria* (Figure 2.9●). Within the Proteobacteria are found many chemoorganotrophic bacteria, such as *Escherichia coli,* the model organism of microbial physiology, biochemistry, and molecular biology. Several phototrophic (Figure 2.10*a*●) and chemolithotrophic (Figure 2.10*b*) species are Proteobacteria, as well. Many chemolithotrophs use hydrogen sulfide (H_2S, the smell from rotten eggs) in their metabolism, pro-

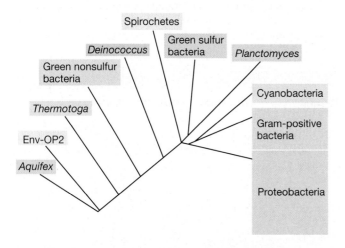

● **Figure 2.9 Detailed phylogenetic tree of *Bacteria*.** Not all known groups of *Bacteria* are depicted on this tree. The relative sizes of the colored boxes are an indication of the number of known genera and species in each of the groups. The Proteobacteria are currently the largest group of *Bacteria* known. The lineage on the tree labeled "Env" (for Environmental) does not represent a cultured organism but instead is a sequence of rRNA genes isolated from an organism in a natural sample (see text). In this example, the closest known relative of Env-OP2 would be *Aquifex*. Although not shown, there are many other such Env's known, located all over the tree.

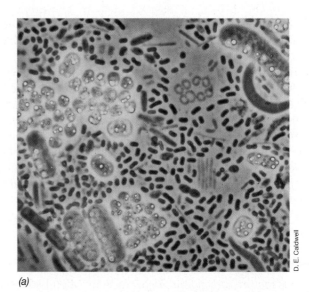

(a)

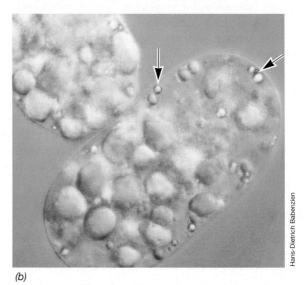

(b)

● **Figure 2.10 Phototrophic and chemolithotrophic Proteobacteria.** (a) The phototrophic purple sulfur bacterium, *Chromatium* (large, red, rod-shaped cells in this photomicrograph of a natural microbial community). A cell is about 10 μm in diameter. (b) The large chemolithotrophic sulfur-oxidizing bacterium, *Achromatium*. A cell is about 20 μm in diameter. Globules of elemental sulfur can be seen in the cells (arrows). Both of these organisms oxidize hydrogen sulfide (H_2S) produced by sulfate-reducing bacteria. Sulfate-reducing bacteria are chemoorganotrophs that oxidize organic compounds or H_2 and couple this to the reduction of sulfate (SO_4^{2-}) to H_2S, thus completing the sulfur cycle (⌒ Section 19.13).

ducing elemental sulfur that is stored within or outside the cell (Figure 2.10*b*). The sulfur is an oxidation product of H_2S and can be further oxidized to sulfate (SO_4^{2-}). The sulfide and sulfur are oxidized to fuel important metabolic functions such as CO_2 fixation (autotrophy) or energy generation (Figure 2.8).

Several other common prokaryotes of soil and water, and species that live in or on plants and animals in both casual and disease-causing ways, are members of the Proteobacteria. These include species of *Pseudomonas*, many of which can degrade complex and otherwise toxic natural and synthetic organic compounds, and *Azotobacter*, a nitrogen-fixing bacterium. A number of key pathogens are Protoeobacteria, including *Salmonella*, *Rickettsia*, *Neisseria*, and many others.

Some bacteria can be distinguished by use of the *Gram-staining* procedure, a technique that stains cells either *gram-positive* or *gram-negative* (staining properties of bacteria will be discussed in Chapter 4). The **Gram-positive lineage** of *Bacteria* contains a number of species united by a common phylogeny and cell wall structure. Here we find the endospore-forming *Bacillus* (discovered by Ferdinand Cohn, ∞ Section 1.5) (Figure 2.11*a*●) and *Clostridium* and related spore-forming prokaryotes such as the antibiotic-producing *Streptomyces*. Also included here are the lactic acid bacteria, common inhabitants of decaying plant material and dairy products, that include such organisms as *Streptococcus* (Figure 2.11*b*) and *Lactobacillus*. Other interesting gram-positive relatives are the mycoplasmas. These prokaryotes lack a cell wall, have very small genomes, and are often pathogenic. *Mycoplasma* is a major genus of organisms in this medically important group (∞ Section 12.21 and Figure 12.62).

The **Cyanobacteria** (Figure 2.12●) are phylogenetic relatives of gram-positive bacteria (Figure 2.9) and are oxygenic phototrophs. Cyanobacteria were critical in the evolution of life, as they were the first oxygenic phototrophs to have evolved on Earth (∞ Figure 1.1*b*). The production of O_2 on an originally anoxic Earth paved the way for the evolution of prokaryotes that could respire using oxygen. The development of "higher organisms," such as the plants and animals, followed billions of years later when Earth had a more oxygen-rich environment.

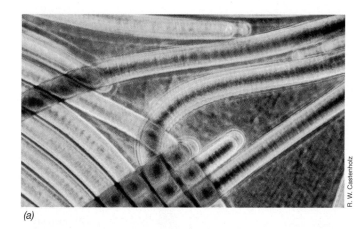

(a)

(b)

● **Figure 2.12 Filamentous cyanobacteria.** (a) *Oscillatoria*, (b) *Spirulina*. Eons ago, cyanobacteria produced the oxygen now present on Earth. Many other morphological forms of cyanobacteria are known, including unicellular, colonial, and heterocystous. The latter contain special structures called *heterocysts* that carry out nitrogen fixation (∞ Sections 12.25 and 17.28).

Several lineages of *Bacteria* contain species with unique morphologies. These include the aquatic **Planctomyces** group, characterized by cells with a distinct stalk that allows the organisms to attach to a solid substratum (Figure 2.13●), and the helically shaped **Spirochetes**

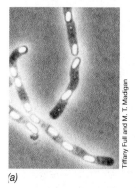

(a)

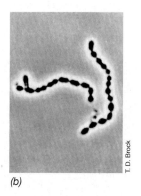

(b)

● **Figure 2.11 Gram-positive bacteria.** (a) The rod-shaped endospore-forming bacterium *Bacillus*, here shown as cells in a chain. Note the presence of endospores (bright refractile structures) inside the cells. Endospores are extremely resistant to heat, chemicals, and radiation. (b) *Streptococcus*, a spherical cell that exists in chains. Streptococci are widespread in dairy products and some are potent pathogens.

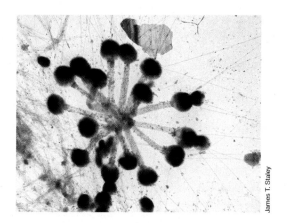

● **Figure 2.13 The morphologically unusual stalked bacterium *Planctomyces.*** Shown are several cells attached by their stalks to form a rosette.

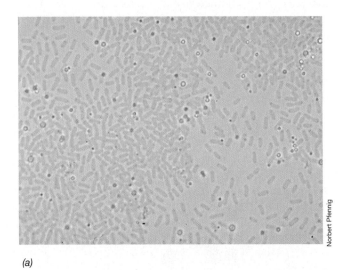

● **Figure 2.14** **Spirochetes.** A cell of *Spirochaeta zuelzerae*. These morphologically distinct prokaryotes are also phylogenetically distinct (see Figure 2.9). Spirochetes are widespread in nature and some cause diseases such as syphilis and Lyme disease.

(Figure 2.14●). Several diseases, most notably syphilis and Lyme disease (∞ Sections 26.12 and 27.4), are caused by spirochetes.

Two other major lineages of *Bacteria* are phototrophic: the **green sulfur bacteria** and the **green nonsulfur bacteria** (*Chloroflexus* group) (Figure 2.15●). Species in both lineages contain similar photosynthetic pigments and can grow as autotrophs. *Chloroflexus* is a filamentous prokaryote that inhabits hot springs and shallow marine bays and is often the dominant organism in stratified microbial mats containing a community of microorganisms. *Chloroflexus* is also noteworthy because it is believed to be an important link in the evolution of photosynthesis (∞ Sections 12.35 and 17.7).

Other major lineages of *Bacteria* include the **Chlamydia** and **Deinococcus** groups (Figure 2.9). The genus *Chlamydia*, most species of which are pathogens, harbors a variety of respiratory and venereal pathogens of humans (∞ Sections 12.27 and 26.13). Chlamydia are *obligate intracellular parasites*, meaning that they live inside the cells of higher organisms, in this case, human cells. Several other pathogenic prokaryotes (for example, species of *Rickettsia*, a member of the Proteobacteria that causes diseases such as typhus and Rocky Mountain spotted fever, or *Mycobacterium tuberculosis*, a gram-positive bacterium that causes tuberculosis) are also intracellular pathogens. The intracellular location of these pathogens provides a means by which they can avoid destruction by the host's immune response.

The *Deinococcus* lineage contains species with unusual cell walls and an innate resistance to high levels of radiation; *Deinococcus radiodurans* (Figure 2.16●) is a major species in this group. This organism can survive doses of radiation many-fold greater than that sufficient to kill animals and can even reassemble its chromosome after it has been shattered by radiation treatment. We learn more about this amazing organism in Section 12.34.

Finally, several lineages of *Bacteria* branch off very early on the phylogenetic tree, very near the root (Figure 2.9). Although phylogenetically distinct from one another, these groups are unified by the common property of growth at very high temperature (*hyperthermophily*). Organisms such as *Aquifex* (Figure 2.17●) and *Thermotoga* grow in environments that are near the boiling point of water. As you might imagine, their habitats are hot springs. The early branching nature of

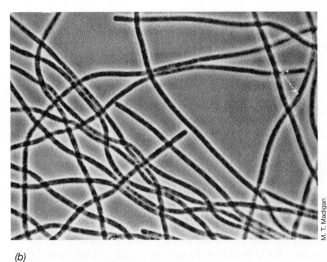

(a)   (b)

● **Figure 2.15** **Phototrophic green bacteria.** (a) *Chlorobium* (green sulfur bacteria); a single cell is about 0.8 μm wide. (b) *Chloroflexus* (green nonsulfur bacteria); a filament is about 1.3 μm wide. Despite sharing many features like pigments and membrane structures in common (∞ Section 17.2), these organisms are phylogenetically quite distinct (Figure 2.9).

● **Figure 2.16 The highly radiation-resistant bacterium *Deinococcus radiodurans*.** This organism can withstand radiation levels far above that sufficient to kill a human. See Section 12.34 for further discussion of this amazing radiation resistance.

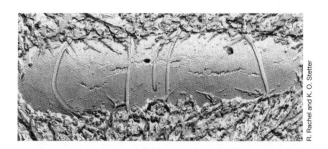

● **Figure 2.17** *Aquifex.* This "early-branching" species of *Bacteria* (see Figure 2.9) is a hyperthermophile, growing optimally above 80°C.

these lineages (Figures 2.7 and 2.9) is consistent with the idea that the early Earth was much hotter than it is today (Section 11.1). If life first evolved on a very warm planet, it makes good sense that the closest extant relatives of early organisms would be hyperthermophiles themselves. Interestingly, the phylogenetic trees of both the *Bacteria* and the *Archaea* show just this (Figures 2.9 and 2.18●). Organisms such as *Aquifex*, *Methanopyrus*, and *Pyrolobus* can therefore be considered modern descendants of very ancient cell lineages.

Archaea

Inspection of the domain *Archaea* (Figure 2.18) shows that two subdivisions exist, the *Euryarchaeota* and the *Crenarchaeota*. Each subdivision forms a major branch on the archaeal tree (Figure 2.8). Many *Archaea* are extremophiles, with species capable of growth at the highest temperatures and extremes of pH of all known organisms (Table 2.1). The organism *Pyrolobus* (Figures 2.18 and 2.19●), for example, is the most heat-loving of all known prokaryotes (Table 2.1).

All *Archaea* are chemotrophic, although *Halobacterium* (to be discussed shortly) can use light to make ATP (but not in the way that phototrophic organisms do). Some *Archaea* use organic compounds in their energy metabolism. Many others are chemolithotrophs, with hydrogen gas (H$_2$) being a widely used energy source (Figure 2.8). Chemolithotrophy is particularly widespread among hyperthermophilic *Archaea*.

The Euryarchaeota branch on the tree of *Archaea* (Figure 2.18) contains three groups of organisms with dramatically different physiologies. Some species require O$_2$ while others are killed by it, and some grow at the upper or lower extremes of pH (Table 2.1). *Methanogens* like *Methanobacterium* are strict anaer-

obes. Their metabolism is unique in that energy is obtained by the *production* of an energy-rich substance, methane (natural gas). Methanogens are ecologically important organisms in the biodegradation of organic matter in nature (Sections 13.4, 17.17, and 19.10), and most if not all of the natural gas found on Earth has accumulated from their metabolism.

The extreme halophiles are relatives of the methanogens (Figure 2.18), but are physiologically distinct from them. Unlike methanogens, which are killed by oxygen, extreme halophiles *require* oxygen and are unified by their requirement for very large amounts of salt (NaCl) for metabolism and reproduction. These organisms are therefore called *halophiles* (salt lovers). In

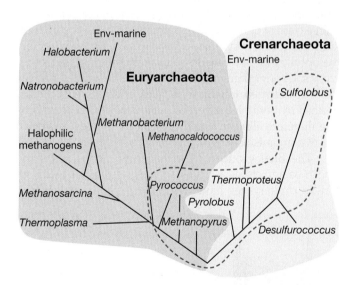

● **Figure 2.18 Detailed phylogenetic tree of *Archaea*.** Not all known groups of *Archaea* are depicted on this tree. There are two major subgroups of *Archaea*, Euryarchaeota and Crenarchaeota. The organisms circled in red are hyperthermophiles, growing at very high temperatures. In pink are shown methanogens and the extreme halophiles and acidophiles. Each major group has its own Env lineages (see legend to Figure 2.9 and text) as well, most of which are marine species. There are roughly the same number of species in both subgroups of *Archaea*, but the total number of cultured *Archaea* is far lower than for *Bacteria*.

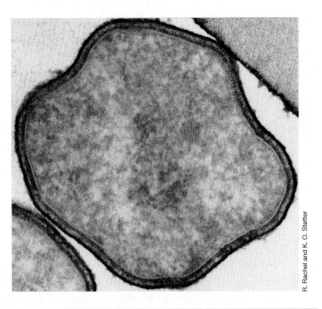

R. Rachel and K. O. Stetter

● **Figure 2.19** *Pyrolobus.* A hyperthermophilic archaeon that grows optimally above the boiling point of water!

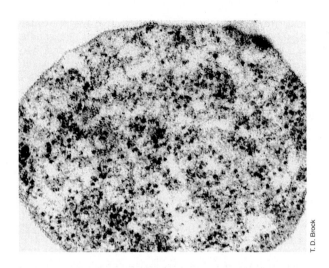

T. D. Brock

● **Figure 2.21** **Extremely acidophilic prokaryotes.** The cell wall-less archaeon, *Thermoplasma*, shown here, and its close relative *Picrophilus* (see Table 2.1) grow at moderately high temperature and extremely low pH. Although members of the *Bacteria*, the genus *Mycoplasma* also contains species lacking a cell wall. Cell wall-less prokaryotes are discussed in Sections 12.21 and 13.5.

fact, organisms like *Halobacterium* are so salt loving that they can actually grow on and within salt crystals (Figure 2.20●).

As previously mentioned (see Section 2.4), many prokaryotes are phototrophic and can generate ATP from light. Although *Halobacterium* species do not produce chlorophyll like true phototrophs, they nevertheless contain a class of light-sensitive pigments that can absorb light and trigger ATP synthesis (👓 Section 13.3). Extremely halophilic *Archaea* inhabit salt lakes, salterns, and other very salty environments. Some extreme halophiles, such as *Natronobacterium*, inhabit soda lakes, environments characterized by both high levels of salt *and* pH. Such organisms are therefore *alkaliphilic* and grow at the highest pH of all known organisms (Table 2.1).

The final group of *Archaea* we will consider is the thermoacidophiles, such as *Thermoplasma* (Figure 2.21●). These prokaryotes have a cell membrane but lack cell

William D. Grant

● **Figure 2.20** **Extremely halophilic *Archaea.*** A vial of brine at the point of NaCl precipitation and containing cells of the extreme halophile, *Halobacterium.* The organism contains pigments that absorb light and lead to ATP production. Cells of *Halobacterium* can also live *within* salt crystals themselves (👓 Microbial Sidebar, Chapter 4, How Long Can an Endospore Survive?).

walls (resembling *Mycoplasma* in this respect) and grow best at moderately high temperatures and extremely low pH. This group includes *Picrophilus*, the most *acidophilic* of all known prokaryotes (Table 2.1).

Phylogenetic Analyses of Natural Microbial Communities

Not all *Archaea* are extremophiles. Many can be found in lakes, soils, and oceans. Unfortunately, however, most of these *Archaea* have thus far defied laboratory culture, and so we know very little about what they are doing in their habitats. Much the same can be said about the legion of uncultured species of *Bacteria*. If these organisms have not been cultured, how do we know that they are there? We know this because it is possible to isolate *ribosomal RNA genes* from cells in a natural sample, such as soil. If a natural sample contains ribosomal RNA, it is because the organisms that made this ribosomal RNA were present in the sample. If we process such a sample for ribosomal RNA genes, isolate and sequence them, we can place them on the phylogenetic tree just as if we had started from cultures of the different organisms (see Figures 2.9 and 2.18).

Using these methods of *molecular microbial ecology* initially devised by Norman Pace, an American microbiologist, it is clear that the extent of prokaryotic diversity is far greater than originally thought. A sampling of virtually any microbial habitat reveals that the vast majority of prokaryotes in that habitat have never been cultured. The challenge now is to try to discover enough about the biology of these uncultured prokaryotes that creative culturing techniques can be devised to grow them. Genomic analyses of uncultured *Archaea* and

Bacteria (environmental genomics, ∞ Section 18.6), will help in this regard. This is because identification of the complement of genes present in these uncultured organisms will reveal their metabolic capacities. Knowledge of the metabolic blueprint of these organisms should then help in the design of specific isolation methods for their culture.

2.5 Concept Check

Several lineages are present in the domains *Bacteria* and *Archaea*, and an enormous diversity of cell morphologies and physiologies are represented there. Retrieval and analysis of ribosomal RNA genes from cells in natural samples have shown that many phylogenetically distinct but as yet uncultured prokaryotes exist in nature.

◆ What important bacterial species that resides in your gut is a member of the Proteobacteria?

◆ Why can it be said that the cyanobacteria prepared Earth for the evolution of higher life forms?

◆ What is unusual about the genus *Halobacterium*?

◆ How do we know a particular lineage of prokaryote exists in a natural habitat without first isolating and growing it in laboratory culture?

2.6 Eukaryotic Microorganisms

Eukaryotic microorganisms are related by their distinct cell structure (Figure 2.1) and phylogenetic history (Figure 2.7). Inspection of the domain *Eukarya* (Figure 2.22●) shows a long branch in the tree of life culminating in the most derived **eukaryotes**, the plants and animals. Interestingly, however, and consistent with their phylogenetic location on the tree, some of the "early-branching" *Eukarya* turn out to be structurally simple eukaryotes, lacking mitochondria and some other key organelles. These cells, such as the diplomonad *Giardia* (Figure 2.22), appear to be modern descendents of primitive eukaryotic cells that did not engage in endosymbiosis or perhaps had, but then lost their endosymbionts (∞ Sections 2.3, 11.3, and 14.4). Most of these early eukaryotes are parasites of humans and other animals, and are unable to live a free and independent existence.

Eukaryotic Microbial Diversity

Like prokaryotes, a diverse array of eukaryotic microorganisms is known. Collectively, microbial eukaryotes are known as the **Protista**, or *protists* for short. Some protists, such as the algae (Figure 2.23*a*●), are phototrophic. Algae contain chlorophyll-rich organelles called *chloroplasts* and can live in environments containing only a few minerals (for example, K, P, Mg, N, S), H_2O, CO_2, and light.

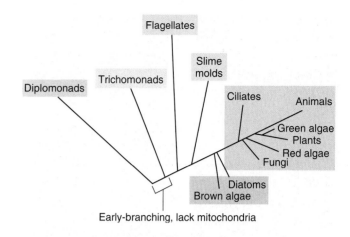

● **Figure 2.22 Detailed phylogenetic tree of *Eukarya*.** Not all known lineages of *Eukarya* are depicted. Some early-branching species of *Eukarya* lack organelles other than the nucleus. Note how "higher organisms" (plants and animals) branch near the apex of the tree.

Algae inhabit both soil and aquatic habitats and are major primary producers in nature. Fungi (Figure 2.23*b*) lack photosynthetic pigments and are either unicellular (yeasts) or filamentous (molds). Fungi are major agents of biodegradation in nature and recycle much of the organic matter in soils and other ecosystems.

Cells of algae and fungi have cell walls, whereas the protozoa (Figure 2.23*c*) do not. Most protozoans are motile, and different species are widespread in nature in aquatic habitats and as pathogens of humans and other animals. Different protozoa are also spread about the phylogenetic tree of *Eukarya*. Some, like the flagellates, are fairly early branching species, while others, like the ciliated species *Paramecium* (Figure 2.23), are phylogenetically more derived (Figure 2.22). The *slime molds* resemble protozoa in that they are motile and lack cell walls. Slime molds differ from protozoa, however, not only in phylogeny but by the fact that their cells undergo a life cycle. During the slime mold life cycle, motile cells aggregate to form a multicellular structure called a *fruiting body* from which spores are produced that yield new motile cells (∞ Section 14.11). Slime molds are the earliest known protists to show such cellular cooperation to form microscopic structures.

Lichens are leaf-like structures often found growing on rocks, trees, and other surfaces (Figure 2.24●). Lichens are an example of microbial *mutualism*, a situation where two organisms live together for mutual benefit. Lichens consist of a fungus and a phototrophic partner, either an alga (eukaryote) or a cyanobacterium (a prokaryote). The phototrophic component is the primary producer while the fungus provides the phototroph with an anchor and protection from the elements. The lichen is thus a dynamic organism that has evolved a successful strategy of mutualistic interaction between two quite different microorganisms.

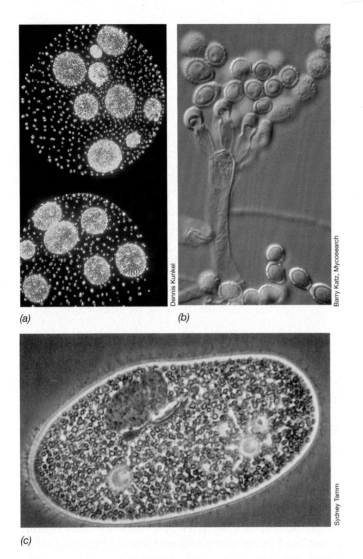

(a)

(b)

(c)

(a)

(b)

● **Figure 2.23** **Microbial *Eukarya*.** (a) Algae; the colonial green alga, *Volvox* (◯◯ Section 14.13). Each spherical-shaped cell contains several chloroplasts, the photosynthetic organelle of phototrophic eukaryotes. (b) Fungi; the spore-bearing structures of a typical mold. Each spore can give rise to a new filamentous fungus (◯◯ Section 14.12). (c) Protozoa; the ciliated protozoan *Paramecium* (◯◯ Section 14.10). Cilia function like oars in a boat, conferring motility on the cell.

● **Figure 2.24** **Lichens.** (a) An orange-pigmented lichen growing on a rock, and (b) a yellow-pigmented lichen growing on a dead tree stump, Yellowstone National Park, USA. The color of the lichen comes from the pigmented (algal) component of the lichen structure. Besides chlorophylls, the algal components of lichens contain carotenoid pigments (◯◯ Section 17.3), which can be yellow, orange, brown, red, green, or purple.

Final Remarks

Our tour of microbial diversity here has only served up an overview of the subject. The story is much larger than this and will continue in Chapters 12–15. Note also that the viruses were intentionally left out of our coverage here. Recall that viruses are not cells, yet require cells for their replication (see Section 2.1). Cells in all domains of life have viral parasites, and we will deal with viral diversity in later chapters (◯◯ Chapters 9 and 16).

Before we can proceed to a more detailed consideration of microbial diversity, however, we must become familiar with the molecular features of cells, especially prokaryotic cells. Along the way we will learn of the great chemical diversity of living organisms, a direct result of

the evolutionary breadth of the tree of life. Then we will be well prepared to revisit the subject of microbial diversity and expand on our coverage here.

 2.6 Concept Check

Microbial eukaryotes are a diverse group that includes algae, protozoa, fungi, and slime molds. Some algae and fungi have developed mutualistic associations called *lichens*.

◆ List at least two ways algae differ from cyanobacteria.

◆ List at least two ways algae differ from protozoa.

◆ How do the components of a lichen benefit each other?

REVIEW QUESTIONS

1. Why does a cell need a cytoplasmic membrane (∞ Section 2.1)?

2. Which domains of life have a prokaryotic cell structure? Can prokaryotic cell structure predict evolutionary relationships (∞ Section 2.1)?

3. How do viruses resemble cells? How do they differ from cells (∞ Section 2.1)?

4. What is meant by the word *genome*? How does the arrangement of the genome of prokaryotes differ from that of eukaryotes (∞ Section 2.2)?

5. For what reason do the processes of **mitosis** and **meiosis** occur in eukaryotic cells (∞ Section 2.2)?

6. How many genes does an organism such as *Escherichia coli* have? How does this compare with the number of genes in one of your cells (∞ Section 2.2)?

7. What is the theory of endosymbiosis (∞ Section 2.3)?

8. Molecular studies have shown many macromolecules in species of *Archaea* resemble their counterparts in various eukaryotes more closely than in species of *Bacteria*. Explain (∞ Section 2.3).

9. How do *chemoorganotrophs* differ from *chemolithotrophs* from the standpoint of energy metabolism? What carbon sources do members of each group use? Are they therefore *heterotrophs* or *autotrophs* (∞ Section 2.4)?

10. What is unusual about the organism *Pyrolobus* (∞ Section 2.5)?

11. What similarities and differences exist between the following three organisms: *Pyrolobus*, *Halobacterium*, and *Thermoplasma* (∞ Section 2.5)?

12. Examine Figure 2.18. What does the lineage "Env-marine" mean (∞ Section 2.5)?

13. How does *Giardia* differ from a human cell, both structurally and phylogenetically (∞ Section 2.6)?

APPLICATION QUESTIONS

1. Prokaryotic cells containing plasmids can often be "cured" of their plasmids (that is, the plasmids can be permanently removed) with no ill effects, while removal of the chromosome would be lethal. Explain.

2. It has been said that knowledge of the evolution of *macroorganisms* greatly preceded that of *microorganisms*. Why do you think that reconstruction of the evolutionary lineage of horses, for example, might have been an easier task than doing the same for any group of prokaryotes?

3. Examine the phylogenetic tree shown in Figure 2.6. Using the sequence data shown, describe why the tree would be incorrect if its branches remained the same but the positions of organisms 2 and 3 on the tree were switched?

4. Microbiologists have cultured a great diversity of microorganisms but know that an even greater diversity exists, despite the fact that they have never seen these organisms or grown them in the laboratory. Explain.

5. Which data from this chapter could you use to convince your friend that extremophiles are not just organisms that were "hanging on" in their respective habitats?

6. Defend this statement: If cyanobacteria had never evolved, life on Earth would have remained strictly microbial.

MACROMOLECULES

The structure of proteins, a key class of macromolecules, dictates their function.

WORKING GLOSSARY

Covalent bond a chemical bond in which electrons are shared between two atoms

Denaturation destruction of the folding properties of a protein leading (usually) to loss of biological activity

Enantiomer one form of a molecule that is the mirror image of another form of the same molecule

Glycosidic bond a type of covalent bond that links sugar units together in a polysaccharide

Hydrogen bond a weak chemical bond between a hydrogen atom and a second, more electronegative element, usually an oxygen or nitrogen atom

Isomers two molecules with the same molecular formula but which differ structurally

Lipid glycerol bonded to fatty acids or other hydrophobic molecules by ester or ether linkage. Often contain other groups, such as phosphate as well

Macromolecule polymer of covalently linked monomeric units

Molecule two or more atoms chemically bonded to one another

Nonpolar possessing hydrophobic (water-repelling) characteristics and not easily dissolved in water

Nucleic acid DNA or RNA

Nucleoside a nucleotide without its phosphate group

Nucleotide a monomer of a nucleic acid containing a nitrogen base (adenine, guanine, cytosine, thymine, or uracil), a molecule of phosphate, and a sugar, either ribose (in RNA) or deoxyribose (in DNA)

Peptide bond a type of covalent bond linking amino acids in a polypeptide

Phosphodiester bond a type of covalent bond linking nucleotides together in a polynucleotide

Polar possessing hydrophilic characteristics and generally water-soluble

Polymer a chemical compound formed by polymerization and consisting of repeating units called monomers

Polynucleotide a polymer of nucleotides bonded to one another by phosphodiester bonds

Polypeptide a polymer of amino acids bonded to one another by peptide bonds

Polysaccharide a polymer of sugar units bonded to one another by glycosidic bonds

Primary structure in an informational macromolecule, such as a polypeptide, the precise sequence of monomeric units

Protein a polypeptide or group of polypeptides that form a molecule of specific biological function

Quaternary structure in proteins, the number and types of individual polypeptides in the final protein molecule

Secondary structure the initial pattern of folding of a polypeptide or a polynucleotide, usually dictated by opportunities for hydrogen bonding

Tertiary structure the final folded structure of a polypeptide that has previously attained secondary structure

CHEMICAL BONDING AND WATER IN LIVING SYSTEMS

To understand how a cell works, an understanding of the molecules present and chemical processes that take place within cells is essential. Molecules, especially **macromolecules**, are the "guts" of the cell and are the subject of this chapter. It is assumed that the reader has some background in elementary chemistry, especially regarding the nature of atoms and atomic bonding. Here we will expand on this background with a primer on relevant biochemical bonds, followed by a discussion of the structure and function of the four classes of macromolecules: **polysaccharides, lipids, nucleic acids**, and **proteins**.

3.1 Strong and Weak Chemical Bonds

The major chemical elements of life include *hydrogen, oxygen, carbon, nitrogen, phosphorus,* and *sulfur*; these elements can bond in various ways to form the molecules of life. A **molecule** is two or more atoms chemically bonded to one another. Thus, two oxygen (O) *atoms* can combine to form a *molecule* of oxygen (O_2). Or, carbon (C), hydrogen (H), and O atoms can combine to form glucose, $C_6H_{12}O_6$, a hexose sugar.

The chemical elements of life are able to form strong bonds in which electrons are shared more or less equally between atoms. These are called **covalent bonds**. To en-

vision a covalent bond, consider the formation of a molecule of water from its constituent elements, O and H:

$$\ddot{O} + 2H\cdot \longrightarrow H\!:\!\ddot{O}\!:\!H$$

Oxygen contains six electrons in its outermost shell, while hydrogen has but a single electron. When they combine to form H_2O, covalent bonds maintain the three atoms in tight association. Depending on the molecule, double and even triple covalent bonds can form, and the strength of these bonds increases dramatically with their number (Figure 3.1●).

The chemical elements of life bond in various combinations to form the individual constituents of macromolecules, called **monomers**. Macromolecules are thus **polymers**, chemical compounds formed from repeating monomeric units. Thousands of distinct monomers are known, but only a relatively small number play important roles in the four classes of macromolecules. To a large extent it is the chemical properties of monomers that, once combined to form macromolecules, give the latter their distinctive structure and function.

Hydrogen Bonding

In addition to covalent bonds, a variety of much weaker chemical bonds also play an important role in biological molecules. Foremost among these are *hydrogen bonds*. **Hydrogen bonds** (Figure 3.2●) form between

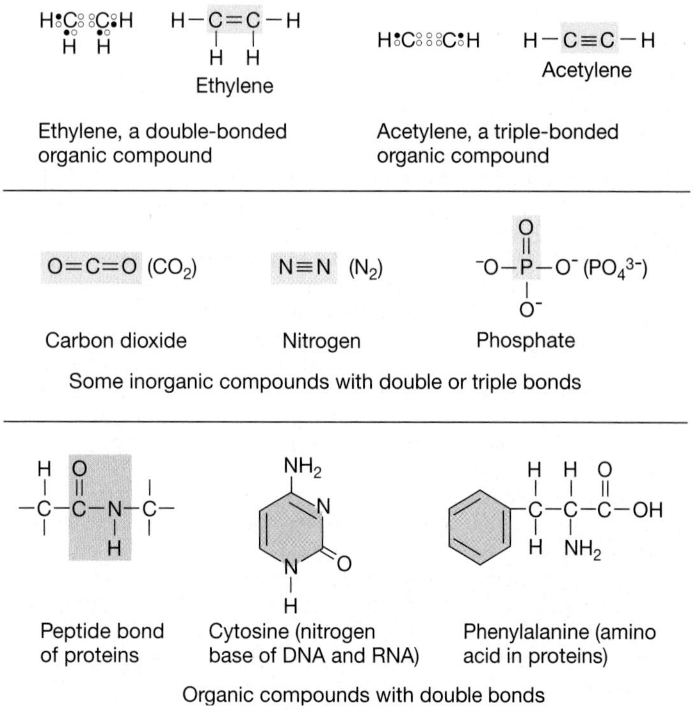

Ethylene

Ethylene, a double-bonded
organic compound

Acetylene

Acetylene, a triple-bonded
organic compound

$O=C=O$ (CO_2) $N\equiv N$ (N_2)

Carbon dioxide   Nitrogen   Phosphate

Some inorganic compounds with double or triple bonds

Peptide bond
of proteins

Cytosine (nitrogen
base of DNA and RNA)

Phenylalanine (amino
acid in proteins)

Organic compounds with double bonds

● **Figure 3.1 Covalent bonding of some molecules containing double or triple bonds.** For acetylene and ethylene, the electronic configuration of the molecules is shown.

hydrogen atoms and more *electronegative* (electron attracting) elements, such as oxygen or nitrogen. For example, an oxygen atom is rather electronegative (electron withdrawing), while a hydrogen atom is not. In the covalent bond between oxygen and hydrogen, the shared electrons therefore orbit slightly nearer to the oxygen nucleus than to the hydrogen nucleus. Because electrons carry a negative charge, this creates a slight charge separation, oxygen negative and hydrogen positive (Figure 3.2a). An individual hydrogen bond by itself is very weak. However, when many hydrogen bonds form within and between molecules, stability of the molecules can increase greatly.

Water molecules readily undergo hydrogen bonding (Figure 3.2a), and this contributes to the distinct *polarity* of water. Because it is **polar**, water molecules readily associate with one another and apart from **nonpolar** (hydrophobic) molecules. As water molecules orient themselves in solution, the slight positive charge on a hydrogen atom can bridge negative charges on two oxygen atoms. This bridge is the hydrogen bond.

Hydrogen bonds also form between atoms in macromolecules (Figure 3.2b and c). When these weak electrical forces accumulate in a large molecule such as a protein, they increase the stability of the molecule and can also affect its overall structure. We will see later that hydrogen bonds play major roles in the biological properties of proteins (Figure 3.2b) and nucleic acids (Figure 3.2c) (see Sections 3.5, 3.7, and 3.8).

Other Weak Bonds

Biomolecules form other types of weak interactions. For instance, **van der Waals forces** are weak attractive forces that occur between atoms when they become closer than about 3–4 angstroms (Å). Van der Waals forces can play significant roles in the binding of substrates to enzymes (◌◌ Section 5.5) and in protein-nucleic acid interactions. **Ionic bonds**, such as that between Na^+ and Cl^- in NaCl, are weak electrostatic interactions that allow for ionization to occur in aqueous solution. Many important biomolecules, such as carboxylic acids and phosphates (see Table 3.1), are ionized at cytoplasmic pH levels (generally about pH 6–8) and can thus be dissolved to high levels in the cytoplasm.

Hydrophobic interactions are also important in biomolecules. Hydrophobic interactions occur because nonpolar molecules or nonpolar regions of molecules tend to associate tightly in a polar environment. Hydrophobic interactions can play a major role in the folding of proteins. Hydrophobic interactions also play important roles in the binding of substrates to enzymes (◌◌ Section 5.5). In addition, hydrophobic interactions often control how differ-

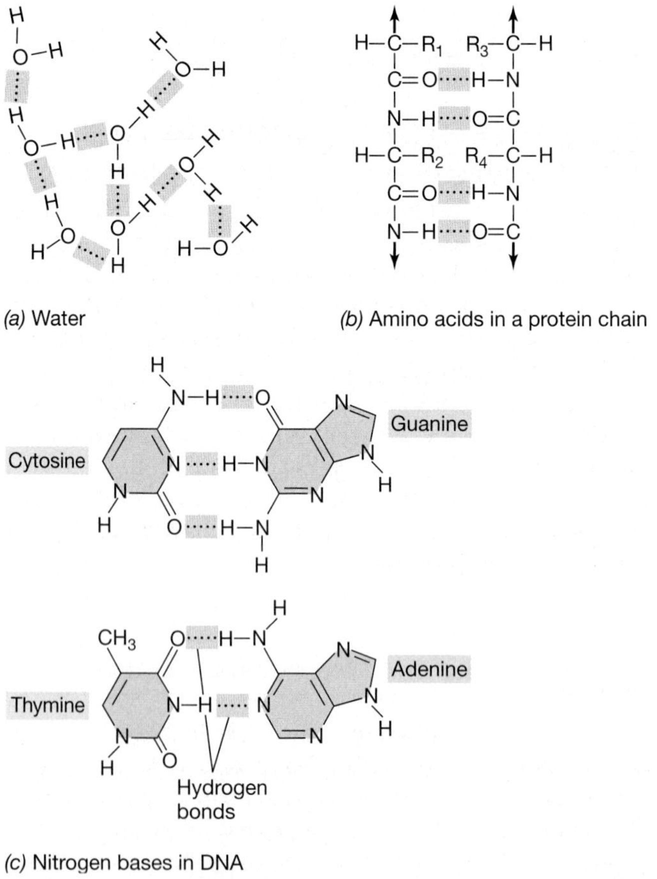

(a) Water

(b) Amino acids in a protein chain

Cytosine   Guanine

CH₃ Thymine   Adenine

Hydrogen
bonds

(c) Nitrogen bases in DNA

● **Figure 3.2 Hydrogen bonding.** In nucleic acids, hydrogen bonds are often depicted as lines rather than dots, with two lines between adenine/thymine pairs and three lines between guanine/cytosine pairs (see Figure 3.11). In (b), the R represents the side chain of each amino acid (see Figure 3.12).

Table 3.1 Some functional groups of biochemical importance

Chemical species	Structure[a]	Biological relevance
Carboxylic acid	$$\overset{\displaystyle O}{\overset{\displaystyle \|}{-C-OH}}$$	Organic, amino, and fatty acids; lipids; proteins
Aldehyde	$$\overset{\displaystyle O}{\overset{\displaystyle \|}{-C-H}}$$	Functional group of reducing sugars such as glucose; polysaccharides
Alcohol	$$\overset{H}{\underset{H}{\overset{\|}{\underset{\|}{-C-OH}}}}$$	Lipids; carbohydrates
Keto	$$\overset{\displaystyle O}{\overset{\displaystyle \|}{-C-}}$$	Pyruvate, citric acid cycle intermediates
Ester	$$\overset{H}{\underset{H}{\overset{\|}{\underset{\|}{-C-O-C-}}}}\overset{O}{\overset{\|}{\,}}$$	Lipids of *Bacteria* and *Eukarya*; amino acid attachment to tRNAs
Phosphate ester	$$\overset{O^-}{\underset{O}{\overset{\|}{\underset{\|}{^-O-P-O-C-}}}}$$	Nucleic acids, DNA and RNA
Thioester	$$\overset{\displaystyle O}{\overset{\displaystyle \|}{-C{\sim}S-}}$$	Energy metabolism; biosynthesis of fatty acids
Ether	$$\overset{H\quad H}{\underset{H\quad H}{\overset{\|\quad\|}{\underset{\|\quad\|}{-C-O-C-}}}}$$	Lipids of *Archaea*; sphingolipids
Acid anhydride	$$\overset{O\quad\; O^-}{\overset{\|\quad\;\|}{-C{\sim}O-P{=}O}}\atop{\underset{O_-}{\underset{\|}{}}}$$	Energy metabolism, for example, acetyl phosphate
Phosphoanhydride	$$\overset{O^-\;\;\; O^-}{\overset{\|\quad\;\;\|}{^-O-P{\sim}O-P-O^-}}\atop{\underset{O\quad\;\; O}{\underset{\|\quad\;\;\|}{}}}$$	Energy metabolism, for example, ATP

[a] A squiggle-type bond depiction (~) indicates an "energy-rich" bond (Section 5.8).

ent subunits in a multisubunit protein associate with one another to form the biologically active molecule and are also important in stabilizing RNA.

Bonding Patterns in Biomolecules

The element **carbon** is a major component of all macromolecules. Carbon can bond not only with itself, but with many other elements as well, to yield large structures of considerable diversity and complexity. In different *organic* (carbon containing) compounds, a variety of bonding patterns are possible. Each of these *functional groups*, as they are called, has unique chemical properties that are important in determining their biological role within the cell. An awareness of these functional groups will make our later discussion of macromolecular structure, cell

physiology, and biosynthesis easier to follow. Table 3.1 lists several functional groups of biochemical importance and examples of molecules and macromolecules that contain them.

3.1 Concept Check

Covalent bonds are strong bonds that bind elements in macromolecules. Weak bonds, such as hydrogen bonds, van der Waals forces, and hydrophobic interactions, also affect macromolecular structure, but through more subtle atomic interactions. A variety of functional groups containing carbon atoms are common in biomolecules.

◆ Why are *covalent* bonds stronger than *hydrogen* bonds?

◆ How can a hydrogen bond play a role in macromolecular structure?

3.2 An Overview of Macromolecules and Water as the Solvent of Life

If you were to chemically analyze a prokaryotic cell like the common intestinal bacterium *Escherichia coli*, what would you find? You would find water as the major constituent, but after removing the water you would find large amounts of macromolecules, much smaller amounts of monomers and a variety of inorganic ions (Table 3.2). About 95% of the dry weight of a cell consists of macromolecules, and of these, *proteins* are by far the most abundant class by weight (Table 3.2).

Proteins are polymers of monomers called *amino acids*. Proteins are found throughout the cell, in both structural as well as catalytic (enzymatic) roles (Figure 3.3*a*●), and an average cell will have many thousands of different types of proteins (Table 3.2).

Nucleic acids are polymers of *nucleotides* and are found in the cell in two forms, **RNA** and **DNA**. After proteins, *ribonucleic acids* (RNAs) are the next most abundant macromolecule in an actively growing cell (Table 3.2 and Figure 3.3*b*). This is because there are thousands of ribosomes (the "machines" that make new proteins) in each cell, and ribosomes are composed of RNA and protein. In addition, smaller amounts of RNA are present in the form of messenger and transfer RNAs, other key players in protein synthesis. In contrast to RNA, *DNA* makes up a relatively insignificant (by weight) fraction of the bacterial cell (Table 3.2). However, although quantitatively minor, DNA is central to cell function as the repository of genetic information.

Lipids have both hydrophobic and hydrophilic properties and play crucial roles in membrane structure and as storage depots for excess carbon (Figure 3.3*d*). **Polysaccharides** are polymers of *sugars* and are present primarily in the cell wall. As with lipids, however, polysaccharides such as glycogen (see later in this chapter) can be major forms of carbon and energy storage in the cell (Figure 3.3*c*).

Water as a Biological Solvent

Macromolecules and all other molecules in cells are bathed in water. Water has several important chemical features that make it an ideal biological solvent. Indeed, water is a necessary prerequisite for life as we know it.

Table 3.2	Chemical composition of a prokaryotic cell[a]	
Molecule	**Percent of dry weight[b]**	**Molecules per cell (different kinds)**
Total macromolecules	96	24,610,000 (~2500)
Protein	55	2,350,000 (~1850)
Polysaccharide	5	4,300 (2)[c]
Lipid	9.1	22,000,000 (4)[d]
Lipopolysaccharide	3.4	1,430,000 (1)
DNA	3.1	2.1 (1)
RNA	20.5	255,500 (~660)
Total monomers	3.0	—[e](~350)
Amino acids and precursors	0.5	—(~100)
Sugars and precursors	2	—(~50)
Nucleotides and precursors	0.5	—(~200)
Inorganic ions	1	—(18)
Total	100%	—

[a] Data from Neidhardt, F.C., et al. (eds.), 1996. *Escherichia coli* and *Salmonella typhimurium—Cellular and Molecular Biology*, 2nd edition. American Society for Microbiology, Washington, DC.

[b] Dry weight of an actively growing cell of *E. coli* $\cong 2.8 \times 10^{-13}$ g; total weight (70% water) = 9.5×10^{-13} g.

[c] Assuming peptidoglycan and glycogen to be the major polysaccharides present.

[d] There are several classes of phospholipids, each of which exists in many kinds because of variability in fatty acid composition between species and because of different growth conditions.

[e] Reliable estimates of monomer and inorganic ion composition are lacking.

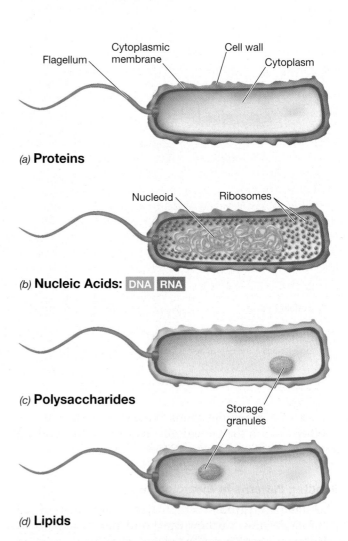

(a) **Proteins**

(b) **Nucleic Acids:** DNA RNA

(c) **Polysaccharides**

(d) **Lipids**

● **Figure 3.3   The locations of macromolecules in the cell.** (a) *Proteins* (brown) are found throughout the cell both as parts of cell structures and as enzymes. The flagellum is a structure involved in swimming motility. (b) *Nucleic acids*. DNA (green) is found in the nucleoid of prokaryotic cells and in the nucleus of eukaryotic cells. RNA (orange) is found in the cytoplasm (mRNA, tRNA) and in ribosomes (rRNA). (c) *Polysaccharides* (yellow) are located in the cell wall and occasionally in internal storage granules. (d) *Lipids* (blue) are found in the cytoplasmic membrane, the cell wall, and in storage granules.

Two key properties of water that make it such a good solvent are its *polarity* and *cohesiveness*.

The polar properties of water are important because many biologically important molecules (Table 3.2) are themselves polar and thus readily dissolve in water. As we will see in Chapter 4, dissolved substances are continually passing into and out of the cell through transport activities of the cytoplasmic membrane (∞ Sections 4.6 and 4.7). These substances include nutrients needed to build new cell material as well as waste products of metabolic processes.

The polar properties of water also promote the aggregation of large molecules, thanks to the increased opportunities for hydrogen bonding. Water forms three-dimensional networks, both with itself (Figure 3.2a) and within macromolecules. By so doing, water molecules help to position atoms within biomolecules for potential interactions. The high polarity of water is also beneficial to the cell because it forces *nonpolar* substances to aggregate and remain together. Membranes, for example, contain large amounts of lipids, which have major nonpolar (hydrophobic) components, and these aggregate in such a way as to prevent the unrestricted flow of polar molecules into and out of the cell.

In addition to hydrogen bonding, the polar nature of water makes it highly *cohesive*. This means that water molecules have a high affinity for one another and form chemically ordered arrangements in which hydrogen bonds (Figure 3.2) are constantly forming, breaking, and re-forming. The cohesive nature of water is responsible for some of its biologically important properties, such as *high surface tension* and *high specific heat* (heat required to raise the temperature 1°C). Also, the fact that water expands on freezing to yield a less dense solid form (ice) has a profound effect on life in temperate and polar aquatic environments. In a lake, for example, ice on the surface insulates the water beneath the ice and prevents it from freezing, thus allowing aquatic organisms to survive under the overlying ice.

Life originated in an aqueous milieu nearly 4 billion years ago, and virtually anywhere on Earth where liquid water exists, microorganisms are likely to be found. With these important properties of water in mind, we now consider the structure of the major macromolecules of life (Table 3.2 and Figure 3.3).

3.2 Concept Check

Proteins are the most abundant class of macromolecule in the cell. Other macromolecules include the nucleic acids (DNA and RNA), lipids, polysaccharides, and lipopolysaccharides. Water is a particularly good solvent for living organisms because of its polarity and cohesiveness.

♦ Why do protein and RNA make up such a large proportion of an actively growing cell?

♦ Why does the high polarity of water make it useful as a biological solvent?

II NONINFORMATIONAL MACROMOLECULES

In this unit we examine the structure and function of *noninformational* macromolecules—polysaccharides and lipids. Although the sequences of monomers in these macromolecules do not carry genetic information, the macromolecules themselves still play important roles in the cell, primarily as structural or reserve materials.

3.3 Polysaccharides

Carbohydrates (sugars) are organic compounds containing carbon, hydrogen, and oxygen in a ratio of 1:2:1. The structural formula for glucose, the most abundant of all sugars, is $C_6H_{12}O_6$ (Figure 3.4●). The most biologically relevant carbohydrates are those containing 4, 5, 6, and 7 carbon atoms (designated as C_4, C_5, C_6, and C_7). C_5 sugars (*pentoses*) are of special significance because of their role as structural backbones of nucleic acids. Likewise, C_6 sugars (*hexoses*) are the monomeric constituents of cell wall

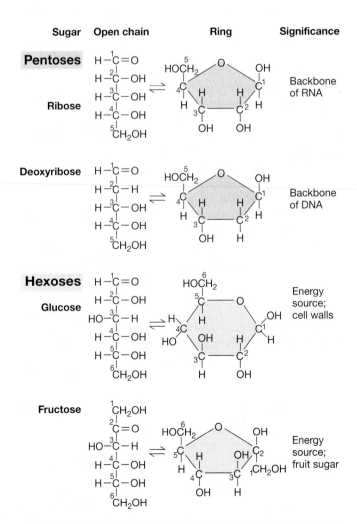

● **Figure 3.4 Structural formulas of a few common sugars.** The formulas can be depicted in two alternate ways, open chain and ring. The open chain is easier to visualize, but the ring form is the commonly used structure. Note the numbering system on the ring.

polymers and energy reserves. Figure 3.4 shows the structural formulas of a few common sugars.

Derivatives of simple carbohydrates can be formed by replacing one or more of the hydroxyl groups by other chemical species. For example, the important bacterial cell wall polymer **peptidoglycan** (⚭ Section 4.8) contains the glucose derivative *N*-acetylglucosamine (Figure 3.5●). Besides sugar derivatives, sugars having the same *structural* formula can differ in their *stereoisomeric properties* (see Section 3.6). Hence, a large number of different sugars are potentially available to the cell for the construction of polysaccharides.

The Glycosidic Bond

Polysaccharides are carbohydrates containing many (sometimes hundreds or even thousands) of monomeric units called *monosaccharides*. The latter are connected by covalent bonds called **glycosidic bonds** (Figure 3.6●). If two monosaccharides are joined by a glycosidic linkage, the resulting molecule is called a *disaccharide*. The addition of one more monosaccharide unit yields a *trisaccharide*, and several more an *oligosaccharide*. An extremely long chain is then a *polysaccharide*.

The glycosidic bond can exist in two different geometric orientations, referred to as alpha (α) and beta (β) (Figure 3.6*a*). Polysaccharides with a repeating structure composed of glucose units bonded between carbons 1 and 4 in the *alpha* orientation (for example, glycogen and starch, Figure 3.6*b*) function as important carbon and energy reserves in bacteria, plants, and animals. Alternatively, glucose units joined by β-1,4 linkages are present in cellulose (Figure 3.6*b*), a stiff plant and algal cell wall component. Thus, even though starch and cellulose are both composed solely of glucose units, their functional properties differ because of the different configurations, α or β, of their glycosidic bonds.

Polysaccharides can also combine with other classes of macromolecules, such as proteins and lipids, to form *complex polysaccharides*—**glycoproteins** and **glycolipids**. These compounds play important roles in cells, in particular as cell surface receptor molecules in cytoplasmic membranes. The compounds typically reside on the external surfaces of the membrane where they are in con-

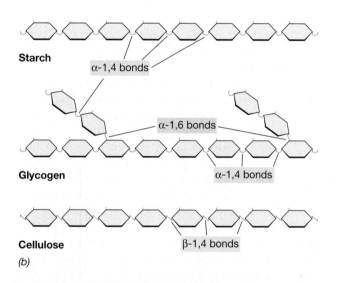

(a)

(b)

● **Figure 3.6 Polysaccharides.** (a) Structure of different glycosidic bonds. Note that both the *linkage* (position on the ring of the carbon atoms bonded) and the *geometry* (α or β) of the linkage can vary about the glycosidic bond. (b) Structures of some common polysaccharides. Compare color coding to (a).

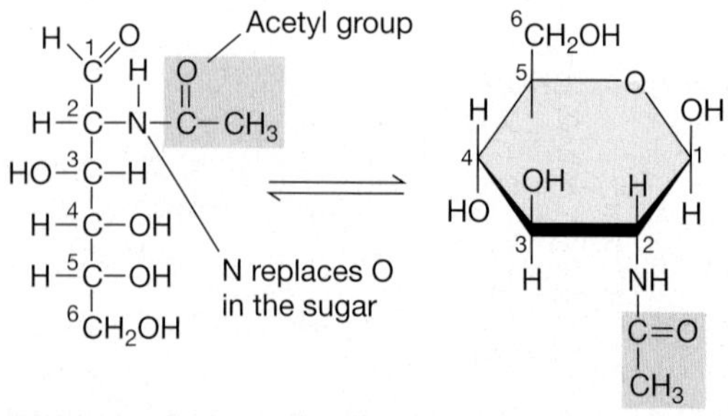

tact with the environment. Glycolipids also constitute a major portion of the cell wall of gram-negative bacteria, and as such, impart a number of unique surface properties to these organisms (⚭ Section 4.9).

3.3 *Concept Check*

Sugars combine into long polymers called polysaccharides. The two different orientations of the glycosidic bonds that link sugar residues impart different properties to the resultant molecules. Polysaccharides can also contain other molecules such as protein or lipid, forming complex polysaccharides.

◆ How can glycogen and cellulose differ so much in their physical properties when they both consist of 100% glucose?

● **Figure 3.5 *N*-acetylglucosamine, a derivative of glucose.** *N*-acetylglucosamine is an important component of the cell wall polysaccharide *peptidoglycan* in species of *Bacteria* (⚭ Section 4.8).

3.4 Lipids

Lipids are essential components of all cells and are *amphipathic* macromolecules, meaning they contain both hydrophilic and hydrophobic properties. Lipid structure varies between the domains of life, and, within a domain, many different lipids are known. *Fatty acids* are major constituents of the lipids of *Bacteria* and *Eukarya*. By contrast, the lipids of *Archaea* are constructed from the hydrophobic molecule *phytane* (∞ Section 4.5).

Fatty acids contain both hydrophobic and hydrophilic components. Palmitate (the ionized form of palmitic acid) (Figure 3.7●), for example, is a 16-carbon fatty acid composed of a chain of 15 saturated (fully hydrogenated and highly hydrophobic) carbon atoms and a single carboxylic acid group (the hydrophilic portion). Other common fatty acids in the lipids of *Bacteria* include saturated or monounsaturated forms, from C_{12} to C_{20} (Figure 3.7).

Triglycerides and Complex Lipids

Simple lipids (fats) consist of fatty acids (or phytanyl units in *Archaea*) bonded to the C_3 alcohol *glycerol* (Figure 3.7). Simple lipids are also called **triglycerides** because three fatty acids are linked to the glycerol molecule.

Complex lipids are simple lipids that contain additional elements such as phosphorus, nitrogen, or sulfur, or small hydrophilic organic compounds such as sugars, ethanolamine (Figure 3.7), serine, or choline. Lipids containing a phosphate group, called **phospholipids**, are an important class of complex lipids because they play a major structural role in the cytoplasmic membrane (∞ Section 4.5).

The amphipathic properties of lipids make them ideal as structural components of membranes. Lipids aggregate to form membranes; the *hydrophilic* (glycerol) portion remains in contact with the cytoplasm and the external environment while the *hydrophobic* portion remains buried away inside the membrane (∞ Section 4.5 and Figure 4.16). Because of this property, membranes are ideal permeability barriers. The inability of polar substances to flow through the hydrophobic region of the lipids renders the membrane impermeable and prevents leakage of cytoplasmic constituents. However, this also means that polar substances necessary for cell function do not leak *in*, either, but we reserve this story for the next chapter (transport, ∞ Section 4.5)

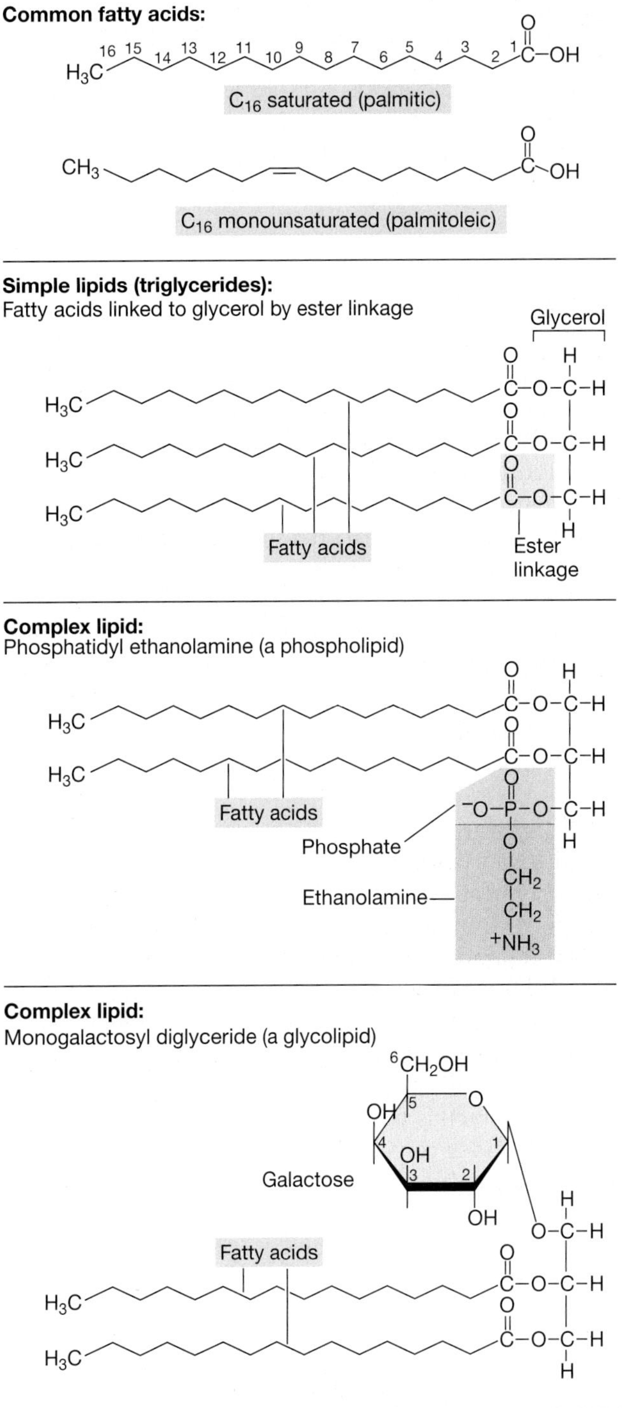

Common fatty acids:

C_{16} saturated (palmitic)

C_{16} monounsaturated (palmitoleic)

Simple lipids (triglycerides):
Fatty acids linked to glycerol by ester linkage

Complex lipid:
Phosphatidyl ethanolamine (a phospholipid)

Complex lipid:
Monogalactosyl diglyceride (a glycolipid)

● **Figure 3.7 Fatty acids, simple lipids (fats), and complex lipids.** Simple lipids are formed by a dehydration reaction between fatty acids and glycerol to yield the ester linkage. The fatty acid composition of a cell varies with growth temperature.

3.4 Concept Check

Lipids contain both hydrophobic and hydrophilic components; their chemical properties make them ideal structural components for cytoplasmic membranes.

◆ What part of a fatty acid molecule is hydrophobic? Hydrophilic?

◆ How does a *phospholipid* differ from a *triglyceride*?

◆ Draw the chemical structure of butyrate, a C_4 fully saturated fatty acid.

III INFORMATIONAL MACROMOLECULES

Unlike polysaccharides and lipids, the sequence of monomers in nucleic acids and proteins carry genetic information. Nucleic acids and proteins are thus the *informational* macromolecules.

3.5 Nucleic Acids

The nucleic acids deoxyribonucleic acid (DNA) and ribonucleic acid (RNA) are macromolecules composed of monomers called *nucleotides*. Therefore, DNA and RNA are **polynucleotides**. As we already know, DNA carries the genetic blueprint for the cell while RNA is the intermediary molecule that converts the blueprint into defined amino acid sequences in proteins (◯◯Figure 1.4).

A nucleotide is composed of three components: a five-carbon sugar, either ribose (in RNA) or deoxyribose (in DNA), a nitrogen base, and a molecule of phosphate, PO_4^{3-}. The general structure of nucleotides of DNA and RNA is very similar (Figure 3.8●).

Nucleotides

The nitrogen bases of nucleic acids belong to either of two chemical classes. *Purine* bases—**adenine** and **guanine**—contain two fused heterocyclic rings (rings containing more than one kind of atom). *Pyrimidine* bases—**thymine**, **cytosine**, and **uracil**—contain a single six-membered heterocyclic ring (Figure 3.9●). Guanine, adenine, and cytosine occur in both DNA and RNA. Thymine is present (with minor exceptions) only in DNA, and uracil is present only in RNA.

Nucleotides consist of a nitrogen base attached to a pentose sugar by a glycosidic linkage between carbon atom 1 of the sugar and a nitrogen atom of the base, either the nitrogen atom labeled 1 (pyrimidine base) or 9 (purine base). Without a phosphate, a base bonded to its

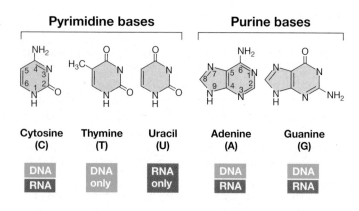

● **Figure 3.9 Structure of the bases of DNA and RNA.** Note the numbering system of the rings. In attaching the base to the 1′ carbon of the sugar phosphate shown in Figure 3.8, pyrimidine bases are bonded through N-1 of the ring and purine bases through N-9 of the ring.

sugar is referred to as a **nucleoside**. **Nucleotides** are thus *nucleosides* containing one or more phosphates (Figure 3.10●).

Nucleotides play other roles in the cell besides their major role as constituents of nucleic acids. Nucleotides, especially adenosine triphosphate (ATP) (Figure 3.10), are key sources of chemical energy, releasing sufficient energy during the cleavage of a phosphate bond to drive energy-requiring reactions in the cell (◯◯ Section 5.8). Other nucleotides or nucleotide derivatives function in oxidation-reduction reactions in the cell (◯◯ Section 5.7) as carriers of sugars in the biosynthesis of polysaccharides (◯◯ Section 5.15) and as regulatory molecules inhibiting or stimulating the activities of certain enzymes or metabolic events. However, we discuss here only the role of nucleotides as building blocks of nucleic acids, the major *informational* function of nucleotides.

Nucleic Acids

The nucleic acid backbone is a polymer of alternating sugar and phosphate molecules (Figure 3.11●). Polynu-

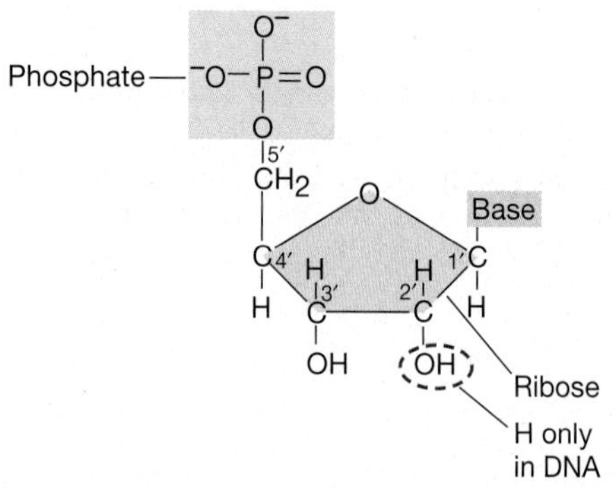

● **Figure 3.8 Nucleotides.** The numbers on the sugar contain a prime (′) after them because the ring structure in the nitrogen base is also numbered (1, 2, 3, etc.) (see Figure 3.9).

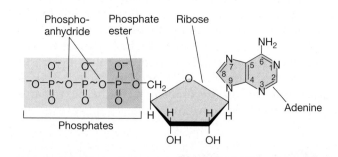

● **Figure 3.10 Components of the important nucleotide, adenosine triphosphate.** The energy of hydrolysis of a phospho-anhydride bond (shown as squiggles) is greater than that of a phosphate ester and will have significance for bioenergetics in Chapter 5 (◯◯ Section 5.8). Adenosine (a nucleoside) consists of a sugar bonded to a nitrogenous base but lacks the phosphate group.

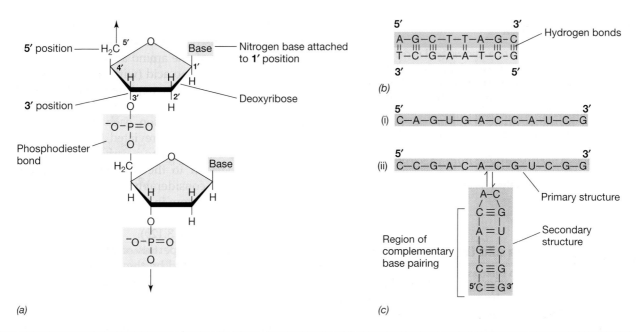

● **Figure 3.11   DNA and RNA.** (a) Structure of part of a DNA chain. The nitrogen bases can be adenine, guanine, cytosine, or thymine. In RNA, an OH group is present on the 2′ carbon of the pentose sugar (see Figure 3.8), and uracil replaces thymine. (b) Simplified structure of DNA in which only the nitrogen bases are shown. Note how the two strands are complementary in base sequence (A = T; G≡C) and bonded by hydrogen bonds. (c) RNA: (i) A sequence showing only primary structure; (ii) A sequence that allows for secondary structure. In RNA, secondary structures form when opportunities for *intra*strand base pairing arise, as shown here. In certain very large RNA molecules, such as ribosomal RNA (∞ Sections 7.15 and 11.5), some parts of the molecule contain only primary structure while others contain both primary and secondary structure. This can lead to highly folded and twisted molecules (∞Figure 11.11c) whose biological function is dependent on their final three-dimensional shape.

cleotides consist of nucleotides covalently bonded via phosphate from carbon 3 [referred to as the 3′ (3 prime) carbon] of one sugar to carbon 5 (5′) of the adjacent sugar (Figure 3.11a). Chemically, the phosphate linkage is a **phosphodiester,** since a single phosphate is connected by ester linkage to two separate sugars (Figure 3.11a).

The *sequence* of nucleotides in a DNA or RNA molecule is referred to as its **primary structure**. As we have discussed, the sequence of bases in a DNA or RNA molecule is informational, encoding the sequence of amino acids in proteins or encoding specific ribosomal or transfer RNAs. The replication of DNA and the synthesis of RNA are key events in the life of a cell (∞ Section 1.2 and Figure 1.4). We will see later that a virtually error-free mechanism is employed to ensure the faithful transfer of genetic traits from one generation to another (∞ Chapter 7).

DNA

In cells, DNA is present in *double-stranded* form. Each chromosome contains two strands of DNA, with each strand containing hundreds of thousands to several million nucleotides linked by phosphodiester bonds. The strands themselves associate with one another by hydrogen bonds that form between the nucleotides of one strand and the complementary nucleotides of the other. When positioned adjacent to one another, purine and pyrimidine bases can undergo hydrogen bonding (see Figure 3.2c).

The most stable hydrogen bonding occurs when guanine (G) forms hydrogen bonds with cytosine (C), and adenine (A) forms hydrogen bonds with thymine (T) (see Figure 3.2c). Specific base pairing, A with T and G with C, means that the two strands of DNA are *complementary* in base sequence. That is, wherever a G is found in one strand, a C is found in the other, and wherever a T is present in one strand, its complementary strand has an A (Figure 3.11b).

RNA

With a few exceptions, all ribonucleic acids are *single-stranded* molecules. However, RNA molecules can fold back upon themselves in regions where complementary base pairing is possible to form folded structures. This pattern of folding in RNA is referred to as its **secondary structure** (Figure 3.11c).

RNA plays three crucial roles in the cell. **Messenger RNA (mRNA)** contains the genetic information of DNA in a single-stranded molecule *complementary* in base sequence to that of DNA. **Transfer RNAs** (tRNAs) are the "adaptor" molecules in protein synthesis. Transfer RNAs convert the genetic information from the language of nucleotides to the language of amino acids, the building blocks of proteins. **Ribosomal RNAs** (rRNAs), of which there are several types, are important structural and catalytic components of the *ribosome,* the protein-synthesizing system of the cell. These various RNA molecules are discussed in detail in Chapters 7 and 11.

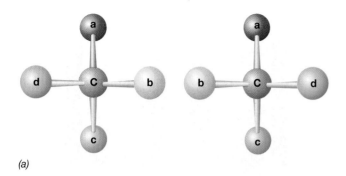

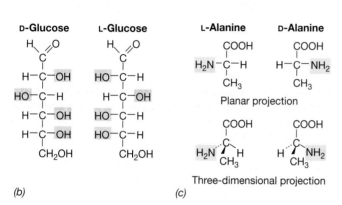

● **Figure 3.15 Isomers.** (a) Ball-and-stick model showing mirror images. (b) Enantiomers of glucose. (c) Enantiomers of the amino acid alanine. Note that no matter how the three-dimensional views are rotated, the L and D forms can never be superimposed. In the three-dimensional projection the arrow should be understood as coming *toward* the viewer whereas the dashed line indicates a plane *away* from the viewer.

3.7 Proteins: Primary and Secondary Structure

Proteins play key roles in cell function. In essence, a cell is what it is and does what it does because of the kinds and amounts of proteins it contains. That is, every different type of cell will have a different complement of proteins. An understanding of protein structure is therefore essential for understanding cells of all types.

Two major classes of proteins are *catalytic* proteins (*enzymes*) and *structural* proteins. **Enzymes** are the catalysts for the wide variety of chemical reactions that occur in cells (∞Chapters 5 and 17). By contrast, structural proteins become integral parts of the cell structures that make up membranes, walls, and cytoplasmic components. However, all proteins show certain structural features, and we discuss these now.

Primary Structure

Proteins are polymers of amino acids covalently bonded by peptide bonds (Figure 3.13). Two amino acids

bonded by peptide linkage constitute a *dipeptide*, three amino acids a *tripeptide*, and so on. When *many* amino acids are covalently linked via peptide bonds, they form a **polypeptide**.

Proteins consist of one or more polypeptides. The number of amino acids in a protein varies from one protein to another. Proteins with as few as 15 and as many as 10,000 amino acids are known. Since proteins may vary in their composition, sequence, and number of amino acids, it is easy to see that enormous variation in protein structure (and thus function) is possible.

The linear array of amino acids in a polypeptide is called its **primary structure**. The primary structure of a polypeptide is very important, because a given primary structure is consistent with only certain types of folding patterns. And it is only the final, folded polypeptide that assumes biological activity.

Secondary Structure

Interactions of the R groups on the amino acids in a polypeptide force the molecule to twist and fold in a specific way. This leads to formation of **secondary structure** (Figure 3.16●). Hydrogen bonds, the weak noncovalent linkages discussed earlier (see Section 3.1), play important roles in polypeptide secondary structure. One common type of secondary structure is the *α-helix*. To envision an *α*-helix, imagine a linear polypeptide wound around a cylinder (Figure 3.16*a*). In this twisted structure, oxygen and nitrogen atoms from different amino acids become positioned close enough to allow for hydrogen bonding. This opportunity for hydrogen bonding gives the *α*-helix its inherent stability (Figure 3.16*a*).

The primary structure of some polypeptides allows for a different type of secondary structure, called a *β-sheet*. In the *β*-sheet, the chain of amino acids in the polypeptide folds back and forth upon itself instead of forming a helix. However, as in the *α*-helix, the folding in a *β*-sheet exposes hydrogen atoms that can undergo hydrogen bonding (Figure 3.16*b*). Typically, a *β*-sheet secondary structure yields a rather rigid polypeptide while *α*-helical secondary structures are more flexible. Thus, an enzyme, for example, whose activity may depend on it being rather flexible, may contain a high degree of *α*-helix secondary structure. By contrast, a structural protein that functions in cellular scaffolding may contain large regions of *β*-sheet secondary structure.

Many polypeptides contain regions of *α*-helix and regions of *β*-sheet secondary structure, the type of folding and its location in the molecule being determined by the available opportunities for hydrogen bonding and hydrophobic interactions (see Figure 3.17). These structural regions, called *domains*, are typically segments of the polypeptide that have specific functions in the final protein molecule.

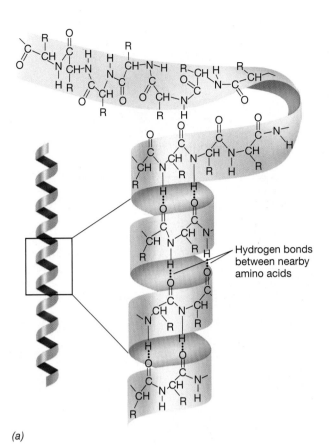

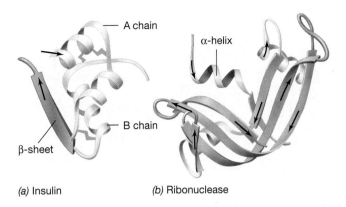

(a) Insulin   (b) Ribonuclease

● **Figure 3.17** **Tertiary structure of polypeptides showing where regions of α-helix or β-sheet secondary structure are located.** (a) *Insulin*, a protein containing two polypeptide chains (◯◯ Section 31.6); note how the B chain contains both α-helix and β-sheet secondary structure and how disulfide linkages (shown in blue) help in dictating folding patterns (tertiary structure). (b) *Ribonuclease*, a large protein with several regions of α-helix and β-sheet secondary structure.

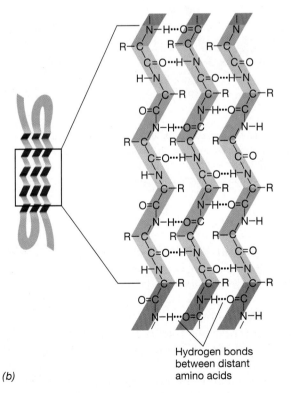

(a)

(b)

Hydrogen bonds between nearby amino acids

Hydrogen bonds between distant amino acids

● **Figure 3.16** **Secondary structure of polypeptides.** (a) α-Helix secondary structure. Note that hydrogen bonding does not involve the R groups but instead occurs between atoms in the peptide bonds. (b) β-Sheet secondary structure.

| 3.8 | **Proteins: Higher Order Structure and Denaturation** |

Once a polypeptide has achieved secondary structure it will continue to fold to form an even more stable molecule. This folding leads to formation of a unique three-dimensional shape, called the *tertiary structure* of the protein.

Like secondary structure, **tertiary structure** is ultimately determined by primary structure. However, tertiary structure is also governed to some extent by the secondary structure of the molecule, because the side chain of each amino acid in the polypeptide is positioned in a specific way (Figure 3.16). If additional hydrogen bonds, covalent bonds, hydrophobic interactions, or other atomic interactions are able to form, the polypeptide will fold to accommodate them (Figure 3.17●). The tertiary folding of the polypeptide ultimately forms exposed regions or grooves in the molecule (Figures 3.17 and 3.18●) that are important for binding other molecules (for example, in the binding of a substrate to an enzyme or the binding of DNA to a specific regulatory protein) (◯◯ Sections 5.5 and 8.4).

Frequently a polypeptide folds in such a way that adjacent sulfhydryl groups of cysteine residues are exposed. These free —SH groups can join covalently to form a *disulfide bond* between the two amino acids. If the two cysteine residues are located in different polypeptides in a protein, the disulfide bond physically links the two molecules (Figure 3.17a). In addition, a single polypeptide can spontaneously fold and bond to itself if a disulfide bond can form within the molecule.

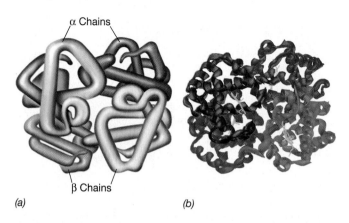

α Chains

β Chains

(a)   (b)

● **Figure 3.18** **Quaternary structure of human hemoglobin.** (a) There are two *kinds* of polypeptide in hemoglobin, α chains (shown in blue and red) and β chains (shown in orange and yellow), but a total of four polypeptides in the final protein molecule (α and β refer here to chain names, not to polypeptide secondary structures). Separate colors are used to distinguish the four distinct chains. (b) Molecular structure of human hemoglobin as determined by X-ray crystallography. In this view each chain (α and β) is in its own color.

If a protein consists of two or more polypeptides, and many proteins do, the number and type of polypeptides that form the final protein molecule are referred to as its **quaternary structure** (Figure 3.18). In proteins showing quaternary structure, each polypeptide, called a *subunit*, contains primary, secondary, and tertiary structure. Some proteins contain multiple copies of a single subunit. A protein containing two identical subunits, for example, would be called a *homodimer*. Other proteins may contain nonidentical subunits, each present in one or more copies (a *heterodimer*, for example, contains one copy each of two different polypeptides). The subunits in multisubunit proteins are held together by noncovalent interactions (hydrogen bonding, van der Waals forces, and hydrophobic interactions) or by covalent linkages, typically intersubunit disulfide bonds.

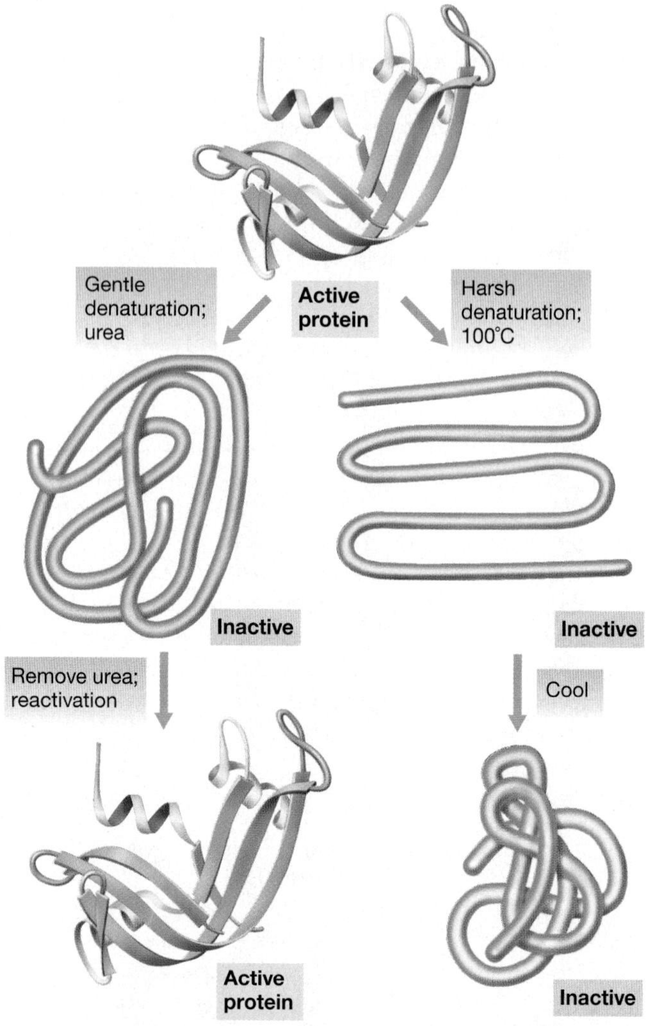

Denaturation

When proteins are exposed to extremes of heat or pH or to certain chemicals or metals that affect their folding, they may undergo **denaturation** (Figure 3.19●). Denaturation causes the polypeptide chain to unfold, destroying the higher order (secondary, tertiary, and quaternary, if relevant) structure of the molecule. Depending on the severity of the denaturing conditions, refolding of the polypeptide may occur after removal of the denaturant (Figure 3.19).

Typically, the biological properties of a protein are lost when it is denatured. Peptide bonds (Figure 3.13) are unaffected, however, and so a denatured molecule *retains its primary structure*. This shows that biological activity is not inherent in the primary structure of a pro-

Gentle denaturation; urea

Active protein

Harsh denaturation; 100°C

Inactive

Inactive

Remove urea; reactivation

Cool

Active protein

Inactive

● **Figure 3.19** **Denaturation of the protein ribonuclease.** The structure of ribonuclease was discussed in Figure 3.17b. Note how harsh denaturation yields a permanently destroyed molecule (from the standpoint of biological function) because of improper folding.

tein, but instead is a function of the uniquely folded form of the molecule as ultimately directed by primary structure. In other words, folding of a polypeptide confers upon it a unique shape that is compatible with a *specific* biological function.

Denaturation of proteins is of more than just academic interest, since this process is a major means of destroying microorganisms. For example, alcohols such as phenol and ethanol are effective disinfectants because they readily penetrate cells and irreversibly denature their proteins. Such chemical agents are thus useful for disinfecting inanimate objects, such as surfaces, and have enormous practical value in household, hospital, and industrial disinfectant applications (∽Chapter 20).

The Road Ahead

Now that we have reviewed the chemistry of life, we can better understand the structural relationships of cells. In the next chapter we will see how the different macromolecules come together to form major structures of the cell, including the membrane, wall, and flagellum. From there we will consider the metabolic properties of cells in Chapter 5. Metabolism, the *machine* function of a cell, drives the biosynthesis and assembly of macromolecules, the *coding* function of a cell (Section 1.2 and Figure 1.4). The combination of these processes results in cell growth (Chapter 6). Cell growth results in *cell populations*, and groups of populations form *microbial communities*. It is at the community level that microorganisms typically have major ecological effects in their habitats.

As we travel through this book, it may be useful to return to Chapter 3 periodically to reinforce the basic principles surrounding the chemistry of life. Although collectively, microorganisms have evolved a nearly limitless chemical diversity of macromolecules, this diversity simply boils down to variations on the four classes of macromolecules discussed in this chapter. And as any successful microbiologist knows, an understanding of the biochemistry of proteins, lipids, nucleic acids, and polysaccharides is essential and pays big dividends in grasping both the basic and more advanced principles of microbiology.

3.7 and 3.8 Concept Check

The primary structure of a protein is determined by its amino acid sequence, but it is the folding (higher order structure) of the polypeptide that determines how the protein functions in the cell.

◆ Define the terms *primary, secondary,* and *tertiary* with respect to protein structure.

◆ How does a *polypeptide* differ from a *protein*?

◆ What secondary structural features tend to make β-sheet proteins more rigid than α-helices?

◆ Describe the *number* and *kinds* of polypeptides present in a homotetrameric protein.

◆ Describe the structural and biological effects of the denaturation of a protein. Of what *practical* value is knowledge of protein denaturation?

REVIEW QUESTIONS

1. Which are the major elements found in living organisms? Why are oxygen and hydrogen particularly abundant in living organisms (Section 3.1)?

2. Define the word *molecule*. How many atoms are in a molecule of hydrogen gas? In a molecule of glucose (Sections 3.1 and 3.3)?

3. Refer to the structure of the nitrogen base *cytosine* shown in Figure 3.1. Draw this structure and then label the positions of all single bonds and double bonds in the cytosine molecule (Section 3.1).

4. Compare and contrast the words *monomer* and *polymer*. Give three examples of biologically important polymers and list the monomers of which they are composed. Which classes of macromolecules are most abundant (by weight) in a cell (Sections 3.1 and 3.2)?

5. List the components that would make up a simple lipid. How does a triglyceride differ from a complex lipid (Section 3.4)?

6. Examine the structures of the triglyceride and of phosphatidyl ethanolamine shown in Figure 3.7. How might the substitution of phosphate and ethanolamine for a fatty acid alter the chemical properties of the lipid (Section 3.4)?

7. RNA and DNA are similar types of macromolecules but show distinct differences as well. List three ways in which RNA differs chemically or physically from DNA. What is the cellular function of DNA and RNA (Section 3.5)?

8. Why are *amino acids* so named? Write a general structure for an amino acid. What is the importance of the R group to final protein structure? Why does the amino acid *cysteine* have special significance for protein structure (Section 3.6)?

9. Chemically, what type of reaction between two amino acids leads to formation of the peptide bond (Section 3.6)?

10. Define the terms *primary, secondary, tertiary,* and *quaternary* as they apply to proteins. Which of these is (are) affected by the denaturation process (Sections 3.7 and 3.8)?

11. Fill in the blanks. A *glycosidic* bond is to a _____ as a _____ bond is to a polypeptide, and a _____ is to a nucleic acid. All of these bonds are examples of _____ bonds, which are chemically much stronger than weak bonds, such as _____, _____, and _____ (Chapter 3).

APPLICATION QUESTIONS

1. Observe the following nucleotide sequences of RNA: (a) GUCAAAGAC, (b) ACGAUAACC. Can either of these RNA molecules have secondary structure? If so, draw the potential secondary structure(s).

2. A few soluble (cytoplasmic) proteins contain a high content of hydrophobic amino acids. How would you predict these proteins would fold as to their tertiary structure and why?

3. Cells of the genus *Halobacterium*, an archaeon that lives in very salty environments, contain over 5 molar (M) potassium (K^+). Because of this high K^+ content, many cytoplasmic proteins of *Halobacterium* cells are enriched in two specific amino acids that are present in much higher proportions in *Halobacterium* proteins than in functionally similar proteins from *Escherichia coli* (which has only very low levels of K^+ in its cytoplasm). Which amino acids are enriched in *Halobacterium* proteins and why? (*Hint:* Which amino acids could best neutralize the *positive* charges due to K^+?)

4. When a culture of the bacterium *Escherichia coli*, an inhabitant of the human gut, is placed in a beaker of boiling water, significant changes in the cells occur almost immediately. However, when a culture of *Pyrodictium*, a hyperthermophilic bacterium that grows optimally in boiling hot springs is put in the same beaker, similar changes do not occur. Explain.

5. Review Figure 3.6*b* and then describe the differences that make each of these polymers unique. If all of the glycosidic bonds in these polymers were hydrolyzed, what single molecule would remain?

6. Review Figure 3.12*b*. Of all the amino acids shown in blue, what is it about their chemistry that unites them as a "family"?

4

CELL STRUCTURE/ FUNCTION

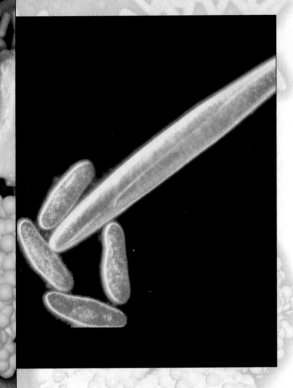

Prokaryotic cells are typically very small, but are occasionally very large. Shown is a photomicrograph of a cell of *Epulopiscium*, a prokaryote, aside four cells of the eukaryote *Paramecium*.

WORKING GLOSSARY

ABC (ATP-Binding Cassette) transporter a membrane transport system consisting of three proteins, one of which hydrolyzes ATP, to transport specific nutrients into the cell

Capsule a polysaccharide or protein outermost layer, usually rather slimy, present on some bacteria

Chemotaxis directed movement of an organism toward (*positive*) or away from (*negative*) a chemical gradient

Cytoplasmic membrane the permeability barrier of the cell, separating the cytoplasm from the environment

Endospore a highly heat-resistant, thick-walled, differentiated structure produced by certain gram-positive *Bacteria*

Flagellum a long, thin cellular appendage capable of rotation in prokaryotic cells and responsible for swimming motility

Gas vesicles gas-filled cytoplasmic structures bounded by protein and conferring buoyancy on cells

Gram-negative a prokaryotic cell whose cell wall contains small amounts of peptidoglycan, and an outer membrane, containing lipopolysaccharide, lipoprotein, and other complex macromolecules

Gram-positive a prokaryotic cell whose cell wall consists chiefly of peptidoglycan and lacks the outer membrane of gram-negative cells

Group translocation an energy-dependent transport process in which the substance transported is chemically modified during the transport process

Lipopolysaccharide (LPS) lipid in combination with polysaccharide and protein, which forms the major portion of the outer membrane in gram-negative *Bacteria*

Magnetosomes particles of magnetite (Fe_3O_4) organized into nonunit membrane-enclosed structures in the cytoplasm of magnetotactic *Bacteria*

Morphology the *shape* of a cell—rod, cocus, spirillum, and the like

Outer membrane a phospholipid and polysaccharide containing unit membrane that lies external to the peptidoglycan layer in cells of gram-negative *Bacteria*

Peptidoglycan a polysaccharide composed of alternating repeats of acetylglucosamine and acetylmuramic acid arranged in adjacent layers and cross-linked by short peptides

Periplasm a gel-like region between the outer surface of the cytoplasmic membrane and the inner surface of the lipopolysaccharide layer of gram-negative *Bacteria*

Peritrichous a pattern of flagellation where flagella are located in many places around the surface of the cell

Phototaxis movement of an organism toward light

Polar in reference to flagellation, having flagella emanating from one or both poles of the cell

Poly-β-hydroxybutyrate (PHB) a common storage material of prokaryotic cells consisting of a polymer of β-hydroxybutyrate or another β-alkanoic acid or mixtures of β-alkanoic acids

Protoplast an osmotically protected cell whose cell wall has been removed

Resolution the ability to distinguish two objects as distinct and separate

S-layer an outermost cell surface layer composed of protein or glycoprotein present on some *Bacteria* and *Archaea*

Sterols hydrophobic heterocyclic ringed molecules that strengthen the cytoplasmic membrane of eukaryotic cells and a few prokaryotes

MICROSCOPY AND CELL MORPHOLOGY

In Chapter 2 we introduced the concept of prokaryotic and eukaryotic cells. In this chapter we examine the structure of the prokaryotic cell, reserving discussion of eukaryotic cell structure for Chapter 14.

The basic design of all cells, *Bacteria*, *Archaea*, and *Eukarya*, share much in common. Like houses, cells are built by connecting building blocks to create more complex structures. Monomers are used to construct macromolecules, and these form into complexes to yield structures with defined functions—ribosomes, membranes, cell walls, and the like. We will see that despite great diversity in the *chemistry* of these macromolecules, cells face similar architectual problems and solve them in common ways.

We begin this chapter with a consideration of microscopy and then look at cell membranes and walls, cell surface structures and inclusions, and locomotion, respectively. We begin with microscopy because, historically, it was the microscope that first revealed the secrets of cell structure, and even today it remains a powerful tool in microbiology.

4.1 Light Microscopy

Visualization of microorganisms requires use of either the *light microscope* or the *electron microscope*. In general, light microscopes are used to look at intact cells at low magnification, while electron microscopes are used to look at internal structure or the details of cell surfaces.

All microscopes employ lenses to magnify the image of a cell such that details of its structure are more apparent. In addition to magnification, however, is the concept of **resolution**, the ability to distinguish two adjacent objects as distinct and separate. Although magnification can be increased virtually without limit, resolution cannot; resolution is dictated by the physical properties of light. It is thus *resolution* and not *magnification* that ultimately dictates what we are able to see with a microscope.

We begin our discussion with the light microscope, for which the limits of resolution are about 0.2 μm [0.2 micrometer or 200 nanometers (nm)]. We then proceed to describe the electron microscope, for which resolution is improved over that of the light microscope by about 1000-fold, allowing the examination of individual molecules.

The Compound Light Microscope

The **light microscope** uses visible light to illuminate cell structures. Several types of light microscopes are commonly used in microbiology: *bright-field, phase-contrast, dark-field,* and *fluorescence.*

With the **bright-field microscope** specimens are visualized because of the differences in contrast (density) that exist between them and the surrounding medium. Contrast differences arise because cells absorb or scatter light to varying degrees. The bright-field microscope is most commonly used in laboratory courses in biology and microbiology and consists of two series of lenses (objective lens and ocular lens), which function together to reveal the image (Figure 4.1●). Many bacterial cells are difficult to see well with the bright-field microscope because of their lack of contrast with the surrounding medium. Pigmented organisms are an exception, because the color of the organism adds contrast, thus improving visualization of the cells (Figure 4.2●).

Magnification and Resolution

The total magnification of a compound microscope is the *product* of the magnification of its objective and ocular lenses (Figure 4.1*b*). Magnifications of about 1500× are the upper limit with a compound light microscope. Above this limit, resolution does not improve. Resolution is a function of the wavelength of light used and a char-

acteristic of the objective lens known as its *numerical aperture* (a measure of light-gathering ability). In general, there is a correspondence between the magnification of a lens and its numerical aperture: Lenses with higher magnification typically have higher numerical apertures (the numerical aperture of a lens is stamped on the lens alongside the magnification). The diameter of the smallest object resolvable by any lens is equal to 0.5λ/numerical aperture, where λ is the wavelength of light used. Based on this formula, resolution is greatest when *blue light* is used to illuminate a specimen and the objective that is used has a *very high* numerical aperture.

As mentioned, the highest resolution possible in a compound light microscope is about 0.2 μm. This means that two objects closer together than 0.2 μm are not resolvable as distinct and separate. Most microscopes used in microbiology have oculars that magnify 10–15× and objectives of 10–100× (Figure 4.1*b*). At 1000×, objects 0.2 μm in diameter can then just be resolved. With the 100× objective, and with certain other objectives of very high numerical aperture, a high-grade optical oil is placed between the specimen and the objective. Lenses on which oil is used are called *oil-immersion lenses.* Immersion oil increases the light-gathering ability of a lens by allowing rays emerging from the specimen at angles that would otherwise be lost to the objective lens to be collected and viewed.

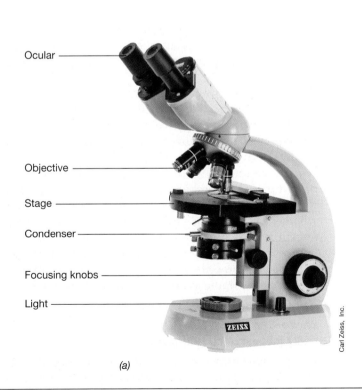

Ocular

Objective

Stage

Condenser

Focusing knobs

Light

Carl Zeiss, Inc.

(a)

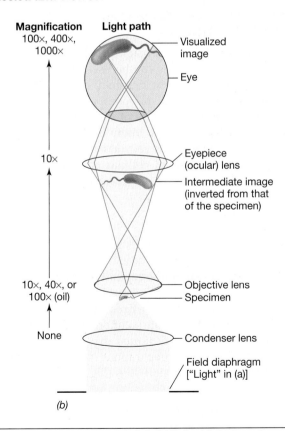

Magnification
100×, 400×, 1000×

Light path

Visualized image

Eye

10×

Eyepiece (ocular) lens

Intermediate image (inverted from that of the specimen)

10×, 40×, or 100× (oil)

Objective lens
Specimen

None

Condenser lens

Field diaphragm ["Light" in (a)]

(b)

● **Figure 4.1   Microscopy.**   (a) A compound light microscope. Various key parts of the microscope are labeled. (b) Path of light through a compound light microscope. Besides 10×, eyepieces (oculars) are available in 15–30×.

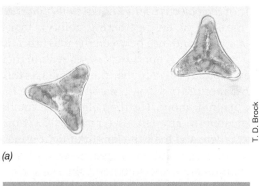

(a)

T. D. Brock

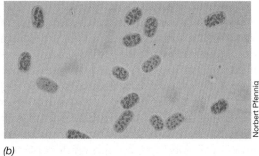

(b)

Norbert Pfennig

● **Figure 4.2 Photomicrographs of pigmented microorganisms by bright-field microscopy.** (a) A green alga (eukaryote). (b) A purple phototrophic bacterium (prokaryote). The algal cells are about 15 μm in diameter, and the bacterial cells are about 5 μm in diameter.

Staining: Increasing Contrast for Bright-Field Microscopy

One of the limitations of bright-field microscopy is insufficient contrast. Dyes can be used to stain cells and increase their contrast so that they can be more easily seen in the bright-field microscope. Dyes are organic compounds, and each class of dye has an affinity for specific cellular materials. Many dyes used in microbiology are positively charged (*basic dyes*) and combine with negatively charged cellular constituents such as nucleic acids and acidic polysaccharides. Examples of basic dyes include *methylene blue, crystal violet,* and *safranin*. Because cell surfaces also tend to be negatively charged, these dyes combine with high affinity to structures on the surfaces of cells and hence are excellent general-purpose stains.

For a **simple stain** one typically uses dried preparations of cell suspensions (Figure 4.3●). A slide containing a dried suspension of heat-fixed cells is flooded for a minute or two with a dilute solution of a dye, rinsed several times in water, and blotted dry. It is typical to observe dried stained preparations of bacteria with a high-power (oil-immersion) lens (Figure 4.3).

Differential Stains: The Gram Stain

Differential stains are so named because they do not stain all kinds of cells the same color. An important differential staining procedure widely used in microbiology is the **Gram stain** (Figure 4.4*a*●). On the basis of their reac-

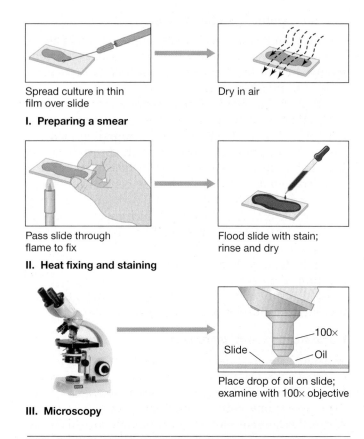

Spread culture in thin film over slide

Dry in air

I. Preparing a smear

Pass slide through flame to fix

Flood slide with stain; rinse and dry

II. Heat fixing and staining

100×
Slide
Oil

Place drop of oil on slide; examine with 100× objective

III. Microscopy

● **Figure 4.3 Staining cells for microscopic observation.** Stains improve the contrast between cells and their background.

tion to the Gram stain, bacteria can be divided into two major groups: *gram-positive* and *gram-negative*. After Gram staining, **gram-positive** bacteria appear purple and **gram-negative** bacteria appear red (Figure 4.4*b*). This difference in reaction to the Gram stain arises because of differences in the cell wall structure of gram-positive and gram-negative cells (as discussed later in this chapter). This leads to ethanol decolorizing gram-negative, but not gram-positive, cells (Figure 4.4).

The Gram stain is one of the most useful staining procedures in microbiology. Typically, one begins the characterization of a new bacterium by determining whether it is gram-positive or gram-negative. If a fluorescent microscope is available (see later discussion of fluorescence microscopy), the Gram stain can be reduced to a one-step procedure where gram-positive and gram-negative cells fluoresce different colors (Figure 4.4*c*).

Phase-Contrast, Dark-Field, and Fluorescence Microscopy

The **phase-contrast microscope** was developed to improve contrast differences between cells and the surrounding medium, making it possible to see cells without staining them (Figure 4.5●). The phase-contrast microscope is widely employed in research because it can be used to observe wet-mount (living) preparations.

The Gram Stain

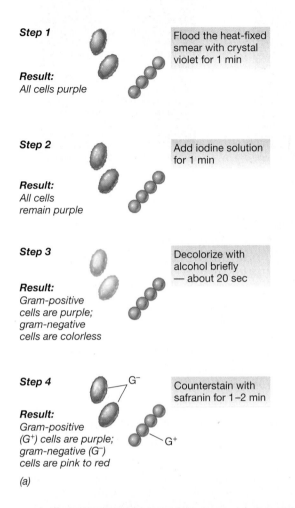

Step 1

Result:
All cells purple

Flood the heat-fixed smear with crystal violet for 1 min

Step 2

Result:
All cells remain purple

Add iodine solution for 1 min

Step 3

Result:
Gram-positive cells are purple; gram-negative cells are colorless

Decolorize with alcohol briefly — about 20 sec

Step 4

Result:
Gram-positive (G⁺) cells are purple; gram-negative (G⁻) cells are pink to red

G⁻

G⁺

Counterstain with safranin for 1–2 min

(a)

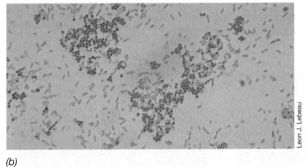

(b)

Leon J. Lebeau

(c)

Molecular Probes, Inc., Eugene, Oregon

Staining, on the other hand, although a widely used procedure in light microscopy, kills cells and can distort their features.

Phase-contrast microscopy was discovered by the Dutch mathematical physicist Frits Zernike in 1936. It is based on the principle that cells slow the speed of light passing through them and thus differ in *refractive index* from their surroundings. This results in a difference in "phase" between the cell itself and its surroundings, and this subtle difference is amplified by a special ring in the objective lens of a phase-contrast microscope. This leads to the formation of a dark image on a light background (Figure 4.5*b*). The ring consists of a "phase plate," the key discovery of Zernike, that amplifies this minute variation in phase. Zernike's discovery of differences in contrast between cells and background stimulated later innovations in microscopy, such as fluorescence and confocal microscopy (see later), where, like phase-contrast, differences in contrast form the heart of the technique. For his discovery of phase-contrast microscopy, Zernike was awarded the 1953 Nobel Prize in Physics.

The **dark-field microscope** is a light microscope in which the lighting system has been modified to reach the specimen from the sides only. The only light reaching the lens is scattered by the specimen, and thus the specimen appears light on a dark background (Figure 4.5*c*). Resolution by dark-field microscopy is somewhat better than by light microscopy, and thus objects can often be resolved by dark-field that are not resolvable in bright-field or even phase-contrast microscopes. Dark-field microscopy is also an excellent way to observe the motility of microorganisms, as bundles of flagella are often resolvable with this technique (see Figure 4.55*a*).

The **fluorescence microscope** is used to visualize specimens that *fluoresce*, that is, emit light of one color when light of another color shines upon them (Figure 4.6●). Fluorescence occurs either because of the presence within cells of naturally fluorescent substances such as chlorophyll or other fluorescing components (*autofluorescence*) (see Figure 4.6*a, b*), or because the cells have been treated with a fluorescent dye (Figures 4.4*c* and 4.6*c*). DAPI (diamidino-2-phenylindole) is a widely used fluorescent dye, staining cells a bright blue (∞Figure 18.6). Using DAPI, cells can be identified in a complex milieu, such as soil, water, food, or a clinical specimen (∞Section 18.3). Fluorescence microscopy is widely used in clinical diagnostic microbiology and also in microbial ecology (∞Chapters 18–20 and 24).

● **Figure 4.4   The Gram stain.** (a) Steps in the Gram stain procedure. (b) Photomicrograph of Gram-stained *Bacteria* that are gram-positive (blue-purple) and gram-negative (pink-red). The species are *Staphylococcus aureus* and *Escherichia coli*, respectively. (c) Photomicrograph of cells of *Pseudomonas aeruginosa* (gram-negative, green) and *Bacillus cereus* (gram-positive, orange) stained with a one-step fluorescent staining method. This method allows for differentiating gram-positive from gram-negative cells in a single staining step.

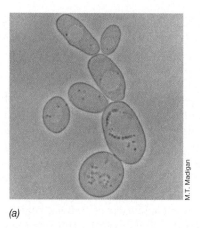

(a)

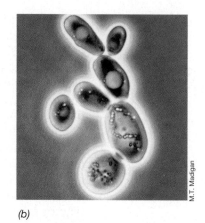

(b)

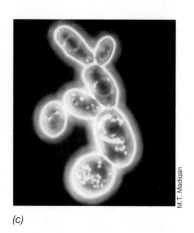

(c)

● **Figure 4.5 Photomicrographs of the same field of cells of the baker's yeast *Saccharomyces cerevisiae,* taken by different types of light microscopy.** (a) Bright-field. (b) Phase contrast. (c) Dark-field. Cells average 8–10 μm in diameter.

 4.1 Concept Check

Microscopes are essential for microbiological studies. Various types of light microscopes exist, including bright-field, dark-field, phase contrast, and fluorescence microscopes. In bright-field microscopy, stains are necessary to increase contrast.

◆ Define the term *resolution*.

◆ What is the upper limit of magnification for a light microscope?

◆ What light microscopic techniques can sometimes improve resolution?

◆ What color would a gram-negative bacterium be after Gram staining by the conventional method?

 4.2 Three-Dimensional Imaging: Interference Contrast, Atomic Force, and Confocal Scanning Laser Microscopy

In the types of light microscopy just considered, the images obtained are essentially two dimensional. How can this limitation be overcome? We will see in the next section that the scanning electron microscope offers one solution to this problem, but so can certain forms of light microscopy.

Differential Interference Contrast Microscopy

Differential interference contrast (DIC) is a type of light microscopy that employs a polarizer to produce polarized light. The polarized light then passes through a prism that generates two distinct beams. These beams traverse the specimen and enter the objective lens where they are recombined into one. Because the two beams pass through different substances with slightly different

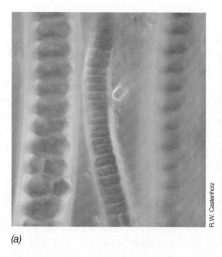

(a)

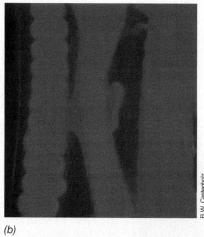

(b)

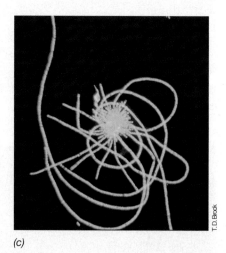

(c)

● **Figure 4.6 Photomicrographs of various microorganisms as visualized by fluorescence microscopy.** (a, b) Cyanobacteria. (a) Cells observed by bright-field microscopy. (b) Same cells observed by fluorescence (cells exposed to light of 546 nm). The red color is due to autofluorescence of chlorophyll *a* and other pigments. (c) Cells of the filamentous bacterium *Leucothrix mucor* stained with the fluorescent dye, acridine orange, which fluoresces green. Cells are 3 μm in diameter and may reach lengths of greater than 100 μm.

refractive indices, the combined beams are not totally in phase but instead create an interference effect. This effect intensifies subtle differences in cell structure. Thus, by DIC microscopy, structures such as the nucleus of eukaryotic cells (Figure 4.7a●), and endospores, vacuoles, and granules of prokaryotic cells, attain a three-dimensional appearance. DIC microscopy is also particularly useful for observing *unstained* cells because it can reveal internal cell structures that are less apparent (or even invisible) by bright-field techniques (compare Figure 4.5a with Figure 4.7a).

Atomic Force Microscopy

Another type of microscopy useful for three-dimensional imaging of biological structures is the **atomic force microscope (AFM)**. In atomic force microscopy, a tiny stylus is positioned extremely close to the specimen such that weak repulsive atomic forces are established between the probe and the specimen. As the specimen is scanned in both the horizontal and vertical directions, the stylus rides up and down the hills and valleys, constantly recording its interactions with the surface. This pattern is processed by a series of detectors that feed the digital information into a computer that generates an image (Figure 4.7b).

Although the images obtained from an atomic force microscope appear similar to those from the scanning electron microscope (compare Figure 4.7b with Figure 4.10b), the AFM has one advantage—no fixatives or coatings are required. The AFM thus allows living and hydrated specimens to be viewed, something that is generally not possible with electron microscopes.

Confocal Scanning Laser Microscopy

Confocal scanning laser microscopy (CSLM) is a computerized microscope that couples a laser source to a light microscope. This technique allows for the generation of three-dimensional digital images of microorganisms and other biological specimens (Figure 4.8●).

In CSLM, a laser beam is bounced off a mirror that directs the beam through a scanning device. Then the laser beam is directed through a pinhole that precisely adjusts the plane of focus of the beam to a given vertical layer within a specimen. By precisely illuminating only a single plane of the specimen, illumination intensity drops off rapidly above and below the plane of focus. Because of this, stray light from other planes of focus is minimized. Thus, in a relatively thick specimen such as a microbial biofilm (Figure 4.8a), not only are cells on the surface of the biofilm apparent, as would be the case with conventional light microscopy, but cells in the various layers can also be observed by adjusting the plane of focus of the laser beam.

Cells in CSLM preparations are frequently stained with fluorescent dyes to make them more distinct

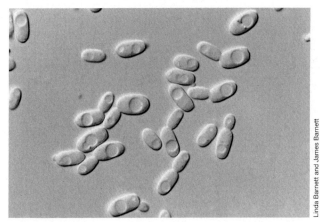

(a)

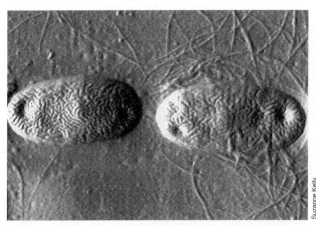

(b)

Linda Barnett and James Barnett

Suzanne Kelly

● **Figure 4.7 Three-dimensional imaging of cells.** (a) Interference contrast microscopy and (b) atomic force microscopy. The yeast cells in (a) are about 8 μm in diameter. Note how the nucleus is clearly visible here (compare Figure 4.7a with Figure 4.5a). The bacterial cells in (b) are about 2.2 μm in length, and the micrograph was taken from a natural biofilm that developed on the surface of a glass slide immersed for 24 h in a dog's water bowl. The slide was air dried before viewing with an atomic force microscope.

(Figure 4.8a). Alternatively, false color images can be generated by adjusting the microscope in such a way as to make different layers take on different colors. The laser confocal microscope is equipped with computer software to assemble digital images for subsequent image processing. Thus, images obtained from different layers can be stored and then digitally overlaid to reconstruct a three-dimensional image of the entire specimen (Figure 4.8a).

CSLM has found widespread use in microbial ecology, especially for identifying phylogenetically distinct populations of cells present in a microbial habitat (see for example, Figure 18.11b), or for resolving the different layered components as in a biofilm (Figure 4.8a; (⮾Figure 19.4b). However, CSLM is useful anywhere thick specimens need to be examined to see how their microbial content varies with depth.

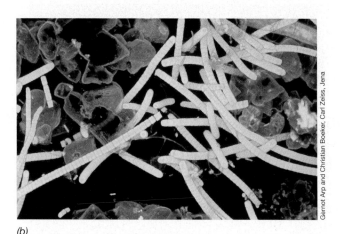

(a)

(b)

● **Figure 4.8   Confocal scanning laser microscopy.**   (a) Confocal image of a mixed microbial biofilm community cultivated in the laboratory. The green, rod-shaped cells are *Pseudomonas aeruginosa* experimentally introduced into the biofilm. Other cells that are different colors are present at different depths in the biofilm. (b) Confocal micrograph of a filamentous cyanobacterium growing in a soda lake.

⬣ 4.2   Concept Check

Interference contrast (DIC) and confocal scanning (CSLM) are forms of light microscopy that allow for greater three-dimensional imaging than other forms of light microscopy, and confocal microscopy allows imaging through thick specimens. The atomic force microscope yields a detailed three-dimensional image of live preparations.

◆   What structure in eukaryotic cells is more easily seen using DIC than bright-field microscopy? (Hint: compare Figures 4.5*a* and 4.7*a*).

◆   How is CSLM able to view different layers in a thick preparation?

4.3   Electron Microscopy

Electron microscopes use electrons instead of photons to image cells or cell structures. In the **transmission electron microscope (TEM)** electromagnets function as lenses, and the whole system operates in a vacuum (Figure 4.9●). Electron microscopes are fitted with cameras to allow a photograph, called an *electron micrograph*, to be taken.

The transmission electron microscope is typically used to examine internal cell structure. The resolving power of the electron microscope is much greater than that of the light microscope, enabling one to view structures at the molecular level (∞Figure 2.4*b*). For example, whereas the resolving power of a good light microscope is about 0.2 *micrometers*, the resolving power of a good TEM is about 0.2 *nanometers*. Thus, individual molecules, such as proteins and nucleic acids, can be visualized in the electron microscope.

Unlike visible light, electron beams do not penetrate very well; even a single cell is too thick to be viewed directly. Consequently, special techniques of *thin sectioning* are needed to prepare specimens for the electron microscope. A single bacterial cell, for instance, is cut into many very thin (20–60 nm) slices, which are then examined individually with the electron microscope (Figure 4.10*a*●). To obtain sufficient contrast, the preparations are treated with stains such as osmic acid, or permanganate, uranium, lanthanum, or lead salts. Because these substances are composed of atoms of high atomic weight, they scatter electrons well and thus improve contrast (Figure 4.10*a*).

Scanning Electron Microscopy

If only the *external* features of an organism need to be observed, thin sections are unnecessary. Intact cells or cell components can be observed directly by TEM with a technique called *negative staining* (see, for example, Figure 4.54). Alternatively, one can use the **scanning electron microscope (SEM)** (Figures 4.9 and 4.10*b*).

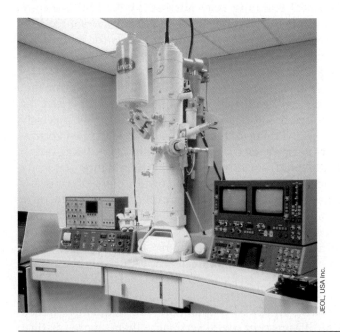

● **Figure 4.9   The electron microscope.**   This instrument encompasses both transmission and scanning electron microscope functions.

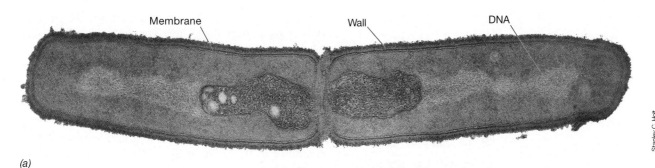

Membrane   Wall   DNA

(a)

Stanley C. Holt

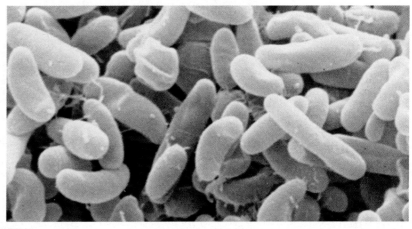

(b)

F. R. Turner

● **Figure 4.10 Electron micrographs of bacterial cells.** (a) Transmission electron micrograph, and (b) scanning electron micrograph. (a) Thin section of a typical gram-positive bacterium, *Bacillus subtilis.* The cell has just divided, and two membrane-containing structures are attached to the cross-wall. Note the light region in the middle (DNA or the *nucleoid*). The cell is about 0.8 μm in diameter. (b) Cells of the phototrophic bacterium *Rhodo-vibrio sodomensis.* A single cell is about 0.75 μm wide. Note how the scanning micrograph allows for great depth of field and thus excellent three-dimensional imaging.

In scanning electron microscopy, the specimen is coated with a thin film of a heavy metal such as gold. An electron beam from the SEM is then directed onto the specimen and scans back and forth across it. Electrons scattered by the metal are collected, and they activate a viewing screen to produce an image (Figure 4.10b). In the SEM, even fairly large specimens can be observed, and the depth of field is extremely good. A wide range of magnifications can be obtained with the SEM, from as low as 15× up to about 100,000×, but only the *surface* of an object can be visualized.

 4.3 Concept Check

Electron microscopes have far greater resolving power than do light microscopes, the limits of resolution being about 0.2 nm. Two major types of electron microscopy are performed: transmission electron microscopy, for observing internal cell structure down to the molecular level, and scanning electron microscopy, useful for three-dimensional imaging and for examining surfaces.

◆ What is an *electron micrograph?* How does an electron micrograph differ from a photomicrograph in terms of how the image is obtained?

◆ Keeping in mind that chemical fixatives are necessary and that electron microscopes must operate in a vacuum, what major *disadvantage* do electron microscopes have compared with light microscopes?

◆ What type of electron microscope would you use to observe the bacterial nucleoid?

4.4 Cell Morphology and the Significance of Being Small

In biology, the term **morphology** refers to cell *shape.* Several morphologies are known among prokaryotes and most have recognized terms to describe them. We explore morphology here and then look at some of the biological benefits of small cells.

Major Cell Morphologies

Schematic examples of typical bacterial shapes along with phase photomicrographs of example organisms are shown in Figure 4.11●. A bacterium that is spherical or ovoid in morphology is called a **coccus** (plural, **cocci**). A bacterium with a cylindrical shape is called a **rod**. Some rods are curved, frequently forming spiral-shaped patterns, and are then called **spirilla**. In many prokaryotes, cells remain together in groups or clusters after division, and the arrangements in these groups are often characteristic of different organisms. For instance, cocci or rods may occur in long chains. Some cocci form sheets of cells, whereas others occur in three-dimensional cubes or irregular cubelike clusters.

Several groups of bacteria are immediately recognizable by their unusual shapes. Examples include **spirochetes**, which are tightly coiled bacteria, **appendaged bacteria**, which possess extensions of their cells as long tubes or stalks, and **filamentous bacteria**, which form long, thin cells or chains of cells (Figure 4.11). The

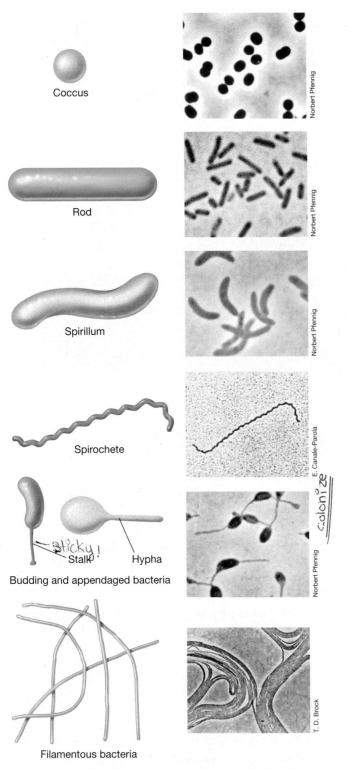

Coccus

Rod

Spirillum

Spirochete

Stalk   Hypha

Budding and appendaged bacteria

Filamentous bacteria

● **Figure 4.11   Representative cell shapes (morphology) in prokaryotes.**   Next to each drawing is a phase photomicrograph showing an example of that morphology. Organisms are coccus, *Thiocapsa roseopersicina* (diameter of a single cell = 1.5 μm); rod, *Desulfuromonas acetoxidans* (diameter = 1 μm); spirillum, *Rhodospirillum rubrum* (diameter = 1 μm); spirochete, *Spirochaeta stenostrepta* (diameter = 0.25 μm); budding and appendaged, *Rhodomicrobium vannielii* (diameter = 1.2 μm); filamentous, *Chloroflexus aurantiacus* (diameter = 0.8 μm).

cell shapes in Figure 4.11 should be studied with the understanding that these are *representative* morphologies. Many variations of these basic morphological types, both subtle and distinct, are known in the microbial world.

The Size of Microbial Cells and the Significance of Being Small

Prokaryotes vary in size from cells smaller than 0.2 μm in diameter to those more than 50 μm in diameter. A few very large prokaryotes, such as the surgeonfish symbiont *Epulopiscium fishelsoni* (Figure 4.12a●), are up to 80 μm in

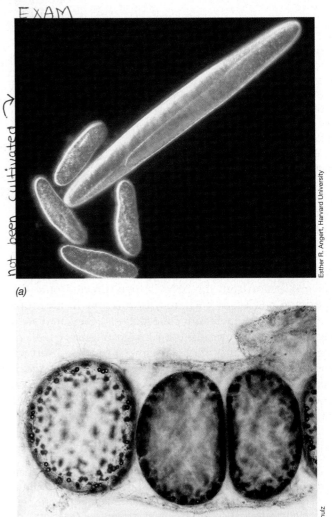

(a)

(b)

● **Figure 4.12   Some very large prokaryotes.**   (a) Dark-field photomicrograph of a giant prokaryote, the surgeonfish symbiont *Epulopiscium fishelsoni*. The rod-shaped *E. fishelsoni* cell in this field is about 600 μm (0.6 mm) long and 75 μm wide and is shown with four cells of the protozoan (eukaryote) *Paramecium*, each of which measures about 150 μm in length. *E. fishelsoni* is a member of the *Bacteria* and phylogenetically related to *Clostridium* species. (b) *Thiomargarita namibiensis*, a large sulfur chemolithotroph (phylum Proteobacteria of the *Bacteria*) and currently the largest known prokaryote. A single coccus-shaped cell is about 400 μm wide. See also Table 4.1 for some large and small prokaryotes.

diameter and can be more than 0.6 millimeters (mm) in length (Table 4.1). The largest known prokaryote in terms of total cell volume, the sulfur chemolithotroph *Thiomargarita* (Figure 4.12*b*) can be 0.75 mm (750 μm) in diameter, nearly visible with the naked eye.

Prokaryotic cells vary dramatically in both size and volume (Table 4.1). Most very large prokaryotes are either sulfur chemolithotrophs or cyanobacteria (∞ Chapter 2 and Table 4.1). Why these cells are so large is not well understood, although for sulfur bacteria large cell size may be a mechanism for storing substrates to high levels. It is thought that limitations in diffusion and metabolism ultimately dictate the upper limit for the size of a prokaryotic cell. The metabolic rate of a cell varies inversely with the square of its size. Thus, for very large cells, diffusion processes may eventually limit metabolism such that the cell is no longer competitive.

Very large cells are not the norm in the prokaryotic world. By contrast to *Thiomargarita* or *Epulopiscium* (Figure 4.12), the dimensions of an average rod-shaped prokaryote, the bacterium *Escherichia coli*, for example, are about 1 × 3 μm and are typical for the vast majority of prokaryotes. For comparison, eukaryotic cells may be 2 μm to more than 200 μm in diameter. In general, prokaryotes are thus very small cells compared with eukaryotes.

The likely reason most prokaryotes are very small is that there are significant advantages to being small. For example, nutrients and waste products pass more readily into and out of a small cell than a large cell, thus accelerating cellular metabolism and growth. This is because relative to cell volume, small cells contain more *surface area* than do large cells. Consider the simple case of a sphere, in which the *volume* is a function of the cube of the radius ($V = 4/3 \pi r^3$), while the *surface area* is a function of the square of the radius ($S = 4\pi r^2$). The surface-to-volume (S/V) ratio of a sphere can thus be expressed as $3/r$ (Figure 4.13●). A cell with a smaller r value therefore has a *higher* S/V ratio than a cell with a larger r value.

Besides the exchange of nutrients, the S/V ratio affects other aspects of a cell's biology. For instance, because growth rate depends to some extent on the rate of nutrient exchange, the higher S/V of small cells typically supports more rapid growth than for larger cells.

Table 4.1	Cell size and volume of prokaryotic cells, from the largest to the smallest		
Organism	**Characteristics**	**Size**[a] (μm)	**Cell volume** (μm^3)
Thiomargarita namibiensis	Spherical sulfur chemolithotroph	750	200,000,000
Epulopiscium fishelsoni	Chemoorganotrophic bacterium	80 × 600	3,000,000
Beggiatoa sp.	Filamentous sulfur chemolithotroph	50 × 160	1,000,000
Achromatium oxaliferum	Ellipsoid sulfur bacterium	35 × 95	80,000
Lyngbya majuscula	Filamentous cyanobacterium	8 × 80	40,000
Prochloron sp.	Prochlorophyte	30	14,000
Thiovulum majus	Spherical sulfur chemolithotroph	18	3,000
Staphylothermus marinus	Hyperthermophile	15	1,800
Titanospirillum velox	Rod-shaped sulfur chemolithotroph	5 × 30	600
Magnetobacterium bavaricum	Magnetotactic bacterium	2 × 10	30
Escherichia coli	Chemoorganotrophic bacterium	1 × 2	2
Mycoplasma pneumoniae	Pathogenic bacterium	0.2	0.005

[a]Where only one number is given, this is the diameter of spherical cells. The values given are for the largest cell size observed in each species. For example, for *T. namibiensis*, an average cell is only about 200 μm in diameter. But on occasion, giant cells of 750 μm are observed. Likewise, an average cell of *S. marinus* is about 1 μm in diameter.

Source: Data obtained from Schulz, H.N., and B.B. Jørgensen. 2001. *Ann. Rev. Microbiol.* 55: 105–137.

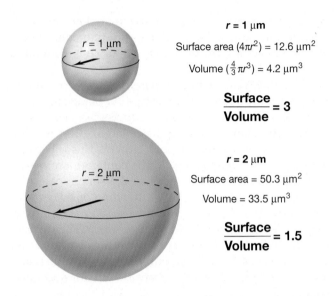

r = 1 μm

Surface area $(4\pi r^2)$ = 12.6 μm²

Volume $(\frac{4}{3}\pi r^3)$ = 4.2 μm³

$$\frac{Surface}{Volume} = 3$$

r = 2 μm

Surface area = 50.3 μm²

Volume = 33.5 μm³

$$\frac{Surface}{Volume} = 1.5$$

● **Figure 4.13   Surface area and volume relationships in cells.**
As a cell *increases* in size, its surface area-to-volume ratio *decreases*.

Moreover, per unit of available resources, small cells will typically develop larger populations than will large cells. And since mutation rates are the same for all organisms, larger cell populations mean more cell divisions and this means more mutations accumulate from spontaneous errors in DNA replication. Mutations are the "raw material" that drives evolutionary change. Coupling this with the fact that prokaryotes are genetically haploid (∞Section 2.2), a condition that allows beneficial mutations to be immediately expressed, the small size of prokaryotes may ultimately explain why these organisms tend to adapt rapidly to changing environmental conditions and easily exploit new habitats.

Lower Limits of Cell Size

From the foregoing it may seem that smaller and smaller bacteria will have greater and greater selective advantages in nature. Obviously there are lower limits to cell size, but what are these? Some microbiologists have proposed that *very* small bacteria exist in nature, cells referred to as *nanobacteria*. The size of putative nanobacteria are on the order of 0.1 μm in diameter for coccus-shaped structures, or even smaller. This is extremely small, even by prokaryotic standards (Table 4.1). Are nanobacteria really cells?

Most reports of nanobacteria have been associated with their supposed formation of precipitates and biofilms (∞Section 19.3) in environments as diverse as mineral surfaces and human tissues. Although some microbiologists firmly believe in the nanobacteria concept, others have claimed that nanobacteria are simply artifacts of chemical or geochemical reactions of nonliving materials and that even the smallest known bacterial cells are significantly larger than reported nanobacteria (Table 4.1). Moreover, if one considers the space needed to house all of the essential biomolecules of life, it is

highly unlikely that these could exist within the volume available to a structure of 0.1 μm or less. Thus, whether nanobacteria are living organisms or not is an unsettled question. If such very small cells actually exist, they would be the smallest known living structures.

Regardless of the status of nanobacteria, many microbial habitats clearly contain very small cells. The open oceans, for example, contain 10^4–10^5 prokaryotic cells per milliliter, and these tend to be very small cells, 0.2–0.4 μm in diameter. We will see that many pathogenic bacteria are very small as well. Thus one can conclude that very small cells do exist in nature, but that structures smaller than about 0.2 μm may not be viable (capable of reproducing) cells.

 4.4   Concept Check

Prokaryotes are typically smaller in size than eukaryotes, and prokaryotic cells can have a wide variety of morphologies. The small size of prokaryotic cells affects their physiology, growth rate, and ecology. Cell-like structures smaller than about 0.2 μm may or may not be living organisms.

◆   List three morphological types of prokaryotes.

◆   What physical property of cells *increases* as cells become smaller?

II CELL MEMBRANES AND CELL WALLS

We now consider two extremely important cell structures in prokaryotes: the cytoplasmic membrane and the cell wall. Each carries out well-defined and critical functions for the cell, including the transport of nutrients (membrane) and the prevention of osmotic lysis (wall).

4.5   Cytoplasmic Membrane: Structure

The **cytoplasmic membrane** is a thin structure that surrounds the cell. Only about 8 nm thick, this vital structure is the barrier separating the inside of the cell (the cytoplasm) from its environment. If the membrane is broken, the integrity of the cell is destroyed, the cytoplasm leaks into the environment, and the cell dies. The cytoplasmic membrane is also a *highly selective permeability barrier*, enabling a cell to concentrate specific metabolites and excrete waste materials.

Chemical Composition of Membranes

The general structure of biological membranes is a **phospholipid bilayer** (Figure 4.14●). As discussed in Section 3.4, phospholipids contain both hydrophobic (fatty acid) and hydrophilic (glycerol-phosphate) components and can exist in many different chemical forms as a result of variation in the groups attached to the glycerol backbone. As phospholipids aggregate in an aqueous solution, they naturally form bilayer structures spontaneously. In a phos-

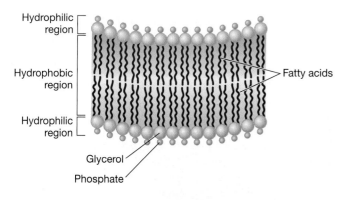

Hydrophilic region

Hydrophobic region

Hydrophilic region

Fatty acids

Glycerol

Phosphate

● **Figure 4.14   Structure of a phospholipid bilayer.** The cytoplasmic membrane is about 8 nm (80 Å) wide. Review Figure 3.7 for the structure of a phospholipid.

pholipid, the fatty acids point inward toward each other to form a hydrophobic environment, while the hydrophilic portions remain exposed to the aqueous external environment (Figure 4.14). The bilayer character of membranes represents the most stable arrangement of lipid molecules in an aqueous environment.

Thin sections of the cytoplasmic membrane can be seen with the electron microscope (Figure 4.15a●). The cytoplasmic membrane appears as two light-colored lines separated by a darker area (Figure 4.15a). This **unit membrane**, as it is called (because each phospholipid leaf forms half of the "unit"), consists of a phospholipid (∞Figure 3.7) bilayer with proteins embedded in it (Figure 4.16●). The major proteins of the cytoplasmic membrane typically have very hydrophobic external surfaces in the regions of the protein that span the membrane, and hydrophilic surfaces that make contact with the environment and the cytoplasm (Figure 4.16). The overall structure of the cytoplasmic membrane is stabilized by hydrogen bonds and hydrophobic interactions (∞Section 3.1). In addition, cations such as Mg^{2+} and Ca^{2+} help stabilize the membrane by combining ionically with negative charges of the phospholipids.

Membrane Proteins

The outer surface of the cytoplasmic membrane faces the environment and in certain bacteria makes contact with a variety of proteins that bind substrates or process large molecules for transport into the cell (periplasmic proteins, see discussion in Sections 4.7 and 4.9). The inner side of the cytoplasmic membrane faces the cytoplasm and interacts with proteins involved in energy-yielding reactions and other important cellular functions.

Many membrane proteins are firmly embedded in the membrane and are called *integral membrane proteins*. Other proteins are not embedded in the membrane but associate quite firmly with one of the membrane surfaces and actually function as if they were membrane-bound proteins. These include proteins in the periplasm (a region between the cytoplasmic membrane and the outer membrane of gram-negative bacteria, see Section 4.9) and some cytoplasmic proteins. Some of these *peripheral membrane proteins*, as they are called, are lipoproteins, which contain a lipid tail on the amino terminus of the protein that anchors the protein in the membrane. These proteins interact directly with integral membrane proteins in important cellular processes such as energy metabolism.

In a diagram, the cytoplasmic membrane may appear somewhat rigid. In reality, however (Figures 4.15 and 4.16), the cytoplasmic membrane is quite fluid—phospholipid and protein molecules have significant freedom to move about within the membrane. Membranes have a viscosity approximating that of a light-grade oil. Thus, integral membrane proteins likely span a highly mobile, yet highly ordered, phospholipid bilayer. We will see in the next section how this *fluid mosaic* structure guides the function of the membrane as well.

Membrane Strengthening Agents: Sterols and Hopanoids

One major difference in chemical composition of membranes between eukaryotic and prokaryotic cells is that eu-

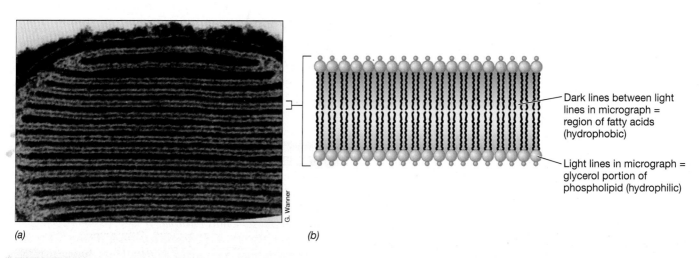

G. Wanner

(a)   (b)

Dark lines between light lines in micrograph = region of fatty acids (hydrophobic)

Light lines in micrograph = glycerol portion of phospholipid (hydrophilic)

● **Figure 4.15   The cytoplasmic membrane.**   (a) Electron micrograph of photosynthetic membrane stacks derived from the cytoplasmic membrane in the phototrophic bacterium *Halorhodospira halochloris*. Note the multiple lipid bilayers. Each bilayer is about 8 nm thick. (b) Enlarged schematic view of a single unit membrane shown in (a).

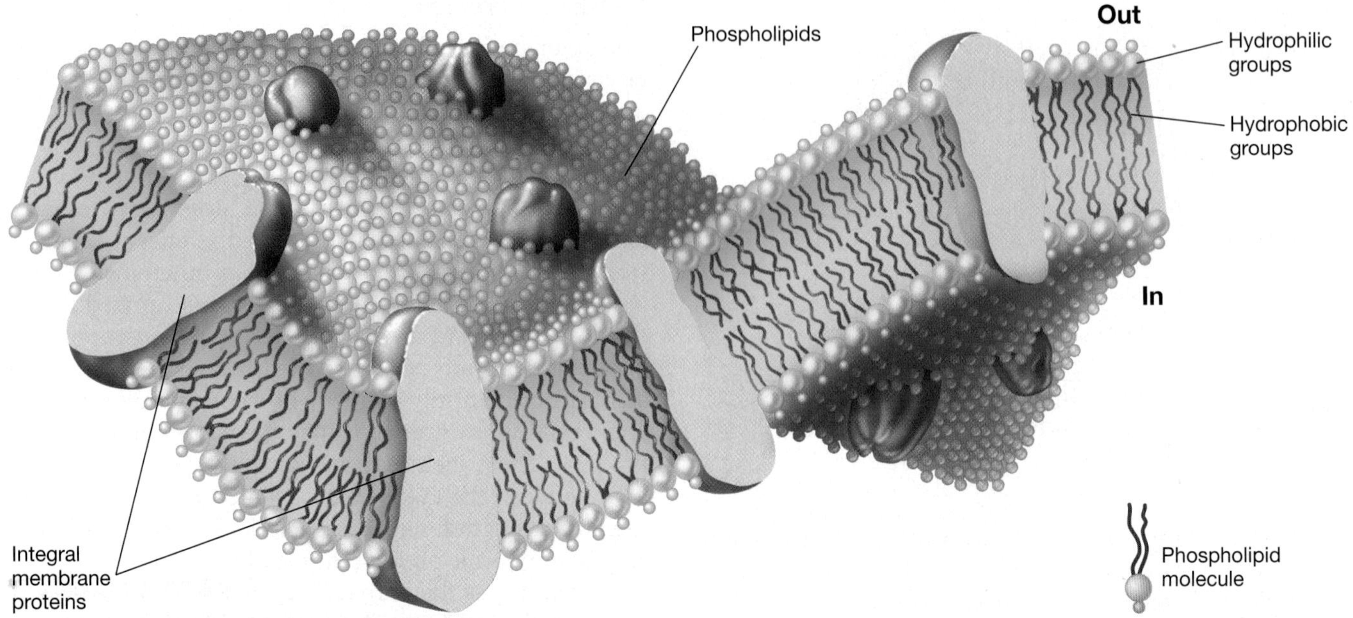

● **Figure 4.16** **Structure of the cytoplasmic membrane.** The inner surface (**In**) faces the cytoplasm and the outer surface (**Out**) faces the environment. The matrix of the cytoplasmic membrane is composed of phospholipids, with the hydrophobic groups directed inward and the hydrophilic groups toward the outside, where they associate with water. Embedded in the matrix are proteins that have considerable hydrophobic character in the region that traverses the fatty acid bilayer. Hydrophilic proteins and other charged substances, such as metal ions, may be attached to the hydrophilic surfaces. Although there are some chemical differences, the overall structure of the cytoplasmic membrane shown is similar in both prokaryotes and eukaryotes (but see an exception to the bilayer design in Figure 4.19*d*).

karyotes have **sterols** in their membranes (Figure 4.17*a*, *b*●). Sterols are absent from the membranes of virtually all prokaryotes (methanotrophic bacteria and the mycoplasmas are exceptions, ∞Sections 12.6 and 12.21). Depend-

● **Figure 4.17** **Sterols and hopanoids.** (a) The structure of cholesterol, a typical sterol. (b) The structure of the hopanoid diploptene. Sterols are found in the membranes of eukaryotes and hopanoids in the membranes of some prokaryotes. The intra-ring labels (1, 2, and 3) are meant to highlight similarities in the parent structure of sterols and hopanoids.

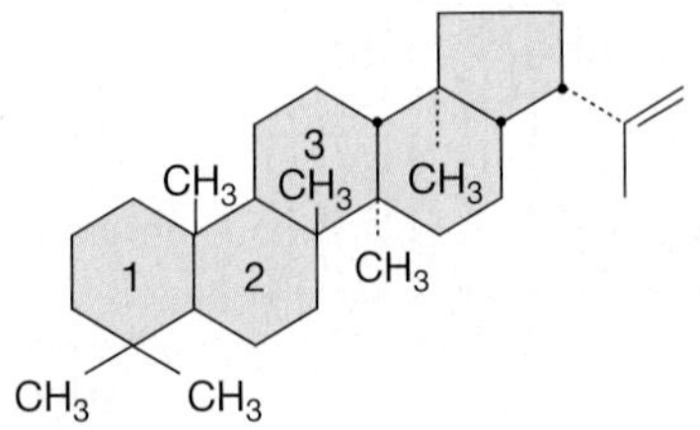

ing on the cell type, sterols can make up from 5 to 25% of the total lipids of eukaryotic membranes.

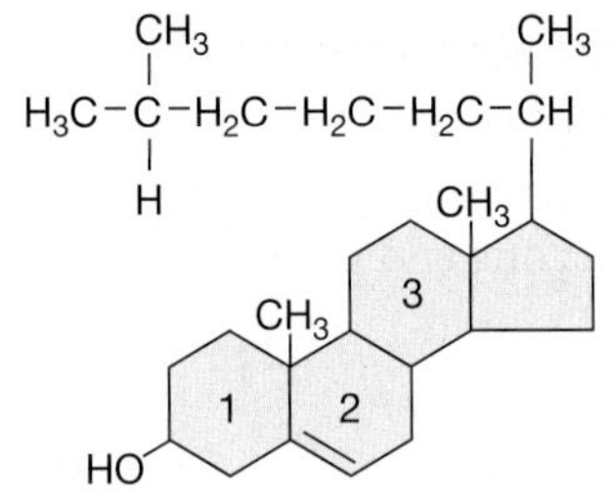

Sterols and related molecules are rigid, planar molecules, whereas fatty acids are flexible. The presence of sterols in a membrane thus stabilizes it and makes it less flexible. Molecules similar to sterols, called *hopanoids*, are present in the membranes of many *Bacteria*, and likely play a role similar to that of sterols in eukaryotic cells. One widely distributed hopanoid is the C_{30} hopanoid *diploptene* (Figure 4.17*b*). As far as is known, hopanoids are not present in species of *Archaea*.

Archaeal Membranes

The lipids of *Archaea* differ from those of other organisms. In contrast to the lipids of *Bacteria* and *Eukarya* in which *ester* linkages bond the fatty acids to glycerol (Figure 4.18*a*●; ∞Section 3.4), the lipids of *Archaea* have *ether* linkages between glycerol and their hydrophobic side chains. In addition, archaeal lipids lack fatty acids. Instead, their side chains are composed of repeating units of the five-carbon hydrocarbon *isoprene* (Figure 4.18*c*). Nevertheless, the overall architecture of the cytoplasmic membrane of *Archaea*, forming inner and outer hydrophilic surfaces with a hydrophobic interior, is the same as in *Bacteria* and *Eukarya*.

Glycerol *diethers* and glycerol *tetraethers* (Figure 4.19*a*, *b*●) are the major lipids present in *Archaea*. In the *tetraether* molecule, the phytanyl (composed of four linked isoprenes)

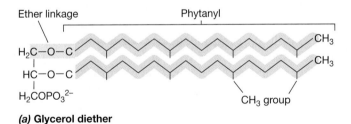

● **Figure 4.18 Chemical bonds in lipids.** (a) The *ester* linkage as found in the lipids of *Bacteria* and *Eukarya*. (b) The *ether* linkage of lipids from *Archaea*. (c) Isoprene, the parent structure of the hydrophobic side chains of archaeal lipids. By contrast, in lipids of *Bacteria* and *Eukarya*, the side chains are composed of fatty acids.

side chains from each glycerol molecule are *covalently bonded* together (Figure 4.19b). This structure therefore yields a lipid *monolayer* instead of a lipid *bilayer* cytoplasmic membrane (Figure 4.19d). Lipid monolayers are quite resistant to peeling apart. This membrane structure is therefore widespread among hyperthermophilic *Archaea*, prokaryotes that grow at very high temperatures (∞Sections 6.10 and 13.5–13.10).

We will encounter many other features that set the prokaryotic organisms *Bacteria* and *Archaea* apart, but the chemistry of their membrane lipids is a major defining feature of each phylogenetic group.

◉ **4.5 Concept Check**

The cytoplasmic membrane is a highly selective permeability barrier constructed of lipids and proteins that forms a bilayer with hydrophilic exteriors and a hydrophobic interior. Other molecules, such as sterols and hopanoids, may strengthen the membrane, and integral proteins involved in transport and other functions traverse it. Unlike *Bacteria* and *Eukarya*, *Archaea* contain ether-linked lipids, and some species have membranes of monolayer instead of bilayer construction.

◆ Draw the basic structure of a lipid bilayer.

◆ Why are compounds like sterols and hopanoids good at stabilizing the cytoplasmic membrane?

◆ Contrast the linkage between glycerol and the hydrophobic portion of lipids in *Bacteria* and *Archaea*.

4.6 **Cytoplasmic Membrane: Function**

The cytoplasmic membrane is more than just a barrier separating the inside from the outside of the cell. The membrane plays several critical roles in cell function. First and foremost, the membrane functions as a *permeability barrier*, preventing the passive leakage of cytoplasmic constituents into or out of the cell (Figure 4.20●). In addition, the membrane is the location of many proteins. Some of these are enzymes involved in bioenergetic functions and others are involved in the transport of substances into and out of the cell.

● **Figure 4.19 Major lipids of *Archaea* and the structure of archaeal membranes.** (a) Glycerol diethers. (b) Diglycerol tetraethers. Note that in both cases, the hydrocarbon is attached to the glycerol by *ether* linkages. Hydrocarbon in (a) phytanyl (C_{20}) and (b) dibiphytanyl (C_{40}). (c, d) Membrane structure in *Archaea*. (c) Lipid bilayer. (d) Lipid monolayer.

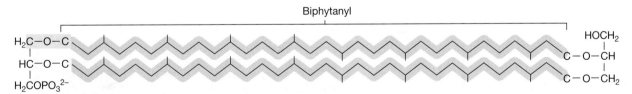

(a) Glycerol diether

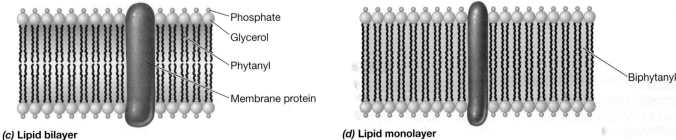

(b) Diglycerol tetraether

(c) Lipid bilayer

(d) Lipid monolayer

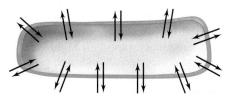

1. Permeability Barrier — Prevents leakage and functions as a gateway for transport of nutrients into and out of the cell

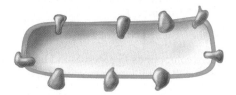

2. Protein Anchor — Site of many proteins involved in transport, bioenergetics, and chemotaxis

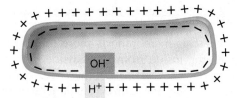

3. Energy Conservation — Site of generation and use of the proton motive force

● **Figure 4.20** **The major functions of the cytoplasmic membrane.** Although structurally weak, the cytoplasmic membrane has many important cellular functions.

We will learn in Chapter 5 that the cytoplasmic membrane is a major site of *energy conservation* in the cell. The membrane can exist in an energetically "charged" form in which a separation of protons (H^+) from hydroxyl ions (OH^-) occurs across its surface (Figure 4.20). This charge separation is a form of energy, analogous to the potential energy present in a charged battery. This energized state of the membrane, called the *proton motive force* (PMF), is responsible for driving many energy-requiring functions in the cell, including some forms of transport, motility, and the biosynthesis of the cell's energy currency, ATP.

The Cytoplasmic Membrane as a Permeability Barrier

The interior of the cell (the cytoplasm) consists of an aqueous solution of salts, sugars, amino acids, nucleotides, vitamins, coenzymes, and a variety of other soluble materials. The hydrophobic nature of the internal portion of the cytoplasmic membrane (Figure 4.16) makes it a tight diffusion barrier. Although some small hydrophobic molecules may pass through the membrane by diffusion, hydrophilic and charged molecules do not pass through but instead must be specifically transported. Because it is charged, even a substance as

small as a hydrogen ion (H^+) cannot diffuse across the cytoplasmic membrane.

One molecule that does freely penetrate the membrane is water, which is sufficiently small to pass between phospholipid molecules in the lipid bilayer (Table 4.2). However, water transport through the membrane can be greatly accelerated by transport proteins called *aquaporins*. These proteins form membrane-spanning channels that specifically transport water into or out of the cytoplasm. Aquaporin AqpZ of *Escherichia coli*, for example, is a well-studied water channel. The synthesis of AqpZ protein is greatly increased under low osmotic conditions. Under these conditions, AqpZ functions as a water *exporter*, preventing the cell from experiencing hypoosmotic shock. Under high osmotic conditions, lower amounts of AqpZ are present, but here the protein functions to transport water *into* the cell to counteract the tendency for water to flow to higher solute conditions.

The relative permeability of a few biologically important substances is shown in Table 4.2. As can be seen, most substances do not passively enter the cell, and thus *transport* is required. The data of Table 4.2 should also be viewed with the understanding that water flow in prokaryotic cells is assisted by aquaporins and is not entirely due to diffusion through the membrane.

The Necessity for Transport Proteins

Transport proteins do more than just ferry things across the membrane—they accumulate solutes *against* the concentration gradient. The necessity for carrier-mediated transport in microorganisms is easy to understand. If diffusion was the only way that solutes could enter the cell, cells would never achieve the intracellular concentrations necessary to carry out biochemical reactions. This is because both the rate of uptake and the intracellular level of diffusible solutes are proportional to their external concentration (Figure 4.21●). The concentration of nutrients in nature, however, is often very low, much lower than typical microbial culture media (the solutions used to grow microorganisms in the laboratory). Hence, cells

Table 4.2	Comparative permeability of membranes to various molecules
Substance	**Rate of permeability** [a]
Water	100
Glycerol	0.1
Tryptophan	0.001
Glucose	0.001
Chloride ion (Cl^-)	0.000001
Potassium ion (K^+)	0.0000001
Sodium ion (Na^+)	0.00000001

[a] Relative scale—permeability with respect to permeability of water given as 100. Permeability of the membrane to water may be affected by aquaporins (see text).

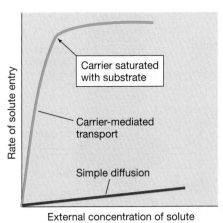

● **Figure 4.21   Relationship between uptake rate and external concentration in diffusion and transport.** Note that in the carrier-mediated process, the uptake rate shows saturation at relatively low external concentrations.

must have mechanisms for accumulating nutrients to levels higher than those in nature, and this is the function of transport systems.

Unlike simple diffusion, carrier-mediated transport systems also show a *saturation effect*. If the concentration of substrate is high enough to saturate the carrier, which typically occurs at even very low substrate concentration, the rate of uptake becomes maximal and the addition of more substrate does not increase the rate (Figure 4.21). This characteristic greatly assists cells in concentrating nutrients in the cytoplasm from an often very dilute nutrient environment.

Another characteristic of carrier-mediated transport processes is the *highly specific* nature of the transport event. Many carrier proteins react only with a single molecule while others show affinities for a related class of molecules. For instance, there are carriers that transport a variety of related sugars or amino acids. This economy in uptake reduces the need for separate transport proteins for every single amino acid or every single sugar.

The synthesis of transport proteins is also *regulated* by the cell such that the specific complement of transporters present in the membrane is a function of both the nutrients present in the environment and their concentration. The latter is an important factor because transport of a particular nutrient often occurs via one type of transporter when the nutrient is present at high concentration and by a different, higher affinity transporter, when present at low concentration.

┌─ ◼ *4.6 Concept Check*

The major function of the cytoplasmic membrane is as a permeability barrier, preventing leakage of cytoplasmic metabolites into the environment. Selective permeability also prevents the diffusion of most solutes. To accumulate nutrients against the concentration gradient, specific transport mechanisms are employed.

◆ Besides permeability, what other functions does the cytoplasmic membrane have?

◆ List two reasons why a cell cannot depend on diffusion as a means of getting nutrients into the cell.

◆ Why is physical damage to the cytoplasmic membrane a more critical problem for the cell than damage to some other cell component?

4.7   Membrane Transport Systems

As just discussed, transport of nutrients and expulsion of wastes are key cellular events. Different mechanisms for transport have evolved in prokaryotes, each with its own unique features. We explore this subject here.

Structure and Function of Membrane Transport Proteins

There are at least three systems for transporting substances in prokaryotes: **simple transport, group translocation**, and the **ABC system**. Simple transporters require only a membrane-spanning protein. Group translocation involves a series of proteins in the transport event. The ABC system involves a *substrate-binding protein*, a *membrane transporter*, and an *ATP-hydrolyzing protein* (Figure 4.22●). All of these transport systems require energy, either in the form of the proton motive force, ATP, or some other energy-rich compound.

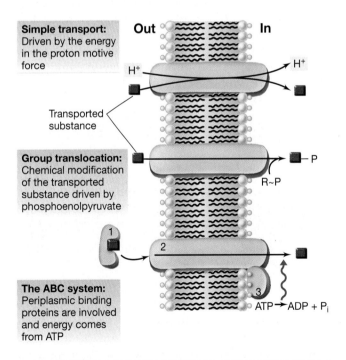

● **Figure 4.22   The three classes of membrane-transporting systems.** Note how simple transporters and the ABC system transport substances *without* chemically modifying them, while group translocation results in the chemical modification (phosphorylation) of the transported substance. The three proteins of the ABC system are labeled 1, 2, and 3.

The membrane-spanning proteins of virtually all bacterial transport systems show significant homologies in both their primary and secondary structure, a testament to their common evolutionary roots. Structurally, these transporters form 12 α-helices (∞Section 3.7) that wind back and forth through the membrane to form a channel through which the transported substance is carried into the cell (Figure 4.23●). The actual transport event involves a conformational change in the protein following binding of its substrate, and this event shuttles the compound across the membrane.

At least three *classes* of transporters are known (Figure 4.24). *Uniporters* transport a molecule in a unidirectional fashion across the membrane. *Symporters* transport a substance *along with* another substance, typically a proton (H⁺). *Antiporters*, as their name implies, transport one substance across the membrane in one direction while simultaneously transporting a second substance in the *opposite* direction (Figure 4.23).

Lactose Uptake in *Escherichia coli*: The Lac Permease

The bacterium *Escherichia coli* metabolizes the disaccharide sugar lactose. Lactose is transported by cells of *E. coli* through the activity of a symporter called the *Lac permease*. Lac permease is a simple transporter. This is shown in Figure 4.24●, where the activity of the lac permease is compared with other simple transporters, including ones that function as uniporters or antiporters.

The lac permease requires energy. Note that as each lactose molecule is transported, the energy in the proton motive force is slowly diminished by the co-transport of protons into the cell. However, the proton motive force is reestablished in the cell through energy-yielding reactions that we will describe in later chapters (∞

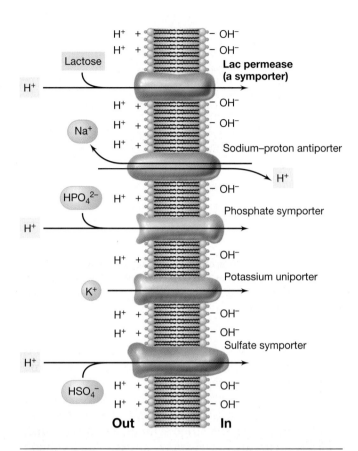

● **Figure 4.24 Function of the Lac permease (a symporter) of *Escherichia coli* and several other well-characterized simple transporters.** Although for simplicity the membrane-spanning proteins are drawn here in globular form, note that their structure is actually as depicted in Figure 4.23. Also note that symporters and antiporters are driven by the energy of the proton motive force. The latter is regenerated by energy-yielding (catabolic) reactions in the cells.

Chapters 5 and 17). The net result of the activity of the lac permease is the accumulation of lactose to a sufficiently high concentration such that its metabolism can yield useful energy for the cell.

Group Translocation: The Phosphotransferase System

Group translocation is a transport mechanism in which the transported substance is *chemically altered* during passage across the membrane. The best-studied group translocation system involves transport of the sugars glucose, mannose, and fructose in *E. coli*. These compounds are *phosphorylated* during transport by the **phosphotransferase system**.

The phosphotransferase system consists of a family of proteins, five of which are necessary to transport any given sugar. Before the sugar is transported, the proteins in the phosphotransferase system are themselves alternately phosphorylated and dephosphorylated in a cascading fashion until the membrane-spanning protein, called *Enzyme II_c*, receives the phosphate group and phosphorylates the sugar in the actual transport event (Figure 4.25●). A small protein called HPr, the enzyme

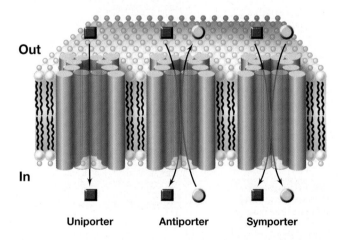

Out

In

| Uniporter | Antiporter | Symporter |

● **Figure 4.23 Structure of membrane-spanning transporters and types of transport events.** In prokaryotes, membrane-spanning transporters typically contain 12 α-helices that align with each other in a circle to form a channel through the membrane. Shown here are three individual transporters, each showing a different type of transport event. For antiporters and symporters, the cotransported molecule is shown in yellow.

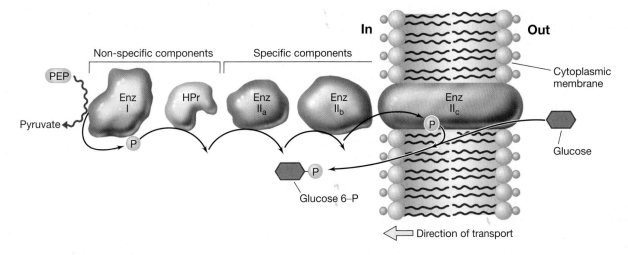

● **Figure 4.25** **Mechanism of the phosphotransferase system of *Escherichia coli*.** For glucose uptake, the system consists of five proteins: Enzyme (Enz) I; Enzymes II_a, II_b, and II_c, and HPr. Sequential phosphate transfer occurs from phosphoenolpyruvate (PEP) through the proteins shown to Enzyme II_c. The latter actually transports (and phosphorylates) the sugar. Proteins HPr and Enz I are nonspecific and involved in the transport of any sugar. The Enz II components are specific for a particular sugar.

that phosphorylates it (Enzyme I), and Enzyme II_a are cytoplasmic proteins. By contrast, Enzyme II_b lies on the inner membrane surface, and Enzyme II_c is an integral membrane protein (Figure 4.25). HPr and Enzyme I are nonspecific components of the phosphotransferase system and participate in the uptake of various sugars, while specific Enzymes II exist for each individual sugar transported (Figure 4.25).

Energy for the phosphotransferase system comes from the energy-rich compound *phosphoenolpyruvate*. However, it should be noted that although energy in the form of one energy-rich phosphate bond is consumed in the process of transporting the glucose molecule (Figure 4.25), the phosphorylation of glucose to glucose-6-P is the first step in its intracellular metabolism anyway (glycolysis, ∞Section 5.10). Thus, the phosphotransferase system prepares glucose for immediate entry into a central metabolic pathway.

Periplasmic Binding Proteins and the ABC System

We will learn a bit later in this chapter (see Section 4.9) that gram-negative bacteria contain a space called the *periplasm* between the cytoplasmic membrane and a lipid-rich outer membrane layer (see Figure 4.35). The periplasm contains various proteins, many of which function in transport. These are called *periplasmic-binding proteins*. Transport systems of this type actually employ three components: (1) periplasmic-binding proteins; (2) membrane-spanning proteins; and (3) ATP-hydrolyzing proteins (kinases). The latter supply the necessary energy.

Transporters of this type have been called *ABC transport systems*, the *ABC* standing for ATP-binding cassette (Figure 4.26●). More than 200 different ABC transport systems have been identified in prokaryotes, and structural studies have shown that they are clearly a *family* of related proteins. ABC transporters exist for organic nu-

trients such as sugars and amino acids, for a variety of inorganic nutrients, such as sulfate and phosphate, and for trace metals.

One of the interesting properties of ABC-type transport systems is the extremely high substrate affinity of the periplasmic-binding proteins. These proteins can

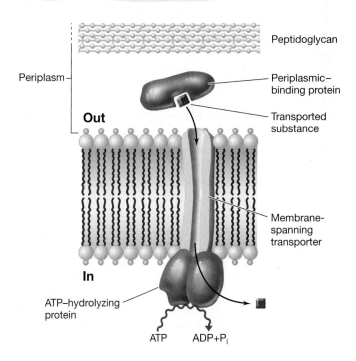

● **Figure 4.26** **Mechanism of an ATP-Binding Cassette (ABC-type) transporter.** The periplasmic binding protein has high affinity for substrate, the membrane-spanning protein is the transport channel, and the cytoplasmic ATP-hydrolyzing protein supplies the energy for the transport event. In *Escherichia coli*, the maltose (a disaccharide sugar) transport system is an example of an ABC system. We discuss the structure of the gram-negative cell wall, including the periplasm, in Section 4.9.

move within the periplasm and bind substrates even when they are present at extremely low concentration. For example, substrate concentrations of 1 micromolar (10^{-6} M) or less can easily be bound and transported by periplasmic-binding proteins. Once trapped, the complex interacts with its respective membrane-spanning component, and the actual transport event occurs driven by the energy of ATP (Figure 4.26).

Interestingly, even though gram-positive bacteria lack a periplasm, binding protein-dependent transport systems have also been found in many of these organisms. However, in gram-positive bacteria, specific binding proteins are not mobile but instead are anchored to the cytoplasmic membrane. Nevertheless, as in gram-negative bacteria, once these binding proteins bind their substrate, they interact with a membrane-spanning component where, at the expense of ATP, transport across the membrane occurs.

Protein Export

Thus far our discussion of transport has focused on small molecules. What about the transport of large molecules, such as proteins? To function properly, many proteins need to be transported *outside* the cytoplasmic membrane or inserted into the membrane in specific ways. Protein translocation occurs in prokaryotic cells through the activities of proteins called *translocases*, a key one being the Sec (for secretory) system.

SecYEG, for example, is a membrane-associated translocase that exports certain proteins while inserting others into the membrane in a specific orientation consistent with their function. Some translocases are very specific in the types of proteins exported, but SecYEG is widely distributed among prokaryotes and can translocate a variety of different proteins. How proteins destined for transport are recognized as such is another story, and we discuss this issue in a later chapter (⬢ Section 7.17).

Protein export is important to bacteria because many bacterial enzymes function outside the cell (exoenzymes). For example, hydrolytic exoenzymes such as amylase or cellulase are excreted directly into the environment where they cleave starch or cellulose (⬢ Figure 3.6*b*), respectively, into glucose. The latter is then used by the cell as a carbon and energy source. Moreover, many pathogenic bacteria excrete protein toxins or other deleterious proteins into the host during infection. All of these large molecules need to move through the cytoplasmic membrane, and translocases like SecYEG assist in these transport events.

4.7 Concept Check

At least three types of transporters are known: simple transporters, phosphotransferase-type transporters, and ABC systems, the latter of which contains three interacting components. Transport requires energy from either the proton motive force, ATP, or some other energy-rich substance.

◆ Contrast the energy requirements of simple transporters, the phosphotransferase system, and the ABC transport system.

◆ Contrast the three classes of transport systems in terms of any chemical alterations that occur in the transported molecule.

◆ Which transport system is best suited for the transport of nutrients present in the environment in extremely low amounts, and why?

◆ How are proteins exported from the cell?

4.8 The Cell Wall of Prokaryotes: Peptidoglycan and Related Molecules

Bacterial cells contain a high concentration of dissolved solutes. This causes a considerable turgor pressure to develop—about 2 atmospheres in a bacterium like *Escherichia coli*. This is roughly the same as the pressure in an automobile tire. To withstand these pressures, bacteria possess **cell walls**, which also function to some extent to give shape and rigidity to the cell.

We have seen that species of *Bacteria* can be divided into two major groups, called **gram-positive** and **gram-negative**. The original distinction between gram-positive and gram-negative was based on the *Gram stain* (see Section 4.1). But differences in cell wall structure are at the heart of the Gram-staining reaction. The appearance of the cell walls of gram-positive and gram-negative cells in the electron microscope differs markedly, as is shown in Figure 4.27●. The gram-negative cell wall is a multilayered structure and quite complex, whereas the gram-positive cell wall primarily consists of a single type of molecule, and is often much thicker.

The focus of this section is on the polysaccharide component of the cell walls of prokaryotes, both *Bacteria* and *Archaea*. These include, in particular, peptidoglycan, but also a variety of related and unrelated polysaccharides found in *Archaea*. In Section 4.9 we describe the special wall components found in gram-negative *Bacteria*.

Peptidoglycan

The cell walls of *Bacteria* have a rigid layer that is primarily responsible for the strength of the wall. In gram-negative *Bacteria*, additional layers are present outside this rigid layer. The rigid layer of both gram-negative and gram-positive *Bacteria* is very similar in chemical composition. Called **peptidoglycan**, this polysaccharide is composed of two sugar derivatives—*N-acetylglucosamine* and *N-acetylmuramic acid*—and a small number of special amino acids including L-alanine, D-alanine, D-glutamic acid, and either lysine or diaminopimelic acid (DAP) (Figure 4.28●). These con-

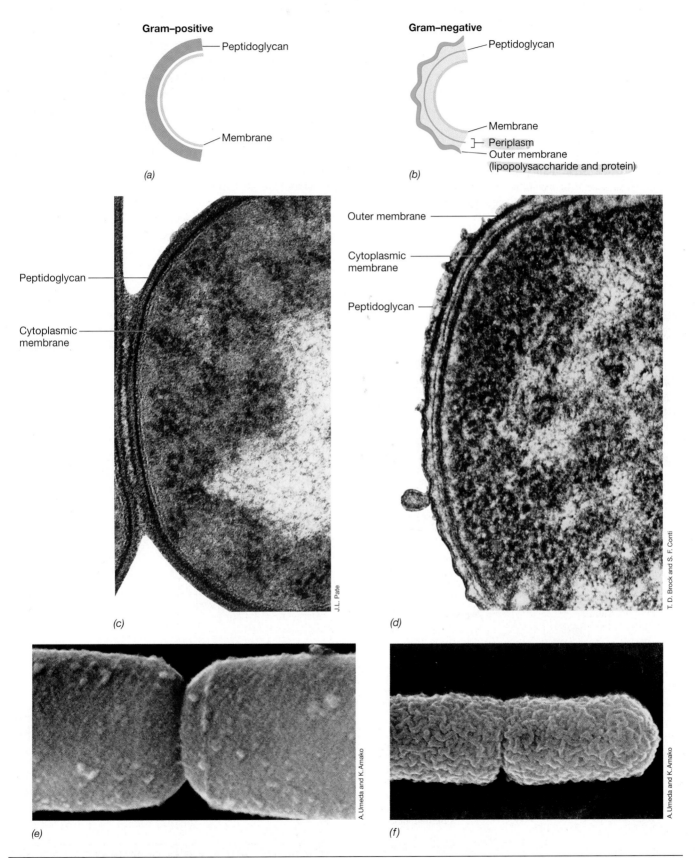

● **Figure 4.27** **Cell walls of *Bacteria*.** (a, b) Schematic diagrams of gram-positive and gram-negative cell walls. (c) Transmission electron micrographs showing the cell wall of a gram-positive bacterium, *Arthrobacter crystallopoietes*, and (d) Gram-negative bacterium, *Leucothrix mucor*. (e, f) Scanning electron micrographs of gram-positive (*Bacillus subtilis*) and gram-negative (*Escherichia coli*) *Bacteria*. Note the surface texture in the cells shown in (e) and (f). A single cell of *B. subtilis* or *E. coli* is about 1 μm in diameter.

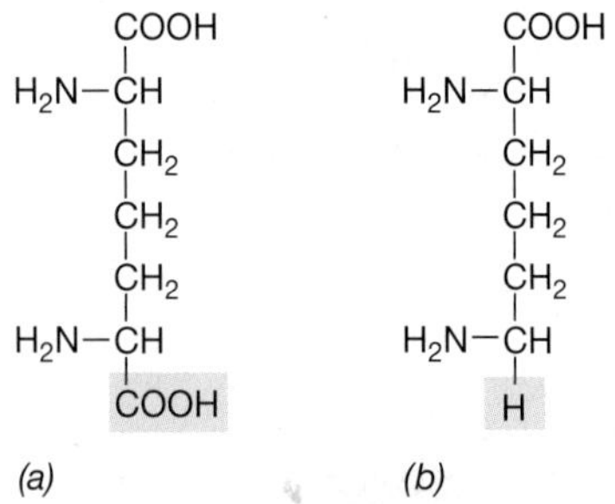

COOH   COOH
| |
H₂N—CH   H₂N—CH
| |
CH₂   CH₂
| |
CH₂   CH₂
| |
CH₂   CH₂
| |
H₂N—CH   H₂N—CH
| |
COOH   H

(a)   (b)

● **Figure 4.28 Cross-linking amino acids in peptidoglycan.** (a) Diaminopimelic acid. (b) Lysine. The only difference in the two molecules is highlighted in color. Besides these two amino acids, several other amino acids are found in peptidoglycan cross-links.

stituents are connected to form a repeating structure, the *glycan tetrapeptide* (Figure 4.29●).

The basic structure of peptidoglycan is a sheet that surrounds the cell formed from individual strands of peptidoglycan lying adjacent to one another. The glycan chains that form the sheet are connected by *tetrapeptide cross-links* formed by the amino acids. The glycosidic bonds connecting the sugars in the glycan strands are very strong, but these chains alone cannot provide rigidity in all directions. The full strength of the peptidoglycan structure is realized only when these chains are cross-linked by amino acids. This cross-linking occurs to different extents in different

Bacteria, with greater rigidity coming from more complete cross-linking.

In *gram-negative Bacteria*, cross-linkage occurs by peptide linkage of the amino group of diaminopimelic acid to the carboxyl group of the terminal D-alanine (Figure 4.30a●). In *gram-positive Bacteria*, cross-linkage occurs by way of a *peptide interbridge*, the kinds and numbers of amino acids varying from organism to organism. For example, in *Staphylococcus aureus*, a well-studied gram-positive bacterium, the interbridge peptide consists of five molecules of glycine (Figure 4.30b). The overall structure of a peptidoglycan molecule is shown in Figure 4.30c.

In gram-positive *Bacteria*, as much as 90% of the cell wall consists of peptidoglycan, although another molecule called *teichoic acid* (discussed later in this section), is usually present in small amounts. And, although some bacteria have only a single layer of peptidoglycan surrounding the cell, many *Bacteria*, especially gram-positive *Bacteria*, have several (up to about 25) sheets of peptidoglycan stacked upon one another. In gram-negative *Bacteria* only about 10% of the wall is peptidoglycan,

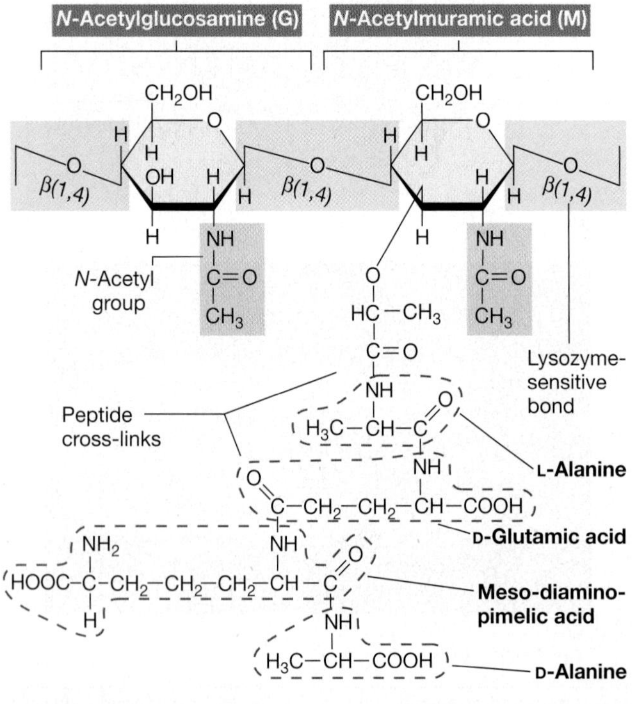

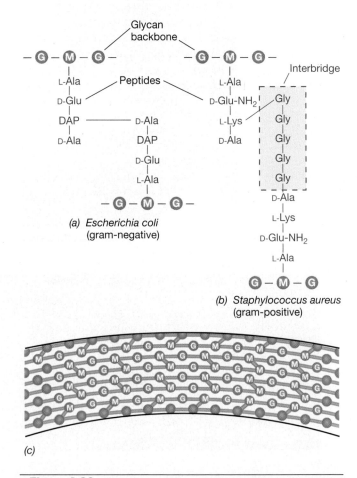

(a) *Escherichia coli*
(gram-negative)

(b) *Staphylococcus aureus*
(gram-positive)

(c)

● **Figure 4.30 How the peptide and glycan units are connected to form the peptidoglycan sheet in *Escherichia coli* and *Staphylococcus aureus*.** (a) No interbridge is present in *E. coli* and other gram-negative *Bacteria*. (b) The glycine interbridge in *S. aureus* (gram-positive). (c) Overall structure of peptidoglycan. G, *N*-acetylglucosamine; M, *N*-acetylmuramic acid.

N-Acetylglucosamine (G) | **N-Acetylmuramic acid (M)**

● **Figure 4.29 Structure of one of the repeating units of the peptidoglycan cell wall structure, the glycan tetrapeptide.** The structure given is that found in *Escherichia coli* and most other gram-negative *Bacteria*. In some *Bacteria*, other amino acids are found.

the majority of the wall consisting of the outer membrane as discussed in Section 4.9. However, the shape of both gram-positive and gram-negative cells is thought to be determined to some extent by the lengths of the peptidoglycan chains and by the manner and extent of cross-linking of the chains.

Diversity in Peptidoglycan

Peptidoglycan is present only in species of *Bacteria*—the sugar *N*-acetylmuramic acid and the amino acid diaminopimelic acid have never been found in the cell walls of *Archaea* or *Eukarya*. However, not all *Bacteria* examined have DAP in their peptidoglycan. This amino acid is present in peptidoglycan in all gram-negative *Bacteria* and in some gram-positive species, but most gram-positive cocci have lysine instead of DAP (Figure 4.30*b*), and a few other gram-positive *Bacteria* have other amino acids. Another unusual feature of peptidoglycan is the presence of two amino acids that have the D configuration, D-alanine and D-glutamic acid. As we saw in Chapter 3, in cellular proteins amino acids are always of the L enantiomeric form (∞ Section 3.6).

More than 100 different peptidoglycan types are known, with the variations occurring in the interbridge. In each different peptidoglycan structure the glycan portion is uniform, with only the sugars *N*-acetylglucosamine and *N*-acetylmuramic acid being present. These sugars are always connected in β-1,4 linkage (Figure 4.29). The tetrapeptide of the repeating unit shows major variation in only one amino acid, the lysine–diaminopimelic acid alternation. However, the D-glutamic acid at position 2 is hydroxylated in some or-

ganisms, whereas substitutions occur in amino acids at positions 1 and 3 in some others. Any of the amino acids present in the tetrapeptide can also occur in the interbridge. But in addition, a number of other amino acids such as glycine, threonine, serine, and aspartic acid can be found in the interbridge. However, branched-chain amino acids, aromatic amino acids, sulfur-containing amino acids, and histidine, arginine, and proline (∞ Figure 3.12) are never found in the interbridge.

Thus, although the peptide chemistry of peptidoglycan can vary, the backbone of peptidoglycan—alternating repeats of glucosamine and muramic acid—is the same in all species of *Bacteria*.

Teichoic Acids and a Summary of the Gram-Positive Wall

Many gram-positive *Bacteria* have acidic substances called **teichoic acids** embedded in their cell wall. The term *teichoic acid* includes all wall, membrane, or capsular polymers containing glycerophosphate or ribitol phosphate residues. These polyalcohols are connected by phosphate esters and usually have other sugars and D-alanine attached (Figure 4.31*a*●). Because they are negatively charged, teichoic acids are partially responsible for the negative charge of the cell surface as a whole. Teichoic acids also function to bind divalent cations such as Ca^{2+} and Mg^{2+}, some of which are transported into the cell. Certain teichoic acids are covalently bound to membrane lipids, and because of this, have been called **lipoteichoic acids.**

Figure 4.31*b* summarizes the structure of the cell wall of gram-positive *Bacteria* and shows how teichoic

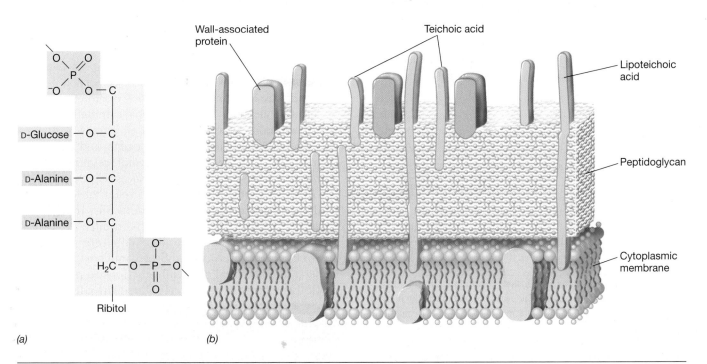

● **Figure 4.31 Teichoic acids and the overall structure of the gram-positive cell wall.** (a) Structure of the ribitol teichoic acid of *Bacillus subtilis*. The teichoic acid is a polymer of the repeating ribitol units shown here. (b) Summary diagram of the gram-positive cell wall.

acids and lipoteichoic acids are arranged in the overall wall structure.

Cells with No Walls

Peptidoglycan—the signature molecule of species of *Bacteria*—can be destroyed by certain agents. One such agent is the enzyme **lysozyme**, a protein that breaks the β-1,4-glycosidic bonds between *N*-acetylglucosamine and *N*-acetylmuramic acid in peptidoglycan (Figure 4.29), thereby weakening the wall. Water then enters the cell, and the cell swells and eventually bursts, a process called **lysis** (Figure 4.32*a*●). Lysozyme is found in animal secretions including tears, saliva, and other body fluids, and presumably functions as a major line of defense against infection by *Bacteria*.

If a solute that does not penetrate the cell, such as sucrose, is added to a cell suspension, the solute concentration outside the cell balances that inside (conditions called *isotonic*). Under these conditions, lysozyme still digests peptidoglycan, but water does not enter the cell and lysis does not occur. Instead, a **protoplast** (a bacterium that has lost its cell wall) is formed (Figure 4.32*b*). If such sucrose-stabilized protoplasts are placed in water, lysis occurs immediately. The word *spheroplast* is often used as a synonym for protoplast, although the two words have slightly different meanings. Protoplasts are cells that are free of residual cell wall material, whereas

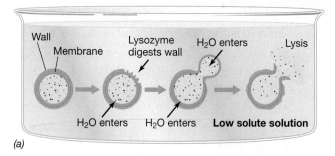

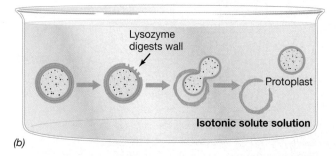

● **Figure 4.32 Protoplasts and their formation.** (a) In dilute solution breakdown of the cell wall releases the protoplast, but it immediately lyses because the cytoplasmic membrane is structurally very weak. (b) In a solution containing an isotonic concentration of a solute such as sucrose, water does not enter the protoplast and it remains stable. Lysozyme breaks the β-1,4 glycosidic bonds in peptidoglycan (see Figure 4.29).

spheroplasts contain pieces of wall material attached to the otherwise membrane-enclosed structure.

Although most prokaryotes cannot survive in nature without their cell walls, some are able to do so. These include the mycoplasmas, a group that causes certain infectious diseases (∞ Section 12.21), and the *Thermoplasma* group, *Archaea* that naturally lack cell walls (∞ Section 13.5). These prokaryotes are free-living protoplasts and are able to survive without cell walls either because they have unusually tough membranes or because they live in osmotically protected habitats, such as the animal body. Certain mycoplasmas have sterols (see Section 4.5) in their cell membranes, which lends strength and rigidity to this structure.

Pseudopeptidoglycan, S-Layers, and Other Cell Walls of *Archaea*

Some species of *Archaea* contain cell walls constructed of a polysaccharide very similar to that of peptidoglycan. This material is called *pseudopeptidoglycan* (Figure 4.33*a*●). The backbone of pseudopeptidoglycan is composed of alternating repeats of *N*-acetylglucosamine and *N*-acetyltalosaminuronic acid (the latter replaces the *N*-acetylmuramic acid of peptidoglycan) (compare Figures 4.29 and 4.33*a*). The backbone of pseudopeptidoglycan also varies from peptidoglycan in that the glycosidic bonds are β-1,3 instead of the β-1,4 bonds found in true peptidoglycan (compare Figures 4.29 and 4.33*a*).

Cell walls of other *Archaea* lack both peptidoglycan and pseudopeptidoglycan and consist of polysaccharide, glycoprotein, or protein. For example, *Methanosarcina* species contain thick polysaccharide walls composed of glucose, glucuronic acid, galactosamine, and acetate. Extremely halophilic (salt-loving) *Archaea* such as *Halococcus* contain cell walls similar to that of *Methanosarcina* but that contain in addition, sulfate (SO_4^{2-}) residues.

The most common cell wall type among *Archaea* is the paracrystalline surface layer (S-layer) (see Section 4.10 for further discussion). The S-layer consists of protein or glycoprotein and generally has a hexagonal symmetry. S-layers have been found among species of all groups of *Archaea*, the extreme halophiles, the methanogens, and the hyperthermophiles (∞ Section 2.5). Several species of *Bacteria* also have S-layers on their outer surfaces (Figure 4.33*b*).

In species of *Archaea* we thus see a variety of cell wall chemistries, varying from molecules that closely resemble peptidoglycan to walls totally lacking a polysaccharide component. But with rare exception, all *Archaea* contain a cell wall of some sort, and as in *Bacteria*, the archaeal cell wall functions to prevent osmotic lysis and to define cell shape. In addition, because they lack peptidoglycan in their cell walls, all *Archaea* are naturally resistant to the action of lysozyme (see earlier) and penicillin, agents that destroy this molecule or prevent its proper synthesis (∞ Section 6.2), respectively.

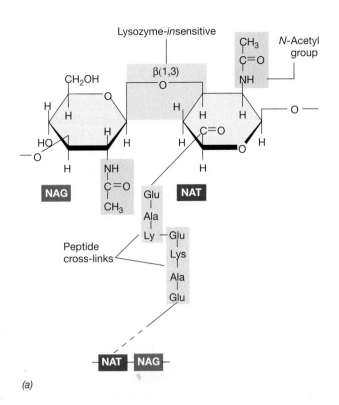

(a)

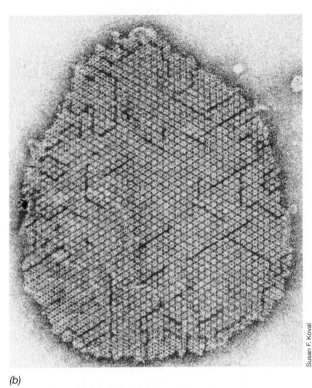

(b)

● **Figure 4.33**   **Pseudopeptidoglycan and S-layers.**   (a) Structure of pseudopeptidoglycan, the cell wall polymer of *Methanobacterium* species. Note the resemblance to the structure of peptidoglycan shown in Figure 4.29, especially the peptide cross-links, in this case between *N*-acetyl-talosaminuronic acid (NAT) residues instead of muramic acid residues. NAG, *N*-Acetylglucosamine. (b) Transmission electron micrograph of a portion of an S-layer showing the paracrystalline nature of this cell wall layer. Shown is the S-layer from the prokaryote *Aquaspirillum serpens* (a member of the *Bacteria*); this S-layer displays hexagonal symmetry as do many of the S-layers found in *Archaea*.

 4.8   Concept Check

The cell walls of *Bacteria* contain a polysaccharide called peptidoglycan. This material consists of strands of alternating repeats of *N*-acetylglucosamine and *N*-acetylmuramic acid, with the latter cross-linked between strands by short peptides. Many sheets of peptidoglycan can be present, depending on the organism. *Archaea* lack peptidoglycan but contain walls made of other polysaccharides or of protein. The enzyme lysozyme destroys peptidoglycan, leading to cell lysis.

◆   List the monomeric components of peptidoglycan.

◆   Why is peptidoglycan such a strong molecule?

◆   How do some cells live without cell walls?

◆   How does pseudopeptidoglycan resemble peptidoglycan? How do the two molecules differ?

4.9   The Outer Membrane of Gram-Negative *Bacteria*

In addition to peptidoglycan, gram-negative *Bacteria* contain an additional wall layer, the **outer membrane**. This layer is effectively a second lipid bilayer, but it is not constructed solely of phospholipid and protein as is the cytoplasmic membrane (Figure 4.16). The outer membrane also contains polysaccharide. The lipid and polysaccharide are linked in the outer membrane to form a *lipopolysaccharide complex*. Because of this, the outer membrane is often called the **lipopolysaccharide layer**, or simply **LPS**.

Chemistry of LPS

Although complex, the chemistry of LPS from several bacteria is now understood. As seen in Figure 4.34●, the polysaccharide portion of LPS consists of two components, the *core polysaccharide* and the *O-polysaccharide*. In *Salmonella* species, where LPS has been best studied, the **core polysaccharide** consists of ketodeoxyoctonate (KDO), seven-carbon sugars (heptoses), glucose, galactose, and *N*-acetylglucosamine. Connected to the core is the **O-specific polysaccharide**, which typically contains galactose, glucose, rhamnose, and mannose (all hexoses), as well as one or more unusual dideoxy sugars such as abequose, colitose, paratose, or tyvelose. These sugars are connected in four- or five-membered sequences, which often are branched. When the sequences repeat, the long *O*-polysaccharide is formed.

The relationship of the *O*-polysaccharide to the rest of the LPS is shown in Figure 4.35●. The lipid portion of the

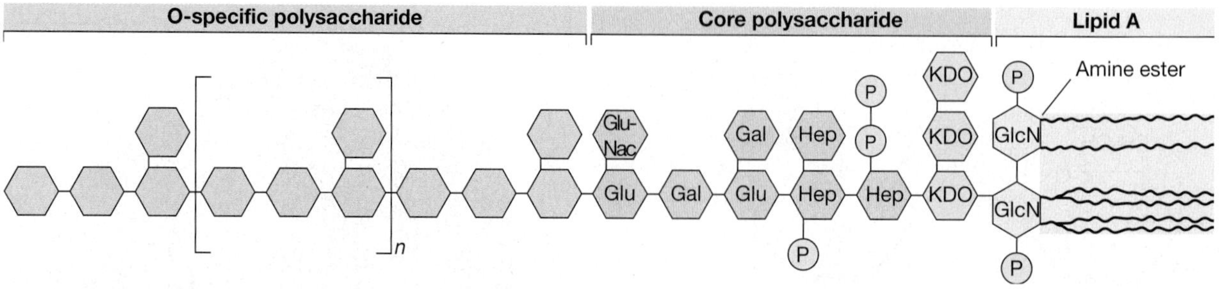

● **Figure 4.34** **Structure of the lipopolysaccharide of gram-negative *Bacteria*.** The precise chemistry of lipid A and the polysaccharide components varies among species of gram-negative *Bacteria*, but the sequence of major components (lipid A–KDO–core–O-specific) is generally uniform. The O-specific polysaccharide varies among species. KDO, ketodeoxyoctonate; Hep, heptose; Glu, glucose; Gal, galactose; GluNac, *N*-acetylglucosamine; GlcN, glucosamine; P, phosphate. Glucosamine and the lipid A fatty acids are linked by an amine ester bond. The lipid A portion of LPS can be toxic to animals and comprises the *endotoxin complex* (∞ Section 21.12). Compare this figure with Figures 4.35 and 4.36, and note the color coding of different portions of the LPS in Figures 4.34 and 4.35.

lipopolysaccharide, referred to as **lipid A** (Figure 4.34), is not a glycerol lipid, but instead the fatty acids are connected by amine ester linkage to a disaccharide composed of *N*-acetylglucosamine phosphate (Figure 4.34). The disaccharide is attached to the core polysaccharide through KDO (Figure 4.34). Fatty acids commonly found in lipid A include caproic, lauric, myristic, palmitic, and stearic acids.

In the outer membrane, LPS associates with several proteins to form the *outer* leaflet of the membrane. A **lipoprotein** complex is found on the *inner* leaflet of the LPS of a number of gram-negative *Bacteria* (Figure 4.35*a*). Lipoprotein functions as an anchor between the outer membrane and peptidoglycan. Finally, in the *outer* leaflet of the outer membrane, LPS replaces phospholipids. The latter are found only in the inner leaf (Figure 4.35*a*). Thus, although the outer membrane can be considered a lipid bilayer, its structure is quite distinct from that of the cytoplasmic membrane (compare Figures 4.16 and 4.35*a*).

Endotoxin

Although the major function of the outer membrane is structural, one of its important biological properties is its *toxicity* to animals. Gram-negative *Bacteria* that are pathogenic for humans and other mammals include members of the genera *Salmonella*, *Shigella*, and *Escherichia*, among others, and some of the symptoms these pathogens elicit in their hosts are due to their toxic outer membrane.

The toxic properties are associated with part of the lipopolysaccharide layer, in particular, *lipid A*. The term **endotoxin** refers to this toxic component of LPS, as we will discuss in Section 21.12. Some endotoxins cause violent symptoms in humans, including severe gastrointestinal distress (gas, diarrhea, vomiting). Endotoxins are responsible for a number of bacterial illnesses, including in particular, *Salmonella* food infection (∞ Section 29.7). Interestingly, LPS from several nonpathogenic bacteria have also been shown to have endotoxin activity. Thus, the organism itself need not be pathogenic to contain toxic outer membrane components.

Porins and the Periplasm

Unlike the cytoplasmic membrane, the outer membrane of gram-negative *Bacteria* is relatively permeable to small molecules even though it is basically a lipid bilayer. This is because proteins called **porins** are present in the outer membrane that function as channels for the entrance and exit of hydrophilic low-molecular-weight substances (Figure 4.35). Several different porins are known, including both specific and nonspecific classes. *Nonspecific porins* form water-filled channels through which any small substance can pass. By contrast, some porins are highly specific because they contain a specific binding site for one or a group of structurally related substances. Structural studies have shown that most porins are proteins that contain *three* identical subunits. Porins are transmembrane proteins (Figure 4.35*a*) and associate to form small holes about 1 nm in diameter in the outer membrane (Figure 4.35*b*).

Although permeable to small molecules, the outer membrane is *not* permeable to enzymes or other large molecules. In fact, one of the major functions of the outer membrane is to keep proteins that are present outside the cytoplasmic membrane from diffusing away from the cell. These enzymes are present in a region called the **periplasm** (see Figures 4.35 and 4.36). This space, located between the outer surface of the cytoplasmic membrane and the inner surface of the outer membrane, is about 12–15 nm wide. The periplasm contents are gel-like in consistency because of the abundance of periplasmic proteins found there (Figure 4.36●).

Depending on the organism, the periplasm can contain several proteins. These include *hydrolytic enzymes*, which function in the initial degradation of food molecules, *binding proteins*, which begin the process of transporting substrates (see Section 4.7), and *chemoreceptors*, which are proteins involved in the chemotaxis response (see Sections 4.16 and 8.12). As previously discussed, most of these proteins reach the periplasm via transport by the SecYEG system (see Section 4.7).

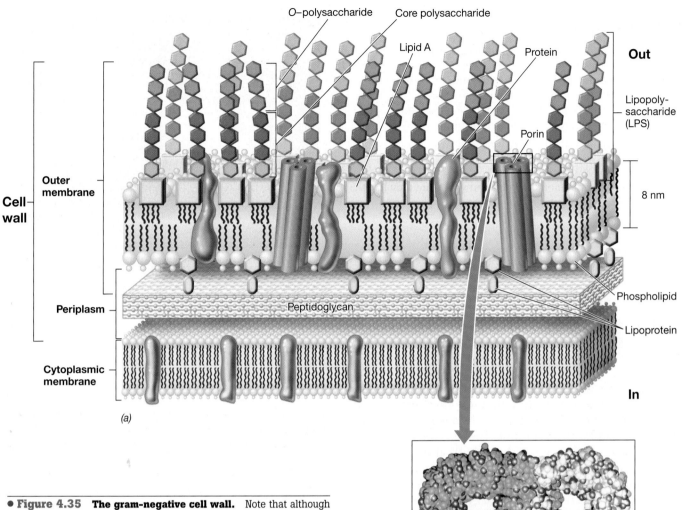

(b)

(a)

● **Figure 4.35 The gram-negative cell wall.** Note that although the outer membrane is often called the "second lipid bilayer," the chemistry and architecture of this layer differ in many ways from that of the cytoplasmic membrane. (a) Arrangement of lipopolysaccharide, lipid A, phospholipid, porins, and lipoprotein in the outer membrane. See Figure 4.34 for details of the structure of LPS. Lipid A can be toxic in humans, and if so, is referred to as endotoxin (⬭⬭ Section 21.12). (b) Molecular model of porin proteins. Note the four pores present, one formed from each of the proteins forming a porin molecule and, a central pore between the porin proteins. The view is perpendicular to the plane of the membrane. Model based on X-ray diffraction studies of *Rhodobacter blasticus* porin.

Relationship of Cell Wall Structure to the Gram Stain

The structural differences between the cell walls of gram-positive and gram-negative *Bacteria* are thought to be responsible for differences in the Gram stain reaction. In the Gram stain (see Section 4.1), an insoluble crystal violet-iodine complex is formed inside the cell. This complex is extracted by alcohol from gram-*negative* but not from gram-*positive Bacteria* (see Figure 4.4). As we have seen, gram-positive *Bacteria* have very thick cell walls consisting of several layers of peptidoglycan. These become dehydrated by the alcohol, causing the pores in the walls to close and preventing the insoluble crystal violet-iodine complex from escaping. By contrast, in gram-negative *Bacteria*, alcohol readily penetrates the lipid-rich outer membrane and extracts the crystal violet-iodine complex from the cell.

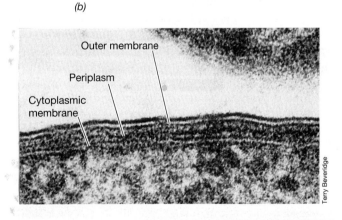

● **Figure 4.36 The cell wall of *Escherichia coli*.** High-magnification thin section of the cell envelope of *E. coli* showing the periplasmic gel bounded by the outer membrane and the cytoplasmic membrane. The large, dark particles in the cytoplasm are ribosomes.

4.9 *Concept Check*

In addition to peptidoglycan, gram-negative *Bacteria* contain an outer membrane consisting of lipopolysaccharide, protein, and lipoprotein. Proteins called porins allow for permeability across the outer membrane. The space between the membranes is the periplasm, which contains various proteins involved in important cellular functions.

◆ What components constitute the LPS layer of gram-negative *Bacteria*?

◆ What is the function of porins and where are they located in a gram-negative cell wall?

◆ What component of the cell has endotoxin properties?

◆ Why does alcohol readily decolorize gram-negative but not gram-positive *Bacteria*?

III SURFACE STRUCTURES AND INCLUSIONS OF PROKARYOTES

In addition to the cell wall and related surface layers just discussed, some prokaryotic cells have other outer layers or structures in contact with the environment. Moreover, most cells can form one or more types of cellular inclusions of one sort or another. The inclusions are then later metabolized as nutrient sources. We examine some of these surface and internal inclusions here.

4.10 Bacterial Cell Surface Structures

Prokaryotes can produce a variety of structures that are attached to or in some way protrude from the cell surface. These include fimbriae, pili, S-layers, capsules, and other slime layers. We examine these structures now.

Fimbriae and Pili

Fimbriae and **pili** are short filamentous structures composed of protein that extend from the surface of a cell. *Fimbriae* (Figure 4.37●) enable organisms to stick to surfaces, including animal tissues in the case of some pathogenic bacteria, or to form pellicles or biofilms (∞ Section 19.3) on surfaces. Notorious among these pathogens include *Salmonella typhimurium* (salmonellosis), *Neisseria gonorrhoeae* (gonorrhea), and *Bordetella pertussis* (whooping cough).

Pili are structurally similar to fimbriae but are typically longer, and only one or a few pili are present on the surface. Because they serve as receptors for certain types of viruses, pili can be seen under the electron microscope when they become coated with virus particles (Figure 4.38●). Although possibly involved in attachment as for fimbriae, pili are clearly involved in the process of conjugation (a form of genetic exchange) in prokaryotes, as will be discussed in Section 10.9.

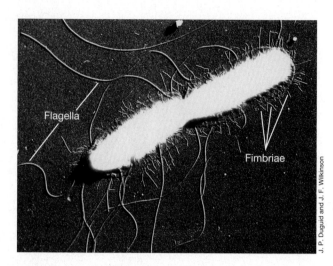

● **Figure 4.37 Fimbriae.** Electron micrograph of a dividing cell of *Salmonella typhi*, showing flagella and fimbriae. A single cell is about 0.9 μm in diameter.

Many classes of fimbriae/pili are known, distinguished by their structure and function. One class, called *type IV fimbriae/pili*, is involved in an unusual form of motility in certain bacteria, called *twitching motility*. Twitching motility is a type of movement on solid surfaces, where it is thought that rapid and reversible extension and retraction of the fimbriae allow the cell to crawl along the surface. Unlike other fimbriae, type IV fimbriae are found only at the poles of the cells, and besides motility, have been implicated as key host colonization factors in a variety of pathogens including *Vibrio cholerae* (cholera), and *Neisseria menigitidis* (bacterial meningitis). Type IV fimbriae are also thought to mediate genetic transfer by the process of transformation (∞ Section 10.7) in a wide variety of bacteria.

Paracrystalline Surface Layers

Many prokaryotes contain a cell surface layer composed of a two-dimensional array of protein. These layers are called **S-layers**. S-layers have been detected in representatives of virtually every phylogenetic grouping of *Bacteria* and are widespread among *Archaea*. In some

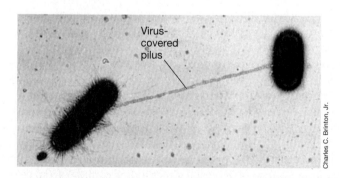

● **Figure 4.38 Pili.** The presence of pili on an *Escherichia coli* cell is revealed by the use of viruses that specifically adhere to the pilus. The cell is about 0.8 μm in diameter.

species of *Archaea* the S-layer is also the cell wall (see Section 4.8). S-layers have a crystalline appearance and show various symmetries, such as hexagonal, tetragonal, or trimeric, depending upon the number and structure of the protein or glycoprotein subunits of which they are composed (see Figure 4.33*b* to view an electron micrograph of an S-layer).

The major function of S-layers is unknown. However, as the interface between the cell and its environment, it is likely that the S-layer functions as a selective sieve, allowing the passage of low-molecular-weight substances while excluding large molecules and structures (such as viruses). The S-layer would then function to retain proteins near the cell, much like the outer membrane does in gram-negative bacteria. Evidence also exists that in pathogenic bacteria that contain S-layers, this structure may provide protection against certain host defense mechanisms.

Capsules and Slime Layers

Many prokaryotic organisms secrete slimy or gummy materials on their surfaces (Figure 4.39●). A variety of these structures consist of polysaccharide, and a few consist of protein. The terms **capsule** and **slime layer** are frequently used to describe these polysaccharide layers. The composition of these layers varies in different organisms but may be thick or thin, rigid or flexible, depending on their chemical nature. The rigid layers are organized in a tight matrix that excludes small particles, such as india ink; this form is called a *capsule* (Figure 4.39*a*). If the layer is more easily deformed, it will not exclude particles and is more difficult to see; this form is called a *slime layer*.

Polysaccharide layers have several functions in bacteria. Surface polysaccharides assist in the *attachment* of microorganisms to solid surfaces. As we will see (∞ Section 21.6), pathogenic microorganisms that enter the animal body by specific routes usually do so

by first binding specifically to surface components of host tissues. This binding is often mediated by surface polysaccharides on the bacterial cell. Many non-pathogenic bacteria also bind to solid surfaces in nature, sometimes forming a thick layer of cells called a **biofilm**. Polysaccharides play a key role in the development of biofilms (∞ Section 19.3).

Slime layers play other roles as well. For example, encapsulated pathogenic bacteria are typically more difficult for phagocytic cells of the immune system (∞ Sections 22.2) to recognize and subsequently destroy. In addition, because outer polysaccharide layers bind a significant amount of water, it is likely that these layers play some role in resistance to desiccation.

⬣ 4.10 Concept Check

Prokaryotic cells often contain various surface structures. These include fimbriae and pili, S-layers, capsules, and slime layers. These structures have several functions, but a key one is in attaching cells to a solid surface.

◆ How do fimbriae differ from pili, both structurally and functionally?

◆ Although they lack lipid, how could S-layers play a role similar to that of the outer membrane?

4.11 Cell Inclusions

Granules or other inclusions are often seen within cells. Their nature differs in different organisms, but they commonly function as energy reserves or as a reservoir of structural building blocks. Inclusions can often be seen directly with the light microscope (see Figure 4.41). Most cellular inclusions are enclosed by a thin *nonunit* membrane consisting of lipid separating the inclusion from the cytoplasm proper.

● **Figure 4.39 Bacterial capsules.** (a) A capsule in *Acinetobacter* species observed by negative staining with india ink and phase-contrast microscopy. The india ink does not penetrate the capsule, and so it is revealed in outline as a light structure on a dark background. (b) Electron micrograph of a thin section of a *Rhizobium trifolii* cell stained with ruthenium red to reveal the capsule. The diameter of the cell proper (not including the capsule) is about 0.7 µm. Although most capsules consist of polysaccharide, some bacteria contain protein capsules. Cells of *Bacillus anthracis*, for example, an organism associated with both animal disease and bioterrorism (∞ Sections 25.12 and 25.13), contain a poly-D-glutamic acid capsule that is effective in preventing cell destruction by host defenses.

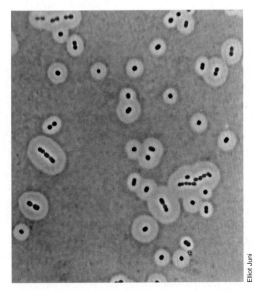
(a)

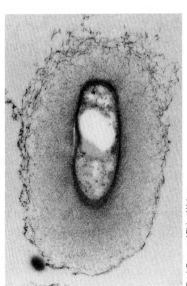
(b)

Carbon Storage Polymers

In prokaryotic organisms, one of the most common inclusion bodies consists of **poly-β-hydroxbutyric acid (PHB)**, a lipid that is formed from β-hydroxbutyric acid units (Figure 4.40a●). The monomers of this acid are connected by ester linkages, forming the long PHB polymer, and these polymers aggregate into granules (Figure 4.40b). The length of the monomer in the polymer can vary considerably, from as short as C_4 to as long as C_{18} in certain organisms. Thus, the more generic term *poly-β-hydroxyalkanoate* (PHA) is used to describe this whole class of carbon/energy storage polymers. PHAs are synthesized when carbon is in excess and are broken down for use as carbon skeletons for biosynthesis or to make ATP when conditions warrant. A wide variety of prokaryotes, including species of both *Bacteria* and *Archaea*, produce PHAs.

Another storage product formed by prokaryotes is **glycogen**, which is a polymer of glucose (∞ Section 3.3 and Figure 3.6). Like PHAs, glycogen is a storage depot for carbon and energy. Glycogen is produced when carbon is in excess in the environment and is consumed when carbon is limiting. Glycogen resembles starch, the major storage reserve of plants, but differs from starch in the way the glucose units are linked together (see Figure 3.6b).

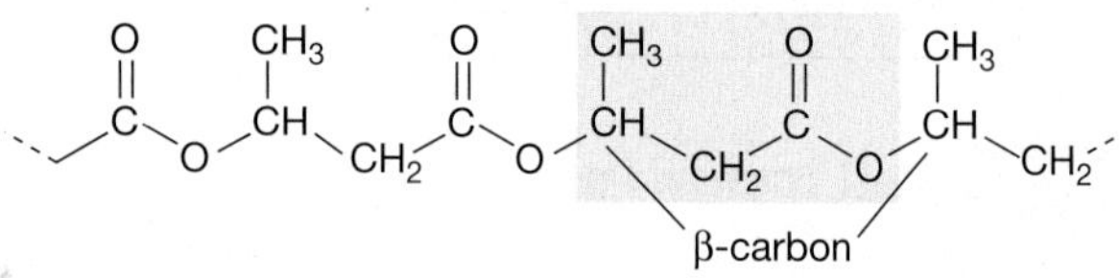

Other Storage Materials and Inclusions

Many microorganisms accumulate inorganic phosphate in the form of granules of **polyphosphate**. These granules can be degraded and used as sources of phosphate for nucleic acid and phospholipid. In addition, many prokaryotes are capable of oxidizing reduced sulfur compounds such as hydrogen sulfide (H_2S). These oxidations are linked to either reactions of energy metabolism (∞ Sections 17.8 and 17.10) or biosynthesis (∞ Section 17.6). In either case, **elemental sulfur** may accumulate inside the cell in readily visible globules (Figure 4.41●). These globules of elemental sulfur remain as long as the source of reduced sulfur is still present. However, as the reduced sulfur source becomes limiting, the sulfur in the granules is oxidized to sulfate (SO_4^{2-}), and the granules slowly disappear as this reaction proceeds.

Interestingly, the sulfur globules themselves actually reside in the *periplasm* rather than the *cytoplasm* (sulfur-oxidizing phototrophic or chemolithotrophic prokaryotes are gram-negative organisms). The periplasm expands outwards to accommodate the globules as H_2S is oxidized ($H_2S \rightarrow S^0$). The periplasm then contracts inward as sulfur is oxidized ($S^0 \rightarrow SO_4^{2-}$). It is considered likely that certain other "intracytoplasmic" inclusions may be periplasmic in gram-negative bacteria, as well. Thus far, it appears that at least poly-β-hydroxyalkanoates (see Figure 4.40) fall into this category.

Magnetosomes are intracellular particles of the iron mineral magnetite—Fe_3O_4 (Figure 4.42●). Magnetosomes impart a magnetic dipole to a cell, allowing it to respond to a magnetic field. Bacteria that produce magnetosomes (Figure 4.42a) exhibit *magnetotaxis*, the

(a)

(b)

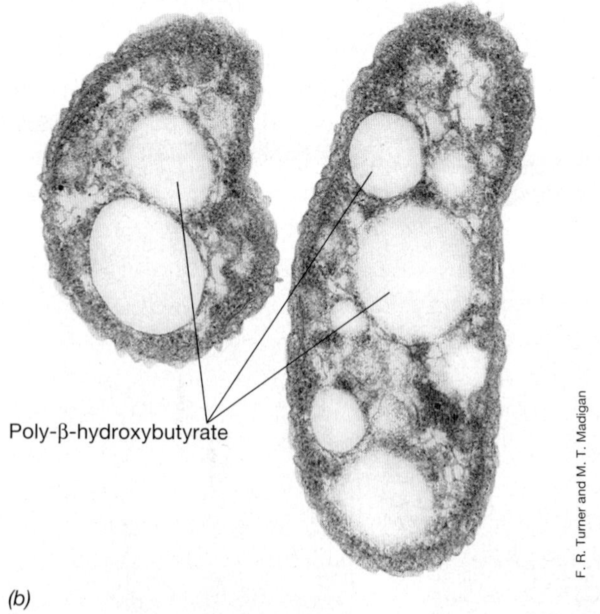

Poly-β-hydroxybutyrate

F. R. Turner and M. T. Madigan

● **Figure 4.40 Poly-β-hydroxybutyrate (PHB).** (a) Chemical structure of PHB, a common poly-β-hydroxyalkanoate. A monomeric unit is shown in color. Other alkanoate polymers are made by substituting longer-chain hydrocarbons for the $-CH_3$ group on the β carbon. (b) Electron micrograph of a thin section of cells of the phototrophic bacterium *Rhodovibrio sodomensis* (see also Figure 4.10b) containing granules of PHB.

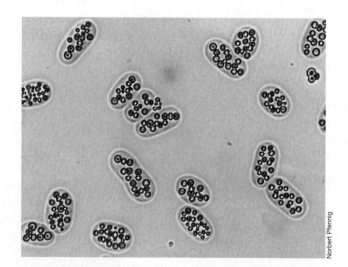

Norbert Pfennig

● **Figure 4.41 Sulfur globules.** Bright-field photomicrograph of cells of the purple sulfur bacterium *Isochromatium buderi*. Note the sulfur globules inside the cell, obtained from the oxidation of hydrogen sulfide (H_2S). A single cell measures about 4×7 μm.

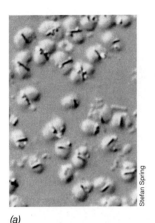
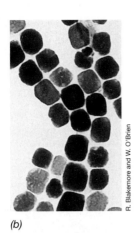

Stefan Spring

(a)

R. Blakemore and W. O'Brien

(b)

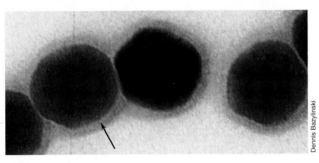

Dennis Bazylinski

(c)

● **Figure 4.42 Magnetotactic bacteria and magnetosomes.** (a) Interference contrast micrograph of coccoid magnetotactic bacteria. Note magnetosomes. A single cell is about 2.2 μm in diameter. (b) Magnetosomes isolated from the magnetotactic bacterium *Magnetospirillum magnetotacticum*. Each particle is about 50 nm in length (⚬⚬ Figure 12.32). (c) Transmission electron micrograph of magnetosomes from a magnetic coccus. The arrow points to the membrane surrounding each magnetosome. A single magnetosome is about 90 nm wide. Although containing lipids and proteins, the magnetosome membrane, like the PHB membrane (see Figure 4.40), is a monolayer and not a true "unit" membrane.

process of orienting and migrating along Earth's magnetic field lines (⚬⚬ Section 12.14). Although the suffix *-taxis* is used in the word *magnetotaxis*, there is no evidence that magnetotactic bacteria employ the sensory systems of chemotactic or phototactic bacteria (see Sections 4.16 and 8.12). Instead, the alignment of magnetosomes in the cell simply imparts magnetic properties to it, which then orient the cell in a particular direction in its environment.

The major function of magnetosomes is unknown. However, magnetosomes have been found in a variety of aquatic *Bacteria* (Figure 4.42*a*) that grow best at low O_2 concentrations. It has thus been hypothesized that a function of magnetosomes is to guide these aquatic cells toward the sediments (since the major magnetic field line points downwards) where O_2 levels are lower.

Magnetosomes are surrounded by a membrane containing phospholipids, proteins, and glycoproteins (Figure 4.42*b, c*). This membrane is not a unit membrane, as is the cytoplasmic membrane (Figure 4.16), but instead is a nonunit membrane, like that surrounding

granules of poly-β-hydroxybutyrate (PHB, Figure 4.40). Magnetosome membrane proteins probably play a role in precipitating Fe^{3+} (brought into the cell in soluble form by chelating agents) as Fe_3O_4 in the developing magnetosome. The morphology of magnetosomes appears to be species-specific, varying in shape from square to rectangular to spike-shaped in certain bacteria, forming into chains inside the cell (Figure 4.42).

4.11 Concept Check

Prokaryotic cells often contain internal granules such as sulfur, polyphosphate, PHAs, and magnetosomes. These substances function as storage materials or in magnetotaxis.

◆ Under what growth conditions would you expect PHAs or glycogen to be produced?

◆ Why would it be impossible for *gram-positive* bacteria to store sulfur like sulfur-oxidizing chemolithotrophs can?

◆ What form of iron is present in magnetosomes?

4.12 Gas Vesicles

A number of prokaryotic organisms are *planktonic*, living a floating existence within the water column of lakes and the oceans. Many planktonic prokaryotes produce **gas vesicles**. These structures confer buoyancy on cells by decreasing their density. Gas vesicles are actually a means of motility, allowing cells to float up and down in a water column in response to environmental factors.

The most dramatic instances of flotation due to gas vesicles are seen in cyanobacteria that form massive accumulations (blooms) in lakes (Figure 4.43●). Gas-vesiculate cells rise to the surface of the lake and are blown by winds into dense masses. Gas vesicles are also present in certain purple and green phototrophic bacteria (⚬⚬ Sections 12.2 and 12.32) and in some nonphototrophic bacteria that live in lakes and ponds. Some species of *Archaea* also contain gas vesicles.

T. D. Brock

● **Figure 4.43 Gas vesicles in nature.** Flotation of gas vesiculate cyanobacteria from a bloom on a nutrient-rich lake, Lake Mendota, Madison, Wisconsin.

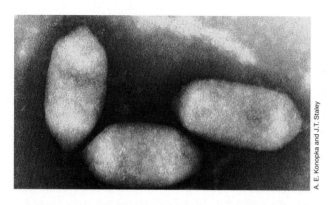

● **Figure 4.44 Isolated gas vesicles.** Transmission electron micrographs of gas vesicles purified from the bacterium *Ancyclobacter aquaticus* and examined in negatively stained preparations. A single gas vesicle is about 100 nm in diameter. [Reproduced with permission from *Archives of Microbiology* 112:133–140 (1977).]

Structure of Gas Vesicles

Gas vesicles are spindle-shaped gas-filled structures made of protein. They are hollow yet rigid and of variable length and diameter (Figure 4.44●). Gas vesicles in different organisms vary in length from about 300 to over 1000 nm and in width from 45 to 120 nm, but the vesicles of any given organism are more or less of constant size.

Gas vesicles may number from a few to hundreds per cell. The gas vesicle membrane is composed of protein, is about 2 nm thick, and is impermeable to water and solutes but permeable to gases. The presence of gas vesicles in cells can be determined either by light microscopy (where clusters of vesicles, called *gas vacuoles*, show up as irregular bright inclusions) or by electron microscopy (Figure 4.45●).

Molecular Structure of Gas Vesicles

Gas vesicles are composed of two different proteins (Figure 4.46●). The major gas vesicle protein, called *GvpA*, is a small, highly hydrophobic and very rigid protein. The rigidity of the gas vesicle membrane is essential for the structure to resist the pressures exerted on it from outside. GvpA makes up the gas vesicle shell and composes 97% of total gas vesicle protein. The minor protein, called *GvpC*, functions to strengthen the shell of the gas vesicle (Figure 4.46).

Gas vesicles are constructed of copies of the GvpA protein aligned as parallel "ribs" forming a watertight shell. The GvpA ribs are strengthened by GvpC, which cross-links GvpA at an angle, binding several GvpA molecules together like a clamp (Figure 4.46). The final shape of the gas vesicle, which can vary in different organisms from long and thin to short and fat (compare Figures 4.44 and 4.45*b*), is a function of how the GvpA and GvpC proteins are arranged to form the intact vesicle.

Because the gas vesicle membrane is freely permeable to gases, the composition and pressure of the gas inside a gas vesicle is the same as that of the gas in which the organism is suspended. And, because the gas

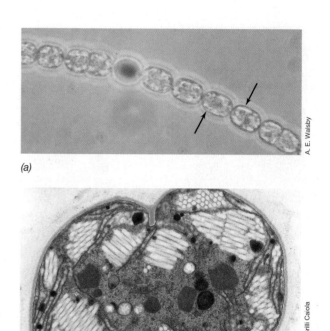

(a)

(b)

● **Figure 4.45 Gas vesicles of the cyanobacteria *Anabaena* and *Microcystis.*** (a) *Anabaena flos-aquae.* The dark cell in the center (a heterocyst) lacks gas vesicles. In the other cells, the vesicles group together as phase-bright gas vacuoles (arrows). (b) Transmission electron micrograph of the cyanobacterium *Microcystis.* Gas vesicles are arranged in bundles, here observable in both longitudinal and cross section.

vesicle attains a density of about 5–20% of that of the cell proper, gas vesicles decrease the density of the cell, thereby increasing its buoyancy. Aquatic phototrophic organisms in particular benefit from this because it allows them to adjust their position vertically in a water column to regions where the light intensity for photosynthesis is optimal.

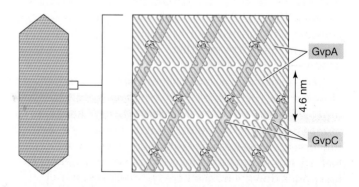

GvpA

4.6 nm

GvpC

● **Figure 4.46 Gas vesicle proteins.** Model of how the two proteins that make up the gas vesicle, GvpA and GvpC, interact to form a watertight but gas-permeable structure. GvpA makes up the rib and is a rigid β-sheet. GvpC is the cross-linker and is of an α-helix structure (∞ Section 3.7 and Figure 3.16).

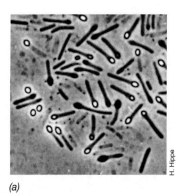

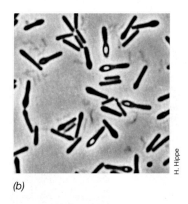

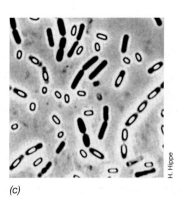

(a) H. Hippe *(b)* H. Hippe *(c)* H. Hippe

● **Figure 4.47** **The bacterial endospore.** Phase contrast photomicrographs illustrating several types of endospore morphologies and intracellular locations. (a) Terminal. (b) Subterminal. (c) Central.

4.12 Concept Check

Gas vesicles are small gas-filled structures made of protein that function to confer buoyancy on cells. Gas vesicles contain two different proteins arranged to form a gas permeable, but watertight structure.

◆ What might be the benefit of gas vesicles to phototrophic cells? (*Hint*: What do phototrophs need to grow?).

◆ How are the two proteins that make up the gas vesicle, GvpA and GvpC, arranged to form such a water-impermeable structure?

4.13 Endospores

Certain *Bacteria* produce structures called **endospores** (Figure 4.47●) during a process called *sporulation* (see Figure 4.50). Endospores (the prefix "endo" means "within") are differentiated cells that are highly resistant to heat and are difficult to destroy, even by harsh chemicals or radiation. The biological function of endospores is undoubtedly to enable the organism to endure difficult times, including extremes of temperature, drying, and nutrient depletion. Endospores are also ideal structures for dispersal by wind, water, or through the animal gut. Endospore-forming bacteria are found most commonly in the soil, and the genera *Bacillus* and *Clostridium* are the best studied of endospore-forming bacteria.

The discovery of bacterial endospores was of immense importance to microbiology because the knowledge of such remarkably heat-resistant forms was essential for the development of adequate methods of sterilization, not only of culture media but also of foods and other perishable products. Although many microorganisms form spores, the bacterial endospore is unique in its degree of heat resistance. Highly resistant bacterial endospores survive heating to temperatures as high as 150°C, although an autoclave operating at 121°C should kill the endospores of most species. Endospores are also resistant to other harmful agents such as drying, ultraviolet radiation, strong acids or bases, and chemical disinfectants, and can remain dormant for extremely long periods (see the Microbial Sidebar, How Long Can an Endospore Survive?).

Endospore Structure

Endospores stand out under the light microscope as strongly refractile structures (see Figure 4.47). Endospores are impermeable to most dyes, so occasionally they are seen as unstained regions within cells that have been stained with basic dyes such as methylene blue. To stain endospores, special staining procedures must be used.

The structure of the endospore as seen with the electron microscope differs distinctly from that of the vegetative cell (Figure 4.48●). In particular, the endospore is

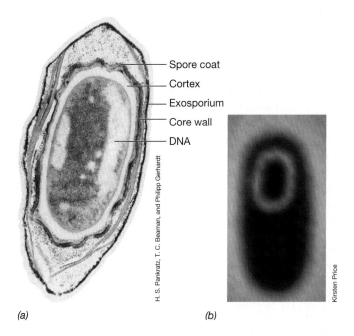

Spore coat
Cortex
Exosporium
Core wall
DNA

(a) H. S. Pankratz, T. C. Beaman, and Philipp Gerhardt

(b) Kirsten Price

● **Figure 4.48** **Dissecting the bacterial endospore.** (a) Transmission electron micrograph of a mature endospore from *Bacillus megaterium.* (b) Fluorescent photomicrograph of a cell of *Bacillus subtilis* undergoing sporulation. The green area is due to a dye that specifically stains a sporulation protein in the spore coat.

Microbial Sidebar ◆ How Long Can an Endospore Survive?

In this chapter we have discussed the dormancy and resistance properties of bacterial endospores and have pointed out that endospores can survive for long periods in a dormant state. But how long is *long*?

Evidence for endospore longevity suggests that these structures can remain viable (that is, capable of germination into vegetative cells) for at least several decades and probably for much longer than that. A suspension of endospores of the bacterium *Clostridium aceticum* (Figure 1*a*) prepared in 1947 was placed in sterile growth medium in 1981, 34 years later, and in less than 12 h growth commenced, leading to a robust culture. *Clostridium aceticum* was originally isolated by the Dutchman K. T. Wieringa in 1940 but was thought to have been lost until this vial of *C. aceticum* endospores was found in a storage room at the University of California at Berkeley and revived.[1] *C. aceticum* is a homoacetogen, capable of making acetate from $CO_2 + H_2$ or from glucose (👓 Section 17.16).

Other more extreme examples of endospore longevity have been documented. Bacteria of the genus *Thermoactinomyces* are thermophilic endospore formers that are widespread in nature in soil, plant litter, and fermenting plant material. Microbiological examination of a Roman archaeological site in the United Kingdom that was dated to over *2000* years ago yielded significant numbers of viable *Thermoactinomyces* endospores in various pieces of debris. Additionally, *Thermoactinomyces* endospores were recovered in fractions of sediment cores from a Minnesota lake known to be over *7000* years old. Although contamination is always a possibility in such studies, samples in both of these cases were processed in such a way as to virtually rule out contamination with "recent" endospores.[2]

What factors could limit the age of an endospore? Cosmic radiation has been considered a major factor because it can introduce

Figure 1 Longevity of endospores. *(a) Photograph of a tube containing endospores from the bacterium* Clostridium aceticum *prepared on May 7, 1947. After remaining dormant for over 30 years, the endospores were suspended in a culture medium after which growth occurred within 12 h.[1] (b) Halophilic bacteria trapped within salt crystals. These crystals (about 1 cm in diameter) were grown in the laboratory in the presence of* Halobacterium *cells (orange) that remain viable in the crystals. Crystals similar to these but of Permian age (≈250 million years old) were reported to contain viable halophilic endosporulating bacteria.*

(a) Gerhard Gottschalk   *(b)* William D. Grant

mutations in DNA.[2] It has been hypothesized that over periods of thousands of years, the cumulative effects of cosmic radiation could introduce so many mutations into the genome of an organism that even highly radiation-resistant structures such as endospores would succumb to the genetic damage. However, extrapolations from actual experimental assessments of the effect of natural radiation on endospores suggest that if suspensions of endospores were partially shielded from cosmic radiation, for example, by being embedded in layers of organic matter, they could retain viability over periods as great as *several hundred thousand years* and perhaps even longer. Amazing, but is this the upper limit?

In 1995 a group of scientists reported the revival of bacterial endospores they claimed were 25–40 million years old.[3] The endospores were allegedly preserved in the gut of an extinct bee trapped in amber of known geological age. The presence of endospore-forming bacteria in these bees was previously suspected from electron microscopic studies of the insect gut that showed endospore-like structures and because *Bacillus*-like DNA was recovered from the insect. Incredibly, samples of bee tissue incubated in a sterile culture medium quickly yielded endospore-forming bacteria. Rigorous precautions were taken to demonstrate that the endospore-forming bacterium revived from the amber-encased bee was not a modern-day contaminant. An even more spectacular claim was made that halophilic (salt-loving) endospore-forming bacteria had been isolated from fluid inclusions in salt crystals of Permian age, over 250 million years old.[4] Presumably these cells were trapped in brines within the crystal (Figure 1*b*) as it formed eons ago, and remained viable for over a quarter billion years!

If these claims of almost unbelievable endospore longevity are supported by repetition of the results in independent laboratories (and such confirmation is crucial for verifying such highly controversial findings), then endospores stored under the proper conditions can likely remain viable indefinitely. This is a remarkable testimony to the endospore, a structure that undoubtedly evolved to maintain viability for relatively short periods but that turned out to be such a well-designed structure that dormancy for hundreds of thousands, if not millions, of years may be possible. ∎

[1]Braun, M., F. Mayer, and G. Gottschalk. 1981. *Clostridium aceticum* (Wieringa), a microorganism producing acetic acid from molecular hydrogen and carbon dioxide. *Arch. Microbiol. 128:* 288–293.
[2]Gest, H., and J. Mandelstam. 1987. Longevity of microorganisms in natural environments. *Microbiol. Sci. 4:* 69–71.
[3]Cano, R. J., and M. K. Borucki. 1995. Revival and identification of bacterial spores in 25- to 40-million-year-old Dominican amber. *Science 268:* 1060–1064.
[4]Vreeland, R.H., W.D. Rosenzweig, and D.W. Powers. 2000. Isolation of a 250 million-year-old halotolerant bacterium from a primary salt crystal. *Nature 407:* 897–900.

structurally more complex in that it has many layers that are absent from the vegetative cell. The outermost layer is the **exosporium**, a thin protein covering. Within this are the **spore coats**, composed of layers of spore-specific proteins (Figure 4.48*b*). Below the spore coat is the **cortex**, which consists of loosely cross-linked peptido-glycan, and inside the cortex is the **core** or **spore proto-plast**, which contains the core wall, cytoplasmic membrane, cytoplasm, nucleoid, ribosomes, and other cellular essentials (∞ Section 2.1 and Figure 2.1*a*). Thus, the endospore differs structurally from the vegeta-tive cell primarily in the kinds of structures found *outside* the core wall.

One chemical substance that is characteristic of en-dospores but absent from vegetative cells is **dipicolinic acid** (Figure 4.49●). This substance has been found in endospores from all endospore-forming bacteria exam-ined and is located in the core. Endospores are also en-riched in calcium ions, most of which are combined with dipicolinic acid. The calcium-dipicolinic acid com-plex of the core represents about 10% of the dry weight of the endospore. The complex functions to reduce water availability within the endospore, thus helping to dehydrate it. In addition, the complex intercalates (in-serts between bases) in DNA, and in so doing, stabi-lizes it to heat denaturation.

Properties of the Endospore Core

The core of a mature endospore differs greatly from the vegetative cell from which it was formed. Besides the high levels of calcium dipicolinate (Figure 4.49), which reduces the water content of the core, the core becomes further dehydrated during the sporulation process. The core of a mature endospore contains only 10–25% of the water content of the vegetative cell, and thus the consis-tency of the core cytoplasm is that of a gel. This dehydra-tion of the core greatly increases the heat resistance of the endospore. However, dehydration has also been shown to confer resistance in the endospore to chemicals, such as hydrogen peroxide (H_2O_2), and causes enzymes re-maining in the core to become inactive.

In addition to the low water content of the en-dospore, the pH of the core is about one unit lower than that of the vegetative cell cytoplasm. Moreover, the core contains high levels of proteins called *small acid-soluble proteins* (SASPs). These are made during the sporulation process and have at least two functions. SASPs bind tightly to DNA in the core and protect it from potential damage from ultraviolet radiation, desiccation, and dry heat. Ultraviolet resistance is conferred when SASPs change the molecular structure of DNA from the normal "B" form to the more compact "A" form. A-form DNA is more resistant to the formation of pyrimidine dimers by UV radiation, a means of mutation (∞ Section 10.4), and to the denaturing effects of dry heat. In addition, SASPs function as a carbon and energy source for the outgrowth of a new vegetative cell from the endospore, a process called *germination* (discussed later in this section).

Endospore Formation

During endospore formation, a vegetative cell is converted to a nongrowing, heat-resistant structure (Figure 4.50●). As previously described and as summarized in Table 4.3, the differences between the endospore and the vegetative cell are profound. Sporulation involves a complex series of events in *cellular differentiation*. Bacterial sporulation does not occur when cells are dividing exponentially, but only when growth ceases owing to the exhaustion of an essen-tial nutrient. Thus, cells of *Bacillus*, a typical endospore-forming bacterium, cease vegetative growth and begin sporulation when a key nutrient such as the carbon or ni-trogen source becomes limiting.

Many genetically directed changes in the cell under-lie the conversion from vegetative growth to sporula-tion. The structural changes occurring in sporulating cells of *Bacillus* are shown in Figure 4.51●. The process of sporulation can be divided into several stages. In *Bacillus subtilis*, where detailed studies have been done, the entire sporulation process takes about 8 hours. Ge-netic studies of mutants of *Bacillus*, each blocked at one

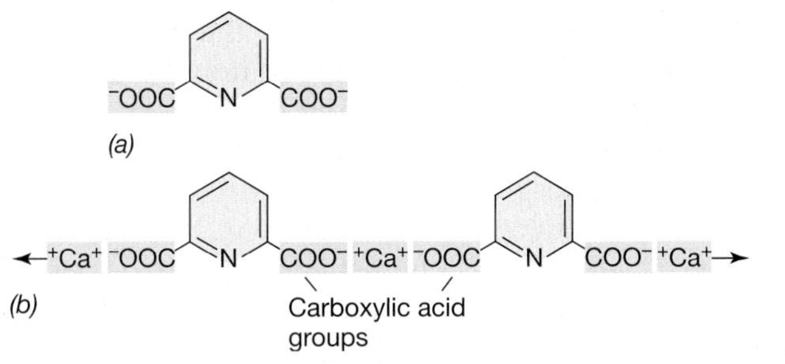

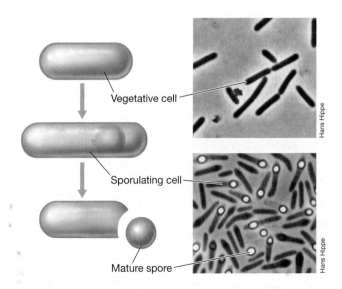

● **Figure 4.49 Dipicolinic acid (DPA).** (a) Structure of DPA. (b) How Ca^{2+} cross-links DPA molecules to form a complex.

● **Figure 4.50 Sporulation.** Phase contrast photomicrographs are of cells of *Clostridium pascui*.

Table 4.3	Differences between endospores and vegetative cells	
Characteristic	**Vegetative cell**	**Endospore**
Structure	Typical gram-positive cell; a few gram-negative cells	Thick spore cortex Spore coat Exosporium
Microscopic appearance	Nonrefractile	Refractile
Calcium content	Low	High
Dipicolinic acid	Absent	Present
Enzymatic activity	High	Low
Metabolism (O_2 uptake)	High	Low or absent
Macromolecular synthesis	Present	Absent
mRNA	Present	Low or absent
Heat resistance	Low	High
Radiation resistance	Low	High
Resistance to chemicals (for example, H_2O_2) and acids	Low	High
Stainability by dyes	Stainable	Stainable only with special methods
Action of lysozyme	Sensitive	Resistant
Water content	High, 80–90%	Low, 10–25% in core
Small acid-soluble proteins (product of *ssp* genes)	Absent	Present
Cytoplasmic pH	About pH 7	About pH 5.5–6.0 (in core)

of the stages of sporulation shown in Figure 4.51, indicate that more than 200 genes are involved in the sporulation process. Sporulation requires that the synthesis of many proteins involved in vegetative cell functions cease and that specific endospore proteins be made (see Figure 4.48*b*). This is accomplished by activation of a variety of endospore-specific genes including *spo, ssp* (which encodes SASPs), and many other genes in response to an environmental trigger to sporulate. The proteins encoded by these genes catalyze the series of events leading from a moist, metabolizing vegetative cell to a relatively dry, metabolically inert but extremely resistant, endospore (Table 4.3 and Figure 4.51).

Germination

An endospore can remain dormant for many years (see the Microbial Sidebar), but it can convert back to a vegetative cell relatively rapidly. This process involves three steps: *activation, germination,* and *outgrowth* (Figure 4.52●).

Activation is most easily accomplished by heating freshly formed endospores for several minutes at an elevated but sublethal temperature. Activated endospores are then conditioned to germinate when placed in the presence of specific nutrients, such as certain amino acids (for example, alanine is a particularly good trigger of endospore germination). *Germination*, usually a rapid process (on the order of several minutes), involves loss of microscopic refractility of the endospore, increased ability to be stained by dyes, and loss of resistance to

heat and chemicals. Loss of calcium dipicolinate and cortex components from the endospores occurs during this stage, and the SASPs are degraded. The final stage, *outgrowth*, involves visible swelling due to water uptake and synthesis of new RNA, proteins, and DNA. The cell emerges from the broken spore coat and eventually begins to divide (Figure 4.52). The cell then remains in vegetative growth until environmental signals once again trigger sporulation .

Diversity and Phylogenetic Aspects of Endospore Formation

How widespread is the process of endospore formation? Nearly 20 genera of *Bacteria* have been shown to form endospores (∞ Table 12.25), although the process has only been studied in detail in a few species of *Bacillus* and *Clostridium*. Nevertheless, many of the secrets to endospore survival, like the formation of calcium dipicolinate complexes (Figure 4.49) and the possession of endospore-specific genes, seem to be universal. It is thus likely that although some of the details may vary, the general principles of endospore construction are the same in all endospore-forming bacteria.

From a phylogenetic perspective, the capacity to produce endospores is uniquely tied to a particular sublineage of the gram-positive bacteria (∞ Sections 2.5 and 12.20). Despite this, the physiologies of endospore-forming bacteria are highly diverse and include anaerobes, aerobes, phototrophs, and chemolithotrophs.

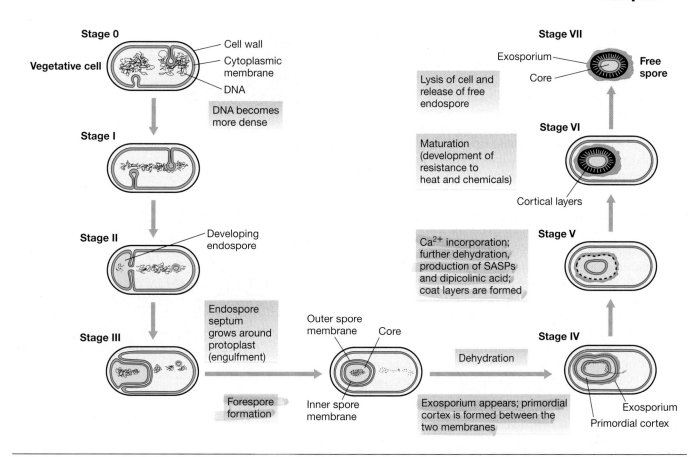

Stage 0

Vegetative cell
- Cell wall
- Cytoplasmic membrane
- DNA

DNA becomes more dense

Stage I

Stage II
- Developing endospore

Stage III

Endospore septum grows around protoplast (engulfment)

Forespore formation

Outer spore membrane — Core

Inner spore membrane

Dehydration

Exosporium appears; primordial cortex is formed between the two membranes

Stage IV
- Exosporium
- Primordial cortex

Ca^{2+} incorporation; further dehydration; production of SASPs and dipicolinic acid; coat layers are formed

Stage V

Cortical layers

Maturation (development of resistance to heat and chemicals)

Stage VI

Lysis of cell and release of free endospore

Stage VII
- Exosporium
- Core
- **Free spore**

● **Figure 4.51   Stages in endospore formation.**   The stages listed (0 through VII) are defined from both genetic studies and microscopic analyses.

Considering this physiological diversity, the actual triggers for endospore formation may vary with different species and could include signals other than simple nutrient star-vation, the major trigger for endospore formation in *Bacillus* and *Clostridium* species. Interestingly, no species of *Archaea* have been shown to sporulate, suggesting that the capacity to produce endospores evolved sometime *after* the major prokaryotic lineages diverged billions of years ago (⚭ Section 2.3 and Figure 2.7).

4.13   Concept Check

The endospore is a highly resistant differentiated bacterial cell produced by certain gram-positive *Bacteria*. Endospore formation leads to a highly dehydrated structure that contains essential macromolecules and a variety of substances such as calcium dipicolinate and small acid-soluble proteins, absent from vegetative cells. Endospores can remain dormant indefinitely but germinate quickly when the appropriate trigger is applied.

♦ What is *dipicolinic acid* and where is it found?

♦ What are *SASPs* and what is their function?

♦ What happens when an endospore germinates?

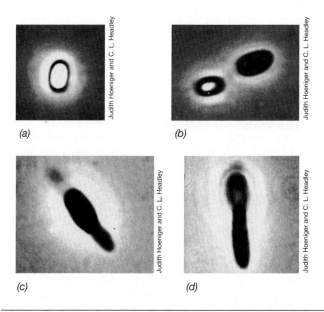

(a)   *(b)*

(c)   *(d)*

Judith Hoeniger and C. L. Headley

● **Figure 4.52   Endospore germination in *Bacillus*.**   Conversion of the mature endospore (a) to a vegetative cell (d); photomicrographs showing the sequence of events starting from a highly refractile mature endospore. In (b) (activation) refractility is being lost, while in (c) and (d) the new vegetative cell is emerging (outgrowth).

IV MICROBIAL LOCOMOTION

We finish our survey of microbial structure and function by examining cell locomotion. Many cells can move under their own power. Motility allows cells to reach different regions of their environment. In the struggle for survival, movement to a new location may offer a cell new resources and growth opportunities and spell the difference between life and death.

In the last few sections of this chapter we examine the different types of cell movement, including swimming and gliding. We then consider how motile cells are able to move in a directed fashion toward or away from particular stimuli (phenomena called *taxes*), and examples of simple behavioral responses.

4.14 Flagella and Motility

Many prokaryotes are motile, and this function is typically due to a special structure, the **flagellum** (plural, **flagella**) (Figure 4.53●). Certain bacterial cells can move along solid surfaces by *gliding* (see Section 4.14), and planktonic microorganisms can regulate their position in a water column by gas-filled structures called gas vesicles (see Sections 4.12 and 12.25). However, the majority of motile prokaryotes move by means of rotating flagella, and so we begin with a detailed consideration of this process.

Bacterial Flagella

Bacterial flagella are long thin appendages free at one end and attached to the cell at the other end. Flagella are so thin (about 20 nm) that a single flagellum can only be seen with the light microscope after staining with special flagella stains that increase their diameter (Figure 4.53). Flagella are easily seen with the electron microscope (Figure 4.54●).

Flagella are arranged differently on different bacteria. In **polar** flagellation, the flagella are attached at one or both ends of the cell (Figures 4.53*b* and 4.54*a*). Occasionally a tuft (group) of flagella may arise at one end of the cell, an arrangement called **lophotrichous** (*lopho* means "tuft"; *trichous* means "hair") (Figure 4.53*c*). Tufts of flagella of this type can be seen in living cells by darkfield microscopy (see Section 4.1 and Figure 4.55*a*●), where the flagella appear light and are attached to lightcolored cells against a dark background. In extremely large prokaryotes, tufts of flagella can also be observed

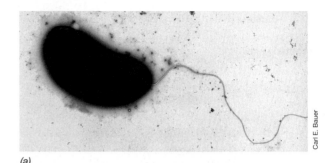

(a)

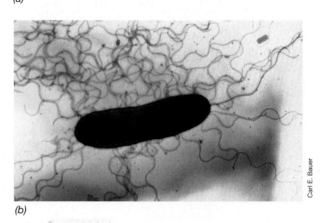

(b)

● **Figure 4.54** **Fine structure of bacterial flagella.** Bacterial flagella as observed by negative staining in the transmission electron microscope. (a) A single polar flagellum. (b) Peritrichous flagella. Both micrographs are of cells of the phototrophic bacterium *Rhodospirillum centenum*. Cells of *R. centenum* are normally polarly flagellated but under certain growth conditions form peritrichously flagellated "swarmer" cells. See also Figure 4.63*b*.

by phase contrast microscopy (Figure 4.55*b*). In **peritrichous** flagellation (Figures 4.53*a* and 4.54*b*), the flagella are inserted at many locations around the cell surface (*peri* means "around"). The type of flagellation, polar or peritrichous, is used as one characteristic in the classification of bacteria.

Flagellar Structure

Flagella are not straight but are helically shaped. When flattened, flagella show a constant distance between adjacent curves, called the *wavelength*, and this wavelength is characteristic for any given species (Figures 4.53–4.55). The filament of bacterial flagella is composed of subunits of a protein called **flagellin**. The shape and wavelength of the flagellum are in part determined by the structure of the flagellin protein and also to some extent by the direction of rotation of the filament. The basic flagellar structure to be described here varies little among species of *Bacteria*. However, in *Archaea*, several different flagellins are known, and the flagellar structure seems to be quite different from that of *Bacteria*, although flagellar function may be quite similar. In species of *Bacteria*, flagellin is highly conserved, suggesting that flagellar motility has deep evolutionary roots within this evolutionary domain.

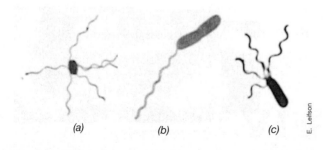

(a)   *(b)*   *(c)*

● **Figure 4.53 Bacterial flagella.** Light photomicrographs of prokaryotes containing different flagellar arrangements. Cells are stained with Leifson flagella stain. (a) Peritrichous. (b) Polar. (c) Lophotrichous.

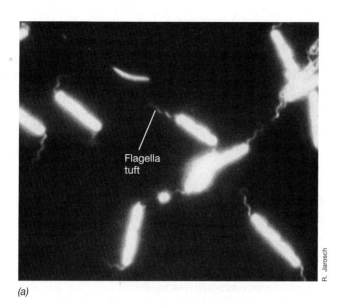

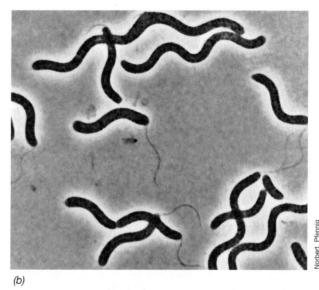

(a) R. Jarosch

(b) Norbert Pfennig

● **Figure 4.55** **Bacterial flagella as observed in living cells.** (a) Dark–field photomicrograph of a group of large rod–shaped bacteria with flagellar tufts at each pole. A single cell is about 2 μm wide. Dark-field microscopy uses horizontal illumination to yield reflected light (see Section 4.1 and Figure 4.5c). (b) Phase contrast photomicrograph of the large phototrophic purple bacterium *Rhodospirillum photometricum*. A single cell measures about 3 × 30 μm. Note the lophotrichous flagella that emanate from one of the poles.

A flagellum consists of several components and functions by *rotation*, much like a propeller in a motor boat. The base of the flagellum is different in structure from that of the filament (Figure 4.56a●). There is a wider region at the base of the flagellum called the *hook*. The hook consists of a single type of protein and functions to connect the filament to the motor portion of the flagellum.

The *motor* is anchored in the cytoplasmic membrane and cell wall. The motor consists of a small central rod that passes through a series of rings. In gram-negative *Bacteria*, an outer ring, called the *L ring*, is anchored in the lipopolysaccharide layer. A second ring, called the *P ring*, is anchored in the peptidoglycan layer of the cell wall. A third set of rings, called the *MS and C rings*, are located within the cytoplasmic membrane and the cytoplasm, respectively (Figure 4.56a). In gram-positive *Bacteria*, which lack an outer membrane, only the inner pair of rings is present. Surrounding the inner ring and anchored in the cytoplasmic membrane are a series of proteins called *Mot proteins* (Figure 4.56a). A final set of proteins, called the *Fli proteins* (Figure 4.56a) function as the motor switch, reversing the direction of rotation of the flagella in response to intracellular signals.

Flagellar Movement

The flagellum is a tiny rotary motor. How does this motor work? Rotary moters contain two components: the rotor and the stator. In the flagellar motor, the *rotor* consists of the C, MS, and P rings. Collectively, these structures make up the **basal body**. The *stator* consists of Mot proteins that surround the MS and C rings and functions to generate torque.

The rotary motion of the flagellum is imparted from the basal body. The energy required for rotation of the flagellum comes from the proton motive force (see Sections 4.6 and 5.12). Proton movement across the cytoplasmic membrane through the Mot complex (Figure 4.56a) drives rotation of the flagellum, and calculations have shown that about 1000 protons must be translocated per single rotation of the flagellum. How this actually occurs is not yet known. However, a "proton turbine" model has been proposed to explain the available experimental data (Figure 4.56b). In this model, protons flowing through channels in the stator exert electrostatic forces on helically arranged charges on the rotor proteins. Attractions between positive and negative charges would cause the basal body to rotate as protons flow though the stator (Figure 4.56b).

Flagellar Synthesis

Several genes are required for flagellar synthesis and subsequent motility. In *Escherichia coli* and *Salmonella typhimurium*, where studies have been most extensive, over 40 genes are necessary for motility. These genes have several functions, including encoding structural proteins of the flagellar apparatus, export of flagellar components through the cytoplasmic membrane to the outside of the cell, and regulation of the many biochemical events surrounding the synthesis of new flagella.

An individual flagellum grows not from its base, as does an animal hair, but from the tip. The MS ring is synthesized first and inserted into the cytoplasmic membrane. Then other anchoring proteins are synthesized along with the hook before filament formation occurs (Figure 4.57●). Flagellin molecules synthesized in the cytoplasm pass up

Web Tutorial 4.2

The Prokaryotic Flagellum

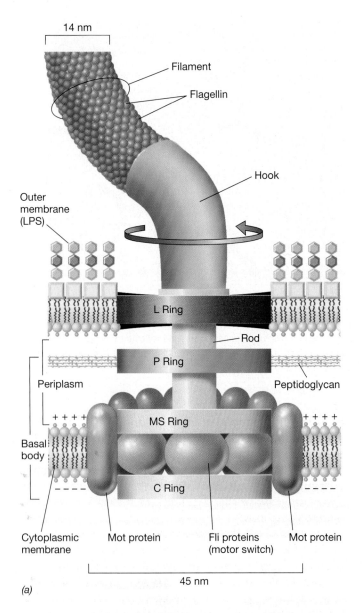

(a)

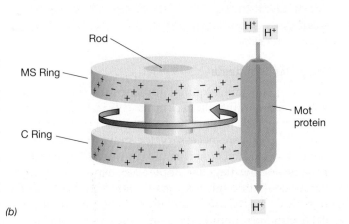

(b)

through a 3-nm channel inside of the filament and add on at the terminus to form the mature flagellum. At the end of the growing flagellum a protein "cap" exists. Cap proteins assist flagellin molecules that have diffused through the channel to organize at the flagellum termini to form new filament (Figure 4.57). Growth of the flagellum occurs more-or-less continuously until the flagellum reaches its final length. Broken flagella still rotate and can be repaired with new flagellin units passed through the filament channel to replace the lost ones.

Cell Speed

Flagella do not rotate at a constant speed but instead can increase or decrease their speed in relation to the strength of the proton motive force. Flagella rotation can move bacteria through liquid media at speeds of up to 60 cell lengths/second (sec). Although this is only about 0.00017 kilometer/hour (km/h), when comparing this speed with that of higher organisms in terms of the *number of lengths* moved per second, it is extremely fast. The fastest animal, the cheetah, moves at a maximum rate of about 110 km/h, but this represents only about 25 body lengths/sec. Thus, when size is accounted for, prokaryotic cells swimming at 50–60 lengths/sec are actually moving much faster than larger organisms!

The motions of polarly and lophotrichously flagellated organisms are different from those of peritrichously flagellated organisms, and these can be differentiated under the microscope. Peritrichously flagellated organisms typically move in a straight line in a slow, stately fashion. Polarly flagellated organisms, on the other hand, move more rapidly, spinning around and dashing from place to place. The different behavior of flagella on polar and peritrichous organisms, including differences in reversibility of the flagellum, is illustrated in Figure 4.58●.

4.14 Concept Check

Motility in most microorganisms is due to flagella. In prokaryotes the flagellum is a complex structure made of several proteins, most of which are anchored in the cell wall and cytoplasmic membrane. The flagellum filament, which is made of a single kind of protein, rotates at the expense of the proton motive force, which drives the flagellar motor.

◆ What is *flagellin* and where is it found?

◆ How does a bacterial flagellum move a cell forward?

◆ How does *polar flagellation* differ from *peritrichous flagellation*?

● **Figure 4.56 Structure and function of the prokaryotic flagellum in gram-negative *Bacteria.*** (a) Structure. The L ring is embedded in the LPS, and the P ring in peptidoglycan. The MS ring is embedded in the cytoplasmic membrane and the C ring in the cytoplasm. A narrow channel exists in the rod and filament through which flagellin molecules diffuse to reach the site of flagellar synthesis. The Mot proteins function as the flagellar motor, whereas the Fli proteins function as the motor switch. The flagellar motor rotates the filament to propel the cell through the medium. (b) Function. A "proton turbine" model has been proposed to explain rotation of the flagellum. Protons, flowing through the Mot proteins, may exert forces on charges present on the C and MS rings, thereby spinning the rotor.

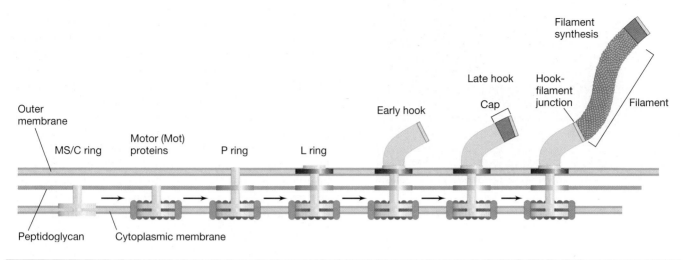

● **Figure 4.57** **Summary of steps in flagella biosynthesis.** Synthesis begins with MS/C ring assembly in the cytoplasmic membrane. This is followed by formation of the other rings, the hook and the cap. At this point, flagellin protein (approximately 20,000 copies are needed to make one filament) flows through the hook to form the filament. Flagellin molecules are guided into position by cap proteins to ensure that the growing filament develops evenly.

4.15 Gliding Motility

Some prokaryotes are motile but lack flagella. These nonswimming yet motile bacteria move across solid surfaces in a process called **gliding**. Unlike flagellar motility, where cells often stop and then start off in a different direction, gliding motility is a smoother form of movement and typically occurs along the long axis of the cell. Gliding motility is widely distributed among *Bacteria* but has been well studied in only a few groups. The gliding movement itself—up to 10 μm/sec in some gliding bacteria—is considerably slower than propulsion by flagella, but still offers the cell a means of moving about its habitat.

Gliding prokaryotes are filamentous or rod-shaped cells (Figure 4.59●), and the gliding process requires that the cells be in contact with a solid surface. The morphology of colonies of a typical gliding bacterium (colonies are masses of bacterial cells that form from successive cell divisions of a single cell, ∞ Section 5.3) are distinctive, since cells glide out and move away from the center of the colony (Figure 4.59c). Perhaps the most well-known gliding bacteria are the filamentous cyanobacteria (Figure 4.59a, b and ∞ Section 12.25), certain gram-negative *Bacteria*, such as *Myxococcus* and other myxobacteria (∞ Section 12.17), and species of *Cytophaga* and *Flavobacterium* (Figure 4.59c, d and ∞ Section 12.31).

Mechanisms of Gliding Motility

Although no gliding mechanism is well understood, it is likely that more than one mechanism occurs. In cyanobacteria (Figure 4.59a, b) it is known that a polysaccharide slime is secreted on the outer surface of the cell

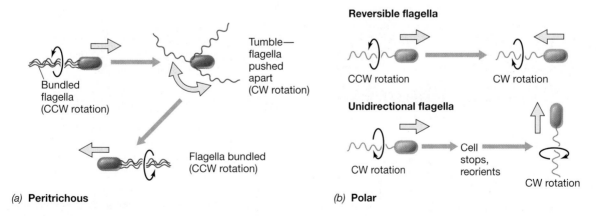

● **Figure 4.58** **Manner of movement in polarly and peritrichously flagellated prokaryotes.** (a) Peritrichous: Forward motion is imparted by all flagella rotating counterclockwise (CCW) in a bundle. Clockwise (CW) rotation causes the cell to tumble, and then a return to counterclockwise rotation leads the cell off in a new direction. (b) Polar: Cells change direction by reversing flagellar rotation (thus pulling instead of pushing the cell), or in unidirectional flagella, by stopping periodically to reorient, and then moving forward by clockwise rotation of its flagella. The yellow arrows show the direction the cell is traveling.

(a)

Richard W. Castenholz

(b)

Richard W. Castenholz

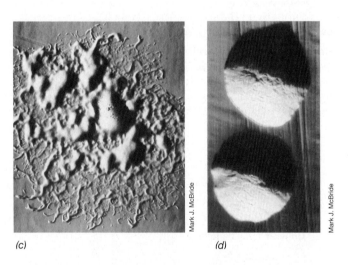

(c)   (d)

Mark J. McBride   Mark J. McBride

● **Figure 4.59 Gliding bacteria.** (a, b) The large filamentous cyanobacterium *Oscillatoria princeps*. (a) Photomicrograph. A cell is about 35 μm wide. (b) Photograph of filaments gliding on an agar surface. Cells can move by gliding against a solid surface or one filament can glide using a second filament as the solid surface. (c, d) The gram-negative gliding bacterium *Flavobacterium johnsoniae*. (c) Masses of cells gliding away from the center of the colony (the colony is about 2.7 mm wide). (d) Nongliding mutant strain showing typical colony morphology of nongliding bacteria (the colonies are 0.7–1 mm in diameter). For the proposed mechanism of gliding in *F. johnsoniae* cells, see Figure 4.60.

as it glides. The slime appears to contact both the cell surface and the solid surface against which the gliding cell moves. As the excreted slime adheres to the surface, the cell is pulled along. This hypothesis is supported by the observation of slime-excreting pores on the cell surface of several filamentous cyanobacteria.

Slime extrusion is not the mechanism of gliding in nonphototrophic gliding bacteria. In *Flavobacterium johnsoniae* (Figure 4.59c), for example, no slime is excreted. Instead, the movement of proteins on the cell surface has been identified as the likely mechanism of gliding. In *F. johnsoniae*, specific motility proteins anchored in the cytoplasmic and outer membranes are thought to propel the cell forward in a type of continuous ratcheting mechanism (Figure 4.60●). The movement of the cytoplasmic membrane proteins is driven by energy released from the proton motive force (∞ Sections 4.6 and 5.12), and they somehow transmit this energy to outer membrane proteins located along the cell surface. It is hypothesized that movement of the proteins against the solid surface literally pulls the cell forward (Figure 4.60).

Like other forms of motility, gliding motility has significant ecological relevance. Gliding allows a cell to exploit new resources or to interact in some beneficial way with other cells. In the latter regard, it is of interest that the myxobacteria, classic examples of gliding bacteria, have a very social and cooperative lifestyle, where gliding motility may play an important role in cell-to-cell interactions (∞ Section 12.17).

⬡ 4.15 Concept Check

Prokaryotes that move by gliding motility do not employ rotating flagella, but instead creep along a solid surface by any of several possible mechanisms.

◆ How does *gliding* motility differ from *swimming* motility in both mechanism and requirements?

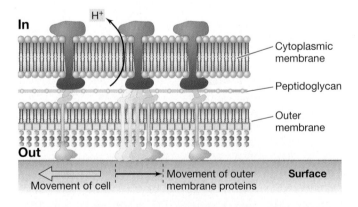

● **Figure 4.60 Model for how gliding motility may occur in *Flavobacterium johnsoniae* and some other gliding bacteria.** Tracks (yellow) are thought to exist in the peptidoglycan that connect cytoplasmic proteins (brown) to outer membrane proteins (orange) and propel them along the solid surface. Note the difference in direction of movement of outer membrane proteins and of the cell proper. Protein propulsion is somehow linked to dissipation of the proton motive force.

◆ Contrast the mechanism of gliding motility in a filamentous cyanobacterium and in *Flavobacterium*.

4.16 Cell Motion as a Behavioral Response: Chemotaxis and Phototaxis

Prokaryotes often encounter *gradients* of physical or chemical agents in nature, and cells have evolved means to respond to these gradients by moving either toward or away from the signal molecule, depending on whether it is a useful or harmful substance. Such directed movements are called *taxes*. **Chemotaxis**, a response to chemicals, and **phototaxis**, a response to light, are two well-known taxes. Here we cover taxes in a general way. We will return in Section 8.13 to cover the details of chemotaxis and its regulation as a model for all prokaryotic taxes.

Chemotaxis as a phenomenon has been well studied in swimming bacteria, and much is known at the genetic level concerning how the chemical status of the environment is communicated to the flagellar assembly. Our discussion here will thus deal solely with swimming bacteria. However, some gliding bacteria have also been shown to be chemotactic, and phototactic movements in filamentous cyanobacteria also occur. Thus, although the mechanisms are unknown, like swimming bacteria, gliding bacteria must in some way be able to communicate with their motility apparatus.

Chemotaxis

To understand chemotaxis let us focus on the behavior of a single bacterial cell faced with a chemical gradient of an attractant (Figure 4.61●). Unlike larger organisms, prokaryotes are too small to sense a gradient along the length of a single cell. Instead, while moving, prokaryotes compare the chemical or physical state of their environment with that sensed a few seconds before. In other words, bacteria respond to the *temporal* (rather than *spatial*) *gradient* of signal molecules as they swim along.

Much research on chemotaxis has been done with the peritrichously flagellated bacterium, *Escherichia coli*. In the absence of a gradient, cells of *E. coli* move in a random fashion that includes **runs**, where the cell is swimming forward in a smooth fashion, and **tumbles**, when the cell stops and jiggles about (Figure 4.61a). Following a tumble, the direction of the next run is random (Figure 4.61a). Thus, by means of runs and tumbles, the cell moves about randomly in its environment but does not really go anywhere.

However, if a gradient of a chemical attractant is present, these random movements become biased. As the organism senses that it is moving toward higher concentrations of the attractant (through periodic sampling of the concentration of the chemical in its environment), runs become longer and tumbles less frequent. The net result of this behavioral response is that the organism moves up the concentration gradient of the attractant (Figure 4.61b).

If the organism is sensing a repellent, the same general mechanism applies, although in this case it is the *decrease* in concentration of the repellent (rather than the *increase* in concentration of an attractant) that promotes runs. Forward movement in a run occurs when the flagellar motor is rotating *counterclockwise*. When the flagella rotate *clockwise*, the bundle pushes apart, forward motion ceases, and the cells tumble (Figures 4.58 and 4.61).

The situation with polarly flagellated cells is somewhat different. Many polarly flagellated bacteria, such as *Pseudomonas* species, can reverse the direction of rotation of their flagella like peritrichously flagellated cells and reverse direction in this fashion (Figure 4.58b). However,

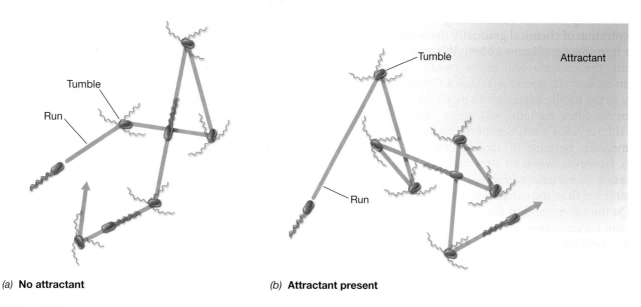

(a) **No attractant**   *(b)* **Attractant present**

● **Figure 4.61   Chemotaxis in a peritrichously flagellated bacterium such as *Escherichia coli*.**   (a) In the absence of a chemical attractant the cell swims randomly in runs, changing direction during tumbles. (b) In the presence of an attractant runs become biased, and the cell moves up the gradient of the attractant.

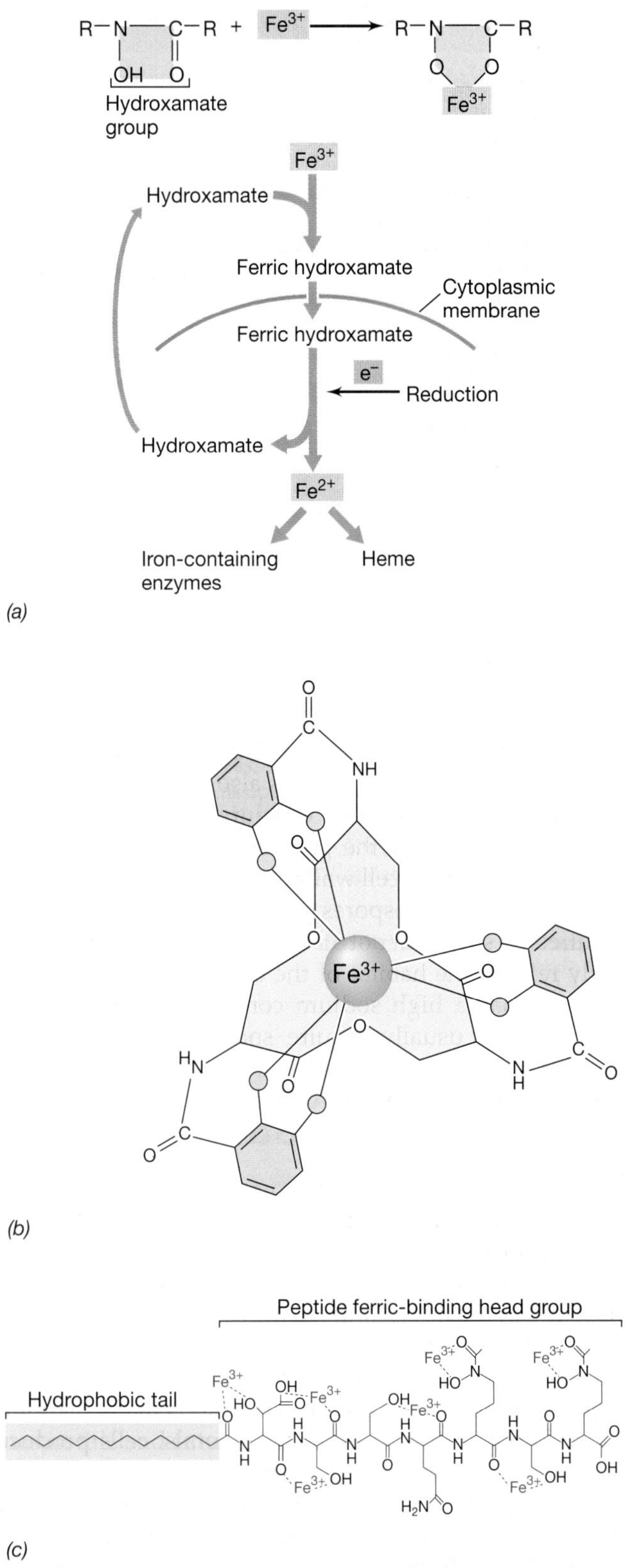

(a)

(b)

(c)

● **Figure 5.1   Iron-chelating agents produced by microorganisms.**
(a) Hydroxamate. Iron is bound as Fe^{3+} and released inside the cell as
Fe^{2+}. The hydroxamate then exits the cell and repeats the cycle. (b) Ferric
enterobactin of *Escherichia coli*. The oxygen atoms of each catechol
molecule are shown in yellow. (c) The peptidic and tailed siderophore
aquachelin, showing Fe^{3+} binding. The hydrophobic tail assists in
threading aquachelin through the membrane into the cell.

iron. Without such tenacious iron-binding agents, many
pathogens would be unable to initiate an infection. Iron
is often in short supply in animal tissues because iron
scavenging systems tie up iron and make it unavailable
to microorganisms (∞Sections 21.7 and 21.8).

In marine waters, iron can be virtually undetectable.
For example, surface ocean waters typically contain only
a few *picograms* (one picogram is 10^{-12} g) of iron per milliliter. To counter this, many marine bacteria produce
structurally complex siderophores that can sequester iron
present in these vanishingly small amounts. These
siderophores contain a peptide head group that complexes Fe^{3+} and a lipid tail that can associate with the cytoplasmic membrane. Once compounds such as *aquachelin*
(Figure 5.1c) bind iron, they aggregate into lipidlike mi-
celles and transport the bound iron into the cell.

Some prokaryotes can grow in the total *absence* of
iron. For example, cells of the bacteria *Lactobacillus plan-
tarum* and *Borrelia burgdorferii* (the latter is the causative
agent of Lyme disease, ∞Section 27.4) do not contain
iron. In these bacteria, Mn^{2+} substitutes for iron as the
metal component of enzymes that normally contain Fe^{2+}.

Micronutrients (Trace Elements)

Although required in just tiny amounts, *micronutrients*
are nevertheless also critical to cell function. Like iron,
micronutrients are metals, and are thus often called *trace
elements*. Micronutrients typically play a role as components of various enzymes, the cells' catalysts. Table 5.2
summarizes the major micronutrients of cells and gives
examples of enzymes in which each plays a role.

For the routine laboratory culture of microorganisms
it is often unnecessary to add trace elements to the culture medium because the requirement for trace elements
is so small. However, if a culture medium contains highly purified chemicals dissolved in high-purity distilled
water, a trace element deficiency can occur. In such cases
a dilute solution of trace metals (Table 5.2) is added to
the medium to make available the necessary metals.

Growth Factors

Growth factors are *organic* compounds that, like micronutrients, are required in very small amounts and
then only by some cells. Growth factors include vitamins, amino acids, purines, and pyrimidines. Although
most microorganisms are able to synthesize all of these
compounds, some microorganisms require one or more
of them preformed from the environment and thus must
be supplied with these compounds in laboratory culture.

Vitamins are the most commonly required growth
factors. Most vitamins function as components of coenzymes (see, for instance, Figures 5.10, 5.12, and 5.15), and
these are summarized in Table 5.3. Vitamin requirements
are highly variable among microorganisms, ranging from
none to several. Lactic acid bacteria, which include the
genera *Streptococcus*, *Lactobacillus*, *Leuconostoc*, among
others (∞Section 12.19), are renowned for their multi-

ple vitamin requirements, which are even more extensive than those of humans (see Table 5.4). The vitamins most commonly required by microorganisms are thiamine (vitamin B_1), biotin, pyridoxine (vitamin B_6), and cobalamin (vitamin B_{12}).

Table 5.2	Micronutrients (trace elements) needed by living organisms[a]
Element	**Cellular function**
Boron (B)	Present in an autoinducer for quorum sensing in bacteria; also found in some polyketide antibiotics
Chromium (Cr)	Required by mammals for glucose metabolism; no known microbial requirement
Cobalt (Co)	Vitamin B_{12}; transcarboxylase (propionic acid bacteria)
Copper (Cu)	Respiration, cytochrome c oxidase; photosynthesis, plastocyanin, some superoxide dismutases
Iron (Fe)[b]	Cytochromes; catalases; peroxidases; iron-sulfur proteins; oxygenases; all nitrogenases
Manganese (Mn)	Activator of many enzymes; present in certain superoxide dismutases and in the water-splitting enzyme in oxygenic phototrophs (Photosystem II)
Molybdenum (Mo)	Certain flavin-containing enzymes; some nitrogenases, nitrate reductases, sulfite oxidases, DMSO-TMAO reductases; some formate dehydrogenases
Nickel (Ni)	Most hydrogenases; coenzyme F_{430} of methanogens; carbon monoxide dehydrogenase; urease
Selenium (Se)	Formate dehydrogenase; some hydrogenases; the amino acid selenocysteine
Tungsten (W)	Some formate dehydrogenases; oxotransferases of hyperthermophiles
Vanadium (V)	Vanadium nitrogenase; bromoperoxidase
Zinc (Zn)	Carbonic anhydrase; alcohol dehydrogenase; RNA and DNA polymerases; and many DNA-binding proteins

[a] Not every micronutrient listed is required by all cells; some metals listed are found in enzymes present in only specific microorganisms.
[b] Needed in greater amounts than other trace metals.

5.1 Concept Check

The hundreds of chemical compounds present inside a living cell are formed from starting materials called nutrients. Elements required in fairly large amounts are called macronutrients, while metals and organic compounds needed in very small amounts are called micronutrients and growth factors, respectively.

◆ What two classes of macromolecules contain the bulk of the *nitrogen* in a cell?

◆ Why is an element like Co^{2+} considered a *micro*nutrient whereas an element like C is considered a *macro*nutrient?

◆ What roles does iron play in cellular metabolism? How do cells sequester iron?

5.2 Culture Media

Culture media are the nutrient solutions used to grow microorganisms in the laboratory. Because laboratory culture is required for the detailed study of a microorganism, careful attention must be paid to both the selection and preparation of media for successful culture to take place.

Classes of Culture Media

Two broad classes of culture media are used in microbiology: *chemically defined* and *undefined (complex)*. **Chemically defined media** (**defined media,** for short) are prepared by adding precise amounts of highly purified inorganic or organic chemicals to distilled water. Therefore, the *exact chemical composition* of a defined medium is known. Of paramount importance in any culture medium is the *carbon source*, since all cells need large amounts of carbon to make new cell material. In a *simple defined medium* (Table 5.4), a *single* carbon source is present. The nature of the carbon source and its concentration depends on the organism to be cultured.

For growing many organisms, knowledge of the exact composition of a medium is not essential. In these

Table 5.3	Growth factors: Vitamins and their functions
Vitamin	**Function**
p-Aminobenzoic acid	Precursor of folic acid
Folic acid	One-carbon metabolism; methyl group transfer
Biotin	Fatty acid biosynthesis; β-decarboxylations; some CO_2 fixation reactions
Cobalamin (B_{12})	Reduction of and transfer of single carbon fragments; synthesis of deoxyribose
Lipoic acid	Transfer of acyl groups in decarboxylation of pyruvate and α-ketoglutarate
Nicotinic acid (niacin)	Precursor of NAD^+ (see Figure 5.10); electron transfer in oxidation-reduction reactions
Pantothenic acid	Precursor of coenzyme A; activation of acetyl and other acyl derivatives
Riboflavin	Precursor of FMN (see Figure 5.15), FAD in flavoproteins involved in electron transport
Thiamine (B_1)	α-Decarboxylations; transketolase
Vitamins B_6 (pyridoxal-pyridoxamine group)	Amino acid and keto acid transformations
Vitamin K group; quinones	Electron transport; synthesis of sphingolipids
Hydroxamates	Iron-binding compounds; solubilization of iron and transport into cell

Table 5.4	Examples of culture media for microorganisms with simple and demanding nutritional requirements[a]	
Defined culture medium for *Escherichia coli*	**Defined culture medium for** *Leuconostoc mesenteroides*	**Complex culture medium for** either *E. coli* or *L. mesenteroides*
K_2HPO_4 7 g	K_2HPO_4 0.6 g	Glucose 15 g
KH_2PO_4 2 g	KH_2PO_4 0.6 g	Yeast extract 5 g
$(NH_4)_2SO_4$ 1 g	NH_4Cl 3 g	Peptone 5 g
$MgSO_4$ 0.1 g	$MgSO_4$ 0.1 g	KH_2PO_4 2 g
$CaCl_2$ 0.02 g	Glucose 25 g	Distilled water 1000 ml
Glucose 4–10 g	Sodium acetate 20 g	pH 7
Trace elements (Fe, Co, Mn, Zn, Cu, Ni, Mo) 2–10 μg each	Amino acids (alanine, arginine, asparagine, aspartate, cysteine, glutamate, glutamine, glycine, histidine, isoleucine, leucine, lysine, methionine, phenylalanine, proline, serine, threonine, tryptophan, tyrosine, valine) 100–200 μg of each	
Distilled water 1000 ml		
pH 7	Purines and pyrimidines (adenine, guanine, uracil, xanthine) 10 mg of each	
	Vitamins (biotin, folate, nicotinic acid, pyridoxal, pyridoxamine, pyridoxine, riboflavin, thiamine, pantothenate, *p*-aminobenzoic acid) 0.01–1 mg of each	
	Trace elements (see first column) 2–10 μg each	
	Distilled water 1000 ml	
	pH 7	

(a)

(b)

[a] The photos are tubes of (a) the defined medium described, and (b) the complex medium described. Note how the complex medium is colored from the various organic extracts and digests that it contains. Photos courtesy of Cheryl L. Broadie and John Vercillo, Southern Illinois University at Carbondale.

instances complex media may suffice and may even be advantageous. **Complex media** employ digests of animal or plant products, such as casein (milk protein), beef, soybeans, yeast cells, or any of a number of other highly nutritious (yet chemically undefined) substances. These digests are commercially available in powdered form and can be quickly weighed out and dissolved in distilled water to yield a medium. However, a major concession in using a complex medium is loss of control of its precise nutrient composition.

In particular situations, especially in clinical microbiology, culture media are often made to be *selective* or *differential* (or both). A **selective medium** contains compounds that selectively inhibit the growth of some microorganisms but not others. By contrast, a **differential medium** is one in which some sort of indicator, typically a dye, is added, that allows for the differentiation of particular chemical reactions occurring during growth. Differential media are quite useful for distinguishing between species of bacteria, some of which may carry out the particular reaction while others do not. Differential and selective media are further discussed in Chapter 24.

Nutritional Requirements and Biosynthetic Capacity

Table 5.4 shows three recipes for culture media, two defined and one complex. The complex medium is easiest to prepare and supports good growth of either organism shown in the table, the enteric bacterium *Escherichia coli* or the lactic acid bacterium *Leuconostoc mesenteroides*, an extremely *fastidious* (nutritionally demanding) bacterium.

The simple defined medium shown in Table 5.4 supports excellent growth of *E. coli* but not of *L. mesenteroides*. Growth of the latter organism in defined medium requires the addition of several organic nutrients and growth factors not needed by *E. coli* (Table 5.4). Which organism, *E. coli* or *L. mesenteroides*, has the greater *biosynthetic* capacity? Obviously, it is *E. coli*, since its ability to grow on a simple defined culture medium means that it can synthesize *all* of its organic constituents from a single carbon compound, in this case glucose (Table 5.4).

In contrast to *E. coli*, *L. mesenteroides* has multiple growth factor and major organic nutrient (for example, amino acid) requirements, indicative of its limited biosynthetic capacity. The complex nutritional needs of *L. mesenteroides* can be satisfied by either preparing a defined medium as shown in Table 5.4 (although such a medium could take hours to prepare) or by using a complex medium (Table 5.4), which can usually be prepared quickly.

Some pathogenic microorganisms have even more stringent nutritional requirements than those shown for *L. mesenteroides*. For culturing these organisms, an *enriched medium* may be needed. An **enriched medium** starts as a complex medium to which additional nutrients, such as serum or whole blood, are added. These added nutrients better mimic conditions in the host and are required for the successful laboratory culture of organisms such as *Streptococcus pyogenes* (strep throat) and *Neisseria gonorrhoeae* (gonorrhea).

A major conclusion to be drawn from Table 5.4 is that *different microorganisms can have vastly different nutri-*

tional requirements*. Thus, for successful culture of a given microorganism it is necessary to understand its nutritional requirements and then supply the nutrients in the proper form and proportions in a culture medium. If care is taken in preparing culture media, it is fairly easy to culture many different types of microorganisms in the laboratory. We discuss the procedures involved in culturing microorganisms now.

 5.2 Concept Check

Culture media supply the nutritional needs of microorganisms and can be either chemically defined or undefined (complex). Selective, differential, and enriched are terms that describe media used for the isolation of particular species or for comparative studies of microorganisms.

◆ Why is the routine culture of *Leuconostoc mesenteroides* easier in a complex medium than in a chemically defined medium?

◆ In which medium, defined or complex (shown in Table 5.4), do you think *Escherichia coli* would grow the fastest? Why?

5.3 Laboratory Culture of Microorganisms

Once a culture medium has been prepared and sterilized to render it free of all microorganisms, it can be *inoculated* (that is, organisms added) and incubated under conditions that will support microbial growth. In a laboratory situation, inoculation will typically be of a **pure culture**, a culture containing only a *single kind* of microorganism.

It is essential that other organisms are prevented from entering a pure culture. Such unwanted organisms, called *contaminants*, are ubiquitous (as Pasteur discovered over 125 years ago, ∞Section 1.6), and microbiological techniques are designed to avoid contaminants. A major method for obtaining pure cultures and for assessing the purity of a culture is the use of solid media, specifically, solid media prepared in the Petri plate, and we consider this now.

Solid versus Liquid Culture Media

Culture media are sometimes prepared in a semisolid form by the addition of a gelling agent to liquid media. Such solid culture media immobilize cells, allowing them to grow and form visible, isolated masses called *colonies* (Figure 5.2●). Bacterial colonies can be of various

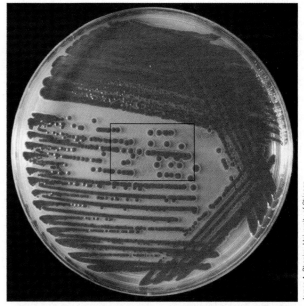

(a)

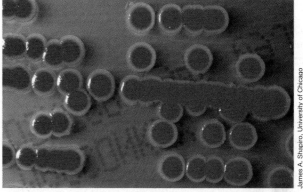

(b)

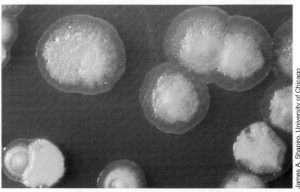

(c)

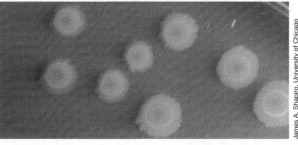

(d)

● **Figure 5.2 Examples of bacterial colonies.** Colonies are visible masses of cells formed from the subsequent division of one or a few cells. The size, shape, texture, and color of a bacterial colony is a function of the organism that made them. Depending on its size and the arrangement of cells within the colony, a colony can have highly variable cell numbers; colonies containing over one *billion* individual cells are not uncommon. (a) *Serratia marcescens,* grown on MacConkey agar. (b) Close-up of colonies outlined in (a). (c) *Pseudomonas aeruginosa,* grown on Trypticase-Soy agar. (d) *Shigella flexneri,* grown on MacConkey agar.

shapes and sizes depending on the organism, the culture conditions, the nutrient supply (including the amount of oxygen present), and several other physiological parameters. Some bacteria produce pigments that cause the colony to be colored (Figure 5.2). Colonies permit the microbiologist to visualize the purity of the culture. Plates that contain more than one colony type are indicative of a contaminated culture. In this way, the Petri plate has been used as one criterion of culture purity for over 100 years (∞Section 1.6).

Solid media are prepared the same way as for liquid media except that before sterilization, *agar* (Microbial Sidebar, Solid Media, the Petri Plate, and Pure Cultures ∞Chapter 1) is added as a gelling agent, usually at a concentration of 1.5%. The agar melts during the sterilization process and the molten medium is then poured into sterile glass or plastic plates and allowed to solidify before use (Figure 5.2).

Aseptic Technique

Because microorganisms are everywhere, culture media must be sterilized before use. For most culture media this is done by heating, typically by moist heat in a large pressure cooker called an *autoclave*. We discuss the operation and principles of the autoclave later, along with other methods of sterilization (∞Section 20.1).

Once a sterile culture medium has been prepared, it is ready to receive an *inoculum* from a previously grown pure culture to start the growth process once again. This manipulation requires the practice of **aseptic technique**, the series of steps used to prevent contamination during manipulations of cultures and sterile culture media (Figures 5.3● and 5.4●). A mastery of aseptic technique is required for success in the microbiology laboratory, and it is one of the first methods learned by the novice microbiologist. Airborne contaminants are the most common problem because laboratory air contains minute dust particles that will have a community of microorganisms on them. When containers are opened, they must be handled in such a way that contaminant-laden air does not enter (Figures 5.3 and 5.4).

Aseptic transfer of a culture from one tube of medium to another is usually accomplished with an inoculating loop or needle that has been sterilized in a flame (Figure 5.3). Cells from liquid cultures can also be transferred to the surface of agar plates (Figure 5.4), where colonies develop from the growth and division of single cells. Picking and restreaking from an isolated colony is a major method of obtaining pure cultures from microbial communities containing many different organisms.

5.3 Concept Check

Microorganisms can be grown in the laboratory in culture media containing the nutrients they require. Successful cultivation and maintenance of pure cultures of microorganisms can only be done if aseptic technique is practiced.

◆ What is meant by the word *sterile*? What would happen if freshly prepared culture media were not sterilized and left at room temperature?

◆ Why is aseptic technique necessary for successful cultivation of pure cultures in the laboratory?

II ENERGETICS AND ENZYMES

We learned in Chapter 2 of the different energy classes of microorganisms—*chemoorganotrophs, chemolithotrophs*, and *phototrophs* (∞Section 2.4). Regardless of how an organism makes a living, however, it must be able to conserve some of the energy it obtains as ATP. Here we discuss the principles of energy conservation, using some simple laws of chemistry and physics to guide our understanding. We then consider enzymes, the cell's biocatalysts.

5.4 Bioenergetics

Energy is defined as the ability to do work. In microbiology, energy is calculated in units of *kilojoules* (kJ), a measure of heat energy. Chemical reactions are accompanied by *changes* in energy. Although in any chemical reaction

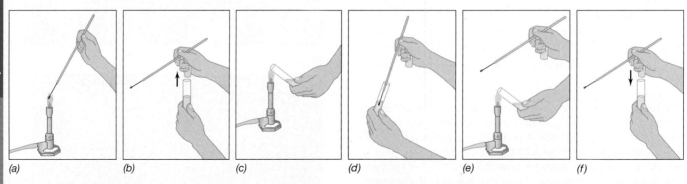

● **Figure 5.3 Aseptic transfer.** (a) Loop is heated until red-hot and cooled in air briefly. (b) Tube is uncapped. (c) Tip of tube is run through the flame. (d) Sample is removed on sterile loop. (e) After removing sample on loop, the tube is reflamed and then the sample is transferred to a sterile medium. (f) The tube is recapped. Loop is reheated before being taken out of service.

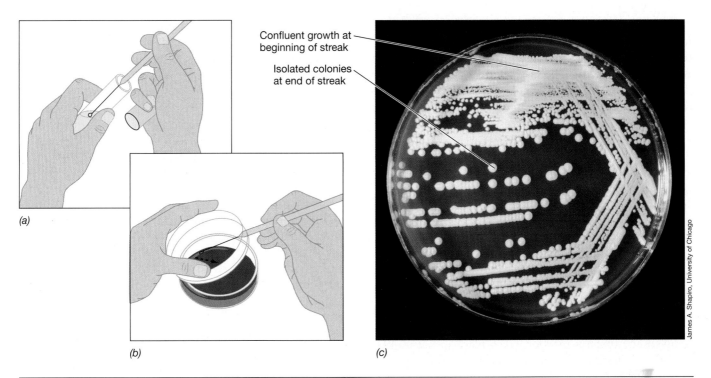

● **Figure 5.4** **Method of making a streak plate to obtain pure cultures.** (a) Loop is sterilized, and then a loopful of inoculum is removed from tube. (b) Streak is made over a sterile agar plate, spreading out the organisms. Following the initial streak, subsequent streaks are made at angles to it, the loop being resterilized between streaks. (c) Appearance of a well-streaked plate after incubation. Colonies of the bacterium *Micrococcus luteus* grown on a blood agar plate. It is from such well-isolated colonies that pure cultures can usually be obtained.

some energy is lost as heat, in microbiology we are interested in **free energy** (abbreviated G), which is defined as the energy released *that is available to do useful work*. The change in free energy during a reaction is expressed as $\Delta G^{0\prime}$, where the symbol Δ should be read "change in." The "0" and "prime" superscripts mean that the free-energy value was obtained under standard conditions: pH 7, 25°C, one atmosphere of pressure, and all reactants and products at 1 M concentration.

Consider the reaction:

$$A + B \rightarrow C + D$$

If $\Delta G^{0\prime}$ for this reaction is *negative,* then the reaction will proceed with the *release* of free energy, energy that the cell may be able to conserve in the form of ATP. Such energy-yielding reactions are called **exergonic**. However, if $\Delta G^{0\prime}$ is *positive,* the reaction *requires* energy in order to proceed. Such reactions are called **endergonic**. Thus, from the standpoint of the microbial cell, exergonic reactions *yield* energy while endergonic reactions *require* energy.

Free Energy of Formation and Calculating $\Delta G^{0\prime}$

To calculate the free energy yield of a reaction, one needs to know the free energy of its reactants and products. This is the *free energy of formation* (G_f^0), the energy released or required during the *formation* of a given molecule from the elements. Table 5.5 gives a few examples of G_f^0. By convention, the free energy of formation of the elements in their elemental form (for instance, C, H_2, N_2) is zero. The G_f^0 of compounds, however, is not zero. If the formation of a compound from its elements proceeds exergonically, then the G_f^0 of the compound is *negative* (energy is released). If the reaction is endergonic (energy is required), then the G_f^0 of the compound is *positive*.

For most compounds G_f^0 is *negative*. This reflects the fact that compounds tend to form spontaneously (that is, with energy being released) from their elements. However, the positive G_f^0 for nitrous oxide (+104.2 kJ/mol, Table 5.5) tells us that this molecule will *not* form spontaneously. Instead, it will decompose spontaneously to nitrogen and oxygen. The free energies of formation of a variety of compounds of microbiological interest are given in Appendix 1.

Table 5.5	Free energy of formation for a few compounds of biological interest
Compound	**Free energy of formation**[a]
Water (H_2O)	−237.2
Carbon dioxide (CO_2)	−394.4
Hydrogen gas (H_2)	0
Oxygen gas (O_2)	0
Ammonium (NH_4^+)	−79.4
Nitrous oxide (N_2O)	+104.2
Acetate ($C_2H_3O_2^-$)	−369.4
Glucose ($C_6H_{12}O_6$)	−917.3
Methane (CH_4)	−50.8
Methanol (CH_3OH)	−175.4

[a] The free energy of formation values (G_f^0) are in *kJ/mol*. See Table A1.1 for a more complete list of free energies of formation.

Using free energies of formation, it is possible to calculate the *change in standard free energy* ($\Delta G^{0\prime}$) that occurs in a given reaction. For the reaction $A + B \rightarrow C + D$, $\Delta G^{0\prime}$ is calculated by subtracting the *sum* of the free energies of formation of the reactants (A and B) from that of the products (C and D). Thus:

$$\Delta G^{0\prime} = G_f^0[C + D] - G_f^0[A + B]$$

The phrase "products minus reactants" is a simple way to recall how to calculate changes in free energy during chemical reactions. However, it is first necessary to balance the reaction chemically before free-energy calculations can be made. Appendix 1 details the steps involved in calculating free energies for any hypothetical reaction.

$\Delta G^{0\prime}$ versus ΔG

Free energy calculations using *standard* conditions are only approximations of the free energy changes that actually occur when a reaction takes place in nature. Although calculations of $\Delta G^{0\prime}$ are usually good estimates of actual free energy changes, under some circumstances they are not. We will see later in this book that the actual concentrations of products and reactants in nature, which are rarely at 1 M levels, can alter the bioenergetics of reactions, and sometimes in significant ways (Sections 17.21, 19.10, and Appendix 1). Thus, what is most relevant to a bioenergetic calculation is not $\Delta G^{0\prime}$, but ΔG, the free energy change that occurs *under the actual conditions* in which the organism is growing. ΔG is a form of the free energy equation that takes into account the actual concentrations of reactants and products in the reaction, and is calculated as:

$$\Delta G = \Delta G^{0\prime} + RT \ln K$$

where R and T are physical constants and K is the equilibrium constant for the reaction in question (Appendix 1). We will pick up the distinction between $\Delta G^{0\prime}$ and ΔG in Chapter 17 where we consider metabolic diversity in more detail. But for now, we will focus on what the expression $\Delta G^{0\prime}$ can tell us about chemical reactions catalyzed by microorganisms.

 5.4 Concept Check

The chemical reactions of the cell are accompanied by changes in energy, expressed in kilojoules. A chemical reaction can occur with the release of free energy (exergonic) or with the consumption of free energy (endergonic).

◆ What is free energy?

◆ In general, are *catabolic* reactions exergonic or endergonic?

◆ Using the data in Table 5.5, calculate $\Delta G^{0\prime}$ for the reaction $CH_4 + \frac{1}{2}O_2 \rightarrow CH_3OH$. How does $\Delta G^{0\prime}$ differ from ΔG?

5.5 Catalysis and Enzymes

A free energy calculation reveals only whether energy is released or required in a given reaction. The number obtained says nothing about the *rate* of the reaction. Consider the formation of water from gaseous oxygen and hydrogen. The energetics of this reaction are quite favorable: $H_2 + \frac{1}{2}O_2 \rightarrow H_2O$, $\Delta G^{0\prime} = -237$ kJ. However, if we were to mix O_2 and H_2 together in a bottle, no measurable formation of water would occur for years. This is because the rearrangement of oxygen and hydrogen atoms to form water first requires that the chemical bonds of the reactants be broken. The breaking of bonds requires energy, and this energy is called **activation energy**.

Activation energy is that required to bring all molecules in a chemical reaction into the reactive state. For a reaction that proceeds with a net release of free energy (that is, an exergonic reaction), the situation is as diagrammed in Figure 5.5●. Although the activation energy barrier is virtually insurmountable in the absence of catalysis, in the presence of the proper catalyst it is much less formidable (Figure 5.5).

Enzymes

The idea of activation energy leads us to the concept of catalysis and enzymes. In a biochemical sense, a **catalyst** is a substance that *lowers* the activation energy of a reaction, thereby *increasing* the reaction rate. Catalysts facilitate reactions but are themselves not consumed or transformed by the reactions. Moreover, catalysts do not affect the energetics or the equilibrium of a reaction; catalysts affect only the *rate* at which reactions proceed.

Most reactions in living organisms would not occur at appreciable rates without catalysis. Biological catalysts

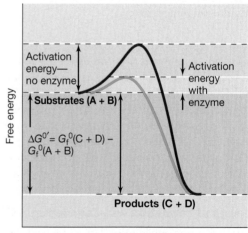

● **Figure 5.5 Progress of a hypothetical exergonic reaction: A + B → C + D, and the concept of activation energy.** Chemical reactions may not proceed spontaneously even though energy would be released, because the reactants must first be activated. Once activation has occurred, the reaction then proceeds spontaneously. Catalysts such as enzymes lower the required activation energy.

are called **enzymes**. Enzymes are proteins (or in a few cases, RNAs) that are highly specific in the reactions they catalyze. That is, each enzyme catalyzes only a *single type* of chemical reaction, or in the case of some enzymes, a *class* of closely related reactions. This specificity is a function of the precise three-dimensional structure of the enzyme molecule. In an enzyme-catalyzed reaction, the enzyme temporarily combines with the reactant, called a **substrate** (S), forming an **enzyme-substrate complex**. Then, as the reaction proceeds, the **product** (P) is released and the enzyme (E) is returned to its original state:

$$E + S \leftrightarrow E—S \leftrightarrow E + P$$

The enzyme is generally much larger than the substrate(s), and the combination of enzyme and substrate(s) typically depends on weak bonds, such as hydrogen bonds, van der Waals forces, and hydrophobic interactions (∞Section 3.1) to join the enzyme to the substrate. The portion of the enzyme to which substrates bind is called the **active site** of the enzyme.

Enzyme Catalysis

The catalytic power of enzymes is impressive. Enzymes typically increase the rate of chemical reactions from 10^8 to 10^{20} times the rate that would occur spontaneously. To catalyze a specific reaction, an enzyme must do two things: (1) bind the correct substrate, and (2) position the substrate relative to the catalytically active amino acids in the enzyme's active site. The enzyme-substrate complex (Figure 5.6●) aligns reactive groups and places strain on specific bonds in the substrate(s). The net result of enzyme-substrate complex formation is a reduction in

the activation energy required to make the reaction proceed from substrate(s) to product(s) (Figure 5.5). These steps are summarized diagrammatically in Figure 5.6 for the key glycolytic enzyme *fructose bisphosphate aldolase* (see Section 5.10).

Note that the reaction depicted in Figure 5.5 is exergonic because the free energy of formation of the substrates is *greater* than that of the products. That is, product formation proceeds with the *release* of free energy. Enzymes can also catalyze reactions that require energy, converting energy-poor substrates to energy-rich products. In these cases, however, not only must an activation energy barrier be overcome, but sufficient free energy must also be put *into* the reaction to raise the energy level of the substrates to that of the products. This is done by *coupling* the energy-requiring reaction to an energy-yielding one, such as the hydrolysis of ATP (see Figure 5.12).

Theoretically all enzymes will catalyze their reaction in either direction. In practice, however, enzymes that catalyze highly exergonic or highly endergonic reactions are essentially unidirectional. If a particularly exergonic or endergonic reaction needs to be reversed during cellular metabolism, a different enzyme is usually involved in the reaction.

Structure and Nomenclature of Enzymes

Most enzymes are proteins, polymers of amino acids (∞Sections 3.6–3.8), and every protein has a specific three-dimensional shape. A given protein thus assumes specific binding and physical properties. The structure of an enzyme may be visualized in a computer-generated space-filling model (Figure 5.7●). In this example of the

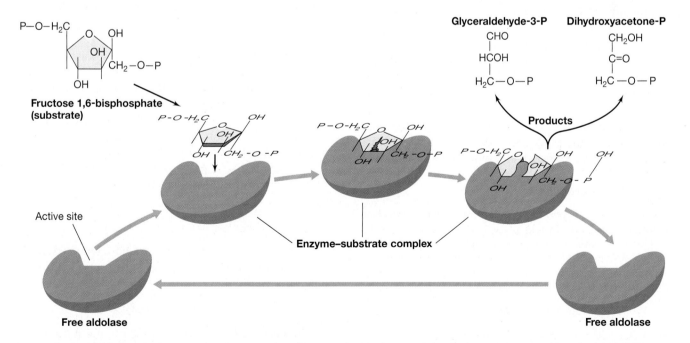

● **Figure 5.6 The catalytic cycle of an enzyme as depicted for the enzyme fructose bisphosphate aldolase.** This enzyme catalyzes the reaction: fructose 1,6-bisphosphate → glyceraldehyde 3-phosphate + dihydroxyacetone phosphate in glycolysis (see Figure 5.14). Following binding of fructose 1,6-bisphosphate in the formation of the enzyme–substrate complex, the conformation of the enzyme is altered, placing strain on certain bonds of the substrate, which break and yield the two products.

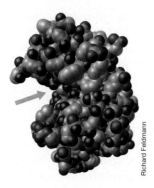

● **Figure 5.7 Computer-generated space-filling model of the enzyme lysozyme.** The substrate (peptidoglycan) binding site (active site) is in the large cleft (arrow) on the left side of the model (∞Section 4.8).

peptidoglycan-cleaving enzyme *lysozyme* (∞Section 4.8), the large cleft is the site where the substrate binds (the active site).

Many enzymes contain small nonprotein molecules that participate in catalysis but are not themselves considered substrates. These small enzyme-associated molecules are divided into two classes based on the way they associate with the enzyme: *prosthetic groups* and *coenzymes*.

Prosthetic groups are bound very tightly to their enzymes, usually covalently and permanently. The heme group present in cytochromes (see Section 5.11) is an example of a prosthetic group. **Coenzymes**, by contrast, are loosely bound to enzymes, and a single coenzyme molecule may associate with a number of different enzymes. Most coenzymes are derivatives of vitamins, and $NAD^+/NADH$ is a good example, being a derivative of the vitamin niacin (Table 5.3).

Enzymes are named either for the substrate they bind or for the chemical reaction they catalyze, by addition of the suffix *-ase*. Thus, cellul*ase* is an enzyme that attacks cellulose, glucose oxid*ase* is an enzyme that catalyzes the oxidation of glucose, and ribonucle*ase* is an enzyme that decomposes ribonucleic acid. A more formal enzyme nomenclature system is employed by biochemists, in which a specific numbering system is used to name enzymes according to the class of chemical reaction that they catalyze. But for the purposes of this book, trivial names ending in "ase" are sufficient and will be used exclusively.

 5.5 Concept Check

The reactants in a chemical reaction must first be activated before the reaction can take place, and this requires a catalyst. Enzymes are catalytic proteins that speed up the rate of biochemical reactions. Enzymes are highly specific in the reactions they catalyze, and this specificity resides in the three-dimensional structure of the polypeptide(s) in the protein.

◆ What is the function of a *catalyst*?

◆ What *class* of macromolecules are enzymes?

◆ Where on an enzyme does the substrate bind?

◆ What is *activation energy*?

III OXIDATION–REDUCTION AND ENERGY–RICH COMPOUNDS

Energy conservation in cells involves **oxidation–reduction (redox) reactions**. The energy released in these reactions is conserved in the production of energy-rich storage compounds, such as ATP. Here we consider oxidation–reduction reactions and the major electron carriers present in both the cytoplasm and the cytoplasmic membrane of a cell. We will then examine the nature of the compounds that actually conserve the energy released in oxidation–reduction reactions.

5.6 Oxidation–Reduction

Chemically, an oxidation is defined as the *removal* of an electron (or electrons) from a substance. A reduction is defined as the *addition* of an electron (or electrons) to a substance. In biochemistry, oxidations and reductions frequently involve the transfer of not just electrons, but both an electron (e^-) plus a proton (H^+). We will, on occasion, need to distinguish between oxidation–reduction reactions involving electrons only or electrons plus protons, but reserve this distinction for the appropriate time (see Sections 5.11 and 5.12).

Electron Donors and Acceptors

Redox reactions occur when electrons removed from an electron donor are taken up by an electron acceptor. For example, the electron donor hydrogen gas (H_2) can release electrons and protons and become oxidized:

$$H_2 \rightarrow 2\,e^- + 2\,H^+$$

However, electrons cannot exist alone in solution; they must be part of atoms or molecules. Thus, the equation as drawn provides chemical information but does not itself represent an independent reaction. The reaction is only a *half reaction,* a term that implies the need for a second half reaction. This is because for any *oxidation* to occur, a subsequent *reduction* must also occur. For example, the oxidation of H_2 could be coupled to the reduction of many different substances, including O_2, in a second reaction:

$$\tfrac{1}{2}O_2 + 2\,e^- + 2\,H^+ \rightarrow H_2O$$

This half reaction, which is a reduction, when coupled to the oxidation of H_2 above, yields the following overall balanced reaction:

$$H_2 + \tfrac{1}{2}O_2 \rightarrow H_2O$$

In reactions of this type, we refer to the substance *oxidized*, in this case H_2, as the **electron donor**, and the substance *reduced,* in this case O_2, as the **electron acceptor** (Figure 5.8●). The key to understanding biological oxidations and reductions is to keep the proper half reactions

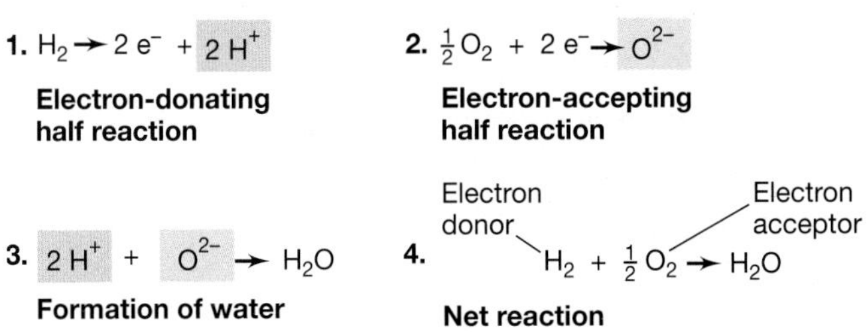

1. $H_2 \rightarrow 2\,e^- +$ 2 H^+

**Electron-donating
half reaction**

2. $\frac{1}{2}O_2 + 2\,e^- \rightarrow$ O^{2-}

**Electron-accepting
half reaction**

3. 2 H^+ + O^{2-} $\rightarrow H_2O$

Formation of water

4.

Electron
donor

Electron
acceptor

$H_2 + \frac{1}{2}O_2 \rightarrow H_2O$

Net reaction

● **Figure 5.8 Example of an oxidation–reduction reaction.** The formation of H_2O from the electron donor H_2 and the electron acceptor O_2.

straight. In a redox reaction there must always be one reaction involving an electron *donor* and another reaction involving an electron *acceptor*.

Reduction Potentials

Substances vary in their tendency to become oxidized or reduced. This tendency is expressed as the **reduction potential** (E_0', standard conditions) of the half reaction. This potential is measured electrically in volts (V) in reference to a standard substance, H_2. By convention, reduction potentials are expressed for half reactions written as *reductions*. If protons are involved in the reaction, as is often the case, then the reduction potential is to some extent influenced by the hydrogen ion concentration (pH). By convention in biology, however, reduction potentials are given for pH 7 because the cytoplasm of most cells is neutral or nearly so. Using these conventions, at pH 7 the E_0' of

$$\frac{1}{2}O_2 + 2\,H^+ + 2\,e^- \rightarrow H_2O$$

is +0.816 volts (V), and the E_0' of

$$2\,H^+ + 2\,e^- \rightarrow H_2$$

is −0.421 V. We will see in the following text that these values of E_0' mean that O_2 is an excellent electron acceptor, while H_2 is an excellent electron donor (Figure 5.8).

Oxidation–Reduction Couples and Complete Redox Reactions

Many molecules can be either electron donors or electron acceptors under different circumstances, depending on the substances with which they react. The chemical constituents on each side of the arrow in half reactions can be thought of as representing a *redox couple*, such as $2\,H^+/H_2$ or $\frac{1}{2}O_2/H_2O$. By convention, when writing a redox couple, the *oxidized* form of the couple is always placed on the left.

In constructing complete oxidation–reduction reactions from their constituent half reactions, the reduced substance of a redox couple whose E_0' is more negative *donates* electrons to the oxidized substance of a redox couple whose E_0' is more positive. Thus, in the couple $2\,H^+/H_2$ (E_0' −0.42 V), H_2 has a greater tendency to

donate electrons than do protons to accept them. On the other hand, in the couple $\frac{1}{2}O_2/H_2O$ (E_0' +0.82 V), H_2O has a very weak tendency to donate electrons, while O_2 has a great tendency to accept them. It follows then that in a reaction of H_2 and O_2, H_2 will be the electron *donor* and become oxidized, and O_2 will be the electron *acceptor* and become reduced (Figure 5.8).

By convention, all half reactions are written as *reductions*. However, in an actual redox reaction, one of the two half reactions will proceed as an oxidation and therefore will be written in the *opposite* direction. Thus, in the reaction shown in Figure 5.8, the oxidation of H_2 to $2\,H^+ + 2\,e^-$ is reversed from the formal half reaction, written as a reduction.

The Electron Tower

A convenient way of viewing electron transfer reactions in biological systems and their importance to bioenergetics is to imagine a vertical tower (Figure 5.9●). The tower represents the range of reduction potentials possible for redox couples in nature, from those with the most negative E_0's on the top to those with the most positive E_0's at the bottom. The *reduced* substance in the redox pair at the top of the tower has the greatest tendency to donate electrons, whereas the *oxidized* substance in the couple at the bottom of the tower has the greatest tendency to accept electrons.

As electrons from the electron donor at the top of the tower fall, they can be "caught" by acceptors at various levels. The difference in reduction potential between two substances on the tower is expressed as $\Delta E_0'$. The farther the electrons drop from a donor before they are caught by an acceptor, the greater the amount of energy released. That is, $\Delta E_0'$ *is proportional to* $\Delta G^{0'}$ (Figure 5.9). Oxygen (O_2), at the bottom of the tower, is the most favorable electron acceptor commonly found in nature. In the middle of the tower, redox couples can be either electron donors or acceptors, depending on which redox couples they react with. For instance, the $2\,H^+/H_2$ couple (−0.42 V) can react with the fumarate/succinate couple (+0.02 V) as follows:

$$H_2 + \text{fumarate}^{2-} \rightarrow \text{succinate}^{2-}$$

By contrast, the oxidation of succinate to fumarate can be coupled to the reduction of NO_3^- or $\frac{1}{2}O_2$:

$$\text{Succinate}^{2-} + NO_3^- \rightarrow \text{fumarate}^{2-} + NO_2^- + H_2O$$

$$\text{Succinate}^{2-} + \frac{1}{2}O_2 \rightarrow \text{fumarate}^{2-} + H_2O$$

Hence, in the presence of H_2 and under anoxic conditions, fumarate can be an electron acceptor (producing succinate). Under other conditions (for example, anoxic in the presence of NO_3^-, or under oxic conditions) succinate can be an electron donor (producing fumarate). Indeed, all the transformations involving fumarate and succinate described here are actually carried out by various microorganisms (including *Escherichia coli*) under certain nutritional and environmental conditions.

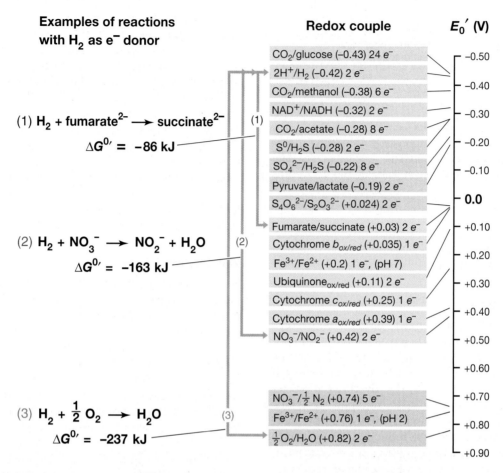

● **Figure 5.9   The electron tower.** Redox couples are arranged from the strongest reductants (negative reduction potential) at the top to the strongest oxidants (positive reduction potentials) at the bottom. As electrons are donated from the top of the tower, they can be "caught" by acceptors at various levels. The farther the electrons fall before they are caught, the greater the difference in reduction potential between electron donor and electron acceptor and the more energy released. As an example of this, on the left is shown the differences in energy released when a single electron donor, H_2, reacts with any of three different electron acceptors, fumarate, nitrate, and oxygen.

Electron Donor ↔ Energy Source

Electron donors are often called **energy sources** because energy is released when they are oxidized. Many potential electron donors exist in nature, including a wide variety of organic and inorganic compounds (◗◗Chapters 17 and 19). However, it is not the electron donor *per se* that contains energy, but the *chemical reaction* in which the electron donor gets oxidized that actually releases energy. The presence of a suitable electron acceptor is thus just as important as an electron donor. Moreover, the amount of energy released in a redox reaction depends on the nature of *both* the electron donor and the electron acceptor. The greater the difference between reduction potentials of the two half reactions, the more energy will be released when they react (Figure 5.9) (see also Appendix 1).

 5.6   *Concept Check*

Oxidation–reduction reactions involve the transfer of electrons from electron donor to electron acceptor. The tendency of a compound to accept or release electrons is expressed quantitatively by its reduction potential, E_0'.

◆ In the reaction $H_2 + \frac{1}{2}O_2 \rightarrow H_2O$, what is the electron *donor* and what is the electron *acceptor*?

◆ What is the E_0' of the $2\,H^+/H_2$ couple? Are protons good electron acceptors? Why or why not?

◆ Why is nitrate (NO_3^-) a better electron acceptor than fumarate?

5.7   NAD as a Redox Electron Carrier

The transfer of electrons in redox reactions typically involves one or more intermediates called *carriers*. When such carriers are used, we refer to the initial donor as the **primary electron donor** and to the final acceptor as the **terminal electron acceptor**. The net energy change of the complete reaction sequence is determined by the *difference* in reduction potentials between the primary donor and the terminal acceptor.

Electron carriers can be divided into two classes: those that are *freely diffusible* (coenzymes) and those that are firmly attached to enzymes in the cytoplasmic membrane (prosthetic groups). The fixed carriers function in membrane-associated electron transport reactions and are discussed in Section 5.11. Common diffusible carriers include the coenzymes nicotinamide-adenine dinucleotide (NAD^+) and NAD-phosphate ($NADP^+$) (Figure 5.10●). NAD^+ and $NADP^+$ are *electron plus H^+ carriers,* transporting $2e^-$ and $2H^+$ at a time.

The reduction potential of the $NAD^+/NADH$ (or $NADP^+/NADPH$) couple is -0.32 V, which places it fairly high on the electron tower. That is, NADH (or NADPH) is a good electron *donor.* However, although the NAD^+ and $NADP^+$ couples have the same reduction potentials, they generally function in different capacities in the cell. $NAD^+/NADH$ is directly involved in energy-generating (catabolic) reactions, whereas $NADP^+/NADPH$ is involved primarily in biosynthetic (anabolic) reactions.

NAD/NADH Cycling

Coenzymes increase the diversity of redox reactions possible in a cell by allowing chemically dissimilar molecules to interact as primary electron donor and terminal electron acceptor, with the coenzyme acting as an intermediary. In the case of $NAD^+/NADH$, for example, electrons removed from one molecule can reduce NAD^+ to NADH, while the latter can be converted back to NAD^+ by donating electrons to a second molecule. Figure 5.11● is a diagram showing the functioning of $NAD^+/NADH$

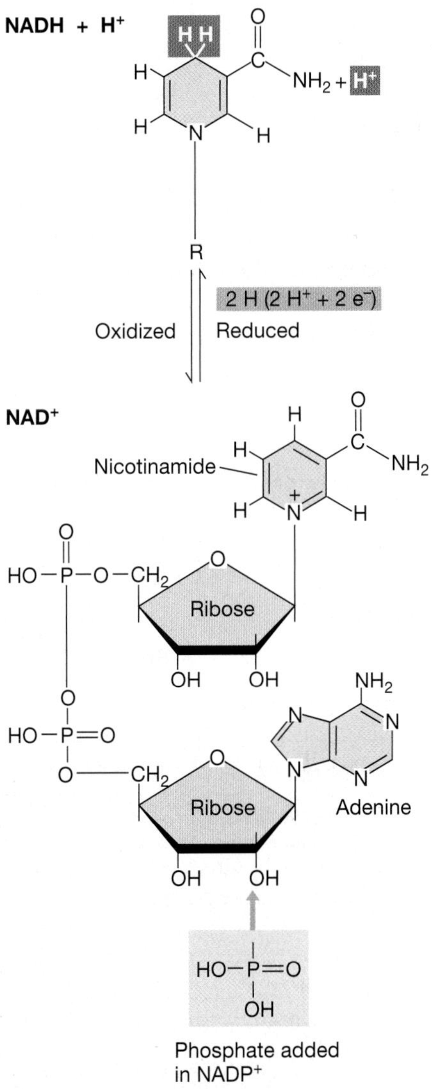

● **Figure 5.10 Structure of the oxidation–reduction coenzyme nicotinamide adenine dinucleotide (NAD^+).** In $NADP^+$, a phosphate group is present, as indicated. Both NAD^+ and $NADP^+$ undergo oxidation–reduction as shown, are freely diffusible, and are $2e^- + 2H^+$ carriers. "R" in the top portion of the art is the adenine dinucleotide portion of NAD^+, shown in full in the bottom portion of the art.

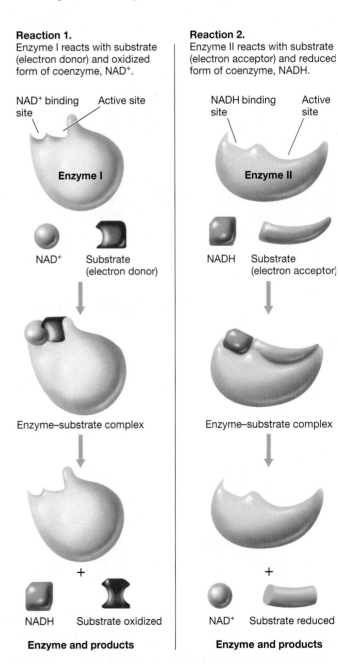

Reaction 1.
Enzyme I reacts with substrate (electron donor) and oxidized form of coenzyme, NAD^+.

Reaction 2.
Enzyme II reacts with substrate (electron acceptor) and reduced form of coenzyme, NADH.

● **Figure 5.11 Schematic example of an oxidation–reduction reaction.** This reaction involves the oxidized and reduced forms of the coenzyme nicotinamide-adenine dinucleotide, NAD^+ and NADH.

in such a two-part reaction. NAD$^+$ and NADH are simply intermediaries in the process, facilitating the reaction but not being consumed in the process. Thus, unlike a primary electron donor or a terminal electron acceptor, where relatively large amounts of each are required, the cell needs only a tiny amount of the coenzymes NAD$^+$/NADH. All that is needed is an amount sufficient to service the redox enzymes in the cell that use these coenzymes in their reaction mechanisms.

 5.7 Concept Check

The transfer of electrons from donor to acceptor in a cell typically involves one or more electron carriers. Some electron carriers are membrane-bound, whereas others, such as NAD$^+$/NADH, are freely diffusible, transferring electrons from one place to another in the cell.

◆ Is NADH a better electron donor than H$_2$? Using the data of Figure 5.9, how can you tell this?

5.8 Energy-Rich Compounds and Energy Storage

Once energy is released from redox reactions it must be conserved by the cell if it is to be used to drive various energy-requiring functions. In living organisms, chemical energy released in redox reactions is conserved primarily in the form of phosphorylated compounds, in particular,

ATP. Such compounds are said to be **energy-rich**. This is because the potential energy released upon hydrolysis of the phosphate bonds is significantly greater than that of the average covalent bond in the cell.

In phosphorylated compounds, phosphate groups are attached via oxygen atoms by *ester* or *anhydride* bonds, as illustrated in Figure 5.12●. However, not all phosphate bonds are energy-rich. The energy of phosphate bonds is expressed in terms of the free energy released when the phosphate is hydrolyzed. As seen in Figure 5.12, the $\Delta G^{0\prime}$ of hydrolysis of the phosphate *ester* bond in glucose 6-phosphate is only -13.8 kJ/mol. By contrast, the $\Delta G^{0\prime}$ of hydrolysis of the phosphate *anhydride* bond in phosphoenolpyruvate is -51.6 kJ/mol, almost four times that of glucose 6-phosphate. Thus, phosphoenolpyruvate, a phosphoanhydride (∞ Table 3.1), is an energy-rich compound, while glucose 6-phosphate, a phosphate ester, is not. Although theoretically either compound could be hydrolyzed to yield energy, cells typically use compounds whose $\Delta G^{0\prime}$ is *greater than* -30 kJ/mol as energy "currencies" in the cell (Figure 5.12 and see Table 17.6).

Adenosine Triphosphate (ATP)

The most important energy-rich phosphate compound in cells is **adenosine triphosphate (ATP)**. ATP consists of the ribonucleoside adenosine to which three phosphate molecules are bonded in series (Figure 5.12). ATP is the

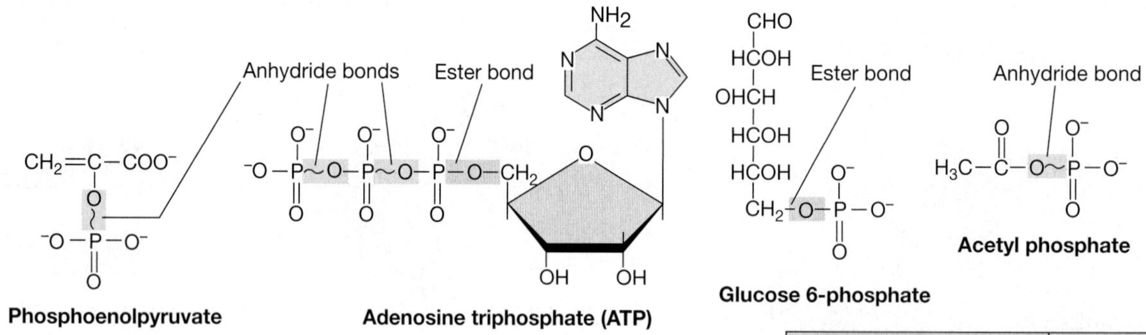

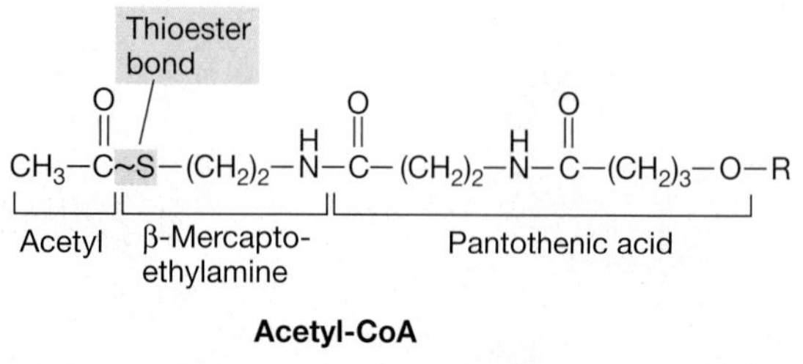

Compound	$G^{0\prime}$ kJ/mol
$\Delta G^{0\prime} > 30$kJ	
Phosphoenolpyruvate	−51.6
1,3-Bisphosphoglycerate	−52.0
Acetyl phosphate	−44.8
ATP	−31.8
ADP	−31.8
Acetyl CoA	−31
$\Delta G^{0\prime} < 30$kJ	
AMP	−14.2
Glucose 6-phosphate	−13.8

● **Figure 5.12 Some compounds important in energy transformations in cells.** The table shows the free energy of hydrolysis of some of the key phosphate esters and anhydrides, indicating that some of the phosphate ester bonds are of higher energy than others. Structures of four of the compounds are given to indicate the position of bonds. ATP contains three phosphates, but only two of them have free energies of hydrolysis >30 kJ. Also shown is the structure of the coenzyme acetyl-CoA. The thioester bond between C and S has a free energy of hydrolysis of >30 kJ. The "R" group of acetyl-CoA is a 3′ phospho ADP group.

prime energy currency in cells, being generated during exergonic reactions and consumed in certain endergonic reactions. From the structure of ATP (Figure 5.12) it can be seen that two of the phosphate bonds are phosphoanhydrides and thus have free energies of hydrolysis greater than 30 kJ. Thus the reactions ATP → ADP + P_i, and ADP → AMP + P_i, each release roughly 32 kJ/mol of energy (Figure 5.12).

Although the energy released in ATP hydrolysis is about −32 kJ, a caveat must be introduced here to more precisely define the energy requirements for the *synthesis* of ATP. In an actively growing cell, the ratio of ATP:ADP is kept high, roughly 1000. This significantly affects the energy requirements for ATP synthesis. In an actively growing cell, the actual energy expenditure (that is, the ΔG, see Section 5.4) for the synthesis of one mole of ATP is on the order of −55 to −60 kJ. Nevertheless, for the purposes of learning the basic principles of bioenergetics, we will remain under "standard conditions" ($\Delta G^{0\prime}$) and take the energy required for synthesis or hydrolysis of ATP to be the same, 32 kJ/mol.

Coenzyme A

In addition to energy-rich *phosphate* compounds, the cell produces certain other compounds that can conserve the energy released in exergonic reactions. These include derivatives of coenzyme A (for example, acetyl-CoA; see structure in Figure 5.12). **Coenzyme A** derivatives contain *thioester* bonds (Figure 5.12, and Table 3.1). These yield sufficient free energy upon hydrolysis to drive the synthesis of an energy-rich phosphate bond. For example, in the reaction:

$$\text{acetyl-S-CoA} + H_2O + ADP + P_i \rightarrow$$
$$\text{acetate}^- + \text{HS-CoA} + ATP + H^+,$$

the energy released in the hydrolysis of coenzyme A is conserved in the synthesis of ATP. Coenzyme A derivatives (acetyl-CoA is just one of many) are especially important to the energetics of *anaerobic* microorganisms, in particular those whose energy metabolism involves fermentation (Sections 5.10, 17.19, and 17.20). We will return to the importance of these compounds in Chapter 17.

Long-Term Energy Storage

ATP is a dynamic entity in the cell; it is continuously being broken down to drive biosynthetic reactions and resynthesized at the expense of catabolic reactions. However, ATP is present in relatively low concentration in an actively growing cell, about 2 millimolar (mM). For longer-term energy *storage,* microorganisms typically produce insoluble polymers that can be oxidized later for the production of ATP.

Examples of long-term energy storage in prokaryotes include the polyglucose polymer glycogen (Figure 3.6), the lipid polymer poly-β-hydroxybutyrate and other polyhydroxyalkanoates

(Figure 4.52), and elemental sulfur, stored by many sulfur chemolithotrophs (Figure 2.10*b*). These polymers are deposited within the cell as large granules that can be seen with the light or electron microscope (Section 4.11). In eukaryotic microorganisms, polyglucose in the form of starch (Figure 3.6), or lipid in the form of simple fats (Figure 3.7), are the major reserve materials. In the absence of an external energy source, cells oxidize these polymers in order to make new cell material or to supply the energy needed to maintain cell integrity, called *maintenance energy*, when in a nongrowing state.

5.8 Concept Check

The energy released in redox reactions is conserved in the formation of certain compounds that contain energy-rich phosphate or sulfur bonds. The most common of these compounds is ATP, the prime energy carrier in the cell. Long-term storage of energy is linked to the formation of polymers, which can be consumed to yield ATP.

◆ How much energy is released per mole of ATP converted to ADP + P_i? Per mole of AMP converted to adenosine and P_i?

◆ Following periods of nutrient abundance, how can cells prepare for periods of nutrient starvation?

IV MAJOR CATABOLIC PATHWAYS, ELECTRON TRANSPORT, AND THE PROTON MOTIVE FORCE

We next look at some common catabolic reactions in microorganisms. First we will compare the processes of fermentation and respiration, followed by a more detailed look at each process. We then move to a discussion of the proton motive force, the key product of respiration and other catabolic processes, including anaerobic respiration, photosynthesis, and chemolithotrophy. Besides yielding ATP, the proton motive force also drives a number of other functions in the cell, including many different transport reactions and both flagellar and gliding motility (Sections 4.7 and 4.14–4.16).

5.9 Energy Conservation: Options

Two mechanisms for energy conservation are known in chemoorganotrophs: **fermentation** and **respiration**. In each mechanism, the synthesis of ATP is driven by the energy released in oxidation–reduction reactions. However, the redox reactions in fermentation and respiration differ significantly. In *fermentation* the redox process occurs in the *absence* of exogenous electron acceptors. By contrast, in *respiration,* molecular oxygen or some other electron acceptor is present as a terminal electron acceptor.

Because every oxidation has to be coupled to a reduction, the oxidation reaction in a fermentation is coupled to the reduction of a compound *derived from the electron donor*. In respiration, on the other hand, an *exogenous* electron acceptor is reduced. This reality of fermentation places constraints on the amount of energy that can be obtained, as we will see when we compare energy yields from fermentation and respiration in a later section.

Substrate-Level Phosphorylation and Oxidative Phosphorylation

In addition to differences in the nature of electron acceptors, fermentation and respiration differ in the *mechanism* by which ATP is synthesized. In fermentation, ATP is produced by **substrate-level phosphorylation**. In this process, ATP is synthesized during steps in the catabolism of an organic compound (Figure 5.13*a*●). This is in contrast to **oxidative phosphorylation,** in which ATP is produced at the expense of the proton motive force

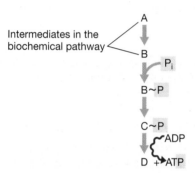

(a) **Substrate-level phosphorylation**

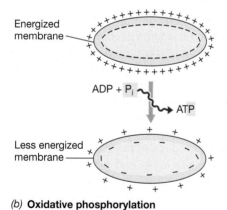

(b) **Oxidative phosphorylation**

● **Figure 5.13 Energy conservation in fermentation and respiration.** (a) In fermentation, ATP synthesis occurs as a result of *substrate-level phosphorylation;* a phosphate group gets added to some intermediate in the biochemical pathway and eventually gets transferred to ADP to form ATP. (b) In respiration, the cytoplasmic membrane, energized by the proton motive force, dissipates some of that energy in the formation of ATP from ADP and inorganic phosphate (P_i) in the process called *oxidative phosphorylation.* The coupling of the proton motive force to ATP synthesis occurs by way of a membrane protein enzyme complex called ATP synthase (ATPase) (see Section 5.12 and Figure 5.21).

(Figure 5.13*b*). A third form of ATP synthesis, **photophosphorylation**, occurs in phototrophic organisms. However, the basic mechanism of photophosphorylation is similar to that of oxidative phosphorylation except that *light* rather than a chemical compound drives the redox reactions that generate the proton motive force (∞Section 17.2).

 5.9 Concept Check

Fermentation and respiration are the two means by which chemoorganotrophs can conserve energy from the oxidation of organic compounds. During these catabolic reactions, ATP synthesis occurs by way of either substrate-level phosphorylation (fermentation) or oxidative phosphorylation (respiration).

◆ Which form of ATP synthesis requires cytoplasmic membrane participation? Why?

◆ How does substrate-level phosphorylation differ from oxidative phosphorylation?

5.10 Glycolysis as an Example of Fermentation

As we mentioned, a fermentation is an *internally balanced* oxidation–reduction reaction. In a fermentation, some atoms of the electron donor become more reduced whereas others become more oxidized, and energy is conserved by substrate-level phosphorylation. A common biochemical pathway for the fermentation of glucose is **glycolysis**, also named the **Embden-Meyerhof pathway** for its major discoverers. Several other fermentations are also known (∞Chapter 17).

Glycolysis is an anoxic process and can be divided into three major stages, each involving a series of individually catalyzed enzymatic reactions (Figure 5.14●). *Stage I* of glycolysis is a series of preparatory reactions that do not involve oxidation–reduction and do not release energy, but that lead to the production of two molecules of the key intermediate, *glyceraldehyde 3-phosphate* from glucose. In *Stage II*, oxidation–reduction reactions occur, energy is conserved in the form of ATP, and two molecules of pyruvate are formed. In *Stage III*, oxidation–reduction reactions occur once again and *fermentation products* are formed (Figure 5.14).

Stages I and II: Preparatory and Redox Reactions

In Stage I, glucose is phosphorylated by ATP, yielding glucose 6-phosphate. The latter is then converted to an isomeric form, fructose 6-phosphate. A second phosphorylation leads to the production of *fructose 1,6-bisphosphate*, a key intermediate of glycolysis. The enzyme **aldolase** splits fructose 1,6-bisphosphate into two three-carbon molecules, *glyceraldehyde 3-phosphate* and its isomer, *dihydroxyacetone phosphate*, which is converted into glyceraldehyde 3-phosphate. Note that thus far,

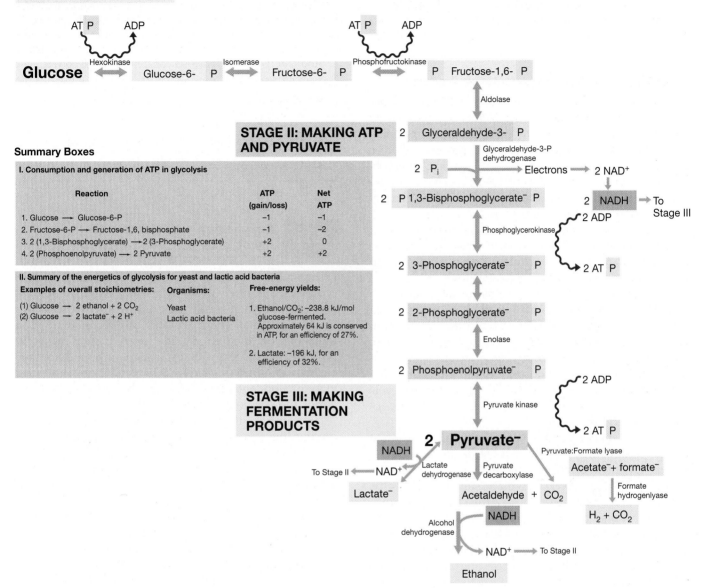

● **Figure 5.14 Embden—Meyerhof pathway (glycolysis).** The sequence of enzymatic reactions in the conversion of glucose to pyruvate and then to fermentation products (enzymes are shown in small type). The product of aldolase is actually glyceraldehyde 3-P and dihydroxyacetone P, but the latter is converted to glyceraldehyde 3-P. Note how pyruvate is the central "hub" of glycolysis—all fermentation products are made from pyruvate, and just a few common examples are given.

all of the reactions, including the consumption of ATP, have proceeded without redox reactions.

The first redox reaction of glycolysis occurs in Stage II during the conversion of glyceraldehyde 3-phosphate to 1,3-bisphosphoglyceric acid. In this reaction (which occurs twice, once for each molecule of glyceraldehyde 3-phosphate), an enzyme containing NAD^+ as a coenzyme accepts $2\,e^- + 2\,H^+$ to form NADH. The enzyme catalyzing this reaction is **glyceraldehyde-3-phosphate dehydrogenase**. Simultaneously, each glyceraldehyde-3-P molecule is phosphorylated by the addition of a molecule of inorganic phosphate. This reaction, in which

inorganic phosphate is converted to organic form, sets the stage for later energy conservation by substrate-level phosphorylation. ATP formation is possible because each of the phosphates on a molecule of 1,3-bisphosphoglyceric acid has a free energy of hydrolysis of >30 kJ (see Figure 5.12). The synthesis of ATP occurs when (1) each molecule of 1,3-bisphosphoglyceric acid is converted to 3-phosphoglyceric acid, and (2) when each molecule of phosphoenolpyruvate is converted to pyruvate (Figure 5.14).

In glycolysis, *two* ATP molecules are consumed in the two phosphorylations of glucose in Stage I, and *four*

ATP molecules are synthesized (two from each 1,3-bis-phosphoglyceric acid converted to pyruvate) in the reactions of Stage II (Figure 5.14). Thus, the net gain to the organism in glycolysis is *two molecules of ATP per molecule of glucose fermented.*

Stage III: Production of Fermentation Products

During the formation of two molecules of 1,3-bisphosphoglyceric acid, two molecules of NAD^+ are reduced to NADH (see Figure 5.14). However, the continued oxidation of glyceraldehyde 3-phosphate can only proceed if there is free NAD^+ available to accept released electrons. This problem is overcome by the oxidation of NADH to NAD^+ by enzymes that reduce pyruvate to any of a variety of **fermentation products** (Figure 5.14).

In the case of yeast, pyruvate is reduced to *ethanol* with the release of CO_2. In lactic acid bacteria, pyruvate is reduced to *lactate*. Many routes of pyruvate reduction in various fermentative prokaryotes are known (∞Chapters 12 and 17), but the net result is the same: NADH is reoxidized to NAD^+ during the process. As a diffusible coenzyme, NADH can diffuse away from glyceraldehyde-3-phosphate dehydrogenase, attach to an enzyme that reduces pyruvate to lactate (lactate dehydrogenase), and diffuse away once again following the oxidation of NADH to NAD^+ to repeat the cycle (Figure 5.14, and see Figure 5.11 for details of NAD^+/NADH shuttling).

In any energy-yielding process, oxidations must balance reductions, and there must be an electron acceptor for each electron removed. In this case, the *reduction* of NAD^+ at one enzymatic step in glycolysis is balanced with its *oxidation* at another. The final product(s) must also be in redox and atomic balance with the starting substrate, glucose. Hence, the fermentation products discussed here, ethanol plus CO_2, or lactate plus protons, are in both electrical and atomic balance with the starting substrate, glucose, as shown here: glucose $(C_6H_{12}O_6)$ = 2 ethanol (C_2H_5OH) + 2 CO_2; glucose $(C_6H_{12}O_6)$ = 2 lactate$^-$ $(C_3H_{10}O_3)$ + 2 H^+.

Glucose Fermentation: Net and Practical Results

The ultimate result of glycolysis is the consumption of glucose, the net synthesis of two ATPs, and the production of fermentation products. For the organism the crucial product is ATP, which is used in a wide variety of energy-requiring reactions, and fermentation products are merely waste products. However, the latter substances are hardly considered waste products by the distiller, the brewer, the cheese-maker, or the baker (see the Microbial Sidebar, The Products of Yeast Fermentation). Thus, fermentation is more than just an energy-yielding process. It is a means of producing natural products useful to humans. We discuss industrial production of fermentation products in more detail in Chapter 30.

 5.10 Concept Check

Glycolysis is a major pathway of fermentation and is a widespread means of anaerobic metabolism. The end result of glycolysis is the release of a small amount of energy that is conserved as ATP and the production of fermentation products. For each glucose consumed in glycolysis, two ATPs are produced.

◆ Which reaction(s) in glycolysis involve oxidations and reductions?

◆ What is the role of NAD^+/NADH in glycolysis?

◆ Why are fermentation products made during glycolysis?

5.11 Respiration and Membrane-Associated Electron Carriers

Fermentation occurs in the absence of external electron acceptors, and it releases only a relatively small amount of energy. Only a few ATP molecules are synthesized as a result. This small energy release can be understood in terms of the principles of oxidation–reduction reactions: (1) the carbon atoms in the starting compound are only partially oxidized (see Figure 5.14), and (2) the difference in reduction potentials between the primary electron donor and terminal electron acceptor is small.

By contrast, if O_2 or some other terminal acceptor is present, the glucose can be oxidized completely to CO_2 and a far higher yield of ATP is possible. Oxidation using O_2 as the terminal electron acceptor is called **aerobic respiration**.

Our discussion of aerobic respiration will deal with both the carbon transformations and redox reactions that occur. Thus, our focus will be on two issues: (1) the way electrons are transferred from the organic compound to the terminal electron acceptor, and (2) the biochemical pathways involved in the transformation of organic carbon to CO_2. During the former, ATP synthesis occurs at the expense of the proton motive force (Figure 5.13b). We begin with a discussion of electron flow.

Electron Transport Carriers

Electron transport systems are *membrane-associated* electron carriers. These systems have two basic functions. First, electron transport systems mediate the transfer of electrons from primary donor to terminal acceptor. And second, these systems conserve some of the energy released during electron transfer for synthesis of ATP.

Several types of oxidation–reduction enzymes are involved in electron transport. These include *NADH dehydrogenases, flavoproteins, iron-sulfur proteins*, and *cytochromes*. In addition, a class of *nonprotein* electron carriers is known, the *quinones*. We now consider each of these classes of electron transport components in slightly more detail.

Microbial Sidebar ♦ The Products of Yeast Fermentation

E very home wine maker or brewer is an amateur microbiologist, perhaps without even realizing it. Indeed, the anaerobic processes of energy generation are at the heart of some of the most striking discoveries of the human race: fermented foods and beverages (Figure 1●).

In the production of breads and most alcoholic beverages, it is the yeast *Saccharomyces cerevisiae* that is exploited to produce ethanol plus CO_2. Found in various sugar-rich environments such as fruit juices and nectar, yeasts can carry out the two opposing modes of chemoorganotrophic metabolism discussed in this chapter, *fermentation* and *respiration*. When oxygen is present, yeasts grow efficiently on various sugars, making yeast cells and CO_2 (the latter from the citric acid cycle, see Section 5.13) in the process. However, when conditions are anoxic, yeasts switch to fermentative metabolism. This results in a reduced cell yield but significant amounts of alcohol plus CO_2.

When grapes are squeezed to make juice, small numbers of yeast cells present on the grapes in the vineyard are transferred to the must. During the first several days of the wine-making process, these yeast cells grow by respiration and consume O_2, making the juice anoxic. As soon as the oxygen is depleted, fermentation can begin, and the process of alcohol formation takes over. This switch from aerobic to anaerobic metabolism is crucial, and special care must be taken to make sure air is kept out of the fermenting vessel.

Wine is only one of many products made with yeast. Others include beer (Microbial

Figure 1 *Major products in which fermentation by the yeast* Saccharomyces cerevisiae *is critical.*

Sidebar, Home Brew, ◌◌Chapter 30) and distilled spirits such as brandy, whisky, vodka, and gin (see Figure 1). In distilled spirits, the ethanol, produced in relatively low amounts (10–15% by volume) by the yeast, is concentrated by distilling to make a beverage containing 40–60% alcohol. Even alcohol for motor fuel is made with yeast in parts of the world where sugar is plentiful but petroleum is in short supply (such as Brazil). In the United States, major production of ethyl alcohol for use as an industrial solvent occurs using corn starch as the fermentable substrate. Yeast also serves as the leavening agent in bread, al-

though here it is not the *alcohol* that is important, as it volatilizes in the baking process, but CO_2, the *other* product of the alcohol fermentation (see Figure 5.14). We discuss yeast and yeast products in some detail in Chapter 30.

We can thus appreciate how the yeast cell, forced to carry out a fermentative lifestyle because the oxygen it needs for respiration is absent, has impacted the lives of humans. Besides being "waste products" of the glycolytic pathway in yeast, ethanol and CO_2 are the key ingredients in the products of the alcoholic beverage and baking industries, respectively. ∎

NADH dehydrogenases are proteins bound to the inside surface of the cytoplasmic membrane. They have an active site that binds NADH and accepts $2 e^- + 2 H^+$ when NADH is converted to NAD^+ (Figure 5.10). The $2 e^- + 2 H^+$ are passed on to flavoproteins.

Flavoproteins are proteins containing a derivative of riboflavin (Figure 5.15●). The flavin portion, which is bound to a protein, is a prosthetic group that is alternately reduced as it accepts hydrogen atoms and oxidized when electrons are passed on. Note that flavoproteins *accept* $2 e^- + 2 H^+$ but *donate* electrons only. We will consider what happens to the two protons later. Two flavins are commonly observed in cells, *flavin mononucleotide*

(FMN) and *flavin-adenine dinucleotide (FAD)*. In the latter, FMN is bonded to ribose and adenine through a second phosphate. Riboflavin, also called vitamin B_2, is a source of the parent flavin molecule in flavoproteins and is a required growth factor for some organisms (see Section 5.2 and Table 5.3).

The **cytochromes** are proteins with iron-containing porphyrin ring (heme) prosthetic groups attached to them (Figure 5.16●). Cytochromes undergo oxidation and reduction through loss or gain of a *single electron* by the iron atom in the heme of the cytochrome:

$$\text{Cytochrome} - Fe^{2+} \longleftrightarrow \text{Cytochrome} - Fe^{3+} + e^-$$

Several classes of cytochromes are known, differing widely in their reduction potentials (see Figure 5.9). Different classes of cytochromes are designated by letters, such as cytochrome *a*, cytochrome *b*, cytochrome *c*, depending upon the type of heme they contain. The cytochromes of one organism may differ slightly from those of another, and so there are designations such as cytochromes a_1, a_2, a_3, and so on among cytochromes of the same class. Occasionally, cytochromes form tight complexes with other cytochromes or with iron-sulfur proteins. An example of such a complex is the cytochrome bc_1 complex, which contains two different *b*-type cytochromes and one *c*-type cytochrome. The cytochrome bc_1 complex plays an important role in energy metabolism, as we will see later (see Section 5.12 and Figure 5.20).

In addition to the cytochromes, in which iron is bound to heme (Figure 5.16), one or more **nonheme iron-proteins** are typically present in electron transport chains. These proteins contain clusters of iron and sulfur atoms, and Fe_2S_2 and Fe_4S_4 clusters are the most common (Figure 5.17●). The iron atoms are bonded to free sulfur and to the protein via sulfur atoms from cysteine residues (Figure 5.17). *Ferredoxin*, a common iron-sulfur protein in biological systems, has an Fe_2S_2 configuration.

The reduction potentials of iron-sulfur proteins vary over a wide range depending on the number of iron and sulfur atoms present and how the iron centers are embedded in the protein. Thus, different iron-sulfur proteins can function at different locations in the electron

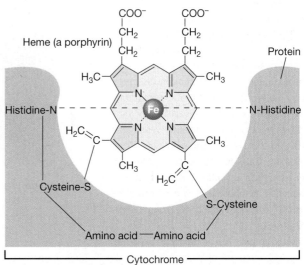

(a) (b) Porphyrin (a tetrapyrrole, $C_{20}H_{14}N_4$)

Heme (a porphyrin) Protein

(c) Cytochrome

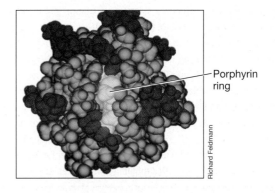

Porphyrin ring

Richard Feldmann

● **Figure 5.16 Cytochrome and its structure.** (a) Structure of the pyrrole ring. (b) Four pyrrole rings are condensed, leading to formation of the porphyrin ring. Various metals can be incorporated into the porphyrin ring system. For example, in chlorophyll pigments, the metal is Mg^{2+} (⚬⚬ Section 17.2 and Figure 17.3); in vitamin B_{12}, it is Co^{2+} (⚬⚬ Section 30.7 and Figure 30.12); and in certain unique porphyrin coenzymes, Ni^{2+} may be the central metal atom (⚬⚬ Section 17.17 and Figure 17.42). (c) In some cytochromes, like cytochrome *c*, the porphyrin ring is covalently linked via disulfide bridges to cysteine molecules in the protein. Note the presence of iron in the center of the ring. (d) Computer-generated model of cytochrome *c*. The protein completely surrounds the porphyrin ring (light color) in the center. Cytochromes carry electrons only; the redox site is the iron atom, which can alternate between the Fe^{2+} and Fe^{3+} oxidation state. The reduction potential of different cytochromes varies widely (see Figure 5.9) depending on the cytochrome type and binding to the parent protein.

Isoalloxazine ring

Ribitol

Oxidized

2 H ($2 e^- + 2 H^+$)

Reduced

● **Figure 5.15 Flavin mononucleotide (FMN) (riboflavin phosphate, a hydrogen atom carrier).** The site of oxidation–reduction is the same in FMN and flavin-adenine dinucleotide (FAD).

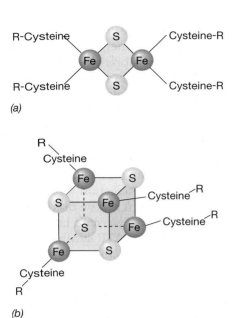

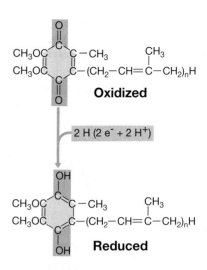

Oxidized

2 H (2 e⁻ + 2 H⁺)

Reduced

● **Figure 5.18** **Structure of oxidized and reduced forms of coenzyme Q, a quinone.** The five-carbon unit in the side chain (an isoprenoid) occurs in a number of multiples. In prokaryotes, the most common number is $n = 6$; in eukaryotes, $n = 10$. Note that oxidized quinone requires $2\ e^-$ and $2\ H^+$ to become fully reduced. An intermediate form, the semiquinone (one H more reduced than oxidized quinone), is formed during the reduction of a quinone.

● **Figure 5.17** **Arrangement of the iron-sulfur centers of nonheme iron-sulfur proteins.** (a) Fe_2S_2 center. (b) Fe_4S_4 center. The cysteine linkages are from the protein portion of the molecule. Iron-sulfur proteins typically carry electrons only.

transport process. Like cytochromes, iron-sulfur proteins carry *electrons only.*

The **quinones** (Figure 5.18●) are highly hydrophobic nonprotein-containing molecules involved in electron transport. Some quinones found in bacteria are related to vitamin K, a growth factor for higher animals. Like flavoproteins, quinones accept $2\ e^- + 2\ H^+$, but pass on only electrons to the next carrier in the chain.

The Proton Motive Force: Chemiosmosis

To understand how electron transport is linked to ATP synthesis, we must first understand how the electron transport system is oriented in the cytoplasmic membrane. The overall structure of the membrane was outlined in Section 4.5 (∞Figure 4.16). Recall that proteins are embedded in the lipid bilayer of membranes such that most have access to both the outside and the inside of the cell. That is, integral membrane proteins are typically *transmembrane* proteins.

Electron transport carriers are oriented so that a *separation* of protons from electrons occurs across the membrane during the transport process. Two electrons plus two protons, removed from substances such as NADH, are transported through the chain by specific carriers. During this process, protons are released into the environment (in gram-negative prokaryotes, protons are released into the periplasm). This extrusion of protons results in a slight acidification of the external surface of the membrane. At the end of the electron transport chain, the electrons are passed to the terminal electron acceptor. In the case of aerobic respiration, O_2 is the terminal acceptor and it is reduced to water.

To reduce O_2 to H_2O, protons from the cytoplasm are needed to complete the reaction. These protons originate from the dissociation of water into H^+ and OH^-. The use of H^+ in the reduction of O_2 to H_2O and the extrusion of H^+ to the environment causes OH^- to accumulate on the *inside* of the membrane. Despite their small size, neither H^+ nor OH^- can diffuse through the membrane because they are charged. Thus, equilibrium cannot be spontaneously restored.

The net result of electron transport is thus the generation of a *pH gradient* and an *electrochemical potential*

 5.11 Concept Check

Electron transport systems consist of a series of membrane-associated electron carriers that function in an integrated fashion to carry electrons from the primary electron donor to oxygen as terminal electron acceptor.

◆ In what major way do quinones differ from other electron carriers in the membrane?

◆ Which electron carriers described in this section accept $2\ e^- + 2\ H^+$? Which accept e^- only?

5.12 Energy Conservation from the Proton Motive Force

The major components of electron transport chains are shown in Figure 5.19● against a backdrop of $E_0{}'$ values. Note that each component has its own unique reduction potential. During electron transport, ATP is produced by the process of *oxidative phosphorylation*. The production of ATP is linked to the establishment of a **proton motive force (pmf)** (∞Section 4.6) across the membrane, and electron transport reactions establish this energized state. We now consider the details of this key cellular process.

the population density may be set by varying the concentration of a single nutrient (the growth-limiting nutrient) in the medium reservoir. Independent control of these two crucial growth parameters is impossible with batch cultures because the batch culture is a closed system where growth conditions are constantly changing with time.

A practical advantage to the chemostat is that a population may be maintained in the exponential growth phase for long periods, for days and even weeks. Because exponential phase cells are usually most desirable for physiological experiments, the experimenter using the chemostat can have such cells available at any time. Thus, experiments can be planned in detail and then performed whenever most convenient. Moreover, repetition of experiments can be done with the knowledge that the cell population will be as close to being the same each time as possible. For some applications, such as the study of a particular enzyme, enzyme activities may be quite lower in stationary phase cells than in exponential phase cells and thus chemostat-grown cultures are ideal.

The chemostat can also be used in microbial ecology. For example, because the chemostat can easily mimic the low substrate concentrations that often prevail in nature, it is possible to study mixed bacterial populations in a chemostat and ask questions about the competitiveness of different organisms at particular nutrient concentrations. Using cultural methods as well as the powerful tools of molecular ecology such as phylogenetic stains and gene tracking (∞Chapter 18), changes in the microbial community can be monitored as a function of chemostat conditions. Such experiments often reveal interactions among individual components of the population that are not obvious from growth studies in batch culture.

Chemostats have also been used for enrichment and isolation of bacteria (∞Sections 1.7, 18.1, and 18.2 for discussion of enrichment and isolation of bacteria). From a mixed inoculum, one can select a stable population under the nutrient and dilution rate conditions chosen and then slowly increase the dilution rate until a single organism remains. In this way, microbiologists recently isolated a bacterium with a 6-minute doubling time—the fastest growing bacterium known.

6.9 Concept Check

Continuous culture devices (chemostats) are a means of maintaining cell populations in exponential growth for long periods. In a chemostat, the rate at which the culture is diluted governs the growth rate, and the population size is governed by the concentration of the growth-limiting nutrient entering the vessel.

◆ How do microorganisms in a *chemostat* differ from microorganisms in a *batch culture*?

◆ What happens in a chemostat if the dilution rate exceeds the maximal growth rate of the organism?

◆ Do pure cultures have to be used in a chemostat?

IV ENVIRONMENTAL EFFECTS ON MICROBIAL GROWTH: TEMPERATURE

Up to now we have described the growth of microorganisms under ideal laboratory conditions. However, the activities of microorganisms are greatly affected by the chemical and physical conditions of their environments. Understanding environmental influences helps explain the distribution of microorganisms in nature and makes it possible to devise methods for controlling or enhancing microbial activities.

Many environmental factors can be considered. However, four key factors play major roles in controlling the growth of all microorganisms: *temperature, pH, water availability*, and *oxygen*. Some other factors can potentially affect the growth of microorganisms, such as pressure and radiation. These more specialized environmental factors will be considered later in this book when we encounter microbial habitats in which they play major roles.

6.10 Effect of Temperature on Growth

Temperature is one of the most, if not *the* most, important environmental factor affecting growth and survival of microorganisms. At either too cold or too hot a temperature, microorganisms will not be able to grow. But the minimum and maximum temperatures vary greatly among different microorganisms, usually reflecting the temperature range and average temperature of their habitats.

Cardinal Temperatures

Temperature can affect living organisms in either of two opposing ways. As the temperature *rises*, chemical and enzymatic reactions in the cell proceed at more rapid rates, and growth becomes faster. However, *above* a certain temperature, particular proteins may be irreversibly denatured. Thus, as the temperature is increased within a given range, growth and metabolic function increase up to a point where denaturation reactions set in. Above this point, cell functions fall sharply to zero.

For every organism there is thus a *minimum* temperature below which growth no longer occurs, an *optimum* temperature at which growth is most rapid, and a *maximum* temperature above which growth is not possible (Figure 6.16●). The optimum temperature is always nearer the maximum than the minimum. These three temperatures, called the **cardinal temperatures**, are characteristic of each organism but are not completely fixed entities, as they can be modified slightly by other factors of the environment, in particular, by the composition of the growth medium.

The cardinal temperatures of different microorganisms differ widely—some organisms have temperature optima as low as 4°C and some higher than 100°C. The temperature range throughout which growth occurs by

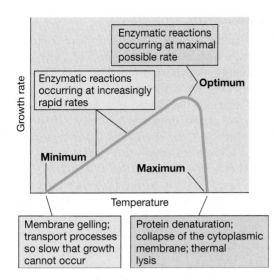

● **Figure 6.16 Effect of temperature on growth rate and the molecular consequences for the cell.** The three cardinal temperatures vary by organism.

various organisms is even wider than this, from below freezing to greater than boiling. However, no single organism can grow over this whole temperature range, and the typical range for any given organism is 30–40 degrees.

The *maximum* growth temperature of an organism most likely reflects the denaturation of one or more essential proteins in the cell. However, the factors controlling an organism's *minimum* growth temperature are not as clear. As mentioned earlier (∞ Section 4.5), the cytoplasmic membrane must be in a fluid state for proper functioning. An organism's minimum temperature may be the result of stiffening of its cytoplasmic membrane such that it no longer functions properly in

nutrient transport or in developing a proton motive force. This explanation is supported by experiments in which the minimum temperature for an organism can be altered to some extent by adjustments in cytoplasmic membrane lipid composition (see Section 6.11). In this regard, the maximum and minimum temperatures supporting growth of an organism are higher and lower, respectively, when tested in complex rather than defined media.

Temperature Classes of Organisms

Although there is a continuum of organisms, from those with very low temperature optima to those with high temperature optima, it is possible to broadly distinguish *four groups* of microorganisms in relation to their temperature optima: **psychrophiles**, with low temperature optima, **mesophiles**, with midrange temperature optima, **thermophiles**, with high temperature optima, and **hyperthermophiles**, with very high temperature optima (Figure 6.17●).

Mesophiles are found in warm-blooded animals and in terrestrial and aquatic environments in temperate and tropical latitudes. Psychrophiles and thermophiles are found in unusually cold and unusually hot environments, respectively. Hyperthermophiles are found in extremely hot habitats such as hot springs, geysers, and deep-sea hydrothermal vents (see Sections 6.12, 13.8, and 19.8).

In *Escherichia coli*, a typical mesophile, a detailed study of growth as a function of temperature has precisely defined its cardinal temperatures. The optimum temperature of *E. coli* in a complex medium (∞ Section 5.2) is 39°C, the maximum is 48°C, and the minimum is 8°C. Thus, the temperature *range* for *E. coli* is 40 degrees, near the high end for the average prokaryote.

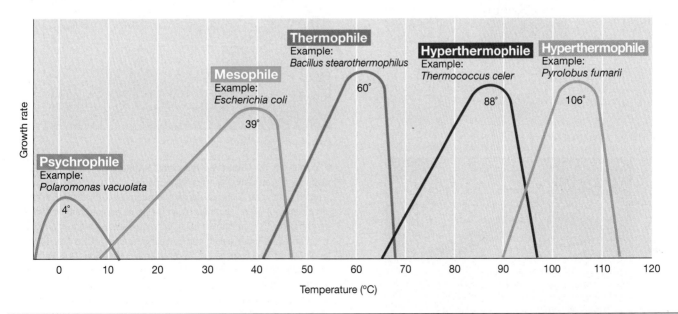

● **Figure 6.17 Relation of temperature to growth rates of a typical psychrophile, a typical mesophile, a typical thermophile, and two different hyperthermophiles.** The temperature optima of the example organisms are shown on the graph. Hyperthermophiles show growth temperature optima above 80°C.

6.10 *Concept Check*

Temperature is a major environmental factor controlling microbial growth. The cardinal temperatures describe the minimum, optimum, and maximum temperatures at which each organism grows. Microorganisms can be grouped by the temperature ranges they require.

◆ What are the cardinal temperatures for *Esherichia coli*? To what temperature class does it belong?

◆ How does a *hyperthermophile* differ from a *psychrophile*?

◆ *Escherichia coli* can grow at a higher temperature in a complex medium than in a defined medium. Why?

6.11 Microbial Growth at Cold Temperatures

Because humans live and work on the surface of Earth where temperatures are generally moderate, it is natural to consider very hot and very cold environments as being "extreme." And they are extreme for human habitation because humans would die quickly if immersed in boiling or freezing water. However, the natural habitats of many microorganisms can be either extremely hot or extremely cold, and the organisms that live there, referred to as **extremophiles** (∞ Section 2.4 and Table 2.1), have evolved to grow *optimally* under these conditions. We consider first organisms that grow at cold temperatures.

Cold Environments

Much of Earth's surface experiences low temperatures. The oceans, which make up over half of Earth's surface, have an average temperature of 5°C, and the depths of the open oceans have constant temperatures of 1–3°C. Vast land areas of the Arctic and Antarctic are permanently frozen or are unfrozen for only a few weeks in summer (Figure 6.18●). These cold environments are not sterile, and some microorganisms can be found alive and growing at any low temperature at which liquid water still exists. Even in many frozen materials there are small pockets of liquid water present where microorganisms can metabolize and grow. Within glaciers, for example, there exists a network of small liquid water channels in which prokaryotes thrive and reproduce.

In considering cold environments, it is important to distinguish between environments that are *constantly* cold and those that are cold *only* in winter. The latter, characteristic of continental temperate climates, may have summer temperatures as high as 40°C and winter temperatures of −20°C or colder. A temperate lake, for example, may have a period of ice cover in the winter, but the time that the water is at 0°C is relatively brief. Such highly variable environments are less favorable for cold-adapted organisms than are the *constantly cold* environments found in polar regions, at high altitudes, and in the depths of the oceans. For example, freshwater lakes in the Antarctic Dry Valleys contain a *permanent* ice

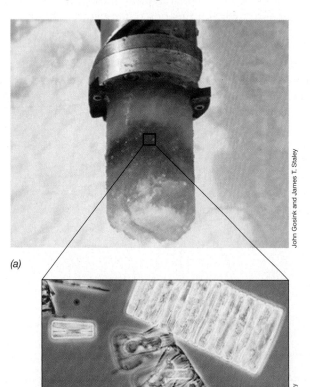

(a)

(b)

John Gosink and James T. Staley

(c)

Deborah Jung and Michael T. Madigan

● **Figure 6.18 Antarctic microbial habitats.** (a) A core of permanently frozen seawater from McMurdo Sound, Antarctica. The core is about 8 cm wide. Note the dense coloration due to pigmented microorganisms. (b) Phase contrast micrograph of phototrophic microorganisms from the core shown in (a). Most organisms are either diatoms or green algae (both eukaryotic microorganisms, ∞ Section 14.13). (c) Photo of the surface of Lake Bonney, McMurdo Dry Valleys, Antarctica. Like many other Antarctic lakes, Lake Bonney, which is about 40 meters deep, remains permanently frozen and has an ice cover of about 5 meters. The water column of Lake Bonney remains near 0°C and contains both oxic and anoxic zones (∞ Section 19.5 and Figure 19.9); thus both aerobic and anaerobic microorganisms are present. However, no higher eukaryotic organisms inhabit Dry Valley lakes, making them unique microbial ecosystems.

cover of several meters in thickness (Figure 6.18*c*). The water column below the ice in these lakes remains at 0°C or colder year round and is thus an ideal habitat for cold-adapted organisms.

Psychrophilic and Psychrotolerant Microorganisms

As noted earlier, organisms with low temperature optima are called **psychrophiles**. A psychrophile can be defined as an organism with an *optimal* growth temperature of *15°C or lower*, a maximum growth temperature below 20°C, and a minimal growth temperature at 0°C or lower. Organisms that grow at 0°C but have optima of 20–40°C are called **psychrotolerant**.

Psychrophiles are found in environments that are constantly cold, such as in polar regions or in marine sediments from polar regions, and they may be rapidly killed by warming to room temperature. For this reason, their laboratory study requires that great care be taken to ensure that they never warm up during sampling, transport to the laboratory, isolation, or other manipulations.

Some of the best-studied psychrophiles have been algae that grow in dense masses within and under the ice in polar regions or other permanent ice fields (Figure 6.18). Psychrophilic algae are also often seen on the surfaces of snowfields and glaciers at such densities that they impart a distinctive red or green coloration to the surface (Figure 6.19*a*●). The most common snow alga is *Chlamydomonas nivalis*; its brilliant red spores are responsible for the red color (Figure 6.19*b*). This green alga grows within the snow as a green-pigmented vegetative cell and then sporulates. As the snow dissipates by melting, erosion, and vaporization, the spores become concentrated on the surface. Snow algae are most commonly seen on melting permanent snowfields in midsummer to late summer and are especially common in sunny, dry areas. In addition to snow algae, several psychrophilic chemoorganotrophic bacteria are known, many from the Antarctic (Figure 6.18). Some of these, particularly isolates from sea ice (frozen seawater that forms seasonally in polar regions), show the lowest maximum growth temperature of all known microorganisms.

Psychrotolerant microorganisms are much more widely distributed than psychrophiles and can be isolated from soils and water in temperate climates as well as from meat, milk and other dairy products, cider, vegetables, and fruit stored at refrigeration temperatures (~4°C). As noted, psychrotolerant microorganisms grow best at a temperature between 20° and 40°C. Because temperate environments warm up in summer, they cannot support the heat-sensitive psychrophiles because the warming provides selection against them. It should be emphasized that although psychrotolerant microorganisms do grow at 0°C, most do not grow very well at this temperature, and one must often wait several weeks before visible growth is seen in culture media. Various genera of *Bacteria*, fungi, algae, and protozoa have members that are psychrotolerant.

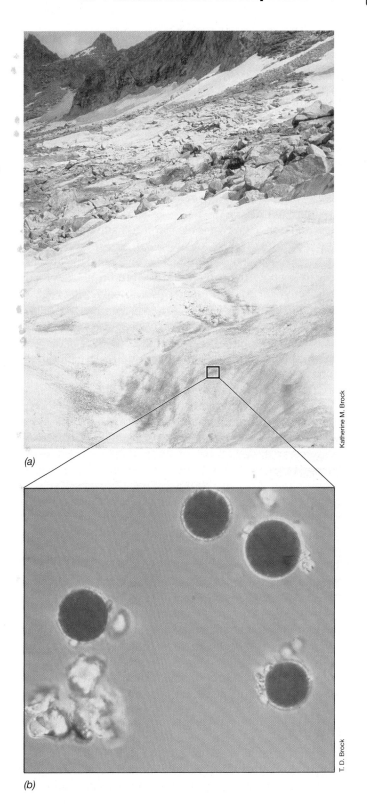

(a)

(b)

Katherine M. Brock

T. D. Brock

● **Figure 6.19 Snow algae.** (a) Snow bank in the Sierra Nevada, California, with red coloration caused by the presence of snow algae (eukaryotic cells, ⌒ Section 14.13). Pink snow such as this is common on summer snow banks at high altitudes throughout the world. (b) Photomicrograph of red-pigmented spores of the snow alga *Chlamydomonas nivalis*. The spores germinate to yield motile green algal cells. Related species of snow algae contain different carotenoid pigments (⌒ Section 17.3), and thus, fields of snow algae can also be green, orange, brown, or purple in color.

Molecular Adaptations to Psychrophily

Psychrophiles produce enzymes that function optimally in the cold and that are often denatured or otherwise inactivated at even very moderate temperatures. The molecular basis for this is not entirely understood, but it is known that cold-active enzymes have greater amounts of α-helix and lesser amounts of β-sheet secondary structure (Section 3.7 and Figure 3.16) than enzymes that are inactive in the cold. Because the β-sheet tends to form a more rigid structure, the greater α-helix content of cold-active enzymes allows these proteins greater flexibility in the cold.

Cold-active enzymes also tend to have greater polar and lesser hydrophobic amino acid contents than their mesophilic and thermophilic counterparts. This also assists in keeping the protein flexible (and thus enzymatically active) at cold temperatures. Moreover, cold-active proteins tend to have decreased levels of weak bonds (Section 3.1) and decreased interactions between their domains compared to proteins from organisms that grow best at higher temperatures. Again, these modifications probably favor protein flexibility.

Another feature of psychrophiles is that compared to mesophiles, transport processes (Section 4.7) occur optimally at low temperature, an indication that the cytoplasmic membranes of psychrophiles are constructed in such a way that low temperatures do not inhibit membrane functions. Cytoplasmic membranes from psychrophiles tend to have a higher content of *unsaturated* fatty acids (Section 5.17). This helps to maintain a semifluid state of the membrane at low temperatures (membranes composed of predominantly saturated fatty acids become waxy and nonfunctional at low temperatures). The lipids of some psychrophilic bacteria also contain *polyunsaturated* fatty acids and long-chain hydrocarbons with multiple double bonds. For example, a hydrocarbon with nine double bonds ($C_{31:9}$) has been identified from the lipids of some Antarctic bacteria, and the bacterium *Psychroflexus* has been shown to contain fatty acids with four and five double bonds. These fatty acids remain more fluid at low temperatures than do saturated or monounsaturated fatty acids.

Freezing

Despite the ability of some organisms to grow at low temperatures, there is a lower limit below which reproduction is impossible. Pure water freezes at 0°C and seawater at −2.5°C, but freezing is not a continuous process. Microscopic pockets of water continue to exist at these and even much lower temperatures. As long as liquid water is available, microbial growth is possible.

Although freezing prevents microbial growth, it does not necessarily cause microbial death. In addition, the medium in which the cells are suspended considerably affects sensitivity to freezing. Water-miscible liquids such as glycerol and dimethylsulfoxide (DMSO), when added at about 10% (final concentration) to the suspending medium, penetrate the cells and protect them by reducing the severity of dehydration effects and preventing ice crystal formation. In fact, the addition of such agents, called *cryoprotectants*, is a common way of *preserving* microbial cultures at very low temperatures (usually at −70 to −196°C). Properly prepared frozen cells can remain viable for long periods (decades or longer).

6.11 Concept Check

Organisms with cold temperature optima are called *psychrophiles*, and the most extreme representatives inhabit permanently cold environments. Psychrophiles have evolved biomolecules that function best at cold temperatures but that can be unusually sensitive to warm temperatures.

◆ How do *psychrotolerant* organisms differ from *psychrophilic* organisms?

◆ What molecular adaptations to the cytoplasmic membrane are seen in psychrophiles, and why are they necessary?

6.12 Microbial Growth at High Temperatures

Microbial life flourishes in high-temperature environments, up to and including boiling water. Above about 65°C, only *prokaryotic* life forms are found, but even here, a huge diversity of both *Bacteria* and *Archaea* exist. We consider some of these hot environments and their microbial life here.

Thermal Environments

Recall that organisms whose growth temperature optimum is *above* 45°C are called **thermophiles**, and those whose optimum is *above* 80° C are called **hyperthermophiles** (Figure 6.17). Temperatures as high as these are found in nature only in certain areas. For example, soils subject to full sunlight are often heated to temperatures above 50°C at midday, and some soils may become warmed to even 70°C, although a few centimeters under the surface the temperature is much lower. Fermenting materials such as compost piles and silage can reach temperatures of 70°C. However, the most extensive and extreme high-temperature environments found in nature are in association with volcanic phenomena. These include, in particular, hot springs.

Many hot springs have temperatures at or near boiling, and steam vents (fumaroles) may reach 150–500°C. Hydrothermal vents in the bottom of the ocean have temperatures of 350°C or greater (Section 19.8). Hot springs occur throughout the world, but are especially concentrated in the western United States, New Zealand, Iceland, Japan, Italy, Indonesia, Central America, and central Africa. The world's largest single concentration of hot springs is in Yellowstone National Park, Wyoming (USA).

Although some hot springs vary in temperature, others are very constant, not varying more than 1–2°C over many years. In addition, different springs have different chemical compositions and pH values, but generally contain sufficient levels of nutrients to support populations, often very large populations, of hyperthermophilic chemoorganotrophs and chemolithotrophs.

Hyperthermophiles in Hot Springs

Many hot springs are at the boiling point for the altitude (92–93°C at Yellowstone, 99–100°C at locations where the springs are close to sea level). In boiling hot springs (Figure 6.20●), a variety of hyperthermophiles are typically present. The growth of such organisms can be studied by immersing microscope slides into the spring and retrieving them after a few days. Microscopic examination of the slides reveals colonies of prokaryotes (Figure 6.20b) that have developed from single bacterial cells that attached to and grew on the glass surface.

Ecological studies of organisms living in boiling springs have shown that growth rates are fairly rapid, and doubling times of as short as 1 h have been recorded. Cultures of many of these prokaryotes have been obtained, and a variety of morphological and physiological

types exist (∞Chapter 13). Phylogenetic studies using ribosomal RNA sequencing (∞Sections 2.3, 11.5, and 11.6) have shown great evolutionary diversity among these hyperthermophiles. Species of both *Bacteria* and *Archaea* are present. Some hyperthermophilic *Archaea* show growth-temperature optima *greater than 100°C* and must be grown in pressurized vessels in the laboratory in order to reach temperatures above the boiling point.

Thermophiles

Many thermophiles (optima 45–80°C) are also present in hot springs as well as other thermal environments. In hot springs, as boiling water overflows the edges of the spring and flows away from the source, it gradually cools, setting up a *thermal gradient*. Along this gradient, various microorganisms grow (Figure 6.21●), with different species growing in the different temperature ranges. By studying the species distribution along such thermal gradients and by examining hot springs and other thermal habitats at different temperatures around the world, it has been possible to determine the upper temperature limits for each type of organism (Table 6.1). From this information we can conclude that (1) prokaryotic organisms in general are able to grow at temperatures higher than those at which eukaryotes can grow; (2) the most thermophilic of all

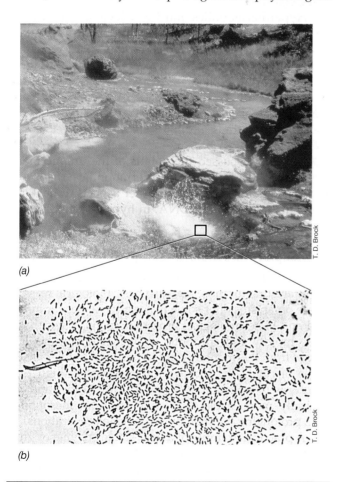

(a)

(b)

● **Figure 6.20   Growth of hyperthermophiles in boiling water.** (a) Boulder Spring, a small boiling spring in Yellowstone National Park. This spring is superheated, having a temperature 1–2°C above the boiling point. The mineral deposits around the spring consist mainly of silica and sulfur. (b) Photomicrograph of a microcolony of prokaryotes that developed on a microscope slide immersed in a boiling spring such as that shown in (a).

● **Figure 6.21   Growth of thermophilic cyanobacteria in a hot spring in Yellowstone National Park.** Characteristic V-shaped pattern formed by cyanobacteria at the upper temperature for phototrophic life, 70–74°C, in the thermal gradient formed from a boiling hot spring. The pattern develops because the water cools more rapidly at the edges than in the center of the channel. The spring flows from the back of the picture toward the foreground. The light-green color is from a high temperature strain of the cyanobacterium *Synechococcus*. As water flows down the gradient, the density of cells increases, less thermophilic strains enter, and the color becomes more intensely green.

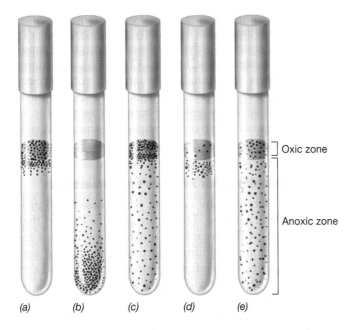

(a) (b) (c) (d) (e)

Oxic zone

Anoxic zone

● **Figure 6.25 Growth versus oxygen concentration.** Aerobic, anaerobic, facultative, microaerophilic, and aerotolerant anaerobe growth, as revealed by the position of microbial colonies (depicted here as black dots) within tubes of thioglycolate broth culture medium. A small amount of agar has been added to keep the liquid from becoming disturbed and the redox dye, resazurin, which is pink when oxidized and colorless when reduced, is added as a redox indicator. (a) Oxygen penetrates only a short distance into the tube, so obligate aerobes grow only at the surface. (b) Anaerobes, being sensitive to oxygen, grow only away from the surface. (c) Facultative aerobes are able to grow in either the presence or the absence of oxygen and thus grow throughout the tube. However, better growth occurs near the surface because these organisms can respire. (d) Microaerophiles grow away from the most oxic zone. (e) Aerotolerant anaerobes grow throughout the tube. However, growth is no better near the surface because these organisms can only ferment.

called *thioglycolate broth,* commonly used to test an organism's requirements for oxygen (Figure 6.25●).

After thioglycolate reacts with oxygen throughout the tube, oxygen can penetrate only near the top of the tube where the medium contacts air. Obligate aerobes grow only at the top of such tubes. Facultative organisms grow throughout the tube but best near the top. Microaerophiles grow near the top but not right at the top. Anaerobes grow only near the bottom of the tube, where oxygen cannot penetrate (Figure 6.25). A redox indicator dye called *resazurin* is added to the medium because the dye changes color in the presence of oxygen (pink when oxygenated, colorless when reduced) and thereby indicates the degree of penetration of oxygen into the medium (Figure 6.25).

To remove all traces of O_2 for the culture of anaerobes, it is possible to place an oxygen-consuming system in a jar holding the tubes or plates. One of the simplest devices for this is an *anoxic jar,* a heavy-walled jar with a gastight seal within which tubes, plates, or other containers to be incubated are placed (Figure 6.26*a*●). The air in the jar is replaced with a mixture of H_2 and CO_2, and in the presence of a chemical catalyst the traces of O_2 left in the jar and culture medium are consumed by the H_2 ($H_2 + O_2 \rightarrow H_2O$), eventually leading to anoxic conditions.

For strict anaerobes, such as the methanogens (∞ Sections 13.4 and 17.17), it is necessary not only to carefully remove all traces of O_2 but also to carry out all manipulations of cultures in an anoxic atmosphere. Strict anaerobes can be killed by even a brief exposure to oxygen. In these cases, a culture medium is first boiled to render it oxygen-free, and then a reducing agent such as H_2S is added, and the mixture is sealed under an oxygen-free gas. All manipulations are carried out under a jet of oxygen-free hydrogen or nitrogen gas that is direct-

(a)

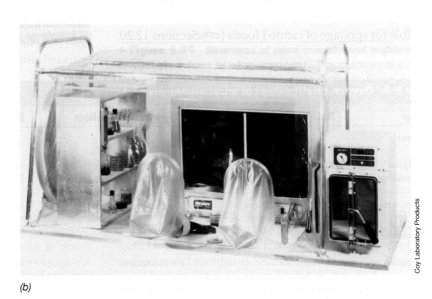

(b)

● **Figure 6.26 Incubation under anoxic conditions.** (a) Anoxic jar. A chemical reaction in the envelope in the jar generates $H_2 + CO_2$. The H_2 reacts with O_2 in the jar on the surface of a palladium catalyst to yield H_2O; the final atmosphere contains N_2, H_2, and CO_2. (b) Anoxic glove bag for manipulating and incubating cultures under anoxic conditions. The airlock on the right, which can be evacuated and filled with O_2-free gas, serves as a port for adding and removing materials to and from the glove bag.

ed into the culture vessel when it is open, thus driving out any oxygen that might enter. For extensive research on anaerobes, special boxes fitted with gloves, called *anoxic glove boxes*, permit work with open cultures in completely anoxic atmospheres (Figure 6.26*b*).

6.15 Concept Check

Aerobes require oxygen to live, whereas anaerobes do not and may even be killed by oxygen. Facultative organisms can live with or without oxygen. Special techniques are needed to grow aerobic and anaerobic microorganisms.

◆ What is a *facultative aerobe*?

◆ How does a reducing agent work?

6.16 Toxic Forms of Oxygen

Oxygen is a powerful oxidant and the best electron acceptor for respiration (∞Section 5.11). But at the same time, oxygen can be a poison to some microorganisms. Why? It turns out that oxygen, *per se*, is not the poison, but instead it is certain *oxygen derivatives* that are toxic to microorganisms. We consider this topic here.

Oxygen Chemistry

Oxygen in its ground state is referred to as **triplet oxygen** ($^3O_2$). However, other electronic configurations of oxygen are possible. One major form of toxic oxygen is called **singlet oxygen** ($^1O_2$), a higher energy form of oxygen in which outer shell electrons surrounding the nucleus become highly reactive and are able to carry out a variety of spontaneous and undesirable oxidations within the cell. Singlet oxygen is produced both photochemically and biochemically, the latter through the action of various peroxidase enzymes. Organisms that frequently encounter singlet oxygen, such as airborne bacteria and phototrophic microorganisms, often contain pigments called **carotenoids**, which function to convert singlet oxygen to nontoxic forms (∞Section 17.3).

Other highly toxic forms of oxygen include **superoxide anion** (O_2^-), **hydrogen peroxide** (H_2O_2), and **hydroxyl radical** ($OH·$). All of these are produced as by-products of the reduction of O_2 to H_2O in respiration (Figure 6.27●). Flavoproteins, quinones, thiols, and iron-sulfur proteins (∞Section 5.11), found in virtually all cells, can also carry out the reduction of O_2 to O_2^-. Thus, whether or not they can use oxygen (Table 6.3), virtually all cells are faced with exposure to toxic oxygen species from time to time.

Superoxide is a strong oxidizing agent and can oxidize virtually any organic compound in the cell, including macromolecules. Peroxides such as H_2O_2 can damage cell components but are generally not as toxic as superoxide or hydroxyl radical. The latter is the most reactive of all toxic oxygen species and can, like superoxide, quickly oxidize any organic substance in the cell.

$$O_2 + e^- \longrightarrow O_2^- \quad \text{Superoxide}$$
$$O_2^- + e^- + 2\,H^+ \longrightarrow H_2O_2 \quad \text{Hydrogen peroxide}$$
$$H_2O_2 + e^- + H^+ \longrightarrow H_2O + OH· \quad \text{Hydroxyl radical}$$
$$OH· + e^- + H^+ \longrightarrow H_2O \quad \text{Water}$$

$$\text{Overall:} \quad O_2 + 4\,e^- + 4\,H^+ \longrightarrow 2\,H_2O$$

● **Figure 6.27 Four-electron reduction of O_2 to H_2O by stepwise addition of electrons.** All the intermediates formed are reactive and toxic to cells except for water, of course.

However, the hydroxyl radical is only a transient species in most cells because its major source is ionizing radiation, a substance to which most cells are rarely exposed. Small amounts of hydroxyl radical can also be produced from H_2O_2 (Figure 6.27). But if peroxides are removed from the cell (through the activity of the enzyme catalase, which is discussed later), this source of hydroxyl radical is virtually eliminated. In later chapters we will see that some toxic oxygen species can be produced by certain immune cells in the animal body and used to kill microbial invaders (∞Section 22.2).

Enzymes That Destroy Toxic Oxygen

With such an array of toxic oxygen derivatives, it is not surprising that organisms have evolved enzymes that destroy these compounds (Figure 6.28●). The most common enzyme in this regard is **catalase**, which attacks hydrogen peroxide. The activity of catalase is illustrated in Figures 6.28*a* and 6.29. Another enzyme that destroys hydrogen peroxide is **peroxidase** (Figure 6.28*b*), which differs from catalase in requiring a reductant for activity, usually NADH, producing H_2O as a product. Superoxide is destroyed by the enzyme **superoxide dismutase** (Figure 6.28*c*), which combines two molecules of superoxide to form

(a) Catalase:
$$H_2O_2 + H_2O_2 \longrightarrow 2\,H_2O + O_2$$

(b) Peroxidase:
$$H_2O_2 + \text{NADH} + H^+ \longrightarrow 2\,H_2O + NAD^+$$

(c) Superoxide dismutase:
$$O_2^- + O_2^- + 2\,H^+ \longrightarrow H_2O_2 + O_2$$

(d) Superoxide dismutase/catalase in combination:
$$4\,O_2^- + 4\,H^+ \longrightarrow 2\,H_2O + 3\,O_2$$

(e) Superoxide reductase:
$$O_2^- + 2\,H^+ + \text{cyt } c_{reduced} \longrightarrow H_2O_2 + \text{cyt } c_{oxidized}$$

● **Figure 6.28 Enzymes that destroy toxic oxygen species.** (a) Catalases and (b) peroxidases are porphyrin-containing proteins, although some flavoproteins (∞Section 5.11) may consume toxic oxygen species as well. (c) Superoxide dismutases are metal-containing proteins, the metals being copper and zinc, manganese, or iron. (d) Combined reaction of superoxide dismutase and catalase. (e) Superoxide reductase catalyzes the one electron reduction of O_2^- to H_2O_2 using reduced cytochrome *c* as the electron donor.

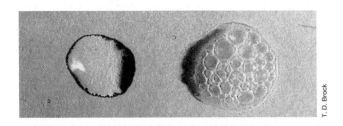

T. D. Brock

● **Figure 6.29 Method for testing a microbial culture for the presence of catalase.** A heavy loopful of cells from an agar culture was mixed on a slide (right) with a drop of 30% hydrogen peroxide. The immediate appearance of bubbles is indicative of the presence of catalase. The bubbles are O_2 produced by the reaction $H_2O_2 + H_2O_2 \rightarrow 2\ H_2O + O_2$.

one molecule of hydrogen peroxide and one molecule of oxygen. Superoxide dismutase and catalase work in tandem to bring about the conversion of superoxide back to oxygen (Figure 6.28*d*).

Aerobes and facultative aerobes typically contain both superoxide dismutase and catalase, although a few obligate aerobes lack catalase. Superoxide dismutase is indispensable to aerobic cells, and the absence of this enzyme in obligate anaerobes was originally thought to be the reason why oxygen is toxic to them (but see next paragraph). Some aerotolerant anaerobes, such as lactic acid bacteria, also lack superoxide dismutase, but they use protein-free manganese (Mn^{2+}) complexes to carry out the dismutation of O_2^- to H_2O_2 and O_2. Such a reaction may have functioned as a primitive form of superoxide dismutase in ancient organisms. This is supported by the fact that all known superoxide dismutases contain a metal cofactor, usually Mn^{2+}, but also Fe^{2+}, or Cu^{2+} plus Zn^{2+}, at the enzyme's active site.

Another means of superoxide disposal is present in certain obligately anaerobic prokaryotes. In *Pyrococcus furiosus* (a member of the *Archaea*), for example, superoxide dismutase is absent, but a unique enzyme,

superoxide reductase, is present and is responsible for superoxide removal. Unlike superoxide dismutase, superoxide reductase reduces superoxide to H_2O_2 *without* the production of O_2 (Figure 6.28*e*), thus avoiding exposure of the organism to O_2. *P. furiosus* also lacks catalase, an enzyme that like superoxide dismutase, also generates O_2 (Figure 6.28*a*). In *P. furiosus*, the H_2O_2 produced by superoxide reductase is removed by the activity of peroxidase-like enzymes that yield H_2O as a final product (Figure 6.28*b*).

Superoxide reductase may be widely distributed among obligate anaerobes, since genomic studies have revealed superoxide reductase-like genes in the genomes of several obligate anaerobes. This means that these organisms, previously thought to be strict anaerobes because of a lack of superoxide dismutase, do indeed have a mechanism to deal with the highly toxic superoxide anion. The sensitivity of these organisms to oxygen in culture media may therefore be for entirely other and as yet unknown reasons.

Interestingly, many obligately anaerobic hyperthermophiles are quite tolerant of cold, oxic conditions. Although they do not grow under such conditions, superoxide reductase presumably prevents their killing. It is thought that this oxygen tolerance is important in the transfer of these organisms from one deep sea hydrothermal system to another (Section 19.8).

6.16 Concept Check

Several toxic forms of oxygen can be formed in the cell, but enzymes are present that can neutralize most of them. Superoxide in particular seems to be a common toxic oxygen species.

◆ How does superoxide dismutase protect a cell from toxic oxygen?

◆ How does the activity of superoxide *dismutase* differ from that of superoxide *reductase*?

REVIEW QUESTIONS

1. Describe the key molecular processes that occur when a cell grows and divides. What proteins assist in this cell division process (⌒Section 6.1)?

2. Describe the role that Fts proteins play in the cell division process (⌒Section 6.2).

3. In what way do derivatives of the rod-shaped bacterium *Escherichia coli* that carry mutations that inactivate the protein MreB look different microscopically from wild-type (unmutated) cells? What is the reason for this (⌒Section 6.2)?

4. How does the antibiotic penicillin kill bacterial cells (⌒Section 6.3)?

5. What is the difference between the specific growth rate (k) of an organism and its generation time (g) (⌒Sections 6.4 and 6.5)?

6. Describe the growth cycle of a population of bacterial cells from the time this population is first inoculated into fresh medium (⌒Section 6.6).

7. Describe one direct and one indirect method by which microbial growth can be measured. Make sure that the methods you choose agree with your definition (⌒Sections 6.7 and 6.8).

8. Briefly describe the process by which a single cell develops into a visible colony on an agar plate. With this explanation as a background, describe the principle behind the viable count method (⌒Section 6.7).

9. How can a chemostat regulate growth rate and cell numbers independently (⌒Section 6.9)?

10. Examine the graph describing the relationship between growth rate and temperature (Figure 6.17). Give an explanation, in biochemical terms, of why the optimum temperature for an organism is usually closer to its maximum than its minimum (⌒Section 6.10).

11. Describe a habitat where you would find a psychrophile. A hyperthermophile (⌒Sections 6.11 and 6.12).

12. Concerning the pH of the environment and of the cell, in what ways are acidophiles and alkaliphiles different? In what ways are they similar (⌒Section 6.13)?

13. Write an explanation in molecular terms for how a cell of a halophile is able to make water molecules flow *into* itself (⌒Section 6.14).

14. List three *chemical classes* of compatible solutes produced by various microorganisms. List at least two things they all have in common (⌒Section 6.14).

15. Contrast an *aerotolerant* and an *obligate anaerobe* in terms of sensitivity and ability to grow in the presence of oxygen (O_2). How does an aerotolerant anaerobe differ from a microaerophile (⌒Section 6.15)?

16. Compare and contrast the enzymes *catalase, superoxide dismutase,* and *superoxide reductase* from the following points of view: substrates, oxygen products, organisms containing them, role in oxygen tolerance of the cell (⌒Section 6.16).

APPLICATION QUESTIONS

1. Starting with four bacterial cells per milliliter in a rich nutrient medium, with a 1-h lag phase and a 20-min generation time, how many cells will there be in 1 liter of this culture after 1 h? After 2 h? After 2 h if one of the initial four cells was dead?

2. Calculate g and k in a growth experiment in which a medium was inoculated with 5×10^6 cells/ml of *Escherichia coli* cells and, following a 1-h lag, grew exponentially for 5 h, after which the population was 5.4×10^9 cells/ml.

3. Return to Chapter 3 and locate a figure that best describes what happens to enzymes from a cell of a mesophile like *Escherichia coli* when placed in a culture medium at 80°C. Contrast this with a figure from Chapter 6 that best describes what would happen if cells of *Pyrolobus fumarii* were placed under the same conditions. Describe why neither organism would grow.

4. Would you expect to find a psychrophilic microorganism alive in a hot spring? Why? It is frequently possible to isolate hyperthermophilic microorganisms from cold-water environments. Supply an explanation for this.

7

ESSENTIALS OF MOLECULAR BIOLOGY

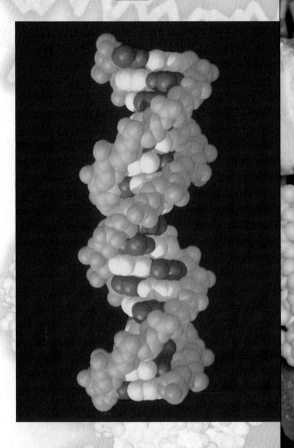

Although many forms of nucleic acid are known in nature, all cells contain double-stranded DNA genomes. The genome directs all of the molecular events in the cell.

WORKING GLOSSARY

Aminoacyl-tRNA synthetase an enzyme that catalyzes attachment of an amino acid to its cognate tRNA

Anticodon a sequence of three bases in a tRNA molecule that base-pairs with a codon during protein synthesis

Antiparallel in reference to double-stranded DNA, one strand runs $5' \rightarrow 3'$ and the complementary strand $3' \rightarrow 5'$

Chromosome a genetic element, usually circular in prokaryotes and linear in eukaryotes, carrying genes essential to cellular function

Codon a sequence of three bases in mRNA that encodes an amino acid

Complementary nucleic acid sequences that can base-pair with each other

DNA gyrase an enzyme found in most prokaryotes that introduces negative supercoils in DNA

DNA polymerase an enzyme that synthesizes a new strand of DNA in the $5' \rightarrow 3'$ direction using an antiparallel DNA strand as a template

Exon the coding DNA sequences in a split gene (contrast with *intron*)

Gene a segment of DNA specifying a protein (via mRNA), a tRNA, or an rRNA

Genome the total complement of genes contained in a cell or virus

Hybridization formation of a duplex nucleic acid with strands derived from different sources by complementary base pairing

Intron the intervening noncoding DNA sequences in a split gene (contrast with *exon*)

Messenger RNA (mRNA) an RNA molecule that contains the genetic information to encode one or more polypeptides

Molecular chaperone a protein that helps other proteins fold or refold from a partially denatured state

Operon a cluster of genes whose expression is controlled by a single operator

Polymerase chain reaction (PCR) a method for the amplification of a specific DNA sequence *in vitro* by repeated cycles of synthesis using specific primers and DNA polymerase

Primary transcript an unprocessed RNA molecule that is the direct product of transcription

Primer an oligonucleotide to which DNA polymerase can attach the first deoxyribonucleotide during DNA replication

Promoter a site on DNA to which RNA polymerase binds to commence transcription

Replication synthesis of DNA using DNA as a template

Restriction enzyme an enzyme that recognizes and breaks DNA at specific sequences

Ribosomal RNA (rRNA) types of RNA found in the ribosome; some participate actively in the process of protein synthesis

Ribosome a cytoplasmic particle composed of ribosomal RNA and protein whose function is to synthesize proteins

RNA polymerase an enzyme that synthesizes RNA in the $5' \rightarrow 3'$ direction using a complementary and antiparallel DNA strand as a template

Semiconservative replication DNA synthesis yielding new double helices, each consisting of one parental and one progeny strand

Transcription the synthesis of RNA using a DNA template

Transfer RNA (tRNA) an adaptor molecule used in translation that has specificity for both a particular amino acid and for one or more codons

Translation the synthesis of protein using the genetic information in messenger RNA as a template

---

In Chapter 1 we characterized cells as *chemical machines* and *coding devices*. As chemical machines, cells accumulate and transform their vast array of macromolecules into new cells. As coding devices, they store, process, and use genes, the genetic information of the cell. Genes and gene processing are the subject of *molecular biology*, the focus of this chapter. Here we will study genes, DNA/RNA structure and function, and DNA replication. We then consider the biosynthesis of proteins, the macromolecules that play important roles in both cell structure and the machine function of the cell.

GENES AND GENE EXPRESSION

7.1 Macromolecules and Genetic Information

The functional unit of genetic information is the **gene**. All microorganisms, indeed all life forms, contain genes, and thus a fundamental understanding of what a gene is and what it does is important for understanding the structure, function, and behavior of cells and viruses. Moreover, because biology is rapidly moving toward

defining cells in terms of their gene complement (the "genomic era"), we must understand the process of biological information flow if we are to understand the biology of microorganisms.

Genes and the Steps in Information Flow

In all cells genes are composed of *deoxyribonucleic acid (DNA)*. The information in the gene is present as the sequence of bases—purine (adenine and guanine) and pyrimidines (thymine and cytosine)—in the DNA. The information stored in DNA is transferred to *ribonucleic acid (RNA)*. RNA can be either an informational intermediate (a messenger), or, in some cases, a more active part of the cell's machinery. Because all three of these molecules, DNA, RNA, and protein, contain genetic information in their sequences, they are called **informational macromolecules** (∞ Section 3.2).

The molecular processes underlying genetic information flow can be divided into three stages (Figure 7.1●).

1. *Replication.* The DNA molecule is a **double helix** of two long chains (∞ Section 3.5). During replication, DNA is duplicated, producing two double helices (Figure 7.1).

2. *Transcription.* DNA participates in protein synthesis through an RNA intermediate. Transfer of the informa-

tion to RNA is called **transcription**, and the RNA molecule that encodes one or more polypeptides is called **messenger RNA** (mRNA). Some genes contain information for other types of RNA, such as **transfer RNA** (tRNA) and **ribosomal RNA** (rRNA). These play roles in protein synthesis but do not themselves encode the genetic information for making proteins.

3. *Translation.* The sequence of amino acids in a polypeptide is determined by a specific sequence of bases in the mRNA. There is a linear correspondence between the base sequence of a gene and the amino acid sequence of a polypeptide (Figure 7.1). *Three* bases on a mRNA molecule encode a single amino acid, and each such triplet of bases is called a **codon**. This **genetic code** is translated into protein by means of the protein-synthesizing system. This system consists of **ribosomes** (which are themselves made up of proteins and rRNA), tRNA, and a number of enzymes.

The three transfer steps shown in Figure 7.1 are those used in all cells. In Chapter 9 we will learn that some viruses violate this **central dogma of molecular biology** (DNA → RNA → protein). This includes instances where RNA is the viral genetic material and either functions as mRNA directly or encodes mRNA, and also a situation in retroviruses such as HIV—the causative agent of AIDS—where its RNA genome encodes the production of DNA (reverse transcription). Note that in both of these cases information transfer remains nucleic acid to nucleic acid, but not in the way defined by the central dogma. RNA viruses are just one of many molecular surprises that await us in the viral world!

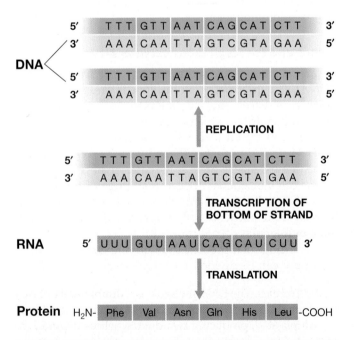

DNA

REPLICATION

TRANSCRIPTION OF BOTTOM OF STRAND

RNA 5′ UUU GUU AAU CAG CAU CUU 3′

TRANSLATION

Protein H₂N- Phe Val Asn Gln His Leu -COOH

● **Figure 7.1 Synthesis of the three types of informational macromolecules.** Note that in any particular region, only one of the two strands of the DNA double helix is transcribed.

Prokaryotic and Eukaryotic Genetics

Replication, transcription, and translation occur in all organisms. However, there are some differences in the mechanisms of these processes in prokaryotes and in eukaryotes. In part this is due to differences in the organization of DNA in these two classes of organisms and to the fact that eukaryotes have a nucleus.

We emphasized in Chapter 2 the basic differences in the organization of DNA in prokaryotes and eukaryotes. To review, the typical prokaryotic genome consists of a single, covalently closed *circular* molecule of DNA present in the cytoplasm of the cell. In contrast, the eukaryotic genome consists of several *linear* pieces of DNA present in individual chromosomes in the cell nucleus. In virtually all prokaryotes there is no membrane separating the chromosome from the cytoplasm (Section 2.2). In eukaryotes, however, the chromosomes are located inside the nucleus while the ribosomes are in the cytoplasm, so transcription and translation are spatially separated processes.

We also emphasized in Chapter 2 how the two domains of prokaryotes—*Bacteria* and *Archaea*—are phylogenetically distinct. This distinction is also obvious in many aspects of the molecular biology of these organisms. In many cases we will see closer parallels between molecular processes in *Archaea* and *Eukarya* than in *Bacteria* and *Archaea*.

In all cells the definition of a gene is the same: a segment of DNA specifying a protein (via mRNA), a tRNA, or an rRNA. However, in eukaryotes, protein-encoding genes are typically split into two or more coding regions, with noncoding regions separating coding regions. The *coding sequences* are called **exons**, and the *intervening noncoding regions*, **introns**. Both introns and exons are transcribed into the **primary transcript**. From here, the mature (functional) mRNA is formed by excision of noncoding regions and transport to the cytoplasm for translation (Section 14.8). A few genes in prokaryotes also contain introns, but the vast majority do not. A summary contrasting molecular events in prokaryotes and eukaryotes is given in Figure 7.2●.

⬣ *7.1 Concept Check*

The three key processes of macromolecular synthesis are (1) DNA replication; (2) transcription (the synthesis of RNA from a DNA template); and (3) translation (the synthesis of proteins using messenger RNA as template). Although the basic processes are the same in both prokaryotes and eukaryotes, the organization of genetic information is more complex in eukaryotes. Most eukaryotic genes have both coding regions (exons) and noncoding regions (introns).

◆ What three informational macromolecules are involved in genetic information flow?

◆ In all cells there are *three* processes involved in genetic information flow. What are they?

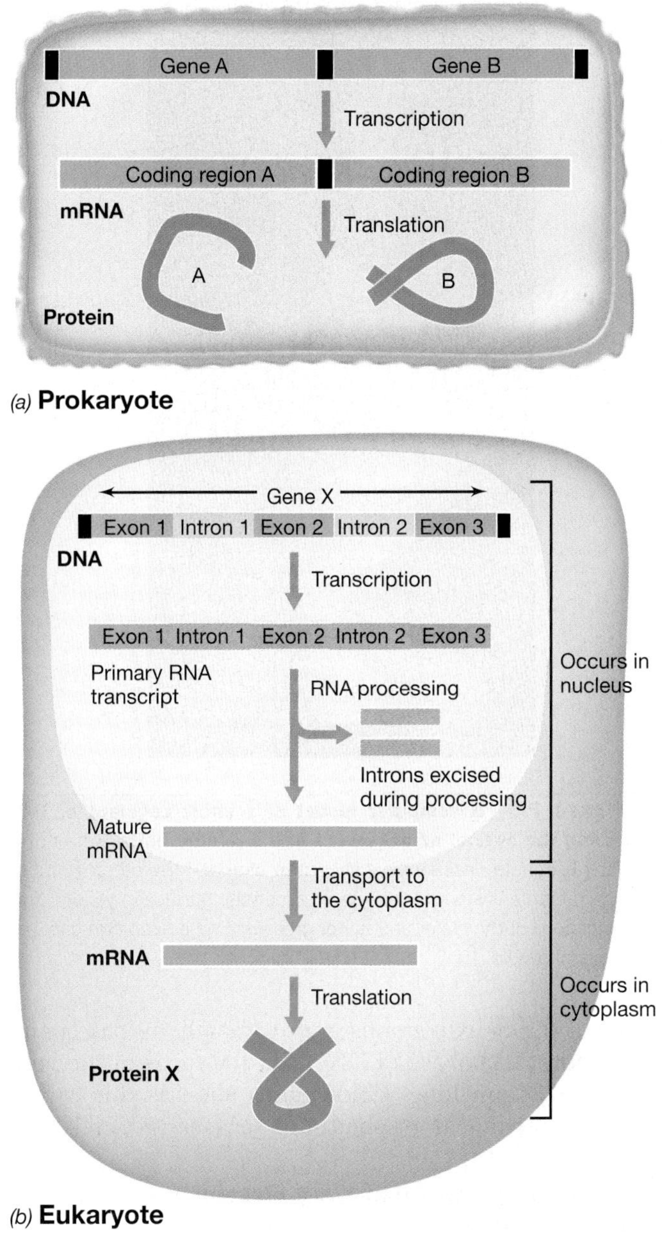

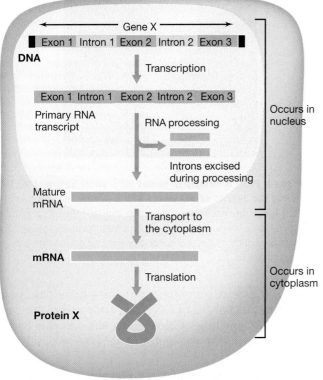

(a) **Prokaryote**

(b) **Eukaryote**

● **Figure 7.2 Contrast of information transfer in prokaryotes and eukaryotes.** (a) Prokaryote. A single mRNA often contains more than one coding region. (b) Eukaryote. Noncoding regions (*introns*) are removed from the primary RNA transcript before translation, leaving *exons* to be joined to form the mature mRNA. For details of RNA processing in eukaryotes, see Section 14.8.

▌▌ DNA STRUCTURE

We dealt with the structure of nucleic acids in a general way in Chapter 3. In the next few sections of this chapter we discuss the details of DNA structure including the types of genetic elements containing DNA that are found in cells. With this information as a foundation, we can then discuss how DNA is replicated, transcribed into RNA, and translated into protein.

7.2 DNA Structure: The Double Helix

Four nucleic acid bases are found in DNA: *adenine* (A), *guanine* (G), *cytosine* (C), and *thymine* (T). As already shown in Figure 3.11, the *backbone* of the DNA chain consists of alternating repeats of phosphate and the pentose sugar *deoxyribose*; connected to each sugar is one of the nucleic acid bases. Recall especially the numbering system for the positions of sugar and base. The phosphate connecting two sugars spans from the 3′-carbon of one sugar to the 5′-carbon of the adjacent sugar (see Figure 7.4). At one end of the DNA molecule the sugar has a phosphate on the 5′-hydroxyl, whereas at the other end the sugar has a free hydroxyl at the 3′-position.

DNA as a Double Helix

As we will discuss in Chapter 9, the chromosomes of some viruses are single-stranded. However, in all *cells*, DNA exists as two polynucleotide strands whose base sequences are **complementary**. The complementarity of DNA arises because of the specific pairing of the purine and pyrimidine bases: Adenine always pairs with thymine, and guanine always pairs with cytosine (Figure 7.3●). The two strands in the resulting *double-stranded* molecule are arranged in an **antiparallel** fashion (distinguished in Figures 7.3 and 7.4● as two shades of green). For example, in Figure 7.4 the strand on the left is arranged 5′ to 3′ (top to bottom), whereas the other strand is 5′ to 3′ (bottom to top).

The two strands are wrapped around each other forming a **double helix** (Figure 7.5●). In the double helix, DNA forms two distinct grooves, the *major groove* and the *minor groove*. Of the many proteins that interact

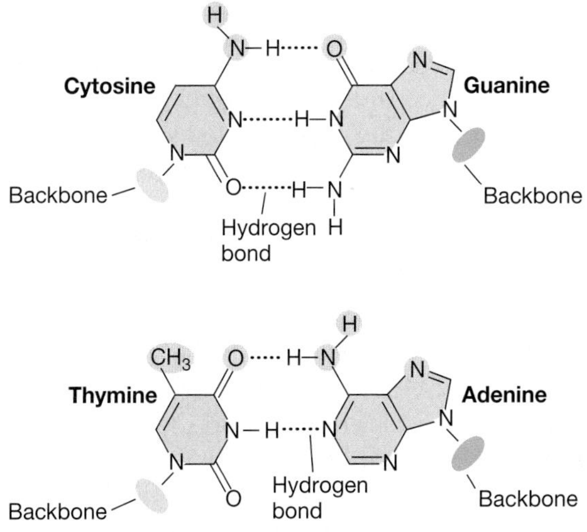

● **Figure 7.3 Specific pairing between adenine (A) and thymine (T) and between guanine (G) and cytosine (C) via hydrogen bonds.** These are the typical base pairs found in double-stranded DNA. Atoms that are found in the major groove of the double helix and that interact with proteins are highlighted in pink. The deoxyribose phosphate backbones of the two strands of DNA are also indicated. Note the different shades of green for the two strands of DNA, a convention used throughout this book.

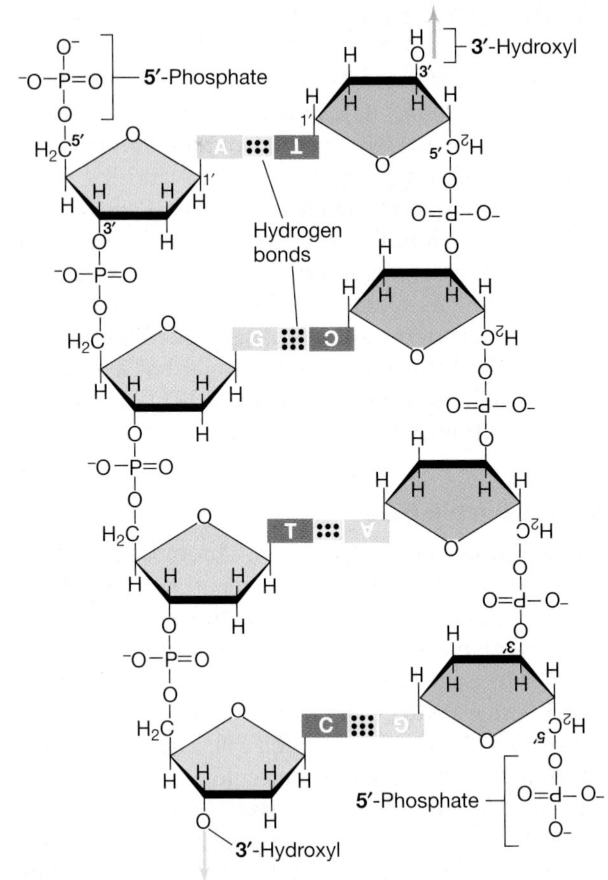

● Figure 7.4 DNA structure. Complementary and antiparallel nature of DNA. Note that one chain ends in a 5′-phosphate group, whereas the other ends in a 3′-hydroxyl. The red bases represent the pyrimidines cytosine (C) and thymine (T), and the yellow bases represent the purines adenine (A) and guanine (G).

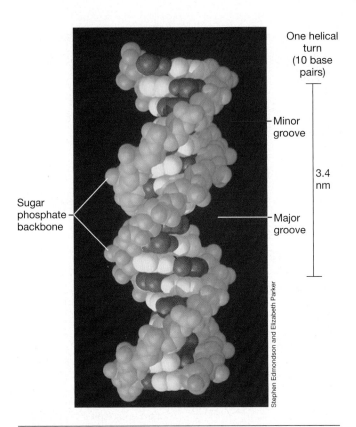

● Figure 7.5 A computer model of a short segment of DNA showing the overall arrangement of the double helix. One of the sugar-phosphate backbones is shown in blue and the other in green. The pyrimidine bases are shown in red and the purines in yellow. Note the locations of the major and minor grooves. One helical turn contains 10 base pairs.

specifically with DNA (as we shall see in Chapter 8), most engage predominantly with the *major* groove, where there is a considerable amount of space. Because of the regularity of the double helix, some atoms of the bases are always exposed in the major groove (and some in the minor groove). Key regions of nucleotides that are known to be important in interactions with proteins are shown in Figure 7.3.

Size of a DNA Molecule

The size of a DNA molecule can be expressed in terms of the *number of thousands* of nucleotide bases or base pairs per molecule. Thus, a DNA molecule with 1000 bases contains 1 *kilobase (kb)* of DNA. If the DNA is a double helix, then one speaks of *kilobase pairs (kbp)*. Thus, a double helix 5000 base pairs in size would be 5 kbp. The bacterium *Escherichia coli* has about 4640 kb pairs of DNA in its chromosome. Given the huge amount of genomic sequence information now becoming available, it is often useful to speak of *millions* of base pairs, or *megabase pairs (Mbp)*. The genome of *E. coli* is 4.64 Mbp.

In terms of actual length, each base pair takes up 0.34 nanometers (nm) in length along the helix, and each

turn of the helix contains approximately 10 base pairs. Therefore, 1 kilobase of DNA has 100 turns of the helix and is 0.34 μm long. Calculations such as this can be very interesting, as we shall soon see (see Section 7.3).

Inverted Repeats, Secondary Structure, and Stem-Loops

Long DNA molecules are quite flexible, but stretches of DNA less than 100 base pairs are more rigid. Some short segments of DNA can be bent by proteins that interact with them. However, certain sequences themselves result in bends in the DNA. The sequences of this *bent DNA* often involve several runs of five or six adenines (in the same strand), each separated by four or five other bases. In addition, some sequences contribute to the ability of DNA to bend when certain proteins interact with the DNA. DNA bending is one mechanism for the *regulation* of gene expression (the turning on and off of genes), as we shall discuss in Chapter 8.

Short, repeated sequences are often found in DNA molecules. Many proteins interact with regions of DNA containing repeated sequences (∞Chapter 8) that are repeated in inverse orientation. This type of repeat is called an **inverted repeat**. Inverted repeats give the DNA sequence a twofold symmetry. As shown in Figure 7.6●, nearby inverted repeats can lead to the formation of **stem-**

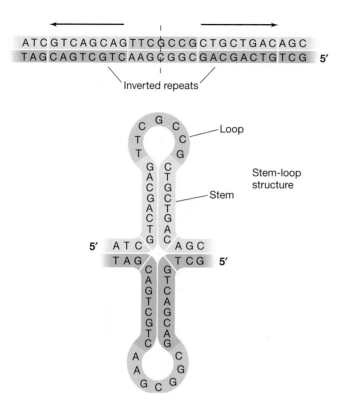

When isolated from cells and kept near room temperature and at physiological salt concentrations, DNA remains in a double-stranded form. However, if the temperature is raised, the hydrogen bonds will break but the covalent bonds holding a chain together will not, and so the two DNA strands will separate. This process is called *denaturation* (*melting*) and can be measured experimentally because single-stranded and double-stranded nucleic acids differ in their ability to absorb ultraviolet radiation at 260 nm (Figure 7.7●).

DNA with a high percentage of GC pairs melts at a *higher* temperature than a similar-sized molecule with more AT pairs. If the heated DNA is allowed to cool slowly, a process called *annealing*, the double-stranded native DNA can reform. Such a process can be used not only to reform native DNA but also to form *hybrid* molecules whose two strands come from different sources. **Hybridization**, the artificial construction of a double-stranded nucleic acid by complementary base pairing of two single-stranded nucleic acids, can be a powerful and useful technique in molecular biology (see Section 7.7).

● **Figure 7.6 Inverted repeats and the formation of a stem-loop structure.** (a) Nearby inverted repeats in DNA. The arrows indicate the symmetry around the imaginary axis (dashed line). (b) Formation of stem-loop structures by pairing of complementary bases on the *same* strand. Note how if RNA was formed from either strand of DNA, the RNA could form a single stem-loop.

7.2 Concept Check

DNA is a double-stranded molecule that forms a helical configuration and is measured in terms of numbers of base pairs. The two strands in the double helix are antiparallel, but inverted repeats allow for the formation of secondary structure. The strands of a double-helical DNA molecule can be denatured by heat and allowed to reassociate following cooling.

loop structures. These are short double-helical regions with normal base pairing and antiparallel strands.

The production of stem-loop structures in DNA may not be a widespread process in cells. However, the production of stem-loop structures in *RNA produced from DNA* following transcription is common. Such *secondary structure* (⬡Section 3.5) formed by base pairing within a single strand of nucleic acid is critical to the functioning of transfer RNA (see Section 7.15) and ribosomal RNA (⬡Section 11.5). Moreover, even if a stem-loop does not form in DNA, inverted repeats in DNA are often binding sites for specific DNA-binding proteins that function as transcriptional regulators (⬡Chapter 8).

The Effect of Temperature on DNA Structure

Although the hydrogen bonds between the base pairs are individually very weak (⬡Section 3.1), many such bonds in a long DNA molecule effectively hold the two strands together. There may be millions or even hundreds of millions of hydrogen bonds in a DNA molecule, depending on the number of base pairs present. Recall from Figure 7.3 that each adenine-thymine base pair has two such bonds, while each guanine-cytosine base pair has three. This makes GC pairs stronger than AT pairs.

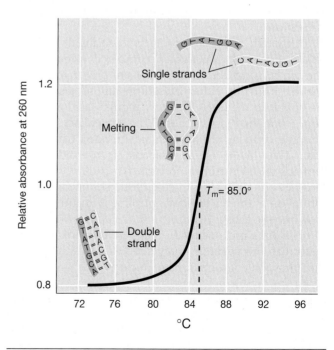

● **Figure 7.7 Thermal denaturation of DNA.** The DNA absorbs more ultraviolet radiation at 260 nm as the double helix is denatured. The transition is quite abrupt, and the temperature of the midpoint, T_m, is proportional to the GC content of the DNA. Although the denatured DNA can be renatured by a slow cooling, the process does not follow a similar curve. Renaturation becomes progressively more complete at temperatures well below the T_m, and then only after a considerable incubation time.

◆ Explain what the word *antiparallel* means as regards the structure of double-stranded DNA.

◆ Define the term *complementary* as it is used to refer to two strands of DNA.

◆ Define the terms, *denaturation*, *reannealing*, and *hybridization* as they apply to nucleic acids.

7.3   DNA Structure: Supercoiling

A *relaxed* DNA molecule is one with exactly the number of turns of the helix one would predict by knowing the number of base pairs. An example is shown in Figure 7.8●. However, if linearized, the *Escherichia coli* chromosome would be more than 1 mm in length, about 400 times longer than the *E. coli* cell itself! How is it possible to pack so much DNA into such a little space? The solution is the imposition of a kind of "higher order" structure on the DNA, in which the double-stranded DNA is further twisted in a process called *supercoiling*.

Figure 7.8 diagrams how supercoiling occurs in a circular DNA duplex. Supercoiling puts the DNA molecule under torsion, much like the added tension to a rubber band that occurs when it is twisted. DNA can be supercoiled in either a *positive* or *negative* direction. **Negative supercoiling** occurs when the DNA is twisted about its axis in the *opposite* direction from that of the right-handed double helix. It is in this form that supercoiled DNA is predominantly found in nature, with some interesting exceptions among the *Archaea* (see below)

Although there are proteins associated with the chromosomes of prokaryotes, compared to eukaryotes one can consider the chromosomes of prokaryotes to be "naked" DNA. In eukaryotes, large amounts of protein are bound to the DNA in a very regular fashion. As has been mentioned (see Section 7.1), each eukaryotic chromosome contains a *linear* double-stranded DNA molecule. This linear DNA molecule is wound around proteins called **histones**, to form structures called **nucleosomes** (Figure 7.9●). In eukaryotic chromosomes, the formation of the nucleosome introduces negative supercoils into the DNA.

Histones are positively charged proteins that can neutralize the negative charge of DNA (resulting from the phosphate groups). The nucleosomes are spaced along the double helix at very regular intervals, but can aggregate to form a fibrous material called *chromatin*. Chromatin itself can be further compacted by folding and looping to eventually form very dense structures called *chromosomes*. It is these chromosomes that are

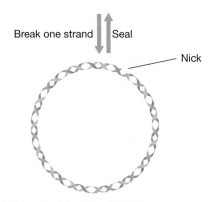

(a) Relaxed, covalently closed circular DNA

Break one strand ⇅ Seal

Nick

(b) Relaxed, nicked circular DNA

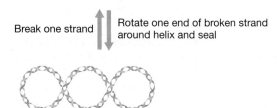

Break one strand ⇅ Rotate one end of broken strand around helix and seal

(c) Supercoiled circular DNA

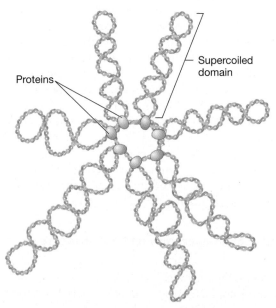

Proteins

Supercoiled domain

(d) Chromosomal DNA with supercoiled domains

● **Figure 7.8   Supercoiled DNA.** Parts (a), (b), and (c) show supercoiled circular DNA and relaxed, nicked circular DNA interconversions. A nick is a break in a phosphodiester bond of one strand. (d) In actuality, the double–stranded DNA in the bacterial chromosome is arranged not in one supercoil but in several *supercoiled domains*, as shown here. In *Escherichia coli* over 50 supercoiled domains are thought to exist, each of which is stabilized by binding to specific proteins.

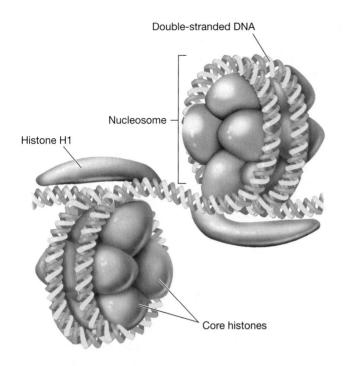

Double-stranded DNA

Nucleosome

Histone H1

Core histones

● **Figure 7.9 Packaging of DNA around a core of histone proteins to form a nucleosome.** Nucleosomes are arranged along the DNA strand somewhat like beads on a string. This arrangement is typical of DNA in eukaryotic cells. In eukaryotes, "core" histones are composed of eight proteins, two copies each of histones H2A, H2B, H3, and H4, and one copy of histone H1. A single nucleosome contains about 200 base pairs of DNA.

most easily visible during cell division in eukaryotic cells (∞Section 14.7).

Topoisomerases: DNA Gyrase

In *Bacteria* and most *Archaea*, there is an enzyme called **DNA gyrase**, which introduces negative supercoils into DNA. The process can be thought to occur in several stages. First, the circular DNA molecule is twisted, then a break occurs where the two chains come together, and then the broken double helix is resealed on the opposite side of the intact strand (Figure 7.10●). DNA gyrase is a type of enzyme called a *topoisomerase*, specifically a topoisomerase II. It is worth noting that some of the antibiotics that affect *Bacteria*, such as the quinolones (e.g., *nalidixic acid*), fluoroquinolones (e.g., *ciprofloxacin*), and *novobiocin*, inhibit the activity of DNA gyrase. Novobiocin is also effective against several species of *Archaea*, where it also seems to inhibit DNA gyrase.

Another enzyme is able to *remove* supercoiling in DNA. This enzyme, called *topoisomerase I*, introduces a single-strand break in the DNA and causes the rotation of one single strand of the double helix around the other. As was shown in Figure 7.8, a break in the backbone (a *nick*) of either strand allows the DNA to return to the relaxed state. However, to prevent the entire bacterial chromosome from becoming relaxed every time a nick is made, the chromosome contains *supercoiled domains* as shown in Figure 7.8*d*. A nick in the DNA in one of these domains does not relax DNA in the others. It is unclear what holds the DNA in these domains, but it is likely to involve specific proteins.

Through the activity of topoisomerases, the DNA molecule can be alternately supercoiled and relaxed. Because supercoiling is necessary for packing the DNA into the cell and relaxing is necessary for DNA to be replicated, these two complementary processes play an important role in the biology of the cell. In most prokaryotes, the extent of negative supercoiling is the result of a balance between the activity of DNA gyrase and topoisomerase I. In addition to replication, however, supercoiling is known to affect *gene expression*. Certain genes are more actively transcribed when DNA is supercoiled, whereas transcription of other genes is inhibited by excessive supercoiling.

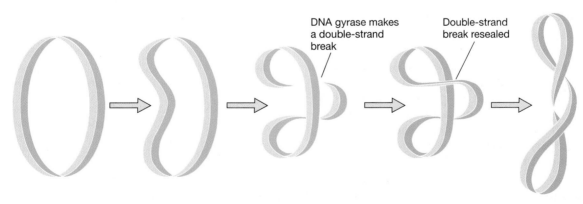

DNA gyrase makes a double-strand break

Double-strand break resealed

1. Relaxed circle

2. One part of circle is laid over the other

3. Result is contact between the helix in two places. Note that no twisting has as yet been introduced.

4. After DNA gyrase action, twisting (a negative supercoil) has been introduced.

5. Supercoiled DNA

● **Figure 7.10 DNA gyrase.** Introduction of supercoiling into a circular DNA by activity of DNA gyrase (topoisomerase II), which makes double-strand breaks. DNA gyrase introduces a *negative* supercoil into DNA.

Archaea and Reverse Gyrase

A few prokaryotes contain an enzyme called *reverse gyrase*. This topoisomerase introduces *positive* supercoils in DNA. The organisms that contain this enzyme are species of *Archaea* that grow at extremely high temperatures (hyperthermophiles, ∞Section 6.12 and Chapter 13). Interestingly, one example, the hyperthermophile *Methanothermus fervidus*, a member of the *Archaea*, also contains histone-like proteins, and forms nucleosome-like structures in which the DNA is *positively* supercoiled (∞Section 13.12).

Positive supercoiling is probably why DNA does not heat denature in hyperthermophiles. Supporting this hypothesis is the fact that with rare exception, reverse gyrase is only present in these very heat-loving organisms. However, as in all cells, in order to be of use, the genetic information in DNA must be accessible to the cell's replication and transcriptional machinery. Therefore, whether or not supercoiling is controlled by the activity of DNA gyrase or reverse gyrase, the structure of DNA within the cell likely remains quite dynamic.

7.3 *Concept Check*

The very long DNA molecule can be packaged into the cell because it is supercoiled. In prokaryotes this supercoiling is brought about by enzymes called topoisomerases. In eukaryotic chromosomes, DNA is wound around proteins called histones, forming structures called nucleosomes. DNA gyrase is a key enzyme in prokaryotes, introducing negative supercoils to the DNA. Reverse gyrase introduces positive supercoiling.

◆ Why is supercoiling important?

◆ What function do topoisomerases serve inside cells?

◆ How do the activities of DNA gyrase and reverse gyrase differ?

7.4 Chromosomes and Other Genetic Elements

Structures containing genetic material (DNA in most organisms but RNA in some viruses) are called *genetic elements*. The **genome** is the total complement of genes in a cell or virus. Although the main genetic element in prokaryotes is the **chromosome**, other genetic elements are known and play important roles in gene function in both prokaryotes and eukaryotes (Table 7.1).

The Chromosome

In Section 2.2 we discussed the fact that a typical prokaryote has a single chromosome containing all (or most) of the genes found inside the cell. Although a single chromosome is the rule among prokaryotes, there are exceptions. A few prokaryotes, both species of *Bacteria* and *Archaea*, contain two chromosomes (Table 7.2). Eukaryotes have multiple chromosomes as a part of their genome. Also, the typical prokaryotic chromosome is a circular DNA molecule, whereas the DNA in all known eukaryotic chromosomes is linear.

In Table 7.2 the number, size, and configuration of chromosomes in a few microorganisms, both prokaryotic and eukaryotic, are given. Note that the chromosome of the bacterium *Borrelia burgdorferi*, the causative agent of Lyme disease (∞Section 27.4), is linear. Although uncommon, linear chromosomes are now known to be present in several other *Bacteria*, including the important antibiotic-producing genus *Streptomyces* (∞Sections 12.24 and 30.5). More detailed information on the sizes and characteristics of microbial genomes is given in Chapter 15.

The prokaryotic examples listed in Table 7.2 are not random. They include species of *Archaea* as well as *Bacteria* and examples of the smallest and some of the largest known prokaryotic chromosomes. Only a few examples of eukaryotic microorganisms are also given in Table 7.2. The linear DNA molecule in a eukaryotic chromosome has a special DNA sequence called a *telomere* at each end and a *centromere* located between the telomeres. Centromeres are important for partitioning the chromosomes during cell division. As we shall see in Chapter 14 (Eukaryotic Microorganisms), telomeres play an important role in the replication of linear DNA molecules (∞Section 14.6 and Figure 14.12).

Eukaryotes typically contain much more DNA than is needed to encode all the proteins required for cell function. For instance, in the human genome only about 3% of the total DNA encodes protein, whereas in *Bacteria* this fraction often exceeds 90%. The "extra" DNA in eu-

Table 7.1	Kinds of genetic elements	
Organism	**Element**	**Description**
Prokaryote	Chromosome	Extremely long, usually circular, double-stranded DNA molecule
	Plasmid	Typically a relatively short, usually circular, double-stranded DNA molecule, which is extrachromosomal
Eukaryote	Chromosome	Extremely long, linear, double-stranded DNA molecule
	Plasmid[a]	Typically a relatively short circular or linear double-stranded DNA molecule, which is extrachromosomal
All Organisms	Transposable elements	Double-stranded DNA molecule always found within another DNA molecule
Mitochondrion or chloroplast	Chromosome	Intermediate-length DNA molecules, usually circular
Virus	Genome	Single- or double-stranded DNA or RNA molecule

[a]Plasmids are uncommon in eukaryotes.

Table 7.2 Sizes, shapes, and numbers of chromosomes in selected microorganisms from each domain of life

Organism	Comments	Chromosome Size (Mbp)[a]	Number	Geometry
Bacteria				
Mycoplasma genitalium	Smallest known cellular genome in Bacteria	0.58	1	
Borrelia burgdorferi	Causes Lyme disease (◯Chapter 27)	0.91[b]	1	
Haemophilus influenzae	Gram-negative, can cause disease (◯Chapter 26)	1.83[c]	1	
Rhodobacter sphaeroides	Gram-negative, phototrophic	4.00[d]	2	
Bacillus subtilis	Gram-positive, genetic model	4.21	1	
Escherichia coli K-12	Gram-negative, genetic model	4.64[e]	1	
Streptomyces coelicolor	Actinomycete, produces antibiotics (◯Chapter 12)	8.66	1	
Archaea				
Nanoarchaeum equitans	Parasite of Ignicoccus (◯Chapters 13 and 15)	0.49	1	
Methanococcus jannaschii	Methanogen, which grows at high temperature (◯Chapters 6 and 13)	1.66	1	
Pyrococcus abyssi	Grows at high temperature (◯Chapters 6 and 13)	1.77	1	
Halobacterium sp. NRC1	Grows in high salt (◯Chapters 6 and 13)	2.57[f]	3	
Sulfolobus solfatarius	Grows at high temperature and high acidity (◯Chapters 6 and 13)	2.99	1	
Eukarya[g]				
Giardia lamblia	Flagellated protozoan that causes acute gastroenteritis (◯Chapter 14)	12.00	4	
Saccharomyces cerevisiae	Yeast, widely used in science and industry (◯Chapters 14 and 30)	12.06[h]	16	
Dictyostelium discoideum	Cellular slime mold, developmental model (◯Chapter 14)	34.0	6	
Tetrahymena thermophila	Ciliated protozoan (◯Chapter 14)	210.0	5	

[a] Mbp is megabase pairs. In the case of the *Eukarya* the genome sizes and chromosome number are for the haploid form. (All the prokaryotic genomes listed have been completely sequenced.)

[b] This is for the linear chromosome. The genome of this organism also contains at least 17 circular and linear plasmids, which themselves have a combined size of more than 0.5 Mbp.

[c] *Haemophilus influenzae* strain Rd was the first cellular organism to have its genome entirely sequenced.

[d] Chromosome I is 3.1 Mbp and chromosome II is 0.90 Mbp. The strain sequenced also contains five plasmids.

[e] The reported sequence does not contain that of the F-plasmid (◯Section 10.9) nor that of bacteriophage lambda (◯Section 9.11), both of which would be present in a typical K-12 strain.

[f] There is one large chromosome of 2.01 Mbp and two minichromosomes of 0.19 Mbp and 0.37 Mbp.

[g] All the organisms listed are single-celled.

[h] *Saccharomyces cerevisiae* was the first eukaryote to have its genome completely sequenced. The number given here does not include the mitochondrial genome and all the copies of some repeated sequences.

karyotes is present as introns (see Figure 7.2) or repetitive sequences, some repeated hundreds of thousands of times. Eukaryotic microorganisms have fewer introns than do "higher" eukaryotes. For example, about 70% of DNA in the yeast *Saccharomyces cerevisiae* encodes protein. Eukaryotes also often have multiple copies of certain genes, such as those that encode transfer RNAs and ribosomal RNAs. These latter may also be found repeated in prokaryotes but usually only a few copies are present at most.

Nonchromosomal Genetic Elements

Besides the chromosome, a number of other genetic elements are known, referred to collectively as *nonchromosomal genetic elements*. These include *viruses, plasmids, organellar genomes,* and *transposable elements.*

Viruses contain genomes, either DNA or RNA, that control their own replication and transfer from cell to cell. The viral genome is also called a *chromosome*. However, it contains genes essential to the virus, not the host cell, and is therefore clearly functionally distinct from cellular chromosomes. Both linear and circular viral chromosomes are known. Viruses are of special interest because they often cause disease. We discuss viruses in Chapters 9 and 16, and viral diseases in Chapters 26 and 27.

Plasmids are genetic elements that exist and replicate separately from the chromosome. The great majority of plasmids are double-stranded DNA, and although most plasmids are circular, some are linear. Plasmids differ from viruses in two ways: (1) they do not cause cellular damage (generally they are beneficial), and (2) plasmids do not have extracellular forms, whereas viruses do. Although plasmids have been recognized in only a few eukaryotes, they have been found in most prokaryotic species and can be of profound importance to the biology of the organism. We discuss plasmids in Chapter 10.

(see Table 7.3) that binds to this region and opens up the helix to expose the strands to the replication machinery. Initiation of DNA replication then begins on the two single strands. As replication proceeds, the site of replication, called the **replication fork**, appears to move down the DNA (Figure 7.14●).

There are five different DNA polymerases in *Escherichia coli*, called *DNA polymerases* I, II, III, IV, and V. It is *DNA polymerase III* (Pol III) that is the primary enzyme of replication at the replicating forks. However, several other non-polymerase enzymes are also involved (Table 7.3). For example, at the replication fork, the DNA double helix is unwound and a small single-stranded region is exposed by the activity of specific proteins called *helicases*. Helicases are ATP-dependent enzymes that move down the helix and separate the strands in advance of the replicating fork (Figures 7.14 and 7.15●). The single-stranded region generated is complexed with a special protein, the *single-strand binding protein*, which stabilizes the single-stranded DNA, preventing the formation of intrastrand hydrogen bonds and reversion to the helix.

Note in Figure 7.14 an important distinction in replication of the two strands, which arises from the fact that DNA replication always proceeds from 5′-phosphate to 3′-hydroxyl (always adding a *new* nucleotide to the 3′-OH of the growing chain). On the strand growing

Table 7.3	Major enzymes involved in DNA replication in *Bacteria*	
Enzyme	**Encoding genes**	**Function**
DNA polymerase III	*polC; dnaE,Q,N,X; holA–E; mutD*	Main polymerizing enzyme
DNA polymerase I	*polA*	Excises RNA primer and fills in gaps
Helicase	*dnaB*	Unwinds helix at the replication fork
Primase	*dnaG*	Primes new strands of DNA
Origin-binding protein	*dnaA*	Binds to origin of replication site; facilitates melting to open the complex
Single-strand binding protein	*ssb*	Prevents opened helix from annealing
DNA ligase	*ligA, lig B*	Seals nicks in DNA

from the 5′-phosphate to the 3′-hydroxyl, called the **leading strand**, DNA synthesis occurs *continuously* because there is always a free 3′-OH at the replication fork to which a new nucleotide can be added. But on the opposite strand, called the **lagging strand**, DNA synthesis occurs *discontinuously* (because there is no 3′-OH at the replication fork to which a new nucleotide can attach). Where is the 3′-OH on this strand? At the *opposite* end, *away* from the replication fork. Therefore, on the lagging strand, an RNA primer must be synthesized by primase

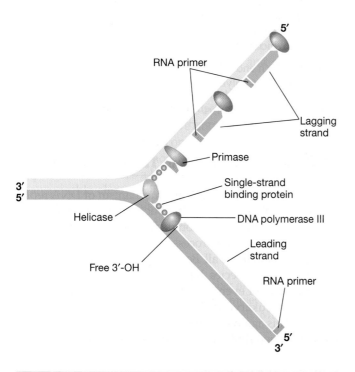

● **Figure 7.14 Events at the DNA replication fork.** Note the polarity and antiparallel nature of the DNA strands. The substrates for primase are ribonucleotide triphosphates, while for DNA polymerase, they are *deoxy*ribonucleotide triphosphates. Single-strand binding protein keeps the template strands from reforming a helix following their separation by helicase. Compare with Figure 7.19, which shows the replisome protein complex.

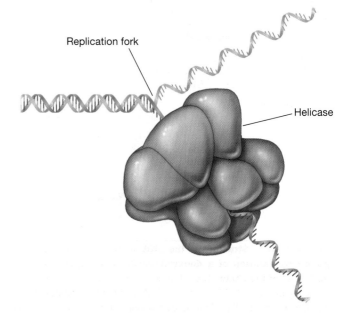

● **Figure 7.15 DNA helicase unwinding a double helix.** In this figure, the protein and DNA molecules are drawn to scale. Simple diagrams are often used to give an idea of molecular events in the cell, but they may give the incorrect impression that most proteins are relatively small compared to DNA. This scale model shows otherwise.

multiple times to provide free 3'-OH groups. By contrast, the leading strand is primed only once, at the origin.

After synthesizing the primer, primase is replaced by Pol III. This enzyme is a complex of ten proteins, including the polymerase core complex itself. Other proteins include those that function as a sliding clamp to maintain the polymerase on each template strand of DNA as well as proteins that assist in replisome (see later) assembly. Pol III adds deoxyribonucleotide triphosphates until it reaches previously synthesized DNA (Figure 7.16●). At this point, Pol III stops. The next enzyme that is involved, *DNA polymerase I* (Pol I), has more than one enzymatic activity. Besides synthesizing DNA, Pol I has a 5' → 3' *exonuclease* activity that removes the RNA primer preceding it (Figure 7.16). When the primer has been removed and replaced with DNA, Pol I is released. The last phosphodiester bond is made by an enzyme called *DNA ligase*.

This enzyme can seal nicks in DNAs that have a 5'-phosphate and 3'-OH, and along with Pol I, is also involved in DNA repair.

Bidirectional Chromosome Replication and the Replisome

In *Escherichia coli*, and probably in all prokaryotes that contain a circular chromosome, replication is *bidirectional* from the origin of replication, as shown in Figures 7.17● and 7.18●. There are thus *two* replication forks on each chromosome, replicating in opposite directions. In circular DNA, bidirectional replication leads to the formation of characteristic structures called *theta structures* (Figure 7.17). Most large DNA molecules, whether from prokaryotes or eukaryotes, have bidirectional replication from fixed origins. In fact, a single eukaryotic chromosome has many origins. Close inspection of the replicating DNA shows that synthesis is occurring in both a leading and lagging manner *on each template strand* (Figure 7.18).

Bidirectional DNA synthesis along a circular chromosome allows DNA replication to occur as rapidly as

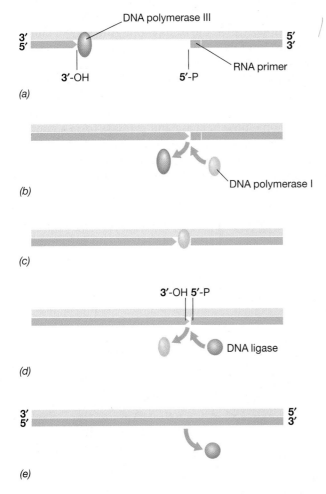

● **Figure 7.16 Sealing two fragments on the lagging strand.** (a) DNA polymerase III is synthesizing DNA in the 5' → 3' direction toward the RNA primer of a previously synthesized fragment on the lagging strand. (b) On reaching the fragment, DNA polymerase I replaces III. (c) DNA polymerase I continues synthesizing DNA while removing the RNA primer from the previous fragment. (d) DNA ligase replaces DNA polymerase I after the primer has been removed. (e) DNA ligase seals the two fragments together.

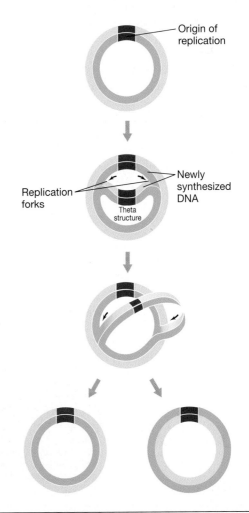

● **Figure 7.17 Replication of circular DNA: the theta structure.** In circular DNA, bidirectional replication from an origin leads to the formation of replication intermediates resembling the Greek letter theta (Θ).

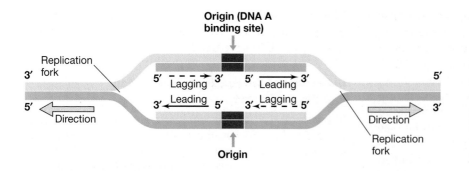

● **Figure 7.18 Dual replication forks in the circular chromosome.** At an origin of replication that directs bidirectional replication, two replication forks must start. Therefore, two leading strands must be primed, one in each direction. In *Escherichia coli*, the origin of replication is identified by the binding of a specific protein, Dna A (see Table 7.3).

possible. Even taking this into account and considering that Pol III can add nucleotides to a growing DNA strand at the rate of about 1000 per second, chromosome replication in *E. coli* still takes about 40 minutes. Under the best growth conditions, *E. coli* can grow with a doubling time of about 20 minutes. However, under these conditions, chromosome replication still takes 40 minutes. The solution to this conundrum is that cells of *E. coli* growing at doubling times shorter than 40 minutes contain *multiple DNA replication forks*. That is, a new round of DNA replication begins before the last round has been completed. Only in this way can a generation time shorter than the chromosome replication time be maintained.

The Replisome

While DNA synthesis is continuing at the replication fork, changes in the coiling of the DNA are occurring, modified by the activities of topoisomerases (see Section 7.3). Unwinding is an essential feature of DNA replication, and because supercoiled DNA is under tension, it unwinds more easily than DNA that is not supercoiled. Thus, by regulating the degree of supercoiling, topoisomerases

regulate the process of replication (and also transcription, as discussed later).

Figure 7.14 shows the differences in replication of the leading and the lagging strands and the various enzymes involved. From such a simplified drawing it would appear that each replication fork contains a host of different proteins all working independently. Actually, this is not the case. The proteins involved in this process are highly dynamic and aggregate to form a large complex called the *replisome* (Figure 7.19●). The lagging strand of DNA loops out to allow the replisome to move smoothly on both strands, and the replisome literally pulls the DNA template through it as replication occurs (Figure 7.19). Therefore, it is the *DNA*, and not DNA polymerase, that moves during replication. Note also in the replisome how helicase and primase form a complex (called the *primosome*) since they must work in close association during the replication process (Figure 7.19).

In summary, the replisome contains, besides DNA polymerase, several key replication proteins: (1) DNA gyrase, to remove supercoils; (2) DNA helicase/primase (the primosome), to unwind and prime the DNA; and

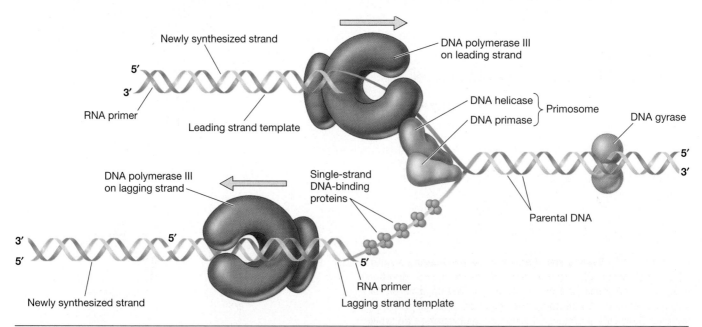

● **Figure 7.19 The replisome.** The replisome consists of two copies of DNA polymerase III, helicase and primase (together forming the primosome), and many copies of single-strand DNA-binding protein. Just upstream of the replisome DNA gyrase is removing supercoils in the DNA to be replicated. Note how the lagging-strand template loops out to keep the components of the replisome away from each other and moving forward smoothly.

(3) Single-strand binding protein, to prevent the separated template strands from re-forming a double helix (Figure 7.19). Table 7.3 summarizes the properties of the essential proteins involved in DNA replication.

Fidelity of DNA Replication: Proofreading

Errors in DNA replication introduce *mutations*, changes in DNA sequence. Mutation rates in cells are remarkably low, between 10^{-8} and 10^{-11} errors per base pair inserted. This accuracy is possible partly because DNA polymerases get *two* chances to incorporate the correct base at a given site. The first chance occurs when complementary bases are inserted by base-pairing rules, A with T and G with C, using the template strand as the pattern. The second chance occurs because of a second enzymatic activity of both Pol I and III, called **proofreading** (Figure 7.20●).

How does proofreading work? In addition to inserting nucleotides in the replicating strand, Pol I and III also contain a $3' \rightarrow 5'$ *exonuclease* (*exo* means "end") activity that can remove a misinserted nucleotide and replace it with the correct nucleotide. Proofreading activity occurs if an incorrect base has been inserted because misinsertion creates a mismatch in base pairing. The polymerase is signaled to this problem by virtue of the fact that a misinserted nucleotide is unable to form the correct (most stable) hydrogen-bonding pattern with its mismatched complement. This proofreading activity gives the polymerase activity a second chance to insert the correct base (Figure 7.20).

Proofreading exonuclease activity is distinct from the $5' \rightarrow 3'$ exonuclease activity of Pol I that is used to remove the RNA primer from both the leading and lagging strands (Figure 7.16). Only Pol I has this latter activity. Exonuclease proofreading occurs in prokaryotes, eukaryotes, and viral DNA replication systems. However, many organisms have other mechanisms for reducing errors made during DNA replication, and we will discuss some of these in Chapter 10.

Termination of Replication

Details about the process of termination of the replicating forks are not completely known. However, it is clear that certain DNA sequences and several proteins are involved in this process. When the replication of the circular molecule is complete (Figure 7.17), the two circular molecules are linked together, much like the links of a chain. These can be unlinked by a topoisomerase. In Chapter 6 we outlined how cell wall synthesis and DNA replication is coupled with cell division (∞Section 6.2). Obviously, it is critical that, after DNA replication, the DNA is partitioned so that each daughter cell has a copy of the chromosome. This process may be assisted by the important cell division protein *FtsZ*, a critical protein that helps orchestrate several key events in the cell division process (∞Section 6.2).

 7.6 Concept Check

DNA synthesis begins at a unique location called the origin of replication. The double helix is unwound by helicase and is stabilized by single-strand binding protein. Extension of the DNA occurs continuously on the leading strand but discontinuously on the lagging strand. Most errors in base pairing are corrected by proofreading functions associated with the activities of DNA polymerases.

◆ Why are there *leading* and *lagging* strands?

◆ What is the *replisome* and what are its components?

◆ Why is proofreading so important to the cell?

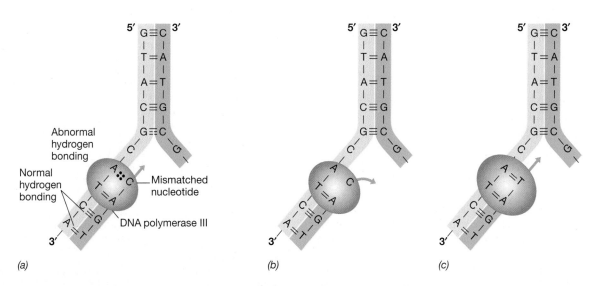

(a)   (b)   (c)

● **Figure 7.20 Proofreading by the 3′ → 5′ exonuclease activity of DNA polymerase III.** (a) A mismatch in base pairing at the terminal base pair causes the polymerase to pause briefly. This is a signal for the proofreading activity (b) to excise the mismatched nucleotide, after which the correct base is incorporated (c) by polymerase activity.

IV TOOLS FOR MANIPULATING DNA

7.7 Restriction Enzymes and Hybridization

Prokaryotic cells typically contain one or more enzymes that can chemically modify DNA. A major class of such enzymes are the *restriction endonucleases*, or **restriction enzymes** for short. Although such systems are widespread among prokaryotes (both *Bacteria* and *Archaea*), they are very rare in eukaryotes. Restriction enzymes recognize certain sequences of DNA and cut the DNA. Besides playing important roles in the cells, restriction enzymes are essential for *in vitro* DNA manipulation, and their discovery gave birth to the field of *genetic engineering* (⚭Chapter 31).

Mechanism of Restriction Enzymes

A major class of restriction endonucleases are the *type II restriction enzymes*. Most of the DNA sequences recognized by type II restriction enzymes exhibit twofold symmetry around a given point. Figure 7.21● shows the sequence that is recognized and cleaved by the restriction endonuclease from *Escherichia coli* called **EcoRI** (this acronym stands for *Escherichia coli*, restriction enzyme **I**). The cleavage sites are indicated by arrows and the axis of symmetry by a dashed line. Note that the two strands have the same sequence if one is read from the left and the other from the right (or, in terms of polynucleotide strands, if both are read $5' \to 3'$ or both read $3' \to 5'$). Such a sequence pattern is called a **palindrome**. (The term *palindrome* is derived from the Greek meaning "to run back again.")

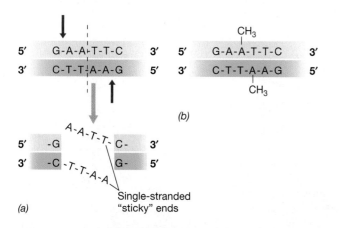

(a)

(b)

CH₃

5' G-A-A-T-T-C 3' 5' G-A-A-T-T-C 3'
3' C-T-T-A-A-G 5' 3' C-T-T-A-A-G 5'

CH₃

5' -G   A-A-T-T- C- 3'
3' -C -T-T-A-A   G- 5'

Single-stranded "sticky" ends

● **Figure 7.21 Restriction and modification of DNA.** (a) (Top panel) The sequence of DNA recognized by the restriction endonuclease *Eco*RI. The red arrows indicate the bonds cleaved by the enzyme. The dashed line indicates the axis of symmetry of the sequence. (Bottom panel) Appearance of DNA after cutting with restriction enzyme *Eco*R1. Note the single-stranded "sticky ends" (⚭Section 10.15). (b) The same sequence after modification by the *Eco*RI methylase. The methyl groups added by this enzyme are shown and these protect the restriction site from cutting by *Eco*R1.

Most restriction enzymes are homodimeric proteins composed of two identical polypeptide subunits, and each subunit recognizes and cuts the DNA on one of the strands. Since the sequences recognized by restriction enzymes are typically short and palindromic, such enzymes typically make *double-stranded* breaks, leaving short, single-stranded regions of DNA free instead of severing the DNA and yielding blunt ends (Figure 7.21). The recognition sequences and cut sites for a few restriction enzymes are given in Table 7.4.

Well over 2000 restriction enzymes with over 200 different specificities are known. Others are being sought because these enzymes are of great importance in genetic research. Consider that the enzyme *Eco*RI can cut *any* double-stranded DNA that has its recognition sequence and cuts only at that sequence. For example, a specific 6 base-pair sequence, such as that cut by the restriction enzyme *Eco*RI (Figure 7.21), should appear in a genome about once every 4096 nucleotides (four bases taken six at a time, 4^6) assuming that the DNA has a "random" sequence. As we will explain in Chapters 10 and 31, such fragments with defined termini have many uses, especially in gene-cloning technologies. Restriction enzymes are such important tools in modern molecular genetic research that they have become widely available commercially.

Table 7.4	Recognition sequences of a few restriction endonucleases	
Organism	**Enzyme designation**[a]	**Recognition sequence**[b]
Bacillus globigii	*Bgl*II	A↓GATCT
Bacillus subtilis	*Bsu*RI	GG↓CC
Brevibacterium albidum	*Bal*I	TGG↓CCA
Escherichia coli	*Eco*RI	G↓AATTC[c]
Haemophilus haemolyticus	*Hha*I	GCG↓C
Haemophilus influenzae	*Hind*II	GTPy↓PuAC
Haemophilus influenzae	*Hind*III	A↓AGCTT
Klebsiella pneumoniae	*Kpn*I	GGTAC↓C
Nocardia otitidiscaviarum	*Not*I	GC↓GGCCGC
Proteus vulgaris	*Pvu*I	CGAT↓CG
Serratia marcescens	*Sma*I	CCC↓GGG
Thermus aquaticus	*Taq*I	T↓CGA

[a] Nomenclature: The first letter of the three letter abbreviation of a restriction endonuclease designates the genus from which the enzyme originates, the second two letters, the species. The roman numeral designates the order of discovery of enzymes in that particular organism, and any additional letters are strain designations.

[b] Arrows indicate the sites of enzymatic attack. Asterisks indicate the site of methylation (modification). G, guanine; C, cytosine; A, adenine; T, thymine; Pu, any purine; Py, any pyrimidine. Only the $5' \to 3'$ sequence is shown.

[c] See Figure 7.21*a*

Modification: Protection from Restriction

The major function of restriction enzymes in prokaryotes is probably to protect the cell from invasion by foreign DNA, for example, viral DNA. If such DNA enters the cell, the cell's restriction enzyme(s) will destroy it. However, a cell must protect its own DNA from inadvertent destruction by its own restriction enzyme(s), and such protection is conferred by a **modification system**. This consists of a modifying enzyme, which chemically **modifies** the specific sequences on its *own* DNA so these sequences are not attacked by the cell's own restriction enzymes. Such modification typically involves *methylation* of specific bases within the recognition sequence so that the restriction nuclease can no longer bind. For example, the sequence recognized by the *Eco*RI restriction enzyme (Figure 7.21*a*) can be modified by methylation of the two most interior adenines (Figure 7.21*b*). The enzyme that performs this modification is called *Eco*RI *methylase*. If even a single strand is modified, the sequence is no longer a substrate for the restriction enzyme *Eco*RI.

Electrophoresis and Restriction Analysis of DNA

Because the base sequences recognized by many restriction enzymes are four to six nucleotides long (Table 7.4), there will generally be only a limited number of such sequences in a single DNA molecule (for example, see Figure 7.23). After cleaving the DNA, the fragments generated can be separated from each other by **gel electrophoresis** and analyzed.

Electrophoresis is a procedure by which charged molecules migrate in an electrical field, the rate of migration being determined by the charge on the molecule and by its size and shape. In gel electrophoresis (Figure 7.22*a*●) the molecules are separated in an agarose gel. When an electrical current is applied, nucleic acids (which are negatively charged) will migrate through the gel, and small or compact molecules migrate more rapidly than large molecules. After a certain period of time the gel can be "stained" with a compound that binds to DNA, such as *ethidium bromide* (◌◌Section 10.3), and the DNA will then fluoresce under ultraviolet light (Figure 7.22*b*).

A typical agarose gel is shown in Figure 7.22*b*. In lane A a restriction enzyme digestion of a standard DNA sample has been applied, and the size of each band is thus known. The other lanes contain a purified source of DNA (in this case a plasmid) cut by one or more restriction enzymes. Because a given restriction enzyme always cuts at the same site, the banding pattern of a given DNA is reproducible. By comparison with the standard, the size of the fragments can therefore be determined. Such a technique can be used to generate what is called a **restriction map** of the DNA. Combining restriction enzyme digestion of a single DNA molecule with fluorescent microscopy allows a restriction map to be made directly

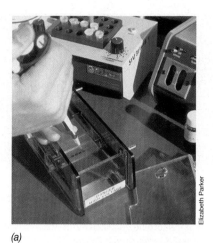

(a)

(b)

● **Figure 7.22 Agarose gel electrophoresis of DNA.** (a) DNA samples are loaded into wells in a submerged agarose gel. (b) A photograph of a stained agarose gel. The DNA has been loaded into wells toward the top of the gel (negative pole) as shown, and the positive pole of the electrical field is at the bottom. The sample in lane A is used as a standard where the size of the fragments was known. Using the standards, one can determine the sizes of the fragments in the other lanes. Bands stain less intensely at the bottom of the gel because the fragments are smaller and chemically there is less DNA to stain.

from the pattern of cuts along the DNA (Figure 7.23●). This has been called *optical mapping*.

Restriction analyses have many uses in microbiology, primarily in comparative studies of two or more DNAs. This includes studies in the classification of microorganisms. For example, the banding patterns generated from restriction analyses of either whole chromosomes or specific genes from a series of organisms can yield an indication of their genetic relationships (◌◌Section 11.11). Multiple restriction digests, either using different enzymes one at a time or in combination, can generate patterns that allow one to readi-

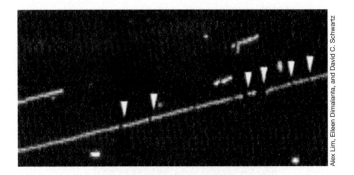

● **Figure 7.23** **Optical mapping of restriction fragments.** The figure shows a digital fluorescence micrograph of a portion of the *Escherichia coli* chromosome digested with the restriction enzyme XhoI (isolated from *Xanthomonas holica*). Sites of cutting by the restriction enzyme are indicated by arrows. Although the DNA strand itself is too small to be seen by light microscopy, *fluorescence* of the DNA is visible. The length of the DNA shown here is about 260 kb pairs. Optical mapping has been used to create restriction maps of the complete chromosomes of *E. coli* and other *Bacteria* such as *Deinococcus radiodurans*.

ly interpret the relative relationships between different DNA molecules.

Nucleic Acid Hybridization and the Southern Blot

DNA fragments can be purified from gels and used for many purposes including *nucleic acid hybridization*, or **hybridization** for short. Recall from Section 7.2 that when DNA is denatured (i.e., made single-stranded), it can be used to form hybrid molecules with other single-stranded DNA (or RNA) molecules having complementary (or almost complementary) base sequences. These latter molecules are called **nucleic acid probes** or, simply, **probes**.

Hybridization can be very useful for finding similar sequences from different genetic elements, or to find the location of a specific gene. Techniques are available for hybridizing probes to DNA fragments that have been separated by gel electrophoresis, and a very common one is the **Southern blot**, named for its inventor, E.M. Southern.

In a Southern blot the DNA fragments in the gel are transferred to a membrane and then denatured to yield single strands. The membrane is then exposed to a labeled probe to allow hybrids to form if the probe is complementary to any of the fragments (the fragments are fixed on the membrane and are thus prevented from reannealing with themselves). Hybridization can be detected by assaying for the label that has become bound to the membrane. Probes can be made radioactive or labeled with agents that yield colored products or that fluoresce. The hybridization procedure when *DNA* is in the gel and RNA or DNA is the probe is called a *Southern blot*. By contrast, when *RNA* is in the gel and DNA or RNA is the probe, the procedure is called a *Northern blot*. Figure 7.24● shows how a Southern blot can be used to identify fragments of DNA containing sequences that hybridize to the probe.

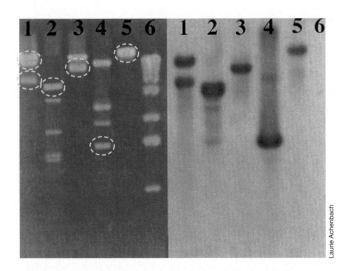

● **Figure 7.24** **Southern blotting.** (Left panel) Agarose gel electrophoresis of DNA molecules. Purified molecules of DNA from several different plasmids were treated with restriction enzymes and then subjected to electrophoresis. (Right panel) Southern blot of the DNA gel shown to the left. After blotting, hybridization with a radioactively labeled probe was carried out. The positions of the bands have been detected by X-ray autoradiography. Note that only some of the DNA fragments (circled in yellow) have sequences complementary to the labeled probe. Lane 6 contained DNA used as a size marker and none of the bands hybridized to the probe.

⬡ **7.7** **Concept Check**

Restriction enzymes recognize specific short sequences in DNA and make breaks in the DNA. The products of restriction enzyme digestion can be separated using gel electrophoresis and complementary sequences detected by hybridization.

◆ Why are *restriction enzymes* useful to the geneticist?

◆ What is a *Southern blot* and what does it tell you?

7.8 **Sequencing and Synthesizing DNA**

In addition to restriction enzyme and hybridization analyses, another common analysis of DNA is **sequencing**, determining the precise order of nucleotides in a DNA molecule. To assist in sequencing and also for hybridization studies, procedures are available to synthesize DNA molecules with defined sequences for use as *primers* and *probes*. Such short fragments of DNA (typically 10–20 nucleotides) are called *oligonucleotides*. Along with restriction enzymes (see Section 7.7), nucleic acid sequencing is a key tool for molecular biology, and we describe the essentials of the sequencing process here.

DNA Sequencing

Most DNA sequencing today is done by the *Sanger dideoxy* method. This method generates DNA fragments that terminate at each of the four bases and that are labeled in some way, for example, with radioactivity. These fragments are then subjected to gel electrophoresis so that molecules with one-nucleotide difference in length are separated on the gel. The procedure requires four separate reactions (and four separate gel lanes) for each sequence determination, one for fragments ending at each of the four bases in DNA: *adenine, guanine, cytosine,* and *thymine.* The positions of the fragments are located by autoradiography (or by fluorescent probes, see later), and the sequences can then be read off the gel.

In the Sanger procedure the sequence is actually determined by making a *copy* of the single-stranded DNA, using the enzyme *DNA polymerase.* As we have previously seen, this enzyme uses deoxyribonucleoside triphosphates as substrates and adds them to a *primer* (∞Section 7.6). In each of the incubation mixtures (four separate test tubes) are small amounts of a different *dideoxy analog* of the deoxyribonucleoside triphosphates (Figure 7.25●). Because the dideoxy sugar lacks the 3'-hydroxyl, when it is inserted, continuation of the chain cannot occur. The dideoxy analog thus acts as a specific *chain-termination reagent.* Oligonucleotide fragments of variable length are obtained, depending on the incubation conditions. Electrophoresis of these fragments is then carried out, and the positions of the bands are determined by exposing X-ray film or by fluores-

cence. By aligning the four dideoxynucleotide lanes and noting the vertical position of each fragment relative to its neighbor, the sequence of the DNA copy can be read directly from the gel (Figure 7.26●).

RNA and Automated Sequencing

The Sanger method can be used to sequence RNA as well as DNA. To sequence RNA, a single-stranded *DNA copy* is first made (using the RNA as the template) by an enzyme found in certain viruses called *reverse transcriptase* (∞Section 9.13). By making the single-stranded DNA in the presence of dideoxynucleotides, various-sized DNA fragments are generated suitable for Sanger-type sequencing. From the sequence of the DNA, the RNA sequence is deduced by base-pairing rules.

The demands of large-scale sequencing projects (∞Chapters 15 and 31) have led to the development of *automated* DNA-sequencing systems. With such systems, the sequencing reactions are still based on the dideoxy method, but fluorescent dye-labeled primers (or nucleotides) are used instead of radioactivity. The products are separated by automated electrophoresis and the bands detected by fluorescence spectroscopy. Each of the four different reactions uses a different fluorescent label, and the lanes are scanned by a fluorescence-detecting laser. In this way, all four reactions can be run on a single lane. The results are analyzed by computer and a sequence generated, with each of the four bases being distinguished by a separate color (Figure 7.26*c*).

Synthetic DNA

Procedures for making **synthetic DNA** are completely automated so that an oligonucleotide of 30–35 bases can easily be made in a few hours, and oligonucleotides of well over 100 bases in length can be made if necessary. For the synthesis of longer polynucleotides, the oligonucleotide fragments can be joined enzymatically using DNA ligase (see Section 7.6).

DNA is synthesized *in vitro* in a *solid-phase procedure* in which the first nucleotide in the chain is fastened to an insoluble porous support (such as silica gel with particles about 50 μm in size). The overall procedure is shown in Figure 7.27●. Several steps are needed for the addition of each nucleotide, but the chemistry need not concern us here. After each step is completed, the reaction mixture is flushed out of the solid support and the series of reactions repeated for the addition of the next nucleotide. Once the desired length is achieved, the oligonucleotide is cleaved from the solid-phase support by a specific reagent and purified to eliminate by-products and contaminants.

Synthetic DNA molecules are widely used for various purposes. In addition to their use as *primers* for DNA sequencing, they are used as primers for the polymerase chain reaction (see Section 7.9). They are also used as *probes* to detect, via nucleic acid hybridization, specific

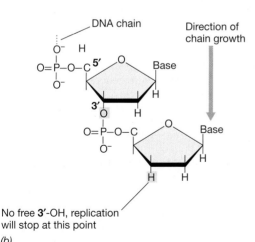

● **Figure 7.25 Dideoxynucleotides and Sanger sequencing.** (a) A normal deoxynucleotide has a hydroxyl group on the 3' carbon and a dideoxynucleotide does not. (b) Chain termination will occur after incorporation of a dideoxynucleotide. Compare this figure with Figure 7.12.

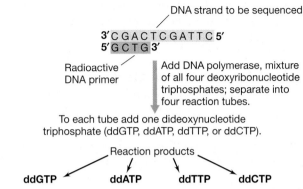

DNA strand to be sequenced

3′ C G A C T C G A T T C 5′
5′ G C T G 3′

Radioactive DNA primer

Add DNA polymerase, mixture of all four deoxyribonucleotide triphosphates; separate into four reaction tubes.

To each tube add one dideoxynucleotide triphosphate (ddGTP, ddATP, ddTTP, or ddCTP).

Reaction products

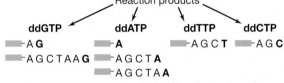

ddGTP	**ddATP**	**ddTTP**	**ddCTP**
▬A **G**	▬ **A**	▬-A G C **T**	▬-A G **C**
▬-A G C T A A **G**	▬-A G C T **A**		
	▬-A G C T A **A**		

Reaction products are separated by electrophoresis on polyacrylamide gel and identified by autoradiography.

G   A   T   C

(a)

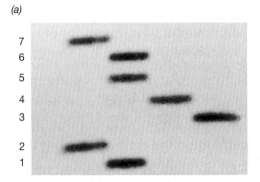

Sequence reads from bottom of gel as A G C T A A G. Sequence of unknown is **3′** T C G A T T C **5′**.

(b)

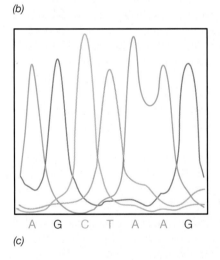

A G C T A A G

(c)

● **Figure 7.26 DNA sequencing using the Sanger method.** (a) The template DNA strand is the DNA whose sequence is to be determined. Note that four different reactions must be run, one with each dideoxynucleotide. Also note that since these reactions are run *in vitro,* the primer for DNA synthesis need not be RNA, and for convenience is DNA. (b) A portion of a gel containing the reaction products from (a). (c) Results of sequencing the same DNA as shown in (a) and (b), but using an automated sequencer and fluorescent labels.

Solid-phase nucleotide support

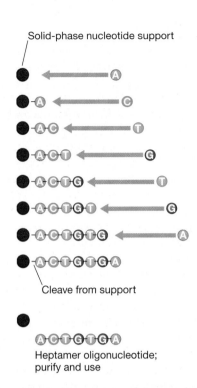

Cleave from support

Heptamer oligonucleotide; purify and use

● **Figure 7.27 Solid-phase procedure for synthesis of a DNA fragment of defined sequence.** Chemical synthesis proceeds by adding one nucleotide at a time to the growing chain.

DNA or RNA sequences in Southern (Figure 7.24) or Northern blots, respectively. We will describe later (see Section 10.18) how synthetic DNA is also used in a procedure called **site-directed mutagenesis** to create mutations of interest at specific locations within a gene.

 7.8 Concept Check

DNA can be sequenced by the Sanger method, which involves copying the DNA to be sequenced in the presence of chain-terminating dideoxynucleotides. The final products are separated by electrophoresis and the sequence read. The short DNA primers required in this method can be synthesized chemically.

◆ Why do the methods involved in DNA sequencing involve the use of *four* different reactions?

◆ What is the term given to a nucleic acid molecule containing several nucleotides?

7.9 Amplifying DNA: The Polymerase Chain Reaction

The development of synthetic DNA and DNA sequencing spawned a new method for the rapid amplification of DNA *in vitro,* the **polymerase chain reaction (PCR).** The polymerase chain reaction can multiply DNA molecules by up to a billionfold in the test tube, yielding large amounts of specific genes for a host of applications in molecular biology. PCR makes use of the enzyme *DNA polymerase,* which copies DNA molecules (∞ Section 7.6).

The steps in PCR amplification of DNA can be summarized as follows:

1. Two DNA oligonucleotide primers flanking the target DNA (Figure 7.28*a*●) are made on an oligonucleotide synthesizer and added in great excess to heat-denatured target DNA.

2. As the mixture cools, the excess of primers relative to the target DNA ensures that most target strands anneal to a primer and not to each other (Figure 7.28*a*).

3. DNA polymerase then extends the primers using the target strands as templates (Figure 7.28*b*).

4. After an appropriate incubation period, the mixture is heated again to separate the strands. The mixture is then cooled to allow the primers to hybridize with complementary regions of newly synthesized DNA, and the whole process is repeated (Figure 7.28*c*).

The secret to PCR is that the *products* of one primer extension serve as the *template* in the next cycle (Figure 7.28). Indeed, the beauty of the PCR technique is that each cycle literally *doubles* the content of the original target DNA. In practice, 20–30 cycles are usually run, yielding a 10^6- to 10^9-fold increase in the target sequence (Figure 7.28*d*).

PCR at High Temperature

The original PCR technique employed *Escherichia coli* Pol III, but because of the high temperatures needed to denature the double-stranded copies of DNA being made, the enzyme was also denatured and had to be replenished every cycle. This problem was solved by employing a thermal stable DNA polymerase isolated from the thermophilic hot spring bacterium *Thermus aquaticus*. DNA polymerase from *T. aquaticus*, known as **Taq polymerase**, is stable to 95°C and thus is unaffected by the denaturation step employed in the PCR reaction. The use of *Taq* DNA polymerase also increased the *specificity* of the PCR reaction because the DNA is copied at 72°C rather than 37°C. At high temperatures, nonspecific hybridization of primers to nontarget DNA rarely occurs, thus making the product of *Taq* PCR more homogeneous than that obtained using the *E. coli* enzyme.

DNA polymerase from the hyperthermophile *Pyrococcus furiosus* (growth temperature optimum, 100°C) (∞Sections 6.12 and 13.6), called *Pfu polymerase* (or "vent polymerase" because *P. furiosus* was isolated from a hydrothermal vent, ∞Section 19.8), is also widely used and is even more thermally stable than *Taq* polymerase. Moreover, unlike *Taq* polymerase, *Pfu* polymerase has proofreading activity (see Section 7.6), making it a particularly good enzyme when high accuracy is crucial.

Because a number of highly repetitive steps are involved in the PCR technique, PCR machines, called *thermocyclers*, have been developed that run through the heating and cooling cycles automatically. Because each cycle requires only about 5 minutes, the automated procedure allows for large amplifications in only a few hours. To supply the demand for thermostable DNA polymerases in the PCR and DNA sequencing markets, the genes for these enzymes have been cloned into *E. coli* and produced commercially in large quantities. As a result, the cost of the PCR method is now just a fraction of what it was when the technique was first introduced.

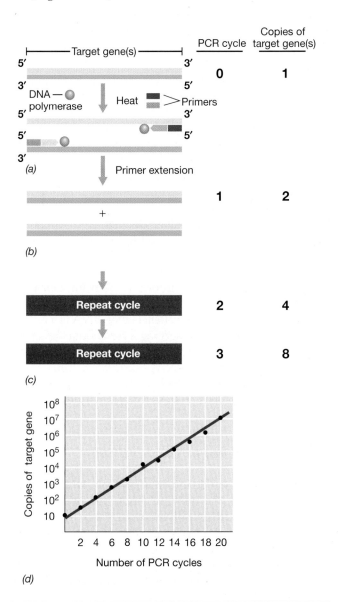

● **Figure 7.28 The polymerase chain reaction (PCR) for amplifying specific DNA sequences.** (a) Target DNA is heated to separate the strands, and a large excess of two oligonucleotide primers, one complementary to the target strand and one to the complementary strand, is added along with DNA polymerase. (b) Following primer annealing, primer extension yields a copy of the original double-stranded DNA. (c) Two additional PCR cycles yield 4 and 8 copies, respectively, of the original DNA sequence. (d) Effect of running 20 PCR cycles on a DNA preparation originally containing 10 copies of a target gene. Note that the graph is semilogarithmic (∞Figure 6.6).

Applications and Sensitivity of PCR

PCR is a powerful tool. It is easy to perform, extremely sensitive and specific, and highly efficient. During each round of amplification the amount of product *doubles*, leading to an exponential increase in the desired DNA (Figure 7.28) This means not only that a large amount of amplified DNA can be produced in just a few hours, but that only a few molecules of target DNA need be present in the sample to start the reaction. The reaction is so specific that, with primers of 15 or so nucleotides and high annealing temperatures, almost no "false priming" occurs, and therefore the PCR product is virtually homogeneous.

PCR is extremely valuable for obtaining DNA for cloning genes or for sequencing purposes because the gene or genes of interest can easily be amplified if flanking sequences are known. PCR is also used routinely in comparative or phylogenetic studies to amplify genes from various sources. In these cases the primers are made to regions of the gene thought to be conserved throughout a wide variety of organisms. For example, because 16S rRNA, a molecule used for phylogenetic analyses (∞Section 2.3), has both highly conserved and highly variable regions, primers specific for the 16S rRNA gene from *Bacteria* or *Archaea* can be synthesized and used to survey the organisms in a habitat for the presence of species of each group. Moreover, if more specific primers are used, only certain subgroups within each domain can be targeted. This technique is in widespread use in microbial ecology and has revealed the enormous diversity of the microbial world, much of it still uncultured (∞Chapters 2, 11–14, 18).

Because it is such a sensitive technique, PCR can be used to amplify very small quantities of DNA. For example, PCR has been used to amplify and clone DNA from sources as varied as mummified human remains and fossilized plants and animals. The ability of PCR to amplify and analyze DNA from cell mixtures has also made it a common tool of diagnostic microbiology. For example, if a clinical sample shows evidence of a gene specific to a particular pathogen, then it can be assumed that the pathogen existed in the sample. With this information, treatment of the patient can begin (∞Section 24.12). In this way, PCR alleviates the need to culture the organism, a time-consuming and often fruitless process. PCR has also been used in DNA *fingerprinting*, a powerful forensic tool that permits identification of individuals, or relationships between individuals, from very small samples of their DNA (Microbial Sidebar DNA Fingerprinting, ∞Chapter 31). PCR can also be coupled with a reverse transcription step in a process called *RT-PCR*. This can be very useful when one wants to make large amounts of DNA using RNA as a template.

7.9 Concept Check

The polymerase chain reaction is a procedure for amplifying DNA *in vitro* and employs a heat-stable DNA polymerase from thermophilic prokaryotes. Heat is used to denature the DNA into two single-stranded molecules, each of which is copied by the polymerase. After each cycle, the newly

formed double strands are separated by heat again, and a new round of copying proceeds. At each cycle, the amount of target DNA doubles.

◆ Why does PCR require primers?

◆ Why is a primer needed at each "end" of the DNA to be amplified?

V RNA SYNTHESIS: TRANSCRIPTION

Ribonucleic acid (RNA) plays a number of important roles in the cell. There are three key differences in the chemistry of RNA and DNA: (1) RNA contains the sugar *ribose* instead of *deoxyribose*; (2) RNA contains the base *uracil* instead of *thymine*; and (3) except in certain viruses, RNA is not double-stranded. A change from *deoxyribose* to *ribose* affects some of the chemical properties of a nucleic acid, and enzymes that affect DNA in general have no effect on RNA, and vice versa. Moreover, the change from *thymine* to *uracil* does not affect base pairing, as the two nucleotide bases pair with adenine equally well.

Three major types of RNA are known: **messenger RNA (mRNA), transfer RNA (tRNA)**, and **ribosomal RNA (rRNA)**. These are all products of the *transcription* of DNA. It should be emphasized that RNA plays a role at two levels, *genetic* and *functional*. At the *genetic* level, RNA carries the genetic information from DNA via mRNA (in the case of RNA viruses, RNA plays a direct genetic function). At the *functional* level, RNA is a macromolecule in its own right, serving a functional and structural role in ribosomes (rRNA) or an amino acid transfer role in protein synthesis (tRNA). Some RNA even has catalytic (enzymatic) activity (ribozymes, ∞Section 14.8). In this section we focus on how RNA is synthesized.

7.10 Overview of Transcription

Transcription of genetic information from DNA to RNA is carried out by the enzyme **RNA polymerase**. Like DNA polymerase, RNA polymerase catalyzes the formation of phosphodiester bonds, but in this case between *ribo*nucleotides rather than *deoxyribo*nucleotides. RNA polymerase requires DNA as a template. The precursors of RNA are the ribonucleoside triphosphates ATP, GTP, UTP, and CTP.

The chemistry of RNA synthesis is much like the chemistry of DNA synthesis (see Figure 7.12). That is, during elongation of an RNA chain, ribonucleotide triphosphates are added to the 3'–OH of the ribose of the preceding nucleotide and polymerized with the release of the two energy-rich phosphate bonds. Thus, as in DNA synthesis, in RNA synthesis the overall direction of chain growth is from the 5'-end to the 3'-end, and the *template* strand is antiparallel to the newly synthesized strand. Unlike DNA polymerase, however, RNA polymerase can initiate synthesis *de novo*; that is, no *primer* is necessary.

RNA Polymerases

The DNA template for RNA polymerase is a double-stranded DNA molecule, but only *one* of the two strands is transcribed for any given gene. Nevertheless, genes are present on either strand of DNA and thus regions of both strands are transcribed at various times during the transcription process. While these principles are true for RNA polymerases from all organisms, RNA polymerase differs markedly among *Bacteria*, *Archaea*, and *Eukarya*. *Bacteria* and *Archaea* each have a single RNA polymerase while in eukaryotes, the nucleus contains three such enzymes: RNA polymerases I, II, and III, each involved in transcribing different types of genes. RNA polymerases from all sources contain some subunits that are evolutionarily conserved (∞Figure 11.16). However, true to their phylogenetic roots, archaeal and eukaryal RNA polymerases are similar and structurally more complex than those of *Bacteria*. The following discussion deals only with RNA polymerase from *Bacteria*, which has the simplest structure and about which the most is known.

All RNA polymerases from species of *Bacteria* are closely related proteins. The enzyme from *Escherichia coli* has four different subunits, designated β, β', α, and σ (sigma), with α present in two copies. The subunits interact to form the active enzyme, called the *RNA polymerase holoenzyme*, but the sigma factor is not as tightly bound as the others and easily dissociates, leading to the formation of what is called the *RNA polymerase core enzyme* ($\alpha_2\beta\beta'$). The core enzyme alone can catalyze the formation of RNA, and the role of sigma is in *recognition* of the appropriate site on the DNA for the initiation of RNA synthesis. The process of RNA synthesis involving RNA polymerase and sigma is illustrated in Figure 7.29●.

Promoters

RNA polymerase is a large protein and makes contact with the DNA over many bases simultaneously. Proteins like RNA polymerase can interact specifically with DNA because portions of the base pairs are exposed in the major groove (see Figure 7.5). However, in order to *start* an RNA chain correctly, RNA polymerase must first recognize the proper regions on the DNA. These important sites are called **promoters**.

Once the RNA polymerase has bound to the promoter, transcription can proceed. In this process, the DNA double helix at the promoter is *opened up* by the RNA polymerase (Figure 7.29). As the polymerase moves, it causes the DNA to unwind in short segments. As a result of this transient unwinding, the template strand is exposed and is copied into the RNA complement. Thus, the promoter can be thought of as *pointing* the RNA polymerase in one direction or the other along the DNA. If a region of DNA has two nearby promoters pointing in opposite directions, then it follows that transcription

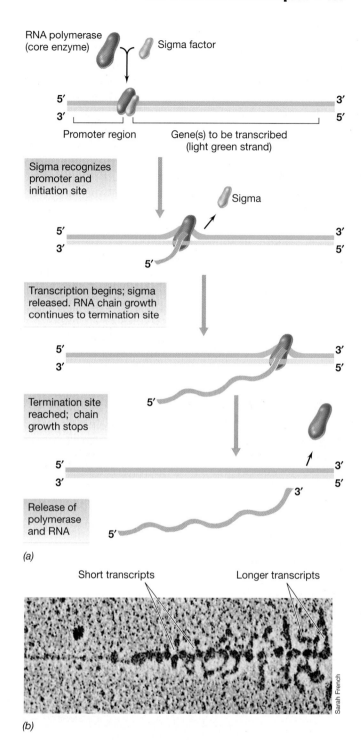

● **Figure 7.29 Transcription.** (a) Steps in RNA synthesis. The initiation and termination sites are specific nucleotide sequences on the DNA. The sigma factor allows RNA polymerase to recognize the initiation site (the promoter). The sigma factor is released during elongation. RNA polymerase moves down the DNA chain, causing temporary opening of the double helix and transcription of one of the DNA strands. When a termination site is reached, chain growth stops, and the mRNA and polymerase are released. (b) Electron micrograph of transcription occurring along a gene on the *Escherichia coli* chromosome. The region of active transcription represents about 2 kb pairs of DNA. Transcription proceeds from left to right, with the shorter transcripts on the left becoming longer transcripts as transcription proceeds.

from one of the promoters will occur in one direction (on one of the strands) while transcription from the other will occur in the opposite direction (on the other strand).

Once a short stretch of RNA has been formed, the sigma factor dissociates. Elongation is then carried out by the core enzyme alone (Figure 7.29). Thus, sigma is only involved in the formation of the initial RNA polymerase-DNA complex at the promoter. As the newly synthesized RNA dissociates from the DNA, the opened DNA closes back into the original double helix. Transcription stops at specific regions called *transcription terminators* (see Section 7.12).

Thus, unlike DNA replication, which involves copying an entire genome, transcription involves much smaller units of DNA, often as little as a single gene. This system allows the cell to transcribe different genes at different frequencies depending on the need within the cell for different proteins. As we shall see in Chapter 8, regulation of transcription in prokaryotes is an important and elaborate process involving many different mechanisms and is very efficient at controlling gene expression and conserving cell resources.

 ### 7.10   Concept Check

The three major types of RNA are messenger RNA (mRNA), transfer RNA (tRNA), and ribosomal RNA (rRNA). Transcription of RNA from DNA involves the enzyme RNA polymerase, which adds bases onto 3'-ends of growing chains. Unlike DNA polymerase, RNA polymerase needs no primer and recognizes a specific start site on the DNA called the promoter.

◆ Which direction (5' → 3' or 3' → 5') *along the template strand* does transcription occur?

◆ What is a *promoter*?

 ## 7.11   Diversity of Sigma Factors, Consensus Sequences, and Other RNA Polymerases

As we have noted, the promoter plays a central role in transcription. Promoters are specific DNA sequences where RNA polymerase binds. The sequences of a large number of promoters from a variety of organisms have been determined, and Figure 7.30● shows the sequence of a few promoters from *Escherichia coli*. It is the *sigma factor*, as part of the RNA polymerase, that is primarily involved in recognition of these promoters.

Sigma Factor Diversity

A single organism can synthesize several different sigma factors. For example, the *Escherichia coli* genome encodes 7 distinct sigma factors while the *Bacillus subtilis* genome encodes 14. These *alternative sigma factors* allow the RNA polymerase to recognize several different promoter sequences. Although in any given organism a single sigma factor is typically involved in the transcription of the bulk of the organism's genes, genes for special or rarely needed functions may require a different sigma factor. This is, of course, one way of controlling expression of these genes. If a needed sigma factor is not present, genes requiring this factor are effectively "turned off."

All the DNA sequences in Figure 7.30 are recognized by the same sigma factor, the major sigma factor in *E. coli*, called σ^{70}. If you carefully examine the sequences, you will see that they are not identical. However, two sequences within the promoter region are *highly conserved* between promoters, and it is these that are recognized by sigma. Both sequences precede (are *upstream* of) the site where

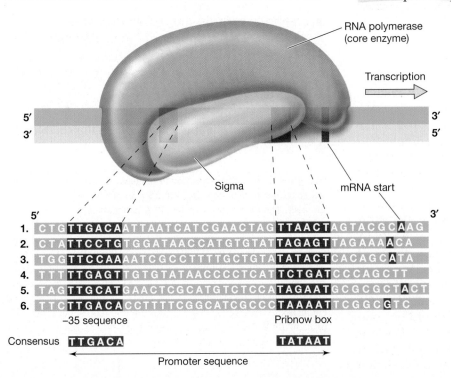

● **Figure 7.30 The interaction of RNA polymerase with the promoter.** Shown below the top of the diagram are six different promoter sequences identified in *Escherichia coli*, a species of *Bacteria*. The contacts of the RNA polymerase with the −35 sequence and the Pribnow box (−10 sequence) are shown. Transcription begins at a unique base just downstream from the Pribnow box. Below the actual sequences at the −35 and Pribnow box regions are consensus sequences derived from comparing many promoters. Note in this example that although the promoter sequences shown are from the 5' → 3' (dark green) strand of DNA, it is the light green strand running 3' → 5' that would actually be transcribed because RNA polymerase works only in a 5' → 3' direction.

transcription starts. One is a region 10 bases before the start of transcription, the −10 region (called the *Pribnow box*). Notice that although each promoter is slightly different, most bases are the same. When comparing the −10 regions of all the promoters recognized by this sigma factor to determine which base occurs most often at each position, one arrives at the *consensus sequence*: TATAAT. In our example, each promoter has from three to five matches for these bases. The second region of conserved sequence is about 35 bases from the start of transcription. The consensus sequence in the −35 region is TTGACA (Figure 7.30). Once again, most of the sequences are not *exactly* the same as the consensus sequence, but are very close.

Note that in Figure 7.30 the promoter sequence on the DNA of only one strand is given. By convention, the strand shown is the one oriented with its 5′-end upstream (therefore, it is *not* the strand used as the template by RNA polymerase, see Figure 7.30). Showing only the sequence of one strand is simply "shorthand" for describing promoters. In reality, promoters are actually double-stranded entities. That is, RNA polymerase recognizes and binds to double-stranded DNA even though transcription itself occurs using only one or the other strand of DNA as template.

Sigma factors in other organisms can sometimes be much more specific as regards binding sequences than the σ^{70} example shown in Figure 7.30. In such cases, very little leeway is allowed in the critical bases that are recognized. In *E. coli*, promoters that are most like the consensus are usually more effective in binding RNA polymerase. These more effective promoters are called *strong promoters* and are very useful in genetic engineering, as will be discussed in Chapter 31.

RNA Polymerases in *Eukarya* and *Archaea*

Recall that the eukaryotic nucleus contains *three* different RNA polymerases. Each of these enzymes recognizes a promoter that is associated with a particular class of gene. **RNA polymerase I** synthesizes most types of rRNA. **RNA polymerase II** synthesizes all the mRNA, and **RNA polymerase III** synthesizes tRNA and one type of rRNA. The basis for this specificity is that each type of RNA polymerase recognizes only those promoters that occur with the particular class of gene. By contrast, in *Bacteria* the promoter for a gene encoding a protein could well be identical to a promoter for a gene encoding a tRNA. Eukaryotic RNA polymerases also require accessory proteins in order to recognize specific promoters but, unlike in *Bacteria*, these eukaryotic initiation factors (and those of the *Archaea*) recognize the promoter elements independently, not as part of a polymerase holoenzyme (Figure 7.31•).

A depiction of RNA polymerase II binding to a promoter on the DNA is shown in Figure 7.31. This promoter contains a *TATA box*, a conserved sequence resembling in some respects the Pribnow box in the promoters of *Bacteria* (see Figure 7.30), as well as an *initiator element* very near the transcription start site. These two regions

are key elements of eukaryotic promoters, although a variety of other sequences may also be involved in any given promoter.

In species of *Archaea*, which contain only a single RNA polymerase most closely resembling that of eukaryotic RNA polymerase II, promoter structure resembles that of eukaryotes. Archaeal promoters that have been characterized contain a 6–8 base pair TATA box 18–27 nucleotides upstream of the transcriptional start site. Archaeal transcription also requires some accessory factors resembling those in eukaryotic transcription (Figure 7.31). Thus, the phylogenetic relationships between *Archaea* and *Eukarya* (∞ Section 2.3) are also reflected in several details of a central molecular process, transcription. We will see many other such molecular relationships between *Archaea* and *Eukarya* as we proceed through this book.

7.11 Concept Check

In *Bacteria*, promoters are recognized by the sigma subunit of RNA polymerase. Promoters recognized by a specific sigma factor have very similar sequences. In the *Eukarya* the major classes of RNA are transcribed by different RNA polymerases, with RNA polymerase II producing most mRNA. The single RNA polymerase of *Archaea* resembles in both structure and function this RNA polymerase II.

◆ What is a *consensus sequence*?

◆ In a eukaryote, which type of RNA polymerase transcribes genes that encode proteins?

7.12 Transcription Terminators

As important as initiation of transcription is, *termination* of transcription is also important to the fidelity of protein synthesis. **Termination** of RNA synthesis occurs at

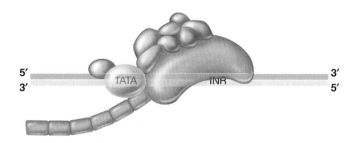

● **Figure 7.31 The interaction of eukaryotic RNA polymerase II with a promoter.** The polymerase itself (brown) is positioned at the initiator element (INR) of the promoter. A TATA box-binding protein (yellow) is shown bound at the TATA box. The polymerase has a repetitive amino acid sequence at one end (shown as a tail-like structure) that can be phosphorylated, affecting activity of the polymerase. The other proteins shown in blue are a few of the large number of accessory factors required for initiation of transcription in eukaryotes.

Microbial Sidebar ◆ Unconventional Amino Acids

The genetic code has codons for 20 amino acids (Table 7.5). However, many proteins contain "other" amino acids. In fact, there are well over 100 different amino acids that have been found in different proteins. It was originally thought that all of these unconventional amino acids were made by modifying one of the standard amino acids *after* it was incorporated into protein, a process called *post-translational modification*. Although this can occur, at least two of these "other" amino acids are inserted into proteins by the translational machinery itself. These exceptions are *seleno-cysteine* and *pyrrolysine*.

Selenocysteine has the same structure as cysteine except that it contains a *selenium* rather than a *sulfur* atom (Figure 1●). It was known for some time that several proteins, including some from humans, contained this unusual amino acid. For example, *Escherichia coli* makes two different formate dehydrogenase enzymes and both contain a selenocysteine residue. When the gene encoding one of these enzymes was sequenced, it was found that the codon encoding the selenocysteine was UGA. Normally, UGA is a nonsense codon in *E. coli* (Table 7.5); but UGA is translated directly as selenocysteine in certain mRNA molecules in *E. coli* and many other prokaryotes and eukaryotes, including humans. Therefore, selenocysteine became the twenty-first genetically encoded amino acid.

Pyrrolysine is a lysine analog that contains an aromatic ring that lysine itself lacks (Figure 1). This amino acid has been found in certain *Archaea* and *Bacteria* but was first discovered in species of *Archaea* called *methanogens*, organisms whose metabolism generates natural

gas (methane, ∞Section 13.4). In certain methanogens the enzyme *methylamine methyltransferase* contains a pyrrolysine residue. When analyzing the gene for this enzyme, it was found that the way in which pyrrolysine is encoded bears a striking resemblance to the selenocysteine story. A nonsense codon, in this case UAG instead of UGA, encodes pyrrolysine.

How can a codon sometimes be a nonsense codon and sometimes a sense codon? The answer in the selenocysteine case is well understood and lies in the *context* of the codon, that is, the sequence of the bases surrounding the UGA codon. This differs from the mitochondrial "alternative code," where UGA encodes tryptophan regardless of sequence context. In certain contexts, the translational machinery decodes UGA as "selenocysteine," while in all other contexts UGA stops translation. Selenocysteine has its own tRNA (as do all the standard amino acids) containing the predicted anticodon (UCA) and also has a special protein factor that brings only this tRNA to the ribosome when called for during the translational process. Pyrrolysine also has its

own tRNA, but differs from the selenocysteine situation in a key way. Selenocysteine is formed by modifying a serine that has been attached to the selenocysteine tRNA by the serine aminoacyl-tRNA synthetase. By contrast, in the case of pyrrolysine, a unique aminoacyl-tRNA synthetase exists that charges the pyrrolysyl tRNA directly with pyrrolysine.

Evolution has obviously allowed for some "play" in the genetic code. The incorporation of selenocysteine and pyrrolysine are excellent examples of this, and one can only wonder what other unusual amino acids may be genetically encoded from either reassigned nonsense codons or perhaps even from reassigned sense codons. The genomic era has simplified the search for examples of the former, as computer inspection of nucleic acid sequences can quickly pinpoint nonsense codons that appear out of place in an otherwise normal open reading frame. Detection of reassigned sense codons, if they occur, is another matter, although computer analysis of correlations between unusual amino acids, their cognate codons, and sequence context, may yield more interesting surprises. ■

Figure 1 *Comparisons of the structures of cysteine/selenocysteine and lysine/pyrrolysine*

7.14 Concept Check

The genetic code is expressed in terms of RNA, and a single amino acid may be encoded by several different but related codons. In addition to the nonsense codons, there is also a specific start codon that signals where the translation process should begin.

◆ Why is it important for the ribosome to read "in frame"?

◆ Describe an *open reading frame*. If you were given a nucleotide sequence, how would you find the open reading frame?

7.15 Transfer RNA

Recall from Section 7.14 that it is the *anticodon* portion of the tRNA that base pairs with the codon. However, a tRNA is much more than simply an anticodon (Figure 7.36●). A tRNA has a specificity for both a codon *and* its cognate amino acid. The tRNA and its specific amino acid are brought together by specific enzymes that ensure that a particular tRNA receives its correct amino acid. These enzymes, called **aminoacyl-tRNA synthetases**, have the important function of recognizing *both* the cognate amino acid *and* the specific tRNA for that amino acid.

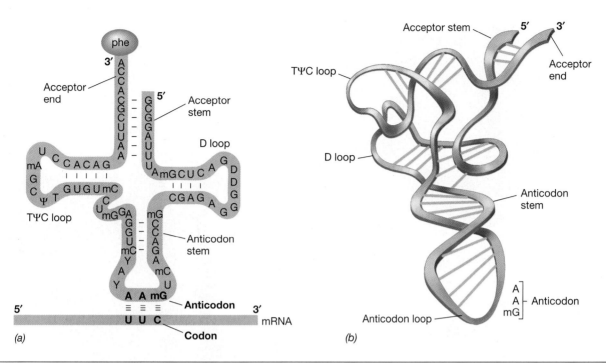

● Figure 7.36 **Structure of a transfer RNA.** (a) The conventional cloverleaf structure of yeast phenylalanine tRNA. The amino acid is attached to the ribose of the terminal A at the acceptor end. A, adenine; C, cytosine; U, uracil; G, guanine; ψ, pseudouracil; D, dihydrouracil; m, methyl; Y, a modified purine. (b) In actuality, the tRNA molecule folds so that the D loop and TψC loops are close together and associate by hydrophobic interactions. Note the unusual bases present in the tRNA and also the presence of an occasional thymine, absent from mRNA.

Structure of tRNAs

There are about 60 different tRNAs in bacterial cells and 100–110 in mammalian cells. Transfer RNA molecules are short, single-stranded molecules that contain extensive secondary structure and have lengths of 73–93 nucleotides. When compared, it has been found that certain bases and secondary structures are constant for all tRNAs, whereas other parts are variable. Transfer RNA molecules also contain some purine and pyrimidine bases differing slightly from the normal bases found in RNA in that they are chemically modified. These modifications are made to the bases after transcription. Some of these unusual bases are *pseudouridine, inosine, dihydrouridine, ribothymidine, methyl guanosine, dimethyl guanosine,* and *methyl inosine.* The final mature and active tRNA not only contains unusual bases, but also extensive double-stranded regions within the molecule as a result of internal base pairing when the molecule folds back on itself (Figure 7.36).

The structure of a tRNA can be drawn in a cloverleaf fashion, as in Figure 7.36*a*. Some regions of tRNA secondary structure are given names having to do either with the bases most often found there (the TψC and D loops, for example) or with specific functions (anticodon loop and acceptor end). The three-dimensional structure of a tRNA is shown in Figure 7.36*b*. Note that bases that appear widely separated in the cloverleaf model are actually closer together when viewed in three dimensions. This means that some of the bases in the "loops" are actually paired with bases in other loops.

One of the key variable parts of the tRNA molecule contains the **anticodon**, the site that recognizes the codon on the mRNA. The anticodon is found in the *anticodon loop* (Figure 7.36). There are *three* nucleotides in the anticodon loop that are specifically involved in the recognition process and that base-pair with the codon (see Section 7.14 and Figure 7.34). Other portions of the tRNA interact with the ribosome (both rRNA and protein components), nonribosomal translation proteins, and the activating synthetase enzyme. At the 3'-end, or **acceptor end**, of all tRNAs, are three unpaired nucleotides. The sequence of these three nucleotides is always cytosine-cytosine-adenine (CCA), and it is to the ribose sugar of the terminal adenine that the amino acid is covalently attached via an ester linkage. From this location on the tRNA, the amino acid is incorporated into the growing polypeptide chain on the ribosome by a mechanism that will be described in the next section.

Recognition, Activation, and Charging of tRNAs

Recognition of the correct tRNA by an aminoacyl-tRNA synthetase involves specific contacts between key regions of the nucleic acid and particular amino acids of its respective synthetase (Figure 7.37●). As might be expected because of the unique sequence in this region, the *anticodon* of the tRNA is important in recognition by the synthetase. However, other contact sites between the tRNA and the synthetase are also important. Studies of tRNA binding to aminoacyl-tRNA synthetases in which specific bases in the tRNA have been changed

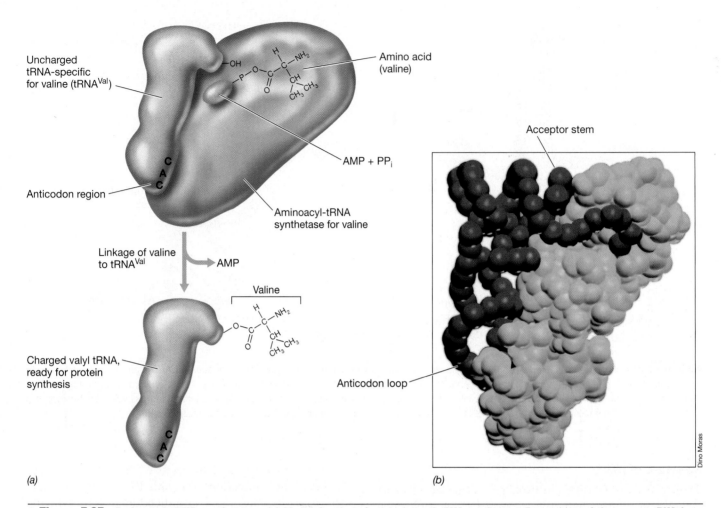

Uncharged tRNA-specific for valine (tRNA^{Val})

Amino acid (valine)

AMP + PP$_i$

Anticodon region

Aminoacyl-tRNA synthetase for valine

Linkage of valine to tRNA^{Val} → AMP

Valine

Charged valyl tRNA, ready for protein synthesis

(a)

Acceptor stem

Anticodon loop

Dino Moras

(b)

● **Figure 7.37 Aminoacyl-tRNA synthetases.** (a) Mode of action of an aminoacyl–tRNA synthetase. Recognition of the correct tRNA by a particular synthetase involves contacts between specific nucleic acid sequences in the D–loop and acceptor stem of the tRNA and specific amino acids of the synthetase. In this diagram, valyl-tRNA synthetase is shown catalyzing the final step of the reaction, where the valine in valyl-AMP is transferred to tRNA. (b) A computer model showing the interaction of glutaminyl-tRNA synthetase (blue) with its tRNA (red). Reprinted with permission from M. Ruff et al. 1991. *Science* 252: 1682–1689. © 1991, AAAS.

by mutation have shown that only a small number of key nucleotides in a tRNA besides the anticodon region are involved in recognition. These other key recognition nucleotides are often part of the acceptor stem or D-loop of the tRNA molecule (see Figure 7.36). It should be emphasized that the fidelity of this recognition process is crucial, for if the wrong amino acid becomes attached to the tRNA, it will be inserted into the polypeptide, likely leading to the synthesis of a faulty protein.

The specific reaction between amino acid and tRNA catalyzed by the aminoacyl-tRNA synthetase first involves *activation* of the amino acid by reaction with ATP:

$$\text{Amino acid} + \text{ATP} \longleftrightarrow \text{aminoacyl—AMP} + \text{P—P}$$

The aminoacyl-AMP intermediate formed normally remains bound to the enzyme until collision with the appropriate tRNA molecule, and, as shown in Figure 7.37*a*, the activated amino acid is then transferred to the tRNA to form a *charged* tRNA:

$$\text{Aminoacyl—AMP} + \text{tRNA} \longleftrightarrow$$

$$\text{aminoacyl—tRNA} + \text{AMP}$$

The pyrophosphate (PP$_i$) formed in the first reaction is split by a pyrophosphatase, forming two molecules of inorganic phosphate. Hence, since *ATP* is used and *AMP* is formed, a total of *two* energy-rich phosphate bonds are required for the activation of an amino acid and its cognate tRNA. Once activation and charging have occurred, the aminoacyl-tRNA leaves the synthetase and travels to the ribosome where polypeptide synthesis occurs. The mechanism of protein synthesis is discussed in the next section.

 7.15 *Concept Check*

One or more transfer RNAs exist for each amino acid found in protein. Enzymes called aminoacyl-tRNA synthetases function to attach an amino acid to a tRNA. Once the correct amino acid is attached to its tRNA, further specificity resides primarily in the codon-anticodon interaction.

◆ What is the function of the *anticodon* of a tRNA?

◆ What is the function of the *acceptor end* of a tRNA?

7.16 Translation: The Process of Protein Synthesis

We learned in Chapter 3 that it is the amino acid *sequence* that determines the structure (and ultimately the function) of a protein. Thus, it is critical to the fidelity of translation that the proper amino acid be inserted at the proper place in the polypeptide chain. This is the role of the protein-synthesizing machinery of the cell, including, in particular, the ribosome.

Ribosomes

Ribosomes are the site of protein synthesis. A cell may have many thousands of ribosomes, the number being positively correlated with growth rate. Each ribosome is constructed of two subunits. In prokaryotes, the ribosome subunits are of 30S and 50S, yielding intact 70S ribosomes. The numbers 30S, 50S, and 70S refer to *Svedberg units*, which are units of sedimentation coefficients of ribosome subunits (30S and 50S) or intact ribosomes (70S) when subjected to centrifugal force in an ultracentrifuge.

Each ribosomal subunit is a ribonucleoprotein complex made up of specific ribosomal RNAs and ribosomal proteins. The 30S subunit contains 16S rRNA and about 21 proteins, while the 50S subunit contains 5S and 23S rRNA and about 34 proteins (Table 7.6 and Figure 7.38a●). In *Escherichia coli*, there are at least 53 distinct ribosomal proteins, most present at one copy per ribosome.

Since the genome of *E. coli* has been completely sequenced and the ribosome has been studied for some time, one might think that we should know exactly how many ribosomal proteins there are overall and in each subunit. However, some proteins are tightly associated with the ribosome, some less so. In addition, some ribosomal proteins are associated only with one subunit, some with both. There are also proteins that are absolutely essential for ribosome function and that interact with the ribosome at various stages in the trans-

lational process but are not considered "ribosomal proteins," per se. Moreover, it should be kept in mind that the ribosome is a very dynamic structure whose parts alternately associate and dissociate and that interact with many other proteins in the cell. Thus, determining the total number of "ribosomal proteins," even in *E. coli*, is not easy.

Steps in Protein Synthesis

The synthesis of a protein involves a complex cycle in which the various ribosomal components play specific roles. Although a continuous process, protein synthesis can be broken down into a number of discrete steps: **initiation**, **elongation**, and **termination–release**. In addition to mRNA, tRNA, and ribosomes, the process involves a number of proteins designated initiation, elongation, and termination factors, while the energy-rich compound guanosine triphosphate provides energy for the process. The key steps in protein synthesis are shown in Figure 7.38.

Initiation

In prokaryotes initiation of protein synthesis always begins with a free 30S ribosome subunit. From this an **initiation complex** forms consisting of a 30S ribosome subunit, mRNA, formylmethionine tRNA, and several initiation proteins called *IF1*, *IF2*, and *IF3*. Guanosine triphosphate is also required for this step. To this initiation complex a 50S ribosomal subunit is added to make the active 70S ribosome. At the end of the translation process, the released ribosome separates again into 30S and 50S subunits.

Just preceding the initiation codon on the mRNA is a sequence of from three to nine nucleotides called the **Shine–Dalgarno sequence** that is involved in the binding of the mRNA to the ribosome. This *ribosome binding site* at the 5′-end of the mRNA is complementary to regions in the 3′-end of the 16S RNA of the ribosome. Base pairing between these two molecules ensures effective formation of the ribosome–mRNA complex *in the correct reading frame*. The presence of the Shine–Dalgarno site on the mRNA and its specific interaction with 16S rRNA allow prokaryotic ribosomes to translate *polycistronic* mRNA because the ribosome can find each initiation site within a message after binding to the Shine-Dalgarno site (see Section 7.13).

Initiation always begins with a special initiator aminoacyl-tRNA binding to the **start codon**, AUG. In *Bacteria* this is **formylmethionine** tRNA. Subsequently, the formyl group at the N-terminal end of the polypeptide is removed; the terminal amino acid of the completed protein is hence methionine. Because the Shine–Dalgarno sites (and other possible interactions between the rRNA and the mRNA) direct the ribosome to the proper start site, prokaryotic messages can use a start codon other than AUG. The most common alternative start codon is GUG. When used in this context, however,

Table 7.6	Ribosome structure[a]		
Property		**Prokaryote**	**Eukaryote**
Overall size		70S	80S
Small subunit		30S	40S
Number of proteins		~21	~30
RNA size (number of bases)		16S (1500)	18S (2300)
Large subunit		50S	60S
Number of proteins		~34	~50
RNA size (number of bases)		23S (2900)	28S (4200)
		5S (120)	5.8S (160)
			5S (120)

[a] Ribosomes of mitochondria and chloroplasts of eukaryotes are similar to prokaryotic ribosomes (∞ Section 14.4).

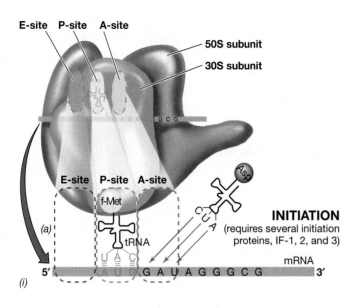

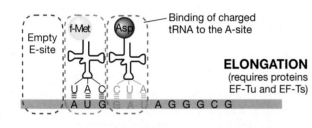

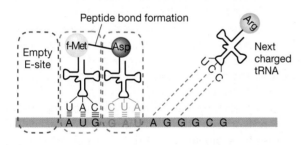

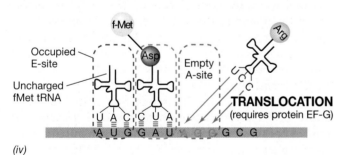

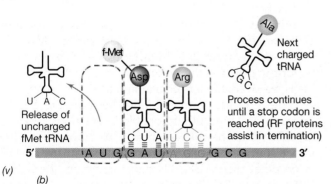

GUG calls for formylmethionine initiator tRNA (and not valine, see Table 7.5). In *Eukarya* and *Archaea*, initiation begins with *methionine* instead of *formylmethionine*. Although all proteins in cells are *initiated* with a methionine or formylmethionine, this amino acid is not always found as the first amino acid in a cell's proteins. Methionine (or formylmethionine in *Bacteria*) is removed from many mature proteins by a specific protease.

Elongation, Translocation, and Termination

The mRNA is threaded through the ribosome primarily bound to the 30S subunit. The ribosome contains other sites where the tRNAs interact. Two of these sites are located primarily on the 50S subunit, and they are termed the *P-site* and the *A-site* (see Figure 7.38b). The A-site, the **acceptor site**, is the site where the new charged tRNA first attaches. Travel and attachment of a tRNA to the A-site is assisted by one of a series of elongation factors (EF) proteins called *EF-Ta*.

The P-site, the **peptide site**, is the site where the growing peptide is held by a tRNA. During peptide bond formation, the peptide moves to the tRNA at the A-site as a new peptide bond is formed. Several nonribosomal proteins are required for **elongation**, especially other proteins of the EF family, such as EF-Tu and EF-Ts, as well as additional molecules of GTP (to simplify Figure 7.38b, the elongation factors are omitted and only a portion of the ribosome is shown). Following elongation, the tRNA that holds the peptide must now be moved (translocated) from the A-site to the P-site, thus opening up the A-site for another charged tRNA (Figure 7.38b).

Translocation requires a specific EF protein called EF-G and one molecule of GTP per each translocation event. At each translocation step the ribosome advances three nucleotides, exposing a new codon at the ribosome A-site. Translocation pushes the now empty tRNA to a third site, called the *E-site*. It is from this **exit site** that the tRNA is actually released from the ribosome (Figure 7.38b). The precision of the translocation step is critical to the accuracy of protein synthesis. The ribosome must move exactly one codon at each step. Although during this process mRNA appears to be moving through the ribosome complex, in reality, the *ribosome* is moving along the mRNA. Thus, the

● **Figure 7.38** **The ribosome and protein synthesis.** (a) Structure of the ribosome, showing the position of the A (acceptor) site, the P (peptide) site, and the E (exit) site. (b) Translation. Initiation and elongation. (i, ii) Interaction between the codon and anticodon brings into the position the correct charged tRNA—in this case the initiator tRNA and the second charged tRNA. (iii) The formation of a peptide bond between amino acids on adjacent tRNA molecules completes elongation. (iv) Translocation of the ribosome from one codon to the next occurs with release of the tRNA from the E-site. (v) The next charged tRNA binds to the A-site.

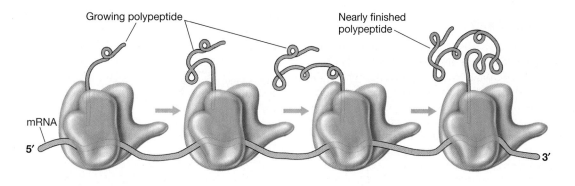

● **Figure 7.39** **Polysomes.** Translation by several ribosomes on a single messenger RNA forms the polysome. Note how the ribosomes nearest the 5'-end of the message are at an earlier stage in the translation process and thus only a portion of the final polypeptide has been made.

three sites on the ribosome that we have identified in Figure 7.38 are not static locations but instead are *moving parts* of a complex biomolecular machine.

Several ribosomes can simultaneously translate a single mRNA molecule forming a complex called a **polysome** (Figure 7.39●). Polysomes increase the speed and efficiency of translation, and because the activity of each ribosome is independent of that of its neighbors, each ribosome in a polysome complex makes a complete polypeptide. Note in Figure 7.39 how ribosomes closest to the 5'-end (the beginning) of the mRNA molecule have short polypeptides attached to them because only a few codons have been read, while ribosomes closest to the 3'-end of the message have nearly finished polypeptides.

The **termination** of protein synthesis occurs when a nonsense codon is reached. No tRNA binds to a nonsense codon. Instead, specific proteins called *release factors* (RF factors) recognize this chain-terminating signal and cleave the attached polypeptide from the terminal tRNA, releasing the finished product. Following this, the ribosome subunits dissociate, and the 30S/50S subunits are then free to form new initiation complexes and repeat the process.

Role of Ribosomal RNA in Protein Synthesis

Ribosomal RNA plays a critical functional role in all stages of protein synthesis, from initiation to termination. The role of the many proteins present in the ribosome, although less clear, may be more as facilitators of RNA function by stabilizing or positioning key sequences in the various ribosomal RNAs.

As already discussed, in prokaryotes it is clear that 16S rRNA is involved in initiation through base pairing with the ribosome binding sequence (the Shine–Dalgarno sequence) on the mRNA. Besides this, other mRNA/rRNA interactions occur during elongation. Ribosomal RNA also plays a role in ribosome subunit association, as well as in tRNA positioning in the A and P sites (see Figure 7.38*b*) on the ribosome. Although charged tRNAs that enter the ribosome recognize the

correct codon by codon:anticodon base pairing, they are also physically attached to the ribosome by interactions of the anticodon stem-loop of the tRNA with specific sequences within 16S rRNA. Moreover, the acceptor end of the tRNA (see Figure 7.36) base pairs with sequences in the 23S rRNA.

In addition to all of this, the *actual formation of peptide bonds* is catalyzed by rRNA. This process that occurs on the 50S subunit of the ribosome and is called the **peptidyl transferase reaction** is not catalyzed by *any* of the multitude of ribosomal or ribosomal-associated proteins, but instead by the activity of 23S rRNA itself. 23S rRNA also plays a role in the translocation process, and in this regard, the EF proteins are known to interact specifically with 23S rRNA. Finally, 16S rRNA is also involved in a catalytic way in termination reactions, probably by interacting with the mRNA or through interactions with release proteins. Thus, besides its role as a structural backbone of the ribosome, ribosomal RNA plays a major catalytic role in the translation process as well.

Effect of Antibiotics on Protein Synthesis

A large number of antibiotics inhibit protein synthesis by interacting with the ribosome. These interactions are quite specific, and many have been shown to involve rRNA. Several of these antibiotics are clinically useful, and several are also effective research tools because they are specific for different steps in protein synthesis. For instance, *streptomycin* inhibits initiation, whereas *puromycin, chloramphenicol, cycloheximide,* and *tetracycline* inhibit elongation.

Adding to their clinical usefulness is the fact that many antibiotics specifically inhibit ribosomes of organisms from only one or two of the phylogenetic domains. Of the antibiotics just listed, for example, chloramphenicol and streptomycin are specific for the ribosomes of *Bacteria* and cycloheximide for ribosomes of *Eukarya*. The mode of action of these and other antibiotics will be discussed in Chapter 20.

7.16 *Concept Check*

The ribosome plays a key role in the translation process, bringing together mRNA and aminoacyl tRNAs. There are three sites on the ribosome: the acceptor site, where the charged tRNA first combines; the peptide site, where the growing polypeptide chain is held; and an exit site. During each step of amino acid addition, the ribosome advances three nucleotides (one codon) along the mRNA and the tRNA moves from the acceptor to the peptide site. Termination of protein synthesis occurs when a nonsense codon, which does not encode an amino acid, is reached.

◆ What are the components of a ribosome?

◆ What functional roles does rRNA play in protein synthesis?

7.17 Folding and Secreting Proteins

We have now discussed how the genetic information present in the sequence of bases in DNA is *replicated* into an identical copy, *transcribed* into a sequence of bases in RNA, and *translated* into a sequence of amino acids in protein. For a protein to function, however, it must be *folded* correctly (∞Sections 3.6–3.8), and it must also end up in the correct location in the cell. Here we briefly deal with these two processes.

Protein Folding

Many polypeptides fold spontaneously into their active form while they are being synthesized (Figure 7.39). However, many others do not and require assistance from other proteins called **molecular chaperones** for proper folding or for assembly into larger complexes. The chaperones themselves do not become part of the assembled proteins, but only assist in the folding process. Indeed, one important function of chaperones is to prevent improper aggregation of proteins.

There are many different kinds of these chaperones and some are even associated with ribosomes. Some are also extremely abundant in the cell, especially under certain growth conditions when protein stability is at risk (for example, at high temperatures). Molecular chaperones seem to be both extremely widespread and their sequences highly conserved.

Four key chaperones in *Escherichia coli* are the proteins *DnaK, DnaJ, GroEL*, and *GroES*. DnaK and DnaJ are ATP-dependent enzymes that bind to newly formed polypeptides and keep them from folding too abruptly, a process that increases the risk of improper folding (Figure 7.40●). Slower folding thus improves the chances of correct folding. If the DnaKJ complex is unable to fold the protein properly, it may transfer the partially folded protein to the two multi-subunit proteins GroEL and GroES. The protein first enters GroEL, a large barrel-shaped protein that, using the energy of ATP hydrolysis, properly folds the protein. GroES assists in this process (Figure 7.40).

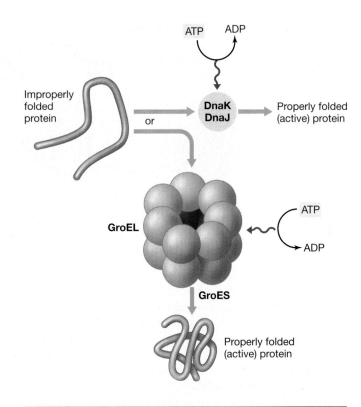

● **Figure 7.40 The activity of molecular chaperones.** An improperly folded protein can be refolded by either the DnaKJ complex or by the GroEL/ES complex. In both cases, energy for refolding comes from ATP.

In addition to folding newly synthesized proteins, chaperones can also refold proteins that have partially denatured in the cell. Such protein denaturation can occur because the organism has temporarily experienced high temperatures in its environment. Chaperones are thus a type of **heat shock protein**, and their synthesis is greatly accelerated when a cell is stressed by excessive heat (∞Section 8.9). The heat shock response is an attempt by the cell to refold its partially denatured proteins for re-use before proteases recognize them as improperly folded and destroy them. Refolding is not always successful, and cells contain a variety of proteases whose function is to specifically target and destroy misfolded proteins in order to generate a pool of amino acids for use in making new proteins.

Protein Secretion and the Signal Recognition Particle

Many proteins carry out their function in the cytoplasmic membrane, in the periplasm of gram-negative cells (∞Section 4.9) or even *outside* the cell proper. Such proteins must get from their site of synthesis on ribosomes into or through the cytoplasmic membrane. How is it possible for a cell to selectively transfer some proteins across a membrane while leaving most proteins in the cytoplasm?

Most proteins that must be transported into or through membranes are synthesized with an extra peptide sequence (about 15–20 amino acids long) at the beginning of the molecule, called the **signal sequence**. Signal sequences are quite variable but typically have a few positively charged residues at the beginning, a central region of hydrophobic residues, and then a more polar region. The signal sequence "signals" the cell's secretory system that this particular protein is to be exported and also helps prevent the protein from completely folding, a process that could interfere with its secretion. Since the signal sequence is the first part of the protein to be synthesized, the early steps in export may begin before the protein is completely synthesized (Figure 7.41●).

A principal role in the identification of proteins to be secreted is played by the **signal recognition particle (SRP)** (Figure 7.41). SRPs are found in all cells. In *Bacteria*, they contain a single protein and a small RNA molecule called 4.5S RNA. This *small RNA* (∞Section 8.14) is not a transfer RNA, a ribosomal RNA, or a messenger RNA. The SRP recognizes the signal sequence-containing protein and delivers it to a special membrane protein complex (for example, to the SecYEG complex, ∞Section 4.7) where the protein is secreted through a pore and into the periplasm or the environment (Figure 7.41). During the transport process the signal sequence is usually removed by a protease, an example of the process of **post-translational modification**.

Less is known about protein translocation through the membrane in species of *Archaea*. However, genomic analyses indicate that *Archaea* produce a SRP and share several aspects of the translocation process with *Bacteria*. Interestingly, and in line with their phylogenetic relationships (∞Section 2.3), *Archaea* share several aspects of the protein translocation process with *Eukarya* as well. In particular, the small RNA in the archael SRP is 7S in size, as it is in eukaryotic cells.

Secretion of Folded Proteins: The TAT System

In the Sec system of protein transport, mediated by the signal recognition particle, the transported proteins are threaded through the cytoplasmic membrane in an unfolded state and are only folded upon reaching the periplasm (in gram-negative cells) (Figure 7.41). However, there are a number of proteins that must be transported outside the cell that contain small cofactors. The latter usually have to be inserted into the protein as it folds into its final form. This occurs in the cytoplasm, and then the folded proteins are transported by a transport system distinct from Sec, called the *Tat protein export system*.

The acronym *Tat* stands for "*t*win *a*rginine *t*ranslocase" because the transported proteins contain a short signal sequence containing a pair of arginine residues. This signal sequence on a folded protein is recognized by the TatBC proteins, and they carry the protein to TatA, the membrane transporter. Energy is required for the actual transport event, and this is sup-

plied by the proton motive force. A wide variety of proteins are transported by the Tat system, especially proteins involved in energy metabolism that function in the periplasm. This includes a variety of iron-sulfur proteins and several other "redox" type proteins. In addition, the Tat pathway transports proteins involved in outer membrane (Section 4.9) biosynthesis and a variety of proteins that do not contain cofactors but that only fold properly within the cytoplasm.

In this chapter we have covered the essentials of the key molecular processes that occur in all cells. We next turn our attention to how cells *regulate* the expression of a particular gene or set of genes, a process as we will see, that is rather diverse and highly elaborate.

🛑 7.17 *Concept Check*

Proteins must be properly folded in order to function correctly. Folding may occur spontaneously but may also involve other proteins called molecular chaperones. Many proteins also need to be transported into or through cell membranes. Such proteins are synthesized with a signal sequence that is recognized by the cellular export apparatus and is removed either during or after export.

◆ What is a *molecular chaperone*?

◆ Why do some proteins have a *signal sequence*?

◆ What is a signal recognition particle?

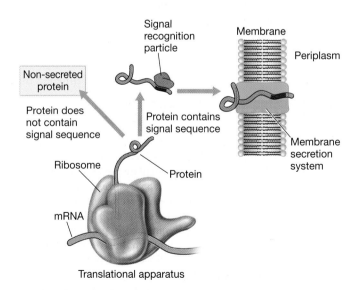

● **Figure 7.41 Secretory proteins and the signal recognition particle (SRP).** The signal sequence is recognized by the SRP and the protein transported to the membrane secretion system. In gram-negative bacteria, the protein is secreted into the periplasm (∞ Section 4.9). In species of *Bacteria*, the SRP is 4.5S in size, while in *Archaea* and *Eukarya*, it is larger, about 7S.

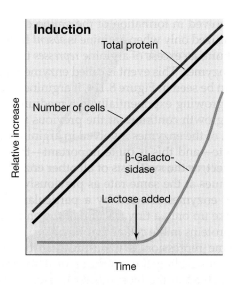

Figure 8.12 Enzyme induction. Induction of the enzyme β-galactosidase on the addition of lactose to the medium. Note that the rate of total protein synthesis remains unchanged.

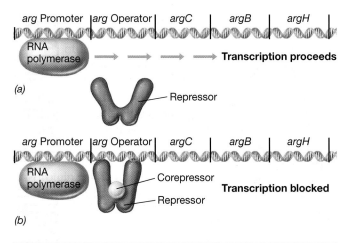

● **Figure 8.13 The process of enzyme repression using the arginine operon as an example.** (a) Transcription of the operon occurs because the repressor is unable to bind to the operator. (b) After a corepressor (small molecule) binds to the repressor, the repressor binds to the operator and blocks transcription; mRNA and the proteins it encodes are not made. In the case of the *argCBH* operon, the repressor would be the arginine repressor and the corepressor would be the amino acid arginine.

lactose. If lactose is *absent* from the medium, the enzyme is not synthesized, but synthesis begins almost immediately after lactose is added. One can immediately see the value to the organism of such a control mechanism, as it provides a means to synthesize specific enzymes *only when they are needed*.

The substance that initiates enzyme induction is called an **inducer,** and a substance that represses enzyme synthesis is called a **corepressor.** These substances, which are always small molecules, are collectively called **effectors.** Not all inducers and corepressors are actual substrates or end products of the enzymes involved. For example, structural *analogs* may induce or repress even though they are not substrates of the enzyme. Isopropyl-thiogalactoside (IPTG), for instance, is an inducer of β-galactosidase even though IPTG cannot be hydrolyzed by the enzyme. In nature, however, inducers and corepressors are probably normal cell metabolites. Interestingly, though, the actual inducer of β-galactosidase, which is one of three proteins encoded by genes in the *lac operon*, is not lactose, but instead the structurally similar compound *allolactose*, a derivative made by the cell from lactose.

Mechanism of Repression and Induction

How can inducers and corepressors affect transcription in such a specific manner? They do this indirectly by combining with specific DNA-binding proteins, which, in turn, affect transcription. In the case of a repressible enzyme (Figure 8.11), the corepressor (in this case, arginine) combines with a specific **repressor protein,** the *arginine repressor,* that is present in the cell (Figure 8.13●). The repressor protein is itself an allosteric protein (see Sections 8.2 and 8.4); that is, its conformation can be altered when the corepressor combines with it.

By binding its effector, the repressor protein becomes *active* and can then combine with a specific region of the DNA near the promoter of the gene, the **operator region.** This region gave its name to the **operon,** a cluster of genes arranged in a linear and consecutive fashion whose expression is under the control of a single operator (∞Section 7.13). All of the genes in an operon are transcribed as a single unit yielding a single mRNA. The operator is located downstream from the promoter where synthesis of mRNA is initiated (Figure 8.13). If the repressor binds to the operator, transcription is physically blocked because RNA polymerase can neither bind nor proceed. Hence, the polypeptide(s) encoded by the genes in the operon cannot be synthesized. If the mRNA is polycistronic (∞Section 7.13), *all* the polypeptides encoded by this mRNA will be repressed.

Enzyme induction may also be controlled by a repressor. The specific repressor protein is *active* in the *absence* of the inducer, completely blocking transcription. When the inducer is added, it combines with the repressor protein and inactivates it. When this occurs, inhibition is overcome and transcription can proceed (Figure 8.14●).

All regulatory systems involving repressors have the same underlying mechanism: *inhibition* of mRNA synthesis by the activity of specific repressor proteins that are themselves under the control of specific small-molecule inducers and corepressors. And, as previously noted, because the repressor's role is inhibitory, regulation involving repressors is called *negative control*.

 8.5 Concept Check

The amount of an enzyme in the cell can be controlled by increasing (induction) or decreasing (repression) the amount of mRNA that encodes the enzyme. This transcriptional regulation involves allosteric regulatory proteins that bind to DNA. For negative control of transcription, the regulatory protein is called a repressor and it functions by inhibiting mRNA synthesis.

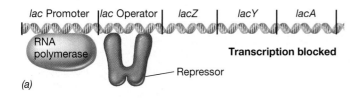

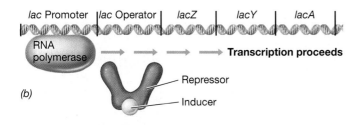

● **Figure 8.14** **The process of enzyme induction using the lactose operon as an example.** (a) A repressor protein binds to the operator region and blocks the action of RNA polymerase. (b) An inducer molecule binds to the repressor and inactivates it so that it no longer can bind to the operator. Transcription by RNA polymerase occurs, and an mRNA for that operon is formed. In the case of the *lac* operon, the repressor would be the *lac* repressor, and the inducer would be allolactose.

◆ Why is "negative control" so named?

◆ How does a repressor inhibit the synthesis of a specific mRNA?

8.6 Positive Control of Transcription

In negative control, the controlling element—the repressor protein—brings about *repression* of mRNA synthesis. By contrast, in **positive control** of transcription, a regulator protein *activates* the binding of RNA polymerase, hence the term *positive*. An excellent example of positive regulation is the catabolism of the disaccharide sugar maltose in *Escherichia coli*.

Maltose Catabolism in *Escherichia coli*

The enzymes for maltose catabolism in *Escherichia coli* are synthesized only after the addition of maltose to the medium. The expression of these enzymes thus follows the pattern shown for β-galactosidase in Figure 8.12 except that maltose rather than lactose is required to affect gene expression. However, control of the synthesis of maltose-degrading enzymes is not under negative control as in the *lac* operon, but under positive control; transcription requires the activity of an **activator protein**.

The *maltose activator protein* cannot bind to the DNA unless this protein first binds maltose, the effector. When the maltose activator protein binds to DNA, it allows RNA polymerase to begin transcription (Figure 8.15●). Like repressors, activators bind specifically to only certain sequences on the DNA. However, the region on the

DNA that is the site of the activator is not called an operator (Figures 8.13 and 8.14), but instead an *activator-binding site* (Figure 8.15). Nevertheless, the genes controlled by this activator-binding site are still called an *operon*.

Binding of Activator Proteins

In negative control, the repressor binds to the operator and blocks transcription. How does an activator protein work? The promoters of positively controlled operons have nucleotide sequences that are not close matches to the consensus sequence (∞Figure 7.27). Thus, even with the correct sigma factor, the RNA polymerase has difficulty recognizing these promoters.

The activator protein, when bound to DNA, helps the RNA polymerase recognize the promoter and begin transcription. For example, the activator protein may cause a change in the structure of the DNA by bending it (Figure 8.16●), allowing the RNA polymerase to make the correct contacts with the promoter to begin transcription. Alternatively, the activator protein may interact directly with the RNA polymerase. This can happen either when the activator binding site is close to the promoter (Figure 8.17*a*●) or when it is several hundred base pairs away from the promoter, a situation where DNA looping is required to make the necessary contacts (Figure 8.17*b*).

Many genes in *Escherichia coli* have promoters under positive control, and many have promoters under negative control. In addition, many operons have a promoter with multiple types of control, while some even have more than one promoter, each with its own control system! Thus, the rather simple picture outlined in Figures 8.13–8.17 is not

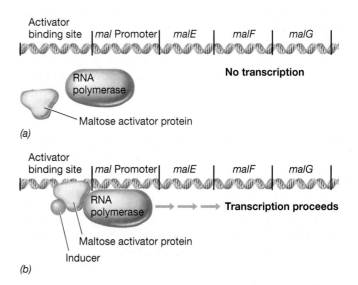

● **Figure 8.15** **Positive control of enzyme induction using the maltose operon as an example.** (a) In the absence of an inducer, neither the activator protein nor the RNA polymerase can bind to the DNA. (b) An inducer molecule binds to the activator protein, which in turn binds to the activator-binding site. This allows RNA polymerase to bind to the promoter and begin transcription. In the case of the *malEFG* operon, the activator protein would be the maltose activator protein, and the inducer would be the sugar maltose.

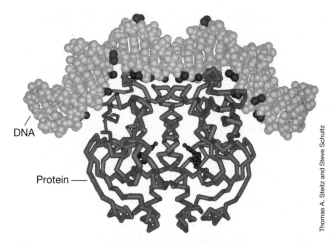

● **Figure 8.16   Computer model of the interaction of a positive regulatory protein with DNA.** This figure shows the cyclic AMP-binding protein (CAP protein; see Section 8.7), a regulatory protein involved in the control of several operons. The α-carbon backbone of this protein is shown in blue and purple. The protein is shown binding to a DNA double helix, which is shown in yellow and light blue. Note that binding of this protein to DNA has caused the DNA to be significantly bent.

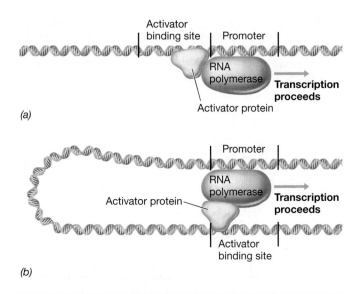

● **Figure 8.17   Activator protein interactions with RNA polymerase.** (a) The activator-binding site is near the promoter. (b) The activator-binding site is several hundred base pairs from the promoter. In this case, the DNA must be looped to allow the activator and the RNA polymerase to contact.

typical of all operons. Multiple control features are common in the operons of virtually all prokaryotes and thus their regulation can be very complex.

Operons versus Regulons

In *Escherichia coli,* the genes required for maltose utilization are spread out over the chromosome in several operons, each of which has an activator-binding site to which a copy of the maltose-activator protein can bind. Therefore, the maltose-activator protein actually controls more than one operon. When more than one operon is under the control of a single regulatory protein, these operons are collectively known as a **regulon.** Therefore, the enzymes for maltose utilization are encoded by the *maltose regulon.*

Regulons are also known for operons under negative control. For example, the arginine biosynthetic enzymes (see Section 8.5) are encoded by the *arginine regulon,* whose operons are all under the control of the arginine repressor protein (only one of the arginine operons was shown in Figure 8.13). However, whether functioning as an activator or a repressor, in regulon control, a specific DNA-binding protein binds *only* at the operons it controls; other operons are not affected.

 8.6   Concept Check

Positive regulators of transcription are called activator proteins. They bind to activator-binding sites on the DNA and stimulate transcription. As in repressors, activator protein activity is modified by effectors. For positive control of enzyme induction, the effector promotes the binding of the activator protein and thus stimulates transcription.

◆ Compare and contrast the activities of an *activator* protein and a *repressor* protein.

◆ Distinguish between an *operon* and a *regulon.*

IV GLOBAL REGULATORY MECHANISMS

An organism often needs to regulate many unrelated genes simultaneously in response to a change in its environment. Regulatory mechanisms that respond to environmental signals by regulating expression of many different genes are called *global control systems.* Both the lactose operon and the maltose regulon respond to global controls. Thus, we begin a consideration of this regulatory system by revisiting the *lac* operon.

8.7   Global Control and the *lac* Operon

Our discussions of negative and positive control did not consider the quite realistic possibility that the environment in which the cells are growing might contain several different carbon sources that the bacteria could use. For example, *Escherichia coli* can use a wide array of carbon sources. When faced with several sugars, including glucose, do cells of *E. coli* use them simultaneously or one at a time? The answer is that *glucose is always used first.* Indeed it would be wasteful to induce enzymes for the metabolism of other sugars if a better carbon source—glucose—was available. One mechanism of global control, *catabolite repression,* ensures that this occurs.

Catabolite Repression

In **catabolite repression** the syntheses of a variety of unrelated, primarily catabolic, enzymes are repressed when cells are grown in a medium that contains *glucose.* Catabolite repression is also called the *glucose effect* because glucose was the first substance shown to initiate it. In some organisms, however, carbon sources other than glucose cause catabolite repression. The key point is that the substrate that catabolite represses the utilization of other substrates, simply needs to be the *better* energy source. In this way, catabolite repression ensures that the organism uses the *best* available carbon and energy source first.

One consequence of catabolite repression is that it leads to two exponential growth phases, a situation called *diauxic growth.* If two utilizable energy sources are available to the cell, in diauxic growth the organism grows first on the best energy source. Then, following a lag period, growth resumes on the other energy source. Diauxic growth is illustrated in Figure 8.18● for growth of *Escherichia coli* on a mixture of glucose and lactose.

As we have seen, the enzyme β-galactosidase, which is required for utilization of lactose, is inducible (Figures 8.12 and 8.14). But, in addition, its synthesis is also subject to catabolite repression. Thus, as long as glucose is present, β-galactosidase is not synthesized and lactose is not used. However, when glucose is exhausted, catabolite repression is abolished, and after a brief lag period, β-galactosidase is synthesized and growth on lactose occurs. Notice that Figure 8.18 shows that the cells grow more rapidly on glucose than on lactose. Even though glucose and lactose are both excellent energy sources for *E. coli,* glucose is the *best* of all possible carbon sources for this organism, and thus growth is fastest on this substrate.

Cyclic AMP and CAP

How does catabolite repression work? Catabolite repression involves control of transcription by an activator protein and is therefore a form of *positive* control (see Section 8.6). In the case of catabolite-repressible enzymes, binding of RNA polymerase to the DNA that encodes them occurs only if another protein, called the **catabolite activator protein (CAP),** has bound first. An allosteric protein itself, CAP binds to DNA only if it has first bound a small molecule called *cyclic adenosine monophosphate* or **cyclic AMP** (see Figure 8.16).

Cyclic AMP (Figure 8.19●) is a key molecule in a variety of metabolic control systems, not only in prokaryotes but in eukaryotes as well. Cyclic AMP is synthesized from ATP by an enzyme called *adenylate cyclase.* However, glucose *inhibits* the synthesis of cyclic AMP and *stimulates* cyclic AMP transport out of the cell. When glucose enters the cell, the cyclic AMP level is lowered, and binding of RNA polymerase to the promoters of CAP-specific operons does not occur. Thus, catabolite repression is really an indirect result of the presence of a better carbon source (glucose). The direct cause of catabolite repression is a *cellular deficiency of cyclic AMP.*

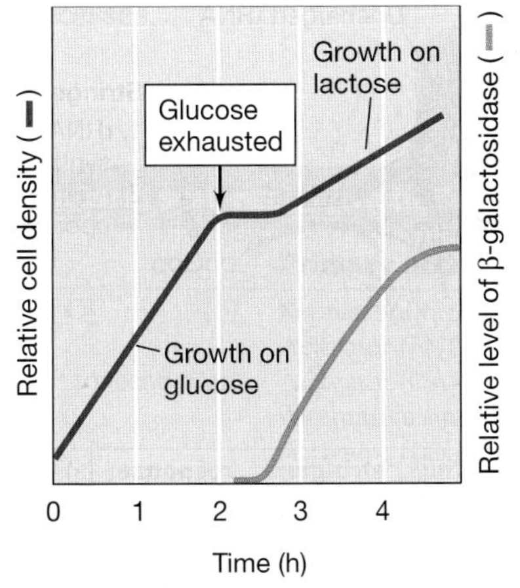

● **Figure 8.18 Diauxic growth on a mixture of glucose and lactose.** Glucose represses the synthesis of β-galactosidase. After glucose is exhausted, a lag occurs until β-galactosidase is synthesized, and then growth can resume on lactose but at a slower rate.

Global Aspects of Catabolite Repression

Why is catabolite repression considered a mechanism of *global* control? In the case of *Escherichia coli* and other organisms in which glucose is the prime energy source, as long as glucose is present, catabolite repression prevents expression of *all other* catabolic operons affected by this control mechanism. This may total dozens of catabolic operons. So, not only is lactose catabolism affected, but also affected are the catabolism of maltose, a host of other sugars, and most of the commonly used carbon and energy sources of *E. coli.*

Let us return to the *lac* operon to put aspects of catabolite repression in context with the entire regulatory picture. To do this, the entire regulatory region of the *lac* operon is shown in Figure 8.20●. For transcription of *lac* genes to occur, two requirements must be met: (1) the level of cyclic AMP must be high enough so that the CAP protein binds to the CAP-binding site (*positive control*), and (2) lactose must be present so that the lactose repressor does not block transcription by binding to the operator (*negative control*). Once these two conditions have been met, the cell is signaled that glucose is absent and lactose is present; then and only then does transcription of the *lac* operon begin.

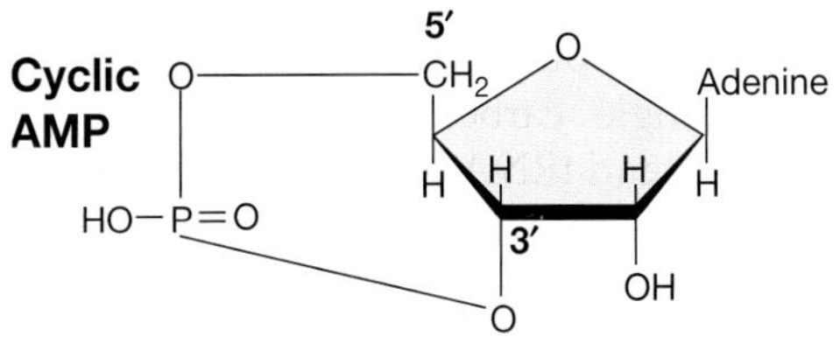

● **Figure 8.19 Cyclic AMP.** Cyclic adenosine monophosphate (cyclic AMP, cAMP) is produced from ATP by the enzyme adenylate cyclase.

unrelated. Examples of these broad global control systems are shown in Table 8.1.

Alternative Sigma Factors

Genes belonging to global control systems do not all use a simple combination of repressors or activators to achieve regulation. Several are controlled by *alternative sigma factors* (Table 8.2). In these cases, regulation is brought about by changing the amount or activity of these sigma factors, since each alternative sigma factor recognizes only a certain subset of genes (for example, particular modulons or stimulons) in the genome.

Recall that sigma is the subunit of RNA polymerase responsible for promoter recognition (∞Section 7.10). Most genes in *Escherichia coli* require the sigma factor σ^{70} (the superscript 70 indicates the size of this protein, 70 kilodaltons) for transcription and have promoters like those that were shown in Figure 7.30. In total there are seven different sigma factors in *E. coli*, and each recognizes different promoter consensus sequences (Table 8.2). Most of these have counterparts in other *Bacteria*. The endospore-forming bacterium *Bacillus subtilis* has 14 sigma factors, with 4 different sigma factors alone dedicated to the transcription of endospore-specific genes (∞Section 4.13) alone.

The key to controlling transcription with alternative sigma factors is to control the synthesis and activity of each sigma factor. The *concentration* of each sigma factor in the cell can be modulated by the transcriptional controls we have previously discussed, or by the rate of their degradation within the cell by specific proteases. By contrast, the *activity* of preformed alternative sigma factors can be controlled by other proteins called *anti-sigma factors*, which can temporarily inactivate one sigma factor or another in response to changes in environmental signals (Table 8.1).

Heat Shock Response

Most proteins are relatively stable. Once made, they continue to perform their functions and are passed along at cell division. However, some proteins are recognized by protease enzymes in the cell as being improperly folded and for this or other reasons are rapidly degraded. For example, in *Escherichia coli*, the sigma factor σ^{32} is normally degraded within a minute or two of its synthesis. However, when cells experience a significant and sudden increase in temperature (a *heat shock*), degradation of σ^{32} is inhibited. This means there will be more σ^{32} in the cell and more transcription of operons that contain its specific promoter (Table 8.2). A major set of genes controlled by σ^{32} in *E. coli* are the *heat shock genes*, which encode the **heat shock proteins** expressed during the **heat shock response.**

Heat shock proteins assist the cell in recovering from stress. They are not only induced by heat, but by several other stress factors that the cell can encounter. These include exposure to high levels of certain chemicals—such as ethanol—or from exposure to high doses of ultraviolet radiation.

In *E. coli* and in most prokaryotes examined, there are three major classes of heat shock protein, *Hsp70*, *Hsp60*, and *Hsp10*. Although not by these names, we have encountered these proteins before. The Hsp70 protein of *E. coli* is *DnaK*, a protein that functions to prevent aggregation of newly synthesized proteins and to stabilize unfolded proteins (∞Section 7.17). DnaK also plays a role in the degradation of σ^{32}, thus serving as a feedback loop for keeping the amount of this sigma factor in check in unstressed cells. During a stress situation, the role of DnaK is primarily in protein salvaging/repair, so σ^{32} becomes more abundant and transcription of heat shock genes occurs at a high rate.

Major representatives of the Hsp60 and Hsp10 families in *E. coli* are the proteins *GroEL* and *GroES*, respectively. These are *molecular chaperones* that catalyze the correct folding of misfolded proteins (∞Section 7.17 and Figure 7.40). Another class of heat shock proteins includes various proteases that function in the cell to remove denatured or irreversibly aggregated proteins. Heat shock proteins are apparently quite ancient and are present in all cells. Molecular sequencing of heat shock proteins, especially Hsp70, has been used to help unravel the phylogeny of eukaryotes (∞Section 14.9).

Table 8.2	Sigma factors in *Escherichia coli*	
Name[a]	**Upstream (−35) Consensus Recognition Sequence**[b]	**Function**
σ^{70}	TTGACA	For most genes, major sigma factor during normal growth
σ^{54}	TTGGCACA	Nitrogen assimilation
σ^{38}	CCGGCG	Major sigma factor during stationary phase, also for genes involved in oxidative and osmotic responses
σ^{32}	TNTCNCCTTGAA[c]	Heat shock response
σ^{28}	TAAA	For genes involved in flagella synthesis
σ^{24}	GAACTT	Response to misfolded proteins in periplasm
σ^{19}	AAGGAAAAT	For certain genes in iron transport

[a] Superscript number in name indicates size of protein in kilodaltons. Most factors also have other names, i.e., σ^{70} is also sometimes called σ^{D}.

[b] For a discussion of consensus sequences, ∞Sections 7.10 and 7.11 and Figure 7.30.

[c]N, any nucleotide.

When the stress situation has passed, for example, upon a temperature downshift, σ^{32} is rapidly inactivated by DnaK, and the synthesis of heat shock proteins is greatly reduced. Since heat shock proteins perform vital functions in the cell, there is always a low level of these proteins present, even when cells are growing under optimal conditions. However, the rapid synthesis of heat shock proteins in stressed cells emphasizes their importance for survival from excessive heat, chemicals, or physical agents. Heat or other stresses can generate large amounts of inactive proteins that need to be refolded (and in the process, reactivated) or degraded to release free amino acids for the biosynthesis of new proteins.

Cold Shock

Cellular stress induced by cold temperatures also induces a number of genes (the *cold shock response*), but these are not Hsp proteins. Instead, the cold shock response is triggered by the effect that cold has on retarding ribosome function. Cold shock proteins include helicases, nucleases, and ribosome-associated proteins that directly or indirectly interact with DNA, RNA, and ribosomes to decrease macromolecular syntheses. Also induced by cold shock are proteins that increase the production of compatible solutes (⦿⦿Section 6.14) in the cell to help defray ice formation in the cytoplasm. Ice nucleation proteins are synthesized if conditions approach freezing; these catalyze the formation of an ice lattice that is less destructive to cell components (such as membranes) than is amorphous ice.

 8.9 *Concept Check*

Cells can control several regulons by employing alternative sigma factors. These recognize only certain promoters and thus allow transcription of a select category of genes. Other global signals include heat and cold shock proteins that function to help the cell overcome temperature stress.

◆ Why do cells have more than one type of sigma factor?

◆ Why might the proteins induced during a heat shock not be needed during a cold shock?

| 8.10 | **Quorum Sensing** |

As previously mentioned, global control systems allow an organism to regulate many different genes at one time in response to signals in its environment. One potential "signal" that prokaryotes can respond to is the presence in their surroundings of other cells *of the same species*. Some prokaryotes have regulatory pathways that are controlled by the *density* of cells of their own kind. This type of control is called **quorum sensing** (the word *quorum* in this sense means "sufficient numbers").

Mechanism of Quorum Sensing

Quorum sensing is a mechanism to ensure that sufficient cell numbers of a given species are present before eliciting a particular biological response. For example, a pathogenic bacterium that secretes a toxin can have no effect as a single cell; production of the toxin by that cell would be a waste of resources. However, if there is a sufficiently high population of cells present, the coordinated expression of the toxin may successfully initiate disease.

Quorum sensing is widespread among gram-negative *Bacteria*. Each species that employs this type of regulation synthesizes a specific acylated *homoserine lactone* (*AHL*) (Figure 8.22*a*●). This molecule is freely diffusible to the outside of the cell. Because of this, the AHL reaches high concentrations in the cell only if there are many cells nearby, each making the same AHL. The specific AHL functions as an inducer that combines with a specific activator protein, thus triggering transcription of specific genes (Figure 8.22*b*).

Quorum sensing was first discovered as the mechanism of regulating bioluminescence in bacteria. Several prokaryotes can emit light in this way, including the marine bacterium *Vibrio fischeri* (⦿⦿Section 12.12 and Figure 12.28). Figure 8.23● shows bioluminescent colonies of *V. fischeri*. The light is a by-product of the activity of a

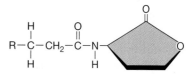

Acyl homoserine lactone (AHL)

(a)

(b)

● **Figure 8.22   Quorum sensing.** (a) General structure of an acyl homoserine lactone (AHL). Different AHLs are structural variants of this parent structure. R, keto group or hydroxyl group. (b) A cell capable of quorum sensing transcribes and translates genes for acyl homoserine lactone synthase at basal levels. This enzyme makes the cells' specific AHL. When numbers of cells of the same species reach a certain level, the concentration of AHL rises sufficiently that it can bind to the activator protein. The latter activates transcription of quorum-specific proteins.

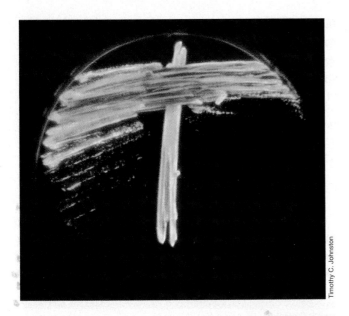

Timothy C. Johnston

● **Figure 8.23** **Bioluminescent bacteria producing the enzyme luciferase** (Section 12.12). Cells of the bacterium *Vibrio fischeri* were streaked on nutrient agar in a Petri dish and allowed to grow overnight. The photograph was taken in a darkened room using only the light generated by the bacteria.

specific enzyme called *luciferase*. The *lux* operons, which encode the proteins involved in bioluminescence, are under control of an activator protein called *LuxR* and are induced when the concentration of the specific *V. fischeri* AHL, *N*-3-oxohexanoyl homoserine lactone, becomes high enough. This AHL is synthesized by the protein encoded by the *luxI* gene.

Examples of Quorum Sensing

A variety of genes are controlled by a quorum-sensing system, including some in disease-causing (pathogenic) bacteria. In *Pseudomonas aeruginosa*, for instance, quorum-sensing triggers the expression of a large number of unrelated genes when the population density becomes sufficiently high. These genes assist cells of *P. aeruginosa* in the transition from growing freely suspended in liquid to growing in a semisolid matrix called a *biofilm* (Section 19.3). The biofilm, formed from specific polysaccharides produced by *P. aeruginosa*, increases the pathogenicity of this organism and prevents the penetration of antibiotics.

The pathogenesis of *Staphylococcus aureus* (Sections 25.7 and 26.9) involves, among many other things, the production and secretion of small cell surface and extracellular peptides that damage host cells or that interfere with the immune system. The genes encoding these virulence factors are under the control of a quorum sensing system that responds to a peptide (rather than an AHL) produced by the organism. The regulation of these genes is quite complex and, in part, involves a *regulatory RNA molecule*. Although we have discussed only regulatory proteins in this chapter, reg-

ulatory RNA molecules also exist (see Section 8.14). The *S. aureus* quorum-sensing system also involves regulatory proteins that are part of a two-component, signal-transducing regulatory system. We discuss this type of regulation in Section 8.12.

Quorum-sensing has been discovered in at least one species of *Archaea*. In the haloalkaliphile *Natronococcus occultus* (an organism that requires both high salt concentrations and high pH, Section 13.3), synthesis of an extracellular protease is under control of a specific AHL. This control ensures that the protease is not produced and excreted until enough cells are present for the level of protease to be sufficient to degrade proteins at a reasonable rate. Although this is the only instance of quorum sensing currently known among the *Archaea*, it would be surprising if other *Archaea* do not employ quorum-sensing control for reactions that can benefit them only when cell densities are high enough to have an impact.

8.10 Concept Check

Quorum sensing allows cells to survey their environment for cells of their own kind. Quorum sensing involves the sharing of specific small molecules. Once a sufficient concentration of the signaling molecule is present, specific gene expression is triggered.

◆ What is the major class of quorum sensing signaling molecules?

◆ In terms of signaling molecules, how does quorum sensing differ in *Pseudomonas* from that in *Staphylococcus*?

V OTHER MECHANISMS OF REGULATION

We have discussed several mechanisms cells use to regulate the transcription of genes: repressors, activators, alternative sigma factors, quorum-sensing molecules. Could there be anything left? A few other important regulatory mechanisms fall outside the scope of our discussion thus far, and we describe them here.

8.11 Attenuation

Some control systems do not employ regulatory proteins binding to DNA. **Attenuation** is such a system. Moreover, in attenuation, the control occurs *after* initiation of transcription but *before* its completion. That is, the number of *completed* transcripts from an operon is reduced, even though the number of *initiated* transcripts is not. The best examples of attenuation involve regulation of genes controlling the biosynthesis of certain amino acids in gram-negative *Bacteria*. The first such system to be described was in the *tryptophan operon* in *Escherichia coli*, and we focus on it here.

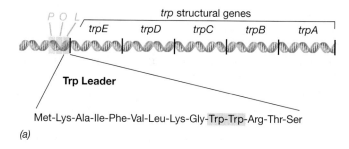

Met-Lys-Ala-Ile-Phe-Val-Leu-Lys-Gly-Trp-Trp-Arg-Thr-Ser

(a)

Threonine	Met-Lys-Arg-Ile-Ser-Thr-Thr-Ile-Thr-Thr-Thr-Ile-Thr-Ile-Thr-Thr-Gly-Asn-Gly-Ala-Gly
Histidine	Met-Thr-Arg-Val-Gln-Phe-Lys-His-His-His-His-His-His-His-Pro-Asp
Phenylalanine	Met-Lys-His-Ile-Pro-Phe-Phe-Phe-Ala-Phe-Phe-Phe-Thr-Phe-Pro

(b)

● **Figure 8.24** **Attenuation and the leader peptide.** Structure of the tryptophan operon and of tryptophan and other leader peptides in *Escherichia coli*. (a) Arrangement of the tryptophan operon. Note that the leader (L) encodes a short peptide containing two tryptophan residues near its terminus (there is a stop codon following the Ser codon). The promoter is labeled P, and the operator is labeled O. The genes labeled *trpE* through *trpA* encode the enzymes involved in tryptophan biosynthesis. (b) Amino acid sequence of leader peptides synthesized in some other amino acid biosynthetic operons. Because isoleucine is made from threonine, it is an important constituent of the threonine leader peptide.

Attenuation and the Tryptophan Operon

The tryptophan operon contains structural genes for five proteins of the tryptophan biosynthetic pathway plus the usual promoter and regulatory sequences at the beginning of the operon (Figure 8.24●). Like many operons, the tryptophan operon has more than one type of regulation. The first enzyme in the pathway, *anthranilate synthase* (a multisubunit enzyme encoded by *trpD* and *trpE*) is subject to feedback inhibition by tryptophan (see Section 8.2). Transcription of the entire tryptophan operon is also under negative control (see Section 8.5).

However, in addition to the promoter (P) and operator (O) regions needed for negative control, there is a sequence in the operon called the **leader sequence**. The leader encodes a polypeptide that contains tandem tryptophan codons near its terminus and functions as an **attenuator** (Figure 8.24).

The basis of control is as follows. If tryptophan is plentiful in the cell, there will be a sufficient pool of charged tryptophan tRNAs and the leader peptide will be synthesized. On the other hand, if tryptophan is in short supply, the tryptophan-rich leader peptide will *not* be synthesized. Synthesis of the leader peptide results in *termination* of transcription of the remainder of the *trp* operon, which includes the structural genes for the biosynthetic enzymes. By contrast, if synthesis of the leader peptide is blocked by tryptophan deficiency, transcription of the rest of the operon occurs.

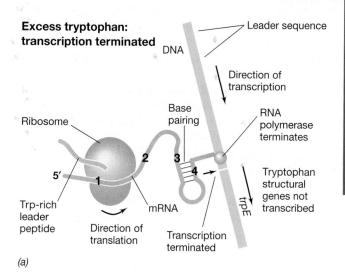

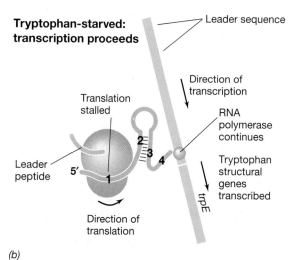

Web Tutorial 8.3 Attenuation and the Tryptophan Operon

Mechanism of Attenuation

How does *translation* of the leader peptide regulate *transcription* of the tryptophan genes downstream? Consider that in prokaryotic cells transcription and translation are simultaneous processes; as mRNA is released from the DNA the ribosome binds to it and translation begins (∞Figure 7.39). That is, while *transcription* of downstream DNA sequences is still proceeding, *translation* of sequences transcribed has already begun (Figure 8.25●).

Attenuation occurs (transcription stops) because a portion of the newly formed mRNA folds into a unique stem-loop that causes cessation of RNA polymerase activity. The

Excess tryptophan: transcription terminated

(a)

Tryptophan-starved: transcription proceeds

(b)

● **Figure 8.25** **Mechanism of attenuation.** Control of transcription of tryptophan operon structural genes by attenuation in *Escherichia coli*. The leader peptide is encoded by regions 1 and 2 of the mRNA. Two regions of the growing mRNA chain are able to form double-stranded loops, shown as 2:3 and 3:4. (a) Under conditions of excess tryptophan, the ribosome translates the complete leader peptide, and so region 2 cannot pair with region 3. Regions 3 and 4 then pair to form a loop that terminates RNA polymerase. (b) If translation is stalled because of tryptophan starvation, loop formation via 2:3 pairing occurs, loop 3:4 does not form, and transcription proceeds past the leader sequence.

stem-loop structure formed by mRNA occurs because two stretches of nucleotides near each other are complementary and can thus base pair. If tryptophan is plentiful, the ribosome will translate the leader sequence until it comes to the leader stop codon. The remainder of the leader RNA then assumes the stem-loop, a *transcription pause site*, which is followed by a uracil-rich sequence that actually causes termination (Figure 8.25).

If tryptophan is in short supply, transcription of genes encoding tryptophan biosynthetic enzymes is obviously desirable. During transcription of the leader, the ribosome pauses at a tryptophan codon because of a shortage of charged tryptophan tRNAs. The presence of the stalled ribosome at this position allows a stem-loop to form that differs from the terminator stem-loop (sites 2 and 3 in Figure 8.25); this alternative stem-loop is *not a* transcription termination signal. Instead, it prevents the terminator stem-loop (sites 3 and 4 in Figure 8.25) from forming. This allows RNA polymerase to move past the termination site and begin transcription of tryptophan structural genes. Thus, in attenuation control, transcription and translation communicate with one another, with the rate of transcription being influenced by the rate of translation.

Attenuation has also been shown to occur in *Escherichia coli* in the biosynthetic pathways for histidine, threonine-isoleucine, phenylalanine, and several other amino acids and essential metabolites as well. As shown in Figure 8.24b, the leader peptide for each of these amino acid biosynthetic operons is rich in that particular amino acid. The *his* operon is dramatic in this regard because its leader contains *seven* histidines in a row near the end of the peptide (Figure 8.23b). This longer stretch of regulatory codons gives attenuation a major effect in regulation, which may compensate for the fact that unlike the *trp* operon, the *his* operon in *E. coli* is not under negative control.

Translation-Independent Attenuation Mechanisms

Gram-positive *Bacteria*, such as *Bacillus*, also use attenuation to regulate certain amino acid biosynthetic operons. As in gram-negative *Bacteria*, the mechanism also involves attenuation of transcription and alternative mRNA secondary structures, which in one configuration lead to termination. However, the mechanism is *translation independent*, and involves an RNA-binding protein.

In the *Bacillus subtilis* tryptophan operon the binding protein is called the *trp attenuation protein*. In the presence of sufficient amounts of the amino acid tryptophan, this regulatory protein binds to the leader mRNA and causes transcription termination. By contrast, if tryptophan is limiting, the protein does not bind to the mRNA. This allows the favorable secondary structure to form and transcription proceeds.

Attenuation also occurs with genes unrelated to amino acid biosynthesis. These mechanisms obviously do not involve the cell assessing its amino acid pools. Some of the operons involved in pyrimidine biosynthesis (the *pyr* operons) in *Escherichia coli* are regulated by attenuation, and the same is true for *Bacillus*. The mechanisms in the two organisms are, however, quite different, although each employs a system to assess the level of pyrimidines in the cell. In *E. coli* the mechanism involves monitoring the rate of *transcription*, not *translation*. If pyrimidines are not limiting, RNA polymerase can move along and transcribe the leader DNA at a normal rate; this allows for a terminator type stem-loop to form in the mRNA. By contrast, if pyrimidines are in short supply, the polymerase pauses at pyrimidine-rich sequences, and this leads to formation of a nonterminator stem-loop that allows for further transcription.

In *Bacillus*, a different mechanism is employed. For *pyr* attenuation, an RNA-binding protein controls the alternative stem-loop structures of the *pyr* mRNA, terminating transcription when pyrimidines are in excess in the cell. In this way the cell can maintain levels of pyrimidines—compounds that require significant cell resources to biosynthesize (∞Section 5.16)—at levels needed to balance biosynthetic needs.

 8.11   Concept Check

Attenuation is a mechanism whereby gene expression (typically at the level of transcription) is controlled *after* initiation of RNA synthesis. Attenuation mechanisms involve a coupling of transcription and translation and can therefore occur only in prokaryotes.

◆   Why can attenuation not occur in *eukaryotes*?

◆   Explain how the formation of one stem-loop in the RNA can block the formation of another.

◆   How does attenuation of the tryptophan operon differ between *Escherichia coli* and *Bacillus subtilis*?

8.12   Signal Transduction and Two-Component Regulatory Systems

Bacteria regulate cell metabolism in response to a wide variety of environmental fluctuations, including temperature changes, changes in pH and oxygen availability, changes in the availability of nutrients, and even changes in the number of cells present. Therefore, there must be mechanisms by which bacteria receive *signals from the environment* and transmit them to the specific target to be regulated.

We have seen in preceding sections that some signals can be small molecules that enter the cell (often by specific uptake mechanisms) and function as *effectors*. However, in many cases the external signal is not transmitted directly to the regulatory protein, but instead to a *sensor* that transmits the signal to the rest of the regulatory machinery, a process called **signal transduction**.

Sensor Kinases and Response Regulators

Most signal transduction systems contain two parts and are thus also called **two-component regulatory systems**. Characteristically, such systems include two different proteins: (1) a specific **sensor kinase protein** located in the cell membrane, and (2) a partner **response regulator protein**.

The sensor kinase is an enzyme that phosphorylates compounds, typically using phosphate from ATP. Sensor kinases detect a signal from the environment and phosphorylate *themselves* (a process called *autophosphorylation*) at a specific histidine residue on their cytoplasmic surface (Figure 8.26●). Sensor kinases are thus also called *histidine kinases*. The phosphoryl group is then transferred to another protein inside the cell, the *response regulator*. The response regulator is typically a DNA-binding protein that regulates transcription, either in a positive or negative fashion, depending on the system (Figure 8.26).

In order to have a balanced regulatory system, there must be a feedback loop; that is, a way to complete the regulatory circuit and terminate the initial signal. This resets the system for another cycle. This feedback loop involves a *phosphatase*, an enzyme that removes the phosphoryl group from the response regulator protein at a constant rate. In some cases this reaction is carried out by the sensor kinase itself, while in other systems there is a separate protein that carries out this reaction (Figure 8.26). Phosphatase activity is typically a slower process than response regulator phosphorylation. However, if phosphorylation ceases due to reduced sensor kinase activity, phosphatase activity eventually returns the response regulator to the fully unphosphorylated state.

Examples of Two-Component Regulatory Systems

Two-component systems regulate a large number of genes in many different bacteria. Interestingly, two-component systems seem to be less widely distributed among species of *Archaea* than *Bacteria*, and are all but absent in parasitic *Bacteria*. A few key examples of two-component systems include those for phosphate assimilation in *Escherichia coli*, nitrogen fixation in *Klebsiella* and *Rhizobium*, and sporulation in *Bacillus*. In *E. coli* it is estimated that almost 50 different two-component systems operate, and a few of these are listed in Table 8.3.

Many two-component regulatory systems also involve global regulatory phenomena, as well (see Table 8.1). For example, the osmolarity of the environment controls which of the two proteins, OmpC or OmpF, is synthesized as part of the *E. coli* outer membrane. OmpC and OmpF are *porins*, proteins that allow metabolites to cross the outer membrane of gram-negative bacteria (⊂⊃Section 4.9 and Figure 4.29). If osmotic pressure is low, OmpF, a porin with a large pore, is synthesized, while if osmotic pressure is higher, OmpC, a porin with a smaller pore, is synthesized. The response regulator of this system is OmpR. When OmpR is phosphorylated, it *activates* transcription of the *ompC* gene and *represses* transcription of the *ompF* gene. The *ompF* gene in *E. coli* is also controlled by antisense RNA (see Section 8.14).

In some cases multiple regulatory elements are involved. For instance, in the *Ntr regulatory system*, which regulates nitrogen assimilation in many *Bacteria* including *E. coli*, the response regulator is an activator protein, NR_I (nitrogen regulator I). NR_I activates transcription from promoters recognized by RNA polymerase using σ^{54}, an alternative sigma factor (see Section 8.9 and Table 8.2). The sensor kinase in the Ntr system, NR_{II} (nitrogen regulator II), fills a dual role, as both kinase and phosphatase. The activity of NR_{II} is in turn regulated by the state of phosphorylation of another protein, P_{II}.

The *Nar regulatory system* (Table 8.3) controls a set of anaerobically induced genes and contains two different sensor kinases and two different response regulators. In addition, all the genes regulated by this system are also under global control of the anaerobically activated transcriptional regulatory protein FNR (see Table 8.1). It is easy to see from all of this that regulation of gene expression in prokaryotes can quickly get very complicated!

Genomic analyses allow for the ready detection of genes encoding two component regulatory systems, since histidine kinases show significant amino acid sequence conservation. Two-component systems closely related to those in a wide variety of *Bacteria* are also present in microbial

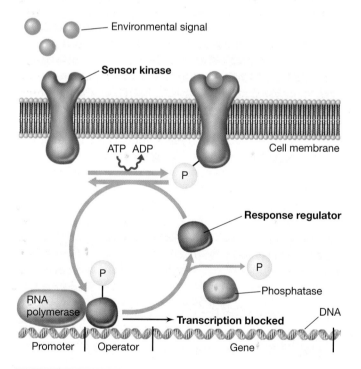

● **Figure 8.26 The control of gene expression by a two-component regulatory system.** The main components of the system include a *sensor kinase* in the cell membrane that phosphorylates itself in response to an environmental signal. The phosphoryl group is then transferred to the other main component, a *response regulator*. In the system diagrammed in this figure the phosphorylated response regulator is a repressor protein. The phosphatase slowly removes phosphate from the response regulator and closes the circuit.

Table 8.3	Examples of two-component regulatory systems that regulate transcription in *Escherichia coli*			
System	**Environmental signal**	**Sensor kinase**	**Response regulator**	**Activity of response regulator**[a]
Arc system	O_2	ArcB	ArcA	Repressor/Activator
Nitrate and nitrite anaerobic regulation (*Nar*)	Nitrate and nitrite	NarX and NarQ	NarL	Activator/Repressor
			NarP	Activator/Repressor
Nitrogen utilization (*Ntr*)	NH_4^+	NR_{II}, the product of *glnL*	NR_I, the product of *glnG*	Activates RNA polymerase at promoters requiring σ^{54}
Pho regulon	Inorganic phosphate	PhoR	PhoB	Activator
Porin regulation	Osmotic pressure	EnvZ	OmpR	Activator/Repressor

[a] Note that several of the response regulator proteins act as both activators and repressors depending on the genes being regulated. Although ArcA can function as either an activator or a repressor, it functions as a repressor on most operons that it regulates.

eukaryotes, such as the yeast *Saccharomyces cerevisiae.* Higher eukaryotes also use phosphorylation as a mechanism of signal transduction in order to respond to environmental changes. Some species of *Archaea*, including in particular the extreme halophiles (*Halobacterium* species, ⟋⟋ Section 13.3) and methanogens (⟋⟋ Section 13.4), employ two-component systems to regulate certain genes, including those encoding chemotaxis proteins (see Section 8.13).

 8.12 Concept Check

Signal transduction systems transmit environmental signals to the cell. In prokaryotes signal transduction typically involves two-component regulatory systems, which include a membrane-integrated sensor kinase and a cytoplasmic response regulator. The activity of the response regulator depends on its state of phosphorylation.

◆ What are *kinases* and what is their role in two-component regulatory systems?

◆ After depletion of an environmental signal, how are the two components reset for a second stimulus response?

8.13 Regulation of Chemotaxis

Not all two-component systems regulate transcription. We have previously seen how prokaryotes can move toward attractants or away from repellants, a process called chemotaxis (⟋⟋ Section 4.12). We noted that prokaryotes are too small to actually sense *spatial* gradients of a chemical, but that they can respond to *temporal* gradients. That is, they can sense the *change in concentration* of a chemical over time rather than the *absolute* concentration of the chemical stimulus. Prokaryotes use a two-component system to sense temporal changes in attractants or repellants and process this information to regulate flagellar rotation.

Step One: Response to Signal

The mechanism of chemotaxis is complex and involves a variety of different proteins. A number of *sensory proteins*

reside in the cell membrane, and these sense the presence of attractants and repellants. These sensor proteins are not themselves sensor kinases but are proteins that interact with cytoplasmic sensor kinases. These sensory proteins allow the cell to monitor the concentration of the substance over time.

The sensory proteins are called **methyl-accepting chemotaxis proteins (MCPs).** In *Escherichia coli*, five different MCPs have been identified, and each is a transmembrane protein (Figure 8.27●). Each MCP can sense a variety of compounds. For example, the *Tar* transducer of *E. coli* can sense the attractants aspartate and maltose as well as repellents, such as the heavy metals cobalt and nickel.

MCPs bind attractants or repellents directly, or in some cases indirectly through interactions with periplasmic binding proteins. Binding of an attractant or repellent initiates a series of interactions with cytoplasmic proteins that eventually affects flagellar rotation. Recall that if rotation of the flagellum is *counterclockwise*, the cell will continue to move in a run, whereas if the flagellum rotates *clockwise*, the cell will tumble, reorient itself, and begin to move in a random direction (⟋⟋ Section 4.14).

MCPs are in contact with the cytoplasmic proteins CheW and CheA (Figure 8.27). CheA is the *sensor kinase* in chemotaxis. When an MCP has bound a chemical, it changes conformation and (with help from CheW) causes CheA to autophosphorylate, forming CheA-P. Attractants *decrease* the rate of autophosphorylation, whereas repellents *increase* this rate. CheA-P then phosphorylates CheY (forming CheY-P); this is the *response regulator*. CheA-P can also phosphorylate CheB, another response regulator, but this is a much slower reaction than the phosphorylation of CheY. We will discuss the activity of CheB-P later.

Step Two: Controlling Flagellar Rotation

CheY is a key protein in the system because it governs the direction of rotation of the flagellum. CheY-P interacts with the flagellar motor to induce clockwise flagellar rotation and tumbling (the motor switch itself consists of proteins encoded by *fla* genes; ⟋⟋ Section 4.14). If unphosphorylated, CheY cannot bind, the flagellar motor

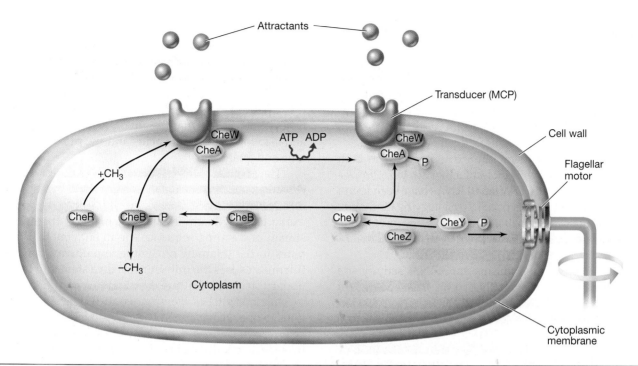

● **Figure 8.27   Interactions of MCPs, chemotaxis (Che) proteins, and the flagellar motor in bacterial chemotaxis.** The MCP forms a complex with the *sensor kinase* CheA and the coupling protein CheW. This combination results in a signal-regulated autophosphorylation of CheA to CheA-P. CheA-P can then phosphorylate the *response regulators* CheB and CheY. Phosphorylated CheY (CheY-P) interacts directly with the flagellar motor switch. CheZ dephosphorylates CheY-P. CheR continually adds methyl groups to the transducer. CheB-P (but not CheB) removes them. The degree of methylation of the MCPs controls their ability to respond to attractants and repellants and leads to adaptation. The structure of the flagellar motor was shown in Figure 4.56.

continues counterclockwise rotation, and the cell continues its run. Another protein, CheZ, dephosphorylates CheY, returning it to a form that allows runs instead of tumbles. Because repellents increase the level of CheY-P, they lead to tumbling, whereas attractants lead to a lower level of CheY-P and smooth swimming.

Step Three: Adaptation

In addition to processing a signal and regulating flagellar rotation, the system must be able to detect a *change* in concentration of the chemotactic agent with time. This problem is solved by yet another aspect of chemotaxis, the process of **adaptation.** Adaptation is the feedback loop necessary to reset the system. This involves covalent modification of the MCPs by CheB, mentioned earlier.

As their name implies, MCPs can be methylated. There is a cytoplasmic protein, CheR (Figure 8.27), that continually adds methyl groups to the MCPs at a slow rate using S-adenosylmethionine as a methyl donor. The phosphorylated form of the response regulator CheB is a *demethylase* that removes methyl groups from the MCPs. The level of methylation of the MCPs affects their conformation and controls adaptation to a sensory signal. This occurs as follows.

If the level of an attractant remains high, the level of phosphorylation of CheA (and, therefore, of CheY and CheB) will remain low, and the cell will swim smoothly. The level of methylation of the MCPs during this period

will increase (because CheB-P is not present to demethylate). However, MCPs no longer respond to the attractant when they are fully methylated. Therefore, even though the level of attractant might remain high, the level of CheA-P (and CheB-P) increases and the cell begins to tumble. However, now the MCPs can be demethylated by CheB-P, and when this happens, the receptors are "reset," and can once again respond to attractants.

The course of events is just the opposite with regard to repellants. Fully methylated MCPs respond best to an increasing gradient of repellants and send a signal for cell tumbling to begin. The cell then moves off in a random direction while MCPs are slowly demethylated. With this mechanism for adaptation, chemotaxis successfully achieves the ability to monitor changing concentrations of both attractants and repellants over time.

 8.13   *Concept Check*

Chemotaxis is under complex regulation involving signal transduction in which regulation occurs in the activity of the proteins involved rather than their synthesis. Adaptation by methylation allows the system to reset itself to the continued presence of a signal.

◆ What is the primary *response regulator* and the primary *sensor kinase* involved in regulating chemotaxis?

◆ Why is adaptation important?

8.14 RNA Regulation and Riboswitches

A major theme in this chapter has been *transcriptional control* of protein synthesis. Less often, genes are controlled at the level of *translation*. Either way, *proteins* are the regulatory elements. However, in some forms of regulation, it is a regulatory *RNA*, not a regulatory *protein*, that plays the key role in controlling gene expression.

RNA regulation is a rapidly expanding area in both prokaryotic and eukaryotic molecular biology. In *Escherichia coli*, for example, a number of **small RNAs** have been found to regulate various aspects of cell physiology by binding to other RNAs or even to small molecules in some cases. Small RNAs (sRNAs) are 40–400 nucleotides long and do not encode proteins, ribosomal RNA, or transfer RNA.

At least three mechanisms for sRNA activity are known. We have seen one sRNA in the *signal recognition particle*, a ribonucleoprotein that identifies newly synthesized proteins that are to be excreted from the cell (∞Section 7.17 and Figure 7.41). A second mechanism of sRNA activity is the binding of the sRNA to a mRNA by complementary base pairing. This ties up the mRNA and prevents its translation (Figure 8.28a●). Small RNAs that show this activity are called *antisense RNA*, so-named because the sRNA has a sequence complementary to the coding "sense" of the mRNA. Transcription of antisense RNA is enhanced by conditions that no longer favor expression of the genes they regulate, and the antisense RNAs are encoded by genes complementary to those that produce the mRNA they bind to. Antisense sRNAs survey the cytoplasm for their mRNA complements and form double-stranded RNAs from them. Double-stranded RNAs are not translatable, and these RNAs are soon degraded by specific ribonucleases (Figure 8.28a).

Riboswitches

A unique form of small RNAs are the **riboswitches**. These are *mRNAs* that contain sequences upstream of the coding sequences that can bind small molecules (Figure 8.28b). Riboswitches are thus far known in a variety of biosynthetic pathways leading to the synthesis of enzymatic cofactors, such as the vitamins thiamine, riboflavin, and B_{12}, and also for a few amino acids and the purine bases adenine and guanine.

The molecule that is bound by a given riboswitch is the substance that would be synthesized by the protein that the mRNA portion of the riboswitch encodes. For example, the thiamine riboswitch binds thiamine upstream of a coding sequence for an enzyme that participates in the thiamine biosynthetic pathway. Binding of thiamine causes the mRNA to attain a stem and loop secondary structure that prevents translation (Figure 8.28b).

In riboswitches we see a control mechanism analogous to one we have seen before. At the beginning of this chapter we discussed the control of preformed enzyme by a mechanism called *feedback inhibition* (see Section 8.1). In that case, conformational changes in an allosteric protein following the binding of a specific metabolite shut down enzymatic activity of the entire pathway (see Figure 8.5). A riboswitch can assume either of two structures as well. In the absence of the metabolite, the Shine–Dalgarno ribosome-binding site on the mRNA can bind to the ribosome and translation can begin. However, when the metabolite binds to the mRNA—a signal that a sufficient amount of the metabolite is present—the secondary structure of the riboswitch prevents mRNA binding to the ribosome, and thus translation is prevented.

How widespread are riboswitches and how did they evolve? Thus far riboswitches have only been found in

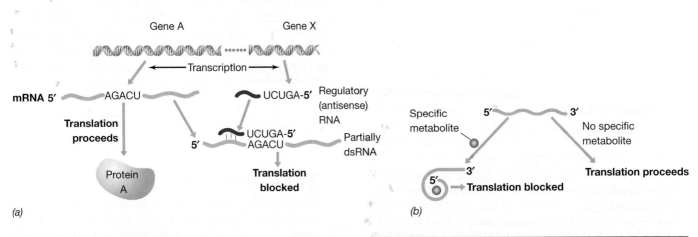

● **Figure 8.28** **Regulatory RNAs.** (a) Antisense RNA. Gene A is transcribed from its promoter (∞Section 7.10) to yield an mRNA that can be translated to form protein A. Gene X is a small gene whose sequence is identical to that of a portion of Gene A but has its promoter at the opposite end. Therefore, if it is transcribed, the resulting RNA will be complementary to the mRNA of gene A. If these two RNAs base pair, translation will be blocked. (b) Riboswitches. Binding of a specific metabolite affects the secondary structure of the mRNA portion of the riboswitch, preventing translation.

some bacteria and a few plants and fungi. Some scientists believe that riboswitches are remnants of the *RNA world*, a period eons ago before cells, DNA, and protein, when it is hypothesized that catalytic RNAs were the only self-replicating life forms (◌◌Section 11.2). In such an environment, riboswitches may have been a primitive mechanism of metabolic control—a simple means by which RNA life forms could have controlled the synthesis of other RNAs. As proteins evolved, riboswitches might have been the first control mechanisms for their synthesis, as well. If true, the riboswitches that remain today are the last vestiges of this simple form of control, since, as we have seen in this chapter, most have been replaced by regulatory proteins.

REVIEW QUESTIONS

1. If an enzyme can be effectively inhibited by feedback inhibition, why would cells also have mechanisms to regulate its synthesis (◌◌Section 8.1)?

2. Contrast DAHP synthase regulation in *Escherichia coli* with that of an organism employing a concerted feedback inhibition. Which employs more proteins? How can the final result be the same in each case (◌◌Section 8.2)?

3. In what ways are the events leading to the production of the protein GyrA in *Mycobacterium leprae* similar and dissimilar to gene splicing in eukaryotes (◌◌Section 8.3)?

4. Describe why a protein that binds to a specific sequence of double-stranded DNA is unlikely to bind to the same sequence if the DNA is single-stranded (◌◌Section 8.4).

5. Most biosynthetic operons need only be under negative control for effective regulation, while most catabolic operons need to be under *both* negative and positive control. Why (◌◌Sections 8.5 and 8.6)?

6. Describe the mechanism by which catabolite activator protein (CAP), the regulatory protein for catabolite repression, functions using the lactose operon as an example. For this operon the CAP protein is not a repressor. Describe the regulatory region of a gene for which the CAP protein *is* a repressor (◌◌Section 8.7).

7. What events trigger the stringent response? Why are the events that occur in the stringent response a logical consequence of the trigger of the response (◌◌Section 8.8)?

8. Describe the proteins produced when cells of *Escherichia coli* experience a heat shock. Of what value are they to the cell (◌◌Section 8.9)?

9. How can quorum sensing be considered a regulatory mechanism for conserving cell resources (◌◌Section 8.10)?

10. Describe how transcriptional attenuation works. What is actually being "attenuated"? Why hasn't the type of attenuation that controls several different amino acid biosynthetic pathways in *Escherichia coli* also been found in eukaryotes (◌◌Section 8.11)?

11. What are the two components that give the name to signal transduction regulation in prokaryotes? What is the function of each of the components (◌◌Section 8.12)?

12. Adaptation allows the regulatory mechanism controlling flagellar rotation to be reset. How is this accomplished (◌◌Section 8.13)?

13. How does regulation by antisense RNA differ from that of riboswitches? Which form of regulation is better suited to control of synthesis of enzymes in a biosynthetic pathway? Why (◌◌Section 8.14)?

14. Although completely different forms of regulations, a form discussed in Section 8.2 shows similarities to a form discussed in Section 8.14. Explain.

APPLICATION QUESTIONS

1. What would happen to regulation from a promoter under negative control if the region where the regulatory protein binds were deleted? What if the promoter were under positive control?

2. Promoters from *Escherichia coli* under positive control are not close matches to the DNA consensus sequence for *E. coli* (◌◌Section 7.9). Why?

3. The attenuation control of some of the pyrimidine biosynthetic pathway genes in *Escherichia coli* actually involves coupled transcription and translation. Can you describe a mechanism whereby the cell could somehow make use of translation to help it measure the level of pyrimidine nucleotides?

4. Most of the regulatory systems described in this chapter involve regulatory proteins. However, regulatory RNA is known to exist. Describe how one could achieve negative control of the *lac* operon using either of two different types of regulatory RNA.

5. Many amino acid biosynthetic operons under attenuation control are also under negative control. Considering that the environment of a bacterium can be highly dynamic, what advantage could be conferred by having attenuation as a second layer of control?

ESSENTIALS OF VIROLOGY

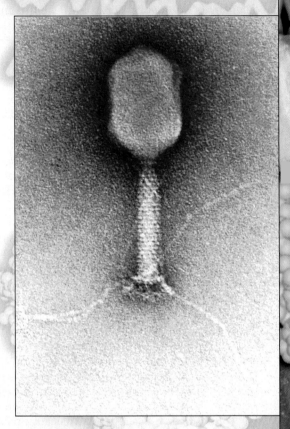

Viruses are infectious agents that require cells for their reproduction. Although the genome of many viruses is double-stranded deoxyribonucleic acid (DNA), for others it is single-stranded DNA and for still others, RNA. The infectious form of a virus is called a *virion*, as shown here for bacteriophage T4.

WORKING GLOSSARY

Bacteriophage a virus that infects prokaryotic cells

Early protein a protein synthesized soon after virus infection

Late protein a protein synthesized toward the end of virus infection

Lysogen a bacterium containing a prophage

Lysogenic pathway a series of steps that, after virus infection, lead to a state (lysogeny) where the viral genome is replicated as a prophage along with that of the host

Lytic pathway a series of steps after virus infection that leads to virus replication and the destruction (lysis) of the host cell

Minus (negative)-strand virus a virus with an RNA genome in which the RNA strand has the opposite sense of (is complementary to) the mRNA of the virus

Nucleocapsid the complex of nucleic acid and proteins of a virus

Oncogene a gene whose expression causes formation of a tumor

Plaque a zone of lysis or cell inhibition caused by virus infection of a lawn of sensitive cells

Plus (positive)-strand virus a virus with an RNA or DNA genome in which the genome has the same complementarity as the mRNA of the virus

Prion an infectious protein whose extracellular form contains no nucleic acid

Provirus (prophage) the genome of a temperate virus when it is replicating with, and usually integrated into, the host chromosome

Retrovirus a virus whose RNA genome has a DNA intermediate as part of its replication cycle

Reverse transcription the process of copying information found in RNA into DNA by the enzyme reverse transcriptase

Temperate virus a virus whose genome is able to replicate along with that of its host and not cause cell death in a state called lysogeny

Transformation in eukaryotes, a process by which a normal cell becomes a cancer cell (but see alternative usage in Chapter 10)

Virion the complete virus particle; the nucleic acid surrounded by a protein coat and in some cases other material

Virulent virus a virus that lyses or kills the host cell after infection; a nontemperate virus

Virus a genetic element containing either RNA or DNA that replicates in cells but is characterized by having an extracellular state

Viroid small, circular, single-stranded RNA that causes various plant diseases

---

Viruses are genetic elements that replicate independently of a cell's chromosome(s) but not independently of cells themselves (∞Section 7.4). However, unlike genetic elements such as plasmids (∞Sections 7.4 and 10.9), viruses have an *extracellular form* that enables them to exist outside the host for long periods and that facilitates transmission from one host to another. In order to multiply, viruses must enter a cell in which they can replicate, a process called **infection**. The infected cell is called the **host**. The study of viruses is a key subdiscipline of microbiology called *virology*, and we introduce the essentials of the field in this chapter.

Viruses, like plasmids and some other genetic elements, exploit the metabolic machinery of the cell. Like these other elements, viruses can confer important new properties on their host cell. These properties will be inherited when the host cell divides if each new cell also inherits the viral genome. These changes are often not harmful and may even be beneficial. Viruses can replicate in a way that is destructive to the host cell, and this accounts for the fact that some viruses are agents of disease. We cover a number of human diseases caused by viruses in later chapters (∞Chapters 26–29).

This chapter is divided into four parts. The first part introduces basic concepts of virus structure. The second part deals with the host cell and how viruses can be quantitated. The third part deals with the basic molecular biology of virus multiplication. The final part provides an overview of a few of the viruses that infect bacteria and animals. Further coverage of viral diversity can be found in Chapter 16. The topics in this chapter also expand on the concepts of macromolecular synthesis and gene regulation covered in Chapters 7 and 8.

Viruses are among the most numerous microorganisms on our planet and infect all types of cellular organisms. Therefore, they are interesting to study in their own right. However, scientists also study viruses for what they can tell us about the genetics and biochemistry of cellular processes and, in the case of many viruses, the development of disease. Furthermore, as we shall see in Chapters 10, 15, and 31, viruses are also important tools for the microbial geneticist and the genetic engineer.

VIRUS AND VIRION

9.1 General Properties of Viruses

Viruses can exist in either extracellular or intracellular forms. In the *extracellular* form, a virus is a minute particle containing nucleic acid surrounded by protein and, occasionally, depending on the specific virus, other macromolecules. In the extracellular form, the virus particle, also called the **virion**, is metabolically inert and does not carry out respiratory or biosynthetic functions. The virion is the structure by which the virus genome moves from the cell in which it has been produced to another cell where the viral nucleic acid can be introduced. Once in the new cell, the *intracellular state* is initiated. In the intracellular state, virus replication occurs: New copies of the virus genome are produced, and the components that make up the virus coat are synthesized.

Viral genomes are very small, and they encode primarily those functions that they cannot adapt from their

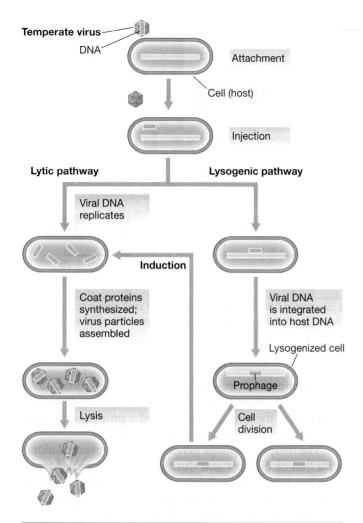

● **Figure 9.16 The consequences of infection by a temperate bacteriophage.** The alternatives upon infection are *replication* and release of mature virus (lysis) or *integration* of the virus DNA into the host DNA (lysogenization). The lysogenic cell can also be induced to produce mature virus and lyse.

Instead, the prophage is integrated into the bacterial chromosome and replicates along with the host cell as long as the genes controlling its lytic pathway are not expressed. Typically this control is maintained by a phage-encoded repressor protein (the gene encoding the repressor protein *is* expressed). The virus repressor protein not only controls the lytic genes on the prophage but also prevents the expression of any incoming genes of the same virus. This results in the lysogens having **immunity** to infection by the same type of virus.

If the phage repressor is inactivated or if its synthesis is prevented, the prophage is induced (Figure 9.16). Induction results in the production of new virions and the lysis of the host cell. In some cases (as we shall see later), induction can be brought about by environmental conditions. If the virus loses the ability to leave the host genome (because of mutation), it becomes a cryptic virus. Genomic studies (∞Chapter 15) have shown that many bacterial chromosomes contain DNA se-

quences that were clearly once part of a viral genome. Thus the establishment and breakdown of the lysogenic state is likely a dynamic process in prokaryotes.

We now focus on a bacteriophage whose lysogenic properties are known in great detail, bacteriophage lambda.

9.10 Concept Check

Lysogeny is a state in which viral genes become integrated into the host chromosome and lytic events are repressed. Viruses capable of inducing the lysogenic state are called temperate viruses. In lysogeny, the virus genome becomes a prophage; however, lytic events can be induced by certain environmental stimuli.

◆ What are the two pathways available to a temperate virus?

◆ What might the advantage be to hosts of temperate viruses?

9.11 Bacteriophage Lambda

One of the best-studied temperate phages is **lambda**, which infects *Escherichia coli*. As with other temperate viruses, both the *lytic* and the *lysogenic* pathways are possible (Figure 9.16). Morphologically, lambda virions look like those of many other tailed bacteriophages, although unlike phage T4, no tail fibers are present (Figure 9.17●; see also Figure 9.12). The genome of lambda consists of a linear *double-stranded* DNA molecule, but at the 5'-terminus of each of the strands is a *single-stranded* tail 12 nucleotides long. These single-stranded ends are complementary (the ends of the DNA are said to be *cohesive*). Thus, when the two ends of the DNA are free in the host cell they associate (the *cos* site), and the DNA is ligated to form a double-stranded circle of 48,502 base pairs. Figure 9.18● shows this process along with the genetic map of lambda.

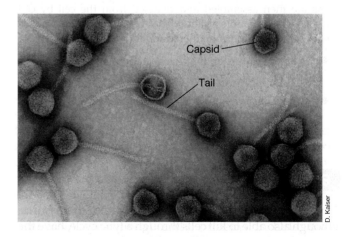

● **Figure 9.17 Bacteriophage lambda.** Electron micrograph by negative staining of phage lambda virions. The head of each virion is about 65 nm in diameter.

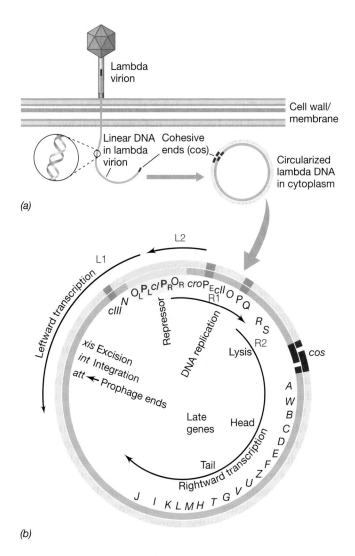

(a)

(b)

● **Figure 9.18** **The lambda genome.** (a) Conversion of lambda DNA from a linear to a circular molecule upon infection. (b) Genetic and molecular map of lambda. The genes are designated by letters; *att*, attachment site for phage to host chromosome. Regulatory genes of special interest: *cI*, repressor protein; O_R, operator right; P_R, promoter right; O_L, operator left; P_L, promoter left; *cro*, gene for second repressor; N, antiterminator protein. The regulatory region of lambda (shown in yellow) is positioned at the top of this circular map. The site created when the cohesive ends of the lambda genome join is called *cos* (shown in red). Early transcription in lambda is primarily leftward (counterclockwise, L1) from P_L and rightward (clockwise, R1) from P_R. The three transcription terminators affected by the N-protein are shown as blue boxes on the DNA. The late rightward transcript, which encodes head and tail proteins and proteins for lytic function, is labeled R2 and begins at a promoter near the Q gene. The transcript labeled L2 is the positively regulated transcript from P_E that encodes the repressor protein.

Lambda Infection and the Lytic Pathway

The lambda virion attaches to a specific receptor protein in the cell wall of *Escherichia coli* (see Section 9.6) and injects its DNA. The DNA circularizes almost immediately (Figure 9.18*a*), and expression of the phage genome begins. The first steps in gene expression are the same whether the final result is lysis or lysogeny.

Production of RNA using host RNA polymerase begins at two key promoters called P_L (*promoter left*) and P_R (*promoter right*) (Figure 9.18*b*). These yield short transcripts that are translated to give the products of the N and the *cro* genes, both proteins being involved in regulatory events. The *Cro* protein regulates whether the lytic or the lysogenic pathway is followed, and we discuss its function later. The N protein is an *antiterminator* that allows the RNA polymerase to transcribe past specific terminators (Figure 9.18*b*), extending the transcripts from P_L and P_R. These longer transcripts can be translated to yield more proteins, including the products of the *cII* and *cIII* genes. The N protein is not completely effective at the terminator before the Q gene, and so only a small amount of the Q protein is made.

The Q protein is also an antiterminator protein. If its concentration becomes high enough, it will allow the transcript from a nearby promoter to synthesize the transcript labeled R2 in Figure 9.18. This transcript is translated to yield the late proteins, all the necessary structural proteins to construct a virion, and the proteins necessary for cell lysis. However, at the same time the Q protein is accumulating, the Cro protein also reaches levels where it can block transcription from both P_L and P_R by binding to both O_L (operator left) and O_R (operator right). Therefore, Cro functions as a repressor protein (∞Section 8.5).

The mechanism of Cro protein binding at O_R is detailed in Figure 9.19●. Note that there are three similar but nonidentical sites at this operator where the Cro protein can bind. It does so first at site 3, and then site 2, and only when those two sites are filled, at site 1. Note that only when bound to site 1 does it block P_R. Note also that once P_L and P_R are blocked, no more cII or cIII proteins can be synthesized. These proteins are needed to enter the lysogenic pathway (see later). Thus, *when Cro is made in high amounts, lambda is irreversibly committed to the lytic pathway.*

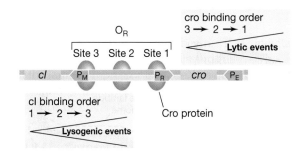

● **Figure 9.19** **Lambda operator right (O_R).** Both regulatory proteins Cro and the lambda repressor bind to O_R to carry out regulatory functions. The Cro protein binds to the three sites in the order site 3, then site 2, and then site 1, turning off cI synthesis first and its own synthesis last. The lambda repressor (cI) binds to these sites in the opposite order, turning off Cro synthesis first and its own synthesis last. Compare this figure with the summary of lambda lytic/lysogenic steps in Figure 9.21.

◆ What events need to happen for lambda to become a prophage?

◆ What regulatory events occur at the O_R site on the lambda chromosome?

9.12 Overview of Animal Viruses

The first few sections of this chapter were devoted to general properties of all viruses, and little was said about animal viruses. Here we will consider some of the attributes of animal viruses, setting the stage for a discussion of the retroviruses, one of which causes *acquired immunodeficiency syndrome* (AIDS).

In our discussions of viral reproduction in Sections 9.5–9.7, we also dealt with the virus "host" in a very general way. However, it is important to remember that the bacteriophage host is a prokaryotic cell while the host of an animal virus is a eukaryotic cell. We will emphasize these important differences in Chapter 16, where we discuss several types of animal viruses in more detail. However, the key points to keep in mind for our discussion here are that (1) unlike in prokaryotes, the entire virus virion enters the animal cell, and (2) eukaryotic cells contain a nucleus, where most viruses have to replicate. In addition, there are details of RNA processing (⌘ Section 14.8) that affect the replication of animal viruses, but again, these are details that need not bother us here.

Classification of Animal Viruses

Various types of animal viruses are illustrated in Figure 9.23●. We discussed the principles by which viruses are classified in Sections 9.1, 9.2, and 9.7. Note that there are many more kinds of enveloped animal viruses than enveloped bacterial viruses (see Section 9.7). This likely relates to the differences in host cell exteriors. Unlike prokaryotic cells, animal cells lack a cell wall, and thus viruses are more easily secreted from the cell. As they are, they remove part of the cell's lipid bilayer as they pass through the membrane.

Most of the animal viruses that have been studied in any detail are those that have been amenable to growth in cell cultures. Animal viruses are classified according to the same schemes as bacterial viruses, including the Baltimore Classification System (see Table 9.2), which classifies viruses on the basis of genome type and reproductive strategy. Animal viruses are known in all replication categories and many of these will be discussed in Chapter 16.

Consequences of Virus Infection in Animal Cells

Viruses can have several different effects on animal cells. **Lytic infection** results in the destruction of the host cell (Figure 9.24●). In the case of enveloped viruses, however, release of virions, which occurs by a kind of budding process, may be slow, and the host cell may not be lysed.

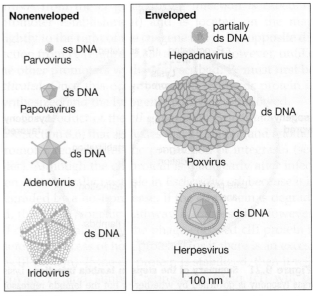

(a) **DNA viruses**

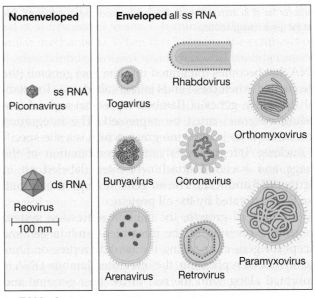

(b) **RNA viruses**

● **Figure 9.23 Diversity of animal viruses.** The shapes and relative sizes of vertebrate viruses of the major taxonomic groups. The hepadnavirus genome has one complete DNA strand and part of its complement (⌘ Section 16.15).

The infected cell may therefore remain alive and continue to produce virus indefinitely. Such infections are called **persistent infections** (Figure 9.24).

Viruses may also cause **latent infection** of a host. In a latent infection, there is a delay between infection by the virus and lytic events. Fever blisters (cold sores), caused by the herpes simplex virus (⌘ Section 16.12), are a typical example of a latent viral infection; the symptoms (the result of lysed cells) reappear sporadically as the virus emerges from latency. The latent stage in viral infection of an animal cell is generally not due to in-

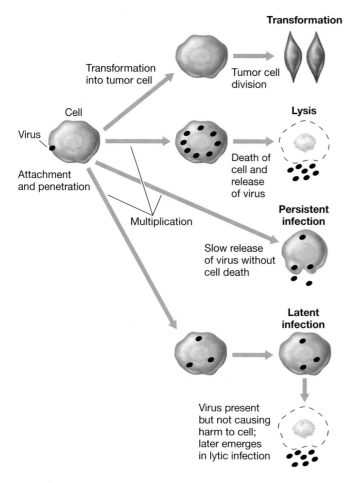

Transformation

Transformation into tumor cell

Tumor cell division

Cell

Virus

Attachment and penetration

Multiplication

Lysis

Death of cell and release of virus

Persistent infection

Slow release of virus without cell death

Latent infection

Virus present but not causing harm to cell; later emerges in lytic infection

● **Figure 9.24 Possible effects that animal viruses may have on cells they infect.** Note that most viruses are lytic and that only a very few are known to cause cancer. Many tumorigenic viruses belong to the Herpesvirus group (∞Section 16.12).

tegration of the viral genome into the genome of the animal cell, as is the case with latent infections by temperate bacteriophages.

Finally, certain animal viruses can catalyze the conversion of a normal cell into a tumor cell, a process called **transformation** (Figure 9.24). We discuss cancer-causing viruses in Section 16.12.

Many different viruses are known. But of all the viruses listed in Figure 9.23, one stands out as having an absolutely unique mode of replication. These are the retroviruses. We explore them here as an example of a complex and highly unusual animal virus with significant medical implications.

 9.12 Concept Check

Animal viruses are known that cover all known modes of viral genome replication. Many animal viruses are enveloped, picking up portions of the host cytoplasmic membrane as they leave the cell. Not all infections of animal host cells result in cell lysis or death; latent or persistent infections are common, and some animal viruses can cause cancer.

◆ Differentiate between a *persistent* and a *latent* viral infection.

◆ Contrast the mechanisms by which animal viruses enter cells with those used by bacterial viruses.

9.13 Retroviruses

Retroviruses contain an RNA genome that is replicated through a DNA intermediate (see Section 9.1 and Figure 9.11). The term *retro* means "backward," and the name of this class of virus is derived from the fact that these viruses transfer information backward from RNA to DNA (∞Section 7.1). Retroviruses employ the enzyme **reverse transcriptase** to carry out this interesting process. The use of reverse transcriptase is not restricted to the retroviruses. Hepatitis B virus (a human virus) and cauliflower mosaic virus (a plant virus) also use **reverse transcription** in their replication processes (∞Section 16.15). But in contrast to the retroviruses, these latter viruses encapsidate *DNA* as their genome rather than *RNA*.

The retroviruses are interesting for several other reasons. For example, they were the first viruses shown to cause *cancer* and have been studied most extensively for their carcinogenic characteristics. Also, one retrovirus, *human immunodeficiency virus* (HIV) causes *acquired immunodeficiency syndrome (AIDS)*. This virus infects a specific kind of T lymphocyte (T-helper cell) in humans that is vital for proper functioning of the immune system. In later chapters we discuss the medical and immunological aspects of AIDS (∞Sections 25.6 and 26.14).

Retroviruses are enveloped viruses (Figure 9.25a●). There are several proteins in the virus coat and typically seven internal proteins, four of which are structural and three enzymatic. The enzymes found in the virus particle are *reverse transcriptase, DNA endonuclease (integrase)*, and a *protease*. The virion also contains specific cellular tRNA molecules used in *replication* (see later in this section).

Features of Retroviral Genomes and Replication

The genome of the retrovirus is unique. It consists of *two* identical single-stranded RNA molecules of plus complementarity. A genetic map of a typical retrovirus genome is shown in Figure 9.25b. Although there are differences between the genetic maps of different retroviruses, all contain the following genes and are arranged in the same order: *gag*, encoding structural proteins; *pol*, encoding reverse transcriptase and integrase; and *env*, encoding envelope proteins. Some retroviruses, such as Rous sarcoma virus, carry a fourth gene downstream from *env* that is involved in cellular transformation and cancer. The terminal repeats shown on the map play an essential role in the replication process (see later).

The overall process of replication of a retrovirus can be summarized in the following steps (Figure 9.26●):

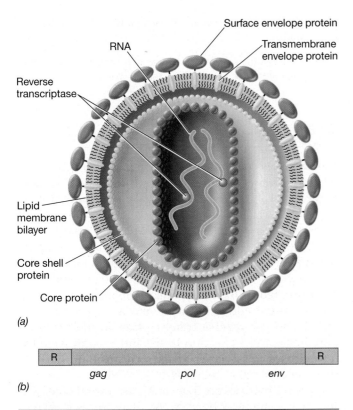

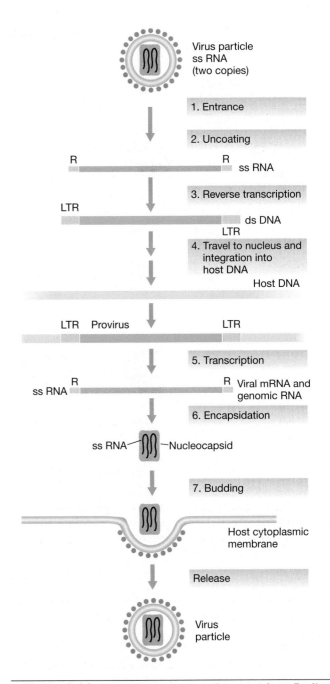

● **Figure 9.25   Retrovirus structure and function.** (a) Structure of a retrovirus. (b) Genetic map of a typical retrovirus genome. Each end of the genomic RNA contains direct repeats (R). See text for more details.

1. **Entrance** into the cell via fusion with cell membrane at sites of specific receptors.

2. **Uncoating of the virion** at the membrane, but such that the genome and enzymes remain in the core.

3. **Reverse transcription** of *one* of the two RNA genomes into a single-stranded DNA that is subsequently converted to a linear double-stranded DNA by reverse transcriptase, which then enters the nucleus.

4. **Integration** of retroviral DNA into the host genome.

5. **Transcription** of retroviral DNA, leading to the formation of viral mRNAs and viral genomic RNA.

6. **Assembly and encapsidation** of the genomic RNA into nucleocapsids in the cytoplasm.

7. **Budding** of enveloped virions at the cytoplasmic membrane and release from the cell.

Activity of Reverse Transcriptase

A very early step after the entry of the RNA genome into the cell is reverse transcription: conversion of RNA into a DNA copy using the enzyme reverse transcriptase present in the virion. The DNA formed is a linear double-stranded molecule and is synthesized in the cytoplasm within an uncoated viral core particle. Details on this process can be found in Section 16.15. Here we note only that reverse transcriptase is a DNA polymerase and, like all DNA polymerases, must have a primer (⌒⌒⌒Section 7.5). The

● **Figure 9.26   Replication process of a retrovirus.** R, direct repeats; LTR, long terminal repeats. For more detail on conversion of RNA to DNA (step 3) refer to Figure 16.25.

primer for retrovirus reverse transcription is a specific *cellular transfer RNA (tRNA)*. The type of tRNA used as primer depends on the virus and is brought into the virion from the previous host cell.

The overall process of reverse transcription results in a product that has long terminal repeats (LTRs, Figure 9.26) that are longer than the terminal repeats on the genome itself (Figure 9.25). This entire double-stranded DNA molecule enters the nucleus along with the integrase protein; here the viral DNA is integrated into the host DNA. The LTRs contain strong promoters of transcription and are also involved in the integration process. The *integration* of the viral DNA into the host genome is analogous to the integration of phage DNA into a bacterial genome to form the lysogenic state (see Section 9.10). Integration of viral DNA can occur anywhere in the cellular DNA, and once integrated, the element, now called a **provirus**, is a stable genetic element and can remain this way indefinitely.

If the promoters in the right LTR are activated, the integrated proviral DNA is transcribed by a cellular RNA polymerase into transcripts. These RNA transcripts either may be encapsidated into virus particles (as the genome) or may be translated into virus proteins. Some virus proteins are made initially as a large primary *gag* protein, which is split by proteolysis into the capsid proteins (∞Figure 16.26). Occasionally, read-through past the *gag* region (involving either inserting an amino acid in response to a stop codon or a shift in reading frame by the ribosome) leads to the translation of *pol*, the reverse transcriptase gene. This allows for reverse transcriptase to be produced for insertion into virions.

When virus proteins have accumulated in sufficient amounts, assembly of nucleocapsids occurs. Encapsidation of the RNA genome leads to the formation of mature nucleocapsids, which move to the cytoplasmic membrane for final assembly into the enveloped virus particles. As nucleocapsids bud through the membrane they are sealed and then released to infect neighboring cells (Figure 9.26).

 9.13 Concept Check

Retroviruses are RNA viruses that replicate through a DNA intermediate. The retrovirus called human immunodeficiency virus causes AIDS. The retrovirus virion contains an enzyme, reverse transcriptase, that copies the information from its RNA genome into DNA. The DNA becomes integrated into the host chromosome in the manner of a temperate virus. The retrovirus DNA can be transcribed to yield mRNA (and new genomic RNA) or may remain in a latent state.

◆ Why are these viruses called *retro*viruses?

◆ How does the life cycle of a temperate bacteriophage differ from that of a retrovirus?

V SUB-VIRAL PARTICLES

If our brief tour of viral diversity, which we expand upon in Chapter 16, didn't exhaust all possibilities for replication schemes and viral idiosyncrasies, we end this chapter on viral essentials by considering two viral-like infectious agents that are not viruses: viroids and prions. Both are best described as sub–viral particles.

9.14 Viroids and Prions

Recall that our definition of a virus is of a genetic element that subverts normal cellular processes for its own

replication and that has a mature infectious extracellular form. There are a few infectious agents that resemble viruses but whose properties are at odds with this definition and are thus not viruses. Two of the most important are *viroids* and *prions*, and we consider these here.

Viroids

Viroids are small, circular, *single-stranded* RNA molecules that are the smallest known pathogens. Viroids range in size from 246 to 399 nucleotides and show a considerable degree of sequence homology, suggesting that they have common evolutionary roots. Viroids cause a number of important plant diseases and can have a severe agricultural impact. A few well-studied viroids include coconut cadang-cadang viroid (246 nucleotides), citrus exocortis viroid (375 nucleotides), and potato spindle tuber viroid (359 nucleotides; see Figure 9.28).

The extracellular form of the viroid is *naked RNA*—there is no protein capsid of any kind. Even more interestingly, *the RNA molecule contains no protein-encoding genes*, and therefore the viroid is totally dependent on host function for its replication. Although the viroid RNA is a single-stranded circle, there is such considerable secondary structure possible that it resembles a short double-stranded molecule with closed ends (Figure 9.27●). The viroid molecule enters the plant through a wound, as from insect or other mechanical damage. The viroid is replicated in the host cell nucleus by one of the plant RNA polymerases.

Viroid-infected plants can be symptomless or develop symptoms that range from mild to lethal, depending on the viroid (Figure 9.28●). Most severe symptoms are growth related, suggesting that viroids are a type of "regulatory RNA" (∞Section 8.14), examples of which are widely known in both plants and animals. Thus, viroids could be regulatory RNAs that have evolved away from constructive roles in the cell to now catalyzing destructive events in the cell. No viroid diseases of animals are known, and the precise mechanisms by which viroids cause plant diseases remain unclear.

Prions

Prions represent the other extreme from viroids. They have a distinct extracellular form, but the extracellular form is *entirely protein*. The prion particle does not contain any nucleic acid. However, it is infectious, and prions are known to cause a variety of diseases in animals, such as scrapie in sheep, *bovine spongiform encephalopathy* in cattle (BSE or "mad cow disease"), chronic wasting disease in

● **Figure 9.27 Viroids.** Structure of viroids, showing how single-stranded circular RNA can form a seemingly double-stranded structure by intrastrand base-pairing.

● **Figure 9.28 Viroids.** Photograph of healthy (left) and potato spindle tuber viroid (PSTV)-infected (right) tomato plants. The host range of most viroids is quite restricted. However, PSTV will infect tomato as well as potato, causing growth stunting, a flat top, and premature plant death (right).

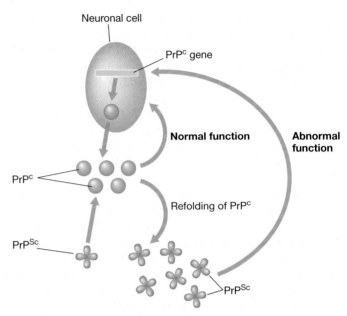

● **Figure 9.29 Mechanism of action of a prion.** Neuronal cells produce prion protein (PrP^c), which carries out some normal function in the cell. Abnormally folded prion protein (PrP^{sc}) can catalyze refolding of PrP^c into PrP^{sc}. The PrP^{sc} form is protease resistant, insoluble, and forms aggregates in neural cells. This eventually leads to destruction of neural tissues and neurological symptoms.

deer and elk, and kuru and a form of Creutzfeldt-Jakob Disease (CJD) in humans. No prion diseases of plants are known. Collectively, prion diseases have come to be known as **transmissible spongiform encephalopathies**, or TSEs. In 1997 the American scientist Stanley B. Pruisner won the Nobel Prize for his pioneering work with these diseases and with the prion proteins.

In 1996 it became clear from disease tracking in England that the prion that causes BSE in cattle can also infect humans, resulting in a variant of CJD, called *new variant CJD (nvCJD)*. Because transmission was from contaminated beef products, in just a few years nvCJD became a worldwide health concern with a major impact on the animal husbandry industry (∞Section 29.11).

Most such instances of BSE occurred in the United Kingdom or other European Union (EU) countries and were linked to improper feeding practices in which protein supplements containing rendered cattle and sheep were used to feed uninfected animals. Since 1994 this practice has been banned in all EU countries, and cases of BSE have dropped dramatically. Thus far, TSE transmission via other domesticated animals, such as swine, chicken, or fish, has not been found.

Although a consensus has not been reached on how prions cause disease, the most likely scenario is as follows (Figure 9.29●). Prion infection results in production of more copies of the prion protein, and this leads to disease symptoms. Symptoms are invariably neurological, and in most cases are due to destruction of brain or related nervous tissue by the accumulation of aggregated prion protein (∞Section 29.10 and Figure 29.10). But if prions lack nucleic acid, how is the protein they consist of encoded?

The host cell contains a gene that encodes a glycoprotein called PrP^C, which is nearly identical to the prion protein. PrP^C is produced in the healthy animal and is found mostly in neurons. The incoming prion, called PrP^{Sc}, is a structurally modified form of the PrP^C protein and can itself modify PrP^C, converting it into PrP^{Sc}. The modification involves an alternative pattern of folding. This causes the protein to lose its normal function, to become partially resistant to proteases, and to become insoluble in the neural cell, leading to aggregation (Figure 9.29). In this state, prion protein accumulates (∞Figure 29.10) and neurological symptoms commence. Therefore, prions do not simply subvert host enzymes in the cell, but convert a normal cell protein into a *self-propagating conformational state*.

Viroids and prions do more than stretch our definition of a virus. They also demonstrate both the unexpected ways that genetic elements can replicate and the unexpected ways they can subvert the host cells. However, prions stand out among all of the genetic elements we have considered in this chapter because the infectious transmissible agent lacks nucleic acid.

 9.14 Concept Check

Viroids are circular single-stranded RNA molecules that do not encode proteins and are completely dependent on host-encoded enzymes. By contrast, prions consist of protein but have no nucleic acid. Collectively, prions and viroids are the smallest known pathogens.

◆ If viroids are circular molecules, why are they drawn as shown in Figure 9.27?

◆ How does a prion induce disease?

REVIEW QUESTIONS

1. In what ways do viral genomes differ from those of cells (∞Section 9.1)?

2. Define *virus*. What are the minimal features needed to fit your definition (∞Section 9.2)?

3. Define the term *host* as it relates to viruses (∞Section 9.3).

4. Describe the events that occur on an agar plate containing a bacterial lawn when a single bacteriophage particle causes the formation of a *bacteriophage plaque* (∞Section 9.4).

5. Under some conditions, it is possible to obtain nucleic acid-free protein coats (*capsids*) of certain viruses. Under the electron microscope these capsids look very similar to complete virions. What does this fact tell you about the role of the virus nucleic acid in the virus assembly process? Would you expect such virions to be infectious? Why (∞Section 9.5)?

6. Describe how a *restriction endonuclease* might play a role in resistance to bacteriophage infection. Why could a restriction endonuclease play such a role whereas a generalized DNase could not (∞Section 9.5)?

7. One can divide the replication process of a virus into five steps. Describe the events associated with each of these steps (∞Sections 9.6 and 9.7).

8. Specifically, why are both the life cycle and the virion of a positive-strand RNA virus likely to be simpler than those of a negative-strand RNA virus (∞Section 9.8)?

9. In terms of structure, how does the genome of bacteriophage T4 resemble and differ from that of *Escherichia coli* (∞Section 9.9)?

10. Many of the viruses we have considered have *early* genes and *late* genes. What is meant by these two classifications? What types of proteins tend to be encoded by early genes? What type of proteins by late genes? For bacteriophages T4 and lambda describe how expression of the late genes is controlled (∞Section 9.9).

11. Define the following: virulent, lysogeny, prophage (∞Section 9.10).

12. A strain of *Escherichia coli* that is missing the outer membrane protein responsible for maltose uptake is *resistant* to lambda infection. A lambda lysogen is *immune* to lambda infection. Describe the functional difference between resistance and immunity. Explain how these conditions are brought about in the examples given (∞Sections 9.10 and 9.11).

13. Describe and differentiate the effects animal virus infection can have on an animal (∞Section 9.12).

14. Typically, transfer RNA is used in translation. However, it also plays a role in the replication of retroviral nucleic acid. Explain this role (∞Section 9.13).

15. What are the similarities and differences between prions, viruses, and viroids (∞Section 9.14)?

APPLICATION QUESTIONS

1. What causes the viral plaques that appear on a bacterial lawn to stop growing larger?

2. The promoters for mRNA encoding *early* proteins in viruses like T4 have a different sequence than the promoters for mRNA encoding *late* proteins in the same virus. Explain why this might be true.

3. One characteristic of *temperate bacteriophages* is that they cause turbid rather than clear plaques on bacterial lawns. Can you think why this might be? (Remember the process by which a bacterial plaque develops.)

4. There are three lambda genes that, when rendered nonfunctional, turn lambda from a temperate to a virulent virus. What are these three genes and how do they normally function?

5. Figure 9.19 shows two promoters, P_R and P_E, on either side of the *cro* gene. Both transcribe through the *cro* gene, but only the transcript from P_R can be translated to yield a functional Cro protein. Explain.

6. Contrast the enzyme(s) present in the virions of a retrovirus and a positive-strand RNA bacteriophage. Why do they differ if each has plus complementarity single-stranded RNA as their genome?

7. Explain why as viral infection leads to more viral particles being formed, that the "growth curve" for viruses shows the pattern in Figure 9.9 rather than that in Figure 6.8.

BACTERIAL GENETICS

10

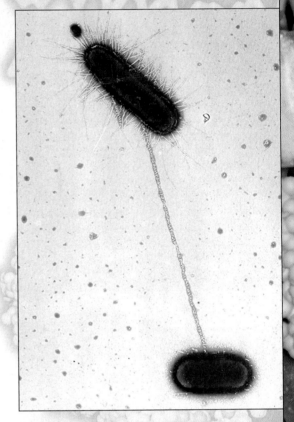

To understand an organism's growth, development, and reproduction, it is necessary to understand its genetics. Genetic analyses contribute to advances in both basic science and applied science. Genetic exchange, as shown here in two conjugating bacteria, is at the heart of bacterial genetics.

WORKING GLOSSARY

Auxotroph an organism that has developed a nutritional requirement through mutation

Cloning vector genetic element into which genes can be recombined and replicated

Conjugation transfer of genes from one prokaryotic cell to another by a mechanism involving cell-to-cell contact and a plasmid

Gene disruption use of genetic techniques to inactivate a gene by inserting within it a DNA fragment containing a selectable marker. The inserted fragment is called a *cassette*, and the process of insertion, *cassette mutagenesis*. Also called *gene knockout*

Genetic map the arrangement of genes on a chromosome

Genotype the precise genetic makeup of an organism; the complete sequence of a cell's chromosome(s) and plasmids if present

Molecular cloning isolation and incorporation of a fragment of DNA into a vector where it can be replicated

Mutagen an agent that causes mutation

Mutant an organism whose genome carries a mutation

Mutation an inheritable change in the base sequence of the genome of an organism

Phenotype the observable characteristics of an organism

Plasmid an extrachromosomal genetic element that has no extracellular form

Point mutation a mutation that involves a single base pair

Recombination the process by which parts or all of the DNA molecules from two separate sources are exchanged or brought together into a single DNA molecule

Screening any of a number of procedures that permit the identification of organisms by phenotype or genotype, but do not inhibit or enhance the growth of particular phenotypes or genotypes

Selection placing organisms under conditions where the growth of those with a particular genotype will be favored

Shotgun cloning making a gene library by random cloning of DNA fragments

Site-directed mutagenesis a technique whereby a gene with a specific mutation can be constructed *in vitro*

Transduction transfer of host genes from one cell to another by a virus

Transformation transfer of bacterial genes involving free DNA (but see alternative usage in Chapter 9)

Transposable element a genetic element that has the ability to move (transpose) from one site on a chromosome to another

Transposon a type of transposable element that carries genes in addition to those involved in transposition

Vector (as in cloning vector) a DNA molecule that, on being replicated in a cell, brings about the replication of other genes inserted into the DNA molecule

Wild type a bacterial strain isolated from nature

In this chapter we present the basic principles and techniques of bacterial genetics. First we will discuss mutation and genetic recombination, and then we consider how genes can be transferred from one organism to another. Next, we will explore some of the basic techniques of gene cloning that have revolutionized the study of genetics of all organisms, from bacteria to humans. We close with a detailed look at the *Escherichia coli* chromosome.

Unlike most eukaryotes, prokaryotes do not reproduce sexually. However, there are mechanisms of genetic exchange in prokaryotes that allow for both gene transfer and recombination. These mechanisms are based on *lateral gene flow* (also called *horizontal gene flow*), because genes are transferred from donor to recipient rather than vertically, from mother cell to daughter cell. To detect genetic exchange between two organisms, it is necessary to employ *genetic markers* whose transfer can be detected. Genetically altered strains are used for this purpose, the alteration(s) being due to one or more mutations in the DNA of the organism. These mutations may involve changes in only one or a few base pairs or even the insertion or deletion of entire genes.

We begin this chapter on microbial genetics with a consideration of the molecular mechanism of mutation and the properties of mutant microorganisms as a prelude to our discussion of genetic exchange.

▮ MUTATION AND ▮ RECOMBINATION

All organisms contain a specific sequence of nucleotides in their **genome**, their genetic blueprint. **Mutation** is an *inherited change* in the nucleotide base sequence of that genome. Mutations can lead to changes—some good, some bad—in an organism. Change can also be brought about by **genetic recombination**. This is the process by which genes contained in two separate genomes are brought together in one molecule. Through this mechanism, new combinations of genes can arise even in the absence of mutation. Whereas mutation usually brings about only a very *small* amount of genetic change in a cell, genetic recombination typically involves much *larger* changes. Entire genes, sets of genes, or even whole chromosomes, can be transferred between organisms. Taken together, mutation and recombination fuel the evolutionary process.

10.1 Mutations and Mutants

As previously mentioned, a *mutation* is a heritable change in the base sequence of the nucleic acid in the genome of an organism. In all cells this nucleic acid is double-stranded DNA (◠◠ Section 7.1). A strain carrying such a change is called a **mutant**. A mutant by definition differs from its parental strain in **genotype**, the nucleotide sequence of the genome. But in addition, the *observable properties* of the mutant—its **phenotype**—may also be altered relative to the parental strain. One refers to this altered phenotype as a *mutant phenotype*. It is common to refer to a strain isolated from nature as a **wild-type strain**. Mutant derivatives can be obtained either from wild-type strains or from a strain derived from the wild type, for example, another mutant.

Genotype Versus Phenotype

Depending on the mutation, a mutant strain may or may not differ in phenotype from its parent. By convention in bacterial genetics, the *genotype* of an organism is designated by three lowercase letters followed by a capital letter (all in italics) indicating the particular gene involved. For example, the *hisC* gene of *Escherichia coli* encodes a protein called the HisC protein, involved in biosynthesis of the amino acid histidine. Mutations in the *hisC* gene would be designated as *hisC1*, *hisC2*, and so on, the numbers referring to the order of isolation of the mutant strains. Each *hisC* mutation would be different and would likely affect the HisC protein in different ways.

The *phenotype* of an organism is designated by a capital letter followed by two lowercase letters, with either a plus or minus superscript to indicate the presence or absence of that property. For example, a His$^+$ strain of *E. coli* is capable of making its own histidine whereas a His$^-$ strain is not. The His$^-$ strain would require a histidine supplement for growth. Returning to our previous example, mutations in the *hisC* gene may lead to a His$^-$ phenotype if they eliminate the function of the HisC protein.

Isolation of Mutants: Screening Versus Selection

Virtually any characteristic of an organism can be changed by mutation. However, some mutations are *selectable*, conferring some type of advantage on organisms possessing them, while others are *nonselectable*, even though they may lead to a very clear change in the phenotype of an organism. A *selectable* mutation confers a clear advantage on the mutant strain under certain environmental conditions, so the progeny of the mutant cell are able to outgrow and replace the parent. A good example of a selectable mutation is drug resistance: An antibiotic-resistant mutant can grow in the presence of antibiotic concentrations that inhibit or kill the parent (Figure 10.1a●) and is thus *selected* for under these conditions. It is relatively easy to detect and isolate selectable mutants by choosing the appropriate environmental conditions. **Selection** is therefore an extremely powerful genetic tool, allowing the isolation of a single mutant from a population containing millions or even billions of parental organisms.

An example of a nonselectable mutation is that of loss of color in a pigmented organism (Figure 10.1b). Nonpigmented cells usually have neither an advantage nor a disadvantage over the pigmented parent cells when grown on agar plates, although there may be a selective advantage for pigmented organisms in nature. We can detect such mutations only by examining large numbers of colonies and looking for the "different" ones, a process called **screening**.

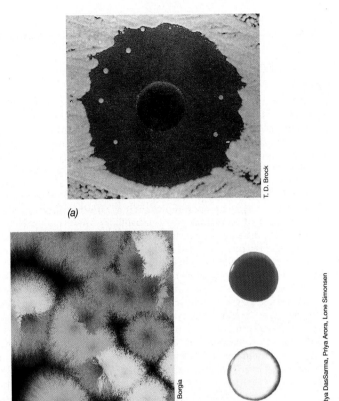

● **Figure 10.1   Selectable and nonselectable mutations.** (a) Development of antibiotic-resistant mutants, a type of easily selectable mutation, within the inhibition zone of an antibiotic assay disc. (b) Nonselectable mutations. Spontaneous pigmented and nonpigmented mutants of the fungus *Aspergillus nidulans*. The wild type has a green pigment. The white or colorless mutants make no pigment, whereas the yellow mutants cannot convert the pigment they do make to the normal (green) color. (c) Colonies of mutants of a species of *Halobacterium*, a member of the *Archaea*. The colonies that appear white are the wild type. The orange/brown colonies are mutants that are missing gas vesicles (⬡ Section 4.12). The gas vesicles scatter light and mask the color of the colony.

Isolation of Nutritional Auxotrophs and Penicillin Selection

Although screening is always more tedious than selection, for certain types of mutations methods are available for screening large numbers of colonies. For instance, nutritionally defective mutants can be detected by the technique of **replica plating** (Figure 10.2●). With the use of sterile velveteen cloth or filter paper, an imprint of colonies from a master plate is made onto an agar plate lacking the nutrient. Colonies of the parental type will grow normally, whereas those of the mutant will not. Thus, the inability of a colony to grow on the replica plate containing minimal medium (Figure 10.2) is a signal that it is a mutant. The colony on the master plate corresponding to the vacant spot on the replica plate (Figure 10.2b) can then be picked, purified, and characterized.

A mutant that has a nutritional requirement for growth is called an **auxotroph**, and the parent from which the auxotroph was derived is called a **prototroph**. (A prototroph may or may not be the wild type. An auxotroph may be derived from the wild type or from a mutant derivative of the wild type.) For instance, mutants of *Escherichia coli* with a His⁻ phenotype are said to be *histidine auxotrophs*. Although of great utility, replica plating is nevertheless a *screening* process, and it can be laborious to isolate mutants by screening.

An ingenious method widely used to isolate auxotrophs is the **penicillin-selection method**. Ordinarily, mutants that require specific nutrients are at a disadvantage in competition with the parent cells, and so there is no direct way of isolating them. For instance, histidine mutants are rare in a mutagenized culture, and it could take a great deal of time to obtain them by replica plating alone. However, penicillin selection can be used to *enrich* a population of mutagenized cells in His⁻ mutants, after which replica plating can be much more effective. How does penicillin selection work?

Penicillin is an antibiotic that kills only *growing* cells. If penicillin is added to a population growing in a medium lacking the nutrient required by the desired mutant, the parent cells will be killed, whereas the (nongrowing) mutant cells will be unaffected. Thus, after preliminary incubation in the absence of the nutrient in a penicillin-containing medium, the population is washed free of the penicillin and transferred to plates containing the nutrient. The colonies that appear include some wild-type cells that escaped penicillin killing, but also include some of the desired mutants. Penicillin selection is thus a kind of *negative selection*; the selection is not *for* the mutant but *against* the parental type. Penicillin selection is often used as a prelude to replica plating to increase the chances of obtaining auxotrophic mutants.

Examples of common classes of mutants and the means by which they are detected are listed in Table 10.1.

 10.1 Concept Check

Mutation is a heritable change in DNA sequence that can lead to a change in phenotype. Selectable mutations are those that give the mutant a growth advantage under certain environmental conditions and are especially useful in genetic research. If selection is not possible, mutants must be identified by screening.

◆ Distinguish between the words *mutation* and *mutant*.

◆ Distinguish between the words *screening* and *selection*.

10.2 Molecular Basis of Mutation

Mutations can be either *spontaneous* or *induced*. **Spontaneous mutations** can occur as a result of exposure to natural radiation (cosmic rays, and so on) that alters the structure of bases in the DNA. Also, oxygen radicals (⚬⚬ Section 6.16) can affect DNA structure by chemically modifying DNA. For example, oxygen radicals can convert guanine into 8-hydroxyguanine, and this causes spontaneous mutations. However, the bulk of spontaneous mutations occur during DNA replication as a result of errors in the pairing of bases.

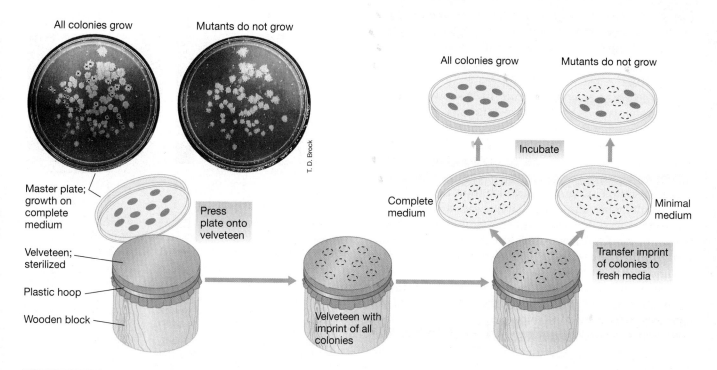

● **Figure 10.2 Screening for nutritional auxotrophs.** The replica plating method can be used for the detection of nutritional mutants. Photos: The photograph on the left shows the master plate. The colonies not appearing on the replica plate are marked with an X. The replica plate (right) lacked one nutrient (leucine) present in the master plate. Therefore, the colonies marked with an X on the master plate are leucine auxotrophs.

Table 10.1	Kinds of mutants	
Phenotype	**Nature of change**	**Detection of mutant**
Auxotroph	Loss of enzyme in biosynthetic pathway	Inability to grow on medium lacking the nutrient
Cold-sensitive	Alteration of an essential protein so it is inactivated at low temperature	Inability to grow at a low temperature (for example, 20°C) that normally supports growth
Drug-resistant	Alteration of permeability to drug or drug target or detoxification of drug	Growth on medium containing a growth-inhibitory concentration of the drug
Nonencapsulated	Loss or modification of surface capsule	Small, rough colonies instead of larger, smooth colonies
Nonmotile	Loss of flagella; nonfunctional flagella	Compact colonies instead of flat, spreading colonies
Pigmentless	Loss of enzyme in biosynthetic pathway leading to loss of one or more pigments	Presence of different color or lack of color
Rough colony	Loss or change in lipopolysaccharide outer layer	Granular, irregular colonies instead of smooth, glistening colonies
Sugar fermentation	Loss of enzyme in degradative pathway	Lack of color change on agar containing sugar and a pH indicator
Temperature-sensitive	Alteration of an essential protein so it is more heat-sensitive	Inability to grow at a temperature normally supporting growth (for example, 40°C) but still growing at a lower temperature (for example, 30°C)
Virus-resistant	Loss of virus receptor	Growth in presence of large amounts of virus

Mutations involving a change in one base pair are referred to as **point mutations**. Point mutations occur from *base-pair substitutions* in the DNA. As is the case with all mutations, the phenotypic change that results from a point mutation depends on exactly where the mutation occurs in the gene, what the nucleotide change is, and what product the gene encodes.

Base-Pair Substitutions

If a point mutation occurs within the coding region of a gene that encodes a polypeptide, any change in the phenotype of the cell is almost certainly the result of a change in the amino acid *sequence* of the polypeptide. The error in the DNA is transcribed into mRNA, and the erroneous mRNA in turn is translated to yield the polypeptide. The triplet code that directs the insertion of an amino acid via a transfer RNA (⌒ Sections 7.15 and 7.16) will thus be incorrect. Figure 10.3● shows the consequences of various base-pair substitutions.

In interpreting the results of a mutation, we must first recall that the genetic code is degenerate (⌒ Section 7.14). Because of this degeneracy, not all mutations in polypeptide-encoding genes result in changes in the polypeptide. This is illustrated in Figure 10.3, which shows several possible results when the DNA that encodes a single tyrosine codon in a polypeptide undergoes mutation. First, a change in the RNA from UAC to UA*U* would have no apparent effect because UAU is also a tyrosine codon. Mutations that give rise to such changes are indeed still mutations, but because they do not affect the primary sequence of the encoded polypeptide, they are called **silent mutations**. Note that silent mutations in coding regions almost always occur in the *third base* of the codon (arginine and leucine can also have silent mutations in the first position).

Second, changes in the first or second base of the triplet more often lead to significant changes in the polypeptide. For instance, a single base change from UAC

to AAC (Figure 10.3) would result in an amino acid change within the polypeptide from tyrosine to asparagine at a specific site. This is referred to as a **missense mutation** because the informational "sense" (sequence of amino acids) in the ensuing polypeptide has changed. If the change occurs at a critical point in the polypeptide chain, the protein could be inactive or have reduced activity. However, not all mutations that cause amino acid substitution necessarily lead to nonfunctional proteins. The outcome depends on where in the polypeptide chain the substitution has occurred and on how it affects its folding and activity. For example, mutations in the active site of

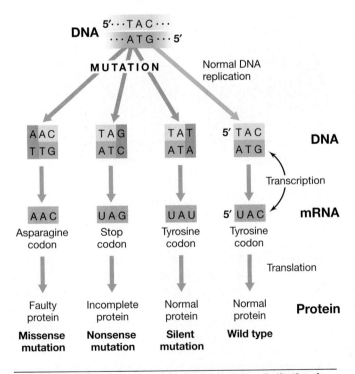

● **Figure 10.3 Possible effects of base-pair substitution in a gene encoding a protein:** Three different protein products are possible from changes in the DNA for a single codon.

an enzyme are more likely to destroy activity than mutations that affect other parts of the protein.

A third possible result of a base pair substitution is the formation of a *nonsense (stop) codon*. This results in premature termination of translation, leading to an incomplete polypeptide that would almost certainly not be functional (Figure 10.3). Mutations of this type are called **nonsense mutations** because the change is from a codon for an amino acid (sense codon) to a nonsense codon (⌘ Table 7.5). Unless the nonsense mutation occurs very near the end of the gene, the incomplete product will be completely inactive.

The terms *transition* and *transversion* are used to describe the nature of the base substitution that occurs in a point mutation. **Transitions** involve mutations in which one purine base (A or G) is substituted for another purine, or one pyrimidine base (C or T) is substituted for another pyrimidine. **Transversions** are point mutations where a purine base is substituted for a pyrimidine base, or vice versa.

Frameshifts and Other Insertions or Deletions

Because the genetic code is read from one end of the nucleic acid in consecutive blocks of three bases (that is, codons), any deletion or insertion of a base pair results in a *reading frameshift*. These **frameshift mutations** can have serious consequences. Single base insertions or deletions change the primary sequence of the encoded polypeptide, typically in a major way (Figure 10.4●). Such microinsertions or microdeletions result from replication errors.

Insertions or deletions can also involve the gain or loss of hundreds or even thousands of base pairs. Such changes inevitably result in complete loss of gene function. Some deletions are so large that they may involve several genes. If any of the deleted genes are essential, the mutation will be lethal. Such deletions cannot be restored through further mutations but only through genetic recombination. Indeed, one way in which large deletions are distinguished from point mutations is that the latter are reversible through further mutations, whereas the former are not. Larger insertions typically arise as a result of errors during genetic recombination. Many insertion mutations are due to the insertion of specific identifiable DNA sequences 700–1400 bp in length called **insertion sequences**, a type of transposable element (⌘ Section 7.4). The behavior of such insertion sequences is discussed in detail in Section 10.14.

Other types of large-scale mutations seem to involve rearrangements brought about by mistakes in recombination. These include **translocations**, in which a large section of chromosomal DNA is moved to a new location (and in eukaryotes often to a different chromosome), and **inversions**, in which the orientation of a particular segment of DNA is reversed with respect to the surrounding DNA.

Transposon and Site-Directed Mutagenesis

Mutations can be introduced through the process of *transposon mutagenesis*. We discuss the details of transpo-

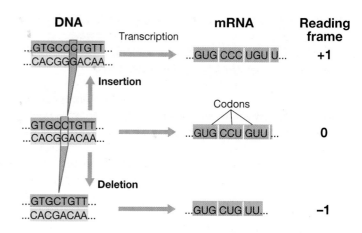

● **Figure 10.4** **Shifts in the reading frame of messenger RNA caused by insertion or deletion mutations in DNA.** The reading frame in mRNA is established by the ribosome that begins at the 5′ end (toward the left in the figure) and proceeds by units of three bases (codons, ⌘ Section 7.14). The normal reading frame is referred to as the 0 frame, that missing a base the −1 frame, and that with an extra base the +1 frame. To determine the effects of a frameshift, translate the codons using Table 7.5.

son mutagenesis later in this chapter (see Section 10.14) but only note here that if insertion of a transposable element occurs *within* a gene, loss of gene function generally results. Because transposable elements can enter the chromosome at various locations, transposons are widely used by microbial geneticists as mutagenic agents.

The mutations that we have been discussing thus far have been random, that is, not directed at any particular gene. However, recombinant DNA technology (see Sections 10.15–10.17) and the use of synthetic DNA (⌘ Section 7.8) make it possible to induce *specific* mutations in *specific* genes. The procedures for carrying out mutagenesis of specific sites in the genome are called *site-directed mutagenesis* and will be discussed later in this chapter (see Section 10.18).

Back Mutations or Reversions

Point mutations are typically reversible by the process of **reversion**. A *revertant* is a strain in which the wild-type (or prototrophic, depending on the strain) phenotype that was lost in the mutant is restored. Revertants can be of two types. In *same-site revertants*, the mutation that restores activity occurs at the same site at which the original mutation occurred. (If the back mutation is not only at the same site but also leads to the wild-type sequence, it is called a *true revertant*.)

In *second-site revertants*, the mutation occurs at a *different* site in the DNA. Second-site mutations cause restoration of a wild-type phenotype because they function as **suppressor** mutations—mutations that compensate for the effect of the original mutation and restore the original phenotype. Several classes of suppressor mutations are known. These include (1) a mutation somewhere else in the same gene that restores enzyme function, such as a second frameshift mutation that occurs near the first and restores the original reading frame; (2) a mutation in another gene that restores the function of the original mutated gene; and (3) a mutation in another gene that results in the production of an enzyme that can replace the mutated one.

Unlike point mutations, large-scale deletion mutations are essentially nonrevertable. By contrast, large-scale insertions can revert as the result of a subsequent deletion that removes the insertion. But typically, frameshift mutations of any magnitude are difficult to restore to the wild type, and mutants carrying frameshift mutations are therefore genetically quite stable. For this reason, geneticists often use them in genetic crosses to avoid inadvertent reversion of mutant strains during the course of a genetic study.

 10.2 Concept Check

Mutations, which can be either spontaneous or induced, arise because of changes in the base sequence of the nucleic acid of an organism's genome. A point mutation, which is due to a change in a single base pair, can lead to a single amino acid change in a polypeptide or to no change at all, depending on the particular codon involved. In a nonsense mutation, the codon becomes a stop codon and an incomplete polypeptide is made. Deletions and insertions cause more dramatic changes in the DNA, including frameshift mutations and often result in complete loss of gene function.

◆ What does it mean to say that point mutations can spontaneously revert?

◆ Do *missense* mutations occur in genes encoding transfer RNAs (tRNAs)?

10.3 Mutation Rates

The *rates* at which various kinds of mutations actually occur vary widely. Some types of mutations occur so rarely that they are almost impossible to detect, whereas others occur so frequently that they present difficulties for an experimenter trying to maintain a genetically stable stock culture.

Errors in DNA replication occur at a frequency of 10^{-7}–10^{-11} per base pair during a single round of replication. A typical gene has about 1000 base pairs. Therefore, the frequency of a mutation occurring in a given gene is in the range of 10^{-4} to 10^{-8} per generation. For instance, in a bacterial culture having 10^8 cells/ml there are likely to be a number of different mutants for a given gene in each milliliter of culture.

When any single-base error occurs during DNA replication, it will more likely lead to *missense* mutations than to *nonsense* mutations because most single-base substitutions yield codons that encode other amino acids (∞ Table 7.5). The next most frequent type of codon change caused by a single-base change would lead to a *silent* mutation. This is because for the most part, alternate codons for a given amino acid differ from each other by a single base change in the "silent" third position (∞ Table 7.5). A given

codon can be changed to any of 27 other codons by a single-base substitution, and on average, about two of these will be silent mutations, about one a nonsense mutation, and the rest will be missense mutations. There are also DNA sequences, usually involving short repeats, which are *hot spots* for mutations because the error frequency of DNA polymerase can be quite high at such sequences. Hot spots are defined by the context of nucleotides in and around the spot that affect normal polymerase function.

Unless a mutation can be selected for, its experimental detection is difficult, and much of the skill of the microbial geneticist involves increasing the efficiency of mutation detection. As we will see in the next section, it is possible to significantly increase the *rate* of mutation by the use of mutagenic treatments. In addition, we will also see that the mutation rate of a gene may change in certain situations. For example, mutations called *adaptive mutations* are strongly selected for under high-stress conditions. Adaptive mutations are generated by mechanisms activated by the organism itself in order to survive a particular stress condition (see Section 10.4).

Mutations in RNA Genomes

Whereas all cells have *DNA* as their genetic material, some viruses have *RNA* genomes (∞ Sections 9.1, 16.1, 16.7–16.10, and 16.15). These genomes can also undergo mutation. Interestingly, however, the mutation rate in RNA genomes is about *1000-fold higher* than in DNA genomes. Why should this be so?

Some RNA polymerases have *proofreading* activities like those of DNA polymerases (∞ Section 7.6), thus limiting the total number of polymerase errors. However, while there are several repair systems for DNA that can correct changes before they become fixed in the genome as mutations (see Section 10.4), comparable RNA repair mechanisms do not exist. This leads to heightened mutation rates for RNA genomes. This high mutation rate in RNA viruses has serious consequences. For example, the RNA genomes of viruses that cause disease can mutate very rapidly, presenting a constantly changing and evolving population of viruses. Such changes are one of many challenges to human medicine posed by the AIDS virus, HIV, an RNA virus with a notorious ability to undergo genetic changes that affect its virulence (∞ Sections 16.15 and 26.14).

 10.3 Concept Check

Different types of mutations can occur at different frequencies. For a typical bacterium, mutation rates of 10^{-7}–10^{-11} per base pair are generally seen. Although RNA and DNA polymerases make errors at about the same rate, RNA genomes typically accumulate mutations at much higher frequencies than DNA genomes.

◆ Which class of mutation, *missense* or *nonsense*, is more common, and why?

◆ Why are RNA viruses genetically unstable?

10.4 Mutagenesis

While the spontaneous rate of mutation is very low, there are a variety of chemical, physical, and biological agents that can increase the mutation rate and are therefore said to *induce* mutations. These agents are called **mutagens**. We discuss some of the major categories of mutagens and their activities here.

Chemical Mutagens

An overview of some of the major chemical mutagens and their modes of action is given in Table 10.2. Several classes of chemical mutagens exist. One class is the **nucleotide base analogs**, molecules that resemble DNA purine and pyrimidine bases in structure yet display faulty pairing properties (Figure 10.5●). When one of these base analogs is incorporated at a site in DNA in place of the natural nucleotide, replication may occur normally most of the time. However, DNA replication errors occur at higher frequencies at these sites, resulting in the more frequent incorporation of the wrong base into the copied strand of DNA and thus introduction of a mutation. During subsequent segregation of this strand in cell division, the mutation is revealed.

Other chemical mutagens induce chemical modifications in one base or another, resulting in faulty base pairing or related changes (Table 10.2). For example, *alkylating agents* (chemicals that react with amino, carboxyl, and hydroxyl groups in proteins and nucleic acids, substituting them with alkyl groups) such as *nitrosoguanidine*, are powerful mutagens and generally induce mutations at higher frequency than base analogs. Alkylating agents differ from the base analogs in that the chemicals are able to introduce changes even in nonreplicating DNA (base analogs have an effect only when incorporated during DNA replication). Both base analogs and alkylating agents tend to induce base-pair substitutions (see Section 10.2).

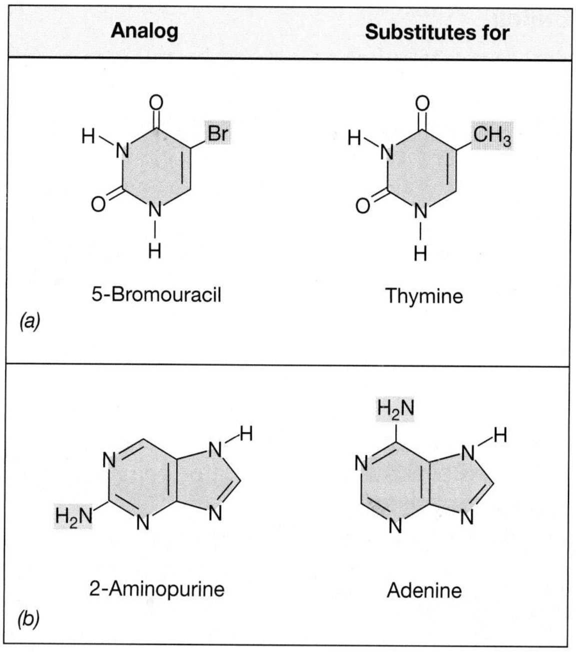

● **Figure 10.5 Nucleotide base analogs.** Structure of two common nucleotide base analogs used to induce mutations and the normal nucleic acid bases they substitute for. (a) 5-Bromouracil can base pair with guanine causing AT to GC substitutions. (b) 2-Aminopurine can base pair with cytosine, causing AT to GC substitutions.

Another group of chemicals, the acridines, are planar molecules that function as *intercalating agents*. These mutagens become inserted between two DNA base pairs and in the process push them apart. During replication, this abnormal conformation can lead to insertions or deletions in acridine-containing DNA. Thus, acridines typically induce *frameshift* mutations (see Section 10.2). Ethidium bromide, which is often used to detect DNA in electrophoresis (see Section 10.12), is also an intercalating agent that functions as a mutagen.

Agent	Action	Result
Base analogs		
5-Bromouracil	Incorporated like T; occasional faulty pairing with G	AT pair → GC pair; occasionally GC → AT
2-Aminopurine	Incorporated like A; faulty pairing with C	AT → GC; occasionally GC → AT
Chemicals reacting with DNA		
Nitrous acid (HNO$_2$)	Deaminates A and C	AT → GC and GC → AT
Hydroxylamine (NH$_2$OH)	Reacts with C	GC → AT
Alkylating agents		
Monofunctional (for example, ethyl methane sulfonate)	Puts methyl on G; faulty pairing with T	GC → AT
Bifunctional (for example, nitrogen mustards, mitomycin, nitrosoguanidine)	Cross-links DNA strands; faulty region excised by DNase	Both point mutations and deletions
Intercalative dyes		
Acridines, ethidium bromide	Inserts between two base pairs	Microinsertions and microdeletions
Radiation		
Ultraviolet	Pyrimidine dimer formation	Repair may lead to error or deletion
Ionizing radiation (for example, X-rays)	Free-radical attack on DNA, breaking chain	Repair may lead to error or deletion

Table 10.2 Chemical and physical mutagens and their modes of action

Radiation

Several forms of radiation are highly mutagenic. We can divide mutagenic radiation into two main categories, *nonionizing* and *ionizing* (electromagnetic) (Figure 10.6●). Although both kinds of radiation are used in microbial genetics to generate mutations, *nonionizing* radiation has the widest use and will be discussed first.

The purine and pyrimidine bases of the nucleic acids absorb ultraviolet (UV) radiation strongly, and the absorption maximum for DNA and RNA is at 260 nm (∞ Figure 7.9). Killing of cells by UV radiation is due primarily to its effect on DNA. Although several effects are known, one well-established effect is the production in DNA of **pyrimidine dimers**. This is a state in which two adjacent pyrimidine bases (cytosine or thymine) on the same strand of DNA become covalently bonded to one another in such a way that during replication, the probability of DNA polymerase misreading the sequence at this point is greatly increased.

The type of UV radiation source most commonly used for mutagenesis is the *germicidal lamp*, which emits large amounts of UV radiation in the 260-nm region. A dose of UV radiation is used that brings about 50–90% killing of the cell population (∞ Section 20.2), and mutants are then selected or screened for among the survivors. If much higher doses of radiation are used, the number of viable cells will be too low. If lower doses are used, insufficient damage to DNA in the cell population will be induced to make recovery of the desired mutant(s) possible. When used at the correct dose, UV radiation is a very useful and convenient tool for isolating mutants and avoids the necessity of handling toxic chemicals.

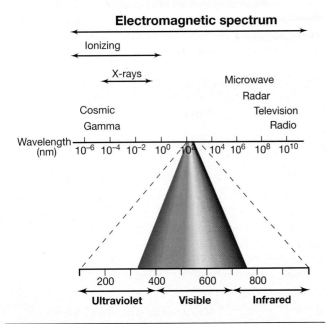

Electromagnetic spectrum

● **Figure 10.6 Wavelengths of radiation.** Note that ultraviolet radiation consists of wavelengths just shorter than visible light. For any electromagnetic radiation, the shorter the wavelength, the higher the energy. DNA absorbs strongly at 260 nm.

Ionizing Radiation

Ionizing radiation is a more powerful form of radiation than UV and includes short-wavelength rays such as X-rays, cosmic rays, and gamma rays (Figure 10.6). These radiations cause water and other substances to ionize, and mutagenic effects are brought about indirectly through this ionization. Among the potent chemical species formed by ionizing radiation are chemical free radicals, of which the most important is the hydroxyl radical, OH· (∞ Section 6.16).

Free radicals react with and inactivate macromolecules in the cell, of which the most important is DNA. At low doses of ionizing radiation only a few "hits" on DNA occur, but at higher doses, multiple hits occur, leading to the death of the cell. In contrast to UV radiation, ionizing radiation penetrates readily through glass and other materials. Because of this, ionizing radiation is used frequently to induce mutations in animals and plants (where its penetrating power makes it possible to reach the gamete-producing cells of these organisms readily). However, because ionizing radiation is more dangerous to use and is less readily available than UV, it finds less use in microbial genetics.

Mutations That Arise from DNA Repair: The SOS System

Recall that a mutation is an *inheritable* change in the genetic material. Therefore, if an error in DNA synthesis can be corrected before the cell divides, no mutation has occurred. Furthermore, some DNA damage cannot be easily replicated. For instance, if a DNA molecule contains pyrimidine dimers, it will not be replicated to give two DNA molecules, each containing pyrimidine dimers. If such DNA damage cannot be repaired, replication can stall and the cell will die. Most cells have a variety of different DNA repair processes to correct mistakes or repair damage. Many of these DNA repair systems are virtually error-free. However, other processes seem to be *error-prone*, and it is the repair process itself that introduces the mutation.

A complex cellular mechanism called the **SOS regulatory system** is activated as a result of some types of DNA damage and initiates a number of DNA repair processes. However, in the SOS system, some DNA repair occurs *in the absence of template instruction*, that is, without base pairing. As you might expect, this results in many errors, hence many mutations.

In the SOS regulatory system, DNA damage serves as a distress signal to the cell, resulting in the coordinate expression of a number of cellular functions involved in DNA repair. The SOS system is normally repressed by a protein called *LexA*. However, this repressor protein is inactivated by a second protein called *RecA*, a protease that is activated as a result of DNA damage (Figure 10.7●). Since one of the DNA repair mechanisms of the SOS system is inherently error-prone, many mutations arise. Thus, through the SOS regulatory system, DNA damage

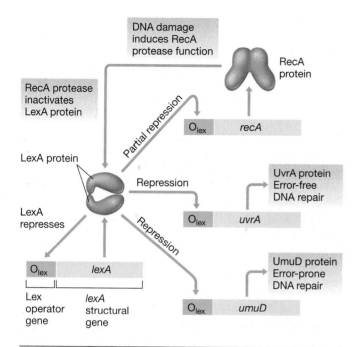

● **Figure 10.7** **Mechanism of the SOS response.** DNA damage results in conversion of RecA protein into a protease that cleaves LexA protein. LexA protein normally represses the activities of the *recA* gene and the DNA repair genes *uvrA* and *umuD* (the UmuD protein is part of DNA polymerase V). However, repression is not complete. Some RecA protein is produced even in the presence of LexA protein. With LexA inactivated, these genes become active.

when cells are exposed to various agents such as chemicals and radiation, leads to mutagenesis. In *Escherichia coli*, where mutagenesis has been studied in great detail, much of this is brought about by *DNA polymerase V*, an enzyme encoded by the *umuCD* genes (see Figure 10.7). The highly error-prone DNA polymerase IV is also induced as part of the SOS response. Collectively, these enzymes are referred to as **DNA mutases**.

Once the DNA damage has been repaired, the SOS system is repressed and further mutagenesis ceases. In addition to its effect on the cellular mutation rate, the SOS regulatory system plays a central role in the regulation of temperate virus replication (⊂⊃ Section 9.10).

It should be emphasized that not all DNA repair occurs in the absence of template instruction. Cells generally have many DNA repair systems that require template instruction and lead to proper DNA repair. These systems apparently work most of the time but are not sufficient to repair the large scale damage done by some very mutagenic chemicals or by large doses of radiation.

Changes in Mutation Rate

High fidelity (low error frequency) in DNA replication is essential if organisms are to enjoy genetic stability. On the other hand, perfect fidelity is impractical, because it would prevent evolution. Therefore, a mutation rate has evolved in cells that is very low yet detectable. This allows organisms to balance the need for genetic stability with that for evolutionary improvement.

The fact that organisms as phylogenetically disparate as hyperthermophilic *Archaea* (⊂⊃ Chapter 13) and *Escherichia coli* have about the same mutation rate might make one believe that evolutionary pressure has selected for organisms with the *lowest possible* mutation rates. However, this is not so. The mutation rate in an organism is subject to change. For example, mutants of some organisms have been selected in the laboratory that are hyperaccurate in DNA replication and repair. However, in these strains, the improved proofreading/repair mechanisms has a significant metabolic cost; thus, a *hyperaccurate* mutant might actually be at a disadvantage in its natural environment. On the other hand, some organisms seem to benefit from a hyperaccurate phenotype that enables them to occupy particular niches in nature. A good example is the bacterium *Deinococcus radiodurans* (⊂⊃ Section 12.34). This organism is 20 times more resistant to UV radiation and 200 times more resistant to ionizing radiation than *E. coli*. This resistance, dependent in part upon redundant DNA repair systems and on a mechanism for exporting damaged nucleotides, allows the organism to survive in environments where other organisms cannot, such as near concentrated sources of radiation or on the surfaces of dust particles exposed to intense sunlight.

In contrast to the hyperaccurate phenotype, some organisms actually benefit from an *increased* mutation rate. DNA repair systems are themselves encoded by genes and thus subject to mutation. For example, the protein subunit of DNA polymerase III involved in proofreading (⊂⊃ Section 7.6) is encoded by the gene *dnaQ*. Certain mutations in this gene lead to mutant strains that are still viable but have an increased rate of mutation. Such a strain is said to be hypermutable or a **mutator**. Mutations leading to a mutator phenotype are known in several other DNA repair systems as well. The mutator phenotype is apparently selected for in complex and changing environments, since strains of bacteria with mutator phenotypes appear to be more abundant under these conditions. Presumably, whatever disadvantage an increased mutation rate may have in such environments is offset by the ability to generate greater numbers of useful mutations. These mutations ultimately increase evolutionary fitness of the population and make the organism more successful in its ecological niche.

As was indicated earlier, a mutator phenotype may be induced in wild-type strains by stress situations. For instance, the SOS response induces error-prone repair. Therefore, when the SOS response is activated, the mutation rate increases. While in some cases this may be a necessary by-product of DNA repair, in other cases the increased mutation rate may itself be of value. That is, the organism may generate *adaptive mutations*. **Adaptive mutations** lead to a phenotype in the mutant that allows it to survive a particular stress. Selection is obviously strong under these conditions because, depending on the stress condition, the only option to potential survival from an increased mutation rate may be death.

10.4 Concept Check

Mutagens are chemical, physical, or biological agents that increase the mutation rate. Mutagens can alter DNA in many different ways. However, alterations in DNA are not mutations unless they can be inherited. Some DNA damage can lead to cell death if not repaired, and both error-prone as well as high-fidelity DNA repair systems exist.

◆ How do mutagens work?

◆ Why might a mutator phenotype be successful in an environment experiencing rapid changes?

10.5 Mutagenesis and Carcinogenesis: The Ames Test

A practical use of bacterial mutants has been developed to identify potentially hazardous chemicals in the environment. Because the sensitivity with which selectable mutants can be detected in large populations of bacteria is very high, bacteria can be used as screening agents for the potential mutagenicity of chemicals. This is relevant because many mutagenic chemicals are also *carcinogenic*, capable of causing cancer in humans or other animals.

The variety of chemicals, both natural and artificial, that humans encounter through agricultural and industrial exposure is enormous. There is good evidence that a large proportion of human cancers have environmental causes, most likely from various chemicals, making the detection of chemical carcinogens an urgent and very important matter. It does not necessarily follow that because a compound is mutagenic it is also carcinogenic. The correlation, however, is quite high, and the knowledge that a compound is mutagenic in a bacterial system serves as a warning of possible danger. The development of bacterial tests for carcinogenic screening was carried out primarily by a group at the University of California in Berkeley under the direction of Bruce Ames, and the mutagenicity test for carcinogens is called the **Ames test** (Figure 10.8●).

Protocol for an Ames Test

The standard way to test chemicals for mutagenesis is to look for an increase in the rate of *back* mutation (reversion) in auxotrophic strains of bacteria in the presence of the suspected mutagen. It is important that the mutation to be examined be a point mutation because reversion will occur in such a strain at the same rate as the forward mutation did. When cells of such an auxotrophic strain are spread on a medium lacking the required nutrient (for example, an amino acid), no growth occurs, and even very large populations of cells can be spread on the plate without formation of visible colonies. However, if back mutants (revertants) are present, those cells will form colonies. Thus, if 10^8 cells are spread on the surface of a single plate, even as few as 10–20 revertants can be detected by the 10–20 colonies they form (Figure 10.8, left photo). If the reversion

rate has been *increased* by addition of a chemical mutagen, the number of revertant colonies will increase. Histidine auxotrophs of *Salmonella enterica* (Figure 10.8) and tryptophan auxotrophs of *Escherichia coli* have been the major tools of the Ames test.

Two additional elements have been introduced in the Ames test to make it much more powerful. The first of these is to use test strains that almost exclusively use error-prone pathways to repair DNA damage; normal repair mechanisms are thus thwarted (see Section 10.3). The second important element in the Ames test is the addition of liver enzyme preparations to convert the chemicals to be tested into their active mutagenic (and potentially carcinogenic) forms. It has been well established that many carcinogens are not directly carcinogenic or mutagenic themselves but undergo modifications in the human body that convert them into active substances. These changes take place primarily in the liver, where enzymes called *mixed-function oxygenases* normally involved in detoxification cause the formation of epoxides or other activated forms of the compounds; these are then highly reactive (and thus mutagenic) with DNA.

In the Ames test, a preparation of enzymes from rat liver is first used to activate the test compound. The activated complex is then taken up on a filter-paper disk, which is placed in the center of a plate on which the proper bacterial strain has been overlaid. After overnight incubation, the mutagenicity of the compound can be detected by looking for a halo of back mutations in the area around the paper disk (Figure 10.8). It is always necessary to carry out this test with several different concentrations of the compound and with appropriate positive (known mutagens) and negative (no mutagen) controls, because compounds vary in their mutagenic activity and may be lethal at higher levels. A wide variety of chemicals have been subjected to the Ames test, and it has become one of the most useful screens for determining the potential carcinogenicity of a compound.

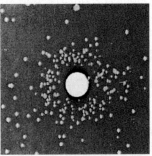

T. D. Brock

● **Figure 10.8 The Ames test is used to evaluate the mutagenicity of a chemical.** Two plates were inoculated with a culture of a histidine–requiring (auxotrophic) mutant of *Salmonella enterica*. The medium does not contain histidine, so only cells that revert back to wild type can grow. Spontaneous revertants appear on both plates, but the chemical on the filter paper disc in the test plate (right) has caused an increase in the mutation rate, as shown by the large number of colonies surrounding the disc. The plate on the left was the negative control, as the disc contained only water. Revertants are not seen very close to the test disc because the concentration of the mutagen is so high there that it is lethal.

10.5 Concept Check

The Ames test employs a sensitive bacterial assay system for detecting chemical mutagens in the environment.

◆ Why does the Ames test measure the rate of *back* mutation rather than the rate of *forward* mutation?

◆ Of what significance is the detection of mutagens to the prevention of cancer?

10.6 Genetic Recombination

Recombination is the physical exchange of genes between genetic elements. In this section, we focus on **homologous recombination**, which results in genetic exchange between *homologous* DNA sequences from two different sources. Homologous DNA sequences are those that have nearly the same sequence; therefore, base pairing can occur over an extended length of the two DNA molecules. This type of recombination is involved in the process referred to as "crossing over" in classical genetics.

Molecular Events in Homologous Recombination

Recombination has been intensively studied in prokaryotes, viruses, and some fungi, especially yeast. In *Bacteria*, homologous recombination involves the participation of the RecA protein, previously mentioned in regard to the error-prone SOS repair system (see Section 10.4). The RecA protein has been shown to be essential in nearly every homologous recombination pathway. RecA-like proteins have been identified in all prokaryotes examined, including the *Archaea*, as well as in yeast and in the higher *Eukarya*.

A molecular mechanism of homologous recombination is shown in Figure 10.9●. The process begins with a *nick* (generated by an endonuclease) in one of the DNA molecules. This nicked strand must be displaced from the other strand by proteins having helicase activity (◌◌ Section 7.6). In some pathways specialized enzymes, such as the RecBCD enzyme of *Escherichia coli*, have both endonuclease and helicase activities. Single-stranded binding protein (◌◌ Section 7.6) then binds to the resulting single-stranded segment. Next, the RecA protein binds to the single-stranded fragment, forming a complex that facilitates annealing with a complementary sequence in the adjacent DNA duplex, simultaneously displacing the resident strand (Figure 10.9). This process is referred to as *strand invasion* and leads to the *pairing* of DNA molecules over long stretches. Following pairing, *exchange* of homologous DNA molecules occurs, leading to the formation of recombination intermediates containing extensive **heteroduplex** regions, where each strand has originated from a different chromosome. This process also requires DNA polymerase and ligase activities. Finally, the linked molecules are *resolved* by nucleases and DNA ligase to form two recombinant DNA molecules.

Effect of Homologous Recombination on Genotype

For new genotypes to arise from homologous recombination, it is essential that the two homologous sequences be genetically distinct. This is obviously the case in a diploid eukaryotic cell (◌◌ Section 14.7), which has two sets of chromosomes, one from each parent. However, in prokaryotes, genetically distinct but homologous DNA molecules are brought together in different ways, but the process of genetic recombination is no less important. Genetic recombination in prokaryotes occurs because *fragments* of homologous DNA from a donor chromosome are transferred to a recipient cell by one of three processes: *transformation*

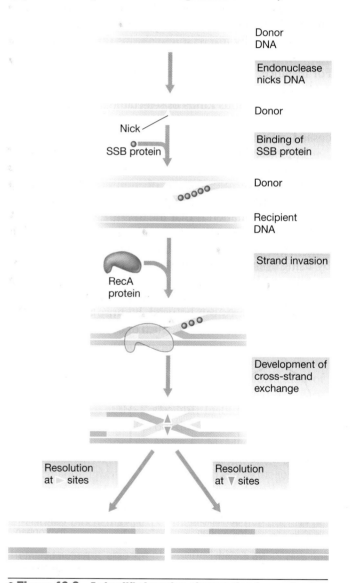

● **Figure 10.9 A simplified version of one molecular mechanism of genetic recombination: homologous recombination.** Homologous DNA molecules pair and exchange DNA segments. The mechanism involves breakage and reunion of paired segments. Two of the proteins involved, a single-stranded binding (SSB) protein and the RecA protein, are shown. The other proteins involved are not shown. The diagram is not to scale: Pairing can occur over hundreds or thousands of bases. Resolution occurs by cutting and ligating the cross-linked DNA molecules. Note that there are two possible outcomes, depending on where strands are cut during the resolution process.

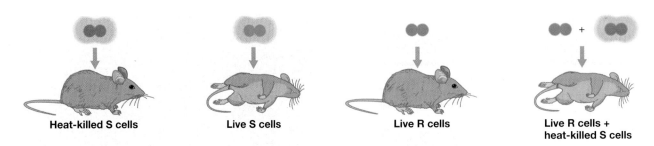

● **Figure 10.13   Griffith's experiments with pneumococcus.** Live smooth (S) cells contain a capsule and kill mice because immune cells cannot kill the encapsulated bacteria; the cells proliferate in the lung and cause a fatal pneumonia. Rough (R) cells have no capsule and are not pathogenic. But a combination of *live* R and *dead* S cells kill mice, and live S type cells can be isolated from the animals. DNA–encoding capsule production is released from dead S cells and taken up by R cells, thus *transforming* them into S-type cells.

material. Thus, three types of studies, the bacteriological ones of Griffith, the biochemical ones of Avery, and the physical-chemical ones of Watson and Crick, solidified the concept of DNA as the genetic material. In subsequent years, this work led to the whole field of molecular biology and molecular genetics.

Competence

Even within transformable genera, only certain strains or species are transformable. A cell that is able to take up DNA and be transformed is said to be **competent**, and the capacity is genetically determined. Competence in most naturally transformable bacteria is regulated, and special proteins play a role in the uptake and processing of DNA. These competence-specific proteins include a membrane-associated DNA-binding protein, a cell wall autolysin, and various nucleases. One pathway to natural competence in *Bacillus subtilis*—an easily transformed species—is part of a quorum-sensing system (a regulatory system that responds to cell density, ⌘ Section 8.10) controlled by a two-component regulatory system (⌘ Section 8.12). Cells produce and excrete a small peptide during growth, and the accumulation of this peptide to high concentrations induces the cells to become competent. In *Bacillus*, about 20% of the cells become competent and stay that way for several hours. However, in *Streptococcus*, 100% of the cells can become competent, but only for a brief period during the growth cycle.

High-efficiency natural transformation is known in only a few *Bacteria*. For example, *Acinetobacter, Azotobacter, Bacillus, Streptococcus, Haemophilus, Neisseria,* and *Thermus* are naturally competent and easily transformable. By contrast, many prokaryotes are poorly transformed, if at all, under natural conditions. Interestingly, *Escherichia coli* and many other gram-negative bacteria fall into this category. However, when cells of *E. coli* are treated with high concentrations of calcium ions and then chilled for several minutes, they become transformable. Cells of *E. coli* treated in this manner take up double-stranded DNA, and therefore transformation of this organism by plasmid DNA is relatively efficient. As we will see later, this discovery was important for getting DNA into *E. coli*—the workhorse of genetic engineering—for applied reasons in the field of biotechnology (⌘ Chapter 31).

Uptake of DNA in Transformation

During transformation, competent bacteria reversibly bind DNA. Soon, however, the binding becomes irreversible. Competent cells bind much more DNA than do noncompetent cells—as much as 1000 times more. As we noted earlier, the sizes of the transforming fragments are much smaller than that of the whole genome, and the fragments are further degraded during the uptake process. In *Streptococcus pneumoniae* each cell can bind only about 10 molecules of double-stranded DNA of 10–15 kbp each. However, as these fragments are taken up, they are converted into single-stranded pieces of about 8 kb, with the complementary strand being degraded. The DNA fragments in the mixture compete with each other for uptake, and if excess DNA that does not contain the genetic marker is added, a decrease in the number of transformants occurs.

In preparations of transforming DNA, typically only about 1 out of 100–300 DNA fragments contains the genetic marker being studied. Thus, at high concentrations of DNA, the competition between DNA molecules results in saturation of the system, so even under the best of conditions it is impossible to transform all the cells in a population for a given marker. The maximum frequency of transformation that has so far been obtained is about 20% of the population; the values usually obtained are between 0.1 and 1.0%. But with the recipient population sizes being very high, even this low frequency is easy to detect. The minimum concentration of DNA yielding detectable transformants is about 0.01 ng/ml, which is so low that it is chemically undetectable.

Interestingly, in transformation in *Haemophilus influenzae* there is a requirement that the DNA fragment have a particular 11-bp sequence for irreversible binding and uptake to occur. This sequence is found at an unexpectedly high frequency in the *Haemophilus* genome, which has been completely sequenced (⌘ Section 15.4). Evidence such as this, and the fact that certain bacteria have been shown to become competent in their natural environment, suggests that transformation is not a laboratory artifact but plays an important role in lateral gene transfer in nature. By promoting new combinations of genes, naturally transformable bacteria increase diversity and fitness of the microbial community as a whole.

Integration of Transforming DNA

Transforming DNA is bound at the cell surface by a DNA-binding protein. Following this, either the entire double-stranded fragment is taken up, or a nuclease degrades one strand and the remaining strand only is taken up, depending on the organism (Figure 10.14●). After uptake, the DNA becomes attached to a competence-specific protein. This prevents the DNA from nuclease attack until it reaches the chromosome, where the RecA protein takes over. The DNA is integrated into the genome of the recipient by recombination (Figure 10.14; see also Figure 10.9). During replication of this heteroduplex DNA, one parental and one recombinant DNA molecule are formed.

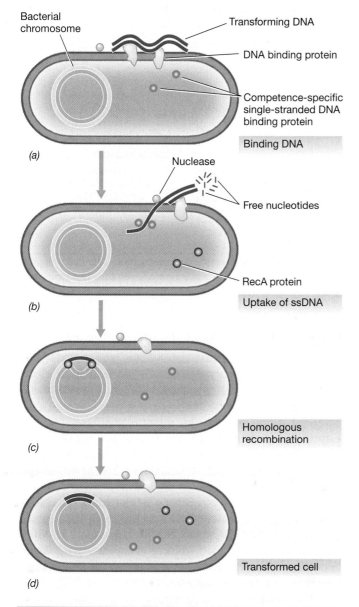

Bacterial chromosome — Transforming DNA

DNA binding protein

Competence-specific single-stranded DNA binding protein

Binding DNA

(a)

Nuclease

Free nucleotides

RecA protein

Uptake of ssDNA

(b)

Homologous recombination

(c)

Transformed cell

(d)

● **Figure 10.14 Mechanism of transformation in a gram-positive bacterium.** (a) Binding of double-stranded DNA by a membrane-bound DNA binding protein. (b) Passage of one of the two strands into the cell while nuclease activity degrades the other strand. (c) The single strand in the cell is bound by specific proteins, and recombination with homologous regions of the bacterial chromosome is mediated by RecA protein. (d) Transformed cell.

On segregation at cell division, the latter is present in the transformed cell, which is now genetically altered as compared to the parental type. The preceding pertains to only small pieces of *linear* DNA. Many naturally transformable *Bacteria* are transformed only poorly by plasmid DNA because the plasmid must remain double-stranded and circular in order to replicate.

Transfection

Bacteria can be transformed with DNA extracted from a bacterial virus rather than from another bacterium. This process is known as **transfection**. If the DNA is from a lytic bacteriophage, transfection leads to virus production and can be measured by the standard phage plaque assay (Section 9.4). Transfection is useful for studying the mechanism of transformation and recombination because the small size of phage genomes allows for the isolation of a nearly homogeneous population of DNA molecules. By contrast, in conventional transformation, the transforming DNA is typically a random assortment of chromosomal DNA of various lengths, and this tends to complicate experiments designed to study the mechanism of transformation.

 10.7 Concept Check

Certain prokaryotes exhibit competence, a state in which cells are able to take up free DNA released by other bacteria. Incorporation of donor DNA into a recipient cell requires the activity of single-stranded binding protein, RecA protein, and several other enzymes. Only competent cells are transformable.

◆ The donor bacterial cell in a transformation is probably dead. Explain.

◆ Even in naturally transformable cells, competency is usually inducible. What does this mean?

10.8 Transduction

In transduction, DNA is transferred from cell to cell by a bacterial virus (bacteriophage). Genetic transfer of host genes by viruses can occur in two ways. In the first, called **generalized transduction**, host DNA derived from virtually any portion of the host genome becomes a part of the DNA of the mature virion in place of the virus genome. The second, called **specialized transduction**, occurs only in some temperate viruses. In specialized transduction, DNA from a specific region of the host chromosome is integrated directly into the virus genome—usually replacing some of the virus genes. The transducing virion in both generalized and specialized transduction is usually noninfective as a virus because bacterial genes have replaced some or all necessary viral genes.

In generalized transduction, if the donor genes do not undergo homologous recombination with the recipient bacterial chromosome, they will be lost. They cannot replicate independently and are not part of a viral genome.

In specialized transduction, homologous recombination may also occur. However, since the donor bacterial DNA is now actually a part of a temperate phage genome, there are two other possibilities: (1) the DNA may be integrated into the host chromosome during lysogenization (⚭ Section 9.10), and (2) the DNA may be replicated in the recipient as part of a lytic infection.

Transduction occurs in a variety of *Bacteria*, including genera of *Desulfovibrio, Escherichia, Pseudomonas, Rhodococcus, Rhodobacter, Salmonella, Staphylococcus,* and *Xanthobacter,* as well as *Methanothermobacter thermoautotrophicus,* a species of *Archaea.* Not all phages can transduce, and not all bacteria are transducible, but the phenomenon is sufficiently widespread that it likely plays an important role in genetic transfer in nature.

Generalized Transduction

In generalized transduction, virtually any gene on the donor chromosome can be transferred to the recipient. Generalized transduction was first discovered and extensively studied in the bacterium *Salmonella enterica* with phage P22 and has also been studied with phage P1 in *Escherichia coli.* An example of how *transducing particles* may be formed is given in Figure 10.15●. When a bacterial cell is infected with a phage, the events of the phage lytic cycle may be initiated. However, during the lytic infection, the enzymes responsible for packaging viral DNA into the bacteriophage sometimes package host DNA accidentally. The resulting virion is called a *transducing particle.* On lysis of the cell, these particles are released along with normal (that is, potentially lytic) virions, and so the lysate contains a mixture of normal virions and transducing particles.

Because transducing particles cannot lead to a normal viral infection (they contain no viral DNA), they are said to be *defective.* When this lysate is used to infect a population of recipient cells, most of the cells become infected with normal virus. However, a small proportion of the population receives transducing particles that inject the DNA they packaged from the previous host bacterium. While this DNA cannot replicate, it can undergo genetic recombination with the DNA of the new host. Because only a small proportion of the particles in the lysate are defective, and each of these contains only a small fragment of donor DNA, the probability of a given transducing particle containing a particular gene is quite low. Typically, only about 1 cell in 10^6 to 10^8 is transduced for a given marker.

Phages that form transducing particles can be either temperate or virulent, the main requirements being that they have a DNA-packaging mechanism that accepts host DNA and that DNA packaging occurs before the host genome is completely degraded. The detection of transduction is most certain when the multiplicity of phage to host is low, so a host cell is infected with only a single phage particle; with multiple infection, the cell will likely be killed by the normal virions in the lysate.

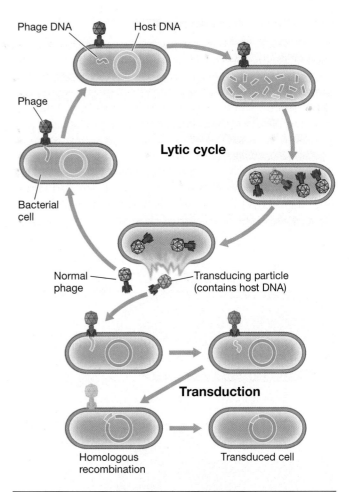

● **Figure 10.15 Generalized transduction.** Note that "normal" virions contain phage genes while a transducing particle contains host genes.

Phage Lambda and Specialized Transduction

Generalized transduction allows the transfer of DNA from one bacterium to another at a low frequency. However, specialized transduction can allow extremely efficient transfer while also allowing a small region of a bacterial chromosome to be selectively transduced. The example we use to discuss specialized transduction was the first to be discovered and involves transduction of the galactose genes by the temperate phage **lambda** of *Escherichia coli.*

As we discussed (⚭ Section 9.10), when a cell is lysogenized by lambda, the phage genome becomes integrated into the host DNA at a *specific site.* The region in which lambda integrates in the *E. coli* chromosome is immediately adjacent to the cluster of genes that encode the enzymes involved in galactose utilization (⚭ Figure 9.21). After insertion, viral DNA replication is under control of the bacterial host chromosome. Upon induction, the viral DNA separates from the host DNA by a process that is the reverse of integration (Figure 10.16●). Ordinarily when the lysogenic cell is induced, the lambda DNA is excised as a unit. Under rare conditions, however, the phage genome is excised incorrectly. Some of the adja-

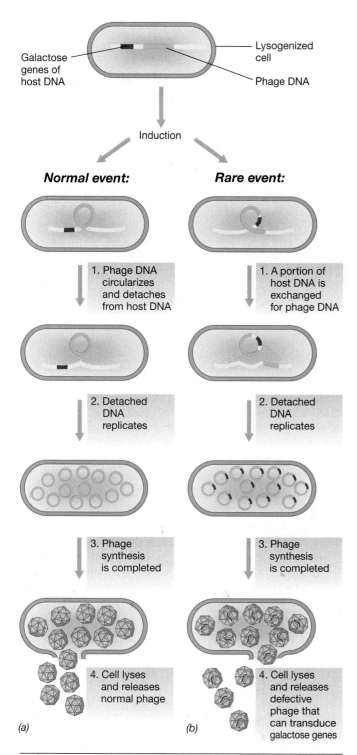

● **Figure 10.16 Specialized transduction.** (a) Normal lytic events, and (b) the production of particles transducing the galactose genes in an *Escherichia coli* cell containing a lambda prophage.

cent bacterial genes (for example, the galactose operon) are excised along with phage DNA. At the same time, some phage genes are left behind (Figure 10.16).

One type of altered phage particle, called **lambda dgal** (λ*dgal*; *dgal* means "defective, galactose"), is defective because of the phage genes lost. It will not make mature phage in a subsequent infection. However, a viable lambda virion, in this case called a **helper phage**, can provide those functions missing in the defective particle. When cells are *coinfected* with λ*dgal* and the helper phage, the culture lysate contains a few λ*dgal* particles mixed in with a large number of normal lambda virions. When a galactose-negative (Gal⁻) bacterial culture is infected at high multiplicity with such a lysate and Gal⁺ transductants selected, many are double lysogens, carrying both lambda and λ*dgal*. When such a double lysogen is induced, the lysate can contain large numbers of λ*dgal* virions and can transduce at high efficiency, although only for the restricted group of *gal* genes.

If a lambda virion is to be viable, there is a limit to the amount of phage DNA that can be replaced with host DNA. Sufficient phage DNA must be retained to provide for production of the phage protein coat and for other phage proteins needed for lysis and lysogenization. However, if a helper phage is used together with the defective phage in a mixed infection, then even fewer phage-specific genes are needed in the defective phage for transduction. Only the *att* (attachment) region, the *cos* site (cohesive ends, for packaging), and the replication origin of the lambda genome are absolutely needed for production of a transducing particle, provided that a helper phage is used (🔗 the genetic map of lambda, Figure 9.18 and Sections 9.10 and 9.11). In this way, specialized transducing phages covering many specific regions of the *E. coli* genome have been isolated. In addition, lambda transducing phages can be constructed by the techniques of genetic engineering to contain genes from any organism (see Section 10.17).

Phage Conversion

When a normal temperate phage (that is, a nondefective phage) lysogenizes a cell and its DNA is converted to the prophage state, the cell is immune to further infection by the same type of phage. This acquisition of immunity can be considered a change in phenotype. However, other phenotypic alterations can often be detected in the lysogenized cell that are unrelated to phage immunity. Such a change, which is brought about through lysogenization by a normal temperate phage, is called **phage conversion**.

Two cases of phage conversion have been especially well studied. One involves a change in structure of a polysaccharide on the cell surface of *Salmonella anatum* on lysogenization with bacteriophage ε[15]. The second involves the conversion of nontoxin-producing strains of *Corynebacterium diphtheriae* (the bacterium that causes the disease diphtheria) to toxin-producing (pathogenic) strains upon lysogenization with phage β (🔗 Section 26.3). In both of these situations, the genes encoding the necessary molecules are an integral part of the phage genome and hence are automatically (and exclusively) transferred upon infection by the phage and lysogenization.

Lysogeny probably carries a strong selective value for the host cell because it confers resistance to infection by viruses of the same type. Phage conversion may also be of considerable evolutionary significance because it results in efficient genetic alteration of host cells. Many bacteria isolated from nature are natural lysogens. It seems reasonable to conclude, therefore, that lysogeny is a common condition and may often be essential for survival of the host in nature.

 10.8 Concept Check

Transduction involves transfer of host genes from one bacterium to another by bacterial viruses. In generalized transduction, defective virus particles randomly incorporate fragments of the cell's chromosomal DNA, but the efficiency is low. In specialized transduction, the DNA of a temperate virus excises incorrectly and takes adjacent host genes along with it; transducing efficiency here may be very high.

◆ What is the major difference between generalized transduction and transformation?

◆ In specialized transduction, the donor DNA can replicate inside the recipient cell without homologous recombination taking place, but this is not true in generalized transduction. Explain.

10.9 Plasmids: General Principles

Before we discuss the third method of genetic transfer, **conjugation**, we must first discuss *plasmids*. **Plasmids** are genetic elements that replicate independently of the host chromosome (∞ Section 7.4). Unlike viruses, plasmids do not have an extracellular form and exist inside cells simply as free, and typically circular, DNA. Plasmids and chromosomes can be differentiated on the basis that plasmids carry only *unessential* (but often very helpful) genes. Essential genes reside on chromosomes. Literally thousands of different plasmids are known. Indeed, over 300 different naturally occurring plasmids have been isolated from strains of *Escherichia coli* alone. In this section we discuss the properties of a few of them.

Physical Nature and Replication of Plasmids

Almost all known plasmids consist of double-stranded DNA. Most plasmids are circular, but many linear plasmids are also known. Naturally occurring plasmids vary in size from approximately 1 kilobase to more than 1 megabase. The typical plasmid is a circular double-stranded DNA molecule less than 5% the size of the chromosome (Figure 10.17●). Most of the plasmid DNA isolated from cells is in the **supercoiled** configuration, which is the most compact form for DNA to exist within the cell (∞ Figure 7.10).

The enzymes involved in actual plasmid replication are normal cell enzymes. Therefore, the genes carried by the plasmid itself are concerned primarily with *control* of the replication initiation process and with apportionment of the replicated plasmids between daughter cells. Also, different plasmids are present in cells in different numbers; this is called the *copy number*. Some plasmids are present in the cell in only 1–3 copies, whereas others

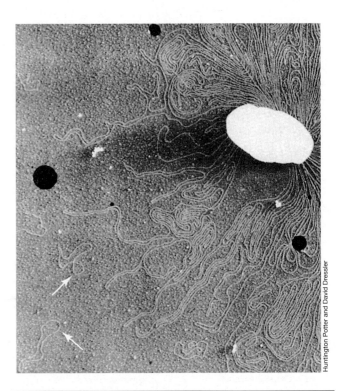

● **Figure 10.17 The bacterial chromosome and bacterial plasmids, as shown in the electron microscope.** The plasmids (arrows) are the circular structures, and are much smaller than the main chromosomal DNA. The cell (large, white structure) was broken gently so the DNA would remain intact.

may be present in over 100 copies. Copy number is controlled by genes on the plasmid and by interactions between the host and the plasmid.

Most plasmids in gram-negative *Bacteria* replicate in a manner similar to that already described for the chromosome (∞ Section 7.6). This involves initiation at an origin of replication and bidirectional replication around the circle, giving a *theta* intermediate (∞ Figure 7.17). However, some plasmids have *unidirectional* replication. Because of the small size of plasmid DNA relative to the chromosome, the whole replication process occurs very quickly, perhaps in a tenth or less of the total time of the cell division cycle.

Most plasmids of gram-positive *Bacteria* replicate by a rolling circle mechanism similar to that used by the phage ϕX174 (∞ Section 16.2 and Figure 16.4). This mechanism gives rise to a single-stranded intermediate, and thus these plasmids are sometimes referred to as *single-stranded DNA plasmids*. Most linear plasmids replicate using a mechanism involving a protein bound to the 5'-end of each strand that is used in priming DNA synthesis (∞ Section 14.6).

Plasmid Incompatibility and Plasmid Curing

Some bacteria may contain several different types of plasmids. For example, *Borrelia burgdorferi* (the Lyme disease pathogen, ∞ Section 27.4) contains 17 different circular and linear plasmids! The ability of two different plasmids to each replicate in the same cell is controlled by plasmid genes involved in regulating DNA replication. When a plasmid is transferred into a cell that already carries another plasmid, a common observation is that the second plasmid may not be maintained and is lost during subsequent cell replication. If this occurs, the two plasmids are said to be **incompatible**. A number of incompatibility (Inc) groups have been recognized. The plasmids of one incompatibility group exclude each other from replicating in the cell but can coexist with plasmids from other groups. Plasmids of an incompatibility group share a common mechanism of regulating their replication and are thus *related* to one another. Therefore, although a bacterial cell may contain different kinds of plasmids, each is genetically distinct.

Some plasmids, called *episomes*, can integrate into the chromosome, and under such conditions their replication comes under control of the chromosome. This situation is remarkably like that of several viruses whose genomes can become incorporated into the host genome (prophages, ➾ Sections 9.10, 16.5, 16.11, and 16.15). Plasmids can sometimes be eliminated from host cells by various treatments. This removal, called **curing**, results from inhibition of plasmid replication without parallel inhibition of chromosome replication. As a result of cell division, the plasmid is diluted out. Curing may occur spontaneously, but it is greatly increased by treatments with certain chemicals such as acridine dyes, which become inserted into DNA, or other treatments that seem to interfere more with plasmid replication than with chromosome replication.

Many of these characteristics are exemplified in a very well-characterized plasmid of *Escherichia coli*, called the *F plasmid*. The F plasmid ("F" stands for "fertility") is a circular DNA molecule of 99,159 bp. Cells containing it can easily be cured with acridine orange. Figure 10.18● shows a genetic map of the F plasmid. One region of the plasmid contains genes involved in regulating DNA replication. It also contains a number of transposable elements (see Section 10.14) involved in its ability to function as an episome. Last, it has a large region of DNA, the *tra* region, containing genes that permit it to be transferred from one cell to another (see later).

Cell-to-Cell Transfer of Plasmids

Because one of the defining characteristics of a plasmid is the lack of a distinct extracellular form, one might imagine that plasmids are transferred only during cell division. Since some prokaryotic cells can take up free DNA from the environment (see Section 10.6), it is possible that lysis of the host, however this may happen, may bring the plasmid in contact with a new host. However, this process occurs naturally in only a few bacterial species and is unlikely to account for much cell-to-cell plasmid transfer. The main mechanism of cell-to-cell transfer is **conjugation**, a function encoded by some plasmids themselves. Conjugation is a replicative process, and both cells end up with copies of the plasmid (Figure 10.19●).

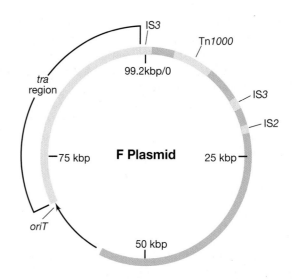

● **Figure 10.18   Genetic map of the F (fertility) plasmid of *Escherichia coli*.** The numbers on the interior show the size of the plasmid in kilobase pairs (the exact size is 99,159 bp). The region shown in dark green at the bottom of the map contains genes primarily responsible for the replication and segregation of the F plasmid in normally growing cells. The light green region, the *tra* region, contains the genes involved in conjugative transfer. The *oriT* sequence is the origin of transfer during conjugation. The arrow indicates the direction of transfer (the *tra* region would be transferred last). The regions shown in yellow on F are transposable elements where integration into identical elements on the bacterial chromosome can occur and lead to the formation of different Hfr strains (see Section 10.12).

Plasmids that govern their own transfer by cell-to-cell contact are called *conjugative*. Not all plasmids are conjugative. Transmissibility by conjugation is controlled by a set of genes within the plasmid called the *tra* (for *transfer*) *region*. The *tra* region contains genes encoding proteins that function in DNA transfer and replication and others that function in mating pair formation. The presence of a *tra* region in a plasmid can have another important consequence if the plasmid becomes integrated into the chromosome. In that case, the plasmid can *mobilize* the transfer of chromosomal DNA from one cell to another. The use of conjugation to transfer host genes is discussed further in the next section.

Some conjugative plasmids from *Pseudomonas* have a *broad host range*. This means that they are transferable to a wide variety of other gram-negative *Bacteria*. Such

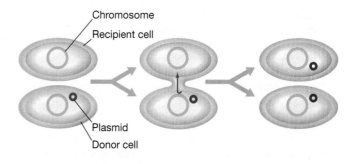

● **Figure 10.19   Plasmid transfer from cell to cell during conjugation.** Note that as in all gene transfer processes in prokaryotes, DNA transfer is *unidirectional*.

plasmids can transfer genetic information between distantly related organisms. Conjugative plasmids have been shown to transfer between gram-negative and gram-positive *Bacteria*, between *Bacteria* and plant cells, and between *Bacteria* and fungi. Even if the plasmid cannot replicate independently in the new host, transfer of the plasmid itself could have important evolutionary consequences if genes from the plasmid recombine with the genome of the new host.

 10.9 Concept Check

Plasmids are small circular or linear DNA molecules that carry any of a variety of unessential genes. Although a cell can contain more than one plasmid, they cannot be genetically closely related. Plasmids can be transferred by lateral transfer in the process of conjugation.

◆ How can a *large* plasmid be differentiated from a *small* chromosome?

◆ What function do the *tra* genes of the F plasmid carry out?

10.10 Types of Plasmids and Their Biological Significance

Clearly all plasmids must carry genes that ensure their own replication. As we have seen, some plasmids also carry genes necessary for conjugation, and they can sometimes be detected biologically, either by the transfer functions themselves or by sensitivities to certain viruses (∞ Section 16.1).

Although plasmids do not carry genes that are essential to the host, the plasmids can carry genes that have a profound influence on the cell's phenotype. For example, in some cases plasmids encode properties considered fundamental to the ecology of the bacterium. For example, the ability of *Rhizobium* to interact and form nitrogen-fixing root nodules with plants requires certain plasmid functions (∞ Section 19.22). Other plasmids have been shown to confer special metabolic properties on a cell, such as for the degradation of toxic pollutants. Indeed, plasmids seem to be a major mechanism for conferring special properties on bacteria, and in many instances for also exporting these properties by horizontal gene flow. The only limitation to the types of genes present on a plasmid is that they do not interfere with their own replication or with the survival of the host. A few of the phenotypes conferred on prokaryotes by plasmids are summarized in Table 10.3.

Resistance Plasmids

Among the most widespread and well-studied groups of plasmids are the *resistance plasmids* (R *plasmids*), which confer resistance to antibiotics and various other growth inhibitors. R plasmids were first discovered in Japan in strains of enteric bacteria that had acquired resistance to a number of antibiotics (multiple resistance) and have since been found throughout the world. The emergence of bacteria resistant to several antibiotics is of considerable medical significance and was correlated with the increasing use of antibiotics for the treatment of infectious diseases. Soon after these resistant strains were isolated it

Table 10.3	**Some phenotypes conferred by plasmids in prokaryotes**
Phenotype class[a]	**Organisms[b]**
Antibiotic production	*Streptomyces*
Conjugation	*Escherichia, Pseudomonas, Rhizobium, Staphylococcus, Streptococcus, Sulfolobus, Vibrio*
Physiological functions	
Degradation of octane, camphor, naphthalene	*Pseudomonas*
Degradation of herbicides	*Alcaligenes*
Formation of acetone and butanol (∞ Section 12.20)	*Clostridium*
Lactose, sucrose or urea utilization and nitrogen fixation	Enteric bacteria
Nodulation and symbiotic nitrogen fixation (∞ Section 19.22)	*Rhizobium*
Pigment production	*Erwinia, Staphylococcus*
Resistance	
Antibiotic resistance (∞ Section 20.12)	*Campylobacter*, Enteric bacteria, *Neisseria, Staphylococcus*
Resistance to cadmium, cobalt, mercury, nickel, and/or zinc (∞ Section 19.16)	*Acidocella, Alcaligenes, Listeria, Pseudomonas, Staphylococcus*
Bacteriocin resistance (and production)	*Bacillus*, Enteric bacteria, *Lactococcus, Propionibacterium*
Virulence	
Host cell invasion	*Salmonella, Shigella, Yersinia*
Coagulase, hemolysin, enterotoxin (∞ Sections 21.9 and 21.11)	*Staphylococcus*
Enterotoxin, K antigen (∞ Sections 12.11 and 21.11)	*Escherichia*
Tumorigenicity in plants (∞ Section 19.21)	*Agrobacterium*

[a] Only a few of the many phenotypes known to be associated with plasmids are given.
[b] Only a few well-characterized examples are given. All of the organisms given in the list are *Bacteria* except for *Sulfolobus*, which is a member of the *Archaea*.

was shown that they could transfer resistance to sensitive strains via cell-to-cell contact. The infectious nature of the conjugative R plasmids permitted their rapid spread through cell populations. Resistance plasmids are now a major problem in clinical medicine (Section 20.12).

Several antibiotic resistance genes can be carried by an R plasmid. In general, these genes encode proteins that either inactivate the antibiotic or prevent its uptake into the cell. Plasmid R100, for example, is a 94.3-kbp plasmid (Figure 10.20●) that carries genes encoding resistance to sulfonamides, streptomycin and spectinomycin, fusidic acid, chloramphenicol, and tetracycline. Plasmid R100 also carries several genes conferring resistance to mercury (Section 19.16). Plasmid R100 can be transferred between enteric bacteria of the genera *Escherichia, Klebsiella, Proteus, Salmonella*, and *Shigella*, but does not transfer to the nonenteric gram-negative bacterium *Pseudomonas*. Different R plasmids with genes for resistance to most antibiotics are known. Many drug-resistant elements on R plasmids, such as those on R100, are also transposable elements (see Section 10.14) and this, plus the fact that many of these plasmids are conjugative, have made them a serious threat to traditional antibiotic therapies.

Plasmids Encoding Toxins and Other Virulence Characteristics

We will discuss in Chapter 21 the characteristics of pathogenic microorganisms that enable them to colonize hosts and establish infections. In the present context, we merely

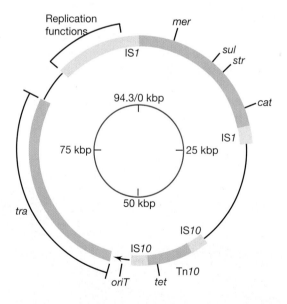

● **Figure 10.20 Genetic map of the resistance plasmid R100.** The inner circle shows the size of the plasmid in kilobase pairs. The outer circle shows the location of major antibiotic resistance genes and other key functions: *cat*, chloramphenicol resistance; *oriT*, origin of conjugative transfer; *mer*, mercuric ion resistance; *sul*, sulfonamide resistance; *str*, streptomycin resistance; *tet*, tetracycline resistance; *tra*, transfer functions. The locations of insertion sequences (IS) and the transposon Tn*10* are also shown. Several genes related to plasmid replication are found in the region from 88–92 kbp.

note the two major characteristics involved in the virulence (disease-causing ability) of pathogens: (1) the ability of the pathogen to attach to and colonize specific host tissues; and (2) the formation of substances (toxins, enzymes, and other molecules) that cause damage to the host.

In several pathogenic bacteria each of these virulence characteristics is carried on plasmids. For example, enteropathogenic strains of *Escherichia coli* are characterized by an ability to colonize the small intestine and to produce a toxin that causes symptoms of diarrhea (Section 21.11). Colonization requires the presence of a cell surface protein called the *colonization factor antigen*, encoded by a plasmid. This protein confers on cells the ability to attach to epithelial cells of the intestine. At least two toxins in enteropathogenic *E. coli* are known to be encoded by a plasmid: the *hemolysin*, which lyses red blood cells, and the *enterotoxin*, which induces extensive secretion of water and salts into the bowel. It is the enterotoxin that is responsible for diarrhea, as will be discussed in Chapter 21.

Some virulence factors are encoded on plasmids, while others are encoded by other types of *mobile genetic elements*, such as transposons and bacteriophages; some virulence factors, of course, are chromosomal. Several examples are known where virulence genes are present on different genetic elements within the same cell. For instance, the genes encoding the virulence determinants of shigatoxin-producing strains of *E. coli* are distributed between the chromosome, a bacteriophage, and a plasmid.

Bacteriocins

Many bacteria produce proteins that inhibit or kill closely related species or even different strains of the same species. These agents, called **bacteriocins** to distinguish them from the antibiotics, have a more narrow spectrum of activity than antibiotics. The genes encoding bacteriocins and the proteins involved with processing and transporting them (and for conferring immunity on the producing organism) are often carried on a plasmid or a transposon. Bacteriocins are named after the species of organism that produces them. Thus, in *Escherichia coli* we have *colicins*, encoded by Col plasmids; *Bacillus subtilis* produces *subtilisin*, and so on.

The Col plasmids of *Escherichia coli* encode various colicins. Colicins released from a cell bind to specific receptors on the surface of susceptible cells. The receptors for colicins are generally entities whose normal function is to transport some substance, frequently a growth factor or micronutrient, through the outer membrane (the lipopolysaccharide layer) of the cell. Colicins kill cells by disrupting some critical cell function. For example, many colicins form channels in the cell membrane that allow potassium ions and protons to leak out, leading to a loss of the cell's energy-forming ability. However, colicin E2 is a DNA endonuclease that can cleave DNA, and colicin E3 is a nuclease that cuts at a specific site in 16S rRNA and inactivates ribosomes. Col plasmids can be either conjugative or nonconjugative.

The bacteriocins or bacteriocin-like agents of *gram-positive* bacteria are quite different from the colicins but are also often encoded by plasmids; some even have commercial value. For instance, lactic acid bacteria produce the bacteriocin Nisin A, which strongly inhibits the growth of a wide range of gram-positive bacteria and is used as a preservative in the food industry.

Engineered Plasmids

The basic tools of genetic engineering, discussed later in this chapter, have made possible the construction in the laboratory of countless new, artificial plasmids. Incorporation into such plasmids of genes from a wide variety of sources allows for the transfer of genetic material across any species barrier. Moreover, genes can be synthesized and introduced into plasmids. The only requirements for artificial plasmids are that (1) they contain genes controlling their own replication; (2) if conjugative, they contain transfer (*tra*) functions that facilitate this process; and (3) they are stably maintained in the host of choice.

We now turn our attention to the details of conjugation and show how certain plasmids can mobilize the bacterial chromosome, allowing transfer of chromosomal genes from donor to recipient.

 10.10 Concept Check

The genetic information that plasmids carry is not essential for cell function under all conditions but may confer a selective growth advantage under certain conditions. Examples include antibiotic resistance, enzymes for degradation of unusual organic compounds, and special metabolic pathways. Virulence factors of many pathogenic bacteria are often plasmid encoded.

◆ How does an *R plasmid* differ from the *F plasmid* discussed previously? How are they similar?

◆ How do *bacteriocins* differ from *antibiotics*?

10.11 Conjugation: Essential Features

Bacterial **conjugation** (mating) is a process of genetic transfer that involves cell-to-cell contact. As we discussed (see Section 10.9), conjugation is a plasmid-encoded mechanism. A conjugative plasmid uses this mechanism to transfer a copy of itself to a new host. However, sometimes other genetic elements can be *mobilized* (meaning *transferred*) during conjugation. These other genetic elements can be other plasmids, or the host chromosome itself. Indeed, conjugation was discovered because the F plasmid of *Escherichia coli* (see Figure 10.17) can mobilize the host chromosome. Mechanisms of conjugative transfer may differ depending on the plasmid involved, but most plasmids in gram-negative *Bacteria* seem to employ a mechanism similar to that used by the F plasmid.

The process of conjugation involves a *donor* cell, which contains a particular type of conjugative plasmid, and a *recipient* cell, which does not. The genes that con-

trol conjugation are contained in the *tra* region of the plasmid (see Section 10.9). Many genes in the *tra* region are involved in mating pair formation, and most of these have to do with the synthesis of a surface structure, the **sex pilus** (Figure 10.21●). Only *donor* cells produce these pili. Different conjugative plasmids may have slightly different *tra* regions, and the pili may also be different. The F plasmid and its relatives encode *F pili*.

Pili allow specific pairing to take place between the donor cell and the recipient cell. All conjugation in gram-negative *Bacteria* is thought to depend on cell pairing brought about by pili. The pili make specific contact with a receptor on the recipient and then retract, pulling the two cells together. The contacts between the donor and recipient cells then become stabilized, probably from fusion of the outer membranes, and DNA is then transferred from one cell to another.

Mechanism of DNA Transfer During Conjugation

DNA synthesis is necessary for DNA transfer to occur during conjugation. A mechanism of DNA synthesis in certain bacteriophages, called **rolling circle replication** (∞ Section 9.11), best explains DNA transfer during conjugation, and this process is described in Figure 10.22●. Conjugation is triggered by cell-to-cell contact, at which time *one strand* of the plasmid DNA circle is nicked and is transferred to the recipient. The nicking enzyme required to initiate the process, TraI, is encoded by the *tra* operon of the F plasmid. This protein also has helicase activity and is thus also involved in unwinding the strand to be transferred. As this transfer occurs, DNA synthesis by the rolling circle mechanism replaces the transferred strand in the donor, while the complementary DNA strand is made in the recipient. Therefore, at the end of the process, both donor and recipient possess completely formed plasmids. For transfer of the F plasmid then, an F-containing cell, which is designated F$^+$, can mate with a cell lacking the plasmid, designated F$^-$, to yield two F$^+$ cells (Figure 10.22).

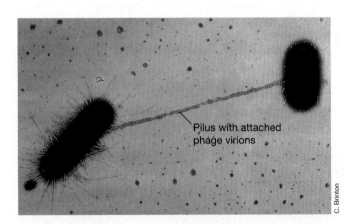

● **Figure 10.21 Direct contact between two conjugating bacteria is first made via a pilus.** The cells are then drawn together to form a mating pair for the actual transfer of DNA. This occurs by retraction (depolymerization) of the pilus within the donor cell. Note the F-specific bacteriophages on the pilus (∞ Section 16.1).

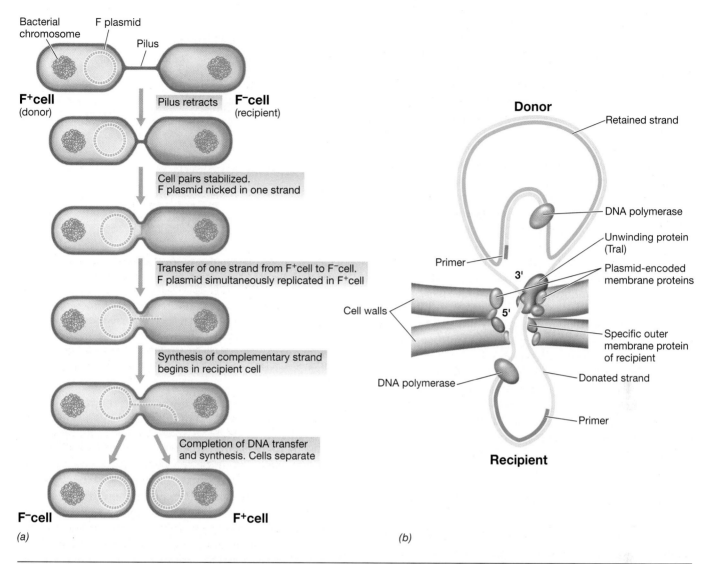

The plasmid DNA transfer process is highly efficient; under appropriate conditions virtually every recipient cell that pairs acquires a plasmid. When the plasmid genes can be expressed in the recipient, the recipient itself becomes a donor and can transfer the plasmid to other recipients. In this fashion, conjugative plasmids can spread rapidly between populations, behaving much like infectious agents. The infectious nature of plasmid transfer is of major ecological significance because a few plasmid-containing cells introduced into an appropriate population of recipients can, if they contain genes that confer a selective advantage, convert the entire recipient population into a plasmid-bearing (and thus donating) population in a short period of time. Plasmids can also be lost from a cell by a process called *curing*. This could happen spontaneously in natural populations when there is no selection pressure to maintain the plasmid. For example, plasmids conferring antibiotic resistance can be lost without affecting the cell's viability if there are no antibiotics in the cell's environment.

10.11 Concept Check

Conjugation is a mechanism of DNA transfer in prokaryotes that requires cell-to-cell contact. Conjugation is controlled by genes carried by certain plasmids (such as the F plasmid) and involves transfer of the plasmid from a donor cell to a recipient cell. Plasmid DNA transfer involves replication via the rolling circle mechanism.

◆ In conjugation, how are donor and recipient cells brought into contact with each other?

◆ How does rolling circle DNA replication differ from the process by which chromosomal DNA replication occurs, described in Sections 7.5 and 7.6?

10.12 The Formation of Hfr Strains and Chromosome Mobilization

In conjugation discussed thus far, no mention was made of the transfer of *chromosomal* genes. However, the F plasmid of *Escherichia coli* (see Section 10.9) can, under certain circumstances, mobilize the chromosome for transfer during cell-to-cell contact. The F plasmid is an episome, a plasmid that can integrate into the host chromosome (see Section 10.9). When the F plasmid is integrated into the chromosome, the chromosome becomes mobilized and can lead to transfer of chromosomal genes. Following genetic recombination between donor and recipient, lateral gene transfer by this mechanism can be very extensive.

Cells possessing an *unintegrated* F plasmid are called F$^+$. Those that have a *chromosome-integrated* F plasmid are called **Hfr** (for *high frequency of recombination*). Both F$^+$ and Hfr cells are donors and are unable to take up a second copy of the F plasmid or genetically related plasmids. But unlike conjugation between an F$^+$ and an F$^-$, conjugation between an F$^-$ and an Hfr donor leads to transfer of genes from the host chromosome because the plasmid *is* part of the chromosome. Following recombination, the F$^-$ recipient cell may express a new phenotype. Plasmid integration is thus *a mechanism for mobilizing the cell's main genetic resources*. The term "high frequency of recombination" refers to the high rates of genetic recombination between genes on the donor chromosome and that of the recipient.

The presence of the F plasmid therefore results in three distinct alterations in the properties of a cell: (1) the ability to synthesize the F pilus, (2) the mobilization of DNA for transfer to another cell, and (3) the alteration of surface receptors so the cell is no longer able to behave as a recipient in conjugation.

Integration of F and Chromosome Mobilization

Integration of the F plasmid into the host chromosome can occur at several specific sites, called *IS* (for *insertion sequences*) sites. These sites are regions of DNA sequence homology between chromosomal and F plasmid DNA (see Section 10.13 for a discussion of insertion sequences). Figure 10.23● shows the integration of an F plasmid at an IS site. Once integrated, the plasmid no longer controls its own replication, but the *tra* operon still functions normally and the strain synthesizes pili. When a recipient is encountered, conjugation is triggered just as in an F$^+$ cell, and DNA transfer is initiated at the *oriT* (*origin of transfer*) site. However, since the plasmid is now part of the chromosome, after part of the plasmid DNA is transferred, *chromosomal genes* begin to be transferred (Figure 10.24●). As is the case of conjugation with just the F plasmid itself (Figure 10.22), chromosomal DNA transfer also involves replication. Thus, after transfer, the Hfr strain remains genetically Hfr because it retains a copy of the transferred genes.

Because a number of distinct insertion sites are present on the chromosome, a number of distinct Hfr strains are possible. A given Hfr strain always donates genes in the same order, beginning with the same position. However, Hfr strains that differ in the position of integration of the F plasmid in the chromosmome transfer genes in different orders (Figure 10.25●). Because breakage of the DNA strand typically occurs during transfer, only *part* of

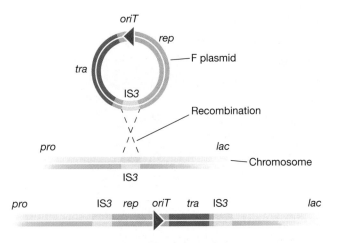

● **Figure 10.23 Integration of an F plasmid into the chromosome with the formation of an Hfr.** The insertion of the F plasmid occurs at a variety of specific sites where IS elements are located, the one here being an IS*3* located between the chromosomal genes *pro* and *lac*. Some of the genes on the F plasmid are shown. The arrow indicates the origin of transfer, *oriT*, with the arrow as the leading end. Thus, in this Hfr *pro* would be the first chromosomal gene to be transferred and *lac* would be among the last (see Figure 10.24).

the donor chromosome is transferred. And, since only part of the chromosome is transferred, it cannot replicate in the recipient cell. Therefore, donor genes normally cannot be detected in the recipient cells unless recombination between the incoming fragment and the recipient chromosome takes place. Finally, although Hfr strains transmit chromosomal genes at high frequency, they generally do not convert F$^-$ cells to F$^+$ or Hfr because the entire F plasmid is only rarely transferred. Instead, an Hfr × F$^-$ cross yields the original Hfr and an F$^-$ cell that

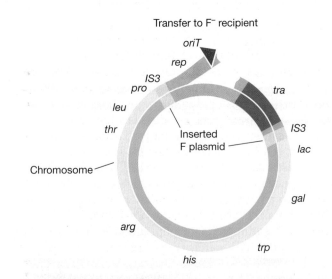

● **Figure 10.24 Breakage of the Hfr chromosome at the origin of transfer and the beginning of DNA transfer to the recipient.** Replication occurs during transfer (see Figure 10.22). The figure is not drawn to scale; the inserted F plasmid is actually less than 3% of the size of the *Escherichia coli* chromosome.

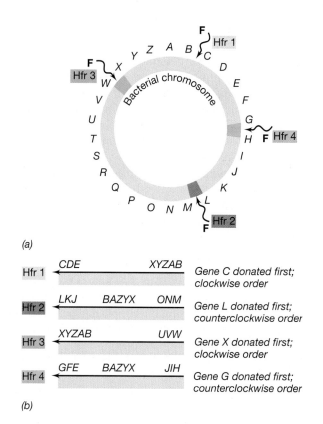

(a)

Hfr 1   CDE   XYZAB   *Gene C donated first; clockwise order*

Hfr 2   LKJ   BAZYX   ONM   *Gene L donated first; counterclockwise order*

Hfr 3   XYZAB   UVW   *Gene X donated first; clockwise order*

Hfr 4   GFE   BAZYX   JIH   *Gene G donated first; counterclockwise order*

(b)

● **Figure 10.25** **Manner of formation of different Hfr strains, which donate genes in different orders and from different origins.** The bacterial chromosome (a) opens at various insertion sequences, at which F plasmids can become inserted. The gene orders are shown in part (b).

now has a new genotype and possibly phenotype. As in transformation and transduction, genetic recombination between Hfr genes and F⁻ genes involves *homologous recombination* in the recipient cell.

At some insertion sites, the F plasmid is integrated with the origin pointing in one direction, whereas at other sites the origin points in the opposite direction. The direction in which the F plasmid is inserted determines which of the chromosomal genes will be transferred into the recipient first (Figure 10.25). By use of various Hfr strains in mating experiments, it was possible to determine the arrangement and orientation of virtually all chromosomal genes in *Escherichia coli* (see Figure 10.42) long before the chromosome of this bacterium was sequenced.

Use of Hfr Strains in Genetic Crosses

As is the case for any system of bacterial gene transfer, one *selects* recombinants from conjugation. However, unlike the situation in transformation and transduction, during conjugation both the donor and recipient cells are viable. It is thus necessary to chose selection conditions where the desired recombinants can grow but where neither of the parental strains can form colonies. Typically a recipient is used that is resistant to an antibiotic but is auxotrophic for some substance, and a donor is used that is sensitive to the antibiotic but is prototrophic for the

same substance. Thus, on minimal medium containing the antibiotic, only recipient cells will grow following the mating event.

For instance, in the experiment shown in Figure 10.26●, an Hfr donor that is sensitive to streptomycin (Strˢ) and is wild-type for synthesis of the amino acids threonine and leucine (Thr⁺ and Leu⁺) and for utilization of lactose (Lac⁺), is mated with a recipient cell that cannot make these amino acids, but that is resistant to streptomycin (Strʳ). The selective medium is a minimal medium containing streptomycin so that only recombinant cells can grow. The composition of each selective medium is varied depending on which genotypic characteristics are desired in the recombinant, as shown in Figure 10.26. The frequency of the process is measured by counting the colonies grown on the selective medium.

The order in which genes are present on the donor chromosome can also be determined by the kinetics of transfer of individual markers. For example, in the process called **interrupted mating**, conjugating cells can be separated by agitation in a mixer or blender. If mixtures of Hfr and F⁻ cells are agitated at various times after mixing and the genetic recombinants scored, it is found that the longer the time between pairing and agitation, the greater the number of genes of the Hfr that will appear in the F⁻ recombinant. As shown in Figure 10.27●, genes present closer to the origin of transfer enter the F⁻ first and are present in a higher percentage of the recombinants than genes that are transferred later. In addition to showing that gene transfer from donor to recipient is a sequential process, experiments of this kind provide a

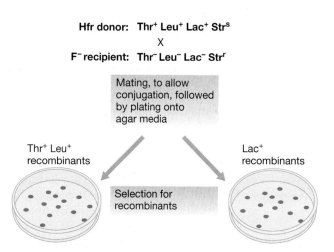

Hfr donor: **Thr⁺ Leu⁺ Lac⁺ Strˢ**

X

F⁻ recipient: **Thr⁻ Leu⁻ Lac⁻ Strʳ**

Mating, to allow conjugation, followed by plating onto agar media

Thr⁺ Leu⁺ recombinants   Lac⁺ recombinants

Selection for recombinants

Agar mimimal medium with streptomycin and glucose; selects for markers Thr⁺ Leu⁺; does not select for Lac

Agar mimimal medium with streptomycin, lactose, threonine, leucine; selects for marker Lac⁺; does not select for Thr or Leu

●**Figure 10.26** **Example experiment for the detection of conjugation.** Thr, threonine; Leu, leucine; Lac, lactose; Str, streptomycin. Note that each medium selects for specific classes of recombinants. The controls for the experiment are to plate samples of the donor and the recipient *before* they are mixed. Neither should be able to grow on the selective media used.

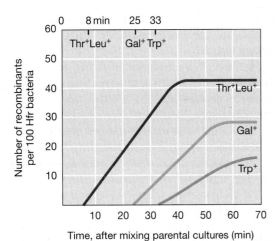

● **Figure 10.27   Rate of formation of recombinants containing different genes after mixing Hfr and F⁻ bacteria by interrupted mating.** The location of the genes along the Hfr chromosome is shown at the upper left. Note that the genes closest to the origin (0 min) are the first ones whose transfer is detected. The experiment is done by mixing Hfr and F⁻ cells under conditions in which essentially all Hfr cells find recipients. At various times, samples of the mixture are shaken violently to separate the mating pairs and plated on a selective medium in which only the recombinants can form colonies.

method of *determining the order* of the genes on the bacterial chromosome (genetic mapping). The arrangement of gene loci on the chromosome is called a **genetic map** (see Section 10.19 and Figure 10.42).

Transfer of Chromosomal Genes to the F Plasmid

Occasionally integrated F plasmids may be excised from the chromosome, and the possibility exists for the incorporation at that time of *chromosomal* genes into the liberated F plasmid. This can happen because both the integrated F plasmid and the chromosome contain a number of identical IS at which recombination can occur (Figure 10.23). Such F plasmids containing chromosomal genes are called F′ (*F prime*) plasmids. F′ plasmids differ from normal F plasmids in that they contain identifiable chromosomal genes, and they transfer these genes at high frequency to recipients. F′-mediated transfer resembles specialized transduction (see Section 10.8) in that only a restricted group of chromosomal genes can be transferred. Transferring a known F′ into a recipient allows one to establish diploids (two copies of each gene) for a limited region of the chromosome. Such partial diploids are called *merodiploids*. This procedure is important for doing complementation tests, as we will see in the next section.

Other Conjugation Systems

Although we have discussed conjugation almost exclusively as it occurs in *Escherichia coli*, conjugative plasmids have been found in many other gram-negative *Bacteria*. Conjugative plasmids of the incompatibility group (see Section 10.8) IncP can be maintained in virtually all gram-negative *Bacteria* and even transferred between different genera. Conjugative plasmids are also

known in gram-positive *Bacteria* (for example, in *Streptococcus*, *Enterococcus*, and *Staphylococcus*).

Several conjugative plasmids have also been found in *Sulfolobus*, a genus of *Archaea*. Little is known about conjugation in *Sulfolobus*, although it is known that cell-pairing occurs before plasmid transfer and that transfer is unidirectional. However, with one exception, the genes involved seem to have little similarity to those in gram-negative *Bacteria*. The exception is a gene similar to *traG*, whose protein product in F plasmid-mediated conjugation seems to be involved in stabilizing mating pairs. It thus seems likely that the mechanism of conjugation in *Archaea* is quite different from that in *Bacteria*.

 10.12   Concept Check

The donor cell chromosome can be mobilized for transfer to a recipient cell. This requires that the F plasmid integrate into the chromosome to form the Hfr phenotype. Transfer of the host chromosome is rarely complete but can be used to map the order of the genes on the chromosome. F′ plasmids are previously integrated F plasmids that have deintegrated and excised some chromosomal genes.

◆   In conjugation involving the F plasmid of *Escherichia coli*, how is the host chromosome mobilized?

◆   Why does an Hfr × F⁻ mating not yield two Hfr cells?

◆   At which sites in the chromosome can the F plasmid integrate?

10.13   Complementation

All methods of bacterial gene transfer involve only a *portion* of the donor chromosome. Therefore, unless recombination takes place with the recipient chromosome, the donor DNA will be lost in the recipient because it cannot replicate independently. In only two instances have we seen that a state of partial diploidy can be stably maintained. One was in the case of a specialized transducing phage where the donor genes are maintained as part of the viral genome (see Section 10.8). The other was the use of F′ plasmids where donor genes have become a part of the F plasmid genome on an episome (see Section 10.12). Since it is possible to create specialized transducing phage or specific plasmids using recombinant DNA techniques (see Sections 10.15–10.17), it is possible to put essentially *any* portion of the bacterial chromosome on a phage or plasmid. This can be quite useful in bacterial genetic analyses, and we consider this now.

Complementation Test and the Cistron

When two mutant strains are genetically crossed (mated), homologous recombination can yield a wild-type recombinant *unless* both of the mutations occur in exactly the same base pairs. For example, if two different Trp⁻ *Escherichia coli* auxotrophs (strains that require the amino acid tryptophan in the medium) are crossed and Trp⁺ recombinants are obtained, it is obvious that the mutations in the two strains were not in the same base pairs. However, this kind of experiment cannot detect whether two mutations are in different regions of *the same gene*. This can be determined by a **complementation test**.

Continuing with our tryptophan auxotroph example, if one of the tryptophan genes has been inactivated in a particular strain by a mutation leading to the Trp⁻ phenotype, then inserting a copy of the wild-type gene on a plasmid or viral genome into the same cell should restore the wild-type phenotype of the cell; that is, the resulting partially diploid cell should be Trp⁺. One would then say that the wild-type gene *complements* the mutation. How can a complementation test be done to look for mutations *in the same gene*?

Even though only a chromosomal fragment is transferred in conjugation, this fragment may contain many genes. Again, from our tryptophan example, recall that the genes for the biosynthesis of tryptophan form an operon (⟳ Figure 8.24), and therefore one can easily transfer the entire operon from the donor cell to a recipient cell. Note then that even if the operon in the donor also has a mutation, it will complement the mutated gene in the recipient *if the two mutations are in different genes*. The two mutations are then said to complement one another. This is shown diagrammatically in Figure 10.28●. If each homologous DNA molecule contributes a different required gene, then the cell will have all the enzymes it requires to synthesize tryptophan.

It is important to note that complementation *does not* involve recombination. To do a complementation test, the mutations must be located in two different DNA molecules: the chromosome and the gene carrier (plasmid or viral DNA). Such mutations are referred to as being in *trans* with respect to one another. If one molecule has both mutations, a condition called *cis*, the second molecule can complement *only* if it is wild-type for the given gene. Therefore, having the mutations in *cis* serves as a positive control in such a complementation experiment. This type of complementation test, called a *cis-trans test*, is used to determine whether two mutations are in the same genetic unit. The genetic unit defined by the cis-trans test is called a **cistron** (a term essentially equivalent to a gene). If two mutations occur in genes encoding different enzymes, or even different subunits of the same enzyme, complementation of the two mutations is possible, and the mutations are therefore not in the same cistron (Figure 10.28).

Although genetic crosses involving complementation analyses are still done in bacterial genetics, in many cases it is simply easier to sequence the gene in question and to analyze the sequence to identify the nature and location of any mutations. This is especially true if the sequence of the wild-type gene is known, since highly specific primers can be used and the DNA quickly sequenced. The term *cistron* is now rarely used in microbial genetics except when describing whether an mRNA has

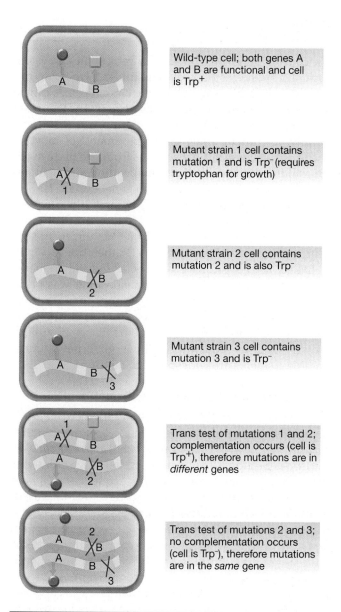

● **Figure 10.28 Complementation analysis.** In this example, the protein products of both genes A and B are required to synthesize tryptophan. Mutations 1, 2, and 3 each lead to the same phenotype, a requirement for tryptophan (Trp⁻). Complementation analysis indicates that mutations 2 and 3 are in the same gene while mutation 1 is in a separate gene.

the genetic information from one gene (*monocistronic* mRNA) or from more than one gene (*polycistronic* mRNA) (⟳ Section 7.13).

 10.13 Concept Check

If a cell is treated so as to to contain two copies for a region of its genome, one can do complementation tests to determine if two mutations are in the same or different genes. This is often necessary because mutations in different genes in the same pathway may give the same phenotype. Complementation tests do not involve recombination.

◆ What is a *merodiploid*?

◆ Complementation tests have been referred to as *cis-trans* tests. Explain.

10.14 Transposons and Insertion Sequences

The order of genes on a bacterial chromosome can be determined by the methods of gene transfer we have considered, just as genes in eukaryotic chromosomes can be mapped by mating experiments. However, the exact arrangement of the genes along a chromosome is not necessarily permanent; some genes are capable of moving. The process by which a gene moves from one place to another in the genome is called **transposition** and is an important process in evolution and in genetic analysis.

Transposition is a *rare* event, occurring at frequencies of 10^{-5}–10^{-7} per generation. Thus, for the most part, genes are relatively stable entities. Moreover, not all genes are capable of transposition. Rather, transposition is linked to the presence of special genetic elements called **transposable elements**. Transposition was originally discovered in corn (maize) and then later in *Bacteria* owing to the extremely sensitive types of genetic analysis available in these organisms. It has now been shown that DNA sequences with the properties of transposable elements are widespread in nature.

Transposons and Transposable Elements

As we discussed earlier (∞ Section 7.4), there are three types of transposable elements in *Bacteria: insertion sequences, transposons,* and some *special viruses* (such as *Mu*) (∞ Section 16.5). In this section we confine our discussion primarily to insertion sequences (IS) and transposons. Both of these elements have two important features in common: They both carry genes encoding a **transposase**, the enzyme necessary for transposition, and they both have short *inverted terminal repeats* at the ends of their DNA (these "ends" are continuous with whatever DNA the element has inserted into). The repeats range in length from fewer than 20 bp in simple IS elements to more than 1 kbp in some transposons, and each different IS has a specific number of base pairs in its terminal repeats. Such inverted terminal repeats are involved in the transposition process. Figure 10.29● shows a genetic map of a common insertion element called IS2 and of the transposon Tn5.

Insertion sequences are the simplest type of transposable element and carry no genes other than those required for them to move to new locations. Insertion sequences are short segments of DNA, about 1000 nucleotides long, that can become integrated at specific sites on the genome. Insertion sequences are found in both chromosomal and plasmid DNA, as well as in certain bacteriophages. Several hundred distinct IS elements have been characterized, and most are designated by a number identifying its type: IS1, IS2, IS3, and so on.

IS elements are scattered about the chromosome, and strains vary in the number and frequency of these elements. For instance, one strain of *Escherichia coli* has five copies of IS2 and five copies of IS3. Many plasmids also carry these insertion sequences (see Figures 10.18 and 10.20), and it is homologous recombination between

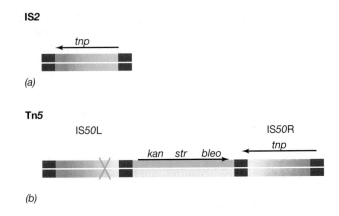

● **Figure 10.29 Maps of the transposable elements IS2 and Tn5.** Inverted repeats are shown in red. The arrows above the maps show the direction of transcription of any genes on the elements. The gene encoding the transposase is *tnp*. (a) IS2 is an insertion sequence of 1327 bp with inverted repeats of 41 bp at its ends. (b) Tn5 is a composite transposon of 5.7 kbp containing the insertion sequences IS50L and IS50R at its left and right ends, respectively. IS50L is not capable of independent transposition because there is a *nonsense mutation* (see Section 10.2) marked by a blue cross in its transposase gene. Otherwise, the two IS50 elements are very nearly identical. The genes *kan, str,* and *bleo* confer resistance to the antibiotics kanamycin (and neomycin), streptomycin, and bleomycin. Tn5 is commonly used to generate mutants in *Escherichia coli* and other gram-negative bacteria.

identical insertion sequences on the F plasmid and the chromosome and not transposition that allows the F plasmid to integrate into the bacterial chromosome and later mobilize it (see Section 10.12 and Figure 10.23). Some species of *Archaea* also have large numbers of IS elements in their chromosomes, suggesting that transposition is characteristic of all prokaryotes.

Transposons are larger than insertion sequences and carry other genes, some of them conferring important properties on the organism carrying them. These often include drug resistance and other easily selectable genes. In addition, there are *conjugative transposons*, transposons that can move between bacterial species by conjugation. These transposons have genes allowing them not only to move from one location on a genome to another, but also *tra* genes necessary to transfer themselves from one bacterium to another (see Section 10.11). Some transposons are actually composite genetic elements, containing a gene or group of genes lying between two identical insertion sequences. The existence of such *composite transposons* indicates that novel transposons likely arise periodically in cells that contain insertion sequences located close to one another.

Transposition and Microbial Evolution

As mentioned previously, the inverted repeats found at the ends of transposable elements and the transposase enzyme are essential for transposition. The transposase recognizes, cuts, and eventually ligates the DNA during transposition (Figure 10.30●). When a transposable ele-

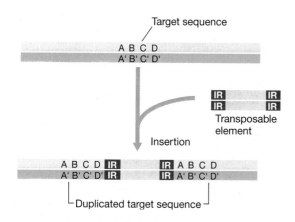

● **Figure 10.30 Transposition.** Insertion of a transposable element generates a duplication of the target sequence. Note the presence of inverted repeats (IRs) at the ends of the transposable element. Figure 10.31 shows more detailed models of the mechanism of transposition.

ment becomes inserted into a target DNA, a short sequence in the target at the site of integration is duplicated during the insertion process. The duplication arises because single-stranded DNA breaks are generated by the transposase. The transposon is then attached to the single-stranded ends that have been generated, and repair of the single-strand portions results in the duplication (see Figure 10.31).

If the duplicated DNA is sufficient to have duplicated an entire gene or group of genes, the organism will contain multiple copies of these particular genes. Such **gene-duplication** events are thought to fuel microbial evolution. This is because mutations occurring in one copy of the gene(s) do not affect the other copy; the function of the faulty protein is still "covered" by the product of the unmutated duplicate gene. However, beneficial mutations in one copy can lead to production of a protein that has superior properties to the normal protein and thereby increase fitness. In this way, evolution can "experiment" with one copy of the gene while the identical copy provides the necessary backup function. Genomic analyses have revealed numerous examples of protein-encoding genes that were clearly derived from gene duplication.

Mechanisms of Transposition

Two mechanisms of transposition are known, called *conservative* and *replicative* (Figure 10.31●). In **conservative transposition**, such as occurs in the transposon Tn5, the transposable element is excised from one location in the chromosome and becomes reinserted at a second location. The copy number of a conservative transposon therefore remains at one. By contrast, in **replicative transposition**, such as occurs in bacteriophage Mu (∞ Section 16.5), a new copy is produced during transposition and is inserted at another location on the chromosome. Thus, after a replicative transposition event, one copy of the transposing element remains at the original site and another copy is found at the new site.

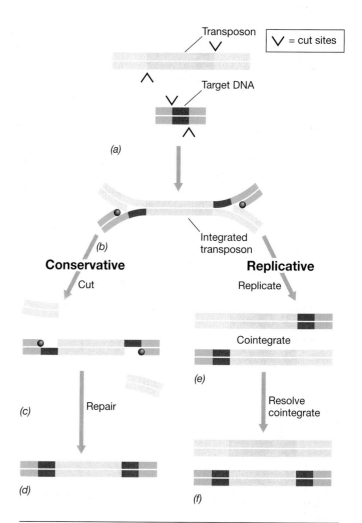

● **Figure 10.31 Mechanisms of transposition.** For simplicity, donor DNA (carrying the transposon) is shown in light green and DNA carrying the target DNA is shown in dark green. (a) In both conservative and replicative transposition the transposase makes cuts (marked with arrows) in the DNA strands at the end of the transposable element (orange) and at the target site (red). The number and location of the cuts may vary depending on the mechanism. (b) The target site becomes ligated to the transposable element. The black balls indicate free 3′ ends of DNA strands at which replication can occur (∞ Sections 7.5 and 7.6). (c) In conservative transposition, further cuts are made before DNA replication/repair occurs, and the transposable element is lost from the donor DNA. (d) Repair leads to duplication of the target site and completion of transposition to the new site. (e) In replicative transposition, replication occurs without the cutting of the transposable element from the donor site leading to two copies of the transposable element as part of a cointegrate. Note, however, this has led to the joining of the donor and target DNA molecules together. (f) These molecules are separated (resolved) in a further reaction. Resolution of cointegrates is shown in more detail in Figure 10.32.

Models for both conservative and replicative transposition are illustrated in Figure 10.31. As seen, single-strand cuts are made at the ends of the transposon (at the sites of the inverted repeats), and staggered single-strand cuts are made at the target site. The transposon is now joined to the target site via the single-stranded ends. In conservative transposition, the donor site is now cut, and replication repair then fills in the single-strand gaps in the target site. This process results in the formation of *direct repeats* in the target site at the ends of the transposon. In replicative transposition, the replication repair takes place while the transposable element is still attached to both the original and the target sites. This leads to the formation of a composite structure called a *cointegrate*. The final event in this pathway is *resolution* of the cointegrate structure, leading to release of the original transposon and the presence of a new copy of the transposon at the target site (Figure 10.32●).

Transposition is essentially a *recombination event*, but one that does not occur between homologous sequences

or use the general recombination system of the cell. It involves *transposase* rather than the RecA protein that is involved in general recombination (Figure 10.9). Because this recombination involves a *specific* base sequence, it is called *site-specific recombination* (in contrast to *homologous recombination* discussed in Section 10.6).

Mutagenesis with Transposable Elements

If the insertion site for a transposable element is *within* a gene, insertion of the transposon will result in mutation (Figure 10.33●). Transposons thus provide a facile means of creating mutants throughout the chromosome. The most convenient element for **transposon mutagenesis** is one containing an antibiotic resistance gene. Clones containing the transposon can then be selected by the isolation of antibiotic-resistant colonies. If the antibiotic-resistant clones are selected on rich medium on which all auxotrophs can grow, they can subsequently be screened on minimal medium supplemented with various growth factors to determine if a growth factor is required.

Transposons are also useful for incorporating an auxotrophic gene marker into a wild-type organism. Normally, auxotrophic recombinants cannot be isolated by positive selection (see Figure 10.2), but if the auxotrophic marker to be introduced contains a transposable element with an antibiotic resistance marker, then one can select for antibiotic-resistant clones, a positive selection procedure, and automatically obtain clones that have incorporated the auxotrophic marker. Two transposons widely used in microbiology for mutagenesis are Tn5 (see Figure 10.30), which confers neomycin and kanamycin resistance, and Tn10, which contains a marker for tetracycline resistance (see Figure 10.20).

Integrons

Integrons are transposons that can capture and express genes from other sources. However, unlike other transposons, integrons are not inserted at random but are highly selective in their insertion site, frequently becoming inserted into plasmids.

Integrons contain a gene that encodes a protein called *integrase*, needed for *site-specific recombination*. Recall that the lambda genome becomes integrated into the

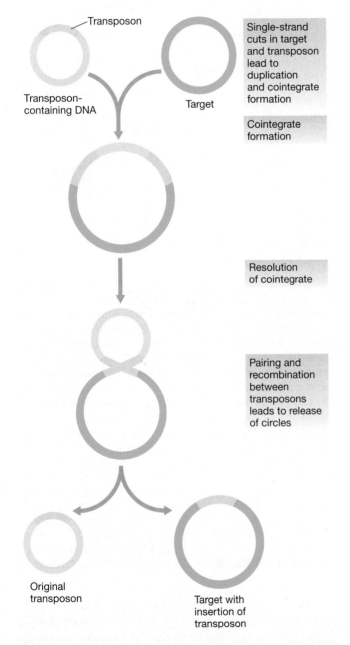

● **Figure 10.32 Replicative transposition.** After the formation of single-strand cuts, a cointegrate structure arises by association of the two molecules (see Figure 10.31). After recombination, resolution of the cointegrate structure leads to the release of the original transposon and duplication of the transposon in the target molecule.

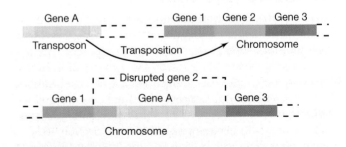

● **Figure 10.33 Transposon mutagenesis.** The transposon moves into the middle of gene 2. Gene 2 is now disrupted by the transposon and is inactivated. Gene A in the transposon will be expressed in both locations.

Escherichia coli genome at a specific site by the activity of the lambda integrase (⌁ Section 9.11). Integrons also contain a specific DNA sequence that allows the integrase to insert groups of genes called *cassettes*, along with a promoter that allows for expression of the newly integrated gene cassette.

Integrons can become part of transposons, plasmids, or even the bacterial chromosome. Some integrons contain as many as five different gene cassettes. Over 40 different antibiotic resistance genes have been identified on such cassettes, as have some genes associated with virulence in certain pathogenic bacteria. Figure 10.34● shows the structure of two integrons from *Pseudomonas aeruginosa*, a potentially serious pathogen. Integrons have been found in various species of *Bacteria*, often in clinical isolates, and thus their selection by horizontal gene transfer in such antibiotic-rich environments as hospitals and clinics is obvious. What is less obvious is the origin of the gene cassettes themselves. These are not simply random genes that can be captured, but genes that are either (1) bounded by specific DNA sequences that are recognized by the integrase, or (2) apparently not expressed until they become part of an integron and can be transcribed from the promoter on the integron.

10.14 Concept Check

Transposons and insertion sequences are genetic elements that can move from one location on a chromosome to another by a process called transposition, a type of site-specific recombination. Transposition can be either replicative or conservative. Transposons often carry genes encoding antibiotic resistance. Transposons can be used as biological mutagens.

◆ What features do *insertion sequences* and *transposons* have in common?

◆ What are integrons and how do they differ from transposons?

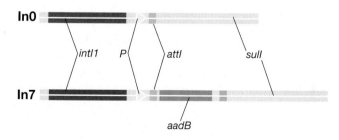

● **Figure 10.34 Structure of two naturally occurring integrons from *Pseudomonas*.** The integron In0 has the basic set of genes: *intI1*, encodes integrase; *attI*, the site where site-specific integration can occur; P, a promoter; and *sulI*, a gene conferring sulfonamide resistance that contains its own promoter. The integron In7 contains all of these genes, but in addition, a gene cassette has been integrated. All cassettes contain a site (blue square) for site-specific recombination. This cassette contains *aadB*, which confers resistance to certain aminoglycoside antibiotics.

III BACTERIAL GENETICS AND GENE CLONING

The three mechanisms of genetic exchange we have discussed, transformation (see Section 10.7), transduction (see Section 10.8), and conjugation (see Sections 10.9–10.11), are used to do bacterial crosses and select recombinants for many reasons. These include making a genetic map of an organism, creating new strains of an organism, or specifically studying the genetics and biochemistry of an organism's metabolism. However, some severe limitations are associated with doing all the crosses *in vivo* using these techniques. Many of these limitations can be overcome by manipulating DNA *in vitro* (in a test tube). The next several sections of this chapter discuss some of these *in vitro* techniques.

The techniques to be described here rely heavily on the molecular biologist's toolkit: (1) restriction enzymes; (2) gel electrophoresis and nucleic acid hybridization; and (3) DNA sequencing, the synthesis of DNA probes, and PCR. These topics were covered in Sections 7.7–7.9, respectively. It might be useful to review these sections before beginning here. We begin with molecular cloning, the capture of specific genes on small molecules of DNA that can be easily manipulated by the experimenter.

10.15 Essentials of Molecular Cloning

Molecular (gene) cloning provides the foundation for genetic-engineering and most molecular genetics procedures and has greatly facilitated the detailed analysis of genomes. The purpose of gene cloning is to isolate multiple copies of specific genes in pure form. Consider the nature of this problem. For a genetically "simple" organism like *Escherichia coli*, an average gene represents 1–2 kbp out of a genome of over 4600 kbp. An average *E. coli* gene is thus less than 0.05% of the total DNA in the cell. In human DNA the problem is even more complicated because the coding regions of average genes are not much larger than in *E. coli*, genes are typically split into pieces, and the genome is almost 1000 times larger! How then can a specific gene be obtained in multiple copies?

If a gene of interest can be isolated onto a fragment of DNA whose replication is under the experimenter's control, making copies of the entire fragment (and thus the gene of interest) is possible. The basic strategy of molecular cloning is thus to move the desired gene(s) from a large, complex genome to a small, simple one (Figure 10.35●). Fortunately, our knowledge of DNA chemistry and enzymology allows us to break and join DNA molecules *in vitro*. This process is known as *in vitro recombination*. Restriction enzymes, DNA ligase, PCR, and synthetic DNA are important tools used for *in vitro* recombination.

Steps in Gene Cloning

Gene cloning can be divided into several steps as summarized here:

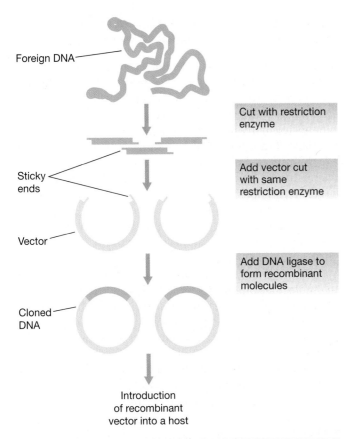

Foreign DNA

Cut with restriction enzyme

Sticky ends

Add vector cut with same restriction enzyme

Vector

Add DNA ligase to form recombinant molecules

Cloned DNA

Introduction of recombinant vector into a host

● **Figure 10.35** **Major steps in gene cloning.** The vector can be a plasmid or a viral genome. By cutting the foreign DNA and the vector DNA with the same restriction enzyme, complementary sticky ends are generated that allow foreign DNA to become part of the vector.

1. **Isolation and fragmentation of the source DNA.** This can be total genomic DNA from an organism of interest, DNA synthesized from an RNA template by reverse transcriptase (⚬ Section 9.13), a gene or genes amplified by the polymerase chain reaction (⚬ Section 7.9), or even totally synthetic DNA made *in vitro* (⚬ Section 7.8). If genomic DNA is the source, it is cut with restriction enzymes first to give a mixture of manageable-sized fragments (Figure 10.35).

2. **Joining the DNA fragments to a cloning vector with DNA ligase.** Cloning vectors are small, independently replicating genetic elements used to replicate genes, and most are derived from plasmids or viruses. Cloning vectors are typically designed to allow *in vitro* insertion of foreign DNA at a restriction site that cuts the vector in a way that does not affect its replication (Figure 10.35). If the source DNA and the vector are cut with the same restriction enzyme, joining can be mediated by annealing of the single-stranded regions called "sticky ends" (Figure 10.35 and see Figure 7.21). "Blunt ends" generated by some restriction enzymes can also be joined using synthetic DNA **linkers** or **adapters**. DNA ligase is required to seal the final phosphodiester bond. The properties of cloning vectors are discussed in Sections 10.15 and 10.16, as well as in Chapter 31.

3. **Introduction and maintenance of the cloned DNA in a host organism.** The recombinant DNA molecule made in a test tube is introduced into a host organism, for example, by the process of transformation (see Section 10.7), where it can then replicate (Figure 10.35). Transfer of the DNA into the host usually yields a mixture of clones. Some cells contain the desired cloned gene, whereas other cells contain other clones generated by joining the source DNA to the vector. Such a mixture is known as a **DNA library** or a **gene library** because many different clones can be purified from the mixture, each containing different cloned DNA segments from the source organism. Making a gene library by cloning *random fragments* of a genome is called **shotgun cloning**, and is a widely practiced technique in gene cloning and genomic analyses (⚬ Sections 15.1 and 15.2).

We now consider in more detail the properties of plasmids and viruses, respectively, used as cloning vectors.

 10.15 *Concept Check*

The isolation of a specific gene or region of a chromosome by molecular cloning is done using a plasmid or virus as the cloning vector. Restriction enzymes and DNA ligase are used in an *in vitro* recombination procedure to produce the hybrid DNA molecule. Once introduced into a suitable host, the target DNA can be produced in large amounts under the control of the cloning vector.

◆ What is the purpose of molecular cloning?

◆ What are the roles of a cloning vector, restriction enzymes, and DNA ligase in molecular cloning?

10.16 **Plasmids as Cloning Vectors**

Plasmids replicate independently of the host chromosome (see Section 10.9). In addition to carrying genes required for their own replication, most plasmids are natural **vectors** because they often carry other genes that confer important properties on their hosts. Plasmids have very useful properties as **cloning vectors**. These include (1) *small size*, which makes the DNA easy to isolate and manipulate; (2) *independent origin of replication*, so plasmid replication in the cell proceeds independently from direct chromosomal control; (3) *multiple copy number*, so they can be present in the cell in several copies, making *amplification* of the DNA possible; and (4) presence of *selectable markers* such as antibiotic resistance genes, making detection and selection of plasmid-containing clones easier.

Although in the natural environment conjugative plasmids are transferred by cell-to-cell contact, plasmid cloning vectors have been genetically modified to prevent their transfer conjugatively in order to achieve biological containment. However, vector transfer in the laboratory can be brought about by transformation or an artificial type of transformation called *electroporation* (see

Sections 10.7 and 31.2). Depending on the host-plasmid system, replication of the plasmid may be under tight cellular control, in which case only a few copies are made, or under relaxed cellular control, in which case a large number of copies is made. Achievement of high copy number is often important in gene cloning, and by proper selection of the host-plasmid system and manipulation of cellular macromolecule synthesis, plasmid copy numbers of several thousand per cell can be obtained.

The Plasmid pBR322

In Section 10.9, we discussed how the F plasmid could obtain genes from the *Escherichia coli* chromosome *in vivo*, becoming an F′ plasmid. Therefore, one might imagine using F as an *in vitro* cloning vector. However, the F plasmid is far too large (almost 100 kb, see Figure 10.18) to be a useful cloning vector, and it has no easily selectable markers. Even so, the very first plasmid cloning vectors used were natural isolates. These were soon replaced, however, by plasmids that were themselves the result of *in vitro* manipulations. An example of one of these modified plasmid cloning vectors is pBR322, which replicates in *Escherichia coli* (Figure 10.36●). Plasmid pBR322 has a number of characteristics that make it suitable as a cloning vehicle:

1. It is relatively small, only 4361 bp.

2. It is stably maintained in its host (*Escherichia coli*) in relatively high copy number, 20–30 copies per cell.

3. It can be amplified to a very high number (1000–3000 copies per cell, about 40% of the genome!) by inhibiting protein synthesis with the antibiotic chloramphenicol.

4. It is easy to isolate in the supercoiled form using a variety of routine techniques.

5. A reasonable amount of foreign DNA can be inserted into it, although inserts of more than 10 kbp lead to plasmid instability.

6. The complete base sequence of this plasmid is known, making it possible to identify all restriction enzyme cut sites.

7. There are *single* cleavage sites for various restriction enzymes such as *Pst*I, *Sal*I, *Eco*RI, *Hind*III, and *Bam*HI. It is crucial that only a single recognition site for a given restriction enzyme exist on the cloning vector so that treatment with that enzyme linearizes the vector but does not cut it into pieces. Individual sites for each of several restriction enzymes increase the versatility and usefulness of the vector.

8. It has genes conferring ampicillin resistance and tetracycline resistance on its host. These permit ready selection of hosts containing the plasmid because such hosts are resistant to both antibiotics. The sites recognized by some of the restriction enzymes are within one or the other of these resistance genes, facilitating the identification of plasmids carrying cloned DNA (see later).

9. It can be inserted into cells easily by transformation or artificial transformation (see Section 10.7).

Cloning Genes into pBR322

The use of plasmid pBR322 in gene cloning is shown in Figure 10.37●. As seen, the *Bam*HI restriction site is within the gene for tetracycline resistance, and the *Pst*I site is within the gene for ampicillin resistance. If foreign DNA is inserted into one of these sites, the antibiotic resistance conferred by the gene containing this site is lost, a phenomenon called **insertional inactivation**. Insertional inactivation is used to detect the presence of foreign DNA within the plasmid. Thus, when pBR322 is digested with *Bam*HI, linked with foreign DNA, and transformed bacterial clones isolated, those clones that are *both* ampicillin resistant and tetracycline resistant *lack* the foreign DNA (the plasmid incorporated into these cells represents vector DNA that had recyclized without picking up foreign DNA). On the other hand, those cells *resistant* to ampicillin but *sensitive* to tetracycline *contain* the plasmid with inserted foreign DNA. Since ampicillin resistance and tetracycline resistance can be determined independently on agar plates using replica plating (Figure 10.2), isolation of bacteria containing the desired clones and elimination of cells not containing the plasmid can readily be accomplished.

Second Generation Plasmid Vectors

Plasmid pBR322 represents an early generation of cloning vectors which were themselves partially constructed by genetic engineering. There are now newer generations of plasmid vectors that have been engineered to have even more useful cloning features and are even simpler to

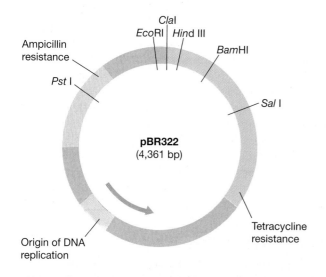

● **Figure 10.36 The structure of plasmid pBR322, a widely used cloning vector.** Shown are the essential features including antibiotic resistance markers and restriction enzyme cut sites. The arrow indicates the direction of DNA replication from the origin.

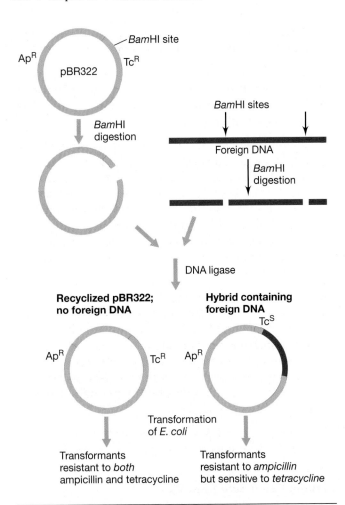

● **Figure 10.37 Cloning with plasmid pBR322.** The use of plasmid pBR322 as a cloning vector, showing how insertion of foreign DNA causes inactivation of the tetracycline resistance gene, permitting easy identification of transformants containing the cloned DNA fragment.

use. These new features, as exemplified by the plasmid pUC19, typically include a **multiple cloning site** (also called a **polylinker**), which is a short segment of DNA containing many different restriction enzyme cut sites but ones that are present only in the polylinker portion of the vector. This polylinker is contained within the coding region of a gene where insertional inactivation is very easy to monitor, such as the gene encoding β-galactosidase, an enzyme involved in the catabolism of lactose (∞ Sections 8.5 and 8.7). When a special reagent is added to the medium, activity of this enzyme leads to a blue-colored product. Thus, cells containing pUC19, which *does not* contain cloned DNA (β-galactosidase is therefore active), form blue colonies, while cells containing the plasmid *with* cloned DNA (β-galactosidase is therefore inactive) are colorless. This allows for screening colonies for cloned DNA by inspection. Such color screening features are also found in bacteriophage cloning vectors, and a specific example is discussed in Section 15.1 and shown in Figure 15.1.

Sometimes insertional inactivation can be detected by *selection*, rather than by screening. For example, in some

vectors, the gene carrying the polylinker normally produces a protein that is lethal to the host cell. Therefore, only cells containing a plasmid in which this gene has been *inactivated* can grow, a very rigorous selection tool indeed!

Cloning using plasmid vectors is a versatile and fairly general procedure widely used in genetic engineering, particularly when the fragment to be cloned is fairly small. Also, plasmids are often used as cloning vectors if *expression* of the cloned gene is desired, since regulatory genes can be engineered into the plasmid to obtain expression of the cloned genes under specific conditions (∞ Section 31.4).

10.16 Concept Check

Plasmids are useful cloning vectors because they are easy to isolate and purify and are able to multiply to high copy numbers in bacterial cells. Antibiotic resistance genes of the plasmid are used to identify bacterial cells containing the plasmid.

◆ Explain why in cloning it is necessary to use a restriction enzyme that cuts the plasmid vector in only one location.

◆ What is *insertional inactivation?*

10.17 Bacteriophage Lambda as a Cloning Vector

Recall that during specialized transduction (see Section 10.8) some host genes become incorporated into a bacteriophage genome. One phage that is used as a specialized transducing phage is bacteriophage lambda (∞ Section 9.11). During specialized transduction lambda acts as a vector, but the recombination occurs in the cell, not in a test tube. Lambda can also be used as a cloning vector for *in vitro* recombination.

Lambda is a particularly useful cloning vector because its biology is well understood, it can hold larger amounts of DNA than most plasmids, and DNA can be efficiently packaged into phage particles *in vitro*. These can be used to infect suitable host cells, and infection is much more efficient than transformation (transfection). Phage lambda has a complex genetic map (∞ Figure 9.18) and a large number of genes. Of particular significance for its use as a cloning vector, however, is the fact that the central third of the lambda genome, the region between genes *J* and *N*, is *unessential* for infectivity and can be replaced with foreign DNA (Figure 10.38●). This allows relatively large DNA fragments, up to about 20 kb, to be cloned into lambda. This is twice the cloning capacity of a small plasmid vector such as pBR322.

Modified Lambda Phages

Wild-type lambda is not suitable as a cloning vector because its genome has too many restriction enzyme sites. To avoid this difficulty, modified lambda phages have been constructed especially for cloning. In one set of modified lambda phages, called *Charon phages*, unwanted restriction

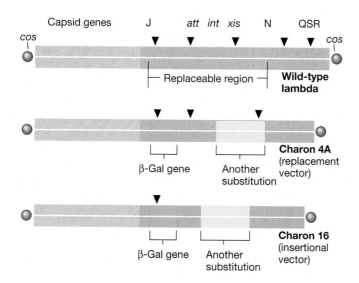

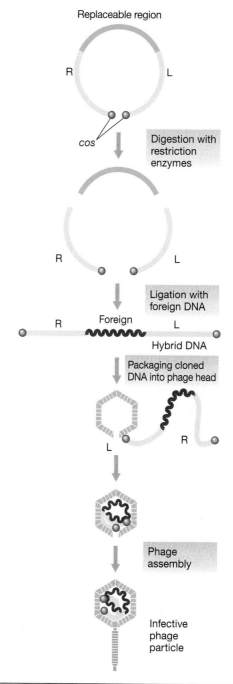

● **Figure 10.38** **Molecular cloning with lambda.** Abbreviated genetic map of bacteriophage lambda showing the cohesive ends as circles (◯◯ Figure 9.18*a*). Charon 4A and 16 are both derivatives of lambda, which have various substitutions and deletions in the nonessential region. One of the substitutions in each case is a gene (*lacZ*) that encodes the enzyme β-galactosidase, which permits detection of clones containing this phage. Whereas the wild-type lambda genome is 48.5 kb pairs, that for Charon 4A is 45.4 and that for Charon 16 is 41.7 kb pairs. The arrowheads shown above the maps of each phage indicate the sites recognized by the restriction enzyme *Eco*RI.

enzyme sites have been removed by mutation. In variants that have only a single restriction site, such as Charon 16, a foreign piece of DNA can be *inserted*. By contrast, in variants with two sites, such as Charon 4A, foreign DNA can *replace* a specific segment of the lambda DNA (Figure 10.38). The latter variants, called **replacement vectors**, are especially useful in cloning large DNA fragments. Charon 4A is used as a replacement vector; the two small interior fragments are cut out and discarded during cloning.

Both vectors are also engineered to contain *reporter genes*, such as the gene for β-galactosidase previously discussed. When the vectors replicate in a lactose-negative (Lac$^-$) strain of *Escherichia coli*, β-galactosidase is synthesized from the phage gene and the presence of lactose-positive (Lac$^+$) plaques can be detected by using a color indicator agar (◯◯ Section 15.1). However, if a foreign gene is inserted *into* the β-galactosidase gene, the Lac$^+$ character is lost. Such Lac$^-$ plaques can be readily detected as colorless plaques among a background of colored plaques (◯◯ Figure 15.1).

Steps in Cloning with Lambda

Cloning with lambda replacement vectors involves the following steps (Figure 10.39●):

1. Isolating the vector DNA from phage particles and digestion with the appropriate restriction enzyme.

2. Connecting the two lambda fragments to fragments of foreign DNA using DNA ligase. Conditions are

● **Figure 10.39** **The use of bacteriophage lambda as a cloning vector.** The maximum size of inserted DNA is about 20 kb pairs.

chosen so molecules are formed of a length suitable for packaging into phage particles.

3. Packaging of the DNA by adding cell extracts containing the head and tail proteins and allowing the formation of viable phage particles to occur spontaneously.

4. Infecting *E. coli* and isolating phage clones by picking plaques on a host strain.

5. Checking recombinant phage for the presence of the desired foreign DNA sequence using nucleic acid hybridization procedures, DNA sequencing, or observation of genetic properties.

Selection of recombinants is less of a problem with lambda replacement vectors (such as Charon 4A) than with plasmids because (1) the efficiency of transfer of recombinant DNA into the cell by lambda is very high, and (2) lambda fragments that have not received new DNA are too small to be incorporated into phage particles; thus, every viable phage virion will contain cloned DNA.

Cosmids

Like replacement vectors, a **cosmid** employs specific lambda genes. Cosmids are plasmid vectors containing foreign DNA plus the *cos* (cohesive end) site from the lambda genome (Section 9.11). These *cos* sites are required for packaging DNA into lambda virions. Cosmids are constructed from plasmids containing cloned DNA by ligating the lambda *cos* region to the plasmid DNA. The modified plasmid can then be packaged into lambda virions *in vitro* as described previously and the phage particles used to transduce *Escherichia coli*. Cosmid construction avoids the necessity of having to transform *E. coli*, which at best is an inefficient process (see Section 10.7).

One additional advantage of cosmids is that they can be used to clone large fragments of DNA, with inserts as large as 50 kbp accepted by the system. Therefore, with bigger inserts, fewer clones are needed to obtain representation of the whole genetic element. Cosmids also permit storage of the DNA in phage particles instead of in plasmids. Phage particles are much more stable than plasmids, so the recombinant DNA can be kept for long periods of time.

 10.17 Concept Check

Bacteriophages such as lambda have been modified to make useful cloning vectors. Larger amounts of foreign DNA can be cloned with lambda than with many plasmids. In addition, the recombinant DNA can be packaged *in vitro* for efficient transfer to a host cell. Plasmid vectors containing the lambda *cos* sites are called cosmids, and they can carry a large fragment of foreign DNA.

◆ Why is the ability to package recombinant DNA in a test tube useful?

◆ What is a *replacement vector*?

10.18 *In Vitro* and Site-Directed Mutagenesis

Methods for *in vitro* DNA manipulation opened up a whole new field of mutagenesis, *in vitro* mutagenesis, better known as **site-directed mutagenesis**. Whereas conventional mutagens (see Section 10.3) introduce mutations *at random* in the intact organism, site-directed mutagenesis makes use of synthetic DNA and DNA cloning techniques to introduce mutations at *precisely determined sites* in genes in a test tube.

Site-Directed Mutagenesis

Site-directed mutagenesis is a powerful tool, as it allows the experimenter to change any base pair in a specific gene. The basic procedure is to synthesize a short DNA oligonucleotide primer (Section 7.8) containing the desired base change and to allow this to base pair with a single-stranded DNA containing the gene of interest. Pairing is complete except for the short region of mismatch. Then the short single-stranded fragment of the synthetic oligonucleotide is extended using DNA polymerase, thus copying the rest of the gene. The double-stranded molecule obtained is then inserted into a host cell by transformation. Mutants are selected by using some sort of positive selection, such as antibiotic resistance (in this case, the modified DNA would also contain an antibiotic resistance marker). The mutant obtained is then used in production of the modified (mutant) protein.

One procedure for site-directed mutagenesis is illustrated in Figure 10.40●. It is convenient to begin the process with the gene of interest cloned into a *single-stranded* DNA vector. The mutagenized DNA can then bind by complementary pairing with the gene of interest (Figure 10.40). A widely used vector for this purpose is *bacteriophage M13*, a single-stranded DNA phage whose genome is easy to manipulate and that replicates in *Escherichia coli* (Section 16.3). Several basic cloning methods are available based on M13, and we discuss it as a vector for cloning DNA for genomics analysis in Chapter 15.

Site-directed mutagenesis can be used to test the efficacy of proteins containing known amino acid substitutions. For example, if one was studying the active site of an enzyme to assess the importance of particular amino acids in this region, site-directed mutagenesis could be used to change a specific amino acid and the modified enzyme assayed and compared to the wild-type enzyme. In such an experiment, the vector containing the mutated DNA would be inserted into a mutant strain unable to make the enzyme in question. In this way, the enzyme activity measured would only be that of the one encoded by the site-directed mutation.

Using site-directed mutagenesis, enzymologists can link virtually any aspect of an enzyme's activity, such as catalysis, resistance or susceptibility to chemical or physical agents, or interactions with other proteins, to *specific amino acids* in the protein. In particular, site-directed mutagenesis has given enzymologists detailed information on which amino acids are critical for an enzyme's function and significant insight into how enzymes catalyze chemical reactions.

Cassette Mutagenesis and Gene Disruption

Because of the large number of restriction enzymes commercially available and therefore the large number of different DNA sequences that can be cut, it is usually possible to find several different restriction sites within a gene of interest. If sites for the appropri-

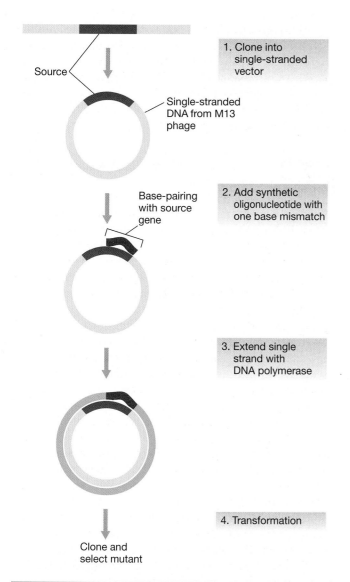

● **Figure 10.40 Site-directed mutagenesis using short synthetic oligodeoxyribonucleotide fragments.** Note that cloning into bacteriophage M13 yields the single-stranded DNA needed for site-directed mutagenesis to work.

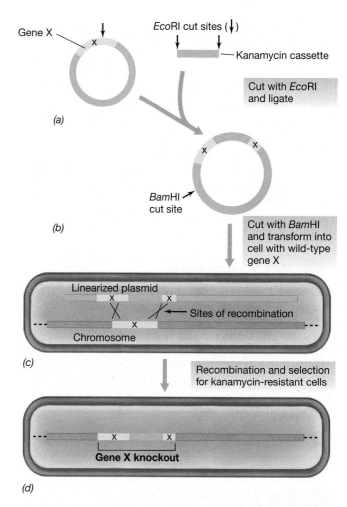

● **Figure 10.41 Gene disruption using cassette mutagenesis.** (a) A plasmid containing a cloned wild-type copy of gene X is cut with the restriction enzyme *Eco*RI and mixed with a DNA fragment (the kanamycin cassette) that contains a gene capable of conferring kanamycin resistance on a cell and that has been obtained using the same restriction enzyme. The cut plasmid and the cassette are ligated. (b) The product of the ligation is a plasmid that now contains the kanamycin cassette as an insertion mutation within gene X. This new plasmid is now cut with a further restriction enzyme, *Bam*HI, and transformed into a cell containing a wild-type gene X. (c) The transformed cell contains the linearized plasmid with a disrupted gene X and its own chromosome with a wild-type copy of the gene. In some cells, homologous recombination (see Figure 10.9) occurs between the wild-type and mutant forms of gene X. (d) Cells that can grow in the presence of kanamycin must have the kanamycin cassette recombined into their chromosome because the linearized plasmid cannot replicate. These cells now have only a single disrupted copy of gene X. This disruption typically abolishes all gene X function.

ate enzyme are not found in the gene, or at the precise location required, they can also be inserted by site-directed mutagenesis (see Figure 10.40). If restriction sites are close together, the intervening DNA fragment can be excised and replaced by a synthetic DNA fragment in which one or more of the nucleotides have been changed. These synthetic fragments are called *cassettes* (or cartridges), and the process is known as **cassette mutagenesis**.

Insertion mutations can also be generated by simply inserting a cassette at a single site. When using cassettes to replace sections of genes, the cassettes are typically the same size as wild-type DNA fragments. However, the cassette used for making insertion mutations can be almost any size and can even be an entire gene. In fact, to ease the selection process, cassettes that encode proteins that confer a particular antibiotic resistance on the host are commonly used. This type of cassette mutagenesis is used in a process called **gene disruption**.

The process of gene disruption is illustrated in Figure 10.41●. In this case, a fragment carrying a gene conferring kanamycin resistance, the *kan* cassette, is inserted at a restriction site in a cloned gene. The vector carrying this mutant gene is then linearized by being cut with a different restriction enzyme, and the linear DNA is transformed into the host with kanamycin resistance selected. The linearized plasmid cannot replicate, and so resistant cells likely arise by homologous recombination (see Section 10.6) between the mutated gene on the plasmid and the wild-type gene on the chromosome.

Note that when a cassette is inserted, the cells have not only gained kanamycin resistance but have also *lost the function* of the gene in which the *kan* cassette was inserted. These mutations are called **knockout mutations**. This process is similar to searching for insertion mutations made by transposons (see Section 10.14), but in this case the experimenter chooses exactly which gene will receive the mutation. Knockout mutations in haploid organisms (such as prokaryotes), yield viable cells only if the disrupted gene is not essential. In fact, gene knockouts are a convenient means of determining whether a given gene is an essential one.

10.18 Concept Check

Synthetic DNA molecules of desired sequence can be made *in vitro* and used to construct a mutated gene directly or to change specific base pairs within a gene via site-directed mutagenesis. Genes can also be disrupted by inserting DNA fragments, called cassettes, into them. The inserted cassette eliminates the function of the wild-type gene while conferring a new, and usually selectable, phenotype on the cell.

◆ How can site-directed mutagenesis be useful to enzymologists?

◆ What are *knockout mutations*?

IV THE BACTERIAL CHROMOSOME

The three mechanisms of genetic exchange described in this chapter—transformation, transduction, and conjugation—together with the powerful tools of gene cloning, can be used to map the locations of genes on a bacterial chromosome. Genes in *Escherichia coli* were initially mapped using conjugation. By using Hfr strains with origins at different sites (Figure 10.25), it was possible to map the entire *E. coli* genome. However, conjugation does not permit ordering genes that are closely linked. Therefore, generalized transduction was used for fine-structure mapping of the *E. coli* chromosome. Bacteriophage P1 can carry small fragments of DNA and proved very useful for detailed mapping of genes in *E. coli*.

The genetic techniques used to map the chromosomes of other prokaryotes were dictated by the efficiency with which genetic transfer occurs in the particular organism. For example, transformation is a very inefficient process in *E. coli* but is a very efficient process in gram-positive bacteria (see Section 10.7). Thus, transformation proved an effective tool for mapping genes in *Bacillus*. In addition to conjugation, transduction was also used to map genes in the *E. coli* chromosome. However, because transduction only involves the transfer of a few genes—many fewer than is possible using conjugation—transduction was only useful for fine structure mapping purposes.

Although all of these "classical" techniques are still useful for the characterization and construction of strains, they have largely been supplanted in constructing genetic maps by molecular cloning and DNA sequencing. Cloning and sequencing has revolutionized the study of genome organization in prokaryotes and eukaryotes, and we consider the field of genomics in Chapter 15.

10.19 Genetic Map of the *Escherichia coli* Chromosome

The genetic map of *Escherichia coli* strain K-12 is shown in Figure 10.42●. The map distances are given in "minutes" of transfer, with the entire chromosome containing 100 minutes and with "zero time" arbitrarily set as that at which the first genetic transfer (the *threonine* operon) can be detected using the first Hfr strain of *E. coli* that was discovered.

The map shown in Figure 10.42 contains only a few of the many thousands of genes present in the *E. coli* chromosome and shows the location of a few restriction enzyme recognition sites. The size of the chromosome is given in both minutes (the original units determined by conjugation studies) and in kilobase pairs. The *E. coli* chromosome, like that of many other prokaryotes, has been completely sequenced. Sequencing of relatively small bacterial chromosomes of 3 or 4 megabases is now done routinely by automated sequencing of random, or *shotgun*, clones (∞ Section 15.2). Because of the enormous amount of genetic information genomic analyses makes available, this information can only be managed through computer databases. This reality has spawned a whole new field of molecular biology called *bioinformatics* (∞ Chapter 15). Database analyses using powerful gene search software has placed the field of bioinformatics in a central position in genomic studies.

Escherichia coli as a Model Prokaryote

Many factors have favored the use of *Escherichia coli* as the workhorse for studies of biochemistry, genetics, and bacterial physiology. As emphasized in Chapters 9 and 16, even *E. coli* viruses have served as model systems of study. Therefore, although the chromosome of *E. coli* was not among the first prokaryotic chromosomes sequenced, this organism remains the best characterized microorganism. In addition to its important role as a model organism, *E. coli* continues to be the organism of choice for both research and applications of genetic engineering (∞ Chapter 31).

The strain of *E. coli* whose chromosome was originally sequenced, *strain MG1655*, is a derivative of *E. coli* K-12, the traditional strain used for genetic studies. The wild-type *E. coli* K-12 is a lysogen of bacteriophage lambda (∞ Section 9.11) and also contains the F plasmid (see Section 10.9). However, strain MG1655 had been "cured" of lambda (by radiation) and of the F plasmid (by acridine treatment, see Section 10.9) before sequencing began. The chromosome of strain MG1655

Deeper /

Analysis
an incred
Since the
ly explore
been sequ
be done r
and bioch
not! Altho
much info
genes, and
necessary
use bioch
their effec
20–40% (c
up in all g
of unknov

This h
certainly h
will expan
yotes alrea
tion, becau

REVIE

1. Write a
   the sam
   organis
   type oc
   without
   example

2. Explain
   otroph a
   E. coli do

3. What ar
   genetic
   affect th

4. Microin
   framesh
   frameshij
   plain ho

5. Explain
   early in
   mutation

6. What is
   rate chan

7. Give an
   one phy
   which ea

8. How car
   any relev

9. How do
   specific r

10. Why is it
    mation o
    genes to

11. Explain
    refers to

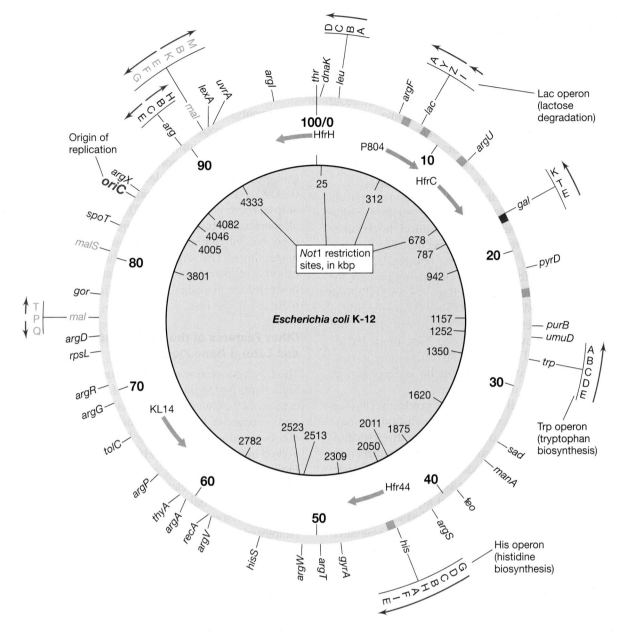

● **Figure 10.42**   **Circular linkage map of the chromosome of *Escherichia coli* strain K-12.** On the outer edge of the map, the locations of a few of the mapped genes are indicated. The actual number of ORFs in the *E. coli* chromosome is 4288 and the chromosome contains 4,639,221 bp. A few operons are also shown, along with the direction in which they are transcribed. Along the inner edge of the map, the numbers from 0 to 100 refer to map position in minutes. The origin of DNA replication is marked *oriC* (84.3 min), and replication proceeds bidirectionally from this point (👁 Section 7.6 and Figures 7.16–7.20). The inner circle shows the locations, in kilobase pairs, of the sites recognized by the restriction enzyme *Not*I. Note that 0 minutes and 0 kbp are both, by convention, at the *thr* locus. The origins and directions of transfer of a few Hfr strains are also shown (arrows). The positions where five copies of the transposable element IS*3* have been located in a particular strain are shown in blue. This element is also found in two copies on the F plasmid and is involved in Hfr formation (see Section 10.12). The position of the site where the bacteriophage lambda prophage integrates is shown in red (👁 Section 9.11). If the prophage was present, it would add an extra 48.5 kbp (slightly over 1 minute) to the map. The genes of the maltose regulon (👁 Section 8.6), which includes several operons, are shown in green. Although most genes in this regulon have an abbreviation beginning with *mal*, note that one of the genes is *lamB*. This gene encodes a membrane protein involved in maltose uptake by the cell, but the protein is also the receptor for bacteriophage lambda. The gene *rpsL* (73 minutes) encodes a ribosomal protein. The gene was once called *str* because mutations in this gene lead to streptomycin resistance.

contains 4,639,221 bp. Analysis of the genomic sequence showed there to be 4288 possibly functional *open reading frames* (ORFs, 👁 Sections 7.13 and 15.2), which account for about 88% of the genome. Approximately 1% of the genome is composed of genes encoding tRNAs and rRNAs, and about 0.5% is noncoding, repetitive sequences. The remaining 10% of the genome contains all of the regulatory sequences: promoters, operators, origin and terminus of DNA replication, and so forth.

Arran
Esche

Early
tion of
chemi
were c
a few
gal ger
min, a
cluster
polycis

Ge
not clu
biosyn
some,
of the
usefulr
examp
often g
prokar
chrom
known
gene. (
mRNA

Wi
the sar
operon
some,
The di
ons is s
ses hav
operon
pressed
that th
DNA r
replicat
on the
ally (
cated a
sequen
operon
the dir
ment fc
lows F
replicat
rection

Alr
teins, h
been co
the E. d
4288 di
this org
of unkr
protein
E. coli p
residues
much la
protein

APPLICATION QUESTIONS

1. A constitutive mutant is a strain that continuously makes a protein that in the wild type is inducible. Describe two ways in which a change in a DNA molecule could lead to the development of a constitutive mutant. How could these two types of constitutive mutants be distinguished genetically?

2. In Chapter 7, we saw that it was critical for the ribosome to translocate with great accuracy in order to maintain the proper reading frame. However, sometimes ribosomes make frameshift errors. Compare the impact on the cell of a ribosome periodically making a frameshift error in the mRNA from a particular gene with the impact of a frameshift mutation in the same gene.

3. Although a large number of mutagenic chemicals are known, no chemical is known to induce mutations in a single gene (gene-specific mutagenesis). From what you know about mutagens, explain why it is unlikely that a gene-specific chemical mutagen will be found. How then is site-specific mutagenesis accomplished?

4. Transposable elements cause mutations when the element is inserted within the gene. These elements disrupt the continuity of a gene. One could also say that introns (Section 7.13) disrupt the continuity of a gene, yet the gene is still functional. Explain why the presence of an intron in a gene does not inactivate that gene but insertion of a transposable element does.

5. When employing the plasmid vector pUC19, DNA fragments made using the restriction enzyme *Bam*HI inactivate the gene encoding β-galactosidase. When employing pBR322, cloning the same fragments inactivates the gene conferring tetracycline resistance. Both vectors contain a gene for ampicillin resistance that is used to select for bacterial transformants after making the recombinant molecules. Explain why it is much more efficient to use pUC19 rather than pBR322 as a cloning vector. (*Hint*: The increased efficiency has to do with finding the cells that contain vectors with cloned DNA. Figure 15.1 will also help you here.)

11

MICROBIAL EVOLUTION AND SYSTEMATICS

The evolutionary history of prokaryotes mirrors the evolutionary history of Earth itself. Of the 4.5 *billion* years that Earth has existed, prokaryotes have thrived for over 4/5 of this time. Evidence for the antiquity of prokaryotes exists in many forms, including oxidized iron formations widespread on Earth.

WORKING GLOSSARY

Allele one form of a given gene

Archaea a group of phylogenetically related prokaryotes distinct from *Bacteria*

Bacteria a group of phylogenetically related prokaryotes distinct from *Archaea*

Binomial system the system devised by Linnaeus for naming living organisms by which an organism is given a genus name and a species epithet

Domain in a taxonomic sense, the highest level of biological classification

Ecotype a population of genetically identical cells sharing a particular resource within an ecological niche

Endosymbiosis the process by which the mitochondrion and chloroplast, which were originally free-living *Bacteria*, established stable residence in primitive eukaryotic cells, eventually yielding the modern eukaryotic cell

Eukarya all eukaryotic cells: algae, protozoa, fungi, slime molds, plant and animal cells

Evolutionary chronometer a molecule, such as ribosomal RNA, whose molecular sequence can be used as a comparative measure of evolutionary divergence

Evolutionary distance in phylogenetic trees, the sum of the physical distance on a tree separating organisms; this distance is inversely proportional to evolutionary relatedness

FAME fatty acid methyl ester

Family in biological classification, an intermediate level of taxonomic hierarchy. Contains one or more genera, each of which consists of one or more species

FISH fluorescent *in-situ* hybridization

GC ratio in DNA (or RNA) from any organism, the percentage of the total nucleic acid that consists of guanine and cytosine bases

Genomic hybridization the experimental determination of genome similarities by measuring the extent of hybridization of DNA from the genome of one organism with that of another

Genus a collection of different species, each sharing one or more (usually several) major properties

Lateral (horizontal) gene transfer the exchange of genes between and among cells in a microbial community

Metazoa multicellular organisms

Multilocus sequence typing a taxonomic tool for classifying organisms on the basis of gene sequence comparisons of 6–7 housekeeping genes

Organelle unit membrane-enclosed structures of bacterial size and specialized in metabolic function found inside eukaryotic cells

Phylogenetic probe an oligonucleotide, sometimes made fluorescent by attachment of a dye, complementary in sequence to some ribosomal RNA signature sequence

Phylogeny the evolutionary history of organisms

Phylum a major lineage of cells in one of the three domains of life

Proteobacteria a large group of phylogenetically related gram-negative *Bacteria*

Ribosomal Database Project (RDP) a large database of small subunit ribosomal RNA sequences that can be retrieved electronically and used in comparative ribosomal RNA sequencing studies

Ribotyping a means of identifying microorganisms from analysis of DNA fragments generated from restriction enzyme digestion of genes encoding their 16S rRNA

RNA life a life form lacking DNA and protein that may have existed on early Earth and in which RNA served both a genetic coding and a catalytic function

Signature sequence short oligonucleotides of defined sequence in 16S or 18S ribosomal RNA characteristic of specific organisms or a group of phylogenetically related organisms and useful for constructing probes

16S ribosomal RNA a large polynucleotide (~1500 bases) that functions as part of the small subunit of the ribosome of prokaryotes and from whose sequence evolutionary information can be obtained; eukaryotic counterpart, 18S rRNA

Small subunit (SSU) RNA ribosomal RNA from the 30S ribosomal subunit of prokaryotes or the 40S ribosomal subunit of eukaryotes

Species in microbiology, a collection of strains that all share the same major properties but differ in one or more significant properties from other collections of strains

Stromatolite a laminated microbial mat, typically built from layers of filamentous prokaryotes and other microorganisms, which can become fossilized

Taxonomy the science of identification, classification, and nomenclature

Universal phylogenetic tree a tree that shows the position of representatives of all domains of living organisms

EARLY EARTH, THE ORIGIN OF LIFE, AND MICROBIAL DIVERSIFICATION

An integrating theme in all of biology is **evolution**. We begin here a unit on microbial diversity as we know it today, the result of nearly 4 billion years of evolution. This chapter focuses on microbial evolution and how the phylogenetic picture of prokaryotes corresponds to their taxonomy and classification. We will see how our understanding of the evolutionary relationships of microorganisms has emerged from molecular biology and how the concept of a "bacterial species" is now beginning to crystallize. This chapter can be considered the background from which we present the organisms themselves and their genomes in the next four chapters.

11.1 Evolution of Earth and Earliest Life Forms

In the first few sections we travel from a time before cells appeared on Earth through to the modern eukaryotic cell. We will learn that prebiotic chemistry could have supplied the necessary prerequisites for the origin of life, that early self-replicating "life forms" may not have been cellular, that early energy sources may have been inorganic, and that the eukaryotic cell is a genetic chimera consisting of at least two distinct organisms.

Origin of Earth

Earth itself is about 4.6 billion years old. Our solar system formed when a large, very hot star exploded, generating a new star (our sun) and the other components of our solar system. Although no rocks dating to the origin of our planet have yet been discovered, presumably because they have been weathered, rocks dating back to nearly 4 *billion* years old (4 gigayears, GY) have been found in several locations on Earth. The oldest rocks discovered thus far are in the Itsaq gneiss complex, in southwestern Greenland, which date to about 3.86 billion years ago. Other rocks of ancient origin include the Warrawoona series, Towers Formation, and Pilbara supergroups in Western Australia, and the Swaziland and Barberton supergroups in southern Africa; all of these rock formations are about 3.5 billion years old (see Figure 11.1).

Ancient rocks are of three types: *sedimentary*, *volcanic*, and *carbonaceous*. The earliest sedimentary rocks are of particular evolutionary interest because we know that the formation of sedimentary rocks requires liquid water. Thus, since sedimentary rocks nearly 4 GY old exist on Earth, it follows that liquid water was also present at that time, probably in the form of oceans or large lakes. The presence of liquid water in turn implies that conditions were compatible with life as we know it.

Evidence for Microbial Life on Early Earth

Evidence for microbial life in ancient rocks rests on the fossilized remains of cells and the isotopically "light" carbon abundant in these rocks (we discuss the use of isotopic analyses of carbon and sulfur as an indication of living processes in Section 18.8). Some ancient rocks contain microfossils that appear very much like bacteria, typically as simple rods or cocci (Figure 11.1●).

In rocks of 3.5 GY or younger, microbial formations called *stromatolites* are abundant. **Stromatolites** are fossilized microbial mats consisting of layers of filamentous prokaryotes and trapped sediment (Figure 11.2●); we discuss some characteristics of microbial mats in Section 18.7. What kind of organisms were these ancient

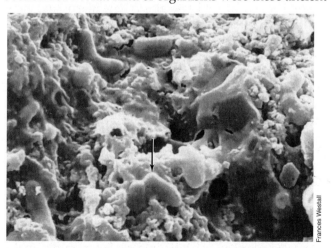

● **Figure 11.1 Ancient microbial life.** Scanning electron micrograph of microfossil prokaryotes from 3.45 billion year old rocks of the Barberton Greenstone Belt, South Africa. Note the rod-shaped bacteria (arrow) attached to particles of mineral matter. The cells are about 0.7 μm in diameter.

● **Figure 11.2 Ancient and modern stromatolites.** (a) The oldest known stromatolite, found in a rock about 3.5 billion years old, from the Warrawoona Group in Western Australia. Shown is a vertical section through the laminated structure preserved in the rock. Arrows point to the laminated layers. (b) Stromatolites of conical shape from 1.6-billion-year-old dolomite rock of the McArthur basin of the Northern Territory of Australia. (c) Modern stromatolites in a warm marine bay, Shark Bay, Western Australia. (d) Modern stromatolites composed of thermophilic cyanobacteria growing in a thermal pool in Yellowstone National Park. Each structure is about 2 cm high. (e) Another view of modern and very large stromatolites from Shark Bay. Individual structures are 0.5–1 m in diameter.

stromatolitic bacteria? By comparing ancient stromatolites with modern stromatolites growing in shallow marine basins (Figure 11.2*c* and *e*) or in hot springs (Figure 11.2*d*; ⚬⚭Figure 18.18*a*), it has been concluded that ancient stromatolites were formed by filamentous phototrophic bacteria, perhaps relatives of the green nonsulfur bacterium *Chloroflexus* (⚬⚭Section 12.35).

Figure 11.3● shows photomicrographs of thin sections of more recent rocks containing cell-like structures remarkably similar to modern filamentous bacteria and green algae (⚬⚭Section 14.13). In the oldest stromatolites these organisms were likely anoxygenic (nonoxygen-evolving) phototrophic bacteria (⚬⚭Sections 12.2, 12.21, 12.32, and 12.35) rather than the O_2-evolving cyanobacteria (⚬⚭Sections 12.25 and 12.26) that dominate modern stromatolites (Figure 11.2*c, e*). Nevertheless, the conclusion is inescapable that prokaryotic microorganisms had evolved an impressive morphological diversity very early in Earth's history.

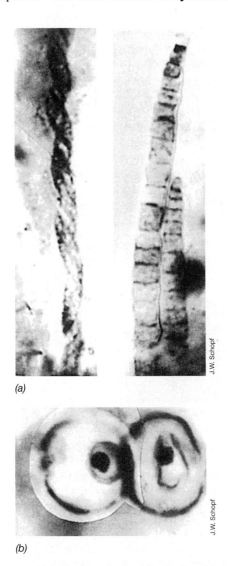

(a)

(b)

J.W. Schopf

● **Figure 11.3 Fossil prokaryotes and eukaryotes from more recent rocks than those shown in Figure 11.1.** The two photographs in (a) show fossil prokaryotic microorganisms from the Bitter Springs Formation, a rock formation in central Australia about 1 billion years old. These forms bear a striking resemblance to modern filamentous cyanobacteria, anoxygenic phototrophs, or filamentous sulfur chemolithotrophs (◌◌Chapter 12). Cell diameters, 5–7 μm. (b) Microfossils of eukaryotic cells from the same rock formation. The cellular structure is remarkably similar to that of certain modern green algae, such as *Chlorella* species (◌◌Section 14.13). Cell diameter, about 15 μm.

Conditions on Early Earth: Hot and Anoxic

The atmosphere of early Earth was devoid of oxygen. Besides water, a variety of gases were present, the most abundant being methane, carbon dioxide, nitrogen, and ammonia. In addition, trace amounts of carbon monoxide and hydrogen likely existed, as well as large reservoirs of sulfide, as a mixture of H_2S and FeS. It is also likely that a considerable amount of hydrogen cyanide, HCN, was produced on early Earth when NH_3 and CH_4 reacted to yield HCN.

Most evidence suggests that the early Earth was hotter than it is today. For its first two hundred million years or so, the surface of Earth may have exceeded 100°C. In addition, the planet was frequently bombarded by meteorites. Under these conditions, free water could not exist. Water only accumulated as Earth cooled. How fast Earth cooled is unknown, but it has been hypothesized that self-replicating entities first appeared at a time when Earth was still fairly hot. Thus, early life forms were likely quite heat tolerant, and may have resembled hyperthermophilic prokaryotes (◌◌Section 6.12 and Chapter 13) in this regard. We will consider the evidence for this hypothesis in Sections 11.8, 11.9, and 13.13.

Origin of Life

The synthesis of biomolecules can occur spontaneously if reducing atmospheres containing the aforementioned gases are subjected to intense energy sources. Of the energy sources available on primitive Earth, the most important was probably ultraviolet (UV) radiation from the sun, as no UV-absorbing ozone layer was present at this time. However, lightning discharges, radioactivity, heat from meteoritic impacts, and thermal energy from volcanic activity were also likely to have been available.

If gaseous mixtures resembling those thought to be present on primitive Earth are irradiated with UV or subjected to electric discharges in the laboratory today, a variety of biomolecules form. These include sugars, amino acids, purines, pyrimidines, various nucleotides, thioesters, and fatty acids. Moreover, under simulated prebiological conditions in the laboratory some of these biochemical building blocks have been shown to *polymerize*, leading to the formation of polypeptides, polynucleotides, and other important macromolecules. We can therefore imagine that by whatever means, a mixture of organic compounds eventually accumulated on the primitive Earth, and in so doing, set the stage for the origin of life. The sterility of Earth at this time meant that these organic compounds were stable and not subject to biodegradation as they would be today.

Catalysis and the Importance of Montmorillonite Clay

To obtain spontaneous synthesis of macromolecules on early Earth, a catalyst would have had to be available. Good possibilities include the surfaces of clays or pyrite (FeS_2). Such surfaces would have provided a stable, relatively dry environment for the synthesis and accumulation of macromolecules into organic films. From these films and over long periods of time, primitive, self-replicating structures could then have emerged.

Experiments with Montmorillonite clay have shown it to be an excellent catalytic surface. For example, in the presence of fatty acids, particles of Montmorillonite clay trigger the formation of small lipid vesicles that appear very much like cells (Figure 11.4●). These vesicles have been shown to self-enlarge by incorporating more fatty acids and to "divide" if passed through a small-pored filter. Such clay particles also bind RNAs and encapsu-

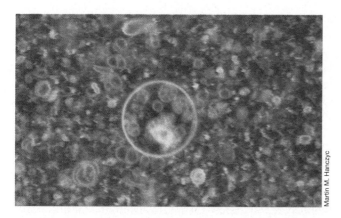

● **Figure 11.4 Lipid vesicles made in the laboratory from the fatty acid myristic acid and RNA.** The vesicle itself stains green and the RNA complexed inside the vesicle stains red. Vesicle synthesis is catalyzed by the surfaces of Montmorillonite clay particles.

late them within the lipid vesicles (Figure 11.4). In addition, if ribonucleotides are added to particles of Montmorillonite clay, the particles catalyze the spontaneous formation of RNA oligonucleotides.

Although clay-catalyzed vesicles are not modern-day cells and RNA oligonucleotides are not modern day genes (although recall that many viruses do have RNA genomes, ∞Chapters 9 and 16), RNA-containing vesicles (Figure 11.4) show many of the properties of cells. If such self-replicating vesicles were formed on early Earth, improved versions containing other macromolecules and a more complex design would eventually have arisen, given the geological time periods available. Such structures could have been the forerunners of the first cellular organisms.

 11.1 Concept Check

Planet Earth is approximately 4.6 billion years old. The first evidence for microbial life can be found in rocks about 3.86 billion years old. Early Earth was anoxic and much hotter than the present. The first biochemical compounds were made by abiotic syntheses that set the stage for the origin of life.

♦   How did the atmosphere of early Earth compare with that of Earth today?

♦   How were early biochemicals formed?

♦   What role could Montmorillonite clay have played in the origin of life?

11.2 Primitive Life: The RNA World and Molecular Coding

What was the first self-replicating entity like? Based on what we know about life today, we can predict that early life forms were simpler than modern cells and that even the simplest self-replicating entities showed at least two

properties: (1) a means of obtaining energy, and (2) some form of heredity. But would these properties have required a *cellular* structure?

It is tempting to extrapolate backward from the present and postulate that early organisms resembled modern cells, differing primarily by having only a few genes and possessing limited transcriptional and translational abilities. However, even a structure like this would have been relatively complex—probably much more complex than the first self-replicating entities. What would the latter have been like then? Following the discovery that certain types of ribonucleic acid (RNA) are catalytic (∞Section 14.8), many scientists now believe that the *earliest* life forms may not have been cells at all, but instead naked RNAs. If true, this would have been the age of "**RNA life**," where RNA was the agent of both catalysis and genetic coding.

RNA Life

In an **RNA world**, if it ever existed, various self-replicating RNAs would have carried out the catalytic reactions necessary for their self-replication (Figure 11.5●). This idea is not far-fetched, because RNAs that do just that are known even today. Various *catalytic RNAs* (∞**ribozymes**, Section 14.8) have been shown to catalyze several different reactions, including self-replication. Moreover, certain ribozymes can synthesize nucleotides from a sugar and a nitrogenous base and polymerize amino acids into polypeptides.

Self-replicating RNA life forms may have evolved into the first *cellular* life forms when they became enclosed within lipoprotein vesicles, perhaps similar to Montmorillonite clay vesicles (Figure 11.4). This step may have occurred to no avail countless times on early Earth, but eventually, the proper constituents and set of circumstances came together and a primitive enclosed structure arose, capable of self-replication (Figures 11.4 and 11.5). Although still lacking DNA and proteins, this protocellular life form would otherwise have resembled a modern cell.

As these structures replicated, natural selection would have pushed for their evolutionary refinement. This would likely have included a selection for more efficient catalysts and a more stable genetic coding process. It is well known that proteins show higher catalytic specificity than do RNAs. Thus, as primitive organisms became more complex and catalytic reactions more biochemically demanding, *proteins* would have replaced *RNAs* as the cells' primary biocatalysts. But this was not an overnight occurrence. It is likely that proteins appeared gradually in cells, perhaps at first complexed with RNA (as ribonucleoprotein complexes, well known in today's cells). However, as evolution selected for more and more precise biochemical catalysts, RNA was eventually replaced almost entirely by proteins as cellular enzymes (Figure 11.5). By this point, the modern cell was clearly on the horizon.

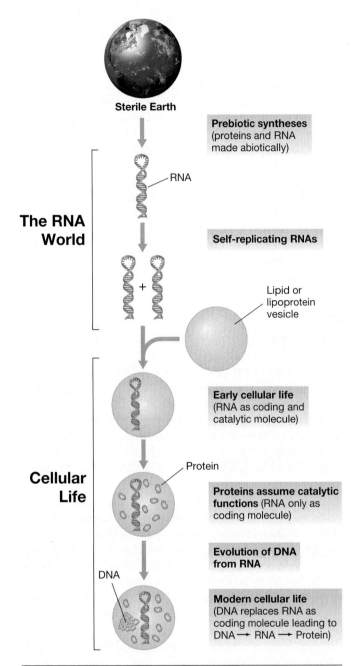

Figure 11.5 **Possible scenario for the evolution of cellular life forms from RNA life forms.** Self-replicating RNAs could have become cellular entities by becoming stably integrated into lipoprotein vesicles. With time, proteins replaced the catalytic functions of RNA, and DNA replaced the coding functions of RNA.

The Modern Cell: DNA → RNA → Protein

Why did the DNA-based cell arise? This may have come about for several reasons. The establishment of DNA as the genome of the cell may have resulted from the need to store genetic information in one place in the cell and keep it there in a stable form. From here, genes could be *expressed* to yield proteins. With further evolutionary input, *selective expression* would have arisen, the latter effecting an energy savings on the cell. Continued evolutionary change in this area would have led to the extensive and highly sophisticated metabolic controls we see in prokaryotes today (⌒⌒ Chapter 8).

The instability of RNA as genetic material might also have led to DNA-based systems. For cells to evolve, there has to be a high degree of genetic stability in the system, otherwise favorable genetic traits will not be faithfully transmitted from generation to generation. This requirement for genetic stability in primitive cells was probably the evolutionary push for the conversion from RNA to DNA as the master genetic molecule in early cells. For reasons probably dealing with the chemistry of their substrates, RNA polymerases are inherently less accurate than DNA polymerases. Thus, to ensure the faithful transfer of genetic information from generation to generation, the evolution of a DNA genome would eventually have been selected. RNAs were retained in protein synthesis, however, where many of their roles (even in modern cells) are catalytic (⌒⌒ Section 7.16).

However it happened, somewhere in the early stages of microbial evolution the three-part system—DNA, RNA, and protein—became fixed in cells as the best solution to biological information processing. That this system was an evolutionary success story is obvious, since without exception, modern cells contain all three types of these macromolecules.

 11.2 **Concept Check**

The first life forms may have been self-replicating RNAs. These were both catalytic and informational. Eventually, DNA became the genetic repository of cells and the three part system, DNA, RNA, and protein, became universal among cells.

◆ What evidence supports the concept of a period of RNA life, and why would RNA life forms likely not have survived to the present?

11.3 **Primitive Life: Energy and Carbon Metabolism**

Life is a highly ordered process. *Energy* is required to assemble an assortment of molecules into a complex biological machine—the cell. How were the energy demands of primitive self-replicating entities met? For catalytic RNAs, the energy required may have come from the exothermic nature of many of the reactions they carried out. But for cells, energy conservation became an issue.

Until the evolution of cyanobacteria, molecular oxygen was unavailable in any significant quantity on Earth (see Figure 11.7). Thus, only energy-generating mechanisms that could occur under *anoxic* conditions could be exploited by primitive cells to fuel their energy needs. However, as we will see in Chapter 17, anoxic conditions place few restrictions on microbial metabolic diversity. A variety of chemoorganotrophic and chemolithotrophic energy-generating mechanisms as well as several variations of photosynthesis occur in the absence of oxygen. All of these mechanisms are rather complex, however, and simpler mechanisms could, and probably did, prevail.

Reactions involving ferrous iron (which was known to have been abundant on early Earth), have often been proposed as energy-yielding reactions for primitive organisms. For example, the reaction:

$$FeS + H_2S \longrightarrow FeS_2 + H_2 \quad \Delta G^{0'} = -42 \text{ kJ/reaction}$$

proceeds exergonically with the release of energy. This reaction yields H_2, and primitive cells could have used this powerful electron donor to generate a proton motive force. The latter could have fueled a primitive ATPase to yield ATP (Figure 11.6●). Other sources of H_2 may also have been available from ferrous iron on the early Earth. For example, Fe^{2+} can reduce protons to hydrogen in the presence of ultraviolet radiation as an energy source (Figure 11.6). Early oceans were thought to be loaded with ferrous iron, and their surface waters were subject to intense UV radiation.

With H_2 as electron *donor*, an electron *acceptor* would also have been required to form a redox pair (◯◯ Section 5.6); this could have been elemental sulfur (S^0). As shown in Figure 11.6, the oxidation of H_2 with the production of H_2S would have required few enzymes. Moreover, because of the abundance of H_2 and sulfur compounds on the early Earth, cells would have had a nearly limitless supply of energy. Interestingly, many hyperthermophilic *Archaea* (which we will see later are, along with hyperthermophilic

Bacteria, the closest extant relatives of Earth's earliest organisms) can oxidize H_2 with S^0 as electron acceptor.

Primitive organisms may have derived carbon from various sources, including organic carbon from abiotic syntheses or even from CO_2, a gas that was abundant on early Earth. Use of carbon dioxide would have had to follow the evolution of autotrophy, the process where CO_2 is converted into all organic components in the cell (◯◯ Sections 17.6 and 17.7). Although originally thought to be unlikely, the hypothesis that autotrophy was an early physiological process has gained momentum from the discovery of autotrophic prokaryotes such as *Aquifex*. This member of the *Bacteria* is a hyperthermophile, contains a very small genome, and branches near to the root of the evolutionary tree of life (see Figure 11.16). All of these are properties one would associate with a "primitive" state. Therefore, it is distinctly possible that early prokaryotes used CO_2 as a major or sole carbon source for growth.

Molecular Oxygen: Banded Iron Formations

A milestone in Earth's history occurred with the evolution of *oxygenic photosynthesis* in the cyanobacteria (Figure 11.7●). Cyanobacteria first appeared on Earth about 2.8 billion years ago, but the O_2 that they produced did not accumulate in the atmosphere because of all of the reducing substances (such as FeS) that were still present in the oceans; these materials reacted sponta-

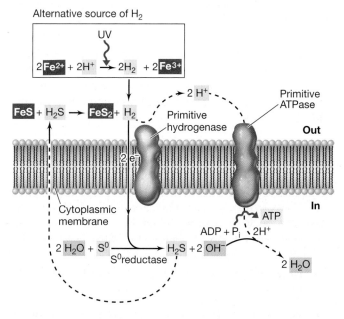

● **Figure 11.6 A possible energy-generating scheme for primitive cells.** Formation of pyrite leads to H_2 production and S^0 reduction, which fuels a primitive ATPase. Note how H_2S plays only a catalytic role; the net substrates would be FeS and S^0. Also note how few different proteins would be required. The $\Delta G^{0'}$ of the reaction $FeS + H_2S \rightarrow FeS_2 + H_2 = -42$ kJ. An alternative source of H_2 could have been the UV-catalyzed reduction of H^+ by Fe^{2+} as shown.

● **Figure 11.7 Banded iron formations.** An exposed cliff about 10 m in height containing layers of iron oxides interspersed with layers containing iron silicates and other silica materials. Brockman Iron Formation, Hammersley Basin, Western Australia. The iron oxides contain iron in the ferric (Fe^{3+}) form produced from ferrous iron (Fe^{2+}) primarily by the oxygen released by cyanobacterial photosynthesis.

neously with O_2 to produce H_2O. In the case of pyrite oxidation, the Fe^{3+} produced became a prominent marker in the geological record. Cyanobacteria almost certainly evolved from anoxygenic phototrophs through the development of a photosystem that could use H_2O as an electron donor for photosynthetic CO_2 reduction; this releases O_2 as a by-product. We discuss the biochemistry of photosynthesis in Chapter 17.

Much of the iron found in rocks of Precambrian origin (>0.5 billion years ago, see Figure 11.8) exists in *banded iron formations* (BIF) (Figure 11.7). These are laminated sedimentary rocks with alternations of iron-rich and silica-rich layers. BIF range in age from 2.7 to 1.9 GY. The evolution of oxygenic photosynthesis (cyanobacteria) is thought to have occurred some 2.75–3 GY ago. The metabolism of cyanobacteria yielded O_2 that oxidized iron present as Fe^{2+} (for example, in FeS_2) to Fe^{3+}. The ferric iron formed various iron oxides that accumulated as BIF (Figure 11.7). Once the abundant Fe^{2+} was consumed, the stage was set for O_2 accumulation in the atmosphere, a major trigger of evolutionary events in both the prokaryotic and eukaryotic worlds (Figure 11.8●).

Oxygenation of the Atmosphere: New Metabolisms and the Ozone Shield

The appearance of oxygenic photosynthesis had enormous consequences for the Earth because as O_2 accumulated, the atmosphere gradually changed from an anoxic to an oxic one (Figure 11.8). With O_2 now available as an electron acceptor, aerobic organisms could evolve. These organisms were able to obtain more energy from the oxidation of organic compounds than anaerobes could (∞Chapter 17), allowing larger populations to develop and increasing the chances for the evolution of new types of organisms and metabolic schemes. There is good evidence from the fossil record that at about the time that Earth's atmosphere became oxidizing there was an enormous burst in the rate of evolution. This triggered the appearance of organelle-containing eukaryotic microorganisms and from them the rapid diversification of multicellular organisms. As evolution proceeded, animals and plants eventually appeared (see Figure 11.16 and Section 11.9).

A major consequence of the appearance of O_2 was the formation of *ozone*, a substance that provides a barrier preventing the intense ultraviolet radiation of the sun from reaching the Earth. When O_2 is subject to short wavelength UV radiation, it is converted to O_3, which strongly absorbs wavelengths up to 300 nm. Until an ozone shield developed, evolution could have continued only in environments protected from direct radiation from the sun, such as under rocks or in the oceans, because the intense UV radiation would have caused lethal DNA damage. However, after the photosynthetic production of O_2 and subsequent development of an ozone shield, organisms could have ranged generally over the

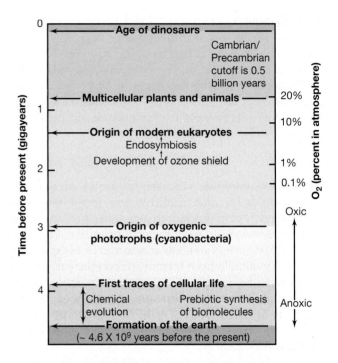

● **Figure 11.8 Major landmarks in biological evolution and Earth's changing geochemistry.** One gigayear (GY) is one billion years. Note how the oxygenation of the atmosphere due to cyanobacterial metabolism was a gradual process, occurring over a period of about 2 billion years. Although full (20%) oxygen levels are required for animals and most other higher organisms, this is not true of prokaryotes, as many are facultative aerobes or microaerophiles (∞ Section 6.15 and Table 6.4). Thus, prokaryotes respiring at reduced O_2 levels may have dominated Earth for the period of a billion years or so before Earth's atmosphere reached current levels of oxygen.

surface of Earth, permitting evolution of a greater diversity of living organisms. Figure 11.8 summarizes some major events in biological evolution and changes in the Earth's geochemistry from a highly reducing to a highly oxidizing planet.

 11.3 Concept Check

Primitive metabolism was anaerobic and likely chemolithotrophic, exploiting the abundant sources of FeS and H_2S present. Carbon metabolism may have included autotrophy. Oxygenic photosynthesis led to development of banded iron formations, an oxic environment, and great bursts of biological evolution.

◆ How could energy have been obtained by early cells from FeS + H_2S?

◆ Why is the advent of cyanobacteria considered a critical step in evolution?

◆ In what form is iron present in BIF?

◆ What role did ozone play in biological evolution, and how did cyanobacteria make the production of ozone possible?

11.4 Eukaryotes and Organelles: Endosymbiosis

Chapter 2 offered a brief tour of eukaryotic microbial diversity (⚬⚬Section 2.6). Chapter 14 will examine eukaryotic microorganisms in more detail. Here, we consider how the modern eukaryotic cell evolved its characteristic internal structure: the membrane-enclosed nucleus and **organelles**.

Origin of the Nucleus

Primitive eukaryotic cells likely consisted of structurally simple cells. Like modern-day prokaryotic cells, they lacked mitochondria, chloroplasts, and a membrane-enclosed nucleus. How then did a nucleus first appear?

As cells in the eukaryotic line of descent became larger, the nucleus and mitotic apparatus evolved, along with the partitioning of DNA into distinct units (chromosomes). Chromosomes may have arisen to ensure the orderly replication and separation of DNA. This is because as primitive genomes enlarged, a point was reached where replication of DNA as a single molecule (as in prokaryotes) was no longer feasible. Development of the nucleus also facilitated handling of the larger genomes needed by the more complex eukaryotic cell and also made possible the recombination of genomes through sexual reproduction (⚬⚬Sections 2.2 and 14.7).

There is no obvious reason why primitive eukaryotes would have needed organelles *other* than the nucleus, especially if the Earth was still anoxic at the time. Hence, the other major organelles likely evolved much later. Their origin, however, is a very interesting story, and we look at that process now.

Endosymbiosis

The modern (organelle-containing) eukaryotic cell did not come about from the gradual partitioning of cellular functions into enclosed structures, the organelles. Instead, organelles originated from the *stable incorporation* of chemoorganotrophic and phototrophic symbionts from the domain *Bacteria* (Figure 11.9●). This process, called **endosymbiosis** (*endo* means "inside"), probably began when an aerobic bacterium established stable residency within the cytoplasm of a primitive eukaryote and supplied the cell with energy in exchange for a protected environment and a ready supply of nutrients (Figure 11.9). This symbiont was the forerunner of the modern mitochondrion. Similarly, the endosymbiotic uptake of an oxygenic phototroph conferred photosynthetic properties on a primitive eukaryote; such a cell was freed from a reliance on organic compounds for energy production and became phototrophic. The phototrophic endosymbiont was the forerunner of the modern day chloroplast (Figure 11.9).

A few eukaryotic cells either never incorporated endosymbionts or, what appears more likely, had them at

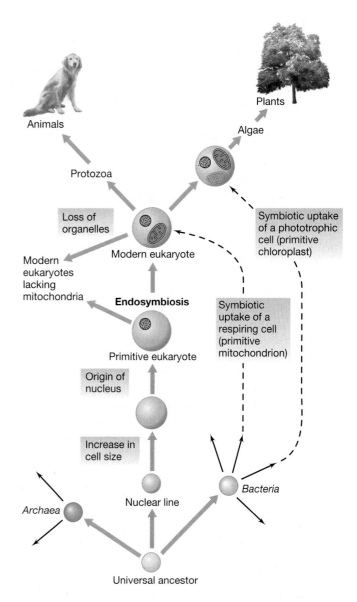

● **Figure 11.9 Origin of the modern eukaryotic cell by endosymbiotic events.** Note how organelles originated from *Bacteria* rather than *Archaea*. Some primitive eukaryotes either never underwent endosymbiotic events or permanently lost their symbionts, but otherwise maintained the basic properties of eukaryotic cells. Compare this figure with Figure 11.16 that shows lineages of *Bacteria* from which organelles originated (⚬⚬Section 14.4).

one time and then disposed of them, possibly because their niche was permanently anoxic. Such organelle-less eukaryotes are known today (⚬⚬Sections 14.2 and 14.9). Most endosymbiotic associations probably increased the fitness of the host cell, and thus today, organelle-containing eukaryotic cells predominate. Microfossil records (see Figure 11.3) suggest that endosymbiotic events began sometime after about 2 GY ago (see Figure 11.10).

Endosymbiosis was probably not a one-time event, but rather a multiple and perhaps even common occurrence between prokaryotic cells and early eukaryotic cells. Unstable or harmful associations were not maintained while beneficial associations were. Selection helped refine the beneficial associations. This included modifications to the structure of the endosymbiont (loss of a cell wall, for example), genomic modifications, and the transfer of genes from the endosymbiont to the nucleus of the host cell (⚭ Sections 14.9 and 15.7). From these associations, the modern eukaryotic cell was born.

Evidence for Endosymbiosis

Many lines of evidence support the endosymbiosis hypothesis. For example, both chloroplasts and mitochondria contain ribosomes that are clearly of the prokaryotic type (⚭ Table 7.4) and show 16S ribosomal RNA sequences (see Section 11.6) characteristic of specific *Bacteria*. Moreover, organellar ribosome function is inhibited by the same antibiotics that affect ribosome function in free-living *Bacteria* (⚭ Section 14.5). Mitochondria and chloroplasts also contain small amounts of DNA arranged in a covalently closed, circular form, typical of prokaryotes (⚭ Section 2.2). Indeed, many telltale signs of a prokaryotic background are present in organelles from modern cells today (⚭ Section 14.4).

Endosymbiotic events occur even today. For example, there are a number of protozoa, especially anaerobic protozoa, that contain symbiotic prokaryotes (⚭ Figure 19.25). Although these symbionts appear to be fully functional cells rather than the highly modified cells that we find in the mitochondria and chloroplasts, they show that eukaryotic cells can and still do take up prokaryotic cells, presumably to the advantage of both partners.

Endosymbiosis was a major boost for the eukaryotic lineage of cells. The presence of respiratory and photosynthetic energy factories conferred such important new properties on eukaryotic cells that the stage was set for an explosion in biological diversity. The period from about 1.5 billion years ago to the present saw the rise and diversification of unicellular eukaryotic microorganisms and the metazoa, culminating in the appearance of the structurally complex higher plants and animals (see Figures 11.8, 11.9, and 11.16).

Biological Evolution and Geological Time Scales

The period from the origin of the first metazoan to the present represents about one-sixth of the total time that life has existed on Earth. Put another way, five-sixths of Earth's biological history was restricted to microbial life, the bulk of this period *solely to prokaryotes*. Prior to this period would have been the age of acellular replicating life forms—the RNA world—if such structures actually existed. Prior to this, planet Earth was sterile, in the truest sense of the word. These eras in the evolution of life are summarized using the geologist's time scale in Figure 11.10●.

Metazoa have left a considerable and highly diverse fossil record. Because of this, our understanding of biological evolution from the time of the metazoa to the pres-

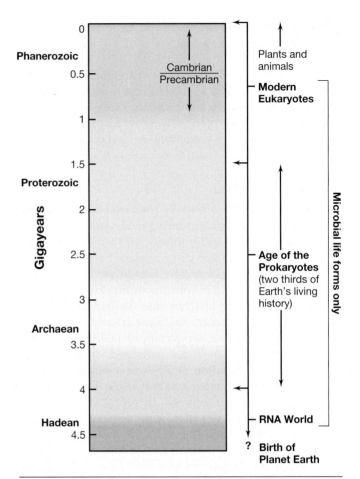

● **Figure 11.10 The eras of self-replicating entities on Earth.** The left scale relates life forms to the major geological time periods.

ent is quite good. The story for microorganisms, however, is quite incomplete, especially with regards to prokaryotes. In fact, the evolutionary history of prokaryotes was a mystery until just a little more than a quarter century ago. The advent of molecular methods for determining evolutionary relationships has opened many new avenues in evolutionary science, especially in the area of microbial evolution. We examine this story now, and see how these methods allow us to place all cells onto a phylogenetic tree of life.

 11.4   Concept Check

The eukaryotic nucleus and mitotic apparatus probably arose as a necessity for ensuring the orderly partitioning of DNA in large-genome organisms. Mitochondria and chloroplasts, the principal energy-producing organelles of eukaryotes, arose from symbiotic association of prokaryotes of the domain *Bacteria* within eukaryotic cells, a process called endosymbiosis. Assuming that an RNA world existed, self-replicating entities have populated Earth for over 4 billion years.

◆ When in geological time did eukaryotes first appear? Metazoa?

◆ What evidence supports the idea that organelles were once free-living species of the domain *Bacteria*?

II METHODS FOR DETERMINING EVOLUTIONARY RELATIONSHIPS

In the next few sections we consider the **phylogeny** of microorganisms—their evolutionary relationships. The major focus here will be on the comparative sequencing of structural RNAs from ribosomes. From such studies, the first revealing picture of microbial phylogeny has emerged. In addition, the techniques employed have yielded new research tools for microbial ecology and clinical diagnostics.

11.5 Evolutionary Chronometers

Certain genes and proteins are **evolutionary chronometers**—measures of evolutionary change. In other words, differences in nucleotide or amino acid sequence of functionally similar (homologous) macromolecules are a function of their *evolutionary distance*. However, in order to measure evolutionary distances by molecular sequencing, it is essential that the correct molecules be chosen for sequencing studies.

Criteria for Molecular Chronometers

A number of criteria define the perfect molecular chronometer. A molecular chronometer should be *universally distributed* across the group chosen for study. This facilitates the comparison of all organisms. Second, the molecule must be *functionally homologous* in each organism; functionally distinct molecules would not be expected to show sequence similarities. Third, it is crucial that the molecule have regions of sequence conservation for *aligning* the sequences for analysis. And finally, the sequence of the molecule chosen should reflect evolutionary change in the organism as a whole. In fact, the greater the phylogenetic distance to be measured, the *slower* must be the rate at which the sequence changes, as too much change over long time periods will scramble the evolutionary signal.

Many genes and proteins have been proposed as molecular chronometers. However, genes encoding *ribosomal RNAs*, central components in the translational system (⚭Sections 7.14–7.16), *ATPase proteins*, the enzyme complex that synthesizes or hydrolyzes ATP (⚭Section 5.12), *RecA*, the enzyme that facilitates genetic recombination (⚭Section 10.6), and certain translational proteins, have provided the most insightful phylogenetic information about microorganisms. All of these molecules were probably essential for even rather primitive cells, and thus their gene sequence variations allow us to look deeply into the evolutionary past. We focus our discussion here on the most widely used of these chronometers, **ribosomal RNA** (Figure 11.11●).

Ribosomal RNAs as Evolutionary Chronometers

Ribosomal RNAs have several features that make them excellent evolutionary chronometers. Ribosomal RNAs are relatively large, functionally constant, universally distributed, and contain several regions in which the nucleotide sequence is conserved in all cells. There are three ribosomal RNA molecules, which in prokaryotes have sizes of 5S, 16S, and 23S (the "S" stands for "Svedberg," a measure of mass based on the sedimentation of a particle in a centrifuge). The focus of sequencing efforts has been on 16S rRNA, or in eukaryotes, the functionally similar, albeit slightly larger, 18S molecule. Because 16S and 18S rRNAs are part of the *small* (30S or 40S) *subunit* of the ribosome (Figure 11.11*b*), the acronym **SSU** (for *small subunit*) sequencing is synonymous with 16S or 18S sequencing.

A huge database of rRNA sequences exists. For example, the **Ribosomal Database Project (RDP)** contains a large collection of such sequences, now numbering over 100,000. The RDP can be accessed electronically (http://rdp.cme.msu.edu/html/) and, besides sequences, contains phylogenetic tutorials, reference citations, previews of new releases of sequences, and a host of other features.

Use of SSU rRNA as a phylogenetic tool was pioneered in the early 1970s by Carl Woese at the University of Illinois, and the method is now widely used throughout biology. For his accomplishments, Woese received the 2003 Crafoord Prize, the highest recognition for scientific achievement in biology, from the Royal Swedish Academy of Sciences.

11.5 Concept Check

Comparisons of sequences of ribosomal RNA can be used to determine the evolutionary relationships between organisms. Phylogenetic trees based on ribosomal RNA have now been prepared for all the major prokaryotic and eukaryotic groups.

◆ Why is ribosomal RNA a good evolutionary chronometer?

◆ What is the *RDP*?

◆ Who is Carl Woese?

11.6 Ribosomal RNA Sequencing as a Tool of Molecular Evolution

The methods for obtaining ribosomal RNA sequences and generating phylogenetic trees are now quite routine, thanks to a combination of molecular biology and computer analyses. Newly generated sequences can be compared with sequences in the RDP and other genetic databases, such as GenBank (USA), DDBS (Japan), or EMBL (Germany). Then, using a treeing algorithm, a phylogenetic tree is produced that best describes the evolutionary information inherent in the sequences.

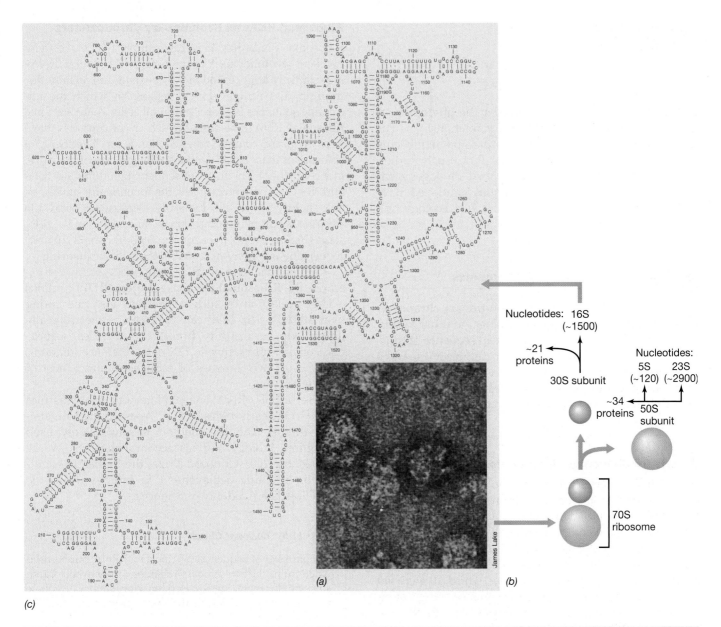

● **Figure 11.11 Ribosomal RNA.** (a) Electron micrograph of 70S ribosomes from *Escherichia coli*. (b) Parts of the ribosome; 5S, 16S, and 23S refer to different forms of RNA in the small subunit of the ribosome. (c) Primary and secondary structure of 16S ribosomal RNA (rRNA). This is the 16S rRNA from *Escherichia coli* (*Bacteria*); 16S rRNA from *Archaea* is similar in overall secondary structure (folding) but has numerous differences in primary structure (sequence). The counterpart to 16S rRNA in eukaryotes is 18S rRNA present in cytoplasmic ribosomes.

Sequencing Methodology

We begin with the assumption that we are working with a pure culture of a microorganism, although, as we will see, pure cultures are not necessary for comparative SSU rRNA sequencing. To begin the process, the polymerase chain reaction (PCR, ∞Section 7.9) is used to amplify the gene encoding 16S ribosomal RNA from genomic DNA. Following this, the PCR product is sequenced by the dideoxy DNA sequencing method (Sanger sequencing; ∞Section 7.8) (Figure 11.12●). This procedure is both rapid and specific. Using PCR primers complementary to conserved sequences in SSU ribosomal RNAs, only a tiny amount of cell material can yield a huge

amount of DNA product for sequencing purposes (Figure 11.12). Once sequencing is done, the sequence data are ready for computer analyses.

Generating Phylogenetic Trees from RNA Sequences

Several different algorithms for sequence analysis and phylogenetic tree formation are available for comparative ribosomal RNA sequencing. However, regardless of which program is used, raw sequence data must first be *aligned* with previously aligned sequences using a sequence editor. Not all rRNAs are exactly the same length. Thus, during alignment, gaps must be inserted

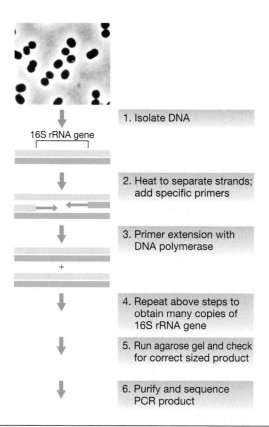

1. Isolate DNA

16S rRNA gene

2. Heat to separate strands; add specific primers

3. Primer extension with DNA polymerase

+

4. Repeat above steps to obtain many copies of 16S rRNA gene

5. Run agarose gel and check for correct sized product

6. Purify and sequence PCR product

● **Figure 11.12   Ribosomal RNA sequencing of a pure culture of a microorganism using the polymerase chain reaction (PCR).** The 16S rRNA gene is amplified and then sequenced by the Sanger method (∞Section 7.8). The primers added are complementary to conserved sequences of 16S rRNA (see Figure 11.11c).

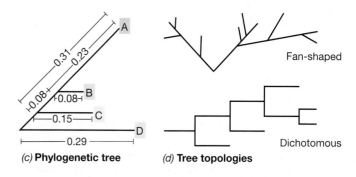

Organism	Sequence	Analysis
A	CGUAGACCUGAC	For A ⟶ B, three differences occur out of a total of twelve; thus $\frac{3}{12} = 0.25$
B	CCUAGAGCUGGC	
C	CCAAGACGUGGC	
D	GCUAGAUGUGCC	

(a) **Sequence alignment and analysis**

Evolutionary distance				Corrected evolutionary distance
E_D	A ⟶ B	0.25		0.30
E_D	A ⟶ C	0.33		0.44
E_D	A ⟶ D	0.42		0.61
E_D	B ⟶ C	0.25		0.30
E_D	B ⟶ D	0.33		0.44
E_D	C ⟶ D	0.33		0.44

(b) **Calculation of evolutionary distance**

A

0.31 0.23

0.08 0.08 B

0.15 C

0.29 D

(c) **Phylogenetic tree**

Fan-shaped

Dichotomous

(d) **Tree topologies**

● **Figure 11.13   Preparing a phylogenetic distance tree from 16S ribosomal RNA sequences.** For illustrative purposes, only short sequences are shown. The evolutionary distance E_D in (b) is calculated as the percentage of *nonidentical* sequences between the RNAs of any two organisms. The corrected E_D is a statistical correction necessary to account for either back mutations to the original genotype or additional forward mutations at the same site that could have occurred. The tree (c) is ultimately generated by computer analysis of the data to give the best fit. (d) Different forms of phylogenetic trees can be generated, including fan-shaped and dichotomous. In either format, the total length of the branches separating any two organisms is proportional to their evolutionary distance.

where necessary in regions where one sequence is shorter than another. The aligned sequences are then imported into a treeing program and comparative analyses done.

Two widely used treeing algorithms are *distance* and *parsimony*. Using distance methods, sequences are aligned and then an **evolutionary distance** (E_D) calculated by having the computer record every position in the dataset in which there is a *difference* in sequence (Figure 11.13●). From these data, a distance matrix can be constructed that shows the E_D between any two sequences in the dataset. Following this, a statistical correction is factored into the E_D that considers the possibility that more than one change may have occurred at any given site (Figure 11.13). Once this is accounted for, a phylogenetic tree is generated in which the lengths of the lines in the tree are proportional to evolutionary distances (Figure 11.13c, d).

Parsimony, another popular method of phylogenetic analysis, generates evolutionary trees based on the assumption that the *minimal amount* of sequence change necessary to diverge two lineages from a common ancestor actually occurred during their evolutionary divergence. Like distance programs, this method requires summing the number of sequence differences in a particular data set, but the algorithm

handles the analysis somewhat differently. Nevertheless, phylogenetic trees based on parsimony appear similar to distance trees, although the branching order within a parsimonious tree can and sometimes does differ from that in a distance tree generated from the exact same data set. Thus, no single phylogenetic tree provides the final word on the evolutionary relationships of the organisms in question. Any tree can only be considered a close approximation to the true phylogeny of a group.

11.6 **Concept Check**

Comparative ribosomal RNA sequencing is now a routine procedure involving the amplification of the gene encoding 16S rRNA, sequencing it, and analyzing the sequence in reference to other sequences. Two major treeing algorithms are distance and parsimony.

◆ What is an *evolutionary distance*?

◆ How is the *polymerase chain reaction (PCR)* of use in comparative ribosomal RNA sequencing?

11.7 Signature Sequences, Phylogenetic Probes, and Microbial Community Analyses

SSU rRNA sequencing has spawned many research tools that employ this technology. These include signature sequences and ribosomal RNA probes for use in microbial ecology and diagnostic medicine.

Signature Sequences

Computer analyses of ribosomal RNA sequences have revealed so-called **signature sequences**, short oligonucleotides unique to certain groups of organisms. For example, signature sequences specific for each of the domains of cellular life are known (Table 11.1). Moreover, signatures defining a specific group within a domain or, in some cases, a particular genus or even a single species, are also known or can be determined by computer inspection of aligned sequences. Because of their exclusivity, signature sequences are very useful. For example, signature sequences can help place newly isolated or previously misclassified organisms into their correct phylogenetic group. But the most common use of signature sequences is for the design of specific nucleic acid probes called **phylogenetic probes**.

Phylogenetic Probes and FISH

Recall that a *probe* is a strand of nucleic acid that can be labeled and used to hybridize to a complementary nucleic acid from a mixture (∽Section 7.7). Probes can be general or specific. For example, universal ribosomal RNA probes can be synthesized that will bind to conserved sequences in the ribosomal RNA of all organisms, regardless of domain. By contrast, specific probes can be designed that will react only with cells of species of the domain *Bacteria* because of unique signatures found in their RNA (Table 11.1). Likewise, archaeal-specific and eukaryl-specific probes will react only with species of *Archaea* or *Eukarya*, respectively. Even major groups within a domain such as genera or families (see Section 11.10) can be targeted with specific probes.

The binding of probes to cellular ribosomes can be seen microscopically when a fluorescent dye is attached to the probe (Figure 11.14●). By treating cells with the appropriate reagents, membranes become permeable and allow penetration of the probe/dye mixture. After hybridization of the probe directly to ribosomal RNA in ribosomes, the cells become uniformly fluorescent and can be observed under a fluorescent microscope (Figure 11.14, see also Figures 11.15● and 11.17). This technique has been nicknamed *FISH*, for *f*luorescent *in*-situ *h*ybridization, and can be applied directly to cells in culture or in a natural environment (the term *in situ* means "in the environment"). In essence, FISH is a *phylogenetic stain*.

FISH technology is widely used in microbial ecology and clinical diagnostics. In ecology, FISH can be

Table 11.1	Signature sequences from 16S or 18S rRNA defining the three domains of life				
			Occurrence among[c]		
Oligonucleotide signatures[a]	Approximate position[b]		*Archaea*	*Bacteria*	*Eukarya*
CACYYG	315		0	>95	0
AAACUCAAA	910		3	100	0
AAACUUAAAG	910		100	0	100
YUYAAUUG	960		100	<1	100
CAACCYYCR	1110		0	>95	0
UCCCUG	1380		>95	0	100
UACACACCG	1400		0	>99	100
CACACACCG	1400		100	0	0

[a] Y, any pyrimidine; R, any purine.
[b] Refer to Figure 11.11c for numbering scheme of 16S rRNA.
[c] Occurrence refers to percentage of organisms examined in any domain that contain that sequence.

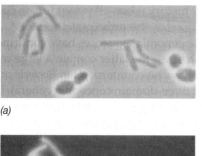

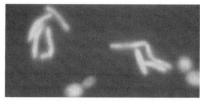

(a)

(b)

(c)

Norman Pace

● **Figure 11.14 Fluorescently labeled ribosomal RNA probes: Phylogenetic stains.** (a) Phase contrast photomicrograph (no probes present) of cells of *Bacillus megaterium* (rod, member of the *Bacteria*) and the yeast *Saccharomyces cerevisiae* (oval-shaped cells, *Eukarya*). (b) Same field, cells stained with a yellow-green universal rRNA probe (this probe reacts with species from any domain). (c) Same field, cells stained with a eukaryal probe (only cells of *S. cerevisiae* react). Cells of *B. megaterium* are about 1.5 μm in diameter and cells of *S. cerevisiae* are about 6 μm in diameter. Section 18.4 discusses genetic stains in more detail, including stains not based on rRNA sequences.

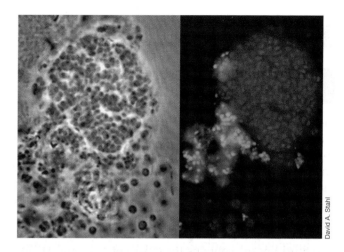

David A. Stahl

● **Figure 11.15 Use of phylogenetic stains to make nitrifying bacteria visible in a granule of activated sewage sludge.** (Left) Phase photomicrograph. (Right) Color photomicrograph of the same field after applying phylogenetic stains. The probe that fluoresces red is specific for a signature sequence (see Table 11.1) in the 16S rRNA of ammonia-oxidizing bacteria, while the probe that fluoresces green is specific for a sequence present only in nitrite-oxidizing bacteria. Both ammonia-oxidizing and nitrite-oxidizing bacteria are phylogenetically closely related members of the *Bacteria* (∞Sections 12.3 and 17.12) and carry out an interdependent series of chemolithotrophic reactions (∞Section 19.12).

used for the microscopic identification and tracking of organisms directly in the environment. FISH also offers a method for assessing the composition of microbial communities directly by microscopy (Figure 11.15 and ∞ Section 18.4). In clinical diagnostics, FISH has been used for the rapid identification of specific pathogens from patient specimens. The technique circumvents the need to grow an organism in culture. Instead, microscopic examination of a specimen can confirm the presence of a specific pathogen. In clinical diagnostics this allows for more rapid diagnosis and treatment of a patient suffering from a microbial infection. By contrast, isolation and identification techniques typically take a minimum of 24 hours to complete and can take much longer.

Microbial Community Analysis

PCR-amplified ribosomal RNA genes do not need to originate from a pure culture grown in the laboratory. Using methods to be described in detail in Chapter 18, a phylogenetic snapshot of a *natural microbial community* can be taken using PCR to amplify the genes encoding SSU ribosomal RNA from all members of that community. Such genes can easily be sorted out, sequenced, and aligned. From the data, a phylogenetic tree can be generated of "environmental" sequences that show the different ribosomal RNAs present in the community. From this tree, specific organisms can be inferred even though none of them were actually cultivated or otherwise identified. Such **microbial community analyses**, as they have come to be called, are a major thrust of microbial ecology research today and have revealed many key features of microbial community structure and microbial interactions (∞Chapter 18 for detailed discussion of this topic).

 ### 11.7 Concept Check

Signature sequences, short oligonucleotides found within a ribosomal RNA molecule, can be highly diagnostic of a particular organism or group of related organisms. Signature sequences can be used to generate specific phylogenetic probes, useful for FISH or microbial community analyses.

◆ Using the sequence data in Table 11.1, choose a specific phylogenetic probe that would allow you to distinguish a cell of *Archaea* from a cell of *Bacteria*.

◆ How can oligonucleotide probes be made visible under the microscope? What is this technology called?

similar to those of hyperthermophilic Crenarchaeota (⌒⌒Sections 13.8–13.10).

Lying basal to the Korarchaeota are the Nanoarchaeota, a group thus far represented by only a single organism, *Nanoarchaeum*. This tiny bacterium is a parasite of another archaeon, called *Ignicoccus*, and contains the smallest genome of all known prokaryotes (⌒⌒Sections 13.11 and 15.4). We consider the Nanoarchaeota and all phyla of *Archaea* in more detail in Chapter 13.

Eukarya

Phylogenetic trees of species in the domain *Eukarya* are generated from comparative sequencing of 18S rRNA, the functional equivalent of 16S rRNA, in eukaryotic cytoplasmic ribosomes. Early eukaryotes may have been similar to present-day microsporidia and diplomonads. These organisms are all obligate parasites that live in association with representatives of various groups of eukaryotes, from microorganisms to humans [for example, the pathogen *Giardia* (⌒⌒Section 28.6) is a member of the diplomonad group]. Interestingly, although these organisms contain a membrane-enclosed nucleus, microsporidia and diplomonads lack mitochondria. In this connection they resemble the type of cell that might have first accepted stable endosymbionts (see Figure 11.9 and ⌒⌒Sections 14.5 and 14.9). (Several problems have emerged in pinpointing the true phylogeny of the microsporidia and these will be discussed in Section 14.9.)

More phylogenetically derived *Eukarya* include the multicellular organisms, culminating in the largest and most structurally complex of the *Eukarya*, the plants and animals (Figure 11.16). When the fossil record is compared with the phylogenetic tree of *Eukarya*, the beginnings of rapid evolutionary radiation can be traced to about 1.5 billion years ago. Geochemical evidence suggests that this is the period in Earth's history in which significant oxygen levels had accumulated in the atmosphere (Figure 11.8). It is thus a good possibility that the onset of oxic conditions and subsequent development of an ozone shield (which would have greatly expanded the number of surface habitats available for colonization) was a major trigger for the rapid diversification of *Eukarya*. We consider the major groups of microbial *Eukarya* and their biology in Chapter 14.

 11.8 Concept Check

Life on Earth evolved along three major lines, called domains, all derived from a common ancestor. Each domain contains several phyla. Two of the domains, *Bacteria* and *Archaea*, remained prokaryotic, whereas the third line, *Eukarya*, evolved into the modern eukaryotic cell.

◆ How does the universal tree support the idea that early Earth was very hot?

◆ How does the universal tree support the hypothesis of endosymbiosis (see Figure 11.9)?

11.9 Characteristics of the Domains of Life

Although the primary domains (*Bacteria*, *Archaea*, and *Eukarya*) were founded on the basis of comparative ribosomal RNA sequencing—genetic criteria—each domain can also be characterized by certain *phenotypic* properties. Some of these characteristics are unique to one domain, whereas others are found in two of three domains. We discuss here an overview of major phenotypic traits of phylogenetic value.

Cell Walls

Virtually all *Bacteria* have cell walls containing *peptidoglycan* (⌒⌒Section 4.8). The only known exceptions are members of the *Planctomyces–Pirella* group (⌒⌒Section 12.28), whose cell walls are composed of protein, and the *Mycoplasma–Chlamydia* groups, which lack cell walls altogether (⌒⌒Sections 12.21 and 12.27). Peptidoglycan can thus be considered a "signature" molecule for species of *Bacteria* (when assaying for peptidoglycan, it is *muramic acid* that is actually detected because this is what is unique to peptidoglycan; see Table 11.3).

Cells of *Eukarya* and *Archaea* lack peptidoglycan. In eukaryotes, if cell walls are present, they are typically made of cellulose or chitin (⌒⌒Chapter 14). In *Archaea* various cell wall chemistries are known, from the peptidoglycan analog *pseudopeptidoglycan* to walls made of polysaccharide, protein, or glycoprotein (⌒⌒Section 4.8). Thus, great diversity exists in the chemistry of microbial cell walls, but it is the presence or absence of *peptidoglycan* that is most useful in distinguishing *Bacteria* from *Archaea*.

Lipids

The chemical nature of membrane lipids is a key feature for differentiating *Archaea* from *Bacteria*. *Bacteria* and *Eukarya* synthesize membrane lipids with a backbone consisting of fatty acids bonded in *ester* linkage to a molecule of glycerol (Figure 11.18●). Although the nature of the fatty acid can be highly variable, the *ester link* to glycerol is the key defining feature. By contrast, archaeal lipids consist of *ether-linked* molecules (Figure 11.18). In ester-linked lipids, the fatty acids are usually straight-chain (linear) molecules, whereas in *Archaea*, long-chain, branched hydrocarbons, either of the phytanyl or biphytanyl type, are present in place of fatty acids and are bonded by ether linkage to glycerol.

A few exceptions to the phylogenetic pattern of lipids are known. For example, the thermophilic sulfate-reducing bacterium *Thermodesulfobacterium* (⌒⌒Section 12.36) and a few other sulfate-reducing bacteria (⌒⌒Section 12.18), as well as *Propionibacterium* species, contain ether-linked lipids. However, the converse is not true. No species of *Archaea* has been shown to contain ester-linked lipids.

In terms of overall membrane structure, the membranes of some *Archaea*, especially hyperthermophilic

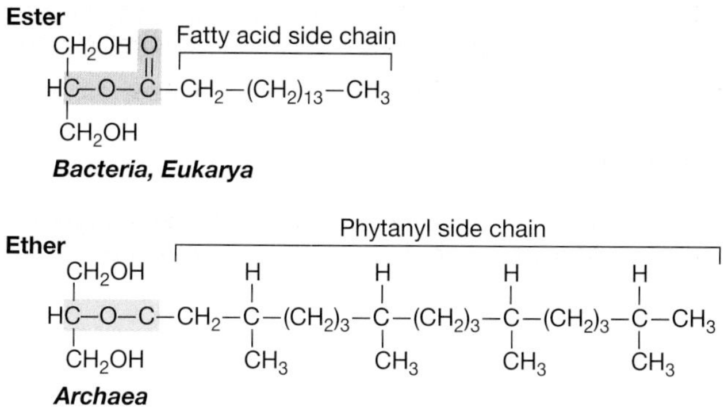

Ester

CH₂OH O   Fatty acid side chain
| ‖
HC—O—C—CH₂—(CH₂)₁₃—CH₃
|
CH₂OH

Bacteria, Eukarya

Ether

Phytanyl side chain

CH₂OH   H   H   H   H
| | | | |
HC—O—C—CH₂—C—(CH₂)₃—C—(CH₂)₃—C—(CH₂)₃—C—CH₃
| | | | |
CH₂OH   CH₃   CH₃   CH₃   CH₃

Archaea

● **Figure 11.18 Lipids in *Bacteria, Eukarya,* and *Archaea*.** In *Bacteria* and *Eukarya*, lipids contain fatty acids (palmitic acid is shown) bonded by *ester* linkages to glycerol. In *Archaea*, the side chains are branched hydrocarbons (phytanyl, C_{20}, is shown) bonded by *ether* linkages to glycerol. Phytanyl is synthesized from isoprene (◌◌Figure 4.18c).

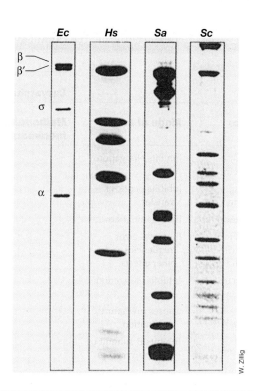

● **Figure 11.19 RNA polymerase and phylogeny.** RNA polymerases from representatives of the three domains: Ec, *Escherichia coli* (*Bacteria*), Hs, *Halobacterium salinarum* (Euryarchaeota, *Archaea*), Sa, *Sulfolobus acidocaldarius* (Crenarchaeota, *Archaea*), and Sc, *Saccharomyces cerevisiae* (*Eukarya*). The purified components of the RNA polymerases have been separated by electrophoresis on a polyacrylamide gel. The largest polypeptide subunits are on the top, and the smallest subunits are on the bottom. Only members of the *Bacteria* contain the simple (four-polypeptide) RNA polymerase.

species, form into a lipid *monolayer* instead of a lipid *bilayer* (◌◌Section 4.5 and Figure 4.20). A lipid monolayer is less likely than a lipid bilayer to denature at the high temperatures at which these organisms grow. The ether linkage between glycerol and the hydrophobic side chains (Figure 11.18) combined with monolayer membrane construction protect the membranes of these extremophilic *Archaea* from the harsh conditions of their habitats.

RNA Polymerase

Transcription is carried out by DNA-dependent RNA polymerases in all organisms; DNA is the template, and RNA is the product. Cells of *Bacteria* contain a single type of RNA polymerase of rather simple quaternary structure. This is the classic RNA polymerase containing *four* polypeptides, α, β, β′, and σ, combined in a ratio of 2:1:1:1, respectively, in the active polymerase (Figure 11.19●) (◌◌also Section 7.10).

Archaeal RNA polymerases are structurally more complex than those of *Bacteria*. The RNA polymerases of species of *Archaea* contain *eight* or more polypeptides, more closely resembling the pattern in *Eukarya* (Figure 11.19). The major RNA polymerase of eukaryotes (there are three) contains 10–12 polypeptides, and the relative sizes of the peptides coincide most closely with those from species of hyperthermophilic *Archaea* (Figure 11.19). Thus, in terms of phylogenetic signatures, the $\alpha_2\beta\beta'\sigma$ polymerase is diagnostic of *Bacteria*; RNA polymerases with more than four subunits are not phylogenetically definitive, except to say that they do not belong to *Bacteria*.

Features of Protein Synthesis

Because of differences in ribosomal RNA sequences and several protein synthesis factors, it is not surprising that certain aspects of the protein-synthesizing machinery differ in representatives of the three domains. Although ribosomes of *Archaea* and *Bacteria* are the same size (70S,

as compared with the 80S ribosomes in the cytoplasm of eukaryotes), several steps in archaeal protein synthesis more strongly resemble those in eukaryotes than in *Bacteria*. Recall that translation always begins at a unique codon, the so-called *start codon*. In *Bacteria* this start codon (AUG) calls for the incorporation of an initiator tRNA containing a modified methionine residue, *formyl-methionine* (◌◌Section 7.16). By contrast, in eukaryotes and in *Archaea*, the initiator tRNA carries an *unmodified* methionine.

The exotoxin produced by *Corynebacterium diphtheriae* is a potent inhibitor of eukaryotic protein synthesis because it ADP-ribosylates (adds ADP-ribose to) an elongation factor required to translocate the ribosome along the mRNA; the modified elongation factor is inactive (◌◌Section 21.10). Diphtheria toxin also inhibits protein synthesis in species of *Archaea* but not in species of *Bacteria*.

Most antibiotics that specifically affect protein synthesis in *Bacteria* do not affect archaeal or eukaryal protein synthesis. The sensitivity of representatives of the three domains to various protein synthesis inhibitors is shown in Table 11.2 (◌◌also Section 7.16), where various antibiotics are grouped according to their modes of action in blocking protein synthesis in various domains of organisms.

Table 11.4 **Some phenotypic characteristics of taxonomic value**

Major category	Components
I. Morphology	Shape; size; Gram reaction; arrangement of flagella, if present
II. Motility	Motile by flagella; motile by gliding; motile by gas vesicles; nonmotile
III. Nutrition and Physiology	Mechanism of energy conservation (phototroph, chemoorganotroph, chemolithotroph); relationship to oxygen; temperature, pH, and salt requirements/tolerances; ability to use various carbon, nitrogen, and sulfur sources; growth factor requirements
IV. Other factors	Pigments; cell inclusions, or surface layers; pathogenicity; antibiotic sensitivity

otes; this is a somewhat broader range than for eukaryotes (Figure 11.21●). Base compositions of DNA have been determined for a wide variety of organisms, and knowledge of an organism's GC ratio can be useful, depending on the situation. For example, two organisms can have identical GC ratios and yet turn out to be quite unrelated (both taxonomically and phylogenetically), because a variety of base *sequences* is possible with DNA of a single base *composition*. In this case, the identical GC ratios are taxonomically meaningless. By contrast, if two organisms' GC ratio differs by greater than about 5%, they will share few DNA sequences in common and are therefore unlikely to be closely related.

 11.10 Concept Check

Conventional bacterial taxonomy places heavy emphasis on analyses of phenotypic properties of the organism. Determining the guanine plus cytosine (GC) base ratio of the DNA of the organism can be part of this process.

◆ List three phenotypic properties that can be used to distinguish bacteria.

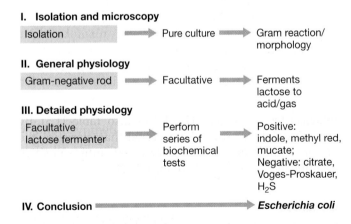

I. Isolation and microscopy

Isolation ⟹ Pure culture ⟹ Gram reaction/ morphology

II. General physiology

Gram-negative rod ⟹ Facultative ⟹ Ferments lactose to acid/gas

III. Detailed physiology

Facultative lactose fermenter ⟹ Perform series of biochemical tests ⟹ Positive: indole, methyl red, mucate; Negative: citrate, Voges-Proskauer, H₂S

IV. Conclusion ⟹ *Escherichia coli*

● **Figure 11.20 Example of methods that would be used for identification of a newly isolated enteric bacterium.** This scheme uses classical microbiological methods (the example given shows the procedures that would be used for identifying *Escherichia coli*). Note that most of the analyses here require that the organisms be grown in pure culture and that solely phenotypic criteria be used in the identification. A description of biochemical tests is presented in Chapter 24 (🔗Section 24.2, Table 24.3, and Figure 24.7).

◆ How is it possible for two organisms to have very similar DNA GC ratios yet share few genes in common?

11.11 Chemotaxonomy

Molecular taxonomy, or **chemotaxonomy** as it is also called, involves molecular analyses of one or more constituents in the cell. Chief among chemotaxonomic methods that have been used routinely are *genomic DNA:DNA hybridization, ribotyping, multilocus sequence typing,* and *lipid profiling.*

DNA:DNA Hybridization

A GC base ratio describes the percentage of each nucleotide present in genomic DNA of a given species but gives absolutely no information on the *sequence* of those nucleotides. Sequences are critical, of course, because if two organisms have many of the same nucleotide sequences in their DNAs, they likely contain many highly similar (if not identical) genes. Two DNAs would thus be expected to *hybridize* to one another in proportion to the *similarities* in their gene sequences. **Genomic hybridization** measures the degree of sequence similarity in two DNAs and is useful for differentiating very closely related organisms where rRNA sequencing may fail to be definitive.

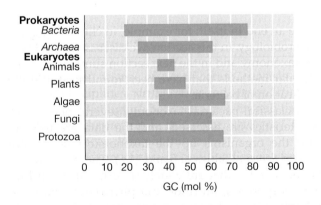

● **Figure 11.21 Ranges of genomic DNA base composition of various organisms.** Note that the greatest range of GC ratios exists with *Bacteria.*

We discussed the theory and methodology of nucleic acid hybridization in Section 7.7. In an actual hybridization experiment, DNA isolated from one organism is made radioactive with ^{32}P or ^3H, sheared to a relatively small size, heated to denature, and mixed with an excess of unlabeled DNA prepared in the same way from a second organism (Figure 11.22●). The DNA mixture is then cooled to allow it to reanneal and double-stranded DNA is separated from any remaining un-

hybridized DNA. Following this, the amount of radioactivity in the hybridized DNA is determined and compared with the control, which is taken as 100% (Figure 11.22). In addition to radioactivity, a variety of nonradioactive DNA labels are available, and these have the advantage that the hybridization experiment generates no radioactive wastes.

There is no fixed convention as to how much hybridization between two DNAs is necessary to assign two organisms to the same taxonomic rank. However, hybridization values of 70% or greater are recommended for considering two isolates to be of the same *species*. By contrast, values of at least 25% are required to argue that two organisms should reside in the same *genus* (see Section 11.12 for working definitions of a bacterial genus and species). Hybridization of DNAs from more distantly related organisms, for example, *Clostridium* (grampositive) and *Salmonella* (gram-negative), will hybridize at only background levels, 10% or less (Figure 11.22).

DNA:DNA hybridization is a sensitive method for revealing subtle differences in the genes of two organisms and is therefore useful for differentiating *closely related* organisms. Indeed, a common application of genomic hybridization in taxonomic studies is to test two organisms that are suspected to be different species even though SSU ribosomal RNA sequencing and phenotypic analyses may have failed to reveal significant differences between them.

Where available, complete genome sequences make hybridization analyses unnecessary. When the genome of a given bacterium is sequenced, its complete assortment of genes can be compared with those of any other sequenced bacterium (∞Chapter 15). At present, however, with only several hundred prokaryotic genomes completely sequenced, such comparisons are not possible in most cases. In future years, microbiologists will have the ultimate tool for molecular taxonomy in complete genome sequences. However, until taxonomy is based on complete genome sequence comparisons, hybridization will remain useful for differentiating close taxonomic relationships.

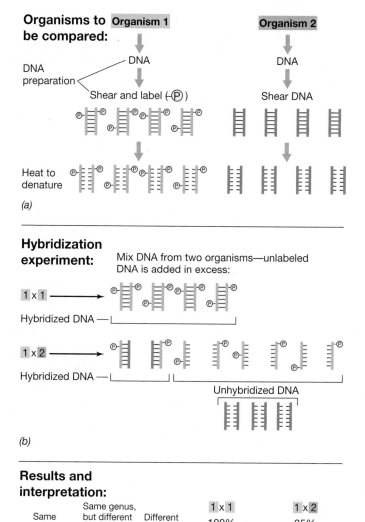

(a)

(b)

(c)

● **Figure 11.22   Genomic hybridization as a taxonomic tool.** (a) DNA is isolated from test organisms. One of the DNAs is labeled (shown here as radioactive phosphate in the DNA of Organism 1). (b) Actual hybridization experiment. Excess unlabeled DNA is added to prevent labeled DNA from reannealing with itself. Following hybridization, hybridized DNA is separated from unhybridized DNA before measuring radioactivity in the hybridized DNA only. (c) Results. Radioactivity in the control (Organism 1 DNA hybridizing to itself) is taken as the 100% hybridization value.

Ribotyping ⟵ restriction enzyme

Ribotyping is a technique for bacterial identification that employs some of the methods previously discussed for ribosomal RNA-based phylogenetic characterizations (see Section 11.6). However, unlike comparative *sequencing* methods, ribotyping does not involve sequencing. Instead, it measures the unique *pattern* that is generated when DNA from an organism is digested by a restriction enzyme and the fragments are separated and probed with a ribosomal RNA probe (Figure 11.23a●).

Because differences in ribosomal RNA sequences between two organisms translate into the presence or absence of specific restriction enzyme cut sites (see

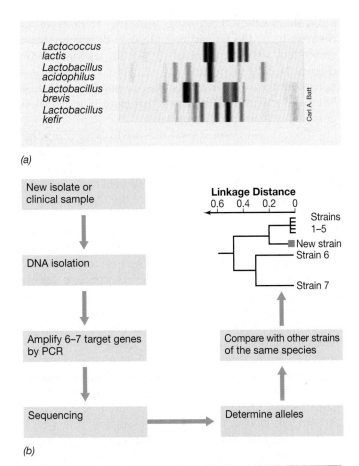

(a)

(b)

● **Figure 11.23 Ribotyping and multilocus sequence typing (MLST).** (a) Ribotype results from four different lactic acid bacteria. The pattern of DNA fragments generated from restriction enzyme digestion of DNA taken from a colony of each bacterium and then probed with 16S rRNA genes is unique to a species or even to strains within a species. The patterns generated on a gel with known organisms are digitized and stored in a database for comparisons in identifying environmental or clinical isolates. Variations in both *position* and *intensity* of the bands are important in identification. (b) Steps in multilocus sequence typing (MLST).

Section 7.7 for a discussion of restriction enzymes), the restriction pattern of a particular bacterial species is unique (Figure 11.23a). In fact, ribotyping is so specific it has been given the nickname "molecular fingerprinting," because a unique series of bands appears for virtually any organism.

In practice, ribotyping starts with DNA from a colony or liquid culture. Using PCR, genes for 16S rRNA and related molecules are amplified, treated with one or more restriction enzymes, separated by electrophoresis, and then probed (Chapter 7 for discussion of these methods). The pattern generated from the fragments of DNA on the gel is then digitized and a computer used to make comparisons of this pattern with patterns from reference organisms available from a database (Figure 11.23a). Ribotyping is both a *rapid* (since it bypasses the actual sequencing, sequence alignment, and analysis requirements of ribosomal RNA sequencing methods) and *specific* method of bacterial identification. For these reasons, ribotyping has found

many applications in clinical diagnostics and for the microbial analyses of food, water, and beverages.

Multilocus Sequence Typing

One of the limitations of both ribosomal RNA sequencing and ribotyping is that analyses focus on only a single gene. **Multilocus sequence typing** (**MLST**) circumvents this problem and is a powerful technique for characterizing strains of organisms within a species.

MLST involves sequencing fragments of six to seven "housekeeping genes" from an organism and comparing these with the same gene set from *different strains* of the same organism. Recall that housekeeping genes encode essential functions in cells and are located on the chromosome rather than on a plasmid (Section 7.4). For each gene, an approximately 450-bp sequence is amplified using PCR and is then sequenced. The comparative sequencing data is then expressed in a dendrogram (Figure 11.23b).

In MLST, strains with *identical sequences* for a given gene are said to have the same **allele** at that gene and are assigned the same number for that gene. A given gene can have many different alleles, and typically, 10–30 alleles exist for a given gene within a group of strains of the same species. Each strain eventually ends up with a series of numbers—its *multilocus sequence type*—characteristic of that strain. The relatedness between each sequence type is then expressed in a dendrogram of linkage distances that vary from 0 (strains are identical) to 1 (strains are only distantly related) (Figure 11.23b).

MLST has sufficient resolving power to distinguish even very closely related strains. It is thus a better instrument than rRNA sequencing for differentiating organisms below the species level. With seven genes in the analysis and 20 alleles per gene it has been calculated that *several billion* distinct genotypes can be resolved with MLST. By contrast, MLST is not useful for comparing organisms above the species level because its resolution is too sensitive to yield a meaningful phylogeny of higher order taxa.

Thus far, MLST has made an impact primarily in clinical microbiology for differentiating strains of a particular pathogen. This is serious business when it is considered that within a species—*Escherichia coli*, for example—some strains, such as strain K-12, may be harmless, whereas others, such as strain O157:H7, can cause serious and even fatal infections (Section 29.8). MLST is also quite useful for epidemiological studies. For example, MLST can track a virulent strain of a bacterial pathogen as it moves through a population and can unequivocally identify the strain even if many closely related strains are also present in the population. In addition, since the technique is PCR based, MLST can be used to test clinical samples without first isolating the organism in laboratory culture. That is, if DNA from a strain of a specific pathogen can be obtained from a clinical sample, MLST analyses can be done.

MLST analyses have revealed some interesting genetic patterns in bacteria. For example, some prokaryotes are essentially clonal, with very little MLST variability apparent. This is typical of strains of *Staphylococcus aureus*, for example. Other organisms, such as *Neisseria meningititis*, are only weakly clonal, and show great variability by MLST analyses. These results indicate that for whatever reason, organisms like *N. meningitidis* have undergone much greater lateral gene flow in the past than has *S. aureus*. These differences undoubtedly reflect differences in the ecology of these organisms and the factors that make them competitively successful. Such information may be useful for developing drug strategies and vaccines based on the genetic stability or instability of a particular pathogen.

Fatty Acid Analyses: FAME

Another popular method of bacterial identification is through characterization of the types and proportions of *fatty acids* present in cytoplasmic membrane and outer membrane (gram-negative bacteria, ∞Section 4.9) lipids of cells. This technique has been nicknamed *FAME*, for *fatty acid methyl ester* and is in widespread use in clinical, public health, and food and water inspection laboratories where the identification of pathogens or other bacterial hazards needs to be done on a routine basis.

The fatty acid composition of prokaryotes can be highly variable, including differences in the fatty acids in terms of their chain length, the presence or absence of double bonds, rings, branched chains, or hydroxy groups (Figure 11.24*a*●). Hence, a fatty acid profile can often identify a particular bacterial species. For actual analyses, fatty acids, extracted from cell hydrolysates of a culture grown under standardized conditions, are chemically derivatized to form their corresponding methyl esters. These now volatile derivatives are then identified by gas chromatography. A chromatogram showing the types and amounts of fatty acids from the unknown bacterium is then compared with a database containing the fatty acid profiles of thousands of reference bacteria grown under the same conditions. The best matches to that of the unknown are then selected (Figure 11.24*b*).

As a chemotaxonomic tool, FAME does have some drawbacks. In particular, FAME analyses require rigid standardization, since fatty acid profiles of an organism can vary as a function of temperature, growth phase (exponential versus stationary), and to a lesser extent, growth medium. Thus, for consistent results, it is necessary to grow the unknown organism on a specific medium and at a specific temperature in order to compare its fatty acid profile with those of organisms from the database that have been grown in the same way. For many organisms, this is impossible, of course, and thus FAME analyses are limited to those organisms that can be grown under the specified conditions.

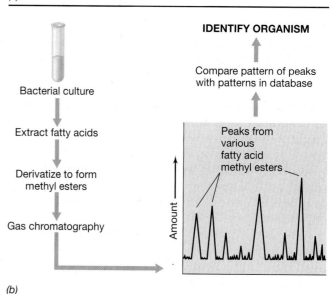

Classes of Fatty Acids in *Bacteria*

Class/Example	Structure of example
I. Saturated: tetradecanoic acid	
II. Unsaturated: omega-7-cis hexadecanoic acid	
III. Cyclopropane: *cis* 7, 8 methylene hexadecanoic acid	
IV. Branched: 13-methyltetradecanoic acid	
V. Hydroxy: 3-hydroxytetradecanoic acid	

(a)

IDENTIFY ORGANISM

Compare pattern of peaks with patterns in database

Bacterial culture

Extract fatty acids

Derivatize to form methyl esters

Gas chromatography

Peaks from various fatty acid methyl esters

Amount

(b)

● **Figure 11.24 Fatty acid methyl ester (FAME) analysis in bacterial identification.** (a) Classes of fatty acids in *Bacteria*. Only a single example is given of each class, but in actuality, more than 200 different fatty acids have been discovered from bacterial sources. A methyl ester contains a methyl group (CH_3) in place of the proton on the carboxylic acid group (COOH) of the fatty acid. (b) Procedure. Each peak from the gas chromatograph is due to one particular fatty acid methyl ester and the peak height is proportional to the amount.

🛑 *11.11 Concept Check*

Molecular taxonomy includes molecular analyses of specific cell components. These include, among others, DNA:DNA hybridization, ribotyping, multilocus sequence typing, and fatty acid analyses.

◆ Hybridization of less than 10% between two organisms' DNA indicates that they are _____.

◆ How does ribotyping differ from 16S ribosomal RNA gene sequencing? From MLST?

◆ What is FAME analysis?

11.12 The Species Concept in Microbiology

In the world of plants and animals, a **species** is defined as a population (1) that can naturally interbreed and produce fertile offspring, and (2) that is reproductively isolated from other species. However, this definition does not hold for prokaryotes. Prokaryotes are haploid and reproduce asexually. Concepts like "the production of fertile offspring" thus have no meaning. Nevertheless, microbiologists traditionally refer to "species" of bacteria and regularly give new isolates of bacteria genus names and species names, the latter called *epithets*.

So what *is* a bacterial species? Although the concept of a prokaryotic species is still evolving, microbiologists today are using SSU ribosomal RNA sequencing, genomic hybridization, and the phenotypic tools discussed in Section 11.11 for discerning prokaryotic species.

Bacterial Species and Higher Taxa

It has been proposed that a prokaryote whose 16S ribosomal RNA sequence differs by more than 3% from that of all other organisms (that is, the sequence is less than 97% identical to any other sequence in the databases), should be considered a new species. This is not an arbitrary number. Support for this proposal includes the observation that genomic DNA from two prokaryotes whose 16S rRNA sequences are less than 97% identical typically hybridize to less than 70%, a minimal value considered evidence for two organisms being of the same species (see Section 11.11 and Figure 11.22). This is depicted in Figure 11.25●.

The data of Figure 11.25 also shows how in some cases 16S sequencing lacks the resolving power of genomic DNA hybridization for differentiating organisms at the species level. Although an organism whose 16S sequence differs from that of all other organisms by more than 3% will likely turn out to be a new species by DNA hybridization criterion as well, some organisms that have very similar (or even identical) ribosomal RNA sequences have genomes that are quite unrelated (Figure 11.25). Thus, in cases where SSU sequencing shows greater than 97% sequence identity, genomic hybridization is an important taxonomic tool for identifying new species.

A new species is usually defined from the characterization of several strains. The species concept is important in microbiology because it gives the collected strains formal taxonomic identity. Groups of species are then collected into genera (singular, **genus**). What constitutes a genus by molecular criteria is more a matter of judgment than for species, but 16S sequence *differences* of more than 5% from all other organisms (this corresponds to less than 95% sequence *identity*) have been taken as support for an organism constituting a new genus. Groups of genera are collected into **families**, families into **orders**, orders into **classes**, and so on up to the high-

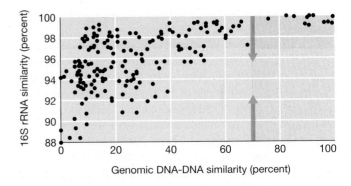

● **Figure 11.25 Relationship between 16S ribosomal RNA sequence similarity and genomic DNA:DNA hybridization between different pairs of organisms.** These data are the results from several independent experiments with various species of the domain *Bacteria*. Points in the orange box represent combinations where 16S sequence similarity and genomic hybridization were both very high; thus, in each case, the two organisms tested were clearly the same species. Points in the green box represent combinations where different species were detected, and both methods show this. The blue box shows examples of organisms that were different species as measured by genomic DNA hybridization but not by 16S sequencing. Note that above 70% DNA hybridization, no 16S similarities were found of less than 97%. Data redrawn from Rosselló-Mora, R., and R. Amann. 2001. *FEMS Microbiol. Revs.* 25:39–67.

est level taxon, the **domain** (Tables 11.4 and 11.5). Almost 6500 species of *Bacteria* and *Archaea* were formally recognized as of 2004 (Table 11.5).

In the identification and naming of a newly isolated organism, it is essential that the organism satisfy all the taxonomic criteria of ranks *above* its species designation. Thus, in the example given in Table 11.4 where the taxonomic hierarchy for the phototrophic bacterium *Allochromatium warmingii* is shown, all species of the genus *Allochromatium* must be rod-shaped purple sulfur gram-negative *Bacteria*; if one of these criteria is not satisfied, the organism is not considered a species of *Allochromatium*. Moreover, the genus *Allochromatium* must meet all of the criteria defining the family Chromatiaceae, and so on up the taxonomic ladder. In other words, as one *descends* the taxonomic hierarchy from the level of domain to that of species, the criteria used to distinguish two organisms become *less general* and *more specific* (Table 11.4).

Bacterial Speciation

How do new bacterial species arise? This is currently a hotly debated point among microbiologists. Nevertheless, models of bacteria speciation have been proposed, and we consider a popular one here. Imagine a population of cells originating from the growth of a single cell and that occupies a particular ecological niche. If cells in this population share a particular resource (for example, a key nutrient), this population of cells can be called an

Table 11.5 Taxonomic hierarchy for the purple sulfur bacterium *Allochromatium warmingii*

Taxonomic division	Name	Properties	Confirmed by
Domain	*Bacteria*	Prokaryotic cells; ribosomal RNA sequences typical of *Bacteria*	Microscopy; 16S ribosomal RNA sequencing; presence of unique biomarkers, for example, peptidoglycan
Phylum	Proteobacteria	Ribosomal RNA sequence typical of Proteobacteria	16S rRNA sequencing
Class	Gammaproteobacteria	Gram-negative bacteria; rRNA sequence typical of Gammaproteobacteria	Gram-staining, microscopy
Order	Chromatiales	Phototrophic purple bacteria	Characterizing pigments (⚬⚬Figure 17.3)
Family	Chromatiaceae	Purple sulfur bacteria	Ability to oxidize H_2S and store S^0 within cells; observe culture microscopically for presence of S^0 (see photo)
Genus	*Allochromatium*	Rod-shaped purple sulfur bacteria	Microscopy (see photo)
Species	*warmingii*	Cells 3.5–4.0 μm × 5–11 μm; store sulfur mainly in poles of cell (see photo)	Measure cells in microscope using a micrometer; look for position of S^0 globules in cells (see photo)

Sulfur (S^0) globules

Photograph of cells of *Allochromatium warmingii*

Norbert Pfennig

ecotype. Different ecotypes can coexist in a habitat, but each is only successful within its niche in the habitat.

Errors in DNA replication (mutations) occur at a low but regular frequency (⚬⚬Section 10.3). Within a given ecotype, the occurrence of a mutation that confers increased fitness on the ecotype (an adaptive mutation), triggers periodic selection. When this occurs, the old ecotype becomes purged by a population of the new ecotype (Figure 11.26●). In this way, populations of cells eventually "move away" from each other in a genetic sense. Repeated rounds of mutation and selection eventually lead to an ecotype that is sufficiently distinct genetically from the original ecotype to be recognized as a new species (Figure 11.26). Note that this series of events *within* an ecotype has no effect on *other* ecotypes, since different ecotypes do not compete for the same resources (Figure 11.26).

The model for speciation shown in Figure 11.26 is based solely on the assumption of *vertical* (mother to daughter) gene flow. However, we know that bacterial speciation is also affected to some degree by **lateral (horizontal) gene transfer**. Lateral flow is the transfer of genes *between* species by conjugation, transduction, and transformation (⚬⚬Chapter 10). Prokaryotes are sexually promiscuous and can exchange genes across broad phylogenetic lines. Thus, a new genetic capability in an ecotype may arise from genes obtained from cells of another ecotype rather than from mutation and selection.

The extent of horizontal gene flow among prokaryotes is variable. Genome sequencing has revealed examples where lateral flow has apparently been rampant and others where it has been scarce (⚬⚬Section 15.8). In addition, multilocus sequence typing (see Section 11.11) has revealed that genetic exchange *within* some species is widespread, while virtually nonexistent within others. Despite the impact of lateral gene flow, *speciation* is thought to be driven primarily by mutation and periodic selection (Figure 11.26) rather than by lateral transfer. This is because the genes obtained by lateral transfers are typically few in number, confer only temporary benefits, and can often be lost if the selective pressure to retain them decreases.

How Many Prokaryotic Species Are There?

The result of nearly 4 billion years of bacterial evolution (see Figures 11.8 and 11.10) is the prokaryotic world we see today. Microbial taxonomists agree that no firm estimate of the number of prokaryotic species can be given at present. However, they also agree that in the final analysis, this number will be very large. Several thou-

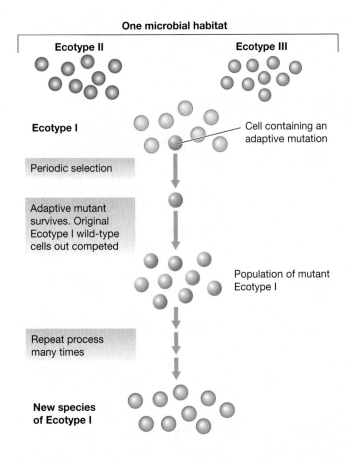

One microbial habitat

Ecotype II   Ecotype III

Ecotype I

 Cell containing an adaptive mutation

Periodic selection

Adaptive mutant survives. Original Ecotype I wild-type cells out competed

Population of mutant Ecotype I

Repeat process many times

New species of Ecotype I

● **Figure 11.26 A model for bacterial speciation.** Several ecotypes can coexist in a single microbial habitat, each occupying their own prime ecological niche. When a beneficial mutation occurs within an ecotype, the cell containing that mutation will eventually form a population that will replace the original ecotype. As this occurs repeatedly within a given ecotype, a genetically distinct population of cells arises that represents a new species. Because other ecotypes do not compete for the same resources, they are unaffected by genetic and selection events that occur outside of their ecotype or habitat.

Table 11.6	Taxonomic ranks and numbers of known prokaryotic species[a]		
Rank	*Bacteria*	*Archaea*	**Total**
Domains	1	1	2
Phyla	25	4[a]	29
Classes	34	9	43
Orders	78	13	91
Families	230	23	243
Genera	1227	79	1306
Species	6740	289	7029

[a] Numbers represent validly named genera and species of *Bacteria* and *Archaea* as of 2005. The phyla category for *Archaea* includes the Korarchaeota and the Nanoarchaeota, not yet officially recognized phyla.

Source: Garrity, G.M., Libum, T.G., and Bell, J.A. 2005. *Bergey's Manual of Systematic Bacteriology*, 2d ed., Vol. 2, part A, pp159–220. Springer-Verlag, New York.

11.12 Concept Check

The species concept applies to prokaryotes as well as eukaryotes, and a similar taxonomic hierarchy exists, with the domain as the highest level taxon. Bacterial speciation may occur from a combination of repeated periodic selection for a favorable trait within an ecotype and lateral gene flow.

◆ If a given genus includes several _____, a _____ includes several genera.

◆ What is an ecotype?

◆ How many species of prokaryotes are already known? How many likely exist?

11.13 Nomenclature and *Bergey's Manual*

Following the **binomial system** of nomenclature used throughout biology, prokaryotes are given genus names and species epithets. The terms used are Latin or Latinized Greek derivations of some descriptive property appropriate for the organism and are set in print in *italics*. For example, over 100 species of the genus *Bacillus* have been described, including *Bacillus (B.) subtilis*, *B. cereus*, and *B. megaterium*. These species epithets mean "slender," "waxen," and "big beast," respectively, and refer to key morphological, physiological, or ecological traits characteristic of each organism.

The nomenclature of prokaryotes, *Bacteria* as well as *Archaea*, is regulated by the rules of the Bacteriological Code—*The International Code of Nomenclature of Bacteria*. The Code presents the formal framework by which prokaryotes are to be officially named, and the procedures by which existing names can be changed, for example, when new data warrants taxonomic rearrangements. There are even rules for rejecting names if errors were made in the original naming process or a name has otherwise become invalid. The *Code* covers the rules for the naming of all species, genera, families, and orders of prokaryotes.

sand prokaryotic species are already known (Table 11.6), and several thousands more, perhaps as many as 100,000–1,000,000 in total (or 10 times this by some estimates) are suspected to exist. By anyone's count, the final number of prokaryotic species will likely be enormous.

Microbial community analyses (see Sections 11.7 and 18.5) indicate that we have only scratched the surface in our ability to *culture* the diversity of prokaryotes in nature. With more exacting tools, both molecular and cultural, for revealing prokaryotic diversity, it is possible that the large numbers of species already predicted will be an underestimate, perhaps by several fold. The reality today is that an accurate estimate of prokaryotic speciation is simply out of reach of current technology. But as with other things in microbiology, this will likely change. Obviously, much exciting work is left for those whose interests include prokaryotic diversity!

Culture Collections and Publication of New Taxa

When a new organism is isolated and thought to be unique, a decision must be made as to whether it is sufficiently different from other species to be described as novel, or perhaps even sufficiently different from all described genera to warrant description as a new genus (in which case a species is automatically created). To achieve formal taxonomic standing as a new genus or species, a detailed description of the isolate and the proposed name is published, and a viable culture of the organism is deposited in two international culture collections. Examples of the latter include the American Type Culture Collection (ATCC, Manassas, Virginia, USA) or the Deutsche Sammlung von Mikroorganismen und Zellkulturen (DSMZ, German Collection for Microorganisms, Braunschweig, Germany) (∞Section 30.1). The deposited strain becomes the *type* strain of the new species and the standard by which other strains thought to be the same can be compared.

Culture collections preserve the deposited culture, typically by freezing at very low temperatures (−80 to −196°C), or by freeze-drying. This practice differs from the botanical or zoological approach to taxonomy. These disciplines employ preserved (dead) specimens (either dried herbarium material or chemically fixed animal specimens) as the basis for comparison with proposed new species. By contrast, microbiologists have always relied on a *living type strain* that can be distributed to the scientific community, grown in different laboratories, and studied and compared. This approach allows for more detailed and reproducible comparisons of the properties of organisms, especially at the molecular level.

If the description of a new organism is published in a journal other than the *International Journal of Systematic and Evolutionary Microbiology (IJSEM)*, the official publication of record for the taxonomy and classification of prokaryotes, a copy of the published paper must be submitted to this journal and the name validated before it is formally accepted as a new taxon of prokaryotes. In each issue the *IJSEM* publishes an approved list of newly created names and serves as the publication of record for research in prokaryotic taxonomy. A website dedicated to tracking new and official names of prokaryotes can be found at (www.bacterio.cict.fr).

Bergey's Manual and *The Prokaryotes*

By validating newly proposed names, *IJSEM* paves the way for their inclusion in *Bergey's Manual of Systematic Bacteriology*, a major taxonomic treatment of prokaryotes (Figure 11.27●). Widely used, *Bergey's Manual* has served the community of microbiologists since 1923 and is a compendium of information on all recognized species of prokaryotes. Each chapter, written by an expert, contains tables, figures, and other systematic information useful for identification purposes. Volume I of the second edition of *Bergey's Manual* appeared in 2001, Volume 2 in 2005, and three additional volumes will appear by 2007.

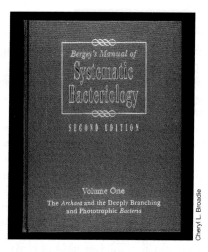

● **Figure 11.27** ***Bergey's Manual of Systematic Bacteriology, second edition.*** In five volumes this source describes the major properties of all known prokaryotes, both *Bacteria* and *Archaea*.

The second edition of *Bergey's Manual* has incorporated many of the concepts that have emerged from ribosomal RNA sequencing and genomic studies and blends this with a wealth of phenotypic information as well.

A second major reference in prokaryotic diversity is *The Prokaryotes*. This work, with more than 4100 pages (four volumes) in its second edition (1992), is now available in a third edition online (http://141.150.157.117:8080/prokPUB/index.htm). The electronic edition incorporates frequent updates to reflect the rapid pace at which new data appear on prokaryotic taxonomy and phylogeny.

Collectively, *Bergey's Manual* and *The Prokaryotes* offer microbiologists the foundations as well as the details of prokaryotic taxonomy and phylogeny as we know it today and are typically the "go-to" sources for microbiologists characterizing newly isolated prokaryotes.

As we will see in Chapter 18, there are many prokaryotes in nature that have thus far eluded laboratory culture. These are probably not "unculturable" prokaryotes, but simply ones that we know too little about to culture at present. However, when these organisms are finally cultured, characterized, and named, *Bergey's Manual* and *The Prokaryotes* will have to expand dramatically to accomodate what will likely be a very large number of prokaryotic species.

 11.13 Concept Check

Prokaryotes are given descriptive genus names and species epithets. Formal recognition of a new prokaryotic species requires depositing a sample of the organism in a culture collection and official publication of the new species name and description. *Bergey's Manual of Systematic Bacteriology* is a major taxonomic compilation of *Bacteria* and *Archaea*.

◆ What is the *IJSEM* and what taxonomic function does it fulfill?

◆ Why might living cell material be of more use in taxonomy than preserved specimens?

WORKING GLOSSARY

Acid-fastness a property of *Mycobacterium* species in which cells stained with the dye basic fuchsin resist decolorization with acidic alcohol

Carboxysome a polyhedral cellular inclusion of crystalline ribulose bisphosphate carboxylase (RubisCO), the key enzyme of the Calvin cycle

Chemolithotroph an organism able to oxidize inorganic compounds (such as H_2, Fe^{2+}, S^0, or NH_4^+) as energy sources (electron donors)

Chlorosome a cigar-shaped structure bounded by a nonunit membrane and containing the light-harvesting bacteriochlorophyll (*c*, *d*, or *e*) in green bacteria and *Chloroflexus*

Consortium two- or more-membered association of prokaryotes, usually living in an intimate symbiotic fashion

Cyanobacteria prokaryotic oxygenic phototrophs that contain chlorophyll *a* and phycobilins but not chlorophyll *b*

Enteric bacteria a large group of gram-negative rod-shaped *Bacteria* characterized by a facultatively aerobic metabolism

Green sulfur bacteria anoxygenic phototrophs containing chlorosomes and bacteriochlorophyll *c*, c_s, *d*, or *e* as light-harvesting chlorophyll

Heliobacteria anoxygenic phototrophs containing bacteriochlorophyll *g*

Heterocyst a differentiated cyanobacterial cell that carries out nitrogen fixation but not oxygenic photosynthesis

Heterofermentative in reference to lactic acid bacteria, capable of making more than one fermentation product

Homofermentative in reference to lactic acid bacteria, producing only lactic acid as a fermentation product

Hyperthermophile an organism with a growth temperature optimum of greater than 80°C

Methanotroph an organism capable of oxidizing methane (CH_4) as an electron donor in energy metabolism

Methylotroph an organism capable of oxidizing organic compounds that do not contain carbon-carbon bonds; if able to oxidize CH_4, also a methanotroph

Mixotroph an organism that can conserve energy from the oxidation of inorganic compounds but requires organic compounds as a carbon source

Nitrifying bacteria chemolithotrophs capable of carrying out the transformation $NH_3 \rightarrow NO_2^-$, or $NO_2^- \rightarrow NO_3^-$

Prochlorophyte a prokaryotic oxygenic phototroph that contains chlorophylls *a* and *b* but lacks phycobilins

Prosthecae an extrusion of cytoplasm, often forming a distinct appendage, bounded by the cell wall

Proteobacteria a major lineage of *Bacteria* that contains a large number of gram-negative rods and cocci

Purple nonsulfur bacteria a group of phototrophic prokaryotes containing bacteriochlorophylls *a* or *b* that grows best as photoheterotrophs and has a relatively low tolerance for H_2S

Purple sulfur bacteria a group of phototrophic prokaryotes containing bacteriochlorophylls *a* or *b* and characterized by the ability to oxidize H_2S and store elemental sulfur inside the cells (or in the genera *Ectothiorhodospira* and *Halorhodospira*, outside the cell)

Spirochete a slender, tightly coiled gram-negative prokaryote characterized by possession of endoflagella used for motility

Stickland reaction fermentation of an amino acid pair in which one amino acid serves as an electron donor and a second serves as an electron acceptor

Sulfate-reducing bacteria a large group of anaerobic *Bacteria* that respire anaerobically with SO_4^{2-} as electron acceptor, producing H_2S

THE PHYLOGENY OF BACTERIA

In Chapter 11 we stressed the evolutionary relationships among microorganisms. In this and the next two chapters we expand on these concepts with a discussion of properties of the major microbial groups themselves. In this chapter we focus on species of _Bacteria_, while in the next two chapters we focus on _Archaea_ and microbial _Eukarya_, respectively.

With nearly 7000 species of prokaryotes known, obviously we will not be able to consider them all. So, using a phylogenetic tree as the focus of our discussion, we will explore some of the best-known cultured species, particularly ones in which much phenotypic information is available. For more detailed information on prokaryotic diversity the student should refer to _Bergey's Manual of Systematic Bacteriology_ and _The Prokaryotes_ (⌀⌀ Section 11.13).

12.1 Phylogenetic Overview of _Bacteria_

At least 18 major lineages (phyla) of _Bacteria_ are known from the study of laboratory cultures, and many others have been identified from retrieval and sequencing of ribosomal RNA genes in natural habitats. Figure 12.1● gives a phylogenetic overview of _Bacteria_. The most phylogenetically ancient (least derived) phylum contains the genus _Aquifex_ and relatives, all of which are hyperthermophilic H$_2$-oxidizing chemolithotrophs. Other "early" phyla such as _Thermodesulfobacterium_, _Thermotoga_, and the green nonsulfur bacteria (the _Chloroflexus_ group), also contain thermophilic species.

Continuing past the green nonsulfur bacteria, we see the deinococci and relatives, the morphologically unique spirochetes, the phototrophic green sulfur bacteria, the chemoorganotrophic _Flavobacterium_ and _Cytophaga_ groups, the budding _Planctomyces-Pirella_ and the _Verrucomicrobium_ groups, the _Chlamydia_, and the genera _Nitrospira_ and _Deferribacter_ (Figure 12.1). Each of these groups is discussed in this chapter.

The remaining phyla of cultured _Bacteria_ constitute the major thrust of this chapter. They include the gram-positive bacteria, the cyanobacteria, and the Proteobacteria. Each of these is a large group containing many genera and are _Bacteria_ about which much phenotypic information is known. The gram-positive bacteria can be separated into two subgroups, called _low_ GC and _high_ GC (Actinobacteria), the terms referring to the fact that the species tend to have DNA GC base ratios (see Section 11.10) either well below or well above 50%, respectively. The gram-positive bacteria are a large group of primarily chemoorganotrophic _Bacteria_ and are discussed in detail

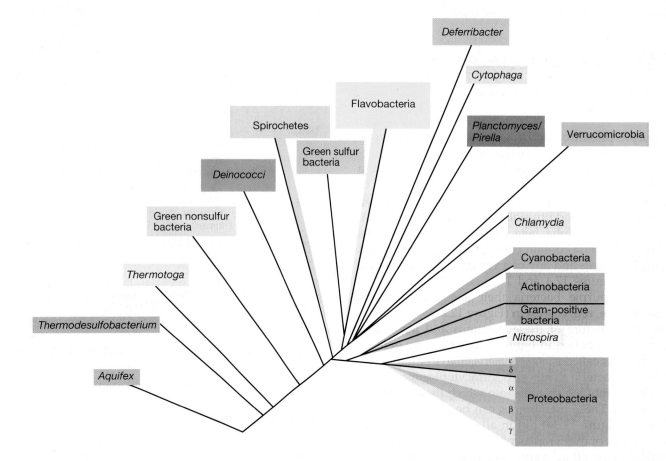

● **Figure 12.1   Detailed phylogenetic tree of the major lineages (phyla) of _Bacteria_ based on 16S ribosomal RNA sequence comparisons.** Including many phyla known only from community sampling (⌀⌀ Section 11.7), over 40 phyla of _Bacteria_ actually exist.

Photobacterium and Bioluminescence

A number of gram-negative, polarly flagellated rods possess the interesting property of emitting light, a process called *bioluminescence*. Most of these bacteria have been classified in the genus *Photobacterium*, but a few *Vibrio* isolates are also bioluminescent (Figure 12.28●). Most **bioluminescent bacteria** are marine, usually found associated with fish. Some fish possess a special organ in which bioluminescent bacteria grow (Figure 12.28c–f). Other bioluminescent marine bacteria live saprophytically on dead fish and occasionally form visible colonies on the fish surface. (To see bioluminescence readily, one should observe the material in a completely dark room after the eyes have become adapted to the dark, Figure 12.28a, b.)

Although *Photobacterium* isolates are facultative aerobes, they are bioluminescent only when O_2 is present. Several components are needed for bioluminescence. These include the enzyme **luciferase** and a long-chain aliphatic aldehyde (for example, *dodecanal*). Reduced flavin mononucleotide ($FMNH_2$) and O_2 are also required for bioluminescence. The primary electron donor is NADH, and the electrons pass through luciferase as follows:

$$FMNH_2 + O_2 + RCHO \xrightarrow{\text{Luciferase}}$$

$$FMN + RCOOH + H_2O + \text{Light}$$

The light-generating system constitutes a bypass route for shunting electrons from $FMNH_2$ to O_2, without involving other electron carriers such as quinones and cytochromes.

Regulation of Bioluminescence

The enzyme luciferase shows a unique kind of regulatory synthesis called **autoinduction**. The luminous bacteria produce a specific organic molecule, the *autoinducer*, which accumulates in the culture medium during growth. When the amount of this substance reaches a critical level, induction of luciferase occurs. The auto inducer in *Vibrio fischeri* has been identified as N-β-ketocaproyl homoserine lactone. Thus, cultures of luminous bacteria at *low* cell density are not luminous but only become luminous when growth reaches a sufficiently *high* density that the autoinducer can accumulate and function. This is a mechanism called **quorum sensing**, because of the density-dependent nature of the phenomenon (∞Section 8.10).

Because the autoinducer cannot accumulate, free-living luminescent bacteria in seawater are not luminous, and luminescence develops only when conditions are favorable for the development of high population densities (Figure 12.28). Although it is not clear why luminescence is density-dependent in free-living bacteria, in symbiotic strains of luminescent bacteria (see

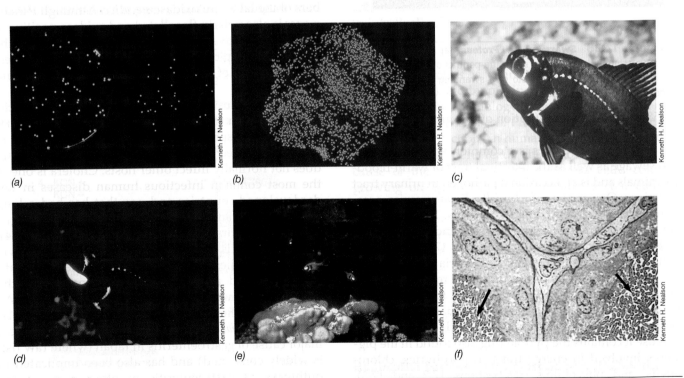

● **Figure 12.28   Bioluminescent bacteria and their role as light organs in the flashlight fish.** (a) Two Petri plates of luminous bacteria photographed by their own light. Note the different colors. Left, *Vibrio fischeri* strain MJ-1, blue light, and right, *V. fischeri* strain Y-1, green light. (b) Colonies of *Photobacterium phosphoreum* photographed by their own light. (c) The flashlight fish *Photoblepharon palpebratus*; the bright area is the light organ containing bioluminescent bacteria. (d) Same fish photographed by its own light. (e) Underwater photograph taken at night of *P. palpebratus* in coral reefs in the Gulf of Eilat. (f) Electron micrograph of a thin section through the light-emitting organ of *P. palpebratus*, showing the dense array of bioluminescent bacteria (arrows).

Figure 12.28*c–f*) the rationale for density-dependent luminescence is clear: luminescence develops only when sufficiently high population densities are reached in the light organ of the fish to allow a visible flash of light. The genetics of quorum sensing using bioluminescence as a model experimental system has been actively explored, and it is now clear that this form of regulation is not limited to bioluminescent bacteria, but instead is a general regulatory feature of a number of different bacteria for phenomena in which a minimum cell density (a "quorum") is required (∞ Section 8.10).

12.11–12.12 Concept Check

The enteric bacteria are a large group of facultative aerobic rods of medical and molecular biological significance. *Vibrio* and *Photobacterium* species are marine organisms; some species are pathogenic while others are bioluminescent.

◆ How is *Escherichia coli* distinguished from *Enterobacter aerogenes* based on physiology?

◆ Describe two major properties of *Proteus* species that distinguish them from other enteric bacteria.

◆ What is necessary for an organism like *Photobacterium* to give off visible light?

12.13 Rickettsias

Key Genera: *Rickettsia, Wolbachia*

The rickettsias are small, gram-negative, coccoid or rod-shaped Proteobacteria in the size range of 0.3–0.7 μm wide and 1–2 μm long. They are, with one exception, *obligate intracellular parasites* (Figure 12.29*a*●) and have not yet been cultivated in the absence of host cells. Rickettsias are the causative agents of several human diseases including typhus fever, Rocky Mountain spotted fever, and Q fever (∞ Section 27.3).

Electron micrographs of thin sections of rickettsias show cells with a normal bacterial morphology (Figure 12.29*b*); both cell wall and cell membrane are clearly present. The cell wall contains muramic acid and diaminopimelic acid. Both RNA and DNA are present, and the rickettsias divide by normal binary fission, with doubling times of about 8 h. The penetration of a host cell by a rickettsial cell is an active process, requiring both host and parasite to be alive and metabolically active. Once inside the host phagocytic cell, the bacteria multiply primarily in the cytoplasm and continue replicating until the host cell is loaded with parasites (see Figures 12.29 and Figure 27.3). The host cell then bursts and liberates the bacteria into the surrounding fluid. Several genera of rickettsias are known, and the properties of four key genera are shown in Table 12.17.

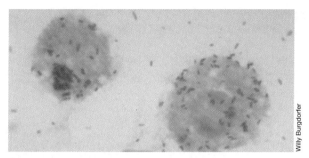

(a)

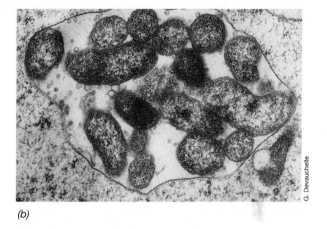

(b)

● **Figure 12.29 Rickettsias growing within host cells.** (a) *Rickettsia rickettsii* in tunica vaginalis cells of the vole, *Microtus pennsylvanicus*. Cells are about 0.3 μm in diameter. (b) Electron micrograph of cells of *Rickettsiella popilliae* within a blood cell of its host, the beetle *Melolontha melolontha*. Notice that the bacteria are growing in a vacuole within the host cell.

Metabolism and Pathogenesis

Much research has been done on the metabolic activities and biochemical pathways of rickettsias in an attempt to explain why they are obligate intracellular parasites. Many rickettsias possess a highly specific energy metabolism: they can oxidize only glutamate or glutamine and cannot oxidize glucose or organic acids. However, *Coxiella burnetii*, the causative agent of the disease Q fever, is able to utilize both glucose and pyruvate as electron donors. Rickettsias possess a respiratory chain complete with cytochromes and are able to carry out electron transport phosphorylation, using NADH as electron donor. They are also able to synthesize at least some of the small molecules needed for macromolecular synthesis and growth, and they obtain the rest of their nutrients from the host cell. Thus, although parasites, rickettsias maintain a number of independent metabolic functions.

Rickettsias do not survive long outside their hosts, and this may explain why they must be transmitted from animal to animal by arthropod vectors. When the arthropod obtains a blood meal from an infected vertebrate, rickettsias present in the blood are inoculated directly into the arthropod, where they penetrate to the

Table 12.17 **Characteristics of rickettsias**

Genus and Species	Rickettsial group	Alternate host	Cellular location	DNA (mol % GC)	Phylogenetic group[a]	DNA hybridization to *R. rickettsii* DNA (%)[b]
Rickettsia						
R. rickettsii	Spotted fever	Tick	Cytoplasm and nucleus	32–33	Alpha	100
R. prowazekii[c]	Typhus	Louse	Cytoplasm	29–30		53
R. typhi	Typhus	Flea	Cytoplasm	29–30		36
Rochalimaea						
R. quintana	Trench fever	Louse	Epicellular	39	Alpha	30
R. vinsonii	—	Vole	Epicellular	39		30
Coxiella						
C. burnetii	Q fever	Tick	Vacuoles	43	Gamma	—
Ehrlichia						
E. chaffensis	Ehrlichiosis (humans)	Tick or	Mononuclear	—	Alpha	—
E. equi	Potomac fever (horses)	domestic animals	leukocytes	—	Alpha	—
Wolbachia[d]						
W. pipientis	—	Arthropods	Cytoplasm	30	Alpha	—

[a] All are Proteobacteria.

[b] For discussion of DNA:DNA hybridization, see Section 11.11.

[c] The genome of this organism has been sequenced and shows several similarities to the mitochondrial genome.

[d] Not a pathogen of humans or other animals.

epithelial cells of the gastrointestinal tract, multiply, and appear later in the feces. When the arthropod feeds on an uninfected individual, it then transmits the rickettsias either directly with its mouthparts or by contaminating the bite with its feces. However, *C. burnetii* (∞Section 27.3) can also be transmitted to the respiratory system by aerosols. *C. burnetii* is the most resistant of the rickettsias to physical damage, probably because it produces a resistant, sporelike form. This probably explains the ability of *C. burnetii* to survive in air.

Rochalimaea is an atypical rickettsia because it can be grown in culture and is thus not an obligate intracellular parasite. In addition, when growing in tissue culture, cells of *Rochalimaea* grow on the *outside surface* of the eukaryotic host cells rather than within the cytoplasm or the nucleus. *Rochalimaea quintana* is the causative agent of *trench fever*, a disease that decimated troops in World War I. Species of the genus *Ehrlichia* cause disease in humans and other animals, two of which, *ehrlichiosis* in humans and *Potomac fever* in horses, can be quite debilitating (∞Section 27.3).

Wolbachia

The genus *Wolbachia* contains species of rod-shaped alpha Proteobacteria that are intracellular parasites of arthropod insects (Figure 12.30●). *Wolbachia* are phylogenetically related to the rickettsias and can have any of several effects on their insect hosts. These include inducing parthenogenesis (development of unfertilized eggs), the killing of males, and feminization (the conversion of male insects into females).

Wolbachia pipientis is the best-studied species in the genus. Cells colonize the insect egg (Figure 12.30), where they multiply in vacuoles of host cells surrounded by a membrane of host origin. *W. pipientis* cells are passed from infected females to her offspring through this egg infection. *Wolbachia*-induced parthenogenesis occurs in a number of species of wasps. In these insects, males normally arise from unfertilized eggs (which contain only one set of chromosomes), while females arise from fertilized eggs (which contain two sets of chromosomes). However, in unfertilized eggs infected with *Wolbachia*, the organism somehow triggers a doubling of the chromosome number, thus yielding only females. Predictably, if female insects are fed antibiotics that kill *Wolbachia*, parthenogenesis ceases.

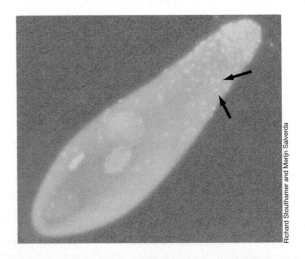

● **Figure 12.30** ***Wolbachia.*** Photomicrograph of a 4',6-diamidine-2' phenylindole dihydrochloride-stained (DAPI, ∞Section 18.3) egg of the parasitoid wasp, *Trichogramma kaykai* infected with *Wolbachia pipientis*, which induces parthenogenesis. The *W. pipientis* cells are primarily located in the narrow end of the egg (arrows).

In certain woodlice (pillbugs), a species of *Wolbachia* causes males to develop as females. Here the *Wolbachia* infection damages the male hormone-producing glands. In other insects, such as certain lady beetles and butterflies, *Wolbachia* infection somehow results in the death of male but not female offspring. In addition to these reproductive anomalies, in some insects *Wolbachia* infection may actually be *essential* for survival. For example, in the nematode worms that cause the diseases *elephantiasis* and *river blindness*, antibiotic treatment kills the worms, apparently by killing their *Wolbachia* symbionts.

Like parasitic prokaryotes in general, *W. pipientis* has a small genome (about 1.5 Mbp), which has now been sequenced. Although this parasite does not cause disease in either vertebrates or its invertebrate hosts, its various reproductive effects raise the interesting question of the extent it has affected evolution and speciation in a major class of insects (arthropods make up 70% of all known insects). However, because the association between *Wolbachia* and insects is so common, it could turn out that the reproductive effects are secondary to a true *Wolbachia*–insect symbiosis. That is, it is likely that by infecting insects, the *Wolbachia* cells benefit from a ready source of nutrients and a protected environment. Moreover, the insects may also benefit from the association in some as yet unknown ways.

 12.13 Concept Check

The rickettsias are obligate intracellular parasites, many of which cause disease. Rickettsias are deficient in many metabolic functions and obtain key metabolites from their hosts.

◆ Name a disease caused by a *Rickettsia* species.

◆ What is meant by the phrase "obligate intracellular parasite"?

◆ What effects can *Wolbachia* have on its insect hosts?

12.14 Spirilla

Key Genera: *Spirillum, Bdellovibrio, Campylobacter*

The **spirilla** are gram-negative, motile, spiral-shaped Proteobacteria that show a wide variety of physiological attributes. Some of the key taxonomic criteria used are cell shape, size, kind of polar flagellation (single or multiple), relation to oxygen (obligately aerobic, microaerophilic, facultative), relationship to plants (as symbionts or plant pathogens) or animals (as pathogens), fermentative ability, and certain other physiological characteristics (nitrogen-fixing ability, halophilic nature, thermophilic nature). The genera to be covered here are given in Table 12.18, where it can be seen that genera of spirilla can be found in each of the five subdivisions of Proteobacteria.

Spirillum, Aquaspirillum, Oceanospirillum, and *Azospirillum*

These helically curved rods are motile by means of polar flagella, usually tufts at both poles (see Figure 12.31b●). The number of turns in the helix may vary from less than one complete turn (in which case the organism looks like a vibrio) to many turns. Spirilla with many turns can superficially resemble spirochetes (see Section 12.33) but differ distinctly from the latter phylogenetically. In addition, spirilla do not have the outer sheath and endoflagella of spirochetes, but instead contain typical bacterial flagella.

Some spirilla are very large bacteria and were seen by early microscopists. It is likely that Antoni van Leeuwenhoek first observed a *Spirillum* species in the 1670s (∞Section 1.5), and the genus was first created by the

---

Table 12.18	**Characteristics of the genera of spiral-shaped bacteria**[a]		
Genus	**Phylogenetic group**[b]	**Characteristics**	**DNA (mol % GC)**
Spirillum	Beta	Cell diameter 1.7 μm; microaerophilic; freshwater	36–38
Aquaspirillum	Alpha or beta	Cell diameter 0.2–1.5 μm; aerobic; freshwater	49–66
Magnetospirillum	Alpha	Vibrio to spirillum-shaped; cell diameter about 0.3 μm; contains magnetosomes; microaerophilic	65
Oceanospirillum	Gamma	Cell diameter 0.3–1.2 μm; aerobic; marine (require 3% NaCl)	42–51
Azospirillum	Alpha	Cell diameter 1 μm; microaerophilic; soil and rhizosphere; fixes N_2	68–70
Herbaspirillum	Beta	Cell diameter 0.6–0.7 μm; microaerophilic; soil and rhizosphere; fixes N_2	66–67
Campylobacter	Epsilon	Cell diameter 0.2–0.8 μm; microaerophilic to anaerobic; pathogenic or commensal in humans and animals; single polar flagellum	30–38
Helicobacter	Epsilon	Cell diameter 0.5–1 μm; tuft of polar flagella; associated with pyloric ulcers in humans	36–38
Bdellovibrio	Delta	Cell diameter 0.25–0.4 μm; aerobic; predatory on other bacteria; single polar sheathed flagellum	33–52
Ancyclobacter	Alpha	Cell diameter 0.5 μm; curved rods forming rings; nonmotile, aerobic; sometimes gas-vesiculate	66–69

[a] All are gram-negative and respiratory but never fermentative.
[b] All are Proteobacteria.

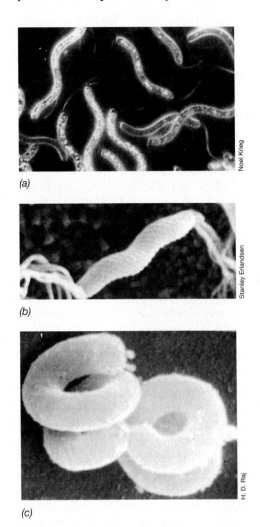

(a)

Noel Krieg

(b)

Stanley Erlandsen

(c)

H. D. Raj

● **Figure 12.31 Spirilla.** (a) *Spirillum volutans*, visualized by dark-field microscopy, showing flagellar bundles and volutin (polyphosphate) granules. Cells are about $1.5 \times 25\ \mu$m. (b) Scanning electron micrograph of an intestinal spirillum. Note the polar flagellar tufts and the spiral structure of the cell surface. (c) Scanning electron micrograph of cells of *Ancyclobacter aquaticus*. Cells are about $0.5\ \mu$m in diameter.

German protozoologist Ehrenberg in 1832. The organism seen by these workers is now called *Spirillum volutans* and is a rather large bacterium (Figure 12.31*a*). A phototrophic organism resembling *S. volutans* is *Thiospirillum* (see Figure 12.4*b*). *S. volutans* is microaerophilic, requiring O_2, but is inhibited by O_2 at normal levels. Another prominent characteristic of cells of *S. volutans* is the formation of granules (volutin granules) consisting of polyphosphate (see Figure 12.31*a* and Section 4.11).

Azospirillum lipoferum is a nitrogen-fixing organism, which was originally described and named *Spirillum lipoferum* by Beijerinck in 1922. *A. lipoferum* is of considerable interest because it enters into a symbiotic relationship with tropical grasses and grain crops. Although not as intimate an association as that between root nodule bacteria and leguminous plants (∞Section 19.22), the *A. lipoferum*/corn association clearly benefits the corn plant from nitrogen fixation.

The small-diameter spirilla (which are not microaerophilic) have been separated into two genera, *Aquaspirillum* and *Oceanospirillum*. The former includes freshwater species while the latter includes species that inhabit seawater and require NaCl for growth (Table 12.18). Numerous species of *Aquaspirillum* and *Oceanospirillum* have been described, the various species being separated on physiological and phylogenetic grounds. These organisms undoubtedly play an important role in the recycling of organic matter in aquatic environments.

Magnetotactic Spirilla

Highly motile microaerophilic magnetic spirilla have been isolated from freshwater habitats. These organisms demonstrate a dramatic directed movement in a magnetic field referred to as **magnetotaxis**. In an artificial magnetic field magnetic spirilla quickly orient their long axis along the north-south magnetic moment of the field. Within the cells are chains of 5–40 magnetic particles called **magnetosomes,** consisting of magnetite (Fe_3O_4) and greigite (Fe_3S_4) (∞Section 4.11 and Figure 4.42). Magnetosomes function as internal magnets that orient the cells along a specific magnetic field.

Magnetic bacteria can have one of two magnetic polarities depending on the orientation of magnetosomes within the cell. Cells in the Northern Hemisphere have the north-seeking pole of their magnetosomes forward with respect to their flagella and thus move in a northward direction. Cells in the Southern Hemisphere have the opposite polarity and move southward. Although the ecological role of bacterial magnets is unclear, the ability to orient in a magnetic field may be of selective advantage in maintaining these microaerophilic organisms in zones of low O_2 concentration near the oxic/anoxic interface. The spirillum *Magnetospirillum magnetotacticum* (Figure 12.32● and Table 12.18) is a major organism in this group.

Bdellovibrio

These small vibroid organisms have the unusual property of preying on other bacteria, using as nutrients the cytoplasmic constituents of their hosts. These bacterial predators are small, highly motile cells that stick to the surfaces of their prey cells. Because of this, they have been given the name *Bdellovibrio* (*bdello* is a prefix meaning "leech"). Other predatory bacteria have been isolated and given such suggestive genus names as *Vampirococcus*. However, *Bdellovibrio* has a unique mode of attack and develops intraperiplasmically.

After attachment of a *Bdellovibrio* cell to its prey, the predator penetrates through the prey wall and replicates in the periplasmic space, eventually forming a spherical structure called a **bdelloplast**. The stages of attachment and penetration are shown in electron micrographs in Figure 12.33● and diagrammatically in Figure 12.34●. A wide variety of gram-negative *Bacteria* can be attacked

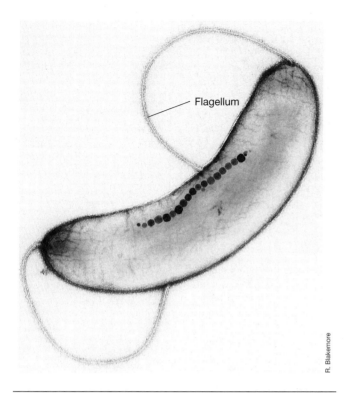

Flagellum

R. Blakemore

● **Figure 12.32 Negatively stained electron micrograph of a magnetotactic spirillum, *Magnetospirillum magnetotacticum*.** A cell measures 0.3×2 μm. This bacterium contains particles of Fe_3O_4 (magnetite), called *magnetosomes* (∞ Figure 4.42), arranged in a chain; the particles align the cell along geomagnetic lines. The organism was isolated from a water treatment plant in Durham, New Hampshire.

by a single *Bdellovibrio* species; gram-positive cells are not attacked.

Bdellovibrio is an obligate aerobe, obtaining its energy from the oxidation of amino acids and acetate. In addition, *Bdellovibrio* assimilates nucleotides, fatty acids, peptides, and even intact proteins, directly from its host without first breaking them down. Thus it is clear that the predatory mode of existence has involved the development in *Bdellovibrio* of interesting and unusual biochemical processes. In addition to a predatory lifestyle, however,

derivatives of predatory strains of *Bdellovibrio* that are prey-independent can be isolated, showing that predation is not obligatory. Prey-independent strains can be grown on complex media containing yeast extract and peptone.

Phylogenetically, bdellovibrios fall into the delta subdivision of Proteobacteria (Figure 12.1) and contain a genome about half the size of that of *E. coli*. Taxonomically, two species of *Bdellovibrio* and one species of the genus *Bacteriovorax* are recognized and supported by the results of genomic DNA:DNA hybridization (∞ Section 11.11).

Members of the genus *Bdellovibrio* are widespread in soil and water, including the marine environment. Their detection and isolation require methods reminiscent of those used in the study of bacterial viruses (∞ Section 9.4). Prey bacteria are spread on the surface of an agar plate to form a lawn, and the surface is inoculated with a small amount of soil suspension that has been filtered through a membrane filter; the latter retains most bacteria but allows the small *Bdellovibrio* cells to pass. On incubation of the agar plate, *plaques* analogous to bacteriophage plaques are formed at locations where *Bdellovibrio* cells are growing. However, unlike phage plaques, which continue to enlarge only as long as the bacterial host is growing, *Bdellovibrio* plaques continue to enlarge even after the prey has stopped growing. This results in ever enlarging plaques on the agar surface. Pure cultures of *Bdellovibrio* can then be isolated from these plaques. *Bdellovibrio* cultures have been obtained from a wide variety of soils and are thus common members of the soil population.

Ancyclobacter

Members of the genus *Ancyclobacter* are ring-shaped, nonmotile, extremely nutritionally diverse chemoorganotrophic bacteria (Figure 12.31*c*). They resemble very tightly coiled vibrios and are widely distributed in aquatic environments. A phototrophic counterpart to *Ancyclobacter* exists in the purple nonsulfur bacterium *Rhodocyclus purpureus* (see Figure 12.6*e*).

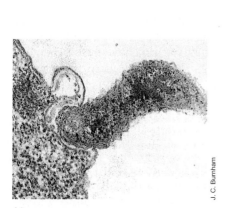

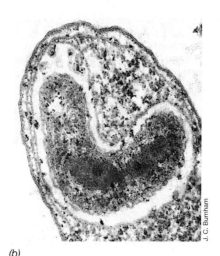

J. C. Burnham

(a)   (b)

● **Figure 12.33 Stages of attachment and penetration of a prey cell by *Bdellovibrio*.** A *Bdellovibrio* cell measures about 0.3 μm in diameter. (a, b) Electron micrographs of thin sections of *Bdellovibrio* attacking *Escherichia coli*; (a) early penetration; (b) complete penetration. The *Bdellovibrio* cell is enclosed in a membranous infolding of the prey cell (the bdelloplast) and replicates in the periplasmic space between the wall and the membrane (see Figure 12.34).

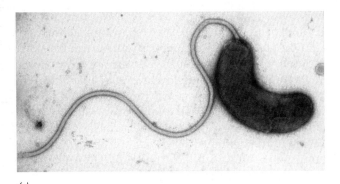

(a)

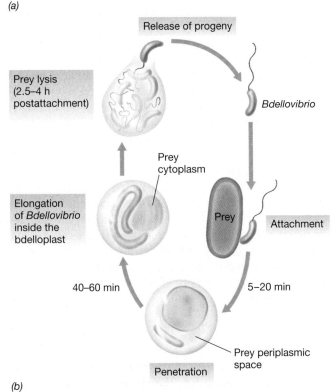

Release of progeny

Prey lysis
(2.5–4 h
postattachment)

Bdellovibrio

Prey
cytoplasm

Elongation
of *Bdellovibrio*
inside the
bdelloplast

Prey

Attachment

40–60 min

5–20 min

Prey periplasmic
space

Penetration

(b)

● **Figure 12.34 Developmental cycle of the bacterial predator *Bdellovibrio bacteriovorus*.** (a) Electron micrograph of a cell of *Bdellovibrio bacteriovorus*. Note the very thick flagellum present on this organism (∞Figure 4.54). (b) Events in predation. Following primary contact between a highly motile *Bdellovibrio* cell and a gram-negative bacterium, attachment and penetration into the prey periplasmic space occurs. Once inside, *Bdellovibrio* elongates and within 4 hours progeny cells are released. The number of progeny cells released varies with the size of the prey bacterium. For example, 5–6 bdellovibrios are released from each infected *Escherichia coli* cell, and 20–30 for a larger cell, such as an *Aquaspirillum* sp.

Campylobacter and Helicobacter

These two genera are key representatives of the *epsilon* subdivision of the Proteobacteria. They are gram-negative, motile spirilla, most species of which are pathogenic to humans or other animals. *Campylobacter* and *Helicobacter* are both microaerophilic and are cultured

from clinical specimens in media incubated at low (3–15%) O_2 and high (3–10%) CO_2.

Campylobacter species cause acute enteritis leading to (usually) bloody diarrhea, and pathogenesis is due to several factors, including an enterotoxin that is related to cholera toxin (∞Section 21.11). *Helicobacter pylori* is closely related to *Campylobacter* species and causes both acute and chronic gastritis, leading to the formation of *peptic ulcers*. We discuss the diseases caused by *Campylobacter* and *Helicobacter*, including modes of transmission of the organisms and clinical symptoms, in more detail in Section 29.9.

⬢ 12.14 Concept Check

Spirilla are spiral-shaped, chemoorganotrophic prokaryotes widespread in the aquatic environment. The genera *Helicobacter* and *Campylobacter* are pathogenic spirilla. Spirilla are distributed among all five subdivisions of the Proteobacteria.

◆ What is a *volutin granule*?

◆ What is unique about the spirilla *Bdellovibrio* and *Magnetospirillum*?

12.15 **Sheathed Proteobacteria: *Sphaerotilus* and *Leptothrix***

Key Genera: *Sphaerotilus, Leptothrix*

Sheathed bacteria are filamentous beta Proteobacteria (Table 12.1) with a unique life cycle involving formation of flagellated swarmer cells within a long tube or sheath. Under certain (generally unfavorable) conditions, the swarmer cells move out and become dispersed to new environments, leaving behind the empty sheath. Under favorable conditions, vegetative growth occurs within the filament, leading to the formation of long, cell-packed sheaths. Sheathed bacteria are common in freshwater habitats that are rich in organic matter, such as polluted streams. They are also abundant in trickling filters and activated sludge digestors in sewage treatment plants (∞Section 28.2), generally being found in flowing waters. In habitats where reduced iron or manganese compounds are present, the sheaths may become coated with ferric hydroxide or manganese oxide (∞Figure 17.29). Iron precipitation is probably due to chemical reactions, but some sheathed bacteria have the biochemical ability to oxidize manganous ions to manganese oxide. Two genera are the major organisms here: *Sphaerotilus*, in which manganese oxidation does not occur, and *Leptothrix*, whose members do oxidize Mn^{2+}.

Sphaerotilus

The *Sphaerotilus* filament is composed of a chain of rod-shaped cells with rounded ends enclosed in a closely fitting sheath. This thin, transparent sheath is difficult to see when it is filled with cells, but when it is partially empty, the sheath can easily be seen by phase contrast microscopy (Figure 12.35*a*●) or by staining. Individual cells are 1–2 μm wide and 3–8 μm long and stain gram negatively.

The cells within the sheath divide by binary fission (Figure 12.35*b*), and the new cells pushed out at the end synthesize new sheath material. Thus, the sheath is always formed at the *tips* of the filaments. Eventually cells are liberated from the sheaths, probably when the nutrient supply is low. These free cells are actively motile, the flagella being arranged lophotrichously (in a bundle at one pole) (Figure 12.35*c*). Probably the flagella are synthesized before the cells leave the sheath and, if so, may even aid in their liberation. It is thought that the swarmer cells then migrate, attach to some solid surface, and begin to

grow, each swarmer being the forerunner of a new filament. The sheath, which is devoid of muramic acid or other components of the peptidoglycan cell wall, is a protein-polysaccharide-lipid complex. However, it differs from the capsules formed by many gram-negative *Bacteria* (∞Section 4.10) in that it forms a linear structure.

Sphaerotilus cultures are nutritionally versatile, able to use a wide variety of simple organic compounds as carbon and energy sources, along with inorganic nitrogen sources. Befitting its habitat in flowing waters, *Sphaerotilus* is an obligate aerobe. *Sphaerotilus* blooms often occur in the fall of the year in streams and brooks when leaf litter causes a temporary increase in the organic content of the water. Its filaments are the main component of a microbial complex that sanitary engineers call "sewage fungus," which is a filamentous slime found on the rocks in streams receiving sewage pollution. In activated sludge works of sewage treatment plants (∞Section 28.2), *Sphaerotilus* growth, like that of *Beggiatoa* (see Section 12.4), is often responsible for a detrimental condition called "bulking." The tangled masses of *Sphaerotilus* filaments so increase the bulk of the sludge that it does not settle properly, thus presenting difficulties in sludge clarification.

Leptothrix

The ability of *Sphaerotilus* and *Leptothrix* to precipitate iron oxides on their sheaths is well established, and such iron-encrusted sheaths are frequently seen in iron-rich waters (Figure 12.36●). Iron precipitation occurs when iron, chelated to organic materials such as humic or tannic acids, is metabolized. The iron gets precipitated on the sheath while the organic constituents get taken up and used as a carbon or energy source.

Besides Fe^{2+} oxidation, *Leptothrix* is capable of oxidizing Mn^{2+} to Mn^{4+}. The reaction is exergonic:

$$Mn^{2+} + \frac{1}{2}O_2 + H_2O \longrightarrow MnO_2 + 2\,H^+$$

$$\Delta G^{0'} = -68\,kJ$$

There is evidence that Mn^{2+} oxidation is coupled to energy-yielding reactions in the cell as well, probably involving electron transport and formation of a proton motive force.

The gene encoding the manganese-oxidizing protein of *Leptothrix* has been isolated, and biochemical studies have shown that the protein resides in the sheath, not inside the cells. Thus the sheath of *Leptothrix* is critical for energy metabolism in this bacterium.

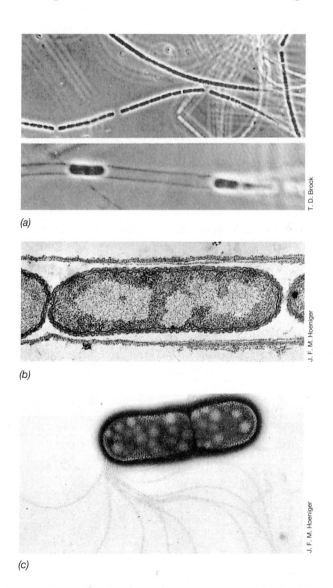

(a)

T. D. Brock

(b)

J. F. M. Hoeniger

(c)

J. F. M. Hoeniger

● **Figure 12.35** *Sphaerotilus natans.* A single cell is about 2 μm wide. (a) Phase contrast photomicrographs of material collected from a polluted stream. Active growth stage (above) and swarmer cells leaving the sheath. (b) Electron micrograph of a thin section through a filament. (c) Electron micrograph of a negatively stained swarmer cell. Notice the polar flagellar tuft.

12.16 Budding and Prosthecate/Stalked *Bacteria*

Key Genera: *Hyphomicrobium, Caulobacter*

This large and heterogeneous group of primarily alpha Proteobacteria contains organisms that form various

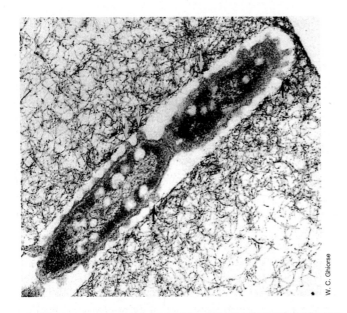

W. C. Ghiorse

● **Figure 12.36** *Leptothrix* **and iron precipitation.** Transmission electron micrograph of a thin section of *Leptothrix* sp. in a sample from a ferromanganese film in a swamp in Ithaca, New York. A single cell measures about 0.9 μm in diameter. Note the protuberances of the cell envelope that contact the sheath (arrows).

kinds of cytoplasmic extrusions: *stalks, hyphae,* or *appendages* (Table 12.19). Extrusions of these kinds, which are smaller in diameter than the mature cell, contain cytoplasm, and are bounded by the cell wall, are collectively called **prosthecae** (Figure 12.37●).

Cell division in budding bacteria occurs as a result of *unequal cell growth.* In contrast to cell division in a typical bacterium, which occurs by *binary fission* and results in the formation of two equivalent cells, cell division in the stalked and budding bacteria involves the formation

of a new daughter cell with the mother cell retaining its identity (Figure 12.38●). The major genera of budding bacteria and a summary of their cell division processes are shown in Figure 12.38.

A fundamental difference between these bacteria and conventional bacteria is not the formation of buds or stalks but the formation of new cell wall material from a *single point* (polar growth) rather than throughout the whole cell (intercalary growth). Several genera not normally considered to be budding bacteria show polar growth without differentiation of cell size (Figure 12.38). An important consequence of **polar growth** is that internal structures, such as membrane complexes, are not involved in the cell division process, thus permitting the formation of more complex internal structures than in cells undergoing intercalary growth. And many budding bacteria, particularly phototrophic species, contain extensive internal membrane systems.

Budding Bacteria: *Hyphomicrobium*

Two well-studied budding bacteria are closely related phylogenetically: *Hyphomicrobium,* which is chemoorganotrophic, and *Rhodomicrobium,* which is phototrophic. These organisms release buds from the ends of long, thin hyphae. The **hypha** is a direct cellular extension of the mother cell (Figures 12.39● and 12.40b●) and contains cell wall, cytoplasmic membrane, ribosomes, and, occasionally, DNA.

Figure 12.39 shows the process of reproduction in *Hyphomicrobium.* The mother cell, which is often attached by its base to a solid substrate, forms a thin outgrowth that lengthens to become a hypha. At the end of the hypha, a bud forms. This bud enlarges, forms a flagellum, breaks loose from the mother cell, and swims

Table 12.19	Characteristics of stalked, appendaged (prosthecate), and budding bacteria		
Characteristics	**Genus**	**Phylogenetic group**[a]	**DNA (mol% GC)**
Stalked bacteria:			
Stalk an extension of the cytoplasm and involved in cell division	*Caulobacter*	Alpha	62–67
Stalked, fusiform-shaped cells	*Prosthecobacter*	Alpha	54–60
Stalked, but stalk an excretory product not containing cytoplasm:			
Stalk depositing iron, cell vibrioid	*Gallionella*	Beta	55
Laterally excreted gelatinous stalk not depositing iron	*Nevskia*	Gamma	60
Appendaged (prosthecate) bacteria:			
Single or double prosthecae	*Asticcacaulis*	Alpha	55–61
Multiple prosthecae			
Short prosthecae, multiply by fission, some with gas vesicles	*Prosthecomicrobium*	Alpha	64–70
Flat, star-shaped cells, some with gas vesicles	*Stella*	Alpha	69–74
Long prosthecae, multiply by budding, some with gas vesicles	*Ancalomicrobium*	Alpha	70–71
Budding bacteria:			
Phototrophic, produce hyphae	*Rhodomicrobium*	Alpha	61–63
Phototrophic, budding without hyphae	*Rhodopseudomonas*	Alpha	64–72
Chemoorganotrophic, rod-shaped cells	*Blastobacter*	Alpha	59–66
Chemoorganotrophic, buds on tips of slender hyphae			
Single hyphae from parent cell	*Hyphomicrobium*	Alpha	59–65
Multiple hyphae from parent cell	*Pedomicrobium*	Alpha	62–67

[a] All are Proteobacteria.

sta[...]
lith[...]
cor[...]

env[...]
for[...]
stru[...]
sur[...]
is o[...]
bind[...]
this[...]
vel[...]
Cau[...]
sior[...]
stal[...]
arat[...]
and[...]
flag[...]
Stal[...]
and[...]
cell[...]
simp[...]
cells[...]
incl[...]

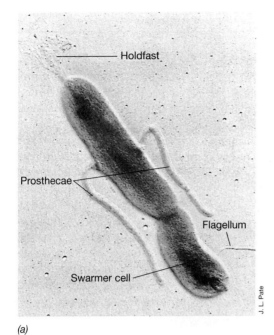

Holdfast

Prosthecae

Flagellum

Swarmer cell

(a)

J. L. Pate

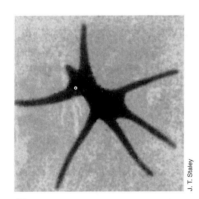

(b)

J. T. Staley

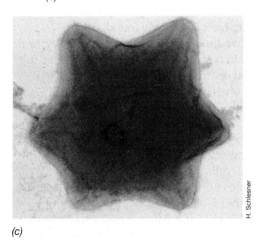

H. Schlesner

(c)

● **Figure 12.37** **Prosthecate bacteria.** (a) Electron micrograph of a shadow-cast preparation of *Asticcacaulis biprosthecum*, illustrating the location and arrangement of the prosthecae. Cells are about 0.6 μm wide. Note also the holdfast material and the swarmer cell in the process of differentiation. (b) Electron micrograph of a negatively stained preparation of a cell of the prosthecate bacterium *Ancalomicrobium adetum*. The appendages are cellular (prosthecae) because they are bounded by the cell wall and contain cytoplasm and are about 0.2 μm in diameter. (c) Electron micrograph of the star-shaped bacterium *Stella*. Cells are about 0.8 μm in diameter.

I. Equal products of cell division:

Binary fission: conventional bacteria

II. Unequal products of cell division:

1. Simple budding: *Pirella, Blastobacter*

2. Budding from hyphae: *Hyphomicrobium, Rhodomicrobium, Pedomicrobium*

3. Cell division of stalked organism: *Caulobacter*

4. Polar growth without differentiation of cell size: *Rhodopseudomonas, Nitrobacter, Methylosinus*

● **Figure 12.38** **Cell division.** Contrast between cell division in conventional bacteria and in budding and stalked bacteria.

away. Later, the daughter cell loses its flagellum and after a period of maturation forms a hypha and buds. Further buds can also form at the hyphal tip of the mother cell leading to arrays of cells connected by hyphae (Figure 12.40*a*). In some cases, a bud begins to form directly from the mother cell without the intervening formation of a hypha, whereas in other cases a single cell forms hyphae from each end (Figure 12.40*a*).

Nucleoid replication events during the budding cycle are of interest (Figure 12.39). First, the DNA located in the mother cell replicates. Then, once the bud has formed, a copy of the circular chromosome is moved down the length of the hypha and into the bud. A cross-septum then forms, separating the still-developing bud from the hypha and mother cell.

Hyphomicrobium is a methylotrophic bacterium (see Section 12.6) and is widespread in freshwater, marine, and terrestrial habitats. Preferred carbon sources are *one-carbon* compounds such as methanol, methylamine, formaldehyde, and formate. Enrichment cultures of *Hyphomicrobium* can be prepared taking advantage of the ability of this organism to grow in very dilute conditions. All that is needed is a mineral salts medium lacking organic carbon and nitrogen, to which an inoculum of soil, mud, or water is added. After several weeks of incubation, a surface film develops and is streaked out on agar medium containing methylamine or methanol as a sole carbon source. Colonies are then checked microscopically for the characteristic *Hyphomicrobium* cellular

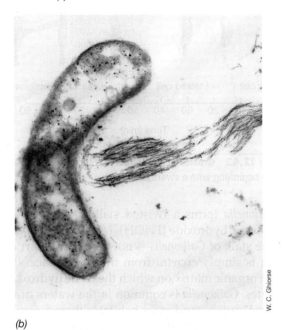

(a)

(b)

W. C. Ghiorse

● **Figure 12.43 The neutrophilic ferrous iron-oxidizer, *Gallionella ferruginea*.** (a) Photomicrograph of cells from an iron seep near Ithaca, New York. Notice the twisted stalk leading to the two bean-shaped cells (arrow). (b) Transmission electron micrograph of a thin section of a cell. Cells are about 0.6 μm wide. Note the twisted stalk of ferric hydroxide emanating from the center of the cell in both photos.

structures called **fruiting bodies** and show complex developmental life cycles involving intercellular communication (see Figure 12.47). Phylogenetically, the gliding myxobacteria are members of the *delta* subdivision of Proteobacteria (Table 12.20).

The fruiting myxobacteria exhibit the most complex behavioral patterns and life cycles of all known prokaryotic organisms. In keeping with this complexity, the chromosome of some myxobacteria is very large. *Myxococcus xanthus*, for example, has a single circular chromosome of 9.2 Mbp, *twice* as large as that of *Escherichia coli* (∞ Sections 15.4 and 10.19). Indeed, this is two-thirds the size of the entire yeast genome, which is contained on 16 chromosomes (∞Section 15.6). The vegetative cells of the fruiting myxobacteria are simple, nonflagellated, gram-negative rods (Figure 12.44*a*●) that glide across surfaces and obtain their nutrients primarily by causing the lysis of other bacteria. Under appropriate conditions, a swarm of vegetative cells aggregate and construct *fruiting bodies*, within which some of the cells become converted to resting structures called **myxospores** (Figure 12.44*b*).

Fruiting Bodies

The fruiting bodies of the myxobacteria vary from simple globular masses of myxospores in loose slime to complex forms with a fruiting body wall and a stalk (Figure 12.45●). The fruiting bodies are often strikingly colored and morphologically elaborate (Figures 12.45 and 12.46●). They can usually be seen with a hand lens or dissecting microscope on moist pieces of decaying wood or plant material. Fruiting bodies of myxobacteria often develop on dung pellets (for example, rabbit pellets) after they have been incubated for a few days in a moist chamber.

Another method for isolating fruiting myxobacteria is to prepare Petri plates of water agar (1.5% agar in distilled water with no added nutrients) onto which is spread a heavy suspension of virtually any bacterium. In the center of the plate a small amount of soil, decaying bark, or other natural material is placed. Myxobacteria in the inoculum lyse the bacterial cells and use their liberated products as nutrients; as they grow, they swarm out across the plate from the inoculum site. After several days to a week, the plates are examined under a dissecting microscope for myxobacterial swarms or fruiting bodies, and pure cultures are obtained by transfer of cells from the fruiting bodies or from the edge of the

Table 12.20	Classification of the fruiting myxobacteria[a]
Characteristics	**Genus/DNA (mol% GC)**
Vegetative cells tapered:	
Spherical or oval myxospores, fruiting bodies usually soft and slimy without well-defined sporangia or stalks	*Myxococcus* (68–71)
Rod-shaped myxospores: Myxospores not contained in sporangia, fruiting bodies without stalks	*Archangium* (67–68)
Myxospores embedded in slime envelope:	
Fruiting bodies without stalks	*Cystobacter* (68)
Stalked fruiting bodies, single sporangia	*Melittangium* (—)
Stalked fruiting bodies, multiple sporangia	*Stigmatella* (68–69)
Fruiting bodies are dark-brown clusters consisting of tiny spherical or disclike sporangia with an outer wall	*Angiococcus* (—)
Vegetative cells not tapered (blunt, rounded ends); myxospores resemble vegetative cells; sporangia always produced:	
Fruiting bodies without stalks; myxospores rod-shaped	*Polyangium* (69)
Fruiting bodies without stalks; myxospores oval; highly cellulolytic	*Sorangium* (—)
Fruiting bodies without stalks; myxospores coccoid	*Nannocystis* (70–72)
Stalked fruiting bodies	*Chondromyces* (69–70)

[a] Phylogenetically, those species examined fall into the delta subdivision of the Proteobacteria (see Table 12.1).

(a) Herbert Voelz

(b) Herbert Voelz

● **Figure 12.44** *Myxococcus.* (a) Electron micrograph of a thin section of a vegetative cell of *Myxococcus xanthus*. A cell measures about 0.75 μm wide. (b) Myxospore of *M. xanthus*, showing the multilayered outer wall. Myxospores measure about 2 μm in diameter.

swarm to organic media. Many myxobacteria can be grown in the laboratory on media containing peptone or casein hydrolysate, which provides organic nutrients in the form of amino acids or small peptides. The organisms are typical aerobes with a complete citric acid cycle (∞Figure 5.22) and respiratory chain.

Life Cycle of a Fruiting Myxobacterium

The life cycle of a typical fruiting myxobacterium is shown in Figure 12.47●. A vegetative cell excretes slime, and as it moves across a solid surface it leaves a slime trail behind (Figure 12.48●). This trail is preferentially used by other cells in the swarm so that a characteristic radiating pattern is soon created, with cells migrating along slime trails (Figure 12.48). The fruiting body ultimately formed (Figures 12.45 and 12.46) is a complex structure produced by the differentiation of cells in the stalk region and in the myxospore-bearing head.

Fruiting body formation does not occur so long as adequate nutrients for vegetative growth are present, but upon nutrient exhaustion, the vegetative swarms begin to fruit. Cells aggregate, likely through a chemotactic response (∞Sections 4.16 and 8.13), with the cells migrating toward each other and forming mounds or heaps (Figure 12.49●); a single fruiting body may have 10^9 or more cells. As the cell mounds become higher, the differentiation of the fruiting body into stalk and head begins (Figure 12.49*b* and *c*). Figure 12.49*d* clearly illustrates the differentiation of the fruiting body into stalk and head. The stalk is composed of slime, within which a few cells may be trapped. The majority of the cells accumulate in the fruiting-body head and undergo differentiation into *myxospores* (Figures 12.44–12.47). In some genera, the myxospores are enclosed in large walled structures called **cysts**.

Compared to the vegetative cell, the myxospore is more resistant to drying, sonic vibration, UV radiation, and heat, but the degree of heat resistance is much less than that of the bacterial endospore (∞Section 4.13). The main function of encysted myxospores is therefore

(a) Hans Reichenbach

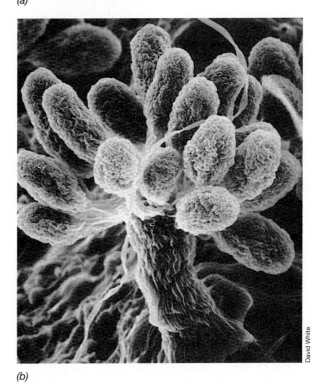

(b) David White

● **Figure 12.45** *Stigmatella aurantiaca.* (a) Color photo of a single fruiting body. The structure is about 150 μm high. (b) Scanning electron micrograph of a fruiting body growing on a piece of wood. Note the individual cells visible in each fruiting structure. The color of the fruiting body shown in (a) is due to the production of carotenoid pigments (∞Section 17.3).

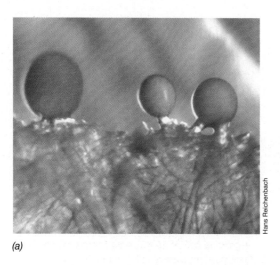
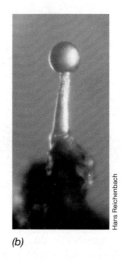
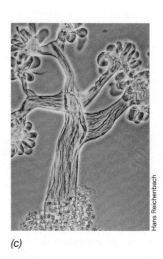

(a)   (b)   (c)

● **Figure 12.46   Fruiting bodies of three species of fruiting myxobacteria.** (a) *Myxococcus fulvus* (125 μm high). (b) *Myxococcus stipitatus* (170 μm high). (c) *Chondromyces crocatus* (560 μm high).

to enable the organism to survive desiccation during dispersal or during drying of the habitat. Upon dissemination to a suitable habitat or restoration of adequate growth conditions, the myxospore eventually germinates by a localized rupture of the capsule, with the growth and emergence of a new vegetative cell.

Myxobacteria are usually colored by carotenoid pigments (see Figures 12.45*a* and 12.46), and the main pigments are carotenoid glycosides. Pigment formation is promoted by light, and at least one function of the pigment is photoprotection. Since in nature the myxobacteria usually form fruiting bodies in the light, the presence of these photoprotective pigments is understandable. In the genus *Stigmatella* (Figure 12.45), light greatly stimulates fruiting body formation and catalyzes production of the lipid pheromone *2,5,8-trimethyl-8-hydroxy-nonane-4-one.* This substance initiates the aggregation step. The fruiting myxo-bacteria are classified on morphological grounds using characteristics of the vegetative cells, the myxo-spores, and fruiting-body structure (Table 12.20), and on phylogenetic grounds, using rRNA sequencing.

● **Figure 12.47   Life cycle of *Myxococcus xanthus.*** Aggregation serves to assemble vegetative cells for fruiting-body formation. Vegetative cells undergo morphogenesis to resting cells called *myxospores.* The latter germinate under favorable nutritional and physical conditions to yield vegetative cells. Vegetative cells can be converted directly to myxospores without fruiting-body formation by certain chemical inducers, notably high concentrations of glycerol. See the photograph of *Myxococcus* fruiting bodies in Figures 12.45 and 12.46.

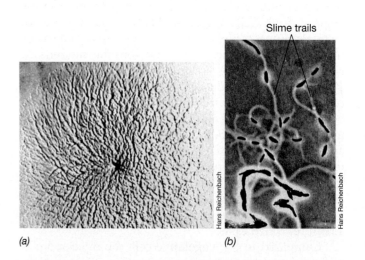

(a)   (b)

● **Figure 12.48   Swarming in *Myxococcus.*** (a) Photomicrograph of a swarming colony (5-mm radius) of *Myxococcus xanthus* on agar. (b) Single cells of *Myxococcus fulvus* from an actively gliding culture, showing the characteristic slime trails on the agar. A cell of *M. fulvus* is about 0.8 μm in diameter.

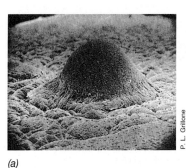
(a)

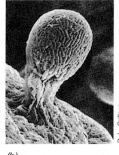
(b)

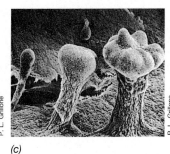
(c)

(d)

● **Figure 12.49 Scanning electron micrographs of fruiting-body formation in *Chondromyces crocatus*.** (a) Early stage, showing aggregation and mound formation. (b) Initial stage of stalk formation. Slime formation in the head has not yet begun, and so the cells of which the head is composed are still visible. (c) Three stages in head formation. Note that the diameter of the stalk also increases. (d) Mature fruiting bodies. The entire fruiting structure is about 600 μm in height (see Figure 12.46c).

 12.17 Concept Check

The fruiting myxobacteria are rod-shaped, gliding bacteria that aggregate to form complex masses of cells called *fruiting bodies*. Myxobacteria are chemoorganotrophic soil bacteria that live by consuming dead organic matter or other bacterial cells.

◆ What environmental conditions trigger fruiting body formation in myxobacteria?

◆ What is a *myxospore* and how does it compare with an *endospore*?

◆ To what specific phylogenetic group do the myxobacteria belong?

12.18 Sulfate- and Sulfur-Reducing Proteobacteria

Key Genera: *Desulfovibrio, Desulfobacter, Desulfuromonas*

Sulfate (SO_4^{2-}) and sulfur (S^0) can be electron acceptors under anoxic conditions for a large group of delta Proteobacteria that utilize organic compounds or H_2 as electron donors. Hydrogen sulfide (H_2S) is the product of both SO_4^{2-} and S^0 reduction. Over 40 genera of these organisms, collectively known as the dissimilative **sulfate-reducing bacteria** and **sulfur-reducing bacteria**, are known, and some of the key ones are shown in Table 12.21.

The genera of sulfate-reducing bacteria in group I, such as *Desulfovibrio* (Figure 12.50a●), *Desulfomonas*, *Desulfotomaculum*, and *Desulfobulbus* (Figure 12.50c), utilize lactate, pyruvate, ethanol, or certain fatty acids as electron donors, reducing sulfate to hydrogen sulfide. The genera in group II, such as *Desulfobacter* (Figure 12.50d), *Desulfococcus*, *Desulfosarcina* (Figure 12.50e), and *Desulfonema* (Figure 12.50b), specialize in the oxidation of fatty acids, particularly *acetate*, reducing sulfate to sulfide. The sulfate-reducing bacteria are for the most part obligate anaerobes, and strict anoxic techniques must be used in their cultivation (Figure 12.50g).

Sulfate-reducing bacteria are widespread in aquatic and terrestrial environments that become anoxic as a result of microbial decomposition processes. The best-

studied genus is *Desulfovibrio* (Figure 12.50a), which is common in aquatic habitats or waterlogged soils containing abundant organic material and sufficient levels of sulfate. *Desulfotomaculum*, phylogenetically a member of the gram-positive *Bacteria*, consists of endospore-forming rods found primarily in soil. Growth and reduction of sulfate by *Desulfotomaculum* in certain canned foods leads to a type of spoilage called *sulfide stinker*. The remaining genera of sulfate reducers are indigenous to anoxic freshwater (group I) or marine environments (groups I and II), and can occasionally be isolated from the mammalian intestine.

Sulfur Reduction

The dissimilative *sulfur-reducing bacteria* can reduce elemental sulfur to sulfide but are unable to reduce sulfate to sulfide. Members of the genus *Desulfuromonas* (Figure 12.50f) can grow anaerobically by coupling the *oxidation* of substrates such as acetate or ethanol to the *reduction* of elemental sulfur to hydrogen sulfide. However, the ability to reduce elemental sulfur, as well as other sulfur compounds such as thiosulfate, sulfite, or dimethyl sulfoxide (DMSO), is a widespread property of a variety of chemoorganotrophic, generally facultatively aerobic bacteria (for example, *Proteus, Campylobacter, Pseudomonas,* and *Salmonella*). *Desulfuromonas* differs from the latter in that it is an obligate anaerobe and utilizes *only* sulfur as an electron acceptor (see Table 12.21). Dissimilative sulfur-reducing bacteria like *Desulfuromonas* reside in many of the same habitats as sulfate-reducing bacteria and often form associations with bacteria that oxidize H_2S to S^0, such as green sulfur bacteria (see Section 12.32). The sulfur produced is then reduced back to H_2S during metabolism of the sulfur-reducer, completing an abbreviated anoxic sulfur cycle (∽Section 19.13).

Physiology of Sulfate-Reducing Bacteria

The range of electron donors used by sulfate-reducing bacteria is fairly broad. Hydrogen, lactate, and pyruvate are almost universally used, and many group I species utilize malate, sulfonates, and certain primary alcohols

Table 12.21	**Characteristics of some key genera of sulfate-and sulfur-reducing bacteria**[a]	
Genus	**Characteristics**	**DNA (mol% GC)**
Group I sulfate reducers: Nonacetate oxidizers		
Desulfovibrio	Polarly flagellated, curved rods, no spores; gram-negative; contain desulfoviridin; one thermophilic	46–61
Desulfomicrobium	Motile rods, no spores; gram-negative; desulfoviridin absent	52–57
Desulfobotulus	Vibrios; gram-negative; motile; desulfoviridin absent	53
Desulfofustis	Motile rods, specializes in the degradation of glycolate and glyoxalate	56
Desulfotomaculum	Straight or curved rods; motile by peritrichous or polar flagellation; gram-negative; desulfoviridin absent; produce endospores; capable of utilizing acetate as energy source	37–46
Desulfomonile	Rod; capable of reductive dechlorination of 3-chlorobenzoate to benzoate (⟳Section 17.18)	49
Desulfobacula	Oval to coccoid cells, marine; can oxidize various aromatic compounds including the aromatic hydrocarbon toluene, to CO_2	42
Archaeoglobus	Archaeon; hyperthermophile, temperature optimum, 83°C; contains some unique coenzymes of methanogenic bacteria, makes small amount of methane during growth; H_2, formate, glucose, lactate, and pyruvate are electron donors, SO_4^{2-}, $S_2O_3^{2-}$, or SO_3^{2-}, electron acceptors (⟳Section 13.7)	41–46
Desulfobulbus	Ovoid or lemon-shaped cells; no spores; gram-negative; desulfoviridin absent; if motile, by single polar flagellum; utilizes propionate as electron donor with acetate + CO_2 as products	59–60
Desulforhopalus	Curved rods, gas vacuolate, psychrophile; uses propionate, lactate, or alcohols as electron donor	48
Thermodesulfobacterium	Small, gram-negative rods; desulfoviridin present; thermophilic, optimum growth at 70°C; a member of the *Bacteria* but contains ether-linked lipids (see Section 12.36)	34
Group II sulfate reducers: Acetate oxidizers		
Desulfobacter	Rods; no spores, gram-negative; desulfoviridin absent; if motile, by single polar flagellum; utilizes only acetate as electron donor and oxidizes it to CO_2 via the citric acid cycle	45–46
Desulfobacterium	Rods, some with gas vesicles, marine; capable of autotrophic growth via the acetyl-CoA pathway	41–59
Desulfococcus	Spherical cells; nonmotile; gram-negative; desulfoviridin present, no spores; utilizes C_1 to C_{14} fatty acids as electron donor with complete oxidation to CO_2; capable of autotrophic growth via the acetyl-CoA pathway	57
Desulfonema	Large, filamentous gliding bacteria; gram-positive, no spores; desulfoviridin present or absent; utilizes C_2 to C_{12} fatty acids as electron donor with complete oxidation to CO_2; capable of autotrophic growth via the acetyl-CoA pathway (H_2 as electron donor)	35–42
Desulfosarcina	Cells in packets (sarcina arrangement); gram-negative; no spores; desulfoviridin absent; utilizes C_2 to C_{14} fatty acids as electron donor with complete oxidation to CO_2; capable of autotrophic growth via the acetyl-CoA pathway (H_2 as electron donor)	51
Desulfoarculus	Vibrios; gram-negative; motile; desulfoviridin absent; utilizes only C_1 to C_{18} fatty acids as electron donor	66
Desulfacinum	Cocci to oval-shaped cells; gram-negative; utilizes C_1 to C_{18} fatty acids, very nutritionally diverse, capable of autotrophic growth; thermophilic	64
Desulforhabdus	Rods; no spores; gram-negative; nonmotile; utilizes fatty acids with complete oxidation to CO_2	52
Thermodesulforhabdus	Gram-negative motile rods; thermophilic; uses fatty acids up to C_{18}	51
Dissimilative sulfur reducers		
Desulfuromonas	Straight rods, single lateral flagellum; no spores; gram-negative; does not reduce sulfate; acetate, succinate, ethanol, or propanol used as electron donor; obligate anaerobe; one species is capable of the reductive dechlorination of trichloroethylene (⟳Section 17.18)	50–63
Desulfurella	Motile short rods; gram-negative; requires acetate; thermophilic	31
Sulfurospirillum	Small vibrios, reduces S^0 with H_2 or formate as electron donors	—
Campylobacter	Curved, vibrio-shaped rods; polar flagella; gram-negative; no spores; unable to reduce sulfate but can reduce sulfur, sulfite, thiosulfate, nitrate, or fumarate anaerobically with acetate or a variety of other carbon or electron donor sources; facultative aerobe	40–42

[a] Phylogenetically, most sulfate- and sulfur-reducing bacteria are delta Proteobacteria.

(for example, ethanol, propanol, and butanol). Some strains of *Desulfotomaculum* utilize glucose, but this is rather rare among sulfate reducers in general. Group I sulfate-reducers oxidize their electron donor to the level of acetate and excrete this fatty acid as an end product.

Group II organisms differ from those in group I by their ability to oxidize fatty acids (including acetate), lactate, succinate, and even benzoate in some species, completely to CO_2. *Desulfosarcina*, *Desulfonema*, *Desulfococcus*, *Desulfobacterium*, *Desulfotomaculum*, and certain species of *Desulfovibrio*, are unique among sulfate-reducers in their ability to grow chemolithotrophically and autotrophically with H_2 as electron donor, sulfate as electron acceptor, and CO_2 as sole carbon source. A few specialized sulfate-reducers can use hydrocarbons, even crude oil itself, as electron donors. This process is noteworthy because until such organisms were recognized, it was thought that hydrocarbons could only be oxidized under oxic conditions (⟳Section 17.23).

In addition to using sulfate as an electron acceptor, many sulfate-reducing bacteria can grow using *nitrate* as an electron acceptor, reducing NO_3^- to NH_3, or reducing sulfonates, such as isethionate ($HO—CH_2—CH_2—SO_3^-$),

(a) *(b)*
(c) *(d)*
(e) *(f)*

(g)

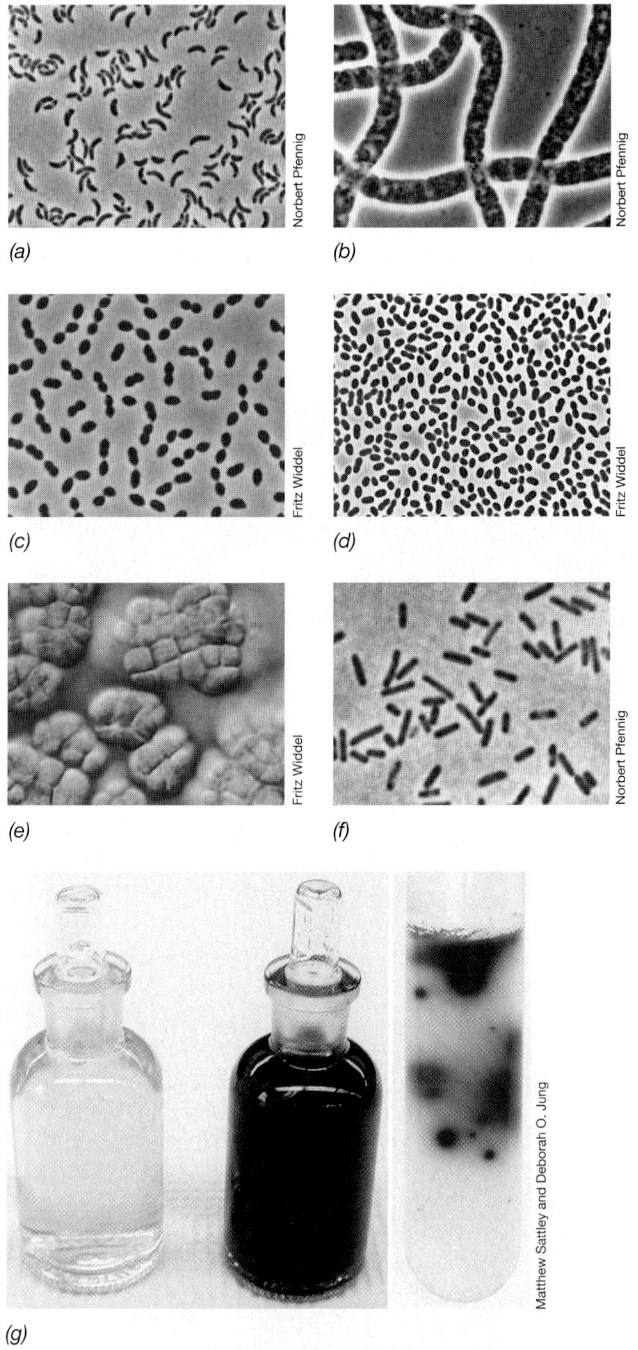

● **Figure 12.50 Phase-contrast photomicrographs of (a–e) representative sulfate-reducing and (f) sulfur-reducing bacteria.** (a) *Desulfovibrio desulfuricans*; cell diameter about 0.7 μm. (b) *Desulfonema limicola*; cell diameter 3 μm. (c) *Desulfobulbus propionicus*; cell diameter about 1.2 μm. (d) *Desulfobacter postgatei*; cell diameter about 1.5 μm. (e) *Desulfosarcina variabilis* (interference contrast microscopy); cell diameter about 1.25 μm. (f) *Desulfuromonas acetoxidans*; cell diameter about 0.6 μm. (g) Enrichment culture of sulfate-reducing bacteria. Left, sterile medium; center, a positive enrichment showing black ferrous sulfide (FeS); right, colonies of sulfate-reducing bacteria in a "shake tube" (∞Section 18.2).

to sulfide. Elemental sulfur can also be reduced to sulfide by most sulfate-reducing bacteria.

Certain organic compounds can be fermented by sulfate-reducing bacteria. The most common fermentable compound is *pyruvate*, which is converted via the phosphoroclastic reaction to acetate, CO_2, and H_2 (∞Figure 17.51). Moreover, although thought to be *obligate* anaerobes, certain isolates of sulfate-reducing bacteria, primarily ones isolated from microbial mats where they coexist with O_2-producing cyanobacteria, are quite oxygen tolerant and can respire with O_2 as electron acceptor. At least one species of *Desulfovibrio*, *D. oxyclinae*, can actually *grow* with O_2 as electron acceptor under microoxic conditions.

Isolation

The enrichment of *Desulfovibrio* is relatively easy on an anoxic lactate-sulfate medium to which ferrous iron is added. A reducing agent, such as thioglycolate or ascorbate, is also added to achieve a lower E_0'. When sulfate-reducing bacteria grow, the sulfide formed from sulfate reduction combines with the ferrous iron to form black, insoluble ferrous sulfide (Figure 12.50*g*). This blackening not only indicates sulfate reduction, but the iron also ties up and detoxifies the sulfide, making possible growth to higher cell yields. After some growth has occurred as evidenced by blackening of the medium, purification is accomplished by streaking onto a tube coated on the inside surface with a thin layer of agar (called *roll tubes*) or on Petri plates in an anoxic glove box (∞Figure 6.26).

Alternatively, agar shake tubes can be used for purification purposes. In the *shake tube method* a small amount of liquid from the original enrichment is added to a tube of molten agar growth medium, mixed thoroughly, and sequentially diluted through a series of molten agar tubes (∞Section 18.2 and Figure 18.3*b*). On solidification, individual cells of sulfate reducers distributed throughout the agar form black colonies (Figure 12.50*g*) that can be removed aseptically, and the whole process is repeated until pure cultures are obtained.

12.18 Concept Check

Sulfate- and sulfur-reducing bacteria are a large group of delta Proteobacteria unified by their physiological process of reducing either SO_4^{2-} or S^0 to H_2S under anoxic conditions. Two physiological subgroups of sulfate-reducing bacteria are known: group I, which is incapable of oxidizing acetate to CO_2, and group II, which is capable of doing so.

◆ What organic substrate would you use to enrich and isolate a *group II* sulfate reducer from nature?

◆ For sulfate-reducing bacteria capable of chemolithotrophic and autotrophic growth: (1) What is the electron donor? (2) What is the electron acceptor? (3) What is the source of cell carbon?

◆ Physiologically, how does *Desulfuromonas* differ from *Desulfovibrio*?

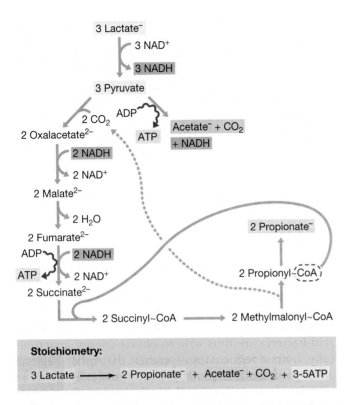

Stoichiometry:

3 Lactate ⟶ 2 Propionate⁻ + Acetate⁻ + CO_2 + 3-5ATP

● **Figure 12.68** **The formation of propionic acid from lactate by *Propionibacterium*.** Note that the four NADH made from the oxidation of 3 lactate to 2 pyruvate/acetate + CO_2 are reoxidized in the reduction of oxalacetate and fumarate. Also, note that the CoA group from propionyl~CoA is transferred to succinate during the formation of propionate in an energy neutral fashion.

12.23 Actinobacteria: *Mycobacterium*

Key Genus: *Mycobacterium*

The genus *Mycobacterium* consists of rod-shaped organisms that at some stage of their growth cycle possess the distinctive staining property called **acid-fastness**. This property is due to the presence on the surface of the mycobacterial cell of unique lipids called **mycolic acids**, found only in the genus *Mycobacterium*. First discovered by Robert Koch during his pioneering investigations on

tuberculosis (∞Section 1.6), acid-fast staining permitted the identification of the organism in tuberculous lesions. It has subsequently proved to be of great taxonomic use in defining the genus *Mycobacterium*.

Acid-Fastness (Ziehl–Neelsen Stain)

A mixture of the dye basic fuchsin and phenol is used in this staining procedure, the stain being driven into the cells by slow heating of the smear on the microscope slide to the steaming point for 2–3 min. The role of the phenol is to enhance penetration of the fuchsin into the lipids. After washing in distilled water, the preparation is decolorized with acid-alcohol. After another wash, a final counterstain of methylene blue is used. Acid-fast organisms in the final preparation appear *red*, whereas the background and non-acid-fast organisms appear *blue* (∞Figure 26.7).

As noted, the key component necessary for acid-fastness is *mycolic acid*. Mycolic acid is actually a group of complex branched-chain hydroxy lipids (the general structure is shown in Figure 12.69*a*●). In the acid-fast stain, the carboxylic acid group of the mycolic acid reacts with the fuchsin dye (Figure 12.69*b*). The mycolic acid is covalently bound to peptidoglycan in the mycobacterial wall, and this complex leaves the cell surface with a waxy, hydrophobic consistency. Mycobacteria are not readily stained by the Gram stain method because of the high surface-lipid content. If the lipoidal portion of the cell is removed with alkaline ethanol, however, the intact cell remaining is non-acid-fast but instead is gram-positive.

Characteristics of Mycobacteria

Mycobacteria are rather pleomorphic and may undergo branching or filamentous growth. However, in contrast to those of the actinomycetes (see Section 12.24), filaments of the mycobacteria become fragmented into rods or coccoid elements upon slight disturbance; a true mycelium is not formed. In general, mycobacteria can be separated into two major groups, *slow growers* and *fast growers* (Table 12.29). *Mycobacterium tuberculosis* is a typical slow grower, and visible colonies are produced from

Species	Growth in 5% NaCl	Nitrate reduction	Growth at 45°C	Human pathogen	Pigmentation
Slow-growing species					
Mycobacterium tuberculosis	−	+	−	+	None
Mycobacterium avium	−	−	−	+	Old colonies pigmented (see Figure 12.70c)
Mycobacterium bovis	−	−	+	+	None
Mycobacterium kansasii	−	+	−	+	Photochromogenic
Fast-growing species					
Mycobacterium smegmatis	+	+	+	−	None
Mycobacterium phlei	+	+	+	−	Pigmented
Mycobacterium chelonae	+	−	−	+	None
Mycobacterium parafortuitum	+	+	−	−	Photochromogenic

Table 12.29 **Some characteristics of representative mycobacteria**

(a) Mycolic acid; R_1 and R_2 are long-chain aliphatic hydrocarbons

(b) Basic fuchsin

● **Figure 12.69 Acid-fast staining.** Structure of (a) mycolic acid and (b) basic fuchsin, the dye used in the acid-fast stain. The fuchsin dye combines with the mycolic acid via ionic bonds between COO^- and NH_2^+.

dilute inoculum only after days to weeks of incubation. (Koch was successful in first isolating *M. tuberculosis* because he waited long enough after inoculating media; ⌘Section 1.6). When growing on solid media, mycobacteria form tight, compact, often wrinkled colonies (Figure 12.70*a*●). This colony morphology is probably due to the high lipid content and hydrophobic nature of the cell surface that make cells stick together.

For the most part, mycobacteria have relatively simple nutritional requirements. Growth often occurs in simple mineral salts medium with ammonium as nitrogen source and glycerol or acetate as sole carbon source and electron donor incubated in air. Growth of *Mycobacterium tuberculosis* is stimulated by lipids and fatty acids, and egg yolk (a good source of lipids) is often added to culture media to achieve more luxuriant growth. A glycerol-whole egg medium (Lowenstein–Jensen medium) is often used in primary isolation of *M. tuberculosis* from pathological materials.

Perhaps because of the high lipid content of its cell walls, *M. tuberculosis* is able to resist such chemical agents as alkali and phenol for considerable periods of time, and this property is used in the selective isolation of the organism from patient sputum and other materials that are grossly contaminated. The sputum is first treated with 1 *N* NaOH for 30 min and then neutralized and streaked onto an isolation medium.

A characteristic of many mycobacteria is their ability to form yellow carotenoid pigments (Figure 12.70*c*). Based on pigmentation, the mycobacteria can be classified into three groups: nonpigmented (including *Mycobacterium tuberculosis*, *M. bovis*); forming pigment only when cultured in the light, a property called *photochromogenesis* (including *M. kansasii*, *M. marinum*); and forming pigment even when cultured in the dark, a property called *scotochromogenesis* (for example, *M. gordonae*; Table 12.29). Photoinduction of carotenoid formation requires short-wavelength (blue) light and occurs only in the presence of O_2. The evidence indicates that the critical event in photoinduction is a light-catalyzed oxidation event, and it appears that one of the early enzymes in carotenoid biosynthesis is photoinduced. As with other carotenoid-containing bacteria, it has been suggested that carotenoids protect mycobacteria against oxidative damage involving singlet oxygen (⌘Section 6.16).

The virulence of *Mycobacterium tuberculosis* cultures has been correlated with the formation of long, cordlike structures (Figure 12.70*b*) on agar or in liquid medium, due to side-to-side aggregation and intertwining of long chains of bacteria. Growth in cords reflects the presence on the cell surface of a characteristic glycolipid, the **cord factor** (Figure 12.71●). The pathogenesis of the disease tuberculosis is discussed in detail in Section 26.5.

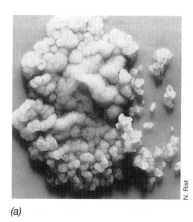

(a)

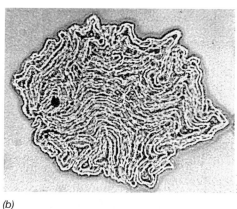

(b)

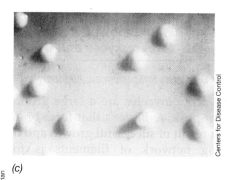

(c)

● **Figure 12.70 Characteristic colony morphology of mycobacteria.** (a) *Mycobacterium tuberculosis*, showing the compact, wrinkled appearance of the colony. The colony is about 7 mm in diameter. (b) A colony of virulent *M. tuberculosis* at an early stage, showing the characteristic cordlike growth. Individual cells are about 0.5 μm in diameter. (See also the historic drawings of *M. tuberculosis* cells made by Robert Koch, ⌘Figure 1.13). (c) Colonies of *Mycobacterium avium* from a strain of this organism isolated as an opportunistic pathogen from an AIDS patient.

(a)

(b)

● **Figure 12.73 Photomicrographs of several spore-bearing structures of actinomycetes.** (a) *Streptomyces*, a monoverticillate type. (b) *Streptomyces*, a closed spiral type. Filaments are about 0.8 μm wide in both cases. Compare these photos with the art in Figure 12.75.

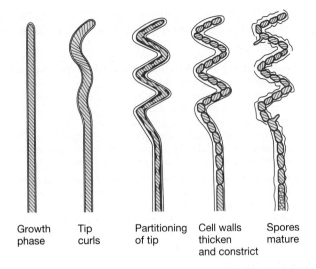

| Growth phase | Tip curls | Partitioning of tip | Cell walls thicken and constrict | Spores mature |

● **Figure 12.74 Spore formation in *Streptomyces*.** Diagram of stages in the conversion of a streptomycete's aerial hypha (sporophores) into spores (conidia).

assortment of inorganic salts to which starch, asparagine, or calcium malate is added as a carbon source and undigested casein or potassium nitrate as a nitrogen source. After incubation for 5–7 days in air, the plates are examined for the presence of the characteristic *Streptomyces* colonies (Figures 12.76 and 12.77●), and spores of interesting colonies can be streaked to isolate pure cultures.

Nutritionally, the streptomycetes are quite versatile. Growth-factor requirements are rare, and a wide variety of carbon sources, such as sugars, alcohols, organic acids, amino acids, and some aromatic compounds, can be utilized. Most isolates produce extracellular hydrolytic enzymes that permit utilization of polysaccharides (starch, cellulose, and hemicellulose), proteins, and fats, and some strains can use hydrocarbons, lignin, tannin, or even rubber. *Streptomyces* can often be obtained by spreading a soil dilution on an alkaline agar medium containing polymers such as casein and starch (Figure 12.76a). A single isolate may be able to break down over 50 distinct carbon sources.

Streptomycetes are strict aerobes whose growth in liquid culture is usually markedly stimulated by forced aeration. Sporulation usually does not take place in liquid culture but only when the organism is growing on the surface of agar or another solid substrate; it can occur, however, when organisms form a pellicle on the surface of an unshaken liquid culture.

Antibiotics of *Streptomyces*

Perhaps the most striking property of the streptomycetes is the extent to which they produce *antibiotics* (Table 12.31). Evidence for antibiotic production is often seen on the agar plates used in the initial isolation of *Streptomyces*: Adjacent colonies of other bacteria show zones of inhibition (Figures 12.76a and 12.77a; ∞Figure 30.7a). About 50% of all *Streptomyces* isolated have proved to be antibiotic producers.

Because of the great economic and medical importance of many streptomycete antibiotics, an enormous amount of work has been done on these producers. Over 500 distinct antibiotic substances have been shown to be produced by streptomycetes, and many more are suspected (∞Sections 30.5 and 30.6); most of these have been identified chemically (Figure 12.77b). Some organisms produce more than one antibiotic, and often the several kinds produced by one organism are not chemically related molecules. The same antibiotic may be formed by different species found in widely scattered parts of the world. And, although an antibiotic-producing organism is resistant to its own antibiotics, it usually remains sensitive to antibiotics produced by other streptomycetes. Many genes are often required to encode the enzymes involved in antibiotic synthesis, and because of this, the genomes of *Streptomyces* species are typically quite large (8 Mbp and larger, ∞Table 15.1).

More than 60 streptomycete antibiotics have found practical application in human and veterinary medicine, agriculture, and industry. Some of the more common antibiotics of *Streptomyces* origin are listed in Table 12.31. They are grouped into classes based on the chemical structure of the parent molecule. The search for new streptomycete antibiotics continues because many infectious diseases are still not adequately controlled by existing antibiotics. Also, the development of antibiotic-resistant pathogens requires the continual discovery of new agents.

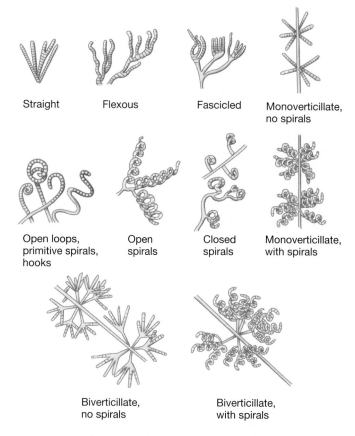

Straight Flexous Fascicled Monoverticillate, no spirals

Open loops, primitive spirals, hooks Open spirals Closed spirals Monoverticillate, with spirals

Biverticillate, no spirals Biverticillate, with spirals

● **Figure 12.75 Various types of sporebearing structures in the streptomycetes.** A given species of *Streptomyces* produces only one morphological type of sporebearing structure. Compare with Figure 12.73.

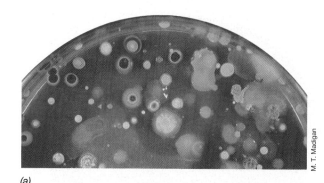

M. T. Madigan

(a)

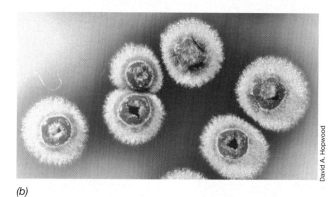

David A. Hopwood

(b)

● **Figure 12.76 *Streptomyces*.** (a) Colonies of *Streptomyces* and other soil bacteria derived from spreading a soil dilution on a casein-starch agar plate. The *Streptomyces* colonies are of various colors (several black *Streptomyces* colonies are in the foreground) but can easily be identified by their opaque, rough, nonspreading morphology. (b) Close-up photo of colonies of *Streptomyces coelicolor*.

Ironically, despite the extensive practical studies on antibiotic-producing streptomycetes by the antibiotic industry and the fact that antibiotics from *Streptomyces* species are a multibillion dollar a year industry, the ecology of *Streptomyces* remains poorly understood. The interactions of these organisms with other prokaryotes and the ecological rationale for why antibiotics are pro-duced in the first place, is still an area where we know very little. One hypothesis for why *Streptomyces* species produce antibiotics is that antibiotic production, which is linked to sporulation (a process itself triggered by nutrient depletion), might be a mechanism to inhibit the growth of other organisms competing with *Streptomyces* cells for limiting nutrients. This would allow the

Table 12.31 Some common antibiotics synthesized by species of *Streptomyces*

Chemical class	Common name	Produced by	Active against[a]
Aminoglycosides	Streptomycin	*S. griseus*[b]	Most gram-negative *Bacteria*
	Spectinomycin	*Streptomyces* spp.	*M. tuberculosis*, penicillinase-producing *N. gonorrhoeae*
	Neomycin	*S. fradiae*	Broad spectrum, usually used in topical applications because of toxicity
Tetracyclines	Tetracycline	*S. aureofaciens*	Broad spectrum, gram-positive and gram-negative *Bacteria*, rickettsias and chlamydias, *Mycoplasma*
	Chlortetracycline	*S. aureofaciens*	As for tetracycline
Macrolides	Erythromycin	*Saccharopolyspora erythraea*	Most gram-positive *Bacteria*, frequently used in place of penicillin, *Legionella*
	Clindamycin	*S. lincolnensis*	Effective against obligate anaerobes, especially *Bacteroides fragilis*
Polyenes	Nystatin	*S. noursei*	Fungi, especially *Candida* infections
	Amphocetin B	*S. nodosus*	Fungi
None	Chloramphenicol	*S. venezuelae*	Broad spectrum; drug of choice for typhoid fever

[a] Most antibiotics are effective against several different *Bacteria*. The entries in this column refer to the common clinical application of a given antibiotic. The structures and mode of action of many of these antibiotics are discussed in Sections 20.7–20.9.
[b] All listings beginning with an "*S.*" are species of the genes *Streptomyces*.

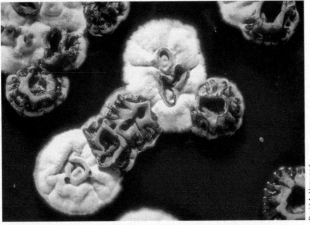

© Eli Lilly & Co. Used with permission.

(a)

David A. Hopwood

(b)

● **Figure 12.77 Antibiotics from _Streptomyces._** (a) Antibiotic action of soil microorganisms on a crowded plate. The smaller colonies surrounded by inhibition zones are streptomycetes; the larger, spreading colonies are _Bacillus_ species. (b) The red-colored antibiotic undecylprodigiosin is being excreted by colonies of _Streptomyces coelicolor_.

Streptomyces to complete the sporulation process and form a dormant structure that would have increased chances of survival.

 12.24 Concept Check

The streptomycetes are a large group of filamentous, gram-positive _Bacteria_ that form spores at the end of aerial filaments. Many clinically useful antibiotics like tetracycline and neomycin have come from _streptomyces_ species.

◆ How do spores and the process of sporulation in a _Streptomyces_ species differ from that in a _Bacillus_ species?

◆ What energy class of organism is a _Streptomyces_ and from what types of compounds do these organisms obtain their energy?

◆ Why might antibiotic production be of advantage to Streptomycetes?

 IV **PHYLUM 4: CYANOBACTERIA AND PROCHLOROPHYTES**

12.25 Cyanobacteria

Key Genera: _Synechococcus, Oscillatoria, Nostoc_

Cyanobacteria comprise a large and morphologically heterogeneous group of phototrophic _Bacteria_. Cyanobacteria differ in fundamental ways from purple and green bacteria (see Sections 12.2 and 12.32), most notably in that they are _oxygenic_ phototrophs. Cyanobacteria represent one of the major phyla of _Bacteria_ and show a distant relationship to gram-positive _Bacteria_ (see Figure 12.1). As we saw in Section 11.1, these organisms were the first oxygen-evolving phototrophic organisms on Earth and were responsible for the conversion of the atmosphere of the Earth from anoxic to oxic.

Structure and Classification of Cyanobacteria

The morphological diversity of the cyanobacteria is impressive (Figure 12.78●). Both unicellular and filamentous forms are known, and considerable variation within these morphological types occurs. Still, cyanobacteria can be divided into just five morphological groups: (1) unicellular dividing by binary fission (Figure 12.78a); (2) unicellular dividing by multiple fission (colonial) (Figure 12.78b); (3) filamentous containing differentiated cells called _heterocysts_ that function in nitrogen fixation (Figure 12.78d and see Figure 12.80); (4) filamentous nonheterocystous forms (Figure 12.78c); and (5) branching filamentous species (Figure 12.78e). Table 12.32 lists the major genera currently recognized in each group. Cyanobacterial cells range in size from those of typical bacteria ($0.5-1$ μm in diameter) to cells as large as 40 μm in diameter (in the species _Oscillatoria princeps_, ∞Figure 4.44a).

The structure of the cell wall of cyanobacteria is similar to that of gram-negative _Bacteria_, and peptidoglycan is present in the walls (Figure 12.79●). Many cyanobacteria produce extensive mucilaginous envelopes, or sheaths, that bind groups of cells or filaments together (see, for example, Figure 12.78a). The photosynthetic membrane system is often complex and multilayered (∞Figure 17.10b), although in some of the simpler cyanobacteria the lamellae are regularly arranged in concentric circles around the periphery of the cytoplasm (Figure 12.79).

Cyanobacteria have only one form of chlorophyll, _chlorophyll a_, and all of them also have characteristic biliprotein pigments, **phycobilins** (∞Figure 17.10a), which function as accessory pigments in photosynthesis. One class of phycobilins, _phycocyanins_, are blue, and together with the green chlorophyll _a_, are responsible for the blue-green color of the bacteria. However,

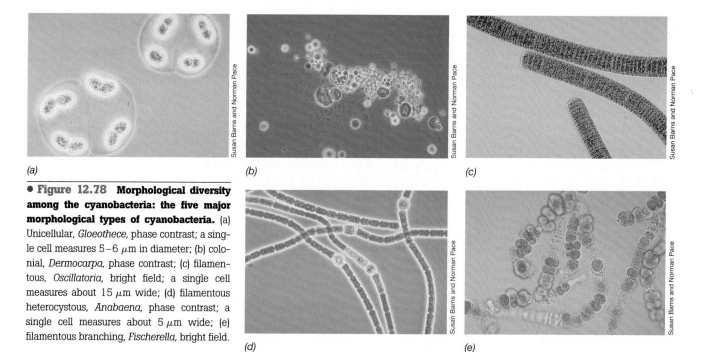

(a)   (b)   (c)

(d)   (e)

● **Figure 12.78 Morphological diversity among the cyanobacteria: the five major morphological types of cyanobacteria.** (a) Unicellular, *Gloeothece*, phase contrast; a single cell measures 5–6 μm in diameter; (b) colonial, *Dermocarpa*, phase contrast; (c) filamentous, *Oscillatoria*, bright field; a single cell measures about 15 μm wide; (d) filamentous heterocystous, *Anabaena*, phase contrast; a single cell measures about 5 μm wide; (e) filamentous branching, *Fischerella*, bright field.

some cyanobacteria produce *phycoerythrin*, a red phycobilin, and species possessing this pigment are red or brown in color.

Structural Variations: Gas Vesicles and Heterocysts

Among the cytoplasmic structures seen in many cyanobacteria are **gas vesicles** (⬡Section 4.12), which are especially common in species that live in open waters (planktonic species). Their function is to regulate cell buoyancy such that cells can remain in a position in the water column where light intensity is optimal for photosynthesis. Some filamentous cyanobacteria form *heterocysts*, which are rounded, seemingly empty cells, usually distributed regularly along a filament or at one end of a filament (Figure 12.80a●). Heterocysts arise from differentiation of vegetative cells and are the sole sites of *nitrogen fixation* (the reduction of N_2 to NH_3, ⬡Section 17.28) in heterocystous cyanobacteria. In *Anabaena*, a

well-studied heterocystous cyanobacterium, complex gene rearrangements occur within the heterocyst to yield a contiguous cluster of *nif* genes that can be expressed as a unit (*nif* genes encode nitrogenase, ⬡Section 17.28).

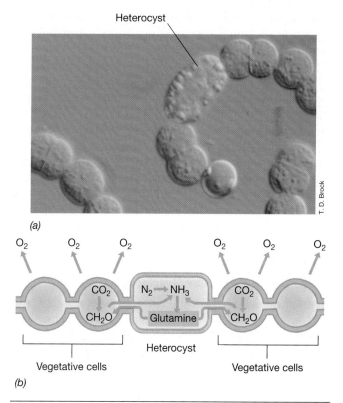

Heterocyst

(a)

O_2 O_2 O_2 O_2 O_2 O_2

CO_2 $N_2 \rightarrow NH_3$ CO_2

CH_2O Glutamine CH_2O

Heterocyst

Vegetative cells   Vegetative cells

(b)

● **Figure 12.80 Heterocysts.** (a) Heterocysts in the cyanobacterium *Anabaena*. Heterocysts are the sole site of nitrogen fixation in heterocystous cyanobacteria. (b) Model for the operation of a heterocyst. The heterocyst lacks oxygen-producing ability (Photosystem II, ⬡Section 17.5) and obtains the needed reductant for nitrogen fixation from organic matter produced by adjacent vegetative cells. Glutamine is the form of fixed nitrogen transported from heterocysts to vegetative cells.

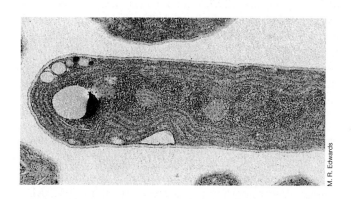

● **Figure 12.79 Thylakoids in cyanobacteria.** Electron micrograph of a thin section of the cyanobacterium *Synechococcus lividus*. A cell is about 5 μm in diameter. Note thylakoid membranes running parallel to the cell wall.

Table 12.32 Genera and grouping of cyanobacteria

Group	Genera	DNA (mol % GC)
Group I. Unicellular: single cells or cell aggregates	*Gloeothece* (Figure 12.78*a*), *Gloeobacter, Synechococcus, Cyanothece, Gloeocapsa, Synechocystis, Chamaesiphon, Merismopedia*	35–71
Group II. Pleurocapsalean: reproduce by formation of small spherical cells called baeocytes produced through multiple fission	*Dermocarpa* (Figure 12.78*b*), *Xenococcus, Dermocarpella, Pleurocapsa, Myxosarcina, Chroococcidiopsis*	40–46
Group III. Oscillatorian: filamentous cells that divide by binary fission in a single plane	*Oscillatoria* (Figure 12.78*c*), *Spirulina, Arthrospira, Lyngbya, Microcoleus, Pseudanabaena*	40–67
Group IV. Nostocalean: filamentous cells that produce heterocysts	*Anabaena* (Figure 12.78*d*), *Nostoc, Calothrix, Nodularia, Cylinodrosperum, Scytonema*	38–46
Group V. Branching: cells divide to form branches	*Fischerella* (Figure 12.78*e*), *Stigonema, Chlorogloeopsis, Hapalosiphon*	42–46

Heterocysts have intercellular connections with adjacent vegetative cells, and there is mutual exchange of materials between these cells. The products of photosynthesis move from vegetative cells to heterocysts, and products of nitrogen fixation move from heterocysts to vegetative cells (Figure 12.80*b*). Heterocysts are low in phycobilin pigments and *lack* photosystem II, the oxygen-evolving photosystem that generates reducing power from H_2O (∞Section 17.5). Without photosystem II heterocysts are unable to fix CO_2 and thus lack the necessary electron donor to reduce N_2 to NH_3; fixed carbon imported to the heterocyst from an adjacent vegetative cell solves this problem (Figure 12.80*b*).

Heterocysts are surrounded by a thickened cell wall containing large amounts of glycolipid, which serves to slow the diffusion of O_2 into the cell. Because of the oxygen lability of the enzyme nitrogenase (∞Section 17.29), the heterocyst maintains an anoxic environment, and by doing so stabilizes the nitrogen-fixing system in organisms that are not only aerobic but also oxygen producing. Indeed, some nonheterocystous filamentous cyanobacteria produce nitrogenase and fix nitrogen in normal vegetative cells if they are grown anaerobically by vigorous bubbling with N_2 to remove O_2.

Cyanophycin and Other Structures

A structure called **cyanophycin** can be seen in electron micrographs of many cyanobacteria. This structure is a copolymer of aspartic acid and arginine:

Asp — Asp — Asp — Asp —Asp —
   | | | | |
Arg Arg Arg Arg Arg

and can constitute up to 10% of the cell mass. Cyanophycin is a nitrogen storage product, and when nitrogen in the environment becomes deficient, this polymer is broken down and used as a cellular nitrogen source. Cyanophycin is also an energy reserve in cyanobacteria. Arginine, derived from cyanophycin, can be hydrolyzed to yield ornithine, with the production of ATP through the activity of the enzyme *arginine dihydrolase*, with carbamyl phosphate (∞Section 17.19) occurring as an intermediate:

$$\text{Arginine} + \text{ADP} + P_i + H_2O \rightarrow$$
$$\text{Ornithine} + 2\,NH_3 + CO_2 + \text{ATP}$$

Arginine dihydrolase is present in many cyanobacteria and may function as a source of ATP for maintenance purposes during dark periods.

Many cyanobacteria exhibit gliding motility; true rotating flagella have never been found. Gliding occurs only when the cell or filament is in contact with a solid surface or with another cell or filament. In some cyanobacteria gliding is not a simple translational movement but is accompanied by rotations, reversals, and flexings of filaments. Most gliding species exhibit directional movement toward light (phototaxis), and chemotaxis (∞Section 4.16) may occur as well.

Among the filamentous cyanobacteria, fragmentation of the filaments often occurs by formation of **hormogonia** (Figure 12.81*a,b*●), which break away from the filaments and glide off. In some species, resting spores or **akinetes** (Figure 12.81*c*) are formed, which protect the organism during periods of darkness, drying, or freezing. These are cells with thickened outer walls; they germinate through the breakdown of the outer wall and outgrowth of a new vegetative filament. However, even the vegetative cells of many cyanobacteria are relatively resistant to drying or low temperatures.

Physiology of Cyanobacteria

The nutrition of cyanobacteria is simple. Vitamins are not required, and nitrate or ammonia is used as nitrogen source. Nitrogen-fixing species are common. Most species tested are obligate phototrophs, being unable to grow in the dark on organic compounds. However, some cyanobacteria can assimilate simple organic compounds such as glucose and acetate if light is present (photoas-

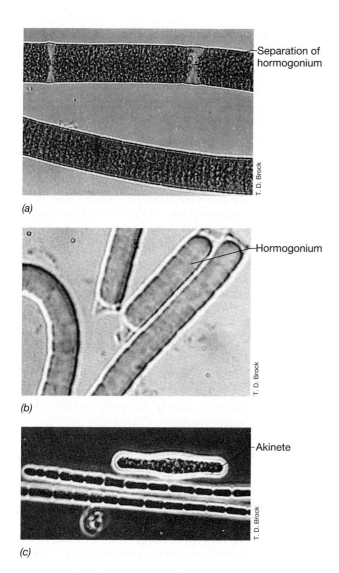

(a)

Separation of
hormogonium

T. D. Brock

(b)

Hormogonium

T. D. Brock

(c)

Akinete

T. D. Brock

● **Figure 12.81** **Structural differentiation in filamentous cyano-bacteria.** (a) Initial stage of hormogonium formation in *Oscillatoria*. Notice the empty spaces where the hormogonium is separating from the filament. (b) Hormogonium of a smaller *Oscillatoria* species. Notice that the cells at both ends are rounded. Nomarski interference contrast microscopy. (c) Akinete (resting spore) of *Anabaena* by phase contrast.

similation). A few cyanobacteria, mainly filamentous species, can actually grow in the dark on glucose or sucrose, using the sugar as both carbon and energy source.

Several metabolic products of cyanobacteria are of considerable practical importance. Many cyanobacteria produce potent neurotoxins, and during water blooms when massive accumulations of cyanobacteria may develop, animals ingesting such water may be killed. Many cyanobacteria are also responsible for the production of earthy odors and flavors in fresh waters, and if such waters are used as drinking water sources, aesthetic problems may arise. The major compound produced is *geosmin* (trans-1,10-dimethyl-trans-9-decalol). This substance is also produced by many actinomycetes (see the discussion in Section 12.24) and is responsible for the distinctive "earthy" odor of moist, freshly turned soil.

Ecology and Phylogeny of Cyanobacteria

Cyanobacteria are widely distributed in nature in terrestrial, freshwater, and marine habitats. In general, they are more tolerant of environmental extremes than are algae and are often the dominant or sole oxygenic phototrophic organisms in hot springs (⊙Table 6.1), saline lakes, and other extreme environments. Many species are found on the surfaces of rocks or soil and occasionally even within rocks themselves (⊙Figure 14.38). In desert soils subject to intense sunlight, cyanobacteria often form extensive crusts over the surface, remaining dormant during most of the year and growing during the brief winter and spring rains. In shallow marine bays, where relatively warm seawater temperatures exist, cyanobacterial mats of considerable thickness may form. Freshwater lakes, especially those that are rich in nutrients, may develop blooms of cyanobacteria (⊙Figure 19.10b). A few cyanobacteria are symbionts of liverworts, ferns, and cycads; a number are found as the phototrophic component of lichens. In the case of the water fern *Azolla* (⊙Sections 17.28 and 19.22), it has been shown that the cyanobacterial endophyte (a species of *Anabaena*) fixes nitrogen that becomes available to the plant.

Base compositions of genomic DNA of a variety of cyanobacteria have been determined. Those of the unicellular forms vary from 35 to 71% GC, a range so wide as to suggest that this group contains many members with little genetic relationship to each other. On the other hand, the values for the heterocyst formers vary much less, from 38 to 46% GC. Phylogenetically, cyanobacteria group along morphological lines in most cases. Filamentous heterocystous and nonheterocystous species form distinct groups, as do the branching forms. However, unicellular cyanobacteria are phylogenetically highly diverse, with different representatives showing phylogenetic relationships to different morphological groups.

12.26 Prochlorophytes and Chloroplasts

Key Genera: *Prochlorococcus, Prochloron, Prochlorothrix*

Prochlorophytes are oxygenic phototrophs that contain chlorophyll *a* and *b* but do *not* contain phycobilins. Prochlorophytes therefore resemble both cyanobacteria (because they are prokaryotic and have chlorophyll *a*) and the green plant/green alga chloroplast (because they contain chlorophyll *b* instead of phycobilins). Phylogenetically, prochlorophytes show specific relationships to cyanobacteria.

Prochloron

Prochloron was the first prochlorophyte discovered. It is found in nature as a symbiont of marine invertebrates (didemnid ascidians), but it has not been cultured in the laboratory. Cells of *Prochloron* expressed from the cavities of didemnid tissue are roughly spherical (Figure 12.82●), and 8–10 μm in diameter.

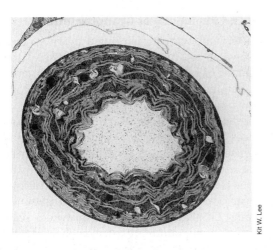

● **Figure 12.82 Electron micrograph of the prochlorophyte *Prochloron*.** Note the extensive intracytoplasmic membranes (thylakoids). Cells are about 10 μm in diameter. Unlike cyanobacteria, prochlorophytes lack phycobiliproteins.

Electron micrographs of thin sections of *Prochloron* (Figure 12.82) show an extensive thylakoid membrane system similar to that observed in the chloroplast (∞Figure 14.6). Further evidence that *Prochloron* is phylogenetically a member of the *Bacteria* is the presence of muramic acid in the cell walls, indicating that peptidoglycan is present (∞Section 4.8). The carotenoids of *Prochloron* are similar to those of cyanobacteria, predominantly β-carotene and zeaxanthin. The GC ratio of genomic DNA from samples of *Prochloron* obtained from different ascidians varies from 31 to 41%, indicating genetic heterogeneity. Thus, different species of *Prochloron* probably exist, but confirmation must await laboratory culture and study of pure cultures.

Prochlorothrix and *Prochlorococcus*

Prochlorothrix is a filamentous prochlorophyte (Figure 12.83●) that can be grown in pure culture. Like *Prochloron*, *Prochlorothrix* contains chlorophylls *a* and *b* and lacks phycobilins, although the thylakoid membranes are less well developed than in *Prochloron* (compare Figures 12.82 and 12.83*b*).

A novel prochlorophyte, *Prochlorococcus*, inhabits the euphotic zone of the open oceans. Cells of these phototrophs are small cocci, measuring less than 1 μm in diameter (∞Figure 19.11*a*). Like other prochlorophytes, cells of *Prochlorococcus* contain chlorophyll *b*. However, *Prochlorococcus* lacks true chlorophyll *a* and produces instead a modified form of chlorophyll *a* called *divinyl chlorophyll a*. Cells of *Prochlorococcus* also contain α-(instead of β-) carotene, a pigment previously unknown in prokaryotes. Because their numbers in the oceans are relatively large ($10^4 - 10^5$ cells/ml), prochlorophytes like *Prochlorococcus* probably have considerable ecological significance as primary producers in open ocean waters. A variety of other prochlorophytes have been isolated including *Acaryochloris* (Figure 12.83*c*), which contains chlorophyll *d* as its major pigment. Chlorophyll is also present in a variety of algae (eukaryotic cells; ∞Section 14.13), some species of which are extremely small (∞Figure 19.11c).

Prochlorophytes, Chloroplasts, and Evolution

Based on our discussion of endosymbiosis (∞Sections 2.6, 11.4, and 14.4), the evolutionary significance of prochlorophytes should be apparent. Until the discovery of prochlorophytes, it was always assumed that the chloroplast originated from endosymbiotic association of *cyanobacteria* with a primitive eukaryotic cell. However,

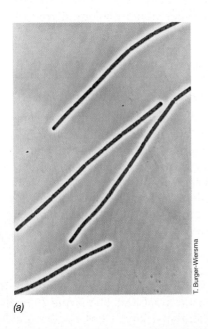

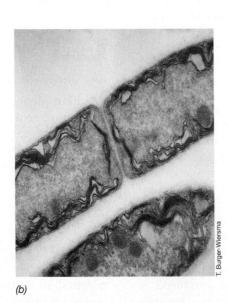

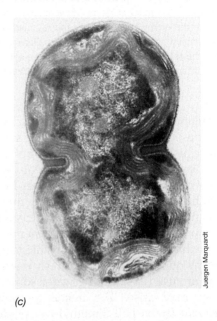

(a)   *(b)*   *(c)*

● **Figure 12.83 Phase and electron micrographs of the filamentous prochlorophytes *Prochlorothrix* and *Acaryochloris*.** (a) Phase contrast. (b) Electron micrograph of thin section showing arrangement of membranes. The diameter of cells is about 2 μm. (c) *Acaryochloris*. This prochlorophyte contains chlorophyll *d* as its main chlorophyll pigment. A cell is about 1.5 μm in diameter.

this hypothesis has been scientifically unsatisfying for at least one major reason: How did the green plant chloroplast evolve the pigment complement it has today if it originated from a cyanobacterial endosymbiont that contained *phycobilins* instead of chlorophyll *b*? The hypothesis that *prochlorophytes* instead of cyanobacteria were the ancestors of the green plant chloroplast eliminates this major point of contention. However, phylogenetic analyses do not show *Prochloron*, *Prochlorococcus*, or *Prochlorothrix* to be the *immediate* ancestors of the green plant chloroplast. Instead, prochlorophytes, cyanobacteria, and the plant chloroplast all *shared* a common ancestor, and remain distinct lineages in the cyanobacterial phylum.

Another hypothesis does account for the pigmentation patterns observed in prokaryotic oxygenic phototrophs. Cyanobacteria and prochlorophytes may have evolved from ancestors that contained phycobilins and chlorophylls other than just chlorophyll *a*, either chlorophyll *b* or *d*. From here, cyanobacterial lineages evolved to lose accessory chlorophylls while prochlorophyte lineages dispensed with phycobilins. The complement of pigments that we find in each group today, then, likely represents the best combination of photosynthetic pigments for fitness in their particular habitat.

12.25–12.26 Concept Check

Cyanobacteria and prochlorophytes are oxygenic phototrophic prokaryotes. Prochlorophytes differ most clearly from cyanobacteria in that prochlorophytes contain chlorophyll *b* or *d* and lack phycobilins. Oxygen in Earth's atmosphere is thought to have originated from cyanobacterial photosynthesis.

◆ Describe at least three ways in which cyanobacteria differ from purple bacteria.

◆ What is a *heterocyst* and what is its function?

◆ How are cyanobacteria, prochlorophytes, and the chloroplasts of corn plants similar and how do they differ?

◆ Of what ecological significance is *Prochlorococcus*?

V PHYLUM 5: CHLAMYDIA

12.27 The Chlamydia

Key Genera: *Chlamydia, Chlamydophila*

Organisms of the genera *Chlamydia* and *Chlamydophila* are obligately parasitic bacteria with poor metabolic capacities that form a distinct phylum of *Bacteria* (see Figure 12.1). Several key species are recognized (Table 12.33): *Chlamydophila psittaci*, the causative agent of the disease *psittacosis*; *Chlamydia trachomatis*, the causative agent of *trachoma* and a variety of other human diseases; and *Chlamydophila pneumoniae*, the cause of a variety of respiratory syndromes (Table 12.33).

Psittacosis is an epidemic disease of birds that is occasionally transmitted to humans and causes pneumonia-like symptoms. **Trachoma** is a debilitating disease of the eye characterized by vascularization and scarring of the cornea. Trachoma is the leading cause of blindness in humans. Other strains of *Chlamydia trachomatis* infect

---

Table 12.33 Differential characteristics of species of the genera *Chlamydia* and *Chlamydophila*

Characteristic	*Chlamydia trachomatis*	*Chlamydophila psittaci*	*Chlamydophila pneumoniae*
Hosts	Humans	Birds, mammals, occasionally humans	Humans
Usual site of infection	Mucous membrane	Multiple sites	Respiratory mucosa
Human-to-human transmission	Common	Rare	Probable
Mol % GC of DNA	42–45	39–43	40
Percent homology to *C. trachomatis* DNA by DNA:DNA hybridization[a]	100	10	10
DNA, kilobase pairs/genome (*Escherichia coli* = 4600)	1000	550	~1000
Human diseases	Trachoma, otitis media, nongonococcal urethritis (males), urethral inflammation (females), lymphogranuloma venereum, cervicitis	Psittacosis	Respiratory syndromes
Domesticated animal diseases	—	Avian chlamydiosis (parrots, parakeets), pneumonia, synovial tissue arthritis, or conjunctivitis (kittens, lambs, calves, piglets, foals)	—

[a] For discussion of DNA:DNA hybridization, see Section 11.11.

the genitourinary tract, and chlamydial infections are currently one of the leading sexually transmitted diseases (⚬Section 26.13). A comparison of the properties of *Chlamydophila psittaci, Chlamydia trachomatis,* and *Chlamydophila pneumoniae* and the diseases they cause is shown in Table 12.33.

Molecular and Metabolic Properties

Besides being disease entities, the chlamydias are intriguing because of the biological, evolutionary, and metabolic problems they pose. Biochemical studies show that the chlamydias have gram-negative-type cell walls, and they have both DNA and RNA, that is, they are clearly cellular. Electron microscopy of thin sections of infected cells shows cells dividing by binary fission (Figure 12.84●).

The biosynthetic capacities of the chlamydias are much more limited than even the rickettsias, the other group of obligate intracellular parasites known among the *Bacteria* (see Section 12.13). Indeed, for some time it was thought that chlamydias were "energy parasites," obtaining not only biosynthetic intermediates from their hosts, as do the rickettsias, but also ATP. However, this hypothesis has been questioned following the sequencing of the genome of *C. trachomatis* (⚬Section 15.3). The approximately 1 Mbp chromosome of *C. trachomatis* contains easily recognizable genes for ATP synthesis and even contains a complement of genes encoding peptidoglycan biosynthetic functions. This suggests that this organism may well contain peptidoglycan even though chemical analyses for this cell-wall polymer have been negative. Nevertheless, the chlamydias still probably have the simplest biochemical capacities of all known *Bacteria*, and a summary of this is shown in Table 12.34.

Interestingly, the *C. trachomatis* genome lacks a gene encoding the protein FtsZ, a key protein involved in septum formation during cell division (⚬Section 6.2). This

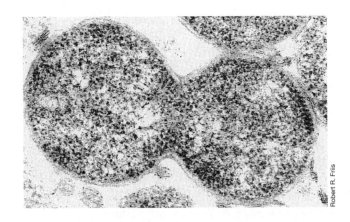

● **Figure 12.84** *Chlamydia.* Electron micrograph of a thin section of a dividing cell (reticulate body, see Figure 12.85) of *Chlamydia psittaci,* a member of the psittacosis group, within a mouse tissue culture cell. A single chlamydial cell is about 1 μm in diameter.

protein was previously thought to be indispensable for growth of all prokaryotes, both *Archaea* and *Bacteria.* Moreover, genes are present in *C. trachomatis* that have a distinct "eukaryotic look" to them, suggesting that *C. trachomatis* has picked up some host genes that may encode functions that assist it in its pathogenic lifestyle (Table 12.33; ⚬Sections 21.7 and 26.13).

Life Cycle of *Chlamydia*

The life cycle of a typical chlamydial species is shown in Figure 12.85●. Two cellular types are seen in the life cycle: (1) a small, dense cell, called an **elementary body**, which is relatively resistant to drying and is the means of *dispersal* of the agent, and (2) a larger, less dense cell, called a **reticulate body**, which divides by binary fission and is the *vegetative* form.

Elementary bodies are nonmultiplying cells specialized for infectious transmission. By contrast, reticulate bodies are noninfectious forms that function only to

Table 12.34 Comparison of obligate intracellular parasites: rickettsias, chlamydias, and viruses			
Property	**Rickettsias**	**Chlamydias**	**Viruses**
Structural			
Nucleic acid	RNA and DNA	RNA and DNA	Either RNA or DNA (single- or double-stranded), never both
Ribosomes	Present	Present	Absent
Cell wall	Peptidoglycan present	Peptidoglycan present[a]	No wall
Structural integrity during multiplication	Maintained	Maintained	Lost
Metabolic capacities			
Macromolecular synthesis	Carried out	Carried out	Only with use of host machinery
ATP-generating system	Present	Present [a]	Absent
Capable of oxidizing glutamate	Yes	No	No
Sensitivity to antibacterial antibiotics	Sensitive	Sensitive (except for penicillin)	Resistant
Phylogeny	Alpha Proteobacteria	Chlamydial phylum	Not cells

[a] The genome of one chlamydial species, *Chlamydia trachomatis,* has been entirely sequenced (⚬Section 15.4), and genes for peptidoglycan synthesis and ATP synthesis are present. However, the lack of penicillin sensitivity of the chlamydia raises doubt whether peptidoglycan is synthesized.

Elementary body

● ▭ ~0.3 μm

Rigid cell wall,
nongrowing,
infectious

Reticulate body

● ▭ ~1μm

Fragile cell wall,
multiplying form,
noninfectious

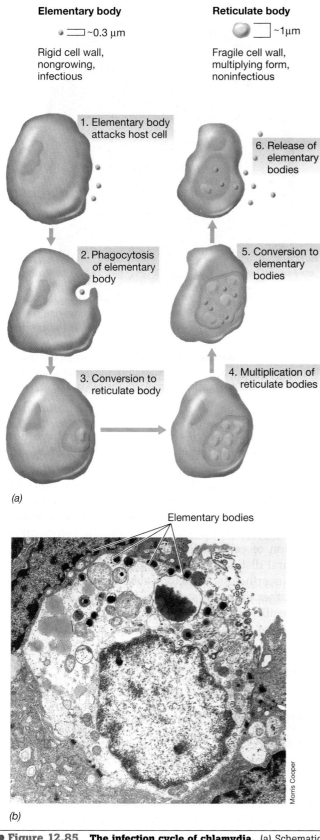

1. Elementary body
   attacks host cell

2. Phagocytosis
   of elementary
   body

3. Conversion to
   reticulate body

4. Multiplication of
   reticulate bodies

5. Conversion to
   elementary
   bodies

6. Release of
   elementary
   bodies

(a)

Elementary bodies

Morris Cooper

(b)

● **Figure 12.85   The infection cycle of chlamydia.** (a) Schematic diagram of the cycle: the whole cycle takes about 48 h. (b) Human chlamydial infection. An infected fallopian tube cell is bursting, releasing mature elementary bodies.

multiply inside host cells to form a large inoculum for transmission. Unlike the rickettsias (see Section 12.13),

the chlamydias are not transmitted by arthropods but are primarily *airborne* invaders of the respiratory system—hence the significance of resistance to drying of the elementary bodies. A dividing reticulate body can be seen in Figure 12.84. After a number of cell divisions, these vegetative cells are converted into elementary bodies that are released when the host cell disintegrates and can then infect other cells. Generation times of 2–3 h have been measured for reticulate bodies, which are considerably faster than those found for the rickettsias.

In sum, the chlamydias appear to have evolved an efficient and effective survival strategy including parasitizing the resources of the host (Table 12.34) and the production of resistant cell forms for transmission. It is thus not surprising that chlamydias have been associated with so many different disease syndromes (Table 12.33; ⚭Section 26.13).

12.27 Concept Check

Chlamydia are extremely small parasitic bacteria that cause a variety of human diseases. Chlamydia contain a very small genome and are apparently deficient in many metabolic functions.

◆ Using the data of Table 12.34 as a guide, how can chlamydias be differentiated from rickettsias? From viruses?

◆ What is the difference between an *elementary* body and a *reticulate* body?

◆ What surprises have emerged from sequencing of the chlamydial genome?

VI PHYLUM 6: PLANCTOMYCES/ PIRELLULA

12.28 *Planctomyces:* A Phylogenetically Unique Stalked Bacterium

Key Genera: *Planctomyces, Pirellula, Gemmata*

This phylum contains a number of morphologically unique bacteria including the genera *Planctomyces, Pirellula, Gemmata,* and *Isosphaera.* The best studied of these has been *Planctomyces* (Figure 12.86●). In Section 12.16 we considered stalked bacteria such as *Caulobacter. Planctomyces* is also a stalked bacterium. However, unlike *Caulobacter,* the stalk of *Planctomyces* is made of protein and does not contain a cell wall or cytoplasm (compare Figure 12.86 with Figure 12.41). The *Planctomyces* stalk presumably functions in attachment but it is a much narrower and finer structure than the prosthecal stalk of *Caulobacter.*

Other Features of the *Planctomyces* Group

Planctomyces and relatives are also of interest because they lack peptidoglycan and their cell walls are of an S-layer type (⚭Sections 4.8 and 4.10), consisting of protein containing

IX PHYLUM 9: THE CYTOPHAGA GROUP

Key Genera: *Cytophaga, Flexibacter, Rhodothermus, Salinibacter*

12.31 *Cytophaga* and Relatives

Organisms of the *Cytophaga* group are long, slender, gram-negative rods, often with pointed ends, that move by gliding (Figure 12.90*a,b*●). The related genus, *Sporocytophaga*, is similar to *Cytophaga* in morphology and physiology, but the cells form resting spherical structures called *microcysts* (Figure 12.90*d*), similar to those produced by some fruiting myxobacteria (see Section 12.17). They are widespread in soil and water, often in great abundance.

Many cytophagas digest polysaccharides like cellulose (Figure 12.90*c*), agar (Figure 12.90*a*), or chitin. The cellulose decomposers can be easily isolated by placing small crumbs of soil on pieces of cellulose filter paper laid on the surface of mineral salts agar. The bacteria attach to and digest the cellulose fibers, forming spreading colonies (Figure 12.90*c*). The cytophagas do not produce soluble, extracellular, cellulose-digesting enzymes (cellulases). Instead, their cellulases remain attached to the cell envelope, which probably explains why the cells must adhere to cellulose fibrils in order to digest them.

In pure culture, *Cytophaga* can be cultured on agar containing embedded cellulose fibers, the presence of the organism being indicated by the clearing that occurs as the cellulose is digested (Figure 12.90*c*; see also Figure 17.62).

Species of *Cytophaga* and *Sporocytophaga* are obligately aerobic and probably account for much of the cellulose digestion that occurs by prokaryotes in oxic environments in nature. A number of *Cytophaga* species are also fish pathogens and can cause serious problems in the cultivated fish business. Two of the most important diseases are *columnaris disease*, caused by *C. columnaris*, and *cold-water disease*, caused by *C. psychrophila*. Both diseases preferentially affect stressed fish, such as those living in waters receiving pollutant discharges or living in high-density confinement situations such as fish hatcheries and aquaculture operations. Infected fish show tissue destruction, frequently around the gills, and this may stem from the fact that *Cytophaga* species isolated from infected fish are typically strongly proteolytic.

The genus *Flexibacter* differs from the cytophagas in that the species usually require complex media for good growth and are not cellulolytic. Cells of some *Flexibacter* species also undergo changes in cell morphology from long, gliding, threadlike filaments lacking cross-walls, to short, nonmotile rods. Many species are pigmented due to carotenoids located in the cytoplasmic membrane, or related pigments called *flexirubins*, located in the gram-negative outer membrane. *Flexibacter* species are com-

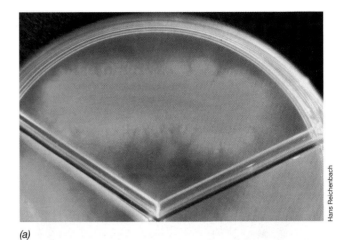
(a)

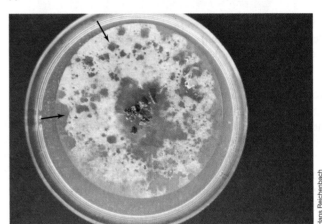
(b)

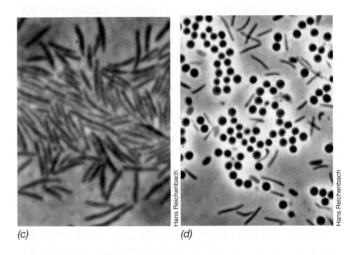
(c)   (d)

● **Figure 12.90** *Cytophaga* **and** *Sporocytophaga*. (a) Streak of an agarolytic marine *Cytophaga* species hydrolyzing agar in the Petri dish. (b) Colonies of *Sporocytophaga* growing on cellulose. Note the clearing zones (arrows) where the cellulose has been degraded. (c) Phase-contrast photomicrograph of cells of *C. hutchinsonii* grown on cellulose filter paper (cells are about 1.5 μm in diameter). (d) Phase-contrast photomicrograph of the rod-shaped cells and spherical microcysts of *Sporocytophaga myxococcoides* (cells are about 0.5 μm and microcysts about 1.5 μm in diameter).

mon soil and freshwater saprophytes, and none have been identified as pathogens.

Rhodothermus/Salinibacter

The genera *Rhodothermus* and *Salinibacter* fall into the *Cytophaga* phylum, but are only distantly relatives. *Rhodothermus* and *Salinibacter* consist of gram-negative, red- or yellow-pigmented, obligatory aerobic chemoorganotrophic bacteria. *Rhodothermus* is thermophilic, with a temperature optimum around 60–65°C. *Rhodothermus* grows best on sugars or simple and complex polysaccharides. The organism inhabits shallow-water submarine hot springs and has also been detected in terrestrial hot springs. *Rhodothermus* produces thermal stable hydrolytic enzymes of biotechnological interest, including an amylase (that degrades starch), a cellulase (cellulose), and a xylanase (that degrades hemicelluloses, abundant in plant cell walls), among many others.

Salinibacter is a genus of extremely halophilic red bacteria that is perhaps the most salt-tolerant and salt-requiring of all *Bacteria*. In fact, its salt requirements rival those of extremely halophilic *Archaea*, such as *Halobacterium* (∞ Section 13.3). *Salinibacter ruber*, the only known species, lives in saltern cystallization ponds and related highly saline environments. *Salinibacter* shares with *Halobacterium* the use of K^+ as a compatible solute, a property that is rarely found among halophiles of the domain *Bacteria* (most halophilic *Bacteria* either synthesize or accumulate organic solutes to maintain water balance in their salty environments, ∞ Section 6.14). In contrast to the highly hydrolytic *Rhodothermus*, *Salinibacter* grows best with amino acids as electron donors, similar to *Halobacterium*. Thus, although phylogenetically unique, we see in *Halobacterium* and *Salinibacter* a likely case of convergent evolution, where physiological strategies have converged along the same pattern, probably because of the unique demands of the extreme environment that these organisms share.

 12.28–12.31 Concept Check

The *Planctomyces* group contains stalked, budding bacteria while the flavobacteria contain a variety of gram-negative *Bacteria* motile by either flagella or by gliding associated with animals, food, and the soil. Members of the Verrucomicrobia are distinguished by their multiple prosthecate cells. The *Cytophaga* group includes a variety of obligately aerobic chemoorganotrophic bacteria in soil, water, hot springs, and thermal environments.

◆ What is unique about the cell wall and arrangement of DNA in *Planctomyces*?

◆ How does the stalk of *Planctomyces* differ from the stalk of *Caulobacter*?

◆ Where might you find large numbers of *Bacteroides* cells in nature?

◆ Describe a method for isolating *Cytophaga* species from nature.

◆ Contrast the habitat and physiology of *Rhodothermus* and *Salinibacter*.

 **PHYLUM 10: GREEN SULFUR BACTERIA**

12.32 *Chlorobium* and Other Green Sulfur Bacteria

Key Genera: *Chlorobium, Chlorobaculum, Prosthecochloris, "Chlorochromatium"*

Green sulfur bacteria are a phylogenetically distinct group of nonmotile anoxygenic phototrophic bacteria that contain only obligately anaerobic and phototrophic species among cultured isolates. The group is morphologically restricted and includes short to long rods (Table 12.35 and Figure 12.91●). Like purple sulfur bacteria they utilize H_2S as an electron donor, oxidizing it first to S^0 and then to SO_4^{2-}. But unlike purple sulfur bacteria, the sulfur produced by green sulfur bacteria resides *outside* the cell (Figure 12.91a; ∞ Figure 17.17b). Most species can also assimilate a few organic compounds in the light (that is, *photoheterotrophy*, ∞ Section 17.4). Strict autotrophy, however, is supported not by the reactions of the Calvin cycle as in purple bacteria, but instead by a reversal of steps in the citric acid cycle (∞ Section 17.7, reverse citric acid cycle), a unique means of autotrophy.

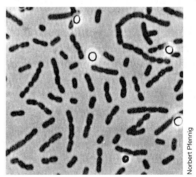

(a)

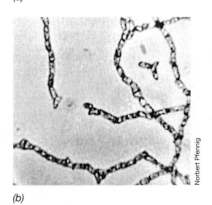

(b)

● **Figure 12.91** **Phototrophic green sulfur bacteria.** (a) *Chlorobium limicola*; cells are about 0.8 μm wide. Note the sulfur granules deposited extracellularly. (b) *Chlorobium clathratiforme*, a bacterium forming a three-dimensional network; cells are about 0.8 μm wide.

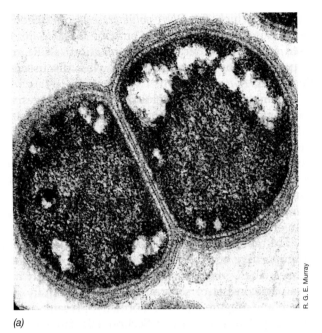

(a)

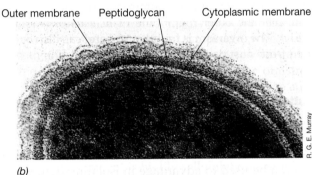

Outer membrane　　Peptidoglycan　　Cytoplasmic membrane

(b)

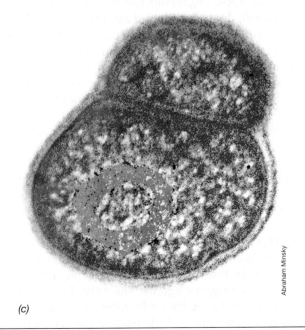

(c)

● **Figure 12.99　The radiation-resistant coccus *Deinococcus radiodurans.*** An individual cell is about 2.5 μm in diameter. (a) Transmission electron micrograph of *D. radiodurans*. Note the outer membrane layer. (b) High-magnification micrograph of wall layer. (c) Transmission electron micrograph of cells of *D. radiodurans* colored to show the toroidal morphology of the nucleoid (green).

to the DNA repair enzyme RecA (∞ Section 10.6), several RecA-independent DNA systems exist in *D. radiodurans* that can repair breaks in single- or double-stranded DNA and excise and repair misincorporated bases. In fact, repair processes are so effective, that the chromosome can even be reassembled from a fragmented state.

It is also thought that the unique arrangement of DNA in *D. radiodurans* cells plays a role in radiation resistance. Cells of *D. radiodurans* always exist as pairs or tetrads (Figure 12.99a). Instead of scattering DNA within the cell as in a typical nucleoid, DNA in *D. radiodurans* is ordered into a toroidal ringlike structure (Figure 12.99c). Repair is then facilitated by the fusion of nucleoids from adjacent compartments, as this provides a platform for homologous recombination. From this extensive recombination, a single repaired chromosome emerges, and the cell containing this chromosome can then grow and divide.

XIII PHYLUM 13: THE GREEN NONSULFUR BACTERIA

This phylum of *Bacteria* is phylogenetically distinct and contains just a few genera, the best known being the anoxygenic phototroph *Chloroflexus*; all cultured representatives are thermophilic. *Thermomicrobium* is a chemotrophic member of this group and is a strictly aerobic, gram-negative rod, growing optimally in complex media at 75°C. Besides its phylogenetic novelty, *Thermomicrobium* is also of interest because of its membrane lipids. Recall that the lipids of *Bacteria* and *Eukarya* contain fatty acids esterified to glycerol (∞ Sections 3.4 and 4.5). By contrast, the lipids of *Thermomicrobium* contain 1,2-dialcohols instead of glycerol, and have neither ester *nor* ether linkages (Figure 12.100●). In addition, cells of *Thermomicrobium* are unusual for species of *Bacteria* in that they lack peptidoglycan, the signature cell-wall polymer of this domain of prokaryotes (∞ Section 11.9).

12.35　*Chloroflexus* and Relatives

Key Genera: *Chloroflexus, Heliothrix, Roseiflexus*

Chloroflexus and most other *green nonsulfur bacteria* (so named because of the physiology of *Chloroflexus*, see later) are filamentous prokaryotes that, along with cyanobacteria, form thick microbial mats in neutral to alkaline hot springs (Figure 12.101●; see also Figure 18.18a). *Chloroflexus*-like organisms have also been found in marine microbial mats. Although an anoxygenic phototroph, *Chloroflexus* is a "hybrid" phototroph in the sense that its photosynthetic mechanism shows

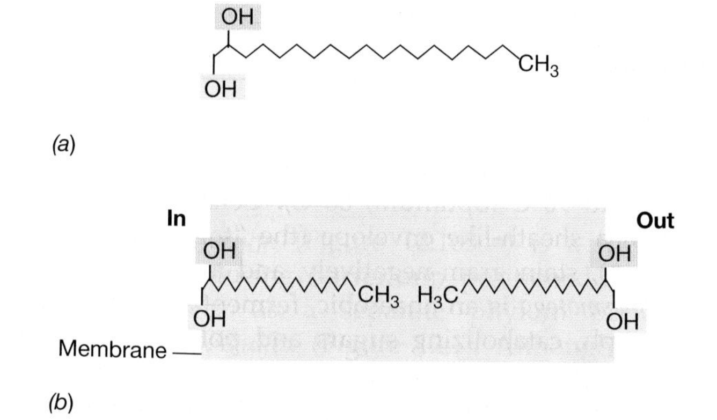

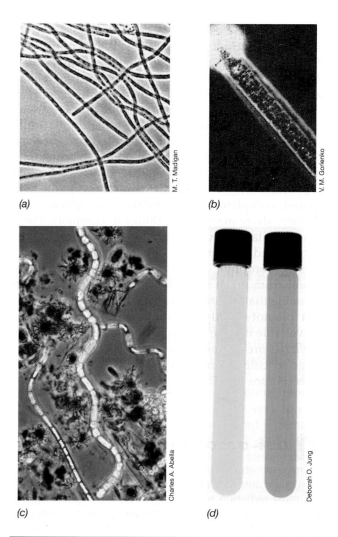

(a)   M. T. Madigan   (b)   V. M. Gorlenko

(c)   Charles A. Abella   (d)   Deborah O. Jung

● Figure 12.101 Green nonsulfur bacteria. (a) Phase photomicrograph of the thermophilic phototroph, *Chloroflexus aurantiacus.* Cells are about 1 μm in diameter. (b) Phase micrograph of the large phototroph *Oscillochloris.* Cells are about 5 μm wide. The brightly contrasting material is a holdfast, used for attachment. (c) Color photomicrograph of filaments of *Chloronema species* growing in a stratified Michigan lake. These cells of *Chloronema* are wavy filaments and about 2.5 μm in diameter. (d) Tube cultures of *Chloroflexus aurantiacus* (right) and *Roseiflexus* (left). *Roseiflexus* lacks bacteriochlorophyll *c* and chlorosomes (∞ Section 17.2 and Figure 17.7) and so lacks the green color of *Chloroflexus.*

**● Figure 12.100 The unusual lipids of *Thermomicrobium.* (a) Membrane lipids from *T. roseum* contain long chain diols like the one shown here (13-methyl-1,2 nonadecanediol). Note that unlike the lipids of other *Bacteria* or of *Archaea* (∞ Section 4.5), neither esternor ether-linked side chains are present. (b) To form a bilayer membrane, dialcohol molecules presumably oppose each other at the methyl groups, with the OH groups being the inner and outer hydrophilic surfaces. Small amounts of the diols have fatty acids esterified to the secondary —OH group (shown in red) while the primary —OH group (shown in green) can bond a hydrophilic molecule like phosphate.

features characteristic of both purple bacteria and green sulfur bacteria. Like the latter, *Chloroflexus* contains bacteriochlorophyll *c* and chlorosomes (see Figure 12.92 for an electron micrograph of chlorosomes). However, the bacteriochlorophyll *a* located in the cytoplasmic membrane of cells of *Chloroflexus* is arranged to form a photosynthetic reaction center structurally similar to those of purple bacteria (by contrast, the reaction center of green sulfur bacteria is structurally quite different, ∞ Figure 17.18).

Because of its phylogeny and unique mechanism of autotrophy (see below), it is possible that *Chloroflexus* may be a vestige of an early form of phototroph that perhaps first evolved a photosynthetic reaction center and then received chlorosome-specific genes later by lateral transfer. From a phylogenetic standpoint, *Chloroflexus* is clearly the earliest known phototrophic bacterium (Figure 12.1).

Physiology of *Chloroflexus*

Physiologically, *Chloroflexus* resembles purple nonsulfur bacteria in that photoautotrophy can occur; however, phototrophic growth occurs best when organic compounds are added as carbon sources (*photoheterotrophy*). *Chloroflexus* also grows well in the dark as a chemoorganotroph by aerobic respiration. Interestingly, and this should be considered in light of the evolutionary position of *Chloroflexus* as the most phylogenetically ancient of anoxygenic phototrophs (see Figure 12.1), autotrophy in *Chloroflexus* is based on a CO_2 incorpora-

tion pathway, the *hydroxypropionate pathway*, unique to this and a few other phylogenetically "ancient" organisms. We consider the biochemistry of this novel autotrophic pathway in Section 17.7.

Other Green Nonsulfur Bacteria

In addition to *Chloroflexus*, other phototrophic green nonsulfur bacteria include the thermophile *Heliothrix* and the large-celled mesophiles *Oscillochloris* (Figure 12.101b) and *Chloronema* (Figure 12.101c). *Oscillochloris* and *Chloronema* are unusual because they are rather

◆ In what ways are *Thermoplasma* and *Picrophilus* similar? In what ways do they differ?

◆ How does *Thermoplasma* strengthen its cell membrane to survive life in the absence of a cell wall?

◆ How does *Ferroplasma* obtain energy for growth?

13.6 Hyperthermophilic Euryarchaeota: Thermococcales and *Methanopyrus*

Key Genera: *Thermococcus, Pyrococcus, Methanopyrus*

A few euryarchaeotes thrive in thermal environments and some are **hyperthermophiles**. We consider here three hyperthermophilic euryarchaeotes that branch on the archaeal tree (Figure 13.1) very near the root. Two of these organisms, *Thermococcus* and *Pyrococcus*, show phenotypic properties very similar to those of hyperthermophilic crenarchaeotes to be discussed in Sections 13.8–13.10 and form a distinct order: the Thermococcales (see Table 13.9). The other, *Methanopyrus*, is a methanogen that closely resembles other methanogens (see Section 13.4 and Table 13.5) in its basic physiology but is unusual in its hyperthermophily and phylogenetic position (Figure 13.1).

Thermococcus and *Pyrococcus*

Thermococcus is a spherical hyperthermophilic euryarchaeote indigenous to anoxic thermal waters in various locations throughout the world. The spherical cells contain a tuft of polar flagella and are thus highly motile (Figure 13.12a●). *Thermococcus* is an obligately anaerobic chemoorganotroph that grows on proteins and other complex organic mixtures (including some sugars) with S^0 as electron acceptor at temperatures from 70 – 95°C.

Pyrococcus is morphologically similar to *Thermococcus* (Figure 13.12b). *Pyrococcus* (the Latin derivation

literally means "fireball") differs from *Thermococcus* primarily by its higher temperature requirements; *Pyrococcus* grows between 70 and 106°C with an optimum of 100°C. *Thermococcus* and *Pyrococcus* are also metabolically quite similar. Proteins, starch, or maltose are oxidized as electron donors and S^0 is the terminal acceptor and is reduced to H_2S. More properties of *Thermococcus* and *Pyrococcus* are described later (see Tables 13.8 and 13.9).

Methanopyrus

Methanopyrus is a rod-shaped hyperthermophilic methanogen (Figure 13.13●). *Methanopyrus* was isolated from sediments near submarine hydrothermal vents and from the walls of "black smoker" hydrothermal vent chimneys (⌀⌀ Sections 13.8 and 19.8 for a discussion of hydrothermal vents). *Methanopyrus* occupies a unique phylogenetic position on the tree of *Archaea* (Figure 13.1). It lies near the base of the archaeal tree and shares phenotypic properties with both the hyperthermophiles (the growth temperature maximum of *Methanopyrus* is 110°C) and the methanogens.

Methanopyrus only produces methane from $H_2 + CO_2$ and grows rapidly for an autotrophic organism (generation time less than 1 h at its temperature optimum of 100°C). Unlike mesophilic methanogens, however, cells of *Methanopyrus* contain large amounts of cyclic 2,3-diphosphoglycerate, a derivative of glycolysis, dissolved in the cytoplasm. This compound, present at more than 1 *molar* concentration in *Methanopyrus*, is

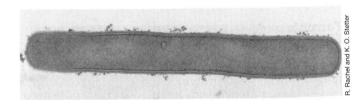

(a)

(b)

● **Figure 13.13 *Methanopyrus*.** *Methanopyrus* grows optimally at 100°C and can make CH_4 only from $CO_2 + H_2$. (a) Electron micrograph of a cell of *Methanopyrus kandleri*, the most thermophilic of all known methanogens (upper temperature limit, 110°C). This cell measures $0.5 \times 8 \ \mu m$. (b) Structure of the novel lipid of *M. kandleri*. This is the normal ether-linked lipid of the *Archaea* (⌀⌀ Section 4.5) with the exception that the side chains are an *unsaturated* form of phytanyl called *geranylgeraniol*. It is thought that this unusual lipid predated the appearance of saturated phytanyl lipids and that *Methanopyrus* is therefore a descendant of a very ancient lineage of *Archaea*; the data of Figure 13.1 support this hypothesis.

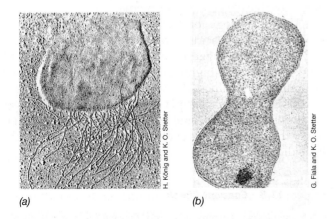

(a)   (b)

● **Figure 13.12 Spherical hyperthermophilic *Archaea* from submarine volcanic areas.** (a) *Thermococcus celer*. Electron micrograph of shadowed cells (note tuft of flagella). (b) Dividing cell of *Pyrococcus furiosus*. Electron micrograph of thin section. Cells of both organisms are about 0.8 μm in diameter.

thought to function as a thermostabilizing agent to prevent denaturation of enzymes and DNA inside the cell (see Section 13.12).

Methanopyrus is also unusual because it contains membrane lipids found in no other known organism. Recall that in the lipids of *Archaea*, the glycerol side chains contain **phytanyl** rather than fatty acids bonded in ether linkage to the glycerol (∞ Section 4.5). In *Methanopyrus*, this ether-linked lipid is an unsaturated form of the otherwise saturated dibiphytanyl tetraethers found in other hyperthermophilic *Archaea* (Figure 13.13*b*; ∞ Section 4.5 and Figure 4.19). Because of the biochemistry of the pathway for how these lipids are biosynthesized, this lipid is thought to be a primitive characteristic. Along with its phylogenic position, hyperthermophily and anaerobic metabolism, the primitive lipids of *Methanopyrus* make this organism an intriguing model for the earliest life forms (∞ Sections 11.1–11.3, and 13.13).

The discovery of *Methanopyrus* offers an explanation for the source of methane in hot oceanic sediments previously thought to be too hot to support biogenic methanogenesis. At the depth at which *Methanopyrus* was found—approximately 2000 m—water remains liquid at temperatures up to 350°C. This leaves open the possibility that other hyperthermophilic methanogens also exist capable of growth at 110°C or at even higher temperatures.

13.7 Hyperthermophilic Euryarchaeota: The Archaeoglobales

Key Genera: *Archaeoglobus, Ferroglobus*

We will see later that a number of hyperthermophilic Crenarchaeota catalyze anaerobic respirations in which elemental sulfur (S^0) is used as an electron acceptor, being reduced to H_2S (see Table 13.8). Curiously, none of these crenarchaeotes are *sulfate*-reducers, that is, capable of reducing SO_4^{2-} to H_2S. However, one hyperthermophilic euryarchaeote, *Archaeoglobus*, is a true sulfate-reducer and forms a phylogenetically distinct lineage within the Euryarchaeota (Figure 13.1).

Archaeoglobus and Its Genome

Archaeoglobus was isolated from hot marine sediments near hydrothermal vents. In its metabolism, *Archaeoglobus* couples the oxidation of H_2, lactate, pyruvate, glucose, or complex organic compounds to the reduction of sulfate to sulfide. Cells of *Archaeoglobus* are irregular cocci (Figure 13.14*a*●) and cultures grow optimally at 83°C.

Archaeoglobus and methanogens have features in common. We will learn in Chapter 17 about the unique biochemistry of methanogenesis. Briefly, this process requires a series of novel coenzymes. With rare exceptions, these coenzymes have only been found in cells of methanogens (∞ Section 17.17). Surprisingly, however, *Archaeoglobus* also contains these coenzymes, and cultures of this organism actually produce small amounts of

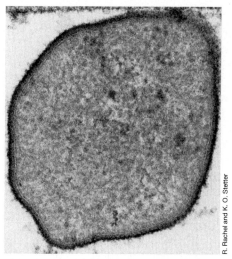

(a)

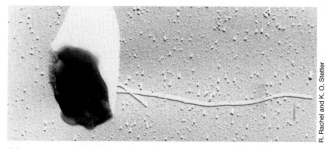

(b)

● **Figure 13.14 Archaeoglobales.** (a) Transmission electron micrograph of the sulfate-reducing hyperthermophile *Archaeoglobus fulgidus*. The cell measures 0.7 μm in diameter. (b) Freeze-etched electron micrograph of *Ferroglobus placidus*, a ferrous iron-oxidizing, nitrate-reducing hyperthermophile. The cell measures about 0.8 μm in diameter.

methane during growth. Thus, *Archaeoglobus*, which also shows a rather close phylogenetic relationship to methanogens (Figure 13.1), may represent a transitional type of organism in a physiological sense, one that bridged the energy-conserving processes of sulfur reduction and methanogenesis.

The genome sequence of *Archaeoglobus* has revealed a conundrum concerning methanogenesis. Although *Archaeoglobus* can make methane, it lacks genes for a key enzyme of methanogenesis, *methyl-CoM reductase* (∞ Section 17.17). Thus, the small amounts of methane that are made by *Archaeoglobus* do not come from the activity of this key enzyme and thus their exact origin is unclear. The genome of *Archaeoglobus* contains about 2400 genes and shares a number of genes with methanogens (see Section 13.4). However, the *Archaeoglobus* genome also contains a number of genes unique to this organism, an example of the broad genetic diversity in *Archaea*. As in species of *Bacteria*, roughly one-third of the genes in every archaeal genome are unique, not present in any other organism (∞ Section 15.3).

Ferroglobus

Ferroglobus (Figure 13.14*b*) is related to *Archaeoglobus* but is not a sulfate-reducing bacterium. Instead, *Ferroglobus* is an iron-oxidizing chemolithotrophic autotroph, conserving energy from the oxidation of Fe^{2+} to Fe^{3+} coupled to the reduction of NO_3^- to NO_2^- plus NO (see Table 13.8). *Ferroglobus* can also use H_2 or H_2S as electron donors in its energy metabolism. *Ferroglobus* was isolated from a shallow marine hydrothermal vent and grows optimally at 85°C.

Ferroglobus is interesting for several reasons, but especially so for its ability to oxidize Fe^{2+} to Fe^{3+} under anoxic conditions. The process might help explain the abundance of Fe^{3+} in ancient rocks, such as the banded iron formations (∞ Section 11.3) previously thought to have originated from the oxidation of Fe^{2+} by O_2 produced by cyanobacteria (oxygenic photosynthesis). The metabolism of *Ferroglobus* thus has implications for dating the origin of cyanobacteria and the subsequent oxygenation of Earth (∞ Sections 11.1 and 11.2). Certain anoxygenic phototrophic bacteria can also oxidize Fe^{2+} under anoxic conditions (∞ Section 17.11). So, several anoxic routes to ancient Fe^{3+} obviously exist, making it a challenge to accurately estimate when cyanobacteria first evolved (∞ Figure 11.8).

 13.6–13.7 Concept Check

Hyperthermophilic euryarchaeota include the Thermococcales, Archaeoglobales, and *Methanopyrus*. All organisms in this group have growth temperature optima above 80°C.

◆ How does the metabolism of *Thermococcus* differ from that of *Methanopyrus*?

◆ Which of the hyperthermophilic Euryarchaeota is a true sulfate reducer?

◆ How does the metabolism of *Ferroglobus* differ from that of *Archaeoglobus*?

III PHYLUM CRENARCHAEOTA

Crenarchaeotes are phylogenetically distinct from Euryarchaeotes (Figure 13.1) and contain representatives that live at both ends of nature's temperature extremes: boiling water and freezing water. However, all cultured crenarchaeotes are *hyperthermophiles* (growth temperature optima above 80°C) and some actually have optima *above* the boiling point of water. We start with an overview of the habitats and energy metabolism of crenarchaeotes and then describe the properties of some key genera.

13.8 Habitats and Energy Metabolism of Crenarchaeotes

A summary of the habitats of Crenarchaeota is shown in Table 13.7. They include very hot and very cold environments. Most hyperthermophilic *Archaea* have been isolated from geothermally heated soils or waters containing elemental sulfur and sulfides (Figure 13.15●) and most species metabolize sulfur in one way or another. In terrestrial environments, sulfur-rich springs, boiling mud, and soils may have temperatures up to 100°C and are mildly to extremely acidic owing to production of sulfuric acid (H_2SO_4) from the biological oxidation of H_2S and S^0 (∞ Sections 17.10 and 19.13). Such hot, sulfur-rich environments, called **solfataras**, are found throughout the world (Figure 13.15), including Italy, Iceland, New Zealand, and Yellowstone National Park in Wyoming (USA).

Depending on the surrounding geology, solfataras can be mildly acidic to slightly alkaline, pH 5–8, or extremely acidic, with pH values below 1. Hyperthermophilic crenarchaeotes have been obtained from both types of environments, but the majority of these organisms inhabit neutral or mildly acidic hot habitats. In addition to these natural habitats, hyperther-

---

Table 13.7 Habitats of Crenarchaeota

Characteristic	Thermal area[a]		Nonthermal area[b]
	Terrestrial	**Marine**	
Locations	Solfataras (hot springs, fumaroles, mudpots, steam-heated soils); geothermal power plants; deep in Earth's crust	Submarine hot springs, hot sediments and vents ("black smokers"); deep oil reservoirs	Planktonic in oceans worldwide; near-shore and deep Antarctic waters; sea ice; symbionts of marine sponges
Temperature	Surface to 100°C; subsurface, above 100°C	Up to 400°C (smokers)	−2 to +4°C
Salinity/pH	Usually less than 1% NaCl; pH 0.5–9	Moderate, about 3% NaCl; pH 5–9	3–8% NaCl; pH 7–9
Gases and other nutrients	CO_2, CO, CH_4, H_2, H_2S, S^0, $S_2O_3^{2-}$, SO_4^{2-}, NH_4^+, N_2	Same as for terrestrial	CO_2, N_2, O_2; for chemolithotrophs, inorganic substrates such as NH_4^+

[a] See Figures 13.10 and 13.15 (∞ Figures 19.19–19.21).

[b] See Figure 13.16. Some species of Euryarchaeota are also present in marine waters.

(a)

(b)

(c)

(d)

● **Figure 13.15** **Habitats of hyperthermophilic *Archaea.*** (a) A typical solfatara in Yellowstone National Park. Steam rich in hydrogen sulfide rises to the surface of the earth. Because of the heat and acidity, only prokaryotes are present. (b) Sulfur-rich hot spring, a habitat containing dense populations of *Sulfolobus.* The acidity in solfataras and sulfur springs comes from the oxidation of H_2S and S^0 to H_2SO_4 (sulfuric acid) by *Sulfolobus* and related prokaryotes. (c) A typical boiling spring of neutral pH in Yellowstone Park; Imperial Geyser. Many different species of hyperthermophilic *Archaea* may reside in such a habitat. (d) An acidic iron-rich geothermal spring, another *Sulfolobus* habitat. Here the oxidation of Fe^{2+} to Fe^{3+} (rather than S^0 to SO_4^{2-}) causes acidic conditions ($Fe^{3+} + 3\,H_2O \rightarrow Fe(OH)_3 + 3\,H^+$).

mophilic *Archaea* also thrive within artificial thermal habitats, in particular the boiling outflows of geothermal power plants.

Hyperthermophilic Crenarchaeota also inhabit undersea hot springs called **hydrothermal vents**. We discuss the geology and microbiology of these habitats in Section 19.8. Here it is only necessary to note that submarine waters can be much hotter than surface waters because the water is under pressure. Indeed, all hyperthermophiles with growth temperature optima above 100°C have come from submarine sources. The latter include both shallow (2–10 meter depth) vents such as those off the coast of Vulcano, Italy, to deep (2000–4000 meter depth) vents near ocean-spreading centers (∞ Section 19.8). Deep vents are the hottest habitats so far known to yield microorganisms.

Cold-Dwelling Crenarchaeotes

In contrast to hyperthermophiles, cold-dwelling crenarchaeotes (Table 13.7) have been identified from community sampling of ribosomal RNA genes from many nonthermal environments. For example, using fluorescent phylogenetic probes (∞ Sections 11.7 and 18.4), crenarchaeotes have been found in marine waters worldwide (Figure 13.16*b*●). In stark contrast to the hyperthermophiles, marine crenarchaeotes thrive even in frigid waters and sea ice, such as those near the Antarctic (Figure 13.16*a*). These organisms are *planktonic* (suspended freely or attached to suspended particles in the water column) and occur in significant numbers ($\sim 10^4$/ml) for waters that are both nutrient poor and very cold (2–4°C in seawater and less than 0°C in sea-ice). Although their physiology is still a mystery, lipid analyses of marine crenarchaeotes filtered from seawater have shown that they contain ether-linked lipids, the hallmark of the *Archaea* (∞ Sections 4.5 and 11.9). The ecology of marine crenarchaeotes is discussed in Section 19.6.

Cold-dwelling species of Euryarchaeota are also present in marine environments. From studies thus far it appears that these organisms are more abundant in temperate surface waters while crenarchaeotes seem to be more abundant in cold (Figure 13.16*a*) waters and in waters of great depth (∞ Section 19.6 and Figure 19.13).

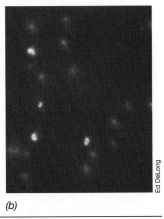

(a)   (b)

● **Figure 13.16 Cold-dwelling Crenarchaeota.** (a) Photo of the Antarctic peninsula taken from shipboard. The frigid waters that lie under the surface ice shown here are habitats for cold-dwelling crenarchaeotes. (b) Fluorescence photomicrograph of seawater treated with a phylogenetic stain composed of a green fluorescing dye attached to an oligonucleotide complementary to a signature sequence in the 16S rRNA of species of Crenarchaeota (⚭ Section 11.7 for a description of the methods used here). Blue cells are stained with DAPI, a stain that stains all cells (⚭ Section 18.3 and Figure 18.6). Cells stained green are thus cold-dwelling Crenarchaeota. See Section 19.6 and Figure 19.13 for a description of *Bacteria* and *Archaea* in open ocean waters.

Energy Metabolism

With a few exceptions, hyperthermophilic Crenarchaeota are obligate anaerobes. Their energy-yielding metabolism is either chemoorganotrophic or chemolithotrophic (or both, for example in *Sulfolobus*) and involves a wide diversity of different electron donors and acceptors (Table 13.8). Fermentation is rare and most bioenergetic strategies are anaerobic respirations (Table 13.8). Energy conservation during these respiratory processes occurs by the same general mechanism widespread in *Bacteria*:

electron transfer within the cytoplasmic membrane leading to the formation of a proton motive force from which ATP is made by way of proton-translocating ATPases (⚭ Section 5.12).

Many hyperthermophilic crenarchaeotes can grow chemolithotrophically under anoxic conditions with H_2 as electron donor and S^0 or NO_3^- as electron acceptor; a few can also oxidize H_2 aerobically (Table 13.8). H_2 respiration with ferric iron (Fe^{3+}) as electron acceptor also occurs in several hyperthermophiles (Table 13.8). Other chemolitho-trophic lifestyles include the oxidation of S^0 and Fe^{2+} aerobically or Fe^{2+} anaerobically with NO_3^- as acceptor (Table 13.8). Only one sulfate-reducing hyperthermophile is known (the euryarchaeote *Archaeoglobus*, see Section 13.7). The only bioenergetic option apparently ruled out is photosynthesis. The most thermophilic phototrophic organism known can grow up to 73°C (⚭ Table 6.1), far too cold for most hyperthermophiles (see Table 13.9).

Armed with a basic understanding of the habitats and energy metabolism of hyperthermophilic crenarchaeotes, let us now look at some representative organisms.

13.9 | **Hyperthermophiles from Terrestrial Volcanic Habitats: Sulfolobales and Thermoproteales**

Key Genera: *Sulfolobus, Acidianus, Thermoproteus, Pyrobaculum*

Volcanic habitats can have temperatures as high as 100°C and are thus suitable for hyperthermophilic *Archaea*. Two phylogenetically related organisms isolat-

Table 13.8	**Energy-yielding reactions of hyperthermophilic** *Archaea*		
Nutritional class	**Energy-yielding reaction**	**Metabolic[a] type**	**Example**
Chemoorganotrophic	Organic compound + $S^0 \rightarrow H_2S + CO_2$	AnR	*Thermoproteus, Thermococcus, Desulfurococcus, Thermofilum, Pyrococcus*
	Organic compound + $SO_4^{2-} \rightarrow H_2S + CO_2$	AnR	*Archaeoglobus*
	Organic compound + $O_2 \rightarrow H_2O + CO_2$	AeR	*Sulfolobus*
	Organic compound $\rightarrow CO_2 + H_2$ + fatty acids	F	*Staphylothermus, Pyrodictium*
	Organic compound + $Fe^{3+} \rightarrow CO_2 + Fe^{2+}$	AnR	*Pyrodictium*
	Pyruvate $\rightarrow CO_2 + H_2$ + acetate	F	*Pyrococcus*
Chemolithotrophic	$H_2 + S^0 \rightarrow H_2S$	AnR	*Acidianus, Pyrodictium, Thermoproteus, Stygiolobus, Ignicoccus*
	$H_2 + NO_3^- \rightarrow NO_2^- + H_2O$ (NO_2^- is reduced to N_2 by some species)	AnR	*Pyrobaculum*
	$4 H_2 + NO_3^- + H^+ \rightarrow NH_4^+ + 2 H_2O + OH^-$	AnR	*Pyrolobus*
	$H_2 + 2 Fe^{3+} \rightarrow 2 Fe^{2+} + 2 H^+$	AnR	*Pyrobaculum, Pyrodictium, Archaeoglobus*
	$2 H_2 + O_2 \rightarrow 2 H_2O$	AeR	*Acidianus, Sulfolobus, Pyrobaculum*
	$2 S^0 + 3 O_2 + 2 H_2O \rightarrow 2 H_2SO_4$	AeR	*Sulfolobus, Acidianus*
	$2 FeS_2 + 7 O_2 + 2 H_2O \rightarrow 2 FeSO_4 + 2 H_2SO_4$	AeR	*Sulfolobus, Acidianus, Metallosphaera*
	$2 FeCO_3 + NO_3^- + 6 H_2O \rightarrow 2 Fe(OH)_3 + NO_2^- + 2 HCO_3^- + 2 H^+ + H_2O$	AnR	*Ferroglobus*
	$4 H_2 + SO_4^{2-} + 2 H^+ \rightarrow 4 H_2O + H_2S$	AnR	*Archaeoglobus*
	$4 H_2 + CO_2 \rightarrow CH_4 + 2 H_2O$	AnR	*Methanopyrus, Methanocaldococcus, Methanothermus*

[a] AnR, anaerobic respiration; AeR, aerobic respiration; F, fermentation

Table 13.9 Properties of some hyperthermophilic Crenarchaeota

Group/Genus[a]	Morphology	Relationship to O$_2$[b]	DNA (mol % GC)	Temperature			Optimum pH
				Minimum	Optimum	Maximum	
Sulfolobales							
Sulfolobus	Lobed coccus	Ae	37	55	75	87	2–3
Acidianus	Coccus	Fac	31	60	88	95	2
Metallosphaera	Coccus	Ae	45	50	75	80	2
Stygiolobus	Lobed coccus	An	38	57	80	89	3
Stetteria	Coccus	An	65	68	95	102	6
Sulfophobococcus	Disc-shaped	An	54–56	70	85	95	7.5
Thermosphaera	Coccus	An	46	67	85	90	7
Thermoproteales							
Thermoproteus	Rod	An	56	60	88	96	6
Thermofilum	Rod	An	57	70	88	95	5.5
Pyrobaculum	Rod	Fac	46	74	100	102	6
Caldivirga	Rod	An	43	60	85	92	4
Thermocladium	Rod	An	52	60	75	80	4.2
Desulfurococcales							
Desulfurococcus	Coccus	An	51	70	85	95	6
Aeropyrum	Coccus	Ae	56	70	95	100	7
Staphylothermus	Cocci in clusters	An	35	65	92	98	6–7
Pyrodictium	Disc-shaped with filaments	An	62	82	105	110[c]	6
Pyrolobus	Lobed coccus	Fac	53	90	106	113	5.5
Thermodiscus	Disc-shaped	An	49	75	90	98	5.5
Ignicoccus	Irregular coccus	An	35	65	90	103	5
Hyperthermus	Irregular coccus	An	56	75	102	108	7
Sulfurisphaera	Coccus	Fac	33	63	84	92	2
Sulfurococcus	Coccus	Ae	43–46	40	75	85	2.5
Archaeoglobales[c]							
Archaeoglobus	Coccus	An	46	64	83	95	7
Ferroglobus	Irregular coccus	An	43	65	85	95	7
Thermococcales[d]							
Thermococcus	Coccus	An	38–57	70	88	98	6–7
Pyrococcus	Coccus	An	38	70	100	106	6–8

[a] The group names ending in "ales" are order names (◯◯● Section 11.10).

[b] Ae, aerobe; An, anaerobe; Fac, facultative.

[c] One species may grow up to 121°C.

[d] Phylogenetically, genera in this order of hyperthermophiles are members of the Euryarchaeota (see Sections 13.6 and 13.7).

---

ed from these environments include *Sulfolobus* and *Acidianus*. These genera form the heart of an order called the Sulfolobales (Table 13.9).

Sulfolobales

Sulfolobus grows in sulfur-rich acidic hot springs (Figure 13.15) at temperatures up to 90°C and at pH values of 1–5.* *Sulfolobus* (Figure 13.17a●) is an aerobic chemolithotroph that oxidizes H_2S or S^0 to H_2SO_4 and

*Historical note: *Sulfolobus* was first discovered by Thomas Brock and colleagues in 1970, and formally described in 1972. The discovery of *Sulfolobus*, along with the previously isolated *Thermus aquaticus* (source of the extremely thermostable *Taq* DNA polymerase, ◯◯● Section 7.9), is generally considered to have launched the field of hyperthermophilic microbiology. *Sulfolobus*, in particular, was the first prokaryote shown to be able to grow above 80°C. Thomas Brock was the senior author of the first seven editions of this book and is now an emeritus professor at the University of Wisconsin in Madison. Since 1980, Karl Stetter and colleagues at the University of Regensburg, Germany, have greatly expanded the field of hyperthermophilic microbiology with the discovery of many new genera and species.

fixes CO_2 as carbon source. *Sulfolobus* can also grow chemoorganotrophically. Cells of *Sulfolobus* are more or less spherical but contain distinct lobes (Figure 13.17a). Cells adhere tightly to sulfur crystals where they can be seen microscopically using fluorescent dyes (Figure 17.26b). Besides the aerobic respiration of sulfur or organic compounds, *Sulfolobus* can also oxidize Fe^{2+} to Fe^{3+}, and this has been applied in the high-temperature leaching of iron and copper ores (◯◯● Section 19.15).

A facultative aerobe resembling *Sulfolobus* also lives in acidic solfataric springs. This organism, named *Acidianus* (Figure 13.17b), differs from *Sulfolobus* most clearly by its ability to grow anaerobically. Remarkably, *Acidianus* is able to use S^0 *both* aerobically and anaerobically. Under aerobic conditions the organism uses S^0 as an electron *donor*, oxidizing S^0 to H_2SO_4, with O_2 as electron acceptor. Anaerobically, *Acidianus* uses S^0 as an electron *acceptor* (with H_2 as electron donor) forming H_2S as

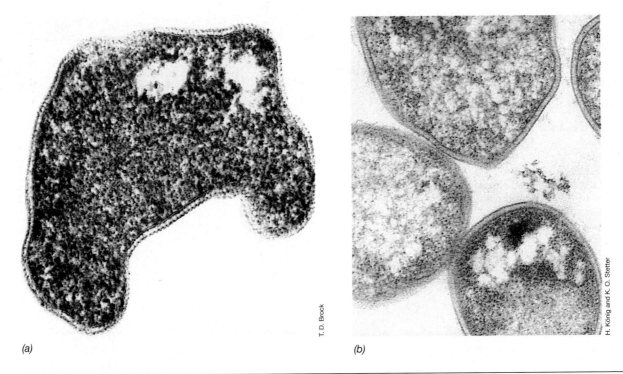

T. D. Brock

H. König and K. O. Stetter

(a)   (b)

● **Figure 13.17** **Acidophilic hyperthermophilic *Archaea*, the Sulfolobales.** (a) *Sulfolobus acidocaldarius*. Electron micrograph of a thin section. (b) *Acidianus infernus*. Electron micrograph of a thin section. Cells of both organisms vary from 0.8 to 2 μm in diameter.

the reduced product. Thus, the metabolic fate of S^0 in cultures of *Acidianus* depends on the presence or absence of O_2. Like *Sulfolobus*, *Acidianus* is roughly spherical in shape but is not as lobed (Figure 13.17*b*). It grows at temperatures from 65°C up to a maximum of 95°C, with an optimum of about 90°C.

Thermoproteales

Key genera within the Thermoproteales are *Thermoproteus*, *Thermofilum*, and *Pyrobaculum*. The genera *Thermoproteus* and *Thermofilum* consist of rod-shaped cells that inhabit neutral or slightly acidic hot springs. Cells of *Thermoproteus* are rigid rods about 0.5 μm in diameter and highly variable in length, ranging from short cells of 1–2 μm (Figure 13.18*a*●) up to filaments 70–80 μm long. Filaments of *Thermofilum* are thinner, some 0.17–0.35 μm wide with filament lengths ranging up to 100 μm (Figure 13.18*b*).

Both *Thermoproteus* and *Thermofilum* are strict anaerobes that carry out a S^0-based anaerobic respiration (Table 13.8). Unlike most hyperthermophiles, the oxygen sensitivity of *Thermoproteus* and *Thermofilum* is extreme, comparable to that of the methanogens (see Section 13.4). Thus, strict precautions must be taken in their culture. Most *Thermoproteus* isolates can grow chemolithotrophically on H_2 or chemoorganotrophically on complex carbon substrates such as yeast extract, small peptides, starch, glucose, ethanol, malate, fumarate, or formate (see Table 13.8). *Thermoproteus* and *Thermofilum* have similar GC base ratios (56–58% GC), but their genomes are distinct by nucleic acid hybridization analyses (∞ Section 11.11).

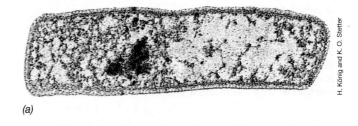

H. König and K. O. Stetter

(a)

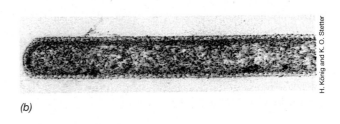

H. König and K. O. Stetter

(b)

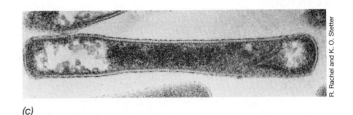

R. Rachel and K. O. Stetter

(c)

● **Figure 13.18** **Rod-shaped hyperthermophilic *Archaea*, the Thermoproteales.** (a) *Thermoproteus neutrophilus*. Electron micrograph of a thin section. A cell is about 0.5 μm in diameter. (b) *Thermofilum librum*. A cell is about 0.25 μm in diameter. (c) *Pyrobaculum aerophilum*. Transmission electron micrograph of a thin section; a cell measures 0.5 × 3.5 μm.

Pyrobaculum (Figure 13.18*c*) is a rod-shaped hyperthermophile but is physiologically unique from other Thermoproteales in at least one sense; some species of *Pyrobaculum* are capable of aerobic respiration. In addition, cultures of *Pyrobaculum* can carry out anaerobic respirations with NO_3^-, Fe^{3+}, or S^0 as electron acceptors and H_2 as electron donor (that is, chemolithotrophic and autotrophic growth). Other species of *Pyrobaculum* can grow anaerobically on organic electron donors, reducing S^0 to H_2S. The growth temperature optimum of *Pyrobaculum* is 100°C, and species of this organism have been isolated from terrestrial hot springs as well as from hydrothermal vents.

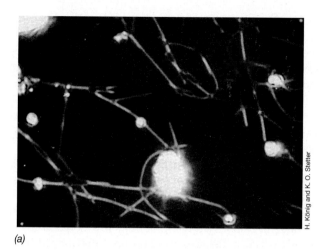

(a)

13.10 Hyperthermophiles from Submarine Volcanic Habitats: Desulfurococcales

Key Genera: *Pyrodictium, Pyrolobus, Ignicoccus, Staphylothermus*

We now turn to the microbiology of *submarine* volcanic habitats, homes to the most thermophilic of all known *Archaea*. These habitats include both shallow water thermal springs and deep-sea hydrothermal vents. We discuss the geology of these fascinating microbial habitats in Section 19.8. The organisms to be described here make up an order-level group of *Archaea* called the Desulfurococcales (Table 13.9).

Pyrodictium and *Pyrolobus*

Pyrodictium and *Pyrolobus* are examples of prokaryotes whose growth temperature optimum lies above 100°C. The optimum for *Pyrodictium* is 105°C and for *Pyrolobus* is 106°C. Cells of *Pyrodictium* are irregularly disc-shaped and grow in culture in a mycelium-like layer attached to crystals of elemental sulfur. The cell mass consists of a network of fibers to which individual cells are attached (Figure 13.19*a, b*•). The fibers are hollow and consist of protein arranged in a fashion similar to that of bacterial flagella (∞ Section 4.14). However, the filaments do not function in motility but instead as organs of attachment (see Figure 13.22). The cell walls of *Pyrodictium* are composed of glycoprotein. Physiologically, *Pyrodictium* is a strict anaerobe that grows chemolithotrophically on H_2 with S^0 as electron acceptor or chemoorganotrophically on complex mixtures of organic compounds (see Table 13.8).

Pyrolobus fumarii (Figure 13.19*c*) currently holds the record for the most thermophilic of all well-characterized

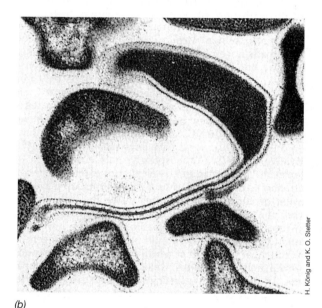

(b)

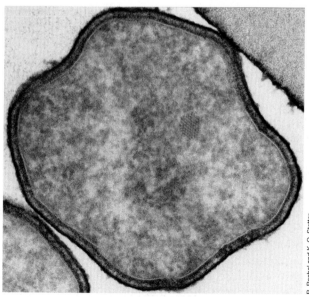

(c)

• **Figure 13.19 Desulfurococcales with growth temperature optima >100°C.** (a) *Pyrodictium occultum* (growth temperature optimum, 105°C), dark-field micrograph. (b) Thin section electron micrograph of *P. occultum.* Cells are highly variable in diameter from 0.3 to 2.5 μm. (c) Thin section of a cell of *Pyrolobus fumarii,* the most thermophilic of all known prokaryotes (growth temperature optimum, 106°C); a single cell is about 1.4 μm in diameter.

prokaryotes. Its growth temperature maximum is 113°C (Table 13.9). *Pyrolobus* lives in the walls of "black smoker" hydrothermal vent chimneys (Section 19.8 and Figure 19.20), where its autotrophic abilities contribute organic carbon to this otherwise inorganic environment. Other species of *Pyrolobus* have been isolated that can apparently grow even above 113°C, so this genus clearly contains the most hyperthermophilic of all known microorganisms.

Pyrolobus cells are coccoid-shaped (Figure 13.19*c*), and the cell wall is composed of protein. The organism is an obligate H_2 chemolithotroph, growing by the oxidation of H_2 coupled to the reduction of NO_3^- (to NH_4^+), $S_2O_3^{2-}$ (to H_2S), or very low concentrations of O_2 (to H_2O). Besides its extremely thermophilic nature, *Pyrolobus* is also resistant to temperatures substantially above its growth temperature maximum. For example, cultures of *P. fumarii* survive autoclaving (121°C) for 1 h, a condition that even bacterial endospores (Section 4.13) cannot withstand.

Desulfurococcus and *Ignicoccus*

Other notable members of the Desulfurococcales include *Desulfurococcus*, the genus for which the order is named (Figure 13.20*a*●), and *Ignicoccus*. *Desulfurococcus* is a strictly anaerobic S^0-reducing bacterium like *Pyrodictium* but differs from this organism in that it is much less thermophilic, growing optimally at about 85°C.

Ignicoccus grows optimally at 90°C, and its metabolism is H_2/S^0 based, as is that of so many hyperthermophilic *Archaea* (Table 13.8). *Ignicoccus* (Figure 13.20*b*) is a novel hyperthermophile because it contains an *outer membrane*, similar to that of *Bacteria* (Section 4.9). The outer membrane of *Ignicoccus* is unusual, however, in that it is present at some distance from the cytoplasm of the cell. This arrangement allows for an unusually large periplasm (Figure 13.20*b*). Indeed, the volume of the *periplasm* of *Ignicoccus* is some 2–3 times that of its *cytoplasm*, in contrast to that of gram-negative *Bacteria*, where periplasmic volume is about 25% that of the cytoplasm. The periplasm of *Ignicoccus* also contains membrane-bound vesicles (Figure 13.20*b*) that may function in exporting substances outside the cell. In addition, however, some *Ignicoccus* species are hosts to a small, parasitic prokaryote, *Nanoarchaeum*, as discussed in the next section.

Staphylothermus

A morphologically unusual member of the order Desulfurococcales is the genus *Staphylothermus* (Figure 13.21●). Cells of *Staphylothermus* are spherical, about 1 μm in diameter, and form aggregates of up to 100 cells, resembling those of *Staphylococcus* (compare Figure 13.21 with Figure 4.4*b*). *Staphylothermus* is not a chemolithotroph like so many of its hyperthermophilic relatives, but in-

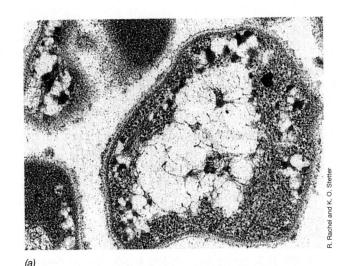

(a)

R. Rachel and K. O. Stetter

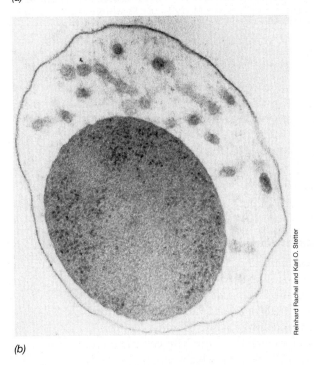

(b)

Reinhard Rachel and Karl O. Stetter

● **Figure 13.20 Examples of Desulfurococcales with growth temperature optima below the boiling point.** (a) Thin section of a cell of *Desulfurococcus saccharovorans*; a cell is 0.7 μm in diameter. (b) Thin section of a cell of *Ignicoccus islandicus*. The cell proper is surrounded by an extremely large periplasm (Section 4.9). The cell itself measures about 1 μm in diameter but cell plus periplasm measures 1.4 μm.

stead is a chemoorganotroph that grows optimally at 92°C. Energy is obtained from the fermentation of peptides, producing the fatty acids acetate and isovalerate as fermentation products (Table 13.8).

Isolates of *Staphylothermus* have been obtained from both shallow marine hydrothermal vents, as well as the very hot black smokers (Section 19.8). This organism is thus widely distributed in submarine thermal areas where it is likely a significant consumer of proteins released from dead organisms.

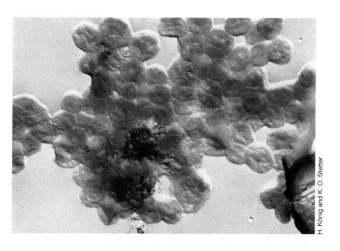

● **Figure 13.21** **The hyperthermophile *Staphylothermus marinus* (growth temperature optimum, 92°C).** Electron micrograph of shadowed cells. A single cell is about 1 μm in diameter.

H. König and K. O. Stetter

 13.8–13.10 Concept Check

Hyperthermophilic Crenarchaeota inhabit the hottest habitats currently known to support life. Cold-dwelling phylogenetic relatives of these organisms are also known. A variety of different morphological types of Crenarchaeota are known and several different metabolic strategies used to support growth.

◆ What are the major differences between the organisms *Sulfolobus* and *Pyrolobus*? *Staphylothermus* and *Ignicoccus*?

◆ What is unusual about the metabolic properties of *Acidianus* regarding elemental sulfur (S^0)?

◆ What energy class of organisms do we call those organisms that use H_2 as electron donor? List the genera of Crenarchaeota that can use H_2 and what they use as electron acceptors.

IV PHYLUM NANOARCHAEOTA

A unique phylum of *Archaea* contains tiny parasitic cells, the **Nanoarchaeota**. These fascinating prokaryotes harbor many surprises, including the smallest genomes of all known prokaryotes.

13.11 *Nanoarchaeum*

Nanoarchaeum is a genus of very small coccoid cells that live as parasites, or possibly as symbionts, of the crenarchaeote *Ignicoccus* (see Section 13.10). Cells of *Nanoarchaeum* are about 0.4 μm in diameter and contain only about 1% of the volume of an *Escherichia coli* cell. Cells of *Nanoarchaeum* replicate only when attached to the surface of *Ignicoccus* cells. *Nanoarchaeum* cells can occur

singly, in pairs, or up to 10 or more cells per *Ignicoccus* cell (Figure 13.22a●). The cell wall of *Nanoarchaeum* consists of an S-layer (∞ Section 4.10) that overlays what appears to be a periplasmic space (Figure 13.22b).

Nanoarchaeum cannot grow in pure culture but only with *Ignicoccus* as a host. Whether *Ignicoccus* receives any benefit from the association is unknown. *Nanoarchaeum* is hyperthermophilic, with optimal growth at about 90°C. The metabolism of *Nanoarchaeum* is unknown, but its host is an autotroph, growing with H_2 as electron donor and elemental sulfur as electron acceptor. Isolates of *Nanoarchaeum* have been obtained from submarine hydrothermal vents as well as terrestrial hot springs, and the organism appears to be distributed worldwide in suitable hot habitats. Only one species, *N. equitans*, is known.

Phylogeny and Genomics of *Nanoarchaeum*

The Nanoarchaeota represent a deep lineage within the *Archaea*, distinct from all other phyla (Figure 13.1). Some difficulty in pinpointing the exact phylogenetic position of the Nanoarchaeota was originally encountered because the 16S ribosomal RNA of the group is so unique. *Nanoarchaeum* 16S rRNA contains many base substitutions, even in regions of the molecule that are highly conserved among other phyla of *Archaea*. In fact, oligonucleotide primers that amplify (using PCR) 16S rRNA genes from all known species of *Archaea*, fail to amplify the gene from *Nanoarchaeum*.

The genome of *Nanoarchaeum* is only 0.49 Mbp. This is the smallest genome of any cell known. Astonishingly, the *Nanoarchaeum* genome lacks identifiable genes for most metabolic functions, including the biosynthesis of monomers such as amino acids, nucleotides, and coenzymes. Also missing in the *Nanoarchaeum* genome are genes encoding proteins for catabolic pathways, such as glycolysis.

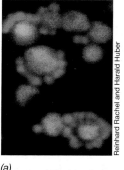
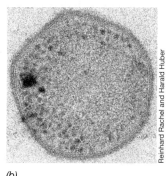

(a)   (b)

Reinhard Rachel and Harald Huber

● **Figure 13.22** **Nanoarchaeum.** (a) Fluorescence micrograph of cells of *Nanoarchaeum* (red) attached to cells of *Ignicoccus* (green). Cells were stained by FISH (∞ Section 11.7) using specific nucleic acid probes targeted to each organism. (b) Transmission electron micrograph of a thin section of a cell of *Nanoarchaeum*. Note the distinct cell wall. Cells of *Nanoarchaeum* are about 0.4 μm in diameter.

Lipid Stability

What about cellular lipids? How do hyperthermophiles prevent their membranes from separating at high temperatures? Virtually all hyperthermophiles produce lipids constructed upon the dibiphytanyl tetraether model (∞ Section 4.5 for the structure of such lipids). Dibiphytanyl tetraether lipids are naturally heat resistant because the covalent bond between phytanyl units forms a lipid *monolayer* membrane structure instead of the normal lipid *bilayer* (∞ Figure 4.19). This structure, supported by covalent bonds, resists the tendency of heat to pull apart a lipid bilayer constructed of fatty acids.

Stability of Monomers

Besides the stability of *macromolecules*, the thermal lability of *monomers* is also an important factor in governing the upper temperature limits for growth of hyperthermophiles. No matter how stable macromolecules are, life is impossible if the basic building blocks themselves are unstable. Interestingly, it is the thermal lability of *small* molecules rather than macromolecules that may actually dictate the upper temperature for life.

Even at temperatures as "low" as 120°C, some important small molecules are destroyed at significant rates. For example, two key molecules in energy metabolism, ATP and NAD^+ (∞ Chapter 5), hydrolyze rapidly at these temperatures. For example, the half-life of ATP or NAD^+ *in vitro* is less than 30 minutes at 120°C and shortens dramatically at temperatures above this. (Note, however, that these are measurements done in water. The stability of these molecules within the cytoplasm could be significantly greater than this due to as yet undiscovered heat protective agents). So, is 120°C the upper temperature for life? Despite the problems cells face with the stability of small molecules, the upper temperature limit for life is likely to lie above 120°C.

The Temperature Limits to Microbial Existence

Because life as we know it depends on liquid water, microbial habitats hotter than 100°C are only found in environments under *pressure*, such as the sea floor. Indeed, all hyperthermophilic *Archaea* capable of growth over 100°C seem to be restricted to these superheated environments (see Table 13.9). How hot can these environments be and still support life? Some evidence already exists for a new *Pyrodictium* species capable of growth at 121°C. Thus, even the remarkable upper temperature limit for growth of *Pyrolobus fumarii* (113°C), is not the upper temperature limit for life.

Deep-sea hydrothermal vent black smokers offer models for superheated natural environments and their microflora. Black smokers emit hydrothermal fluid at 250–350°C and form upright metallic structures called *chimneys* from metal sulfides in the fluid

(∞ Figure 19.19). Several hyperthermophiles have been isolated from smoker chimney walls, including *Pyrodictium*, *Methanopyrus*, and *Pyrolobus*. The walls show a gradient in temperature from 250 to 350°C inside to 2°C outside, and the organisms probably remain in regions of the wall nearest their temperature optima. Studies of the superheated water itself show it to be sterile, consistent with measurements that have shown that both small molecules and macromolecules would be instantly destroyed at such high temperatures.

Laboratory experiments on the stability of biomolecules point to 140–150°C as the most likely upper temperature limit for life as we know it. Above this, organisms would not be able to overcome the heat lability of the essential biomolecules of life. For example, ATP is instantly degraded at 150°C. Thus, if organisms exist that are capable of growth at temperatures even as "low" as 150°C, their energy economy would almost certainly have to be based on something other than ATP. Is this territory that evolution cannot overcome? It is too early to tell for sure. However, if "super" hyperthermophiles capable of growth at or above 150°C are someday discovered, they will likely reveal fundamental new principles in biology and biochemistry.

13.13 Hyperthermophilic *Archaea*, H_2, and Microbial Evolution

Why do so many *Archaea* seem to inhabit extreme environments? Although molecular probing of nonextreme environments (such as soil and water) indicate that *Archaea* are all around us (and even *in* us, as methanogens in our large intestine), a common theme of cultured *Archaea* is *adaptation to environmental extremes*. Until we have cultured the thus far elusive marine and soil *Archaea*, we cannot say for sure that extremophily is a hallmark of *Archaea*. But let's say for the sake of argument that it is. What would this say about Earth and evolution in general?

Extreme environments of various types existed on early Earth just as they do today, and it is within such environments that life may first have flourished. At the time that cellular life evolved nearly 4 billion years ago, it is almost certain that Earth was far hotter than it is today (∞ Section 11.1), and probably suitable only for hyperthermophiles. So, are extant hyperthermophiles relics of Earth's earliest life forms? Let's see what the phylogenetic tree of *Archaea* tells us about hyperthermophiles.

Hyperthermophiles and Their Slow Evolutionary Clocks

Molecular sequencing of rRNA genes suggests that hyperthermophilic *Archaea* evolved at slower rates than other organisms (Figure 13.1). Much the same can be said about hyperthermophilic *Bacteria* such as *Thermotoga* and *Aquifex*

(◌ Figure 12.1). This conclusion emerges from inspection of the evolutionary trees—lines to the hyperthermophiles are typically rather short and branch near the root (◌ Figures 12.1 and 13.1).

Thermal habitats themselves may explain why hyperthermophiles have such slow evolutionary clocks. Organisms living at very high temperatures may be under unusually strong constraints to maintain the characteristics critical to life there. Any change in amino acid sequence of a protein resulting from a spontaneous mutation must be a change that does not decrease its heat stability. If it does, the mutation could well be lethal. For organisms living in nonextreme environments, such pressures would not be a significant controlling factor in the diversification of their proteins. In other words, beyond a certain point, survival of hyperthermophiles may not be amenable to additional evolutionary change, since attempts to increase fitness must be balanced by the need for all cellular molecules to remain heat stable. If life originated on a hot planet Earth, as most evolutionary scenarios predict, then hyperthermophilic *Archaea* and *Bacteria* are likely the closest living relatives to early life forms that remain today. Therefore the biology of these hyperthermophiles is not only interesting, but may offer us a window into the past.

Hydrogen (H₂) as a Primitive Energy Source

Before we leave this chapter we must point out a final clue to the biology of early life forms. Note how often *H₂ metabolism* enters into the metabolic picture of hyperthermophilic prokaryotes, especially those growing at the highest temperatures (Figure 13.25●). Many hyperthermophiles grow anaerobically with H₂ as an electron donor and one or more of the electron acceptors S^0, NO_3^-, Fe^{3+}, or O_2 (see Table 13.8). H₂ metabolism is likely a physiological relic of ancient metabolic schemes.

These schemes would have evolved in primitive organisms because of the ready availability of H₂ and suitable inorganic electron acceptors in their primordial environments, and because the catabolism of H₂ can be a relatively simple process biochemically (◌ Figure 11.6).

Figure 13.25 shows the known upper temperature limits for growth of prokaryotes showing each of the three energy-conserving processes known in biology. Photosynthesis is obviously the most susceptible to heat, and one can only speculate that this is due to the instability of the large multi-protein complexes that must interact in the process of photophosphorylation (◌ Figure 17.15). Chemoorganotrophy occurs up to at least 110°C, as this is the upper temperature limit for growth of *Pyrodictium occultum*, an organism that can ferment certain organic compounds as well as grow chemolithotrophically on H₂ with S^0 as electron acceptor (Table 13.9). Above 110°C, only H₂-oxidizing *Archaea* are known, including the genera *Pyrolobus* and a *Pyrodictium* species that couples H₂ oxidation to the reduction of Fe^{3+} (Table 13.9 and Figure 13.25). Although there are indications from analyses of hydrothermal vent samples that *Archaea* exist in waters above 120°C, cultures of such organisms have not yet been obtained.

The diversity of H₂-oxidizing hyperthermophilic *Archaea* (and *Bacteria*, ◌ Sections 12.5 and 12.37) known today, attests to the evolution of H₂ oxidation as a metabolic success story. Indeed, even though we know little about the metabolism of the parasitic *Nanoarchaeum* (see Section 13.11), it is known that H₂ stimulates its growth in laboratory culture. Could *Nanoarchaeum* be the genetically simplest of all H₂-oxidizing prokaryotes? The use of H₂ by so many different physiological groups of prokaryotes, including even those living deep in the Earth (◌ Section 19.4), points to the antiquity of this source of energy.

Both hydrothermal vents, the habitat of the most extreme hyperthermophiles, and Earth's hot, deep subsurface have often been suggested as environments where life could have first arisen. These environments contain H₂ and

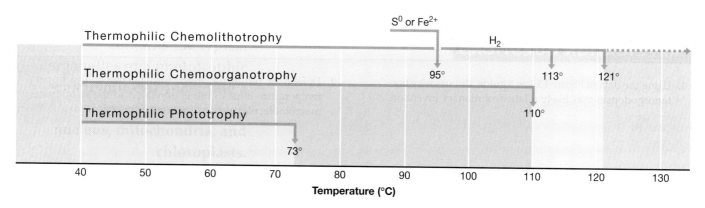

● **Figure 13.25 Upper temperature limits for energy metabolism.** Phototrophy, *Synechococcus lividus* (*Bacteria*, cyanobacteria); chemoorganotrophy, *Pyrodictium occultum* (*Archaea*); chemolithotrophy-S^0 as electron donor, *Acidianus infernus* (*Archaea*); chemolithotrophy-Fe^{2+} as electron donor, *Ferroglobus placidus* (*Archaea*); chemolithotrophy-H₂ as electron donor, *Pyrolobus fumarii* (*Archaea*, 113°C); *Pyrodictium* sp., strain 121 (*Archaea*, 121°C). Strain 121 also grows on formate but this compound may be metabolized through $H_2 + CO_2$.

14.3 *Concept Check*

Chloroplasts are the site of photosynthetic energy production and CO_2 fixation in eukaryotic phototrophs.

◆ Differentiate the *stroma* from *thylakoids*.

◆ What is the function of RubisCO and where is it found?

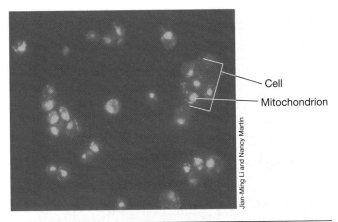

● **Figure 14.7 Cells of the yeast *Saccharomyces cerevisiae.*** The cells have been stained with 4′6-diamidine-2′-phenylindole dihydrochloride (DAPI) (∞ Section 18.3) to show mitochondrial DNA. Each mitochondrion has two to four circular chromosomes that stain blue with the fluorescent dye used.

14.4 Endosymbiosis: Relationships of Mitochondria and Chloroplasts to Bacteria

On the basis of their relative autonomy, size, and morphological resemblance to prokaryotes, it was suggested long ago that mitochondria and chloroplasts were descendants of ancient bacteria. Establishment of these organelles was postulated to have been by the process of **endosymbiosis** (*endo* means "within"), the engulfment of a prokaryotic cell by a larger cell ("the host") and conversion of the engulfed cell into an organelle (∞ Section 11.4 and Figure 11.9). Several lines of molecular evidence support the endosymbiotic theory, and we summarize these here:

1. **Mitochondria and chloroplasts contain DNA.**

   Although most of their functions are encoded by nuclear DNA, a few organellar components are encoded by a small genome present within the organelle itself. These include ribosomal RNA, transfer RNAs, and certain proteins of the respiratory chain (mitochondria) and photosynthetic apparatus (chloroplast). Thus, nonphototrophic eukaryotic cells are genetic chimeras containing DNA from *two* different sources, the *endosymbiont* and the *host cell nucleus*. Phototrophic eukaryotes—algae and higher plants—contain DNA from *three* different sources, the mitochondrial and chloroplast endosymbionts, and the nucleus. Most mitochondrial DNA and all chloroplast DNA exists in a covalently closed *circular* form, just as it does in most prokaryotes (∞ Sections 2.2, 7.2, and 7.3). Mitochondrial DNA can be seen in cells by using special staining methods (Figure 14.7●). We discuss many other very interesting features of organellar genomes in Section 15.7.

2. **The eukaryotic nucleus contains bacterially derived genes.**

   Genomic sequencing (∞ Chapter 15) and other genetic studies have clearly shown that several nuclear genes encode properties unique to organelles. Because the sequences of these genes more closely resemble those of *Bacteria* than those of *Archaea* or eukaryotes, it is concluded that these genes were transferred to the nucleus from bacterial symbionts during the transition from an initially engulfed cell to a modern day organelle.

3. **Mitochondria and chloroplasts contain their own ribosomes.**

   Ribosomes, cell structures involved in protein synthesis (∞ Section 7.16), exist in either a large form [80 Svedberg (S) units], typical of the cytoplasm of eukaryotic cells, or in a smaller form (70S), present in prokaryotes. Mitochondria and chloroplasts also contain ribosomes, and they are 70S in size, the same as those of prokaryotes.

4. **Antibiotic specificity.**

   Several antibiotics (streptomycin is one example) kill or inhibit *Bacteria* by specifically interfering with 70S-ribosome function. These same antibiotics also inhibit protein synthesis in mitochondria and chloroplasts.

5. **Molecular phylogeny.**

   Phylogenetic studies using comparative ribosomal RNA sequencing methods (∞ Sections 11.4–11.8) and organellar genome studies (∞ Section 15.7) have shown convincingly that the chloroplast and mitochondrion originated from the *Bacteria*. The modern eukaryotic cell thus arose from an association of at least two quite distinct organisms by the process of endosymbiosis (∞ Section 11.3 and Figure 11.7).

The evidence that *hydrogenosomes* are endosymbionts is also strong. For example, in the obligately anaerobic ciliated protozoan *Nyctotherus ovalis* that lives in the hindgut of termites (∞ Section 19.10), mitochondria are absent and hydrogenosomes have been identified that contain DNA and ribosomes. Moreover, the nucleus of hydrogenosome-containing eukaryotes contains genes encoding proteins of bacterial origin. Thus, the hydrogenosome originated by endosymbiosis, just as the mitochondrion did. In fact, hydrogenosomes are thought to be metabolically degenerate mitochondria that dispensed with respiration to exploit pyruvate fermentation as a means of energy conservation in a host cell with an anaerobic lifestyle. Although not rivaling the energy available from respiration in the mitochondrion, acetate production in the hydrogenosome supplies cells with more energy than if the fermentation products were lactate or ethanol (Figure 14.4*b*; ∞ Sections 5.10 and 17.19). The mitochondrion and the hydrogenosome can thus be viewed as functionally related organelles that specialize in different metabolic strategies for making ATP. Other structures called *mitosomes* are present in some eukaryotic cells and are probably even more degenerate mitochondria, having lost virtually all energy-related functions altogether (see Section 14.9).

As we can see, many forms of evidence point to organelles as having arisen from the endosymbiotic uptake of free-living *Bacteria* by eukaryotic host cells. In this cozy arrangement, host cells obtained permanent partners specializing in energy generation, while the symbionts received a stable and supportive growth environment. That endosymbiosis was an evolutionary success story can be attested to by the presence of mitochondria, hydrogenosomes, or chloroplasts in virtually all eukaryotic cells today.

 14.4 Concept Check

Key metabolic organelles of eukaryotes are the chloroplast, involved in photosynthesis, and the mitochondrion or hydrogenosome, involved in respiration or fermentation. These organelles were originally *Bacteria* that established permanent residence inside other cells (endosymbiosis).

◆ Summarize the molecular evidence that supports the relationship of organelles to *Bacteria*.

◆ Why might *Nyctotherus* be better off with hydrogenosomes than with mitochondria?

14.5 Other Organelles and Eukaryotic Cell Structures

A variety of other cytoplasmic structures are typically present in eukaryotic cells. These include the *endoplasmic reticulum, Golgi apparatus, lysosomes, peroxisomes,* and the organelles of motility—*flagella* and *cilia*. In contrast to mitochondria and chloroplasts, however, these structures lack DNA and ribosomes, and are not of endosymbiotic origin.

Endoplasmic Reticulum and Golgi Complex

The **endoplasmic reticulum** (ER) is a network of membranes continuous with the nuclear membrane. Two types of endoplasmic reticulum are recognized: *rough*, which contains attached ribosomes, and *smooth*, which does not (Figure 14.1). Smooth ER participates in the synthesis of lipids and in some aspects of carbohydrate metabolism. Rough ER, through the activity of its ribosomes, is a major producer of glycoproteins and also

produces new membrane material that is transported throughout the cell to enlarge the various cell membrane systems (see Figure 14.1) before cell division.

The **Golgi complex** consists of a stack of membranes distinct from the ER (Figures 14.1 and 14.8●), but which functions in concert with the ER. In the Golgi complex products of the ER are chemically modified and sorted into those destined to be secreted, for example, hormones or digestive enzymes, and those that function in other membranous structures in the cell. Golgi arise from the division of preexisting Golgi and contain various enzymes that modify secretory and membrane proteins differently, depending on their final destination in the cell. Many of the modifications involve *glycosylation* (adding sugar residues) to convert the proteins into specific glycoproteins.

Lysosomes and Peroxisomes

Lysosomes (Figure 14.1) are membrane-enclosed structures that contain various *digestive enzymes* that the cell uses to digest macromolecules such as proteins, fats, and polysaccharides. In doing so, the cell recycles these materials for new biosyntheses. The internal pH of the lysosome is about 5, two units lower than that of the cytoplasm, and the hydrolytic enzymes within the lysosome function optimally at this pH. These hydrolytic enzymes are nonspecific in their activity and could potentially destroy key cellular macromolecules if not contained. Thus, the lysosome allows for lytic activities to be partitioned away from the cytoplasm proper. Following hydrolysis of macromolecules in the lysosome, the resulting monomers pass from the lysosome into the cytoplasm as nutrients for the cell.

The **peroxisome** is a membrane-enclosed structure (Figure 14.1) whose function is to produce hydrogen peroxide (H_2O_2) from the reduction of O_2 by various hydrogen

● **Figure 14.8 The Golgi complex.** Transmission electron micrograph of a portion of a cell of the protozoan *Toxoplasma gondii*. The Golgi is colored in red. Note the multiple folded membranes that make up the Golgi complex (the membrane stacks are 0.5–1.0 μm in diameter). Other structures such as cytoplasmic granules are shown in other colors. *T. gondii* is a model system for the study of the Golgi complex because each cell contains only one such structure.

donors, including alcohols and long chain fatty acids. The H_2O_2 produced in the peroxisome is degraded to H_2O and O_2 by the enzyme *catalase* (∞ Section 6.16). Peroxisomes play other roles as well, such as synthesizing bile salts that aid in the absorption and digestion of fats. Peroxisomes originate in the cell by incorporating their proteins and lipids from the cytoplasm, eventually becoming a membrane-enclosed entity that can enlarge and divide in synchrony with the cell.

Microfilaments and Microtubules

Just as houses are supported by structural reinforcement, the large size of eukaryotic cells and their ability to move requires structural reinforcement. This internal structural network comes from proteins that form filamentous structures called **microfilaments** and **microtubules**. Together, these structures form the **cell cytoskeleton**.

Microfilaments are about 8 nm in diameter and are polymers of the protein *actin*. These fibers form scaffolds throughout the cell, defining and maintaining the shape of the cell (Figures 14.1 and 14.9●). *Microtubules* are larger filaments, about 25 nm in diameter, and are composed of the protein *tubulin*. Microtubules assist microfilaments in maintaining cell structure. They also play an important role in cell motility, both in the movement of internal cell structures (for example, in the separation of chromosomes during cell division) and in movement of the organism itself (for example, in movement of the flagellum in flagellated eukaryotic cells, see Figures 14.1 and 14.10).

We saw in an earlier chapter that prokaryotic cells produce structural and functional homologs of actin and tubulin in the form of the proteins MreB and FtsZ, respectively (∞ Section 6.1). This indicates that although prokaryotes are typically much smaller than eukaryotes, they still require some minimal level of internal scaffolding. The existence of the proteins MreB and FtsZ in prokaryotes also shows that the eukaryotic cytoskeleton has deep evolutionary roots.

Flagella and Cilia

Flagella and **cilia** are present on species of many groups of eukaryotic microorganisms. Flagella and cilia are organelles of motility, allowing cells to move by swimming motility. Cilia are essentially short flagella that beat in synchrony to propel the cell—usually quite rapidly—through the medium. Flagella are long appendages present singly or in groups that push the cell along—typically more slowly than by cilia—through a whip-like motion (Figure 14.10*a*●).

The flagella of eukaryotic cells are structurally quite distinct from bacterial flagella (∞ Section 4.14) and do not rotate. In cross-section, cilia and flagella are very similar. Each contains a bundle of 9 pairs of microtubules surrounding a central pair of microtubules called the *axoneme* (Figure 14.10*b*). Microtubules are composed of the protein *tubulin*. A second protein, called *dynein* is attached to the tubulin and functions as an ATPase, hydrolyzing ATP to yield the energy necessary to drive motility. Movement of flagella and cilia is similar. In both cases, movement involves the coordinated sliding of axonemal microtubules (Figure 14.10*b*) against one another in a direction toward or away from the base of the cell. This movement confers the whip-like action on the flagellum or cilium that results in cell propulsion.

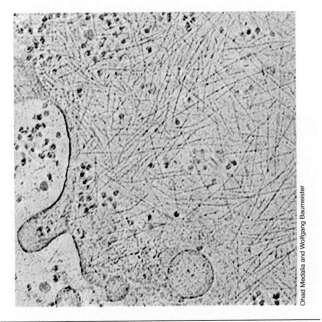

● **Figure 14.9   Microfilaments and eukaryotic cell architecture.** An electron tomographic image of a cell of the cellular slime mold *Dictyostelium discoideum* (see also Figure 14.28) showing the network of actin microfilaments that functions along with microtubules as the cell *cytoskeleton*. Microfilaments are about 8 nm in diameter. Electron tomography is a method for three-dimensional reconstruction of cells from a series of images taken with a transmission electron microscope.

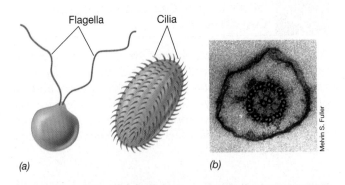

Flagella   Cilia

(a)   *(b)*

Ohad Medalia and Wolfgang Baumeister

Melvin S. Fuller

● **Figure 14.10   Flagella and cilia, motility organelles in eukaryotic cells.** (a) Flagella can be present as single or multiple filaments. Cilia are structurally very similar to flagella but much shorter. Eukaryotic flagella move in a whip-like motion, unlike the flagella of prokaryotes, which propel the cell by rotating, much like a propeller on a motor boat (∞Section 4.14 and Figures 4.53–4.58). (b) Cross section through a flagellum of the fungus *Blastocladiella emersonii* showing the outer sheath, the outer nine pairs of microtubules, and the central pair of single microtubules.

A clear **distinction** can thus be made between the prokaryotic **and** eukaryotic flagellum. The filament of the prokaryotic flagellum is made from a helical array of a single **protein**, flagellin, and the structure itself is firmly anchored in a rotary motor complex embedded in the cell **wall** and membrane (∞ Figure 4.56). In addition, the **bacterial** flagellum functions as a propeller, being driven **by** the energy of the proton motive force (∞ Sections 4.14 and 5.12). The eukaryotic flagellum, by contrast, **propels** the cell by the whip-like motion of sliding microtubules driven by the energy of ATP. Nevertheless, in **all** motile cells, motility likely has survival value; the **ability** to move allows motile organisms to explore new **habitats** and exploit their resources.

 14.5 *Concept Check*

Besides the **major** organelles of eukaryotes, several other structures **with** defined functions are present in the cytoplasm. These **include** the endoplasmic reticulum, the site of ribosomes **and** cellular lipid syntheses; the Golgi apparatus involved in **protein** modification and secretion; lysosomes, which play **a** role in macromolecular digestion; and the peroxisome, an **organelle** involved in H_2O_2 production. In addition, proteinaceous tubes called microfilaments and microtubules are **present**, forming the cell's cytoskeleton. Flagella and cilia **are** organelles of motility that have extensive microtubular **structure.**

♦ How **does** *smooth* ER differ from *rough* ER?

♦ Why **are** the activities that occur in the lysosome best partitioned away from the cytoplasm proper?

♦ Besides **scaffolding**, what other functions do microtubules have?

ESSENTIALS OF EUKARYOTIC GENETICS AND MOLECULAR BIOLOGY

Several aspects of the genetics and molecular biology of eukaryotes **lack** counterparts in most prokaryotic cells. These **include** (1) the replication of linear (as opposed to circular) genetic elements; (2) mitosis and meiosis; and (3) processing of messenger RNAs. We briefly explore each of these topics here.

14.6 Replication of Linear DNA

Most prokaryotic chromosomes are circular, as are most plasmids and some viruses, and the genomes of most organelles (∞ Section 15.7). In contrast, eukaryotic cells contain *linear DNA*. Almost all the steps in DNA replication are identical whether the chromosome is linear or circular. However, there is a key problem with the replication of linear genetic elements that is not an issue with circular ones: *replication of DNA at the extreme 5′–end of each strand.*

To understand the nature of the problem, first review Figure 7.13. Imagine that the left end of the DNA in this diagram is actually one end of a linear chromosome. Even if the RNA primer is very short and there is a special enzyme to remove it, no DNA polymerase can replace it with DNA since *all* known DNA polymerases require a primer. Therefore, if nothing is done, the DNA molecule will become shorter each time it is replicated. The replication of linear DNA thus requires special attention and there are at least two solutions to the problem.

Replication of Linear DNA Using a Protein Primer

Viruses that contain linear DNA genomes (this includes most viruses that infect eukaryotes) and many linear plasmids solve the problem of replicating linear DNA by using a *protein* primer instead of an *RNA* primer. Although all DNA polymerases must add each nucleotide to a free hydroxyl (OH) group, some DNA polymerases can add the first base onto an hydroxyl group present on specific proteins that bind to the ends of linear chromosomes (Figure 14.11●). These proteins are encoded by the plasmid or virus and they function to recognize the ends of the chromosomes. These protein primers are not removed, so these plasmids and viruses have proteins permanently attached to the 5′-ends of their DNA. Protein primers are also the means by which some linear chromosomes of *Bacteria* (∞ Section 15.4) are replicated.

Telomeres and Telomerase

A special method is used to replicate the ends of eukaryotic chromosomes, which are called *telomeres*. Telomeres

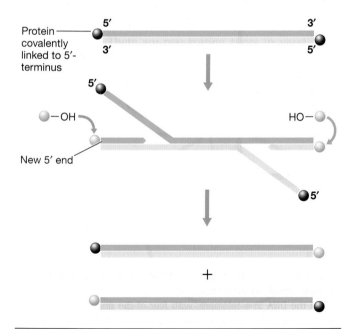

● **Figure 14.11   Replication of linear DNA using protein primers.** New strands of DNA are primed by proteins covalently attached to their 5′ ends. Note the free OH group on the protein. DNA polymerase III can add a nucleotide to this OH group.

REVIEW QUESTIONS

1. List at least three features of eukaryotic cells that clearly differentiate them from prokaryotic cells (◌◌ Sections 4.1–4.5).

2. How are the mitochondrion and the hydrogenosome similar structurally? How do they differ? How do they differ metabolically (◌◌ Section 14.2)?

3. What major physiological processes (energy related and carbon related) occur in a chloroplast (◌◌ Section 14.3)?

4. What does knowing that an antibiotic like streptomycin blocks protein synthesis in organelles tell us about their relationship to *Bacteria* (◌◌ Section 14.4)?

5. What are the chemical differences between *microfilaments* and *microtubules*? Are such structures really absent from prokaryotic cells? (You will want to review Section 6.2 before answering this question.) (◌◌ Section 14.5).

6. What are the functions of the following eukaryotic cell structures: endoplasmic reticulum, Golgi complex, lysosome, peroxisome (◌◌ Section 14.5).

7. Contrast the action of flagella in prokaryotes and eukaryotes (◌◌ Sections 14.4 and 14.5).

8. Discuss how in the absence of telomerase activity, genes would eventually be lost from eukaryotic DNA (◌◌ Section 14.6).

9. Compare and contrast the processes of *mitosis* and *meiosis*. Which process is absolutely necessary for growth of *Saccharomyces cerevisiae* and why? How is the mating type of a yeast cell determined (◌◌ Section 14.7)?

10. Why do eukaryotic mRNAs have to be "processed," while most prokaryotic RNAs do not (◌◌ Section 14.8)?

11. The phylogeny of microbial eukaryotes is unsettled. Discuss the major disagreements between the two phylogenetic trees shown in Figure 14.20 (◌◌ Section 14.9).

12. What are the common names of some major groups of protozoa (◌◌ Section 14.10)?

13 What major differences exist between the organisms *Physarum* and *Dictyostelium* (◌◌ Section 14.11)?

14. Compare and contrast protozoa with fungi and algae, listing at least two ways you could differentiate a member of one group from the other. Which of these groups most closely resemble slime molds and why (◌◌ Sections 14.12 and 14.13)?

15. In what ways can different groups of algae be distinguished (◌◌ Section 14.13)?

16. What is a *frustule* and which group of algae produce them (◌◌ Section 14.13)?

APPLICATION QUESTIONS

1. If organelles such as the mitochondrion and chloroplast had originally been free-living *eukaryotic* cells, how would the molecular properties of organelles listed in Section 14.4 differ?

2. In the next chapter you will learn that eukaryotic cells have, in general, much larger genomes than prokaryotic cells. List three reasons why this is not surprising.

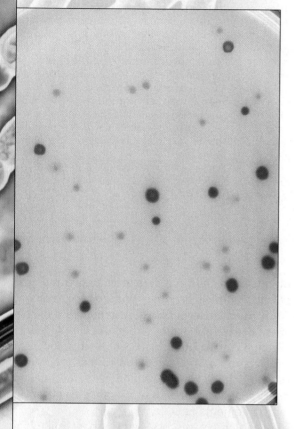

15

MICROBIAL GENOMICS

Genome sequencing reveals the genetic blueprint of an organism. Any genomics project begins with cloning a genome into managable-sized pieces, such as within a bacteriophage, as shown here. White plaques of bacteriophage M13 that have replicated in cells of *Escherichia coli* contain cloned DNA.

WORKING GLOSSARY

Annotation in reference to genomics, the conversion of raw sequence data into a list of genes present in the organism (the "assignment" of genes)

Artificial chromosome a cloning vector that can carry very large inserts of foreign DNA and exist in the cell very much like a cellular chromosome. The most widely used are bacterial artificial chromosomes (BACs) and yeast artificial chromosomes (YACs)

Bioinformatics the use of computational tools to acquire, analyze, store, and access DNA and protein sequences

Gene family genes that are related in sequence to other genes within the organism

Genome the total complement of genes of a cell or a virus

Genomics the discipline involving mapping, sequencing, analyzing, and comparing genomes

Homologue a gene that shares evolutionary ancestry with one or more other genes

Horizontal gene transfer the presence of a gene in an organism that originated in another organism and was transferred by genetic exchange

Open reading frame (ORF) a sequence of DNA that contains no in-frame stop codons and, if transcribed, could be translated to yield a protein of known length and composition. A *functional* ORF is one that actually encodes a protein in the cell.

Ortholog a gene found in one organism that is similar to that in another organism but differs because of speciation (see also *Paralog*)

Paralog a gene within an organism whose similarity to one or more other genes in the same organism is the result of gene duplication (see also *Ortholog*)

Proteome the total complement of proteins present in a cell, tissue, or organism at any one time

Proteomics the large-scale or genome-wide study of the structure, function, and regulation of the proteins of an organism

RNA editing modification of the RNA transcript of a protein-encoding gene by a process other than splicing, to achieve the required coding properties

Shotgun sequencing sequencing of previously cloned small fragments of a genome in a random fashion followed by computational methods to reconstruct the entire genome

Transcriptome the complement of mRNAs produced in an organism under a specific set of conditions

A n organism's **genome** is its entire complement of genes. Knowledge of the genome sequence of an organism reveals not only its genes, but also yields important clues to how the organism functions and its evolutionary history. Genomic sequences also facilitate the study of *gene expression*, the transcription and translation of the genetic information in the genome. The traditional approach for studying gene expression was to focus on a single gene or group of related genes. Today, expression of the *entire complement* of an organism's genes can be examined in a single experiment. The word **genomics** refers to the discipline of mapping, sequencing, analyzing, and comparing genomes. This chapter is focused on microbial genomics.

The first genome to be sequenced was the 3569-nucleotide RNA genome of the *virus* MS2 (∞ Section 16.1) in 1976. The first DNA genome to be sequenced was the 5386-nucleotide sequence of the small, single-stranded DNA virus, ϕX174 (∞ Section 16.2) in 1977. This feat was accomplished by a group led by Frederick Sanger and debuted the Sanger dideoxynucleotide technique for DNA sequencing (∞ Section 7.8). The first *cellular* genome to be sequenced was the 1,830,137-bp chromosome of *Haemophilus influenzae* described in 1995 by Hamilton O. Smith, J. Craig Venter, and their colleagues at The Institute for Genomic Research (TIGR), Rockville, MD (USA). By spring of the year 2000 a "rough draft" had been established of the sequence of the haploid human genome, which is about 3 billion bp. At present, several hundred genomes have been sequenced and many others are in progress. We are clearly in the "Age of Genomics," and the fallout from this new era in biology promises new insight on topics as diverse as clinical medicine and microbial evolution.

It would have been impossible to sequence and analyze large and complex genomes without a corresponding improvement in technology. A major part of this improvement has been the automation of DNA sequencing (∞ Section 7.8). Another has been the development of powerful computational methods to analyze, store, and access DNA and protein sequences, a field called **bioinformatics**. Yet another major advance has been in cloning large fragments of DNA.

In this chapter we discuss some model microbial genomes, some of the techniques used to analyze these genomes, and what microbial genomics has shown us thus far. We begin with cloning, since success at this stage is necessary before a genomics project can really begin.

GENOMIC CLONING TECHNIQUES

Most of the techniques we will discuss in this section are modifications or expansions of the *in vitro* techniques discussed in Chapter 10. The principles for these *in vitro* technologies remain the same, but now the emphasis is on studying the *entire genome* of an organism, rather than a single gene or set of genes.

15.1 Vectors for Genomic Cloning and Sequencing

We have previously seen some plasmid and viral vectors used for molecular cloning (∞ Sections 10.15–10.17). Phage pBR322 and the Charon derivatives of bacteriophage lambda, for example, have been extensively used for cloning and sequencing, including genomic analysis. Other, more specialized vectors include bacteriophage M13 and bacterial and yeast artificial chromosomes, and we focus on these here.

Vectors Derived from Bacteriophage M13

M13 is a filamentous bacteriophage that replicates without killing its host (Section 16.3). Mature particles of M13 are released from host cells without lysing by a budding process, and infected cultures can provide continuous sources of phage DNA. Most of the genome of phage M13 contains genes essential for virus replication. However, a small region called the *intergenic sequence* can be used as a cloning site. Variable lengths of foreign DNA, up to about 5 kbp, can be cloned without affecting phage viability. As the genome gets larger, the virion simply grows longer.

Phage M13mp18 is a derivative of M13 in which the intergenic region has been modified to facilitate cloning (Figure 15.1a ●). One useful modification is the insertion of a functional fragment of *lacZ*, the *Escherichia coli* gene that encodes the enzyme β-*galactosidase*. Cells infected with M13mp18 can be detected easily by their color on indicator plates (besides cleaving lactose into glucose and galactose, β-galactosidase cleaves a compound that yields a blue color, Figure 15.1b). This *lacZ* gene has itself been modified to contain a 54-bp DNA fragment called a *polylinker*. The polylinker contains several restriction enzyme cut sites absent from the M13 genome and can therefore be used for cloning. The polylinker is inserted into the beginning of the coding portion of the *lacZ* gene and does not affect enzyme activity. However, insertion of additional DNA into the polylinker during cloning inactivates the gene. Phages that contain additional DNA inserts give rise

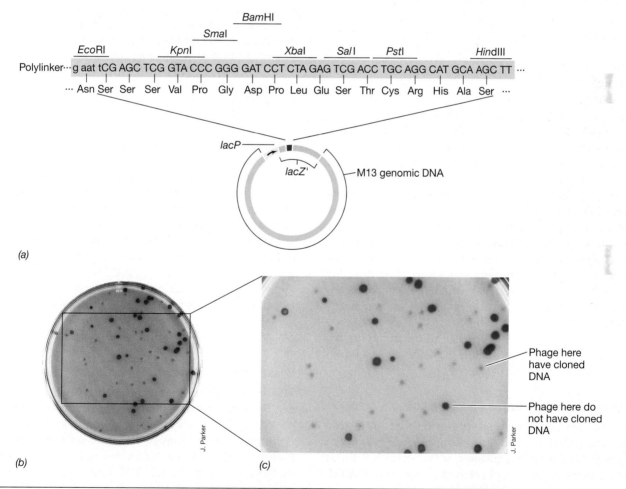

(a)

(b)   (c)

● **Figure 15.1   Cloning using bacteriophage M13.** (a) A partial map of M13mp18, a derivative of M13 constructed for use as a cloning vector. The vector contains the *lac* promoter *(lacP)* and a gene, *lacZ'*, which encodes a functional part of β-galactosidase. At the beginning of this gene is a polylinker that contains several restriction sites but maintains the proper reading frame. The amino acids encoded by the polylinker are shown. Most DNA fragments cloned into the polylinker disrupt the *lacZ'* gene and abolish β-galactosidase activity. (b) A plate with plaques formed by M13mp18 and by clones made using this vector on a lawn of sensitive bacteria plated on a medium containing the chemical 5-bromo-4-chloro-3-indolyl-β-D-galactopyranoside, called X-gal. When β-galactosidase hydrolyzes X-gal, it releases a relatively insoluble blue dye. Many plaques on this plate are blue, indicating the presence of vector *without* cloned DNA. Many other plaques are colorless, indicating that foreign DNA has been inserted into the vector and the *lacZ'* gene has been disrupted. In terms of genetic techniques, cloning using bacteriophage M13 combines aspects of both *selection* and *screening*. Using blue/white color differences, phage containing cloned DNA can be *selected*. However, unless PCR product is cloned, one must still *screen* the white plaques to find the ones containing the cloned gene of interest. (c) An enlargement of a portion of this plate.

to colorless plaques (no β-galactosidase activity), and it is therefore very simple to identify clones (Figure 15.1*b, c*). Similar constructs have been used in lambda cloning vectors and plasmid cloning vectors to allow identification of plaques or colonies containing cloned DNA.

Use of M13 in Molecular Cloning

To clone DNA in M13 vectors, replicative double-stranded DNA (🔗 Section 16.3) is isolated from the infected host and treated with a restriction enzyme. The foreign DNA is then treated with the same restriction enzyme. On ligation, double-stranded M13 molecules are obtained that contain the foreign DNA. When these molecules are introduced into the cell by transformation (🔗 Section 10.7), they are replicated and produce single-stranded DNA bacteriophage particles containing the cloned DNA.

The single-stranded M13 DNA produced can then be used directly in DNA sequencing. Since the base sequence where the foreign DNA is inserted is known (based on the specificity of the restriction enzyme used), it is possible to construct an oligonucleotide primer complementary to this region and with this determine the sequence of the whole DNA downstream from this point using Sanger sequencing. In this way, M13 derivatives have proven extremely useful in sequencing foreign DNA, even rather long molecules, and have featured prominently in the sequencing of several genomes.

Vectors like M13, or plasmid vectors that hold about 2 kb of cloned DNA, are adequate for making *gene libraries* (🔗 Section 10.15) for sequencing of prokaryotic genomes. Bacteriophage lambda vectors, which hold 20 kb or more (🔗 Section 10.17), are also widely used in genomics projects. However, as the size of the genome to be sequenced increases, so will the number of clones needed to obtain a complete sequence. Therefore, for making libraries of DNA from eukaryotic microorganisms or from higher eukaryotes, such as from humans, it is useful to have vectors that can carry very large segments of DNA. This allows the size of the initial library to be manageable. Such vectors have been developed and are called *artificial chromosomes*.

Bacterial Artificial Chromosomes: BACs

Many bacteria contain large plasmids that are stably replicated within the cell. For example, the *F plasmid* of *Escherichia coli* is such a plasmid (🔗 Sections 7.4 and 10.10). F is very stably replicated in *E. coli*, and naturally occurring derivatives, called F′ plasmids, are known that can carry large amounts of chromosomal DNA (🔗 Section 10.12). Because of these desirable properties, the F plasmid has been used to construct cloning vectors called **bacterial artificial chromosomes**, or **BACs**.

Figure 15.2● shows the structure of a BAC based on the F plasmid. The vector is only 6.7 kb compared to the 99.2 kb of F itself. The BAC contains only a few genes from F, including *oriS* and *repE*, which are necessary for replication, and *sopA* and *sopB*, which keep the copy num-

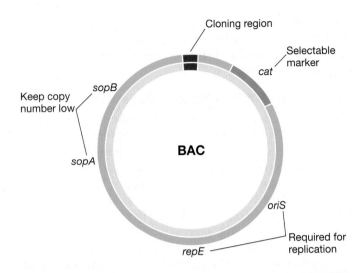

● **Figure 15.2 Genetic map of a bacterial artificial chromosome.** BACs are derivatives of the F plasmid of *Escherichia coli*. The BAC diagrammed is 6.7 kb. At the top of the map is a cloning region that contains several unique restriction enzyme sites. The *cat* gene confers resistance to the antibiotic chloramphenicol. The other genes shown are involved in plasmid replication. In the 99.2-kb F plasmid itself (🔗Figure 10.18) all these genes are located in a relatively compact replication region. Therefore, BACs contain only a small fraction of the entire F plasmid.

ber very low (🔗 Section 10.9). Inserted into the plasmid is the gene *cat*, which confers chloramphenicol resistance on the host, and a cloning region that includes several restriction sites for cloning DNA. Foreign DNA of *over 300 kb* can be inserted and stably maintained in a BAC vector such as this. The host for a BAC is typically a mutant strain of *E. coli* that is missing the normal restriction and modification systems of the wild type (🔗 Sections 7.7 and 9.6). This prevents the BAC from being destroyed. Typically, the host strain will be defective in some recombination pathways (🔗 Section 10.6), as well. This prevents recombination and rearrangements of the cloned DNA from the BAC into the host's chromosomes.

Yeast Artificial Chromosomes: YACs

Historically, the term **artificial chromosome** originated not with BACs but with **yeast artificial chromosomes**, or **YACs**. These vectors replicate in yeast like normal chromosomes, but they have sites where very large fragments of DNA can be inserted. To function like normal eukaryotic chromosomes, YACs must have (1) an *origin of DNA replication*, (2) *telomeres* for replicating DNA at the ends of the chromosome (🔗 Section 14.6), and (3) a *centromere* (the section of the chromosome required for segregation during mitosis). They must also contain a cloning site and a gene that can be used for selection after transformation into the host, typically the yeast *Saccharomyces cerevisiae*. Figure 15.3● shows a diagram of a YAC vector into which foreign DNA has been cloned. YAC vectors are themselves only about 10 kbp, but they can have 200–800 kbp of cloned DNA inserted.

● **Figure 15.3** **Diagram of a yeast artificial chromosome (YAC) containing foreign DNA.** The foreign DNA was cloned into the vector at a *NotI* restriction site. The telomeres are labeled TEL and the centromere CEN. The origin of replication is labeled ARS (for autonomous replication sequence). The gene used for selection is called URA3. The host into which the clone is transformed has a mutation in that gene so that it normally requires uracil for growth (Ura⁻). Host cells containing this YAC become Ura⁺. The diagram is not drawn to scale; the inserted DNA would normally be 200–800 kbp long and the vector only about 10 kbp.

After confirming that a particular fragment of DNA has been cloned on a BAC or a YAC, this region can be *subcloned* into a plasmid or bacteriophage vector for more detailed analysis or sequencing. Although YACs can hold larger DNA inserts than BACs, there is a greater problem with recombination and rearrangement of the cloned DNA within yeast than within *E. coli*. For this reason BACs are now more widely used in genomic cloning than are YACs.

 15.1 Concept Check

Specialized cloning vectors have been constructed that are useful for the sequence and assembly of genomes. Some, like the M13 derivatives, are useful for both cloning and for direct DNA sequencing. Others, like artificial chromosomes, are useful for cloning very large fragments of DNA, fragments approaching a megabase in size.

◆ Compare the capacity for cloning foreign DNA in M13, lambda, BACs, and YACs.

◆ The yeast artificial chromosome behaves like a chromosome in a yeast cell. What makes this possible?

15.2 Sequencing the Genome

The analysis of a genome begins with the formation of a genomic library—the molecular cloning of all of the DNA fragments of the genome generated from restriction enzyme cuts. Once this is done, actual sequencing begins.

Shotgun Sequencing

Virtually all genomic sequencing projects today employ **shotgun sequencing**. This technique is made possible by high-throughput sequencing capacity, robotics, and powerful computational capacities. Shotgun sequencing has been used on chromosomes from prokaryotes as well as eukaryotes, including for the privately funded version of the human genome project.

Whole genome shotgun techniques involve cloning the entire genome in a random fashion and then sequencing the resultant clones. That is, the clones are sequenced without knowing the order or orientation of any of the cloned DNA. The sequences are then analyzed by a computer that searches for overlapping sequences.

The overlaps allow the computer to assemble the sequenced fragments in the correct order.

Much of the sequencing in the shotgun method is redundant. To ensure that all sequences are obtained it is necessary to sequence a very large number of clones, many of which will be identical or nearly identical. Typically there will be 7–10 sequences over any given part of the genome. This seven- to tenfold *coverage*, as it is referred to, greatly reduces errors in the sequence because the redundancy in sequencing allows for a consensus nucleotide to be selected at any point in the sequence where ambiguity may exist.

For shotgun sequencing to work effectively, it is essential that the cloning itself be efficient (one needs a large number of clones) and, in so far as possible, the DNA cloned should be randomly generated. Restriction sites are not random, but by cutting the genomic DNA with an enzyme that recognizes a short sequence that occurs commonly in the DNA, one can approach random digestion. To obtain more truly random fragments, DNA can be mechanically sheared by forcing DNA through a *nebulizer*. This is a device with a small opening similar to a nozzle that reduces the DNA solution to a spray; in the process, the DNA is sheared. The DNA fragments can be purified by size using gel electrophoresis (Section 7.7) and then cloned into a vector and transformed into a host.

Genome Assembly

Once shotgun sequencing is completed, all of the fragments must be ordered, a process called **assembly**. Assembly of a circular genome from a prokaryote, for example, involves putting all of the fragments in the correct order and eliminating overlaps and then generating a genome suitable for *annotation*, the process of identifying genes and other functional regions (see Section 15.3).

Sometimes shotgun sequencing and assembly will not yield a complete genome sequence; that is, sometimes there will be *gaps* left in the sequence. In such situations, clones can be sought that are predicted to cover the gap. One method of doing this is to perform PCR reactions using specific primers complementary to sequences that are already known and that flank a given gap. Note that these additional clones are targeted, not random, as in the shotgun method. However, the sequences in these clones are likely to contain overlapping sequences sufficient to close the gap.

Some genome projects have the goal of obtaining a *closed genome*, meaning that the entire genome sequence is determined. Other projects stop at the *draft* stage, dispensing with the sequencing of the small gaps. Since shotgun sequencing and assembly are heavily automated procedures while gap closure is not, a closed genome is considerably more expensive to generate than a draft genome sequence.

 15.2 Concept Check

Shotgun techniques employ random cloning and sequencing of relatively small genome fragments followed by computer-generated assembly of the genome using overlaps as a guide to the final sequence.

◆ Why is shotgun sequencing considered to give a very *accurate* genome sequence?

◆ What is done during genome *assembly*?

15.3 Annotating the Genome

It is not the sequence of the genome, *per se*, that is the final goal in genome sequencing projects, but instead, determining the *genes* that the sequence contains. Once sequencing and assembly is completed, the next step in genomic analysis is **annotation**, the conversion of raw sequence data into a list of the genes present in the genome.

Locating Putative Genes: Identifying ORFs

The great majority of genes of any organism encode proteins, and in most microbial genomes, the great majority of the genome consists of coding sequences. Because the genomes of microbial eukaryotes typically have fewer introns than plant and animal genomes, and prokaryotes have almost none, microbial genomes essentially consist of hundreds to thousands of **open reading frames (ORFs)** (Section 7.13) separated by regulatory regions and transcriptional terminators. A *functional ORF* is one that actually encodes a protein in the cell. Thus, the simplest way to locate protein-encoding genes, or potential protein-encoding genes, is to have a computer search the sequence of the genome for ORFs.

How Does the Computer Find an ORF?

In the cell, ribosomes establish a reading frame by initiating translation at a *start codon*, usually an AUG. The ribosome then proceeds until it reaches an in-frame *stop codon* (Section 7.13). The first step in finding an ORF is to look for these signals in the sequence. However, the identification of possible *functional* ORFs is typically more complex than just searching for in-frame start and stop codons, since these will appear randomly with reasonable frequency. Thus, other clues are sought.

One hint that an ORF is functional will be its *size*. The majority of cellular proteins contain 100 or more

amino acids, so most functional ORFs are longer than 100 codons (300 nucleotides). However, simply programming the computer to ignore ORFs shorter than 100 codons will miss some functional genes. So other factors must be considered. Because most organisms show preferences among synonymous codons (Section 7.14), *codon bias* can also give a clue as to whether an ORF is functional. If the codon usage in a given ORF is considerably different from the consensus codon usage, the ORF may not be functional or may be functional but obtained by lateral gene transfer (see Section 15.8). It should also be remembered that prokaryotic ribosomes start translation not at the first (most 5′) possible start codon, but at one immediately downstream of a *Shine–Dalgarno sequence* on the mRNA (Section 7.16). Therefore, searching the DNA sequence of a prokaryotic genome for potential Shine–Dalgarno sequences can help establish both whether an ORF is functional and which start codon is actually used.

Figure 15.4● shows a region of DNA with the characteristics of an ORF that a computer would recognize as a "hit." Although any given gene is always transcribed from a single strand, in all but the smallest plasmid or viral genomes, both strands are transcribed in some part of the genome. Thus, computer inspection of *both strands* of DNA is required. Figure 15.4 shows a diagram of a region of a genome in which one gene is transcribed from one strand and one from the other. Of course, some genes encode transfer RNAs and ribosomal RNAs, and these are not recognized by programs that search only for ORFs. However,

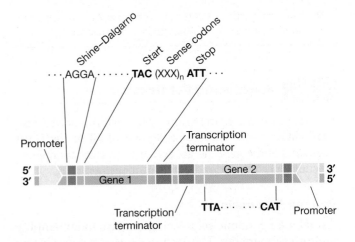

● **Figure 15.4 Locating possible functional open reading frames.** The sequence shown for Gene 1 from one strand includes bases for a potential Shine–Dalgarno sequence (AGGA) separated from a start codon (ATG) by about 8 bases. The bases encoding the start codon are followed by about 100 sense codons and then a stop (nonsense) codon (UAA is the most commonly used stop codon). In the DNA molecule with the ORF described as gene 1, note that upstream of this ORF is a promoter and downstream a transcription terminator. Also shown is Gene 2, which has the same components, but that would be transcribed in the opposite direction. Shine–Dalgarno sequences, start codons, sense codons, and stop codons function only at the level of RNA. Here we are showing DNA bases (base pairs) that encode these functional sequences.

genes for tRNAs and rRNAs can usually be located rather easily because the sequences of these RNAs are very highly conserved (⌀ Sections 7.15 and 11.5–11.7).

Locating Putative Genes: Genomic Comparisons

An ORF is also likely to be functional if its sequence is similar to sequences of ORFs obtained from the genomes of other organisms (regardless of whether they encode known proteins) or if some part of the ORF has a sequence known to encode a protein functional domain. This is because proteins that carry out the same function in different cells tend to be *homologous*; that is, they are proteins that are related in an evolutionary sense and typically share sequence and structural features. The computer can search for such sequence similarities in the annotation process.

It would be almost impossible to assemble even a small prokaryotic genome and to locate genes and other important functional DNA sequences without the availability of sophisticated computational tools to handle large databases. The use of computers to do sophisticated analyses of this type is called working *in silico*, a term analogous to the terms *in vivo* and *in vitro* (*in silico* refers to the silicon processor chips present in the computer). Figure 15.5● shows a genetic map constructed by computer from shotgun sequencing of the 4.4 Mbp genome of *Mycobacterium tuberculosis*, the causative agent of tuberculosis (⌀ Section 26.5). Although a highly detailed figure, all ORFs are indicated, as well as genes encoding tRNAs, rRNAs, and a few other sequences.

15.3 Concept Check

After major sequencing is through, computers search for ORFs and genes encoding protein homologues as part of the annotation process.

◆ What is an *open reading frame (ORF)*?

◆ How can protein homology assist in the annotation process?

MICROBIAL GENOMES

Over 250 microbial genomes, mostly prokaryotic, have now been sequenced, and hundreds more projects are currently ongoing. Here we will discuss only a few of them and explore what analysis of these genomes tells us.

15.4 Prokaryotic Genomes: Sizes and ORF Contents

Several hundred prokaryotic genomes are now available in public databases (for an up-to-date list of genome sequencing projects search the URL *http://www.genomesonline.org/*). These include many species of *Bacteria* and *Archaea* and representatives containing circular as well as linear ge-

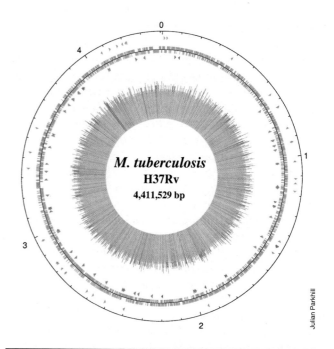

● **Figure 15.5** **Chromosome map of *Mycobacterium tuberculosis*.** The outer circle shows the size of the chromosome in Mbp with 0 placed at the point of origin of DNA replication. Inside this is a ring showing the locations of tRNA genes (blue) and the single rRNA operon (orange). The next ring shows ORFs transcribed clockwise (dark green) and counterclockwise (light green). Beneath this is a ring showing the location of some repetitive DNA elements, including insertion sequences (orange). The histogram in the center shows regional GC content (⌀ Section 11.10) with yellow rays showing regions below 65% GC and orange rays showing regions above 65% GC. For additional discussion of the organism *M. tuberculosis*, see Section 12.23, and for the disease tuberculosis, Section 26.5. The figure was generated with software from DNASTAR and is reprinted by permission from *Nature 393*: 537–544 (1998). © Macmillan Magazines Ltd.

nomes. Table 15.1 lists a few representative examples. Although the sequencing of prokaryotic genomes is becoming more rapid and routine, sequencing and analysis of a complete genome is still a major undertaking, requiring both considerable financial as well as human efforts. Thus, one still needs important scientific and/or societal reasons and considerable financial resources before initiating a genome sequence.

Note that the list of prokaryotic genomes in Table 15.1 contains several pathogens. Naturally, such organisms would have high priority for sequencing resources. The hyperthermophiles on the list may have important uses in biotechnology since the enzymes in these organisms will be heat stable (⌀ Sections 6.10, 6.12, 13.12 and 30.9). Indeed, the needs of the biomedical and biotechnology industries are important considerations for generating the interest and the funding to pursue genomic sequencing. However, the list in Table 15.1 also includes organisms such as *Bacillus subtilis*, *Escherichia coli*, and *Pseudomonas aeruginosa*, all of which remain widely studied genetic model systems, and *Caulobacter crescentus*, a model system for cell differentiation (⌀ Section 12.16).

Sizes of Prokaryotic Genomes

Based on analyses of prokaryotic genomes of species of *Bacteria* and *Archaea*, there is a strong correlation between a genome's size and its ORF content (Figure 15.6●). Indeed, regardless of the organism, each *megabase* of prokaryotic DNA encodes about *1000* ORFs (Figure 15.6). This shows that unlike in eukaryotes, where noncoding DNA can make up a large fraction of the genome (especially in large genome organisms, see Table 15.3), in prokaryotes, as genomes increase in size, they also increase proportionally in gene number.

Analyzing genomic sequences allows us to answer some fundamental biological questions. For example, genomic analysis of the small genomes of *Mycoplasma genitalium* and *Mycoplasma pneumoniae* shows that all 470 ORFs found in *M. genitalium* are also present in *M. pneumoniae*. By seeking homologous genes in other organisms and by determining which genes are unessential using such techniques as transposon mutagenesis (∞ Section 10.14), scientists have deduced that about 300 protein-encoding genes may be sufficient to maintain a cellular existence. No known organism has these few genes. However, some remarkable organisms do have relatively small genomes. For example, *Methanocaldococcus jannaschii* (*Archaea*) is an *autotroph* (∞ Sections 13.4 and 17.17) and contains only 1738 ORFs. This is enough, however, to enable it to be not only free living, but also to synthesize all of its organic cellular components from CO_2 (∞ Section 13.4). *Aquifex aeolicus* (*Bacteria*) is also an autotroph and contains the smallest genome of any known autotroph at just 1.5 Mbp (Table 15.1). Both *Methanocaldococcus* and *Aquifex* are also hyperthermophiles. One can thus conclude, rather surprisingly, that an autotrophic lifestyle in near boiling water does not require a huge genome to support it.

It is no coincidence that Table 15.1 lists several relatively small genomes. Sequencing strategies are simpler for small genomes, so there was some emphasis on sequencing small genomes in early sequencing efforts. Note that the genome of *Mycoplasma genitalium* is smaller than that of some viruses, such as the *Chlorella* viruses (∞ Section 16.6) and bacteriophage G (∞ Section 9.2). The smallest known prokaryotic genome, one even smaller than that of *M. genitalium*, is that of *Nanoarchaeum equitans* (*Archaea*). This genome is some 90 kbp smaller that that of *M. genitalium* (Table 15.1). *N. equitans* is a hyperthermophile and a parasite of a second hyperthermophile, *Ignicoccus* (∞ Section 13.11). Analyses of the gene content of *N. equitans* shows it to be free of virtually all genes that encode proteins involved in anabolism or catabolism. Interestingly, however, the genome of *N. equitans* actually contains more ORFs than the larger genome of *M. genitalium* (Table 15.1). This is because the *N. equitans* genome is extremely compact and contains virtually no noncoding DNA.

Some prokaryotes have very large genomes. *Bradyrhizobium japonicum*, the organism that forms nitrogen-fixing root nodules on soybeans (∞ Section 19.22), for example, has 2800 more ORFs than the yeast *Saccharomyces cerevisiae*, a eukaryote (see Section 15.6 and

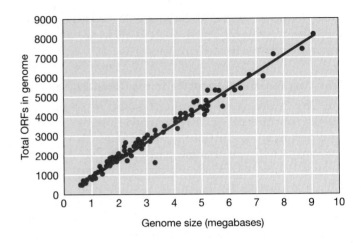

● **Figure 15.6 Correlation between genome size and ORF content in prokaryotes.** These data are from analyses of 115 completed prokaryotic genomes and include species of both *Bacteria* and *Archaea*. Data from *Proc. Natl. Acad. Sci. (USA)* 101: 3160–3165 (2004).

Table 15.1). The bacterium *Streptomyces coelicolor*, whose linear chromosome of over 8 Mbp is typical of *Streptomyces* species, has 7846 ORFs, almost 1800 more than *S. cerevisiae*. Some even larger prokaryotic genomes are known, but their sequence is not yet complete. These include the 9.2-Mbp chromosome of *Myxococcus xanthus* and the 12.3-Mbp chromosome of *Sorangium cellulosum*, both gliding bacteria (∞ Section 12.17). These are the largest prokaryotic genomes known. Genomes of comparable size have not yet been found in *Archaea* (Table 15.1). Prokaryotic genomes thus range in size from those of large viruses to those of small eukaryotes.

15.4 Concept Check

Sequenced prokaryotic genomes range in size from 0.49 Mbp to 9.1 Mbp. The smallest prokaryotic genomes are the size of the largest viruses, and the largest prokaryotic genomes have more genes than some eukaryotes. In prokaryotes, ORF content is proportional to genome size.

◆ How many more genes does *Bradyrhizobium* have than *Nanoarcheaum*? How does this affect the lifestyles of these two organisms?

◆ Approximately how many ORFs will a genome of 4 Mbp contain?

15.5 Prokaryotic Genomes: Bioinformatic Analyses and Gene Distributions

Following assembly and annotation, a typical genomic analysis proceeds to the *gene comparison stage*. The basic question here is "how does the complement of genes in this organism compare with that of other organisms?" Such analyses put the just-sequenced genome in perspective with the genomes of other organisms, similar as well as more distant. These activities are a major part of the field of **bioinformatics**.

Table 15.1	Select prokaryotic chromosomes[a]		
Organism	**Size (base pairs)**	**ORFs[b]**	**Comments**
Bacteria			
Mycoplasma genitalium	580,070	470	Smallest known bacterial genome (⚬⚬ Section 12.21)
Mycoplasma pneumoniae	816,394	677	Causes pneumonia (⚬⚬ Section 12.21)
Borrelia burgdorferi	910,725	853	Spirochete, has linear chromosome, causes Lyme disease (⚬⚬ Sections 12.33 and 27.4)
Chlamydia trachomatis	1,042,519	894	Obligate intracellular parasite, common human pathogen (⚬⚬ Sections 12.27 and 26.13)
Rickettsia prowazekii	1,111,523	834	Obligate intracellular parasite, causes epidemic typhus (⚬⚬ Sections 12.13 and 27.3)
Treponema pallidum	1,138,006	1041	Spirochete, causes syphilis (⚬⚬ Sections 12.33 and 26.12)
Aquifex aeolicus	1,551,335	1544	Hyperthermophile, autotroph (⚬⚬ Section 12.37)
Prochlorococcus marinus	1,657,990	1716	Most abundant phototroph in oceans (⚬⚬ Section 12.26 and 19. 6)
Helicobacter pylori	1,667,867	1590	Causes peptic ulcers (⚬⚬ Section 26.10)
Streptococcus pyogenes	1,852,442	1752	Pathogen, causes step throat and scarlet fever (⚬⚬ Section 26.2)
Thermotoga maritima	1,860,725	1877	Hyperthermophile (⚬⚬ Section 12.36). See also Figure 15.7
Chlorobium tepidum	2,154,946	2288	Model phototrophic bacterium (⚬⚬ Section 12.32)
Staphylococcus aureus	2,814,816	2593	Major cause of nosocomial infections (⚬⚬ Sections 12.19 and 25.7)
Deinococcus radiodurans	3,284,156	2185	Radiation resistant, multiple chromosomes (⚬⚬ Section 12.34)
Synechocystis sp.	3,573,470	3168	Cyanobacterium (⚬⚬ Section 12.25)
Bdellovibrio bacteriovorus	3,782,950	3584	Predator of other prokaryotes (⚬⚬ Section 12.14)
Geobacter sulfurreducens	3,814,139	3467	Model for bioremediation (⚬⚬ Section 17.18)
Caulobacter crescentus	4,016,942	3767	Complex life cycle (⚬⚬ Section 12.16)
Bacillus subtilis	4,214,810	4100	Gram-positive genetic model (⚬⚬ Section 12.20)
Mycobacterium tuberculosis	4,411,529	3924	Causes tuberculosis (⚬⚬ Sections 12.23 and 26.5)
Escherichia coli	4,639,221	4288	Gram-negative genetic model (⚬⚬ Section 12.11)
Bacillus anthracis	5,227,293	5738	Pathogen, biowarfare agent (⚬⚬ Section 12.20)
Rhodopseudomonas palustris	5,459,213	4836	Metabolically versatile anoxygenic phototroph (⚬⚬ Section 12.2)
Pseudomonas aeruginosa	6,264,403	5570	Metabolically versatile opportunistic pathogen (⚬⚬ Section 12.7)
Streptomyces coelicolor	8,667,507	7825	Has linear chromosome, produces antibiotics (⚬⚬ Section 12.24)
Bradyrhizobium japonicum	9,105,828	8317	N_2 fixation, nodulates soybeans (⚬⚬ Section 19.22)
Archaea			
Nanoarchaeum equitans	490,885	552	Smallest known genome (⚬⚬ Section 13.11)
Thermoplasma acidophilum	1,564,905	1509	Thermophilic, acidophilic (⚬⚬ Section 13.5)
Methanocaldococcus jannaschii	1,664,976	1738	Methanogen (⚬⚬ Section 13.4)
Aeropyrum pernix	1,669,695	1841	Hyperthermophile (⚬⚬ Section 13.9)
Pyrococcus horikoshii	1,738,505	2061	Hyperthermophile (⚬⚬ Section 13.6)
Methanothermobacter thermoautotrophicus	1,751,377	1855	Methanogen (⚬⚬ Section 13.4)
Archaeoglobus fulgidus	2,178,400	2436	Hyperthermophile (⚬⚬ Section 13.7)
Halobacterium salinarum	2,571,010	2630	Extreme halophile, bacteriorhodopsin (⚬⚬ Section 13.3)
Sulfolobus solfataricus	2,992,245	2977	Hyperthermophile, sulfur chemolithotroph (⚬⚬ Section 13.9)

[a] Information on these and hundreds of other prokaryotic genomes can be found in the TIGR Database (www.tigr.org/tdb), a web site maintained by The Institute for Genomic Research (TIGR), Rockville, MD, a not-for-profit research institute, and at *http://www.genomesonline.org*. Links are listed there to other relevant web sites.

[b] Open reading frames. The purpose of reporting ORFs is to predict the total number of proteins that an organism might encode. Of course, genes encoding known proteins are included, as are all ORF's that could encode a protein greater than 100 amino acid residues. Smaller ORFs are typically not included unless they show similarity to a gene from another organism or unless the codon usage is typical of the organism being studied.

Gene Content of Prokaryotic Genomes

In many ways, the complement of genes in a particular organism defines its biology. In other words, *genomes are molded by an organism's lifestyle.* One might imagine, for instance, that obligately parasitic organisms such as *Treponema pallidum* (⚬⚬ Section 26.12) might require relatively few genes for amino acid biosynthesis, since the amino acids they need can be supplied by their hosts. In fact, this is the case, as computer inspection of the *T. pallidum* genome showed it to lack recognizable genes for amino acid biosynthesis, although it encodes several proteases that can convert peptides taken up from the host into free amino acids. In contrast, the nonparasitic *Escherichia coli* has 131 genes involved in amino acid biosynthesis and metabolism, and *Bacillus subtilis* has over 200.

Comparative analyses such as these are useful in the search for genes that encode enzymes that almost certainly must exist because of the known properties of an organism. *Thermotoga maritima* (*Bacteria*) for example, is a hyperthermophile found in heated marine sediments and is known to catabolize a large number of sugars. Figure 15.7● summarizes some of the metabolic pathways and transport systems of *T. maritima* that have been derived from analysis of its genome. As might be expected, the genome is particularly rich in genes involved in transport, particularly for carbohydrates and amino acids. A large portion of its genome (7%) is also involved in metabolism of simple and complex sugars. All this suggests that *T. maritima* exists in an environment rich in organic material.

Some analysis on the division of genes and their activities in prokaryotes is given in Table 15.2. Similar data are assembled when each new finished and annotated genome is published. Thus far it is clear that a distinct pattern exists in gene distribution in prokaryotes. Metabolic genes, for example, are typically the most abundant class of gene in prokaryotic genomes, although genes involved in translation overtake metabolic genes on a percentage basis as genome size decreases (Table 15.2, and see Figure 15.8). Surprisingly, as important as these genes are to the biology of a cell, genes involved in DNA replication and transcription make up

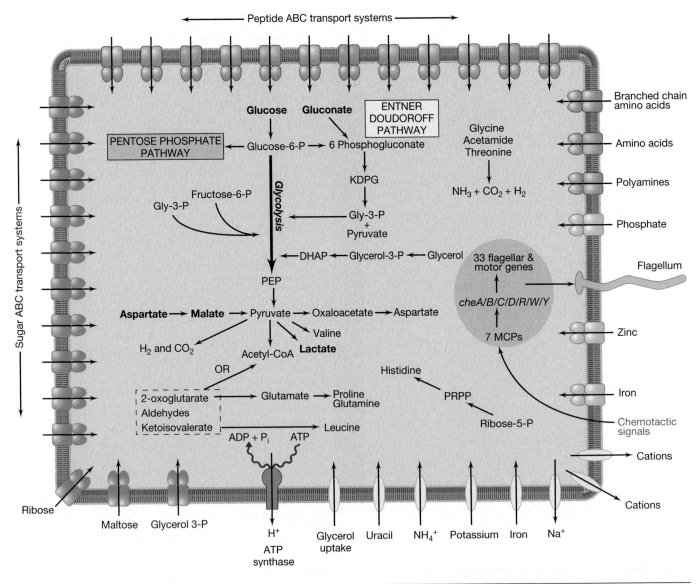

● **Figure 15.7** **Overview of some metabolic pathways and transport systems of *Thermotoga maritima*, a member of the domain *Bacteria*.** The figure shows a blueprint for the metabolic capabilities of this organism. These include some of the pathways for energy production and the metabolism of certain organic compounds, as well as proteins involved in transport, that were identified from analysis of the genomic sequence. Gene names are not shown. The genome contains several ABC transport systems (◯◯ Section 4.7). There are 12 for carbohydrates, 14 for peptides and amino acids, and still others for ions. These are shown as multisubunit structures in the figure. Other types of transport proteins have also been identified and are shown as simple ovals. The flagellum is shown, and this organism has seven transducers (MCPs) and several chemotaxis (*che*) genes (◯◯ Section 8.13) as well as several genes involved in flagellar assembly (◯◯ Section 4.14). A few aspects of sugar metabolism are also shown. This figure is adapted from work published by The Institute for Genomic Research (TIGR).

Table 15.2	Gene function in bacterial genomes		
	Percentage of genes on chromosome in that category		
Functional categories	**Escherichia coli (4.64 Mbp)[a]**	**Haemophilus influenzae (1.83 Mbp)[a]**	**Mycoplasma genitalium (0.58 Mbp)[a]**
Metabolism	21.0	19.0	14.6
Structural	5.5	4.7	3.6
Transport	10.0	7.0	7.3
Regulation	8.5	6.6	6.0
Translation	4.5	8.0	21.6
Transcription	1.3	1.5	2.6
Replication	2.7	4.9	6.8
Other, known	8.5	5.2	5.8
Unknown	38.1	43.0	32.0

[a] Chromosome size. Each organism listed contains only a single circular chromosome.

only a minor fraction of the typical prokaryotic genome (Table 15.2).

Unidentified ORFs

Although there are differences from organism to organism, in most cases the number of genes that can actually be identified in a given genome is 70% or less of the total number of ORFs detected. Unidentified ORFs are said to encode *hypothetical proteins*, proteins that likely exist although we do not yet know their function. An unidentified ORF contains all of the signs of a protein-encoding gene—a ribosome binding site (or a location in an operon downstream from a Shine–Dalgarno site), a start codon and a stop codon—but lacks sufficient amino acid sequence homology with any known protein to be readily identified as such.

Unidentified ORFs reflect the fact that there is still much we don't know about prokaryotic genes. On the other hand, as gene functions become identified in one organism, homologous unidentified ORFs in other organisms will likely be found to encode the same protein. It is interesting to note that even in the world's best-understood organism, *Escherichia coli*, functions have been assigned to only about 2700 of its almost 4300 genes. However, most of the genes involved in macromolecular syntheses and central metabolism essential for growth of *E. coli* have been identified. Therefore, as the functions of more unidentified ORFs are identified, it is likely that most of them will encode nonessential or redundant proteins. As progress is made in identifying hypothetical proteins, the percentage of *E. coli* genes involved in macromolecular synthesis or central metabolism will therefore likely decrease (Table 15.2). It has been predicted that many of the unidentified genes in *Escherichia coli* encode proteins that carry out regulatory functions, are those needed only under special conditions or for the catabolism of unusual substrates, or are those that encode redundant proteins as "backup" systems for key metabolic reactions.

Gene Categories as a Function of Genome Size

As the data of Table 15.2 show, the *percentage* of an organism's genes devoted to one or another cell function is to some degree a function of genome size. This is summarized for a large number of genomes from species of *Bacteria* in Figure 15.8●. In this figure it can be seen that the *percentage* of genes devoted to protein synthesis, for example, rises dramatically in *small* genome organisms. By contrast, genes devoted to transcription and to regulation involving two-component regulatory systems (signal transduction, ∞ Section 8.12), increase significantly in *large* genome organisms. Other core cellular processes, such as DNA replication and energy production, show only minor variations in gene number with genome size (Figure 15.8).

The genomic analyses summarized in Figure 15.8 suggest that although many genes can be dispensed with, genes that encode the protein-synthesizing apparatus cannot. The smaller the genome, the greater the percentage of genes devoted to translational processes (Table 15.2). Large genome organisms, on the other hand, contain more genes involved in transcriptional and other types of regulation than small genome organisms. These regulatory systems likely allow the cell to better respond to its available resources with specific gene expression. In small genome organisms these regulatory processes are dispensable, probably because small genome prokaryotes are often parasitic and can obtain much of what they need from their hosts. Large genome organisms, by contrast, have the coding capacity to include many such regulatory and special metabolic genes in their genome. This likely makes these organisms more competitive in their habitats, which for very large genome prokaryotes is often soil. Soil is a habitat where carbon and energy sources are often scarce or available only intermittently and are of a wide variety of different types (∞ Sections 19.2 and 19.4). A

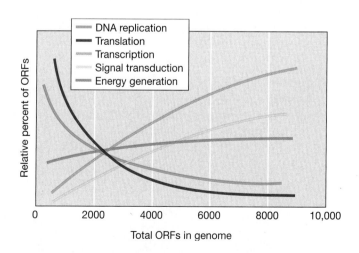

● **Figure 15.8 Functional category of genes as a relative percent of the genome versus total ORFs in the genome.** Note how genes encoding translational or DNA replication increase in percentage in *small* genome organisms, while transcriptional regulatory genes increase in percentage in *large* genome organisms. Data from *Proc. Natl. Acad. Sci. (USA)* 101: 3160–3165 (2004).

large genome encoding several metabolic options would thus be strongly selected for in such a habitat. Interestingly, all of the prokaryotes listed in Table 15.1 whose genomes are in excess of 6 Mbp inhabit soil.

Gene Distribution in *Bacteria* and *Archaea*

Analyses of gene categories have been done on several prokaryotes beyond the three species of *Bacteria* shown in Table 15.2, and the results are compared in Figure 15.9●. Note that in this figure the results are the *averages* of the gene content of several individual genomes and include both *Bacteria* and *Archaea*. On average, species of *Archaea* seem to devote a higher percentage of their genomes to energy and coenzyme production than do *Bacteria* (these results are undoubtedly skewed a bit due to the large number of novel coenzymes produced by methanogenic *Archaea*, ⌒⌒ Sections 13.4 and 17.17). On the other hand, *Archaea* tend to contain fewer genes devoted to carbohydrate metabolism or to cell membrane functions, such as transport and membrane biosynthesis, than do *Bacteria*.

Both groups of prokaryotes have relatively large numbers of genes whose function is either unknown or that encode only hypothetical proteins, although in both categories, more uncertainty exists among species of *Archaea* than species of *Bacteria* (Figure 15.9). Thus, in addition to controls on gene complement that are related to genome size (Figure 15.8), additional controls appear to operate at the level of cellular domain, as well (Figure 15.9). Data of these types are a likely reflection of lifestyle differences between *Bacteria* and *Archaea* that remain to be elucidated.

 15.5 Concept Check

Many genes can be identified by their sequence similarity to genes found in other organisms. However, a significant percentage of sequenced genes are of unknown function. On average, the gene complement of *Bacteria* and *Archaea* are related but distinct. Bioinformatics plays an important role in genomic analyses.

◆ Does every functional ORF encode a protein?

◆ What is a hypothetical protein?

◆ What category of genes do prokaryotes contain the most of on a percentage basis?

15.6 Eukaryotic Microbial Genomes

A large number of microbial eukaryotes are known, and to date, a variety of microbial and higher eukaryotic genomes have been sequenced (Table 15.3). The yeast *Saccharomyces cerevisiae* is an extremely important eukaryote because of its widespread use in industry (⌒⌒ Chapter 30) and its use as a model organism for cell biology, and so we focus on it here.

The Yeast Genome

The haploid yeast genome contains 16 chromosomes ranging in size from 220 kbp to about 2352 kbp. The total yeast nuclear genome (excluding the mitochondria and some plasmid and viruslike genetic elements) is approximately 13,392 kbp. You may wonder why the words *about* and *approximately* are used when this genome has been completely sequenced. Yeast, like many other eukaryotes, has a large amount of repetitive DNA (⌒⌒ Section 7.4). When the yeast genome was published in 1997, not all of the "identical" repeats had been sequenced. It is difficult to sequence a very long run of identical or nearly identical sequences and then assemble the data into a coherent framework. For example, yeast chromosome XII contains a stretch of approximately 1260 kb containing 100–200 repeats of yeast rRNA genes. Another repeated sequence follows this long series of rRNA gene repeats. Because this entire region has not actually been completely sequenced, the precise size of the yeast genome is not known.

In addition to having 100–200 identical copies of the rRNA operons, the yeast nuclear genome has 275 genes

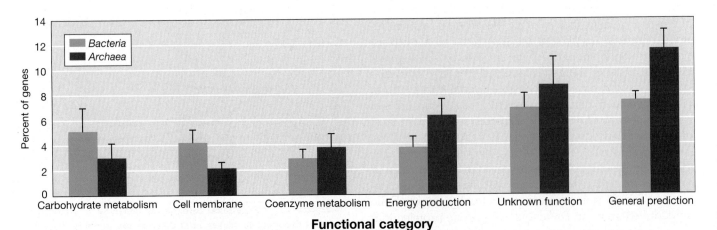

● **Figure 15.9 Variations in gene category in *Bacteria* and *Archaea*.** Data are averages from 34 species of *Bacteria* and 12 species of *Archaea*. "Unknown function" represents genes known to encode proteins but whose functions are unknown. "General prediction" encode hypothetical proteins that may or may not exist. Data from *Proc. Natl. Acad. Sci. (USA)* 101: 3160–3165 (2004).

for tRNAs (only a few are identical) and 80 genes for other types of noncoding RNAs (∞ Section 7.16). There were 6340 possible ORFs originally predicted in yeast, but more detailed analysis has reduced this number to 5570, less than that of some prokaryotes (Table 15.1). Of these, almost 3400 encode proteins whose functions are known. The wide variety of genetic and biochemical techniques available for studying this organism have resulted in significant advances in the understanding of the function of the remaining proteins as well (see Section 15.11).

Minimal Gene Complement of Yeast and the Presence of Introns

Of all the known genes in yeast, which are absolutely essential? A mechanism for exploring this question in yeast or in any microorganism exists and involves *knockout mutations* (∞ Section 10.14). Knockouts are mutations in which a gene has been rendered nonfunctional. Ordinarily such mutations cannot be obtained in a gene essential for cell viability in a haploid organism. However, yeast can also be grown in a diploid state (∞ Section 14.7). By obtaining such mutations in a diploid and then investigating whether they can also exist in a haploid, it is possible to determine whether a particular gene is essential for a cell's viability. Using such techniques, it has been shown, surprisingly, that at least 877 of the yeast ORFs are essential, while 3121 clearly are not. Note that this number of essential genes is considerably greater than the approximately 300 that are predicted to be the minimal number required for cellular existence in prokaryotes (see Section 15.3). However, since eukaryotes are more complex organisms than prokaryotes, a larger minimal gene complement would be expected.

Since yeast is a eukaryote, it contains introns (∞ Section 7.1). However, in the protein-encoding genes of yeast there are only a total of 225 introns. Most yeast genes that have introns have a single small intron near the 5′ end of the gene. This situation is much different than in the genomes of higher organisms shown in Table 15.3. In the worm *Caenorhabditis elegans*, for example, the average gene has 5 introns, and in the fruitfly *Drosophila*, the average gene has 4 introns. Introns are also very common in the mustard plant *Arabidopsis*, which also averages 5 introns per gene, and over 75% of *Arabidopsis* genes have introns. In humans almost all protein-encoding genes have introns, and it is not uncommon for a single gene to have 10 or more. In these organisms, the introns are typically much larger than the exons, as well.

Other Eukaryotic Microorganisms

The genomes of several other notable eukaryotic microorganisms have been sequenced. *Plasmodium falciparum* is the parasite that causes malaria (∞ Section 27.5); the sequencing of its genome was an international effort. The 27-Mbp genome of *P. falciparum* consists of 14 chromosomes ranging in size from 0.7 to 3.4 Mbp. *Encephalitozoon cuniculi* is an intracellular pathogen of humans and other animals that causes lung infections. *E. cuniculi* lacks mitochondria (∞ Section 14.9), and although its haploid genome contains 11 chromosomes, the genome size is only 2.9 Mbp (Table 15.3); this is smaller than that of many prokaryotic genomes (see Table 15.1). *Ustilago maydis* is a plant pathogenic fungus with a genome of approximately 20 Mbp. It causes smut disease in corn (maize), a disease with a large economic impact.

Table 15.3 Some eukaryotic nuclear genomes[a]

Organism	Comments	Genome size	Chromosome number	Protein-encoding genes[b]
Encephalitozoon cuniculi	Very small genome; human pathogen	2.9 Mbp	11	1997
Saccharomyces cerevisiae	This yeast is an industrially important organism (∞ Sections 30.10 and 30.13) that is also a model for biochemical and genetic studies	13 Mbp	16	5,570
Caenorhabditis elegans	This roundworm is an important model for studying animal development	97 Mbp	6	19,099
Drosophila melanogaster	The fruit fly is an intensively studied model organism	180 Mbp	4	13,601
Arabidopsis thaliana	This plant serves as a model organism for genetic studies	125 Mbp	5	25,498
Mus musculus	Mouse, a model mammalian system	2500 Mbp	23	30,000
Homo sapiens	The human genome is available only in draft form	3000 Mbp	23	25,000–35,000[c]

[a] All data are for the haploid nuclear genomes of these organisms.

[b] The number of protein-encoding genes is in all cases an estimate based on the number of known genes and sequences that seem likely to encode functional proteins.

[c] There is still debate over the number of genes in the human genome, despite the release of "draft" sequences by two different groups. The number of genes could be as high as 60,000.

Considerable progress has been made in sequencing the genomes of other eukaryotic microorganisms, primarily pathogens, including *Leishmania major* (leishmaniaisis), *Candida albicans* (various yeast infections; ∞ Section 27.8), *Entamoeba histolytica* (amebic dysentery; ∞ Sections 14.9 and 28.8), *Giardia lamblia* (giardiasis; ∞ Section 28.6), and *Pneumocystis carinii* (AIDS-associated pneumonia; ∞ Section 26.14). In addition, the complete sequence of the mouse and rat genomes are known. The human genome has been sequenced of course, but not yet fully annotated (Table 15.3).

15.6 Concept Check

The complete genomic sequence of the yeast *Saccharomyces cerevisiae* and that of many other microbial eukaryotes has been determined. Yeast may encode up to 5570 proteins of which only 877 appear essential for viability. Relatively few of the protein-encoding genes of yeast contain introns.

◆ How might you show that a gene is essential?

◆ What is unusual about the genome of the eukaryote *Encephalitozoon*?

III OTHER GENOMES AND THE EVOLUTION OF GENOMES

Besides cells, other structures have genomes, including organelles and viruses. We consider here organellar genomes and then some issues of genome evolution. We end with a consideration of how genomes can be "mined" for genes of interest.

15.7 Genomes of Organelles

The major organelles, mitochondria and chloroplasts, contain their own DNA (∞ Section 14.4). Both mitochondria and chloroplasts originated from *Bacteria* by endosymbiosis (∞ Sections 11.4 and 14.4). Therefore, it is not surprising that the proteins that organellar DNA encode are more closely related to those of *Bacteria* than to those of *Eukarya* or *Archaea*. We will see that this is for the most part true; however, mitochondrial genomes are somewhat enigmatic. We begin with the photosynthetic organelle, the chloroplast.

Chloroplast Genomes

Known chloroplast genomes are all *circular* DNA molecules. Moreover, although there are several copies of the genome in each chloroplast, each is identical. The typical chloroplast genome is about 120–160 kbp and contains two inverted repeats of 6–76 kb (Figure 15.10●). Several chloroplast genomes have been completely sequenced and a few of these are summarized in Table 15.4. The flagellated protozoan *Mesostigma viride* belongs to the

earliest diverging green plant lineage. Its chloroplast contains more protein-encoding genes and tRNA genes than any other so far known and has the typical genome structure illustrated in Figure 15.10.

Many of the genes in the chloroplast genome encode proteins involved in photosynthesis and autotrophy. However, the chloroplast genome also encodes rRNAs used in chloroplast ribosomes, tRNAs used by the translational apparatus, and a few of the proteins used in transcription and translation, as well as some other proteins. Some proteins that function in the chloroplast are encoded by genes in the nucleus, presumably genes that migrated there as the chloroplast evolved from a free-living photosynthetic endosymbiont. Unlike free-living prokaryotes, introns are common in chloroplast genes, and they are primarily of the self-splicing type (∞ Section 14.8).

Analyses of chloroplast genomes have firmly supported the endosymbiotic hypothesis (∞ Section 11.4). For example, chloroplast genomes have been shown to contain genes that are homologues of those in *Escherichia coli*, cyanobacteria, and other *Bacteria*. These include, among others, genes encoding proteins involved in cell division (∞ Section 6.2), suggesting that the mechanism of chloroplast division is similar to that of *Bacteria*. In addition, chloroplast genes involved in protein transport through membranes are highly related to those of *Bacteria*.

Mitochondrial Genomes

Mitochondria are involved in energy production and are found in most eukaryotic organisms. Mitochondrial genomes primarily encode proteins involved in oxidative phosphorylation and, as is the case with chloroplast genomes, also encode rRNAs, tRNAs, and translational proteins needed for protein synthesis. However, most mitochondrial genomes encode many fewer proteins than do those of chloroplasts.

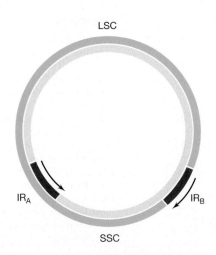

● **Figure 15.10 The map of a typical chloroplast genome.** The genomes of chloroplasts are circular double-stranded DNA molecules. Most contain two inverted repeat regions (IR$_A$ and IR$_B$), which form the borders of a small single copy region (SSC) and a large single copy region (LSC).

Table 15.4   Some chloroplast genomes[a]

| Organism | | Size (bp) | Genes encoding | | | Inverted repeats[d] |
			Proteins[b]	tRNA	rRNA[c]	
Chlorella vulgaris	Green alga	150,613	77	31	1	Absent
Euglena gracilis	Protozoan	143,170	67	27	3	Absent
Mesostigma viride	Protozoan	118,360	92	37	2	Present
Pinus thunbergii	Black pine	119,707	72	32	1	Present[e]
Oryza sativa	Rice	134,525	70	30	2	Present
Zea mays	Corn	140,387	70	30	2	Present

[a] All chloroplast genomes are circular, double-stranded DNA (⚬⚬ Section 14.4).
[b] These include genes encoding proteins of known function and ORFs that might be functional.
[c] Each unit is an rRNA operon, containing genes for each of the rRNAs (⚬⚬ Section 7.13).
[d] See Figure 15.10.
[e] Although the inverted repeats are present, they are greatly truncated.

Well over 200 mitochondrial genomes have been sequenced. The largest mitochondrial genome has 62 protein-encoding genes, while others encode as few as 3 proteins. The mitochondria of almost all animals including humans encode only 13 proteins (plus 22 tRNAs and 2 rRNAs) and that of the yeast *Saccharomyces cerevisiae* only 8 proteins.

Whereas chloroplasts use the "universal" genetic code, mitochondria use slightly different, simplified codes (⚬⚬ Section 7.14). These seem to have arisen from the selection pressures for smaller genomes. The standard set of 22 mitochondrial tRNAs is insufficient to read the entire genetic code, even with the "standard" wobble pairing taken into consideration. Therefore, base-pairing between the anticodon and the codon (⚬⚬ Figure 7.34) is even more flexible in mitochondria than it is in cells.

Unlike chloroplast genomes, which are all single, circular DNA molecules, the genomes of mitochondria are quite diverse. For example, some mitochondrial genomes are linear, including the mitochondria of some species of algae, protozoans, and fungi. In other cases, such as in the yeast *Saccharomyces cerevisiae*, although genetic and restriction mapping have shown the mitochondrial genome to be circular, it seems that the major *in vivo* form is linear. (Recall that bacteriophage T4 has a genetically circular genome structure but is physically linear, ⚬⚬ Section 9.9.) Figure 15.11● shows a map of the 16,569-bp (13 gene) human mitochondrial genome. The yeast mitochondrial genome is larger (85,779 bp) but has only 8 protein-encoding genes. Outside of the genes encoding the RNAs and proteins, the genome of yeast mitochondria contains large stretches of extremely AT–rich DNA that serve no apparent function.

Finally, it should be noted that in the mitochondria of several organisms, small plasmids exist, making mitochondrial genome analysis even more complicated. An additional complication in analyzing some mitochondrial and chloroplast genomes is that it is sometimes difficult to find the gene for a particular protein even when the sequence of both the protein and the organellar DNA

are both known. This is because of **RNA editing** (see the Microbial Sidebar).

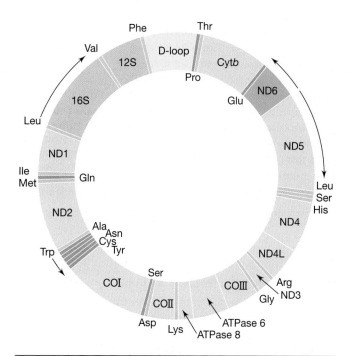

● **Figure 15.11   Map of the human mitochondrial genome.** The circular genome of the human mitochondrion contains 16,569 bp. The genome encodes the 16S and 12S rRNA (corresponding to the prokaryotic 23S and 16S rRNA) and 22 tRNAs. These genes are indicated in two shades of orange, a darker orange for genes that are transcribed counterclockwise from the map as drawn and those in lighter orange that are transcribed clockwise. (The amino acid designations for the tRNA are also on the outside for counterclockwise-transcribed genes and on the inside of the map for clockwise-transcribed genes.) The 13 protein-encoding genes are shown in green (once again with darker green indicating those transcribed counterclockwise and lighter green for those transcribed clockwise). The genes encode: Cyt*b*, cytochrome *b*; ND1–6, components of the NADH dehydrogenase complex; COI–III, subunits of the cytochrome oxidase complex; ATPase 6 and 8, polypeptides of the mitochondrial ATPase complex. The two promoters are in the region called the D-loop, a region also involved in DNA replication.

Microbial Sidebar ◆ RNA Editing

In Chapters 7 and 14 we saw that certain genes have coding regions that are split by noncoding regions called *introns*. Typically, introns are removed after transcription to form a mature mRNA, a process called *splicing* (⌘ Section 14.8). Interestingly, there is a phenomenon found in the genomes of organelles that is almost the opposite of splicing: *RNA editing*.

RNA editing involves either the insertion or deletion of nucleotides *into the final mRNA* that were not present in the DNA transcribed. Editing can also involve the chemical *modification* of a base in the mRNA that changes it from one base to another. In either case, RNA editing can alter codons in such a way that one or more different amino acids are inserted in a polypeptide than those encoded by its gene.

In the mitochondria of trypanosomes and related protozoa (⌘ Section 14.10) some mitochondrial transcripts are edited such that large numbers (hundreds in some cases) of uridylates are added or, more rarely, deleted. An example of this type of RNA editing is shown in Figure 1 ●. RNA editing is precisely controlled by short sequences present in the mRNA that "guide" the enzymes involved in their specific edits. Obviously this process must be very precisely controlled. Inserting too many or too few bases would yield a frameshift product that would likely be nonfunctional.

The other type of RNA editing, the changing of one base into another, is common in the mitochondria and chloroplasts of higher plants. At specific sites in some mRNAs, a C will be converted to a U by oxidative deamination (the opposite modification is more rare). There are at least 25 sites of C to U conversion in the maize chloroplast. Although mostly found in organellar genomes, an example of the programmed conversion of a C to a U is also known for a mammalian nuclear gene. Depending on the location of the edit, a new codon may be formed, leading to formation of a protein sequence not predictable from the gene that encodes it.

RNA editing, although a curious phenomenon, was not a significant obstacle in analyzing *organellar* genomes. This is because the number of proteins they encode is small and the proteins highly conserved. By contrast, had RNA editing been a widespread phenomenon in *cells*, genomic annotations and the identification of orthologous genes in different organisms could have been an even more formidable challenge than it has been to date.

The function and origin of RNA editing is unknown. But some scientists have pointed out that this process may be yet another remnant, along with ribozymes (⌘ Section 14.8) and other catalytic RNAs, of the RNA World (⌘ Section 11.2).   ■

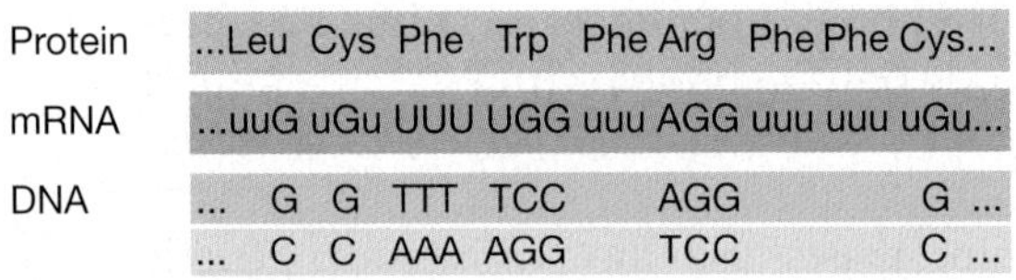

Protein	...Leu Cys Phe Trp Phe Arg Phe Phe Cys...
mRNA	...uuG uGu UUU UGG uuu AGG uuu uuu uGu...
DNA	... G G TTT TCC AGG G ...
	... C C AAA AGG TCC C ...

Figure 1 RNA editing. *The upper part of the figure shows a portion of the amino acid sequence of subunit III of the enzyme cytochrome oxidase from the protozoan* Trypanosoma brucei *(⌘ Section 14.10). This protein is encoded by mitochondria. Beneath the amino acid sequence, the sequence of the messenger RNA (mRNA) for this region is shown. The bases in uppercase letters are those transcribed from the gene, which is shown below. The bases in the mRNA in lowercase have been inserted into the transcript by RNA editing. Although the DNA has many informational gaps, there are no actual gaps in the molecule itself. The spaces between the base pairs are simply to aid in visualization.*

Organelles and the Nuclear Genome

Chloroplasts and mitochondria require more proteins than they encode. For example, far more proteins are involved in organellar translation alone than are encoded by them. Thus, many organellar functions are orchestrated by nuclear genes.

It is estimated that the yeast mitochondrion contains over 400 different proteins, and as we have seen, only 8 of them are encoded by the yeast mitochondrion. Hence, almost all proteins required by the yeast mitochondrion are encoded by genes *in the nucleus*. Although it might seem reasonable that the genes (and proteins) that function in specific processes in the eukaryotic nucleus/cytoplasm could be put to the same use in organelles, this is not the case. Although the genes for many organellar proteins are present in the nucleus, transcribed there and translated in the cell's cytoplasm, the gene products are used specifically by the organelles and must be transported there.

Among the first mitochondrial proteins to be studied were those involved in translation and it was clear that these nuclear-encoded mitochondrial proteins were closely related to counterparts in *Bacteria*, not to those in *Eukarya*. Thus, it appeared at first that most genes encoding mitochondrial proteins had been transferred from the mitochondrion to the nucleus. However, in the absence of genomics approaches, it was difficult to gain further insight into this problem.

What is required to analyze this problem is a genome sequence of a eukaryote, that of a species of *Bacteria* phylogenetically closely related to the mitochondrial genome, and genomic sequences of other *Bacteria* for comparative purposes. Knowledge of the mitochondrial genome of the eukaryote was actually not required (since it encodes so few proteins), but a knowledge of the proteins that function in the mitochondria was. All these requirements were met in the case of the yeast *Saccharomyces cerevisiae*, and the analysis has been revealing.

Surprisingly, of the 400 nuclear genes encoding mitochondrial proteins, only about 50 were closely related to the phylogenetic lineage in *Bacteria* that led to mitochondria (α Proteobacteria, Section 12.1). Another 150 were clearly related to proteins of *Bacteria*, but not necessarily α Proteobacteria. However, the remaining 200 proteins were encoded by genes that have no identifiable homologues among known genes of *Bacteria*. The *Bacteria*-like proteins were mostly involved in energy conversions, translation, and biosynthesis, whereas the other proteins were mostly involved in membranes, regulation, and transport. Thus, although the mitochondrion has many of the telltale signs of having originated from an ancient endosymbiotic event, genomic analyses have shown that its genetic history is more complicated than previously thought.

15.7 Concept Check

Chloroplasts and mitochondria have small genomes independent of nuclear genomes. These genomes encode rRNAs, tRNAs, and a few proteins involved in energy metabolism. Although the genomes of the organelles are independent of the nuclear genome, the organelles themselves are not. Many genes in the nucleus encode proteins required for organellar function. These genes have various phylogenetic histories.

◆ How is genome size and gene content correlated in yeast and human mitochondria?

◆ What is unusual about the genes that encode mitochondrial functions in yeast?

◆ What is RNA *editing*? How does it differ from RNA *processing*?

15.8 Evolution and Gene Families

The first priority of genomics is, of course, to determine the number, sequence, and function of genes in an organism. However, there is more to genomics than just sequencing and annotating genes and then interpreting the results in terms of how an organism might interact with its environment. Comparative genomics also helps us to understand the *evolutionary relationships* between organisms. Reconstructing evolutionary relationships from genome sequences helps to distinguish primitive from derived characteristics and can resolve ambiguities in phylogenetic trees based on analyses of a single gene (for example, 16S rRNA, Sections 11.5 and 14.9). Such knowledge helps us better understand early life forms and, in time, may answer the most basic question in biology: How did life first arise?

Gene Duplication and Gene Families: Paralogs and Orthologs

Genomes from both prokaryotic and eukaryotic sources contain **gene families**—genes that are related to other genes *within* the organism. Although large gene families are not the rule, comparative genomics has shown that many genes have arisen by *duplication* of other genes. Such genes are called **paralogs**, genes whose similarity is the result of gene duplication at some time in the evolution of an organism. Genes found in one organism that are similar to genes *in another organism* but which differ because of speciation are called **orthologs**. An example of paralogous genes would be genes encoding lactate dehydrogenase (LDH) isoenzymes in humans. These enzymes are structurally distinct yet all highly related and carry out the same enzymatic reaction. By contrast, orthologs of LDH exist between the LDH from *Escherichia coli*, for example, and an LDH isoenzyme from humans.

The study of genes and gene families is a major task in comparative genomics. Because chromosomes from many different microorganisms have already been sequenced, such comparisons can be easily done and the results are often surprising. For instance, genes in *Archaea* involved in DNA replication, transcription, and translation, are more similar to those in *Eukarya* than to those in *Bacteria*. Unexpectedly, however, many other genes in *Archaea*, for example, those encoding metabolic functions other than information processing, are more similar to those in *Bacteria* than those in *Eukarya*. The powerful analytical tools of bioinformatics allow us to deduce such genetic relationships between domains of life very quickly, at either the single gene, groups of genes, or entire genome level. The results obtained thus far lend further support to the phylogenetic picture of life deduced originally by comparative ribosomal RNA sequence analysis (Section 11.5) and suggest that many genes in all organisms have common evolutionary roots. However, such analyses have also revealed instances of horizontal gene flow, an important issue to which we now turn.

Horizontal Gene Transfer

Evolution is premised on the transfer of genetic traits from one generation to the next. However, *horizontal (lateral) gene transfer* also occurs, and it can complicate evolutionary studies, especially of entire genomes. **Horizontal gene transfer** occurs whenever genes are transferred from one cell to another *other than* by the usual inheritance process, from mother cell to daughter cell. In prokaryotes, at least three methods for horizontal gene transfer are known: *transformation, transduction,* and *conjugation* (Chapter 10).

Horizontal gene flow may be extensive in nature and is a process that can even cross domains. However, to be detectable by comparative genomics, the difference between the organisms must be rather large. For example, several genes with eukaryotic origins have been found in *Chlamydia* (∞ Sections 15.3 and 26.13) and *Rickettsia* (∞ Sections 12.13 and 27.3), both human pathogens. Moreover, genomic analyses have shown that *Thermotoga maritima* (a species of *Bacteria*) contains over 400 genes (greater than 20% of its genome) that are clearly of archaeal origin. Of these genes, 81 are found in discrete clusters. This strongly suggests that they were obtained by horizontal gene transfer—presumably from thermophilic *Archaea* that share the hot environments inhabited by *Thermotoga*.

Horizontal gene transfers can be spotted in genomes once the genes have been annotated. The presence of genes that encode proteins typically found only in distantly related species is one signal that the genes originated from horizontal transfer. However, another clue to horizontally transferred genes is the presence of an ORF or ORFs whose GC content (∞ Section 11.10) or codon bias differs significantly from that of the remaining genome. Using these clues many likely examples of horizontal transfer have been documented in the genomes of various prokaryotes. Horizontally transferred genes typically encode metabolic functions *other than* the core molecular processes of DNA replication, transcription, and translation and may account for the previously mentioned similarities of metabolic genes in *Archaea* and *Bacteria*. In addition, there have been several examples of virulence genes of pathogens (∞ Sections 21.8 and 21.9) having been transferred by horizontal means. Obviously, prokaryotes are exchanging genes in nature, and the process likely functions to "fine-tune" an organism's genome to a particular situation or habitat.

The *Chlamydia trachomatis* genome revealed potential horizontal transfer from a eukaryotic source, possibly even its human host. This is the opposite of the situation where genes in the eukaryotic nucleus have been transferred there from the ancestor of the mitochondrion (see Section 15.7). If these *C. trachomatis* genes were actually obtained from humans, it shows that horizontal gene transfer is not limited by the prokaryotic/eukaryotic boundary.

Evolutionary analyses of genomes also have a practical component. The ability to identify genes that are clearly prokaryotic or, in particular, genes characteristic of pathogenic bacteria, has helped develop powerful and extremely specific diagnostic tools and therapeutic agents that target the products of these genes for use in clinical medicine.

 15.8 Concept Check

Genomics can be used to study the evolutionary history of an organism. Organisms contain gene families, genes with related sequences. If these arose because of gene duplication, the genes are said to be paralogs; if they arose by speciation, they are called orthologs. Organisms may acquire genes from other organisms in their environment by a process called horizontal gene transfer.

◆ Contrast gene *paralogs* with gene *orthologs*.

◆ Describe a mechanism by which horizontal gene transfer might occur in a prokaryote.

15.9 Genomic Mining

As we have discussed, genomic analysis provides an organism's genetic blueprint. From genomic analyses we can gain valuable insight into the metabolism and ecology of an organism. Often, however, genomic analyses indicate that one or more of the genes usually associated with a given pathway or process seem to be missing. This could mean that the organism uses a novel pathway or that at least one novel reaction is involved. In such cases it would be of interest to search for a gene encoding a protein that can carry out the missing function. This can easily be done by computer. Searching through a database containing the complete sequence of an organism's genome to find such a new gene is referred to as **genomic mining**. The search for the DNA polymerase of the cyanobacterium *Synechocystis* is a good example (Figure 15.12●).

DNA Polymerase of *Synechocystis*: Inteins and the Fruits of Genomic Mining

All species of *Bacteria* contain a DNA polymerase similar to DNA polymerase III of *Escherichia coli*, which is used as the primary polymerase in DNA replication (∞ Section 7.6). This is a complex enzyme containing many different

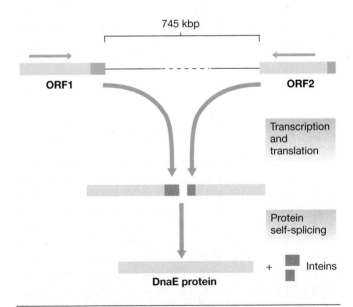

● **Figure 15.12 Inteins: The split *dnaE* gene of *Synechocystis*.** The two ORFs are transcribed in opposite directions, as shown by the orange arrows. The protein products of these ORFs each contain part of an intein (the protein product of ORF2 has been drawn in the standard amino terminal to carboxyl terminal orientation). Several DNA replication enzymes from *Archaea* also show inteins, so the phenomenon is not limited to *Bacteria*.

subunits, and the subunit that catalyzes the actual polymerase reaction is DnaE, a product of the *dnaE* gene. Because DNA polymerases are highly conserved proteins and of prime importance to all cells, inspection of the genome of *any* member of the *Bacteria* will quickly yield a gene with sequence similarity (an *orthologous* gene) to the *dnaE* gene of *E. coli*. Interestingly, however, the original search of the genome of *Synechocystis* (see Table 15.1) failed to identify such a gene. Instead, what was found were two ORFs of unknown function, which, if combined, formed a gene with high similarity to *dnaE*. But these two ORFs were over 700 kbp apart on the *Synechocystis* chromosome and located on opposite DNA strands, indicating that they were transcribed in opposite directions. Could these two ORFs really be part of *dnaE*?

Careful sequence analysis of the two ORFs revealed that this was indeed the case. The genes encoded complementary halves of an **intein**, a self-splicing protein (👁 Section 8.3). The two *Synechocystis* ORFs were then cloned. When transcribed and translated, this intein catalyzed a splicing reaction between the two halves of DnaE to form a complete and enzymatically active DnaE protein (Figure 15.12).

Why has *Synechocystis* arranged its *dnaE* gene in this split fashion? This is unknown, but evolution would tell us that whatever the reason, it is the best solution for expressing this gene in this particular organism. But *Synechocystis* is not unique in this regard. Several archaeal DNA replication genes also contain inteins, including those of *Nanoarchaeum equitans*, the parasitic archaeon with the tiny genome (Table 15.1). It is difficult to imagine how these highly unusual mechanisms would have been discovered without the ability to scan and compare entire genome sequences, as can be done routinely today. Mining genomes is thus likely to yield lots of surprises and may also lead to discoveries that have practical applications as well.

 15.9 Concept Check

Often it is necessary to search carefully through a genomic database to find a particular gene, a process called genomic mining. This can be done to find novel genes or to find genes that one predicts must be present.

◆ How does an *intein* differ from a *self-splicing intron*?

 **IV GENE FUNCTION AND REGULATION**

15.10 Proteomics

An additional aim of genomic studies is to determine which genes are *expressed* to yield protein products and to determine the function of these products. The term **proteome** refers to the collection of proteins present in a cell *at any one time*. Note that the proteome is not simply "all the proteins encoded by an organism's genome." Although the genes of an organism are constantly present, gene expression is regulated (👁 Chapter 8). The number and types of proteins present in a cell constantly change in response to an organism's environment or other factors, such as developmental cycles. The genomewide study of the structure, function, and regulation of an organism's proteins is called **proteomics**, or **functional genomics**.

Two-Dimensional Gel Electrophoresis

One approach to proteomics has been the use of *two-dimensional polyacrylamide gel electrophoresis*. This technique can separate, identify, and measure all the proteins present in a sample of cells. A two-dimensional (2-D) gel separating proteins from *Escherichia coli* is shown in Figure 15.13●. In the first dimension (the horizontal dimension in the figure) the proteins are separated by their *isoelectric points*, the pH at which the net charge on each protein reaches 0. In the second dimension, the proteins are denatured in a way that gives each amino

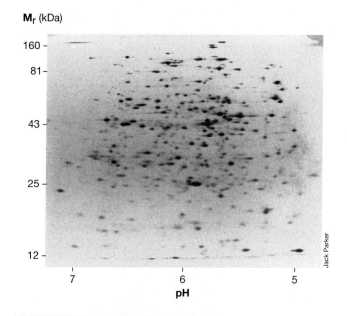

● **Figure 15.13 Proteomics: Two-dimensional polyacrylamide gel electrophoresis of proteins.** The figure shows an autoradiogram of the proteins of a culture of *Escherichia coli*. Each spot on the gel is a particular protein. The proteins were labeled with radioactive methionine during growth to allow for visualization and quantitation. The proteins were first separated by isoelectric focusing under denaturing conditions, concentrating on the pH range 5–7 where most *E. coli* proteins are found (the basic ribosomal proteins are missing from this separation). The second dimension separates denatured proteins by their mass (M$_r$; given here in kilodaltons), with the largest proteins being toward the top of the gel.

acid residue a fixed charge. The proteins are then separated by *size* (in much the same way DNA molecules are separated by size; ∞ Figure 7.22).

In the case of *E. coli* and a few other organisms, hundreds of proteins resolved in 2-D gels have been identified by biochemical or genetic means and their regulation studied under a variety of conditions. Using 2-D gels, the appearance or disappearance of a particular protein under different growth conditions can be used to assess the proteome as a function of environmental conditions. One method of connecting an unknown protein with a particular gene using the 2-D gel system is to elute the protein from the gel and sequence the N-terminus region of the protein. This sequence information may be sufficient to design oligonucleotide probes that amplify the gene encoding the protein from genomic DNA by the polymerase chain reaction (∞ Section 7.9). Then, following sequencing and comparative genomics, it is possible that the gene's function can be determined.

Functional and Structural Genomics

Although proteomics often requires intensive experimentation (Figure 15.13), *in silico* techniques can also be quite useful. After obtaining the sequence of an organism's genome, initial analysis can compare its sequence to that of other organisms to locate and identify genes that are similar to ones already known. The "sequence" here that is most important is the *amino acid* sequence of the protein. Because of degeneracy of the genetic code (∞ Section 7.14), differences in the DNA sequence may not necessarily lead to differences in the amino acid sequence (Figure 15.14●).

Proteins with greater than 50% sequence identity frequently have similar functions. Proteins with identities above 70% are all but certain to have similar functions. One important clue to the possible function of a protein is the identification of regions in the coding sequence that are present in other proteins and that encode domains of the protein having known functions. The latter would include regions such as metal-binding domains, nucleotide-binding domains, or binding domains for some other cofactor. For example, a protein with an NAD^+-binding domain, for example, is almost certainly a protein involved in a redox reaction (∞ Section 5.7) of some type.

Specific protein domains have specific sequences, but what makes them *functional* is their three-dimensional structure (∞ Section 3.8). **Structural genomics** is the determination of the three-dimensional structures of proteins representative of the range of protein structure and function found in an organism. The ultimate aim is to build a body of structural information that will allow *in silico* techniques to predict the probable structure and potential function for any protein from knowledge of its primary structure. Although this goal is not yet a reality, as more and more sequence data are generated and analyzed, it is one whose time will come. The link between genomics and proteomics will then be even stronger.

Coupling proteomics with genomics is yielding important clues to how gene expression in different organisms correlates with environmental stimuli. Not only does such information have important basic science benefits, but it also has potential applications. These include advances in medicine, the environment, and agriculture. In all of these areas, understanding the link between the genome and the proteome and how it is regulated could give humans unprecedented control over major aspects of disease, pollution, and agricultural productivity.

 15.10 Concept Check

The proteome encompasses all the proteins present in an organism at any one time. The aim of proteomics is to study these proteins to learn their structure, function, and regulation.

◆ Can an organism have more than one proteome?

◆ What is *functional* genomics?

15.11 Microarrays and the Transcriptome

One major aim of proteomics is to study gene expression. But there are other ways to study this process. Recall that gene expression is often regulated at the level of transcription (∞ Chapter 8). For a gene to be expressed, it must first be transcribed. Knowing the conditions under which a gene is transcribed may also give information about a gene's function. In analogy to the proteome, the entire complement of *mRNAs* produced under a given set of conditions is called a **transcriptome**. And a powerful technique exists for transcriptome analysis: **microarrays**.

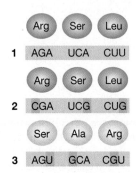

● **Figure 15.14 Comparison of nucleic acid and amino acid sequence similarities.** Three different nucleotide sequences are shown (for convenience RNA is shown). Both sequence 2 and sequence 3 differ from sequence 1 in only three positions. However, the amino acid sequence encoded by 1 and 2 are identical, whereas that encoded by sequence 3 is unrelated to the other two.

Microarrays and the DNA Silica Chip

In Chapter 7 we discussed how nucleic acid hybridization techniques help to locate genes on specific fragments of DNA (∞ Section 7.7). Hybridization techniques can also be used in conjunction with genomic sequence data to measure the expression of genes by hybridizing mRNA to specific DNA fragments. This technique has been radically enhanced with the development of **microarrays**, or **gene chips** as they are also called.

Microarrays are small solid-state supports to which genes or portions of genes representing the entire genome of an organism are affixed and spatially arrayed in a known pattern (see Figure 15.16a). The genes are synthesized by PCR, or, alternatively, oligonucleotides are designed for each gene based on the genomic sequence. Once attached to the solid support, these genes (or gene fragments) can then be hybridized with mRNA from cells grown under a specific condition and scanned and analyzed by a computer. Hybridization between a specific mRNA and the DNA on the chip confirms expression of that gene.

A method for making and using microarrays is shown in Figure 15.15●. The same process used to produce computer chips (photolithography) has been adapted to produce 1 to 2-cm silica microarray chips, each of which can hold thousands of different DNA fragments (Figures 15.15 and 15.16●). For example, one company markets a single array human genome chip that contains the entire human genome on it (Figure 15.16a). The chip can analyze over 47,000 transcripts, plus has room for 6500 additional oligonucleotides for use in clinical medicine! Figure 15.16b shows a part of a chip used to assay expression of the *Saccharomyces cerevisiae* genome.

The gene chip shown in Figure 15.16b easily holds the 5600 protein-encoding genes of *S. cerevisiae*, so that global gene expression in this organism can be measured in a single experiment. To do an experiment, the chip is hybridized with mRNA obtained from yeast cells grown under specific conditions and then tagged with a fluorescent dye. A particular tagged mRNA binds only to the DNA on the chip that is *complementary* to its sequence. To assay hybridization, the chip is scanned with a laser fluorescence detector and the signals analyzed by computer. A distinct pattern of hybridization is typically observed, depending upon which complementary DNA sequences on the chip show hybridization with mRNAs (Figures 15.15 and 15.16b). Although hybridization indicates *expression* of the gene, a qualitative result, the *intensity* of the binding is a quantitative measure of gene expression (Figure 15.16b). Following hybridization and laser scanning, the computer makes a list of which genes were expressed and to what extent. Thus, using gene chips, one knows in an instant the transcriptome of the organism of interest grown under specified conditions.

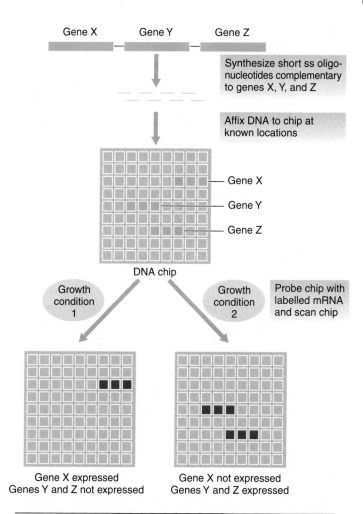

● **Figure 15.15 Measuring the transcriptome: Making and using DNA chips.** Short single-stranded (ss) oligonucleotides corresponding to all the genes of an organism are synthesized individually (∞ Section 7.8) and affixed at known locations to make a DNA chip (microarray). The DNA chip is assayed by hybridizing labeled mRNA obtained from cells grown under a specific condition to the DNA probes on the chip and then scanning the chip with a laser.

Applications of DNA Chips: Gene Expression

Using gene chips one can ask general or very specific types of scientific questions, depending on the genes affixed to the chip. For instance, one can assay an organism's global gene expression by using the entire population of mRNA as a probe (Figure 15.16b). By contrast, one could compare expression of different genes, singly or in sets, under different conditions. The ability to relatively quickly analyze the simultaneous expression of thousands of genes both qualitatively and quantitatively has tremendous potential for sorting through the complexity of metabolism and regulation. This is true for both organisms as "simple" as prokaryotes, with their 400–8000 or more genes (Table 15.1), or as complex as humans, with our approximately 30,000 different genes.

For example, the *S. cerevisiae* gene chip (Figure 15.16b) was used in a detailed study of metabolic con-

Web Tutorial 15.1 DNA Chips

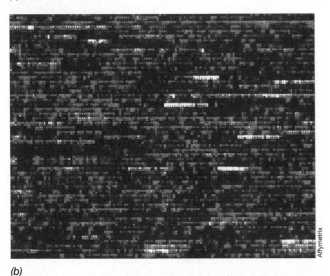

(a)

(b)

● **Figure 15.16 Measuring the transcriptome: DNA chips and their use to assay gene expression.** (a) The human genome chip. This chip has over 40,000 gene fragments on it. (b) A hybridized chip. The photo shows fragments from one-fourth of the entire genome of the yeast, *Saccharomyces cerevisiae* affixed to a portion of a silica gene chip. Each gene is present in several copies and has been probed with fluorescently labeled mRNA obtained from yeast cells grown under a specific condition. The background of the chip is blue. Locations where the RNA has hybridized to the DNA are indicated by a gradation of colors up to maximum hybridization, which shows as white. Because the location of different genes on the chip is known, once the chip is scanned it will reveal which specific genes were expressed.

trol in this organism. Yeast can grow by fermentation and by respiration. Using transcriptome analysis it was possible to see which genes were shut down and which were turned on when yeast cells were switched from fermentative (anoxic) to respiratory metabolism. Transcriptome analyses of gene expression in such an experiment showed that yeast undergo a major metabolic "reprogramming" during the switch to aerobic

growth. A whole suite of genes that control ethanol (a key fermentation product) production were strongly repressed while citric acid cycle functions (needed for aerobic growth) were strongly activated by the switch. Overall, over 700 genes were "turned on" and over 1000 "turned off" during this metabolic transition. Moreover, by using a microarray, the expression pattern of several yeast genes of unknown function could be monitored as well during the fermentative to respiratory switch, yielding clues to their possible function. At present, there is no other method available that can come close to giving as much information about gene expression as can microarrays. Clearly, microarrays offer a powerful approach for exploring gene expression patterns on a genomic scale.

Applications in Identification

Besides their use in probing gene expression, microarrays can be used to *identify* microorganisms. For example, one can use fragmented genomic DNA from a particular organism as a probe and differentiate between closely related strains by differences in their hybridization patterns. This allows for very rapid identification of pathogenic viruses or bacteria from clinical samples or detection of specific strains of these organisms in various substances, such as food. Such chips have been used in the food industry to detect particular pathogens, for example *Escherichia coli* O157:H7 (∞ Section 29.8).

DNA chips are also available that will identify *macroorganisms*. A commercially available chip called the *FoodExpert-ID* contains 88,000 gene fragments from a variety of vertebrate animals and is used in the food industry to ensure food purity. For example, the chip can confirm the presence of the meat listed on a food label and can also detect foreign animal meats that may have been added as supplements to or substitutes for the food product listed on the label. The eventual goal is to have each meat product receive an "identity card" listing all the animal species whose tissues were detected in it. This is intended to give consumers more confidence in the wholesomeness of their food products. The FoodExpert-ID can also be used to detect vertebrate byproducts in animal feed, a growing concern with the advent of prion-mediated diseases such as "mad cow disease" (∞ Sections 9.14 and 29.11).

Applications in Environmental Genomics

DNA chip technology is also useful for identifying and assessing the activities of microbial communities in the environment, an important goal of the microbial ecologist (∞ Chapter 19). Indeed a whole new field in microbial ecology is opening up called **environmental genomics** (∞ Section 18.6). Because of the capacity and power of whole genome shotgun sequencing and gene expression microarrays, it is now possible to sequence the collective genomes of the microorganisms present in a habitat, such as a soil or lake water sample, and identify their patterns of gene expression. The collective genomes, called the *metagenome*, define the genetic potential of an ecosystem. Then, using microarrays, one can explore these natural microbial communities for their patterns of gene expression. These techniques have empowered microbial ecologists to ask new and very complex questions about how microbial ecosystems function and to measure the contributions of specific members of a microbial community to the overall activities in an ecosystem. We discuss environmental genomics along with specific examples of the technology in Section 18.6.

 15.11 Concept Check

Microarrays are genes or gene fragments attached to a solid support in a known pattern. These arrays can be used to hybridize to mRNA and analyzed to determine patterns of gene expression. The arrays are large enough and dense enough that the transcription pattern of the entire genome (the transcriptome) can be analyzed.

◆ What do microarrays tell you that studying gene expression by assaying a particular enzyme, for example, cannot?

◆ Why might it be useful to know how gene expression of the entire genome responds to a particular condition?

REVIEW QUESTIONS

1. What is meant when it is said that cloned genes in phage M13 can be detected "visually" (∞ Section 15.1)?

2. Compare and contrast BACs with YACs with respect to: size, features necessary for replication, and amount of cloned DNA held (∞ Section 15.1).

3. Describe how "shotgun sequencing" methods can lead to the sequence of a complete microbial genome when the sequencing is essentially random (∞ Section 15.2)?

4. What is meant by "annotating" the genome? How is this done? How does *annotation* differ from *assembly*? Which process has to occur first (∞ Section 15.3)?

5. The organisms in Table 15.1 are listed in ascending order of chromosome size. What is the relationship between genome size and ORF content (∞ Section 15.4)?

6. How much larger is your genome than that of *Nanoarchaeum*? How many more genes do you have than *Nanoarchaeum* (∞ Sections 15.4 and 15.6)?

7. As a proportion of the total genome, which class of genes predominates in small genome organisms? In large genome organisms (∞ Section 15.5)?

8. In *Bacteria* and *Archaea* the acronym ORF is almost a synonym for the word "gene." However, in eukaryotes this is not, strictly speaking, true. Explain (∞ Section 15.6).

9. Whose genomes are larger, those of chloroplasts or those of mitochondria? Describe one unusual feature about a chloroplast and a mitochondrial genome (∞ Section 15.7).

10. The gene encoding the beta (β) subunit of RNA polymerase from *Escherichia coli* is said to be *orthologous* to the *rpoB* gene of *Bacillus subtilis*. What does that mean about the relationship between the two genes? What protein do you suppose the *rpoB* gene of *Bacillus subtilis* encodes? The genes for the different sigma factors (∞ see Table 8.2) of *Escherichia coli* are *paralogous*. What does that say about the relationship between these genes (∞ Section 15.8)?

11. Explain how horizontally transferred genes can be detected in a genome (∞ Section 15.8).

12. What is an intein? How can inteins be discovered (∞ Section 15.9)?

13. What does a 2-D protein gel show? How can the results of such a gel be tied to protein function (∞ Section 15.10)?

14. Distinguish between the terms *genome, genomics, proteome, proteomics,* and *transcriptome* (∞ Section 15.11).

APPLICATION QUESTIONS

1. When using shotgun sequencing, it is possible that there may be "gaps" left in the sequence. Describe how it is possible to close these gaps and complete the sequence.

2. The sequence of the yeast nuclear genome was published, but the entire sequence was never actually completely determined. The yeast mitochondrial genome proved very difficult to sequence accurately. Describe in both cases the practical difficulties that were encountered in the sequencing.

3. Describe how one might determine which *proteins* in *Escherichia coli* are repressed (∞ Section 8.5) when a culture is shifted from a *minimal* medium (which contains only a single carbon source) to a *rich* medium, containing a large number of amino acids, bases, and vitamins? Describe how one might study which *genes* are expressed during each growth condition.

WORKING GLOSSARY

Anaerobic respiration respiration in which some substance, such as SO_4^{2-} or NO_3^-, is used as a terminal electron acceptor instead of O_2

Anammox anoxic ammonia oxidation

Anoxic oxygen-free

Anoxygenic photosynthesis photosynthesis in which O_2 is not produced

Antenna in reference to photocomplexes, light harvesting pigment molecules that funnel energy to the reaction center

Autotroph an organism that can use CO_2 as sole carbon source

Bacteriochlorophyll the chlorophyll pigment of anoxygenic phototrophs

Calvin cycle the biochemical route of CO_2 fixation in many autotrophic organisms

Carboxysomes crystalline inclusions of RubisCo

Carotenoids a hydrophobic accessory pigment present along with chlorophyll in photosynthetic membranes

Chemolithotroph a microorganism capable of oxidizing inorganic compounds as energy sources

Chlorophyll the light-sensitive, Mg-containing porphyrin of photosynthetic organisms that initiates the process of photophosphorylation

Chlorosome cigar-shaped structures present in the periphery of cells of green sulfur and green nonsulfur bacteria and that contain the antenna bacteriochlorophylls (*c, d,* or *e*)

Denitrification anaerobic respiration in which NO_3^- is reduced to gaseous nitrogen compounds, primarily N_2

Disproportionation splitting of a compound into two new compounds, one more oxidized and one more reduced than the original compound

Fermentation anaerobic catabolism of an organic compound in which the compound serves as both an electron donor and an electron acceptor and in which ATP is produced by substrate-level phosphorylation

Homoacetogenesis (acetogenesis) energy metabolism involving the production of acetate from either H_2 plus CO_2 or from organic compounds

Hydrogenase an enzyme, widely distributed in anaerobic microorganisms, capable of taking up or evolving H_2

Hydroxypropionate pathway an autotrophic pathway found in *Chloroflexus* and a few *Archaea*

Methanogenesis the biological production of methane (CH_4)

Methanotroph an organism that can oxidize methane

Methylotroph an organism capable of growth on compounds containing no C—C bonds

Mixotrophic a nutritional state in which an inorganic compound serves as energy source (electron donor) and organic compounds serve as carbon source

Monooxygenase an enzyme that catalyzes the incorporation of one atom of O_2 into a substrate while the other atom is reduced to H_2O

Nitrification the microbial conversion of NH_3 to NO_3^-

Nitrogenase an enzyme capable of reducing N_2 to NH_3 in the process of nitrogen fixation

Nitrogen fixation the biological reduction of N_2 to NH_3 by nitrogenase

Oxygenic photosynthesis photosynthesis carried out by cyanobacteria and green plants in which O_2 is evolved

Photophosphorylation the production of ATP in photosynthesis

Photosynthesis the series of reactions in which ATP is synthesized by light-driven reactions and CO_2 is fixed into cell material

Phototroph an organism that can use light as an energy source

Phycobiliprotein the accessory pigment complex in cyanobacteria that contains a molecule of phycocyanin or phycoerythrin coupled to proteins

Reaction center a photosynthetic complex containing chlorophyll (or bacteriochlorophyll) and several other components within which occurs the initial electron transfer reactions of photosynthetic electron flow

Reductive dechlorination an anaerobic respiration where a chlorinated organic compound is used as an electron acceptor, usually with release of Cl^-

Reverse electron transport the energy-dependent movement of electrons against the thermodynamic gradient to form a strong reductant from a weaker electron donor

RubisCO the acronym for ribulose bisphosphate carboxylase, a key enzyme of the Calvin cycle

Syntrophy a process whereby two or more microorganisms cooperate to degrade a substance neither can degrade alone

The preceding chapters considered the great *phylogenetic* diversity of microbial life on Earth. Now we focus on their *metabolic* diversity, with special emphasis on the biochemical processes behind this diversity. We can then study microbial ecology—the interactions of microorganisms with each other and their environments. Metabolic diversity and microbial ecology go hand in hand; the metabolic diversity to be described in this chapter fuels the nutrient cycles and other microbial activities that we will discuss in Chapters 18 and 19. Indeed, as we will see, microorganisms and their metabolic reactions play key roles in maintaining the web of life on Earth and are of crucial importance in agriculture and other aspects of human affairs (∞ Section 1.4).

I THE PHOTOTROPHIC WAY OF LIFE

Phototrophy, the use of *light* as an energy source, is widespread in the microbial world. In the first five sections we consider the major forms of phototrophy, including that which forms the very oxygen we breathe. We then proceed to see how most phototrophs satisfy all of their carbon needs from CO_2.

17.1 Photosynthesis

The most important biological process on Earth is **photosynthesis**, the conversion of light energy to chemical energy. Organisms that can carry out photosynthesis are called **phototrophs** (Figure 17.1●). Most phototrophic organisms are also **autotrophs**, capable of growing with CO_2 as sole carbon source. Energy from light is used in the reduction of CO_2 to organic compounds (*photoautotrophy*). However, some phototrophs can use organic carbon as their carbon source; this lifestyle is called *photoheterotrophy* (Figure 17.1).

The ability to photosynthesize requires light-sensitive pigments, the *chlorophylls*, found in plants, algae, and several groups of prokaryotes. Light reaches pho-

PHOTOTROPHS (use light as energy source)

Photoautotrophs (Use CO_2)   **Photoheterotrophs** (Use Organic Carbon)

● **Figure 17.1 Classification of phototrophic organisms in terms of energy and carbon sources.** In nature, phototrophic organisms switch their mode of metabolism depending on the resources available in their habitat. Photoheterotrophy typically represses photoautotrophy.

totrophic organisms in distinct units of energy called *quanta*. Absorption of light quanta by chlorophylls begins the process of photosynthetic energy conversion, and the net result is ATP.

Energy Production and CO₂ Assimilation

The growth of a photoautotroph can be characterized by two distinct sets of reactions: (1) ATP production, and (2) CO_2 reduction to organic compounds. For autotrophic growth, energy is supplied from ATP, while electrons for the reduction of CO_2 come from NADH (or NADPH). These are produced by the reduction of NAD^+ (or $NADP^+$) by electrons originating from various electron donors to be discussed later.

To drive autotrophic reactions, some phototrophic bacteria obtain reducing power from electron donors in their environment, typically reduced sulfur sources (H_2S, S^0, $S_2O_3^{2-}$) or H_2. By contrast, green plants, algae, and cyanobacteria use H_2O, an inherently poor electron donor (∞ Figure 5.9), as a source of reducing power to reduce $NADP^+$ to NADPH. The oxidation of H_2O produces molecular oxygen (O_2) as a by-product (Figure 17.2●). Because oxygen is produced, photosyn-

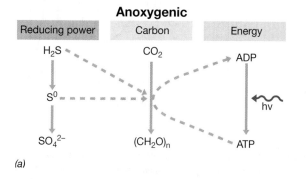
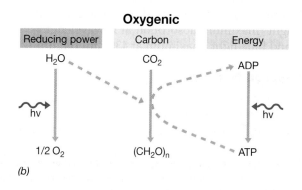

● **Figure 17.2 Patterns of photosynthesis.** Energy and reducing power synthesis in (a) anoxygenic versus (b) oxygenic phototrophs. Although both types of phototrophs obtain their energy from light (hv), in oxygenic phototrophs light also drives the oxidation of water to oxygen.

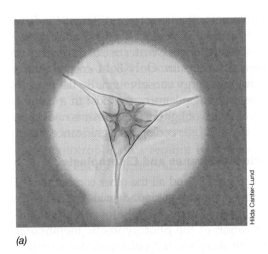

(a)

Hilda Canter-Lund

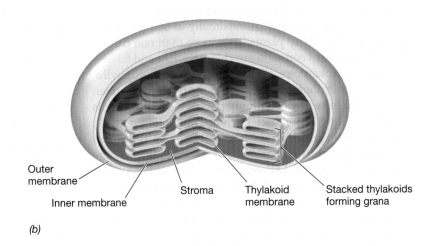

Outer membrane
Inner membrane
Stroma
Thylakoid membrane
Stacked thylakoids forming grana

(b)

● **Figure 17.5** **The chloroplast.** (a) Photomicrograph of an algal cell showing chloroplasts. (b) Details of chloroplast structure, showing how the convolutions of the thylakoid membranes define an inner space called the stroma and form membrane stacks called grana.

Section 14.3). The chlorophyll pigments are attached to sheetlike (lamellar) membrane structures of the chloroplast (Figure 17.5b). These photosynthetic membrane systems are called **thylakoids**; stacks of thylakoids are called *grana* (Figure 17.5b). The thylakoids are so arranged that the chloroplast is divided into two regions, the *matrix space* that surrounds the thylakoids, and the *inner space* within the thylakoid array (Figure 17.5b). This arrangement makes possible the development of a light-driven proton motive force that can be used to synthesize ATP, as will be described in Section 17.5.

In prokaryotes, chloroplasts are absent. Here photosynthetic pigments are integrated into internal membrane systems. These systems arise (1) from invagination of the cytoplasmic membrane (purple bacteria) (for example, ∞ Figure 12.3; see also Figure 17.12), (2) from the cytoplasmic membrane itself (heliobacteria; ∞ Section 12.20), (3) in both the cytoplasmic membrane and specialized non-unit membrane-enclosed structures called *chlorosomes* (green bacteria; see Figure 17.7 and ∞ Section 12.32), or (4) in thylakoid membranes (cyanobacteria, ∞ Section 12.25).

Reaction Centers and Antenna Pigments

Within a photosynthetic membrane chlorophyll or bacteriochlorophyll molecules are associated with proteins to form *complexes* consisting of anywhere from 50 to 300 molecules (Figure 17.6●). Only a very small number of these pigment molecules, called **reaction centers**, participate directly in the conversion of light energy to ATP (Figure 17.6). Reaction center chlorophylls or bacteriochlorophylls are surrounded by the more numerous **light-harvesting** or **antenna** chlorophylls or bacteriochlorophylls. The antenna pigments function to harvest light and funnel the energy of light to the reaction center (Figure 17.6). At the low light intensities that often prevail in nature, this arrangement of pigment molecules allows for the capture and utilization of photons that would otherwise be insufficient to drive reaction center photochemistry by themselves.

The ultimate in low-light efficiency is found in the **chlorosome** of green sulfur bacteria and *Chloroflexus* (Figure 17.7●). This structure is a giant antenna system. But unlike the antenna of other phototrophs, bacteriochlorophyll molecules in the chlorosome are not bound to proteins and instead function much like a solid state circuit. Chlorosome bacteriochlorophylls (bacteriochlorophylls *c*, *d*, and *e*, see Figure 17.4) absorb light and transfer the energy to bacteriochlorophyll *a* located in the reaction center in the cytoplasmic membrane (Figure 17.7). This arrangement is highly efficient for absorbing light at low

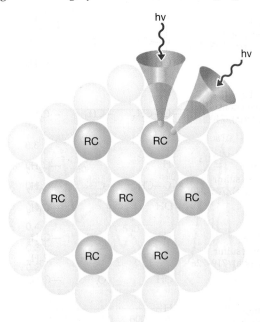

hv
hv

RC RC
RC RC RC
RC RC

● **Figure 17.6** **Model for the arrangement of light-harvesting chlorophylls/bacteriochlorophylls versus reaction centers within a photosynthetic membrane.** Light energy, absorbed by light-harvesting molecules (light green), is transferred to the reaction centers (dark green, RC) where photosynthetic electron transport reactions begin. All pigment molecules are held in place within the membrane by specific pigment-binding proteins. Compare this figure to Figures 17.13 and 17.15.

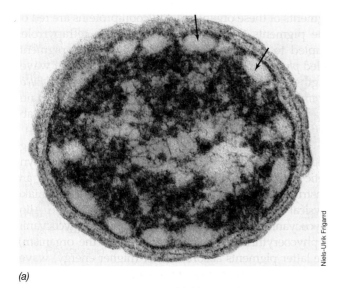

(a)

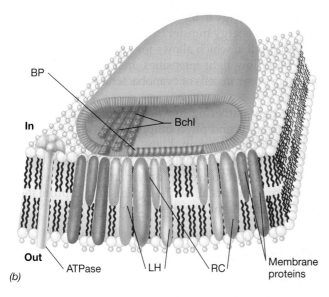

(b)

● **Figure 17.7 The chlorosome of green sulfur and green non-sulfur bacteria.** (a) Electron micrograph of a cell of the green sulfur bacterium *Chlorobium tepidum*. Note the chlorosomes (arrows). (b) Model of chlorosome structure. The chlorosome (green) lies appressed to the inside surface of the cytoplasmic membrane. Antenna bacteriochlorophyll (Bchl) molecules (Bchls *c*, *d*, or *e*) are arranged in tubelike arrays inside the chlorosome, and energy is transferred from these bacteriochlorophylls through light-harvesting Bchl *a* molecules (LH) to reaction center (RC) Bchl *a* in the cytoplasmic membrane (blue). Base plate (BP) proteins function as connectors between the chlorosome and the cytoplasmic membrane.

intensities. In this regard, it has been shown that green sulfur bacteria can grow at the lowest light intensities of all known phototrophs.

 17.2 Concept Check

The central pigment of photosynthesis is chlorophyll (or bacteriochlorophyll). Chlorophylls are located in photosynthetic membranes where the light reactions of photosynthesis are carried out. Antenna chlorophyll molecules harvest light energy and transfer it on to reaction center chlorophylls.

◆ Considering their function, why is it necessary for chlorophyll pigments to be located in membranes?

◆ What is the difference between the numbers of *antenna* and *reaction center* chlorophyll molecules in a photosynthetic complex, and why?

◆ What pigments are found within the *chlorosome*?

17.3 Carotenoids and Phycobilins

Although chlorophyll or bacteriochlorophyll is obligatory for photosynthesis, phototrophic organisms contain various accessory pigments involved in the capture and processing of light energy. These include the *carotenoids* and the *phycobilins*. These pigments primarily play a photoprotective role (carotenoids) or function in light-harvesting (phycobilins).

Carotenoids

The most widespread accessory pigments are the **carotenoids**, which are always found in phototrophic organisms. Carotenoids are hydrophobic pigments that are firmly embedded in the membrane. Figure 17.8● shows the structure of a typical carotenoid, *β-carotene*. Carotenoids have long hydrocarbon chains with alternating C—C and C=C bonds, an arrangement called a *conjugated* double-bond system. Carotenoids are typically yellow, red, brown, or green in color (∞ Figure 12.2) and absorb light in the blue region of the spectrum (see Figure 17.3). The types and structures of carotenoids of various phototrophs have been well studied, and the major carotenoids of anoxygenic phototrophs are shown in Figure 17.9●. These pigments are responsible for the brilliant colors of red, purple, pink, green, yellow, or brown that are observed in different species of anoxygenic phototrophs (∞ Figures 12.2 and 12.5).

Carotenoids are closely associated with chlorophyll or bacteriochlorophyll in photosynthetic pigment complexes but do not function directly in ATP synthesis. They can, however, *transfer* energy to the reaction center, and this transferred energy may be used to make ATP in the same way as light energy captured directly by chlorophyll. Carotenoids also function as photoprotective agents. Bright light can be harmful to cells in that it catalyzes photooxidation reactions that can lead to the production of toxic forms of oxygen, such as singlet oxygen ($^1O_2$) (∞ Section 6.16). The latter can oxidize and thereby damage components of the photosynthetic apparatus itself. Carotenoids quench toxic oxygen species and absorb much of this harmful light. Because

● **Figure 17.8 Structure of β-carotene, a typical carotenoid.** The conjugated double-bond system is highlighted in orange.

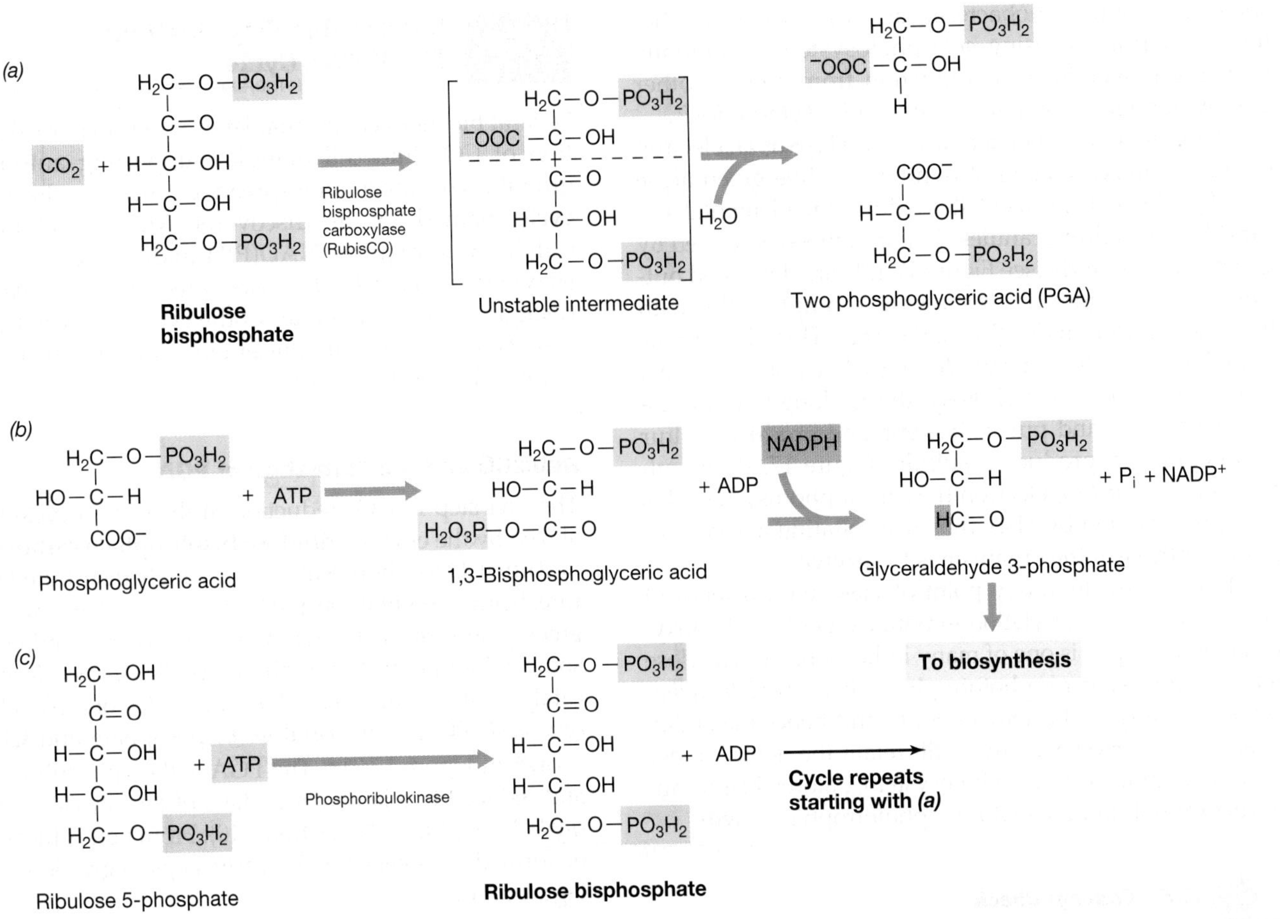

(a)

Ribulose
bisphosphate

CO$_2$ + Ribulose → Unstable intermediate → Two phosphoglyceric acid (PGA)

Ribulose bisphosphate carboxylase (RubisCO)

H$_2$O

(b)

Phosphoglyceric acid + ATP → 1,3-Bisphosphoglyceric acid + ADP → NADPH → Glyceraldehyde 3-phosphate + P$_i$ + NADP$^+$

To biosynthesis

(c)

Ribulose 5-phosphate + ATP → **Ribulose bisphosphate** + ADP → Cycle repeats starting with (a)

Phosphoribulokinase

● **Figure 17.21   Key reactions of the Calvin cycle.** (a) Reaction of the enzyme *ribulose bisphosphate carboxylase.* (b) Steps in the conversion of 3-phosphoglyceric acid (PGA) to glyceraldehyde 3-phosphate. Note that both ATP and NADPH are required. (c) Conversion of ribulose 5-phosphate to the CO_2 acceptor molecule ribulose bisphosphate by the enzyme *phosphoribulokinase.*

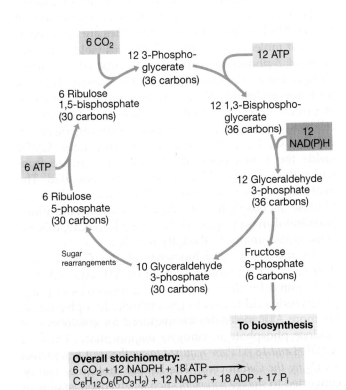

Overall stoichiometry:
6 CO$_2$ + 12 NADPH + 18 ATP ⟶
C$_6$H$_{12}$O$_6$(PO$_3$H$_2$) + 12 NADP$^+$ + 18 ADP + 17 P$_i$

poly-β-hydroxyalkanoates (∞ Section 4.11) and used later to build new cell material.

Carboxysomes

Several autotrophic prokaryotes that use the Calvin cycle for CO_2 fixation produce polyhedral cell inclusions called **carboxysomes**. The inclusions, about 100 nm in diameter, are surrounded by a thin, nonunit membrane and consist of a tightly packed crystalline array of RubisCO molecules (Figure 17.23●; ∞ Figure 12.10a). It is thought that carboxysomes are a mechanism to increase the amount of RubisCO in the cell to allow for more rapid CO_2 fixation without affecting the osmolarity of the cytoplasm (osmotic pressure is not affected because the carboxysome is insoluble).

● **Figure 17.22   The Calvin cycle.** Shown is the production of one hexose molecule from CO_2. For each *six* molecules of CO_2 incorporated, *one* fructose 6-phosphate is produced. In phototrophs, ATP comes from photophosphorylation and NAD(P)H from light or reverse electron flow. Use the color coding here to follow the biochemical reactions occurring in Figure 17.21.

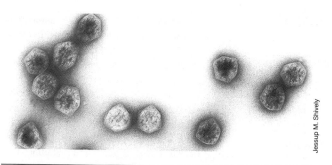

● **Figure 17.23   Crystalline Calvin cycle enzymes: Carboxysomes.** Electron micrograph of carboxysomes purified from the chemolithotrophic sulfur oxidizer *Thiobacillus neapolitanus*. The structures are about 100 nm in diameter. Carboxysomes are present in a wide variety of obligately autotrophic prokaryotes.

Carboxysomes have been found in obligately chemolithotrophic sulfur-oxidizing bacteria, the nitrifying bacteria, and cyanobacteria and prochlorophytes (⚉ Sections 12.3, 12.4, 12.26, and 12.27, respectively). They are not present in *facultative* autotrophs (organisms that can grow either as autotrophs or as heterotrophs) such as purple anoxygenic phototrophs. Thus, the carboxysome appears to be an evolutionary adaptation to life under *strictly autotrophic conditions*.

 17.6   Concept Check

The fixation of CO$_2$ by most phototrophic and other autotrophic organisms occurs via the Calvin cycle, in which the enzyme ribulose bisphosphate carboxylase (RubisCO) plays a key role. The Calvin cycle is an energy-demanding process in which CO$_2$ is converted into sugar.

◆ What reaction does the enzyme *ribulose bisphosphate carboxylase* carry out?

◆ Why is reducing power needed for autotrophic growth?

◆ What is a *carboxysome*?

17.7   Autotrophic CO$_2$ Fixation: Reverse Citric Acid Cycle and the Hydroxypropionate Cycle

Alternative mechanisms of autotrophic CO$_2$ fixation are present in green sulfur bacteria and green nonsulfur bacteria. In the green sulfur bacterium *Chlorobium* (see Figures 17.7a and 17.17b), CO$_2$ fixation occurs by a reversal of steps in the citric acid cycle (⚉ Figure 5.22), a pathway called the *reverse citric acid cycle* (Figure 17.24a●). *Chlorobium* contains two ferredoxin-linked enzymes that catalyze the reductive fixation of CO$_2$ into intermediates of the citric acid cycle. The two ferredoxin-linked reactions involve the carboxylation of succinyl-CoA to α-ketoglutarate and the carboxylation of acetyl-CoA to pyruvate (Figure 17.24a). Most of the other reactions of the reverse citric acid cycle are cat-

alyzed by enzymes working in reverse of the normal oxidative direction of the cycle. One exception is *citrate lyase*, an ATP-dependent enzyme that cleaves citrate into acetyl-CoA and oxaloacetate in green sulfur bacteria. In the oxidative direction of the cycle, citrate is produced from these same precursors by the enzyme *citrate synthase* (⚉ Figure 5.22).

The reverse citric acid cycle as a mechanism of autotrophy has also been found in certain nonphototrophic hyperthermophiles, including *Archaea* such as *Thermoproteus* (⚉ Section 13.9), and in *Aquifex*, a very early branching autotroph on the phylogenetic tree of *Bacteria* (⚉ Figure 12.1 and Section 12.37). These findings, of course, leave open the possibility that this pathway is more widespread than previously thought and suggest that it may have been an "early" evolutionary form of autotrophy.

Autotrophy in *Chloroflexus*

The green nonsulfur phototroph *Chloroflexus* (⚉ Section 12.35) grows autotrophically with either H$_2$ or H$_2$S as electron donors. However, neither the Calvin cycle nor the reverse citric acid cycle operates in this organism. Instead, two molecules of CO$_2$ are reduced to glyoxylate by a unique cyclic pathway, the **hydroxypropionate pathway** (Figure 17.24b). This pathway leads to the synthesis of hydroxypropionate as a key intermediate.

In phototrophic bacteria, the hydroxypropionate pathway has been confirmed only in *Chloroflexus*, the earliest branching anoxygenic phototroph on the tree of *Bacteria* (⚉ Figure 12.1). This suggests that the hydroxypropionate pathway may have been the first attempt at autotrophy in anoxygenic phototrophs. If *Chloroflexus* evolved long before other phototrophic organisms, as the tree of *Bacteria* suggests (Figure 12.1), the hydroxypropionate pathway was perhaps the first autotrophic pathway in *any* phototrophic organism. But in addition to *Chloroflexus*, the hydroxypropionate pathway functions in several hyperthermophilic *Archaea*, including *Metallosphaera*, *Acidianus*, and *Sulfolobus* (⚉ Section 13.9). These are all nonphototrophic prokaryotes and lie near the base of the archaeal domain. The roots of the hydroxypropionate pathway may thus be very deep; it was perhaps nature's first attempt at autotrophy.

 17.7   Concept Check

The reverse citric acid cycle and the hydroxypropionate cycle are pathways of CO$_2$ fixation found in green sulfur and green nonsulfur bacteria, respectively.

◆ Including the route of CO$_2$ fixation, discuss at least three ways that you could distinguish a purple sulfur bacterium from a green sulfur bacterium.

◆ Including the route of CO$_2$ fixation, what similarities and differences exist between green *sulfur* and green *nonsulfur* bacteria?

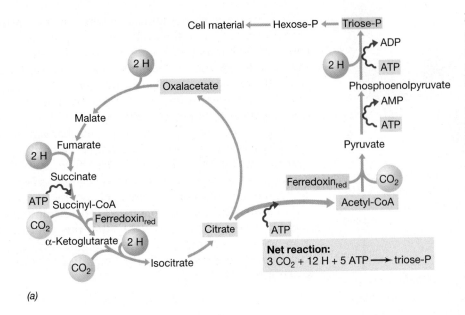

● **Figure 17.24 Unique autotrophic pathways in phototrophic green bacteria.** (a) The reverse citric acid cycle is the mechanism of CO_2 fixation in the green sulfur bacterium *Chlorobium* (see also Figure 17.17*b*; ⊂⊃ Section 12.32). Ferredoxin$_{red}$ indicates carboxylation reactions requiring reduced ferredoxin (2 H each). Reduced ferredoxin is generated in *Chlorobium* by light-driven reactions (see Figure 17.18). Starting from oxalacetate, each turn of the cycle results in three molecules of CO_2 being incorporated and pyruvate as the product. The cleavage of citrate by the ATP-dependent enzyme citrate lyase regenerates the C_4 acceptor oxalacetate and produces acetyl-CoA for biosynthesis. The conversion of pyruvate to phosphoenolpyruvate consumes two high-energy phosphate bond equivalents. (b) The hydroxypropionate pathway is the means of autotrophy in the green nonsulfur bacterium *Chloroflexus* (⊂⊃ Section 12.35). Acetyl-CoA is carboxylated twice to yield methylmalonyl-CoA. This intermediate is rearranged to yield acetyl-CoA and glyoxylate. The latter is converted to cell material probably through a serine or glycine intermediate. The source of reducing power in the hydroxypropionate pathway is NADPH.

III CHEMOLITHOTROPHY: ENERGY FROM THE OXIDATION OF INORGANIC ELECTRON DONORS

Chapter 5 focused on microorganisms whose energy metabolism was based on *chemoorganotrophy*, the use of organic compounds as energy sources. The next four sections focus on the *chemolithotrophs*, highlighting the strategies, problems, and advantages of a lifestyle geared toward the use of *inorganic* chemicals as energy sources.

17.8 Inorganic Electron Donors and Energetics

Organisms that obtain energy from the oxidation of *inorganic* compounds are called **chemolithotrophs**. Most chemolithotrophic bacteria can also obtain their carbon from CO_2, so they are also autotrophs. As we have noted, for growth on CO_2 as sole carbon source an or-

ganism needs (1) energy (ATP), and (2) reducing power. Some chemolithotrophs are **mixotrophic**, meaning that although they are able to obtain energy from the oxidation of an inorganic compound, they require an *organic* compound as carbon source.

In chemolithotrophs, ATP generation is in principle similar to that in chemoorganotrophs, except that the electron donor is *inorganic* rather than organic. ATP synthesis is coupled to *oxidation* of the electron donor. Reducing power in chemolithotrophs is obtained either directly from the inorganic compound, if it has a sufficiently low reduction potential, or by *reverse electron transport* reactions, as discussed in Section 17.5 for phototrophic purple bacteria.

Sources of Inorganic Electron Donors

Chemolithotrophs have many sources of inorganic electron donors, which may be geological, biological, or anthropogenic in nature. Volcanic activity is a major source of reduced sulfur compounds, primarily H_2S. Agricultural and mining operations add inorganic electron donors to the environment, especially nitrogen and iron compounds, as does the burning of fossil fuels and the input of industrial wastes. Biological sources are also quite extensive, especially the production of H_2S, H_2, and NH_3. The ecological success (∞ Chapter 19) and metabolic diversity (this chapter) of chemolithotrophs underscores the abundance of sources and supplies of inorganic electron donors in nature.

Energetics of Chemolithotrophy

A review of reduction potentials listed in Table A1.2 reveals that a number of inorganic compounds can provide sufficient energy for ATP synthesis when oxidized with O_2 as electron acceptor. Recall from Chapter 5 that the further apart two half reactions are in terms of E_0', the greater the amount of energy released. For instance, the difference in reduction potential between the H^+/H_2 couple and the $\frac{1}{2}O_2/H_2O$ couple is -1.23 V, which is equivalent to a free-energy yield of -237 kJ/mol (see Appendix 1 for calculations). On the other hand, the potential difference between the H^+/H_2 couple and the NO_3^-/NO_2^- couple is less, -0.84 V, equivalent to a free-energy yield of -163 kJ/mol. This is still quite sufficient for the production of ATP (the energy-rich phosphate bond of ATP has a free energy of -31.8 kJ/mol, see Table 17.6). However, a similar calculation will show that there is insufficient energy available from, for example, the oxidation of H_2S using CO_2 as electron acceptor.

Such energy calculations make it possible to predict the kinds of chemolithotrophs that should exist in nature. Since organisms must obey the laws of thermodynamics, only reactions that are thermodynamically favorable are potential energy-yielding reactions (see also Section 17.21). Table 17.1 summarizes energy yields for some reactions known to be carried out by chemolithotrophic microorganisms. The organisms themselves were considered in Chapters 12 and 13. We also examine ecological aspects of chemolithotrophy in Chapter 19,

where we will see that chemolithotrophic reactions are at the heart of most nutrient cycles.

 17.8 Concept Check

Chemolithotrophs are able to oxidize inorganic chemicals as their sole sources of energy and reducing power. Most chemolithotrophs are also able to grow autotrophically.

◆ For what *two* purposes are inorganic compounds used by chemolithotrophs?

◆ Why does the oxidation of H_2 yield more energy with O_2 as electron acceptor than with SO_4^{2-} as electron acceptor?

17.9 Hydrogen Oxidation

Hydrogen is a common product of microbial metabolism, and a number of chemolithotrophs are able to use it as an electron donor in energy metabolism. A wide variety of anaerobic H_2-oxidizing *Bacteria* and *Archaea* are known, differing in the electron acceptor they use (for example, nitrate, sulfate, ferric iron, and others), and these organisms are discussed later in this chapter (see Sections 17.14–17.18). Here we consider only the *aerobic* H_2-oxidizing bacteria.

Energetics of H_2 Oxidation

Generation of ATP during H_2 oxidation results from the oxidation of H_2 by O_2 leading to formation of a proton motive force. The overall reaction:

$$H_2 + \tfrac{1}{2}O_2 \longrightarrow H_2O \qquad \Delta G^{0\prime} = -237 \text{ kJ}$$

is highly exergonic and can support the synthesis of at least one ATP. The reaction is catalyzed by the enzyme **hydrogenase**, the electrons from H_2 initially being transferred to a quinone acceptor. From here electrons pass through a series of cytochromes to eventually reduce O_2

Table 17.1	Energy yields from the oxidation of various inorganic electron donors[a]					
Electron donor	**Reaction**	**Type of chemolithotroph**	**E_0' of couple (V)**	**$\Delta G^{0\prime}$ (kJ/reaction)**	**Number of electrons**	**$\Delta G^{0\prime}$ (kJ/2e⁻)**
Phosphite[b]	$4\,HPO_3^{2-} + SO_4^{2-} + H^+ \rightarrow 4\,HPO_4^{2-} + HS^-$	Phosphite bacteria	-0.69	-91	2	-91
Hydrogen	$H_2 + \tfrac{1}{2}O_2 \rightarrow H_2O$	Hydrogen bacteria	-0.42	-237.2	2	-237.2
Sulfide	$HS^- + H^+ + \tfrac{1}{2}O_2 \rightarrow S^0 + H_2O$	Sulfur bacteria	-0.27	-209.4	2	-209.4
Sulfur	$S^0 + 1\tfrac{1}{2}O_2 + H_2O \rightarrow SO_4^{2-} + 2\,H^+$	Sulfur bacteria	-0.20	-587.1	6	-195.7
Ammonium[c]	$NH_4^+ + 1\tfrac{1}{2}O_2 \rightarrow NO_2^- + 2\,H^+ + H_2O$	Nitrifying bacteria	$+0.34$	-274.7	6	-91.6
Nitrite	$NO_2^- + \tfrac{1}{2}O_2 \rightarrow NO_3^-$	Nitrifying bacteria	$+0.43$	-74.1	2	-74.1
Ferrous iron	$Fe^{2+} + H^+ + \tfrac{1}{4}O_2 \rightarrow Fe^{3+} + \tfrac{1}{2}H_2O$	Iron bacteria	$+0.77$	-32.9	1	-65.8

[a] Data calculated from values in Appendix 1; values for Fe^{2+} are for pH 2, and others are for pH 7. At pH 7 the Fe^{3+}/Fe^{2+} couple is about $+0.2$ V.

[b] Except for phosphite, all reactions are shown coupled to O_2 as electron acceptor. The only known phosphite oxidizer couples to SO_4^{2-} as electron acceptor.

[c] Ammonium can also be oxidized with NO_2^- as electron acceptor by anammox organisms (see Section 17.12).

to water (Figure 17.25●). Some hydrogen bacteria have *two* hydrogenases, one soluble and one membrane-bound (∞ Table 12.6). In this case, the membrane-bound enzyme is involved in energetics while the soluble hydrogenase takes up H_2 and reduces NAD^+ to NADH directly (the reduction potential of H_2, -0.42 V, is so low that reverse electron flow as a means of making reducing power is unnecessary) (Figure 17.25). The organism *Ralstonia eutropha* has been a model for studying aerobic H_2 oxidation, and we discussed some of the properties of this organism in Section 12.5.

Autotrophy in H_2 Bacteria

Although most hydrogen bacteria can also grow as chemoorganotrophs, when growing chemolithotrophically they fix CO_2 by the Calvin cycle (see Section 17.6). The stoichiometry observed here is:

$$6 H_2 + 2 O_2 + CO_2 \longrightarrow (CH_2O) + 5 H_2O$$

where (CH_2O) represents cell material. However, when readily useable organic compounds such as glucose are present, synthesis of Calvin cycle and hydrogenase enzymes in typical aerobic H_2-oxidizing bacteria is repressed. In nature, H_2 levels in oxic environments are transient and low at best. It is thus likely that aerobic hydrogen bacteria closely regulate their catabolic enzymes. These organisms probably shift between chemoorganotrophy and chemolithotrophy often, depending on levels of useable organic compounds and H_2 in their habitats. Moreover, many aerobic H_2 bacteria grow best microaerobically. It is therefore likely that these organ-

isms are most successful as H_2 chemolithotrophs in *oxic/anoxic interfaces*, where H_2 from fermentative metabolism would be in greater and more continuous supply than in highly oxic habitats.

17.10 Oxidation of Reduced Sulfur Compounds

Many reduced sulfur compounds can be used as electron donors by a variety of sulfur bacteria, called *colorless* to distinguish them from the bacteriochlorophyll-containing (pigmented) green and purple sulfur bacteria discussed earlier in this chapter (see Figure 17.17). Indeed, the very concept of chemolithotrophy emerged from studies of the sulfur bacteria in the work of the great Russian microbiologist Sergei Winogradsky (see the Microbial Sidebar, Winogradsky and Chemolithoautotrophy, and ∞ Section 1.7).

Energetics of Sulfur Oxidation

The most common sulfur compounds used as electron donors are hydrogen sulfide (H_2S), elemental sulfur (S^0), and thiosulfate ($S_2O_3^{2-}$). The final product of sulfur oxidation in most cases is sulfate (SO_4^{2-}), and the total number of electrons involved between H_2S (oxidation state, -2) and sulfate (oxidation state, $+6$) is eight (see Table 17.3 for a summary of sulfur oxidation states). Less energy is available when one of the intermediate sulfur oxidation states is used:

$$H_2S + 2 O_2 \longrightarrow SO_4^{2-} + 2 H^+$$
$$\text{(final product, sulfate)} \quad \Delta G^{0\prime} = -798.2 \text{ kJ/reaction}$$
$$HS^- + \tfrac{1}{2} O_2 + H^+ \longrightarrow S^0 + H_2O$$
$$\text{(final product, sulfur)} \quad -209.4 \text{ kJ/reaction}$$
$$S^0 + H_2O + 1\tfrac{1}{2} O_2 \longrightarrow SO_4^{2-} + 2 H^+$$
$$\text{(final product, sulfate)} \quad -587.1 \text{ kJ/reaction}$$
$$S_2O_3^{2-} + H_2O + 2 O_2 \longrightarrow 2 SO_4^{2-} + 2 H^+$$
$$\text{(final product, sulfate)} \quad -818.3 \text{ kJ/reaction}$$
$$(-409.1 \text{ kJ/S atom oxidized})$$

The oxidation of the most reduced sulfur compound, H_2S, occurs in stages, and the first oxidation step results in the formation of elemental sulfur, S^0. Some H_2S-oxidizing bacteria deposit this elemental sulfur inside the cell (Figure 17.26a●). The sulfur deposited as a result of the initial oxidation is an energy reserve. When the supply of H_2S has been depleted, additional energy can be obtained from the oxidation of sulfur to sulfate.

When elemental sulfur is provided externally as an electron donor, the organism must grow attached to the sulfur particle because of the extreme insolubility of elemental sulfur (Figure 17.26b). By adhering to the particle, the organism can remove sulfur atoms for oxidation to sulfate. This is thought to occur through the action of membrane or periplasmic proteins that solubilize the

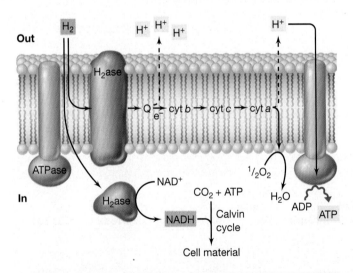

● **Figure 17.25 Bioenergetics and function of the two hydrogenases of aerobic H_2 bacteria.** In *Ralstonia eutropha*, where two hydrogenases are present, the membrane-bound hydrogenase is involved in energetics while the cytoplasmic hydrogenase makes NADH for the Calvin cycle. Note how the membrane-bound hydrogenase begins the flow of electrons leading to formation of a proton motive force. Some H_2 bacteria have only the membrane-bound hydrogenase, and in these organisms reducing power synthesis occurs from reverse electron flow. H_2ase, hydrogenase; cyt, cytochrome; Q, quinone.

Microbial Sidebar ◆ Winogradsky and Chemolithoautotrophy

The concept of chemolithotrophic autotrophy was first conceived by the great Russian microbiologist Sergei Winogradsky (Section 1.7). Winogradsky studied sulfur bacteria because certain colorless sulfur bacteria (*Beggiatoa, Thiothrix*) are very large (Figure 1) and easy to investigate, even in the absence of pure cultures. Springs with waters rich in H_2S are fairly common around the world. In the Bernese Oberland district of Switzerland, vast populations of *Beggiatoa* and *Thiothrix* develop in the outflow channels of sulfur springs. It was here that Winogradsky found suitable material for microscopic and physiological studies by merely lifting up the white filamentous masses of cells (Section 12.4 and Figures 12.11 and 12.13) and performing experiments directly in the field.

Winogradsky first showed that colorless sulfur bacteria are present only in water containing H_2S. As the water flows away from the source, the H_2S gradually dissipates, and sulfur bacteria were no longer present. This suggested to him that their development depends on the presence of H_2S. Winogradsky then showed that when *Beggiatoa* filaments are starved for sulfide, they lost their sulfur granules. He found, however, that the granules are rapidly restored if a small amount of H_2S was added (Figure 1). He concluded that H_2S was being oxidized to elemental sulfur.

But what happens to the sulfur granules when the filaments are starved of H_2S? Winogradsky showed by some clever microchemical tests that when the sulfur granules disappeared, sulfate appeared in the medium. He concluded that *Beggiatoa* (and by inference other colorless sulfur bacteria) oxidize

H_2S to elemental sulfur and subsequently to sulfate ($H_2S \rightarrow S^0 \rightarrow SO_4^{2-}$). Because this organism seemed to require H_2S for development in the springs, he further postulated that this oxidation was the principal source of *energy* for these organisms.

Winogradsky's studies on *Beggiatoa* thus provided the first evidence that an organism could oxidize an *inorganic* substance as an energy source. This was the origin of the concept of chemolithotrophy. From these beginnings, Winogradsky turned to a study of the nitrifying bacteria, and it was with this group that he first showed that autotrophic fixation of CO_2 is coupled to the oxidation of an inorganic compound. The process of nitrification had been known before Winogradsky's work from studies on the fate of sewage when added to soil. For example, ammonia-rich sewage passed through a soil column showed conversion to nitrate. Winogradsky proceeded from here to isolate nitrifying bacteria using completely mineral media in which CO_2 was the sole carbon source and ammonia was the sole electron donor. Because ammonia is chemically stable, it was easy to show that the oxidation of ammonia to nitrite, and subsequently to nitrate, is a strictly bacterial process.

In fact, Winogradsky further showed that nitrification is a *two-step* process, with one group of organisms converting NH_4^+ to NO_2^- and a second, NO_2^- to NO_3^- (see Section 17.12). Because no organic materials were present in the medium, it was also possible to show that organic matter (the bacterial cell material) was formed from CO_2. When the ammonia or nitrite was left out of the medium, no growth occurred. Careful chemical

analyses showed that the amount of organic matter formed by the bacteria was proportional to the amount of ammonia or nitrite they oxidize. Winogradsky concluded, "This [process] is contradictory to that fundamental doctrine of physiology which states that a complete synthesis of organic matter cannot take place in nature except through chlorophyll-containing plants by the action of light." The concept of chemolithoautotrophy was thus born.

We now know that at least in one way, autotrophy in most chemolithotrophs and phototrophs is similar. In both groups, the pathway of CO_2 fixation follows the same biochemical steps (the Calvin cycle) involving the enzyme ribulose bisphosphate carboxylase (see Section 17.6).■

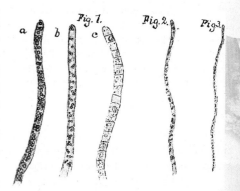

Figure 1 Beggiatoa and chemolithotrophy. *Drawings made by Winogradsky of Beggiatoa and translation (from the French) of the legend accompanying these figures. "Fig. 1. The tip of a filament of Beggiatoa alba: (a) in sulfurous [sulfide-containing] water, (b) after 24 h in water nearly depleted in H_2S, (c) after 48 h in water without H_2S [note depletion of sulfur globules with time]. Fig. 2. The tip of a filament of Beggiatoa media. Fig. 3. The tip of a filament of Beggiatoa minima."* From Winogradsky, S. 1949. *Microbiologie du Sol.* Masson, Paris.

---

sulfur, probably by reduction of S^0 to HS^-, from which it is transported into the cell and enters chemolithotrophic metabolism(Figure 17.27●).

One of the products of the sulfur oxidation reactions is H^+. Production of protons lowers the pH. Consequently, one result of the oxidation of reduced sulfur compounds is the acidification of the medium. The acid formed by the sulfur bacteria is a stronger acid, *sulfuric acid*—H_2SO_4—and sulfur bacteria are often able to bring about a marked reduction in the pH of the medium.

Biochemistry and Energetics of Sulfur Oxidation

The biochemical steps in the oxidation of various sulfur compounds are summarized in Figure 17.27. Starting with sulfide, sulfite (SO_3^{2-}) is produced, a six-electron oxidation. If S^0 is the starting substrate, sulfite is also produced, although the S^0 must first be reduced to sulfide (Figure 17.27a). There are two ways in which SO_3^{2-} can be oxidized to SO_4^{2-}. The most widespread system is that employing the enzyme *sulfite oxidase*. Sulfite oxidase transfers electrons from SO_3^{2-} directly to cytochrome *c*,

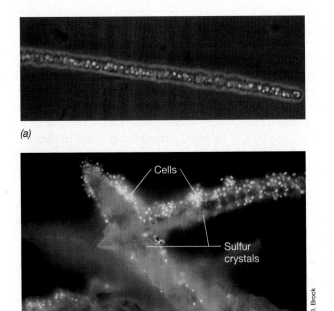

(a)

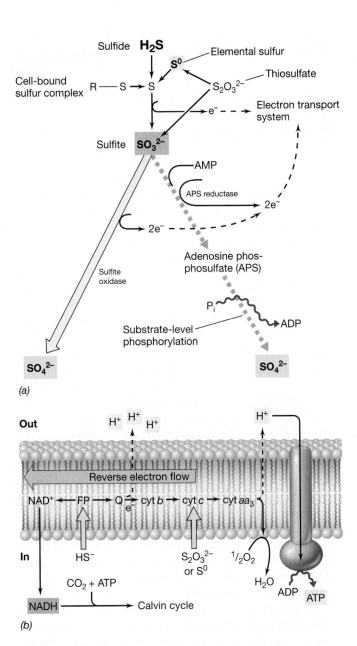

(a)

(b)

● **Figure 17.26 Sulfur bacteria.** (a) Deposition of internal sulfur granules by *Beggiatoa*. (b) Attachment of the sulfur-oxidizing archaeon *Sulfolobus acidocaldarius* to a crystal of elemental sulfur. Cells are visualized by fluorescence microscopy after staining them with the dye acridine orange. The sulfur crystal does not fluoresce. See Figure 17.27 for how sulfide and sulfur are oxidized to yield ATP.

● **Figure 17.27 Oxidation of reduced sulfur compounds by sulfur chemolithotrophs.** (a) Steps in the oxidation of different compounds. The sulfite oxidase pathway accounts for the majority of sulfite oxidized. (b) Electrons from sulfur compounds feed into the electron transport chain to drive a proton motive force; electrons from thiosulfate and elemental sulfur enter at the level of cytochrome *c*. NADH must be made by energy-consuming reactions of reverse electron flow since the electron donors have a more electropositive E_0' than does $NAD^+/NADH$. Cyt, cytochrome; FP, flavoprotein; Q, quinone. For the structure of APS, see Figure 17.38.

and ATP is made from this during electron transport and proton motive force formation (Figure 17.27*b*).

In addition to sulfite oxidase, a few sulfur chemolithotrophs oxidize SO_3^{2-} to SO_4^{2-} via a reversal of the activity of *adenosine phosphosulfate (APS) reductase*, an enzyme essential for the metabolism of sulfate-reducing bacteria (compare Figures 17.27*a* and 17.38). This reaction, run in the direction of SO_4^{2-} *production* by sulfur chemolithotrophs, yields one high-energy phosphate bond when AMP is converted to ADP (Figure 17.27*a*). When thiosulfate is the electron donor for sulfur chemolithotrophs, it is split into S^0 and SO_3^{2-}, both of which are eventually oxidized to SO_4^{2-}.

All the electrons from reduced sulfur compounds eventually reach the electron transport system as shown in Figure 17.27*b*. Depending on the E_0' of the couple, electrons enter at either the flavoprotein ($E_0' = \sim -0.2$) or cytochrome *c* ($E_0' = +0.3$) level and are shuttled to O_2, generating a proton motive force that leads to ATP synthesis by ATPase. Electrons for autotrophic CO_2 fixation come from reverse electron flow (see Section 17.5), eventually yielding NADH, and CO_2 is fixed via the Calvin cycle (Figure 17.27*b*). Although the sulfur chemolithotrophs are primarily an aerobic group (Section 12.4), some species can grow anaerobically using nitrate as an electron acceptor; *Thiobacillus denitrificans* is a classic example of this lifestyle.

17.9–17.10 Concept Checks

Hydrogen (H_2) and reduced sulfur compounds such as H_2S and S^0 are excellent electron donors for chemolithotrophs. These compounds can be oxidized by the hydrogen bacteria or the sulfur bacteria, respectively, thereby generating a proton motive force and ATP synthesis. These chemolithotrophs are also autotrophs and fix CO_2 by the Calvin cycle.

◆ What special enzyme is needed for growth on H_2?

◆ How many electrons are available from the oxidation of H_2S if S^0 is the final product? If SO_4^{2-} is the final product?

17.11 Iron Oxidation

The aerobic oxidation of iron from the ferrous (Fe^{2+}) to the ferric (Fe^{3+}) state is an energy-yielding reaction for some prokaryotes. Only a small amount of energy is available from this oxidation (see Table 17.1), and for this reason the iron bacteria must oxidize large amounts of iron in order to grow. The ferric iron produced forms insoluble ferric hydroxide [$Fe(OH)_3$] precipitates in water (Figure 17.28●). This is in part because at neutral pH ferrous iron rapidly oxidizes nonbiologically to the ferric state. It is thus stable for long periods only under anoxic conditions. At acid pH, however, ferrous iron is stable under oxic conditions. This explains why most iron-oxidizing bacteria are obligately acidophilic.

● **Figure 17.29 Iron bacteria at neutral pH: *Sphaerotilus*.** Phase contrast photomicrograph of empty iron-encrusted sheaths of *Sphaerotilus* collected from seepage at the edge of a small swamp. *Sphaerotilus* is a typical interface organism, living in habitats where anoxic ferrous-containing water meets oxic zones.

The best-known iron-oxidizing bacteria, *Acidithiobacillus ferrooxidans* and *Leptospirillum ferrooxidans*, can both grow autotrophically using ferrous iron (Figure 17.28b) as electron donor. Both organisms can grow at pH values below pH 1. These organisms are very common in acid-polluted environments such as coal-mining dumps (Figure 17.28a). *Ferroplasma*, a member of the *Archaea*, is an extremely acidophilic iron oxidizer and is even capable of growing at pH values below 0 (⬯ Section 13.5). We will discuss the role of all of these organisms in acid-mine pollution and mineral oxidation in Sections 19.14 and 19.15.

Despite the instability of Fe^{2+} at neutral pH, there are a number of iron-oxidizing bacteria that thrive in such environments, but these are situations where ferrous iron is moving from anoxic to oxic conditions. At interfaces between these zones iron bacteria can oxidize Fe^{2+} as it comes from an anoxic source before the Fe^{2+} oxidizes spontaneously. *Gallionella ferruginea* and *Sphaerotilus natans* are examples of organisms that live at these interfaces. They are typically seen mixed in with the characteristic deposits they form (Figure 17.29●; also ⬯ Figure 12.43a).

(a)

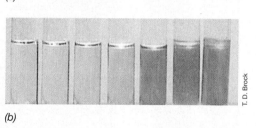

(b)

● **Figure 17.28 Iron-oxidizing bacteria.** (a) Acid mine drainage, showing the confluence of a normal river and a creek draining a coal-mining area. The acidic creek is very high in ferrous iron (Fe^{2+}). At low pH values, ferrous iron does not oxidize spontaneously in air, but *Acidithiobacillus ferrooxidans* carries out the oxidation. Insoluble ferric hydroxide and complex ferric salts precipitate, forming the precipitate called "yellow boy" by coal miners. (b) Cultures of *A. ferrooxidans*. Shown is a dilution series, with no growth in the tube on the left and increasing amounts of growth from left to right. Growth is evident from the production of Fe^{3+} from Fe^{2+}, which readily forms $Fe(OH)_3$ and acidity, leading to the yellow-orange color ($Fe^{3+}+3H_2O \rightarrow Fe(OH)_3+3H^+$).

Energy from Ferrous Iron Oxidation

The bioenergetics of iron oxidation by *Acidithiobacillus ferrooxidans* is of interest because of the very electropositive reduction potential of the Fe^{3+}/Fe^{2+} couple (+0.77 V at pH 2). The respiratory chain of *A. ferrooxidans* contains cytochromes of the *c* and *a* types and a periplasmic copper-containing protein called *rusticyanin* (Figure 17.30●). Because the reduction potential of the Fe^{3+}/Fe^{2+} couple is so high, the route of electron transport to oxygen ($\frac{1}{2}O_2/H_2O$, $E_0' = +0.82$ V) can only be very short.

Ferrous iron oxidation begins in the periplasm, where rusticyanin oxidizes Fe^{2+} to Fe^{3+}, a one-electron

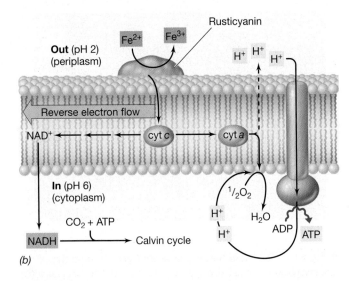

(b)

● **Figure 17.30 Electron flow during Fe^{2+} oxidation by the acidophile *Acidithiobacillus ferrooxidans*.** The periplasmic copper-containing protein rusticyanin is the immediate acceptor of electrons from Fe^{2+}. From here, electrons travel a short electron transport chain resulting in the reduction of O_2 to H_2O. Reducing power to drive the Calvin cycle comes from reactions of reverse electron flow. Note the steep pH gradient (~4 units) across the membrane.

transition. This protein then reduces cytochrome *c*, and this subsequently reduces cytochrome *a*. The latter interacts directly with O_2 to form H_2O (Figure 17.30). ATP is then synthesized from proton-translocating ATPases in the membrane; ATP yields are typically low because of the high potential of the electron donor.

Because of the large natural gradient of protons across the *A. ferrooxidans* membrane (the periplasm is pH 1–2 while the cytoplasm is pH 5.5–6), protons entering the cytoplasm via the ATPase must be consumed in order to maintain the internal pH within acceptable limits (Figure 17.30). The protons are consumed during the production of H_2O, but this reaction also requires electrons; these come from Fe^{2+} as follows:

$$2 Fe^{2+} + \tfrac{1}{2}O_2 + 2 H^+ \longrightarrow 2 Fe^{3+} + H_2O$$

Thus, as long as *A. ferrooxidans* has Fe^{2+} available, ATP synthesis can occur at the expense of the natural proton motive force that exists across the cytoplasmic membrane (Figure 17.30).

Autotrophy in *A. ferrooxidans* is driven by the Calvin cycle, and because of the high potential of the electron donor, Fe^{2+}, much energy is consumed in reverse electron flow reactions to obtain the reducing power (NADH) necessary to drive CO_2 fixation. Thus, a relatively poor energetic yield coupled with large energetic demands in biosynthesis means that *A. ferrooxidans* must oxidize large amounts of Fe^{2+} in order to produce even a very small amount of cell material. Because of this, in environments where acidophilic Fe^{2+}-oxidizing bacteria thrive, their presence is signaled not by the formation of much cell material but by the presence of large amounts of ferric iron

precipitates (Figures 17.28; ∞ Figure 19.37). We consider the important ecological processes connected with the iron-oxidizing bacteria in Sections 19.14 and 19.15.

Ferrous Iron Oxidation by Anoxygenic Phototrophs

Ferrous iron can be oxidized under *anoxic* conditions by certain anoxygenic phototrophic bacteria (Figure 17.31●). The ferrous iron is used in this case not as an electron donor in energy metabolism, but as an electron donor for CO_2 reduction (autotrophy). At neutral pH where these organisms thrive, the Fe^{3+}/Fe^{2+} couple (about 0.2 V) is much less electropositive than at pH 2. Thus, electrons from Fe^{2+} can reduce cytochrome *c* in the photosystem of purple bacteria (see Section 17.5 for a discussion of anoxygenic photosynthesis). The organisms involved, which are species of phototrophic purple bacteria (Figure 17.31*b*), can also use FeS as electron donor; under these conditions both Fe^{2+} and S^{2-} are oxidized as electron donors.

Certain phototrophic green sulfur bacteria (genus *Chlorobium*; ∞ Section 12.32) can also use Fe^{2+} as a photosynthetic electron donor. Moreover, various chemotrophic denitrifying bacteria have been isolated that can couple the oxidation of Fe^{2+} to the reduction of NO_3^- to N_2 and grow under anoxic conditions. However, like the aerobic iron bacteria, in these organisms iron is an electron donor for *both* energy and reducing power needs.

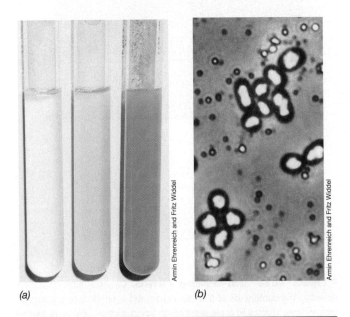

(a)   *(b)*

● **Figure 17.31 Ferrous iron oxidation by anoxygenic phototrophic bacteria.** (a) Fe^{2+} oxidation in anoxic tube cultures. Left to right: Sterile medium, inoculated medium, a growing culture. The brown–red color is mainly due to $Fe(OH)_3$ precipitate. (b) Phase contrast photomicrograph of an iron-oxidizing purple bacterium. The bright refractile areas within cells are gas vesicles (∞ Section 4.12). The granules outside the cells are iron precipitates. This organism is phylogenetically related to the purple sulfur bacterium *Chromatium* (∞ Section 12.2).

The discovery of Fe^{2+}-oxidizing phototrophs has important implications for both understanding the evolution of photosynthesis and explaining the large deposits of ferric iron found in ancient sediments. Such ferric iron was previously thought to have been formed from the oxidation of Fe^{2+} by O_2 produced by oxygenic phototrophs (⚬⚬ Section 11.1). However, because of the age of these sediments, it is more likely that the ferric iron was formed by anoxygenic phototrophs oxidizing Fe^{2+} in anoxic environments.

 17.11 Concept Check

The iron bacteria are chemolithotrophs able to use ferrous iron (Fe^{2+}) as sole energy source. Most iron bacteria grow only at acid pH and are often associated with acid pollution from mineral and coal mining. Some phototrophic purple bacteria can oxidize Fe^{2+} to Fe^{3+} anaerobically.

◆ Why is only a very small amount of energy available from the oxidation of Fe^{2+} to Fe^{3+} at acidic pH?

◆ What is the function of *rusticyanin* and where is it found in the cell?

◆ How can Fe^{2+} be oxidized anoxically?

17.12 Nitrification and Anammox

The most common *inorganic nitrogen compounds* used as electron donors are ammonia (NH_3) and nitrite (NO_2^-). These compounds are oxidized aerobically by the chemolithotrophic **nitrifying bacteria** (⚬⚬ Section 12.3) in the process of **nitrification**. The nitrifying bacteria are widely distributed in soil and water. One group, the *nitrosifyers* (*Nitrosomonas* is one genus), oxidizes ammonia to nitrite, and another group (*Nitrobacter*) oxidizes nitrite to nitrate. The complete oxidation of ammonia to nitrate, an eight-electron transfer, is thus carried out by two groups of organisms acting in concert (see the Microbial Sidebar).

Bioenergetics and Enzymology of Nitrification

The electrons from nitrogen compounds enter an electron transport chain, and electron flow establishes a proton motive force linked to ATP synthesis. However, because of the reduction potential of their electron donors, nitrifying bacteria are faced with bioenergetic problems similar to those of the sulfur chemolithotrophs. The E_0' of the NO_2^-/NH_3 couple (the first step in the oxidation of NH_3) is $+0.34$ V. The E_0' of the NO_3^-/NO_2^- couple is even higher, about $+0.43$ V. These relatively high reduction potentials mean that nitrifying bacteria must donate electrons to their electron transport chains at rather late steps in the overall process. This effectively limits the energy released and the amount of ATP that can be produced from each pair of electrons.

Several key enzymes are involved in oxidizing reduced nitrogen compounds. In ammonia-oxidizing bacteria, NH_3 is oxidized by **ammonia monooxygenase** (see Section 17.22 for a discussion of monooxygenase enzymes) that produces NH_2OH and H_2O (Figure 17.32). *Hydroxylamine oxidoreductase* then oxidizes NH_2OH to NO_2^-, removing *four* electrons in the process. Ammonia monooxygenase is an integral membrane protein, whereas hydroxylamine oxidoreductase is periplasmic (Figure 17.32●). In the reaction carried out by ammonia monooxygenase,

$$NH_3 + O_2 + 2\,H^+ + 2\,e^- \longrightarrow NH_2OH + H_2O$$

there is a need for two exogenously supplied electrons plus two protons to reduce one atom of oxygen to water. These electrons originate from the oxidation of hydroxylamine and are supplied to ammonia monooxygenase from hydroxylamine oxidoreductase via cytochrome c and ubiquinone (Figure 17.32). Thus, for every *four* electrons generated from the oxidation of NH_3 to NO_2^-, only *two* actually reach the terminal oxidase (cytochrome aa_3, Figure 17.32).

Nitrite-oxidizing bacteria employ the enzyme *nitrite oxidoreductase* to oxidize nitrite to nitrate, with electrons traveling a very short electron transport chain (because of the high potential of the NO_3^-/NO_2^- couple) to the terminal oxidase (Figure 17.33●). Cytochromes of the a and c types are present in the electron transport chain of

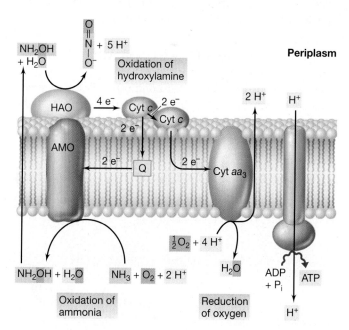

● **Figure 17.32 Oxidation of ammonia and electron flow in ammonia-oxidizing bacteria.** The reactants and the products of this reaction series are highlighted. The cytochrome c (cyt c) in the periplasm is a different form of cyt c than that in the membrane. AMO, ammonia monooxygenase; HAO, hydroxylamine oxidoreductase; Q, ubiquinone.

nitrite oxidizers, and generation of a proton motive force (which ultimately drives ATP synthesis) occurs through the activity of cytochromes aa_3 (Figure 17.33). Like the situation with iron oxidation (see Section 17.11), only small amounts of energy are available in this reaction. Thus, growth yields of nitrifying bacteria are low.

Carbon Metabolism in Nitrifying Bacteria

Like sulfur- and iron-oxidizing chemolithotrophs, aerobic nitrifying bacteria employ the Calvin cycle for CO_2 fixation, and the ATP and reducing power requirements of this process place additional burdens on an already relatively low-yielding energy-generating system (NADH to drive the Calvin cycle is formed by reverse electron flow). The energetic constraints are particularly severe for nitrite oxidizers, and it is perhaps for this reason that most nitrite oxidizers can also grow chemo-organotrophically on glucose and certain other organic substrates (∞ Section 12.3).

Anoxic Ammonia Oxidation: Anammox

Although classical nitrifying bacteria are *strict aerobes*, at least when growing on their reduced nitrogen substrates, ammonia can also be oxidized under **anoxic** conditions. This process, known as **anammox** (for *anoxic ammonia oxidation*), is highly exergonic and linked to energy metabolism of the organisms involved. Anammox involves the oxidation of ammonia with nitrite as the electron acceptor to yield gaseous nitrogen:

$$NH_4^+ + NO_2^- \longrightarrow N_2 + 2\,H_2O \quad \Delta G^{0'} = -357\ \text{kJ}$$

The organism that catalyzes anammox, *Brocadia anammoxidans*, is a phylogenetically distinct member of the Planctomycetes phylum of *Bacteria* (∞ Section 12.28) (Figure 17.34●). Planctomycetes are unusual members of the *Bacteria*, lacking peptidoglycan and containing membrane-enclosed compartments inside the cell, including a structure analogous to the nucleus of eukaryotic cells (∞ Section 12.28). In *B. anammoxidans* a major compartment is present called the *anammoxosome*; this is the location of the anammox reaction (Figure 17.34b).

The anammoxosome is a unit membrane-enclosed structure; in other words, an *organelle* (Figure 17.34b). Yet, *Brocadia* is clearly a prokaryote. The anammoxosome is another type of membrane-enclosed compartment beside the aforementioned "nucleus." Anammoxosome membrane lipids are unique as well. The lipids consist of fatty acids containing multiple cyclobutane rings (∞ Figure 11.24) that are connected to glycerol by both ester and ether bonds. The lipids aggregate to form an unusually dense membrane structure, highly resistant to diffusion. The strong anammoxosome membrane likely

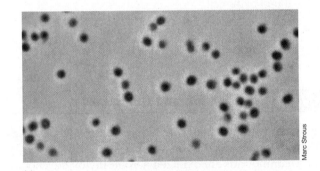

(a)

Marc Strous

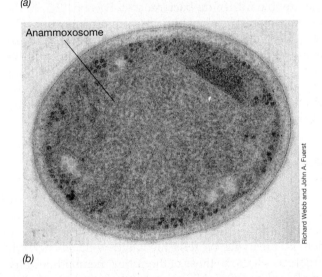

Anammoxosome

(b)

Richard Webb and John A. Fuerst

● **Figure 17.34 The anammox organism, *Brocadia anammoxidans.*** (a) Phase-contrast photomicrograph. A single cell is about 1 μm in diameter. (b) Transmission electron micrograph of a cell. Notice the membrane enclosed compartments including the large fibrillar anammoxosome in the center portion of the cell (arrow). *Brocadia* is phylogenetically related to the *Planctomyces*, organisms that contain several different types of membrane–enclosed compartments within their cytoplasm (∞ Section 12.28 and Figure 12.87).

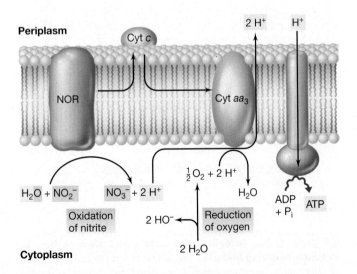

● **Figure 17.33 Oxidation of nitrite to nitrate by nitrifying bacteria.** The reactants and products of this reaction series are highlighted. NOR, nitrite oxidoreductase.

protects the cell from the toxic intermediates produced during the anammox reaction.

The source of NO_2^- in the anammox reaction is the product of ammonia oxidation by aerobic nitrifying bacteria (Figure 17.32). The two groups of nitrifyers, aerobic (for example, *Nitrosomonas*) and anaerobic (*Brocadia*), live together in ammonia-rich habitats such as sewage and other wastewaters. In these environments suspended particles are present that contain both oxic and anoxic zones where the two groups of ammonia oxidizers can coexist. In mixed laboratory cultures, high levels of oxygen inhibit anammox and favor classic nitrification, and thus it is likely that in nature the extent of anammox-mediated ammonia oxidation is governed by the concentration of O_2 in the system.

Like classical nitrifying bacteria, *Brocadia anammoxidans* is also an autotroph. The anammox organism can grow with CO_2 as sole carbon source and uses nitrite as electron donor to produce cell material:

$$CO_2 + 2 NO_2^- + H_2O \longrightarrow CH_2O + 2 NO_3^-$$

Although an autotrophic nitrifyer, it appears that *B. anammoxidans* lacks Calvin cycle enzymes, and the mechanism of CO_2 fixation is unclear. However, the use of nitrite as electron donor for CO_2 fixation along with the production of nitrate is exactly the same reaction carried out by aerobic nitrifyers like *Nitrobacter* (see Figure 17.33). Thus, although phylogenetically distinct, aerobic nitrifyers and the anammox organism share common substrates and ecology.

Before anammox was understood, it was thought that ammonia was stable in anoxic environments. Now, however, it is clear that ammonia can be oxidized in the absence of O_2. From an environmental standpoint, anammox is a very beneficial process in the treatment of anoxic wastewaters. The removal of ammonia and related amines helps reduce the fixed nitrogen pollution of rivers and streams, thereby maintaining high water quality. Ecological studies of anammox have shown that this process also occurs in marine sediments by organisms similar to *Brocadia*. This discovery has helped account for a significant fraction of ammonia known to be lost from marine environments but that has been unaccounted for as to mechanism.

 17.12 Concept Check

Ammonia and nitrite can be used as electron donors by the nitrifying bacteria. The ammonia-oxidizing bacteria produce nitrite, which is then oxidized by the nitrite-oxidizing bacteria to nitrate. Anoxic NH_3 oxidation is coupled to both N_2 and NO_3^- production in the anammoxosome.

◆ What is the inorganic electron donor for *Nitrosomonas*? For *Nitrobacter*?

◆ What are the substrates for the enzyme *ammonia monooxygenase*?

◆ What do nitrifying bacteria use as a carbon source?

◆ What is the *anammox* reaction and how does it differ from aerobic nitrification?

III THE ANAEROBIC WAY OF LIFE: ANAEROBIC RESPIRATIONS

In eukaryotes, anaerobic growth is rare. In prokaryotes, by contrast, anaerobic growth is common, and the mechanisms of anaerobic metabolism are highly diverse. In the next several sections we survey anaerobic respiration and see the many ways by which prokaryotes can conserve energy using electron acceptors *other* than O_2.

17.13 Anaerobic Respiration

We examined the process of *aerobic* respiration in some detail in Chapter 5. As we noted there, molecular oxygen functions as a terminal electron acceptor, accepting electrons from electron carriers by way of an electron transport chain. However, we also noted that a variety of other electron acceptors can be used instead of O_2, in which case the process is called **anaerobic respiration**. Here we consider some of these processes.

The bacteria carrying out anaerobic respiration possess electron transport systems containing cytochromes, quinones, iron-sulfur proteins, and other typical electron transport proteins. Their respiratory systems are thus analogous to those of conventional aerobes. In some cases, such as with the denitrifying bacteria, the anaerobic respiration process competes in the same organism with an aerobic one. In such cases, if O_2 is present, aerobic respiration is favored, and when O_2 is depleted from the environment, the alternate electron acceptor is reduced. Other organisms carrying out anaerobic respiration are *obligate* anaerobes and are unable to use O_2.

Alternative Electron Acceptors and the Electron Tower

The energy released from the oxidation of an electron donor using O_2 as electron acceptor is higher than if the same compound is oxidized with an alternate electron acceptor (∞ Figure 5.9). These energy differences are apparent if the reduction potentials of each acceptor are examined (Figure 17.35●). Because the O_2/H_2O couple is the most electropositive, more energy is available when O_2 is used than when another electron acceptor is used. Other electron acceptors that are near the O_2/H_2O couple are Fe^{3+}, NO_3^-, and NO_2^-. More electronegative acceptors are SO_4^{2-}, S^0, and CO_2. A summary of the most common types of anaerobic respiration is given in Figure 17.35.

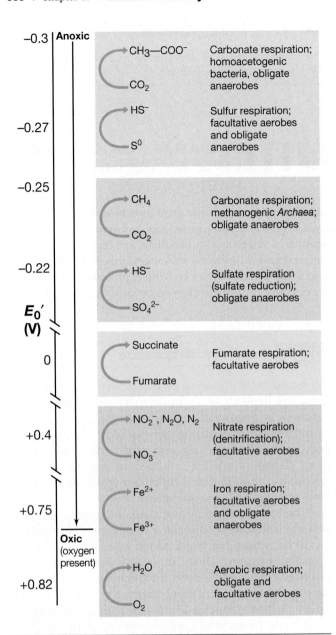

● **Figure 17.35 Examples of anaerobic respirations.** The couples are arranged in order from most electronegative E_0' (top) to most electropositive E_0' (bottom). Compare with Figure 5.9 to see how the energy yields of these anaerobic respirations vary.

Assimilative and Dissimilative Metabolism

Inorganic compounds such as NO_3^-, SO_4^{2-}, and CO_2 are reduced by many organisms as sources of cellular nitrogen, sulfur, and carbon, respectively. The end products of such reductions are primarily amino groups ($-NH_2$), sulfhydryl groups ($-SH$), and organic carbon compounds, respectively. When an inorganic compound such as NO_3^-, SO_4^{2-}, or CO_2 is reduced for use in biosynthesis, it is said to be *assimilated*, and the reduction process is called **assimilative metabolism**. Assimilative metabolism of NO_3^-, SO_4^{2-}, and CO_2 is conceptually quite different from their use as electron acceptors for *energy* metabolism in anaerobic respiration. To distinguish these two kinds of reduction processes, the use of

these compounds as electron acceptors in energy metabolism is called **dissimilative metabolism**.

Assimilative and dissimilative metabolism differ markedly. In *assimilative* metabolism, only enough of the compound (NO_3^-, SO_4^{2-}, or CO_2) is reduced to satisfy the needs for cell growth. The products are eventually converted to cell material in the form of macromolecules. In *dissimilative* metabolism, a large amount of the electron acceptor is reduced, and the reduced product is *excreted* into the environment. Many organisms carry out assimilative metabolism of compounds such as NO_3^-, SO_4^{2-}, and CO_2 (for example, many *Bacteria, Archaea*, fungi, algae, and higher plants), whereas only a restricted variety of organisms, primarily prokaryotes, carry out dissimilative metabolism.

⬡ *17.13 Concept Check*

Although oxygen is the most widely used electron acceptor in energy-yielding metabolism, a number of other compounds can be used as electron acceptors. This process of anaerobic respiration is less energy efficient but makes it possible for respiration to occur in environments where oxygen is absent.

◆ What is anaerobic respiration?

◆ With H_2 as electron donor, why is the reduction of NO_3^- a more favorable reaction than the reduction of S^0?

17.14 Nitrate Reduction and Denitrification

Inorganic nitrogen compounds are some of the most common electron acceptors in anaerobic respiration. Table 17.2 summarizes the various inorganic nitrogen species with their oxidation states. The most widespread inorganic nitrogen compounds in nature are *ammonia* and *nitrate*, both of which are formed in the atmosphere by inorganic chemical processes, and also nitrogen gas— N_2—the most stable form of nitrogen in nature. We discuss *nitrogen fixation*, the use of N_2 as a biosynthetic nitrogen source, later in this chapter (see Section 17.28).

One of the most common alternative electron acceptors is nitrate, NO_3^-, which is reduced to N_2O, NO, and N_2. Because these products of nitrate reduction are all gaseous, they can easily be lost from the environment, a process called **denitrification** (Figure 17.36●). The pro-

Table 17.2	Oxidation states of key nitrogen compounds
Compound	**Oxidation state**
Organic N ($R-NH_2$)	−3
Ammonia (NH_3)	−3
Nitrogen gas (N_2)	0
Nitrous oxide (N_2O)	+1 (average per N)
Nitrogen oxide (NO)	+2
Nitrite (NO_2^-)	+3
Nitrogen dioxide (NO_2)	+4
Nitrate (NO_3^-)	+5

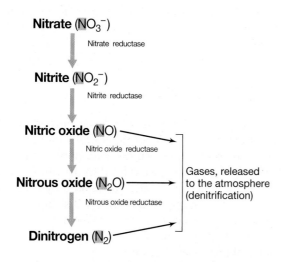

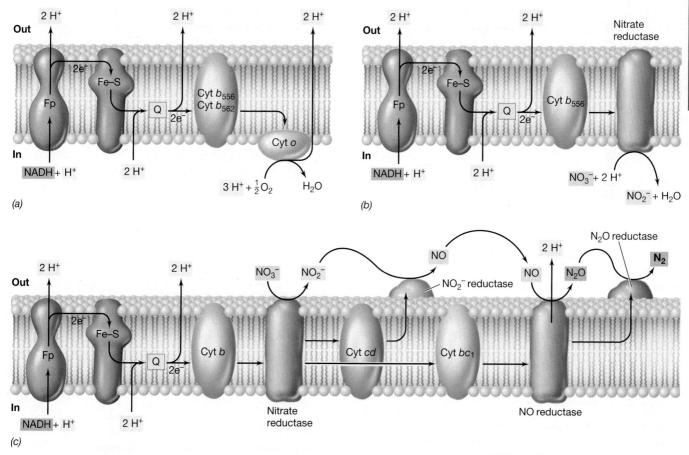

cess is the main means by which gaseous N_2 is formed biologically. N_2 is much less available to organisms than nitrate as a source of nitrogen, so for agricultural purposes, at least, denitrification is a detrimental process. For sewage treatment (∞ Section 28.2), however, denitrification is beneficial because it converts NO_3^- to N_2. This effectively decreases the load of fixed nitrogen in the sewage treatment effluent that can stimulate algal growth (∞ Section 19.5).

Biochemistry of Dissimilative Nitrate Reduction

The enzyme involved in the first step of dissimilative nitrate reduction, *nitrate reductase*, is a molybdenum-containing membrane-integrated enzyme whose synthesis is repressed by molecular oxygen. All subsequent enzymes of the pathway (Figure 17.37●) are coordinately regulated and thus also repressed by O_2, but in addition to anoxic conditions, nitrate must also be present before these enzymes are fully expressed.

The first product of nitrate reduction is nitrite (NO_2^-), and the enzyme *nitrite reductase* reduces it to

● **Figure 17.36   Steps in the dissimilative reduction of nitrate.** Some organisms, for example *Escherichia coli*, can carry out only the first step. All enzymes involved are derepressed by anoxic conditions. Also, some prokaryotes are known that can reduce NO_3^- to NH_4^+ in dissimilative metabolism.

● **Figure 17.37   Respiration.** Electron transport processes in the membrane of *Escherichia coli* when (a) O_2 or (b) NO_3^- is used as an electron acceptor and NADH is the electron donor. Fp, flavoprotein; Q, ubiquinone. Under high oxygen conditions, the sequence of carriers is cyt $b_{562} \rightarrow$ cyt $o \rightarrow O_2$. However, under low oxygen conditions (not shown), the sequence is cyt $b_{568} \rightarrow$ cyt $d \rightarrow O_2$. Note how more protons are translocated per two electrons oxidized aerobically during electron transport reactions than anaerobically with nitrate as electron acceptor, because the aerobic terminal oxidase (cyt *o*) can pump one proton. (c) Possible scheme for electron transport in membranes of *Pseudomonas stutzeri* during denitrification. Nitrate and nitric oxide (NO) reductases are located in the cytoplasmic membrane whereas nitrite (NO_2^-) and nitrous oxide (N_2O) reductases are periplasmic. The immediate electron donors to the various reductases, with the exception of nitrate reductase, have not been definitively identified.

nitric oxide (NO) (Figure 17.37c). Some organisms can reduce NO_2^- to ammonia (NH_3) in a dissimilative process, but the production of gaseous products—*denitrification*—is of greatest global significance. This is because it consumes a fixed form of nitrogen (NO_3^-) readily available to plants and produces gaseous nitrogen compounds, some of which are of environmental significance. For example, N_2O can be converted to NO by sunlight, and NO reacts with ozone in the upper atmosphere to form nitrite (NO_2^-), which returns to Earth as acid rain (nitrous acid, HNO_2). The remaining biochemical steps in denitrification are shown in Figure 17.37c.

The biochemistry of dissimilative nitrate reduction has been studied in detail in several organisms including *Escherichia coli*, where NO_3^- is reduced only to NO_2^- (Figure 17.37b), and *Paracoccus denitrificans* and *Pseudomonas stutzeri*, where true denitrification occurs (Figure 17.37c). The *E. coli* nitrate reductase accepts electrons from a b-type cytochrome, and a comparison of the electron transport chains in aerobic versus anaerobic respiration in *E. coli* is shown in Figure 17.37a, b. As seen here, because of the reduction potential of the NO_3^-/NO_2^- couple (+0.43 V), only two proton translocating steps occur during nitrate reduction while three can occur in aerobic respiration $\left(\frac{1}{2}O_2/H_2O, +0.82\ V\right)$. In *P. denitrificans* and *P. stutzeri* nitrogen oxides are formed from nitrite by *nitrite reductase, nitric oxide reductase*, and *nitrous oxide reductase*, as summarized in Figure 17.37c. During electron transport, a proton motive force is established, and ATP is produced by ATPase in the usual fashion. Additional ATP is available when NO_3^- is reduced to N_2 because the NO reductase is linked to proton extrusion (Figure 17.37c).

Other Properties of Denitrifying Prokaryotes

Most denitrifying prokaryotes are phylogenetically members of the Proteobacteria (∞ Sections 12.2–12.18) and are facultative aerobes. Aerobic respiration occurs when air is present, even if nitrate is also present in the medium. Many denitrifying bacteria will also reduce other electron acceptors anaerobically, such as ferric iron (Fe^{3+}) and certain organic electron acceptors (see Section 17.18). In addition, many denitrifying bacteria can grow by fermentation. Thus, the denitrifying bacteria are quite metabolically diverse in terms of alternative energy-generating mechanisms.

 17.14 Concept Check

Nitrate is a commonly used electron acceptor in anaerobic respiration. Its use requires the enzyme nitrate reductase that reduces nitrate to nitrite. Many bacteria that use nitrate in anaerobic respiration eventually produce N_2, a process called *denitrification*.

◆ For *Escherichia coli*, why is more energy released in aerobic respiration than during NO_3^- reduction?

◆ Where is the dissimilative nitrate reductase found in the cell? What unusual metal does it contain?

◆ Why does an organism like *Pseudomonas stutzeri* derive more energy from NO_3^- respiration than does *Escherichia coli*?

17.15 Sulfate Reduction

Several inorganic *sulfur* compounds are important electron acceptors in anaerobic respiration. A summary of the oxidation states of the key sulfur compounds is given in Table 17.3. Sulfate, the most oxidized form of sulfur, is one of the major anions in seawater and is used by the *sulfate-reducing bacteria*, a group that is widely distributed in nature. The end product of sulfate reduction is H_2S, an important natural product that participates in many biogeochemical processes (∞ Section 19.13).

Again, as with nitrogen, it is important to distinguish between *assimilative* and *dissimilative* metabolism. Many organisms, including higher plants, algae, fungi, and most prokaryotes, use sulfate as a sulfur source for biosynthesis. The ability to use sulfate as an *electron acceptor* for energy-generating processes, however, involves the large-scale reduction of SO_4^{2-} and is restricted to the sulfate-reducing bacteria. In assimilative sulfate reduction, the H_2S formed is immediately converted into organic sulfur in the form of amino acids, and so on, but in dissimilative sulfate reduction, the H_2S is excreted.

As the reduction potentials in Table A1.2 and Figure 17.35 show, sulfate is a much less favorable electron acceptor than either O_2 or NO_3^-. However, sufficient energy to make ATP is available when an electron donor that yields NADH or FADH is used. Because of the less favorable energetics, growth yields are lower for an organism growing on SO_4^{2-} than for one growing on O_2 or NO_3^-. Table 17.3 lists some of the electron donors used by sulfate-reducing bacteria. H_2, lactate, and pyruvate are used by a wide vari-

Table 17.3	Sulfur compounds and electron donors for sulfate reduction	
Compound		**Oxidation state**
Oxidation states of key sulfur compounds		
Organic S (R—SH)		−2
Sulfide (H_2S)		−2
Elemental sulfur (S^0)		0
Thiosulfate ($S_2O_3^{2-}$)		+2 (average per S)
Sulfur dioxide (SO_2)		+4
Sulfite (SO_3^{2-})		+4
Sulfate (SO_4^{2-})		+6
Some electron donors used for sulfate reduction		
H_2		Acetate
Lactate		Propionate
Pyruvate		Butyrate
Ethanol and other alcohols		Long-chain fatty acids
Fumarate		Benzoate
Malate		Indole
Choline		Hexadecane

ety of sulfate-reducing bacteria, while the others have more restricted use. However, a large variety of morphological and physiological types of sulfate-reducing bacteria are known (⬭ Sections 12.18 and 13.17). With the exception of *Archaeoglobus*, a member of the *Archaea*, all known sulfate-reducing prokaryotes are *Bacteria*.

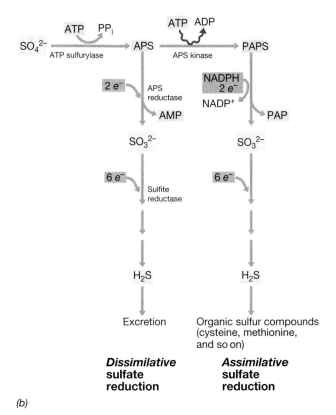

(a)

(b)

● **Figure 17.38** **Biochemistry of sulfate reduction.** (a) Two forms of *active sulfate* can be made, adenosine 5′-phosphosulfate (APS) and phosphoadenosine 5′-phosphosulfate (PAPS). Both are derivatives of adenosine diphosphate (ADP), with the second phosphate of ADP being replaced by sulfate. (b) Schemes of assimilative and dissimilative sulfate reduction.

Biochemistry and Energetics of Sulfate Reduction

The reduction of SO_4^{2-} to H_2S, an eight-electron reduction (Table 17.3), proceeds through a number of intermediate stages. The sulfate ion is stable and cannot be reduced without first being activated. Sulfate is activated by means of ATP. The enzyme *ATP sulfurylase* catalyzes the attachment of the sulfate ion to a phosphate of ATP, leading to the formation of **adenosine phosphosulfate (APS)** as shown in Figure 17.38●. In dissimilative sulfate reduction, the sulfate in APS is reduced directly to sulfite (SO_3^{2-}) by the enzyme *APS reductase* with the release of AMP. In assimilative reduction, another P is added to APS to form *phosphoadenosine phosphosulfate (PAPS)* (Figure 17.38*a*), and only then is the sulfate reduced. In both cases, the first product of sulfate reduction is *sulfite*, SO_3^{2-}. Once SO_3^{2-} is formed, sulfide is formed by the enzyme *sulfite reductase* (Figure 17.38*b*).

In the process of dissimilative sulfate reduction, electron transport reactions occur leading to proton motive force formation, and this drives ATP synthesis by an ATPase. A major electron carrier is cytochrome c_3, a periplasmic low potential cytochrome (Figure 17.39●). Cytochrome c_3 accepts electrons from a periplasmically located hydrogenase (see below) and transfers these electrons to a membrane-associated protein complex.

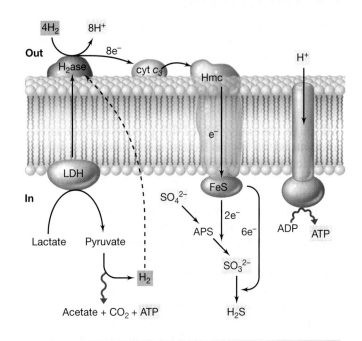

● **Figure 17.39** **Electron transport and energy conservation in sulfate-reducing bacteria.** In addition to external hydrogen (H_2), H_2 originating from the catabolism of organic compounds such as lactate and pyruvate can fuel hydrogenase. The enzymes hydrogenase (H_2ase), cytochrome c_3, and a cytochrome complex (Hmc) are periplasmic proteins. A separate protein functions to shuttle electrons across the cytoplasmic membrane from Hmc to a cytoplasmic iron–sulfur protein that supplies electrons to APS reductase (forming SO_3^{2-}) and sulfite reductase (forming H_2S see Figure 17.38*b*). LDH, lactate dehydrogenase.

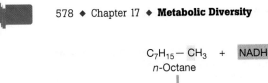

$C_7H_{15}-CH_3$ + NADH + O:O
n-Octane (O₂)

↓ Monooxygenase

$C_7H_{15}CH_2OH$ + NAD⁺ + H₂O
n-Octanol

NAD⁺
NADH

H
|
$C_7H_{15}C=O$
n-Octanal

H₂O — NAD⁺
NADH

OH
|
$C_7H_{15}C=O$
n-Octanoic acid

ATP — CoA
AMP + PP

β-Oxidation
to acetyl-CoA (see Figure 17.69)

● **Figure 17.55 Monooxygenase activity.** Steps in oxidation of an aliphatic hydrocarbon, the first of which is catalyzed by a monooxygenase. Some sulfate-reducing and denitrifying bacteria can degrade aliphatic hydrocarbons under anoxic conditions.

Earth when life first evolved and became available only after the evolution of cyanobacteria (⚭ Chapter 11).

 17.22 Concept Check

In addition to its role as an electron acceptor, oxygen is also a chemical reactant in certain biochemical processes. Enzymes called oxygenases introduce O₂ into a biochemical compound.

◆ How do *monooxygenases* differ in function from *dioxygenases*?

17.23 Hydrocarbon Oxidation

Hydrocarbons are organic compounds containing only carbon and hydrogen and are highly insoluble in water. Low-molecular-weight hydrocarbons are gases, whereas those of higher molecular weight are liquids or solids at room temperature. In *aliphatic* hydrocarbons, the carbon atoms are joined in open chains. There is a tremendous variation among aliphatic hydrocarbons in chain length, degree of branching, and number of double bonds. Aromatic hydrocarbons contain the aromatic ring and can be viewed as derivatives of benzene (see Figure 17.56).

Aliphatic Hydrocarbon Metabolism

A number of bacteria and several molds and yeasts can use hydrocarbons as electron donors to support growth under aerobic conditions. The initial oxidation step of saturated aliphatic hydrocarbons involves molecular oxygen (O₂) as a reactant, and one of the atoms of the oxygen molecule is incorporated into the oxidized hydrocarbon, typically at the terminal carbon atom. This reaction is carried out by a monooxygenase (see Section 17.22), and a typical reaction sequence is that shown in Figure 17.55. The end product of the reaction sequence is acetyl-CoA, and this is further catabolized in the citric acid cycle.

Aliphatic hydrocarbons, both saturated and unsaturated, can be oxidized to CO₂ under anoxic conditions as well. This is done by a restricted group of denitrifying and sulfate-reducing bacteria. Obviously monooxygenases cannot be involved here. The mechanism of degradation involves *carboxylation* (addition of CO₂) of the hydrocarbon to form a fatty acid. Following this, further degradation occurs by β-oxidation, a process that does not require O₂ (see Figure 17.69).

Aromatic Hydrocarbons

Many aromatic hydrocarbons can be used as electron donors aerobically by microorganisms, of which bacteria of the genus *Pseudomonas* have been the best studied. The metabolism of these compounds, some of which are quite large molecules, typically has as its initial stage the formation of *catechol* or a structurally related compound, as shown in Figure 17.56a●.

These single-ring compounds are referred to as *starting substrates* because oxidative catabolism proceeds only after the large aromatic molecules have been converted to these more simple forms. Protocatechuate and catechol may then be further degraded to compounds that can enter the citric acid cycle: succinate, acetyl-CoA, and pyruvate (⚭ Figure 5.22). Several steps in the catabolism of aromatic hydrocarbons usually require oxygenases. Figures 17.56a, b, and c show three different oxygenase-catalyzed reactions, one using a monooxygenase and two using a dioxygenase.

Anoxic Degradation of Aromatic Compounds

Aromatic compounds can also be degraded anaerobically by facultatively aerobic bacteria if the aromatic compound already contains an atom of oxygen. Such phenolic compounds are degraded by certain denitrifying, phototrophic, ferric iron-reducing, and sulfate-reducing bacteria. From a biochemical standpoint, the anoxic catabolism of aromatic compounds proceeds by *reductive* rather than oxidative ring cleavage (Figure 17.57●). This involves *ring reduction* followed by *ring cleavage* to yield a straight-chain fatty acid or dicarboxylic acid. These intermediates can be converted to acetyl-CoA and used for both biosynthetic and energy-yielding purposes. Benzoate and benzoate derivatives are common natural products and are readily degraded anaerobically.

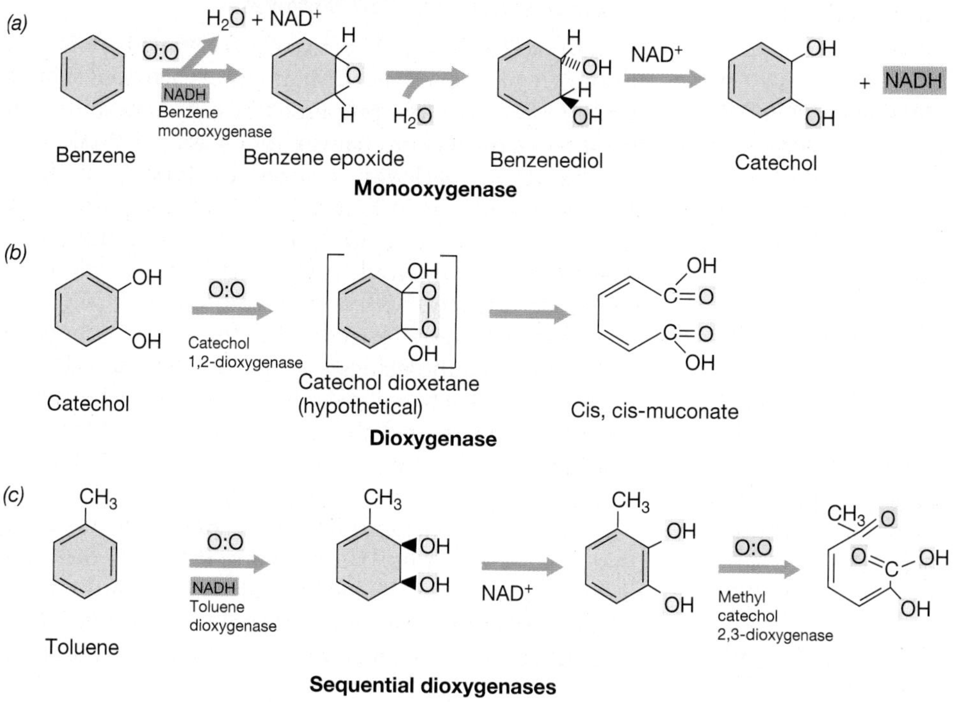

(a)

Benzene — Benzene monooxygenase (O:O, NADH → H₂O + NAD⁺) — Benzene epoxide (H₂O) — Benzenediol (NAD⁺ → NADH) — Catechol

Monooxygenase

(b)

Catechol — Catechol 1,2-dioxygenase (O:O) — Catechol dioxetane (hypothetical) — Cis, cis-muconate

Dioxygenase

(c)

Toluene — Toluene dioxygenase (O:O, NADH) — (NAD⁺) — Methyl catechol 2,3-dioxygenase (O:O) —

Sequential dioxygenases

● **Figure 17.56 Roles of oxygenases in catabolism of aromatic compounds.** (a) Hydroxylation of benzene to catechol by a monooxygenase in which NADH is an electron donor. (b) Cleavage of catechol to *cis,cis*-muconate by a dioxygenase. Reactant oxygen atoms are shown in color in both reactions to demonstrate the different mechanisms. (c) The activity of toluene dioxygenase and methyl catechol 2,3-dioxygenase in the degradation of toluene. The oxygen atoms that each enzyme introduces are distinguished by different colors. Catechol and proto-catechuate are common intermediates in aerobic aromatic catabolism. Proto-catechuate contains a carboxyl group (COO⁻) two carbon atoms removed from one of the OH groups of catechol.

Anoxic degradation of benzene and toluene, aromatic compounds lacking an oxygen atom (see Figure 17.56 for structures), also occurs. Catabolism of benzene occurs by certain denitrifying bacteria and toluene by certain ferric iron-reducing, phototrophic purple, and denitrifying bacteria. Biochemically, toluene is eventually converted to the benzoate derivative benzoyl-CoA and then presumably further catabolized via ring reduction as shown in Figure 17.57.

⬢ 17.23 Concept Check

Many microorganisms can degrade aliphatic and aromatic hydrocarbons. Catabolism aerobically involves the activity of oxygenase enzymes. Anoxic aromatic degradation proceeds by reductive rather than oxidative pathways.

♦ Draw the chemical structure for benzene. Do the same for benzoate. Are these compounds aliphatic or aromatic?

♦ What fundamental difference exists in the *anaerobic* degradation of an aromatic compound compared with its *aerobic* metabolism? Give an example of this.

17.24 Methanotrophy and Methylotrophy

Section 12.6 considered the unique situation, relative to carbon metabolism, of *methanotrophs* and *methylotrophs*. Although not autotrophs, **methylotrophs** use C_1 compounds (or other organic compounds lacking C—C bonds) for energy metabolism and biosynthesis. We focus here on the physiology of this process.

Biochemistry of Methane Oxidation

The individual steps in methane oxidation to CO_2 can be summarized as:

$$CH_4 \rightarrow CH_3OH \rightarrow CH_2O \rightarrow HCOO^- \rightarrow CO_2$$

Methanotrophs are those methylotrophs that can use CH_4 (⌘ Section 12.6). Methanotrophs assimilate either all or one-half of their carbon (depending on the pathway they use) at the level of *formaldehyde* (CH_2O). We will see that this affects a major energy savings compared with autotrophs,

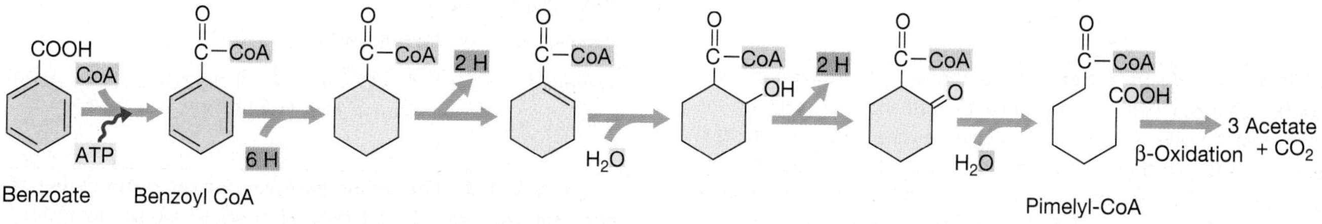

Benzoate — Benzoyl CoA (CoA, ATP) — (6 H) — (2 H, H₂O) — (H₂O) — (2 H) — Pimelyl-CoA — β-Oxidation → 3 Acetate + CO₂

● **Figure 17.57 Anoxic degradation of benzoate by reductive ring cleavage.** Note that all intermediates of the pathway are bound to coenzyme A. The acetate produced is further catabolized in the citric acid cycle (⌘ Section 5.13 and Figure 5.22).

where carbon is assimilated from the more oxidized CO_2. But here our focus is on energy conservation.

The initial step in the oxidation of methane under oxic conditions involves an enzyme called **methane monooxygenase**. As we discussed in Section 17.22, oxygenase enzymes catalyze the incorporation of oxygen from O_2 into carbon compounds (and some nitrogen compounds, see Section 17.12) and seem to be widely involved in the metabolism of hydrocarbons. Monooxygenases incorporate one atom of O_2 into the substrate while the second atom is reduced to H_2O. In the methanotroph *Methylosinus*, where the methane oxidation process has been thoroughly studied, the electrons needed for the oxidation of CH_4 to CH_3OH come from cytochrome *c* (Figure 17.58●.) This need for reducing power precludes synthesis of ATP during the oxidation of methane to methanol. In agreement with this, growth yields (grams of cells produced per mole of substrate consumed) of methanotrophs are the same whether methane or methanol is used as the growth substrate.

The other oxidation steps from CH_3OH to CO_2 use typical redox cofactors, such as quinones or NADH, which enter the electron transport chain. From these, ATP is made from the resulting proton motive force generated (Figure 17.58).

In addition to the *aerobic* catabolism of CH_4, anoxic oxidation of this hydrocarbon can occur. This is a syntrophic process involving the cooperation of sulfate-reducing bacteria and methanogens, both strict anaerobes, as partner organisms. We explore this process in Section 19.10.

C_1 Assimilation into Cell Material

As was noted in Section 12.6, two classes of methanotrophs are known, *type I* and *type II*. Members of each class share a number of phenotypic properties in common, and each also shows a distinct mechanism for C_1 assimilation.

The **serine pathway**, utilized by type II methanotrophs, is outlined in Figure 17.59●. In this pathway, a two-carbon unit, *acetyl-CoA*, is synthesized from one molecule of formaldehyde (produced from the oxidation of CH_4, see Figure 17.58) and one molecule of CO_2. The serine pathway requires reducing power and energy in the form of two molecules each of NADH and ATP, respectively, for each acetyl-CoA synthesized. The serine pathway employs a number of enzymes of the citric acid cycle and one enzyme, *serine transhydroxymethylase*, unique to the pathway (Figure 17.59).

The **ribulose monophosphate pathway**, used by type I methanotrophs, is outlined in Figure 17.60●. It is more efficient than the serine pathway because *all* of the carbon atoms for cell material are derived from formaldehyde. And, since formaldehyde is at the same

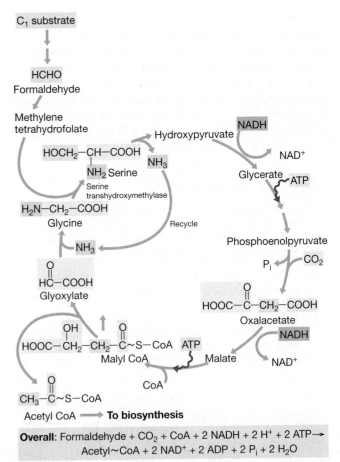

Overall: Formaldehyde + CO_2 + CoA + 2 NADH + 2 H$^+$ + 2 ATP→ Acetyl~CoA + 2 NAD$^+$ + 2 ADP + 2 P$_i$ + 2 H$_2$O

● **Figure 17.59 The serine pathway for the assimilation of C_1 units into cell material by type II methylotrophic bacteria.** The product of the pathway, acetyl-CoA, is used as the starting point for making new cell material. The key enzyme of the pathway is *serine transhydroxymethylase*.

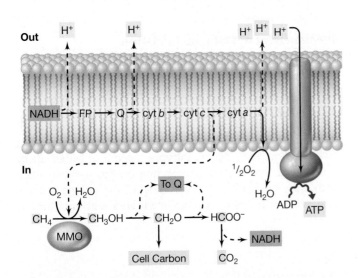

● **Figure 17.58 Oxidation of methane by methanotrophic bacteria.** Methane (CH_4) is converted to methanol (CH_3OH) by the enzyme *methane monooxygenase*. The electrons needed to drive this first step come from cytochrome *c*, and no energy is conserved in this reaction. A proton motive force is established from electron flow in the membrane, and this fuels ATPase. Note how carbon for biosynthesis comes primarily from formaldehyde (CH_2O). MMO, methane monooxygenase; FP, flavoprotein; cyt, cytochrome; Q, quinone. Although not depicted as such, MMO is actually a membrane-associated enzyme.

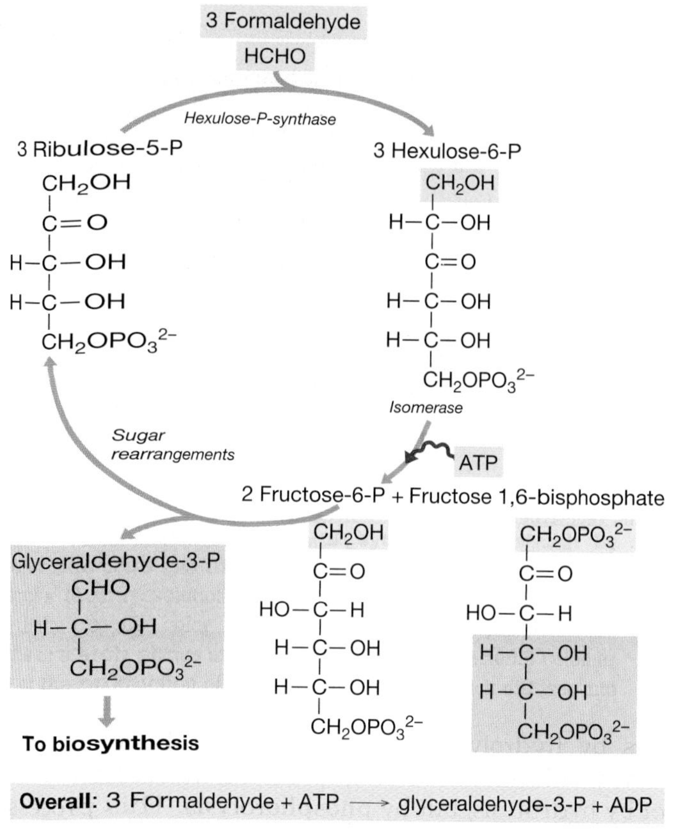

3 Formaldehyde

HCHO

Hexulose-P-synthase

3 Ribulose-5-P

CH_2OH

$C=O$

$H-C-OH$

$H-C-OH$

$CH_2OPO_3^{2-}$

3 Hexulose-6-P

CH_2OH

$H-C-OH$

$C=O$

$H-C-OH$

$H-C-OH$

$CH_2OPO_3^{2-}$

Isomerase

Sugar rearrangements

ATP

2 Fructose-6-P + Fructose 1,6-bisphosphate

CH_2OH

$C=O$

$HO-C-H$

$H-C-OH$

$H-C-OH$

$CH_2OPO_3^{2-}$

$CH_2OPO_3^{2-}$

$C=O$

$HO-C-H$

$H-C-OH$

$H-C-OH$

$CH_2OPO_3^{2-}$

Glyceraldehyde-3-P

CHO

$H-C-OH$

$CH_2OPO_3^{2-}$

To biosynthesis

Overall: 3 Formaldehyde + ATP $\longrightarrow$ glyceraldehyde-3-P + ADP

● **Figure 17.60 The ribulose monophosphate pathway for assimilation of one-carbon compounds, as found in type I methylotrophic bacteria.** The complete name of the hexulose sugar is D-erythro-L-glycero-3-hexulose 6-phosphate. Three formaldehydes are needed to carry the cycle to completion, with the net result being one molecule of glyceraldehyde-3-P. The key enzyme of this pathway is *hexulose-P-synthase*. The "sugar rearrangements" involve enzymes of the pentose phosphate pathway (see Figure 17.65).

oxidation level as cell material, *no reducing power is needed.* The ribulose monophosphate pathway requires one molecule of ATP for each molecule of glyceraldehyde-3-phosphate synthesized (Figure 17.60). Consistent with the lower energy requirements of the ribulose monophosphate pathway, the cell yield (grams of cells produced per mole of CH_4 oxidized) of type I methanotrophs is higher than for type II methanotrophs.

The enzymes *hexulosephosphate synthase*, which condenses one molecule of formaldehyde with one molecule of ribulose-5-phosphate, and *hexulose-6-P isomerase* (Figure 17.60) are unique to the ribulose monophosphate pathway. The remaining enzymes of this pathway are involved in sugar rearrangements in many different organisms. It should also be noted that the substrate for the initial reaction in this pathway, *ribulose-5-P*, is almost identical to the C_1 acceptor in the Calvin cycle, *ribulose 1,5-bisphosphate* (∞ Section 17.6), an indication that these two cycles probably share evolutionary roots.

17.24 Concept Check

Methanotrophy is the use of CH_4 as a carbon and energy source. The enzyme *methane monooxygenase* is a key enzyme in the catabolism of methane. C_1 units get assimilated into cell material at the level of formaldehyde in methanotrophs by either the ribulose monophosphate pathway or the serine pathway.

◆ What are the energy and reducing power requirements for the ribulose monophosphate pathway? For the serine pathway? Why do they differ?

◆ Why does the oxidation of CH_4 to CH_3OH require reducing power?

◆ Which pathway, the Calvin cycle or the ribulose monophosphate pathway, requires the greater energy input? Why?

17.25 Hexose, Pentose, and Polysaccharide Metabolism

We complete our discussion of chemoorganotrophic metabolism with consideration of a few special aspects of the catabolism of organic compounds, especially the use of *polymeric* substances that must first be hydrolyzed to monomeric units before energy-generating mechanisms can be employed. We also examine a special pathway employed by cells to produce and catabolize pentose sugars. We begin with the microbial degradation of polysaccharides.

Hexose and Polysaccharide Utilization

Sugars with six carbon atoms, called **hexoses**, are the most important electron donors for many chemoorganotrophs and are also important structural components of microbial cell walls, capsules, slime layers, and storage products. The most common hexose sources in nature are listed in Table 17.9, from which it can be seen that most are polysaccharides, although a few are disaccharides. *Cellulose* and *starch* are two of the most important natural polysaccharides.

Although both starch and cellulose are composed of glucose units, they are bonded differently (Table 17.9, ∞ Figure 3.6), and this profoundly affects their properties. Cellulose is much more insoluble than starch and is usually less rapidly digested. Cellulose forms long fibrils, and organisms that digest cellulose are often found closely associated with them (Figure 17.61●). Many fungi are able to digest cellulose, and these are mainly responsible for decomposition of plant materials on the forest floor. Among bacteria, however, cellulose digestion is restricted to relatively few groups, of which the gliding bacteria such as *Sporocytophaga* and *Cytophaga* (Figures 17.61 and 17.62●; ∞ Figure 12.90), clostridia, and actinomycetes are among the most common.

Anoxic digestion of cellulose is carried out by a few *Clostridium* species, which are common in lake sediments, animal intestinal tracts, and systems for anoxic

18.7 Radioisotopes and Microelectrodes

In many situations, direct chemical measurements of microbial transformations are sufficient for assessing microbial activities in an environment. For example, the fate of lactate oxidation by sulfate-reducing bacteria in a sediment sample can easily be followed. Lactate is consumed and SO_4^{2-} is reduced to H_2S (Figure 18.16a●). All of these substances are easily measured chemically. However, when very high sensitivity is required, turnover rates need to be determined, or the fate of portions of a molecule needs to be followed, *radioisotopes* are very useful. For instance, if photoautotrophy is to be measured, the light-dependent uptake of $^{14}CO_2$ into microbial cells can be measured (Figure 18.16b). If sulfate reduction is of interest, the rate of conversion of $^{35}SO_4^{2-}$ to $H_2^{35}S$ can be assessed (Figure 18.16c). Chemoorganotrophic activity can easily be assessed by tracking the release of $^{14}CO_2$ from ^{14}C-labeled organic compounds (Figure 18.16d), and so on.

Isotope methods are widely used in microbial ecology. These must, however, employ proper controls, because some transformation of labeled compound might be due to a strictly chemical (rather than microbial) process. The *killed cell control* is key here. It is absolutely essential to know that the transformation being measured is prevented by chemical agents or by heat treatments that either block microbial action or actually kill the organisms. Formalin at a final concentration of 4% is commonly used as a chemical sterilant in microbial ecology studies. This level of formalin kills all cells, and transformations of radiolabeled materials that occur in its presence can be ascribed to nonliving processes (Figure 18.16d).

Radioisotopes in Combination with FISH: FISH–MAR

Radioisotopes can be used as measures of microbial activity in a microscopic technique called microautoradiography (MAR). In this method, cells, either of a pure culture or from a microbial community, are exposed to a radioisotope, such as an organic compound for chemoorganotrophs or CO_2 for an autotroph. Following exposure, cells are then affixed to a slide and the slide dipped in photographic emulsion. When left in darkness for a period, radioactive decay from the incorporated substrate exposes silver grains in the emulsion. These appear as black dots within and around the cells (Figure 18.17a●).

MAR can be combined with FISH (see Section 18.4) in **FISH-MAR**, a powerful technique for combining identification and activity measurements. FISH-MAR allows one to both identify which organisms in a natural sample are metabolizing a particular radiolabeled substance (by MAR) and to identify the organism(s) doing so (by FISH) (Figure 18.17b). FISH-MAR thus goes a major step beyond simple identification. The technique reveals valuable physiological information on the components as well. Such information is useful not only in understanding the activity of the microbial ecosystem, but also helps in guiding enrichment cultures. For example, knowledge of which organisms are metabolizing which substrates can be used for the design of specific enrichment protocols.

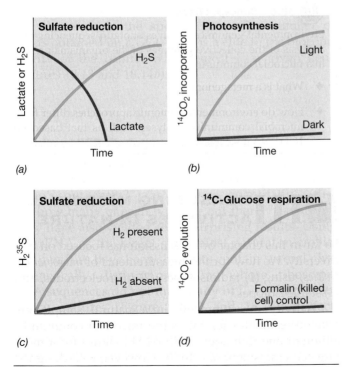

● **Figure 18.16 Microbial activity measurements.** (a) Chemical assay for lactate and H_2S during sulfate reduction. (b–d) Use of radioisotopes. (b) Photosynthesis measured in natural seawater with $^{14}CO_2$. (c) Sulfate reduction in mud measured with $^{35}SO_4^{2-}$. (d) Production of $^{14}CO_2$ from ^{14}C-glucose.

Microelectrodes

Microbial ecologists have used small glass electrodes, called **microelectrodes**, to study the activity of microorganisms in nature. The most useful microelectrodes have been those that measure pH, oxygen, N_2O, CO_2, H_2, or H_2S. As the name implies, microelectrodes are very small, the tips of the electrode ranging in diameter from 2 to 100 μm (Figure 18.18a●). The electrodes must be carefully inserted into the habitat in increments of a millimeter or less in order to follow microbial activities at the microenvironmental level (Figure 18.18b).

Microelectrodes have been used extensively in the study of microbial activities, including photosynthesis in microbial mats. **Microbial mats** are layered microbial communities typically containing cyanobacteria in the uppermost layers, anoxygenic phototrophic bacteria in subsequent layers (until the mat becomes light-limited)

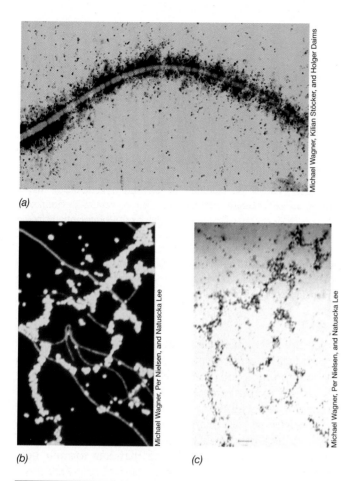

(a)

(b)   *(c)*

● **Figure 18.17 FISH-MAR:** Fluorescence *in situ* hybridization (FISH) combined with microautoradiography (MAR). (a) An uncultured filamentous cell belonging to the gamma Proteobacteria (by FISH analysis) and also an autotroph (by MAR-measured uptake of $^{14}CO_2$). Radioactive decay exposes the photographic emulsion, generating the black dots. (b) Uptake of ^{14}C-glucose by a mixed culture of *Escherichia coli* (yellow cells) and *Herpetosiphon aurantiacus* (filamentous, green cells). (c) MAR of the same field of cells shown in (b). Incorporated radioactivity exposes the film and shows that glucose was assimilated by the cells of *E. coli* but not by the filaments of *H. aurantiacus*.

and chemoorganotrophic, especially sulfate-reducing, bacteria in the lower layers (Figures 18.18*b* and 18.19●). Microbial mats develop in a variety of environments, especially in hot springs (Figure 18.19) and marine intertidal zones.

With the use of O_2 microelectrodes, oxygen concentrations in microbial mats (Figure 18.19*b*) or soil particles (∞ Figure 19.2) can be very accurately measured over extremely fine intervals. A *micromanipulator* is used to immerse electrodes gradually through a sample such that readings can be taken every 0.05–0.1 mm (Figures 18.18*b* and 18.19*b*). Using a bank of microelectrodes, each sensitive to a different chemical, simultaneous measurements of several microbial transformations can be made at the same time. For example, oxygen and sulfide measurements are often taken together because gradients of both form in many microbial environments as a result of photosynthesis and sulfate re-

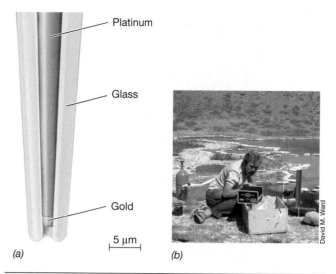

(a)   *(b)*

● **Figure 18.18 Microelectrodes.** (a) Schematic drawing of an oxygen microelectrode. The platinum rod functions as a cathode, and when voltage is applied, O_2 is reduced to H_2O, generating a current. The current resulting from the reduction of the O_2 at the gold surface of the cathode is proportional to the oxygen concentration in the sample. Note the scale of the electrode. (b) Photo of microelectrodes being used in a microbial mat (see Figure 18.19*a*).

duction, respectively (Figure 18.18*b*). Near the zone where O_2 and H_2S begin to mix, intense activity by phototrophic and chemolithotrophic sulfur bacteria (∞ Sections 12.2 and 12.4) may consume both of these substrates, at least to the limits of detection of the microelectrodes (Figure 18.19*b*).

Using microelectrodes as assay tools one can perturb the habitat while the electrodes are immersed, for instance by adding an organic compound or shading the mat surface, and measure what effects these perturbations have on the production or consumption of specific compounds in real time. From studies such as these, it is often possible to deduce the kinds of microbial interactions taking place in an environment. It is also possible to ask specific questions approachable by experiment. In this way, it may be possible to model the *in situ* metabolism of a microbial environment.

18.7 Concept Check

The activity of microorganisms in natural samples can be assessed very sensitively using radioisotopes and/or microelectrodes. In most cases, measurements are of the net activity of a microbial community rather than of a population of a single species.

◆ Why are radioisotopes so useful in measuring microbial activities?

◆ What does the acronym *MAR* stand for, and how are MAR reactions detected?

◆ Explain how a microelectrode works.

(a)

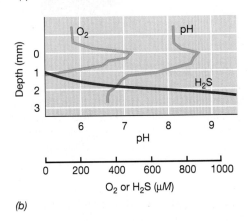

(b)

● **Figure 18.19 Microbial mats and the use of microelectrodes to study them.** (a) Photograph of a core taken through the kind of hot spring microbial mat used in the experiment shown in part (b). Upper layer (dark green) contains cyanobacteria (⌒⌒ Section 12.25), beneath which are several layers of anoxygenic phototrophic bacteria (orange), primarily *Chloroflexus* (⌒⌒ Section 12.35). The whole thickness of the mat is about 2 cm. (b) Oxygen, sulfide, and pH microprofiles in a hot spring microbial mat. Note the millimeter scale on the ordinate.

18.8 Stable Isotopes

For many elements several *isotopes* exist, differing in the number of neutrons present. Certain isotopes are unstable and break down as a result of radioactive decay. Others, called **stable isotopes**, are not radioactive but are metabolized differently by microorganisms. Stable isotopes can be used to study various microbial transformations in nature (Figure 18.20●).

The two elements that have proven most useful for stable isotope studies in microbial ecology are *carbon*, especially CO_2, and *sulfur*, especially SO_4^{2-} and H_2S. Carbon exists in nature primarily as ^{12}C, but a small amount (about 5%) exists as ^{13}C. Likewise, sulfur exists primarily as ^{32}S, although some sulfur is found as ^{34}S. The natural abundance of these isotopes in a given compound changes when C or S is metabolized by organisms because biochemical reactions

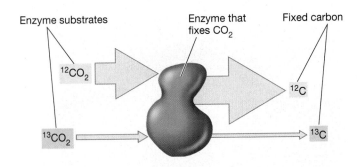

● **Figure 18.20 Mechanism of isotopic fractionation using carbon as an example.** Although the ratio of natural abundance of $^{12}CO_2$ to $^{13}CO_2$ is about 19:1, enzymes that fix CO_2 preferentially fix the lighter isotope (^{12}C). This results in fixed carbon being enriched in ^{12}C and depleted in ^{13}C relative to the starting substrate. The degree of ^{13}C depletion is calculated as an isotopic fractionation (see legend to Figure 18.21 for calculation). The size of the arrows indicates the relative abundance of each isotope of carbon.

favor the *lighter* isotope. That is, the heavier isotope is *discriminated against* relative to the lighter isotope when the compounds are metabolized by enzymes (Figure 18.20). For example, when CO_2 is fed to a phototrophic organism, the cellular carbon becomes *enriched* in ^{12}C and *depleted* in ^{13}C, relative to a carbon standard. Likewise, sulfide produced from the bacterial reduction of sulfate is "lighter" than sulfide that has formed geochemically. This process, known as **isotopic fractionation** (Figure 18.20), is a uniquely biological phenomenon and can be used as an assay for whether a particular transformation is microbiological or not.

Use of Isotopic Fractionation in Microbial Ecology

The isotopic composition of a sample contains a record of its past biological activity. In the case of carbon (Figure 18.21●), plant material and petroleum (which is derived from plant material) have similar isotopic compositions. Carbon from both sources is isotopically lighter than a nonbiological standard because it was fixed as CO_2 by a pathway that discriminated against $^{13}CO_2$ (Figure 18.21). Moreover, methane of microbiological origin is extremely light, indicating that methanogenic *Archaea* (⌒⌒ Section 13.4) discriminate against ^{13}C in a major way. By contrast, marine carbonates are clearly of geological origin (Figure 18.21). Because of the differences in the proportion of ^{12}C and ^{13}C in carbon of biological and geological origin, the $^{13}C/^{12}C$ isotopic ratio of geological strata has been used to detect the onset of living processes in ancient rocks. Interestingly, organic carbon in rocks as old as 3.5 billion years shows some fractionation (Figure 18.21), supporting the idea that life existed at this time.

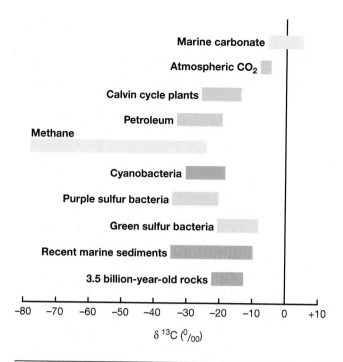

● **Figure 18.21 Carbon isotopic compositions of various substances.** The values are given in parts per thousand (‰) and were calculated using the formula:

$$\frac{(^{13}C/^{12}C \text{ sample}) - (^{13}C/^{12}C \text{ standard})}{(^{13}C/^{12}C \text{ standard})} \times 1000$$

The standard is a belemnite sample from the PeeDee rock formation. Note that carbon fixed by autotrophic organisms is *enriched* in ^{12}C and *depleted* in ^{13}C.

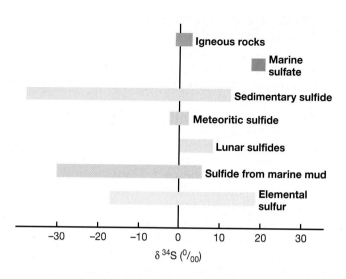

● **Figure 18.22 Summary of the isotope geochemistry of sulfur, indicating the range of values for ^{34}S and ^{32}S in various sulfur-containing substances.** The values are given in parts per thousand ‰ and were calculated using the formula:

$$\frac{(^{34}S/^{32}S \text{ sample}) - (^{34}S/^{32}S \text{ standard})}{(^{34}S/^{32}S \text{ standard})} \times 1000$$

The standard is an iron sulfide mineral from the Canyon Diablo meteorite. Note that sulfide and sulfur of *biogenic* origin tend to be *depleted* in ^{34}S (enriched in ^{32}S).

present for only the past 750 million years or so (Section 11.3 and Figure 11.8).

Epilogue

With an understanding of some of the techniques in the microbial ecologist's toolbox, we move on to Chapter 19 to explore the discipline of microbial ecology itself. In that chapter we will consider the composition of microbial communities and their major activities in nature. In many cases we will use one or more of the tools described in this chapter to help illustrate important principles and to identify the most active microbial species in highly diverse microbial communities.

The activities of sulfate-reducing bacteria in nature are easy to recognize from the fractionation of $^{34}S/^{32}S$ (Figure 18.22●). As compared with a sulfide standard, sedimentary sulfide is highly enriched in ^{32}S, while non-biogenic sulfide (as, for instance, in igneous rocks from volcanic deposits) is significantly heavier (Figure 18.22). Fractionation allows one to monitor these important links in the sulfur cycle, and it has been widely used to trace the activities of sulfur-cycling prokaryotes through geological strata. Sulfur isotope analyses have also been used as evidence for the lack of life on the Moon. For example, the data in Figure 18.22 show that the isotopic composition of sulfides in lunar rocks closely approximates that of the sulfide standard and not that of biogenic sulfide.

Oxygen isotopic analyses ($^{18}O/^{16}O$) have been used to trace Earth's transition from an anoxic to an oxic environment, since Earth's molecular oxygen originated from oxygenic photosynthesis by cyanobacteria (Section 11.1). Using this tool it has been possible to show that Earth was oxygenated only very gradually and that current oxygen levels have been

18.8 Concept Check

Isotopic fractionation can reveal the biological origin of various substances. Fractionation is a result of the activity of enzymes that discriminate against the heavier form of an element when binding their substrates.

◆ How can the $^{12}C/^{13}C$ composition of a substance reveal its possible biological origin?

◆ What is the simplest explanation for why lunar sulfides are isotopically heavy?

REVIEW QUESTIONS

1. What is the basis of the *enrichment culture technique*? Why is an enrichment medium usually suitable for the enrichment of only a certain group or groups of bacteria (⌀ Section 18.1)?

2. What is the principle of the Winogradsky column and what types of organisms does it serve to enrich? How might a Winogradsky column be used to enrich organisms present in an extreme environment, like a hot spring microbial mat (⌀ Section 18.1)?

3. Describe the principle of MPN for enumerating bacteria from a natural sample (⌀ Section 18.2).

4. Why would the laser tweezers be a superior method to dilution and liquid enrichment for obtaining an organism present in a sample in *low* numbers (⌀ Section 18.2)?

5. Compare and contrast the use of fluorescent dyes and fluorescent antibodies for use in enumerating bacteria in natural environments. What advantages and limitations do each of these methods have (⌀ Section 18.3)?

6. Can nucleic acid probes in microbial ecology be as sensitive as culturing methods? What advantages do nucleic acid methods have over culture methods? What disadvantages (⌀ Section 18.4)?

7. How does *chromosomal painting* work? Why might it be a good method for enumerating microorganisms in a habitat capable of carrying out some specific metabolic process (⌀ Section 18.4)?

8. What is the *green fluorescent protein*? In what ways does a green fluorescing cell differ from a cell fluorescing from, for example, staining with a phylogenetic strain (⌀ Section 18.4)?

9. How can a phylogenetic picture of a microbial community be obtained without culturing the inhabitants (⌀ Section 18.5)?

10. After PCR amplification of total community DNA using a specific primer set, why is it usually necessary to either clone or run DGGE on the products before sequencing them (⌀ Section 18.5)?

11. Give an example of how environmental genomics has discovered a known metabolism in a new organism (⌀ Section 18.6).

12. What are the major advantages of *radioisotopic methods* in the study of microbial ecology? What type of controls (discuss at least two) would you include in a radioisotopic experiment to show $^{14}CO_2$ incorporation by phototrophic bacteria or to show $^{35}SO_4^{2-}$ reduction by sulfate-reducing bacteria (⌀ Section 18.7)?

13. What can FISH-MAR tell you that FISH alone cannot (⌀ Section 18.7)?

14. Will autotrophic organisms contain *more* or *less* ^{12}C than the CO_2 that feeds them (⌀ Section 18.8)?

APPLICATION QUESTIONS

1. Why is the enrichment for nitrifying bacteria as described in Table 18.1? Why are each of the resources and conditions necessary? (You might want to review material in Sections 12.3 and 17.12 before answering.)

2. Design an experiment for measuring the activity of sulfur-oxidizing bacteria in soil. If only certain species of the sulfur-oxidizers present were metabolically active, how could you tell this? How would you prove that your activity measurement was due to biological activity?

3. You wish to know whether *Archaea* exist in a lake water sample but are unsuccessful in culturing any. Using techniques described in this chapter, how could you determine whether *Archaea* existed in the sample and if they did, what proportion of the cells in the lake water were *Archaea*?

4. You wish to identify whether organisms capable of autotrophic growth using the Calvin cycle exist in various soil samples (⌀ Section 17.6). This pathway requires a unique enzyme, *ribulose bisphosphate carboxylase* (RubisCO). Describe two ways you could do this, one using a microscopic method and another using a method that does not involve either culturing or microscopic methods.

5. Design an experiment to solve the following problem. Determine the rate of methanogenesis ($CO_2 + 4 H_2 \rightarrow CH_4 + H_2O$) in anoxic lake sediments and whether or not it is H_2-limited. Also, determine the morphology of the dominant methanogen (recall that these are *Archaea*, ⌀ Section 13.4). Finally, calculate what percentage the dominant methanogen is of the *total archaeal* and *total prokaryotic* populations in the sediments.

19

MICROBIAL ECOLOGY

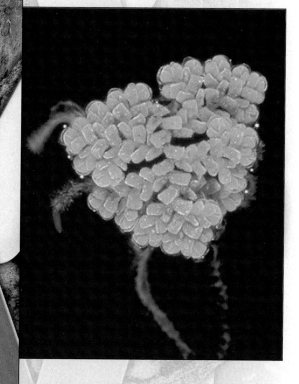

Microorganisms interact in nature with other microorganisms and with plants and animals. A beneficial association is seen here between the water fern *Azolla* and nitrogen-fixing symbiotic cyanobacteria of the genus *Anabaena*, cells of which live in cavities within the leaves of the plant.

WORKING GLOSSARY

Acid mine drainage acidic water containing H_2SO_4 derived from the microbial oxidation of iron sulfide minerals

Anoxic an oxygen-free environment, usually also highly reducing (low E_0')

Bacteroid morphologically misshapen *Rhizobium* cells inside a leguminous plant root nodule; can fix N_2

Barophilic an organism that grows best when placed under a pressure greater than 1 atm

Barotolerant an organism that can grow under elevated pressures but that grows best at atmospheric pressure

Biochemical oxygen demand (BOD) the microbial oxygen-consuming properties of a water sample

Biofilm colonies of microbial cells encased in a porous slime and attached to a surface

Biogeochemistry study of biologically mediated chemical transformations

Bioremediation the cleanup of oil, toxic chemicals, and other pollutants by microorganisms

Black smoker an extremely hot (250–350° C) deep-sea hot spring emitting both hot water and various minerals

Cometabolism metabolism of a compound in the presence of a second organic compound, which is used as the primary energy source

Denitrification the microbial process by which NO_3^- is reduced to gaseous nitrogenous compounds

Ecosystem a community of organisms and their natural environment

Guild a population of metabolically related microorganisms

Hydrothermal vent a deep-sea warm or hot spring

Infection thread in the formation of root nodules, a cellulosic tube through which *Rhizobium* cells can travel to reach and infect root cells

Leghemoglobin an O_2-binding protein found in root nodules

Lichen a fungus and an alga (or cyanobacterium) living in symbiotic association

Microbial leaching the removal of valuable metals such as copper from sulfide ores by microbial activities

Microenvironment the immediate environmental surroundings of a microbial cell or group of cells

Mycorrhizae a symbiotic association between a fungus and the roots of a plant

Nitrification the process by which NH_3 is oxidized to NO_3^-

Nitrogen fixation the microbiological reduction of N_2 to NH_3

Oxic an oxygen-containing environment, typically with a high E_0'

Primary producer an organism that uses light to synthesize new organic material from CO_2

Proteorhodopsin a light-sensitive protein present in some open ocean *Bacteria* that catalyzes ATP formation

Pyrite a common iron-containing ore, FeS_2

Reductive dechlorination removal of Cl as Cl^- from an organic compound by reducing the carbon atom from C—Cl to C—H

Rhizosphere the region immediately adjacent to plant roots

Root nodule a tumorlike growth on plant roots that contains symbiotic nitrogen-fixing bacteria

Rumen the first vessel in the multichambered stomach of ruminant animals in which cellulose digestion occurs

Syntrophy a metabolic process in which two or more microorganisms cooperate to carry out a process that could not be carried out by a single organism

Ti plasmid a conjugative plasmid present in the bacterium *Agrobacterium tumefaciens* that can transfer genes into plants

Volatile fatty acids (VFAs) the major fatty acids (acetate, propionate, and butyrate) produced during the rumen fermentation

Xenobiotic a totally synthetic product not naturally occurring in nature

---

Microorganisms are not alone in nature. Each microorganism in an ecosystem interacts with both its surroundings and with other organisms. These interactions can result in significant chemical and physical changes in the environment that may be beneficial or detrimental to other organisms. Whether harmful, beneficial, or neutral, the collective activities of microorganisms to a great extent control the workings of the biosphere.

In this chapter we will examine these key microbial activities. We will look first at microbial ecosystems in a general way and then explore some different environments for microorganisms—soil, freshwater, and the oceans. We will then examine some key reactions in microbial ecology, focusing on the nutrient cycles that help

sustain plant life on Earth. We finish up with coverage of some of the special relationships microorganisms have with plants and animals.

MICROBIAL ECOSYSTEMS

19.1   Populations, Guilds, and Communities

In nature, individual microbial cells grow to form **populations**. Metabolically related populations are called **guilds**, and sets of guilds interact in **microbial communities** (Figure 19.1●). Microbial communities in turn interact with communities of macroorganisms and the environment to define the entire **ecosystem**.

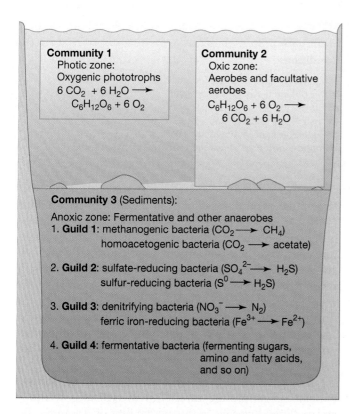

Community 1
Photic zone:
Oxygenic phototrophs
$6 CO_2 + 6 H_2O \longrightarrow$
$C_6H_{12}O_6 + 6 O_2$

Community 2
Oxic zone:
Aerobes and facultative
aerobes
$C_6H_{12}O_6 + 6 O_2 \longrightarrow$
$6 CO_2 + 6 H_2O$

Community 3 (Sediments):

Anoxic zone: Fermentative and other anaerobes
1. **Guild 1**: methanogenic bacteria ($CO_2 \longrightarrow CH_4$)
    homoacetogenic bacteria ($CO_2 \longrightarrow$ acetate)

2. **Guild 2**: sulfate-reducing bacteria ($SO_4^{2-} \longrightarrow H_2S$)
    sulfur-reducing bacteria ($S^0 \longrightarrow H_2S$)

3. **Guild 3**: denitrifying bacteria ($NO_3^- \longrightarrow N_2$)
    ferric iron-reducing bacteria ($Fe^{3+} \longrightarrow Fe^{2+}$)

4. **Guild 4**: fermentative bacteria (fermenting sugars,
    amino and fatty acids,
    and so on)

● **Figure 19.1 Populations, guilds, and communities—an example of microbial community structure in a lake ecosystem.** Microbial communities consist of populations of cells of various species. In a lake ecosystem various communities would exist as shown here. For the sediment community, major guilds are indicated. The reduction of CO_2, SO_4^{2-}, S^0, NO_3^-, and Fe^{3+} are examples of anaerobic respirations, discussed in Sections 17.13–17.18.

Energy enters ecosystems in the form of sunlight, organic carbon, and reduced inorganic substances. Light is used by phototrophic organisms (◯ Sections 17.1–17.5) to synthesize new organic matter (Figure 19.1●) containing not only carbon but also nitrogen, sulfur, phosphorus, iron, and other elements. This newly synthesized organic material, along with organic matter that enters the ecosystem from the outside (called *allochthonous* organic matter) drives the metabolic activities of chemoorganotrophic organisms. Chemolithotrophs, by contrast, obtain their energy from inorganic electron donors, such as H_2, Fe^{2+}, S^0, or NH_3 (◯ Chapter 17).

Biogeochemical Cycles

Microorganisms play an important role in the recycling of several elements in the environment, in particular carbon, sulfur, nitrogen, and iron. The study of these chemical transformations is called **biogeochemistry**. A **biogeochemical cycle** describes the transformations catalyzed by both biological and chemical agents during the cycling of these key elements of living systems. Often these cycles involve oxidation–reduction reactions (◯ Section 5.6) as the element moves through the ecosystem (Figure 19.1). Sulfur, for example, in the

form of hydrogen sulfide (H_2S), can be oxidized by a variety of microorganisms, both phototrophic and chemotrophic, to sulfur (S^0) and sulfate (SO_4^{2-}). Sulfate is a key nutrient for plants. Sulfate, in turn, can be reduced to sulfide by activities of the sulfate-reducing bacteria. This closes the biogeochemical cycle for sulfur by regenerating sulfide (see Figure 19.29).

Microorganisms of various types are intimately involved in biogeochemical cycling and in many instances are the only biological agents capable of regenerating forms of the elements needed by other organisms, particularly plants. We will discuss many biogeochemical cycles in this chapter, including those of carbon, nitrogen, sulfur, and iron.

 19.1 Concept Check

Microbial communities consist of guilds of metabolically related organisms. Microorganisms play major roles in energy transformations and biogeochemical processes that result in the recycling of elements essential to living systems.

◆ How does a microbial *guild* differ from a microbial *community*?

◆ What is a *biogeochemical cycle*?

19.2 Environments and Microenvironments

Microorganisms inhabit every conceivable habitat that will sustain life. Microbial habitats are exceedingly diverse and often rapidly changing. Besides the common habitats of soil and water, microorganisms inhabit the surfaces of higher organisms and in many cases actually live *within* them. Microorganisms frequently reach large numbers in such habitats and may benefit the plant or animal in a nutritionally significant way.

The Microorganism and the Microenvironment

The growth of microorganisms in nature, as in laboratory culture, depends on the *resources* (nutrients) available and on the *growth conditions* (Table 19.1). Differences in the type and quantity of resources and the physiochemical conditions of a habitat define the **niche** for each particular microorganism. Ecological theory states that for every organism there exists at least one niche, the *prime niche*, in which that organism is most successful. The organism dominates the prime niche but may also inhabit other niches; in these it is less ecologically successful than in its prime niche. Countless microbial niches exist on Earth and are in part responsible for evolution of the great metabolic diversity (◯ Chapter 17) and biodiversity (◯ Chapters 12–14) of microorganisms we see today.

Because microorganisms are so small, their habitats are also small. Microbial ecologists use the term **microenvironment** to describe where a microorganism actually lives

Table 19.1	**Major resources and conditions that govern microbial growth in nature**
Resources	**Conditions**
Carbon (organic, CO_2)	Temperature: cold → warm → hot
Nitrogen (organic, inorganic)	Water potential: dry → moist → wet
Other macronutrients (S, P, K, Mg)	pH: 0 → 7 → 14
Micronutrients (Fe, Mn, Co, Cu, Zn, Mn, Ni)	O_2: oxic → microoxic → anoxic
O_2 and other electron acceptors (NO_3^-, SO_4^{2-}, Fe^{3+}, etc.)	Light: bright light → dim light → dark
Inorganic electron donors (H_2, H_2S, Fe^{2+}, NH_4^+, NO_2^-, etc.)	Osmotic conditions: freshwater → marine → hypersaline

and metabolizes within its habitat. Microbiologists must thus learn to "think small" when considering the microorganism in its environment. For example, for a typical 3-μm rod-shaped bacterium, a distance of 3 mm is huge. It is equivalent to that which a human experiences over a distance of 2 km! Across the 3-mm distance, chemical and physical gradients may exist that can greatly affect the activities of the organism.

In a 3-mm particle of soil, for example, several different microenvironments could exist, differing chemically and physically in many ways. Visualize, for example, the distribution of an important microbial nutrient like oxygen in a soil particle. Microelectrodes (∞ Section 18.7) can measure oxygen concentrations throughout small soil particles. As shown in data from an actual experiment (Figure 19.2●), soil particles are not homogeneous in terms of their oxygen content. The outer zones of a small soil particle may be fully **oxic**, meaning that O_2 is present, whereas the center, only a very short distance away, can remain completely **anoxic** (O_2-free) (Figure 19.2).

We thus see that different niches can exist across a very small spatial dimension, and this is why various physiological types of microorganisms can coexist in a single "soil particle," such as that shown in Figure 19.2. Anaerobic organisms can be active near the center of the particle, microaerophiles (aerobes that require very low oxygen levels) active further out, and obligately aerobic organisms can metabolize in the outermost and fully oxic region of the particle. Facultatively aerobic bacteria can be distributed throughout the particle (∞ Section 6.15).

Physiochemical conditions in a microenvironment can change rapidly, both temporally and spatially. For example, the oxygen concentrations shown in Figure 19.2 represent "instantaneous" measurements. Measurements taken following a period of microbial respiration or after an increase in soil water content could show a drastically different gradient of oxygen across the microenvironment. It can thus be concluded that microenvironments are *heterogeneous* and that conditions in a given microenvironment can change rapidly. In this way microenvironments promote high microbial diversity in a relatively small physical space.

Nutrient Levels and Growth Rates

Nutrients (or in ecological terms, *resources*, see Table 19.1) typically enter an ecosystem in intermittent fashion. A

large pulse of nutrients—for example, an input of leaf litter or the carcass of a dead fish or animal—may be followed by a period of nutrient deprivation. Microorganisms in nature often face "feast-or-famine," and thus many microorganisms produce *storage polymers* as reserve materials when resources permit. Reserve polymers store excess nutrients present under favorable growth conditions for use during periods of nutrient deprivation. Examples of storage materials are poly-β-hydroxybutyrate and other alkanoates, polysaccharides, polyphosphate, and so on (∞ Section 4.11).

Extended periods of exponential growth of microorganisms in nature are rare. Growth typically occurs in spurts, linked closely to the availability of resources. Because physiochemical conditions in nature are rarely optimal all at the same time, growth rates of microorganisms in nature are thus well below the maximum growth rates recorded in the laboratory most of the time. For instance,

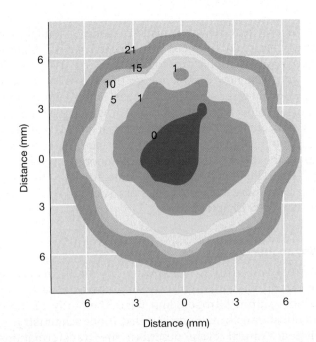

● **Figure 19.2 Contour map of O_2 concentrations in a soil particle.** The axes show the dimensions of the particle. The numbers on the contours are O_2 concentrations (in percentage, air is 21% O_2). In terms of oxygen relationships for microorganisms, each zone can be considered a different microenvironment.

the generation time of *Escherichia coli* in the intestinal tract is about 12 hours (two doublings per day), whereas in pure culture it grows much faster, with a minimum generation time of 20 minutes under the best of conditions.

Moreover, estimates have shown that typical soil bacteria grow in nature at less than 1% of the maximal growth rate measured in the laboratory. On average, these slow growth rates reflect the fact that (1) nutrients (resources, Table 19.1) are frequently in low supply and/or growth conditions suboptimal; (2) the distribution of nutrients throughout the microbial habitat is not uniform; and (3) except for rare instances, microorganisms grow in mixed populations rather than pure culture in natural environments. Thus an organism that grows rapidly in pure culture may do poorly in a natural environment where it must compete with other organisms better suited to the resources and growth conditions available.

Microbial Competition and Cooperation

Competition among microorganisms for available resources may be intense, with the outcome dependent on several factors, including rates of nutrient uptake, inherent metabolic rates and, ultimately, growth rates. A typical habitat will contain a mixture of different microorganisms (Figure 19.1), with the density of each population dependent on how closely the habitat resembles its prime niche.

Instead of directly competing for the same resources, some microorganisms work together to carry out transformations that neither could accomplish alone. These types of microbial partnerships, called **syntrophy** (∞ Section 17.21), are crucial to the success of certain anaerobic bacteria, as will be described in Section 19.10. Metabolic cooperation can also be seen in the activities of organisms that carry out *complementary* metabolisms. For example, in Chapter 17 we discussed metabolic transformations that involved two distinct groups of organisms, such as those of the *nitrosifying* and the *nitrifying* bacteria. Together, these organisms can oxidize NH_3 to NO_3^-, although neither group is capable of doing this alone (∞ Section 17.12). Because the product of the nitrosifying bacteria (NO_2^-) is the substrate for the nitrifying bacteria, such organisms typically live in nature in tight association (∞ Figure 18.12*a*).

19.2 Concept Check

Microorganisms are very small and their habitats are likewise small. The microenvironment is the place in which the microorganism actually lives. Microorganisms in nature often live a feast-or-famine existence such that only the best-adapted species thrive in a given niche. Cooperation among microorganisms is also important in many microbial interrelationships.

◆ What aspects define the niche of a particular microorganism?

◆ Why can many different physiological groups of organisms live in a single habitat?

19.3 Microbial Growth on Surfaces and Biofilms

Surfaces are important microbial habitats because nutrients can adsorb to them. In the microenvironment of a surface, nutrient levels may be much higher than they are in the bulk solution. As a consequence, microbial numbers and activities are usually much greater on surfaces than in water.

Microscope slides can be used as experimental surfaces to which organisms can attach and grow. Typically, a slide is immersed in a microbial habitat, left for a period of time, and then retrieved and examined by microscopy (Figure 19.3*a*●). Microcolonies readily develop on such surfaces, much as they do on natural surfaces in nature. In fact, periodic microscopic examination of immersed microscope slides has been used to measure growth rates of attached organisms in nature (∞ Figure 6.20*b*).

A surface may itself also be a nutrient, such as a particle of organic matter, where attached microorganisms catabolize nutrients directly from the surface of the particle. Plant materials, for example, are rapidly colonized by microorganisms in soil, and simple staining techniques can detect microbial populations attached to the solid surface (Figure 19.3*b*).

Biofilms

Oftentimes microorganisms grow on surfaces enclosed in **biofilms**—assemblages of bacterial cells attached to a surface and enclosed in adhesive polysaccharides excreted by the cells (Figure 19.4●). Biofilms trap nutrients for

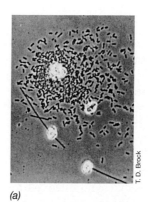
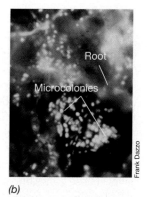

(a)   *(b)*

● **Figure 19.3 Microorganisms on surfaces.** (a) Bacterial microcolonies developing on a microscope slide immersed in a small river. The bright particles are mineral matter. The short, rod-shaped cells are about 3 μm long. (b) Fluorescence photomicrograph of a natural microbial community colonizing plant roots in soil. Note microcolony development. The preparation has been stained with acridine orange.

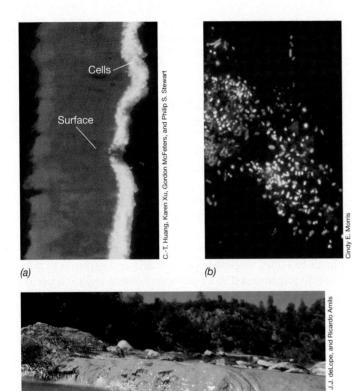

(a) *(b)*

(c)

● **Figure 19.4   Microbial biofilms.** (a) A cross-sectional view of an experimental biofilm made up of cells of *Pseudomonas aeruginosa*. The yellow–green layer (about 15 μm in depth) contains cells and is stained by an enzyme activity stain for the enzyme alkaline phosphatase. The red area is the surface. (b) Confocal laser scanning microscopy (∞ Section 4.2) of a natural biofilm (top view) that developed on a leaf surface. The color of the cells indicates their depth in the biofilm: red, cells on the surface; green, 9 μm depth; blue, 18 μm depth. (c) A biofilm of iron-oxidizing prokaryotes on the surface of rocks in the iron-rich Rio Tinto, Spain. As Fe^{2+} rich water passes over and through the biofilm, iron-oxidizing microorganisms convert Fe^{2+} to Fe^{3+} to obtain energy (∞ Section 17.11).

growth of the microbial population and help prevent detachment of cells on surfaces present in flowing systems (Figure 19.5●). Biofilms typically contain many porous layers, and microscopic examination of the microorganisms in each layer can be done using scanning laser confocal microscopy (∞ Section 4.2) (Figure 19.4).

Cell-to-cell communication is critical in the development and maintenance of a biofilm. Attachment of a cell to a surface is a signal for the expression of *biofilm-specific genes*. These genes encode proteins that synthesize cell-to-cell signaling molecules and begin polysaccharide formation (Figure 19.5a). In *Pseudomonas aeruginosa*, a notorious biofilm former (Figure 19.4a), the major signaling molecules are compounds called *homoserine lactones*. As these molecules accumulate, they function as chemotactic agents to recruit nearby *P. aeruginosa* cells (a mechanism called *quorum sensing*, ∞ Section 8.10), and the biofilm

develops (Figure 19.4a). *P. aeruginosa* has been implicated in the disease *cystic fibrosis*, where a tenacious biofilm forms in the lungs, leading to symptoms of pneumonia (∞ Sections 8.10 and 31.7).

Why Do Bacteria Form Biofilms?

Bacteria form biofilms because the biofilm mode of growth undoubtedly improves survival and growth of the organisms. At least four reasons underlie the formation of biofilms. First, biofilms are a type of *defense*. Biofilms resist physical forces that could sweep unattached cells away, phagocytosis by cells of the immune system, or the penetration of toxic molecules, such as antibiotics. Second, biofilm formation *allows cells to remain in a favorable niche*. Biofilms attached to nutrient-rich surfaces, such as animal tissues, or to surfaces in flowing systems, such as a rock in a stream (Figure 19.4c), fix bacterial cells in a location where nutrients are often more

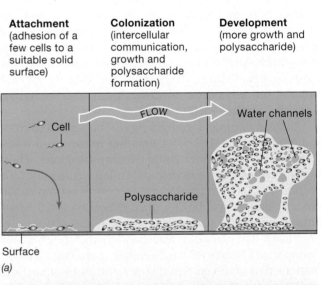

Attachment (adhesion of a few cells to a suitable solid surface)

Colonization (intercellular communication, growth and polysaccharide formation)

Development (more growth and polysaccharide)

(a)

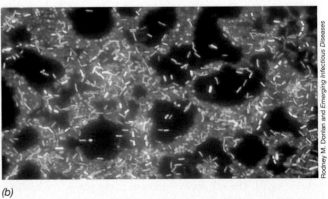

(b)

● **Figure 19.5   Biofilms.** (a) Formation of a bacterial biofilm. The biofilm begins with attachment of a few cells, after which growth and intercellular communication occurs. Polysaccharide formation occurs and becomes more extensive as the biofilm grows. (b) Photomicrograph of a DAPI-stained (∞ Section 18.3) biofilm that developed on a stainless steel pipe. Note the water channels.

abundant or are constantly replenished. Third, biofilms form because they *allow bacterial cells to live in close association* with each other. As we have already seen in the case of *Pseudomonas aeruginosa*, this facilitates intercellular communication via sensory molecules, which may ultimately benefit the cells. Moreover, when cells are in close proximity, opportunities for genetic exchange improve. And fourth, biofilms may be the *typical way bacterial cells grow in nature*, where nutrient concentrations are not nearly what they are in culture media. In other words, the biofilm may be the "default" mode of growth for prokaryotes in nature versus the more artificial rich medium liquid culture approach taken in the laboratory.

Biofilm Control

Biofilms have significant implications in human medicine and commerce. In the body, bacterial cells within a biofilm are protected from attack by the immune system, and antibiotics and other antimicrobial agents often fail to pierce the biofilm. Biofilms have been implicated in a variety of medical and dental conditions besides cystic fibrosis, including periodontal disease, kidney stones, tuberculosis, Legionnaire's disease, and *Staphylococcus* infections. Medical implants are excellent surfaces for biofilm development. These include both short-term devices, such as a urinary catheter, as well as long-term implants, such as artificial joints. Estimates are that as many as 10 million people a year in the United States alone experience biofilm infections from implants or intrusive medical procedures. Biofilms explain why routine oral hygiene is so important. Dental plaque is a typical biofilm and contains acid-producing bacteria responsible for dental caries (∞ Section 21.3).

In industrial situations biofilms can slow the flow of water, oil, or other liquids through pipelines and can accelerate corrosion of the pipes themselves. Biofilms also initiate degradation of submerged objects, such as structural components of offshore oil rigs, boats, and shoreline installations. The safety of drinking water may be compromised by biofilms that develop in water distribution pipes, many of which in the United States are nearly 100 years old. Although water pipe biofilms mostly contain harmless bacteria, if pathogens successfully colonize the biofilm, standard chlorination practices may be insufficient to kill them. Periodic releases of cells can then lead to outbreaks of disease. There is some concern that *Vibrio cholerae*, the causative agent of cholera (∞ Section 28.5), may be propagated in this manner.

Biofilm control is a big business and thus far only a limited repertoire of tools is available. Collectively, industries spend billions of dollars each year treating pipes and other surfaces to keep them free of biofilms. New antibiotics that can penetrate biofilms and other drugs that prevent biofilm formation by interfering with intercellular communication are being developed. A class of chemicals called *furanones*, for example, has

shown promise as biofilm preventatives in tests on abiotic surfaces. Because furanones are stable and nontoxic in humans, they may also have applications as antibiofilm agents in human medicine.

 19.3 Concept Check

Biofilms are bacterial assemblages encased in slime that form on surfaces. Biofilms can lead to the destruction of inert as well as living surfaces from the excretory products of the bacterial cells. Biofilm formation is a complex process involving cell-to-cell communication.

◆ What is the chemical nature of the biofilm matrix?

◆ Why might a biofilm be a good habitat for bacterial cells living in a flowing system?

◆ Give an example of a medically relevant biofilm that most humans likely harbor.

SOIL AND FRESHWATER MICROBIAL HABITATS

Among the major habitats of microorganisms are soils and freshwater, the latter including lakes, ponds, and streams. Soils and waters vary in their physical structure, nutrient composition, temperature, and water potential. All of these factors combine to influence the diversity and numbers of microorganisms present.

19.4 Terrestrial Environments

The term *soil* refers to the loose outer material of Earth's surface, a layer distinct from bedrock that lies underneath (Figure 19.6●). Soil development involves complex interactions among the parent material (rock, sand, glacial drift, and so on), the topography, climate, and living organisms. Soils can be divided into two broad groups— **mineral soils** and **organic soils**—depending on whether they derive initially from the weathering of rock and other inorganic material or from sedimentation in bogs and marshes, respectively. Our discussion will concentrate on *mineral soils*, the predominant soil in most terrestrial environments.

Soil Formation

Soil forms as a result of combined physical, chemical, and biological processes. An examination of almost any exposed rock reveals the presence of algae, lichens (see Section 19.20), or mosses. These organisms are dormant or nearly so on dry rock and then grow when moisture is present. They are phototrophic and produce organic matter, which supports the growth of chemoorganotrophic bacteria and fungi.

O horizon
Layer of undecomposed plant materials

A horizon
Surface soil (high in organic matter, dark in color, is tilled for agriculture; plants and large numbers of microorganisms grow here; microbial activity high)

B horizon
Subsoil (minerals, humus, and so on, leached from soil surface accumulate here; little organic matter; microbial activity detectable but lower than at A horizon)

C horizon
Soil base (develops directly from underlying bedrock; microbial activity generally very low)

(a)

(b)

Michael T. Madigan

● **Figure 19.6   Soil.** (a) Profile of a mature soil. The soil horizons are soil zones as defined by soil scientists. (b) Photo of a soil profile showing O, A, and B horizons. This soil from Carbondale, Illinois, is rich in clay, and thus contains very small particles and is very compact. Such soils are not as well drained as those containing sand as a major component.

The numbers of chemoorganotrophs increase directly with the degree of plant cover. Carbon dioxide produced during respiration by chemoorganotrophs is converted to carbonic acid ($CO_2 + H_2O \rightleftharpoons H_2CO_3$), which is important in the dissolution of rocks, especially limestone. Many chemoorganotrophs also excrete organic acids, which further promote the dissolution of rock into smaller particles.

Freezing, thawing, and other physical processes lead to cracks in the rocks. A raw soil forms in these crevices, and pioneering plants can develop there. The plant roots penetrate farther into crevices and increase the fragmentation of the rock, and their excretions promote the development of a microflora in the **rhizosphere** (the soil that surrounds plant roots, see Figure 19.3*b*). When the plants die, their remains are added to the soil and become nutrients for an even more extensive microbial development. Minerals are further rendered soluble, and as water percolates, it carries some of these chemical substances deeper.

As weathering proceeds, the soil increases in depth, thus permitting the development of larger plants and trees. Soil animals become established and play an important role in keeping the upper layers of the soil mixed and aerated. Eventually, the movement of materials downward results in the formation of layers, called a *soil profile* (Figure 19.6). The rate of development of a typical soil profile depends on climatic and other factors, but it typically takes hundreds of years.

Soil as a Microbial Habitat

The most extensive microbial growth takes place on the *surfaces* of soil particles, usually within the rhizosphere (see Figures 19.2, 19.3*b*, and 19.6*b*). As pointed out in Section 19.2, even a small soil aggregate can have many different microenvironments (compare Figures 19.2 and 19.7●), and thus, several different types of microorganisms may be present. To examine soil particles directly for microorganisms, fluorescence microscopes are often used, the organisms in the soil being stained with a dye that fluoresces (see Figure 19.3*b*). To observe a *specific* microorganism in a soil particle, fluorescent-antibody staining (∞ Figure 18.8) or phylogenetic stains (∞ Section 18.4) can be used. Microorganisms can also be seen on soil surfaces by means of the scanning electron microscope (Figure 19.8●). The scanning electron microscope gives excellent information on the morphology of

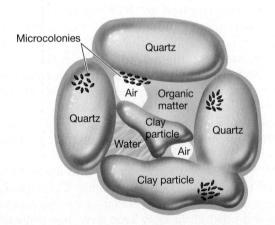

Microcolonies

Quartz
Air
Organic matter
Quartz
Clay particle
Quartz
Water
Air
Clay particle

● **Figure 19.7   A soil aggregate composed of mineral and organic components, showing the localization of soil microorganisms.** Very few microorganisms are found free in the soil solution; most of them occur in microcolonies attached to the soil particles.

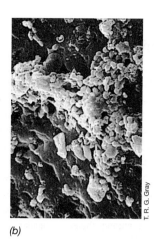
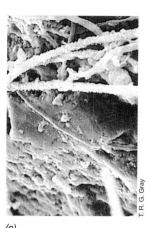

(a) *(b)* *(c)*

● **Figure 19.8 Visualization of microorganisms on the surface of soil particles by use of the scanning electron microscope.** (a) A microcolony of short, rod-shaped bacteria. (b) Actinomycete spores (Section 12.24). The cells in (a) and the spores in (b) are about $1-2$ μm wide. (c) Fungal hyphae. The fungal hyphae are about 4 μm wide and are coated with mineral matter.

soil bacteria and can also be used to enumerate cells on soil particle surfaces.

One of the major factors affecting microbial activity in soil is the availability of *water*, and we have previously discussed the importance of water to microbial growth (Section 6.14). Water is a highly variable component of soil, depending on soil composition, rainfall, drainage, and plant cover. Water is held in the soil in two ways—by *adsorption* onto surfaces or as *free water* in thin sheets or films between soil particles.

The water present in soils has a variety of materials dissolved in it, and the mixture is called the *soil solution*. In well-drained soils, air penetrates readily, and the oxygen concentration of the soil solution can be high. In waterlogged soils, however, the only oxygen present is that dissolved in the water, and this is soon consumed by microorganisms. Such soils quickly become anoxic, showing profound changes in their biological properties. We discussed oxygen relationships in soil microenvironments in Section 19.2 and Figure 19.2.

The *nutrient status* (resources, Table 19.1) of a soil is the other major factor affecting microbial activity. The greatest microbial activity is in the organic-rich surface layers, especially in and around the rhizosphere (see Figure 19.6*b*). The numbers and activity of soil microorganisms depend to a great extent on the balance of nutrients present. In some soils carbon is not the limiting nutrient, but instead the availability of *inorganic* nutrients such as phosphorus and nitrogen limit microbial productivity.

Deep Subsurface Microbiology

Interest in the chemistry of groundwater and the potential leaching of pollutants and their transfer in groundwater aquifers has led to studies of the role microorganisms play in the *deep subsurface* terrestrial environment. The deep soil subsurface, which can extend for several hundred meters below the soil surface, is not a biological wasteland. Although cell numbers are much lower than in the A soil horizon (Figure 19.6), a variety of microorganisms, primarily bacteria, are present in deep subsurface soils. For example, in samples collected aseptically from bore holes drilled down to 300 m, a diverse array of prokaryotes has been found, including anaerobes such as sulfate-reducing bacteria, methanogens, and homoacetogens (Sections 17.15–17.17), as well as various aerobes and facultative aerobes.

Microorganisms in the deep subsurface presumably have access to nutrients because groundwater flows through their habitats, but activity measurements suggest that metabolic rates of these bacteria are rather low in their natural habitats (see the Microbial Sidebar, Microbial Life Deep Underground). Compared to microorganisms in the upper layers of soil, the biogeochemical significance of deep subsurface microorganisms may thus be minimal. However, over very long periods, the metabolic activities of these buried microorganisms may be responsible for the slow mineralization of organic compounds and the release of products into groundwater.

The potential for *in situ* bioremediation (see Section 19.18) of toxic substances (for example, benzenes and agricultural chemicals) leached from soil into groundwater by deep subsurface microorganisms is of particular current interest. The resulting addition of inorganic nutrients may stimulate biodegradation of these chemicals, and the introduction of specific microorganisms into contaminated aquifers may promote biodegradation.

⬡ **19.4 *Concept Check***

The soil is a complex habitat with numerous microenvironments and niches. Microorganisms are present in the soil primarily attached to soil particles. The most important factor influencing microbial activity in surface soil is the availability of water, whereas in deep soil (the subsurface environment) nutrient availability plays a major role.

◆ What is the difference between *mineral* soils and *organic* soils?

◆ What factors govern the extent and type of microbial activity in soils?

◆ Which region of soil is the most microbially active?

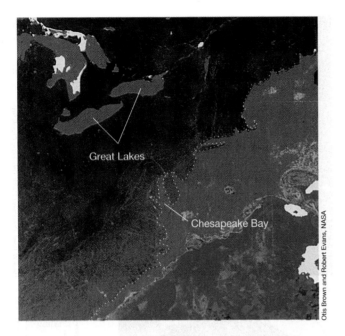

● **Figure 19.12 Distribution of chlorophyll in the western North Atlantic Ocean as recorded by satellite.** The east coast of the United States from mid-Florida to northern Maine is shown in dotted outline. Near the center of the photograph is Chesapeake Bay; the Great Lakes are at the upper left. Areas rich in phytoplankton are shown in red (>1 mg chlorophyll/m^3); blue and purple areas have lower chlorophyll concentrations (<0.01 mg/m^3). Note the high primary productivity of coastal areas and the Great Lakes.

typical purple and green bacteria (⥁ Sections 12.2 and 12.33) whose anoxygenic photosynthesis occurs under *anoxic* conditions. Instead, these are the so-called **aerobic anoxygenic phototrophs**—organisms that carry out anoxygenic photosynthesis but only under *aerobic* conditions. These include organisms like *Erythrobacter* and *Roseobacter*, genera of Proteobacteria (⥁ Chapter 12), that can carry out photosynthesis and presumably use the ATP produced to support their metabolism and growth. There appears to be a great diversity of such organisms in marine waters, especially in in-shore regions. Oligotrophic freshwater lakes also appear to be populated with aerobic phototrophs.

In addition to these phototrophs, many prokaryotes in the ocean's photic zone (top 300 meters) contain a form of the visual pigment rhodopsin. The cells use this pigment to convert light energy into ATP. In Section 13.3 we discussed the well-studied case of *bacteriorhodopsin*, present in the extreme halophile *Halobacterium* (a species of *Archaea*), and how this molecule was involved in ATP synthesis. The form of rhodopsin found in open ocean prokaryotes is very similar to bacteriorhodopsin but is present in cells that are phylogenetically *Bacteria*, not *Archaea*. Called **proteorhodopsin**, because the organisms that contain it are *Proteobacteria* (⥁ Section 12.1), this molecule is presumably the

basis of energy metabolism for these marine bacteria, releasing them from a dependence on scarce organic carbon for an energy source.

Environmental genomics (⥁ 18.6) has shown that genes encoding a diverse array of rhodopsin-like proteins exist in many phyla of *Bacteria* other than Proteobacteria in the open oceans. Thus, this form of metabolism in particular seems to be widely exploited among *Bacteria* and may have evolved (or have been obtained by lateral gene transfer) because of the very low concentrations of organic or inorganic electron donors available in marine waters.

Archaea/Bacteria Distribution

Numbers of prokaryotes in the open oceans *decrease* with depth. As previously mentioned, in surface waters cell numbers average between 10^5–10^6 cells/ml. Below 1000 m, however, total numbers fall to between 10^3–10^5/ml. Using phylogenetic stains to differentiate cells of each phylogenetic domain (⥁ Section 18.4), an interesting distribution of *Bacteria* and *Archaea* with depth has been found in the open oceans. In general, species of *Bacteria* predominate in upper waters (<1000), while numbers are about equal or show a slight predominance of

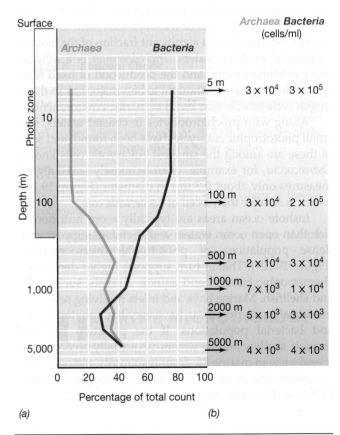

● **Figure 19.13 Percentage of total prokaryotes belonging to either the *Archaea* or the *Bacteria* in North Pacific Ocean water.** (a) Distribution of *Archaea* and *Bacteria* with depth. (b) Absolute numbers of *Archaea* and *Bacteria* with depth (per milliliter). Adapted from *Nature* 409:507–510 (2000).

Archaea in lower waters (Figure 19.13●). The *Archaea* present in lower waters are almost exclusively species of the Crenarchaeota, a phylum of *Archaea* that includes the hyperthermophiles (∞ Sections 13.8–13.11). Extrapolating from the data of Figure 19.13, it has been estimated that 1.3×10^{28} and 3.1×10^{28} cells of *Archaea* and *Bacteria*, respectively, exist in the world's oceans. This indicates that the oceans contain the largest microbial biomass on the surface of the Earth (∞ Section 1.3).

 19.6 Concept Check

Marine waters are more nutrient deficient than many freshwaters, yet substantial numbers of microorganisms exist there. Many of these use light to drive ATP synthesis. In terms of prokaryotes, species of the domain *Bacteria* tend to predominate in oceanic surface waters whereas *Archaea* are more prevalent in deeper waters.

◆ How does the organism *Prochlorococcus* contribute to both the carbon and oxygen cycles in the oceans?

◆ What is proteorhodopsin and why was it so named? How did the discovery of proteorhodopsin help solve the mystery of how large numbers of bacterial cells could exist in open ocean waters essentially devoid of dissolved organic matter?

◆ How are *Bacteria* and *Archaea* differentially distributed in ocean waters?

19.7 Deep-Sea Microbiology

Visible light penetrates no further than about 300 m in open ocean waters; this illuminated region is called the **photic zone** (Figure 19.13). Beneath the photic zone down to a depth of about 1000 m, considerable biological activity still occurs as a result of the action of animals and chemoorganotrophic microorganisms. Water at depths greater than 1000 m is, by comparison, relatively biologically inactive and has come to be known as the **deep sea**. Greater than 75% of all ocean water is deep-sea water, primarily at depths between 1000 and 6000 m.

Conditions in the Deep Sea

Organisms that inhabit the deep sea are faced with three major environmental extremes: *low temperature, high pressure*, and *low nutrient levels*. Below depths of about 100 m ocean water stays a constant 2–3° C. We discussed the responses of microorganisms to changes in temperature in Section 6.10. As would be expected, bacteria isolated from depths below 100 m are *psychrophilic* (cold-loving). Some are *extreme* psychrophiles, growing only in a narrow range near the *in situ* temperature. Deep-sea microorganisms must also be able to withstand the enormous hydrostatic pressures associated with great

depths. Pressure increases by *1 atm* for every *10 m* depth. Thus, an organism growing at a depth of 5000 m must be able to withstand pressures of 500 atm.

Barotolerant and Barophilic Bacteria

Different responses to pressure are evident in different deep-sea prokaryotes. Some organisms simply *tolerate* high pressure; they are called **barotolerant**. By contrast, others actually *grow best* under pressure; these are called **barophilic**.

Organisms isolated from depths down to about 3000 m and tested for growth or metabolic activity as a function of pressure are typically *barotolerant*. In barotolerant organisms, higher metabolic rates are observed at 1 atm than at 300 atm, although growth rates at the two pressures are about the same (Figure 19.14●). However, barotolerant isolates do not grow at pressures above 500 atm. By contrast, cultures derived from samples taken at greater depths, 4000–6000 m, are typically *barophilic*, growing optimally at pressures of about 400 atm (Figure 19.14). However, although barophiles grow best under pressure, they retain the ability to grow at 1 atm (Figure 19.14).

In even deeper water (10,000 m) **extreme (obligate) barophiles** can be found. These organisms have major pressure requirements for growth. For example, the bacterium *Moritella*, isolated from the Mariana Trench (Pacific Ocean, >10,000 m depth), grows fastest at a pressure of 700–800 atm and nearly as well at 1035 atm, the pressure it experiences in its natural habitat (Figures 19.14 and 19.15●).

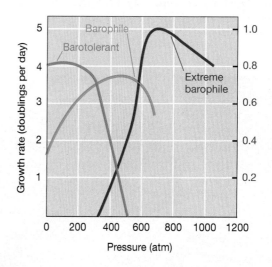

● **Figure 19.14 Growth of barotolerant, barophilic, and extremely barophilic bacteria.** The extreme barophile (*Moritella*) was isolated from the Mariana Trench (see Figure 19.15). Note the much slower growth rate (at any pressure) of the extreme barophile (right ordinate) as compared to the barotolerant and barophilic bacteria (left ordinate). Note also the inability of the extreme barophile to grow at low pressures.

But unlike barotolerant or barophilic prokaryotes, *Moritella* will not grow at pressures of less than about 400 atm (Figure 19.14). Interestingly, however, *Moritella* is not killed by decompression because it can tolerate moderate periods of decompression. However, viability is lost when the culture is left for several hours in a decompressed state. *Moritella* is also sensitive to temperature. Its optimal growth temperature is its environmental temperature (2°C), and temperatures above 10° C significantly reduce viability.

Molecular Effects of High Pressure

Pressure affects cellular physiology and biochemistry in part because it decreases the binding capacity of enzymes for their substrates. Thus, the enzymes of extreme barophiles must fold so as to minimize these pressure-related effects. Other potential pressure-sensitive events include protein synthesis and membrane activities, such as transport. An organism grown under high pressure has a higher proportion of unsaturated fatty acids in its cytoplasmic membrane. This change is presumably of adaptive significance because it makes the membrane less likely to gel at high pressures. The rather slow growth rates of extreme barophiles such as *Moritella* (see Figure 19.14) are probably due to a combination of pressure effects on cellular biochemistry and low temperatures, where reaction rates decrease considerably to begin with (∞ Figure 6.16).

In barophiles capable of growth up to 500–600 atm (Figure 19.14), it has been shown that growth at high pressure is accompanied by changes in the protein composition of the cell wall. In one barophile studied in detail, a specific outer membrane protein (∞ Section 4.9) called *OmpH* (*outer membrane protein H*) is synthesized in cells grown under pressure but not in cells grown at 1

atm. The OmpH protein is a type of *porin*. Porins are structural proteins that form channels for the diffusion of organic molecules through the outer membrane and into the periplasm (∞ Section 4.9 and Figure 4.35*b*). Presumably, the porin present in cells of the barophile grown at low pressures cannot function properly at high pressure, and thus a new type of porin molecule must be synthesized. Curiously, relatively few proteins seem to be controlled by pressure in barophiles, because many proteins are the same in cells grown at both high and low pressure. Cell wall and related structural proteins, and transport proteins seem to vary the most in response to pressure, suggesting that they are likely the keys to a barophilic lifestyle.

19.7 *Concept Check*

The deep sea is a cold, dark habitat where high hydrostatic pressure and low nutrient availability prevails. Barophiles grow best under pressure, and extreme barophiles, obtained from the greatest depths, require high pressure for growth.

◆ How does pressure change with depth?

◆ What molecular adaptations occur in barophiles that allow them to grow optimally under pressure?

19.8 Hydrothermal Vents

Although we think of the deep sea as a remote, low-temperature, high-pressure environment capable of supporting the slow growth of barotolerant and barophilic bacteria, there are some amazing exceptions. Dense, thriving *animal* communities, supported by the activities of microorganisms, cluster about thermal springs in deep-sea waters throughout the world. Geologically, these springs are associated with *sea floor spreading centers* (*rifts*), regions where hot basalt and magma near the sea floor cause the floor to slowly drift apart. Seawater seeping into these cracked regions mixes with hot minerals and is emitted from the springs (Figure 19.16●). These underwater hot springs are known as **hydrothermal vents**.

Two major types of vents have been found. *Warm vents* emit hydrothermal fluid at temperatures of 6–23° C (into seawater at 2° C). *Hot vents*, referred to as **black smokers** because the mineral-rich hot water forms a dark cloud of precipitated material upon mixing with seawater, emit hydrothermal fluid at 270–380° C (Figure 19.16, and see Figure 19.19). Warm vents emit fluid at 0.5–2 cm/s, whereas hot vents have higher flow rates, about 1–2 m/s.

Animals Living at Hydrothermal Vents

Using small pressurized submarines, it is possible to visit hydrothermal vents and study the organisms associated with them. Thriving invertebrate communities are present near hydrothermal vents, including *tube worms* over

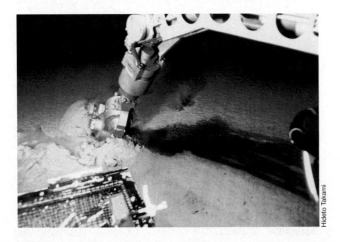

● **Figure 19.15 Sampling the deep sea.** The sampling arm of the unmanned submersible *Kaiko* inserting a sampling tube into sediment on the sea floor of the Mariana Trench (off the Philippines, Pacific Ocean) at a depth of 10,897 m and collecting a sample. The tubes of sediment are then retrieved and used for enrichment and isolation of barophilic bacteria.

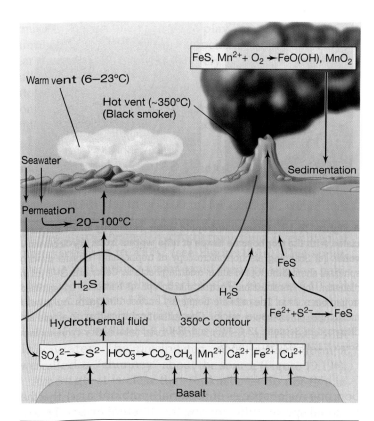

(a)

Dudley Foster, Woods Hole Oceanographic Institution

(b)

James Aguiar, Woods Hole Oceanographic Institution

● **Figure 19.16 Deep sea hot springs: Hydrothermal vents.** Schematic diagram showing the geological formations and major chemical species occurring at warm vents and black smokers. At warm vents (see Figure 19.17), the hot hydrothermal fluid is cooled by cold 2–3° C seawater permeating the sediments. In black smokers (see Figure 19.19), hot hydrothermal fluid near 350° C reaches the sea floor directly. Warm vents and black smokers are typically found at about 2000 in depth. Such depths can be explored by humans in small submersibles such as *Alvin*, used by researchers at the Woods Hole Oceanographic Institution, Woods Hole, MA.

2 m in length and large numbers of giant clams and mussels (Figure 19.17●). Considering that other locations in the deep sea are so biologically unproductive, how do dense animal communities exist in the absence of phototrophic primary producers? What are their energy source(s)?

Chemical analyses of hydrothermal fluid show large amounts of reduced inorganic materials, including H_2S, Mn^{2+}, H_2, and CO. Some vents contain little H_2S but have high levels of NH_4^+. Organic matter is not present in the fluid emitted from the hydrothermal vents. It is clear that the animals are dependent on the activities of *chemolithotrophic* bacteria (⟁ Sections 12.3–12.5 and 17.8–17.12), which grow at the expense of the inorganic energy sources emitted from the vents. Carbon dioxide, abundant in seawater as CO_3^{2-} and HCO_3^-, is fixed into organic carbon by the chemolithotrophs, and some of it is used to feed the vent animals.

Microorganisms in Hydrothermal Vents

Large numbers of sulfur-oxidizing chemolithotrophs are present in and around sulfide-emitting vents. Samples collected near such vents have yielded cultures of *Thiobacillus*,

(c)

Carl Wirsen, Woods Hole Oceanographic Institution

● **Figure 19.17 Invertebrates from habitats near deep-sea thermal vents.** (a) Tube worms (family *Pogonophora*), showing the sheath (white) and plume (red) of the worm bodies. (b) Close-up photograph showing worm plume. The plume collects nutrients from hydrothermal vents, including in particular, H_2S. The latter is transported to symbiotic sulfide-oxidizing bacteria that live within the worm. See text for details. (c) Mussel bed in vicinity of a warm vent. Note yellow deposition of elemental sulfur (from the oxidation of H_2S by chemolithotrophic symbionts) in and around mussels. See Table 19.2 for a list of chemolithotrophic prokaryotes found near hydrothermal vents. Some vent animals may rely on nitrifying bacteria (⟁Sections 12.3 and 17.12) or methylotrophic bacteria (⟁ Sections 12.6 and 17.24) for their nutrients.

(a)

(b)

● **Figure 19.43 Environmental consequences of large oil spills in the marine environment and the effect of bioremediation.** (a) A contaminated beach along the coast of Alaska containing oil from the *Exxon Valdez* spill of 1989. (b) The center rectangular plot (arrow) was treated with inorganic nutrients to stimulate bioremediation of spilled oil by microorganisms, whereas areas to the left and right were untreated.

nonvolatile components in oil are oxidized by bacteria within a year of the spill. Certain oil fractions, such as branched-chain and polycyclic hydrocarbons, however, remain in the environment much longer. Spilled oil that travels to the sediments is only slowly degraded, and it can have a significant long-term impact on fisheries and related activities that depend on unpolluted waters for productive yields.

Interfaces where oil and water meet often occur on a large scale. For example, it is virtually impossible to keep moisture from accumulating in bulk-fuel storage tanks, because water forms in a layer beneath the petroleum. Gasoline and crude oil storage tanks (Figure 19.44●) are thus potential habitats for hydrocarbon-oxidizing microorganisms. If sufficient sulfate is present in the water, sulfate-reducing bacteria can grow in the tanks, consuming hydrocarbons under anoxic conditions (Section 17.23). The sulfide (H_2S) produced is highly corrosive and causes pitting and subsequent leakage of the tanks.

Petroleum Production

Although microbial degradation of petroleum can be quite extensive, microbial *production* of hydrocarbons

● **Figure 19.44 Oil biodegradation.** Bulk fuel-storage tanks, where massive microbial growth may occur at oil-water interfaces.

also occurs, particularly in certain green algae. For example, in the colonial alga *Botryococcus braunii*, growth is accompanied by the excretion of long-chain hydrocarbons C_{30}–C_{36} that have the consistency of oil (Figure 19.45●). In *B. braunii* about 30% of the cell dry weight is petroleum, and there has been interest in using this and other oil-producing algae as renewable sources of petroleum. There is even evidence that oil in certain types of oil shale formations originated from green algae like *B. braunii* that grew in lake beds in ancient times.

19.17 Concept Check

Hydrocarbons are subject to microbial attack. Hydrocarbon-oxidizing microorganisms are used for bioremediation of spilled oil and their activities are assisted by addition of inorganic nutrients to balance the influx of carbon from the oil. Some algae can produce hydrocarbons.

◆ What is *bioremediation*?

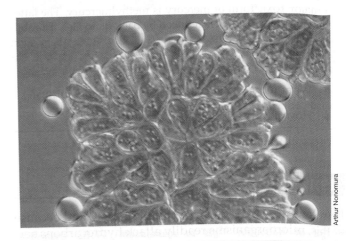

● **Figure 19.45 Phototrophic oil production.** Photomicrograph by Nomarski interference contrast of cells of the green alga *Botryococcus braunii*. Note oil droplets, produced and excreted by this alga, at the margin of the cells.

◆ Why might the addition of inorganic nutrients stimulate oil degradation while the addition of glucose might not?

19.18 Biodegradation of Xenobiotics

Xenobiotics are synthetic chemicals that are not naturally occurring substances. Xenobiotics include pesticides, polychlorinated biphenyls (PCBs, used in the electric-generating and related industries), munitions, dyes, and chlorinated solvents, among other things. Many xenobiotics are structurally related to natural compounds, and can thus be slowly degraded by enzymes that already exist to degrade these natural compounds. However, some xenobiotics differ chemically in such a major way from anything organisms have naturally experienced that they degrade extremely slowly, if at all. We focus here on pesticides as an example of the potential of microbial degradation.

Pesticide Catabolism

Some of the most widely distributed xenobiotics are the *pesticides*, which are common components of toxic wastes. Over 1000 pesticides have been marketed for chemical pest control purposes. Pesticides include *herbicides, insecticides,* and *fungicides*. Pesticides are of a wide variety of chemical types, including chlorinated compounds, aromatic rings, nitrogen- and phosphorus-containing compounds, and others (Figure 19.46●). Some of these substances are suitable as carbon sources and electron donors for certain soil microorganisms, whereas others are not.

If a substance can be attacked by microorganisms, it will eventually disappear from the soil. Such degradation in the soil is desirable because toxic accumulations of the compound are avoided. However, even closely related compounds may differ remarkably in their degradability, as is shown in Table 19.7. Thus some compounds persist relatively unaltered for years, while others are significantly degraded in only weeks or months.

The relative persistence rates of xenobiotics are only approximate because a variety of environmental factors—temperature, pH, aeration, and organic matter content of the soil—also influence decomposition. Some of the chlorinated insecticides (Figure 19.46) are so recalcitrant that they can persist for over 10 years. However, even complete disappearance of a pesticide from an ecosystem does not necessarily mean that it was degraded by microorganisms; pesticide loss can also occur by volatilization, leaching, or spontaneous chemical breakdown. *Bioavailability* to some extent also governs microbial attack on xenobiotic compounds. Many xenobiotics are quite hydrophobic and thus not very soluble in water, and adsorption of these compounds to organic matter and clay in soils and sediments prevents access to the organism. Addition of surfactants or emulsifiers often increases bioavailability, and ultimately biodegradation, of the xenobiotic compound.

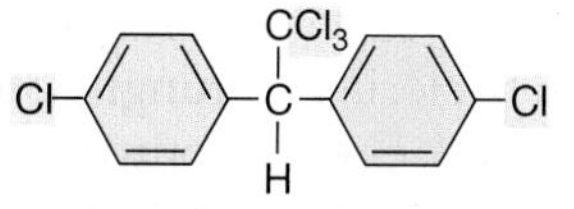

DDT; dichlorodiphenyltrichloroethane (an organochlorine)

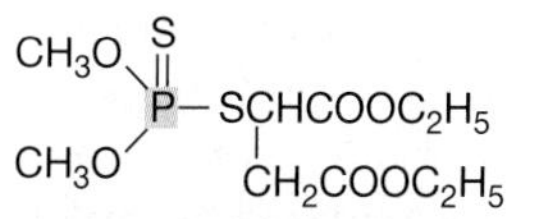

Malathion; mercaptosuccinic acid diethyl ester (an organophosphate)

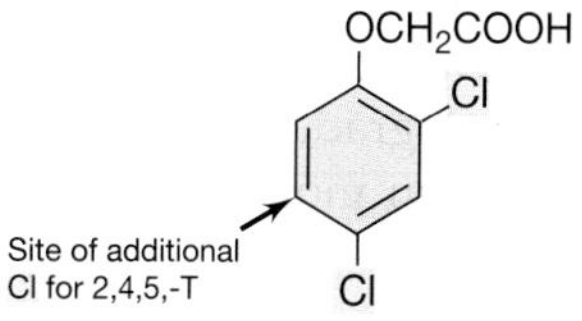

2,4-D; 2,4-dichlorophenoxy acetic acid (a chlorophenoxy acetic acid derivative)

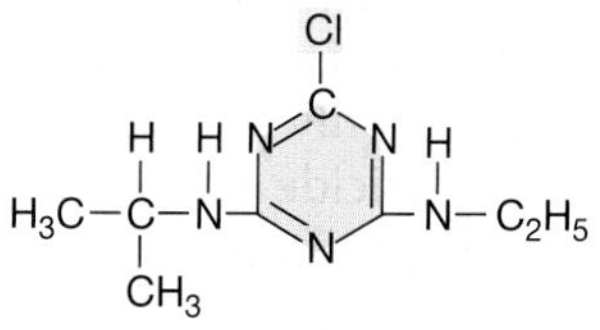

Atrazine, 2-chloro-4-ethylamino-6-isopropylaminotriazine (a triazine derivative)

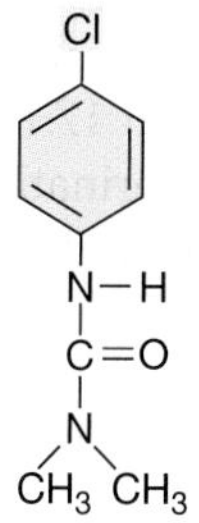

Monuron; 3-(4-chlorophenyl)-1 1-dimethylurea (a substituted urea)

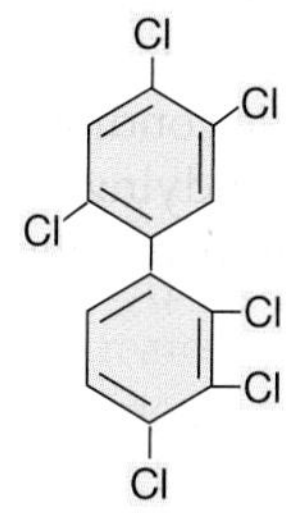

Chlorinated biphenyl (PCB); shown is 2, 3, 4, 2′, 4′, 5′-hexachlorobiphenyl

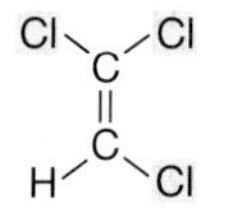

Trichloroethylene

● **Figure 19.46 Examples of xenobiotic compounds.** Although none of these compounds exist naturally, various microorganisms exist that will break them down (see persistence data in Table 19.7 and the pathway for biodegradation of 2,4,5-T in Figure 19.47).

Table 19.7	Persistence of herbicides and insecticides in soils
Substance	**Time for 75–100% disappearance**
Chlorinated insecticides	
DDT [1,1,1-trichloro-2,2-bis-(*p*-chlorophenyl)ethane][a]	4 years
Aldrin	3 years
Chlordane	5 years
Heptachlor	2 years
Lindane (hexachlorocyclohexane)	3 years
Organophosphate insecticides	
Diazinon	12 weeks
Malathion[a]	1 week
Parathion	1 week
Herbicides	
2,4-D (2,4-dichlorophenoxyacetic acid)	4 weeks
2,4,5-T (2,4,5-trichlorophenoxyacetic acid)	20 weeks
Dalapon	8 weeks
Atrazine[a]	40 weeks
Simazine	48 weeks
Propazine	1.5 years

[a] Structure shown in Figure 19.46.

19.19 The Plant Environment

As microbial habitats, plants are clearly vastly different from animals. Compared with warm-blooded animals, plants vary greatly in temperature, both diurnally and throughout the year. Compared with the complex circulatory system of animals, the internal communication system of the plant is only poorly developed, and so transfer of microorganisms within the plant is relatively inefficient.

The above-ground parts of the plant, especially the leaves and stems, are subjected to intermittent drying. For this reason many plants contain waxy coatings on their leaves that retain moisture and keep out microorganisms. The roots, on the other hand, are a main area of microbial activity. Here moisture is less variable and nutrient concentrations are higher.

Rhizosphere and Phyllosphere

The **rhizosphere** is the region immediately outside the root (see Figure 19.6b); it is a zone where microbial activity is usually high. The **rhizoplane** is the actual root surface. The bacterial count is almost always higher in the rhizosphere/rhizoplane than it is in regions of the soil devoid of roots, often many times higher (Figure 19.50●). This is because roots excrete significant amounts of sugars, amino acids, hormones, and vitamins. Bacteria and fungi thus often form microcolonies on the root surface.

The **phyllosphere** is the surface of the plant leaf. Under conditions of high humidity, as in wet forests in tropical and temperate zones, the microflora of leaves may be quite high and include fungi (Figure 19.50a) as well as bacteria. Many bacteria on leaves fix nitrogen (∞ Section 17.28, and see Section 19.22), while the leaves in turn provide carbohydrates and other nutrients to the bacteria.

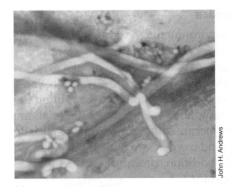

(a)

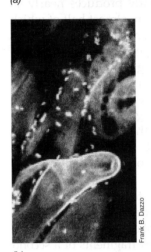

(b)

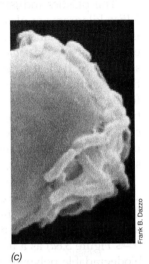

(c)

● **Figure 19.50 Examples of phyllosphere and rhizosphere microbial life.** (a) Fluorescent micrograph of the fungus *Aureobasidium pullulans* on the phyloplane (leaf surface) of apple. Cells were stained by FISH (∞ Section 18.4), and both fungal filaments and spores show the green stain. A fungal filament is about 7 μm in diameter. (b) Laser-scanning confocal (∞ Section 4.2) micrograph of bacterial cells on the rhizosphere/rhizoplane of clover. The association of nitrogen-fixing bacteria with the roots of leguminous plants is discussed in Section 19.22. (c) Scanning electron micrograph of cells of *Rhizobium leguminosarum* biovar *trifolii* attached to the root hair of clover. The bacterial cells in (b) and (c) are about 1 μm in diameter.

 19.19 Concept Check

Key microbial habitats on plants include the rhizoplane/rhizosphere and the phyllosphere.

◆ Why might the roots of plants be a more attractive environment for bacteria than in free soil away from plants?

19.20 Lichens and Mycorrhizae

Lichens are leafy or encrusting growths often found growing on bare rocks, tree trunks, house roofs, and the surfaces of bare soils (Figure 19.51●). Lichens consist of a symbiosis of two organisms, a fungus and an alga or cyanobacterium. However, a given fungus can establish the lichen symbiosis with several different phototrophs, and vice versa.

The alga is phototrophic and able to produce organic matter, which is then used for nutrition of the fungus. The fungus, unable to carry out photosynthesis, provides a firm anchor within which the alga can grow protected from erosion by rain or wind. In addition, the fungus facilitates the uptake of water. From the rock or other substrate on which the lichen is living, the fungus also absorbs inorganic nutrients essential for the growth of the alga. Lichens are usually found on surfaces where other organisms do not grow, and their success in colonizing such environments is due to the mutual interrelationships between the alga and fungus partners.

Lichen Structure and Ecology

Lichens consist of a tight association of many fungal cells within which the phototroph cells are embedded (Figure 19.52●). The shape of the lichen is determined primarily by the fungal partner, and a wide variety of fungi are able to form lichen associations. The diversity of algal types is much smaller, and many different kinds

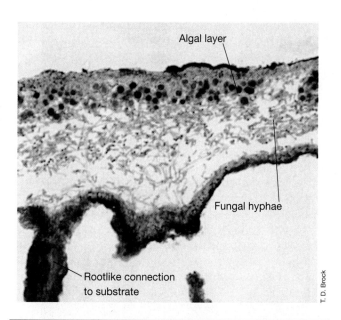

● Figure 19.52 Lichen structure. Photomicrograph of a cross section through a lichen. The algal layer is positioned within the lichen structure so as to receive the most sunlight.

● Figure 19.51 Lichens. (a) A lichen growing on a branch of a dead tree. (b) Lichens coating the surface of a large rock.

of lichens may have the same algal component. Some lichens contain cyanobacteria, frequently N_2-fixing species, instead of algae as the phototrophic component. The algae or cyanobacteria are usually present in defined layers or clumps within the lichen structure (Figure 19.52).

The fungus clearly benefits from associating with the alga, but how does the alga benefit? *Lichen acids*, complex organic compounds excreted by the fungus, promote the dissolution and chelation of inorganic nutrients needed by the alga. Another role of the fungus is to protect the phototroph from drying; most of the habitats in which lichens live are dry (rock, bare soil, roof tops) (Figure 19.52), and fungi are, in general, much better able to tolerate dry conditions than are algae.

Most lichens grow extremely slowly. For example, a 2-cm lichen observed on the surface of a rock may actually be several years old. Measurements of lichen growth vary from 1 mm or less per year to over 3 cm/year, depending on the organisms composing the symbiosis, and the amount of rainfall and sunlight received.

Mycorrhizae

Mycorrhizae (literally, "root fungus") refers to the symbiotic association between plant roots and fungi. The roots of the majority of terrestrial plants are mycorrhizal. There are two classes of mycorrhizae. In **ectomycorrhizae** (Figure 19.53●), fungal cells form an extensive sheath around the outside of the root with only little penetration into the root tissue itself. In **endomycorrhizae**, the fungal mycelium becomes deeply embedded within the root tissue.

Ectomycorrhizae are found mainly in forest trees, especially conifers, beeches, and oaks, and are most highly developed in boreal and temperate forests. In such forests, almost every root of every tree is mycorrhizal. The root system of a mycorrhizal tree such as pine (genus *Pinus*) is composed of both long and short roots. The short roots, which are characteristically dichotomously branched in *Pinus* (Figure 19.53*a*), show typical fungal colonization, and long roots are also frequently colonized. Endomycorrhizae are even more common than ectomycorrhizae. *Arbuscular mycorrhizae*, a type of endomycorrhizae, are found in the roots of over 80% of all terrestrial plant species so far examined.

Most mycorrhizal fungi do not catabolize cellulose and leaf litter, but instead use simple carbohydrates for growth and usually have one or more vitamin requirements. They obtain their carbon from root secretions but get inorganic minerals from the soil. Mycorrhizal fungi are rarely found in nature except in association with roots and hence most can be considered obligate symbionts. Mycorrhizal fungi produce plant growth substances that induce morphological alterations in the roots, stimulating formation of the mycorrhizal state. However, despite the close relationship between fungus and root, a single species of pine can form a mycorrhizal association with over 40 species of fungi.

The beneficial effect of the mycorrhizal fungus on the plant is best observed in poor soils, where trees that are mycorrhizal thrive but nonmycorrhizal ones do not. For example, when trees are planted in prairie soils, which ordinarily lack a suitable fungal inoculum, trees that are artificially inoculated at the time of planting grow much more rapidly than uninoculated trees (Figure 19.54●). The mycorrhizal plant can absorb nutrients

(a)

• **Figure 19.53 Mycorrhizae.** (a) Typical ectomycorrhizal root of the pine, *Pinus rigida*, with rhizomorphs of the fungus *Thelophora terrestis*. (b) Seedling of *Pinus contorta* (lodgepole pine), showing extensive development of the absorptive mycelium of its fungal associate *Suillus bovinus*. This grows in a fanlike formation from the ectomycorrhizal roots to capture nutrients from the soil.

(b)

• **Figure 19.54 Effect of mycorrhizal fungi on plant growth.** Six-month-old seedlings of Monterey pine (*Pinus radiata*) growing in prairie soil: left, nonmycorrhizal; right, mycorrhizal.

from its environment more efficiently, and thus has a competitive advantage. This improved nutrient absorption is due to the greater surface area provided by the fungal mycelium. For example, in the pine seedling shown in Figure 19.53b, the ectomycorrhizal fungal mycelium makes up the overwhelming part of the absorptive area of the plant root system.

In addition to simply helping plants absorb nutrients, mycorrhizae also appear to play a significant role in controlling plant diversity. Indeed, field experiments have shown that there is a positive correlation between (1) the abundance and diversity of mycorrhizae in a soil, and (2) the extent of the plant diversity that develops in it. Thus, mycorrhizae are a prime example of a plant-microorganism symbiosis that benefits both partners: the mycorrhizal plant is better able to function physiologically and compete successfully in a species-rich plant community, while the fungus benefits from a steady supply of organic nutrients.

19.20 Concept Check

Lichens are symbiotic associations between a fungus and an alga or cyanobacterium. Mycorrhizae are formed from fungi that associate with plant roots and improve their ability to absorb nutrients. Mycorrhizae have a great beneficial effect on plant health and competitiveness.

◆ How do *endomycorrhizae* differ from *ectomycorrhizae*?

◆ Why are mycorrhizal associations with plants considered a type of symbiosis?

19.21 *Agrobacterium* and Crown Gall Disease

Some microorganisms are plant pathogens; that is, they cause plant disease. The genus *Agrobacterium*, a close relative of the root nodule bacterium *Rhizobium* (see next section), comprises organisms that cause the formation of tumorous growths on a wide variety of plants. The two species most widely studied are *A. tumefaciens*, which causes **crown gall disease**, and *A. rhizogenes*, which causes **hairy root disease**.

Although plants often form a benign accumulation of tissue, called a *callus*, when wounded, the growth induced by *A. tumefaciens* (Figure 19.55●) is different in that the callus shows uncontrolled growth. It thus resembles tumor growth in animals. Once induced, these tumors continue to grow in the absence of *Agrobacterium* cells. That is, once *Agrobacterium* has brought about the induction of the tumorous condition, its presence is no longer necessary.

A large plasmid called the **Ti** (*tumor induction*) **plasmid** (Figure 19.56●) must be present in the *Agrobacterium* cells if they are to induce tumor formation (Figure 19.55). In *Agrobacterium rhizogenes*, a similar plasmid called the *Ri plasmid* is necessary for induction of hairy root. Following infection, a part of the Ti plasmid called the *transfer DNA* (T-DNA), is integrated into the plant's genome. T-DNA carries the genes for tumor formation and also for the production of a number of modified amino acids called **opines**. *Octopine* [N^2-(1,3-dicarboxyethyl)-*L*-arginine] and *nopaline* [N^2-(1,3-dicarboxypropyl)-*L*-arginine] are two common opines. Opines are produced by plant cells transformed by T-DNA and are a source of carbon and nitrogen for *Agrobacterium* cells.

Recognition and T-DNA Transfer

To initiate the tumorous state, cells of *Agrobacterium* must first attach to a wound site on the plant. The recognition of *Agrobacterium* by plant tissue involves complementary receptor molecules on the surfaces of the bacterial and plant cells. It is thought that the plant receptor molecule is a type of *pectin* (a complex polysaccharide) and that the bacterial receptor is a type of polysaccharide containing β-glucans, embedded in the cell wall lipopolysaccharide.

Studies with nontumorigenic mutants of *Agrobacterium tumefaciens* have shown that most functions necessary for attachment of the bacterium to the plant are borne on the bacterial chromosome. Following attachment, rapid synthesis of cellulose microfibrils by the bacterium anchors the cells to the wound site and literally entraps the bacterial cells, forming large bacterial aggregates on the plant cell surface. This sets the stage for plasmid transfer from bacterium to plant.

The structure of the Ti plasmid is given in Figure 19.56. Although a number of genes are needed for infectivity, only a small region of the Ti plasmid, the *T-DNA* (Figure 19.56), is actually transferred to the plant. The T-DNA con-

● **Figure 19.55 Crown gall.** Photograph of a crown gall tumor on a tobacco plant caused by crown gall bacteria of the genus *Agrobacterium*. *A. tumefaciens* causes these brown, woody-like growths on a wide variety of plants. Crown gall is a particular problem on fruit trees and various ornamental flowers, such as roses and daisies. The disease usually does not kill the plant but may weaken it and make it more susceptible to drought and other diseases. Detached galls serve as a reservoir of *A. tumefaciens* cells to maintain the infection.

tains genes that induce tumorigenesis. The *vir* genes that reside on the Ti plasmid encode proteins that are essential for T-DNA transfer (Figure 19.56). Expression of *vir* genes is induced by plant signal molecules synthesized by wounded plant tissues. Examples of inducers include the

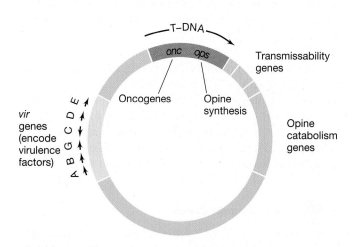

● **Figure 19.56 Structure of the Ti plasmid of *Agrobacterium tumefaciens*.** T-DNA is the region actually transferred to the plant. Arrows indicate the direction of transcription of each gene. The entire Ti plasmid is about 200 kbp of DNA, and the T-DNA is about 20 kbp (⌦ Section 31.10 for further discussion of Ti).

phenolic compounds acetosyringone, *p*-hydroxybenzoic acid, and vanillin.

The *vir* genes are the key to T-DNA transfer. The *virA* gene encodes a protein kinase (VirA) that interacts with inducer molecules and then phosphorylates the product of the *virG* gene (Figure 19.57●). The latter is activated by phosphorylation and functions to activate other *vir* genes. The product of the *virD* gene (VirD) has endonuclease activity and nicks DNA in the Ti plasmid in a region adjacent to the T-DNA (Figure 19.57). The product of the *virE* gene is a DNA-binding protein that binds the *single strand* of T-DNA generated from endonuclease activity and transports this small fragment of DNA into the plant cell. VirB, located in the bacterial membrane, mediates transfer of the single strand of DNA between bacterium and plant.

T-DNA transfer (Figure 19.57) thus resembles bacterial conjugation (Figure 10.11). The T-DNA becomes inserted into the nuclear genome of the plant and integration occurs wherever specific inverted or direct tandem repeats are present. The tumorigenesis (*onc*) genes of the Ti plasmid (Figure 19.56) encode enzymes involved in plant hormone production, as well as at least one key enzyme of opine biosynthesis. Expression of these genes leads to tumor formation. The Ri plasmid involved in hairy root disease also contains *onc* genes. However, in this case the genes confer increased auxin responsiveness to the plant; this leads to overproduction of root tissue, resulting in the symptoms of the disease. The Ri plasmid also encodes several opine biosynthetic enzymes.

Genetic Engineering with the Ti Plasmid

From the standpoint of microbiology and plant pathology, both crown gall and hairy root disease involve a unique type of interaction in which bacterial DNA is physically transferred to plant cells. However, once the mechanism of DNA transmission was understood, it quickly became clear that the Ti system could be used as a vector to introduce genetically engineered DNA into plants. In other words, Ti is a *natural plant transformation system*. Thus, the focus of the Ti/crown gall system has shifted away from the disease itself to new applications in biotechnology.

Through the power of genetic engineering a host of modified ("disarmed") Ti plasmids that lack disease genes are now available for the production of transgenic plants. Success stories are already recorded in the areas of herbicide and insect resistance, and many other areas remain to be explored. We discuss the use of the Ti plasmid as a vector in plant biotechnology in more detail in Section 31.10.

 19.21 Concept Check

The crown gall bacterium *Agrobacterium* enters into a unique relationship with higher plants. A plasmid in the bacterium (the Ti plasmid) is able to transfer part of itself into the genome of the plant, in this way bringing about the production of crown gall disease. The crown gall plasmid has also found extensive use in the genetic engineering of crop plants.

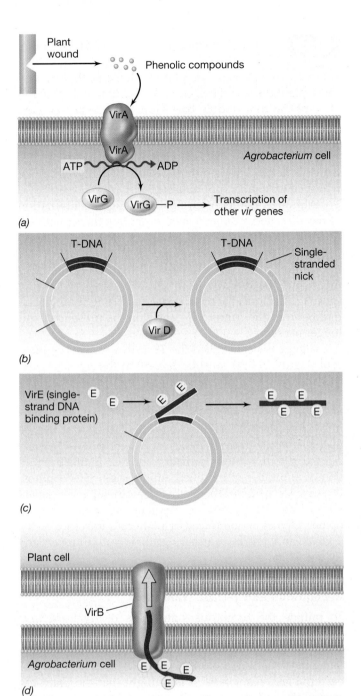

● **Figure 19.57 Mechanism of transfer of T-DNA to the plant cell by *Agrobacterium tumefaciens.*** (a) VirA activates VirG by phosphorylation, and VirG activates transcription of other *vir* genes. (b) VirD is an endonuclease. (c) VirE is a single-strand DNA-binding protein. (d) VirB functions as a conjugation bridge between *Agrobacterium* and the plant cell. Plant DNA polymerase produces the complementary strand to the single strand of T-DNA that is transferred before the DNA gets integrated into the plant genome.

◆ What are *opines* and why are they produced?

◆ How do the *vir* genes differ from *T-DNA* in the Ti plasmid?

◆ How has an understanding of crown gall disease benefited the area of plant molecular biology?

19.22 Root Nodule Bacteria and Symbiosis with Legumes

One of the most important plant bacterial interactions is that between leguminous plants and certain gram-negative *nitrogen-fixing* bacteria. Legumes are defined as plants that bear seeds in pods and are the third largest family of flowering plants. This large group includes such agriculturally important plants as *soybeans, clover, alfalfa, beans,* and *peas.* These plants are key commodities for the soy-processing industry and the feeding of domesticated animals, as well as major vegetables for human nutrition. Several agricultural industries revolve around leguminous crops, and the ability of legumes to grow without nitrogen fertilizer saves farmers millions of dollars in fertilizer costs yearly.

Rhizobium, Bradyrhizobium, Sinorhizobium, Mesorhizobium, Azorhizobium, and *Photorhizobium* are gram-negative motile rod-shaped Proteobacteria that can grow free-living in soil or can infect leguminous plants and establish a symbiotic existence. Infection of the roots of a legume with the appropriate species of one of these genera leads to the formation of **root nodules** (Figure 19.58●) that fix nitrogen (∞ Section 17.28). Nitrogen fixation by these symbioses are of considerable agricultural importance, as it leads to significant increases in combined nitrogen in the soil. Because nitrogen deficiencies often occur in unfertilized bare soils, nodulated legumes can grow well in areas where other plants cannot (Figure 19.59●).

Leghemoglobin and Cross-Inoculation Groups

Under normal conditions, neither the legume plant nor its bacterial symbiont can fix nitrogen. How then does interaction between the two lead to nitrogen fixation? In pure culture, rhizobia are able to fix N_2 alone when grown under *microaerophilic* conditions. Apparently the rhizobia need some O_2 to generate energy for N_2 fixation, but their nitrogenases (like those of other nitrogen-fixing organisms, ∞ Section 17.28) are inactivated by O_2.

In the nodule, precise O_2 levels are controlled by the O_2-binding protein **leghemoglobin**. This red, iron-containing protein is present in healthy N_2-fixing nodules (Figure 19.60●) and is induced through the interaction of the plant host and the bacterial symbiont. Leghemoglobin functions as an "oxygen buffer," cycling between the oxidized (Fe^{3+}) and reduced (Fe^{2+}) forms to keep free O_2 levels within the nodule low. The ratio of leghemoglobin-bound O_2 to free O_2 in the root nodule is on the order of 10,000:1.

About 90% of all leguminous plant species are capable of nodulation. However, there is a marked specificity between species of legume and rhizobial species. A single rhizobial species is generally able to infect certain species of legumes and not others. A species or group of rhizobia able to infect a group of related

● **Figure 19.58 Soybean root nodules.** The nodules develop by infection with *Bradyrhizobium japonicum.* The main stem of this soybean plant is about 0.5 cm in diameter.

legumes is called a **cross-inoculation group** (Table 19.8). However, even if a rhizobial strain can infect a certain legume, it is not always able to bring about the production of nitrogen-fixing nodules. If the strain is genetically *ineffective,* the nodules formed will be small, greenish-white, and incapable of fixing nitrogen. By contrast, if the strain is *effective,* the nodule will be large, reddish (Figure 19.60), and nitrogen-fixing. Effectiveness is determined by genes in the bacterium (see the discussion of *nod* genes later in this section) and can be measured by acetylene reduction (∞ Section 17.28).

● **Figure 19.59 Effect of nodulation on plant growth.** A field of unnodulated (left) and nodulated (right) soybean plants growing in nitrogen-poor soil.

(a)

(c)

Functions and Products of Intestinal Flora

The intestinal flora carry out a wide variety of essential metabolic reactions (Table 21.2) resulting in a variety of products. For all of these metabolic products, the composition of the intestinal flora and the diet influence the type and amount of compounds produced.

Among these products are vitamins B_{12} and K. These essential vitamins are not made by humans, but are made by the indigenous microbial flora and absorbed from the gut. Steroids, produced in the liver and released into the intestine from the gall bladder as bile acids, are modified in the intestine by the microbial flora. The modified active steroid compounds are also absorbed from the gut.

Other products generated by the activities of fermentative and methanogenic microorganisms include gas (*flatus*) and the odor-producing substances listed in Table 21.2. The gas results from the metabolism of fermentative and methanogenic microorganisms. Normal adults expel several hundred milliliters of gas, of which about half is N_2 from swallowed air, from the intestines every day. Some foods metabolized by fermentative bacteria in the intestines result in the production of hydrogen (H_2) and carbon dioxide (CO_2). Methanogens (∞Section 13.4), found in the intestines of over one-third of normal adults, convert H_2 and CO_2 produced by the other intestinal microorganisms to methane (CH_4). The methanogens in the rumen of cattle produce significant amounts of methane, up to a quarter of the total global production (∞Section 19.10).

During the passage of food through the gastrointestinal tract, water is absorbed from the digested material, which gradually becomes more concentrated and is converted to feces. *Bacteria* make up about one-third of the weight of fecal matter. Organisms living in the lumen of the large intestine are continuously displaced downward by the flow of material, and bacteria that are lost are continuously replaced by new growth. Thus, the large intestine resembles a chemostat (∞Section 6.9). The time needed for passage of material through the complete gastrointestinal tract is about 24 hours in humans; the growth rate of bacteria in the lumen is one to two doublings per day.

When an antibiotic is given orally, it inhibits the growth of the normal flora as well as pathogens, leading to the loss of antibiotic-susceptible bacteria in the intestinal tract. In the absence of the full complement of normal flora, opportunistic microorganisms such as antibiotic-resistant *Staphylococcus, Proteus, Clostridium difficile*, or the yeast *Candida albicans* occasionally become established. Establishment of opportunistic pathogens can lead to a harmful alteration in digestive function or even to disease. For example, antibiotic treatment allows less susceptible microorganisms such as *C. difficile* to flourish, causing infection and colitis. After antibiotic therapy stops, the normal flora eventually becomes reestablished, but often only after a considerable period.

 21.4 Concept Check

The stomach is very acidic and is a barrier to most microbial growth. The intestinal tract is slightly acidic to neutral and supports a diverse population of microorganisms in a variety of nutritional and environmental conditions.

◆ Why might the small intestine be more suitable for growth of facultative aerobes than the large intestine?

◆ Identify several essential compounds made by indigenous intestinal microorganisms. What would happen if the microorganisms were completely eliminated from the body by the use of antibiotics?

21.5 Normal Microbial Flora of Other Body Regions

Each individual mucous membrane supports the growth of a specialized group of microorganisms. These organisms are part of the normal local environment and are characteristic of healthy tissue. In many cases, pathogenic microorganisms cannot colonize mucous membranes because of the presence of the normal resident population of microorganisms. Here we discuss two mucosal environments and their resident microorganisms.

Respiratory Tract

The anatomy of the respiratory tract is shown in Figure 21.10● (Figure 26.2). In the **upper respiratory tract** (nasopharynx, oral cavity, and throat) microorganisms live in areas bathed with the secretions of the mucous membranes. Bacteria enter the upper respiratory tract from the air during breathing, but most of them are trapped in the nasal passages and expelled again with the nasal secretions. The resident organisms most commonly found are staphylococci, streptococci, diphtheroid bacilli, and gram-negative cocci. Potentially harmful bacteria such as *Staphylococcus aureus* and *Streptococcus pneumoniae* are often part of the normal flora of the nasopharynx of healthy individuals (Table 21.1). These individuals are *carriers* of the

Table 21.2	Biochemical/metabolic contributions of intestinal microorganisms
Vitamin synthesis	Product: thiamine, riboflavin, pyridoxine, B_{12}, K
Gas production	Product: CO_2, CH_4, H_2
Odor production	Product: H_2S, NH_3, amines, indole, skatole, butyric acid
Organic acid production	Product: acetic, propionic, butyric acids
Glycosidase reactions	Enzyme: β-glucuronidase, β-galactosidase, β-glucosidase, α-glucosidase, α-galactosidase
Steroid metabolism (bile acids)	Process: esterification, dehydroxylation, oxidation, reduction, inversion

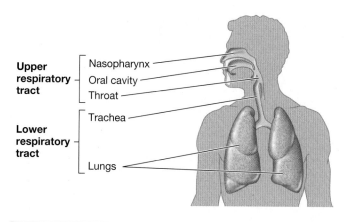

● **Figure 21.10   The respiratory tract.** The upper respiratory tract is populated with a large variety and number of microorganisms, but the lower respiratory tract has relatively few microbial inhabitants, unless there is an ongoing active infection (⚬⚬Figure 26.2).

and components of the adaptive immune system such as IgA (⚬⚬Section 22.9) are particularly active at mucosal surfaces and may also inhibit the growth of pathogens.

The normal **lower respiratory tract** (trachea, bronchi, and lungs) has no resident microflora, despite the large numbers of organisms potentially able to reach this region during breathing. Dust particles, which are fairly large, settle out in the upper respiratory tract. As the air passes into the lower respiratory tract, the flow rate decreases markedly, and organisms settle onto the walls of the passages. The walls of the entire respiratory tract are lined with ciliated epithelium, and the cilia, beating upward, push bacteria and other particulate matter toward the upper respiratory tract where they are then expelled in the saliva and nasal secretions. Only particles smaller than about 10 μm in diameter reach the lungs.

pathogens but do not normally acquire disease, presumably because the other resident microorganisms compete successfully for resources and limit pathogen growth. The innate immune system (⚬⚬Section 22.2 and Section 23.1)

Urogenital Tract

In the male and female urogenital tracts (Figure 21.11a●), the bladder itself is typically sterile, but the epithelial cells lining the urethra are colonized by facultatively aerobic gram-negative rods and cocci (Table 21.1). Microorganisms

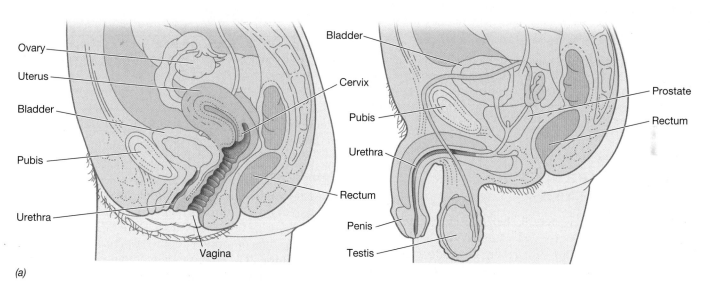

(a)

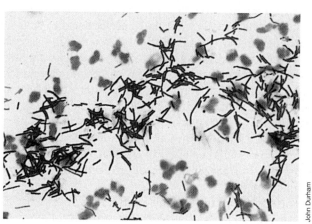

(b)

● **Figure 21.11   Microbial growth in the genitourinary tract.** (a) The genitourinary tracts of the human female and male, showing regions (red) where microorganisms often grow. Note that the upper regions of the genitourinary tracts of both males and females are sterile in normal individuals. (b) Gram stain of *Lactobacillus acidophilus*, the predominant organism in the vagina of women between the onset of puberty and the end of menopause. Individual rods are 3–4 μm long. The genitourinary tract of older and younger women is less acidic and is populated by a much more heterogeneous group of microorganisms that grow at neutral to slightly alkaline pH.

such as *Escherichia coli*, *Proteus mirabilis*, and others, normally present in the body or in the local environment, influenced by local factors such as pH changes, can multiply and become pathogenic. Such organisms are a frequent cause of urinary tract infections, especially in women.

The vagina of the adult female is weakly acidic and contains significant amounts of the polysaccharide glycogen. *Lactobacillus acidophilus*, a resident organism in the vagina, ferments glycogen to produce lactic acid and maintain the acidic conditions (Figure 21.11*b*) (∞ Section 12.19). Other organisms such as yeasts (*Torulopsis* and *Candida* species), streptococci, and *E. coli* may also be present. Before puberty, the female vagina is alkaline and does not produce glycogen, *L. acidophilus* is absent, and the flora consists predominantly of staphylococci, streptococci, diphtheroids, and *E. coli*. After menopause, glycogen production ceases, the pH rises, and the flora again resembles that found before puberty.

 21.5 Concept Check

The presence of a population of normal nonpathogenic microorganisms in the respiratory and urogenital tracts is essential for normal organ function and often prevents the colonization of pathogens.

◆ Potential pathogens are often found in the normal flora of the upper respiratory tract. Why do they not cause disease in most cases?

◆ Why is *Lactobacillus* found in the urogenital tract of normal adult women? Why is it not found in postmenopausal women?

 **II | HARMFUL MICROBIAL INTERACTIONS WITH HUMANS**

Microbial interactions may be harmful to the host and cause disease. Here we will examine mechanisms of pathogenesis, the ability of microorganisms to cause disease. Microbial pathogenesis begins with exposure and adherence of the microorganisms to host cells, followed by invasion, colonization, and growth. Unchecked growth of the pathogen then results in host damage. Disease-producing microorganisms use several different strategies to establish **virulence**, the relative ability of a pathogen to cause disease (Figure 21.12●). Here we sequentially consider the factors responsible for establishing virulence.

21.6 Entry of the Pathogen into the Host

A pathogen must usually gain access to host tissues and multiply before damage can be done. In most cases, this requires that the organisms penetrate the skin, mucous

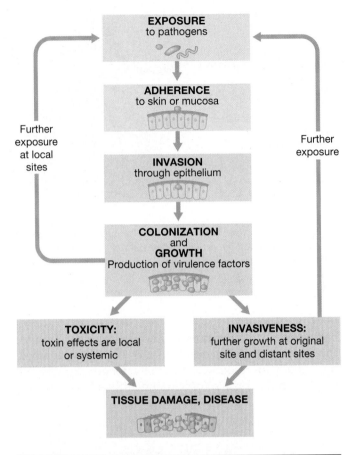

● **Figure 21.12 Microorganisms and pathogenesis.** The presence and even the growth of microorganisms on the host does not necessarily lead to disease.

membranes, or intestinal epithelium, surfaces that normally act as microbial barriers.

Specific Adherence

Most microbial infections begin at breaks or wounds in the skin or on the mucous membranes of the respiratory, digestive, or genitourinary tract. Bacteria or viruses able to initiate infection often adhere *specifically* to epithelial cells (Figure 21.13●) through macromolecular interactions on the surfaces of the pathogen and the host cell.

An infecting microorganism does not adhere to all epithelial cells equally but selectively adheres to cells in a particular region of the body. For example, *Neisseria gonorrhoeae*, the causative agent of the sexually transmitted disease gonorrhea (∞ Section 26.12), adheres much more strongly to urogenital epithelia than to other tissues through a surface protein called *Opa* (*o*pacity *a*ssociated *proteins*). Host cells bind specifically to Opa with a protein called CD66, a protein found only on the surface of human epithelial cells. Thus, *N. gonorrhoeae* interacts exclusively with target cells through a cell surface receptor–ligand pair.

The species of the host also influences specificity. In many cases, a bacterial strain that normally infects hu-

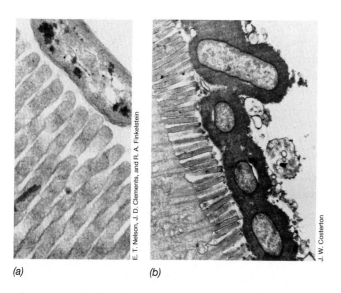

(a)   *(b)*

● **Figure 21.13 Adherence of pathogens to animal tissues.** (a) Transmission electron micrograph of a thin section of *Vibrio cholerae* adhering to the brush border of rabbit villi. Note the absence of a capsule. (b) Enteropathogenic *Escherichia coli* in a fatal model of infection in the newborn calf. The bacterial cells are attached to the brush border of calf villi via a well-defined capsule. The rods are about 0.5 μm in diameter.

mans adheres more strongly to the appropriate human cells than to similar cells in another animal (for example, the rat), and vice versa.

Some macromolecules responsible for bacterial adherence are not covalently attached to the bacteria. These are usually polysaccharides, proteins, or protein-carbohydrate mixtures synthesized and secreted by the bacteria (∞Section 4.10). A loose network of polymer fibers extending outward from a cell is known as a **slime layer** (Figure 21.4*b*). A polymer coat consisting of a dense, well-defined layer surrounding the cell is known as a **capsule** (Figures 21.13 and 21.14●).These structures may be important for adherence not only to host tissues, but also between other bacteria. In addition, these structures can protect bacteria from host defense mechanisms such as phagocytosis (∞Section 22.2).

Fimbriae and *pili* (∞Section 4.10) are bacterial cell surface protein structures that may also function in the attachment process. For instance, the pili of *Neisseria gonorrhoeae* play a key role in the attachment of this organism to urogenital epithelium, and fimbriated strains of *Escherichia coli* (Figure 21.15●) are much more frequent causes of urinary tract infections than strains lacking fimbriae. Among the best-characterized fimbriae are the so-called *type I fimbriae* of enteric bacteria (*Escherichia, Klebsiella, Salmonella, Shigella*). Type I fimbriae are uniformly distributed on the surface of cells. Pili are generally longer than fimbriae, with fewer pili found on the cell surface. Both pili and fimbriae function by binding host cell glycoproteins, initiating the attachment event. Flagella (∞Section 4.14) also increase adherence to host cells and can lead to enhanced virulence (see Figure 21.17).

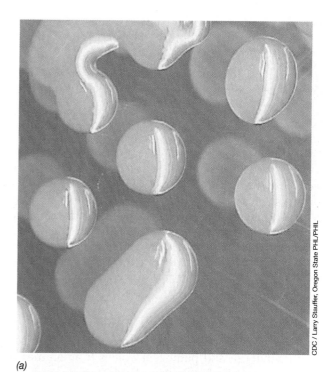

(a)

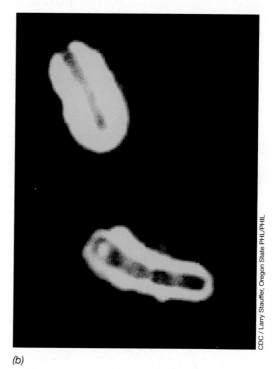

(b)

● **Figure 21.14 *Bacillus anthracis* capsules.** (a) Capsules of *B. anthracis* on bicarbonate agar media. Encapsulated colonies are typically very large and mucoid in appearance. The individual encapsulated colonies are 0.5 cm in diameter. (b) Direct immunofluorescent stain of *B. anthracis* capsules. Antibodies coupled to fluorescein isothiocyanate (FITC) (∞Section 24.9) stain the capsule bright green, indicating that the capsule extends about 1 μm from the cell, which is about 0.5 μm in diameter. Further coverage of *B. anthracis*, the disease anthrax, and the use of *B. anthracis* as a biological weapon can be found in Sections 25.12 and 25.13.

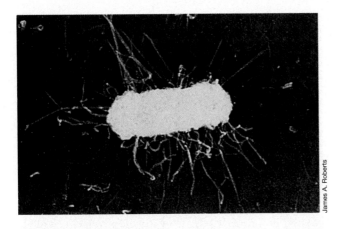

James A. Roberts

● **Figure 21.15 Fimbriae.** Shadow-cast electron micrograph of the bacterium *Escherichia coli* showing type P fimbriae. Type P fimbriae resemble type I fimbriae but are somewhat longer. The cell shown is about 0.5 μm wide.

Evidence for specific interactions between the mucosal epithelium and pathogens comes from studies of diarrhea caused by *Escherichia coli*. Most strains of *E. coli* are nonpathogenic and are part of the normal flora, inhabiting the cecum and the colon (see Figure 21.8). Several strains of *E. coli* are usually present in the body at the same time, and large numbers of these nonpathogens routinely pass through the body and are eliminated in feces. However, *enterotoxic* strains of *E. coli* express fimbrial proteins called *CFA (colonization factor antigens)* that adhere specifically to cells in the *small intestine*. From here, they can colonize and produce enterotoxins (see Section 21.11) that cause diarrhea as well as other illnesses (∞Section 29.8). The nonpathogenic strains of *E. coli* seldom have the CFA proteins. Some major factors important in microbial adherence are shown in Table 21.3.

Invasion

A few microorganisms are pathogenic solely because of the toxins they produce. These organisms do not need to gain access to host tissues, and we will discuss them separately (see Sections 21.10 and 21.11). However, most pathogens penetrate the epithelium to initiate pathogenicity, a process called *invasion*. At the point of entry, usually at small breaks or lesions in the skin or in mucosal surfaces, growth is established. Growth may also begin on intact mucosal surfaces, especially if the normal flora is altered or eliminated, for example, by antimicrobial chemotherapy. Pathogens may then more readily colonize the tissue and begin the invasion process. Pathogen growth may also be established at sites distant from the original point of entry. Access to distant, usually interior, sites is through the blood or lymphatic circulatory system (∞Section 22.1).

 21.6 *Concept Check*

Pathogens gain access to host tissues by adherence at mucosal surfaces through interactions between pathogen and host macromolecules. Pathogen invasion starts at the site of adherence and may spread throughout the host via the circulatory systems.

◆ How do CFA molecules on *Escherichia coli* and Opa proteins on *Neisseria gonorrhoeae* influence adherence to mucosal tissues?

◆ How does adherence initiate invasion?

21.7 Colonization and Growth

If a pathogen gains access to tissues, it may multiply, a process called **colonization**. Because the initial inoculum of a pathogen is too small to cause damage, a pathogen must find appropriate nutrients and environmental conditions to grow in the host. Temperature, pH, and presence or absence of oxygen affect pathogen growth, but the availability of microbial nutrients is most important.

Although a vertebrate host might seem to be a nutritional paradise for microorganisms, not all nutrients are plentiful. Soluble nutrients such as sugars, amino acids, and organic acids are limited, and so organisms able to use complex nutrients such as glycogen may be favored. Not all vitamins and growth factors are in adequate sup-

Table 21.3	Major adherence factors used to facilitate attachment of microbial pathogens to host tissues[a]
Factor	**Example**
Capsule/ slime layer (∞Section 4.10, Figure 21.4, Figure 21.13, Figure 21.14)	Pathogenic *Escherichia coli*—capsule promotes adherence to the brush border of intestinal villi *Streptococcus mutans*—dextran slime layer promotes binding to tooth surfaces
Adherence proteins	*Streptococcus pyogenes*—M protein on the cell binds to receptors on respiratory mucosa *Neisseria gonorrhoeae*—Opa protein on the cell binds to receptors on urogenital epithelium
Lipoteichoic acid (∞Section 4.8 and Figure 4.31)	*Streptococcus pyogenes*—facilitates binding to respiratory mucosal receptor (along with M protein)
Fimbriae (pili) (∞Section 4.10 and Figure 21.15)	*Neisseria gonorrhoeae*—pili facilitate binding to urogenital epithelium *Salmonella* species—type I fimbriae facilitate binding to epithelium of small intestine Pathogenic *Escherichia coli*— colonization factor antigens (CFAs) facilitate binding to epithelium of small intestine

[a] Most receptor sites on host tissues are glycoproteins or complex lipids such as gangliosides or globosides.

ply in all tissues at all times. *Brucella abortus*, for example, grows very slowly in most tissues of infected cattle but grows very rapidly in the placenta, where it causes abortion. This is due to the elevated concentration of erythritol, a nutrient that enhances growth of *B. abortus* when it infects the placenta (see Table 21.6).

Trace elements may also be in short supply and can influence establishment of the pathogen. For example, the concentration of *iron* greatly influences microbial growth (∞Section 5.1). Specific proteins called *transferrin* and *lactoferrin*, present in animals, bind iron tightly and transfer it through the body. These proteins have such high affinity for iron that microbial iron deficiency may be common; an iron salt solution given to an infected animal greatly increases the virulence of some pathogens.

As we noted in Section 5.1, many bacteria produce iron-chelating compounds (*siderophores*) that help them obtain iron, a growth-limiting micronutrient, from the environment. Some iron chelators isolated from pathogenic bacteria are so efficient that they can remove iron from animal iron-binding proteins. For example, a siderophore called *aerobactin*, produced by certain strains of *Escherichia coli* and encoded by the Col V plasmid (∞Section 10.9), readily removes iron bound to transferrin. Likewise, *Neisseria* species produce a transferrin-specific receptor and can remove iron from the bound transferrin.

Localization in the Body

After initial entry, the organism often remains localized and multiplies, producing a small focus of infection such as the boil that may arise from *Staphylococcus* skin infections (∞Section 26.9). Alternatively, the organisms may enter the lymphatic vessels and be deposited in lymph nodes. If an organism reaches the blood, it will be distributed to distant parts of the body, usually concentrating in the liver or spleen. Spread of the pathogen through the blood and lymph systems can result in a generalized (systemic) infection of the body, with the organism growing in a variety of tissues. If extensive bacterial growth in tissues occurs, some of the organisms are usually shed into the bloodstream in large numbers, a condition called **bacteremia**. Widespread infections of this type almost always start as a localized infection in a specific organ such as the kidney, intestine, or lung.

 21.7 Concept Check

A pathogen must gain access to nutrients and appropriate growth conditions before it can colonize and grow in substantial numbers in host tissue. Organisms may grow locally at the site of invasion, or may spread through the body.

◆ Why are colonization and growth necessary for the success of most pathogens?

◆ What host factors limit or accelerate colonization and growth of a microorganism at a local site?

21.8 Virulence

Virulence is the relative ability of a parasite to cause disease, and here we discuss some basic methods used to measure it. We will then provide specific examples of particularly virulent organisms, highlighting the strategies used to enhance virulence.

Measuring Virulence

The virulence of a pathogen can be estimated from experimental studies of the LD_{50} (lethal dose$_{50}$), the dose of an agent that kills 50% of the animals in a test group. Highly virulent pathogens frequently show little difference in the number of cells required to kill 100% of the population as compared with the number required to kill 50% of the population. This is illustrated in Figure 21.16● for experimental *Streptococcus* and *Salmonella* infections in mice. Only a few cells of *Streptococcus pneumoniae* are required to establish a fatal infection and kill all members of a test population of mice once the virulence of a particular strain has been established. In fact, the LD_{50} for *S. pneumoniae* is hard to determine because so few organisms are needed to produce a lethal infection. By contrast, the LD_{50} for *Salmonella typhimurium*, a much less virulent pathogen, is much higher than for *S. pneumoniae*. The number of cells of *Salmonella typhimurium* cells required to kill 100% of the population is more than 100 times greater than the number of cells needed to achieve the LD_{50}.

When pathogens are kept in laboratory culture and not passed through animals, their virulence is often decreased or even lost completely. Such organisms are said to be *attenuated*. **Attenuation** probably occurs because nonvirulent mutants may grow faster and, after successive transfers to fresh media, such mutants are selectively

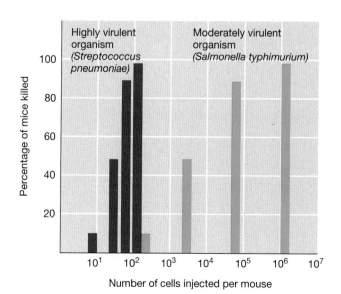

● **Figure 21.16 Microbial virulence.** Comparison of differences in microbial virulence based on the number of cells of *Streptococcus pneumoniae* or *Salmonella typhimurium* required to kill mice.

favored. Attenuation often occurs more readily when culture conditions are not optimal for the species. If an attenuated culture is reinoculated into an animal, organisms sometimes regain virulence, but in many cases loss of virulence is permanent. Attenuated strains are often used for production of vaccines, especially viral vaccines (∞Section 22.13). Measles, mumps, and rubella vaccines, for example, contain attenuated live viruses.

Toxicity and Invasiveness

Virulence is due to the ability of a pathogen to cause host damage through toxicity and invasiveness. Each pathogen uses these properties to cause disease.

Toxicity is the ability of an organism to cause disease by means of a preformed toxin that inhibits host cell function or kills host cells. For example, the disease *tetanus* is caused by a potent exotoxin produced by *Clostridium tetani* (see Section 21.10). *C. tetani* cells rarely leave the wound where they were first introduced, growing relatively slowly at the wound site. Yet *C. tetani* is able to bring about disease because it produces tetanus toxin that moves to distant parts of the body and initiates irreversible muscle contraction and often death of the host.

Invasiveness is the ability of an organism to grow in host tissue in such large numbers that the pathogen inhibits host function. A microorganism may still be able to produce disease through invasiveness even if it produces no toxin. For example, the major virulence factor for *Streptococcus pneumoniae* is the polysaccharide capsule that prevents the phagocytosis of pathogenic strains (see Section 21.6, ∞Section 26.2, and Figure 26.3), defeating a major defense mechanism used by the host to prevent invasion. Encapsulated strains of *S. pneumoniae* are able to cause extensive host damage because they are highly invasive. They grow in lung tissues in enormous numbers where they initiate host responses that lead to pneumonia (∞Section 26.2). Nonencapsulated strains are quickly and efficiently taken up and destroyed by phagocytes.

Clostridium tetani and *Streptococcus pneumoniae* exemplify the extremes of *toxicity* and *invasiveness*, respectively. Most pathogens fall between these extremes and use a combination of toxins and invasiveness to cause disease.

Virulence in *Salmonella*

Salmonella species employ a mixture of toxins, invasiveness, and other virulence factors to enhance pathogenicity (Figure 21.17●). First, several toxins contribute to the virulence of *Salmonella*, and at least three toxins are produced: *enterotoxin* (Table 21.4), *endotoxin* (∞Section 4.9 and Figure 4.35), and *cytotoxin*. Cytotoxin acts by inhibiting host cell protein synthesis and allowing Ca^{2+} to escape from host cells. A number of virulence factors are involved in invasion by *Salmonella*. The cell surface polysaccharide O antigen (∞Figure 4.35), the flagellar H antigen, and fimbriae enhance adherence. The capsular Vi polysaccharide inhibits complement binding and antibody-mediated killing (∞Section 22.11). The *inv* (invasion) genes of *Salmonella* encode at least 10 different proteins involved in invasion. For example, *invH* encodes a surface adhesin protein, while *invC*, *invG*, *invI*, and *invJ* encode proteins involved in assembly of the special *surface appendages* used for host-cell binding.

Salmonella readily establish infections through intracellular parasitism and grow inside the cells that line the intestine as well as in macrophages, white blood cells that normally ingest and kill bacteria (∞Section 22.2). The toxic oxygen products of macrophages are neutralized by proteins induced by the *Salmonella oxyR* gene.

● **Figure 21.17 Virulence factors in *Salmonella* pathogenesis.** Structural elements known to be important in pathogenesis are shown. Protein products of the *pho* gene system have not been identified, but are known to neutralize the effects of the detergent-like, macrophage-produced defensins.

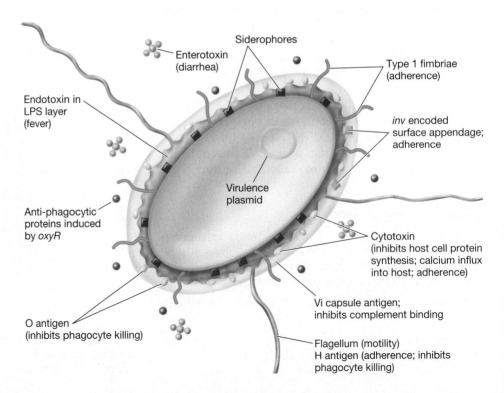

Siderophores

Enterotoxin (diarrhea)

Type 1 fimbriae (adherence)

Endotoxin in LPS layer (fever)

inv encoded surface appendage; adherence

Virulence plasmid

Anti-phagocytic proteins induced by *oxyR*

Cytotoxin (inhibits host cell protein synthesis; calcium influx into host; adherence)

Vi capsule antigen; inhibits complement binding

O antigen (inhibits phagocyte killing)

Flagellum (motility) H antigen (adherence; inhibits phagocyte killing)

Table 21.4 Exotoxins and extracellular virulence factors produced by human pathogens

Organism	Disease	Toxin or factor[a]	Action
Bacillus anthracis	Anthrax	Lethal factor (LF) Edema factor (EF) Protective antigen (PA) (AB)	PA is the cell-binding B component, EF causes edema, LF causes cell death
Bacillus cereus	Food poisoning	Enterotoxin complex	Induces fluid loss from intestinal cells
Bordetella pertussis	Whooping cough	Pertussis toxin (AB)	Blocks G protein signal transduction, kills cells
Clostridium botulinum	Botulism	Neurotoxin (AB)	Flaccid paralysis (see Figure 21.20)
Clostridium tetani	Tetanus	Neurotoxin (AB)	Spastic paralysis (see Figure 21.21)
Clostridium perfringens	Gas gangrene, food poisoning	α-Toxin (CT)	Hemolysis (lecithinase, see Figure 21.18b)
		β-Toxin (CT)	Hemolysis
		γ-Toxin (CT)	Hemolysis
		δ-Toxin (CT)	Hemolysis (cardiotoxin)
		κ-Toxin (E)	Collagenase
		λ-Toxin (E)	Protease
		Enterotoxin (CT)	Alters permeability of intestinal epithelium
Corynebacterium diphtheriae	Diphtheria	Diphtheria toxin (AB)	Inhibits protein synthesis in eukaryotes (Figure 21.19)
Escherichia coli (enterotoxigenic strains only)	Gastroenteritis	Enterotoxin (AB)	Induces fluid loss from intestinal cells
Haemophilus ducreyi	Chancroid	Cytolethal distending toxin[b] (CDT) (AB)	Genotoxin (DNA lesions cause apoptosis in host cells)
Pseudomonas aeruginosa	*P. aeruginosa* infections	Exotoxin A (AB)	Inhibits protein synthesis
Salmonella sp.	Salmonellosis, typhoid fever, paratyphoid fever	Enterotoxin (AB)	Inhibits protein synthesis and lyses host cells
		Cytotoxin (CT)	Induces fluid loss from intestinal cells
Shigella dysenteriae	Bacterial dysentery	Shiga toxin (AB)	Inhibits protein synthesis
Staphylococcus aureus	Pyogenic (pus-forming) infections (boils, and so on), respiratory infections, food poisoning, toxic shock syndrome, scalded skin syndrome	α-Toxin (CT)	Hemolysis
		Toxic shock syndrome toxin (SA)	Systemic shock
		Exfoliating toxin A and B (SA)	Peeling of skin, shock
		Leukocidin (CT)	Destroys leukocytes
		β-Toxin (CT)	Hemolysis
		γ-Toxin (CT)	Kills cells
		δ-Toxin (CT)	Hemolysis, leukolysis
		Enterotoxin A, B, C, D, and E (SA)	Induce vomiting, diarrhea, shock
		Coagulase (E)	Induces fibrin clotting
Streptococcus pyogenes	Pyogenic infections, tonsillitis, scarlet fever	Streptolysin O (CT)	Hemolysin
		Streptolysin S (CT)	Hemolysin (see Figure 21.18a)
		Erythrogenic toxin (SA)	Causes scarlet fever rash
		Streptokinase (E)	Dissolves fibrin clots
		Hyaluronidase (E)	Dissolves hyaluronic acid in connective tissue
Vibrio cholerae	Cholera	Enterotoxin (AB)	Induces fluid loss from intestinal cells (Figure 21.22)

[a] AB, A-B toxin; CT, cytolytic toxin; E, enzymatic virulence factor; SA, superantigen toxin.
[b] Cytolethal distending toxin is found in other gram-negative pathogens including *Actinobacillus actinomycetemcomitans*, *Campylobacter* sp., *Escherichia coli*, *Helicobacter* sp., *Salmonella typhi*, and *Shigella dysenteriae*.

Macrophage-produced antibacterial molecules called *defensins* are neutralized by the products of the *phoP* and *phoQ* genes. Thus, the *oxy* and *pho* gene products of *Salmonella* enhance pathogenicity by fostering methods of intracellular invasion, neutralizing host defenses that normally inhibit intracellular bacterial growth.

Several plasmid-borne virulence factors are also involved in persistence and spread in most *Salmonella* species. For example, antibiotic resistance is encoded on a plasmid. Finally, *Salmonella* also produce *siderophores*, bacterial iron-chelating proteins (∞Section 5.1 and see Section 21.7). Thus, *Salmonella*, and probably most other pathogens, use several factors to establish virulence and initiate disease. These virulence factors are shared between and among various members of the enteric bacteria (∞Section 12.11).

21.8 Concept Check

Virulence is determined by invasiveness, toxigenicity, and other factors produced by a pathogen. In most pathogens, a number of factors contribute to virulence. Attenuation is loss of virulence. *Salmonella* displays a wide variety of traits that enhance virulence.

◆ Distinguish between *toxicity* and *invasiveness*. Give examples of organisms that rely almost exclusively on *toxicity* or *invasiveness* to promote virulence.

◆ Explain how an organism may become attenuated. Discuss the role of attenuated organisms in vaccine production.

III VIRULENCE FACTORS AND TOXINS

Extracellular virulence factors and toxins produced by microorganisms promote pathogenesis. A variety of these proteins are produced by different pathogens, but many share molecular characteristics and modes of action. Here we examine representative examples of several virulence factors and toxins with an eye toward common mechanisms.

21.9 Virulence Factors

Pathogen-produced extracellular proteins that aid in the establishment and maintenance of disease are called *virulence factors*. Most virulence factors are *enzymes* that enhance pathogen colonization and growth. For example, streptococci, staphylococci, and certain clostridia produce **hyaluronidase** (Table 21.4), an enzyme that promotes spreading of organisms in tissues by breaking down hyaluronic acid, a polysaccharide that functions as intercellular cement. Hyaluronidase digestion of the intercellular matrix enables these organisms to spread from an initial site. Streptococci and staphylococci also produce an array of proteases, nucleases, and lipases that depolymerize host proteins, nucleic acids, and lipids. Similarly, clostridia that cause gas gangrene produce collagenase, or κ-toxin (Table 21.4), which breaks down the tissue-supporting collagen network, enabling these organisms to spread through the body.

Fibrin, Clots, and Virulence

Fibrin clots are often formed at a site of microbial invasion by the host. The clotting mechanism, triggered by tissue injury, isolates the pathogens, limiting infection to a local

region. Some organisms produce fibrinolytic enzymes that dissolve the clots and make further invasion possible. One fibrinolytic substance produced by *Streptococcus pyogenes* is known as **streptokinase** (Table 21.4).

By contrast, other organisms produce enzymes that *promote* the formation of fibrin clots, causing localization and protection of the organism rather than spread of the organism. The best-studied fibrin-clotting enzyme is **coagulase** (Table 21.4), produced by pathogenic *Staphylococcus aureus*. Coagulase causes fibrin to be deposited on *S. aureus* cells and protects the coated bacteria from attack by host cells. The fibrin matrix produced as a result of coagulase activity probably accounts for the extremely localized nature of many staphylococcal infections, as in boils and pimples (∞Figure 26.21). Coagulase-positive *Staphylococcus aureus* is considered to be more virulent than coagulase-negative strains.

21.9 Concept Check

Pathogens produce a variety of enzymes that enhance virulence by breaking down or altering host tissue to provide access and nutrients. Still other pathogen-produced virulence factors provide protection to the pathogen by interfering with normal host defense mechanisms. These factors enhance colonization and growth of the pathogen.

◆ What advantage does the pathogen gain by producing enzymes that digest structural components of host tissues?

◆ How can the activity of coagulase work to enhance the growth of *Staphylococcus aureus*?

21.10 Exotoxins

Exotoxins are toxic proteins released extracellularly as the organism grows. These toxins may travel from a limited focus of infection and cause damage at distant sites. Table 21.4 provides a summary of the properties and actions of some of the best-known exotoxins as well as other extracellular virulence factors.

Most exotoxins fall into one of three categories; the *cytolytic toxins*, the *A-B toxins*, or the *superantigen toxins*. The cytolytic toxins work by enzymatically attacking cell constituents, causing lysis. The A-B toxins consist of two covalently bonded subunits, A and B. The B component generally binds to a cell surface receptor, allowing the transfer of the A subunit across the targeted cell membrane, where it functions to damage the cell. The superantigens work by stimulating large numbers of immune response cells, resulting in extensive inflammatory reactions (∞Section 22.16).

Cytolytic Toxins

Various pathogens produce proteins that damage the host cytoplasmic membrane, causing cell lysis and death. Because the activity of these toxins is most easily

● **Figure 21.18   Hemolysis.** (a) Zones of hemolysis around colonies of *Streptococcus pyogenes* growing on a blood agar plate. (b) Action of lecithinase, a phospholipase, around colonies of *Clostridium perfringens* growing on an agar medium containing egg yolk, a source of lecithin. Lecithinase dissolves the membranes of red blood cells, leading to the clearing zones shown.

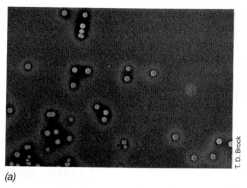
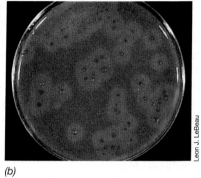

(a) (b)

detected with red blood cells (erythrocytes), they are called **hemolysins** (Table 21.4). However, they also work on cells other than erythrocytes. The production of hemolysins is demonstrated in the laboratory by streaking the organism on a *blood agar plate*. During growth of the colonies, hemolysin is released and lyses the surrounding red blood cells, typically creating a zone of hemolysis (Figure 21.18*a*●).

Some hemolysins attack the phospholipid of the host cytoplasmic membrane. Because the phospholipid lecithin (phosphatidylcholine) is often used as a substrate, these enzymes are called **lecithinases** or **phospholipases**. An example is the α-toxin of *Clostridium perfringens*, a lecithinase that dissolves membrane lipids, resulting in cell lysis (Figure 21.18*b*) (Table 21.4). Since the cytoplasmic membranes of all organisms contain phospholipids, phospholipases sometimes destroy bacterial as well as animal cytoplasmic membranes.

Some hemolysins, however, are not phospholipases. Streptolysin O, a hemolysin produced by streptococci, affects the sterols of the host cytoplasmic membrane. *Leukocidins* (Table 21.4) are lytic agents capable of lysing white blood cells and may decrease host resistance (●Section 22.2).

Diphtheria Toxin

Diphtheria toxin, produced by *Corynebacterium diphtheriae*, is an important virulence factor in the pathogenesis of diphtheria (●Section 26.3). Rats and mice are relatively resistant to diphtheria toxin, but human, rabbit, guinea pig, and bird cells are very susceptible, with only a single toxin molecule required to kill each cell. Diphtheria toxin is an A-B toxin secreted by cells of *C. diphtheriae* as a single polypeptide. Fragment B promotes specific binding of the toxin to a host cell receptor (Figure 21.19●). After binding, proteolytic cleavage between Fragment A and B

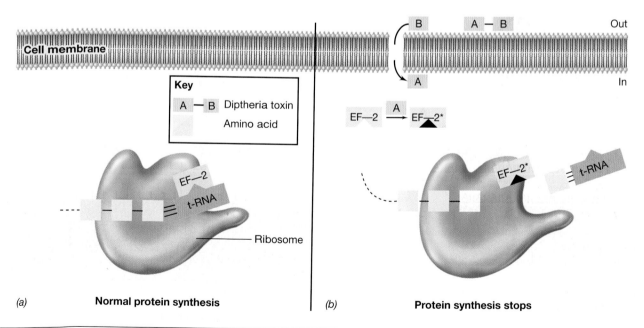

(a) **Normal protein synthesis**   (b) **Protein synthesis stops**

● **Figure 21.19   The action of diphtheria toxin from *Corynebacterium diphtheriae*.** (a) Elongation factor 2 (EF-2) normally binds to the ribosome and brings an amino acid–charged t-RNA to the ribosome, causing protein elongation. (b) Diphtheria toxin binds to the cell membrane, where it is cleaved and the A peptide is internalized. The A peptide catalyzes the ADP-ribosylation of elongation factor 2 (EF-2*). The modified elongation factor no longer aids transfer of amino acids to the growing polypeptide chain, resulting in shutdown of protein synthesis and death of the cell.

Web Tutorial 21.1   Diphtheria and Cholera Toxins

MHC proteins function primarily as antigen-presenting molecules that interact with both the antigen and the TCR.

MHC genes encode both *class I* and *class II* MHC proteins. **Class I MHC proteins** are found on the surfaces of *all* nucleated cells. **Class II MHC proteins** are found only on the surface of B lymphocytes, macrophages, and dendritic cells, all of which are antigen-presenting cells (APCs). The differential cellular distribution of these molecules is linked to their individual functions.

Class I MHC proteins consist of two polypeptides (Figure 22.11a●), a membrane-embedded alpha (α) chain encoded in the MHC gene region, and a smaller protein called *beta-2 microglobulin* ($\beta_2 m$) encoded by a non-MHC gene. Class II MHC proteins consist of two noncovalently linked polypeptides called α and β. Like class I α chains, these polypeptides are embedded in the cytoplasmic membrane and project outward from the cell surface (Figure 22.10b).

MHC proteins are not structurally identical, even within a given species. Different individuals often show subtle differences in the amino acid sequence of their MHC proteins. These limited sequence variations are called *polymorphisms*. There are several hundred different MHC genes in humans. These MHC polymorphisms are the major antigenic barrier for tissue transplantation from one individual to another. As a result, tissue transplants not matched for MHC identity are recognized as nonself and are therefore rejected. The detailed molecular structure and genetic organization of the MHC genes and proteins are presented in Chapter 23.

Antigen Presentation

MHC proteins function as molecular reference points that permit T cells to identify foreign antigens. T cells continually sample the molecular landscape on the surface of other cells to identify cells carrying nonself antigens. The TCR on a given T cell binds only to MHC molecules having foreign antigens embedded in the MHC structure; a T cell does not recognize a foreign antigen unless it is presented in the context of an MHC protein.

How does this happen? A variety of host cells can acquire nonself antigens through infection or phagocytosis. The host cells then degrade (process) the antigens and load the processed antigen peptides into the MHC protein. The MHC–peptide complex is passed through the cytoplasmic membrane and presented on the surface of the cell to be recognized by T cells. Two distinct antigen-processing schemes are known, one for MHC I antigen presentation and one for MHC II antigen presentation (Figure 22.12●).

MHC I proteins present peptide antigens derived from pathogen proteins found in the cytoplasm of non-phagocytic cells (Figure 22.12a). The foreign pathogen proteins result from intracellular infections by viruses and other intracellular pathogens. Proteins derived, for example, from infecting viruses, are taken up and digested in the cytoplasm in a structure called the *proteosome*. Peptide antigens of about 10 amino acids in length are transported via an energy-dependent reaction into the endoplasmic reticulum (ER) through a pore formed by two proteins, called the *transporters associated with antigen processing (TAP)*. Once the peptides have entered the ER, they are bound by the MHC I protein, held in place near the TAP site by a group of *chaperone* proteins. These chaperone proteins direct the MHC I molecule to the TAP site, holding it in place until a peptide is bound. The MHC I–peptide complex is then released from the chaperones and moves to the cell surface where it integrates into the membrane and can be recognized by T_C cells.

Thus, the MHC proteins act as a *platform* on which the foreign viral antigen is bound. Next, the TCR on the surface of a T cell interacts with both antigen (nonself) and MHC protein (self) on the surface of the target cell. This T cell–target cell interaction induces specialized T-cytotoxic (T_C) cells to produce cytotoxic proteins called *perforins* (see Section 22.6) that kill the virus-infected target cell.

A second antigen presentation scheme involves the MHC class II proteins (Figure 22.12b). MHC II proteins are expressed exclusively in the phagocytic APCs, where they function to present peptides from engulfed extracellular pathogens such as bacteria. MHC class II proteins are initially assembled in the ER, much like MHC I proteins. However, unlike MHC I proteins, a chaperone protein called *Ii*, or *invariant chain*, binds to the MHC II protein, blocking the MHC II peptide-binding site and preventing the MHC II proteins from binding peptides inside the ER.

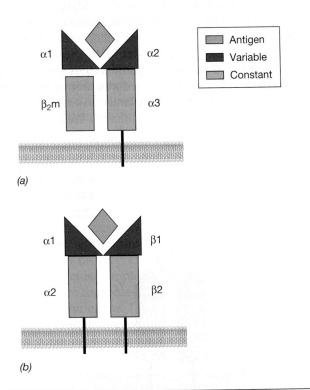

(a)

(b)

● **Figure 22.11 Structures of the MHC proteins.** (a) Class I MHC protein. The $\alpha 1$ and $\alpha 2$ domains interact to form the peptide antigen-binding site. (b) Class II MHC protein. The $\alpha 1$ and $\beta 1$ domains combine to form the peptide antigen-binding site

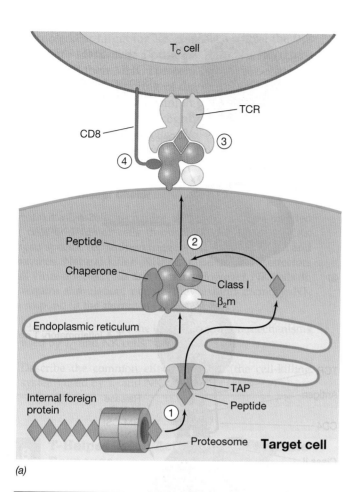

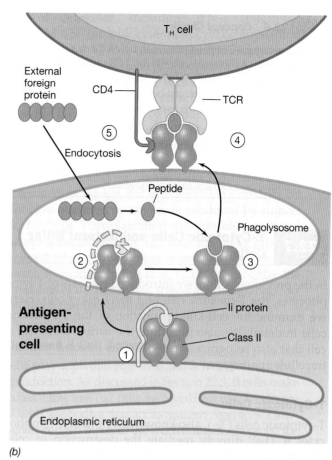

(a)

(b)

● **Figure 22.12   Antigen presentation by MHC I and MHC II proteins.** (a) In the MHC I antigen presentation pathway, the membrane-bound MHC I proteins are made and assembled in the endoplasmic reticulum. Chaperone proteins stabilize MHC I until antigen is bound. ① Protein antigens manufactured within the cell, for instance from viruses, are degraded by the proteosome in the cytoplasm and transported across the endoplasmic reticulum membrane through a pore formed by the TAP proteins. ② The peptides then bind to MHC I, are transported to the cell surface, and ③ interact with T-cell receptors (TCRs) on the surface of T_C cells. ④ The CD8 coreceptor on the T_C cell engages the class I MHC, resulting in a stronger complex. The T_C cells then release cytokines and cytotoxins, killing the target cell. Any nucleated cell can act as a target cell for T cells recognizing peptide-MHC I complexes. (b) In the MHC II antigen presentation pathway, ① MHC II proteins are produced in the endoplasmic reticulum and are assembled with a blocking protein, Ii (invariant chain), preventing MHC II from complexing with peptides found in the endoplasmic reticulum. ② Lysosomes containing MHC II then fuse with phagosomes, forming phagolysosomes where the Ii and foreign proteins, imported from outside the cell by endocytosis, are digested. ③ The MHC II protein then binds to the digested foreign peptides, and the complex is transported to the cell surface ④ where it interacts with TCRs and ⑤ the CD4 coreceptor on T_H cells. The T_H cells then release cytokines that act on other cells to activate an immune response. Only APCs can be targets for T cells recognizing the peptide–MHC II complex. The APCs are macrophages, dendritic cells, and B cells.

These MHC II-Ii complexes are transported from the ER to the cell vacuoles (lysosomes) (see Section 22.2). The phagosome containing the foreign antigen fuses with the lysosome to form a phagolysosome, and the foreign antigens, as well as the Ii peptide, are digested by lysosomal enzymes. The foreign peptides, generally about 11 to 15 amino acids in length (slightly larger than MHC I-binding peptides), are then bound in the newly opened MHC II antigen-binding site. The complex is then transported to the cytoplasmic membrane, where it is presented to specialized T-helper (T_H) cells. The T_H cells, through the TCR, recognize the MHC II-peptide complex. This interaction activates the T_H cells to secrete cytokines that either stimulate antibody production by specific B-cell clones or cause inflammation (see Sections 22.8 and 22.10).

CD4 and CD8 Coreceptors

In addition to the TCR, each T cell expresses a unique cell surface protein that functions as a *coreceptor*. T_H cells express a CD4 protein coreceptor, while T_C cells express a CD8 protein coreceptor (Figure 22.12). When the TCR binds to the peptide–MHC complex, the coreceptor on the T cell also binds to the MHC protein, strengthening the molecular interactions and enhancing activation of the T cell. CD4 binds only to the class II protein, strengthening T_H cell interaction with APCs that express MHC II protein. Likewise, CD8 binds only to the MHC I protein, enhancing the binding of T_C cells to MHC I-bearing target cells. The CD4 and CD8 proteins are used *in vitro* as markers to differentiate T_H cells from T_C cells (∞ Section 24.9).

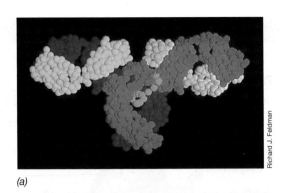

(a)

Richard J. Feldman

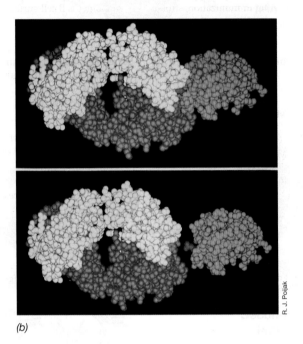

(b)

R. J. Poljak

● **Figure 22.16 Immunoglobulin structure and the antigen-binding site.** (a) Three-dimensional view of an IgG molecule. The heavy chains are shown in red and dark blue. The light chains are in green and light blue. (b) Structure of the combining site of an antigen and Ig. The antigen (lysozyme) is in green. The variable region of the Ig heavy chain is shown in blue, and the light chain in yellow. The amino acid shown in red is a glutamine residue of lysozyme. The glutamine residue fits into a pocket on the Ig molecule, but the overall antigen-antibody recognition involves contacts made between several other amino acids on both the Ig and the antigen. Reprinted with permission from *Science* 233:747 (1986) ©AAAS.

silon (ε). The constant domain sequences constitute the carboxy-terminal three-fourths of the heavy chains of Igs of IgG, IgA, and IgD, respectively, and four-fifths of the heavy chains of IgM and IgE (Figure 22.17●). Each antibody of the IgM class, for example, contains amino acids in its heavy chain constant domains that constitute the *mu* sequence.

In a typical immune response, two Igs of different classes, often IgM and IgG, may bind the same epitope. In this case, the *variable* (antigen-binding) domains of the heavy and light chains from both IgM and IgG may be identical, but the class-determining *constant* domains of the

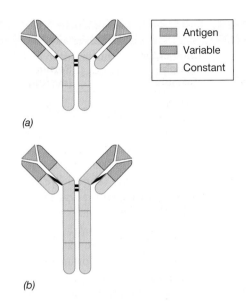

(a)

(b)

● **Figure 22.17 Immunoglobulin classes.** All classes of Igs have V_H and V_L domains (red) that bind antigen (brown). (a) IgG, IgA, and IgD have three constant domains (blue). (b) IgM and IgE each have a fourth constant domain.

heavy chains are different, with IgM expressing a mu (μ) heavy chain and IgG expressing a gamma (γ) heavy chain.

The structure of IgM is shown in Figure 22.18●. IgM is usually found as aggregate of five immunoglobulin molecules attached by at least one *J chain*. Each heavy chain of IgM contains a fourth constant domain

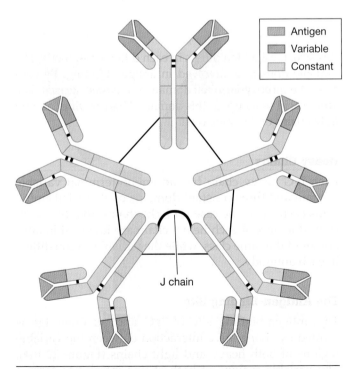

J chain

● **Figure 22.18 Immunoglobulin M.** IgM is found in serum as a pentameric protein consisting of five IgM proteins covalently linked to one another via disulfide bonds and a J chain protein. Because it is a pentamer, IgM can bind up to 10 antigens, as shown.

(C_H4). IgM is the first class of Ig made in a typical immune response to a bacterial infection, but IgMs are generally of low affinity. Antigen-binding strength is enhanced to some degree, however, by the high *valence* of the pentameric IgM molecule; 10 binding sites are available for interaction with antigen (Table 22.2 and Figure 22.18). The term *avidity* is used to describe the *strength of binding* by multivalent antigen-binding molecules. Thus, IgM has *low* affinity but *high* avidity for antigen. About 10% of serum antibodies are IgM. IgM monomers are also found on the surface of B cells, where they bind antigen.

IgA, in the dimeric form (Figure 22.19●), is present in body fluids such as saliva, tears, breast milk colostrum, and mucosal secretions from the gastrointestinal, respiratory, and genitourinary tracts. All of these mucosal surfaces are associated with mucosal-associated lymphoid tissue (MALT) (Figure 22.2) that produces IgA. These mucosal surfaces total about 400 m^2, and large amounts of secretory IgA are produced—about 10 g per day. By contrast, the serum IgG produced in an individual is about 5 g per day. Thus the total amount of *secretory* IgA produced by the body is higher than the amount of *serum* IgG. The secretory form consists of two IgA molecules covalently linked by a J (joining) chain peptide and a protein *secretory component* that aids in transport of IgA across membranes (Figure 22.19). IgA is also present in serum as a monomer (Table 22.2).

IgE is found in serum in extremely small amounts (about 1 of every 50,000 serum Ig molecules is IgE). Despite its low concentration, IgE is important because immediate-type hypersensitivities (allergies) are mediated by IgE (see Section 22.15). The molecular weight of an IgE molecule is significantly higher than most other Igs (Table 22.2) because, like IgM, IgE has a fourth constant domain (Figure 22.17). This additional constant region functions to bind IgE to mast cell surfaces (see Figure 22.25), an important prerequisite for allergic reactions.

IgD, present in serum in low concentrations, has no known function. However, IgD, with IgM, is abundant on the surfaces of B cells, and IgD is especially abundant on memory B cells. Thus, IgD surface Igs may be important for the secondary antibody response.

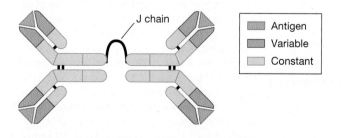

● **Figure 22.19 Immunoglobulin A.** Secretory IgA (sIgA) is often found in body secretions as a dimer consisting of two IgA proteins covalently linked to one another via a J chain protein. A secretory protein component, not shown, aids in transport of IgA across membranes.

 22.9 Concept Check

Ig (antibody) proteins consist of four chains, two heavy and two light. The antigen-binding site is formed by the interaction of variable regions of heavy and light chains. Each class of Ig has different structural and functional characteristics.

◆ Identify the Ig domains involved in antigen binding.

◆ Differentiate between Ig classes using functional and structural characteristics.

22.10 Antibody Production

In this section, we will examine a typical antibody response, first reviewing the complex cell interactions necessary to produce effective antigen-specific immunity. We will then look briefly at the unique genetic mechanism used to generate antibody diversity. Finally, we will review the salient features leading up to and including antibody production in response to antigen exposure.

T Cell–B Cell Interactions

The production of immunoglobulins in response to antigen involves interactions between T cells and B cells through their respective antigen-specific cell surface molecules, the TCR on the T cell, and the surface Ig on the B cell (Figure 22.14). Each step in the antibody production mechanism is *highly* specific. Upon initial antigen exposure, the antigen is bound by a B cell with an antigen-specific surface Ig. The antigen-Ig complex is then internalized, and the antigen is processed and loaded onto MHC II proteins. Thus, the B cell first functions as an antigen-presenting cell, using its surface Ig to capture antigen. The MHC–antigen complex then moves to the cell surface, displaying the antigen for interaction with a T_H2 cell having an antigen-specific TCR on its surface (see Figure 22.12b). Binding of the B-cell MHC–antigen complex activates the antigen-specific T_H2 cell, stimulating cytokine production. The released cytokines stimulate the nearby antigen-specific B cells, activating them to produce and secrete the antigen-specific antibody.

The Genetic Mechanism of Antibody and TCR Diversity

Each individual is capable of producing billions of different antibodies. How does the immune system choose the correct antibody for each antigen? Early theories postulated that individual genes encoded each antibody, but we now know that only a small number of genes are used to encode this immense antibody diversity. Antibody production starts with stepwise rearrangements of the genes necessary to encode an Ig. During development of lymphocytes in the bone marrow, both heavy and light chain gene rearrangements occur in B cells. The genes are reassorted—individual gene pieces are mixed and matched in a variety of combinations—by gene splicing and rearrangements in the maturing lymphocytes, a process known as *somatic recombination*. Figure 22.20● shows a typical rearrangement and expression pattern for the human κ light chain. The heavy chain genes rearrange in an analogous, but more complex fashion. In each mature B cell, the final result is a single functional heavy chain gene and a single functional light chain gene. Each of these rearranged genes is transcribed and expressed as an Ig antigen receptor protein on the surface of the B cell. Upon antigen exposure, as we have seen, soluble Ig is produced by the B cell. In addition, antigen induces genetic hypermutation in the Ig genes, further modifying and diversifying the expressed antibodies. The large number of possible gene rearrangements, coupled with the somatic mutations after antigen exposure, allow almost unlimited diversity in Igs (⚬⚬Section 23.6).

Similar rearrangements also occur during T-cell development and result in the generation of considerable diversity in T-cell receptors. However, T cells do not use the hypermutation mechanism to expand diversity (⚬⚬Section 23.8).

Antibody Production and Immune Memory

Starting with a B cell that expresses a surface Ig, antibody production begins with antigen exposure and culminates with the production and secretion of an antigen-specific antibody according to the following sequence:

1. Antigens are spread via the lymphatic and blood circulatory systems to neighboring secondary lymphoid organs such as lymph nodes, spleen, or mucosal-associated lymphoid tissue (MALT) (Section 22.1 and Figure 22.2). The route of antigen exposure influences the nature of the antibodies produced. Intravenously injected antigen travels via the blood to the spleen, where mostly IgM and IgG and serum IgA antibodies are formed. If antigen is introduced subcutaneously, intradermally, topically, or intraperitoneally, the lymphatic system carries antigen to the nearest lymph nodes, again stimulating production of mostly IgM, IgG, and serum IgA. Antigen introduced to mucosal surfaces is delivered to the nearest MALT. For example, antigen delivered by mouth is delivered to the MALT in the intestinal tract, preferentially stimulating specific antibody production in the gut, with a primary IgM response and a secondary antigen-specific secretory IgA response.

2. Following the initial antigen exposure, each antigen-stimulated B cell multiplies and differentiates to form both antibody-secreting **plasma cells** and **memory cells** (Figure 22.14). *Plasma cells* are relatively short-lived (less than 1 week), but produce and secrete large amounts of IgM antibody in this **primary antibody response** (Figure 22.21●). There is a latent period before specific antibody appears in the blood, followed by a gradual increase in *antibody titer* (antibody quantity), and then a slow fall in the primary antibody response.

3. The **memory B cells** generated by the initial exposure to antigen may live for years. If reexposure to the immunizing antigen occurs at a later time, memory cells need no T-cell activation; they quickly transform to plasma cells and begin producing IgG. Upon reexposure, the antibody titer rises rapidly to a level 10–100 times greater than the titer achieved following the first exposure. This rise in antibody titer is referred to as the **secondary antibody response** (Figure 22.21). The secondary response is the result of *immunologic memory* and produces a more rapid, more abundant antibody response when compared with the primary response. The secondary response is also characterized by a switch from IgM to IgG production, a phenomenon called *class switching* (Figure 22.21).

4. Over time the titer slowly decreases, but subsequent exposure to the same antigen can cause another secondary response. The secondary response is the basis for the immunization procedure known as a "booster shot" (for example, the yearly rabies shot given to do-

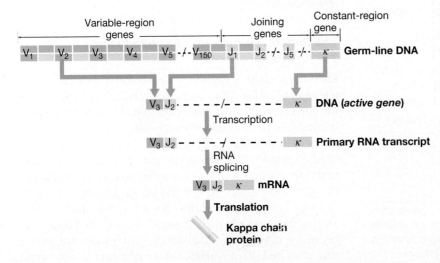

● Figure 22.20 Immunoglobulin kappa chain gene rearrangement in human B cells. The gene segments are arranged in tandem in the kappa (κ) light-chain complex on chromosome 2. DNA rearrangements are completed in the maturing B cell. Any one of the 150 V (variable) sequences may combine with any one of the 5 J sequences. Thus, 750 (150 V × 5) recombinations are possible, encoding 750 distinct kappa chains, but only one rearrangement actually occurs in each cell. An analogous process occurs in the heavy chain genes in each B cell. The heavy chain gene complex has even more gene segments, allowing greater numbers of recombinations and potential heavy chains, but again only one heavy chain rearrangement is found in each mature B cell. The mature Ig protein is made by combining two light chain proteins with two heavy chain proteins after translation (⚬⚬Section 23.6).

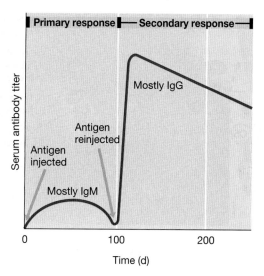

● **Figure 22.21 Primary and secondary antibody responses in serum.** The antigen injected at day 0 and day 100 must be identical to induce the secondary response. The secondary response, also called a booster response, may be more than tenfold greater than the primary response. Note the class switch from IgM production in the primary response to IgG production in the secondary response.

mestic animals). Periodic reimmunization maintains high levels of circulating antibody specific for a certain antigen and provides long-term active protection against individual infectious diseases (see Section 22.12).

 22.10 Concept Check

Antibody production is initiated by antigen contact with an antigen-specific B cell that processes the antigen and presents it to an antigen-specific T_H2 cell. The activated T_H2 cell then signals the antigen-specific B cell to produce antibody. Activated B cells live for years as memory cells and can rapidly produce large quantities (high titers) of antibodies upon re-exposure to antigen.

◆ How do B cells act as APCs?

◆ How do T_H2 cells activate antigen-specific B cells?

◆ Explain the rationale for a "booster" immunization.

22.11 **Complement, Antibodies, and Pathogen Destruction**

A related group of sequentially interacting proteins called **complement** play an important role as an effector in both innate and adaptive immunity. Complement proteins are often activated by interactions with antigen–antibody complexes, but can also be activated by innate mechanisms in the absence of specific antibodies. Complement proteins can react sequentially to cause *lysis* of bacterial cells or can label antigen-expressing cells for

enhanced recognition by phagocytes, accelerating destruction of the antigens.

Classical Complement Activation and Cell Damage

Complement is made up of a number of proteins, many with enzymatic activity. These proteins are activated in a sequential, ordered fashion by antigen–antibody complexes on bacterial cells. The result is cell membrane damage, leakage of contents, and possibly lysis of the bacterial cell. Complement proteins are found at comparable levels in the serum of all individuals. A major function of antibody is to recognize invading cells and activate the complement system for attack. Thus, many antibodies, each specific for a single antigen, can recruit the ubiquitous complement proteins; the immune system does not need individual enzymes to attack each invading agent.

The individual proteins of complement are designated C1, C2, C3, and so on. Classical activation of complement occurs only with antibodies of the IgG and IgM classes (see Table 22.2). When IgG or IgM antibodies bind antigens, especially on cell surfaces, the antibodies fix (bind) the ever-present complement proteins (Figure 22.22●). The complement proteins react in a sequential cascade, with activation of one complement component leading to activation of the next, and so on. The key steps start with (1) binding of antibody to antigen (initiation); (2) binding of C1 components (C1q, C1r, and C1s) to the antibody–antigen complex, leading to C4–C2 binding at an adjacent membrane site, and activation and binding of C3; (3) membrane-bound C3 catalyzes formation of a C5-C6-C7 complex at a second membrane site, and C8 and C9 are then deposited with the C5–C6–C7 complex, resulting in membrane damage and cell lysis (Figure 22.23●). The C5-9 components are known as the *membrane attack complex (MAC)* because they interact at a single membrane site to form a pore.

By-products of complement activation include chemoattractants called *anaphylatoxins*. They cause inflammatory reactions at the site of complement deposition (see Section 22.15). Reactions involving C3 result in chemotactic attraction and activation of phagocytes, resulting in increased phagocytosis. Reactions involving C5 lead to T-cell attraction and cytokine release.

Complement is a bactericidal and lytic agent for many gram-negative *Bacteria* when used in conjunction with specific antibodies. However, gram-positive *Bacteria* are not killed by complement, even in the presence of specific antibodies. Gram-positive *Bacteria* can, however, be destroyed through opsonization.

Opsonization

When antibody binds antigen on the surface of a cell, the cell is much more likely to be phagocytized. When complement binds an antibody–antigen complex on a cell surface, the cell is even more likely to be phagocytized. This is because most phagocytes, including macrophages and B cells, have antibody receptors (*FcR*) as well as C3 receptors

vaccines, periodic reimmunization is necessary to maintain effective immunity.

The importance of immunization in controlling infectious diseases is well established. Introduction of an effective vaccine into a population often dramatically reduces the incidence of the disease (∞Figures 26.13 and 26.23). The degree of immunity obtained by vaccination, however, varies greatly with the individual as well as with the quality and quantity of the vaccine. Lifelong immunity is rarely achieved by means of a single injection, or even a series of injections, and the immune cells induced by immunization gradually disappear from the body. On the other hand, antigenic stimulation often occurs even in the absence of artificial immunization through natural infections. A natural infection induces a secondary response, leading to an increase in production of antibody. In the complete absence of antigenic stimulation, the length of effective immunity varies considerably with different antigens. For example, effective immunity to tetanus toxoid may last many years, but immunity induced by a particular influenza virus vaccine may disappear within a year or two in the absence of a secondary stimulation.

Immunizations not only benefit the individual but are effective public health measures because disease spreads poorly through a population with a large proportion of immune individuals (∞Section 25.5).

Passive Immunity

Passive immunity is introduced by injecting preformed antibodies. The antibody-containing preparation is known as a *serum*, an *antiserum*, or an *antitoxin* (antibodies directed against a toxin). Antisera are obtained from immunized animals, such as horses, or from humans with high antibody titers. These individuals are said to be hyperimmune. The antiserum or antitoxin is standardized to contain a known antibody titer; a sufficient number of units of antiserum must be injected to neutralize any antigen that might be present in the body. The immunoglobulin fraction separated from serum pooled from a number of individuals is also used for passive immunization. Pooled sera contain a variety of antibodies induced by artificial or natural exposure to various antigens.

22.13 Concept Check

Immunity to infectious disease can be either passive or active, natural or artificial. Immunization, a form of artificial active immunity, is widely employed to prevent infectious diseases. Most agents used for immunization are either attenuated or inactivated pathogens or inactivated forms of natural microbial products.

◆ Provide an example of natural passive immunity. How does natural passive immunity benefit the immunized individual?

◆ Provide an example of artificial active immunity. How does artificial active immunity benefit the immunized individual?

22.14 New Immunization Strategies

Most immunization preparations are produced from whole organisms or toxoids, as described in the previous section. However, there are several other methods for producing antigens suitable for immunization.

Synthetic and Genetically Engineered Immunizing Agents

The simplest alternate approach to vaccine development is the use of *synthetic peptides*. To make a vaccine, a peptide can be synthesized that corresponds to a known epitope on an infectious agent. For example, the structure of the protein antigen responsible for immunity to foot-and-mouth virus, an important animal pathogen, is known. A synthetic peptide of 20 amino acids constituting an important antigenic determinant of the protein has been made and attached to suitable carrier molecules. This synthetic vaccine evokes an excellent neutralizing antibody response to foot-and-mouth virus. However, as a general method, this approach has one major problem: The entire sequence, representing the complete antigenic profile of the protein, must be known to make an effective vaccine. This condition has been met for foot-and-mouth virus, and a variety of other pathogens have now been sequenced, providing the potential for identifying the specific antigenic profile necessary to design an effective vaccine (∞Section 15.3).

Using information derived from genomics of pathogens, molecular biology techniques can be used to make vaccines. For example, genes that encode antigens from virtually any virus can be cloned into the vaccinia virus genome and expressed. Inoculation with the *genetically engineered* vaccinia virus can then be used to induce immunity to the product of the cloned gene. Such a preparation is a *recombinant-vector vaccine*. This method depends on the identification and cloning of the gene that encodes the antigen and also on the ability of the vaccinia virus to express the cloned gene as an antigenic protein. An effective recombinant vaccinia-rabies vaccine has been developed for use in animals. Recombinant DNA methods to develop vaccines will be discussed in Section 31.6.

Another immunization strategy involves the production of recombinant DNA proteins as immunogens. First, a pathogen gene must be cloned in a suitable microbial host. The host, chosen for its ability to express protein encoded by the cloned gene, will then express the pathogen protein. The pathogen protein can then be harvested and used as a vaccine. This is known as a *recombinant antigen vaccine*. For example, the current hepatitis B virus vaccine is a major hepatitis surface protein antigen (HbsAg) expressed by yeast cells.

DNA Vaccines

A novel method for immunization is based on expression of cloned genes in host cells. Bacterial plasmids containing cloned DNA are injected intramuscularly into a host animal. After several weeks, the host responds with T_C cells, T_H1 cells, and antibodies directed to the protein encoded by the cloned DNA. Taken up by host cells, the DNA is transcribed and translated to produce immunogenic proteins, triggering a conventional immune response. These plasmids are called DNA vaccines.

DNA vaccines provide considerable advantages over many conventional immunization strategies. For instance, since only a single foreign pathogen gene is normally cloned and injected, there is no chance of an infection as there might be with an attenuated vaccine. Second, genes for individual antigens such as a tumor-specific antigen, or even a single antigenic determinant can be cloned, targeting the immune response to a particular cell component. The response can also be targeted directly to APCs by including a MHC class II promoter in the gene construct. The promoter assures selective expression in dendritic cells, B cells, and macrophages, the only cells that express the MHC II protein. The expressed and processed antigen can then be presented on both MHC I and MHC II proteins. Thus, a single bioengineered plasmid can encode an antigen and elicit a complete immune response, inducing immune T cells and antibodies.

■ 22.14 Concept Check

Alternate immunization strategies using bioengineered molecules eliminate exposure to microorganisms and, in some cases, even to protein antigen. Application of these strategies may provide safer and more targeted vaccines.

◆ Provide two examples of alternate immunization strategies.

◆ What is the advantage of each alternative immunization strategy over current immunization procedures?

 V IMMUNE RESPONSE DISEASES

Immune reactions can cause host cell damage and disease. **Hypersensitivity** responses are inappropriate immune re-

sponses that result in host damage. Hypersensitivity diseases are categorized according to the antigens and effector mechanisms that produce disease (Table 22.5). **Superantigens** are proteins produced by certain bacteria and viruses that cause widespread stimulation of immune cells, resulting in host damage by activating a massive inflammatory response.

22.15 | **Allergy, Hypersensitivity, and Autoimmunity**

Antibody-mediated *immediate hypersensitivity* is commonly called *allergy*. Cell-mediated reactions also cause disease in the form of *delayed-type hypersensitivity*. Autoimmune diseases are directed against self antigens.

Immediate Hypersensitivity (Type I Hypersensitivity)

Immediate hypersensitivity or Type I hypersensitivity is mediated by release of vasoactive products from IgE antibody-coated mast cells (Table 22.5). Immediate hypersensitivity reactions, commonly called *allergies*, occur within minutes after exposure to antigen. Depending on the individual and the antigen, immediate hypersensitivity reactions can be very mild or can cause extremely severe life-threatening reactions, a condition know as *anaphylaxis*. Antigens that cause Type I hypersensitivities are called *allergens*.

About 20% of the population suffers from immediate hypersensitivities involving allergic (*anaphylactic*) reactions to specific allergens such as pollens, animal dander, certain foods, and a variety of other agents (Table 22.6). Almost all allergens are delivered at the surface of mucous membranes such as the lungs or the gut. There they stimulate T_H2 cells to produce cytokines that preferentially induce B cells to make IgE antibodies. Rather than circulating like IgG or IgM, the allergen-specific IgE antibodies bind via a cell-membrane-binding constant domain (see Section 22.8) to IgE receptors on mast cells (Figure 22.25●). Mast cells are nonmotile granulocytes (see Section 22.2) associated with

---

Table 22.5	Hypersensitivity			
Classification	**Description**	**Immune mechanism**	**Time of latency**	**Examples**
Type I	Immediate	IgE sensitization of mast cells	Minutes	Reaction to bee venom (sting) Hay fever
Type II	Cytotoxic[a]	IgG interaction with cell surface antigen	Hours	Drug reactions (penicillin)
Type III	Immune complex	IgG interaction with soluble or circulating antigen	Hours	Systemic lupus erythematosis (SLE)
Type IV	Delayed type	T_H1 inflammatory cells	Days	Poison ivy Tuberculin test

[a] Autoimmune diseases may be caused by Type II, Type III, or Type IV reactions.

Table 22.6	Immediate hypersensitivity allergens

Pollen and fungal spores (hay fever)
Insect venoms (bee sting)
Certain foods
Animal dander
Mites in house dust

the connective tissue adjacent to capillaries throughout the body. With any subsequent exposure to the immunizing allergen, the cell-bound IgE molecules bind the antigen. Cross-linking of two or more IgEs by an antigen triggers the release of several soluble allergic mediators from the mast cells, a process called *degranulation*. These mediators cause allergic symptoms within minutes of antigen exposure. In general, these symptoms are relatively short-lived, but once initially sensitized by an allergen, an individual can respond to each subsequent exposure to the antigen.

The primary chemical mediators released from mast cells are histamine and serotonin, modified amino acids that cause rapid dilation of blood vessels and contraction of smooth muscle, initiating the symptoms of systemic anaphylaxis. These symptoms include vasodilation (causing a sharp drop in blood pressure), severe respiratory distress caused by bronchial edema, flushed skin, mucus production, sneezing, and itchy, watery eyes. If severe cases of anaphylaxis are not treated immediately with adrenalin to counter smooth muscle contraction, increase blood pressure, and promote breathing, death can

occur due to anaphylactic shock. Fortunately, the magnitude of most allergic reactions is limited to mild local anaphylaxis involving symptoms such as itchy, watery eyes. Less serious allergic symptoms are treated with drugs called *antihistamines* that neutralize the histamine mediators. Treatment for more serious symptoms may also include general antiinflammatory drugs such as steroids. Finally, immunization with escalating doses of the allergen may be done to shift antibody production from IgE to IgG. The IgG interacts with antigen, preventing interactions with the relatively scarce IgE, stopping allergic symptoms and inhibiting production of more IgE. This procedure is called *desensitization*.

Delayed-Type Hypersensitivity (Type IV Hypersensitivity)

Type IV hypersensitivity or delayed-type hypersensitivity (DTH) is cell-mediated hypersensitivity characterized by tissue damage due to inflammatory responses produced by T_H1 inflammatory cells (Table 22.5). Symptoms begin to appear several hours after secondary exposure to the eliciting antigen, with a *maximal* response usually occurring in 24–48 hours. Typical antigens include certain microorganisms, a few self-antigens (Table 22.7), and several chemicals that covalently bind to the skin, creating new antigens. Hypersensitivity to these newly created antigens is known as *contact dermatitis* and results in skin reactions to poison ivy (Figure 22.26●), jewelry, cosmetics, latex, and other chemicals. Within several hours after exposure to the agent, the skin feels itchy at the site of contact, and reddening and

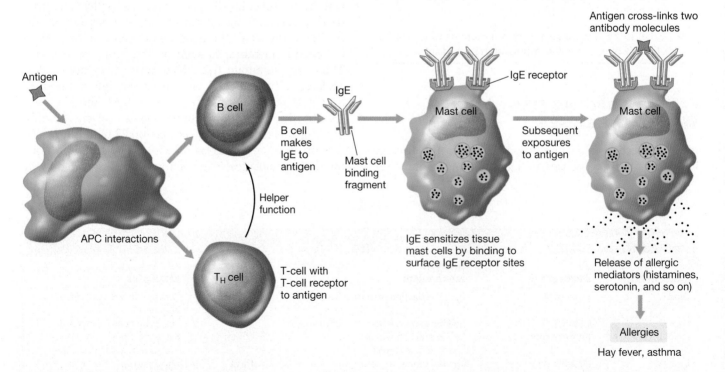

●**Figure 22.25 Immediate hypersensitivity.** IgE binds to mast cells by means of a high-affinity surface receptor for the C_H4 domain. Binding arms the mast cell. Antigen contact and cross-linking of the IgE proteins initiates release of vasoactive mediators, resulting in systemic symptoms ranging from mild allergies to life-threatening anaphylaxis.

Table 22.7 Autoimmune diseases of humans

Disease	Organ or area affected	Mechanism (hypersensitivity type)
Juvenile diabetes (insulin-dependent diabetes mellitus)	Pancreas	Cell-mediated immunity and autoantibodies against surface and cytoplasmic antigens of islets of Langerhans (II and IV)
Myasthenia gravis	Skeletal muscle	Autoantibodies against acetylcholine receptors on skeletal muscle (II)
Goodpasture's syndrome	Kidney	Autoantibodies against basement membrane of kidney glomeruli (II)
Rheumatoid arthritis	Cartilage	Autoantibodies against self IgG antibodies, which form complexes deposited in joint tissue, causing inflammation and cartilage destruction (III)
Hashimoto's disease (hypothyroidism)	Thyroid	Autoantibodies to thyroid surface antigens (II)
Male infertility (some cases)	Sperm cells	Autoantibodies agglutinate host sperm cells (II)
Pernicious anemia	Intrinsic factor	Autoantibodies prevent absorption of vitamin B_{12} (III)
Systemic lupus erythematosis	DNA, cardiolipin, nucleoprotein, blood clotting proteins	Massive autoantibody response to various cellular constituents results in immune complex formation (III)
Addison's disease	Adrenal glands	Autoantibodies to adrenal cell antigens (II)
Allergic encephalitis	Brain	Cell-mediated response against brain tissue (IV)
Multiple sclerosis	Brain	Cell-mediated and autoantibody response against central nervous system (II and IV)

swelling appear, indicative of a general inflammatory response. Localized tissue destruction, often in the form of blistering, occurs as a result of the activities of immune cells, again with a maximal reaction in about 24–48 hours.

An example of delayed-type hypersensitivity is the development of immunity to the causal agent of tuberculosis, *Mycobacterium tuberculosis* (Figure 22.24). This protective cellular immune response was discovered by Robert Koch in his classical studies on tuberculosis (∞ Section 1.6) and has been widely studied. Antigens derived from the bacterium, when injected subcutaneously into an animal previously infected with *M. tuberculosis*, elicit a characteristic skin reaction that develops fully only after a period of 24–48 hours. By contrast, skin reactions to IgE-mediated immediate hypersensitivity responses develop within minutes after antigen injection. In the region of the introduced antigen, T_H1 cells become stimulated by the antigen and release cytokines that attract and activate large numbers of macrophages, which in turn produce a skin reaction at the site of injection. The reaction is characterized by induration (hardening), edema (swelling), erythema (reddening), pain, and localized heating, common symptoms of an inflammatory response (see Section 22.3). The activated macrophages then ingest and destroy the invading antigen. This skin response is the basis for the tuberculin test used to determine prior exposure to *M. tuberculosis* (Figure 22.26).

A number of microbial infections elicit delayed-type hypersensitivity reactions. In addition to tuberculosis, these include leprosy, brucellosis, psittacosis (all caused by species of *Bacteria*), mumps (caused by a virus), and coccidioidomycosis, histoplasmosis, and blastomycosis (caused by fungi). In all of these cellular immune reactions, visible antigen-specific skin responses resembling

the tuberculin reaction (Figure 22.26) occur after injection of antigens derived from the pathogens, indicating prior exposure to the pathogen.

In addition, T_H1-mediated delayed hypersensitivity can also be involved in autoimmune responses directed against self-antigens. For example, T_H1 cells are involved in the pathogenesis of allergic encephalitis and Type I (juvenile) diabetes mellitus (Table 22.6). However, many autoimmune diseases are mediated by antibodies, as we will now discuss.

Autoimmune Diseases (Type II and Type III Hypersensitivities)

T and B cells destined to react with self-antigens are normally eliminated during the process of lymphocyte maturation. However, in some individuals, T and B lymphocytes can be activated to produce immune reactions against self-proteins, leading to autoimmune disease (Table 22.7).

Depending on the specific disorder, autoimmune disease may involve **autoantibodies**, antibodies that interact with self-antigens, or a cellular immune response to self constituents. Certain autoimmune diseases are highly organ-specific. For example, in *Hashimoto's disease*, autoantibodies are made against thyroid gland products such as thyroglobulin. The disease affects thyroid function and is classified as a Type II hypersensitivity: Antibodies interact with antigens found on the surface of host cells and cause destruction of the tissue (Table 22.5). In this case, antibodies to thyroglobulin fix complement and cause destruction of the thyroid gland cells. In *juvenile diabetes* (insulin-dependent diabetes mellitus), autoantibodies against the insulin-producing cells in the pancreas are observed, but tissue destruction occurs primarily through inflammatory reactions mediated by T_H1 cells.

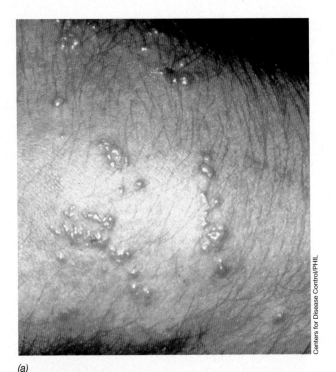

(a)

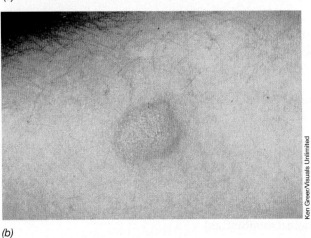

(b)

● **Figure 22.26 Cell-mediated immunity.** (a) Poison ivy blisters on an arm. The raised rash appears maximally in sensitized individuals about 24–48 hours after exposure to plants of the genus *Rhus.* (b) A positive tuberculin test, typical for delayed hypersensitivity. The raised area of inflammation on the forearm is 1.5 cm in diameter. Both reactions result from the action of antigen-specific T_H1 cells.

Systemic lupus erythematosis (SLE) is a clear example of a Type III hypersensitivity: Antibodies bind to soluble proteins, producing insoluble immune complexes, leading to complement deposition and inflammation. Thus, Type III hypersensitivity is an *immune complex disorder* (Table 22.5). This disease and others like it involve a large-scale production of autoantibodies against soluble, circulating self-antigens. In the case of SLE, the antigens include nucleoproteins and DNA. The disease symptoms are induced by circulating antigen–antibody complexes that deposit in different body tissues such as the kidney,

lungs, and spleen. Complement fixation and the resulting lytic and inflammatory responses cause local, often severe, cell damage at the site of the complex deposition.

Organ-specific autoimmune diseases are sometimes more easily controlled clinically because the product of organ function, such as thyroxin in autoimmune hypothyroidism or insulin in juvenile diabetes, can often be supplied in pure form from another source. SLE and other diseases that affect multiple organs can be controlled only by general immunosuppressive therapy, such as the use of steroid drugs. However, steroid therapy carries its own risks. General immunosuppression significantly increases chances of opportunistic infections.

Heredity has an important influence on the incidence, type, and severity of autoimmune diseases. Many autoimmune diseases correlate strongly with the presence of certain major histocompatibility complex (MHC) antigens (see Section 22.6; ⟳Section 23.3). Studies of model autoimmune diseases in mice support such a genetic link, but the precise conditions necessary for developing autoimmunity may also depend on other factors, such as gender. Women, for example, are up to 20 times more likely to develop SLE than are men.

 22.15 Concept Check

Hypersensitivity results when foreign antigens induce cellular or antibody immune responses, leading to host tissue damage. Autoimmunity occurs when the immune response is directed against self-antigens, resulting in host tissue damage.

◆ Discriminate between *immediate hypersensitivity* and *delayed hypersensitivity* with respect to antigens and effectors.

◆ Identify the two main categories of autoimmune disease with respect to antigens and immune effectors.

22.16 Superantigens

We discussed the mechanisms of action for several different categories of bacterial toxins in Chapter 21. Most toxins interact directly with host cells to cause tissue damage. Endotoxins, for example, interact directly with a large variety of cell types, causing release of endogenous pyrogens and other soluble mediators, and producing fever and general inflammation (see Section 22.2). Most exotoxins also interact directly with cells to cause cell damage (⟳Sections 21.10 and 21.11). However, certain exotoxins, the superantigens, act *indirectly* on host cells, using a novel immune mechanism to cause extensive host tissue damage.

Superantigen Activation of T Cells

Superantigens are proteins capable of eliciting a very strong response because they activate more T cells than a normal immune response. The superantigens differ from conventional antigens in their mode of binding to the TCR. Most superantigens are produced by viruses and bacteria. Streptococci and staphylococci, in particular, produce a wide variety of different and very potent superantigens (⚬Table 21.4).

Conventional foreign antigens bind to the TCR at the variable (V) domain antigen-binding site. In a typical immune response, less than 0.01% of all available T cells normally interact with a conventional foreign antigen (see Section 22.5). However, superantigens bind to a site on the Vβ domain of the TCR that is *outside* the antigen-specific TCR binding site. Superantigen binds to *all* TCRs with a shared structure, and many different TCRs share the same structure *outside* the antigen-binding site. In some cases, superantigen exposure can result in the binding and activation of 5% to 25% of all T cells. The superantigens also bind to class II MHC molecules on APCs, again at a specific site outside the normal peptide binding site (Figure 22.27●). These cell surface interactions, mimicking conventional antigen presentation (see Section 22.7), activate large numbers of T cells. The activated T cells, in turn, produce cytokines. The cytokines then stimulate other cells, in particular, macrophages and other phagocytes. Because of the extensive cytokine production, this cell-mediated response is characterized by *systemic* inflammatory reactions, often resulting in fever, diarrhea, vomiting, mucus production, and systemic shock. In extreme cases, exposure to superantigens can be fatal. Superantigen shock is virtually indistinguishable from septic shock as discussed in Section 22.3.

Several diseases can be attributed to superantigens and were mentioned previously (⚬Section 21.10 and Table 21.4). Notable superantigen diseases include *Staphylococcus aureus* food poisoning, characterized by fever, vomiting, and diarrhea, caused by one of several superantigens that act as staphylococcal enterotoxins. *Staphylococcus aureus* also produces the superantigen responsible for *toxic shock syndrome*. *Streptococcus pyogenes*

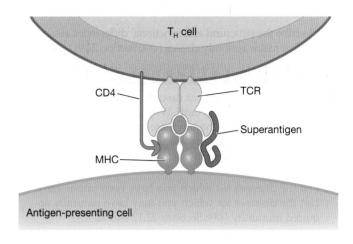

● **Figure 22.27 Bacterial superantigens.** Superantigens act by binding to both the MHC protein and the TCR at positions outside the normal binding site. Because the superantigen binds to conserved regions of MHC and TCR proteins, the superantigen can interact with large numbers of cells, stimulating massive T-cell activation, cytokine release, and even systemic inflammation.

produces erythrogenic toxin, the superantigen responsible for scarlet fever (⚬Figure 26.5).

 22.16 Concept Check

Superantigens bind and activate large numbers of T cells in a novel fashion. Superantigen-activated T cells are capable of producing systemic diseases characterized by massive inflammatory reactions.

◆ Discriminate between normal and superantigen activation of T cells.

◆ Identify the binding site for superantigens on both T cells and APCs.

REVIEW QUESTIONS

1. What is the origin of the phagocytic and antigen-specific cells involved in the immune response? Track the maturation of B cells and T cells (⚬Section 22.1).

2. What are some pathogen-associated molecular patterns (PAMPs) that are recognized by pattern recognition molecules (PRMs)? What is the significance of the interactions between these molecules (⚬Section 22.2)?

3. Explain how phagocytes engulf and kill microorganisms, with particular attention to oxygen-dependent mechanisms (⚬Section 22.3).

4. Identify the most important features of the adaptive immune response (⚬Section 22.4).

5. What molecules induce immune responses? What properties are necessary for a molecule to induce an immune response (⚬Section 22.5)?

6. Describe the basic structure of class I and class II major histocompatibility complex (MHC) proteins. In what functional ways do they differ (⚬Section 22.6)?

7. Differentiate between T$_C$ cells and NK cells. What is the activation signal for each cell type (⚬Section 22.7)?

8. How do T$_H$ cells differ from T$_C$ cells? Differentiate between the functional roles of T$_H$1 and T$_H$2 cells (⚬Section 22.8).

contains about 150 tandemly arranged light-chain V genes and 5 distinct J genes. About 200 tandem V genes, 50 D genes, and 4 J genes exist for the heavy chains. In addition, the heavy-chain constant domain (C_H) genes and the light-chain constant domain genes (C_L) are present. The V, D, J, and C genes are separated by noncoding sequences (introns) typical of gene arrangements in eukaryotes (Section 7.1 and 14.8). During maturation of B lymphocytes, genetic recombination occurs in each B cell. Randomly selected V, D, and J segments are fused, resulting in an active heavy-chain gene and an active light-chain gene. The active gene (containing an intervening sequence between the VDJ gene segment and the C gene segment) is transcribed, and the resulting primary RNA transcript is spliced to yield the final messenger mRNA. The mRNA is then translated to make the heavy and light chains of the Ig molecule.

Reassortment and VDJ (VJ) Joining

Up to this point, all Ig diversity has been generated from recombination of existing genes. The final light chain and heavy chain genes expressed by a given B cell is due to *chance reassortment* of these rearranged heavy chains and light chains. For example, based on the numbers of genes at the kappa (κ) light-chain loci, there are $150\,V \times 5\,J$ possible rearrangements or 750 possible light chains (Section 22.10 and Figure 22.20). At the heavy-chain (H-chain) loci, there are approximately $200\,V \times 50\,D \times 4\,J$, or 4000 possible heavy chains. Assuming that each heavy chain and light chain has an equal chance to be expressed in each cell, there are 750×4000, or 3,000,000 possible antibodies that can be expressed (Figure 23.5).

Another diversity-generating mechanism involves the DNA-joining mechanism. DNA joining of the V-D-J or V-J genes is imprecise and frequently varies the site of VDJ (or VJ) fusions by a few nucleotides. This genetic imprecision is sufficient to change an amino acid or two and leads to even greater antibody diversity. In each B cell, only a single protein-producing rearrangement occurs in the heavy- and light-chain genes. Called *allelic exclusion*, this mechanism ensures that *each B cell produces only one Ig.*

Hypermutation

Additional antibody diversity is generated in B cells by **somatic hypermutation,** the mutation of Ig genes at higher rates than the mutation rates observed in other genes. Somatic hypermutation of Ig genes is usually evident after a second exposure to the immunizing antigen. As we saw, a second exposure to antigen results in a change in the predominant antibody class produced, with a switch from IgM to IgG production (Section 22.10). In addition, class switching is accompanied by an increase in antigen-binding strength (affinity). This *affinity maturation* is one of the factors responsible for the dramatically stronger secondary response in immunity (Section 22.10 and Figure 22.21). Affinity maturation adds virtually unlimited possibilities to the generation of Ig diversity, producing at least 10^{12} possible Igs.

23.6 Concept Check

Recombination allows shuffling of various pieces of the final Ig genes. Random reassortment of the heavy- and light-chain genes maximizes genetically encoded diversity. Imprecise joining of VDJ and VJ segments as well as hypermutation and affinity maturation also contribute to virtually unlimited immunoglobulin diversity.

◆ Describe the recombination events that produce a mature heavy-chain gene.

◆ Describe the somatic mutation events that further enhance antibody diversity.

◆ How many possible Ig-binding sites can be produced?

IV T-CELL RECEPTORS

TCRs are cell-surface antigen receptors on T lymphocytes that recognize peptide antigens embedded in MHC proteins (Section 22.6). In this section, we look at the structure, antigen-binding function, and genetic organization of the TCRs.

23.7 TCR Proteins and Antigen Binding

The TCR is a membrane-integrated protein found only on the surface of T cells. TCRs consist of two polypeptides, the alpha (α) chain and the beta (β) chain. The α:β TCR specifically binds foreign peptides that are embedded in MHC molecules on the surface of APCs or target cells (Section 22.6). Thus, *the TCR binds both a self-MHC protein and a foreign peptide.* The TCRs accomplish this dual binding function through a binding site composed of the V domains of the α chain and β chain. As with Igs, the α-chain and β-chain V domains of TCRs contain CDR1, CDR2, and CDR3 segments that bind to antigen.

The three-dimensional structure of the TCR-peptide-MHC protein complex is shown in Figure 23.6●. Both TCR and MHC proteins bind directly to peptide antigen. The MHC protein binds one face of the peptide, the **agretope,** whereas the TCR interacts with the other peptide face, the **epitope.** The CDR regions are directly involved in the binding of the MHC-peptide complex, and each CDR has a specific binding function. The CDR3 regions of the TCR α chain and β chain make contact with the peptide epitope. The CDR1 and CDR2 regions of the TCR α and β chains contact the peptide-binding α-helices of the MHC proteins.

23.7 Concept Check

The TCR is a protein that binds to peptide antigens presented by MHC proteins. The CDR3 regions of both the α chain and the β chain bind to the peptide epitope, whereas the CDR1 and CDR2 regions bind to the MHC protein.

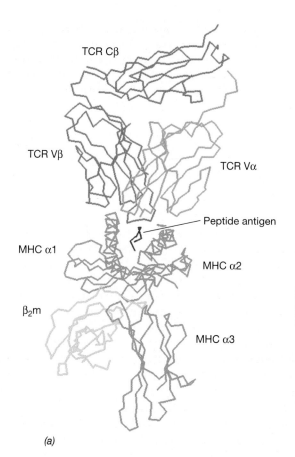

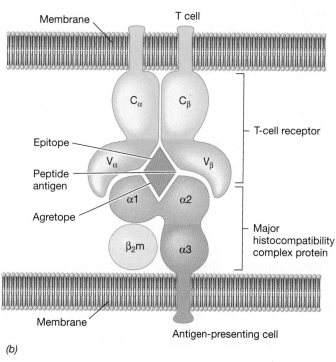

(b)

● **Figure 23.6 The TCR-peptide-MHC I protein complex.** (a) The three-dimensional structure, showing the orientation of TCR, peptide (brown), and MHC. This structure was derived from data deposited in the Protein Data Bank. (b) A diagram-matic interpretation of the MHC–peptide–TCR structure.

◆ Distinguish among the functions of the TCR CDR1, CDR2, and CDR3 segments.

◆ Distinguish between an *agretope* and an *epitope*.

23.8 TCR Genes and Diversity

The TCR α and β chains are encoded by distinct gene segments for constant and variable domains. Much like Ig genes, TCR V region genes are arranged as a number of tandemly arranged genes. The α chain has about 80 variable (V) genes and 61 joining (J) segments, whereas the β chain has 50 variable (V) genes, 2 diversity (D) genes, and 13 joining (J) segments (Figure 23.7●). The β chain V, D, and J genes and the α chain V and J genes undergo recombination to form functional V-region genes. Additional diversity called *N-region diversity* is generated by random additions of one to six nucleotides between V and D gene segments and between D and J gene segments. Finally, the D region of the β chain can be transcribed in all three reading frames, leading to production of three separate transcripts from each D-region gene and creating greater diversity than would be expected from the D gene segments alone. As we discussed for reassortment of Ig H and L chains, individual α and β chains are produced by each T cell at random and joined to form a complete α:β heterodimer. Somatic hypermutation mechanisms are not involved in generating

TCR diversity. However, the potential diversity is still enormous, and theoretically on the order of 10^{15} different TCRs can be generated.

23.8 Concept Check

The V domain of the β chain of the TCR is encoded by V, D, and J gene segments. The V domain of the α chain of the TCR is encoded by V and J gene segments. Diversity generated by recombination, reassortment of gene products, transcription of D regions in three reading frames, and random N nucleotide addition ensure practically unlimited antigen-binding TCRs.

◆ Explain the contributions to diversity for reading of the D region in all three reading frames.

◆ Explain the contributions of the N region additions to TCR diversity.

V MOLECULAR SIGNALS IN IMMUNITY

Antigen recognition through Igs and TCRs is the *first signal* and the most important molecular interaction for activating antigen-specific T and B lymphocytes. Other molecular signals, however, need to be transmitted to lymphocytes before they can respond to antigens. Here we examine some of the other molecular signals necessary to activate the immune response. First, we explore

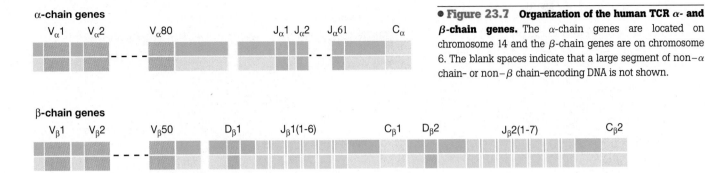

α-chain genes

$V_\alpha 1$ $V_\alpha 2$ $V_\alpha 80$ $J_\alpha 1$ $J_\alpha 2$ $J_\alpha 61$ C_α

β-chain genes

$V_\beta 1$ $V_\beta 2$ $V_\beta 50$ $D_\beta 1$ $J_\beta 1(1\text{-}6)$ $C_\beta 1$ $D_\beta 2$ $J_\beta 2(1\text{-}7)$ $C_\beta 2$

● **Figure 23.7 Organization of the human TCR α- and β-chain genes.** The α-chain genes are located on chromosome 14 and the β-chain genes are on chromosome 6. The blank spaces indicate that a large segment of non–α chain- or non–β chain-encoding DNA is not shown.

the mechanisms that *select* immune cells to react with foreign antigens and to ignore self proteins. Next, we examine the molecular *second signals* that are responsible for activating immune T cells. Finally, we introduce *cytokines* and *chemokines*, soluble proteins capable of activating other cells in the immune response.

23.9 Clonal Selection and Tolerance

For an effective immune response, T cells must discriminate between the *dangerous non-self antigens* and the *nondangerous self antigens* that compose our body tissue. Thus, T cells must achieve **tolerance,** or specific unresponsiveness to self antigens. As we achieve tolerance, we select immune cells that interact only with dangerous non-self antigens.

Clonal Selection

Clonal selection is a hypothesis stating that each antigen-reactive B cell or T cell has a cell-surface receptor for a single antigen. When stimulated by interaction with that antigen, each cell can replicate, and antigen-stimulated B and T cells grow and differentiate. As a result, antigen-stimulated cells divide and produce a pool of cells that express the same antigen-specific receptors. These copies of the original antigen-reactive cell are known as *clones* (Figure 23.8●). Cells that have not interacted with antigen do not grow.

To respond to the seemingly infinite variety of antigens, a large number of antigen-reactive cells are needed in the body. However, antigen-reactive cells must not provoke immune reactions against self antigens in the host. To meet these special requirements, the immune system selects clones that may be useful against non-self antigens and eliminates or suppresses self-reactive clones.

T-Cell Selection and Tolerance

T cells undergo *selection for* antigen-reactive T cells and *selection against* clones that react with self antigens. Selection against self-reactive clones results in the development of *tolerance*, or specific immune unresponsiveness to self antigens. Previously, we discussed the failure of tolerance and the subsequent development of autoimmunity (∞Section 22.15).

Lymphocytes that will become T cells leave the bone marrow and enter the thymus, a primary lymphoid organ, via the lymphatic ducts (Figure 23.9●, ∞Figure 22.2). In the first T-cell maturation stage in the thymus, called **positive selection**, immature T cells interact with the thymic epithelium, using their newly developed TCRs to bind to MHC proteins on the thymus. The T cells that do not bind MHC proteins are programmed to die, a process called **apoptosis**; the T cells that bind thymic MHC proteins continue to grow. Thus, positive selection retains T cells that recognize self-MHC proteins and deletes T cells that do not recognize self-MHC proteins.

In the second stage of T-cell maturation, **negative selection**, the positively selected T-cells continue to interact with thymic MHC proteins, which are complexed with antigens. The antigens in the thymus are mostly of self origin. T cells that react with the thymic self antigens are potentially dangerous because they can react with self tissue antigens (autoimmunity). Therefore, these self-reactive T cells must be eliminated. The self-reactive T cells bind tightly to thymus cells and eventually die. However, the T cells that are destined to interact with non-self antigens do not bind as tightly, presumably because the non-self antigens are not found in the thymus. These T cells do not die, but leave the thymus and migrate to the spleen and lymph nodes where they can contact foreign antigens presented by B lymphocytes and other APCs.

This two-stage mechanism for selecting for antigen-reactive T cells while inducing tolerance is called **clonal deletion**. Precursors of T-cell clones that are either useless or harmful die, and more than 99% of all T cells that enter the thymus do not survive the selection process. T cells that survive positive and negative selection then leave the thymus and travel to the secondary lymphoid organs, where they can participate in the immune response to foreign antigens.

B-Cell Tolerance

The acquisition of immune tolerance in B cells is also necessary because antibodies produced by self-reactive B cells (autoantibodies) may damage host tissue (∞ Section 22.15). B cells also undergo clonal deletion. Many self-reactive B cells are eliminated during development in the bone marrow, the primary lymphoid organ responsible for B-cell development in humans (∞Section 22.1).

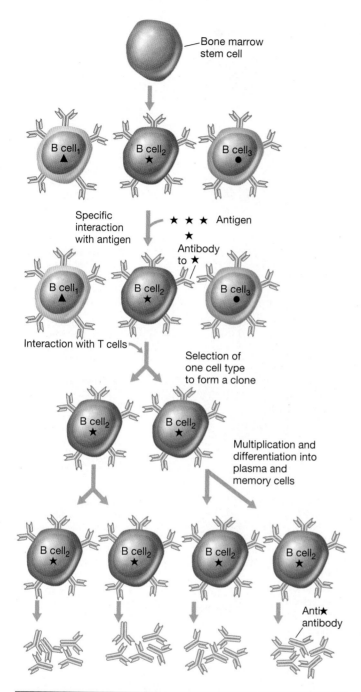

● **Figure 23.8** **Clonal selection.** Individual B cells, specific for a single antigen, proliferate and expand to form a clone after interaction with the specific antigen. The antigen acts as the selection agent, driving selection and then proliferation of the individual antigen-specific B cell. Clonal copies of the antigen-reactive cell all have the same antigen-specific surface antibody. Continued exposure to antigen results in continued expansion of the clone. An analogous situation exists for T cells.

In addition to the clonal deletion events, **clonal anergy** (unresponsiveness) also plays a role in final selection. Immature B cells that are self-reactive do not develop, even when exposed to high concentrations of self antigens in the bone marrow. Although self antigens may occupy some B-cell Ig receptors, the B cell is dependent on T cells for a second activation signal, as we

shall see (see Section 23.10). If no antigen-reactive T cell is available to provide help because of elimination during T-cell selection, the B cell remains in a state of anergy.

 23.9 **Concept Check**

The thymus is a primary lymphoid organ that provides an environment for the maturation of antigen-reactive T cells. Immature T cells that do not interact with MHC protein (positive selection) or react strongly with self antigens (negative selection) are eliminated by *clonal deletion* in the thymus. T cells that survive positive and negative selection leave the thymus and can participate in an effective immune response. B cell reactivity to self antigens is controlled through clonal deletion, selection, and anergy.

◆ Provide an example of a self protein to which you are tolerant and another human protein to which you would *not* be tolerant.

◆ Distinguish between *positive* and *negative* T-cell selection.

◆ For B cells, distinguish between clonal deletion and clonal anergy.

23.10 Second Signals

As we have seen, T cells selected for reactivity against dangerous non-self antigens in the thymus are also selected for *tolerance*, or unresponsiveness to self antigens, and T cells that react with self antigens are deleted in the thymus. However, many self antigens are not expressed in the thymus. T-cell clones responsive to these non-thymus antigens can avoid clonal deletion, but may become unresponsive to self antigen through interactions in the secondary lymphoid organs, developing *clonal anergy*. Analogous to the situation discussed above for B cells, the key to inducing clonal anergy in T cells lies in the two-signal mechanism used to activate T cells.

T-Cell Activation and the Second Signal

When selected mature T cells leave the thymus, they travel to the secondary lymphoid organs (lymph nodes, spleen, and mucosal-associated lymphoid tissue, or MALT), where they take up residence. These T cells have not yet been exposed to antigen and are therefore called naive or uncommitted T cells. Uncommitted T cells must be activated by APC (∞ Section 22.6) before they are fully competent as effector cells.

The first step in activation of uncommitted T cells is binding of the peptide: MHC protein complex on the APC by the TCR (Figure 23.10●). This is *signal 1* and is absolutely required for activation. Without signal 1, a T cell cannot be activated.

The next step in activation of the T cell involves the interaction of two additional proteins, one found on the APC called *B7* and one found only on T cells, called *CD28*. The interaction and binding of B7 to CD28 is

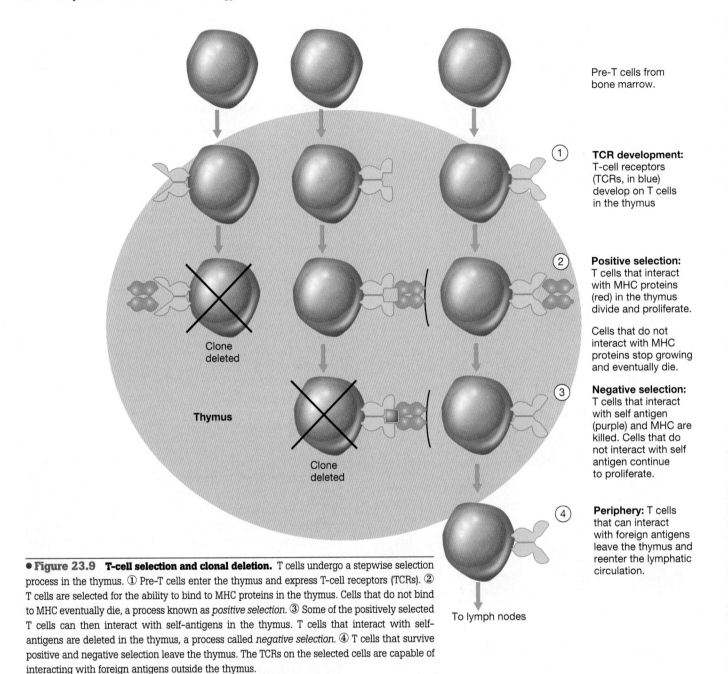

Pre-T cells from bone marrow.

① **TCR development:** T-cell receptors (TCRs, in blue) develop on T cells in the thymus

② **Positive selection:** T cells that interact with MHC proteins (red) in the thymus divide and proliferate.

Cells that do not interact with MHC proteins stop growing and eventually die.

③ **Negative selection:** T cells that interact with self antigen (purple) and MHC are killed. Cells that do not interact with self antigen continue to proliferate.

④ **Periphery:** T cells that can interact with foreign antigens leave the thymus and reenter the lymphatic circulation.

To lymph nodes

● **Figure 23.9** **T-cell selection and clonal deletion.** T cells undergo a stepwise selection process in the thymus. ① Pre-T cells enter the thymus and express T-cell receptors (TCRs). ② T cells are selected for the ability to bind to MHC proteins in the thymus. Cells that do not bind to MHC eventually die, a process known as *positive selection*. ③ Some of the positively selected T cells can then interact with self-antigens in the thymus. T cells that interact with self-antigens are deleted in the thymus, a process called *negative selection*. ④ T cells that survive positive and negative selection leave the thymus. The TCRs on the selected cells are capable of interacting with foreign antigens outside the thymus.

signal 2 and activates the T cell. In the absence of signal 2, the T cell cannot be activated (Figure 23.11●). The activated T cell then becomes an effector cell. A T_C cell that is activated will kill any target cell that displays antigen, even those target cells that do not display CD28. Activated effector cells need only signal 1 (peptide-MHC) to induce effector activity (Figure 23.10).

This requirement for a second activation signal has major implications for establishing and maintaining clonal anergy. For example, an uncommitted T_C cell that interacts with a self antigen on a non-APC will receive only signal 1, since non-APCs do not display the B7 protein necessary to complete signal 2. In the absence of signal 2, this T_C is anergized and cannot be activated (Figure 23.10). Thus, the B7:CD28 second signal is absolutely required for activation. Absence of signal 2 in the presence of signal 1 induces per-

manent anergy. Uncommitted T_H1 and T_H2 lymphocytes are activated in the same way, also using the B7-CD28 co-receptor second signal.

A different second signal is used to activate B cells and other immune response effectors, as we will see in the next section.

 23.10 *Concept Check*

Many self-reactive T cells are deleted during development and maturation in the thymus. Uncommitted T cells are activated in the secondary lymphoid organs by first binding peptide:MHC with their TCRs (signal 1), followed by binding of the B7 APC protein to the CD28 T-cell protein (signal 2). Uncommitted self-reactive T cells are anergized in the secondary lymphoid organs if they interact with signal 1 in the absence of signal 2.

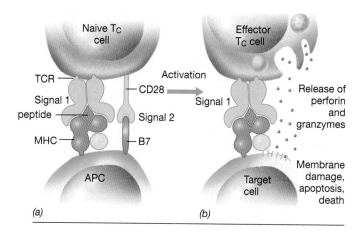

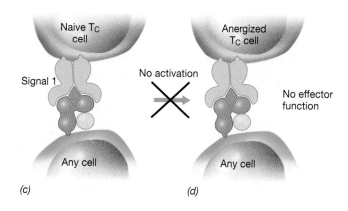

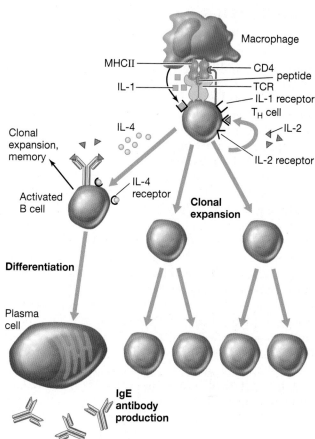

● **Figure 23.10** **Signal 1 and signal 2 are required to activate naive T cells.** (a) A naive T_C cell interacts via TCR with the peptide-MHC complex on an APC. This is signal 1. The T_C cell also has a CD28 protein that interacts with a B7 protein on the APC. This is signal 2. The simultaneous interactions of the T_C cell and APC via signal 1 and signal 2 activate the naive T cell. (b) This permanently activated T_C cell is then capable of killing any target cell as long as signal 1 interactions take place. (c) A naive T_C cell interacts via the TCR with the peptide-MHC complex on any cell. Although the conditions for signal 1 (interactions via TCR with the peptide-MHC complex) are met, signal 2 cannot be generated because only APCs display the B7 protein. (d) In the absence of signal 2, the T_C cell becomes permanently unresponsive, or anergized.

♦ Define signal 1 and signal 2 for an *uncommitted* T cell.

♦ Identify the signal(s) necessary to induce effector function in an *activated* T cell.

23.11 Cytokines and Chemokines

Intercellular communication, necessary for activation in the immune system, is accomplished through a family of soluble proteins known as **cytokines.** Cytokines, produced by leukocytes, regulate a variety of cellular functions in immune cells. Some cytokines have very broad effects, activating a variety of cell types. However, the

● **Figure 23.11** **Cytokines and antibody production.** Some major effects and interactions of IL-1, IL-2, and IL-4 cytokines in the antibody response. Cytokines provide signals for macrophages, T cells, and B cells. Cytokines stimulate enhanced antigen uptake by macrophages, as well as activation, differentiation, and expansion of both T and B lymphocytes.

unifying feature of these heterogeneous proteins is that they are all produced by leukocytes. The cytokines produced by lymphocytes are often called *lymphokines.* The cytokines called *interleukins (IL)* mediate interactions between leukocytes.

In general, cytokines are secreted from one cell and bind specific receptors on a target cell. Some cytokines bind to receptors on the cell that produced them. Thus, these cytokines have *autocrine* (self-stimulatory) abilities. Binding the cytokine receptors generally activates a signal transduction pathway (Section 8.12), relaying information across the cell membrane to control activities such as protein synthesis and cell division. These signals can ultimately result in cell growth, differentiation, and clonal proliferation.

Table 23.2 lists some important cytokines, their major producer cells, their most common target cells, and their biological effects. In all, there are nearly 40 known cytokines, most of which are produced by either T_H cells or monocytes and macrophages. We now examine the activity of three cytokines involved in the induction of an antigen-specific antibody-mediated immune response.

IL-1, IL-2, and IL-4

As we have seen, macrophages are responsible for antigen uptake, processing, and presentation (⊂∞ Section 22.16). In addition, they secrete a cytokine known as IL-1 that affects several different cell types (Table 23.2 and Figure 23.11). IL-1 is a key component of the immune response because T_H cells are activated by IL-1. Because of the proximity of T_H cells and macrophages in the lymph nodes, nearby T_H2 cells are very likely to be activated. IL-1 binding by the IL-1 receptors (IL-1R) on the T_H2 cell acts as an activation signal. The activated T_H2 cell, in turn, responds by producing IL-2, which is secreted and bound by the IL-2R on the surface of the T_H2 cells. Thus, IL-2 can activate the same cell that secreted it. Under the influence of IL-2, the cell divides, making clonal copies. In the process, the T_H2 cell also makes other cytokines, in this case IL-4, which then binds to the IL-4R on B cells. The IL-4:IL-4R complex stimulates the B cells to differentiate into plasma cells, which ultimately produce antibodies (⊂∞ Section 22.10). Thus, IL-1, IL-2, and IL-4 cytokines are soluble mediators and activators for macrophages, T lymphocytes, and B cells, the cells that interact to produce the antibody-mediated immune response.

Other Cytokines

Table 23.2 shows the activity of several other cytokines. Many of these proteins affect cells involved in specific immunity. For example, IL-4 primarily affects B cells. However, several cytokines do not affect T or B lymphocytes, but act on other cells; the cytokine-activated cells in turn serve as important modulators of innate host responses. For example, interferons (IFN-α and IFN-γ) are produced by leukocytes and inhibit viral replication in virtually any cell in the body. Tumor necrosis factors TNF-α and TNF-β can kill a variety of tumors if the TNF-producing cells have access to the tumor. TNF-α is also a critical activator of inflammation (⊂∞ Section 22.3). Interferons and TNFs appear to have no target cell specificity, but are produced by T cells as well as phagocytes and can amplify the effects of immune cells.

Chemokines

Chemokines are a group of small proteins that function as chemoattractants for phagocytic cells and T cells. They are produced by lymphocytes and a wide variety of other cells in response to bacterial products, viruses, and other agents that cause damage to host cells. Chemokines attract phagocytes and T cells to the site of injury, stimulating an inflammatory response as well as potentiating a specific immune response.

About 40 chemokines are known. Perhaps the best-studied chemokines are IL-8 and *macrophage chemoattractant protein-1 (MCP-1)* (Table 23.2). IL-8 is produced by a wide variety of cell types including monocytes, macrophages, fibroblasts (connective tissue cells), and keratinocytes (skin cells) in response to tissue injury or contact

---

Table 23.2 Properties of some major cytokines and chemokines

Cytokine	Producers	Major targets	Effect
IL-1[a]	Monocytes	T_H	Activation
IL-2	Activated T cells	T cells	Growth, differentiation
IL-3	T_H1	Hematopoietic stem cells	Growth factor
IL-4	T_H2	B cells	IgG1 and IgE synthesis
IL-5	T_H2	B cells	IgA synthesis
IL-10	T_H2	T_H1	Inhibits T_H1
IL-12	Macrophages, dendritic cells	T_H1, NK cells	Differentiation, activation
IFN-α[b]	Leukocytes	Normal cells	Anti-viral
IFN-γ	T_H1	Macrophages	Activation
GM-CSF[c]	T_H1	Myeloid stem cells	Differentiation to granulocytes, monocytes
TGF-β[d]	T_H1 and T_H2	Macrophages	Inhibits activation
TNF-α[e]	T_H1, macrophages, NK cells	Macrophages	Activation
TNF-β	T_H1	Macrophages	Activation
Chemokines			
IL-8	Macrophages, fibroblasts, keratinocytes	Neutrophils, T cells	Attractant and activator
MCP-1[f]	Macrophages, fibroblasts, keratinocytes	Macrophages, T cells	Attractant and activator

[a]IL, interleukin; [b]IFN, interferon; [c]GM-CSF, granulocyte, monocyte-colony stimulating factor; [d]TGF, T-cell growth factor; [e]TNF, tumor necrosis factor, [f]MCP, macrophage chemoattractant protein.

with pathogens. IL-8 is secreted by the affected cells and binds to the surrounding tissue, where it is a chemoattractant for T cells and neutrophils (∞Section 22.1). This results in a neutrophil-mediated inflammatory response followed by a specific immune response mediated by the attracted T cells. As is the case for the cytokine receptors, engaged chemokine receptors on the target cells act through signal transduction pathways (∞Section 8.12) to induce activation of the target phagocytes or T cells.

MCP-1 is also produced by a variety of cells and attracts macrophages and T cells, again stimulating production of inflammatory mediators and potentially organizing an antigen-specific immune response. Thus, chemokines are potent initiators of nonspecific inflammatory reactions that lead to recruitment of T cells and antigen-specific immune reactions.

 23.11 Concept Check

Cytokines are soluble mediators produced by leukocytes that regulate interactions between cells. Several cytokines such as IL-1, IL-2, and IL-4 affect leukocytes and are critical components in the generation of specific immune responses. Other cytokines such as IFN and TNF affect a wide variety of cell types. *Chemokines* are produced by a variety of cell types in response to injury and are potent attractants for nonspecific inflammatory cells and T cells.

◆ Compare the target cells for IL-1, IL-4, IL-12, IFN-γ, and TNF-β.

◆ How do cytokines and chemokines differ with respect to their cell sources? Their cell targets?

◆ What events stimulate cytokine production? What events stimulate chemokine production?

REVIEW QUESTIONS

1. Identify the pattern recognition molecules (PRMs) and their interacting pathogen-associated molecular patterns (PAMPs) (∞Section 23.1).

2. Define the criteria used to assign a gene and its encoded protein to the Ig gene superfamily (∞Section 23.2).

3. Identify the major structural features of class I and class II MHC proteins (∞Section 23.3).

4. Polymorphism implies that each different MHC protein binds a different peptide motif. For the HLA class I polymorphisms, how many different HLA proteins are expressed in an individual? By the entire human population (∞Section 23.4)?

5. Which Ig chains are used to construct a complete antigen-binding site? Which domains? Which CDRs (∞Section 23.5)?

6. Calculate the total number of V_H and V_L domains that can be constructed from the available Ig genes. How many complete Ig proteins can be produced from the reassortment of all possible heavy chains and light chains (∞Section 23.6)?

7. Describe the interaction of the TCR with peptide antigen and MHC protein. Be sure to identify the roles of the CDRs in the TCR (∞Section 23.7).

8. In TCRs, diversity can be generated by recombination and reassortment events such as in Igs. However, additional diversity is accomplished by the use of N-region nucleotide additions and reading of the D (diversity) segment in all three reading frames. Explain how these diversity-generating mechanisms work (∞Section 23.8).

9. Explain *positive* and *negative* selection of T cells (∞Section 23.9).

10. What molecular interactions are necessary for activation of naive T cells? For activation of effector cells (∞Section 23.10)?

11. What are the chief effects of cytokines and chemokines? What are the chief differences between cytokines and chemokines (∞Section 23.11)?

APPLICATION QUESTIONS

1. Identify the consequences of a genetic mutation that eliminates a PRM by predicting the outcome for the host. Do this for each PRM.

2. Construct a table that lists the common features of proteins encoded by members of the Ig gene superfamily. For Igs, TCRs, and MHC proteins, identify the structural components that fit these common features.

3. Polymorphism implies that each different MHC protein binds a different peptide motif. However, for the class I proteins, only 6 peptide motifs can be recognized in an individual, while over 350 motifs can be recognized by the entire human population. What advantage does this have for the population? For the individual?

4. While genetic recombination events are important for generating significant diversity in the antigen-binding site of Igs, postrecombination somatic events may be even more important in achieving overall Ig diversity. Do you agree or disagree with this statement? Explain.

5. What would happen to the T-cell repertoire in the absence of positive selection? In the absence of negative selection?

6. What would be the result of activation of all peripheral T cells that contact antigen? How does the second signal scheme prevent this from happening?

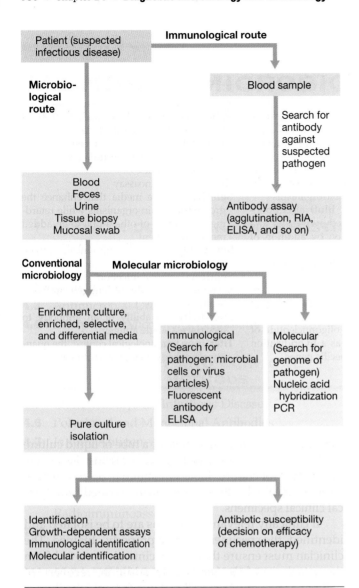

● **Figure 24.1 Isolation and identification of pathogens.** Clinical and diagnostic methods used for isolation and identification of infectious pathogens.

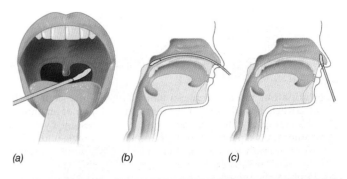

● **Figure 24.2 Methods for obtaining specimens from the upper respiratory tract.** (a) Throat swab. (b) Nasopharyngeal swab passed through the nose. (c) Swabbing the inside of the nose.

Certain media are **selective media**, media that enhance the growth of certain organisms while retarding the growth of others due to an added media component (Table 24.1). Finally, **differential media** are specialized media that allow identification of organisms based on their growth and appearance on the media (see Figure 24.7).

Blood Cultures

Bacteremia is the presence of bacteria in the blood (∞ Section 21.7). Bacteremia is extremely uncommon in healthy individuals, normally occurring only transiently in response to invasive procedures like dental work or trauma. The prolonged presence of bacteria in the blood is generally indicative of systemic infection. The most common pathogens found in blood include *Pseudomonas aeruginosa*, enteric bacteria, especially *Escherichia coli* and

Klebsiella pneumoniae, and the gram-positive cocci *Staphylococcus aureus* and *Streptococcus pyogenes*.

Septicemia is a blood infection resulting from the growth of a virulent organism entering the blood from a focus of infection, multiplying, and traveling to various body tissues to initiate new infections. Septicemia results in severe systemic symptoms, including fever and chills, followed by prostration. Severe cases of septicemia may result in *septic shock*, a life-threatening systemic condition characterized by severe reduction in blood pressure and multiple organ failures, including heart, kidneys, and lungs. Blood cultures provide the only immediate way of isolating and identifying the causal agent, and diagnosis therefore depends on careful and proper blood culture.

The standard blood culture procedure is to draw 20 ml of blood aseptically from a vein and inject it into two blood culture bottles containing an anticoagulant and an all-purpose culture medium. One bottle is incubated aerobically and one anaerobically. Blood culture bottles are incubated at 35°C and examined several times each hour for up to five days in most automated systems. Most clinically significant bacteria are recovered within two days, while more fastidious organisms may be recovered in three to five days. Some blood culture systems employ a chemical that lyses red and white blood cells, releasing intracellular pathogens. Microorganisms from blood cultures are commonly detected by indicators of microbial growth in automated systems, microscopic examination, and subculture. Automated blood culture systems detect growth by monitoring carbon dioxide production and turbidity as often as every 10 minutes.

A certain amount of contamination from the normal flora of the skin is unavoidable when blood is drawn, and a contamination rate of 2–3% can be expected. Certain organisms commonly found on the skin, such as *Staphylococcus epidermidis*, coryneform bacteria, or propionibacteria, are common contaminants. However, these organisms can occasionally cause infection of the heart (subacute bacterial endocarditis) or colonization of intravascular devices such as artificial heart valves. Thus, considerable microbiological and clinical experience is necessary when interpreting blood culture results.

Table 24.1	**Recommended enriched and selective media for primary isolation of pathogens**					
		Media[a]				
Specimen	**Blood agar**	**Enteric agar**	**CA**	**MTM**	**ANA**
Fluids from chest, abdomen, pericardium, joint	+	+	+	−	+
Feces: rectal or enteric transport swabs[b]	+	+	+	−	−
Surgical tissue biopsies	+	+	−	−	+
Throat, sputum, tonsil, nasopharynx, lung, lymph nodes	+	+	+	−	−
Urethra, vagina, cervix	+	+	+	+	−
Urine	+	+	−	−	−
Blood[c]	+	+	+	−	+
Wounds, abscesses, exudates	+	+	+	−	+

SOURCE: Adapted from Murray, P. R., E. J. Baron, J. H. Jorgenson, M. A. Pfaller, and R. H. Yolken. 2003. *Manual of Clinical Microbiology*, 8th edition. American Society for Microbiology, Washington, DC..

[a] Blood agar, 5% whole sheep blood added to trypticase soy agar; enteric agar, either eosin-methylene blue (EMB) agar or MacConkey agar; CA, chocolate (heated blood) agar; MTM, modified Thayer-Martin agar; ANA, anaerobic agar, thioglycolate-containing blood agar or supplemented thioglycolate agar incubated anaerobically.

[b] Special enteric pathogen media, SMAC (MacConkey agar with sorbitol), is also used to culture fecal and enteric samples. SMAC is a selective and differential medium used for the isolation and identification of sorbitol-negative enteric pathogens such as enteropathogenic *Escherichia coli*.

[c] Blood is cultured initially in broth. Depending on the Gram stain characteristics of isolates, subculturing is done on MacConkey agar (gram-negative) or chocolate agar (gram-positive)

Urine Cultures

Urinary tract infections are very common, especially in females. Because the causal agents are often members of the normal flora (for example, *Escherichia coli*), considerable care must be taken in the bacteriological analysis of urine. In most cases, urinary tract infection occurs as a result of an organism ascending the urethra from the outside, infecting the bladder. Urinary tract infections are also the most common form of *nosocomial* (hospital-acquired) infection (⚮Section 25.7).

Significant urinary infection generally results in bacterial counts of 10^5 or more organisms per milliliter of a clean-voided midstream specimen. In the absence of infection, contamination of the urine from the external genitalia (almost unavoidable to some extent) results in less than 10^3 organisms per milliliter. The most common urinary tract pathogens are members of the enteric bacteria, with *E. coli* accounting for about 90% of the cases. Other urinary tract pathogens include *Klebsiella, Enterobacter, Proteus, Pseudomonas, Staphylococcus saprophyticus,* and *Enterococcus faecalis. Neisseria gonorrhoeae*, the causal agent of gonorrhea, does not grow in the urine itself, but on the urethral epithelium, and is diagnosed by different methods (see later).

Direct microscopic examination of urine may be used to indicate *bacteriuria*, the presence of abnormal numbers of bacteria in the urine. However, because nearly all urine contains some level of bacterial growth, bacteriuria is most commonly monitored by using commercially available dipstick tests. For example, one dipstick test monitors the reduction of nitrate by detecting the reduction product, nitrite. A positive test is indicated by a color change on the dipstick (Figure 24.3●). Since significant nitrite production occurs in urine only when large numbers ($>10^5$ per milliliter) of enteric organisms

are present, the method is a very rapid check for urinary tract infections. Other dipstick tests for urinary tract infections, often used in conjunction with nitrate reduction, detect esterase (produced by leukocytes, ⚮ Section 22.1) and peroxidase (produced by a variety of bacteria, ⚮Sections 6.15 and 17.22). Positive dipstick tests are followed by a urine culture.

A Gram stain (⚮Section 4.1) may also be done directly on urine samples exhibiting bacteriuria to identify characteristic morphotypes of potential urinary tract pathogens. These include gram-negative rods (enteric *Bacteria,* ⚮Section 12.11), gram-negative cocci (*Neisseria,* ⚮ Section 12.10), and gram-positive cocci (*Enterococcus,* ⚮ Section 12.19). The Gram stain and other direct staining methods are also useful for direct detection of bacteria in other body fluids such as sputum (⚮Section 26.5).

● **Figure 24.3 Urinalysis dipstick test.** A control strip is shown underneath the test strip. From left to right, the strip measures abnormal levels of glucose, bilirubin, ketones, specific gravity, blood, pH, protein, urobilinogen, nitrite, and leukocytes (esterase) in a urine sample. Abnormal readings for esterase (trace positive, far right) and nitrite (strong positive, second from right) indicate bacteriuria. Subsequent culture of this sample indicated the presence of *Escherichia coli*.

To culture potential urinary tract pathogens, two media are normally used: (1) blood agar as a nonselective general medium, and (2) a medium selective for enteric bacteria, such as MacConkey or eosin-methylene blue agar (EMB) (see Section 24.2 and Figure 24.4●). These specialized enteric media permit the initial differentiation of lactose fermenters from nonfermenters, while the growth of gram-positive organisms such as *Staphylococcus* spp. (common skin contaminants) is inhibited. Experienced clinical microbiologists may make a tentative identification of an isolate by observing the color and morphology of colonies of the suspected pathogen growth on various media as described in Table 24.2. Such an identification must be followed with more detailed tests, but clinical microbiologists use this information in conjunction with further test results, discussed throughout the remainder of this chapter, to make a positive identification.

Urine cultures can be done quantitatively by counting colonies on blood agar or a selective agar medium, using a calibrated amount of urine, usually 1 μl, as the inoculum for a plate.

Finally, if no bacterial growth is obtained despite persistent urinary tract infection symptoms, a clinician may request direct cultures for fastidious organisms such as *Neisseria gonorrhoeae, Chlamydia trachomatis, Branhamella* species, mycoplasma, or several anaerobic organisms.

Fecal Cultures

Proper collection and preservation of feces is important in the isolation of intestinal pathogens. During storage,

● **Figure 24.4 An eosin-methylene blue (EMB) agar plate.** The plate shows a lactose fermenter, *Escherichia coli* (left), and a nonlactose fermenter, *Pseudomonas aeruginosa* (right). Note the green metallic sheen of the *E. coli* colonies.

fecal acidity increases so extended delay between sampling and sample processing must be avoided. This is especially critical for the isolation of *Shigella* and *Salmonella* species, both of which are sensitive to acid pH.

Freshly collected fecal samples are placed in a vial containing phosphate buffer for transport to the lab. Bloody or pus-containing stools as well as stools from patients with suspected foodborne or waterborne infections are inoculated into a variety of selective media (see Section 24.2) for isolation of individual bacteria. Intestinal parasites are identified by observing cysts microscopically in the stool sample or through antigen-detection assays (see Section 24.5 through 24.11) rather than by culture methods. Many laboratories also use a variety of selective and differential media and incubation conditions to identify *Escherichia coli* O157:H7 and *Campylobacter*, two important intestinal pathogens generally acquired from contaminated food or water (∞Sections 29.8 and 29.9).

Wounds and Abscesses

Infections associated with traumatic injuries such as animal or human bites, burns, cuts, or the penetration of foreign objects must be carefully sampled to recover the relevant pathogen, and results must be interpreted carefully. Wound infections and abscesses are frequently contaminated with normal flora, and swab samples from such lesions are frequently misleading. For abscesses and other purulent lesions, the best sampling method is to aspirate pus with a sterile syringe and needle following disinfection of the skin surface. Internal purulent lesions are sampled by biopsy or from tissues removed in surgery.

A variety of pathogens are associated with wound infections. Since some of these pathogens are anaerobes, proper evaluation requires that samples be obtained, transported, and cultured under anaerobic as well as aerobic conditions. For example, potential pathogens commonly associated with purulent discharges from wound infections are *Staphylococcus aureus*, enteric bacteria, *Pseudomonas aeruginosa*, and anaerobes from the genera *Bacteroides* and *Clostridium*. The major isolation media are blood agar, several selective media for enteric bacteria (Tables 24.1 and 24.2), and blood agar containing additional supplements and reducing agents for obligate anaerobes. Gram stains from such specimens are examined directly by microscopy.

Genital Specimens and the Laboratory Diagnosis of Gonorrhea

In males, a purulent urethral discharge is the classic symptom of the sexually transmitted disease *gonorrhea* (∞Section 26.12). If no discharge is present, a sample can be obtained using a sterile narrow-diameter cotton swab that is inserted into the anterior urethra, left in place a few seconds to absorb any exudate, and then removed for identification of *Neisseria gonorrhoeae*, the causative agent of gonorrhea. Alternatively, a sample of the first early morning urine of an infected individual

Table 24.2	Colony characteristics of frequently isolated gram-negative rods cultured on various clinically useful media					
				Agar media[a]		
Organism	**EMB**	**MC**	**SS**	**BS**	**HE**
Escherichia coli	Dark center with greenish metallic sheen (see Figure 24.4)	Red or pink	Red to pink	Mostly inhibited	Yellow-pink
Enterobacter	Similar to *E. coli*, but colonies are larger	Red or pink	White or beige	Mucoid colonies with silver sheen	Yellow-pink
Klebsiella	Large, mucoid, brownish	Pink	Red to pink	Mostly inhibited	Yellow-pink
Proteus	Translucent, colorless	Transparent, colorless	Black center, clear periphery	Green	Clear
Pseudomonas	Translucent, colorless to gold (see Figure 24.4)	Transparent, colorless	Mostly inhibited	No growth	Clear
Salmonella	Translucent, colorless to gold	Translucent, colorless	Opaque	Black to dark green	Green or transparent with black centers
Shigella	Translucent, colorless to gold	Transparent, colorless	Opaque	Brown or inhibited	Green or transparent

SOURCE: Adapted from Murray, P. R., E. J. Baron, J. H. Jorgenson, M. A. Pfaller, and R. H. Yolken. 2003. *Manual of Clinical Microbiology*, 8th edition. American Society for Microbiology, Washington, DC.

[a] BS, Bismuth sulfite agar; EMB, eosin-methylene blue agar; MC, MacConkey agar; SS, *Salmonella-Shigella* agar; HE, Hektoen enteric agar.

usually contains viable cells of *N. gonorrhoeae*. In females suspected of having gonorrhea or other genital infections, samples are usually obtained by swab from the cervix and the urethra.

Clinical microbiology procedures are central to the diagnosis of gonorrhea. *N. gonorrhoeae* (referred to clinically as *gonococcus*) colonizes mucosal surfaces of the urethra, uterine cervix, anal canal, throat, and conjunctiva. The organism is sensitive to drying and therefore is transmitted almost exclusively by direct person-to-person contact, usually by sexual intercourse. Public health measures to control gonorrhea involve identification of asymptomatic carriers, and this requires microbiological analysis.

Neisseria gonorrhoeae is usually found as a gram-negative diplococcus, but can be quite pleomorphic. No similar microorganisms are observed among the normal flora of the urogenital tract. Thus, direct microscopy of a gram-stained vaginal or cervical smear showing gram-negative diplococci is a probable indication of gonorrhea. In acute gonorrhea, microscopic examination of purulent discharges usually reveals phagocytized gram-negative diplococci in the neutrophils (∞Section 22.2) (Figure 24.5a●).

Most laboratory testing of urogenital samples for *N. gonorrhoeae* (and the often-associated *Chlamydia trachomatis*, ∞Section 26.13) is done using nucleic acid probe or

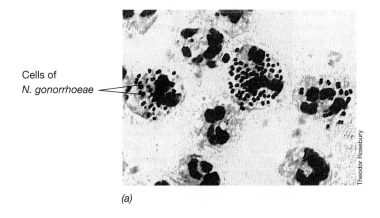

Cells of *N. gonorrhoeae*

(a)

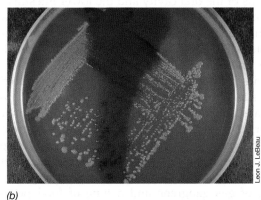

(b)

● **Figure 24.5** **Identification of *Neisseria gonorrhoeae*.** (a) Photomicrograph of *Neisseria gonorrhoeae* within human polymorphonuclear leukocytes from a urethral exudate. Note the paired diplococci (leader). (b) *N. gonorrhoeae* growing on Thayer-Martin agar. The plate has been stained in the middle with a reagent that turns colonies blue if cells contain cytochrome *c* (the oxidase test). *N. gonorrhoeae* cells are oxidase-positive.

(Figure 24.7a). Production of hydrogen and/or carbon dioxide during sugar fermentation is assayed by observing gas production either in gas collection vials or in agar (Figure 24.7a,b). Hydrogen sulfide (H_2S) production is indicated following growth in a medium containing ferric iron. If sulfide is produced, ferric iron reacts with H_2S to form a black precipitate of iron sulfide (Figure 24.7b). Utilization of citric acid, an organic acid with three carboxylic acid groups, is accompanied by a pH increase, and a specific dye in the citric acid test medium changes color as conditions become alkaline (Figure 24.7c). Hundreds of differential tests have been developed for clinical use, but only about 20 are used routinely (Figure 24.7d).

The biochemical patterns for various pathogens are stored in a computer databank. The results of the differential tests on an unknown pathogen are entered, and the computer makes the best identification by comparing the characteristics of the unknown to metabolic patterns in the pathogen database. For many organisms, as few as three or four key tests are sufficient to make an unambiguous identification. In cases of a dubious match, however, more sophisticated identification procedures may be required.

Clinical Diagnosis

Growth-dependent rapid identification systems are often used to identify enteric bacteria because these organisms are frequently implicated in routine urinary tract and intestinal infections (Figure 24.7) (see Section 24.1). Other growth-dependent rapid identification kits are available for other bacterial groups or even for single bacterial species. For example, kits containing a battery of tests have been developed for *Staphylococcus aureus*, *Streptococcus pyogenes*, *Neisseria gonorrhoeae*, *Haemophilus influenzae*, and *Mycobacterium tuberculosis*. Other kits are available for identification of the pathogenic fungi *Candida albicans* and *Cryptococcus neoformans* (∞Section 27.8).

The decision to use a specific diagnostic test is usually made by the clinical microbiologist who must consider the nature of the clinical specimen, the basic characteristics (morphology and Gram stain) of pure cultures obtained, and previous experience with similar cases.

 24.2 Concept Check

Traditional methods for identifying pathogens depend on observing metabolic changes induced as a result of growth. These growth-dependent methods provide rapid and accurate pathogen identification.

◆ Distinguish between *selective* and *differential* media. Give an example of a medium used for each purpose.

◆ Suggest appropriate general, selective, and differential media for preliminary isolation of pathogens from (1) a urine culture, and (2) a blood culture.

24.3 Antimicrobial Drug Susceptibility Testing

Identification of microbial pathogens isolated from clinical specimens is used to confirm medical diagnoses and to guide antimicrobial therapy. Determination of the susceptibility of microbial isolates to antimicrobial agents is a critical health care issue.

We discussed some principles for the measurement of antimicrobial activity in Chapter 20. The susceptibility of a culture can be most easily determined by an agar diffusion method or by using a tube dilution technique to determine the minimum inhibitory concentration (MIC) of an agent that is necessary to inhibit growth (∞ Section 20.4). Food and Drug Administration (FDA) regulations control the automated instruments used for susceptibility testing in the United States. Procedures and standards, including experimental endpoints for each organism and antibiotic, are constantly updated by the National Committee for Clinical Laboratory Standards (NCCLS), a nonprofit, standards-developing organization that establishes voluntary consensus standards for antibiotic testing as well as other healthcare technologies (http://www.nccls.org).

The MIC procedure for antibiotic susceptibility testing involves an *antibiotic dilution assay*, either in culture tubes (∞Figure 20.10) or in the wells of a microtiter plate (Figure 24.8e●). Wells containing serial dilutions of antibiotics are inoculated with a standard amount of the test organism. Growth in the presence of the various antibiotics is observed by measuring turbidity. Antibiotic susceptibility is usually expressed as the *highest dilution* (lowest concentration) of antibiotic that completely inhibits growth. This defines the value of the MIC.

The standard procedure that assesses antimicrobial activity is called the *Kirby–Bauer method* (Figure 24.8). Agar media are inoculated by evenly spreading a defined density of a suspension of the pure culture on the agar surface. Filter paper disks containing a defined quantity (μg/disk) of the antimicrobial agents are then placed on the inoculated agar. After a specified period of incubation, the diameter of the inhibition zone around each disk is measured. Table 24.4 presents zone sizes for several antibiotics. Inhibition zone diameters are then interpreted into susceptibility catagories based on zone size. Interpretive criteria for the various agents for different bacteria are provided either by the FDA or the NCCLS.

Etest® (AB BIODISK, Solna, Sweden) is a non-diffusion based technique that employs a preformed and predefined gradient of an antimicrobial agent immobilized on a plastic strip. The concentration gradient covers a MIC range across 15 two-fold dilutions of the conventional method. When applied to the surface of an inoculated agar plate, the gradient is transferred from the strip to the agar and remains stable for a period that covers the wide variation of critical times associated with the growth characteristics of different microorganisms. After overnight incubation or longer, an elliptical zone of inhi-

Table 24.4 Standards for dilution for some antimicrobial disk diffusion susceptibility tests[a]

Antibiotic	Amount on disk	Inhibition zone diameter (mm)[b]		
		Resistant	Intermediate	Susceptible
Ampicillin[c]	10 μg	13 or less	14–16	17 or more
Ampicillin[d]	10 μg	28 or less	—	29 or more
Ceftriaxone	30 μg	13 or less	14–20	21 or more
Chloramphenicol	30 μg	12 or less	13–17	18 or more
Clindamycin	2 μg	14 or less	15–20	21 or more
Erythromycin	15 μg	13 or less	14–22	23 or more
Gentamicin	10 μg	12 or less	13–14	15 or more
Methicillin[c]	5 μg	9 or less	10–13	14 or more
Nitrofurantoin	300 μg	14 or less	15–16	17 or more
Penicillin G[e]	10 units	28 or less	—	29 or more
Penicillin G[f]	10 units	14 or less	—	15 or more
Streptomycin	10 μg	6 or less	7–9	10 or more
Sulfonamide	10 μg	12 or less	13–16	17 or more
Tetracycline	30 μg	14 or less	15–18	19 or more
Trimethoprim-sulfamethoxazole	1.25/23.75 μg	10 or less	11–15	16 or more
Tobramycin	10 μg	12 or less	13–14	15 or more
Vancomycin[g]	10 μg	14 or less	15–16	17 or more
Vancomycin[h]	10 μg	14 or less (test for MIC)	—	15 or more

[a] Standards are defined and updated by NCCLS, an international nonprofit organization that develops voluntary consensus standards for antibiotic testing and other healthcare technologies (http://www.nccls.org)

[b] See Figure 24.8 for illustrations of standard susceptibility testing methods.

[c] For Enterobacteriaceae.

[d] For staphylococci and highly penicillin-sensitive organisms.

[e] For staphylococci.

[f] For organisms such as enterococci that may cause some systemic infections treatable with high doses of penicillin G.

[g] For enterococci.

[h] For staphylococci.

bition centered along the axis of the strip develops. The MIC value in μg/ml can be read at the point where the ellipse edge intersects the precalibrated Etest strip, providing a precise MIC. This value can then be interpreted using FDA or NCCLS MIC data (Figure 24.8f).

Because of the widespread occurrence of antimicrobial drug resistance (Section 20.12), an antibiotic susceptibility test is recommended for most pathogens isolated from clinical specimens. Data such as those in Table 24.4 are useful for choosing the best antibiotic for a specific bacterial infection. Although most community-acquired pathogens are susceptible to a number of different antibiotics, some pathogens, for example, *Pseudomonas aeruginosa*, are susceptible to very few antimicrobial drugs.

Nosocomial pathogens are those acquired in hospitals. Many nosocomial pathogens have developed antibiotic resistance, and a few are resistant to all known antibiotics (Section 20.12). Thus, antibiotic susceptibility testing for hospital-acquired pathogens is essential for effective chemotherapy. Using the drug susceptibility information gathered in this fashion, the hospital infection control microbiologist generates periodic reports called **antibiograms** that indicate the susceptibility of clinically isolated organisms to the antibiotics in current local use. Antibiograms are valuable for monitoring control of known pathogens and for tracking the emergence of new antibiotic-resistant strains of pathogens, especially at the local level.

 24.3 Concept Check

Antimicrobial drugs are widely used for the treatment of infectious diseases. Pathogens should be tested for susceptibility to individual antibiotics to ensure appropriate chemotherapy. This rigorous approach to antimicrobial drug treatment is usually applied only in health care settings.

♦ Describe the Kirby–Bauer technique. For an individual organism and antimicrobial agent, what do the results indicate?

♦ What is the value of antimicrobial drug susceptibility testing for the microbiologist, the physician, and the patient, both for community-based and nosocomial infections?

24.4 Safety in the Microbiology Laboratory

Microbiology laboratories, and especially clinical laboratories, pose significant potential for infectious biological hazards to workers and those who come into contact with workers. Hence, a defined protocol of standard laboratory practices for handling clinical samples has been established to avoid laboratory accidents. In the United States, every clinical and research institution that deals with human or primate tissue is required by law to have an occupational exposure control plan for the handling of all

BSL-4 laboratories are designed for *maximum containment of life-threatening pathogens that have a high probability of transmission by aerosols and for which there is no effective immunization or cure.* Physical containment in BSL-4 facilities must include some form of total isolation of the organism, such as manipulation through gloves in a sealed biological safety cabinet or by personnel wearing full-body, positive pressure suits with air supplies (Figure 24.9●). Some examples of pathogens that must be manipulated in a BSL-4 facility include hemorrhagic fever viruses (Lassa, Marburg, Ebola, ∞Section 25.11), variola virus (smallpox, ∞Section 25.12), and drug-resistant *Mycobacterium tuberculosis* (∞Section 26.5). BSL-4 laboratories are found only in research and development facilities that are specifically designed and equipped to handle the most dangerous pathogens.

 24.4 *Concept Check*

Safety in the clinical laboratory requires effective training, planning, and care to prevent the infection of laboratory workers with pathogens. Materials such as live cultures, inoculated culture media, used hypodermic needles, and patient specimens require specific precautions for safe handling.

◆ What are the major precautions necessary to prevent spread of a bloodborne pathogen to laboratory personnel?

◆ What are the major causes of laboratory infections?

◆ Identify the basic features of BSL containment laboratories, from BSL-1 to BSL- 4.

 **II** **IMMUNOLOGY AND CLINICAL DIAGNOSTIC METHODS**

Immunoassays are widely used in clinical, reference, and research laboratories for the detection of specific pathogens or pathogen products. They are used to confirm infections when the agent is not or cannot be cultured. When culture methods for pathogens are not routinely available or are prohibitively difficult to perform, as is the case with most viral infections, including infection with the human immunodeficiency virus (HIV), immunoassays often provide an effective and relatively simple means of identifying a specific pathogen.

24.5 Immunoassays for Infectious Disease

Specific immune antibodies to pathogens are utilized *in vitro* to detect infectious diseases. In some cases, the antibody or cellular immune response of a patient is used to indicate infection by a pathogen. In other cases, antibodies are used in tests to identify pathogens *in vitro*.

Innate Immunity

The immune response was discussed in Chapter 22. The major aspects of immunity are summarized in Figure

● **Figure 24.9 A worker in a BSL-4 (biological safety level-4) laboratory.** BSL-4 is the highest level of biological control, affording maximum worker protection and pathogen containment. The worker has a whole–body sealed suit with an outside air supply and ventilation system. Air locks control all access to the laboratory. All material leaving the laboratory is autoclaved or chemically decontaminated.

24.10●. For a pathogen that the body has never before encountered, the pathogen must first be recognized, usually by cells called phagocytes (∞Section 22.1). A group of specialized receptors on the phagocytes recognize one or more repeating macromolecular patterns, such as the repetitive chains of saccharide groups on the cell wall (∞Section 22.2). Recognition of macromolecular patterns shared by a number of pathogen types or species activates the phagocyte to ingest and destroy the targeted pathogens, a process called *phagocytosis*. Pattern recognition followed by destruction of pathogens is called *innate immunity* or *nonspecific immunity* and is the cornerstone for the ability to resist invasive pathogens.

Antigen-Specific Immunity and Antibodies

In the second phase of immunity, the phagocytes present pathogen-derived *antigens* (proteins obtained from the destroyed pathogen) to antigen-specific lymphocytes known as T cells (Figure 24.10 and ∞Section 22.1). T cells known as *T helper* (T_H) cells do not act directly on the pathogen but recruit and stimulate (help) other cells. T_H2 cells, a particular T_H subset, activate other antigen-specific lymphocytes, the *B cells*. The B cells then respond by producing soluble, antigen-binding proteins known as *antibodies* (Figure 24.10 and ∞Sections 22.9 and 22.10). A *primary antibody response* generally occurs within five days, but antibodies do not reach peak quantities for several weeks. The antibody proteins are pathogen-specific, exclusively recognizing individual antigens from particular pathogens.

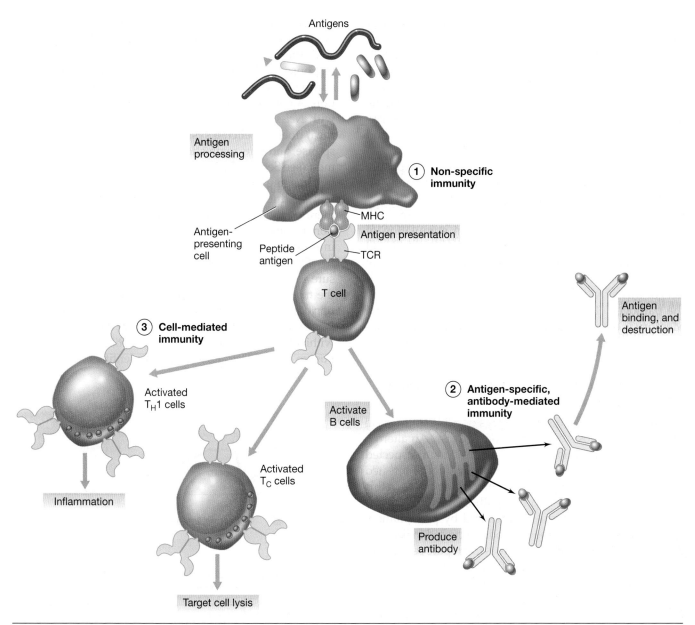

● **Figure 24.10** **The immune response.** Exposure to pathogens stimulates ① innate immunity, mediated by non–specific phagocyte cells, and antigen-specific immunity. Specific immunity is geared to recognize individual pathogen antigens and is mediated by ② antigen-reactive antibodies or ③ by antigen-reactive T cells, including T inflammatory cells (T_H1) and T cytotoxic cells (T_C).

The specific antibodies interact with the antigen on target cells, but cannot kill the cells. A group of proteins, known collectively as complement (◯◯Section 22.11), may attach to antibodies bound to the pathogen and lyse all cells with attached antibody. The complement–antibody interaction affects only those cells that have been targeted by antibodies. For example, antibodies specific for cell surface proteins of *Salmonella* sp. interact only with *Salmonella*: Complement may cause lysis of the antibody-sensitized *Salmonella* cell, but not of a neighboring *Escherichia coli* cell that is not antibody-sensitized. Thus, the immune response is *specific* for individual antigens by virtue of specific antibodies, but the final effect may occur by means of nonspecific mechanisms such as complement.

Cell-Mediated Immunity and Immune Memory

In many cases, antibody-mediated immunity is not an effective mechanism for controlling the spread of infection. Some infectious agents parasitize the body from *within* cells. For example, animal viruses reproduce using host cell systems and, therefore, spend a large portion of their life cycle within the host cells (◯◯Section 9.12). Likewise, bacteria such as *Mycobacterium tuberculosis*, the causative agent of tuberculosis, live in phagocytes (◯◯Sections 22.2 and 26.5). Because antibodies are geared to recognize pathogens in the blood or at mucosal cell surfaces, the infected host cells must be identified and destroyed by other means, usually involving the cell-to-cell interactions of the *cell-mediated immunity*. One type of T_H cell, the

◆ How are fluorescent antibodies used to identify specific cells in complex mixtures such as blood?

24.10 Enzyme-Linked Immunosorbent Assay and Radioimmunoassay

It is desirable to use immunoassays that have high *sensitivity*, the ability to detect very low quantities of antigen–antibody complexes (see Section 24.5 and Table 24.8). Because of their high sensitivity, **enzyme-linked immunosorbent assay (ELISA)** and **radioimmunoassay (RIA)** methods are widely used. ELISA and RIA employ enzymes and radioisotopes, respectively, to label antibody molecules used for antigen detection. The covalent attachment of radioactive molecules or enzymes to antibody molecules decreases the amount of antigen–antibody complex required to detect a reaction. This increased sensitivity has been used in clinical diagnostics and research and has allowed the development of a variety of new immunological tests.

ELISA

The covalent attachment of enzymes to antibody molecules creates an immunological tool possessing both high specificity and high sensitivity. The technique, called *ELISA*, makes use of antibodies to which enzymes have been covalently bound such that the enzyme's catalytic properties and the antibody's specificity are unaltered. Typical bound enzymes include peroxidase, alkaline phosphatase, and β-galactosidase, all of which catalyze reactions that develop colored products that can be detected in very low amounts with a spectrophotometer.

Two basic ELISA methodologies have been developed, one for detecting antigen (*direct* ELISA) and the other for detecting antibodies (*indirect* ELISA). For detecting antigens such as virus particles from a blood or fecal sample, the direct ELISA method is used. In this procedure the antigen is "trapped" between two layers of antibodies (Figure 24.23●). Thus, this method is sometimes called the *sandwich ELISA*. The specimen is added to the wells of a microtiter plate previously coated with antibodies specific for the antigen to be detected. If an antigen (virus particle) is present in a sample, it will be trapped by the antigen-binding sites on the antibodies. After washing unbound material away, a second antibody containing a conjugated enzyme is added. The second antibody is also specific for the antigen: It binds to other exposed determinants. Following a wash, the enzyme activity of the bound material in each microtiter well is determined by adding the substrate for the enzyme. The color produced is proportional to the amount of antigen present (Figure 24.23).

To detect *antibodies* in human serum, an indirect ELISA is employed. An indirect ELISA is widely used to detect antibodies to human immunodeficiency virus (HIV) in human body fluids, and we will discuss this test in detail because the principles involved apply to all indirect ELISA tests (Figure 24.24●).

Modified rapid ELISA procedures use reagents adsorbed to a fixed support material such as paper strips, nitrocellulose or plastic membranes, or plastic "dipsticks." These tests cause a color change on the strip or stick in a very short time. These rapid "point of care" tests are used for diagnosis of infectious diseases such as HIV-AIDS (∞Section 26.14) or "strep throat" (pharyngeal infection with *Streptococcus pyogenes*, ∞Section 26.2).

Nondisease applications for rapid tests include pregnancy testing and drug testing (Figure 24.25●). In most of these tests, a body fluid, generally urine or blood, is applied to the reagent-support matrix. After reacting with the body fluids, the support is washed and developed with a second reagent that identifies antigen (direct test) or antibody (indirect test such as HIV-AIDS) bound to the matrix. These tests are especially valuable where a small number of samples are to be analyzed at the site where the urine or blood is collected (for instance, away from a clinical laboratory) and can be performed by individuals lacking certified laboratory skills. Results can be reported at the site, avoiding the need for delays in patient care or for follow-up visits to obtain test results. The drawback to these tests, however, is that they tend to be less specific than more elaborate tests. As a result, these point-of-care tests often need to be confirmed by standard laboratory tests.

The HIV ELISA

The virus that causes HIV-AIDS, the *human immunodeficiency virus (HIV)* (∞Section 26.14), is transmitted by body fluids including blood. Sensitive, specific, rapid, and cost-effective screening tools are needed to test blood samples from individuals exposed to HIV and to ensure that HIV is not being inadvertently transmitted during blood transfusions or through the transfer of blood products. An ELISA test is used for the routine screening of blood to test for infection with HIV.

The HIV ELISA test is an *indirect* ELISA designed to measure *antibodies* to HIV present in serum. Initial infection with HIV leads to the production of antibodies to several HIV antigens, in particular, those of the HIV envelope. These antibodies can be detected by the HIV ELISA test (Figure 24.24).

To carry out an HIV ELISA test, microtiter plates are first coated with a preparation of disrupted HIV particles; about 200 ng of disrupted HIV is required in each well. A diluted patient serum sample is then added, and the mixture is incubated to allow HIV-specific antibodies to bind to HIV antigens. To detect the presence of antigen–antibody complexes, a second antibody is then added. This second antibody is an enzyme-conjugated anti-human IgG preparation. The anti-human IgG antibodies bind to any HIV-specific IgG antibodies now bound to the HIV antigen preparation. Following addi-

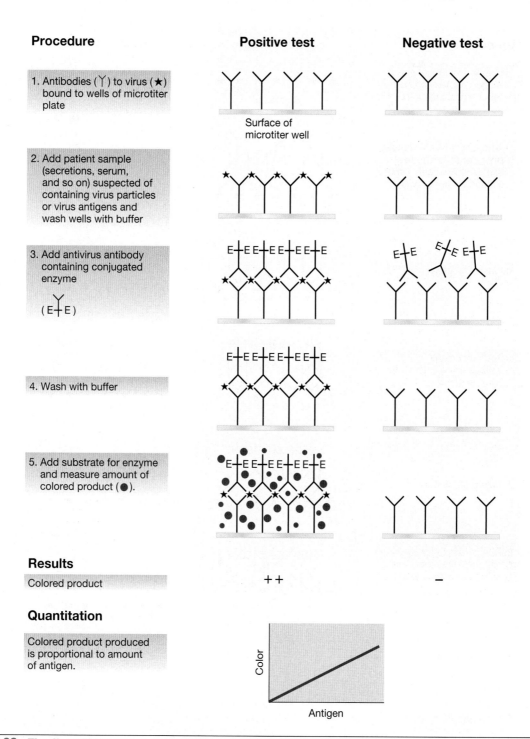

● **Figure 24.23**   **The direct ELISA test.** A direct ELISA test can be used to detect antigenic pathogen components or antigenic metabolites in blood, urine, and other body fluids.

tion of the second antibody, the enzyme substrate is added and the enzyme activity is assayed. The color obtained in the enzyme assay is proportional to the amount of anti-human IgG antibody bound (Figure 24.24). The binding of the second antibody is an indication that antibodies from the patient's serum recognized the HIV antigens and that the patient has antibodies to HIV and therefore has been infected with HIV. Control sera (known to be HIV-positive or HIV-negative) are assayed in parallel with patient samples to establish specificity

(positive control) and measure the extent of background absorbance in the assay (negative control).

The HIV ELISA test is a rapid, highly sensitive, specific method for detecting exposure to HIV. Since ELISAs in general are highly adaptable to mass screening and automation, the HIV ELISA test is used as a standard screening method for blood. However, this test method can give erroneous results under certain circumstances. For example, the test occasionally gives *false-positive* results. Because a number of factors can contribute to these results,

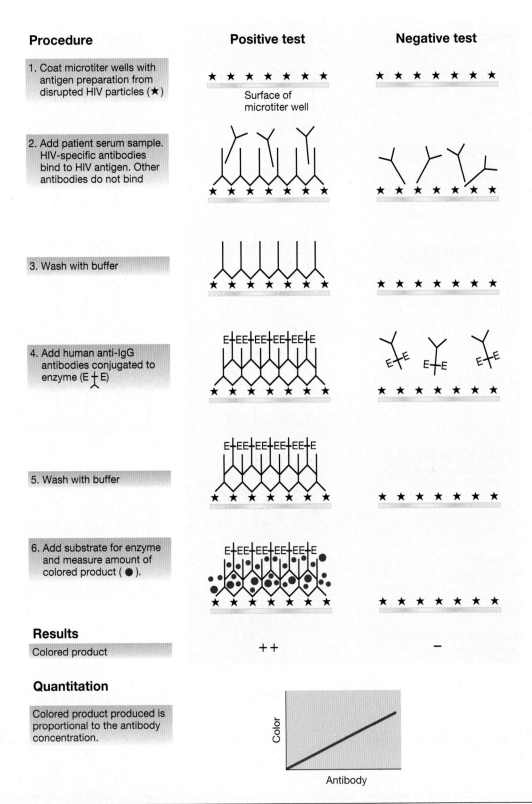

● **Figure 24.24   Indirect ELISA test.** An indirect test is used for detecting antibodies to human immunodeficiency virus (HIV), the causal agent of acquired immunodeficiency syndrome (AIDS).

none of which are related to exposure to HIV, all positive HIV ELISA tests *must* be confirmed by another independent test, usually the Western blot (immunoblot) test (see Section 24.11). A positive HIV Western blot test after a positive HIV ELISA test is considered proof of HIV infection.

The other drawback to the HIV ELISA test is the possibility of obtaining *false-negative* results. The immune system develops a detectable antibody response titer only several weeks after antigen exposure (see Section 24.5). In the case of HIV infection, this lag time is esti-

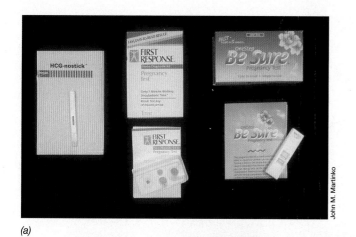

(a)

John M. Martinko

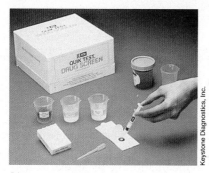

(b)

Keystone Diagnostics, Inc.

● **Figure 24.25 Rapid ELISA-based assay kits.** Pregnancy test kits (a) and the drug-testing kit (b) are available for point-of-care testing for these applications. Other kits are used for diagnosis of infectious diseases including streptococcal sore throat and HIV-AIDS.

mated to be six weeks to a year. Therefore, individuals who have been recently infected with HIV may not yet be producing detectable amounts of antibody when they are tested. Another reason for a false-negative result in the HIV ELISA test is the total destruction of the immune system seen in advanced cases of AIDS; if no immune cells are left in the body, no antibodies can be made and the ELISA test is not useful. However, at this stage of disease, an AIDS diagnosis can be based on clinical information (∞∞Section 26.14), and the ELISA test is useful only as a confirmatory indicator.

Other ELISA Tests of Clinical Importance

Besides the ELISA test for HIV, literally hundreds of clinically useful ELISAs have been developed. Some of these are direct ELISAs for detecting antigens, including bacterial toxins such as cholera toxin, enteropathogenic *Escherichia coli* toxin, and *Staphylococcus aureus* enterotoxin. Viruses currently detected using direct ELISA techniques include rotavirus, hepatitis viruses, rubella virus, bunyavirus, measles virus, mumps virus, and parainfluenza virus.

Indirect ELISAs have been developed for detecting antibodies to a variety of clinically important bacteria.

Although not meant to be a complete list, ELISAs for detecting serum antibodies to *Salmonella* (gastrointestinal diseases), *Yersinia* (plague), *Brucella* (brucellosis), a variety of rickettsias (Rocky Mountain spotted fever, typhus, Q fever), *Vibrio cholerae* (cholera), *Mycobacterium tuberculosis* (tuberculosis), *Mycobacterium leprae* (leprosy), *Legionella pneumophila* (legionellosis), *Borrelia burgdorferi* (Lyme disease), and *Treponema pallidum* (syphilis) have been developed. ELISAs have also been developed for detecting antibodies to *Candida* (yeast) and a variety of parasites, including those causing amebiasis, Chagas' disease, schistosomiasis, toxoplasmosis, and malaria.

The speed, low cost, lack of radioactive waste, and long shelf life make ELISA tests particularly attractive for many laboratories. But it is the high *sensitivity* that makes the ELISA a very important immunodiagnostic tool.

Radioimmunoassay

Radioimmunoassay (RIA) employs radioisotopes as antibody or antigen conjugates, in contrast to the enzymes used in ELISA. The isotope iodine-125 (^{125}I) is commonly used as the conjugate because antibodies or antigens can be readily iodinated without disrupting their immune specificity. RIA is used clinically to measure rare serum proteins such as human growth hormone, glucagon, vasopressin, testosterone, and insulin present in humans in extremely small amounts (Figure 24.26●). RIAs are also used in some tests for illegal drugs.

In most cases, a *direct* RIA is employed. The direct assay is a two-step procedure. First, a series of microtiter wells are prepared containing known concentrations of pure antigen (such as a hormone). Radiolabeled antigen-specific antibodies are added to the wells, and the radioactivity bound in each of these standard wells is measured. These data establish a standard curve for antigen concentrations. Next, the antigen sample from a patient is allowed to bind to another well, and radioactive antibodies are added and measured, as before. The amount of radioactivity bound by the patient sample is then compared to a standard plot generated from the binding data obtained using the pure antigen, and the concentration of antigen in the patient serum is interpolated from the standard plot (Figure 24.26).

RIA has the same sensitivity range as ELISA (Table 24.8) and can also be performed very rapidly. However, the instruments used to detect radioactivity are quite specialized and expensive. RIA generates a considerable amount of radioactive waste, and the radioactive decay time (half life) of the radioisotopes used for detection may limit the useful life of a test kit. As a result, RIA is often used only when ELISA is not sufficiently accurate or sensitive. For example, RIA is often more useful than ELISA for detecting serum protein levels (as described earlier) because serum components inhibit some ELISA enzyme–substrate

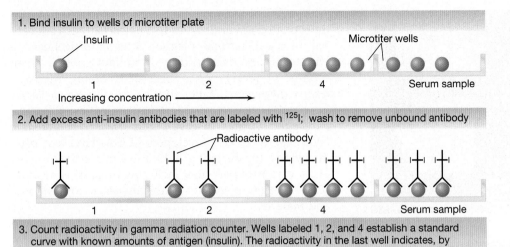

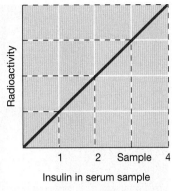

● **Figure 24.26 Radioimmunoassay (RIA).** Using RIA to detect insulin levels in human serum. Following establishment of a standard curve, the insulin concentration in a serum sample can be estimated.

or antigen–antibody interactions. Thus, for certain applications, each test system has clear advantages.

 24.10 Concept Check

ELISA and RIA methods are the most sensitive immunoassay techniques. Both involve linking a detection system, either an enzyme or a radioactive molecule, to an antibody or antigen, significantly enhancing sensitivity. ELISA and RIA are used for clinical and research work; tests have been designed to detect either antibody or antigen in many applications.

◆ Why are ELISA and RIA techniques more sensitive than immunoassays such as precipitation and agglutination?

◆ Compare ELISA and RIA with respect to their relative uses, advantages, and disadvantages.

24.11 Immunoblot Procedures

Antibodies can also be used in the **immunoblot** method to identify individual specific *proteins* associated with specific pathogens, even in complex mixtures such as cell lysates or blood. The immunoblot method employs three techniques discussed previously: (1) the separation of proteins on polyacrylamide gels, (2) the transfer (blotting) of proteins from gels to a nitrocellulose or nylon membrane, and (3) identification of the proteins by specific antibodies. Protein blotting and the subsequent identification of the proteins by specific antibodies is also called the *Western blot* technique to distinguish it from the (DNA) *Southern blot* technique (∞Section 7.7).

In the first step of an immunoblot, a protein mixture is subjected to electrophoresis on a polyacrylamide gel. This separates the proteins into several distinct bands,

each of which represents a single protein of specific molecular weight (Figure 24.27●). The proteins are then transferred to the membrane by an electrophoretic transfer process that elutes the proteins from the gel onto the membrane. At this point, antibodies raised against the pathogen components are added to the blot. Following a short incubation period to allow the antibodies to bind, a radioactive marker that binds antigen–antibody complexes is added. A common radioactive marker is *Staphylococcus* protein A iodinated with radioactive iodine, ^{125}I. Protein A has a strong affinity for antibody and binds firmly. Once the radioactive marker has bound, its location on the blot can be detected by exposing the membrane to X-ray film; the gamma rays emitted by the ^{125}I expose the film only in the region where the radioactive antibody has bound to antigen–antibody complexes (Figure 24.27).

For many clinical applications, immunoblots employ enzyme-linked immunosorbent assay (ELISA) technology (see Section 24.10) for detection of bound antigen–antibody complexes. Following treatment of the blotted proteins with specific antibody, the membrane is washed and then treated with a second antibody, which binds to the first. Covalently attached to this second antibody is an enzyme. The original antigen–antibody complexes are visualized when the enzyme is exposed to substrate: The product of the enzyme reaction leaves a colored product on the membrane at any spot where the enzyme-labeled secondary antibodies are bound to the antigen-reactive antibodies. By comparing the location of the color bands on the blot with the position of colored bands from control samples, a protein associated with a given pathogen can be positively identified.

The immunoblot procedure can be used to detect either antigen (*direct* evidence for pathogen presence) or antibody (*indirect* evidence for pathogen exposure).

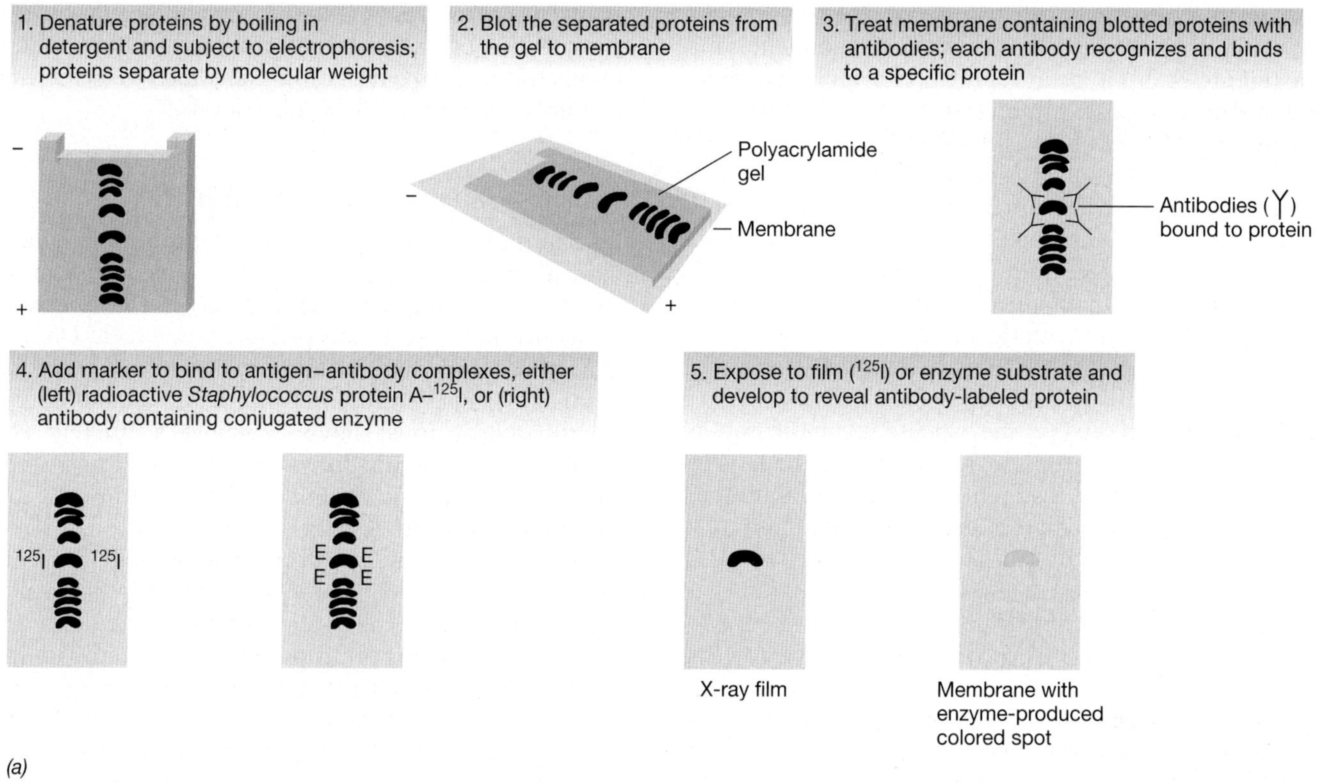

1. Denature proteins by boiling in detergent and subject to electrophoresis; proteins separate by molecular weight

2. Blot the separated proteins from the gel to membrane

3. Treat membrane containing blotted proteins with antibodies; each antibody recognizes and binds to a specific protein

Polyacrylamide gel

Membrane

Antibodies ($\curlyvee$) bound to protein

4. Add marker to bind to antigen–antibody complexes, either (left) radioactive *Staphylococcus* protein A–^{125}I, or (right) antibody containing conjugated enzyme

^{125}I ^{125}I

E   E
E   E

5. Expose to film (^{125}I) or enzyme substrate and develop to reveal antibody-labeled protein

X-ray film

Membrane with enzyme-produced colored spot

(a)

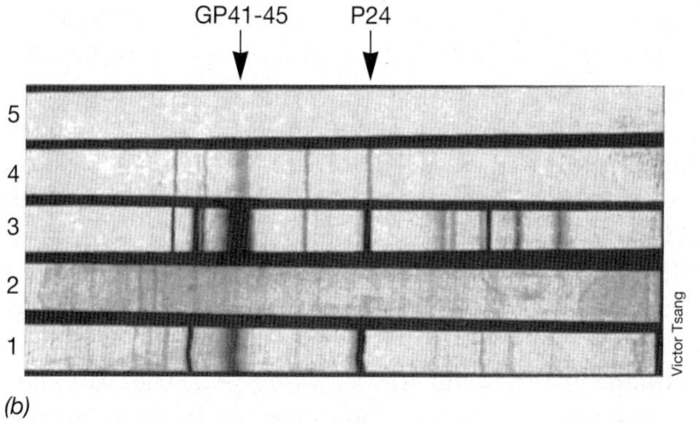

GP41-45   P24

5
4
3
2
1

(b)

Victor Tsang

● **Figure 24.27 The Western blot (immunoblot) and its use in the diagnosis of human immunodeficiency virus (HIV) infection.** (a) Protocol for an immunoblot. (b) Developed HIV immunoblot. The proteins P24 and GP41–45 are coat proteins of the virus and are diagnostic for HIV. Lane 1, Positive control serum (from known AIDS patients); lane 2, negative control serum (from healthy volunteer); lane 3, strong positive from patient sample; lane 4, weak positive from patient sample; lane 5, reagent blank to check for background binding.

The HIV Immunoblot

Immunoblots have had a significant clinical impact on the diagnosis and confirmation of HIV exposure. Because an immunoblot is more laborious, more time-consuming, less sensitive, and more costly than the ELISA test, HIV-ELISA tests have been widely used for screening purposes. However, the HIV-ELISA test occasionally yields false-positive results. Thus, the highly specific immunoblot is used to confirm positive ELISA results.

Like the HIV ELISA, the HIV immunoblot is designed to detect the presence of *antibodies* to HIV in a serum sample. To perform the immunoblot, a purified preparation of HIV is treated with sodium dodecyl sulfate (SDS), a detergent that solubilizes HIV proteins and also inactivates the virus. The HIV proteins are then resolved by polyacrylamide gel electrophoresis and blotted from the gel onto membranes (Figure 24.26). At least seven major HIV proteins are resolved by electrophoresis, and two of them, designated P24 and GP41-45, are used as specific diagnostic proteins in the HIV immunoblot. Protein P24 is the HIV core protein, and proteins GP41-45 are HIV coat proteins (∞Section 16.15).

Following blotting of the proteins, the membrane strips are incubated with the test serum sample. If the sample is HIV-positive, antibodies against HIV proteins will be present and will bind to the HIV proteins on the membrane (Figure 24.27). To detect whether antibodies from the serum sample have bound to HIV antigens, a *detecting antibody*, anti-human IgG conjugated to the enzyme peroxidase, is added to the strips. If the detecting antibody binds, the activity of the conjugated enzyme, after addition of substrate, will form a brown band on the strip at the site of antibody binding. The patient is confirmed as HIV-positive if the position of the bands in the patient serum and a positive control serum are identical; negative control serum is also analyzed in parallel and must show no bands (Figure 24.27).

Although the intensity of the bands obtained in the HIV immunoblot varies somewhat from sample to sample (Figure 24.27 *b*), the interpretation of an immunoblot is generally unequivocal, and thus the test is valuable for confirming HIV-ELISA positives and eliminating false positives. To make the HIV immunoblot clinically accessible, membrane strips containing inactivated HIV antigens (previously separated by electrophoresis) are available commercially. Separate strips can be incubated directly with the patient and control serum samples and subsequently treated with the detecting antibody.

The immunoblot technique is also used to confirm the specificity of antibody tests for Lyme disease (see Table 24.5). However, because of the expense, technical requirements, lower *sensitivity* (but higher *specificity*), and time involved, immunoblot tests are less useful than the rapid, low-cost ELISA methods for general screening purposes.

24.11 Concept Check

Immunoblot procedures are used to detect antibodies to specific antigens or to detect the presence of the antigens themselves. The antigens are separated by electrophoresis, transferred (blotted) to a membrane, and exposed to antibody. Immune complexes are visualized with enzyme-labeled or radioactive secondary antibodies. Immunoblots are extremely specific, but procedures are complex and time-consuming.

◆ What advantage does the immunoblot have over immunoassays such as ELISA and RIA?

◆ Why is the immunoblot not used for general screening for HIV exposure?

III MOLECULAR AND VISUAL DIAGNOSTIC METHODS

Extremely sensitive methods using molecular or electron microscopy methods are available for detecting pathogens. These methods do not depend on pathogen isolation or growth, or on the detection of an immune response to the pathogen. Since it is often difficult and sometimes impossible to grow and identify pathogens in culture, growth-independent identification methods based on molecular or physical methods are widely used.

24.12 Nucleic Acid Methods

Molecular methods are widely used in diagnostic microbiology. These methods use *genotypic* rather than *phenotypic* characteristics to identify specific pathogens. The success of genetic or DNA-based diagnostic procedures is based on several principles: (1) Nucleic acids can be readily isolated from infected tissues; (2) nucleic acids can be readily visualized and measured; (3) the nucleic acid sequence of an individual pathogen genome is unique, and so nucleic acid analysis can be used for unequivocal identification; and (4) nucleic acid sequences can be amplified to increase the amount of material available for analysis.

Nucleic Acid Probes in Diagnostics

One of the most powerful analytical tools available to clinical microbiologists is *nucleic acid hybridization*. Instead of detecting a whole organism or its products, hybridization detects the presence of *specific DNA sequences* associated with a specific organism. To identify a microorganism through DNA analysis, the clinical microbiologist must have available a **nucleic acid probe** for that microorganism. Nucleic acid probes normally consist of a *single strand* of DNA with a sequence unique to the gene of interest. A typical probe may be several kilobases in length. If a microorganism from a clinical specimen contains DNA or RNA sequences complementary to the probe, the probe sequence can hybridize (following appropriate sample preparation to yield single-stranded DNA from the microorganism), forming a *double-stranded* molecule (Figure 24.28●). To detect a reaction, the probe is labeled with a *reporter molecule*, a radioisotope, an enzyme, or a fluorescent compound that can be detected following hybridization. Depending on the reporter (radioisotopes are the most sensitive), as little as 0.25 μg of DNA per sample can be used for identification.

Nucleic acid probes offer many advantages over immunological assays. Nucleic acids are much more stable than proteins at high temperatures and at high pH and are more resistant to organic solvents and other chemicals. Because of the relative chemical stability of the target nucleic acids, nucleic acid probe technology can even be used to positively identify organisms that are no longer alive. Additionally, some nucleic acid probes may be more specific than antibodies and can detect single-nucleotide differences between DNA sequences.

Clinical Laboratory Probes

In most clinical probe assays, colonies from plates or pieces of infected tissue are treated with strong alkali, usually NaOH, to lyse the cells and partially denature the DNA, forming single-stranded molecules (Figure 24.28a). This mixture is then affixed to a matrix (filter or dipstick) or left in solution and the labeled probe added. Hybridization is allowed to occur at a temperature at which sequence homology between target DNA and probe DNA is necessary to form a stable duplex (the actual temperature used in a given probe assay is governed by the length and nucleic acid composition of the probe and target DNA).

Following a wash to remove unhybridized probe DNA, the extent of hybridization is measured using the reporter molecule attached to the probe. Depending on how the probe was labeled, this involves measurement of radioactivity, enzyme activity, or fluorescence. Nucleic acid probes have been marketed for the identification of several major microbial pathogens and are in widespread use for the detection of *Neisseria gonorrhoeae* and *Chlamydia trachomatis* (Table 24.9 and ⚭Sections 26.12 and 26.13).

In addition to their use in clinical diagnostics, molecular probes are finding widespread application in food industries and in food regulatory agencies. Probe detection systems can be used to monitor foods for their content of important pathogens such as *Salmonella* and *Staphylococcus*. In probe assays of food, an enrichment period is usually employed to allow low numbers of cells in the food to multiply to a sufficient number to be detectable by the probe. Probes designed for use in the food industry employ dipsticks coated with pathogen-specific DNA to hybridize with pathogen DNA from solution. Two-component probes are used here, one serving as a *reporter probe* and the other as a *capture probe* (Figure 24.28 *b*). Following hybridization of the reporter or capture probe to target DNA, the dipstick, which contains a sequence complementary to the capture probe (usually poly dT to capture poly dA on the probe) is inserted into the hybridization solution, and it traps hybridized DNA for removal and measurement.

Advances in bacterial phylogenetics based on 16S ribosomal RNA (rRNA) sequences (⚭Section 11.6) have allowed the construction of species-specific and even strain-specific nucleic acid probes. Typing based on differences in rRNA sequences is called *ribotyping*. Ribotyping visualizes the unique DNA restriction patterns of the rRNA genes when DNA from a particular organism is digested by restriction endonucleases (⚭Section 7.7). The digested DNA is separated on an agarose gel, and a labeled rRNA probe is used to visualize the unique restriction patterns of the genes encoding rRNA (⚭Section 11.11 and Figure 11.23). Within a *species*, and especially within a *strain*, the restriction pattern is highly conserved and is a molecular fingerprint for the organism. Because all organisms have ribosomal RNA genes, ribotyping has been applied in a wide variety of clinical as well as phylogenetic studies.

Nucleic acid probes detecting less than 1 μg of nucleic acid per sample can identify DNA extracted from about 10^6 bacterial cells or about 10^8 virus particles. Although molecular probes are not as sensitive as direct

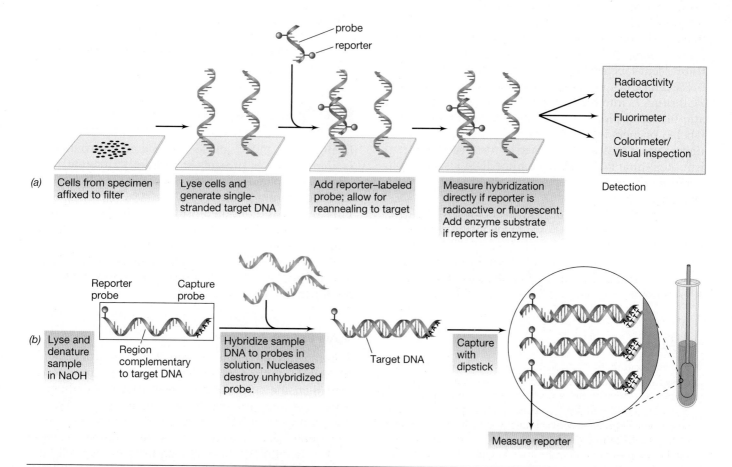

● **Figure 24.28   Nucleic acid probe methodology in clinical diagnostics.** (a) Membrane filter assay. The detecting system (reporter) can be a radioisotope, a fluorescent dye, or an enzyme. (b) Dipstick assay. In the dipstick assay a dual reporter or capture probe is used. The capture probe contains a poly-dA tail that hybridizes to a poly-dT oligonucleotide affixed to the dipstick, binding the oligonucleotide–target–reporter complex. The complex can be detected as in (a) above.

culture (where as few as 1–10 cells per sample can be detected), probe methods are useful in situations where culture of organisms is difficult or even impossible. However, other applications of DNA technology rival the sensitivity of the culture method.

The Polymerase Chain Reaction in Diagnostics

Extremely short sequence-specific nucleic acid oligonucleotides (usually 15–24 nucleotides in length) can act as primers for the polymerase chain reaction (PCR) amplification of DNA from specific pathogens. PCR is widely used in biology research and is also a major clinical diagnostic tool (∞Section 7.9). A typical PCR procedure is designed to amplify DNA about 1 billion-fold. Theoretically, this allows the visualization of a single DNA molecule obtained from a single bacterial cell. For example, sequence-specific primers for a pathogen-specific gene might be used to examine DNA derived from suspected infected tissue, even in the absence of an observable, culturable pathogen. These methods are particularly useful for identifying viral and intracellular infections. The presence of an appropriate amplified gene segment (Figure 24.29●) confirms the presence of the pathogen.

Several specific organisms for which either hybridization or PCR diagnostic methods are in use are listed in Table 24.9. Various PCR applications are used in clinical laboratories. Currently, many PCR tests employ *real-time PCR*. Real-time PCR involves the use of fluorescent-labeled PCR primers. PCR machines with precision optics monitor incorporation of the labeled PCR primers into the PCR products at every cycle. An amplification plot monitors the incorporation of the labeled primers into the amplified product. An amplified product indicates presence of the targeted organism. Real-time PCR yields an immediate visual result, avoiding the need for further detection methods such as electrophoresis.

Another application of PCR technology, *RT-PCR* (reverse-transcriptase PCR) is used to follow the progression of HIV-AIDS and is discussed in Section 26.14. This method incorporates production of cDNA directly from samples containing pathogen-specific RNA, as in the case of RNA retroviruses such as HIV (∞Section 9.13).

24.12 Concept Check

Nucleic acid hybridization is a powerful laboratory tool used for identification of microorganisms. A nucleic acid sequence specific for the microorganism of interest must be available in order to design a probe. Perhaps the most widespread use of probe-based technology is in the application of gene amplification (PCR) methods. Various DNA-based methodologies are currently used in clinical, food, and research laboratories.

◆ What advantage does nucleic acid hybridization have over standard culture methods for identification of microorganisms? What disadvantages?

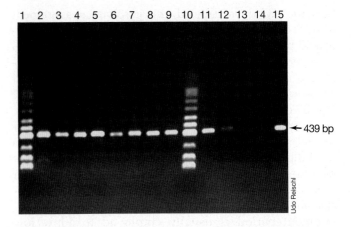

● **Figure 24.29 Polymerase chain reaction (PCR) analysis of patient sputum for *Mycobacterium tuberculosis* in the diagnosis of tuberculosis.** Sputum samples from patients were used as a source of DNA. Amplification was initiated with a primer pair, which produced the indicated 439-base pair product when a pure culture of *M. tuberculosis* was used as the DNA source (lane 15). Lanes 2–9, 11 and 12 are from sputums positive for *M. tuberculosis* (lane 12 is a weak positive). Lanes 13 and 14 are from *M. tuberculosis*-negative sputum samples. Lanes 1 and 10 are molecular weight reference markers. For a description of PCR technology, see Section 10.17.

◆ Cite an example where information about a microorganism can be obtained with a nucleic acid probe or PCR assay in the absence of standard growth-dependent assays.

24.13 Diagnostic Virology

Identification and diagnosis of infections with pathogenic animal viruses presents significantly different problems as compared with bacterial pathogens. Viruses are all obligate intracellular parasites (∞Chapter 9). Since viruses cannot replicate outside the cell, they cannot be directly cultured on artificial media, but must be grown in mammalian cells. However, viruses or their cytopathic effects can be observed directly, and a variety of direct and indirect *in vitro* test procedures are available for identification of viral pathogens.

Virus Growth *In Vitro*

Laboratory cultivation of animal viruses from clinical materials is more difficult, time-consuming, and specialized than the cultivation of most bacterial pathogens. This is because viruses grow only in living cells. We discussed the use of cell cultures for the growth of viruses in Section 9.3, and such cultures are commonly used in diagnostic virology. Several long-lived human cell lines are available for propagation of viruses. These cell lines, often termed *immortal cell lines*, grow rapidly and divide indefinitely in cell culture media and some are readily infected by mammalian viruses.

Table 24.9	Pathogens identified with nucleic acid probe and PCR methods

Pathogen	Diseases
Bacteria	
Campylobacter sp.	Food infections
Chlamydia trachomatis	Venereal syndromes; trachoma
Enterococcus sp.	Nosocomial infections
Escherichia coli (enteropathogenic strains)	Gastrointestinal disease
Haemophilus influenzae	Infectious meningitis
Legionella pneumophila	Pneumonia
Listeria monocytogenes	Listeriosis
Mycobacterium avium	Tuberculosis
Mycobacterium tuberculosis	Tuberculosis
Mycoplasma hominis	Urinary tract infection; pelvic inflammatory disease
Mycoplasma pneumoniae	Pneumonia
Neisseria gonorrhoeae	Gonorrhea
Neisseria meningitidis	Meningitis
Rickettsia sp.	Typhus, hemorrhagic fever, etc.
Salmonella sp.	Gastrointestinal disease
Shigella sp.	Gastrointestinal disease
Staphylococcus aureus	Purulent discharges (boils, blisters, pus-forming skin infections)
Streptococcus pyogenes	Scarlet fever; rheumatic fever; strep throat
Streptococcus pneumoniae	Pneumonia
Treponema pallidum	Syphilis
Fungi	
Blastomyces dermatitidis	Blastomycosis
Candida sp.	Candidiasis, thrush
Coccidioides immitis	Coccidioidomycosis
Histoplasma capsulatum	Histoplasmosis
Viruses	
Cytomegalovirus	Congenital viral infections
Epstein-Barr virus	Burkitt's lymphoma; mononucleosis
Hepatitis viruses A, B, C, D, E	Hepatitis
Herpes virus (types I and II)	Cold sores; genital herpes
Human immunodeficiency virus (HIV)	Acquired immunodeficiency syndrome (AIDS)
Human papilloma virus	Genital warts; cervical cancer
Influenza	Respiratory disease
Polyoma virus	Neurological disease
Rotavirus	Gastrointestinal disease
Protozoa	
Leishmania donovani	Leishmaniasis
Plasmodium sp.	Malaria
Pneumocystis carinii	Pneumonia
Trichomonas vaginalis	Trichomoniasis
Trypanosoma sp.	Trypanosomiasis

In addition to these immortal cell lines, viruses can also be grown in Rhesus monkey kidney cell lines. Monkey kidney cell lines are called *primary* cell lines because they die after a limited number of cell divisions and cannot be maintained indefinitely in the laboratory. Primary monkey kidney cells support growth of a number of pathogenic viruses and are valuable for the initial isolation of unknown viruses.

Electron Microscopy

Diagnostic virology can also be done by electron microscopy. Because many viruses have distinctive morphologies (∞Chapter 9), they can often be detected in clinical samples by observing the sample with an elec-

tron microscope (Figure 24.30●). In most specimens, the virus particles must first be concentrated and separated from human tissues, and a variety of techniques, generally employing centrifugation and filtration methods, are used to obtain a virus-enriched sample. Although not as reliable as immunologic or nucleic acid probe methods, the observation of virus particles of a specific morphology in a particular type of human tissue is presumptive evidence for infection.

Antibodies directed against particular viruses can be used to increase the sensitivity and specificity of this method. Agglutinating antibodies may cause the viral particles to aggregate, making them easier to distinguish from cellular debris under the electron microscope. Viruses can also be visualized by treatment of

The **Centers for Disease Control and Prevention (CDC)** in the United States, through the National Center for Infectious Diseases (NCID), operates a number of surveillance programs, as shown in Table 25.6. In many cases, diseases are reportable to more than one surveillance network. While redundant reporting may at first seem unnecessary, a number of diseases fall into several categories that may affect health-care plans and policies. Cross-referencing and reporting of data from infections, for example, of vancomycin-resistant staphylococci with the National Nosocomial Infections Surveillance System (NNIS), and with CDC as a notifiable disease (Table 25.5) provides the hospital infection control team with a national data base. Using this information, health-care providers can formulate and implement rational plans for isolation, diagnosis, and drug-sensitivity testing of staphylococcal infections to identify antibiotic resistant strains and to begin appropriate treatment.

Pathogen Eradication

Disease eradication can be accomplished in specific cases and has been successful in the eradication of naturally occurring smallpox. As we mentioned earlier in this section, smallpox was a disease with a reservoir consisting solely of the individuals with acute smallpox infections, and transmission of smallpox was exclusively person-to-person. Infected individuals transmitted the disease through

Table 25.6	**National Center for Infectious Diseases (NCID) surveillance systems for infectious disease notification and tracking, United States, 2004[a]**
Surveillance System (acronym)	**Disease Surveillance Responsibility**
121 Cities Mortality Reporting System	Influenza, pneumonia, all deaths
Active Bacterial Core Surveillance	Invasive bacterial diseases
BaCon Study	Bacterial contamination associated with blood transfusion
Border Infectious Disease Surveillance Project (BIDS)	Infectious disease along the U.S.-Mexican border
Dialysis Survey Network (DSN)	Vascular access infections and bacterial resistance in hemodialysis patients
Electronic Foodborne Outbreak Investigation and reporting System (EFORS)	Foodborne outbreaks
EMERGEncy ID NET	Emerging infectious diseases
Foodborne Diseases Active Surveillance Network (FOODNET)	Foodborne disease
Global Emerging Infections Sentinel Network (GeoSentinel)	Global emerging diseases
Gonococcal Isolate Surveillance Project (GISP)	Antimicrobial resistance in *Neisseria gonorrhoeae*
Integrated Disease Surveillance and Response (IDSR)	World Health Organization (WHO/AFRO) initiative for infectious diseases in Africa
Intensive Care Antimicrobial Resistance Epidemiology (ICARE)	Antimicrobial resistance and antimicrobial use in health-care settings
International Network for the Study and Prevention of Emerging Antimicrobial Resistance (INSPEAR)	Global emerging of drug-resistant organisms
Measles Laboratory Network	Measles in the Americas and the Caribbean
National Antimicrobial Resistance Monitoring System: Enteric Bacteria (NARMS)	Antimicrobial resistance in human nontyphoid *Salmonella*, *Escherichia coli* O157:H7, and *Campylobacter* isolates from agricultural and food sources
National Malaria Surveillance	Malaria in the United States
National Molecular Subtyping Network for Foodborne Disease Surveillance (PulseNet)	Molecular fingerprinting of foodborne bacteria
National Nosocomial Infections Surveillance System (NNIS)	Hospital-acquired infections
National Notifiable Diseases Surveillance System (NNDSS)	Notifiable diseases (see Table 25.5)
National Respiratory and Enteric Virus Surveillance System (NREVSS)	Respiratory syncytial virus (RSV), human parainfluenza viruses, respiratory and enteric adenoviruses, and rotavirus
National Surveillance System for Health Care Workers (NaSH)	Health-care worker occupational infections
National Tuberculosis Genotyping and Surveillance Network	Tuberculosis genotyping repository
National West Nile Virus Surveillance System	West Nile virus
Public Health Laboratory Information System (PHLIS)	Notifiable diseases
Surveillance for Emerging Antimicrobial Resistance Connected to Healthcare (SEARCH)	Emerging antimicrobial resistance in health-care settings
Unexplained Deaths and Critical Illnesses Surveillance System	Emerging infectious diseases worldwide
United States Influenza Sentinel Physicians Surveillance Network	260 clinical sites that report incidence and prevalence of influenza infections
Viral Hepatitis Surveillance Program (VHSP)	Viral hepatitis
Waterborne-Disease Outbreak Surveillance System	Waterborne diseases

[a]Contact information for these surveillance systems is available at: *http://www.cdc.gov/ncidod/osr/site/surv_resources/surv_sys.htm.*

direct contact with previously unexposed members of the population. Although smallpox, a viral disease, cannot be treated, immunization practices were very effective: Vaccination with a related viral strain conferred virtually complete immunity. In 1967, the World Health Organization (WHO) developed a plan to eradicate smallpox. Because of the success of vaccination programs worldwide, endemic smallpox was then confined to Africa, the Middle East, and the Indian subcontinent. After a preliminary program to vaccinate everyone in remaining endemic areas, every suspected smallpox outbreak was targeted by a team of WHO personnel who traveled to the outbreak site, quarantined individuals with active disease, and vaccinated all contacts. To break the chain of possible infection, they then immunized everyone who had contact with the contacts. This aggressive policy resulted in the elimination of the active natural disease within a decade, and the WHO announced the eradication of smallpox in 1980.

Polio, another viral disease with a very effective immunization program, is also targeted for eradication (endemic polio has been eradicated from the Western Hemisphere). Using much the same strategy to target polio as was used for smallpox, in the 1990s the WHO undertook a massive immunization program directed toward remaining endemic areas. By 2004, known *endemic* polio was restricted to Nigeria (with spread to nearby western African nations in 2003), India, and Pakistan. Recent sporadic outbreaks have also occurred in Afghanistan, and a single case was reported in Egypt in 2003.

Leprosy, another disease restricted to humans, is also targeted for eradication. Active cases of leprosy can now be effectively treated with a multidrug therapy that cures the patient and also prevents spread of *Mycobacterium leprae*, the causal agent (⚬⚬Section 26.5).

Other diseases that are being targeted for eradication are Chagas' disease (treat active cases and destroy the insect vector of the *Trypanosoma cruzi* parasite in the American tropics) and dracunculiasis (treat drinking water to prevent transmission of *Dracunculus medinensis*, the Guinea helminth parasite in Africa, Arabia, Pakistan, and other places in Asia). Candidates for eradication also include syphilis (⚬⚬Section 26.12) and rabies (⚬⚬ Section 27.1).

 25.9 Concept Check

Food and water purity regulations, vector control, immunization, quarantine, disease surveillance, and pathogen eradication are public health measures that play a major role in reduction of disease incidence.

◆ Compare public measures for controlling infectious disease caused by insect reservoirs and by human carriers.

◆ Identify public health methods used to halt the spread of an epidemic disease.

25.10 Global Health Considerations

The World Health Organization has divided the world into six geographic regions for the purpose of collecting and reporting health information, such as reports of morbidity and mortality. These geographic regions are Africa, the Americas (North America, the Caribbean, Central America, and South America), the eastern Mediterranean, Europe, Southeast Asia, and the Western Pacific. Here we compare mortality data from a developed region, the Americas, to that from a developing region, Africa.

Infectious Disease in the Americas and Africa: A Comparison

About 853 million people live in the Americas. Each year there are about 6 million deaths, or about 7 deaths per 1000 inhabitants per year. In Africa, there are about 672 million people and about 10.7 million annual deaths, or about 15.9 deaths per 1000 inhabitants per year. Although these statistics alone are cause for concern, examination of the causes of mortality in these regions is even more disturbing. Figure 25.10● indicates that most African

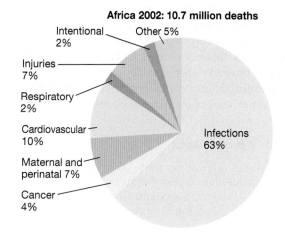

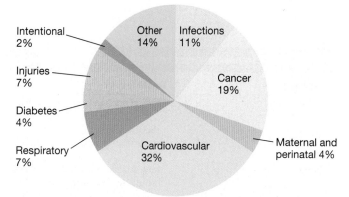

● **Figure 25.10 Causes of death in the Americas and Africa, 2002, by percentage of cause.** There were 10.7 million deaths in Africa, 6.7 million due to infectious diseases. There were 6 million deaths in the Americas, 623,000 due to infectious diseases. Intentional deaths include murder, suicide, and war.

deaths are due to infectious diseases, while in the Americas, cancer and cardiovascular diseases are the leading causes of mortality. In Africa, there are over 10 times as many deaths due to infectious diseases as compared to the Americas. Based on experiences in developed countries over the last century (∞Figure 1.7), these differences in death rates from infection are due to differences in public health services. Lack of resources in developing regions limits access to health care, medicines, safe food and water, and immunization.

Travel to Endemic Areas

The high incidence of disease in many parts of the world is a concern for people traveling to such areas. However, travelers can be immunized against many of the diseases that are endemic in foreign countries. Some typical recommendations for immunization for those traveling abroad are shown in Table 25.7. Many foreign countries currently require immunization certificates for yellow fever, but most other nonstandard immunizations are recommended only for people who are expected to be at high risk. In many parts of the world, travelers may be exposed to diseases for which there are no effective immunizations (for example, AIDS, Ebola hemorrhagic fever, dengue fever, amebiasis, encephalitis, malaria, and typhus). Travelers should take reasonable precautions such as avoiding unprotected sex, avoiding insect and animal bites, drinking only water that has been properly treated, eating properly stored and prepared food, and undergoing antibiotic and chemotherapeutic prophylactic programs or when exposure is suspected.

25.10 *Concept Check*

Infectious diseases account for nearly 30% of all worldwide mortality. Most infectious diseases occur in developing countries. Travelers to endemic disease areas should be immunized when possible and should take appropriate precautions to prevent infection.

◆ Contrast mortality due to infectious diseases in Africa and the Americas.

◆ List a series of infectious diseases for which you have *not* been immunized and with which you could come into contact next year.

25.11 Emerging and Reemerging Infectious Diseases

Infectious diseases are *global*, dynamic health problems. Here we examine some recent patterns of infectious disease, some reasons for the changing patterns, and the methods used by epidemiologists to identify and deal with new threats to public health.

The worldwide distribution of diseases can change dramatically and rapidly. Alterations in the pathogen, the environment, or the host population contribute to the spread of new diseases, with potential for high morbidity and mortality. Diseases that suddenly become prevalent are **emerging** diseases. Emerging infections are not limited to "new" diseases but also include **reemergence** of diseases thought to be controlled, especially when antibiotics become less effective and public health systems fail. Recent, dramatic examples of global emerging and reemerging disease are shown in Figure 25.11●. Diseases with potential for emergence or reemergence are described in Table 25.8. In addition, the epidemic diseases listed in Table 25.2 have the potential to emerge or reemerge as widespread epidemics and pandemics.

The phenomenon of suddenly emerging epidemic diseases is not new. Some of the diseases that suddenly emerged into prominence in the past were syphilis (caused by *Treponema pallidum*) (∞Section 26.12) and plague (caused by *Yersinia pestis*) (∞Section 27.7). In the Middle Ages, up to one-third of all living humans were killed by the plague epidemics that swept Europe, Asia, and Africa. Influenza caused a devastating worldwide epidemic in 1918–1919 (Figure 25.1 and ∞Section 26.8). In the 1980s, legionellosis (caused by *Legionella pneumophila*) (∞Section 28.7), acquired immunodeficiency syndrome (AIDS) (∞Section 26.14), and Lyme disease (∞Section 27.4) emerged as major new diseases. Current

Table 25.7	**Immunizations required or recommended for international travel**[a]	
Disease	**Destination**	**Recommendation**[b]
Yellow fever	Tropical and subtropical countries, especially in sub-Saharan Africa and South America	*Immunization required* for entry
Rabies	Rural, mountainous, and upland areas	*Immunization recommended* if direct contact with wild carnivores is anticipated
Typhoid fever	Many African, Asian, Central, and South American countries	*Immunization recommended* in areas endemic for typhoid fever

[a]*National Center for Infectious Diseases Travelers' Health*, U.S. Department of Health and Human Services, *http://www.cdc.gov/travel*.
[b]Vaccinations are generally recommended for diphtheria, pertussis, hepatitis A, hepatitis B, tetanus, polio, measles, mumps, rubella, and influenza as appropriate for the age of the traveler (∞Section 22.13). Most U.S. citizens are immunized against these diseases through normal immunization practices. Requirements and recommendations for specific vaccinations for each country are found at the website. Recommendations are also made for other appropriate infectious disease prevention measures, such as prophylactic drug therapy for malaria and plague prevention when visiting endemic areas. There are no current requirements for reentry to the United States.

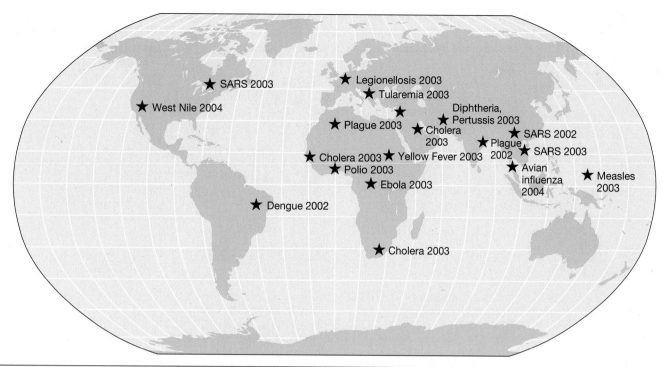

● **Figure 25.11** **Recent outbreaks of emerging and reemerging infectious diseases.** Emerging and reemerging diseases are first recognized as local epidemics. Recent epidemic outbreaks of significant but rare diseases are recorded here. All of these diseases are capable of producing widespread epidemics and even pandemics. The distribution of established pandemic diseases such HIV/AIDS and annual predictable epidemic diseases such as human influenza is not shown here.

emerging pathogens in the United States include West Nile virus (∞Section 27.6). On a worldwide scale, the rapid emergence of severe acute respiratory syndrome (SARS) has the potential to create a disastrous pandemic, especially considering the ease of spread of this highly virulent disease (Section 25.8).

Emergence Factors

Some factors responsible for emergence of new pathogens are (1) human demographics and behavior; (2) technology and industry; (3) economic development and land use; (4) international travel and commerce; (5) microbial adaptation and change; (6) breakdown of public health measures; and (7) abnormal natural occurrences that upset the usual host-pathogen balance.

The demographics of human populations have changed dramatically in the last two centuries. In 1800, less than 2% of the world's population lived in urban areas. By contrast, today nearly one-half of the world's population lives in cities. The numbers, sizes, and population densities of modern urban centers makes disease transmission much easier. For example, dengue fever (Table 25.8) is now recognized as a serious hemorrhagic disease in tropical cities, largely due to the spread of dengue virus in the mosquito *Aedes aegypti*. The disease now spreads as an epidemic in tropical urban areas. Prior to 1950, dengue fever was rare, presumably because the virus was not easily spread among a more dispersed, smaller population.

Human behavior, especially in large population centers, also contributes to disease spread. For example, sex-

ually promiscuous practices in population centers have been a major contributing factor to the spread of hepatitis and AIDS (Table 25.6; ∞Sections 26.11 and 26.14).

Technological advances and industrial development have had a generally positive impact on living standards worldwide, but in some cases these advances have contributed to the spread of diseases. For example, while tremendous technological advances have been made in health care during the twentieth century, there has been a dramatic increase in nosocomial infections (see Section 25.7). Antibiotic resistance in microorganisms is another negative outcome of modern health care practices. For example, vancomycin-resistant enterococci and multiple drug-resistant *Streptococcus pneumoniae* are important emerging diseases in developed countries.

Transportation, bulk processing, and central distribution methods have become increasingly important for quality assurance and economy in the food industry. However, these same factors can increase the potential for common-source epidemics when sanitation measures fail. For example, a single meat-processing plant spread *Escherichia coli* O157:H7 (Table 25.8) to at least 500 individuals in four states in the United States. The contaminated food source, ground beef, was recalled and the epidemic was eventually stopped, but not before several people died (∞Section 29.8). SARS spread rapidly to 32 countries through international travel from the point of origin of the disease, near the southern Chinese city of Guangzhou (see Section 25.8).

Economic development and changes in land use also can potentially promote disease spread. For example,

Table 25.8 **Emerging and reemerging epidemic infectious diseases** (continued on pages 839 and 840)

Agent	Disease and symptoms	Mode of transmission	Cause(s) of emergence
Bacteria, Rickettsias, and Chlamydias			
Bacillus anthracis	Anthrax: respiratory distress, hemorrhage	Inhalation or contact with endospores	Bioterrorism
Borrelia burgdorferi	Lyme disease: rash, fever, neurological and cardiac abnormalities, arthritis	Bite of infective *Ixodes* tick	Increase in deer and human populations in wooded areas
Campylobacter jejuni	Campylobacter enteritis: abdominal pain, diarrhea, fever	Ingestion of contaminated food, water, or milk; fecal-oral spread from infected person or animal	Increased recognition; consumption of undercooked poultry
Chlamydia trachomatis	Trachoma, genital infections, conjunctivitis, infant pneumonia	Sexual intercourse	Increased sexual activity; changes in sanitation
Escherichia coli O157:H7	Hemorrhagic colitis; thrombocytopenia; hemolytic uremic syndrome	Ingestion of contaminated food, especially undercooked beef and raw milk	Development of a new pathogen
Haemophilus influenzae biogroup *aegyptus*	Brazilian purpuric fever; purulent conjunctivitis, fever, vomiting	Discharges of infected persons; flies are suspected vectors	Possible increase in virulence due to mutation
Helicobacter pylori	Gastritis, peptic ulcers, possibly stomach cancer	Contaminated food or water, especially unpasteurized milk; contact with infected pets	Increased recognition
Legionella pneumophila	Legionnaires' disease: malaise, myalgia, fever, headache, respiratory illness	Air-cooling systems, water supplies	Recognition in an epidemic situation
Mycobacterium tuberculosis	Tuberculosis: cough, weight loss, lung lesions; infection can spread to other organ systems	Sputum droplets (exhaled through a cough or sneeze) of a person with active disease	Immunosuppression, immunodeficiency
Neisseria meningitidis	Bacterial meningitis	Person-to-person contact	Urbanization, breakdown or lack of local public health surveillance
Staphylococcus aureus	Abscesses, pneumonia, endocarditis, toxic shock	Contact with the organism in a purulent lesion or on the hands	Recognition in an epidemic situation; possibly mutation
Streptococcus pyogenes	Scarlet fever, rheumatic fever, toxic shock	Direct contact with infected persons or carriers; ingestion of contaminated foods	Change in virulence of the bacteria; possibly mutation
Vibrio cholerae	Cholera: severe diarrhea, rapid dehydration	Water contaminated with the feces of infected persons; food exposed to contaminated water	Poor sanitation and hygiene; possibly introduced via bilge water from cargo ships
Viruses			
Dengue	Hemorrhagic fever	Bite of an infected mosquito (primarily *Aedes aegypti*)	Poor mosquito control; increased urbanization in tropics; increased air travel
Filoviruses (Marburg, Ebola)	Fulminant, high mortality, hemorrhagic fever	Direct contact with infected blood, organs, secretions, and semen	Unknown; in Europe and the United States, virus-infected monkeys shipped from developing countries via air
Hendravirus	Respiratory and neurological disease in horses and humans	Contact with infected bats, horses	Human intrusion into natural environment
Hantaviruses	Abdominal pain, vomiting, hemorrhagic fever	Inhalation of aerosolized rodent urine and feces	Human intrusion into virus or rodent ecological niche
Hepatitis B	Nausea, vomiting, jaundice; chronic infection leads to hepatocellular carcinoma and cirrhosis	Contact with saliva, semen, blood, or vaginal fluids of an infected person; mode of transmission to children not known	Probably increased sexual activity and intravenous drug abuse; transfusion (before 1978)
Hepatitis C	Nausea, vomiting, jaundice; chronic infection leads to hepatocellular carcinoma and cirrhosis	Exposure (percutaneous) to contaminated blood or plasma; sexual transmission	Recognition through molecular virology applications; blood transfusion practices, especially in Japan
Hepatitis E	Fever, abdominal pain, jaundice	Contaminated water	Newly recognized

Table 25.8	Emerging and reemerging epidemic infectious diseases (continued)		
Agent	**Disease and symptoms**	**Mode of transmission**	**Cause(s) of emergence**
Viruses			
Human immuno-deficiency viruses: HIV-1 and HIV-2	HIV disease, including AIDS: severe immune system dysfunction, opportunistic infections	Sexual contact with or exposure to blood or tissues of an infected person; vertical transmission	Urbanization; changes in lifestyle or mores; increased intravenous drug use; international travel; medical technology (transfusions and transplants)
Human papillomavirus	Skin and mucous membrane lesions (often, warts); strongly linked to cancer of the cervix and penis	Direct contact (sexual contact or contact with contaminated surfaces)	Newly recognized; perhaps changes in sexual lifestyle
Human T-cell lymphotrophic viruses (HTLV-I and HTLV-II)	Leukemias and lymphomas	Vertical transmission through blood or breast milk; exposure to contaminated blood products; sexual transmission	Increased intravenous drug abuse; medical technology (transfusion and transplantation)
Influenza	Fever, headache, cough, pneumonia	Airborne; especially in crowded, enclosed spaces	Animal-human virus reassortment; antigenic shift
Lassa	Fever, headache, sore throat, nausea	Contact with urine or feces of infected rodents	Urbanization and conditions favoring infestation by rodents
Measles	Fever, conjunctivitis, cough, red blotchy rash	Airborne; direct contact with respiratory secretions of infected persons	Deterioration of public health infrastructure supporting immunization
Monkey pox	Rash, lymphadenopathy, pulmonary distress	Direct contact with infected primates and other hosts	Travel to endemic areas, consumption and handling of infected primates and other hosts
Nipah virus	Hemorrhagic fever	Close contact with bats and pigs in Malaysia	Exposure to infected animals
Norwalk and Norwalk-like agents	Gastroenteritis; epidemic diarrhea	Most likely fecal-oral; vehicles may include drinking and swimming water, and uncooked foods	Increased recognition
Rabies	Acute viral encephalomyelitis	Bite of a rabid animal; contact with infected neural tissue	Introduction of infected host reservoir to new areas
Rift Valley	Febrile illness	Bite of an infective mosquito	Importation of infected mosquitoes and/or animals; development (dams, irrigation)
Rotavirus	Enteritis: diarrhea, vomiting, dehydration, and low grade fever	Primarily fecal-oral; fecal-respiratory transmission can also occur	Increased recognition
Severe acute respiratory syndrome coronavirus (SARS-CoV)	Respiratory infection and pneumonia	Original zoonotic infection now spread person to person via infected droplets	Zoonotic spread from captured exotic animals (civet)
Venezuelan equine encephalitis	Encephalitis	Bite of an infective mosquito	Movement of mosquitoes and hosts (horses)
West Nile virus	Meningitis, encephalitis	*Culex pipiens* mosquito and avian hosts	Agricultural development, increase in mosquito breeding areas, rapid spread to nonimmune populations
Yellow fever	Fever, headache, muscle pain, nausea, vomiting	Bite of an infective mosquito (*Aedes aegypti*)	Lack of effective mosquito control and widespread vaccination; urbanization in tropics; increased air travel
Protozoa and Fungi			
Candida	Candidiasis: fungal infections of the gastrointestinal tract, vagina, and oral cavity	Endogenous flora; contact with secretions or excretions from infected persons	Immunosuppression; medical management (catheters); antibiotic use
Cryptococcus	Meningitis; sometimes infections of the lungs, kidneys, prostate, liver	Inhalation	Immunosuppression
Cryptosporidium	Cryptosporidiosis: infection of epithelial cells in the gastrointestinal and respiratory tracts	Fecal-oral, person to person, waterborne	Development near watershed areas; immunosuppression

Table 25.8	Emerging and reemerging epidemic infectious diseases (continued)		
Agent	**Disease and symptoms**	**Mode of transmission**	**Cause(s) of emergence**
Protozoa and Fungi (cont.)			
Giardia lamblia	Giardiasis; infection of the upper small intestine, diarrhea, bloating	Ingestion of fecally contaminated food or water	Inadequate control in some water supply systems; immunosuppression; international travel
Microsporidia	Gastrointestinal illness, diarrhea; wasting in immunosuppressed persons	Unknown; probably ingestion of fecally contaminated food or water	Immunosuppression; recognition
Plasmodium	Malaria	Bite of an infective *Anopheles* mosquito	Urbanization; changing parasite biology; environmental changes; drug resistance; air travel
Pneumocystis carinii	Acute pneumonia	Unknown; possibly reactivation of latent infection	Immunosuppression
Toxoplasma gondii	Toxoplasmosis; fever, lymphadenopathy, lymphocytosis	Exposure to feces of cats carrying the protozoan; sometimes foodborne	Immunosuppression; increase in cats as pets
Other Agents			
Bovine prions	Bovine spongiform encephalitis (BSE, animal) and variant Creutzfeld-Jacob disease (vCJD, human)	Foodborne	Consumption of contaminated beef

Rift Valley fever, a mosquito-borne viral infection, has been on the increase since the completion of the Aswan High Dam in Egypt in 1970. The dam flooded 2 million acres, and the enlarged shoreline increased breeding grounds for mosquitoes at the edge of the new reservoir. The first major epidemic of Rift Valley fever occurred in Egypt in 1977, when an estimated 200,000 people became ill and 598 died. Several epidemic outbreaks have occurred in the area since then, and the disease has become endemic near the reservoir.

Lyme disease, the most common vectorborne disease in the United States, is on the rise largely due to changes in land use patterns. Reforestation and the resulting increase in the numbers of deer and mice (the natural reservoirs for the disease-producing *Borrelia burgdorferi*) have resulted in greater numbers of infected ticks, the arthropod vector (∞Section 27.4). In addition, larger numbers of homes and recreational areas in and near forests increase contact between the infected ticks and humans, consequently increasing disease incidence.

International travel and commerce also affect the spread of pathogens. For example, filoviruses (*Filoviridae*), a group of RNA viruses, cause fevers culminating in hemorrhagic disease in infected hosts. These untreatable viral diseases generally have a mortality rate of greater than 20%. Most outbreaks have been restricted to equatorial central Africa, where the primate natural hosts and other vectors live. Travel of potential hosts to or from endemic areas is usually implicated in disease transmission. For example, one of the filoviruses was imported into Marburg, Germany, with a shipment of African green monkeys used for laboratory work. The virus quickly spread from the primate vector to some of the human handlers. Twenty-five people were initially infected, and six more developed disease as a result of contact with the human cases. Seven people died in this outbreak of what became known as the *Marburg virus*. Another shipment of laboratory monkeys brought a different filovirus to Reston, Virginia, in the United States. Fortunately, the virus was not pathogenic for humans, but due to its respiratory transmission mode, the Reston virus infected and killed most of the monkeys at the Reston facility within days. These two filoviruses are closely related to the Ebola virus (Table 25.8).

Sporadic Ebola outbreaks in central Africa, often characterized by mortality rates greater than 50%, highlight a group of pathogens for which there is no immunity or therapy. These pathogens could potentially be spread via air travel throughout the world in a matter of days. A highly contagious respiratory agent like the Reston virus that also possesses the high mortality potential of the Ebola virus could devastate population centers worldwide in a matter of weeks.

Microbial adaptation and change can contribute to pathogen emergence. For example, nearly all RNA viruses, including influenza and HIV, undergo rapid, unpredictable genetic mutations. RNA viruses lack correction mechanisms for replication steps, and so they incorporate mutations in their genome at an extremely high rate compared with most of the DNA viruses. The RNA viruses are considered to be major epidemiological problems because of their constantly changing genomes.

Bacterial genetic mechanisms are capable of enhancing virulence and promoting emergence of new epidemics. One group of virulence-enhancing mechanisms are the mobile genetic elements, bacteriophages, plasmids and transposons (∞Sections 16.1–16.5, 10.9, and 10.11). Table 25.9 lists some virulence factors carried on these mobile genetic elements that contribute to pathogen emergence.

Antibiotic resistance is another factor in bacterial pathogen resurgence (∞Section 20.12) and in virus emergence. Although several drugs are effective against certain viral diseases (∞Section 20.10), resistance to these drugs is very common, especially among the RNA viruses. For example, many strains of HIV develop resistance to azidothymidine (AZT) unless it is used in combination with other drugs (∞Section 26.14).

A breakdown of public health measures is sometimes responsible for the emergence or resurgence of diseases. For instance, cholera (caused by *Vibrio cholerae*, ∞Section 28.5) can be adequately controlled, even in endemic areas, by providing proper sewage disposal and water treatment. However, in 1991 an outbreak of cholera due to contaminated municipal water supplies in Peru was one of the first indications that the current cholera pandemic had reached the Americas (∞Section 28.5). In 1993, the municipal water supply of Milwaukee, Wisconsin, was contaminated with the chlorine-resistant protozoan *Cryptosporidium*, resulting in over 400,000 cases of intestinal disease, 4000 of which required hospitalization. Enhanced filtration systems were required to rid the water supply of the pathogen (∞Section 28.6).

Inadequate public vaccination programs can lead to the resurgence of previously controlled diseases. For example, recent outbreaks of diphtheria (caused by *Corynebacterium diphtheriae*) (∞Section 26.3) in the former Soviet Union result from inadequate immunization of susceptible children due to the breakdown in public health infrastructures. Pertussis, another vaccine-preventable childhood respiratory disease (caused by *Bordetella pertussis*) (∞Section 26.4), has increased recently in eastern Europe and in the United States due to inadequate immunization.

Finally, abnormal natural occurrences sometimes upset the usual host-pathogen balance. For example, hantavirus is a well-known human pathogen that occurs in many rodent populations, even in laboratory animals (∞Section 27.2). A number of lethal cases of hantavirus infection and disease were reported in 1993 in the American Southwest and were linked to exposure to wild animal droppings. The likelihood of exposure to mice and droppings was increased due to a larger than normal wild mouse population resulting from near-record rainfall, a long growing season, and a mild winter.

Addressing Emerging Diseases

Many of the emerging diseases we have discussed are absent from the official notifiable disease list for the United States (Table 25.5). How then do public health officials define emerging diseases and prevent major epidemics? The keys for addressing emerging diseases are *recognition* of the disease and *intervention* to prevent disease transmission.

The first step in disease recognition is *surveillance*. Epidemic diseases that exhibit particular *clinical syndromes* warrant intensive public health surveillance. These syndromes are (1) acute respiratory diseases, (2) encephalitis and aseptic meningitis, (3) hemorrhagic fever, (4) acute diarrhea, (5) clusterings of high fever cases, (6) unusual clusterings of any disease or deaths, and (7) resistance to common drugs or treatment. Thus, new diseases are recognized because of their epidemic incidence, clusterings, and syndromes. As the prevalence and pathology of an emerging disease are recognized, the disease is added to the notifiable disease list.

Genetic element	Organism	Virulence factors
Table 25.9	**Virulence factors encoded by bacteriophages, plasmids, and transposons**[a]	
Bacteriophage	*Streptococcus pyogenes*	Erythrogenic toxin
	Escherichia coli	Shiga-like toxin
	Staphylococcus aureus	Enterotoxins A, D, E, staphylokinase, toxic shock syndrome toxin-1 (TSST-1)
	Clostridium botulinum	Neurotoxins C, D, E
	Corynebacterium diphtheriae	Diphtheria toxin
Plasmid	*Escherichia coli*	Enterotoxins, pili colonization factor, hemolysin, urease, serum resistance factor, adherence factors, cell invasion factors
	Bacillus anthracis	Edema factor, lethal factor, protective antigen, poly-D-glutamic acid capsule
	Yersinia pestis	Coagulase, fibrinolysin, murine toxin
Transposon	*Escherichia coli*	Heat-stable enterotoxins, aerobactin siderophores, hemolysin and pili operons
	Shigella dysenteriae	Shiga toxin
	Vibrio cholerae	Cholera toxin

[a] For discussion of bacteriophages, plasmids, and transposons, see Sections 9.8–9.11 and 16.1–16.5, 10.9, and 10.14, respectively.

For example, AIDS was recognized as a disease in 1981 and was added to the notifiable disease list (see Table 25.5) in 1984. Lyme disease was first recognized as a separate clinical disease in the 1980s and added to the notifiable disease list in 1991. Likewise, outbreaks of gastrointestinal disease due to enteropathogenic *Escherichia coli* O157:H7 have been increasing in recent years, and the strain was added to the notifiable disease list in 1995.

Intervention to prevent spread of emerging infections must be a public health response involving a variety of methods. Disease-specific intervention is the key to controlling individual outbreaks. Methods such as quarantine, immunization, and drug treatment must be applied to contain and isolate outbreaks of specific diseases. Finally, for vectorborne and zoonotic diseases, the nonhuman host or vector must be identified to allow intervention in the life cycle of the pathogen and interrupt transfer to humans.

 25.11 Concept Check

Changes in host, vector, or pathogen conditions, whether natural or artificial, can result in conditions that encourage the explosive emergence or reemergence of certain infectious diseases. Global surveillance and intervention programs must be developed to prevent new epidemics and pandemics.

◆ What factors are important in the emergence or reemergence of potential pathogens?

◆ Indicate general and specific methods that would be useful for dealing with emerging infectious diseases.

25.12 Biological Warfare and Biological Weapons

Biological warfare is the use of biological agents to incapacitate or kill a military or civilian population in an act of war or terrorism. Biological weapons have been used against targets in the United States, and weapons-making facilities are suspected to be in the hands of several governments as well as extremist groups.

Characteristics of Biological Weapons

Biological weapons must be organisms or toxins that are (1) easy to produce and deliver, (2) safe for use by the offensive soldiers, and (3) able to incapacitate or kill individuals under attack in a reproducible and consistent manner. Many organisms or biological toxins fit these rather general criteria, and we will discuss several of these below.

Although bioweapons are potentially useful in the hands of conventional military forces, the greatest likelihood of bioweapons use is probably by terrorist groups. This is in part due to the availibility and low cost of producing and propagating many of the organisms useful for biological warfare. Biological weapons are accessible to nearly every government and even to well-financed private organizations.

Candidate Biological Weapons

Virtually all pathogenic bacteria or viruses are potentially useful for biological warfare, and several of the most likely candidate organisms are relatively simple to grow and disseminate. Commonly considered biological weapons agents are listed in Table 25.10. The most commonly mentioned candidate bioweapon is *Bacillus anthracis*, the causal agent of anthrax. We will discuss anthrax separately.

Other important bacterial bioweapons candidates include *Yersinia pestis*, the organism responsible for plague (∞Section 27.7), *Brucella abortus* (fever and bacteremia; brucellosis), *Francisella tularensis* ("rabbit fever"), and *Salmonella* (foodborne and waterborne illnesses) (∞ Section 29.7). Viral pathogens with bioweapons potential include hemorrhagic fever viruses and encephalitis viruses. These agents cause diseases associated with significant morbidity, and some have very high mortality rates.

Table 25.10 Potential bioterrorism agents and diseases

Bacteria and rickettsias

Bacillus anthracis (anthrax)
Brucella sp. (brucellosis)
Burkholderia mallei (glanders)
Burkholderia pseudomallei (melioidosis)
Chlamydia psittaci (psittacosis)
Vibrio cholerae (cholera)
Clostridium botulinum toxin (botulism[a])
Clostridium perfringens (Epsilon toxin[a])
Coxiella burnetii (Q fever)
Escherichia coli O157:H7 (gastrointestinal disease)
Francisella tularensis (tularemia)
Yersinia pestis (plague)
Staphylococcus aureus enterotoxin B[a]
Salmonella Typhi (typhoid fever)
Salmonella sp. (salmonellosis)
Shigella (shigellosis)
Rickettsia prowazekii (typhus)

Viral agents

Variola major (smallpox)
Alphaviruses (viral encephalitis)
Venezuelan equine encephalitis virus
Eastern equine encephalitis virus
Western equine encephalitis virus
Nipah virus
Viral hemorrhagic fevers viruses
    filoviruses; Ebola, Marburg
    arenaviruses, Lassa, Machupo
    hantaviruses

Protozoa

Cryptosporidium parvum (waterborne gastroenteritis)

Plants

Ricinus communis (ricin toxin from castor bean[a])

[a]Preformed toxin; all other agents require infection.

Source: Information is from the Centers for Disease Control and Prevention, Atlanta, GA, USA.

Bacterial toxins such as the botulinum toxin of *Clostridium botulinum* are also possible bioweapons (Table 25.8) (∞Sections 21.10 and 29.5). Large amounts of the preformed toxin delivered to a population through a common vehicle such as drinking water could have devastating consequences: The lethal dose of botulinum toxin for a human is 2 μg or less.

Smallpox

Smallpox virus (∞Section 16.13) is an intimidating potential biological warfare agent because it can be easily spread by contact or aerosol and it has a mortality rate of 30% or more. Its potential for use as a biowarfare agent is considered low, partly because the only *known* stocks of smallpox virus are in guarded repositories in the United States and Russia. However, a finite possibility remains for terrorist groups or military forces to gain access to smallpox virus. Because of this possibility, the U.S. government has made provisions to immunize front-line health-care and public safety personnel for smallpox. Although an extremely effective smallpox vaccine exists, this live *vaccinia virus* vaccine has not been in general use for almost 30 years because wild smallpox was eradicated worldwide by 1977 and because the vaccine can have serious side effects. As a result, over 90% of the current worldwide population is now inadequately vaccinated and susceptible to the disease. Preparations for a potential smallpox attack in the United States have included controversial recommendations for immunization of selected individuals: persons having close contact with smallpox patients; workers involved in the direct evaluation, care, or transportation of smallpox patients; laboratory personnel handling clinical specimens from smallpox patients; and other persons such as housekeeping personnel who would have increased contact with infectious materials from smallpox patients.

Although vaccinia immunization is very effective, it carries significant risk. Normal vaccine reactions include formation of a pustule, with a scab that falls off in two to three weeks, leaving a small scar. Mild adverse reactions occur in a large number of individuals and commonly include fevers and rashes. Vaccination is not currently recommended for persons with eczema or other chronic or acute skin conditions, heart disease, pregnant women, and those with reduced immune competence, such as individuals using anti-inflammatory steroid medications and those with HIV/AIDS.

About 1 in 1000 vaccinated individuals develop serious complications from the vaccine. Serious reactions include myocarditis and *erythema multiforme*, a toxic or allergic response to the vaccine. *Generalized vaccinia* (systemic vaccinia infection) occasionally occurs in individuals with skin conditions such as eczema. Very serious, life-threatening *progressive vaccinia* sometimes occurs, especially in individuals who are immunosuppressed due to therapy or disease. About one to two individuals per million will die from this vaccination.

Delivery of Biological Weapons

Most organisms suitable for bioweapons use can be spread in an aerosolized form, providing simple, rapid, widespread dissemination and infection. Examples of several exposures involving aerosols are instructive.

In 1962, one of the last outbreaks of smallpox in a developed country occurred in Germany. A German worker developed smallpox after returning from Pakistan, a country with endemic smallpox at the time. The individual was immediately hospitalized and quarantined, but the patient had a cough, and the aerosolized virus caused illness in 19 *vaccinated* individuals; at least one individual died from the resulting infection.

Planned bioterrorist attacks occurred even before the anthrax attacks of 2001 (see Section 25.12). In 1984 in The Dalles, Oregon (United States), cultists inoculated a salad bar with a *Salmonella typhimurium* culture in aerosol form at 10 local restaurants, causing 751 cases of foodborne salmonellosis in a region that usually has less than 10 cases per year (∞Section 29.7). In 1995, a radical political group released Saran nerve gas into a Tokyo subway, killing several people and injuring scores of others. Although this was a chemical weapon, this group also possessed anthrax cultures, bacteriological media, drone airplanes, and spray tanks.

In terms of strategic warfare, delivery of preformed bacterial toxins such as botulinum toxin or staphylococcal enterotoxin (Table 25.5) to large populations is somewhat impractical because most potent exotoxins are proteins that would lose effectiveness as they are diluted or are destroyed in common sources such as drinking water. However, delivery of toxins could be aimed at selected individuals, small groups, or even at random to instigate panic in a population.

Prevention and Response to Biological Weapons

Proactive measures against bioweapons have already begun with periodic planned efforts to update the international agreements of the 1972 Biological and Toxic Weapons Convention. The fifth and most recent update was in 2002. At the practical level, governments are now supporting the large-scale production and distribution of vaccines and the development of strategic and tactical plans to prevent and contain the effects of bioweapons.

The U.S. government, through the CDC, has devised and enhanced the Select Agent Program surveillance systems to monitor possession and use of potential bioterrorism agents. The CDC Laboratory Response Network and the Health Alert Network have been upgraded to enhance diagnostic capabilities and increase the reporting abilities of local and regional health-care centers to more quickly identify bioterrorism events as well as emerging diseases.

25.12 Concept Check

Bioterrorism is a threat in a world of rapid international travel and easily accessible technical information. Biological agents can be used as weapons by contemporary military forces or by terrorist groups. Aerosols or common sources such as food and water are the most likely mode of inoculation. Prevention and containment measures rely on a well-prepared public health infrastructure.

◆ What characteristics make a pathogen or its products particularly useful as a bioweapon?

◆ Identify two infectious agents that could be effective bioweapons. How could the agents be disseminated?

25.13 Anthrax as a Biological Weapon

Bacillus anthracis is a preferred agent for biowarfare and bioterrorism. Here we discuss its unique properties, the diseases it causes, and methods for prevention, diagnosis, and treatment.

Biology and Growth

Bacillus anthracis is a ubiquitous saprophytic soil inhabitant. It grows as an aerobic gram-positive rod, 1 μm in diameter by 3–4 μm in length. As with other members of the genus *Bacillus*, it produces endospores resistant to heat and drying (∞Sections 4.13 and 12.20) (Figure 25.12a●). Viable endospores are sometimes recovered from contaminated animal products such as hides and fur. Growth on blood agar results in large colonies with a characteristic "ground glass" appearance (Figure 25.12b). Strains having a poly-D-glutamic acid capsule are resistant to phagocytosis (∞Section 21.6).

Infection and Pathogenesis

Bacillus anthracis endospores are the standard means of acquiring anthrax. The disease usually affects domestic animals, especially ungulates—cows, sheep, and goats. The number of infections in animals, while considerable, is not known. The animals acquire the disease from plants or soil in pastures. In humans and animals, three forms of the disease can occur. *Cutaneous anthrax* results when abraded skin is contaminated by *B. anthracis* endospores (Figure 25.13a●). *Gastrointestinal anthrax* occurs due to consumption of endospore-contaminated plants and, potentially, from anthrax-infected carcasses. Human gastrointestinal anthrax is rarely seen. *Pulmonary anthrax* occurs when the endospores are inhaled. Inhalation of the endospores or the live bacteria results in pulmonary infections characterized by pulmonary and cerebral hemorrhage (Figure 25.13b). Untreated pulmonary anthrax infections have a mortality rate of nearly 100%. Cutaneous anthrax cases are rare in the

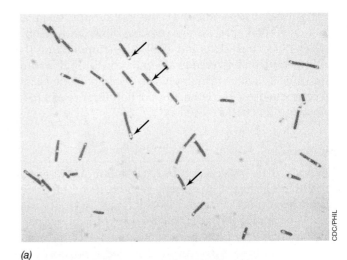

(a)

(b)

● **Figure 25.12 Bacillus anthracis.** (a) *B. anthracis* is a gram–positive endospore-forming rod (∞Section 12.20) approximately 1 μm in diameter and 3–4 μm in length. Note the formation of endospores (arrows). Endospore formation (∞Section 4.13) enhances the ability to disseminate *B. anthracis* in aerosols. (b) *B. anthracis* colonies on blood agar. The nonhemolytic colonies take on a characteristic "ground glass" appearance.

United States, and pulmonary anthrax cases, even in agricultural workers, are extremely rare. The last natural pulmonary anthrax case occurred in 1976. However, several cases of pulmonary anthrax were identified in the United States in 2001 due to bioterrorism events.

Pathogenesis results from inhalation of 8000–50,000 spores of an encapsulated toxigenic strain. Pathogenic *B. anthracis* produces three proteins—*protective antigen (PA)*, *lethal factor (LF)*, and *edema factor (EF)*. PA and LF form *lethal toxin*, while PA and EF form *edema toxin*. PA is the cell-binding B component of these A-B toxins (∞ Section 21.10 and Table 21.4). EF cause edema and LF causes cell death. Growth of *B. anthracis* in the lymph nodes and lymphatic tissues draining the lungs leads to edema and cell death, culminating in tissue destruction, shock, and death.

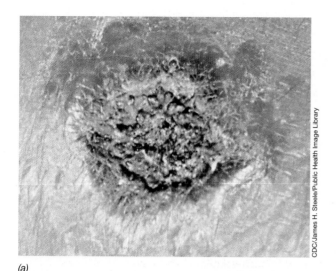

(a)

(b)

CDC/James H. Steele/Public Health Image Library

CDC/Public Health Image Library

● **Figure 25.13 Anthrax.** (a) Cutaneous anthrax. The blackened lesion on the forearm of a patient, about 2 cm in diameter, results from tissue necrosis. Cutaneous anthrax, even when untreated, usually is a localized, nonlethal infection. (b) Inhalation anthrax. Fixed and sectioned brain showing hemorrhagic meningitis (dark coloration) due to a fatal case of inhalation anthrax.

Clinical symptoms can start with sore throat, fever, and muscle aches. After several days, symptoms progress to include difficulty in breathing, followed by systemic shock. Fatality rates can approach 90% even when exposure is suspected, and can be nearly 100% for cases where treatment is not started at or before the onset of symptoms.

Weaponized Anthrax

The term *weaponized* is applied to strains and preparations of *B. anthracis*, usually in endospore form, that exhibit properties that enhance dissemination and use as biological weapons. Such strains and preparations were developed by several governments in the post-World War II era, but overt development of new biological weapons was halted by international treaty in 1972. The physical characteristics of the weaponized anthrax preparations generally include a small particle size, usually interspersed with a very fine particulate agent such as talc. This small-particle, powdery form ensures that the endospores will spread easily by air currents. Thus, opening an envelope containing endospores or releasing the powder–spore mixture into a ventilation system or other air current has the potential to contaminate surrounding areas and personnel.

A weaponized form of anthrax was used as a vehicle for a series of bioterrorism attacks in the United States in 2001. These incidents were attributed to envelopes or packages containing weaponized anthrax endospores. The attacks were apparently directed at the news media (Florida), and the government (Washington, D.C. area). A third focus of attack, the Pennsylvania-New Jersey-New York area, had no defined single target, but disrupted mail service in the Northeast; some anthrax-contaminated mail facilities were still not in use in 2003. In all, there were 22 anthrax infections. Eleven were cutaneous anthrax. Of eleven cases of inhalation anthrax, five cases resulted in death. The bioterrorists were never identified.

The incidents in the United States were not the first or the most serious anthrax bioweapons infections. In a previous incident, *B. anthracis* spores were inadvertently released into the atmosphere from a bioweapons facility in Sverdlovsk, Russia in 1979. Less than 1 g of spores was released, and everyone in the area surrounding the facility was immunized and given prophylactic antibiotic therapy as soon as the first anthrax case was diagnosed. However, 77 individuals outside the facility contracted pulmonary anthrax and 66 died.

Vaccination, Prophylaxis, Treatment, and Diagnosis

Vaccination for anthrax has been restricted to individuals who are considered at risk. This includes agricultural animal workers and military personnel. The current vaccine, called *AVA (anthrax vaccine adsorbed)*, is prepared from a cell-free *B. anthracis* culture filtrate.

Treatment of infection, which seems to have a minimum incubation time of about eight days, is usually done with antibiotics. Ciprofloxacin, a broad-spectrum quinolone antibiotic (⟅Figure 20.19) is used for strains that are penicillin-resistant, as are many known laboratory and bioweapons strains. Ciprofloxacin is also used as a prophylactic measure to treat potentially exposed individuals.

A number of rapid diagnostic tests are available to detect microbial endospores. However, positive identification of *B. anthracis* relies on culture techniques and direct observation of either infected tissues or cultured organisms. The characteristic ground-glass appearance on blood agar, coupled with isolation of gram-positive rods growing in extended chains is presumptive evidence for *B. anthracis* (Figure 25.12).

25.13 Concept Check

Bacillus anthracis has emerged as an important pathogen because of its use as a bioweapon. Highly infective weaponized endospore preparations have been used as bioterror agents. Inhalation anthrax has a fatality rate of about 90% in untreated individuals. Effective treatment relies on timely observation and diagnosis of symptoms. Treatment does not guarantee survival for inhalation anthrax.

◆ What factors contribute to the preferred use of *B. anthracis* as a bioweapon?

◆ Indicate the steps you would use to define and treat a bioterror attack using *B. anthracis*.

REVIEW QUESTIONS

1. List the five most common causes of mortality due to infectious diseases throughout the world. Are any of these diseases preventable by immunization (◑Section 25.1)?

2. Distinguish between *mortality* and *morbidity*, *prevalence* and *incidence*, and *epidemic* and *pandemic*, as these terms relate to infectious disease (◑Section 25.2).

3. Explain the difference between a *chronic* carrier and an *acute* carrier of an infectious disease (◑Section 25.3).

4. Give examples of host-to-host transmission of disease via direct contact. Also give examples of indirect host-to-host transmission of disease via vector agents and fomites (◑Section 25.4).

5. How can immunity to a pathogen by a large proportion of the population protect the nonimmune members of the population from acquiring a disease? Will this herd immunity work for diseases that have a common source, such as water? Why or why not (◑Section 25.5)?

6. Identify the major risk factors for acquiring human immunodeficiency virus (HIV) infection in the United States. Does this pattern hold for all geographic regions (◑Section 25.6)?

7. Hospital environments are conducive to the spread of infectious diseases. Review the reasons for the enhanced spread of infection in hospitals. What are the sources of most nosocomial infections (◑Section 25.7)?

8. Describe the source, the pathogen, and the treatment for severe acute respiratory syndrome (SARS). Does SARS have the potential to become a major epidemic (◑ Section 25.8)?

9. Describe the major medical and public health measures developed in the twentieth century that were instrumental for controlling the spread of infectious diseases in developed countries (◑Section 25.9).

10. Compare the role of infectious diseases on mortality in developed and developing countries (◑Section 25.10).

11. Review the major reasons for the emergence of new infectious diseases. What methods are available for identifying and controlling the emergence of new infectious diseases (◑Section 25.11)?

12. Describe the general properties of an effective biological warfare agent. How does *smallpox* meet these criteria? Identify other organisms that meet the basic requirements for a bioweapon (◑Section 25.12).

13. Describe the use of *Bacillus anthracis* as a bioweapon. Devise a plan to protect yourself against a *B. anthracis* attack (◑Section 25.13).

APPLICATION QUESTIONS

1. Smallpox, a disease that was limited to humans, was eradicated. Plague, a disease with a zoonotic reservoir in rodents (Table 25.2) can never be eradicated. Explain this statement and why you agree or disagree with the possibility of eradicating plague on a global scale. Devise a plan to eradicate plague in a limited environment such as a town or city. Be sure to use methods that involve the reservoir, the pathogen, and the host.

2. Acquired immunodeficiency syndrome (AIDS) is a candidate for a disease that can be eliminated because it is propagated by person-to-person contact and there are no known animal reservoirs. Design a program for eliminating AIDS in a developed country and in a developing country. How would these programs differ from one another? What factors would work against the success of your program, both in terms of human behavior and in terms of the AIDS disease itself? Why are the numbers of HIV-infected and AIDS patients continuing to grow, especially in developing countries?

3. Travel to developing countries involves some exposure to infectious diseases. What general precautions should you take before, during, and after visits to developing countries? Where can you obtain information on the infectious disease status in a specific foreign country? When you return from a foreign country, are you a disease risk to your family or your associates? Explain.

4. Identify a specific pathogen that would be a suitable agent for effective biological warfare. Describe the properties of the pathogen in the context of its use as a bioweapon. Describe conditions for growing large amounts of the pathogen. Identify a suitable delivery method. Since you will propagate and deliver the pathogen, describe the precautions you will take to protect yourself. Now reverse your role. As a public health official in a large city, describe how you would recognize and diagnose the disease caused by the agent. Indicate the measures you would take to treat the illnesses caused by the agent. How could you best limit the damage? Would quarantine and isolation methods be useful? What about immunization and antibiotics?

26

PERSON-TO-PERSON MICROBIAL DISEASES

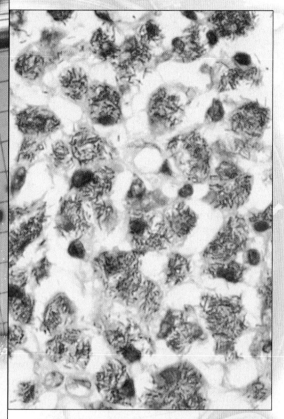

Mycobacterial infections are characterized and identified by the red (acid-fast) staining properties of the pathogen in tissue and sputum samples.

WORKING GLOSSARY

Antigenic drift minor changes in antigens due to gene mutation in influenza virus

Antigenic shift major changes in antigens due to gene reassortment in influenza virus

Cirrhosis breakdown of the normal liver architecture resulting in fibrosis

Congenital syphilis syphilis contracted by an infant from its mother during birth

Fusion inhibitor a synthetic polypeptide that binds to viral glycoproteins, inhibiting fusion of viral and host cell membranes

Hepatitis a liver inflammation commonly caused by an infectious agent

Jaundice production and release of excess bilirubin in the liver due to destruction of liver cells, resulting in yellowing of the skin and whites of the eye

Meningitis inflammation of the meninges (brain tissue), sometimes caused by *Neisseria meningitidis* and characterized by sudden onset of headache, vomiting, and stiff neck, often progressing to coma within hours

Meningococcemia fulminant disease caused by *Neisseria meningitidis* and characterized by septicemia, intravascular coagulation, and shock

Non-nucleoside reverse transcriptase inhibitor (*NNRTI*) a non-nucleoside compound that inhibits the action of viral reverse transcriptase by binding directly to the catalytic site

Nucleoside reverse transcriptase inhibitor (*NRTI*) a nucleoside analog compound that inhibits the action of viral reverse transcriptase by competing with nucleosides

Opportunistic infection an infection usually observed only in an individual with a dysfunctional immune system

Protease inhibitor a compound that inhibits the action of viral protease by binding directly to the catalytic site, preventing viral protein processing

Rheumatic fever an inflammatory autoimmune disease triggered by an immune response to infection by *Streptococcus pyogenes*

Scarlet fever characteristic reddish rash resulting from an exotoxin produced by *Streptococcus pyogenes*

Sexually transmitted infection (*STI*) an infection that is usually transmitted by sexual contact

Toxic shock syndrome (*TSS*) acute systemic shock resulting from a host response to an exotoxin produced by *Staphylococcus aureus*

Tuberculin test a skin test for previous infection with *Mycobacterium tuberculosis*

Viral load a quantitative assessment of the amount of virus in a host organism, usually in the blood

More than 500,000 microbial species exist in nature (Chapter 11), but only a few hundred species cause disease. Most microorganisms carry out essential activities independent of interactions with other organisms, and many microorganisms are closely associated with plants or animals in stable, beneficial relationships (Chapter 19). However, pathogenic species have profoundly negative effects on host organisms. In the next four chapters, representative human pathogens and their biology are examined, as well as the pathology, diagnosis, treatment, and prevention of the diseases they cause. This discussion is organized based on the pathogen's *mode of transmission* and presents infectious disease in relation to the ecology of the pathogen. In this chapter disease transmission via direct *person-to-person* interactions is considered. In Chapters 27 through 29, diseases whose modes of transmission involve animal or arthropod vectors or common sources such as soil, water, or food are examined.

Connections between and among seemingly unrelated organisms can be made by examining pathogens according to their modes of transmission and the diseases they cause. For example, influenza virus and streptococci produce diseases with overlapping symptoms, although the causal agents, one viral and one bacterial, are markedly different. Here these pathogens are discussed together because they are spread person-to-person via a respiratory route. Using this approach, connections can be established between biologically diverse but ecologically and pathogenically related agents.

AIRBORNE TRANSMISSION OF DISEASES

Aerosols, such as those generated by the sneeze shown in Figure 26.1●, are important for person-to-person transmission of many infectious diseases. Most respira-

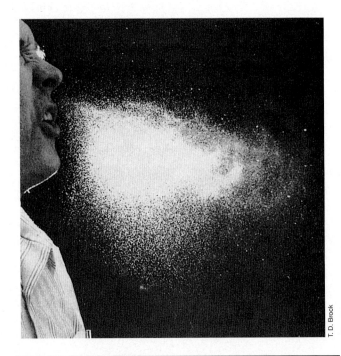

● **Figure 26.1   High speed photograph of an unstifled sneeze.**

tory diseases are spread almost exclusively in this fashion. For example, *Mycobacterium tuberculosis* has successfully used this strategy to infect at least one-third of the world's population (Section 25.1). Influenza and cold viruses are passed by respiratory routes so commonly that virtually everyone acquires more than one cold or case of influenza every year.

26.1 Airborne Pathogens

Air is not a growth medium for microorganisms. Thus, microorganisms found in air are derived from soil, water, plants, animals, people, or other sources. In outdoor air, soil organisms predominate. Indoors, the concentration of microorganisms is considerably higher than outdoors, especially for those originating in the human respiratory tract.

Most microorganisms survive poorly in air, and so effective transmittal to another human occurs only over short distances. However, certain human pathogens (*Staphylococcus, Streptococcus*) survive under dry conditions fairly well and remain alive in dust for long periods of time. Gram-positive *Bacteria* are in general more resistant to drying than gram-negative *Bacteria* because of their thick, rigid cell wall. Likewise, the waxy layer of mycobacterial cell walls (see Section 26.5) resists drying. The endospores of endospore-forming *Bacteria* are extremely resistant to drying but are not generally passed from human to human in the endospore form.

An enormous number of moisture droplets are expelled during sneezing (Figure 26.1), and a considerable number are expelled during coughing or talking. Each infectious droplet has a diameter of about 10 μm and contains one or two microbial cells or virions. The initial speed of the droplet movement is about 100 m/sec (more than 200 mi/h) in a sneeze and ranges from 16 to 48 m/sec during coughing or shouting. The number of bacteria in a single sneeze varies from 10,000 to 100,000. Because of their small size, the moisture droplets evaporate quickly in the air, leaving behind a nucleus of organic matter and mucus to which bacterial cells are attached.

Respiratory Infections

Humans breathe about 500 million liters of air in a lifetime, much of it containing microorganism-laden dust, a potential source of inoculum for respiratory infections. The speed at which air moves through the respiratory tract varies, and in the lower respiratory tract the rate is quite slow. As air slows down, particles in it stop moving and settle. Large particles settle first and the smaller ones later, and only particles smaller than 3 μm travel as far as the bronchioles in the lower respiratory tract (Figure 26.2●). Of course, most pathogens are much smaller than this and different organisms characteristically colonize the respiratory tract at different levels. This is due to the unique metabolic requirements and virulence factors associated with each pathogen. The upper and lower respiratory tracts offer decidedly different environments, favoring certain microorganisms.

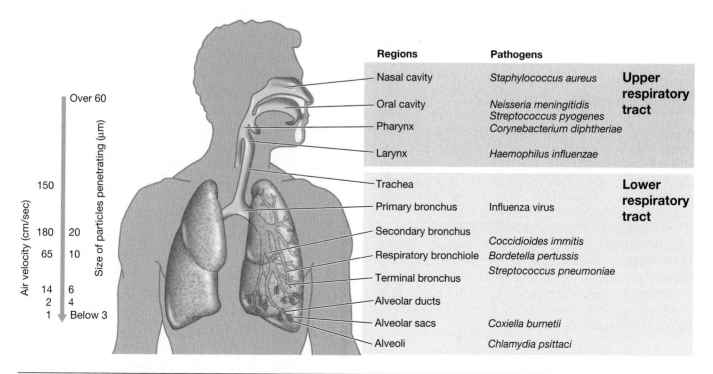

● **Figure 26.2** **The respiratory system of humans.** Pathogenic microorganisms generally initiate infections at the indicated locations.

Bacterial Respiratory Pathogens

We begin here with a consideration of some common bacterial respiratory pathogens. We then examine viral respiratory pathogens and the much less treatable and preventable diseases they cause.

Most bacterial pathogens affecting the respiratory tract inhabit only humans and are normally transmitted from person to person. Since humans are the only reservoir for these pathogens, their survival is dependent on person-to-person transmission. A few respiratory pathogens such as *Legionella pneumophila*, transmitted primarily from water or soil do not require person-to-person propagation, and we discuss these later (see Chapter 28). As we explained above, many person-to-person respiratory pathogens are gram-positive *Bacteria*. Bacterial respiratory infections, while serious by themselves, often initiate secondary problems that can be life-threatening. Thus, accurate and rapid diagnosis and treatment of bacterial respiratory infections is necessary to limit host damage. Fortunately, most respiratory bacterial pathogens respond readily to antibiotic therapy, and many can also be controlled by immunization.

26.1 Concept Check

Many respiratory pathogens are gram-positive *Bacteria*. Because gram-positive *Bacteria* are resistant to drying, they are easily transmitted in air. Most respiratory pathogens are transferred from person to person via respiratory aerosols generated by coughing, sneezing, talking, or breathing.

◆ What physical features of gram-positive *Bacteria* allow them to survive for long periods in air and dust?

◆ Identify pathogens more commonly found in the upper respiratory tract. Identify pathogens more commonly found in the lower respiratory tract.

26.2 Streptococcal Diseases

Streptococcus pyogenes and *Streptococcus pneumoniae* are important human respiratory pathogens. *S. pyogenes* is transmitted by the respiratory route. *S. pneumoniae* is found in the respiratory flora of up to 40% of healthy individuals, and endogenous strains can cause severe respiratory disease in compromised individuals.

Streptococci are nonsporulating, homofermentative, aerotolerant, anaerobic gram-positive cocci (⌾ Section 12.19). Cells of *Streptococcus pyogenes* typically grow in elongated chains (⌾ Figure 12.54). Pathogenic strains of *Streptococcus pneumoniae* typically grow in pairs or short chains and virulent strains typically produce an extensive polysaccharide capsule (Figure 26.3●).

Streptococcus pyogenes: Epidemiology and Pathogenesis

Streptococcus pyogenes is frequently isolated from the upper respiratory tract of healthy adults. Although num-

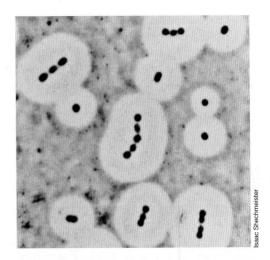

● **Figure 26.3 India ink negatively stained preparation of cells of *Streptococcus pneumoniae*.** Note the extensive capsule surrounding the cells. The cells are about 0.5 μm in diameter.

bers of endogenous *S. pyogenes* are usually low, if host defenses are weakened or a new, highly virulent strain is introduced, acute streptococcal infections are possible. *S. pyogenes* is the cause of streptococcal pharyngitis, also called *strep throat* (Figure 26.2). Most isolates from clinical cases of strep throat produce a toxin that lyses red blood cells, a condition called β-*hemolysis* (⌾ Figure 21.18). Streptococcal pharyngitis is characterized by a severe sore throat, enlarged tonsils, tonsillar exudate, tender cervical lymph nodes, a mild fever, and general malaise. *S. pyogenes* can also cause related infections of the inner ear (*otitis media*), the mammary glands (*mastitis*), and infections of the superficial layers of the skin, a condition referred to as *impetigo* (impetigo can also be caused by *Staphylococcus aureus*) (Figure 26.4●).

About half of the clinical cases of severe sore throat are due to *Streptococcus pyogenes*, with the remainder of viral origin. An accurate, prompt diagnosis is important because if the sore throat is due to a virus, treatment with antibacterial drugs (antibiotics) will be useless, whereas if the sore throat is due to *S. pyogenes*, immediate antibacterial therapy is indicated. Rapid, complete treatment of streptococcal sore throat is important because it can occasionally lead to more serious streptococcal syndromes such as scarlet fever, rheumatic fever, acute glomerulonephritis, and streptococcal toxic shock syndrome.

Certain strains of *Streptococcus pyogenes* carry a lysogenic bacteriophage that encodes production of streptococcal pyrogenic exotoxin A, an exotoxin responsible for most of the symptoms of streptococcal toxic shock syndrome and **scarlet fever**. Exotoxin A is a *superantigen* and acts by recruiting massive numbers of T cells to the infected tissues (⌾ Section 22.16). The T cells then secrete *cytokines*, which activate large numbers of effector cells, resulting in systemic inflammation and tissue destruction. Streptococcal toxic shock syndrome is the term given to this systemic, often life-threatening condition. Exotoxin A

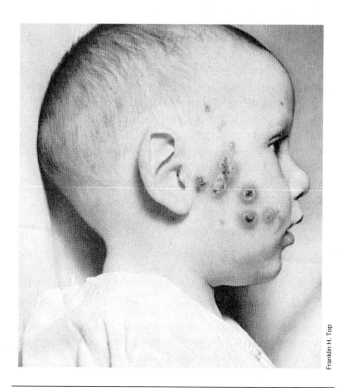

● **Figure 26.4 Typical lesions of impetigo.** Impetigo is commonly caused by *Streptococcus pyogenes* or *Staphylococcus aureus.*

and the related exotoxin C may also cause the pink-red rash of scarlet fever to develop (Figure 26.5●) and also act to damage small blood vessels and initiate fever.

Occasionally, *Streptococcus pyogenes* causes fulminant systemic infection, often marked by *necrotizing fasciitis,* a rapid and progressive disease resulting in extensive destruction of subcutaneous tissue. These infections are responsible for the dramatic, but fortunately rare, reports of "flesh-eating bacteria." In these cases, streptococcal pyrogenic exotoxins A and B and the surface M protein act as superantigens. The resulting inflammation and extensive tissue destruction cause death in up to 30% of the cases. In all of these cases, successful treatment of the streptococcal infection stops production of the superantigen and its effects.

Untreated or insufficiently treated cases of *Streptococcus pyogenes* infection may lead to severe *delayed sequelae* or follow-up diseases. **Rheumatic fever**, one of these delayed sequelae, is caused by *rheumatogenic* strains of *S. pyogenes* containing cell-surface antigens that are similar to certain human cell-surface antigens. The immune response to the invading pathogen produces antibodies that cross-react with host tissue antigens on the heart, joints, and kidneys, resulting in damage to these tissues. Rheumatic fever is a type of *autoimmune disease,* with antibodies reacting with self antigens (∞ Section 22.15). Damage may be permanent, and it is often accelerated by later streptococcal infections resulting in recurrent bouts of rheumatic fever.

Another delayed sequela of *Streptococcus pyogenes* infection is *acute glomerulonephritis,* a painful kidney disease.

This immune-complex disease (∞ Section 22.15) results from the formation of streptococcal antigen-antibody complexes in the blood. The immune complexes lodge in the *glomeruli,* or filtration membranes of the kidney, causing inflammation of the kidney (*nephritis*) accompanied by severe pain. Within several days, these complexes are usually dissolved and the patient quickly returns to normal. Unfortunately, even timely antibacterial treatment may not prevent glomerulonephritis. However, only a few strains of *S. pyogenes*—so-called *nephritogenic* strains— produce this painful disease.

Since infection induces strain-specific immunity, reinfection by a particular *S. pyogenes* strain is rare. However, there are over 60 different strains defined by distinct cell surface M proteins. Thus, an individual can be infected multiple times by different *S. pyogenes* strains. There are no available vaccines to prevent *S. pyogenes* infections.

Diagnosis of *Streptococcus pyogenes*

Because serious host damage following a streptococcal sore throat can occur, several rapid antigen detection (RAD) systems have been developed for identification of *S. pyogenes.* Surface antigens are first extracted by enzymatic or chemical means directly from a swab of the patient's throat. Immunological methods such as latex bead agglutination, enzyme-linked immunoassay (ELISA), or fluorescent antibody staining (∞ Sections 24.8, 24.10, and 24.9) using antibodies specific for surface proteins unique to *S. pyogenes* are employed. Specimens are taken directly from a patient throat swab and are processed and

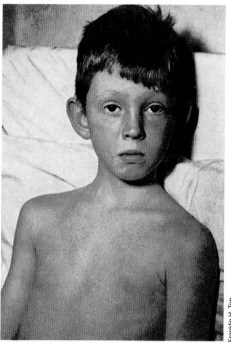

● **Figure 26.5 Scarlet fever.** The typical rash of scarlet fever results from the action of the erythrogenic toxin produced by *Streptococcus pyogenes.*

analyzed in minutes. These rapid diagnostic procedures allow the physician to immediately initiate appropriate antibiotic therapy in order to avoid complications such as rheumatic fever.

A more accurate confirmation of infection by pathogenic streptococci is a positive *S. pyogenes* culture from the throat done on sheep blood agar (∞ Figure 21.18). While the RAD tests are nearly as *specific* as throat cultures, they can be up to 40% less *sensitive*, leading to false negative reports (∞ Section 24.7). Throat cultures take up to two days to process, hence the popularity of the RAD tests. The most sensitive methods for identifying recent streptococcal infections are serology tests, where patients are examined for the presence or increase (rise in titer) of antibodies to various streptococcal antigens (∞ Section 24.7). The presence of new antibodies or an increase in the quantity of an existing antibody confirms a very recent streptococcal infection.

Streptococcus pneumoniae

The other major pathogenic streptococcal species, *Streptococcus pneumoniae*, causes lung infections that often develop as secondary infections to other respiratory disorders. Capsulated strains of *S. pneumoniae* are particularly pathogenic because they are potentially very invasive. Cells invade alveolar tissues (lower respiratory tract) of the lung where the capsule enables the cells to resist phagocytosis and elicit a strong host inflammatory response. Reduced lung function (pneumonia) can result from accumulation of recruited phagocytic cells and fluid. The *S. pneumoniae* cells can then spread from the focus of infection as a bacteremia, sometimes resulting in bone infections, inner ear infections, and endocarditis. Pneumococcal pneumonia is a serious infection and untreated cases have a mortality rate of about 30%. Even with aggressive antimicrobial treatment, individuals hospitalized with pneumococcal pneumonia have up to 10% mortality.

Laboratory diagnosis of *S. pneumoniae* involves the culture of gram-positive diplococci from either patient sputum or blood. There are over 90 different serotypes (antigenic capsule variants), and, as for *S. pyogenes*, infection induces immunity to only the infecting serotype of *S. pneumoniae*.

Prevention and Treatment

An effective multivalent vaccine is available for prevention of infection by at least two-thirds of the 90 known strains of *Streptococcus pneumoniae*, including all common pathogenic strains. The vaccine consists of a mixture of the capsular polysaccharides from the most prevalent pathogenic strains. The vaccine is recommended for the elderly, health-care providers, individuals with compromised immunity, and others at high risk for respiratory infections (∞ Section 22.13).

Penicillin and its semisynthetic derivatives (∞ Sections 20.8 and 30.6) are the agents of choice for treating *S. pyogenes* infections. Erythromycin and other antibacterial drugs are used in individuals who have acquired penicillin allergies (∞ Section 22.15).

Most strains of *S. pneumoniae* respond to penicillin therapy. However, there are penicillin-resistant strains, especially among strains causing hospital-acquired infections (∞ Section 25.7). Thus individual isolates must be tested for penicillin susceptibility (∞ Section 24.3). Erythromycin is the drug of choice for penicillin-resistant organisms, but cephalosporin, fluoroquinolone, ceftriaxone, cefotaxime, or vancomycin (∞ Sections 20.6–20.9) may also be used. However, some strains have acquired resistance to each of these drugs, and some strains have acquired multiple drug resistance, underscoring the need to test each isolate individually.

⬟ 26.2 Concept Check

Diseases caused by streptococci include streptococcal sore throat and pneumococcal pneumonia. Occasionally, *Streptococcus pyogenes* infections develop from pharyngitis into serious conditions such as scarlet fever and rheumatic fever. Pneumonia caused by *Streptococcus pneumoniae* is a serious disease with high mortality. Definitive diagnosis for both pathogens is by culture. Infections with both pathogens are treatable with antimicrobial drugs, but drug-resistant strains are known, especially for *Streptococcus pneumoniae*.

◆ How does *Streptococcus pyogenes* infection cause rheumatic fever?

◆ What is the primary virulence factor for *Streptococcus pneumoniae*?

26.3 Corynebacterium and Diphtheria

Corynebacterium diphtheriae causes diphtheria, a severe respiratory disease that usually infects children. Diphtheria is preventable and treatable. *C. diphtheriae* is a gram-positive, nonmotile, aerobic bacterium that forms irregular rods that may appear as club-shaped cells during growth (Figure 26.6a⬤) (∞ Section 12.22).

Epidemiology and Pathology

Corynebacterium diphtheriae enters the body via the respiratory route with cells lodging in the throat and tonsils. Infection is usually spread from healthy carriers or infected individuals to susceptible individuals by airborne droplets. Previous infection or immunization (see below) provides resistance to the effects of the potent exotoxin. Although limited information is available concerning the mechanism of adherence of *C. diphtheriae* to these tissues, the organism produces a neuraminidase capable of splitting *N*-acetylneuraminic acid (a component of glycoproteins found on animal cell surfaces), and this may enhance the invasion process. The inflammatory response of throat tissues to *C. diphtheriae* infection results in formation of a characteristic lesion called a *pseudomembrane* (Figure 26.6b), which consists of damaged host cells and cells of *C. diphtheriae*. As described in Section 10.9, certain strains of *C. diphtheriae* are lysogenized by bacteriophage β. These lysogenized toxin-producing strains are pathogenic because they produce a powerful exotoxin, the *diphtheria*

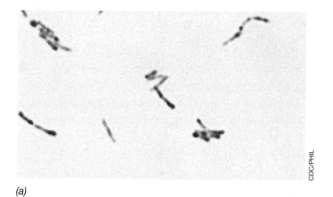

(a)

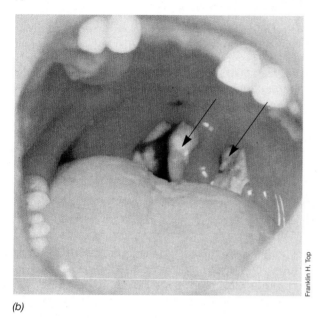

(b)

CDC/PHIL

Franklin H. Top

● **Figure 26.6 Diphtheria.** (a) Cells of *Corynebacterium diphtheriae* showing typical club-shaped appearance. The cells are 0.5 to 1.0 μm in diameter and may be several micrometers in length. (b) Pseudomembrane (arrows) in an active case of diphtheria caused by the bacterium *C. diphtheriae.*

toxin. Diphtheria toxin inhibits eukaryotic protein synthesis and thus kills cells (∞ Section 21.10).

Tissue death due to absorption of the toxin causes the appearance of a pseudomembrane in the patient's throat. The pseudomembrane may block the passage of air, and death from diphtheria is usually due to a combination of the effects of partial suffocation and tissue destruction by exotoxin. In untreated infections, the toxin can cause systemic damage to the heart (about 25% of diphtheria patients develop myocarditis), kidneys, liver, and adrenal glands.

Although diphtheria was once a major childhood disease, it is now rarely encountered because an effective vaccine is available. In developed countries like the United States, the disease is virtually unknown. Worldwide, there are still more than 50,000 cases of diphtheria per year, largely because of a lack of effective immunization programs. For example, diphtheria outbreaks have recently occurred in refugee camps in Afghanistan in individuals lacking immunization.

Diagnosis, Prevention, and Treatment

Corynebacterium diphtheriae isolated from the throat is diagnostic for diphtheria. Nasal or throat swabs are used to inoculate blood agar, tellurite medium, and the selective Loeffler's medium that inhibits the growth of most other respiratory pathogens.

Prevention of diphtheria is accomplished with a highly effective vaccine. The vaccine is made by treating the diphtheria exotoxin with formalin to yield an immunogenic toxoid preparation. Diphtheria toxoid is part of the *DTaP* (*d*iphtheria, *t*etanus, *a*cellular *p*ertussis) vaccine (∞ Section 22.13).

A patient diagnosed with diphtheria is treated simultaneously with antibiotics and diphtheria antitoxin (an antitoxin contains neutralizing antibodies formed in another animal) (∞ Sections 24.7 and 22.13). Penicillin, erythromycin, and gentamicin are generally effective for stopping *C. diphtheriae* growth and further toxin production, but do not alter the effects of preformed toxin. Early administration of both antibiotics and antitoxin is necessary for effective treatment of the disease.

 26.3 Concept Check

Diphtheria is an acute respiratory disease caused by the gram-positive bacterium *Corynebacterium diphtheriae.* Early childhood immunization (DTP) is very effective for preventing this very serious respiratory disease.

◆ Is the pathogenesis of diphtheria due to infection?

◆ How can the spread of diphtheria be prevented?

26.4 *Bordetella* and Whooping Cough

Whooping cough is a potentially serious childhood respiratory disease caused by infection with *Bordetella pertussis. B. pertussis* is a small, gram-negative, aerobic coccobacillus and was linked to whooping cough by Bordet and Gengou in 1906.

Epidemiology and Pathology

Whooping cough or **pertussis** is an acute, highly infectious respiratory disease observed most often in children under 5 years of age. *Bordetella pertussis* attaches to cells of the upper respiratory tract by producing a specific adherence factor called *filamentous hemagglutinin antigen,* which recognizes a complementary molecule on the surface of host cells. Once attached, *B. pertussis* grows and produces pertussis exotoxin that induces synthesis of cyclic adenosine monophosphate (cyclic AMP) (∞ Section 8.7), which is at least partially responsible for the events that lead to host tissue damage. *B. pertussis* also produces an endotoxin, which also may induce some of the symptoms of whooping cough. Clinically, whooping cough is characterized by a recurrent, violent cough that can last up to 6 weeks. The spasmodic coughing gives the disease its name, for a whooping sound results from the patient inhaling in deep breaths to obtain sufficient air. Worldwide, there are up to 50 million cases and 350,000 deaths due to pertussis each year.

In recent years, there has been an alarming upward trend in the cases of pertussis in the United States. Starting in the 1980s, there has been a consistent upward trend of *Bordetella pertussis* infections and disease, reversing a general downward trend that started with the introduction of an effective pertussis vaccine. In 1976, the year of lowest prevalence and incidence, there were only 1010 reported cases of pertussis. In 2003, there were 8489 cases. Up to 60%, or over 4000 cases per year, are in individuals over 5 years of age, including many adolescents and adults who lack appropriate immunity. About 24% of cases were in children less than six months of age who had not yet received all of the recommended doses of pertussis vaccine. The threat of this very communicable disease remains high, as illustrated by recent sporadic outbreaks in cities in the United States, and epidemic outbreaks in refugee camps in Afghanistan. Up to 32% of coughs lasting one to two weeks or longer may be caused by *B. pertussis*. Worldwide, research indicates that immunization programs should still be targeted to children, but immunization of adolescents and adults should also be a priority to build herd immunity (Section 25.4) and prevent serious infections by this endemic respiratory pathogen.

Diagnosis, Prevention, and Treatment

Diagnosis of whooping cough can be made by fluorescent antibody staining of a nasopharyngeal swab specimen or by culture of the organism. For best recovery of *Bordetella pertussis*, a nasopharyngeal aspirate is inoculated directly onto a blood–glycerol–potato extract agar plate (although not selective, this medium supports good recovery of *B. pertussis*). β-hemolytic colonies containing small gram-negative coccobacilli are tested for *B. pertussis* by a latex bead agglutination test or are stained with an anti-*B. pertussis* fluorescent antibody for positive identification (Sections 24.8 and 24.9). If available, a polymerase chain reaction test (PCR) is considered the most sensitive and preferred diagnostic test (Section 24.12). Improved diagnostic and reporting techniques may be one reason for the recent observed increase in pertussis cases in the United States.

A vaccine consisting of proteins derived from *B. pertussis* is part of the routinely administered DTaP (diphtheria, tetanus, acellular pertussis) vaccine. This vaccine is effective only when given to children at appropriate intervals beginning soon after birth (Section 22.13). In the United States, up to 50% of children who acquire whooping cough have not been properly immunized, and current immunization preparations are only 60–90% effective.

Undesirable side effects of pertussis vaccination include local swelling, redness, and fever, and occur in up to one in four vaccinated individuals. Occasional more serious problems such as seizures (1 in 14,000), uncontrollable crying (1 in 1000), or very high fever (1 in 16,000) also occur. Overall, the risk of infection and serious consequences of acquiring whooping cough far outweigh the risks of immunization. Currently, about eight

deaths occur per year. No deaths have been attributed to the acellular pertussis vaccine.

Cultures of *B. pertussis* are killed by ampicillin, tetracycline, and erythromycin, although antibiotics alone do not seem to be sufficient to kill the pathogen *in vivo*. Because a patient with whooping cough remains infectious for up to 2 weeks following commencement of antibiotic therapy, the immune response may be as important, if not more so, than antibiotics in the elimination of *B. pertussis* from the body.

26.4 Concept Check

In the United States, there has been an increase in the number of annual cases of whooping cough. From an average of less than 2000 pertussis cases per year in the 1970s, the number of cases has now risen to over 8,000 per year. Inadequately immunized children are at high risk for acquiring pertussis.

◆   What measures can be taken to decrease the current incidence of pertussis in a population?

◆   Indicate potential problems with the use of pertussis vaccines.

26.5 *Mycobacterium*, Tuberculosis, and Leprosy

Tuberculosis is caused by the gram-positive, acid-fast bacillus *Mycobacterium tuberculosis* (Section 12.23). The famous German microbiologist Robert Koch isolated and described the causative agent of tuberculosis, *M. tuberculosis*, in 1882 (Section 1.6). The acid-fast properties of all mycobacteria result from the waxy mycolic acid constituent of the cell wall. Shared by all members of the genus, mycolic acid allows these organisms to retain carbolfuchsin, a red dye, even after washing in 3% hydrochloric acid in alcohol (Figure 26.7●).

Epidemiology

Mycobacterium tuberculosis is transmitted by the respiratory route, and even the simple act of talking can spread the organism from person to person. At one time, tuberculosis was the most important infectious disease of humans and accounted for one-seventh of all deaths worldwide. Presently, over 11,000 new cases of tuberculosis and about 750 deaths occur each year in the United States. Worldwide, tuberculosis still accounts for almost *1.6 million deaths per year*, about 11% of all deaths due to infectious disease. Up to one-third of the world's population have been infected with *M. tuberculosis* (Table 25.1). In recent years, many of the new tuberculosis cases in the United States have occurred in acquired immunodeficiency syndrome (AIDS) patients.

Pathology

The interaction of the human host and *Mycobacterium tuberculosis* is extremely complex, determined both by the virulence of the strain and the resistance of the host. Cell-

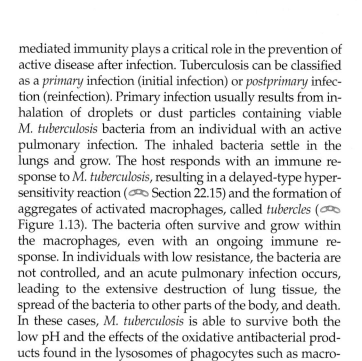

● **Figure 26.7** *Mycobacterium avium* **in an acid-fast stained lymph node biopsy from a patient with AIDS.** Multiple bacilli are evident inside each cell. The individual rods, stained red with carbol fuchsin (arrows), are about 0.4 μm in diameter and up to 4 μm in length.

mediated immunity plays a critical role in the prevention of active disease after infection. Tuberculosis can be classified as a *primary* infection (initial infection) or *postprimary* infection (reinfection). Primary infection usually results from inhalation of droplets or dust particles containing viable *M. tuberculosis* bacteria from an individual with an active pulmonary infection. The inhaled bacteria settle in the lungs and grow. The host responds with an immune response to *M. tuberculosis*, resulting in a delayed-type hypersensitivity reaction (∞ Section 22.15) and the formation of aggregates of activated macrophages, called *tubercles* (∞ Figure 1.13). The bacteria often survive and grow within the macrophages, even with an ongoing immune response. In individuals with low resistance, the bacteria are not controlled, and an acute pulmonary infection occurs, leading to the extensive destruction of lung tissue, the spread of the bacteria to other parts of the body, and death. In these cases, *M. tuberculosis* is able to survive both the low pH and the effects of the oxidative antibacterial products found in the lysosomes of phagocytes such as macrophages (∞ Section 22.2).

In most cases of tuberculosis, however, acute infection does not occur. The infection remains localized, is usually inapparent, and appears to end. But this initial infection hypersensitizes the individual to the bacteria or their products and consequently alters the response of the individual to

subsequent *M. tuberculosis* exposures. A diagnostic test, called the **tuberculin test**, can be used to measure this hypersensitivity. When *tuberculin*, a protein fraction extracted from *M. tuberculosis*, is injected intradermally into a hypersensitive individual, it elicits a localized immune reaction within 1–3 days at the site of injection. The reaction is characterized by *induration* (hardening) and *edema* (swelling) (∞ Figure 22.26). An individual exhibiting this reaction is said to be *tuberculin-positive*, and many healthy adults show positive reactions as a result of previous inapparent infections. A positive tuberculin test does not indicate active disease but only that the individual has been exposed to the organism in the past and has generated a cell-mediated immune response.

For most individuals, this immunity is protective and lifelong. However, some tuberculin-positive patients develop postprimary tuberculosis through reinfection from outside sources or as a result of reactivation of bacteria that have remained alive but dormant in lung macrophages, often for years. Because of the latent nature of tuberculosis infection, individuals who have a positive tuberculin test are generally treated with antibiotics (see below). Factors such as aging, malnutrition, overcrowding, stress, and hormonal changes may reduce effective immunity in untreated individuals and allow reactivation of dormant infections.

Secondary pulmonary infections often progress to chronic infections that result in destruction of lung tissue, followed by partial healing and calcification at the infection site. Thus, chronic postprimary tuberculosis often results in a gradual spread of tubercular lesions in the lungs. Bacteria are found in the sputum in individuals with active disease and areas of destroyed tissue can be seen in X-rays (Figure 26.8●).

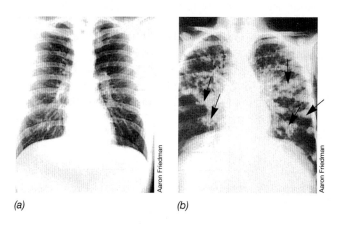

● **Figure 26.8 Tuberculosis and diagnostic X-ray.** (a) Normal chest X-ray. The faint white lines are arteries and other blood vessels. The heart is visible as a white bulge in the lower right quadrant. (b) An advanced case of pulmonary tuberculosis; white patches (arrows) indicate areas of disease. These patches, or tubercles, may contain live *Mycobacterium tuberculosis*. Lung tissue and function is permanently destroyed by these lesions.

Prevention and Treatment

Individuals who have active cases of tuberculosis may spread the disease simply by coughing or speaking near uninfected individuals. Because tuberculosis is highly contagious, the United States Occupational Safety and Health Administration has stringent requirements for the protection of health-care workers who are responsible for tuberculosis patient care. For example, patients with infectious tuberculosis must be hospitalized in negative-pressure rooms. In addition, health-care workers who have patient contact must be provided with personally fitted face masks with high-efficiency particulate air (HEPA) filters. These special filters prevent the passage of *Mycobacterium tuberculosis* in sputum or on dust particles.

Chemotherapy of tuberculosis has been a major factor in control of the disease. Initial success in chemotherapy occurred with the introduction of streptomycin, but the real revolution in tuberculosis treatment came with the discovery of *isonicotinic acid hydrazide (isoniazid or INH)* (Figure 26.9●). This drug, virtually specific for mycobacteria, is effective, inexpensive, and is readily absorbed when given orally. Although the mode of action of isoniazid is not completely understood, it affects the synthesis of mycolic acid by *Mycobacterium* (mycolic acid is a lipid that complexes with peptidoglycan in the mycobacterial cell wall) (∞ Section 12.23). Isoniazid may mimic the activity of a structurally related molecule, nicotinamide (Figure 26.9), becoming incorporated in place of nicotinamide and thus inactivating enzymes requiring this compound for activity.

Treatment of mycobacteria with very small amounts of isoniazid (as little as 5 picomoles [pmol] per 10^9 cells) results in complete inhibition of mycolic acid synthesis, and continued incubation results in loss of outer membrane areas of the cell, a loss of cellular integrity, and death. Following treatment with isoniazid, mycobacteria lose their acid-alcohol fastness, in keeping with the role of mycolic acid in this staining property (∞ Section 12.23). However, mycobacterial resistance to isoniazid and other drugs is increasing at an alarming rate, especially in AIDS patients (see Section 26.14).

Treatment typically involves daily doses of isoniazid and rifampin for 2 months, followed by biweekly doses for a total of 9 treatment months, to eradicate the tubercle bacilli and prevent emergence of antibiotic-resistant organisms.

Failure to complete the entire prescribed treatment plan may allow the infection to be reactivated, and the reactivated organisms often have acquired resistance to the original treatment drugs. Inadequate treatment encourages antibiotic resistance because a high number of mutations spontaneously occur in *M. tuberculosis*, rapidly conferring resistance to single antibiotics. To ensure treatment and thus discourage development of antibiotic resistant organisms, direct observation of treatment (DOT) may be necessary for noncompliant individuals.

In populations such as hospitals and nursing homes where resistant strains are most likely to be present, patients are routinely treated with up to 4 antimycobacterial drugs for 2 months, followed by rifampin-isoniazid treatment for a total of 6 months. Multiple drug therapy reduces the possibility that strains will emerge having resistance to more than one drug.

In many countries, immunization with an attenuated strain of *M. bovis*, the *bacillus Calmette-Guerin (BCG)* strain, is routine for prevention of tuberculosis. However, in the United States and other countries where the prevalence of tuberculosis is low, immunization with BCG is usually discouraged. The live BCG vaccine induces a delayed-type hypersensitivity response (∞ Section 22.15), and all individuals who receive it develop a positive tuberculin test, neutralizing the value of the tuberculin test as a diagnostic and epidemiologic indicator for the spread of *M. tuberculosis* infection.

Mycobacterium leprae and Hansen's Disease (Leprosy)

Mycobacterium leprae, discovered by G. A. Hansen in 1873, is the causative agent of *Hansen's disease*, or *leprosy. M. leprae* is the only *Mycobacterium* species that has not been grown on artificial media. The only experimental animal that has been successfully used to grow *M. leprae* and reproduce a similar disease is the armadillo.

The most serious form of leprosy is characterized by folded, bulblike lesions on the body, especially on the face and extremities (Figure 26.10●), due to growth of *M. leprae* cells in the skin. The lesions contain up to 10^9 bacterial cells per gram of tissue. Like other mycobacteria, *M. leprae* from the lesions stain deep red with carbol fuchsin in the acid-fast staining procedure, providing a rapid, definitive demonstration of active infection (∞ Section 12.23). This *lepromatous* form of leprosy has a very poor prognosis. In severe cases the disfiguring lesions lead to destruction of peripheral nerves and loss of motor function. Many patients exhibit less pronounced lesions from which no bacterial cells can be recovered. These individuals have the *tuberculoid* form of the disease. *Tuberculoid leprosy* is characterized by a vigorous delayed-type hypersensitivity response (∞ Section 22.15) and a good prognosis for spontaneous recovery. Hansen's disease of either form, and the continuum of intermediate forms, is treated using a multiple drug therapy (MDT) protocol, which includes some combina-

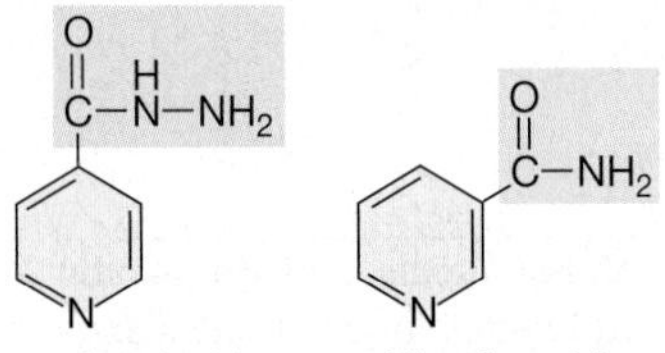

● **Figure 26.9   Structure of isoniazid (isonicotinic acid hydrazide).** Isoniazid is an effective chemotherapeutic agent for tuberculosis. Note the structural similarity to nicotinamide.

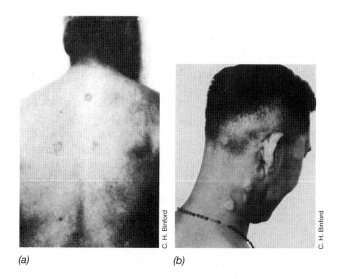

● **Figure 26.10 Lepromatous leprosy lesions on the skin.** Lepromatous leprosy is due to infection with *Mycobacterium leprae*. The lesions can contain up to 10^9 bacterial cells per gram of tissue, indicating an active uncontrolled infection with a poor prognosis.

tion of dapsone (4, 4′-sulfonylbisbenzeneamine), rifampin, and clofazimine. As in the case of tuberculosis, drug-resistant organisms have appeared, especially after treatment with single drugs or inadequate treatment. Extended drug therapy of up to 1 year with a MDT protocol is required for eradication of the organism.

The pathogenicity of *M. leprae* is due to a combination of delayed hypersensitivity (∞ Section 22.15) and the invasiveness of the organism. Transmission involves both direct contact and respiratory routes, and incubation time varies from several weeks to years or even decades. *M. leprae* grows within macrophages, causing an intracellular infection that can result in the large numbers of bacteria within the skin.

In many areas of the world, the incidence of Hansen's disease is very low. Worldwide, there were 763,917 new cases of leprosy reported in 2002, but only 96 cases in the United States. Ninety percent of all cases occur in Madagascar, Mozambique, Tanzania, and Nepal. Up to 2 million people are permanently disabled as a result of leprosy, but, because of the chronic nature and long latent period of the disease, leprosy may go unrecognized and unreported in as many as 12 million people.

Other Pathogenic *Mycobacterium* Species

A common pathogen of dairy cattle, *Mycobacterium bovis*, is pathogenic for humans as well as other animals. *M. bovis* enters humans via the intestinal tract, typically from the ingestion of raw milk. After a localized intestinal infection, the organism eventually spreads to the respiratory tract and initiates the classic symptoms of tuberculosis. *M. bovis* is a different organism from *M. tuberculosis*, but the two organisms are highly homologous at the DNA level. Al-

though *M. bovis* has several gene deletions, there is no observed difference in their infectivity and pathogenesis in humans. Pasteurization of milk and elimination of diseased cattle have eradicated bovine-to-human transmission of tuberculosis in developed countries.

A number of other *Mycobacterium* species are also occasional human pathogens. For example, *M. kansasii*, *M. scrofulaceum*, *M. chelonae*, and other members of the genus (∞ Section 12.23) cause disease. Respiratory disease due to any one of the *M. avium* complex (MAC) group of organisms is particularly pathogenic in AIDS patients as compared with members of the normal population (Figure 26.7) (see Section 26.14).

 26.5 Concept Check

Tuberculosis is one of the most prevalent and dangerous single diseases in the world. Its incidence is on the increase in developed countries, in part because of the emergence of drug-resistant strains. The pathology of tuberculosis and leprosy is influenced by the cellular immune response.

◆ Why is *Mycobacterium tuberculosis* such a widespread respiratory pathogen?

◆ Describe factors that contribute to the incidence of drug resistance in mycobacterial infections.

◆ Identify other species of the genus *Mycobacterium* that can cause human disease.

26.6 *Neisseria meningitidis*, Meningitis, and Meningococcemia

Meningitis is an inflammation of the *meninges*, or the membranes that line the central nervous system, especially the spinal cord and brain. Meningitis can be caused by viral, bacterial, fungal, or protozoan infections. Here we will deal with infectious bacterial meningitis caused by *Neisseria meningitidis* and a related infection, **meningococcemia**.

Neisseria meningitidis, often called meningococcus, is a gram-negative, nonsporulating, obligately aerobic, oxidase positive, encapsulated diplococcus (Figure 26.11●) of about 0.6–1.0 μm in diameter. At least 13 different pathogenic strains of *N. meningitidis* are recognized, based on antigenic differences in the capsular polysaccharides.

Epidemiology and Pathology

Meningococcal meningitis often occurs in epidemics, usually in closed populations such as military installations and college campuses. It normally affects older school-age children and young adults. Up to 30% of individuals normally carry *Neisseria meningitidis* in the nasopharynx with no apparent harmful effects. In epidemic situations, the prevalence of carriers may rise to 80%. The trigger for conversion from the asymptomatic carrier state to pathogenic acute infection is not known.

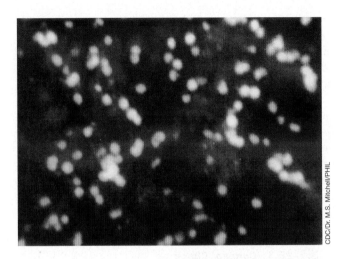

● **Figure 26.11 Fluorescent antibody stain of *Neisseria meningitidis.*** The organism causes adult meningitis and meningococcemia, as observed in cerebrospinal fluid of an infected patient. The individual cocci are about 0.6–1.0 μm in diameter.

In an acute meningococcus infection, the bacterium is transmitted to the host, usually via the airborne route. There it attaches to the cells of the nasopharynx and gains access to the bloodstream, causing bacteremia and simple upper respiratory tract symptoms. The bacteremia sometimes leads to fulminant meningococcemia, characterized by septicemia, intravascular coagulation, shock, and death in over 10% of cases. Meningitis is another possible serious outcome of infection. Meningitis is characterized by sudden onset of headache, vomiting, and stiff neck, and can progress to coma and death in a matter of hours. Death occurs in up to 3% of acute meningococcal meningitis victims.

In the United States, there were 1814 cases of serious meningococcal disease in 2002, the lowest number of cases since 1977. The pattern of decreased incidence indicates the success of widespread vaccination in susceptible populations. However, the mortality rate was almost 12%.

Diagnosis, Prevention, and Treatment

Specimens isolated from nasopharyngeal swabs, blood, or cerebrospinal fluid are inoculated onto modified Thayer-Martin medium (MTM) (∞ Section 24.1), a selective medium that suppresses the growth of most normal flora, but allows the growth of *Neisseria meningitidis* and *Neisseria gonorrhoeae*. Colonies showing a gram-negative diplococcus coupled with a positive oxidase test (∞ Table 24.3) are presumptively identified as *Neisseria*. However, due to the rapid onset of life-threatening symptoms, preliminary diagnosis is often based on clinical symptoms and treatment is started before culture tests confirm infection with *N. meningitidis.*

Penicillin G is the drug of choice for treatment of *Neisseria meningitidis* infections. However, resistant strains

have been reported. Chloramphenicol is the accepted alternative agent for treatment of infections in penicillin-sensitive individuals. A number of broad-spectrum cephalosporins are also effective (∞ Section 20.8).

Naturally occurring strain-specific antibodies acquired by subclinical infections are effective for preventing infections in most adults. Unfortunately, there are no long-term vaccines available for the prevention of meningitis or meningococcemia. However, vaccines consisting of purified polysaccharides from some of the most prevalent pathogenic strains are available and are used to immunize close contacts when epidemics occur. The vaccine is also used to prevent infection in certain susceptible populations such as the military and students living in dormitories. In addition, rifampin is often used as a chemoprophylactic antibiotic to prevent disease in close contacts and family members of infected individuals.

Other Causes of Meninigitis

A number of other organisms can also cause meningitis. Acute meningitis is usually caused by one of the pyogenic bacteria such as *Staphylococcus, Streptococcus,* or *Haemophilus influenzae*. *H. influenzae* primarily infects young children. An effective vaccine for preventing *H. influenzae* meningitis is available and is required in the United States for school-age children (∞ Section 22.13).

Several viruses also cause meningitis. Among these are herpes simplex virus (HSV), lymphocytic choriomeningitis virus (LCM), mumps virus, and the enteroviruses. In general, viral meningitis is less severe than bacterial meningitis.

26.6 Concept Check

Neisseria meningitidis is a common cause of meningococcemia and meningitis in young adults and occasionally occurs in epidemics in closed populations. Bacterial meningitis and meningococcemia are serious diseases with very high mortality rates. Treatment and prevention strategies are in place to deal with epidemic outbreaks, but an effective universal vaccine is not yet available.

◆ Describe the infection by *Neisseria meningitidis* and the resulting development of meningococcemia.

◆ Can meningococcemia be prevented? Explain.

26.7 Viruses and Respiratory Infections

As we discussed (∞ Section 20.10), viruses are less easily controlled by chemotherapeutic means than bacteria or other microorganisms because the growth of viruses is intimately tied to host cell functions. Most chemotherapeutic agents that specifically attack viruses cause at least some harm to host cells as well. Not surprisingly, therefore, the most prevalent infectious diseases, especially in developed countries, are caused by viruses. Most viral diseases are acute, self-limiting infections, but some can be problematic in normal healthy adults. In addition, serious viral diseases such as smallpox and rabies have been effectively controlled by immunization. We begin here by describing measles, mumps, rubella, and chickenpox, all viral diseases transmitted in infectious droplets by an airborne route.

Measles

Measles (*rubeola* or *7-day measles*) is an acute highly infectious childhood disease characterized by nasal discharges, redness of the eyes, cough, and fever. The measles virus is a *paramyxovirus* (∞ Section 16.9) that enters the nose and throat by airborne transmission, quickly leading to systemic viremia. As the disease progresses, fever and cough appear and rapidly intensify, and a rash appears (Figure 26.12●); in most cases measles symptoms last for 7–10 days. Circulating antibodies to measles virus are measurable about 5 days after initiation of infection, and both serum antibodies and cytotoxic T lymphocytes (∞ Sections 22.9 and 22.7) combine to eliminate the virus from the system. A variety of complications may occur due to measles infection, including inner ear infection, pneumonia, and, in rare cases, measles encephalomyelitis. Encephalomyelitis can cause neurological disorders and a form of epilepsy, and has a mortality rate of nearly 20%.

Although once a common childhood illness, measles generally occurs nowadays in rather isolated outbreaks because of widespread immunization programs begun in the mid-1960s (Figure 26.13a●). Because of the highly infectious nature of the disease, all public school systems in the United States require proof of immunization before children can enroll. Active immunization is done with the MMR (measles, mumps, and rubella) vaccine (∞ Section 22.13). A childhood case of measles generally confers lifelong immunity to reinfection.

Mumps

Mumps, like measles, is caused by a paramyxovirus and is also highly infectious. Mumps is spread by airborne droplets, and the disease is characterized by inflammation of the salivary glands leading to swelling of the jaws and neck (Figure 26.14●). The virus spreads through the bloodstream and may infect other organs including the brain, testes, and pancreas. Severe complications may include encephalitis and, very rarely, sterility. The host immune response produces antibodies to mumps virus surface proteins, and this generally leads to a quick recovery. An attenuated vaccine is highly effective for preventing mumps (Figure 26.13b). Hence, the prevalence of mumps in developed countries has been greatly reduced in the last three decades, with disease usually restricted to individuals who did not receive the MMR vaccine.

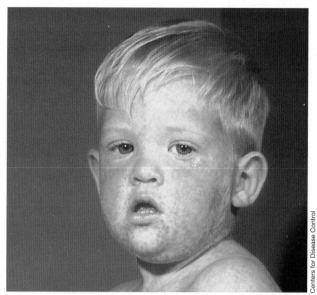

(a)

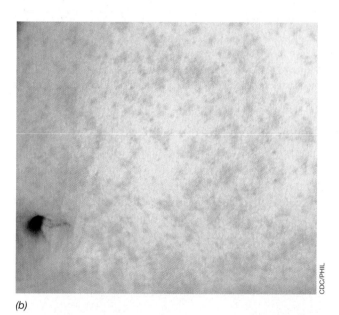

(b)

● **Figure 26.12 Measles in children.** (a) The light pink rash starts on the head and neck, and (b) spreads to the chest, trunk, and limbs. Discrete papules coalesce into blotches as the rash progresses for several days.

Rubella

Rubella (*German measles* or *3-day measles*) is caused by a single-stranded positive sense RNA virus of the togavirus group (∞ Section 16.8). The symptoms of the disease resemble those of measles but are generally milder. Rubella is less contagious than true measles, and thus a good proportion of the population has never been infected. However, during the first 3 months of pregnancy, rubella virus can infect the fetus by placental transmission and cause serious fetal abnormalities.

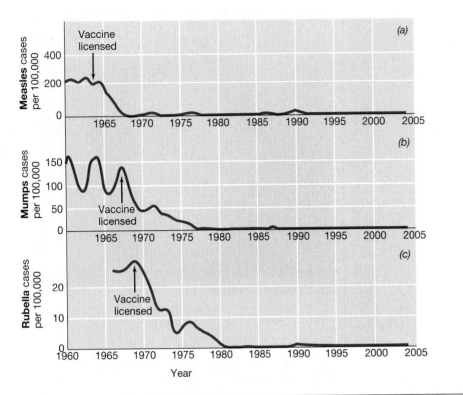

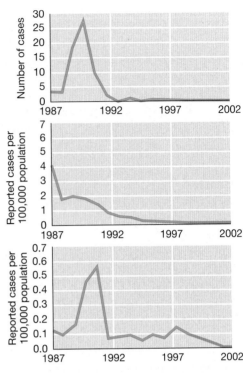

● **Figure 26.13** **Viral diseases and vaccines.** Major childhood viral diseases are now controlled by the MMR (measles, mumps, rubella) vaccine in the United States. (a) Measles. (b) Mumps. (c) Rubella. The insets show a more recent picture of these diseases. Data were obtained from the Centers for Disease Control and Prevention, Atlanta, GA, USA.

Rubella can cause stillbirth, deafness, heart and eye defects, and brain damage in live births. Thus, pregnant women should not be immunized with the rubella vaccine or contract rubella during this period. For this reason, routine childhood immunization against rubella should be practiced. An attenuated virus is adminis-

tered as part of the MMR vaccine (see Figure 26.13c and Section 22.13).

Chickenpox and Shingles

Chickenpox (varicella) is a common childhood disease caused by the Varicella-zoster virus (VZV), a herpesvirus (∞ Section 16.12). VZV is highly contagious and is transmitted by infectious droplets, especially when susceptible individuals are in close contact. In school children, for example, close confinement during the winter months leads to the spread of chickenpox through airborne droplets from infected classmates and through contact with contaminated fomites. The virus enters the respiratory tract, multiplies, and is quickly disseminated via the bloodstream, resulting in a systemic papular rash that quickly heals, rarely leaving disfiguring marks (Figure 26.15●). An attenuated virus vaccine is now recommended for use in the United States (∞ Section 22.13). The current reported annual incidence of chickenpox is now about one-third of the number of cases reported prior to 1994, the year immunization became widespread.

The VZV virus can remain dormant in nerve cells for years with no apparent symptoms. The virus occasionally migrates from this reservoir to the skin surface, causing a painful skin eruption referred to as *shingles* (zoster). Shingles most commonly strikes immunosuppressed individuals or the elderly. Studies with human volunteers suggest that T cells are important in destroying the virus.

Centers for Disease Control

● **Figure 26.14** **Mumps.** Glandular swelling characterizes infection with the mumps virus.

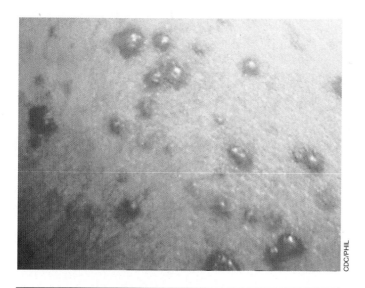

● **Figure 26.15** **Chickenpox.** Mild papular rash associated with the infection by Varicella–zoster virus (VZV), the herpesvirus that causes chickenpox.

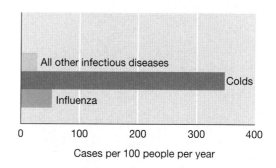

● **Figure 26.16** **Colds and influenza.** These viral diseases are the leading causes of acute infectious disease in the United States. These data are typical for recent years.

The prophylactic use of human hyperimmune globulin prepared against the virus is useful for preventing the onset of symptoms of shingles. Such therapy is advised only for patients where secondary infections occasionally associated with shingles, such as pneumonia or encephalitis, may be life-threatening.

 26.7 **Concept Check**

Viral respiratory diseases are highly infectious and may cause serious health problems. However, the common childhood viral diseases measles, mumps, rubella, and chickenpox are all controllable with appropriate immunization procedures.

◆ Describe the potential serious outcomes of infection by each of these viruses.

◆ Identify the date immunization was made available for measles, mumps, rubella, and chickenpox.

26.8 Colds and Influenza

Colds and influenza are the most common infectious diseases. As shown in Figure 26.16●, there are about two cases of influenza and about 15 colds for every other infectious disease. These viral infections are transmitted via droplets spread from person to person in coughs, sneezes, and respiratory secretions.

Symptoms of a common cold and symptoms of "the flu" (influenza) often seem similar, but the two diseases are symptomatically distinct and caused by quite different viruses. A typical common cold, caused by a rhinovirus, is associated with nasal discharges, cough, chills, and perhaps a sore throat. Influenza, caused by an orthomyxovirus, is generally associated with a different set of symptoms. Although either condition may cause illness, colds are usually of shorter duration and the symptoms are milder. Table 26.1 compares the symptoms of colds and influenza.

The Common Cold

Each person probably averages more than three colds per year throughout his or her lifetime (Figure 26.16). Cold symptoms include rhinitis (inflammation of the nasal region, especially the mucous membranes), nasal obstruction, watery nasal discharges, and a general feeling of malaise, usually without an accompanying fever. *Rhinoviruses*, single-stranded RNA viruses of the *picornavirus* group (see Figure 26.17a● and Section 16.8), are the most common causes of colds. At least 115 different serotypes of rhino-viruses have been identified. Another group of single-stranded RNA viruses, the *coronaviruses* (Figure 26.17b), are responsible for about 15% of all colds in adults. A variety of other viruses including adenoviruses, coxsackie viruses, respiratory syncytial viruses (RSV), and orthomyxoviruses, are responsible for about 10% of common colds. Colds generally induce a specific, local, neutralizing IgA

Table 26.1	Colds and influenza	
Symptoms	**Common cold**	**Influenza**
Fever	Rare	Common (39–40°C) sudden onset
Headache	Rare	Common
General malaise	Slight	Common; often quite severe; can last several weeks
Nasal discharge	Common and abundant	Less common; usually not abundant
Sore throat	Common	Less common
Vomiting and/or diarrhea	Rare	Common in children

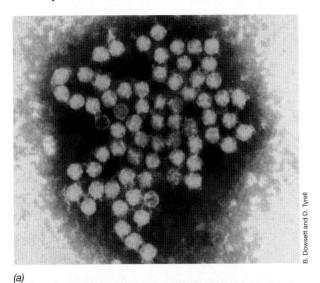

(a)

B. Dowsett and D. Tyrell

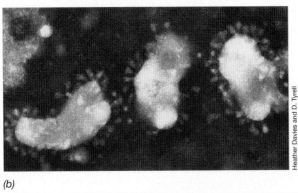

(b)

Heather Davies and D. Tyrell

● **Figure 26.17 Electron micrographs of some common cold viruses.** (a) Human rhinovirus. (b) Human coronavirus. Each rhinovirus virion is about 30 nm in diameter. Each coronavirus virion is about 60 nm in diameter.

response (⌘ Section 22.9). However, the number of potential infectious agents makes immunity via immunization or previous exposure to a given virus very unlikely.

Aerosol transmission of the virus is probably the major means of spreading colds, although experiments with volunteers suggest that direct contact and/or fomite contact is also an important method of transmission. Most antiviral drugs are ineffective, but a pyrazidine derivative (Figure 26.18*a*●) has shown promise for preventing colds after virus exposure. In addition, new experimental antiviral drugs are being designed based on information derived from three-dimensional structures. For example, the antirhinovirus drug WIN 52084 (Figure 26.18*b*) binds to the virus, changing its three-dimensional surface configuration and disrupting rhinovirus binding to the host cell receptor, *ICAM-1 (intercellular adhesion molecule-1)*, thus preventing infection. Interferon-α, a cytokine (⌘ Section 23.11), is also effective in preventing the onset of colds. Thus, there are several experimental possibilities for cold prevention and treatment, although none are widely accepted as effective and safe. Because colds are generally brief and self-limiting, the usual treatment is aimed at symptoms, especially nasal discharges, with a variety of antihistamine and decongestant drugs.

(a)

(b)

● **Figure 26.18 Experimental antirhinovirus drugs.** (a) The structure of 3-methoxy-6-[4-(3-methylphenyl)]-1-piper-azinyl. (b) The structure of WIN 52084, a receptor-blocking drug.

Influenza

Influenza is caused by an RNA virus of the orthomyxovirus group (⌘ Section 16.9). Influenza virus is a single-stranded, negative-sense, helical RNA genome surrounded by an envelope made up of protein, a lipid bilayer, and external glycoproteins (Figure 26.19● and ⌘ Figure 16.17). Three different types of influenza viruses exist: influenza A, influenza B, and influenza C. We limit our discussion here to influenza A since it is the most important human pathogen.

The genome of influenza A virus is single-stranded RNA that is arranged in a highly unusual manner. As discussed in Section 16.9, the influenza virus genome is *segmented*, with genes found on each of eight distinct fragments of its single-stranded RNA (⌘Figure 16.17). This arrangement allows the reassortment of gene fragments between different strains of influenza virus *if* more than one strain of influenza infects a cell at one time. Thus, in a cell infected with two strains of virus, packaging of the RNA segments will occur in a random

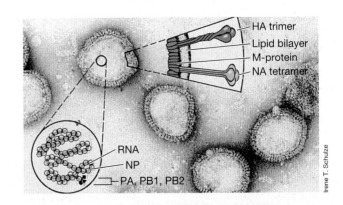

Irene T. Schulze

● **Figure 26.19 Electron micrograph of the influenza virus.** The photo shows the location of the major viral coat proteins and the nucleic acid. Each virion is about 100 nm in diameter. HA, hemagglutinin (three copies make up the HA coat spike); NA, neuraminidase (four copies make up the NA coat spike); M, coat protein; NP, nucleoprotein; PA, PB1, PB2, other internal proteins, some of which may have enzymatic functions.

pattern, with each complete virus containing a random assortment of gene segments from each virus. *The unique reassorted viruses are new viral strains.* Reassortment results in the phenomenon called **antigenic shift**. Antigenic shift refers to *major* modifications in the protein coat of the virions, especially to the two proteins important in the attachment and eventual release of virus from host cells, hemagglutinin (HA or H) and neuraminidase (NA or N) (👓 Figures 16.17 and 26.19). The H and N protein antigens can also acquire minor antigenic changes due to genetic variations such as point mutations in a gene segment, altering one or more amino acids in the H or N protein. This phenomenon is known as **antigenic drift**.

Influenza Epidemiology

Human influenza virus is transmitted from person to person through the air, primarily in droplets expelled during coughing and sneezing. The virus infects the mucous membranes of the upper respiratory tract and occasionally invades the lungs. Symptoms include a low-grade fever lasting for 3–7 days, chills, fatigue, headache, and general aching (Table 26.1). Recovery is usually spontaneous and rapid. Most of the serious consequences of influenza infection occur from bacterial secondary infections in persons whose resistance has been lowered by the influenza infection. Especially in infants and elderly people, influenza is often followed by bacterial pneumonia; death, if it occurs, is usually due to the bacterial infection. Annually, influenza causes 3–5 million cases of severe illness and 250,000–500,000 deaths worldwide.

Most infected individuals develop immunity to the infecting virus, and it is impossible for a strain of similar antigenic type to cause an epidemic for about 2 to 3 years. Immunity is largely dependent on the production of secretory antibody (IgA) (👓 Section 22.9), especially to antigenic determinants of the hemagglutinin (H) and neuraminidase (N) proteins.

Influenza exists in human populations as an *endemic viral disease* and widespread outbreaks occur every year from late autumn through the winter months. Antigenic *drift* results in a reduction of immunity in the population and is responsible for the recurrence of *epidemics*, severe localized influenza outbreaks, occurring in a 2- to 3-year cycle. *Pandemics*, worldwide epidemics, occur much less frequently, being from 10 to 40 years apart, and are the result of antigenic *shift* (👓 Section 25.3).

The 1957 outbreak of the so-called Asian flu provided an opportunity to study the development of a pandemic (Figure 26.20●). The pandemic probably arose when a virulent mutant virus strain, differing antigenically from all previous strains, appeared in the population. Since immunity to this strain was not present, the virus spread rapidly throughout the world. It first appeared in the interior of China in February 1957 and by April had spread to Hong Kong. From Hong Kong, the virus was spread by naval ships to San Diego, California. In May, an outbreak occurred in Newport, Rhode Island, on a naval vessel. From that time, outbreaks continuously occurred in various parts of the United States. Peak incidence occurred in October, when 22 million new cases developed. In the pandemic of 1918, the influenza A "Spanish Flu" killed 230 million people

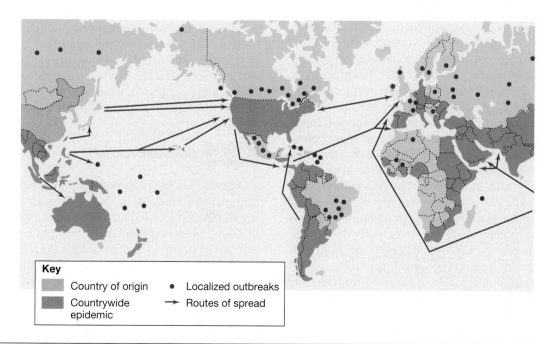

Key
- Country of origin
- Countrywide epidemic
- • Localized outbreaks
- → Routes of spread

● **Figure 26.20   An influenza pandemic.** The spread of the Asian influenza pandemic of 1957 is shown. The original epidemic focus of this pandemic was probably in China. Evidence suggests that agricultural practices involving poultry and swine, and human interactions with these animals, allowed the reassortment of influenza viral genomes from the three host species. New strains for which there is no immune memory in humans are sometimes formed, leading to rapid and uncontrollable spread and pandemic disease.

worldwide (Figure 25.1). Although pandemics occur periodically, none have been as catastrophic as the 1918 flu. The basis for the virulence of the 1918 pandemic is not understood.

Evidence links the pandemics of 1918 and 1957 to the reassortment of avian influenza viruses and human viruses in swine. Swine cells have receptors for both avian and human influenza orthomyxoviruses and can bind and propagate both avian and human influenza strains. If swine are infected with both human and avian strains at the same time, the two unrelated viruses can reassort (antigenic shift), and a unique influenza can be produced that will infect humans and for which there has been no previous antigen exposure. This reassortment from animal strains and infection into humans occurs periodically, but unpredictably, constantly presenting the possibility of a rapidly emerging, highly virulent influenza strain for which there is no preexisting immunity in the human population.

A new avian influenza strain appeared in Hong Kong in 1997, apparently jumping directly from the avian host to humans. The resulting strain, the Influenza A H5N1 strain, also called *avian influenza*, caused numerous infections and several deaths and reemerged in Viet Nam in 2004. This strain is still spread directly from an avian host to humans and can only be spread among humans after prolonged close contact. However, there is some indication that H5N1 has infected pigs, setting the stage for reassortment with human influenza strains that also infect pigs. This reassortment could create a new virus, initiating a new influenza pandemic. Vaccines are being developed to provide appropriate immunity to this potential emergent strain.

Influenza Prevention and Treatment

Influenza epidemics can be controlled by immunization. However, the choice of appropriate vaccines is complicated by the large number of existing strains and the continuing ability of these strains to undergo antigenic drift or antigenic shift. When new strains evolve, vaccines are not immediately available, but through careful worldwide surveillance (Section 25.9), samples of the major emerging strains of influenza virus are usually obtained before epidemic outbreaks occur. In the United States, candidate strains are chosen at the end of each influenza season, are grown in eggs, and inactivated. The inactivated viral strains are mixed to prepare a polyvalent vaccine used for immunization prior to the next influenza season. Influenza immunization is recommended for those individuals most likely to succumb or be exposed to serious secondary illnesses. Influenza immunization is recommended for everyone over 50 years of age, for those suffering from chronic debilitating diseases (e.g., AIDS patients, see Section 26.14), and for health-care workers. Effective artificial immunity from the inactivated influenza vaccine lasts only a few years, and is strain-specific. Therefore, immunization preparations are updated annually.

Influenza A may also be controlled by use of *amantadine* and *rimantadine*, synthetic amines that inhibit viral re-

plication. Both influenza A and B virus release are blocked by the neuraminidase inhibitors *oseltamivir* and *zanavir* (Section 20.10 and Table 20.5). These drugs are all used to treat ongoing influenza and shorten the course and severity of infection. They are most effective when given very early in the course of the infection. Amantadine, rimantadine, and oseltamivir are also effective chemoprophylactic agents that prevent the onset and spread of influenza infection.

Treatment of influenza symptoms with aspirin, especially in children, is not recommended. There is evidence of a link between aspirin treatment of influenza and *Reye's syndrome*, a rare but occasionally fatal complication involving the central nervous system.

26.8 Concept Check

Colds and influenza, or flu, are the most common infectious diseases. While they are not usually life-threatening diseases by themselves, they can lower resistance and allow serious secondary bacterial infections. Influenza outbreaks occur annually and more serious epidemics and pandemics occur periodically.

◆ Define and compare the cause and symptoms of influenza and of common colds, and discuss the possibilities for effective immunization programs for each.

◆ Distinguish between *antigenic drift* and *antigenic shift*.

DIRECT CONTACT TRANSMISSION OF DISEASES

A number of diseases are spread primarily by direct contact with an infected person or by contact with blood or excreta from the infected person. Many of the respiratory diseases we have discussed can also be spread by direct contact. Here we discuss staphylococcal infections, ulcers, and hepatitis, diseases spread primarily person to person through direct contact with infected individuals.

26.9 Staphylococcus

The genus *Staphylococcus* contains common pathogens of humans and animals, occasionally causing life-threatening disease. Staphylococci commonly infect the skin and wounds. Most staphylococcal infections result from the transfer of staphylococci in normal flora from an infected but asymptomatic individual to a susceptible individual.

Staphylococci are gram-positive cocci of about 0.8–1.0 μm in diameter that divide in several planes to form irregular clumps (Section 12.19 and Figure 12.51). Staphylococci are nonsporulating but are resistant to drying and are readily dispersed in dust particles through the air and on surfaces. In humans, two species are important: *Staphylococcus epidermidis*, a nonpigmented form usually found on the skin or mucous membranes, and *Staphylo-*

coccus aureus, a yellow-pigmented form. While both species are potential pathogens, *S. aureus* is more commonly associated with human disease. Both species frequently occur in the normal microbial flora of the upper respiratory tract or on the skin (Figure 26.2).

Epidemiology and Pathogenesis

Staphylococci cause a variety of diseases including acne, boils (Figure 26.21●), pimples, impetigo (Figure 26.4), pneumonia, osteomyelitis, carditis, meningitis, and arthritis. Many of these diseases cause the production of pus, so they are said to be *pyogenic* (pus-forming). The most common habitats of *Staphylococcus aureus* are the upper respiratory tract, especially the nose and throat, and the surface of the skin. Healthy individuals are often carriers, and the resident staphylococci do not cause disease. Infants often become infected during the first week of life from the mother or from another close human contact. Serious staphylococcal infections often occur when the resistance of the host is low because of hormonal changes, debilitating illness, wounds, or treatment with steroids or other drugs that compromise the immune system.

Those strains of *S. aureus* most frequently causing human disease produce a number of extracellular enzymes or toxins (∞ Section 21.9). At least four different *hemolysins* have been recognized, and a single strain often produces several. The production of hemolysins is responsible for the hemolysis seen around colonies on blood agar plates. *S. aureus* is also capable of producing an *enterotoxin* associated with foodborne illness (∞ Sections 21.11, 22.16, and 29.5). All staphylococci also produce *catalase*, an enzyme that converts H_2O_2 to H_2O and O_2. Catalase is not considered a virulence factor, but the catalase test is used to distinguish staphylococci from streptococci, which do not produce catalase.

Another substance produced by *S. aureus* is *coagulase*, an enzyme that causes fibrin to coagulate and form a clot (∞ Section 21.9). The production of coagulase is generally associated with pathogenicity. Clotting induced by coagulase results in the accumulation of fibrin around the bacterial cells making it difficult for host defense agents to come into contact with the bacteria and preventing phagocytosis (Figure 26.21). Most *S. aureus* strains also produce *leukocidin*, which causes the destruction of leukocytes. Production of leukocidin in skin lesions such as boils and pimples results in considerable host cell destruction and is one of the factors responsible for pus formation (Figure 26.21). Some strains of *S. aureus* also produce other extracellular virulence factors, including proteolytic enzymes, hyaluronidase, fibrinolysin, lipase, ribonuclease, and deoxyribonuclease (∞ Table 21.4).

Certain strains of *S. aureus* have been implicated as the agents responsible for **toxic shock syndrome (TSS)**, a serious outcome of staphylococcal infection characterized by high fever, rash, vomiting, diarrhea, and occasionally death. Toxic shock was first recognized in menstruating women and was associated with use of highly absorbent tampons. In menstruating females, blood and mucus in the vagina can become colonized by *S. aureus* from the skin, and the presence of a tampon concentrates this material, creating ideal microbial growth conditions. Largely through education and alterations in materials used in tampons, toxic shock due to tampon use is now relatively rare. However, toxic shock syndrome is still seen in both men and women as a result of staphylococcal infections following surgery.

The symptoms of TSS result indirectly from an exotoxin called *toxic shock syndrome toxin (TSST)*. TSST is a superantigen (∞ Section 22.16). TSST is released by the growing staphylococci, causing a massive T-cell reaction

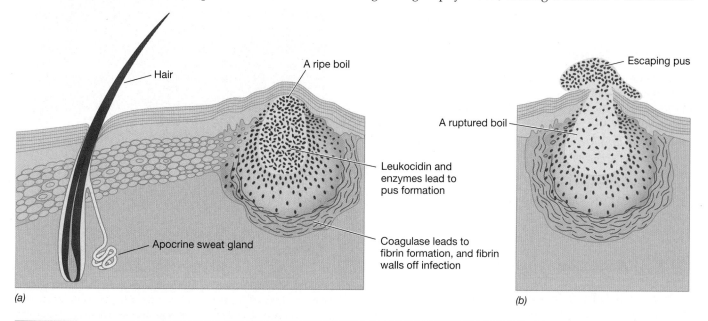

(a)

(b)

● **Figure 26.21   The structure of a boil.** (a) Staphylococci initiate a localized infection of the skin and become walled off by coagulated blood and fibrin through the action of coagulase. (b) The rupture of the boil releases pus and bacteria.

resulting in the inflammatory response characteristic of superantigen reactions. A TSS-like disease may also be caused by different superantigens from other pathogenic *Bacteria*, including *Streptococcus pyogenes* (see Section 26.2).

Staphylococcal enterotoxin A, another superantigen, causes a form of food poisoning. After ingestion of toxin-contaminated food, the toxin stimulates T cells localized along the intestine, resulting in a massive T-cell response, release of mediators, and increased permeability of the intestine. The final outcome is the severe but short-lived diarrhea and vomiting associated with staphylococcal food poisoning (∞ Section 29.5).

Treatment and Prevention

Extensive use of antibiotics has resulted in the selection of resistant strains of *Staphylococcus aureus* and *Staphylococcus epidermidis*. Hospital (nosocomial) infections with antibiotic-resistant staphylococci often occur in patients whose resistance is lowered due to other diseases, surgical procedures, or drug therapy (∞ Section 25.7). Patients often acquire staphylococci from hospital personnel who are asymptomatic carriers of drug-resistant strains. As a result, appropriate antimicrobial drug therapy for *S. aureus* infections is a major problem in hospital environments. Although some community-acquired staphylococcal infections are treatable with penicillin, disease-producing isolates of *S. aureus* must be individually checked for antibiotic susceptibility (∞ Section 24.3).

Prevention of staphylococcal infections is problematic because most individuals are asymptomatic carriers, and diseases such as acne and impetigo can be transmitted by simple contact with contaminated fingers. In hospital environments such as surgical wards and nurseries, carriers of known pathogenic strains must either be excluded or be treated with topical or systemic antimicrobial drugs to eradicate the carrier state.

 26.9 Concept Check

Although staphylococci are usually harmless inhabitants of the upper respiratory tract and skin, several serious diseases can result from pyogenic infection, including some caused by staphylococcal superantigens.

◆ What is the normal habitat of *Staphylococcus aureus*? How is *S. aureus* spread from person to person?

◆ Identify the mechanisms of staphylococcal food poisoning and toxic shock syndrome.

26.10 *Helicobacter pylori* and Gastric Ulcers

Helicobacter pylori was first identified in human intestinal biopsies in 1983. This organism is a pathogen associated with gastritis, ulcers, and gastric cancers. *H. pylori* is a gram-negative, highly motile, spiral-shaped bacterium

that is related to *Campylobacter* (Figure 26.22●) (∞ Section 29.9). It is 2.5–3.5 μm long and 0.5–1.0 μm in diameter and has one to six polar flagella at one end. *H. pylori* colonizes the non-acid-secreting mucosa of the stomach and the upper intestinal tract, including the duodenum (∞ Section 21.4).

Epidemiology

Up to 80% of gastric ulcer patients have concomitant *Helicobacter pylori* infections, and up to 50% of asymptomatic adults in developing countries are chronically infected. Person-to-person contact and ingestion of contaminated food or water are the probable transmission methods for *H. pylori*. Although there is no known nonhuman reservoir of *H. pylori*, the organism has occasionally been recovered from cats kept as household pets, indicating that it can be spread to or from animals in close contact with humans. Infection occurs in high incidence in certain families, and the overall prevalence in the population increases with age. These factors suggest a host-to-host type of transmission (∞ Sections 25.3, 25.4, and 25.5). However, infections with *H. pylori* sometimes also occur in epidemic clusters, suggesting that a common source such as food or water may sometimes be involved.

Pathology, Diagnosis, and Treatment

Although the causal relationship between *Helicobacter pylori* infection and ulcers has not been unequivocally

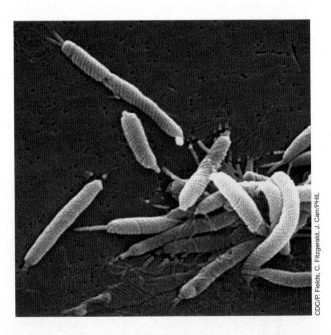

● **Figure 26.22 Scanning electron micrograph of *Helicobacter*.** Members of the genus range in size from 1.5–10 μm in length and 0.3–1 μm in diameter. *Helicobacter* spp. are found in the intestinal and hepatic tracts of mammals and birds. Note the sheathed flagella. The organism in the lower left is 3.2 μm in length.

established, current evidence indicates that *H. pylori* is a major preventable and treatable cause of many gastric ulcers. The bacterium is slightly invasive and colonizes the surfaces of the gastric mucosa, where it is protected from the effects of stomach acids by the gastric mucus layer. After mucosal colonization, a combination of pathogen products and host responses causes inflammation, tissue destruction, and ulceration. Pathogen products such as *vacA* (a cytotoxin; Table 21.4), *urease*, and *lipopolysaccharide* may contribute to localized tissue destruction and ulceration. Antibodies to *H. pylori* are usually present in infected individuals, but are not protective and do not prevent colonization. Individuals who acquire *H. pylori* tend to have chronic, long-term infections unless they are treated with antibiotics. Chronic gastritis due to untreated *H. pylori* infection may lead to the development of gastric cancers.

Clinical signs of *H. pylori* infection include belching and stomach (epigastric) pain. Definitive diagnosis involves the recovery and culture or observation of *Helicobacter pylori* from a gastric ulcer biopsy. Serum antibodies indicate *H. pylori* infection, but since infections seem to be chronic and antibodies may persist for months after a given infection (Section 24.7), *H. pylori* antibodies are not reliable indicators of acute, active disease. A simple *in vivo* test, the *urease test*, is used as a metabolic test for *H. pylori*. In this test, the patient ingests ^{13}C or ^{14}C-labeled urea. The presence of labeled CO_2 in the patient's exhaled breath indicates the presence of urease, produced almost exclusively by *H. pylori*. Recovery of *H. pylori* organisms or antigens from the stool is also indicative of infection.

Evidence for a causal association between *Helicobacter pylori* and gastric ulcers comes from antibiotic treatments for the disease. Long-term treatment of ulcers with antacid preparations has seldom been successful. Most patients relapse within 1 year. However, by treating ulcers as an infectious disease, permanent cures are often obtained. Treatment of *H. pylori* infection usually consists of a combination of drugs including metranidazole, a second antibiotic such as tetracycline or amoxycillin, and a bismuth-containing antacid preparation. The combination treatment, administered for 14 days, abolishes the *H. pylori* infection and provides a long-term cure.

26.10 Concept Check

Helicobacter pylori infection appears to be the most common cause of gastric ulcers. Treatment of gastric ulcers now involves antibiotics, which seem to promote a permanent cure.

◆ Describe the infection by *H. pylori* and the resulting development of an ulcer.

◆ Describe evidence indicating that *H. pylori* infections are spread from person to person and also from common sources.

26.11 Hepatitis Viruses

Hepatitis is a liver inflammation commonly caused by an infectious agent. Hepatitis sometimes results in acute illness followed by destruction of functional liver anatomy and cells, a condition known as **cirrhosis**. Hepatitis due to an infection can cause chronic or acute disease and some forms lead to liver cancer. Although many viruses and a few bacteria can cause hepatitis, a restricted group of viruses is often associated with liver disease. Hepatitis viruses are diverse, and none of these viruses are genetically related, but all infect cells in the liver, causing hepatitis. Table 26.2 defines the five known hepatitis viruses.

Epidemiology

Hepatitis A virus (HAV) is transmitted from person to person or by ingestion of fecally contaminated food or water. Often called *infectious hepatitis*, the virus often causes mild, even subclinical infections, but rare cases of severe liver disease can occur. The most significant food vehicles for hepatitis A are shellfish, usually oysters and clams harvested from water polluted by human fecal material. In recent years however, HAV has also been transmitted in fresh produce. In 2003, a significant outbreak in the eastern United States was traced to eating uncooked or undercooked green onions. The general trend for numbers of HAV infections has moved downward (Figure 26.23●), partly due to the availability of an effective vaccine. HAV causes more cases of viral hepatitis than any other virus, and over 30% of individuals in the United States have antibodies to HAV, indicating they have been infected at some time in their lives.

Disease	Virus and genome	Vaccine	Disease	Transmission route
Hepatitis A	*Hepatovirus* (HAV) ss RNA	Yes	Acute	Enteric
Hepatitis B	*Orthohepadnavirus* (HBV) ds DNA	Yes	Acute, chronic, oncogenic	Parenteral, sexual
Hepatitis C	*Hepacivirus* (HCV) ss RNA	No	Chronic, oncogenic	Parenteral
Hepatitis D	*Deltavirus* (HDV) ss RNA	No	Fulminant, only with HBV	Parenteral
Hepatitis E	Calciviridae family (HEV) ss RNA	No	Fulminant disease in pregnant women	Enteric
Hepatitis G	Flaviviridae family (HGV) ss RNA	No	Asymptomatic	Parenteral

Table 26.2 Hepatitis viruses

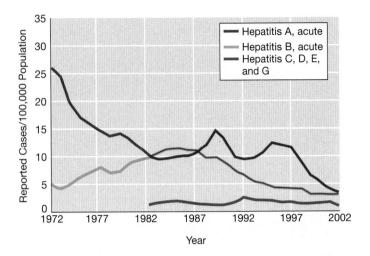

● **Figure 26.23   Hepatitis in the United States.** The prevalence of hepatitis is shown by viral agent. In 2002 there were 8795 cases of hepatitis A, 7996 cases of hepatitis B, and 1835 cases of other hepatitis, mostly caused by HCV. Data were obtained from the Centers for Disease Control and Prevention, Atlanta, GA, USA.

Infection due to hepatitis B virus (HBV) is often called *serum hepatitis*. HBV is a hepadnavirus (⚭ Section 16.15), a partially double-stranded DNA virus. The mature virus particle containing the viral genome is called a Dane particle (Figure 26.24●). HBV causes acute, often severe disease that can lead to liver failure and death. Chronic HBV infection can lead to cirrhosis and liver cancer. HBV is usually transmitted by a *parenteral* (outside the gut) route, such as blood transfusion or through shared hypodermic needles contaminated with infected blood. HBV may also be transmitted through exchanges of body fluids, as in sexual intercourse. The numbers of new HBV infections is decreasing, again due to an effective vaccine. However, approximately 5,000

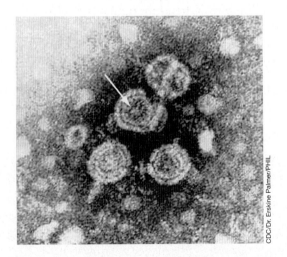

● **Figure 26.24   Hepatitis B virus (HBV).** The arrow indicates a complete HBV particle, which is about 42 nm in diameter and is called a Dane particle (⚭Section 16.15 and Figure 16.27).

people die each year due to complications such as cancers generated by chronic HBV infection.

Hepatitis D virus (HDV) is a *defective virus* that lacks genes for its own protein coat (⚭ Section 16.15). HDV is also transmitted by parenteral routes, but since it is a defective virus, it cannot replicate and express a complete virus unless the cell is also infected with HBV: The HDV genome replicates independently but uses the protein coat of HBV for expression. Thus, HDV infections are always seen as coinfections with HBV.

Hepatitis C virus (HCV) is also transmitted parenterally. HCV generally produces a mild or even asymptomatic disease at first, but up to 85% of individuals develop chronic hepatitis, with up to 20% leading to chronic liver disease and cirrhosis. Chronic infection leads to *hepatocarcinoma* (liver cancer) in 3–5% of infected individuals each year. The latency period for development of cancer can be several decades after the primary infection. The *reported* new cases of HCV in the United States (Figure 26.23) are only a fraction of the approximately 25,000 new infections annually. Large numbers of HCV-related deaths occur annually due to chronic HCV infections that develop into liver cancer. HCV-induced liver disease is the most common liver disease currently seen in clinical settings in the United States and accounts for up to 10,000 of the 25,000 annual deaths due to liver cancer, other chronic liver disease, and cirrhosis.

Hepatitis E virus (HEV) transmits hepatitis via an enteric route. HEV causes an acute, self-limiting hepatitis that varies in severity from case to case, but is often the cause of rapid fulminant disease in pregnant women. HEV is endemic in Mexico as well as in tropical and subtropical regions of Africa and Asia.

Hepatitis G virus (HGV) is commonly found in the blood of patients with other forms of acute hepatitis, but HGV alone seems to cause very mild disease or is completely asymptomatic. Several different samples of volunteer blood donors tested as high as 8.1% positive for HGV, but since HGV is not associated with demonstrable clinical disease, the significance of these findings is not clear.

Pathology and Diagnosis

Hepatitis is an acute disease of the liver. Symptoms include fever, **jaundice** (production and release of excess bilirubin by the liver due to destruction of liver cells, resulting in yellowing of the skin), *icterus* (yellowing of the whites of the eyes), *hepatomegaly* (liver enlargement), and *cirrhosis* (breakdown of the normal liver tissue architecture, including fibrosis). Mild hepatitis may be characterized by relatively minor elevation of liver enzymes such as alanine aminotransferase (ALT). Fulminant disease is characterized by rapid onset of severe symptoms such as jaundice and cirrhosis and is often a life-threatening condition. The various hepatitis viruses cause similar acute clinical disease and cannot be readily distinguished based on the clinical findings alone. Chronic hepatitis infections, usually caused by HBV or HCV, are often asymptomatic or produce very mild symptoms but can cause serious liver disease, even in the absence of hepatocarcinoma.

Diagnosis of hepatitis is based primarily on clinical findings and laboratory tests that determine liver function problems. Cirrhosis is diagnosed by visual examination of biopsied liver tissue. A number of virus-specific assays are also used to confirm diagnosis, identify the infectious agent, and determine a course of treatment. Direct culture of hepatitis viruses is usually not used for identification purposes, and HCV and HGV have not been successfully cultured.

Some of the most widely used methods for determining hepatitis identity are enzyme-linked immuno-assay (ELISA) tests (∞ Section 24.10). Most ELISA tests are designed to identify viral proteins in blood specimens. However, tests are available that indicate the presence of IgM or IgG *antibodies* to HBV. IgM is associated with the primary immune response to HBV and IgG is associated with the secondary response to HBV. Therefore, identification of the antibody class can determine whether or not the HBV infection is a new infection (IgM) or whether the antibody is due to a secondary response to a chronic or latent HBV infection (IgG) (∞ Section 22.9). Other immune-based tests used for the detection of hepatitis viruses include immunoblots (∞ Section 24.11), immunoelectron microscopy (IEM) (∞ Section 24.13), and immunofluorescence (∞ Section 24.9).

PCR-based tests and dot-blot DNA hybridization tests (∞ Section 24.12) are also used for the detection of the viral genome in blood or in liver tissue obtained by biopsy.

Prevention and Treatment

Infection with HAV or HBV can be prevented with effective vaccines. HBV vaccination is recommended and in most cases is required for school-age children in the United States (∞ Section 22.13). No effective vaccines are available for the other hepatitis viruses.

"Universal precautions" for prevention of disease spread by bloodborne pathogens are mandated by law primarily to protect against HBV infection and the human immunodeficiency virus (HIV) (∞ Section 24.4). These standards mandate precautions for personnel when handling infectious waste and body fluids and are designed to prevent infection by all parenterally transmitted hepatitis viruses (HBV, HCV, HDV, and HGV). The precautions prescribe a high level of vigilance and aseptic handling and containment procedures to deal with patients, body fluids, and infected waste materials.

Hepatitis A can be spread through contamination of common sources such as food and water, and epidemic outbreaks of hepatitis A can be prevented by maintaining pathogen-free food and water supplies (∞ Chapter 28 and Chapter 29).

Postexposure treatment of hepatitis is sometimes successful. Pooled human immune gamma globulin can be used to prevent HAV infection if given soon after exposure.

For postexposure prevention of HBV infection, specific hepatitis B immune globulin, coupled with administration of the HBV vaccine, has been effective (∞ Section 22.13).

Most treatment of hepatitis is supportive, providing rest and time to allow liver damage to resolve and be repaired. In some cases, some antiviral drugs are effective for treatment. Interferon α is effective against HCV when combined with ribavirin in some patients. HBV can be treated with the antiviral drugs foscarnet, ribavirin, lamivudine, or ganciclovir (∞ Section 20.10).

⬡ *26.11 Concept Check*

Hepatitis caused by viruses can cause cirrhosis, an acute liver disease. HBV and HCV can cause chronic infections leading to liver cancer. Vaccines are available for HAV and HBV. The overall prevalence of hepatitis has decreased significantly in the last 20 years in the United States, but viral hepatitis is still a major public health problem because of the high infectivity of the viruses.

◆ Describe the mode of transmission for hepatitis A virus, hepatitis B virus, and hepatitis C virus.

◆ Describe potential prevention and treatment methods for hepatitis A virus and hepatitis B virus.

III SEXUALLY TRANSMITTED INFECTIONS

Several important human pathogens are transmitted almost exclusively by sexual contact. These pathogens cause **sexually transmitted infections,** or **STIs.**

STIs, also called *sexually transmitted diseases (STDs)* or *venereal diseases,* are caused by a wide variety of bacteria, viruses, protozoa, and even fungi (Table 26.3). Unlike respiratory pathogens that are shed constantly in large numbers by an infected individual, sexually transmitted pathogens are generally found only in body fluids from the genitourinary tract that are exchanged during sexual activity. This is because sexually transmitted pathogens are very sensitive to drying and other environmental stresses such as heat and light. Their habitat, the human genitourinary tract, is a protected, moist environment. Thus, these organisms preferentially and sometimes exclusively colonize the genitourinary tract.

Effectively diagnosing and treating STIs is very difficult for a number of social and biological reasons. First, up to one-third of all STIs involve teenagers with multiple sex partners, making it difficult to identify the infection source and stop its spread. Second, many STIs have minor symptoms and infected individuals do not seek treatment. Third, social stigmas still attached to STIs prevent many individuals from seeking prompt treatment. However, prompt effective treatment of STIs is desirable for a number of reasons. First, most STIs are curable and virtually all are controllable with appropriate medical intervention.

Table 26.3	Sexually transmitted diseases and treatment guidelines	
Disease	**Causative organism(s)^a**	**Recommended treatment^b**
Gonorrhea	*Neisseria gonorrhoeae* (B)	Cefixime or ceftriaxone, *and* azithromycin or doxycycline
Syphilis	*Treponema pallidum* (B)	Benzathine penicillin G
Chlamydia trachomatis infections	*Chlamydia trachomatis* (B)	Doxycycline or azithromycin
Nongonococcal urethritis	*C. trachomatis* (B) or *Ureaplasma urealyticum* (B) or *Mycoplasma genitalium* (B) or *Trichomonas vaginalis* (P)	Azithromycin
Lymphogranuloma venereum	*C. trachomatis* (B)	Doxycycline
Chancroid	*Haemophilus ducreyi* (B)	Azithromycin
Genital herpes	Herpes simplex type 2 (V)	No known cure; symptoms can be controlled with topical application of acyclovir (Figure 26.31).
Genital warts	Papilloma virus (certain strains)	No known cure; symptomatic warts can be removed surgically, chemically, or by cryotherapy.
Trichomoniasis	*Trichomonas vaginalis* (P)	Metronidazole
Acquired immunodeficiency syndrome (AIDS)	Human immunodeficiency virus (HIV)	No known cure; nucleotide base analogs, protease inhibitors, fusion inhibitors, and nonnucleoside reverse transcriptase inhibitors are clinically useful in some treatments (see Table 26.4).
Pelvic inflammatory disease	*N. gonorrhoeae* (B) or *C. trachomatis* (B)	Cefotetan
Vulvovaginal candidiasis	*Candida albicans* (F)	Butoconazole

^{a} B, bacterium; V, virus; P, protozoan; F, fungus.
^{b} Recommendations of the U.S. Department of Health and Human Services, Public Health Service. For many drugs, there are a number of possible alternatives.

Second, delay or lack of treatment can lead to long-term problems such as infertility, cancer, heart disease, degenerative nerve disease, birth defects, or stillbirth.

Because transmission of STIs is limited to intimate physical contact, generally during sexual intercourse, the spread of venereal diseases can be controlled by sexual abstinence (no exchange of body fluids) or by the use of barriers such as condoms that stop the exchange of body fluids during sexual activity.

STIs are very common and continue to pose social as well as medical problems. We deal here with a discussion of a number of prevalent STIs.

26.12 Gonorrhea and Syphilis

Gonorrhea and syphilis are preventable, treatable bacterial STIs. Because of differences in their symptoms, the overall pattern of disease prevalence is quite different. Gonorrhea is very prevalent, and often asymptomatic, especially in women. As a result, the disease is often unrecognized and remains untreated. Syphilis, on the other hand, now has a low prevalence. This is partly because the primary stage of syphilis exhibits very obvious symptoms and infected individuals usually seek immediate treatment (Figure 26.25●).

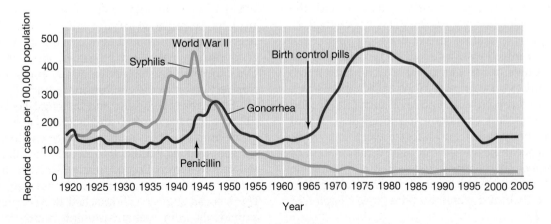

● **Figure 26.25** **Reported cases of gonorrhea and syphilis in the United States.** Note the downward trend in disease incidence after the introduction of antibiotics and the upward trend in the incidence of gonorrhea after the introduction of birth control pills. In 2002 there were 351,852 cases of gonorrhea and only 6862 cases of syphilis in the United States.

Gonorrhea

Neisseria gonorrhoeae, often called *gonococcus*, causes gonorrhea. *N. gonorrhoeae* is a gram-negative, nonsporulating, obligately aerobic, oxidase-positive diplococcus (∞ Section 12.10) related biochemically and phylogenetically to *Neisseria meningitidis* (see Section 26.6). *N. gonorrhoeae* is very sensitive to drying and normally does not survive away from the mucus membranes of the genitourinary tract (Figure 26.26●). Gonococci are killed rapidly by drying, sunlight, and ultraviolet light. Because of its extreme sensitivity to environmental conditions, *N. gonorrhoeae* can only be transmitted by intimate person-to-person contact. The pathogen enters the body by way of the mucus membranes of the genitourinary tract.

The symptoms of gonorrhea are quite different in the male and female. In females, gonorrhea is characterized by a mild vaginitis that is difficult to distinguish from vaginal infections caused by other organisms, and thus, the infection may easily go unnoticed. Complications from untreated gonorrhea in females can lead, however, to a condition known as *pelvic inflammatory disease (PID)*. PID is a chronic inflammatory disease that can lead to long-term complications such as sterility. In the male, the organism causes a painful infection of the urethral canal (∞ Figure 21.11). Complications from untreated gonorrhea affecting both males and females include damage to heart valves and joint tissues due to immune complex deposition (∞ Section 22.15).

In addition to gonorrhea, the organism also causes eye infections in newborns. Infants born of infected mothers may acquire eye infections during birth. There-fore, prophylactic treatment of the eyes of all newborns with an ointment containing erythromycin is generally mandatory to prevent gonococcal infection in infants. We discussed the clinical microbiology and diagnosis of gonorrhea in Section 24.1.

Treatment of gonorrhea with penicillin has been successful in the past. However, strains of *N. gonorrhoeae* resistant to penicillin arose in the 1980s and are now widespread. Fortunately, the penicillinase-producing strains respond to alternative antibiotic therapy, with a single dose of the β-lactam antibiotics cefixime or ceftriaxone. The quinolones ciprofloxacin, oflaxacin, or levofloxacin can also be used. By 2001, however, 14.3% of *N. gonorrhoeae* strains isolated from Hawaii were resistant to quinolones, underscoring the need to test isolates for antibiotic sensitivity. An antichlamydial agent, normally azithromycin or doxycycline, is often given at the same time because nearly 50% of gonorrhea patients are also infected with the hard-to-diagnose *Chlamydia trachomatis* (Table 26.3).

The incidence of gonococcus infection remains relatively high for the following reasons: (1) Acquired immunity does not exist; hence repeated reinfection is possible. Although antibodies are produced, they are either not protective, or they are strain-specific and provide no cross immunization. Within a single *Neisseria gonorrhoeae* strain, antigenic switches can occur, changing *opacity protein antigens (Opa)* and *surface pilin antigens*. These changes create new serotypes and apparently prevent effective immunity. (2) The use of oral contraceptives alters the local mucosal environment in favor of the pathogen. Oral contraceptives induce the body to mimic pregnancy, which results, among other things, in a lack of glycogen production in the vagina and a rise in the vaginal pH. Lactic acid bacteria normally found in the adult vagina fail to develop under such circumstances, allowing colonization by *N. gonorrhoeae* transmitted from an infected partner (∞ Section 21.5). (3) Symptoms in the female are so mild that the disease may be unrecognized, and a promiscuous infected female can infect many males. The disease can be controlled if the sexual contacts of infected persons are quickly identified and treated, but it is often difficult to obtain this information and even more difficult to arrange treatment.

Syphilis

Syphilis is caused by a spirochete, *Treponema pallidum*. *T. pallidum* is about 10–15 μm in length and about 0.15 μm in diameter (Figure 26.27●). The spirochete is extremely sensitive to environmental stress and, therefore, syphilis is normally transmitted from person to person by intimate sexual contact.

Syphilis, often transmitted with *N. gonorrhoeae*, is potentially much more serious than gonorrhea. Largely because of differences in the symptoms and pathobiology of the two diseases, the incidence of syphilis in the United States is much lower than the incidence of gonorrhea (Figure 26.25).

● **Figure 26.26 The causative agent of gonorrhea, *Neisseria gonorrhoeae*.** The scanning electron micrograph of the microvilli of human fallopian tube mucosa shows how cells of *N. gonorrhoeae* attach to the surface of epithelial cells. Note the distinct diplococcus morphology. Cells of *N. gonorrhoeae* are about 0.8 μm in diameter.

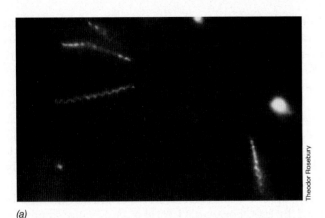

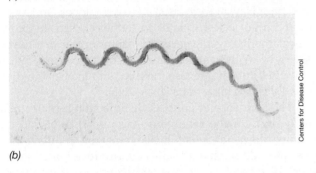

● **Figure 26.27   The syphilis spirochete, *Treponema pallidum.*** (a) Dark-field microscopy of an exudate. *Treponema pallidum* cells measure 0.15 μm wide and 10–15 μm long. (b) Shadow-cast electron micrograph of a cell of *T. pallidum*. Note the endoflagella, typical of spirochetes (◯◯ Section 12.33).

Syphilis does not pass through unbroken skin, and initial infection most probably takes place through tiny breaks in the epidermal layer. In the male, initial infection is usually on the penis; in the female it is most often in the vagina, cervix, or perineal region. In about 10% of cases, infection is extragenital, usually in the oral region. During pregnancy, the organism can be transmitted from an infected woman to the fetus; the disease acquired by the infant is called **congenital syphilis**.

Syphilis is an extremely complex disease and, in an individual patient, may progress into any of three stages, but the disease always begins with a localized infection called *primary syphilis*. In primary syphilis, *T. pallidum* multiplies at the initial site of entry, and a characteristic *primary* lesion known as a *chancre* (Figure 26.28●) is formed within 2 weeks to 2 months. Darkfield microscopy of the exudate from syphilitic chancres often reveals the actively motile spirochetes (Figure 26.27a). In most cases the chancre heals spontaneously and the organisms disappear from the site. Some cells, however, spread from the initial site to various parts of the body, such as the mucous membranes, the eyes, joints, bones, or central nervous system, and extensive multiplication occurs. A hypersensitivity reaction to the treponeme often takes place, revealed by the development of a generalized skin rash; this rash is the key symptom of *secondary syphilis*. At first, the patient may be highly infectious, but eventually *T. pallidum* is cleared from the secondary lesions and infectivity is reduced.

The subsequent course of the disease in the absence of treatment is highly variable. About one-fourth of infected individuals appear to undergo a spontaneous cure as demonstrated by a decrease in antibody titer, and another one-fourth do not exhibit any further symptoms, although static or elevated antibody titers indicate a persistent, active infection. About half of these patients develop *tertiary syphilis*, with symptoms ranging from relatively mild infections of the skin and bone to serious and even fatal infections of the cardiovascular system or central nervous system. Involvement of the nervous system can cause generalized paralysis or other severe neurological damage. Relatively low numbers of organisms are present in individuals with tertiary syphilis, and most of the symptoms probably result from inflammation due to delayed hypersensitivity reactions to the spirochetes (◯◯ Sections 22.3 and 22.15).

The clinical immunology and microbiology as well as the laboratory diagnosis of syphilis were discussed in Section 24.1. The single most important physical sign of a primary syphilis infection, the chancre, is also diag-

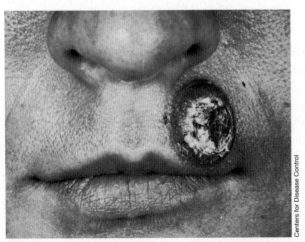

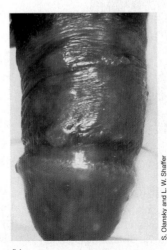

● **Figure 26.28   Primary syphilis lesions.** (a) Chancre on lip. (b) Several chancres on penis. The chancre is the characteristic lesion of primary syphilis at the site of infection by *Treponema pallidum*. Patients who acquire such lesions generally seek medical intervention, and the obvious chancre hastens diagnosis and treatment. Syphilis is treatable and curable with a single injection of penicillin G (◯◯ Table 26.3). The biology of the genus *Treponema* is discussed in Section 12.33.

(a)

(b)

nostic for the disease. Infected individuals generally seek treatment for syphilis because of the highly visible chancre.

Penicillin is highly effective in syphilis therapy, and the primary and secondary stages of the disease can usually be controlled by a single injection of benzathine penicillin G. In tertiary syphilis, penicillin treatment must be extended for longer periods of time. The incidence of primary and secondary syphilis in the United States has decreased significantly over the last two decades and is now at the lowest level since record keeping began.

 26.12 Concept Check

Gonorrhea and syphilis, caused by *Neisseria gonorrhoeae* and *Treponema pallidum*, respectively, are STIs with potential serious consequences if not treated. Although the incidence of these diseases has generally declined in recent years, there are still over 350,000 cases of gonorrhea and 6000 cases of syphilis annually in the United States.

◆ Explain at least one potential reason for the high incidence of gonorrhea as compared with syphilis.

◆ Describe the progression of untreated gonorrhea and untreated syphilis. Do treatments produce a cure for each disease?

26.13 Chlamydia, Herpes, and Trichomoniasis

Chlamydia, herpes, and trichomoniasis are important STIs transmitted by a bacterium, a virus, and a protozoan, respectively. These diseases are very prevalent in the population and are much more difficult to diagnose and treat than are syphilis and gonorrhea.

Chlamydia

A number of sexually transmitted diseases can be ascribed to infection by the obligate intracellular bacterium *Chlamydia trachomatis* (Figure 26.29● and ⟳ Section 12.27). The total incidence of sexually transmitted *C. trachomatis* infections (a reportable disease, ⟳ Table 25.3) probably greatly outnumbers the incidence of gonorrhea. There are over 600,000 *reported* cases and there may be up to 3 million new *C. trachomatis* sexually transmitted infections every year, making this organism the most prevalent cause of venereal disease. *C. trachomatis* also causes a serious eye infection called *trachoma* (⟳ Section 12.27), but the strains of *C. trachomatis* responsible for venereal infections are distinct from those causing trachoma. Chlamydial infections may also be transmitted congenitally to the newborn in the birth canal, causing newborn conjunctivitis and pneumonia.

Nongonococcal urethritis (NGU) due to *C. trachomatis* is one of the most frequently observed sexually transmitted diseases in males and females, but the infections are often inapparent. In a small percentage of cases, chlamydial NGU leads to serious acute complications, including testicular swelling and prostate inflammation in men, and cervicitis, pelvic inflammatory disease, and fallopian tube damage in women; cells of *C. trachomatis* attach to microvilli of fallopian tube cells, enter, multiply, and eventually lyse the cells (Figure 26.29). Untreated NGU can cause infertility.

Chlamydia NGU is relatively difficult to diagnose by traditional isolation and identification methods. To expedite

(a)

(b)

● **Figure 26.29 Cells of *Chlamydia trachomatis* (arrows) attached to human fallopian tube tissues.** (a) Cells attached to the microvilli of a fallopian tube. (b) A damaged fallopian tube containing a cell of *C. trachomatis* (arrow) in the lesion.

diagnoses, a variety of tests have been developed for identifying *C. trachomatis* from a vaginal or pelvic swab or from discharges. These clinical tests include nucleic acid probe tests as well as fluorescent monoclonal antibodies and various enzyme-linked immunosorbent assay (ELISA) tests for detecting specific *C. trachomatis* antigens. If a chlamydial infection is suspected, treatment is initiated with the antibiotics azithromycin or doxycycline (Table 26.3).

Chlamydial NGU is frequently observed as a secondary infection following gonorrhea. If both *Neisseria gonorrhoeae* and *Chlamydia trachomatis* are transmitted to a new host in a single event, treatment of gonorrhea with cefixime or ceftriaxone is usually successful but does not eliminate the chlamydia. Although cured of gonorrhea, such patients, still infected with chlamydia, eventually experience an apparent recurrence of gonorrhea that is instead a case of chlamydial NGU. Thus, patients treated for gonorrhea are also given azithromycin or doxycycline to treat the potential coinfection with the usually undiagnosed *C. trachomatis*.

Lymphogranuloma venereum is a sexually transmitted disease caused by distinct strains of *C. trachomatis* (LGV 1, 2, or 3). The disease occurs most frequently in males and consists of a swelling of the lymph nodes in and about the groin. From the infected lymph nodes, chlamydial cells may travel to the rectum and cause a painful inflammation of rectal tissues called *proctitis*. Because of the potential for regional lymph node damage and the complications of proctitis, lymphogranuloma venereum is also a serious sexually transmitted chlamydial infection. It is the only chlamydial infection that invades beyond the epithelial cell layer. Other *Chlamydia* infect and complete their development in the epithelial cells.

Herpes

Herpesviruses are a large group of complex double stranded DNA viruses (⟂ Section 16.12), many of which are human pathogens. A subgroup of herpesviruses, the *herpes simplex viruses*, are responsible for cold sores and genital infections.

Herpes simplex virus type 1 (HSV-1) infects the epithelial cells around the mouth and lips, causing cold sores (fever blisters) (Figure 26.30*a*●). HSV-1 may, however, occasionally infect other body sites including the anogenital regions. HSV-1 is spread via direct contact or through saliva. The incubation period of HSV-1 infections is short (3–5 days), and the lesions heal without treatment in 2–3 weeks. Relapses of HSV-1 infections are relatively common, and it is thought that the virus is spread primarily by contact with infectious lesions. Latent herpes infections are apparently quite common, with the virus persisting in low numbers in nerve tissue. Recurrent acute herpes infections are due to a periodic triggering of virus activity by unknown or indeterminant causes such as coinfections and stress. Oral herpes caused by HSV-1 is quite common and apparently has no harmful effects on the host beyond the discomfort of the oral blisters.

Herpes simplex virus 2 (HSV-2) infections are associated primarily with the anogenital region, where the virus causes painful blisters on the penis of males or on the cervix, vulva, or vagina of females (Figure 26.30*b*). HSV-2 infections are generally transmitted by direct sexual contact, and the disease is most easily transmitted during the active blister stage rather than during periods of inapparent (presumably latent) infection. HSV-2 occasionally infects other body sites such as the mucous membranes of the mouth. HSV-2 can also be transmitted to a newborn at birth by contact with herpetic lesions in the birth canal. The disease in the newborn varies from latent infections with no apparent damage to systemic disease resulting in brain damage or death. To avoid herpes infections in newborns, delivery by caesarean section is advised for pregnant women with genital herpes infections. The long-term effects of genital herpes infections are not yet fully understood. However, studies have indicated a significant correla-

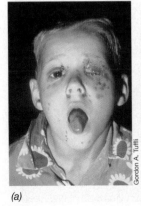
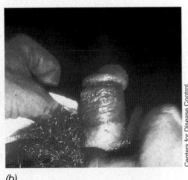
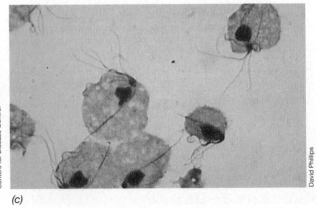

(a)   (b)   (c)

● **Figure 26.30 Herpesvirus and *Trichomonas*.** (a) A severe case of herpes blisters on the face due to infection with herpes simplex virus type 1. (b) Genital herpes due to infection on the penis with herpes simplex virus type 2. (c) Cells of the flagellated protozoan *Trichomonas vaginalis*.

tion between *genital* herpes infections and cervical cancer in females.

Genital herpes infections are presently incurable, although a limited number of drugs have been successful in controlling the infectious blister stages. The guanine analog acyclovir (Figure 26.31●), given orally and also applied topically, is particularly effective in limiting the shed of active virus from blisters and promoting the healing of blistering lesions. Acyclovir specifically interferes with herpesvirus DNA polymerase, inhibiting viral DNA replication.

Trichomoniasis

Nongonococcal urethritis may also be caused by infections with the protozoan *Trichomonas vaginalis* (Figure 26.30c). The biology of *Trichomonas* is discussed in Sections 14.2, 14.9, and 14.10. *T. vaginalis* does not produce the resting cells or *cysts* important to the life cycle of many protozoans. As a result, transmission is usually from person to person, generally by sexual intercourse. However, cells of *T. vaginalis* can survive for 1–2 hours on moist surfaces, 30–40 minutes in water, and up to 24 hours in urine or semen. Thus, transmission of *T. vaginalis* by contaminated toilet seats, sauna benches, and paper towels occasionally occurs. *T. vaginalis* infects the vagina in women, the prostate and seminal vesicles of men, and the urethra of both males and females.

Many cases of trichomoniasis are totally asymptomatic in males. In women trichomoniasis is characterized by a vaginal discharge, vaginitis, and painful urination. The infection is more common in females; surveys indicate that 25–50% of sexually active women are infected, while only about 5% of men are infected because of the killing action of prostatic fluids. The male partner of an infected female should be examined for *T. vaginalis* and treated if necessary because promiscuous asymptomatic males can transmit the infection to several females. Trichomoniasis is diagnosed by observation of the motile protozoan in a wet mount of fluid discharged from the patient (Figure 26.30c). The antiprotozoal drug *metronidazole* is particularly effective in treating trichomoniasis (Table 26.3).

 26.13 Concept Check

Chlamydia, the most prevalent STI, is caused by infection with the bacterium *Chlamydia trachomatis*. Untreated chlamydial nongonococcal urethritis causes serious complications in males and females. Herpes lesions can also be transmitted sexually and are caused by herpes simplex virus type 1 and herpes simplex virus type 2. HSV-2 is generally associated with sexual transmission and infection of the anogenital regions. *Trichomonas vaginalis* is a protozoan responsible for trichomoniasis, another STI. In general, these STIs are widespread and are more difficult to diagnose and treat than gonorrhea or syphilis. There is no cure for herpes.

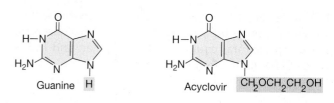

● **Figure 26.31 Guanine and the guanine analog acyclovir.** Acyclovir has been used therapeutically to control genital herpes (HSV-2) blisters.

◆ Describe pertinent clinical features and treatment protocols for chlamydia, herpes, and trichomoniasis.

◆ Why are these diseases more difficult to diagnose than gonorrhea or syphilis?

26.14 Acquired Immunodeficiency Syndrome: AIDS and HIV

Acquired immunodeficiency syndrome (AIDS) was recognized as a distinct disease in 1981. More than 900,000 cases of AIDS have been reported since then in the United States alone, and more than 500,000 people have died (⌀ Section 25.6). Over 1.4 million in the United States have been infected with HIV and about 1 million are living with the infection. Worldwide, the outlook is even more serious, with more than 50 million people already infected with the human immunodeficiency virus (HIV), the causative agent of AIDS. Over 16 million people have already died from AIDS.

HIV is divided into two major types, HIV-1 and HIV-2. HIV-1 is genetically similar, but distinct from HIV-2. HIV-2, discovered in West Africa in 1985, has reduced virulence as compared with HIV-1, and causes a milder AIDS-like disease. Currently, more than 99% of global AIDS cases are due to HIV-1, and our discussion of AIDS will center on infections with HIV-1.

AIDS is caused by the human immunodeficiency virus (HIV-1), a retrovirus (⌀ Section 16.15). The genome contains 9749 nucleotides in each of its two identical single-stranded RNA genomes. The *reverse transcriptase* enzyme present in the intact virion catalyzes formation of a complementary single-stranded DNA molecule using RNA as a template. The enzyme then converts the complementary DNA (cDNA) into double-stranded DNA, which can integrate into the host cell genome.

The numbers of HIV-infected individuals will continue to rise unless effective treatment or prevention methods are discovered. In the United States, the numbers of newly diagnosed HIV infections are increasing, and declines in HIV morbidity and mortality attributable to combination antiretroviral therapy have ended, reversing the trends of decreasing AIDS prevalence and severity seen in the 1990s. We have already discussed the epidemiology of AIDS (⌀ Section 25.6) and the clinical diagnostic methods for identifying and tracking

HIV infection (◠◠ Sections 24.10–24.11). In this section, we will focus on the pathogenesis of AIDS, concentrating on the natural course of HIV infection and the effects of HIV on the immune system.

A Definition of AIDS

AIDS was first suspected of being a disease affecting the immune system because a large number of opportunistic infections were observed in certain populations (◠◠ Section 25.6). **Opportunistic infections** are infections usually observed only in individuals with a dysfunctional immune system. This definition was adopted in 1993 by the Centers for Disease Control and Prevention (Atlanta, Georgia, USA) and is used to define AIDS in the United States.

The current case definition for acquired immunodeficiency syndrome (AIDS) includes those who (a) test positive for human immunodeficiency virus (HIV) and (b) meet one of the two following criteria.

1. They have a CD4 T-cell number of less than $200/mm^3$ of whole blood (the normal count is $600–1000/mm^3$) or a CD4 T-cell/total lymphocytes percentage of less than 14%; or

2. They may have a CD4 T-cell number of more than $200/mm^3$ *and* any of the following conditions: fungal diseases including candidiasis (Figure 26.32a●), coccidiomycosis, cryptococcosis (Figure 26.32b), histoplasmosis (Figure 26.32c), isosporiasis, *Pneumocystis carinii* pneumonia (Figure 26.32d), cryptosporidiosis (Figure 26.32e), or toxoplasmosis (Figure 26.32f) of the brain; bacterial diseases including pulmonary tuberculosis or other *Mycobacterium* spp. infections (Figure 26.32g), or recurrent *Salmonella* septicemia; viral diseases including cytomegalovirus infection, HIV-related encephalopathy, HIV wasting syndrome, chronic ulcers, or bronchitis due to *herpes simplex* (◠◠ Section 16.12); or progressive multifocal leukoencephalopathy, malignant diseases such as invasive cervical cancer, Kaposi's sarcoma (Figure 26.33●), Burkitt's lymphoma, primary lymphoma of the brain, or immunoblastic lymphoma; recurrent pneumonia due to any agent.

The most common opportunistic infection in AIDS patients is by the protozoan *Pneumocystis carinii* pneumonia. A frequent non-microbial disease in AIDS patients is Kaposi's sarcoma, an atypical cancer observed in high frequency in HIV-infected homosexual males. Kaposi's sarcoma is a cancer of the cells lining the blood vessels and is

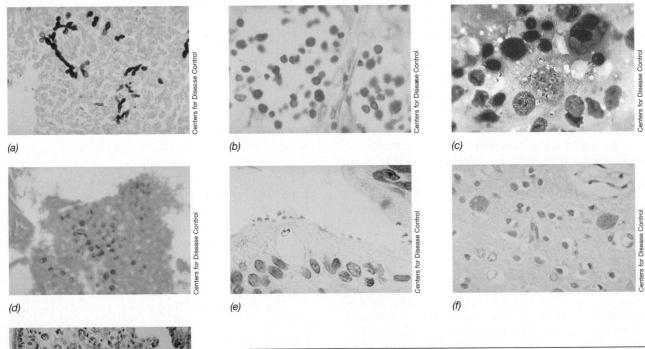

(a) *(b)* *(c)*

(d) *(e)* *(f)*

(g)

● **Figure 26.32 Opportunistic pathogens associated with cases of acquired immunodeficiency syndrome (AIDS).** (a) *Candida albicans*, from heart tissue of patient with systemic *Candida* infection. (b) *Cryptococcus neoformans*, from liver tissue of a patient with cryptococcosis. (c) *Histoplasma capsulatum*, from liver tissue of patient with histoplasmosis. (d) *Pneumocystis carinii*, from patient with pulmonary pneumocytosis. (e) *Cryptosporidium* sp. from small intestine of a patient with cryptosporidiosis. (f) *Toxoplasma gondii*, from brain tissue of patient with toxoplasmosis. (g) *Mycobacterium* spp. infection of small bowel, acid-fast stain.

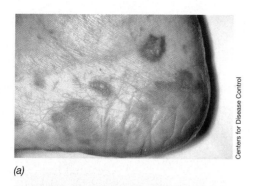

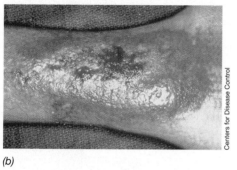

(a)

(b)

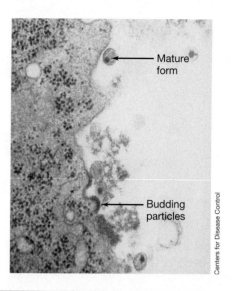

Mature form

Budding particles

● **Figure 26.34 Transmission electron micrograph of a thin section of a lymphocyte releasing human immunodeficiency virus (HIV).** Cells were from a hemophiliac patient who developed AIDS. HIV particles are 90–120 nm in diameter. Compare with Figure 9.25.

● **Figure 26.33 Kaposi's sarcoma.** Lesions are shown as they appear on (a) the heel and lateral foot, and (b) the distal leg and ankle.

characterized by purple patches on the surface of the skin, especially in the extremities (Figure 26.33). Kaposi's sarcoma is caused by coinfection of HIV and human herpesvirus 8 (HHV-8) and is 20,000-fold more prevalent in AIDS patients than in the general population.

In summary, an individual has AIDS if he or she (1) tests positive for HIV or HIV antibodies *and* (2) has a drastically reduced T-helper lymphocyte count, *or* (3) has at least one of a number of opportunistic infections or atypical cancers.

HIV Pathogenesis

HIV has the ability to infect cells displaying the CD4 cell-surface protein. The two cell types most commonly infected are macrophages and T-helper (T$_H$) cells, both of which are important components of the immune system (⚭ Sections 22.1 and 22.8). Infected macrophages and T cells produce and release large numbers of HIV particles, which in turn infect other cells that display CD4 (Figure 26.34●).

In addition to CD4, HIV must interact with coreceptors on target cells. HIV infection normally occurs first in macrophages, a type of antigen-presenting cell (APC) that has a very low level of CD4 on its surface (Figure 26.35●). At the cell surface, the macrophage CD4 molecule binds to the gp120 protein of HIV. The viral gp120 protein then interacts with another macrophage protein, the membrane-spanning chemokine receptor CCR5 (⚭ Section 23.11). CCR5 is a coreceptor for HIV and, together with CD4, forms the docking site

where the HIV envelope fuses with the host cell membrane; this allows the insertion of the viral nucleocapsid into the cell. The CCR5 coreceptor is required for HIV binding to macrophages. Individuals who express a variant CCR5 protein do not bind HIV and do not acquire HIV infections. After HIV has infected the macrophage APCs, a different form of gp120 is made, which in turn binds to a different coreceptor, the CXCR4 chemokine receptor on T cells.

HIV then enters and destroys the CD4 T-helper lymphocytes, the T$_H$1 and T$_H$2 cells that are responsible for cell-mediated inflammatory responses and B-cell help, respectively (⚭ Section 22.8). Thus, HIV starts as a macrophage infection and progresses to a T-cell infection. The result of HIV infection is the systematic destruction of macrophages and T cells, leading to a catastrophic breakdown of immunity. Specific knowledge of the coreceptors involved in HIV infection in macrophages and T cells may be used to design HIV-specific chemokine-receptor blocking agents to prevent the attachment of HIV to either the macrophage or the T cell, thus preventing infection.

In cases of AIDS, CD4 lymphocytes are greatly reduced in number. However, HIV infection does not immediately kill the host cell. HIV can exist as a provirus and not an infectious virion after reverse transcription to produce DNA from the RNA genome. The reverse-transcribed HIV genome is integrated as viral cDNA into host chromosomal DNA, The cell may show no outward sign of infection, and HIV DNA can remain in this latent state for long periods, replicating as the host DNA replicates. Eventually, virus synthesis occurs and new HIV particles are produced and released from the cell. T cells producing HIV no longer divide and eventually die.

Web Tutorial 26.1 **HIV Replication**

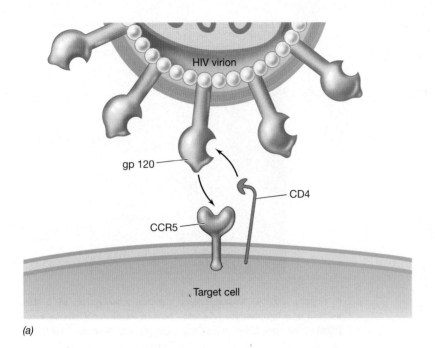

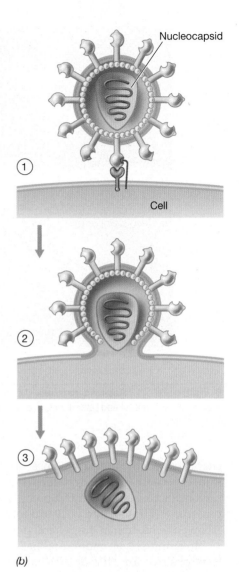

● **Figure 26.35 Infection of a CD4 target cell with the human immunodeficiency virus (HIV).** (a) Interaction of HIV with a target cell through specific binding of HIV gp120 to the CD4 receptor and a coreceptor on target cells such as macrophages or T cells. The coreceptor shown is the membrane-spanning chemokine receptor CCR5 found on macrophages. A similar coreceptor, CXCR4, is found on CD4 T cells. (b) Fusion of the HIV envelope with the host cell begins with ① the interaction of the virus with a receptor/coreceptor pair on the host cell. Cells lacking either the receptor or the coreceptor cannot bind HIV and are not infected. Following binding at the cell surface, ② the viral envelope and host cell membrane coalesce. Finally, ③ the nucleocapsid is inserted into the host cell, beginning the active viral infection. For coverage of the biology and replication of retroviruses such as HIV, see Sections 9.13 and 16.15 and Figures 9.25, 9.26, 16.25, and 16.26.

Accelerated destruction of CD4 cells occurs following the processing of HIV antigens by infected T cells. Such cells insert molecules of gp120 from HIV particles into their cell surfaces. The embedded gp120 protein on the infected cells then sticks to uninfected T cells by binding to the CD4 molecule. The infected and uninfected cells can then fuse to produce multinucleate giant cells called *syncytia*. One HIV-infected T cell may eventually bind and fuse with up to 50 uninfected T cells. Shortly after syncytia formation occurs, the fused cells lose immune function and die.

The HIV infection results in a progressive decline in CD4 cell number. In a normal human, CD4 cells constitute about 70% of the total T-cell pool; in AIDS patients, the number of CD4 cells steadily decreases, and by the time opportunistic infections become established, CD4 cells may be almost absent (☜ Figure 24.22 and Figure 26.36●).

The dramatic reduction of CD4 T cells has serious health consequences in AIDS patients. As CD4 cells decline in number, cytokine production falls, leading to the gradual reduction of uninfected T cells as well as all other lymphocytes. In the process, both the humoral and cellular immune system in AIDS patients are effectively destroyed. Systemic infections by fungi and mycobacteria (Figure 26.30) point to a loss in T_H1 cellular immunity (☜ Section 22.8). Opportunistic viral and bacterial infections associated with AIDS indicate a decline in antibody production due to the loss of the T_H2 cells necessary to stimulate antibody production by B cells (☜ Section 22.8).

The progression of untreated HIV infection to AIDS follows a predictable pattern. During the clinical latency period, a very active infection is in progress. First, there is an intense immune response to HIV: About 1 billion virions are destroyed each day and HIV numbers drop. However, this means that HIV is replicating at a very high rate, and this replication results in the corresponding destruction of about 100 million CD4 T cells each day. Eventually, the immune response is simply overwhelmed, HIV levels increase, and the T cells are completely destroyed, crippling the immune response and

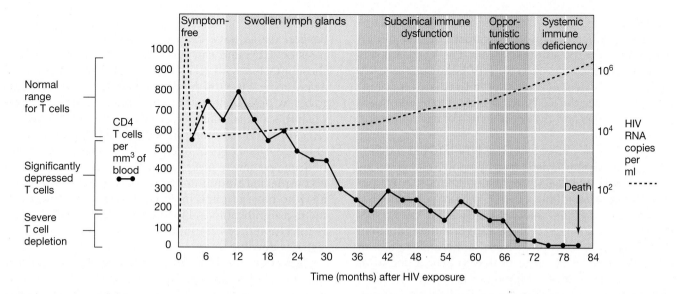

● **Figure 26.36** **Decline of CD4 T lymphocytes and progress of HIV infection.** During the typical progression of untreated AIDS, there is a gradual loss in the number and functional ability of the CD4 T cells, while the *viral load*, measured as HIV-specific RNA copies per milliliter of blood, gradually increases after an initial decline.

allowing the emergence of opportunistic infections. The example in Figure 26.36 documents T-cell destruction and the increase in HIV over a typical time course for untreated HIV infection culminating in AIDS.

Diagnosis

HIV infection can be diagnosed by immunological means. Both radioimmunoassay (RIA) and enzyme-linked immunosorbent assay (ELISA) tests (◌◌ Section 24.10) have been developed for screening blood samples to detect anti-HIV antibodies. The ELISA test has proven particularly valuable for large-scale screening of donated blood to prevent transfusion-associated HIV. Statistics have shown that about 0.25% (2–3 per 1000) of all blood donated by volunteer donors in the United States tests HIV-positive in the ELISA assay. A positive HIV-ELISA test must be confirmed by a second procedure called an immunoblot (Western blot), a technique that combines the analytical tools of protein purification and immunology (◌◌ Section 24.11).

A number of other rapid tests are being developed and marketed to identify individuals infected with HIV. One test uses a single drop of patient blood and a single reagent. The reagent is a bioengineered antibody in which one binding site is directed to a red blood cell antigen, while the other site is directed to the gp41 HIV surface antigen. In a positive test, the bifunctional antibody cross-links the red blood cells to the HIV, resulting in a visible agglutination (◌◌ Section 24.8). In another test, saliva is used as a source of secretory antibody to HIV. The saliva is expelled onto a cartridge containing immobilized HIV antigens. A second antibody, reactive

with the bound antibody and conjugated to an enzyme, is then added. After addition of the enzyme substrate, a positive reaction shows a colored product, as in the ELISA methods (◌◌ Section 24.10). The rapid tests are designed to provide maximum convenience, speed (*minutes* instead of the hours or days required for ELISA or immunoblot), extended shelf life, portability, and ease of use and interpretation. However, in general, the rapid tests are not as sensitive or accurate as the standard HIV-ELISA and HIV-immunoblot tests (◌◌ Section 24.10 and 24.11).

All tests, no matter how sensitive or accurate, fail to detect HIV-positive individuals who have recently acquired the virus but have not yet made a detectable antibody response. This period may be more than 6 weeks after exposure to HIV. Despite this drawback, these tests ensure the general safety of the blood supply, and statistics indicate that the risk of contracting HIV through contaminated blood or blood products is now very low. Sexual promiscuity and group intravenous drug use are the major routes of HIV infection today (◌◌ Figure 25.8).

The presence of HIV antibodies in the blood is the usual indicator for HIV exposure and infection (◌◌ Sections 24.10 and 24.11). In addition, several laboratory tests have been developed, based on a virus-specific *reverse transcriptase-polymerase chain reaction (RT-PCR)* (◌◌ Section 18.5). These tests identify HIV RNA directly and quantitatively from blood samples (◌◌ Section 24.12). The RT-PCR estimates the number of viruses present in the blood, or the **viral load**. The RT-PCR test indicates the magnitude of HIV replication and correlates with the rate of CD4 T cell destruction. The CD4 test indicates the extent of HIV-induced immune damage already suffered, a direct indicator of the magnitude of destruction of the immune system. The RT-PCR test is not routinely used to screen for HIV because it is costly and technically demanding. After initial discovery of infection, however, the RT-PCR test is used to monitor progression of AIDS and the effectiveness of chemotherapy (Figure 26.37●).

The prognosis of an *untreated* HIV-infected individual is poor. Opportunistic pathogens or malignancies (Figures 26.32 and 26.33) eventually kill most AIDS patients. Long-term studies of AIDS patients indicate that the average person infected with HIV progresses through several stages of decreasing immune function, with CD4 cells dropping from a normal range of 600–1000/mm³ of blood to near zero over a period of 5–7 years (Figure 26.36). Although the *rate* of decline in immune function varies from one HIV-infected individual to another, it is rare for an HIV-positive individual to live for more than 10 years without some form of chemotherapeutic intervention.

Treatment

Several drugs have been identified that delay symptoms of AIDS and can significantly prolong the life of those infected with HIV (Table 26.4). Therapy is aimed at reducing the viral load of HIV-infected individuals to below detectable levels. The strategy used to accomplish this aim is called *highly active antiretroviral therapy (HAART)* and is accomplished by giving two or more antiretroviral drugs at once to inhibit the development of drug resistance. A *cure* for HIV infection has not been accomplished through drug therapy. Even in individuals who have no detectable viral load after drug treatment, a significant viral load returns if HAART therapy is interrupted or discontinued, or if multiple drug resistance develops.

Effective anti-HIV drugs fall into four categories. The first of these are a group of *nucleoside analogs* that act as *reverse transcriptase inhibitors*. *Reverse transcriptase* is the enzyme that converts the single-stranded RNA genetic information into complementary DNA (∞ Section 16.15). The oldest effective anti-HIV drug, *azido-*

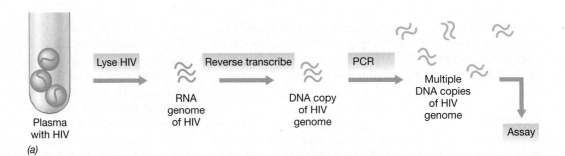

● **Figure 26.37 Monitoring of HIV load.** (a) Detection of HIV by the RT-PCR test (reverse transcription-polymerase chain reaction). The HIV copies obtained are compared quantitatively with DNA copies from a control template that is amplified in the same PCR amplification. HIV load is expressed as the number of HIV copies per milliliter of patient plasma. For a discussion of reverse transcription and the PCR technique, see Sections 16.15 and 7.9, respectively. (b) Time course for HIV infection as monitored by HIV RT-PCR. Progression of infection is estimated based on viral load at successive times after infection. CD4 T cell counts are measured in cells per cubic millimeter (see Figure 24.22). In the upper panel, a viral load of greater than 10⁴ copies per milliliter correlates with below normal CD4 cell numbers (normal = 600–1500/mm³), indicating a poor prognosis and early death of the patient. In the lower panel, a viral load of less than 10⁴ copies per milliliter correlates with normal CD4 cell numbers, indicating a good prognosis and extended survival of the patient. This test is used to monitor the course of HIV infection and is particularly useful for tracking the efficacy of drug treatment protocols. Data are adapted from the Centers for Disease Control and Prevention, Atlanta, GA, USA.

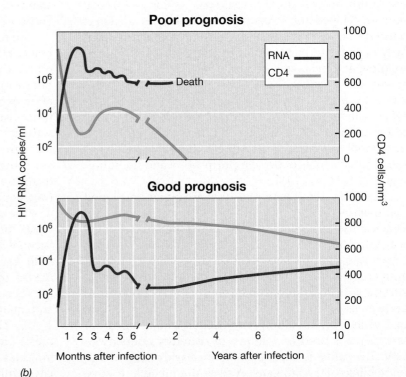

Table 26.4	Chemotherapeutic agents approved for HIV/AIDS treatment
Drug	**Mechanism of activity**
Azidothymidine (AZT, ZDV, or Zidovudine) (Figure 26.38a)	*Nucleoside analog;* reverse transcriptase inhibitor; nucleotide chain synthesis terminator; increases survival time and reduces incidence of opportunistic infection in AIDS patients; toxic to bone marrow cells; may be used in combination with other drugs in multiple drug treatment protocols.
Dideoxycytidine (ddC or zalcitabine) Dideoxyinosine (ddI or didanosine) Stavudine (d4T) Lamivudine (3TC)	*Nucleoside analogs;* reverse transcriptase inhibitors; mechanism of action and effects are the same as AZT; may have less toxicity than AZT in some patients; may be used in combination with other drugs in multiple drug treatment protocols.
Efavirenz Nevirapine (Figure 26.38b) Delavirdine	*Nonnucleoside reverse transcriptase inhibitors* (NNRTIs); bind directly to reverse transcriptase and disrupt the catalytic site; do not compete with nucleosides; may be used in combination with other drugs in multiple-drug treatment protocols.
Indinavir Nelfinavir Saquinavir (Figure 26.38c)	*Protease inhibitors;* computer-designed peptide analogs designed to bind to the active site of HIV protease, inhibiting processing of viral polypeptides and virus maturation; may be used in combination with other drugs in multiple drug treatment protocols.
Enfuvirtide	*Fusion inhibitor;* synthetic polypeptide that binds to the gp41 protein and inhibits the fusion of HIV membranes with host cell membranes.

thymidine (AZT), is an effective inhibitor of HIV replication because it closely resembles the nucleoside thymidine but lacks the correct attachment site for the next base in a replicating nucleotide chain, resulting in termination of the growing DNA chain. Thus, AZT and the other nucleoside analogs are **nucleoside reverse transcriptase inhibitors (NRTIs)**, and they stop HIV replication (Figure 26.38a●, Table 26.4). When these drugs are given to HIV-infected individuals, the result is a rapid decrease in viral load. However, within several weeks, drug resistant strains of HIV arise in each patient as a result of mutation and selection. Although this process seems very rapid, HIV replicates very quickly (see above) and as little as four single-nucleotide mutations result in resistance to a given nucleoside analog.

The second category of anti-HIV drugs are the **nonnucleoside reverse transcriptase inhibitors (NNRTIs)** (Figure 26.38b; Table 26.4). These compounds directly inhibit the action of reverse transcriptase by interacting with the protein and altering the conformation of the catalytic site. Unfortunately, a single mutation in the reverse transcriptase gene is often sufficient to reduce the effectiveness of these drugs.

Another category of anti-HIV drugs are the **protease inhibitors** (Figure 26.38c; Table 26.4). The protease inhibitors are computer-designed peptide analogs that inhibit processing of viral polypeptides by binding to the active site of the processing enzyme, HIV protease (∞ Section 20.13). This action inhibits virus maturation. However, as with the other enzyme-targeted chemotherapy strategy (reverse transcriptase; see above), a single mutation in the HIV protease gene is capable of rendering these drugs nonfunctional.

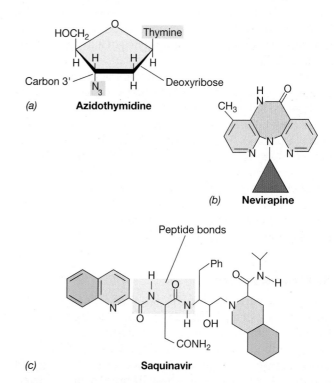

● **Figure 26.38 Examples of HIV/AIDS chemotherapeutic drugs.** (a) Azidothymidine (AZT), a nucleoside analog. The missing OH group on the 3′ carbon causes nucleotide chain elongation to terminate when this analog is incorporated, inhibiting virus replication. (b) Nevirapine, a non-nucleoside reverse transcriptase inhibitor, binds directly to the catalytic site, also inhibiting elongation of the nucleotide chain. (c) Saquinavir, a protease inhibitor, was designed by computer modeling to fit the active site of the HIV protease (∞ Section 20.13). Saquinavir is a peptide analog: The tan highlighted area shows the region analogous to peptide bonds. Blocking the activity of HIV protease prevents the processing of HIV proteins and maturation of the virus (∞ Figure 20.28).

A final category of anti-HIV drugs is represented by a single drug, *enfuvirtide*, a **fusion inhibitor** composed of a 36–amino acid synthetic peptide that acts by binding to the gp41 membrane protein of HIV (Table 26.4). Binding of the protein stops the conformational changes necessary for the fusion of viral and CD4 cell membranes. Enfuvirtide is prescribed to HIV-positive individuals who have developed antiretroviral drug resistance or increased viral load after conventional HAART therapy.

Because of the problems with drug resistance, a typical recommended protocol for treatment of an individual with established HIV infection includes at least one protease inhibitor *plus* a combination of two nucleoside analogs (Table 26.4). Multiple drug therapy reduces the possibility that a drug-resistant virus could emerge because the virus would need to develop resistance to three drugs simultaneously. This combination therapy is then monitored by RT-PCR to track changes in viral load. An effective protocol reduces viral load to nondetectable levels (less than 500 copies of HIV per milliliter of blood) within several days (Figure 26.37). The therapy is continued and monitored for viral load indefinitely. If the viral load again reaches detectable limits, the drug cocktail is changed because an increase in viral load indicates the emergence of a drug-resistant strain of HIV.

In addition to drug resistance, some of the antiviral drugs are toxic to the host. In many cases, nucleoside analogs are not well tolerated by patients, presumably because they interfere with host functions such as cell division (Table 26.4). In general, the NNRTIs and the protease inhibitors are better tolerated because they interfere with only virus-specific functions. However, drug resistance coupled with host toxicity are major problems in HIV therapy, and new chemotherapeutic agents and drug protocols are constantly being developed and tailored to the needs of individual patients.

AIDS Immunization

The genetic variability of HIV has thus far hampered the development of an AIDS vaccine. One vaccine strategy is to induce antibodies to the envelope protein, gp120. These antibodies would then block CD4-gp120 interactions (Figure 26.35) and thus block infection. A gp120 subunit vaccine is now in phase III human clinical trials, the step before general public use. The vaccine consists of two gp120 variants known to be predominant in the local population in Thailand, where it is being tested. Unfortunately, this approach probably will not be successful for a universal vaccine because the gene encoding gp120 mutates frequently, forming antigenic variants of the protein that are not recognized by antibodies made to a different form.

Clinical immunization trials are proceeding with *subunit vaccines* (⊂⊃ Section 31.8). In these vaccines, genes for several HIV envelope proteins have been engineered into vaccinia virus or adenovirus particles. Using these harmless viruses as expression vectors and vehicles for delivery of HIV antigens, several vaccines elicit a potent humoral and cellular immune response to HIV.

Other potential immunization candidates include *killed intact HIV*. These inactivated vaccines are restricted to use in HIV-infected individuals because inactivation procedures may not deactivate 100% of the HIV; it would be unethical to expose uninfected individuals to even a small risk of HIV infection. Next, some laboratories are exploring the possibilities of producing *live attenuated virus* for use as an immunizing agent. This strategy is bolstered by the finding that individuals infected with HIV-2, a related virus that causes a milder form of AIDS with a very long latent period, prevents infection with HIV-1, the strain responsible for severe AIDS. However, there are serious potential risks. For example, integrated virus could cause cancer, mutations might reactivate virulence, and so on.

In all, there are about 20 different AIDS immunization protocols undergoing development, but no immunization has yet been found to be protective and safe. Additionally, AIDS immunization would most probably not be useful for *treating* most patients that already have the disease because these individuals lack significant immune function and could not respond to a vaccine. Thus, despite considerable knowledge of the structural and functional mechanisms of HIV infection and a clear understanding of the AIDS disease process, there is no proven medical intervention strategy for the prevention or cure of HIV/AIDS. Public education and avoidance of high risk behavior remain the major tools used to prevent HIV/AIDS (see the Microbial Sidebar, Sexual Activity and AIDS).

 26.14 Concept Check

AIDS is now one of the most prevalent infectious diseases in the human population. HIV destroys the immune system, and opportunistic pathogens then kill the host. There is still no effective vaccine for HIV. However, several antiviral drugs slow the progress of AIDS. The only prevention for the spread of HIV infection is through avoidance of behavior such as intravenous drug use (needle sharing) and unsafe sexual practices.

◆ Review the definition of AIDS. What diagnostic features are shared by all AIDS patients?

◆ What is the current treatment for HIV infection? Is it effective? What are the side effects?

Microbial Sidebar ◆ Sexual Activity and AIDS

Sexual promiscuity has always been associated with sexually transmitted diseases, but the acquired immunodeficiency syndrome (AIDS) epidemic discussed in this chapter and elsewhere in this book has focused attention on the dangers of multiple sex partners and on the high risk associated with certain sex practices. HIV/AIDS is only one type of sexually transmitted disease. Others include gonorrhea, syphilis, herpes simplex, nonspecific urethritis (caused by *Chlamydia*), protozoal vaginitis (caused by *Trichomonas vaginalis*), fungal vaginitis (caused by *Candida albicans*), and venereal warts (caused by the human papilloma virus). Some of these sexually transmitted diseases have been associated with human society for all of recorded history. However, AIDS is unique. There are no drugs or immunizations to cure or prevent AIDS. Drugs used for treatment are costly and available only in developed countries, so 90% of the 50 million or more HIV-infected individuals do not have access to therapy. AIDS kills more than 2 million people every year, and this number will grow as people who have harbored and spread HIV for up to 10 years develop full-blown AIDS.

Because AIDS is linked to certain sex practices, prevention means avoidance of these sex practices. The United States Surgeon General has issued a report that makes specific recommendations that individuals can follow if they wish to reduce the likelihood of AIDS infection. Among the recommendations are

1. Avoid mouth contact with penis, vagina, or rectum.
2. Avoid all sexual activities that could cause cuts or tears in the linings of the rectum, vagina, or penis.
3. Avoid sexual activities with individuals from high-risk groups. These include prostitutes (both male and female), promiscuous homosexual men, bisexual individuals, and intravenous drug users.
4. If a person has had sex with a member of one of the high-risk groups, a blood test should be done to determine if infection with HIV has occurred. Keep in mind that the blood test should be repeated at intervals for a year or more, because of the lag time in the immune response. If the test is positive, the sexual partners of the HIV-positive individual must be protected by use of a condom during sexual intercourse.

AIDS is not just a disease of male homosexuals. In certain cultures, AIDS is as common in women as in men, and in the United States the fastest growing method of transmission is between heterosexual partners. The disease is linked to promiscuous sexual activities and other activities that involve exchange of body fluids, which include not only male homosexuality but also female prostitution and intravenous drug use.

Is it possible to have sex without incurring the risk of AIDS? Certain sex practices are inherently much safer than others. Safe sex practices include dry kissing (mouths closed), mutual masturbation (in the absence of breaks in the skin), and intercourse protected by a condom. Dangerous sex practices include wet kissing (mouths open), masturbation where breaks in the skin occur, oral sex (either male or female), and unprotected sexual intercourse (either vaginal or anal). The U.S. Surgeon General has recommended that if the health status of the partner is unknown, a condom be used for all sex practices in which exchange of body fluids occurs.

The AIDS epidemic has focused new attention on the condom (see photo). Condoms have always played two roles in sexual activity: disease protection and prevention of pregnancy. Although the best way to avoid AIDS is to avoid dangerous sex practices, a latex condom should be used when having sexual intercourse with any individual whose infection status is unknown. The U.S. Surgeon General strongly recommends the use of condoms for all extramarital sexual activity. In certain countries, widespread advertising campaigns very explicitly promote the use of condoms.

Statements and programs advocating monogamy, abstinence, avoidance of sexual activity outside of matrimony, and even warnings about the potential fatal consequences of HIV infection will not by themselves control the AIDS epidemic. Studies of other sexually transmitted diseases indicate that fear of disease is not sufficient to prevent risky sexual activities. The sex drive in some individuals is so strong that it will suppress the fear of even a disease like AIDS. Every individual must therefore take the responsibility for protecting himself or herself from infection.

For more information about AIDS, the Public Health Service has a toll-free telephone resource, the CDC National STD and AIDS Hotline at 800-342-2437. The Website is at *http://www.ashastd.org/nah/.* ∎

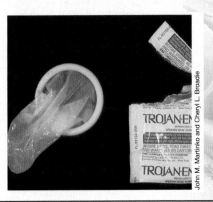

REVIEW QUESTIONS

1. Why do gram-*positive* bacteria cause respiratory diseases more frequently than gram-*negative* bacteria (Section 26.1)?
2. What are the typical symptoms of a streptococcal respiratory infection? Why should streptococcal infections be treated promptly (Section 26.2)?
3. Describe the causal agent and the symptoms of diphtheria. Are the disease symptoms due to infection or to toxicity? Explain. How is diphtheria prevented (Section 26.3)?
4. Describe the causal agent and the symptoms of whooping cough. Describe the changes in vaccine technology that have led to safer pertussis vaccines (Section 26.4).

5. Describe the process of infection by *Mycobacterium tuberculosis*. Does infection always lead to active tuberculosis? Why or why not? Why are individuals in the United States not vaccinated with the BCG vaccine (⚭ Section 26.5)?

6. Describe the symptoms of meningococcemia and meningitis. How are these diseases treated? What is the prognosis for each (⚭ Section 26.6)?

7. Compare and contrast measles, mumps, and rubella. Include in your discussion a description of the pathogen, major symptoms encountered, and any potential consequences of these infections. Why is it important that women be vaccinated against rubella *before* puberty (⚭ Section 26.7)?

8. Why are colds and influenza such common respiratory diseases? Give at least two reasons for the high incidence of each of these diseases (⚭ Section 26.8).

9. Distinguish between pathogenic staphylococci and those that are part of the normal flora (⚭ Section 26.9).

10. Describe the evidence linking *Helicobacter pylori* to gastric ulcers. How would you treat an ulcer patient (⚭ Section 26.10)?

11. Describe the major pathogenic hepatitis viruses. How are they related to one another? How are they spread (⚭ Section 26.11)?

12. Why did the incidence of gonorrhea rise dramatically in the mid-1960s, while the incidence of syphilis actually decreased at the same time (⚭ Section 26.12)?

13. For the sexually transmitted diseases chlamydia, herpes, and trichomoniasis, describe the methods of treating each infection. In each case, is the treatment an effective cure? Why or why not (⚭ Section 26.13)?

14. Describe how human immunodeficiency virus (HIV) effectively shuts down both humoral immunity and cell-mediated immunity. Why are vaccines to HIV so difficult to develop (⚭ Section 26.14)?

APPLICATION QUESTIONS

1. How can an epidemic of whooping cough be controlled? How can it be prevented? Since the incidence of this disease is no longer decreasing, apply your prevention methods to the current "mini-epidemic" (Section 26.4). Would your methods be cost-effective?

2. Why does tuberculosis often lead to a permanent reduction in lung capacity, whereas most other respiratory diseases cause only temporary respiratory problems?

3. Your college roommate goes home for the weekend, becomes extremely ill, and is diagnosed with bacterial meningitis at a local hospital. Because he was away, university officials are not aware of his illness. What should you do to protect yourself against meningitis? Should you notify university health officials?

4. Measles, mumps, and rubella were once very common childhood diseases. However, outbreaks of these diseases are now regarded as serious incidents requiring immediate attention from public health officials. Explain this shift in attitudes in the context of disease prevalence, availability of vaccines, and the potential health consequences of each disease.

5. Discuss the molecular biology of *antigenic shift* in influenza viruses and comment on the immunologic consequences for the host. Why does antigenic shift prevent the production of a single universally effective vaccine for influenza control? Next, compare antigenic *shift* to antigenic *drift*. Which mechanism is more important for the evolution of the influenza virus? Which causes the greatest antigenic change? Which creates the biggest problems for vaccine developers? Why?

6. Arrange the hepatitis viruses in order of disease severity, both in the short term and in the long term.

7. As the director of your dormitory's public health advisory group, you are charged to present information on chlamydia, herpes, and trichomoniasis, all STIs. Besides this textbook, where can you get reliable information about STIs? Present information on prevention, symptoms, and treatment for each STI. Will your program for each disease overlap? For each of the diseases, discuss the social, legal, and public health issues that must be considered for identifying the sexual partners of infected individuals.

27

ANIMAL-TRANSMITTED, ARTHROPOD-TRANSMITTED, AND SOILBORNE MICROBIAL DISEASES

The mosquito transmits malaria, the biggest infectious disease killer of all time, and West Nile virus, an emerging pathogen.

WORKING GLOSSARY

Ehrlichiosis one of a group of emerging tick-transmitted diseases caused by rickettsias of the *Ehrlichia* genus

Hantavirus pulmonary syndrome (HPS) an emerging acute viral disease characterized by respiratory pneumonia, obtained by transmission of hantavirus from rodents

Lyme disease an emerging tick-transmitted disease caused by the spirochete *Borrelia burgdorferi*

Malaria an insect-transmitted disease characterized by recurrent episodes of fever and anemia caused by the protozoan *Plasmodium* spp., usually transmitted between mammals through the bite of the *Anopheles* mosquito

Mycosis infection caused by a fungus

Plague an endemic disease in rodents caused by *Yersinia pestis* that is occasionally transferred to humans through the bite of a flea

Rabies a usually fatal neurological disease caused by the rabies virus that is usually transmitted by the bite or saliva of an infected animal

Rickettsia obligate intracellular parasite genus responsible for diseases including typhus, Rocky Mountain spotted fever, and ehrlichiosis

Rocky Mountain spotted fever a tick-transmitted disease caused by *Rickettsia rickettsii*, causing fever, headache, rash, and gastrointestinal symptoms

Sickle cell anemia a genetic trait that confers resistance to malaria but causes a reduction in the efficiency of red blood

cells by reducing the oxygen-binding affinity of hemoglobin

Thalasemia a genetic trait that confers resistance to malaria but causes a reduction in the efficiency of red blood cells by altering a red blood cell enzyme

Tetanus a disease involving rigid paralysis of the voluntary muscles, caused by an exotoxin produced by *Clostridium tetani*

Typhus a louse-transmitted disease caused by *Rickettsia prowazekii*, causing fever, headache, weakness, rash, and damage to the central nervous system and internal organs

West Nile fever a neurological disease caused by West Nile virus and transmitted by mosquito from birds to humans

Zoonosis an animal disease transmitted to humans

Through the ages, insect-transmitted diseases such as plague and malaria have killed millions, changing the course of human history and evolution. Today, vectorborne diseases such as Lyme disease, hantavirus pulmonary syndrome, and West Nile fever are key emerging diseases, even in highly developed regions of the world. Soilborne diseases also present major problems because organisms in soil cannot be effectively controlled. Tetanus, for example, while totally preventable, remains a serious, often life-threatening disease.

All of these are diseases with *nonhuman* reservoirs. In this chapter, we look at animal-transmitted, vectorborne, and soilborne pathogens and the diseases they cause. The natural host for the *animal-transmitted pathogens* is a nonhuman vertebrate. When infected animal populations come in contact with humans, the result is often human infection. Vectorborne pathogens are spread to new hosts via the bite of an arthropod vector that last fed on an infected host. Humans are often accidental hosts in the life cycle of these pathogens, but they may also act as a disease reservoir, as in the case of malaria. Soilborne pathogens include a variety of fungi, as well as the members of the genus *Clostridium*.

ANIMAL-TRANSMITTED DISEASES

A **zoonosis** is an animal disease transmissible to humans, generally by direct contact, aerosols, or bites. Immunization and veterinary care prevent most infectious diseases in domesticated animals, preventing much zoonotic dis-

ease transfer to humans. However, *feral* (wild) animals cannot be immunized, nor do they receive veterinary care. Diseases in these populations often occur on a periodic, cyclic basis. In the next sections, we will look at two important examples of animal-transmitted diseases—rabies and hantavirus pulmonary syndrome.

27.1 Rabies

Rabies is one of a handful of zoonotic diseases that occurs primarily in animals but is spread to humans under certain conditions. The major reservoir of rabies in the United States is in wild animals, primarily raccoons, skunks, coyotes, foxes, and bats. However, a small but significant number of cases are also seen in domestic animals (Figure 27.1●).

Epidemiology

Rabies is a major preventable infectious disease in humans worldwide. Nevertheless, about 52,000 people die every year from rabies, primarily in developing countries where it is endemic in domestic animals such as dogs. Annually, about 1 million people worldwide receive rabies treatment for animal bites.

Rabies is caused by a member of the Rhabdovirus family (a negative-stranded RNA virus) (∞ Section 16.9) that infects cells in the central nervous system of most warm-blooded animals, almost invariably leading to death if not treated. The virus, present in the saliva of rabid animals, enters the body through a wound from a bite or through contamination of mucous membranes by the infected saliva.

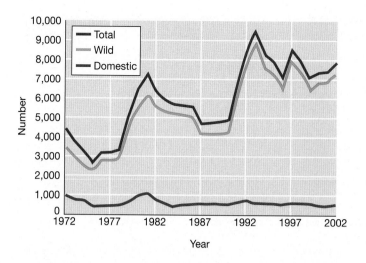

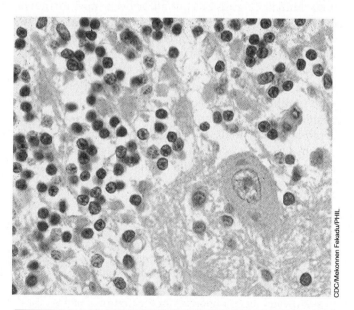

● **Figure 27.1 Rabies cases in wild and domestic animals in the United States.** Since 1990, there have been 30 human rabies cases and eight deaths. Rabies is endemic in wild animal populations, especially in raccoons in the eastern United States. The periodic rise in reported cases following several years of decline is the result of fluctuations in the host population due to deaths from rabies. High case numbers indicate a large infected population. The population is decimated by the infection, and for several years the lower host numbers prevent rabies spread, and case numbers are reduced. As the population recovers, rabies reemerges, case numbers increase, and the host population is again decimated, continuing the cycle. Over 500 cases of rabies occur annually in domestic animals, nearly all acquired from contact with wild animals. Data are from the Centers for Disease Control and Prevention, Atlanta, GA, USA.

Rabies virus multiplies at the site of inoculation and then travels to the central nervous system. The incubation period for the onset of symptoms is highly variable, depending on the animal, the size, location, and depth of the wound, and the number of viral particles transmitted in the bite. In dogs, the incubation period averages 10–14 days. In humans, nine months or more may pass before rabies symptoms become apparent.

The virus proliferates in the brain (especially in the thalamus and hypothalamus), leading to fever, excitation, dilation of the pupils, excessive salivation, and anxiety. A fear of swallowing (hydrophobia) develops from uncontrollable spasms of the throat muscles. Death eventually results from respiratory paralysis. In humans, an *untreated* rabies infection progressing to the symptomatic stage is nearly always fatal.

Diagnosis, Treatment, and Prevention

Rabies is diagnosed in the laboratory by examining tissue samples. Fluorescent antibody tests (⟳ Section 24.9) or immunoperoxidase tests using monoclonal antibodies that recognize rabies virus-infected brain or corneal tissue are used for confirming a clinical diagnosis of rabies, either in a potentially rabid animal or in postmortem examination of a human case. In addition, characteristic virus inclusion bodies in the cytoplasm of nerve cells,

called *Negri bodies* (Figure 27.2●), obtained by biopsy or postmortem sampling, are taken as confirmation of rabies. Reverse transcriptase–PCR testing and sequencing can also be done to confirm the presence a of particular rabies strain in a clinical specimen.

Because of the lethal nature of rabies, any contact with potentially rabid animals must be taken seriously. A wild animal suspected of being rabid should be captured, sacrificed, and immediately examined for evidence of rabies. If a domestic animal, generally a dog, cat, or ferret, bites a human, especially if the bite is unprovoked, the animal is typically held 10 days to check for clinical signs of rabies. If the animal is wild, exhibits rabies symptoms, or a determination cannot be made after 10 days, the patient will be passively immunized with rabies immune globulin (purified anti-rabies virus antibodies obtained from a hyperimmune individual) (⟳ Section 22.13), injected at both the site of the bite and intramuscularly. The patient will also be immunized with an inactivated rabies virus preparation. A summary of guidelines for treating possible human exposure to rabies is shown in Table 27.1. Because of the very slow progression of rabies in humans, this combination immune therapy is nearly 100% effective, stopping the onset of the active disease.

Rabies is prevented largely through immunization. Inactivated rabies vaccines are used in the United States for both human and domestic animal immunizations, and a variety of inactivated and attenuated virus preparations are also used worldwide. Because of the long incubation period, passive and active immunization of potentially exposed individuals is sufficient to prevent disease. Therefore, prophylactic rabies immunization is

● **Figure 27.2 Tissue section from the brain of a human rabies victim, stained with hematoxylin and eosin.** The rabies virus causes characteristic cytoplasmic inclusions known as Negri bodies. Here they are seen as dark-stained, sharply differentiated, roughly spherical masses of about 2–10 μm in diameter, distributed throughout the cytoplasm. The Negri bodies contain rabies virus antigens.

Table 27.1	**Guidelines for treating possible human exposure to rabies virus**

Unprovoked bite by a domestic animal

Animal suspected of rabies
1. Sacrifice animal and test for rabies.
2. Begin treatment of human immediately.[a]

Animal not suspected of rabies
1. Hold for 10 days. If no symptoms, do not treat human.
2. If symptoms develop, treat human immediately.[a]

Bite by wild carnivore (for example, skunk, bat, fox, raccoon, coyote)

Regard animal as rabid
1. Sacrifice animal and test for rabies.
2. Begin treatment of human immediately.[a]

Bite by wild rodent, squirrel, livestock, rabbit

Consult local or state public health officials about possible recent cases of rabies transmitted by these animals (these animals rarely transmit rabies). If no reports, do not treat human.

[a] All bites should be thoroughly cleansed with soap and water. Treatment is generally a combination of rabies immunoglobulin and human diploid cell rabies vaccine (five injections intramuscularly).

generally not recommended for humans, except for individuals at high risk, such as veterinarians and animal control personnel.

This rabies prevention strategy has been extremely successful, and fewer than three cases of human rabies are reported in the United States each year, nearly always the result of bites by wild animals. Since domestic animals often have exposure to wild animals, all dogs and cats should be vaccinated beginning at 3 months of age, and booster inoculations should be given yearly or triannually. Other domestic animals, including large farm animals, are also often immunized with rabies vaccines. However, the key to effective rabies prevention and possible eradication, at least in the United States, lies in control of the disease in the large rabies virus reservoir in wild animals (Figure 27.1). If all or even most members of the disease reservoir are immune, the disease can be stopped and possibly even eradicated. Currently, subunit vaccines consisting of rabies virus genes that encode rabies coat proteins expressed in vaccinia virus (∞ Section 31.8) are available. Because effective doses can be given orally, subunit vaccines can be included in food "baits" and used to immunize local populations of susceptible wild animals, reducing the incidence and spread of rabies. Such vaccines are a completely safe means of controlling rabies in the wild animal reservoir and could lead to eventual eradication of the disease.

27.1 Concept Check

Rabies occurs primarily in wild animals and is an important endemic zoonotic disease in developing countries. In the U.S. rabies can be transmitted from the wild animal reservoir to domestic animals or, very rarely, to humans. Vaccination of dogs and cats is important for the control of rabies.

◆ What is the procedure for treating a human bitten by an animal *if* the animal cannot be found?

◆ What advantages might an oral vaccine have over a parenteral vaccine for rabies control in wild animals?

27.2	**Hantavirus Pulmonary Syndrome**

Hantavirus is the cause of an acute respiratory syndrome, **hantavirus pulmonary syndrome** (HPS). Hantavirus is named for Hantaan, Korea, the site of a hemorrhagic fever outbreak where the virus was first recognized as a human pathogen. The 1993 hantavirus outbreak in the United States occurred in the Four Corners region of Arizona, Colorado, New Mexico, and Utah, and was eventually traced to the deer mouse (*Peromyscus maniculatis*) population in the area. The outbreak caused 32 deaths in 53 infected adults, and it underscored the potential danger of outbreaks due to diseases that are directly transmitted from a variety of different animal reservoirs, sometimes under new or unusual circumstances.

Hantaviruses are related to hemorrhagic fever viruses such as Lassa fever virus and Ebola virus (∞ Section 25.11), and all are occasionally transmitted to humans from animal reservoirs. The genus *Hantavirus* is a member of the Bunyaviridae, a family of enveloped single-stranded RNA viruses (∞ Section 16.8) (Figure 27.3●). The family includes a number of viruses that cause either *hantavirus pulmonary syndrome (HPS)* or *hemorrhagic fever with renal syndrome (HFRS)*.

Epidemiology

Hantaviruses are usually found in rodents, including mice and rats of several species, lemmings, and voles, and are occasionally found in other animals. The HFRS strains are more common in the Eastern Hemisphere, and have been implicated in a number of HFRS outbreaks in recent years. Up to 200,000 cases per year are recognized, chiefly in China, Korea, and Russia. The HPS strains are more prevalent in the Western Hemisphere, and continued investigation of the ecology of these viruses is likely to uncover a number of other pathogenic strains.

Hantaviruses are most commonly transmitted by inhalation of virus-contaminated rodent excreta. Hu-

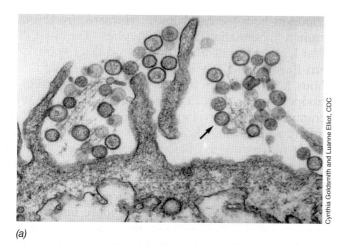

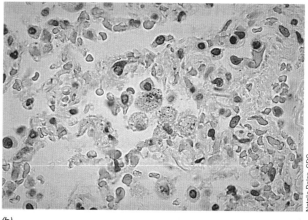

(a)   *(b)*

● **Figure 27.3 Hantavirus.** (a) An electron micrograph of the Sin Nombre hantavirus. The arrow indicates one of several virions. The virus is approximately 0.1 μm in diameter. (b) Immunostaining of Andes hantavirus antigens in alveolar macrophages. Each granular dark blue stained area indicates cellular infection of an individual macrophage (approximately 15 μm in diameter). Hantaviruses belong to the Bunyaviridae and, like many phylogenetically similar viruses (for example, Rift Valley virus and Ebola virus), cause hemorrhagic fevers with very high human mortality rates.

mans seem to be an accidental host and are infected only when they come into contact with rodents and their feces. For example, the individuals who acquired HPS in the Four Corners outbreak had been exposed to mice or their droppings, the result of a warm winter and an unusual increase in numbers of rodents in 1993 (∞ Section 25.11). The *aerosol route* of infection is most common, with most aerosols being in the form of dust generated from mouse droppings or dried urine. However, there are rare reports of person-to-person transmission, as well as a few incidents where HPS or HFRS was spread by a rodent bite.

Pathology, Diagnosis, Treatment, and Prevention

Hantavirus pulmonary syndrome is characterized by a sudden onset of fever, myalgia (muscle pain), thrombocytopenia (reduction in the number of blood platelets), leukocytosis (an increase in the number of circulating leukocytes), and pulmonary capillary leakage. Death occurs in a matter of several days in about 50% of cases, usually due to shock and cardiac complications precipitated by pulmonary edema (leakage of fluid into the lungs, causing suffocation). These symptoms are typical of the Sin Nombre hantavirus, which caused the Four Corners outbreak, but a variety of other symptoms may be evident, depending on the strain of virus causing the disease. For example, the Bayou strain common in rodents in the southeastern United States also causes kidney failure.

If hantavirus from candidate infections can be grown in tissue culture (∞ Section 9.3), the strain can be identified by serological techniques including a virus plaque-reduction neutralization assay. In this assay, patient serum is tested for antibodies that inhibit the formation of viral plaques in tissue culture (∞

Section 9.3). More commonly, ELISAs (enzyme-linked immunoassays) are performed on patient blood to identify antibodies, indicating exposure and an immune response (∞ Section 24.10). The presence of the viral genome, indicating infection, is detected based on a polymerase chain reaction (PCR) assay using patient tissue or blood specimens (∞ Section 24.12).

There is no virus-specific treatment or vaccine for any of the hantaviruses. However, the disease can be prevented by avoiding contact with rodents and rodent habitat, since the mode of transmission is usually through exposure to rodent excreta. Reduction in exposure can be accomplished by destroying mouse habitat, restricting food supplies (for example, keeping food in sealed containers), and aggressive rodent extermination measures. The long-term prognosis for disease eradication is poor because a considerable percentage of the rodent population in a given geographical area is generally infected with the local hantavirus strain. For example, retrospective serological testing of the deer mice in the Four Corners area in 1993 indicated that 30% of the local mouse population carried the Sin Nombre hantavirus.

 27.2 Concept Check

Hantaviruses occur worldwide in rodent populations and cause serious diseases such as hantavirus pulmonary syndrome (HPS) when accidentally spread to humans. In the United States, hantavirus infections were not recognized until 1993.

◆ Why were hantaviruses not recognized as a major public health problem until 1993 in the United States?

◆ Describe the spread of hantaviruses to humans. What are some effective measures for preventing infection by hantaviruses?

on polymerase chain reaction (PCR) tests of whole blood or serum to detect the presence of *Ehrlichia* DNA (∞ Section 24.12).

Ehrlichiosis is spread by the bites of infected ticks. The mammalian reservoirs include deer and possibly rodents, in addition to the human hosts. Retrospective serological analyses in areas with relatively high incidence of tickborne disease indicate that HGE may be a more prevalent disease than Rocky Mountain spotted fever. Many infections are not properly identified because of the variable nature of the symptoms. However, since 1998, ehrlichiosis has been a reportable disease in the United States (∞ Section 25.9). On average, about 500 cases are reported each year. HGE comprises about 60% of total cases while HME is responsible for 40%. Ehrlichiosis will be reported more frequently as physicians become more familiar with this emerging tickborne disease.

As with other tickborne illnesses, outdoor activities in tick-infested habitat are the major predisposing factors in acquiring ehrlichiosis. Golfers and hikers are particularly prone to infection. Prevention of ehrlichiosis involves reducing exposure to ticks and tick bites by avoiding tick habitat, wearing tick-proof clothing, and applying appropriate insect repellents such as those containing diethyl-*m*-toluamide (DEET). At the community level, tick densities can be successfully reduced through areawide application of *acaricides* (chemicals specifically toxic for ticks and related arthropods) and removal of tick habitat such as leaves and brush. Doxycycline, a semisynthetic tetracycline derivative, is the antibiotic of choice for the treatment of ehrlichiosis.

Other Rickettsial Diseases

Q fever is a pneumonia-like infection caused by an obligate intracellular parasite, *Coxiella burnetii*, related to the rickettsias (∞ Section 12.13). Although not transmitted to *humans* directly by an insect bite, the agent of Q fever is transmitted to *animals* by insect bites, and various arthropod species serve as a reservoir of infection. Domestic animals generally have inapparent infections, but may shed large quantities of *C. burnetii* cells in their urine, feces, milk, and other body fluids. Contact with the animals or animal products serves as a source of infection for humans. The resulting influenza-like illness may progress to include prolonged fever, headache, chills, chest pains, pneumonia, and endocarditis.

Laboratory diagnosis of infection with *C. burnetti* can be made by a variety of immunologic tests designed to measure host antibodies to the pathogen. An immunofluorescence antibody test (IFA) is the serological test of choice (∞ Section 24.9). *C. burnetii* infections respond to tetracycline, and therapy is usually begun quickly in any suspected human case of Q fever in order to prevent heart damage. Finally, Q fever is one of the infectious diseases that has been studied as a possible agent for biological warfare (∞ Section 25.12).

Scrub typhus, or *tsutsugamushi disease*, is restricted to Asia, the Indian subcontinent, and Australia, and is caused by *Orientia tsutsugamushi*. Although the disease is similar to typhus, it is transmitted by *mites* to its normal rodent hosts.

Diagnosis and Control

In the past, rickettsial infections have been difficult to diagnose because the characteristic rash associated with many rickettsial diseases may be mistaken for measles, scarlet fever, or adverse drug reactions. Clinical confirmation of rickettsial diseases has now been greatly aided by the introduction of specific immunological and molecular biology reagents. These include antibody-based tests that detect rickettsial surface antigens by latex bead agglutination assays, immunofluorescence assays, ELISA analyses (∞ Sections 24.8, 24.9, and 24.10), and PCR assays (∞ Section 24.12). Control of most rickettsial diseases requires control of the vectors: lice, fleas, and ticks. For humans traveling in wooded or grassy areas, the use of insect repellants containing diethyl-*m*-toluamide (DEET) usually prevents tick attachment. Firmly attached ticks should be removed gently with forceps, care being taken to remove all the mouth parts. A solvent such as ethanol applied to a tick with a saturated swab usually expedites removal. Although a vaccine is available for the prevention of typhus, the few cases reported do not warrant its general administration in the United States. No vaccines are currently available for the prevention of Rocky Mountain spotted fever or ehrlichiosis.

 27.3 Concept Check

Rickettsias are obligate intracellular parasitic *Bacteria* that are transmitted by arthropods. Most rickettsial infections are readily controlled by antibiotic therapy, but prompt recognition and diagnosis of these diseases is still difficult.

◆ What are the arthropod vectors and animal hosts for typhus, Rocky Mountain spotted fever, and ehrlichiosis?

◆ What precautions can be taken to prevent rickettsial infections?

27.4 Lyme Disease

Lyme disease is an emerging tickborne disease that affects humans and other animals. Lyme disease was named for Old Lyme, Connecticut, where cases were first recognized, and is the most prevalent tickborne disease in the United States. Lyme disease is caused by a spirochete, *Borrelia burgdorferi* (Figure 27.6●; ∞ Section 12.33), which is spread primarily by the deer tick, *Ixodes scapularis* (Figure 27.7●). The ticks that carry *B. burgdorferi* feed on the blood of birds, domesticated animals, various wild animals, and occasionally humans.

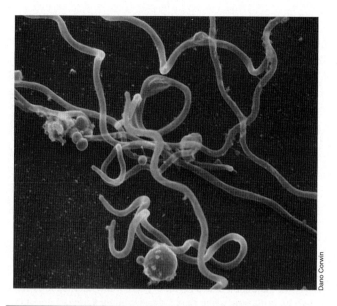

Dano Corwin

● **Figure 27.6 Electron micrograph of the Lyme spirochete,** *Borrelia burgdorferi.* The diameter of a single cell is approximately 0.4 μm.

Epidemiology

Deer and the white-footed field mouse are prime mammalian reservoirs of *B. burgdorferi* in the northeastern United States. However, in other parts of the country, different species of rodents and ticks are involved in the transmission of Lyme disease. In the western United States, *Ixodes pacificus* and the wood rat are common vectors and hosts.

Lyme disease has also been identified in Europe and Asia. In Europe, the tick vector is *Ixodes ricinus*, which may also harbor *Borrelia garinii*, another organism that causes Lyme diseaselike symptoms. In Asian countries, *Borrelia afzelii* is transmitted by *Ixodes persulcatus* and causes Lyme disease. In all cases, different local rodent reservoirs have been identified. Thus, Lyme disease seems to

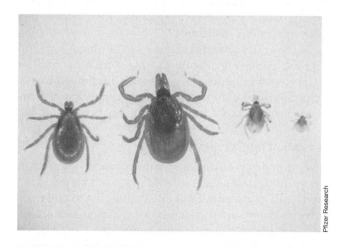

Pfizer Research

● **Figure 27.7 Deer ticks (*Ixodes scapularis*), the major vectors of Lyme disease.** Left to right, male and female adult ticks, nymph, and larval forms. The length of an adult female is about 3 mm. All forms feed on humans and are capable of transmitting *Borrelia burgdorferi.*

have a broad geographic distribution and is transmitted to humans by closely related *Borrelia* and tick species that have a variety of rodent and other mammalian reservoirs.

The deer tick and other members of the genus are smaller than many other species of ticks, making them easy to overlook (Figure 27.7). Unlike the case with other tickborne diseases, a very high percentage of deer ticks (up to 50% in certain regions of the Northeast) carry *B. burgdorferi* cells. Extended contact with the infected tick vector increases the probability of disease transmission.

In the United States most cases of Lyme disease have been reported from the Northeast and upper Midwest, but cases have been observed in nearly every state, and Lyme disease is spreading west and south. Figure 27.8● shows the spread of Lyme disease across the continental United States and the rapid rise in the total number of annual cases.

Pathology

Cells of *Borrelia burgdorferi* are transmitted to humans while the tick is obtaining a blood meal (Figure 27.9*a*●). A systemic infection develops, leading to the main symptoms of Lyme disease, which include headache, backache, chills, and fatigue. In about 75% of all cases, a large rash, known as *erythema migrans (EM)*, is observed at the site of the tick bite (Figure 27.9*b*). At this time, Lyme disease is treatable with tetracycline or penicillin. Untreated Lyme disease may progress to a chronic stage beginning weeks to months after the initial tick bite. Chronic Lyme disease is characterized by arthritis in 40–60% of patients. Neurological involvement such as palsy, weakness in the limbs, and facial ticks occurs in 15–20% of patients. Cardiac damage occurs in about 8% of all cases. In untreated cases, cells of *B. burgdorferi* infecting the central nervous system may lie dormant for long periods before eliciting a variety of additional chronic symptoms, including visual disturbances, facial paralysis, and seizures.

No toxins or other virulence factors have yet been identified in Lyme disease pathogenesis. In many respects the latent symptoms of Lyme disease resemble those of syphilis, caused by a different spirochete, *Treponema pallidum.* Indeed, some of the neurological symptoms of Lyme disease resemble those of chronic syphilis (⊙⊙ Section 26.12). However, unlike syphilis, Lyme disease is not spread by sexual intercourse or other types of human contact. Small numbers of *Borrelia burgdorferi* cells are shed in the urine of infected individuals, and there is some indication that Lyme disease can spread from domestic animal populations, particularly cattle, by infected urine.

Diagnosis

Serological tests have been developed for detection of antibodies to *Borrelia burgdorferi.* Antibodies appear about 4–6 weeks after infection and can be detected by an indirect enzyme-linked immunosorbent assay (ELISA) or a

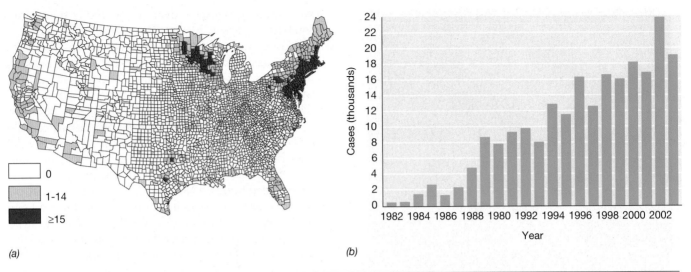

(a)

(b)

● **Figure 27.8 Lyme disease in the United States.** (a) Incidence of Lyme disease in the United States in 2002. Each county that reported Lyme disease in 2002 is shaded, with counties reporting more than 15 cases shown in red. Lyme disease, while most prevalent in the Northeast and upper Midwest, is found throughout the United States. (b) Number of reported cases of Lyme disease by year in the United States. In 2002, there were 23,763 cases, more than in any previous year. Lyme disease is reported through the National Notifiable Diseases Surveillance System of the Centers for Disease Control and Prevention.

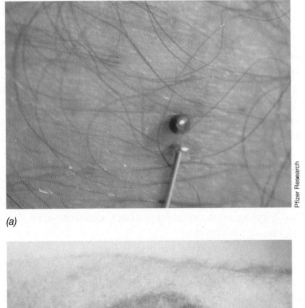

(a)

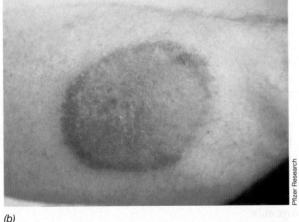

(b)

● **Figure 27.9 Lyme disease infection.** (a) Deer tick obtaining a blood meal from a human. (b) Characteristic circular rash associated with Lyme disease. The rash, known as *erythema migrans* (EM), typically starts at the site of the bite and grows in a circular fashion over a period of several days. This typical EM example is about 5 cm in diameter.

fluorescent antibody assay (∞ Sections 24.10 and 24.9). However, the most definitive serological test for Lyme disease is a Western blot (∞ Section 24.11). Because antibodies to the Lyme spirochete antigens persist for years after infection, the presence of antibodies does not necessarily confer immunity to the disease or indicate recent infection.

A polymerase chain reaction (PCR) assay (∞ Section 24.12) has also been developed for the detection of *Borrelia burgdorferi* from many body fluids and tissues. While they are rapid and sensitive, the PCR methods cannot differentiate between live *B. burgdorferi* in active disease and dead *B. burgdorferi* found in treated or inactive disease. *B. burgdorferi* can also be cultured from nearly 80% of the original erythema migrans lesions (Figure 27.9*b*), but culture is usually not done because of the long latent period before the organism grows on a highly specialized medium.

In the end, Lyme disease is usually diagnosed clinically. If a patient has Lyme disease symptoms and has had a recent tick exposure, especially if followed by erythema migrans, then a presumptive diagnosis of Lyme disease is made and antibiotic treatment is initiated.

Prevention and Treatment

Prevention of Lyme disease requires proper precautions to prevent tick attachment. In tick-infested areas such as woods, tall grass, and brush, it is advisable to wear protective clothing such as shoes, long pants, and a long-sleeved shirt with a snug collar and cuffs. Tucking the pants into tight-fitting socks worn with boots forms an effective barrier to tick attachment. After spending time in a tick-infested environment, individuals should check themselves carefully for ticks and gently remove any attached ticks (including the head). Insect repellants containing diethyl-*m*-toluamide (DEET) are very effective if applied to both skin and clothing. An effective human Lyme disease vaccine was withdrawn from the market in 2001 due to poor sales and possible adverse reactions. Lyme disease vaccines are available for veterinary use.

Treatment of acute Lyme disease can be with doxycycline (a tetracycline derivative, Section 20.9) or with amoxicillin (a β-lactam antibiotic, Section 20.8) for 20–30 days. Established Lyme arthritis or other symptoms indicating chronic *B. burgdorferi* infection are treated with high doses of penicillin or ceftriaxone, a β-lactam antibiotic that crosses the blood-brain barrier and attacks spirochetes residing in the central nervous system. Long-term Lyme arthritis is treated with doxycycline or amoxicillin plus probenicid, an agent that helps retain high serum levels of the antibiotic, for 30 days or more.

27.4 Concept Check

Lyme disease is now the most prevalent arthropod-borne disease in the United States. It is transmitted from several mammalian host vectors to humans by ticks. Prevention and treatment of Lyme disease are straightforward, but accurate and timely diagnosis is a major problem.

◆ What are the primary symptoms of Lyme disease?

◆ What antibiotics can be used to treat Lyme disease?

27.5 Malaria

Malaria is a disease caused by a protozoan, a member of the Sporozoa group (Section 14.10). The malaria parasite is one of the most important human pathogens and

has played an important role in the development and spread of human culture. As we will see, malaria has even affected human genetics and evolution. Malaria is still a significant human disease even though there are several effective treatments available. Over *100 million* people worldwide have malaria, and each year over 1 million of these will die (Section 25.1). The mammalian reservoir for malaria is humans. Four species of sporozoa infect humans. The most widespread is *Plasmodium vivax* and the most serious is *Plasmodium falciparum*. This parasite carries out part of its life cycle in human reservoir and part in the mosquito vector, which spreads the parasite from person to person. Only female mosquitoes of the genus *Anopheles* transmit malaria (see Figure 27.12a).

Epidemiology

Anopheles mosquitoes inhabit warmer parts of the world; therefore, malaria occurs predominantly in the tropics and subtropics. Malaria did not exist in the northern regions of North America prior to settlement by Europeans but was a major problem in areas such as the southern United States, where appropriate mosquito habitat existed. The disease is associated with wet low-lying areas, and the name *malaria* is derived from the Italian words for "bad air."

The life cycle of the malaria parasite is complex (Figure 27.10●). First, the human host is infected by plasmodial *sporozoites*, small, elongated cells produced in the mosquito, which localize in the salivary gland of the insect. The mosquito injects saliva (containing an anticoagulant) along with the sporozoites. The sporozoites travel

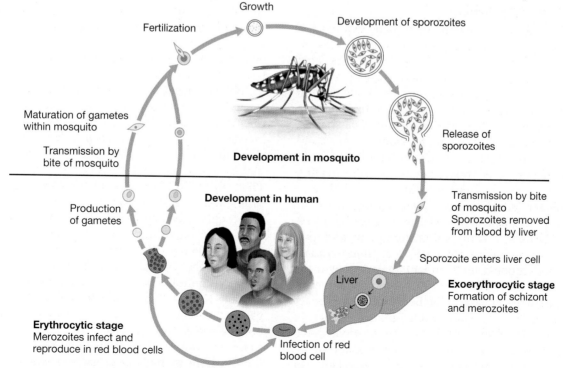

● **Figure 27.10** **Life cycle of the malaria parasite, *Plasmodium vivax*.** The *Plasmodium* genera of protozoans (Section 14.10) comprises the malarial pathogens, all of which have a life cycle dependent on growth in both a warm-blooded host and the mosquito vector. Transmission of the protozoan to and from the warm-blooded host is done by the bite of a mosquito.

through the bloodstream to the liver, where they may remain quiescent, or they may replicate and become enlarged in a stage known as a *schizont*. The schizonts then segment into a number of small cells called *merozoites*, and these cells are liberated from the liver into the bloodstream. Some of the merozoites then infect red blood cells (erythrocytes). The cycle in erythrocytes proceeds as in the liver and, in the case of *P. vivax*, usually repeats at synchronized intervals of 48 hours. During this 48- hour period, the defining clinical symptoms of malaria occur, characterized by *chills* followed by *fever* of up to 40°C (104°F). The chill–fever pattern occurs with the release of *P. vivax* cells from the erythrocytes during the synchronized asexual reproduction cycle. Vomiting and severe headache may accompany the chill–fever cycles, and asymptomatic periods generally alternate with periods in which the characteristic symptoms are present. Because of the loss of red blood cells, malaria generally causes anemia and some enlargement of the spleen.

Not all protozoal cells liberated from red blood cells are able to infect other erythrocytes. The protozoal cells that cannot infect erythrocytes are called *gametocytes* and are infective only for the mosquito. These gametocytes are ingested when another *Anopheles* mosquito bites the infected person; they mature within the mosquito into *gametes*. Two gametes fuse, and a zygote is formed. The zygote then migrates by ameboid motility to the outer wall of the insect's intestine where it enlarges and forms a number of sporozoites. These are released and some reach the salivary gland of the mosquito, from where they can be inoculated into another human, and the cycle begins again.

Diagnosis and Treatment

Conclusive diagnosis of malaria in humans requires the identification of *Plasmodium*-infected erythrocytes in blood smears (Figure 27.11●). Fluorescent nucleic acid stains, nucleic acid probes, polymerase chain reaction (PCR) methods, and antigen-detection methods may all be used to verify *Plasmodium* infections, or to differentiate between infections with various *Plasmodium* species.

Prophylaxis (when traveling to endemic areas) and treatment of malaria are usually accomplished with *chloroquine*. Chloroquine is the drug of choice for treating parasites within red cells, but does not kill malarial parasites outside the cells. The closely related drug *primaquine* eliminates sporozoites, merozoites, and gametes outside the cells. Treatment with both chloroquine and primaquine produces a cure. However, even in individuals who have undergone drug treatment, malaria may recur years after the primary infection. Apparently, small numbers of sporozoites survive in the liver and can reinitiate malaria months or years later by releasing merozoites.

In many parts of the world *Plasmodium* strains have developed resistance to chloroquine or primaquine or both, and some strains have developed resistance to other drugs as well. For use in areas with known drug-resistant strains, mefloquin or doxycycline is prescribed for prophylaxis and malarone, a combination of ato-

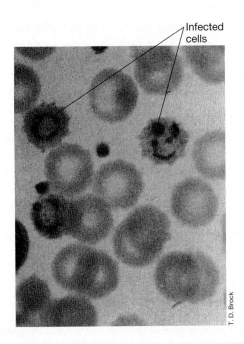

Infected cells

● **Figure 27.11** *Plasmodium vivax,* **the causative agent of malaria.** The parasite is shown growing inside human red blood cells.

vaquone and proguanil, is recommended for both treatment and prophylaxis. A new category of antimalarial drugs are synthetic derivatives of *artemisinin*, a natural compound containing reactive peroxide groups that form free radicals. These compounds are anti-parasitic agents *in vivo* and are now undergoing clinical trials.

Prevention

Antimalarial drug treatment is an expensive and short-term solution to malaria prevention and control, and drug-resistant strains of the parasite complicate matters even further. The most effective control measure is to interrupt the life cycle of the parasite by eliminating one of the obligate hosts, the *Anopheles* mosquito.

Two approaches to mosquito control are possible: (1) *elimination of habitat* by drainage of swamps and similar breeding areas, or (2) *elimination of the mosquito* by insecticides, followed by treatment of patients with primaquine, thereby breaking the *Plasmodium* life cycle. During the 1930s, about 33,000 miles of ditches were constructed in 16 southern states in the United States, removing 544,000 acres of mosquito breeding area. Millions of gallons of oil were also spread on swamps to reduce the oxygen supply to mosquito larvae. With the discovery of the insecticide dichlorodiphenyltrichloroethane (DDT) (∞ Figure 19.49), chemical control of both larvae and adult mosquitoes was possible. During World War II, the Public Health Service organized an Office of Mosquito Control in War Areas, and because many U.S. military bases were in the southern states, this organization carried out an extensive eradication program in the United States as well as overseas. In 1946, a year in which there were 48,610 cases of malaria in the United States, Congress established a five-year malaria eradication program. In endemic areas, the program involved drug prophylaxis and treatment for individuals, along with DDT treatment of mosquito infestations. By 1953 there were only 1310 malaria cases. In 1935 there were about 4000 deaths from malaria; in 1952 there were only 25 deaths.

Although the overall public health threat from malaria in the United States is now minimal, very low numbers of *endemic* malaria cases have resurfaced in recent years as far north as New York City. Significant periodic increases in malaria incidence also occur due to cases imported by soldiers or immigrants from malaria-endemic areas. On average, about 1,500 cases of malaria and five deaths occur in the United States each year.

In other parts of the world, eradication has been much slower, but the same control measures are used and are still effective. Reduction of mosquito habitat, control of mosquitoes by insecticides, and treatment of infected individuals with drugs both for cure and prophylaxis are still the major strategies for controlling malaria. Several malaria vaccines are in development, including synthetic peptide vaccines, recombinant particle vaccines, and DNA vaccines (Section 22.13 and 31.8).

Malaria and Human Evolution

Malaria has undoubtedly been endemic in Africa for thousands of years. In West Africans, resistance to malaria caused by *Plasmodium falciparum* is associated with an altered red blood cell protein, hemoglobin S, which differs from normal hemoglobin A at only a single amino acid. In the beta chain of hemoglobin S, the neutral amino acid *valine* is substituted for the *glutamic acid* of hemoglobin A. As a result, hemoglobin S binds oxygen less efficiently than hemoglobin A. Under conditions of low oxygen concentration, hemoglobin S forms long, thin aggregates that cause the red cell to change from a biconcave round cell (Figure 22.4) to an elongated C-shaped cell. Because of the shape of the cell, this condition is known as a *sickle cell*. Individuals who are *homozygous* for the sickle cell trait are particularly susceptible to changes in oxygen concentrations and suffer from **sickle cell anemia**.

Individuals who are heterozygous for hemoglobin S have the *sickle cell trait*, but have increased resistance to malaria. With the sickle cell trait, hemoglobin S can still produce sickled cells, but not as readily as in the case of the homozygotes. However, the growth of *P. falciparum* inside the red cell causes the heterozygous cells to sickle more easily than uninfected cells. The aggregated hemoglobin S in sickled cells apparently disrupts the membrane of red cells, allowing potassium to diffuse from the cell. *P. falciparum* cannot grow in the low-potassium environment of the disrupted cell. Thus, persons with the sickle cell trait are resistant to malaria.

In certain Mediterranean regions where malaria is endemic, resistance to *P. falciparum* is associated with a deficiency in the red blood cells of the enzyme *glucose-6-phosphate dehydrogenase (G6PD)*, an enzyme that acts as an intracellular antioxidant (reducing) compound. The faulty G6PD

leads to higher levels of intracellular oxidants such as the H_2O_2 produced inside the red cell by the growing *P. falciparum*. The increased levels of oxidants, normally removed by the activity of functional G6PD, damage parasite membranes and limit parasite growth.

In many Mediterranean populations, a diverse group of genetic abnormalities affect hemoglobin production and efficiency. These are known collectively as the **thalassemias**. The thalassemias are also statistically and geographically associated with increased resistance to malaria, and, like the G6PD deficiency, are associated with decreased levels of antioxidants in red cells.

Hemoglobin S, G6PD deficiency, and thalassemias are the result of genetic mutations that confer resistance to malaria infections and thus are selected in the population, although the mutations also confer red blood cell abnormalities and oxygen-processing deficiencies.

Another case in which the malaria parasite influences evolution involves the major histocompatibility complex (MHC) and the immune system (Sections 23.3 and 23.4). As discussed previously, the MHC class I and class II proteins present antigens to T cells for initiation of an immune response. In malaria-prone equatorial West Africa, individuals are very likely to have one particular MHC class I gene and one particular set of class II genes. These selected MHC genes are more common in the West African population and are virtually unknown in other human population groups. Individuals who express these genes have as much resistance to severe fatal malaria infections as those with the hemoglobin S trait. The MHC proteins encoded by these selected genes are exceptionally good antigen-presenting molecules for certain malarial antigens and initiate a strong protective immune response to *Plasmodium* spp. infection. As is the case for selection of the hemoglobin variants, the parasite is a selection factor for MHC genes that enhance host survival. Individuals with selected MHC genes that confer malaria resistance have a measurable survival advantage and are more likely to live and pass the resistance-conferring genes to their descendents.

Thus, in several ways malaria has been a selective agent in human evolution. Other pathogens, such as *Mycobacterium tuberculosis* (tuberculosis, Section 26.5) and *Yersinia pestis* (plague, see also Section 27.7), may also have promoted selective changes in humans, but in no case is the evidence as clear as it is for malaria.

 27.5 Concept Check

Malaria is a widespread, mosquito-borne infectious disease occurring mainly in tropical and subtropical regions of the world. It is a major cause of morbidity and mortality in developing countries and has been a selection factor for several resistance genes in humans. The disease is preventable with a combination of public health and chemotherapy measures.

◆ What is the natural reservoir for *Plasmodium* species? How can malaria be prevented or eradicated?

◆ Review genetic mechanisms responsible for malaria resistance. Why are antimalarial genes not found in all humans?

27.6 West Nile Virus

West Nile virus (**WNV**) causes **West Nile fever**, a rapidly emerging human viral disease transmitted through the bite of a mosquito. WNV is a member of the flavivirus group and has a symmetrical, enveloped icosahedral capsid with a positive-sense, single-stranded RNA genome of about 11,000 nucleotides (⌘ Section 16.8). The icosahedron is about 40–60 nm in diameter, and the virus can invade the nervous system of its warm-blooded host (Figure 27.12●).

Epidemiology

WNV infection in humans was first identified in Uganda in 1937. By the 1950s, the virus had spread to Egypt and Israel. In the 1990s, WNV outbreaks occurred in horses, birds, and humans in a number of African and European countries. In 1999, the first cases were reported in the United States, and early cases were centered in the Northeast, around New York. From 1999 through 2001, there were 149 confirmed cases of human WNV disease, including 18 deaths. Moving with the seasonal appearance and disappearance of the mosquito vectors, by 2002 this emerging disease had shifted from the East Coast to the Midwest, with a peak reported incidence of 884 in Illinois and total nationwide case totals of 4156. By 2003, there were 9186 confirmed cases, now centered in the upper Midwest and West. Colorado had the highest incidence of 2477 cases (Figure 27.13●). Illinois had only 53 cases. Deaths due to WNV infections *decreased* from 284 in 2002 to 231 in 2003, even with over twice as many diagnosed cases! This was probably due to increased recognition and testing for WNV infections in individuals displaying even mild symptoms (see below).

WNV normally causes active disease in birds, and is transferred to susceptible hosts by the bite of an infected mosquito. A number of mosquito species are known vectors, and at least 130 species of birds are known host reservoirs. The infected birds develop a viremia lasting 1–4 days, and survivors develop life-long immunity. Mosquitoes feeding on viremic birds are infected and can then infect susceptible birds, renewing the cycle. The incidence of disease in the avian population in a given area decreases as susceptible avian hosts die or recover and develop immunity. However, the mosquito vectors transmit the WNV to new susceptible hosts in new areas, moving the epidemic in a wavelike fashion across the continent. Incomplete data from 2004 indicate that the epidemic has already shifted to the West Coast and Southwest. The highest incidence of human disease in 2004 occurred in California and Arizona, with significantly lower numbers

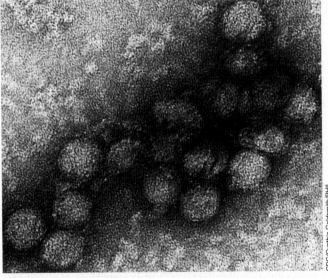

(a)

(b)

● **Figure 27.12 West Nile virus.** (a) The mosquito *Culex quinquefasciatus*, shown here engorged with human blood, is a West Nile virus vector. (b) An electron micrograph of the West Nile virus. The icosohedral virion is about 40–60 nm in diameter.

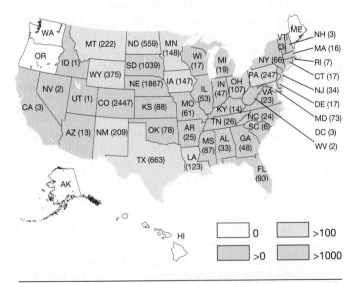

● **Figure 27.13 West Nile virus in the United States in 2003.** The virus caused 9186 cases of human disease, including 231 deaths. From 1999 through 2003, the peak incidence of the disease in the United States shifted from the East Coast, through the Midwest, and to the upper Midwest and West. Each year, the focus of cases moved as the virus infected new, susceptible host populations. Data are from the Centers for Disease Control and Prevention, Atlanta, GA, USA

of cases in the upper Midwest and Colorado. Within the next several years, the disease will cease to be an epidemic in the contiguous United States, and WNV will be an endemic pathogen in North American bird populations.

Humans and other animals are terminal hosts, since they do not develop the viremia necessary to infect mosquitoes. The human mortality rate for *diagnosed* infections is very significant at 2–3%, and horses have mortality rates of up to 40%. Most human infections are asymptomatic or very mild. About 20% of infected individuals develop a mild illness called *West Nile fever*, with an incubation period of 3–14 days and a duration of 3–6 days. Fever may be accompanied by headache, nausea, myalgia, rash, lymphadenopathy (swelling of lymph nodes), and malaise. Less than 1% of infected individuals develop serious neurological diseases such as West Nile encephalitis or meningitis. Adults over age 50 appear to be more susceptible to neurological complications than do others. Diagnosis of WNV disease includes assessment of clinical symptoms followed by confirmation with a positive ELISA test for WNV antibodies in serum or cerebrospinal fluid (Section 24.10).

Prevention and Control

Like St. Louis encephalitis virus and other mosquito-borne viruses that cause encephalitis (Figure 25.3), transmission of WNV is seasonal in the United States, and is dependent on exposure to the mosquito population. The primary means of control for WNV spread is by limiting exposure to the disease vector. This can be accomplished individually by avoiding mosquito habitat, remaining indoors between dusk and dawn (the prime hours for mosquito activity), wearing appropriate mosquito-resistant clothing, and application of repellents containing DEET, as for Lyme disease (see Section 27.4). At the community level, controlling the mosquito vector population by destroying mosquito habitat and applying appropriate insecticides are important public health measures to control WNV. There is no effective human vaccine, although several candidates are in development. Veterinary vaccines are in use, but their efficacy is unknown. Treatment, as for most viral illnesses, is centered around supportive therapy such as rest, fluids, and symptomatic relief of fever and pain. There are no antiviral drugs known to be effective *in vivo*.

27.6 *Concept Check*

West Nile fever is an emerging mosquito-borne viral disease. The natural cycle of the disease involves infections of birds by the bite of an infected mosquito. Humans and other vertebrates are occasional terminal hosts. While most human infections are asymptomatic and undiagnosed, complications in diagnosed infections can cause up to 3% mortality due to encephalitis and meningitis.

◆ Identify the vector and reservoir for West Nile virus.

◆ Trace the progress of West Nile virus in the United States since 1999.

27.7 Plague

Pandemic occurrences of **plague** have been directly responsible for more human deaths than any other infectious disease except malaria and tuberculosis. Plague killed between 25% and 33% of Europe's population in individual epidemics in the Middle Ages.

Plague is caused by *Yersinia pestis*, a gram-negative facultatively aerobic rod that is a member of the enteric bacteria group (Section 12.11) (Figure 27.14●). Plague is a natural disease of domestic and wild rodents; rats are the primary disease reservoir. Humans are only accidental hosts and are not critical for the maintenance of the disease. Fleas are intermediate hosts and act as vectors, spreading plague between the mammalian hosts (Figure 27.15●). Most infected rats die soon after symptoms appear, but a low proportion survive and develop a chronic infection, providing a persistent reservoir of virulent *Y. pestis*.

Epidemiology

The majority of cases of human plague in the United States occur in the southwestern states, where the disease, called *sylvatic plague*, is endemic among wild rodents. Plague is transmitted by the rat flea (*Xenopsylla cheopis*), which ingests *Y. pestis* cells by sucking blood from an infected animal. Cells multiply in the flea's intestine and can be transmitted to a healthy animal in the next bite. As the disease spreads, rat mortality becomes so great that infected fleas seek new hosts, including humans. Once in humans, cells of *Y. pestis* usually travel to the lymph nodes, where they cause the formation of swollen areas referred to as *buboes*. For this reason the disease is frequently referred to as *bubonic plague* (Figure 27.14*b*). The buboes become filled with *Y. pestis* and the capsule on cells of *Y. pestis* prevents phagocytosis by cells of the immune system (Section 22.2). Secondary buboes form in peripheral lymph nodes, and cells eventually enter the bloodstream, causing a generalized septicemia. Multiple hemorrhages produce dark splotches on the skin giving plague its historical name, the "Black Death" (Figure 27.14*c*). If not treated prior to the septicemic stage, the symptoms of plague (lymph node swelling and pain, prostration, shock, and delirium) progress and usually cause death within 3–5 days.

Pathology

The pathogenesis of plague is not clearly understood, but cells of *Yersinia pestis* produce a number of virulence factors, including toxins, that contribute to the disease process. The V and W antigens of *Y. pestis* cell walls are protein-lipoprotein complexes that inhibit phagocytosis. Other envelope proteins are also present. An exotoxin called *murine toxin*, because of its extreme toxicity for mice, is produced by virulent strains of *Y. pestis*. Murine

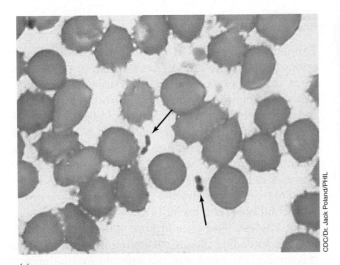

(a)

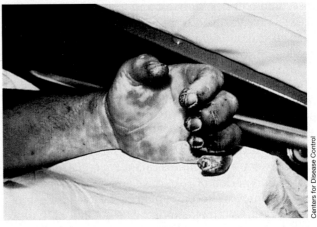

(c)

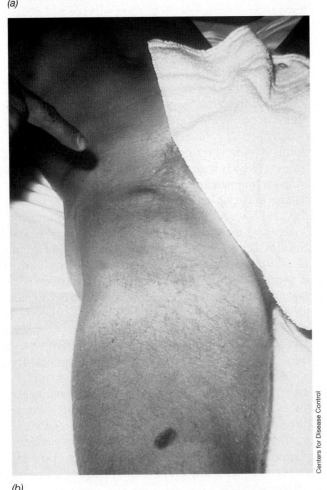

(b)

● **Figure 27.14   Plague in humans.** (a) *Yersinia pestis*, the causative agent of plague is a gram-negative rod, about 2 μm in length and up to 1 μm in diameter. The organisms in this blood smear (arrows) show the characteristic bipolar staining pattern. (b) A bubo formed in the groin. (c) Gangrene and sloughing of skin in the hand of a plague victim. Bubonic plague can be controlled by antibiotic therapy if diagnosed early in the course of the disease. Pneumonic plague and septicemic plague cannot be controlled by antibiotics, and so mortality is very high.

Pneumonic plague occurs when cells of *Yersinia pestis* are either inhaled directly or reach the lungs during bubonic plague. Symptoms are usually absent until the last day or two of the disease when large amounts of bloody sputum are produced. Untreated individuals rarely survive more than 2 days. Pneumonic plague is highly contagious and can spread rapidly via the person-to-person respiratory route if infected individuals are not immediately quarantined. *Septicemic plague* involves the rapid spread of *Y. pestis* throughout the body via the bloodstream without the formation of buboes and usually causes death before a diagnosis can be made.

Treatment and Control

Bubonic plague can be successfully treated if rapidly diagnosed. *Yersinia pestis* infection is treated with streptomycin or gentamycin, given parenterally. Alternatively, doxycycline, ciprofloxacin, or chloramphenicol may be given intravenously. If treatment is started promptly, mortality from bubonic plague can be reduced to 1–5% of those infected. Pneumonic and septicemic plague can also be treated, but these forms progress so rapidly that antibiotic therapy, even if begun when symptoms first appear, is usually too late. Although potentially a devastating disease, there have been only 97 cases of plague in the United States since 1990. Unfortunately, eight of these patients died. Worldwide, there are usually fewer than 1500 confirmed cases and 300 deaths per year. *Y. pestis* is an organism that could be used for a bioterrorism attack (⌦ Section 25.12), and oral doxycycline and ciprofloxacin are recommended as prophylactic antibiotics in this setting.

toxin is a respiratory inhibitor that blocks mitochondrial electron transport reactions at the point of coenzyme Q (⌦ Section 5.11). Although it is not clear that murine toxin is involved in the pathogenesis of human plague (murine toxin is highly toxic for certain animal species but not for others), it produces systemic shock, liver damage, and respiratory distress in mice. These symptoms are all seen in human plague as well. *Y. pestis* also produces a highly immunogenic *endotoxin* that may also play a role in the disease process.

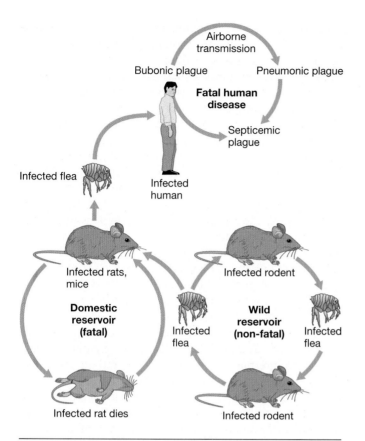

● **Figure 27.15** **The epidemiology of plague due to *Yersinia pestis*.** Plague in most wild rodents is generally a mild, self-limiting infection. Plague in rats and humans is frequently fatal. Infected fleas desert the dead host and look for other hosts, such as humans, an accidental host.

Plague control is accomplished through surveillance of infected animals, vectors, and human contacts. Plague-infected animal populations must be destroyed when identified. Undoubtedly, improved public health practices and the overall control of rodent populations have limited human exposure to plague.

 27.7 Concept Check

Plague is largely confined to individuals who come into contact with rodent populations that are endemic reservoirs for *Yersinia pestis*. A disseminated systemic infection or a pneumonic infection leads to rapid death, but the bubonic form is treatable with antibiotics if rapidly diagnosed.

◆ Distinguish among *sylvatic, bubonic, septicemic,* and *pneumonic* plague.

◆ What is the insect reservoir, the natural host, and the treatment for plague?

III SOILBORNE DISEASES

We now investigate several diseases that are normally spread through soil. Fungi are common and ubiquitous soil microorganisms found worldwide, and a few can be pathogens. Some bacteria are also important soilborne pathogens. In contrast to many person-to-person or vector-borne pathogens, soilborne pathogens are accidental agents of infection, with no life-cycle dependency on the accidental host. Soil is an unlimited reservoir of these pathogens and, thus, these pathogens cannot be eliminated.

27.8 The Pathogenic Fungi

Fungi in some form grow in nearly every ecological niche, but are most commonly found in nature as free-living saprophytes (Section 14.12). However, a few fungi are occasionally found as accidental, often opportunistic, pathogens.

The fungi include the eukaryotic organisms commonly known as *yeasts*, which normally grow as single cells (Figure 27.16*a*●), and *molds*, which grow in branching

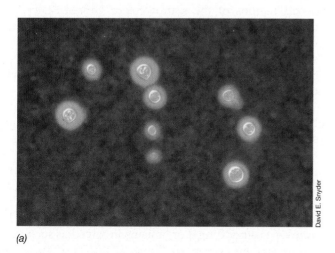

(a)

David E. Snyder

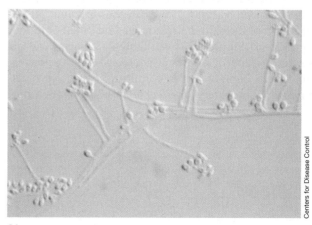

(b)

Centers for Disease Control

● **Figure 27.16** **Typical forms of pathogenic fungi.** (a) Yeast form of *Cryptococcus neoformans*, stained with India ink to show the capsule. The cells are from 4 to 20 μm in diameter (b) *Sporothrix schenckii*, showing the branching, or hyphae, characteristic of the mold form of fungi. The round basidiospores are about 2 μm in diameter. Fungi are discussed in Section 14.12.

chains called *hyphae* (Figure 27.16*b*). The taxonomy and biological diversity of these organisms were discussed in Section 14.12. Fortunately, most fungi are harmless to humans. Only about 50 species cause human disease, and the overall incidence of serious fungal infections is rather low, although certain superficial fungal infections are quite common.

Epidemiology

Fungi cause disease through three major mechanisms. First, some fungi trigger immune responses that result in allergic (hypersensitivity) reactions following exposure to specific fungal antigens (∞ Section 22.15). Reexposure to the same fungi, whether growing on the host or in the environment, may cause allergic symptoms. For example, *Aspergillus* spp. (∞ Figure 14.31), a common saprophyte often found in nature as a leaf mold, produces potent allergens, often causing asthma and other hypersensitivity reactions. *Aspergillus* also has other mechanisms for producing disease.

A second fungal disease-producing mechanism involves the production and activity of *mycotoxins*, a large, diverse group of fungal exotoxins. The best-known examples of mycotoxins are produced by *Aspergillus flavus*, a species that commonly grows on improperly stored food such as grain. The toxins produced by *A. flavus* are known as *aflatoxins* (Figure 27.17●). Aflatoxins are highly toxic and induce tumors in some animals, especially in birds that feed on contaminated grain. Their direct role in human disease is not well defined.

The third fungal disease-producing mechanism is through infections called *mycoses*.

Mycosis

The growth of a fungus on or in the body is called a **mycosis** (plural, **mycoses**). Mycoses are fungal infections that range in severity from relatively innocuous, superficial lesions to serious, life-threatening diseases.

Mycoses fall into three categories. The first of these are the *superficial mycoses*. These diseases involve fungal colonization of the skin, hair, or nails, and infect only the surface layers (Figure 27.18*a*●). Table 27.2 lists some of the common superficial fungi. In general, these diseases are relatively benign and self-limiting. Some, such as *Trichophyton* infections of the feet (*athlete's foot*), are quite

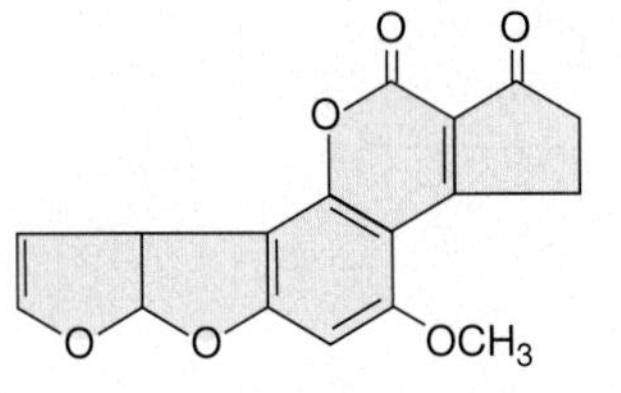

● **Figure 27.17 Structure of aflatoxin B1.** This toxin is one of a group of related compounds produced by *Aspergillus flavus*.

common. Spread is by personal contact with an infected person or by contact with contaminated surfaces such as bathtubs, shower stalls, or floors, or by contact with contaminated shared articles such as towels or bed linens. Treatment for severe cases is with topical application of miconazole nitrate or griseofulvin. Griseofulvin can also be administered orally. After entering the bloodstream, it passes to the skin where it can inhibit fungal growth.

The *subcutaneous mycoses* are a second category of fungal infections. They involve deeper layers of skin (Figure 27.18*b*) and a different group of organisms (Table 27.2). One disease in this category is *sporotrichosis*, an occupational hazard of agricultural workers, miners, and others who come into contact with the soil. The causative organism is found as a ubiquitous saprophyte on wood and in soil. The lesions are usually initiated by fungal infection of a small wound or abrasion. *Sporothrix schenckii* can readily be isolated from the lesion and cultured *in vitro*. Treatment is with oral potassium iodide or oral ketoconazole.

The *systemic mycoses* are the third and most serious category of fungal infections. They involve fungal growth in internal organs of the body and are subclassified as *primary* or *secondary* infections. A primary infection is one resulting directly from the fungal pathogen in otherwise normal, healthy individuals. A secondary infection involves infection in hosts with a predisposing condition such as antibiotic therapy or immunosuppression. In the United States, the most widespread primary fungal infections are *histoplasmosis*, caused by *Histoplasma capsulatum*, and *coccidioidomycosis* (San Joaquin Valley fever), caused by *Coccidioides immitis*. Both of these organisms normally live in soil. Both are respiratory diseases in which the host becomes infected by inhaling airborne spores, which germinate and grow in the lungs. Histoplasmosis is primarily a disease of rural areas in the midwestern United States, especially in the Ohio and Mississippi River valleys. Most cases are mild and are often mistaken for more common respiratory infections. San Joaquin Valley fever is generally restricted to the desert regions of the southwestern United States. The fungus lives in desert soils, and the spores are disseminated on dry, windblown particles that are inhaled. In some areas in the southwestern United States, as many as 80% of the inhabitants may be infected, although most individuals suffer no apparent ill effects.

A number of systemic fungal infections, including histoplasmosis and coccidioidomycosis, are especially serious and common in individuals whose immune systems have been impaired, for example, by acquired immunodeficiency syndrome (AIDS) (∞ Section 26.14) or by immunosuppressive drugs. These are *secondary* fungal diseases because normal individuals either do not get the disease or generally have a less severe form. Examples of other fungal organisms involved as secondary pathogens are given in Table 27.2. These fungi are known as *opportunistic* pathogens because of their particular ability to cause serious infections only in individu-

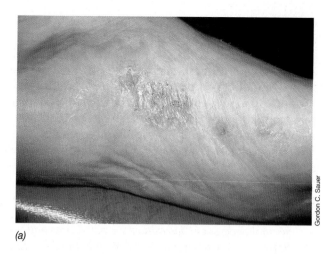

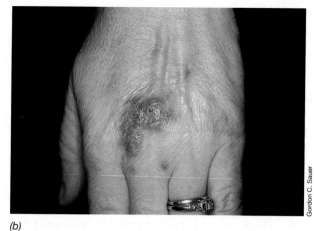

Gordon C. Sauer

(a) *(b)*

● **Figure 27.18 Fungal infections.** (a) Superficial mycosis of the foot (athlete's foot) due to infection with *Trichophyton rubrum*. (b) Sporotrichosis, a subcutaneous infection due to *Sporothrix schenckii*.

als with impaired defense mechanisms (in particular AIDS patients) (∞ Section 26.14).

Treatment and Control

Effective chemotherapy against systemic fungal infections is very difficult (∞ Section 20.11). Most antibiotics that inhibit fungi also harm other eukaryotic organisms, including the human host. One of the most effective antibiotics, amphotericin B, is widely used to treat systemic fungal infections of humans, but serious side effects such as kidney toxicity may occur.

Control of infections by elimination of fungal pathogens from the environment is impractical. As with many common-source pathogens, control of fungal growth is very difficult because there is a limitless reservoir. Exposure to fungi cannot be eliminated, but risks can be reduced by decontamination and air filtration in restricted local environments.

⬡ 27.8 Concept Check

A variety of soilborne fungi produce disease in humans. Superficial, subcutaneous, and systemic mycoses are infections that are difficult to control because of a lack of specific antifungal drugs and the ubiquitous nature of the pathogens. Fungal infections may cause serious systemic disease, often in individuals with impaired immunity, such as in AIDS patients.

◆ Describe *superficial*, *subcutaneous*, and *systemic* mycoses.

◆ Distinguish between a *primary* and a *secondary* fungal disease.

27.9 Tetanus

Tetanus is a serious, often life-threatening disease. Although tetanus is preventable through immunization (∞ Section 22.13), 473 individuals have acquired tetanus in the United States within the last decade and 68 have died.

Disease	**Causal organism**	**Main disease foci**
Table 27.2 Some pathogenic fungi and the diseases they cause		
Superficial mycoses (dermatomycoses)		
Ringworm	*Microsporum*	Scalp of children
Favus	*Trichophyton*	Scalp
Athlete's foot	*Epidermophyton, Trichophyton*	Between toes, skin
Jock itch	*Trichophyton, Epidermophyton*	Genital region
Subcutaneous mycoses		
Sporotrichosis	*Sporothrix schenckii*	Arms, hands
Chromoblastomycosis	Several fungal genera	Legs, feet
Systemic mycoses		
Cryptococcosis[a]	*Cryptococcus neoformans*	Lungs, meninges
Coccidioidomycosis[a]	*Coccidioides immitis*	Lungs
Histoplasmosis[a]	*Histoplasma capsulatum*	Lungs
Blastomycosis	*Blastomyces dermatitidis*	Lungs, skin
Candidiasis[a]	*Candida albicans*	Oral cavity, intestinal tract

[a] Considered opportunistic pathogens and have been implicated in the pathogenesis of AIDS (∞ Section 26.14).

Biology and Epidemiology

Tetanus is caused by an exotoxin produced by *Clostridium tetani*, a motile, anaerobic, endospore-forming rod (∞ Section 12.20). The natural reservoir of *C. tetani* is soil, where it is a ubiquitous resident, although it is occasionally found in the gut of mammals, as are other *Clostridium* species. *C. tetani* normally gains access to the body through a soil-contaminated wound, typically a deep puncture. In the wound, anoxic conditions allow germination of endospores and growth of the organism. *C. tetani* produces a potent exotoxin, the *tetanus toxin*. The organism is noninvasive, and so its sole method of causing disease is through the direct action of tetanus toxin on host cells. The incubation time is variable and may take from 4 days to several weeks, depending on the number of endospores inoculated at the time of injury. Tetanus is not transmitted from person to person.

Pathogenesis

We have already examined the activity of tetanus toxin at the cellular and molecular level (∞ Section 21.10). The toxin directly affects the release of inhibitory signaling molecules in the nervous system. These inhibitory signals control the "relaxation" phase of muscle contraction. The overall result is rigid paralysis of the voluntary muscles, often called *lockjaw* because it is observed first in the muscles of the jaw and face (Figure 27.19●). Death is usually due to respiratory failure, and the mortality rate is very high (15% over the last decade in the United States).

Diagnosis, Control, Prevention, and Treatment

Diagnosis of tetanus is based on exposure, clinical symptoms, and, rarely, identification of the toxin in the blood or tissues of the patient. The organism may also be cultured from the wound, but success is highly variable.

The natural reservoir of *Clostridium tetani* is the soil. Since *C. tetani* is an accidental pathogen in humans and is not dependent on humans or other animals for its propagation, there is no possibility for eradication. Therefore, control measures must focus on methods of prevention.

Tetanus is a preventable disease. The existing toxoid vaccine (∞ Section 22.13) is completely effective for disease prevention. Virtually all tetanus cases occur in individuals who were inadequately immunized. The fastest growing age group for contracting tetanus is, surprisingly, individuals from 25–59 years of age, presumably because public health immunization programs target infants, school-age individuals, and seniors 60 years of age and older.

Appropriate treatment of serious cuts, lacerations, and punctures includes administration of a "booster" tetanus toxoid immunization. If the wound is severe and is contaminated by soil, treatment should also include administration of an antitoxin preparation, especially if the patient's immunization status is unknown or is out of date. The antitoxin is a preparation of antibodies to tetanus, usually made in horses, that neutralizes the tetanus toxin as it is released (∞ Section 22.13). These measures prevent active tetanus from occurring.

Treatment of acute symptomatic tetanus involves administration of antibiotics, usually penicillin, to stop growth and toxin production by *C. tetani*; administration of antitoxin to prevent binding of newly released toxin to cells; and supportive therapy such as sedation, muscle relaxants, and mechanical respiration to control the effects of paralysis. Treatment at this level cannot provide a reversal of symptoms, since toxin that is already bound to tissues cannot be neutralized. Even with antitoxin, antibiotics, and supportive therapy, tetanus patients have significant morbidity and mortality.

Several species of *Clostridium* are pathogens and virtually all exist as normal members of the microbial community in soil, making their eradication impossible. All cause disease because of their production of potent exotoxins. *Clostridium tetani* is found almost exclusively in soil, but *C. botulinum* and *C. difficile* are also occasionally found in the gut of humans and other animals as part of the normal microbial flora (∞ Section 21.4). *C. perfringens* and *C. botulinum*, as we shall see in Chapter 29, are important potential pathogens in common-source foodborne diseases. In each case, the source of the *Clostridium* is food contaminated with soil containing clostridial endospores. The endospores then germinate because of inadequate decontamination and food preservation methods, and produce potent exotoxins that cause disease symptoms.

 27.9 Concept Check

Clostridium tetani is a ubiquitous soilborne microorganism that can cause *tetanus*, a disease characterized by toxin production and rigid paralysis. Tetanus is preventable with appropriate immunization. Treatment for acute tetanus is generally unsatisfactory, with significant morbidity and mortality.

◆ Describe infection by *C. tetani* and the elaboration of tetanus toxin.

◆ Describe the overall effect of tetanus toxin on the host and outline the steps taken to prevent tetanus in an individual who has sustained a puncture wound.

● **Figure 27.19 A soldier dying from tetanus.** Note the rigid paralysis. The painting by Charles Bell is in the Royal College of Surgeons, Edinburgh, Scotland.

REVIEW QUESTIONS

1. Identify the animals most likely to carry rabies in the United States. Which immunization programs are in place for the treatment of rabies? For the prevention of rabies (⌒⌒ Section 27.1)?

2. Why has hantavirus pulmonary syndrome (HPS) emerged as a human pathogen in the United States? How can HPS be prevented (⌒⌒ Section 27.2)?

3. What are the three major categories of organisms that cause rickettsial diseases? For typhus, Rocky Mountain spotted fever, and ehrlichiosis, identify the most common reservoir and vector (⌒⌒ Section 27.3).

4. Identify the most common reservoir and vector for Lyme disease in the United States. How can the spread of Lyme disease be controlled? How can Lyme disease be treated (⌒⌒ Section 27.4)?

5. Malaria involves severe long-term fever followed by chills. These symptoms are related to activities of the pathogen *Plasmodium vivax* or *Plasmodium falciparum*. Describe the growth stages of this pathogen in the human host and relate them to the fever–chill pattern.

Why might a person of western European descent be more susceptible to malaria than a person of African or Mediterranean descent (⌒⌒ Section 27.5)?

6. Explain the life cycle and the role of the mosquito in spreading West Nile virus. What animals are the primary hosts? Are humans productive alternative hosts? Explain (⌒⌒ Section 27.6).

7. For a potentially serious disease like bubonic plague, why aren't vaccines provided for the general population? What public health measures are used to control this disease (⌒⌒ Section 27.7)?

8. What is the reservoir for most fungal pathogens? How can fungal exposure be controlled? What particular problems, especially in terms of therapy, do fungi pose for the clinician (⌒⌒ Section 27.8)?

9. Describe the invasiveness and toxicity of *Clostridium tetani*. Discuss the major mechanism of pathogenesis for tetanus and define measures for prevention and treatment of tetanus (⌒⌒ Section 27.9).

APPLICATION QUESTIONS

1. Describe the sequence of events you would take if a child received a bite (provoked or unprovoked) from a stray dog with no record of rabies immunization. Present one scenario where you were able to capture and detain the dog and another for a dog that escaped. How would these procedures differ from a situation in which the child was bitten by a dog that had documented, up-to-date rabies immunizations?

2. Discuss at least three common properties of the disease agents and review the disease process for Rocky Mountain spotted fever, typhus, and ehrlichiosis. Why is ehrlichiosis emerging as an important rickettsial disease? Compare its emergence to that of Lyme disease.

3. Malaria eradication has been a goal of public health programs for at least 100 years. What factors preclude our ability to eradicate malaria? If an effective vaccine was developed, could malaria be eradicated? Compare this possibility to the possibility of eradicating plague.

4. Devise a plan to prevent the spread of West Nile virus to humans in your community. Identify the costs involved in such a plan, both at the individual level and at the community level. Find out if a mosquito abatement program is active in your community. What methods are used for the reduction of mosquito populations?

28

WASTEWATER TREATMENT, WATER PURIFICATION, AND WATERBORNE MICROBIAL DISEASES

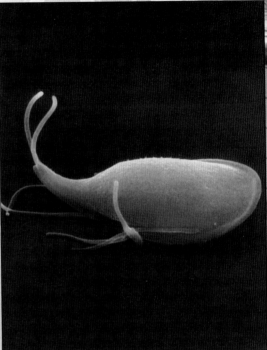

Water is the most important common source of infectious disease. Water treatment standards are designed to prevent infections from bacteria, viruses, and protozoans such as the cell of the protozoan *Giardia* shown in the photo.

WORKING GLOSSARY

Aerobic secondary wastewater treatment digestive reactions carried out by microorganisms under aerobic conditions to treat wastewater containing low levels of organic materials

Anoxic secondary wastewater treatment digestive and fermentative reactions carried out by microorganisms under anoxic conditions to treat wastewater containing high levels of insoluble organic materials

Biochemical oxygen demand (*BOD*) the relative amount of dissolved oxygen consumed by microorganisms for complete oxidation of organic and inorganic material in a water sample

Chloramine a chemical manufactured on site by combining chlorine and ammonia at precise ratios

Chlorine a chemical fed in its gaseous state to disinfect water. A residual level is maintained throughout the distribution system

Clarifier (coagulation basin) a reservoir in which the suspended solids of raw water are coagulated and removed

Coagulation the formation of large insoluble particles from much smaller, colloidal particles by the addition of aluminum sulfate and anionic polymers

Coliform facultatively aerobic, gram-negative, nonspore-forming lactose-fermenting bacterium

Cyst an infectious form of a protozoan parasite that is encased in a thick-walled, chemically and physically resistant coating

Distribution system water pipes, storage reservoirs, tanks, and other means used to deliver drinking water to consumers or store it before delivery

Effluent water treated wastewater discharged from a wastewater treatment facility

Filtration the removal of suspended particles from water by passing it through one or more permeable membranes or media (e.g., sand, anthracite, or diatomaceous earth)

Finished water water delivered to the distribution system after treatment

Flocculation the water-treatment process after coagulation that uses gentle stirring to cause suspended particles to form larger, aggregated masses (floc)

Meningoencephalitis invasion, inflammation, and destruction of brain tissue, by the ameba *Naegleria fowleri* or a variety of other pathogens

Polymer in water purification, a chemical in liquid form used as a coagulant in the clarification process to produce flocculation

Potable drinkable; safe for human consumption

Primary wastewater treatment physical separation of wastewater contaminants, usually by separation and settling

Raw water surface water or groundwater that has not been treated in any way (also called untreated water)

Sediment soil, sand, minerals, and other large particles found in raw water

Sewage liquid effluents contaminated with human or animal fecal material

Suspended solid small particle of solid pollutant that resists separation by ordinary physical means

Tertiary wastewater treatment physico-chemical processing of wastewater to reduce levels of inorganic nutrients

Turbidity a measurement of suspended solids in water

Untreated water surface water or groundwater that has not been treated in any way (also called raw water)

Wastewater liquid derived from domestic sewage or industrial sources, which cannot be discarded in untreated form into lakes or streams

---

Wastewater is liquid derived from domestic sewage or industrial sources that cannot be discarded in untreated form into lakes or streams due to public health, economic, environmental, and aesthetic considerations. Clean, pure water is absolutely essential to public health, and procedures to monitor, assess, and remediate water quality are thus necessary. However, water quality is sometimes compromised, even in publicly operated wastewater and drinking water systems. Lapses in water quality can promote dramatic and even life-threatening spread of infectious disease. This chapter examines standard methods of water treatment and purification and a variety of waterborne diseases.

▌WASTEWATER MICROBIOLOGY AND WATER PURIFICATION

Water is the most important potential *common source* of infectious diseases, and water purification is the most important single measure available for ensuring public health. The methods commonly used to assess water quality depend on standardized microbiology techniques. In addition to physical and chemical purification procedures, water treatment and purification schemes use a variety of standardized methods that use microorganisms to identify, remove, or degrade pollutants.

28.1 Public Health and Water Quality

How can we routinely ensure that drinking water is safe? Even water that looks clear and clean may be contaminated with pathogenic microorganisms and may pose a serious health hazard. Unfortunately, it is impractical to screen water for every pathogenic organism that may be present, and a few nonpathogenic microorganisms are generally tolerable, and even unavoidable, in a water supply. It is, however, possible to sample water supplies for the presence of certain groups of microorganisms that may include pathogens. We discuss here the general methods used to identify potentially harmful microorganisms in water.

Coliforms

The presence of specific *indicator organisms* signals that a given water source might be contaminated with pathogens. The most widely used indicator for microbial water contamination is the **coliform** group of microorganisms. Coliforms are used as indicators of water contamination because many of them inhabit the intestinal tract of humans and other animals in large numbers. Thus, their presence in water indicates fecal contamination. Coliforms are defined as facultatively aerobic, gram-negative, nonspore-forming, rod-shaped *Bacteria* that ferment lactose with gas formation within 48 hours at 35° C. However, this is an operational rather than a taxonomic definition, and the coliform group actually includes a variety of organisms. Most coliforms are members of the enteric bacterial group (∞ Section 12.11). For example, the coliform group includes the organism *Escherichia coli*, a common intestinal organism, and the organism *Klebsiella pneumoniae*, a less common pathogenic intestinal inhabitant. However, *Enterobacter aerogenes*, an organism not found in the enteric group or in the intestine, is also classified as a coliform because of its fermentative properties.

In general, we assume that the presence of coliform organisms in a water sample indicates fecal contamination and makes the water unsafe for human consumption. When excreted into water, the coliforms eventually die, but they do not die as quickly as some pathogens. The coliforms and the pathogens behave similarly during water purification.

The Coliform Test

Two procedures are commonly used to test for coliforms in water samples. These are the *most-probable-number (MPN)* procedure and the *membrane filter (MF)* procedure. The MPN procedure employs liquid culture medium in test tubes in which samples of drinking water are added to the media. Growth in the culture vessels indicates microbial contamination of the water supply. For the more common MF procedure, at least 100 ml of the water sample is passed through a sterile membrane filter, which removes the bacteria (∞ Section 20.3). The filter is placed on a plate of eosin-methylene blue (EMB) culture medium, which is highly selective for coliform organisms (Figure 28.1●; ∞ Section 24.2). Following incubation, coliform colonies are counted, and from this value the number of coliform *Bacteria* in the original water sample can be calculated. In well-regulated water supply systems, coliform tests should be negative. If coliform tests are positive, a breakdown has occurred in the purification or distribution systems.

Drinking water standards in the United States are specified under the Safe Drinking Water Act, which provides a framework for the development of drinking water standards. For the membrane filter (MF) technique, 100 ml samples are filtered. To be considered safe, the number of coliform bacteria in drinking water samples cannot exceed any of the following levels: (1) 1 per 100 ml as the arithmetic mean of all samples examined per month; (2) 4 per 100 ml in more than one sample when fewer than 20 are examined per month; or (3) 4 per 100 ml in more than 5% of the samples when 20 or more samples are examined per month. Water utilities report their results to the United States Environmental Protection Agency, and if they do not meet the prescribed stan-

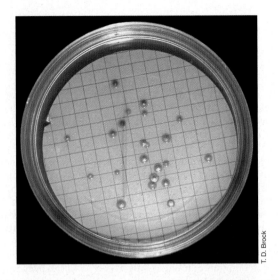

● **Figure 28.1 Coliform colonies growing on a membrane filter.** A drinking water sample was passed through the filter. The filter was then placed on eosin-methylene blue (EMB) medium that is both selective and differential for lactose-fermenting bacteria (coliforms) (∞ Section 24.2). The dark color of the colonies is characteristic of coliforms. Each colony developed from one viable coliform cell present in the original sample.

dards, the utilities must notify the public and take steps to correct the problem. Many smaller communities and even large cities sometimes fail to meet these standards.

Public Health and Drinking Water Purification

Today the incidence of waterborne disease in developed countries is so low that it is difficult to directly measure the effectiveness of treatment practices using drinking water standards. Most intestinal infections in developed countries are no longer transmitted by water but by food (∞ Chapter 29). Highly effective water treatment practices, however, were not in place until the twentieth century.

Microbial culture methods for evaluating the health significance of polluted drinking water were not practiced until coliform-counting procedures were developed and adapted in about 1905. Until then, water purification was limited to *filtration* to reduce turbidity. Although filtration significantly decreased the microbial load of water, many microorganisms still passed through the filters. In about 1910, **chlorine** was discovered to be an extremely efficient disinfectant for large water supplies. Chlorine gas was so effective and so inexpensive that its use as a general disinfectant for drinking water spread widely and was of major significance in reducing the incidence of waterborne disease. Figure 28.2● illustrates the dramatic drop in incidence of typhoid fever (infection with *Salmonella typhi*) in a major American city after purification procedures using filtration and chlorination were introduced. Similar results were obtained in other cities.

The first dramatic improvements in public health in the United States, starting near the beginning of the twentieth century, were largely due to the establishment

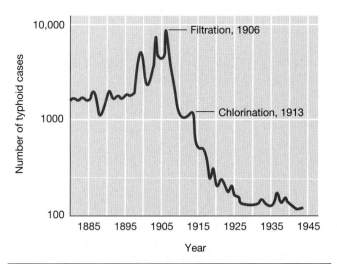

● **Figure 28.2   The effect of water purification on the incidence of waterborne disease.** The graph shows the incidence of typhoid fever in Philadelphia (Pennsylvania, USA) before and after the introduction of filtration and chlorination processes for drinking water supplies. Note the dramatic reduction in the incidence of typhoid fever after the introduction of both filtration and chlorination. This result parallels the results obtained for reduction of other waterborne diseases.

of water treatment procedures employed in large-scale, publicly operated, wastewater and drinking water treatment facilities. The effectiveness of filtration and chlorination could not have been assessed if standard methods for determining the coliform content of drinking water had not been developed. Thus, public works engineering, microbiology, and public health moved forward together.

 28.1   Concept Check

Drinking water quality is determined by counting coliform bacteria. Strict adherence to uniform microbiologic standards makes this method a reliable and reproducible indicator of fecal contamination in all public water supplies in the United States. Filtration and chlorination of water supplies significantly decreases microbial load. Application of water purification methods to drinking water is the most important public health measure ever devised.

◆   Why do the bacterial colonies recovered from drinking water and grown on EMB media indicate fecal contamination of the water supply?

◆   What general procedures are used to reduce microbial load in water supplies?

28.2   Wastewater and Sewage Treatment

Wastewater and sewage treatment involves a large-scale use of microorganisms for bioconversion on an industrial scale. Wastewater enters a treatment plant and, following treatment, the **effluent water**—treated wastewater discharged from the wastewater treatment facility—is

suitable for release into surface waters such as lakes and streams or to drinking water purification facilities.

Wastewater and Sewage

Wastewater from domestic sewage or industrial sources cannot be discarded in untreated form into lakes or streams. **Sewage** is liquid effluent contaminated with human or animal fecal materials. Wastewater commonly contains potentially harmful inorganic and organic compounds as well as pathogenic microorganisms. Complete wastewater treatment involves physical, chemical, and biological (microbiological) treatments to remove or neutralize contaminants.

On average, individuals in the United States use 100–200 gallons of water per day for washing, cooking, drinking, and sanitary needs. The wastewater must then be treated to remove contaminants before it can be released into surface waters. About 16,000 publicly owned treatment works (POTW) operate in the United States. Most POTWs are fairly small, treating 1 million gallons (3.8 million liters) or less of wastewater per day. Collectively, however, these plants treat about 32 *billion* gallons of wastewater daily. Wastewater plants are usually constructed to handle both domestic and industrial wastes. *Domestic wastewater* is made up of sewage, "gray water" (the water resulting from washing, bathing, and cooking), and wastewater from food processing.

Industrial wastewater includes liquid discharged from the petrochemical, pesticide, food and dairy, plastics, pharmaceutical, and metallurgical industries. Industrial wastewater may contain toxic substances and some industrial facilities are required by the United States Environmental Protection Agency (EPA) to pretreat toxic or heavily contaminated discharges before they enter POTWs. Pretreatment may simply involve mechanical processes in which debris that could clog equipment in the wastewater treatment plant is removed. However, certain wastewaters are pretreated biologically or chemically to remove highly toxic substances such as cyanide, heavy metals such as arsenic, lead and mercury, or organic materials such as acrylamide, atrazine (a herbicide), and benzene. These substances are converted to less toxic forms by treatment with chemicals or microorganisms capable of neutralizing, oxidizing, precipitating, or volatilizing these wastes. The pretreated wastewater can then be released to the POTW.

Wastewater Treatment and Biochemical Oxygen Demand

The goal of a wastewater treatment facility is to reduce organic and inorganic materials in wastewater to a level that no longer supports microbial growth and to eliminate other potentially toxic materials. The efficiency of treatment is expressed in terms of a reduction in the **biochemical oxygen demand (BOD)**, the relative amount of dissolved oxygen consumed by microorganisms to completely oxidize all organic and inorganic matter in a water sample (∞ Section 19.5). Higher levels of oxidizable organic and inorganic materials in the wastewater result in a higher BOD.

Typical values for domestic wastewater, including sewage, are approximately 200 BOD units. For industrial wastewater, for example, from sources such as dairy plants, the values can be as high as 1500 BOD units. An efficient wastewater treatment facility reduces levels to less than 5 BOD units in the water released from the treatment plant. Wastewater facilities must be designed to treat both low-BOD sewage and high-BOD industrial wastes.

Treatment is a multistep operation employing a number of independent physical and biological processes (Figure 28.3●). Primary, secondary, and sometimes tertiary treatments are employed to reduce fecal and chemical contamination in the incoming water. Each level of treatment employs more complex and more expensive technologies.

Primary Wastewater Treatment

Primary wastewater treatment uses only *physical separation* methods to separate solid and particulate organic and inorganic materials from wastewater. Wastewater

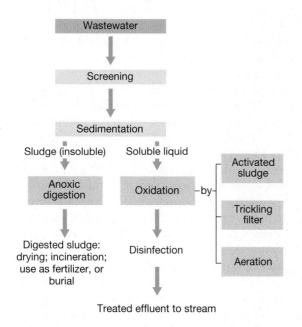

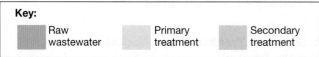

● **Figure 28.3 Wastewater treatment processes.** Effective water treatment plants use the primary and secondary treatment methods shown here. Wastewater can contain from 200 BOD units for sewage and domestic waste to 1500 or more BOD units for industrial waste. Effluent from the treatment plant usually contains 5 BOD units or less. Tertiary treatment of the effluent water to further reduce BOD is rarely done because it is not cost effective.

entering the treatment plant is passed through a series of grates and screens that remove large objects. The effluent is left to settle for several hours to allow solids to settle to the bottom of the separation reservoir (Figure 28.4●).

Municipalities that provide only primary treatment suffer from extremely polluted water when the effluent is discharged into adjacent waterways because high levels of soluble and suspended organic matter and other nutrients remain in water following primary treatment. These nutrients can propagate undesirable microbial growth (∞ Figure 19.10), further reducing water quality. Therefore, most treatment plants employ secondary and even tertiary treatments to reduce the organic content of the wastewater before release to natural waterways. Secondary treatment processes use microbial digestion both in the presence and absence of oxygen to further reduce organic nutrients in wastewater.

Anoxic Secondary Wastewater Treatment

Anoxic secondary wastewater treatment involves a series of digestive and fermentative reactions carried out by a number of bacterial and archaeal species under anoxic conditions. Anoxic treatment is typically used to treat wastewater containing large quantities of insoluble organic matter (and, hence, very high BOD), such as fiber and cellulose waste from food- and dairy-processing plants. The anoxic degradation process itself is carried out in large enclosed tanks called *sludge digesters* or *bioreactors* (Figure 28.5*a* and *b*●). The process requires the collective activities of many different types of microorganisms. The major reactions are summarized in Figure 28.5*c*.

Through the action of the resident anoxic microorganisms, the macromolecular waste components are first digested by polysaccharases, proteases, and lipases into soluble components. These soluble components are then fermented to yield a mixture of fatty acids, H_2,

● **Figure 28.4 Primary treatment of wastewater.** Wastewater is pumped into the reservoir (left) where settling of solids occurs. As the water level rises, the water spills through the grates to successively lower levels. Water at the lowest level, now free of solids, enters the spillway (arrow) and is pumped to a secondary treatment facility.

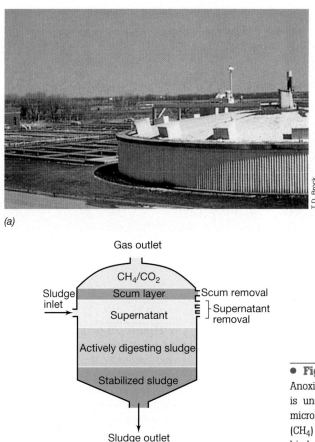

(a)

(b)

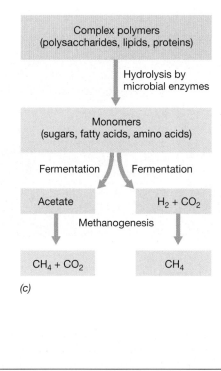

(c)

● **Figure 28.5 Anoxic secondary wastewater treatment.** (a) Anoxic sludge digester. Only the top of the tank is shown; the remainder is underground. (b) Inner workings of a sludge digestor. (c) Major microbial processes occurring during anoxic sludge digestion. Methane (CH_4) and carbon dioxide (CO_2) are the major products of anaerobic biodegradation (⚮ Section 17.17). Figure 19.23 shows the major microbial groups involved in anoxic decomposition as well as the interactions among them that lead to methanogenesis.

and CO_2; the fatty acids are further fermented to acetate, CO_2, and H_2. These products are then used as substrates by methanogenic *Archaea* (⚮ Sections 12.6 and 17.17), capable of carrying out the reactions $CH_3COOH \rightarrow CH_4 + CO_2$ and $4\,H_2 + CO_2 \rightarrow CH_4 + 2\,H_2O$ (Figure 28.5c). Thus, the major products of anoxic sewage treatment are CH_4 (methane) and CO_2. The methane can be collected and either burned off or used as fuel to heat and power the treatment plant.

Aerobic Secondary Wastewater Treatment

Aerobic secondary wastewater treatment uses digestive reactions carried out by microorganisms under aerobic conditions to treat wastewater containing low levels of organic materials. In general, nonindustrial wastewater can be treated efficiently using only aerobic secondary treatment. Several kinds of aerobic decomposition processes are used for wastewater treatment, but the trickling filter and activated sludge methods are the most common. A trickling filter (Figure 28.6a●) is a bed of crushed rocks, about 2 m thick, on top of which the wastewater is sprayed. The liquid slowly passes through the bed, the organic matter adsorbs to the rocks, and microbial growth occurs on the large exposed surface of the rocks. The complete mineralization of organic matter to

carbon dioxide, ammonia, nitrate, sulfate, and phosphate takes place in the microbial biofilm on the rocks.

The most common aerobic treatment system is the activated sludge process. Here, the wastewater to be treated is mixed and aerated in a large tank (Figure 28.6b). Slime-forming bacteria, including *Zoogloea ramigera*, among others, grow and form *flocs* (larger, aggregated masses) (Figure 28.7●). These flocs form a substratum to which protozoa, small animals, and sometimes filamentous bacteria and fungi attach. The basic process of oxidation is similar to the trickling filter. The aerated effluent containing the flocs is pumped into a holding tank or *clarifier* where the flocs settle. Some of the floc material (called *activated sludge*) is then returned to the aerator to serve as inoculum, and the rest is pumped to the anoxic sludge digestor (Figure 28.5) or is removed, dried, and burned or used for fertilizer.

Wastewater normally stays in an activated sludge tank for 5 to 10 hours, a time too short for complete oxidation of all organic matter. However, during this time much of the soluble organic matter is adsorbed to the floc and incorporated by the microbial cells. The BOD of the liquid effluent is considerably reduced (up to 95%) by this process, with most of the BOD now contained in the settled flocs. Nearly complete BOD reduction can occur if the flocs are then transferred to the anoxic sludge digestor for conversion to CO_2 and CH_4.

(a)

(b)

(c)

● **Figure 28.6 Aerobic secondary wastewater treatment processes.** A treatment facility for a small city, Carbondale, Illinois. (a) Trickling filter. The booms rotate, distributing wastewater slowly and evenly on the rock bed. The rocks are 10–15 cm in diameter, and the bed is 2 m deep. (b–c) The activated sludge process. (b) Aeration tank of an activated sludge installation in a metropolitan wastewater treatment plant. The tank is 30 m long, 10 m wide, and 5 m deep. (c) Wastewater flow through an activated sludge installation. Recirculation of activated sludge to the aeration tank introduces microorganisms responsible for oxic digestion of the organic components of the wastewater. The anoxic sludge digestor is detailed in Figure 28.5 and the major microbial interactions are described in Figure 19.23.

(diagram labels: Wastewater from primary treatment; Aeration tank; Settling tank; Clear effluent; Air; Activated sludge; Anoxic sludge digestor; Activated sludge return; Excess sludge)

Tertiary Wastewater Treatment

Tertiary wastewater treatment is any physicochemical or biological process employing bioreactors, precipitation, filtration, or chlorination procedures similar to those employed for drinking water purification (see Section 28.3). The goal is to sharply reduce the levels of inorganic nutrients, especially phosphate, nitrite, and nitrate, from the final effluent.

Most treatment plants now chlorinate the effluent at the end of the secondary treatment procedures (to further reduce the possibility of biological contamination) and discharge the treated water into streams or lakes. In the eastern United States, many wastewater treatment facilities are using ultraviolet (UV) radiation to disinfect effluent water. Ozone (O_3), a strong oxidizing agent that is an effective bactericide and viricide, is also used at over 40 plants in the United States for wastewater disinfection.

Wastewater receiving proper tertiary treatment contains so few nutrients that it cannot support extensive microbial growth. Tertiary treatment is the most complete method of treating sewage but has not been widely adopted due to expense.

🛑 *28.2 Concept Check*

Wastewater treatment is primarily concerned with treating sewage and industrial wastes to reduce the BOD (biochemical oxygen demand) to acceptable levels. Primary, secondary, and tertiary treatment of water involve physical, biological, and physicochemical processes. After secondary or optional tertiary treatment, water may be suitable for release directly to a water purification plant.

◆ What is biochemical oxygen demand (BOD)? Why is BOD reduction necessary in wastewater treatment?

◆ Identify the processes in primary, secondary (anoxic and oxic), and tertiary sewage treatment.

◆ Identify the final products of wastewater treatment. How might these end products be used?

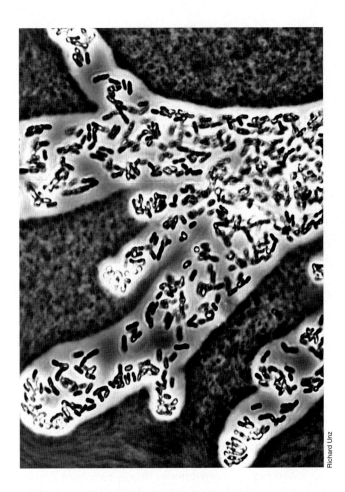

Richard Unz

● **Figure 28.7** **A floc formed by the bacterium *Zoogloea ramigera*.**
The floc found in the activated sludge process consists of a large number of
small, rod-shaped cells of *Z. ramigera* surrounded by a polysaccharide
slime layer, arranged in characteristic fingerlike projections. Negative stain
using India ink. The biology of *Zoogloea* is discussed in Section 12.7.

28.3 Drinking Water Purification

Wastewaters treated by secondary methods are generally
of such quality that they can be directly discharged into
rivers and streams. However, such water is not **potable**,
or safe for human consumption. The production of potable water requires further treatment to remove potentially pathogenic microorganisms, eliminate taste and odor,
reduce nuisance chemicals such as iron and manganese,
and decrease **turbidity** (a measure of *suspended solids*).
Suspended solids are small particles of solid pollutants
that resist separation by ordinary physical means.

Physical and Chemical Purification

A typical drinking water treatment installation for a
medium-sized city is shown in Figure 28.8*a*●. Figure
28.8*b* traces the flow of **raw water (untreated water)**
through a typical treatment scheme. Raw water is first
pumped from the source, in this case a river, to a *sedi-*

mentation basin where anionic **polymers**, alum (aluminum sulfate), and chlorine are added. **Sediment**, soil,
sand, mineral particles, and other large particles, settle
out. The sediment-free water is then pumped to a
clarifier or **coagulation basin**, a large holding tank
where **coagulation** takes place. The alum and anionic
polymers form large particles from the much smaller
suspended solids. After mixing, the particles continue to interact, forming large, aggregated masses, a process known
as **flocculation**. The large, aggregated particles, called
floc, settle out by gravity, trapping microorganisms and
absorbing suspended organic matter and sediment.

After coagulation and flocculation, the clarified
water undergoes **filtration**. The water is passed through
a series of filters designed to remove any remaining organic or inorganic solutes, as well as any suspended particles and microorganisms. The filters usually consist of
thick layers of sand, activated charcoal, and ion exchange filtration media. When combined with previous
purification steps, the filtered water is free of all particulate matter, most organic and inorganic chemicals, and
nearly all microorganisms.

Disinfection

Clarified, filtered water must then be disinfected before
it is released to the supply system as pure, potable
finished water. Chlorination is the most common
method of disinfection. In sufficient doses, chlorine kills
microorganisms within 30 minutes (certain pathogenic
protozoa such as *Cryptosporidium* are not easily killed by
chlorine treatment and thus can be important waterborne pathogens; see Section 28.6). In addition to killing
microorganisms, chlorine reacts with organic compounds, oxidizing and effectively neutralizing them.
Therefore, since most taste- and odor-producing compounds are organic in nature, chlorine treatment also improves water taste and smell. Chlorine is added to water
either from a concentrated solution of sodium hypochlorite or calcium hypochlorite or as a gas from pressurized
tanks. The latter method is used most commonly in large
water treatment plants because it is most amenable to
automatic control.

Chlorine is consumed when it reacts with organic
materials. Therefore, sufficient quantities of chlorine
must be added to water containing organic materials so
that a small amount, called the *chlorine residual*, remains
to react with the microorganisms after all reactions with
organic materials have occurred. The water plant operator performs chlorine analyses on the treated water to determine the residual level of chlorine. A residual chlorine
level of 0.2–0.6 μg/ml is suitable for most water supplies. After chlorine treatment, the now potable water is
pumped to storage tanks from which it flows by gravity
or pumps through a **distribution system** of storage tanks
and supply lines to the consumer. Residual chlorine levels ensure that the finished water will reach the

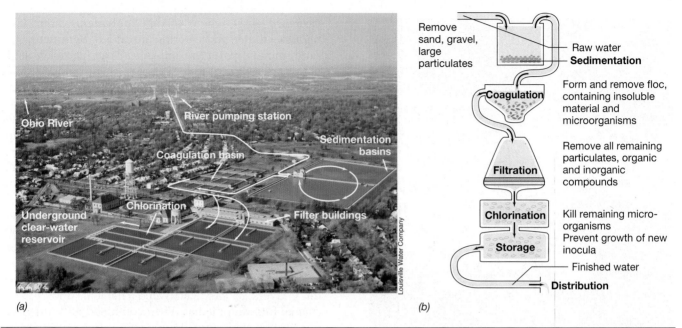

(a)   (b)

● **Figure 28.8 Water purification plant.** (a) Aerial view of a water treatment plant in Louisville, Kentucky, USA. The arrows indicate direction of flow of water through the plant. (b) Schematic overview of a typical community water purification system.

consumer without becoming contaminated (assuming that there is no catastrophic failure, such as a broken pipe, in the distribution system). Chlorine gas, even when dissolved in water, is extremely volatile and can dissipate within hours from treated water. To further ensure that residual chlorine levels are maintained throughout the distribution system, most municipal water treatment plants also introduce ammonia gas with the chlorine to form the stable, nonvolatile chlorine-containing compound **chloramine**, $HOCl + NH_3 \rightarrow NH_2Cl + H_2O$.

Ultraviolet (UV) radiation is also used as an effective means of disinfection. As we discussed in Section 22.2, UV radiation is used to treat secondarily treated effluent from water treatment plants. In Europe, UV irradiation is commonly used for drinking water applications and is being considered for use in the United States. For disinfection purpose, UV light is generated from mercury vapor lamps. Their major energy output is at 253.7 nm, a wavelength that is bactericidal and viricidal (⚭ Section 10.4). However, standard application of UV radiation does not kill cysts and oocysts of protozoans like *Giardia* and *Cryptosporidium* (see Section 28.6).

The advantages of using UV radiation over chemical disinfection procedures like chlorination are several. First, UV irradiation is a physical process that introduces no chemicals to the water. Second, short contact times allow it to be used in existing flow systems, keeping capital costs very low. Third, numerous studies indicate that no disinfection by-products are formed. Especially in smaller systems where finished water is not pumped long distances or held for long periods (reducing the need for residual chlorine), UV disinfection may be preferable to chlorination.

 28.3 Concept Check

Drinking water plants employ industrial-scale physical and chemical systems that remove or neutralize biological, inorganic, and organic contaminants from a variety of natural, community, and industrial sources. Water purification plants employ clarification, filtration, and disinfection processes to produce potable water. Finished potable water is free of chemical and biological contamination.

◆ Trace the treatment of water through a drinking water treatment plant, from the inlet to the final distribution point (faucet).

◆ What specific purposes do *sedimentation, coagulation, filtration,* and *disinfection* accomplish in the drinking water treatment process?

▌▌WATERBORNE MICROBIAL DISEASES

Common-source infectious diseases are caused by microbial contamination of materials shared by a large number of individuals. The most important common source of infectious disease is contaminated water. This is because the failure of a single step in drinking water treatment may result in exposure of thousands or even millions of individuals to an infectious agent.

Common-source waterborne diseases are a very significant source of morbidity and mortality, especially in developing countries. A variety of bacteria, viruses, and protozoa cause waterborne infectious diseases. Waterborne diseases begin as infections. Water may cause infection even if only a small number of microorganisms are present. The exact numbers of pathogens necessary to cause disease are functions of the virulence of the pathogen and the general ability of the host to resist infection (∞ Sections 21.8 and 21.14).

28.4 Sources of Waterborne Infection

Human pathogens can be transmitted through improperly treated water used for drinking and cooking. Another very common source of disease transmission is through pathogen-contaminated water used for swimming and bathing.

Potable Water

Because everyone consumes water through drinking and cooking, water is a common source of pathogen dissemination and has a very high potential for the catastrophic spread of epidemic disease. As discussed previously, water supplies in developed countries usually meet rigid quality standards, effectively limiting the sp-read of waterborne diseases. There are, however, occasional waterborne disease outbreaks in developing countries due to lapses in water quality. Isolated outbreaks affecting low numbers of individuals also occur from consumption of contaminated water from nonregulated sources such as private wells or from untreated water from streams or lakes. These sources may be contaminated by fecal material from humans or animals.

Microorganisms transmitted in water generally grow in the intestines and leave the body in feces. Fecal pollution of water supplies may then occur. In addition, if the contamination is not identified (for example, by the coliform test; see Section 28.1) and eliminated by disinfection (see Section 28.3), then a new host may consume the water and the pathogen may colonize the intestine and cause disease. In the United States, a number of different bacterial and protozoan pathogens are occasionally transmitted in drinking water (Table 28.1). We will discuss giardiasis and cryptosporidiosis in Section 28.7, but we will save our discussions of salmonellosis and *Escherichia coli* O157:H7 infections for Chapter 29, where we introduce these pathogens as agents of foodborne infections, their most common mode of transmission.

Recreational Water

Recreational water includes freshwater recreational areas such as ponds, streams, and lakes, as well as public swimming and wading pools. The operation, disinfection, and filtration of public swimming and wading pools are regulated by state and local health departments. In the United States, the Environmental Protection Agency has established guidelines for recreational fresh water (monthly geometric mean of ≤33/100 ml for enterococci or ≤126/100 ml for *Escherichia coli*), although local and state authorities can set standards above or below the guidelines. Private and therefore unregulated swimming pools, spas, and

Table 28.1	Infectious disease outbreaks associated with drinking water in the United States[a]		
Disease	**Agent**	**Outbreaks**	**Cases**
Salmonellosis	*Salmonella* species	2	208
Giardiasis	*Giardia intestinalis*	6	52
Cryptosporidiosis	*Cryptosporidium parvum*	1	5
Acute gastro-intestinal illness	*Escherichia coli* O157:H7	4	60
	Campylobacter jejuni	2	117
	E. coli O157:H7 and *C. jejuni*	1	781
	Small round virus	1	70
	Norwalk-like viruses	3	356
	Unknown	17	416

[a] Compiled from data provided by the Centers for Disease Control and Prevention for 1999–2000. There were a total of 37 outbreaks and 2065 cases of infectious disease due to drinking water contamination by infectious agents. Regulated community-owned water systems were responsible for 237 cases (11.5%). Noncommunity water systems such as those in some factories, schools, and on cruise ships accounted for 1425 cases (69%). Individual water supply systems such as wells, springs, and streams accounted for 403 cases (19.5%).

hot tubs are also occasional sources of outbreaks of waterborne diseases.

Over the last 12 years for which data are complete, an average of 13 waterborne disease outbreaks have occurred annually from recreational waters in the United States. Table 28.2 categorizes these outbreaks according to the diseases produced.

Waterborne Infections in Developing Countries

Worldwide, waterborne infections are a much larger problem than in the United States and other developed countries. Developing countries often have inadequate water and sewage treatment facilities, and access to safe, potable water is limited. As a result, diseases such as cholera (Section 28.5), typhoid fever, and amebiasis (Section 28.8) are important public health problems worldwide.

 28.4 Concept Check

Drinking water and recreational water may both be sources of waterborne pathogens. In the United States, the number of disease outbreaks due to either of these sources is relatively small in relation to the large number of exposures to water. Worldwide, lack of adequate water treatment facilities and access to clean water contribute significantly to the spread of infectious diseases.

◆ Identify the microorganism most commonly responsible for disease outbreaks due to drinking water contamination.

◆ Identify the microorganism most commonly responsible for disease incidence (numbers of infections) due to drinking water contamination.

◆ Identify the organisms most likely to cause an outbreak of disease due to contaminated recreational water.

Table 28.2	Infectious disease outbreaks associated with recreational water in the United States[a]	
Disease	**Number of outbreaks**	**Percent**
Gastroenteritis[b]	74	46.8
Dermatitis/keratitis[c]	50	31.6
Meningoencephalitis[d]	22	13.9
Other[e]	12	7.6

[a] Compiled from data provided by the Centers for Disease Control and Prevention for 1989–2000.
There were 158 outbreaks of recreational waterborne disease, or about 13 outbreaks per year.
[b] Most cases of gastroenteritis were due to *Cryptosporidium parvum* (Section 28.6), *Escherichia coli* O157:H7 (∞ Section 29.8), or a Norwalk-like virus (Section 28.8).
[c] Most cases of dermatitis were caused by *Pseudomonas aeruginosa*.
[d] Meningoencephalitis was caused by the ameba *Naegleria fowleri* (Section 28.8).
[e] Other diseases include leptospirosis caused by *Leptospira interrogans*, Pontiac fever due to infection by *Legionella* (Section 28.7), and acute respiratory infections of unknown cause.

28.5 Cholera

Cholera is a severe diarrheal disease that is now largely restricted to the developing parts of the world. Cholera is an example of a major waterborne disease that can be controlled by application of appropriate water treatment measures.

Biology and Epidemiology

Cholera is caused by *Vibrio cholerae*, a gram-negative, curved rod (∞ Section 12.12) transmitted through ingestion of contaminated water. However, as with many waterborne diseases, cholera is also associated with food consumption. For example, in the Americas, consumption of raw shellfish and raw vegetables has been associated with cholera. Presumably, the vegetables were washed in contaminated water and the shellfish beds were contaminated by untreated sewage.

Since 1817, cholera has swept the world in seven major pandemics. Two distinct strains of *V. cholerae* are recognized, known as the *classic* and the *El Tor* biotypes. The *V. cholerae* O1 El Tor biotype started the seventh pandemic in Indonesia in 1961, and its spread continues to the present. In 1992, a genetic variant, the *V. cholerae* O139 Bengal serotype, arose in Bangladesh and caused an extensive epidemic. In 2001, there were 184,311 reported cases and 2728 deaths worldwide, with over 94% of disease occurring in Africa. The 1961 pandemic has caused over 5 million cases of cholera and more than 250,000 deaths.

Cholera is endemic in Africa, Southeast Asia, the Indian subcontinent, and Central and South America. Epidemic cholera occurs frequently in areas where sewage treatment is either inadequate or absent. Even in developed countries, the disease is a threat. Sporadic endemic outbreaks of cholera (fewer than 250 total cases) have been reported in the United States, mostly along the Gulf Coast, over the last decade. Currently less than five cases are reported each year in the United States, and the cases are rarely caused by drinking water. Raw shellfish seems to be the most common vehicle, presumably because *V. cholerae* appears to be endemic and free-living in coastal waters, adhering to normal flora (Figure 28.9●).

Pathogenesis

Following ingestion of a substantial inoculum, the *Vibrio cholerae* cells take up residence in the small intestine. Studies in human volunteers have shown that stomach acidity is responsible for the large inoculum needed to initiate cholera. The ingestion of $10^8 - 10^9$ cholera vibrios is generally required to cause disease, but human volunteers given bicarbonate to neutralize gastric acidity developed cholera when given as few as 10^4 cells. Even lower cell numbers can initiate infection if *V. cholerae* is ingested with food, presumably because the food protects the vibrios from stomach acidity.

In the small intestine *V. cholerae* attaches to epithelial cells, where it grows and releases enterotoxin (∞ Section 21.11). Cholera enterotoxin causes severe diarrhea that can result in dehydration and death unless the patient is given fluid and electrolyte therapy. The enterotoxin causes fluid losses of up to 20 liters (20 kg or 44 lb) per day. The mortality rate from *untreated* cholera is generally 25 to 50%, but can be much greater under conditions of severe crowding and malnutrition.

Diagnosis, Prevention, and Treatment

Cholera is diagnosed by the presence of the gram-negative comma-shaped *Vibrio cholerae* bacilli in the "rice water" stools of patients with severe diarrhea (Figure 28.10●).

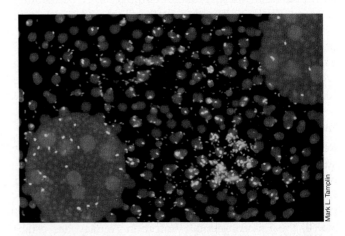

● **Figure 28.9** **Cells of *Vibrio cholerae* attached to the surface of *Volvox*, a freshwater alga.** The isolate was from a cholera-endemic area in Bangladesh. The *V. cholerae* cells are stained green by a monoclonal antibody to bacterial cell surface proteins. The red color is due to the fluorescence of chlorophyll *a* in the algae.

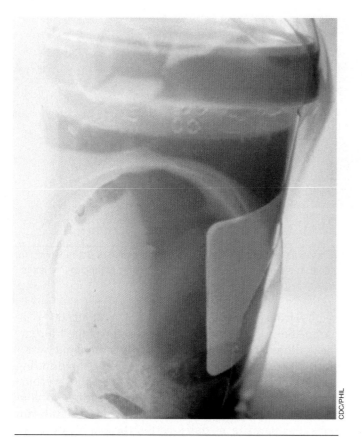

● **Figure 28.10** **A stool sample from a cholera patient.** The "rice-water" stool is nearly liquid. The solid material that has settled in a bottom layer is mucus. The stool from cholera patients is essentially isotonic with blood, containing high amounts of Na^+, K^+, and HCO_3^- (bicarbonate) ions, as well as large numbers of *Vibrio cholerae* cells. The biology and metabolism of *Vibrio cholerae* is discussed in Section 12.12, and the action of the cholera toxin is shown in Figure 21.22.

Immunization is not normally recommended for cholera prevention. Natural infection and the current vaccines offer only limited, short-duration immunity. No vaccine protects against the *V. cholerae* O139 serotype. Public health measures such as adequate sewage treatment and a reliable source of safe drinking water are the most important measures for preventing cholera. *V. cholerae* is completely eliminated from wastewater during proper sewage treatment and drinking water purification procedures. For individuals traveling in cholera-endemic areas, untreated water, raw foods, fish, and shellfish should be avoided.

Intravenous or oral liquid and electrolyte replacement therapy (20 g glucose, 4.2 g NaCl, 4.0 g NaHCO$_3$, 1.8 g KCl, dissolved in one liter of water) is the most effective means of cholera treatment. Oral treatment is preferred since no special equipment or sterile precautions are necessary. Effective treatment reduces the mortality rate to about 1%. Streptomycin or tetracycline may shorten the course of cholera, but antibiotics are of little benefit without simultaneous fluid and electrolyte replacement.

28.5 Concept Check

Vibrio cholerae is a pathogen that causes cholera, an acute diarrheal disease resulting in severe dehydration. Cholera occurs in pandemics. The current pandemic has endemic foci in the Americas, the Indian subcontinent, Asia, and Africa. In endemic areas, appropriate precautions to avoid contaminated water and food are reasonable preventative measures. Treatment with oral rehydration and electrolytes is the most efficient and effective way to treat the disease, reducing overall mortality to about 1%.

◆ Identify measures for preventing cholera in endemic areas.

◆ Define the most likely methods for acquiring cholera.

◆ Identify specific measures for treating cholera.

28.6 Giardiasis and Cryptosporidiosis

Giardiasis and cryptosporidiosis are diseases caused by the protozoans *Giardia intestinalis* and *Cryptosporidium parvum*, respectively. These organisms continue to be problematic even in well-regulated water supplies because they are found in nearly all surface waters and are highly resistant to chlorine.

Giardiasis

Giardia intestinalis is a flagellated protozoan (∞ Section 14.10) that is usually transmitted to humans in fecally contaminated water, although foodborne and sexual transmission of giardiasis has also been documented. Giardiasis is an acute gastroenteritis caused by the protozoan parasite (Figure 28.11●). The protozoal cells, called trophozoites (Figure 28.11*a*) produce a resting stage called a **cyst** (Figure 28.11*b*). The cyst has a thick protective wall that allows the pathogen to resist drying and chemical disinfection. After ingestion in contaminated water, the cysts germinate, attach to the intestinal wall, and cause the symptoms of giardiasis: an explosive, foul-smelling, watery diarrhea, intestinal cramps, flatulence, nausea, weight loss, and malaise. Symptoms may be acute or chronic. The foul-smelling diarrhea and the absence of blood or mucus in the stool are diagnostically helpful in distinguishing giardiasis from diarrhea of bacterial or viral origin. Many individuals who are infected exhibit no symptoms, simply acting as infected carriers.

Giardia intestinalis was implicated in 6 of the 37 recent drinking water infectious disease outbreaks in the United States (Table 28.1). Giardiasis can also occur after accidental ingestion of water from infected swimming pools or lakes. *Giardia* cysts have been found in 97% of surface water sources (lakes, ponds, and streams) in the United States. The thick-walled cysts are resistant to chlorine and ultraviolet radiation, and many outbreaks have been associated with water systems that use only chlorination as a means of water purification. Water

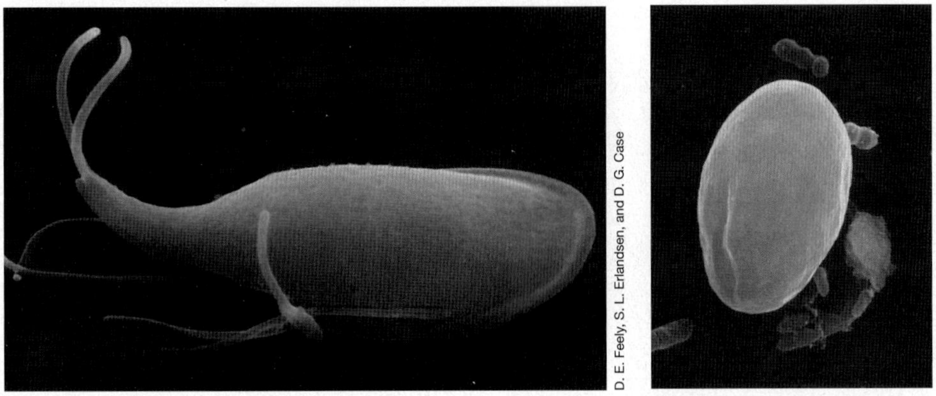

D. E. Feely, S. L. Erlandsen, and D. G. Case

S. L. Erlandsen

(a)   *(b)*

● **Figure 28.11**   **Scanning electron micrographs of the parasite *Giardia*.** (a) Motile trophozoite. The trophozoite is about 15 μm in length. (b) Cyst. The cyst is about 11 μm in length. The interesting phylogeny of this protozoan is discussed in Section 14.9. The biology of protozoa in general is discussed in Section 14.10.

subjected to proper clarification and filtration followed by chlorination or other disinfection (see Section 28.3) is generally free of *Giardia* cysts.

Isolated cases of giardiasis have also been associated with untreated drinking water in wilderness areas. Beavers and muskrats are frequent carriers of *Giardia* and may transmit cells or cysts to water supplies, causing human infection when ingested. As a safety precaution, all water consumed from rivers and streams, for example, during a camping or hiking trip, should be filtered *and* treated with iodine or chlorine, or filtered *and* boiled. Boiling is the preferred method to assure that water is free of pathogens.

Laboratory diagnostic methods include the demonstration of *Giardia* cysts in the stool or the demonstration of *Giardia* antigens in the stool using a direct ELISA (enzyme-linked immunosorbent assay; ∞ Section 24.10). The drugs quinacrine, furazolidone, and metronidazole are useful in treating acute disease.

Cryptosporidiosis

The protozoan *Cryptosporidium parvum* occurs in nature as a parasite in a variety of warm-blooded animals. The parasites are small 2–5 μm round *coccidia* that invade and grow intracellularly in mucosal epithelial cells of the stomach and intestine. The protozoan produces thick-walled, chlorine-resistant, infective oocysts, which are shed into water in high numbers in the feces of infected warm-blooded animals (Figure 28.12●). The infection is passed on when other animals consume the fecally contaminated water. *Cryptosporidium* cysts are highly resistant to chlorine (up to 14 times as resistant as the chlorine-resistant *Giardia*) and ultraviolet radiation disinfection, and therefore sedimentation and filtration methods must be used to remove *Cryptosporidium* from water supplies.

Cryptosporidium parvum was responsible for the largest single common-source outbreak of a waterborne disease ever recorded in the United States. In Milwau-

kee, Wisconsin, in the spring of 1993, over 403,000 people in the population of 1.6 million developed a diarrheal illness that was traced to the municipal water supply. Spring rains and runoff from surrounding farmland had drained into Lake Michigan and overburdened the water purification system, leading to contamination by *C. parvum*. The protozoan is a significant intestinal parasite in dairy cattle, the likely source of the outbreak.

Cryptosporidiosis is usually a self-limiting mild diarrhea that subsides in 2 weeks or less in normal individuals. However, individuals with impaired immunity such as acquired immunodeficiency syndrome (AIDS) patients or the very young or old can develop serious complications. In the Milwaukee outbreak, about 4400 people

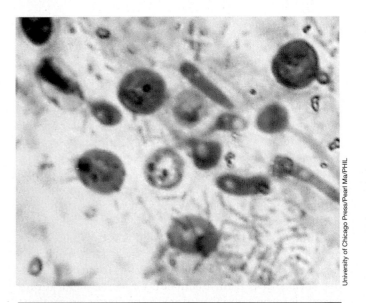

● **Figure 28.12 An acid-fast stained fecal smear showing *Cryptosporidium* oocysts.** The thick-walled oocysts, stained red, are about 3–5 μm in diameter. *Cryptospordium*, an obligate intracellular parrasite, is a member of the Sprorozoa, a group of protozoans discussed in Section 14.10.

University of Chicago Press/Pearl Ma/PHIL

required hospital care, and several died of complications from the disease, including severe dehydration.

The Milwaukee outbreak highlights the vulnerability of water purification systems, the need for constant water monitoring and surveillance, and the catastrophic consequences of the failure of a large water supply system. From an economic point of view, the epidemic cost the inhabitants of Milwaukee over $96 million in medical costs and lost productivity.

Laboratory diagnostic methods for cryptosporidiosis include the demonstration of *Cryptosporidium* oocysts in the stool (Figure 28.12). Treatment is unnecessary for those with normal immunity. For individuals undergoing immuno-suppressive therapy (e.g., prednisone), discontinuation of immunosuppressive drugs is recommended. Immunocompromised individuals should be given supportive therapy (for example, intravenous fluids and electrolytes).

 28.6 Concept Check

Giardiasis and cryptosporidiosis are spread by the chlorine-resistant cysts of *Giardia intestinalis* and *Cryptosporidium parvum*, respectively, in drinking water and recreational water contaminated by the feces of infected humans or animals. Infection with either parasite causes diarrhea and may lead to more serious disease in compromised individuals.

◆ Explain the importance of cysts in the survival and infectivity of both *Giardia intestinalis* and *Cryptosporidium parvum*.

◆ Why are protozoans often associated with waterborne diseases, even in developed countries? Outline steps to reduce their impact.

28.7 Legionellosis (Legionnaires' Disease)

Legionella pneumophila, the bacterium that causes legionellosis, is an important waterborne pathogen normally transmitted in aerosols rather than through drinking water or recreational water (⌒⌒ Section 25.11).

Biology and Epidemiology

Legionella pneumophila was first discovered as the pathogen that caused an outbreak of pneumonia during an American Legion convention in Philadelphia (USA) in the summer of 1976. *L. pneumophila* is a thin, gram-negative obligately aerobic rod (Figure 28.13●) with complex nutritional requirements, including an unusually high iron requirement. *L. pneumophila* can be detected by immunofluorescence techniques (⌒⌒ Section 24.9) and can be isolated from terrestrial and aquatic habitats as well as from patients suffering from legionellosis.

L. pneumophila is present in small numbers in lakes, streams, and soil. It is relatively resistant to heating and chlorination, so it can spread through water distribution

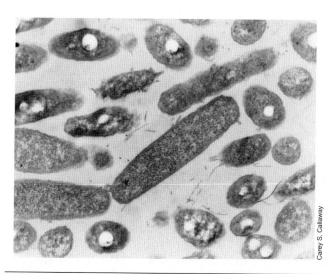

● **Figure 28.13** *Legionella pneumophila* **and Legionnaires' disease.** Transmission electron micrograph of a thin section of *Legionella pneumophila* cells. Cells are 0.3–0.6 μm in diameter and up to 2 μm in length. The clear areas in the cells are polymers of β-hydroxybutyrate, a carbon storage polymer (⌒⌒ Section 4.11).

systems. It is commonly found in large numbers in cooling towers and evaporative condensers of large air conditioning systems. The pathogen grows in the water and is disseminated in humidified aerosols. Human infection is via airborne droplets, but the infection is not spread person to person. Multiple outbreaks of legionellosis tend to peak in mid-to late summer months when air conditioners are extensively used.

L. pneumophila has also been found in hot water tanks and whirlpool spas where it grows to high numbers in warm (35–45°C), stagnant water. Epidemiological studies indicate that *L. pneumophila* infections occur at all times of the year, primarily as a result of aerosols generated by heating/cooling systems and common practices such as showering or bathing. Overall, the incidence of *reported* cases of legionellosis has been about 4–6 cases per million in the United States, but up to 90% of actual cases are probably not diagnosed or properly reported. Prevention of legionellosis can be accomplished by improving the maintenance and design of water-dependent cooling and heating systems and water delivery systems. The pathogen can be eliminated from water supplies by hyperchlorination or by heating to >63°C.

Pathogenesis

L. pneumophila is an intracellular parasite that invades and grows in alveolar macrophages and monocytes (⌒⌒ Section 22.2). Infections are often asymptomatic or produce a mild cough, sore throat, mild headache, and fever. These mild, self-limiting cases, called *Pontiac fever*, need no treatment, and resolve in 2–5 days. However, more serious infections resulting in cases of pneumonia often occur in elderly individuals whose resistance has been previously compromised. Certain serotypes of *L. pneumophila* (more than 10 are known) are strongly associated with the pneumonic form of the infection. Prior to the onset of pneumonia, intestinal disorders are common, followed by high fever, chills, and muscle aches. These symptoms precede the dry cough and chest and abdominal pains typical of legionellosis. Death occurs in up to 10% of cases and is usually due to respiratory failure.

Diagnosis and Treatment

Clinical detection of *L. pneumophila* is usually done by culture from bronchial washings, pleural fluid, or other body fluids. Serological (antibody) tests are used as retrospective evidence for *Legionella* infection (∞ Section 24.7). As an aid in diagnosis, *L. pneumophila* antigens can sometimes be detected in patient urine. *Legionella pneumophila* is sensitive to the antibiotics rifampin and erythromycin. Intravenous administration of erythromycin is the treatment of choice in most cases.

 28.7 Concept Check

Legionella pneumophila is a respiratory pathogen that causes legionellosis and Pontiac fever. *L. pneumophila* grows to high numbers in warm water and is spread via aerosols. The prevalence of legionellosis is not decreasing and infections are underreported.

◆ Indicate the source of *Legionella pneumophila*.

◆ Identify specific measures for control of *Legionella pneumophila*.

28.8 Typhoid Fever and Other Waterborne Diseases

A variety of bacteria, viruses, and protozoa can transmit common-source waterborne diseases. These diseases are a significant source of morbidity, especially in developing countries. Here we briefly discuss several other important waterborne diseases.

Typhoid Fever

On a global scale, probably the most important pathogenic *Bacteria* transmitted by the water route are *Salmonella typhi*, the organism causing typhoid fever, and *Vibrio cholerae*, the organism causing cholera, as discussed above. Although *Salmonella typhi* may also be transmitted by contaminated food (∞ Section 29.7) and by direct contact from infected individuals (∞ Section 25.4), the most common and serious means of transmission worldwide is through water. Fortunately, typhoid fever has been virtually eliminated in developed countries, primarily due to effective water treatment procedures (see Figure 28.2). In the United States, there are fewer than 400 cases in most years, but, as was discussed in Section 28.1, typhoid fever was a major public health threat before drinking water was routinely filtered and chlorinated. However, breakdown of water treatment methods, contamination of water during floods, earthquakes, and other disasters, or cross-contamination of water supply pipes from leaking sewer lines, can propagate epidemics of typhoid fever, even in developed countries.

Viruses

Viruses can also be transmitted in water and cause human disease. Quite commonly, enteroviruses such as poliovirus (∞ Section 16.8), Norwalk-like virus (∞ Table 25.8), and hepatitis A virus (∞ Section 26.11) are shed into the water in fecal material. The most serious of these is poliovirus, but wild poliovirus has been eliminated from the Western Hemisphere (∞ Section 25.9). Although viruses can survive in water for relatively long periods, they are inactivated by water disinfection agents such as chlorine. The maintenance of 0.6 parts per million (ppm) of chlorine in water (Section 28.3) ensures the neutralization of viruses in the water supply.

Amebiasis

A number of amoebae inhabit the tissues of humans and other vertebrates, usually in the oral cavity or intestinal tract, and some of these are pathogenic. We discussed the general properties of ameboid protozoa in Section 14.10.

Worldwide, *Entamoeba histolytica* is a common pathogenic protozoan transmitted to humans, primarily by contaminated water and occasionally through contaminated food (Figure 28.14●). *E. histolytica* is an *anaerobic* ameba; the trophozoites lack mitochondria. Like *Giardia*, the trophozoites of *E. histolytica* produce cysts. Ingested cysts germinate in the intestine, where amebic cells grow both on and in intestinal mucosal cells. Many infections are asymptomatic, but continued growth may lead to invasion and ulceration of the intestinal mucosa, causing diarrhea and severe intestinal cramps. Diarrhea can progress to invasion of the intestinal wall, a condition known as *dysentery*, characterized by intestinal inflammation, fever, and the passage of intestinal exudates, including blood and mucus. If not treated, invasive trophozoites of *E. histolytica* can invade the liver, and occasionally the lung and brain. Growth in these tissues can cause severe abscesses and even death. Worldwide, up to 100,000 individuals die each year from invasive amebic dysentery. The disease is extremely common in tropical and subtropical countries worldwide, with at least 50 million people developing symptomatic diarrhea annual-

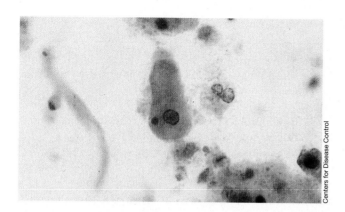

● **Figure 28.14** **The trophozoite of *Entamoeba histolytica*, the ameba that causes amebiasis.** Note the discrete, darkly stained nucleus. The small red structures are red blood cells. The trophozoites range from 12–60 μm in length. The biology of the amoebae is discussed in Section 14.10.

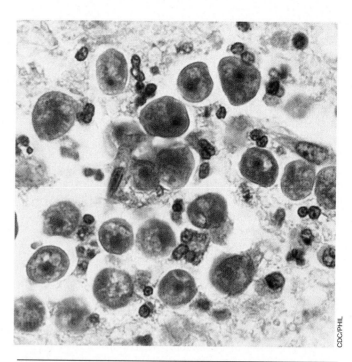

● **Figure 28.15** **Trophozoites of *Naegleria fowleri*, an ameba that causes meningoencephalitis, in brain tissue.** A large number of oval to round and ameboid (irregularly shaped) trophozoites are present as dark-stained structures with densely stained nuclei. The photograph shows extensive destruction of the surrounding brain tissue. Individual trophozoites are 10–35 μm long. The biology of the amoebae is discussed in Section 14.10.

ly and up to tenfold more having asymptomatic disease. In the United States, there are several hundred cases per year, with most occurring near international borders in the Southwest.

E. histolytica amebiasis can be treated with the drug *dehydroemetine* for invasive disease and diloxanide furoate for certain asymptomatic cases, such as in immune-compromised individuals, but amebicidal drugs are not universally effective. Spontaneous cures do occur, suggesting that the host immune system plays some role in ending the infection. However, protective immunity is not an outcome of primary infection, and reinfection is common. The disease occurs at very low incidence in regions that practice adequate sewage treatment. Ineffective sewage treatment and use of untreated surface waters for drinking purposes are the usual scenarios for cases of amebiasis.

Laboratory diagnosis involves the demonstration of *E. histolytica* cysts in the stool, trophozoites in tissue, or the demonstration of antibodies to *E. histolytica* in the blood using an ELISA (enzyme-linked immunosorbent assay; ∞ Section 24.10).

Naegleria fowleri can also cause amebiasis, but in a very different form. *N. fowleri* is a free-living ameba found in soil and in water runoff. *N. fowleri* infections usually result from swimming or bathing in warm, soil-contaminated water sources such as hot springs or lakes and streams in the summer. This free-living ameba enters the body through the nose and burrows directly into the brain. Here, the organism propagates, causing extensive hemorrhage and brain damage (Figure 28.15●). This condition is called **meningoencephalitis**. Death usually results within a week. In the past 12 years, there have been 22 outbreaks of meningoencephalitis from recreational waters in the United States (Table 28.2). From 1999 through 2000, there were

4 outbreaks, each involving a single individual who was infected by swimming or wading in a lake, pond, or stream in summer.

Diagnosis of *N. fowleri* infection requires observation of the amoebae in the cerebrospinal fluid. If a definitive diagnosis can be done quickly, the drug amphotericin B can be successfully used to treat infections.

 ### 28.8 Concept Check

Typhoid fever, viral infections, and amebiasis are important waterborne diseases. Waterborne typhoid fever and viral illnesses, while still common diseases in developing countries, have been controlled by effective water treatment in developed countries. Amebic dysentery caused by *Entamoeba histolytica* is a worldwide problem that affects millions of people. Meningoencephalitis is a rare but serious condition caused by *Naegleria* amebiasis.

◆ Explain the impact of effective water hygiene on the spread of human diseases such as typhoid fever and polio spread by fecal contamination of water supplies.

◆ Describe public health measures that could be used to eliminate or reduce the number of cases of amebiasis due to *Entamoeba histolytica* or meningoencephalitis due to *Naegleria fowleri*.

REVIEW QUESTIONS

1. Define the term *coliform* and explain the coliform test. Why is the coliform test used to assess the purity of drinking water (◯◯ Section 28.1)?

2. Trace the purification of wastewater in a typical treatment plant. What is the overall reduction in the BOD (◯◯ Section 28.2)?

3. Identify (stepwise) the methods used to process drinking water. What important contaminants are targeted by each step in the process (◯◯ Section 28.3)?

4. Identify the major sources of waterborne infection. Why are common sources of infection so dangerous to public health (◯◯ Section 28.4)?

5. Why are antibiotics ineffective for the treatment of cholera? What methods are commonly used to treat cholera victims (◯◯ Section 28.5)?

6. Giardiasis and cyptosporidiosis remain significant public health problems even in areas with stringent water-quality standards. Explain (◯◯ Section 28.6).

7. Describe the main features of legionellosis. What distinguishes this disease from other waterborne diseases (◯◯ Section 28.7)?

8. Indicate the methods that have been used to control salmonellosis, Norwalk virus, and amebiasis in water systems in developed countries (◯◯ Section 28.8).

APPLICATION QUESTIONS

1. Why is reduction in the BOD of wastewater the primary goal of wastewater treatment? What occurs if the BOD of wastewater is not significantly reduced before it is distributed to local water sources such as lakes or streams?

2. In the United States, the federal government has defined a strict set of drinking water standards. These standards are enforced by law. However, recreational water standards are not regulated as tightly. Explain why recreational water standards are more flexible and devise recreational water standards for the area in which you live.

3. Cholera has a long and complex history of interaction with humans, and remains today as a very serious waterborne disease. Worldwide, we are in the midst of an ongoing pandemic. Using sources such as the World Health Organization and the Centers for Disease Control and Prevention, define the status of the current pandemic with regard to its geographic distribution and most recent outbreaks. Comment on methods that could be used to decrease the spread of cholera and control the annual epidemic outbreaks that occur in endemic areas. Can cholera be eradicated?

4. As a visitor to a country in which cholera is an endemic disease, what specific steps would you take to reduce your risk of cholera exposure? Will these precautions also prevent you from contracting other waterborne diseases? Which ones? Identify waterborne diseases for which your precautions will not prevent infection. Present further recommendations for preventing those diseases.

5. Why are surface waters contaminated with the cysts of various protozoans? What steps might public health officials take to remedy this problem?

6. Discuss the wastewater and drinking water treatment schemes that must be in place to control such diseases as typhoid fever. Is it possible to eliminate *Salmonella typhi* as has been effectively done for poliovirus? Would it be possible, or even prudent, to eliminate *Naegleria fowleri* or *Entamoeba histolytica*? Explain.

29

FOOD PRESERVATION AND FOODBORNE MICROBIAL DISEASES

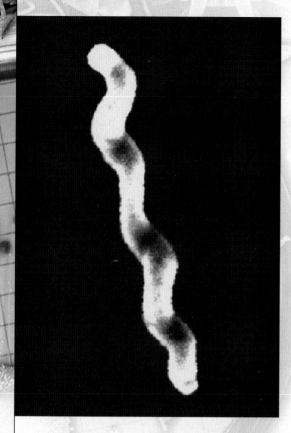

Food infections and food poisoning are important causes of morbidity and mortality worldwide. Food infections with *Campylobacter jejuni*, shown in this electron micrograph, are due largely to contaminated raw and ready-to-eat foods.

WORKING GLOSSARY

Botulism food poisoning due to ingestion of food containing botulinum toxin produced by *Clostridium botulinum*

Canning the process of sealing food in a closed container and heating to destroy living organisms

Fermentation the anaerobic catabolism of organic compounds, generally carbohydrates, in the absence of an external electron acceptor

Food infection disease caused by active infection resulting from ingestion of pathogen-contaminated food

Food poisoning (food intoxication) disease caused by the ingestion of food that contains preformed microbial toxins

Food spoilage a change in the appearance, smell, or taste of a food that makes it unacceptable to the consumer

Irradiation the exposure of food to ionizing radiation for the purpose of inhibiting growth of microorganisms and insect pests or to retard ripening

Listeriosis gastrointestinal food infection caused by *Listeria monocytogenes* that may lead to bacteremia and meningitis

Lyophilization (freeze-drying) the process of removing all water from frozen food under vacuum

Nonperishable (stable) food food of low water activity that has an extended shelf life and is resistant to spoilage by microorganisms

Perishable food fresh food generally of high water activity that has a very short shelf life due to potential for spoilage by growth of microorganisms

Pickling the process of acidifying food to prevent microbial growth and spoilage

Salmonellosis enterocolitis caused by any of over 2000 variants of *Salmonella* spp.

Semiperishable food food of intermediate water activity that has a limited shelf life due to potential for spoilage by growth of microorganisms

Water activity (a_w) the availability of water for use in metabolic processes

---

Foodborne disease is an emerging area of concern in microbiology. Microorganisms are important factors in our food supply. Microorganisms are ubiquitous in our environment and are found in water, air, and even in food. Fresh foods, most prepared foods, and even some preserved foods are contaminated with microorganisms. However, microbial activity is highly desirable in certain foods. A variety of foods are produced or enhanced by microbial action. For example, dairy products such as cheese, buttermilk, sour cream, and yogurt are all produced by microbial fermentation. Sauerkraut is a fermented vegetable food. Meat products including certain sausages, pates, and liver spreads are produced using microbial fermentation techniques. Cider vinegar is produced by lactic acid bacteria (Section 30.11), and alcoholic beverages are produced by fermentation processes using yeast (Section 30.13). We discuss the use of microorganisms to produce edible foods on an industrial scale in Chapter 30.

In this chapter, however, we will concentrate on the negative aspects of microbial growth in food. Uncontrolled and unwanted microbial growth destroys vast quantities of food, causing significant economic loss as well as a tremendous loss of nutrients. Consumption of food contaminated with particular microorganisms or microbial products can also cause food infections or food poisoning.

FOOD PRESERVATION AND MICROBIAL GROWTH

First we discuss microorganisms that are important spoilage agents in foods. We then present a variety of methods used for preserving foods by controlling unwanted microbial growth.

29.1 Microbial Growth and Food Spoilage

A wide variety of microorganisms, including a few human pathogens, colonize and grow on common foods. Most foods provide a suitable medium for the growth of some microorganisms, and microbial growth usually reduces food quality and availability.

Food Spoilage

Food spoilage is any change in the appearance, smell, or taste of a food product that makes it unacceptable to the consumer. Spoiled food is not necessarily unsafe to eat, but spoiled food is generally regarded as unpalatable and will not be purchased or consumed. Food spoilage causes economic loss to producers, distributors, and consumers in the form of reduced quality and quantity as well as higher prices.

Since foods are organic material, they provide nutrients for the growth of a wide variety of chemoorganotrophic bacteria. The physical and chemical characteristics of the food determine its degree of susceptibility to microbial activity. With respect to spoilage, foods can be classified into three major categories: (1) **perishable food**, including many fresh food items; (2) **semiperishable food**, such as potatoes and nuts; and (3) **stable** or **nonperishable food**, such as flour and sugar (Table 29.1).

These food categories differ largely with regard to *moisture content*, which is related to **water activity**, a_w (Section 6.14). Water activity is the availability of water for use in metabolic processes. *Nonperishable foods* have *low water activity* and can generally be stored for considerable lengths of time without spoilage. *Perishable and semiperishable foods* are those with *higher water activity*. These foods must be stored under conditions that slow or stop microbial growth.

Table 29.1 Food classification by storage potential

Food classification	Examples
Perishable	Meats, fish, poultry, eggs, milk, most fruits and vegetables
Semiperishable	Potatoes, some apples, and nuts
Nonperishable	Sugar, flour, rice, and dry beans

Fresh foods are spoiled by a variety of bacteria and fungi, and each type of fresh food is typically colonized by particular microorganisms (Table 29.2). Because the chemical properties of foods vary widely, different foods are colonized by the indigenous spoilage organisms that are best able to use the available nutrients.

For example, enteric bacteria such as *Salmonella, Shigella*, and *Escherichia*, all potential pathogens that live in the gut of animals, are rarely implicated in fruit or vegetable spoilage, but often contaminate and spoil meat. At slaughter, intestinal contents, including live bacteria, can leak and contaminate the meat. Likewise, lactic acid bacteria, the most common microorganisms in dairy products, are the major spoilers of milk and milk products. *Pseudomonas* species are found in both soil and animals and are thus widely involved in the spoilage of fresh foods.

Microbial growth in foods follows the normal pattern for bacterial growth (◯◯ Sections 6.4–6.6). The *lag phase* may be of variable duration in a food, depending on the contaminating organism and its previous growth history. The time required for the population density to reach a significant level in a given food product depends on both the size of the initial inoculum and the rate of growth during the *exponential phase*. The rate of growth during the exponential phase depends on the *temperature*, the *nutrient value of the food*, and other conditions of growth. Only when the microbial population density reaches a substantial level are spoilage effects usually observed. Throughout much of the exponential growth phase, population densities may be so low that no effect can be observed, and only the last few doublings lead to observable spoi-

lage (◯◯ Section 6.6). Thus, for much of the period of microbial growth in a food, there is no visible or easily detectable change in food quality.

 29.1 Concept Check

Foods often spoil due to growth of contaminating microorganisms. Foods vary considerably in their sensitivity to microbial growth, depending on their nutrient value and water content. Perishable and semiperishable foods have limited shelf life due to spoilage. A variety of microorganisms induce spoilage, and some food spoilage microorganisms are also potential pathogens.

◆ List the major categories of food with respect to water availability. Suggest storage conditions for each food type.

◆ Identify at least three bacterial *genera* that cause both food spoilage and human disease.

29.2 Food Preservation

We now examine food storage and preservation processes. In many cases these methods slow the growth of microorganisms that produce food spoilage or are foodborne human pathogens. In other cases, methods of preservation are used to kill all microorganisms.

Cold

Besides moisture, one of the most crucial factors affecting microbial growth is temperature (◯◯ Section 6.10). In general, a *lower* storage temperature results in less microbial growth and slower spoilage. A number of *psychrotolerant* (cold tolerant) microorganisms (◯◯ Section 6.11), however, can survive and grow at refrigerator temperatures. Therefore, storage of perishable food products for long periods of time (greater than several days) is possible only at temperatures below freezing. Freezing and thawing alter the physical structure of many foods. Therefore, freezing is not an acceptable preservation method for many fresh foods, but is widely used for the

Table 29.2 Microbial spoilage of fresh food[a]

Food product	Type of microorganism	Common spoilage organisms, by genus
Fruits and vegetables	Bacteria	*Erwinia,* **Pseudomonas, Corynebacterium** (mainly vegetable pathogens; rarely spoil fruit)
	Fungi	*Aspergillus, Botrytis, Geotrichum, Rhizopus, Penicillium, Cladosporium, Alternaria, Phytophthora,* various yeasts
Fresh meat, poultry, and seafood	Bacteria	*Acinetobacter, Aeromonas,* **Pseudomonas,** *Micrococcus, Achromobacter, Flavobacterium,* **Proteus, Salmonella, Escherichia, Campylobacter, Listeria**
	Fungi	*Cladosporium, Mucor, Rhizopus, Penicillium, Geotrichium,* **Sporotrichium, Candida,** *Torula, Rhodotorula*
Milk	Bacteria	**Streptococcus,** *Leuconostoc, Lactococcus, Lactobacillus,* **Pseudomonas, Proteus**
High-sugar foods	Bacteria	**Clostridium, Bacillus,** *Flavobacterium*
	Fungi	*Saccharomyces, Torula, Penicillium*

[a] The organisms listed are the most commonly observed spoilage agents of fresh, perishable foods. Genera in bold face include possible human pathogens.

preservation of meats and many fruits and vegetables. Freezers providing a temperature of −20°C are most commonly used. At −20°C, storage for weeks or months is possible, but microbial growth may still occur in pockets of liquid water trapped within the frozen mass. For long-term storage, temperatures such as −80°C (dry ice temperature) are necessary. Maintenance of such low temperatures is expensive and consequently is not used for routine food storage.

Pickling and Acidity

Another major factor affecting microbial growth in food is pH or acidity. Foods vary somewhat in pH, but most are neutral or acidic. Microorganisms differ in their ability to grow under acidic conditions, but conditions of pH 5 or less inhibit the growth of most spoilage organisms (∞ Section 6.13). Therefore, acid is often used in food preservation, a process called **pickling**. Vinegar, which is dilute acetic acid, is usually added in the pickling process (vinegar is a fermentation product of the acetic acid bacteria; its industrial production will be discussed in Section 30.11). In addition to vinegar, pickling methods usually include the addition of large amounts of salt or sugar to decrease water availability (see below) and further inhibit microbial growth. Common pickled foods include cucumbers (sweet, sour, and dill pickles), peppers, meats, fish, and fruits.

Drying and Dehydration

Water activity, or a_w, is a measure of the availability of water for use by microorganisms in metabolic processes. In pure water, the a_w is 1.00 and the molecules are loosely ordered and rearrange freely. When solute is added, the a_w is decreased. As water molecules reorder around the solute, the free rearrangement of the solute-bound water molecules becomes energetically unfavorable. The microbial cells must then compete with solute for the reduced amount of free water. In general, bacteria are poor competitors for the remaining free water, but fungi are good competitors. In practice, this means that the addition of high concentrations of solutes such as sugars or salts reduces a_w and bacterial growth is inhibited. For example, nearly all bacterial growth is inhibited at a concentration of 7.5% NaCl (a_w of 0.957) with the exception of some of the gram-positive cocci such as *Staphylococcus* species. On the other hand, molds compete well for free water under conditions of low a_w and often grow well in foods such as sugary syrups.

A number of commercially important foods are preserved by addition of salt or sugar. Foods preserved by addition of sugar are mainly fruits (jams, jellies, and preserves). Salted products are primarily meats and fish. Sausage and ham are preserved by various curing salts, including NaCl, and some also undergo a smoking process. Individual products vary widely in a_w, depending on how much salt is added and how much the meat

has been dried. Some cured meat products, such as "country ham," can be kept at room temperature for extended periods of time. Others with higher a_w still require refrigeration for long-term storage

Microbial growth can also be controlled by lowering the available water content of the food by *drying*. Milk, meat, fish, vegetables, fruit, eggs, and other economically important foods are commonly preserved by some form of drying. The least damaging physical method used to dry foods is the process of **lyophilization (freeze-drying)**, where foods are frozen and water is removed under vacuum. This method is very expensive, however, and is used mainly for specialized applications like preparation of military rations, where long-term storage under adverse conditions may be necessary.

Heating

The application of heat to reduce the bacterial load or to kill all bacteria is also employed for preservation of liquids and wet foods. *Pasteurization*, a process in which liquids are heated to a specified temperature for a precise time, was discussed in Section 20.1. Pasteurization does not sterilize liquids, but reduces the bacterial load of both spoilage organisms and pathogens and significantly extends the shelf life of the liquid.

Spray drying is the process of spraying, or atomizing, liquids such as milk in a heated atmosphere. The atomization reduces surface area on the droplets, allowing rapid drying under controlled heating conditions without destroying the food. This technology is widely used in the production of powdered milk, certain concentrated liquid dairy products, and food ingredients such as flavorings.

Canning is a process in which food is sealed in a container such as a can or glass jar and heated. The process kills all living organisms or at least ensures that there will be no growth of residual organisms. When the can is properly sealed and heated, the food should remain stable and unspoiled indefinitely, even when stored at ambient temperatures.

The temperature–time relationships for canning depend on the type of food, its pH, the size of the container, and the consistency or density of the food. Because heat must penetrate completely to the center of the food within the can, heating times must be longer for large cans or very dense foods. Acid foods can often be canned effectively by heating just to boiling, 100°C, whereas nonacid foods must be heated to autoclave temperatures (121°C). Unfortunately, heating times long enough to guarantee absolute *sterility* of every can (for example, with the use of an autoclave, ∞ Section 20.1) would change the food so greatly that it would likely be unpalatable and lose nutritional value. Therefore, even properly canned foods may not be sterile.

Microbial growth in a sealed can of food is frequently the result of organisms that produce extensive amounts of gas. This can build pressure inside the can,

● **Figure 29.1   Changes in sealed tin cans as a result of microbial spoilage.** (a) Normal can; the top of the can is slightly indented due to negative pressure (vacuum) inside. (b) Swelling resulting from minimal gas production. Note that the top is slightly distended. (c) Severe swelling due to extensive gas production. (d) The can shown in (c) was dropped and the gas pressure resulted in a violent explosion, tearing the lid apart.

resulting in bulges or, in severe cases, explosion of the can (Figure 29.1●). The environment inside a can is anoxic, and some of the anoxic bacteria that grow in canned foods are toxin producers of the genus *Clostridium* (∞ Sections 12.20, 21.10 and 29.6). Therefore, food from a can that is visibly altered should never be eaten. On the other hand, the lack of obvious gas production is not an absolute guarantee that the food is safe to consume.

Chemical Preservation

Over 3000 different compounds are used as food additives. These chemical additives are classified by the U.S. Food and Drug Administration as *"generally recognized as safe" (GRAS)* and find wide application in the food industry for enhancing or preserving texture, color, freshness, or flavor. A small number of these compounds are used to control microbial growth in food (Table 29.3). Many of these microbial growth inhibitors, like *sodium*

propionate, have been used for many years with no evidence of human toxicity. Others, like *nitrites* (carcinogen precursor), *ethylene oxide* and *propylene oxide* (mutagens; ∞ Sections 10.4 and 10.5), are more controversial food additives because of statistical evidence that these compounds may negatively affect human health. However, the use of spoilage-retarding additives significantly extends the useful shelf life of finished foods. Thus, chemical food additives contribute significantly to an increase in quantity and in the perceived quality of available food items.

Because of lengthy and costly testing programs for any new chemical proposed as a food preservative, very few new compounds will likely be added to the list of safe and approved chemical food preservatives in Table 29.3 in the near future.

Irradiation

Irradiation of food with ionizing irradiation is now a standard method for reducing contamination by bacteria, fungi, and even insects (∞ Section 20.2). Table 29.4 lists foods for which radiation treatment has been approved. Foods such as spices are routinely irradiated. In the United States, fresh meat products such as hamburger and poultry can now be irradiated to limit contamination by *Escherichia coli* O157:H7 and other enteric pathogens (hamburger) and *Campylobacter jejuni* (poultry). For food irradiation, gamma rays are generated from ^{60}Co or ^{137}Cs sources, or from high-energy electrons produced by linear accelerators. The food products receive a controlled radiation dose. This dose varies considerably by each food category and purpose. For example, a dose of 44 kilograys (kGys) is used to sterilize meat products used on United States NASA space flights and is nearly 10

Table 29.3	Chemical food preservatives
Chemical	**Foods**
Sodium or calcium propionate	Bread
Sodium benzoate	Carbonated beverages, fruit, fruit juices, pickles, margarine, preserves
Sorbic acid	Citrus products, cheese, pickles, salads
Sulfur dioxide, sulfites, bisulfites	Dried fruits and vegetables, wine
Formaldehyde (from food-smoking process)	Meat, fish
Ethylene and propylene oxides	Spices, dried fruits, nuts
Sodium nitrite	Smoked ham, bacon

Table 29.4	Irradiated foods by category and purpose[a]
Food category	**Purpose for irradiation**
Fresh pork	Control of *Trichinella spiralis* parasite
Fresh fruits and vegetables	Inhibition of growth and maturation (ripening)
Dried spices, herbs, and flavoring mixtures	Microbial disinfection
Refrigerated or frozen uncooked meat products, including ground meat	Control of foodborne pathogens
	Extension of shelf life
Packaged frozen meats used in the National Aeronautics and Space Administration (NASA) flight program	Sterilization
Dry or dehydrated enzyme preparations (e.g., meat tenderizer)	Microbial disinfection
Frozen, uncooked poultry and poultry products	Control of foodborne pathogens

[a] Consumer labeling laws in the United States require that all irradiated foods must be marked with the radura (Figure 29.2) and be conspicuously labeled "Treated with radiation" or "Treated by irradiation" in addition to information required by other regulations.

times higher than the dose of 4.5 kGys used for control of pathogens in hamburger (Table 29.4). In the United States, a consumer product information label must be affixed to foods that are irradiated (Figure 29.2●).

29.2 Concept Check

Food microbiology deals with methods for limiting spoilage and the growth of disease-causing microorganisms in food during processing and storage. Foods vary considerably in their sensitivity to microbial growth, depending on their nutrient content, water availability, and pH. The growth of microorganisms in perishable foods can be controlled by refrigeration, freezing, canning, pickling, dehydration, chemical preservation, or irradiation.

◆ Outline at least four methods of food preservation. How does each method limit growth of microorganisms?

◆ Identify food spoilage microorganisms that are also pathogens.

● **Figure 29.2 The radura, the international symbol for radiation.** Packaging of foods treated with radiation must be labeled with the radura as well as the statement "treated by irradiation" or "treated with radiation."

29.3 Fermented Foods

Here we consider foods that are preserved, produced, or enhanced through the actions of microorganisms. Microorganisms are instrumental in the preparation of a wide variety of common foods and beverages. Microbial processes can produce significant alterations in raw products, and the product is called a *fermented food*. **Fermentation** is the anaerobic catabolism of organic compounds, generally carbohydrates, in the absence of an external electron acceptor (∞ Sections 5.9 and 5.10). The microorganisms involved in food fermentations include the lactic acid bacteria (∞ Section 12.19), the acetic acid bacteria (∞ Section 12.8), and the propionic acid bacteria (∞ Section 12.22) (Table 29.5). These bacteria do not grow below about pH 4, so food fermentation is a self-limiting process .

One of the most common fermented foods is yeast bread, in which the fermentation of simple sugars and grain carbohydrates by the yeast *Saccharomyces cerevisiae* (Table 29.5) produces carbon dioxide, raising the bread and producing the holes in the finished loaf (Figure 29.3●). Fermented beverage products such as wine, beer, and whiskey will be discussed in Chapter 30, as large-scale industrial applications of microbial processes (∞ Microbial Sidebar, Chapter 5).

Dairy Products

Fermented dairy products were originally developed as methods to preserve milk, an economically important fresh food that normally undergoes rapid spoilage (Section 29.2). Dairy products include cheese and other fermented milk products such as yogurt, buttermilk, and sour cream (Figure 29.3). Milk contains the disaccharide *lactose*. Lactose can be hydrolyzed by the enzyme *lactase* into glucose and galactose. These monosaccharides are fermented to the final product of lactic acid (∞ Figure 5.14). The result of this fermentation is a significant decrease in pH from neutral or slightly basic in raw milk to less than 5.3 in cheeses and less than 4.6 in other fermented milk products. The organisms that accomplish this fermentation are members of the lactic acid bacteria, and all produce lactic acid as a major fermentation product (∞ Section 12.19) (Table 29.5).

Starter cultures of various lactic acid bacteria are introduced into the raw milk and fermentation proceeds for various times, depending on the desired product (Table 29.5). For some cheeses, a second inoculum may be introduced to produce another fermentation. For example, Swiss-type cheeses are inoculated with *Propionibacterium*. The secondary fermenter catabolizes lactic acid to propionic acid, acetic acid, and CO_2 (∞ Section 12.22 and Figure 12.68). The carbon dioxide produces the large holes or eyes that characterize Swiss-type cheeses. The secondary addition of *Lactobacillus* and the mold *Penicillium roqueforti* produce the blue veins and distinctive taste and aroma of blue cheese. Each cheese type is produced under carefully controlled conditions. Time of fermentation, temperature, the extent of aging, and the types of fermentative microorganisms must be rigidly controlled to ensure a distinctive and reproducible product.

● **Figure 29.3** **Fermented foods.** Bread, sausage meats, cheeses, other dairy products, and vegetables, are all food products that are produced or enhanced by fermentation reactions catalyzed by microorganisms.

Table 29.5	Fermented foods and primary fermentation microorganisms[a]
Food category	**Primary fermenting microorganism**
Dairy foods	
Cheeses	*Lactococcus*
	Lactobacillus
	Streptococcus thermophilus
Fermented milk products	
Buttermilk	*Lactococcus*
Sour cream	*Lactococcus*
Yogurt	*Lactobacillus*
	Streptococcus thermophilus
Alcoholic beverages	*Zymomonas*
	Saccharomyces[b]
Yeast breads	*Saccharomyces cerevisiae*[c]
Meat products	
Dry sausages (pepperoni, salami)	*Pediococcus*
and semidry sausages	*Lactobacillus*
(summer sausage, bologna)	*Micrococcus*
	Staphylococcus
Vegetables	
Cabbage (sauerkraut)	*Leuconostoc*
	Lactobacillus
Cucumbers (pickles)	Lactic acid bacteria
Soy sauce	*Aspergillus*
	Tetragenococcus halophilus
	Yeasts

[a] Unless otherwise noted, these are all members of the lactic acid bacteria (∞ Section 12.19). *Zymomonas* is related to *Pseudomonas* (∞ Section 12.7).
[b] Yeast (∞ Section 30.13). A variety of *Saccharomyces* species are used in alcohol fermentations.
[c] Baker's yeast.

Meat Products

Meat products fall into several categories. *Sausages* are generally made from pork or beef and, occasionally, from poultry meats. The most common are the dry sausages, such as salami and pepperoni, and the semidry sausages such as bolognas and summer sausages (Figure 29.3).

Sausages are made using a uniformly blended mixture of meat, salt, and seasonings. A starter culture of lactic acid bacteria is added, and fermentation proceeds to reduce the pH in the sausages to below 5. After fermentation, sausages are often smoked and then dried. The moisture content of dry sausages is about 30%. After final processing, dry sausages can be held at room temperature for extended periods of time. The semidry sausages have a final moisture content of about 50% and are less resistant to spoilage if not refrigerated.

Fish, often mixed with rice, shrimp, and spices, are also fermented to make a wide variety of fish pastes and fish-flavoring products.

Vegetables and Vegetable Products

A number of vegetable food products are made using fermentation processes, and the variety of specialty fermented vegetable foods is practically endless. The most economically important fermented vegetable foods are sauerkraut (fermented cabbage) and some pickles (fermented cucumbers). Olives, onions, tomatoes, and peppers, as well as a variety of fruits, are also fermented.

Vegetables are often fermented in salt brine to aid in preservation and enhance flavor. Fermentation may also improve digestibility by breaking down plant tissues. For example, fermented legume products (peas, beans, lentils) have a marked reduction in the flatulence-producing oligosaccharides that characterize fresh legumes.

Soy sauce is a complex fermentation product made by fermentation of soy beans and wheat. Cooked soybeans are mixed with an enzyme preparation derived from a culture of *Aspergillus* (Table 29.5) on rice or other cereals. This starter culture is spread on a wheat-soybean mixture and is then grown for 2–3 days. This preparation, called *koji*, is mixed with a brine (17–19% NaCl) and fermentation is allowed to proceed for 2–4 months or more. Various microbial species, including members of the *Lactobacillus* and *Pediococcus* genera, as well as several yeasts and molds, produce fermentation products that contribute to the desirable characteristics of the final product. After fermentation, the liquid sauce is decanted, filtered, pasteurized (Section 20.1), and bottled as soy sauce.

 29.3 Concept Check

Microbial fermentation is an important process used for preserving and enhancing a number of foods, including breads, dairy products, meats, and vegetables.

◆ Identify important dairy, meat, and vegetable products that are produced or enhanced by microbial fermentation processes.

◆ Identify the microbial group or groups that are most important for food fermentations.

II MICROBIAL SAMPLING AND FOOD POISONING

Failure to adequately decontaminate and preserve food may allow the growth of pathogens, resulting in diseases with significant morbidity and mortality. Like waterborne diseases, foodborne illnesses are usually common-source diseases. A single contaminated food source at a food-processing plant or a restaurant, for instance, may affect a large number of individuals. Each year in the United States, there are nearly 25,000 *reported* foodborne disease outbreaks, causing an estimated 13 million foodborne illnesses. Most of these reported outbreaks affect very small numbers of individuals, occurring in isolated instances in the home. These outbreaks are usually the result of improper food handling and preparation at the consumer level. Most foodborne outbreaks and diseases remain unreported because the connection between food and illness is not made.

29.4 Foodborne Diseases and Microbial Sampling

Table 29.6 summarizes the most prevalent foodborne diseases in the United States and the microorganisms that cause them. These common diseases can be separated into two categories, *food poisoning* and *food infection*. Specialized microbial sampling techniques are necessary to isolate the pathogens responsible for foodborne diseases.

Foodborne Diseases

Food poisoning or **food intoxication** is disease that results from ingestion of foods containing preformed microbial toxin. The microorganisms that produced the toxins do not have to grow in the host and are often not alive at the time the contaminated food is consumed. The illness is due to ingestion and subsequent action of the preformed bioactive toxin. We previously discussed some of these toxins, notably the exotoxin of *Clostridium botulinum* (Section 21.10) and the superantigen toxins of *Staphylococcus aureus* (Section 22.16). **Food infection** is an infection resulting from ingestion of pathogen-contaminated food. We will specifically discuss major foodborne infections in Sections 29.7–29.11.

Microbial Sampling for Foodborne Disease

As discussed in Section 29.1, microorganisms are always present in fresh foods. Because pathogens may be present along with many harmless organisms, rapid non–culture-based methods have been developed to detect important pathogens such as *Escherichia coli* O157:H7, *Salmonella*, *Staphylococcus*, and *Clostridium botulinum*. A number of molecular and immunology-based methods are in use to test for both toxins and pathogen contamination of foods and other commodities such as drugs and cosmetics. We discussed in Section 24.12 the use of nucleic acid probes and PCR for the detection of specific pathogens, including foodborne pathogens.

The presence of a potential foodborne pathogen or toxin is, however, often not enough to link particular foods or pathogens to a specific foodborne disease outbreak. To complete investigations of foodborne illness outbreaks, it is desirable to isolate and positively identify the causal organism. Isolation and growth of pathogens from nonliquid foods usually requires preliminary treatment to suspend microorganisms embedded or entrapped within the food. The most suitable method is with a specialized blender called a *stomacher*. The stomacher blends and homogenizes using paddles to crush and extrude the food sample. The food should be examined as soon after sampling as possible; if examination cannot begin within 1 hour of sampling, the food should be refrigerated. A frozen food should be thawed in its original container in a refrigerator and examined or cultured as soon as thawing is complete. In addition to recovering agents from the food, it is also necessary to recover foodborne pathogens from the patients to establish a cause and effect relationship between the pathogen and the illness. In many cases, fecal samples can be cultured to recover organisms for testing (Section 24.1).

Food samples or patient samples can be inoculated onto enriched media, followed by transfer to differential or selective media for isolation and identification, as we described for human pathogens (Section 24.2). Final identification of foodborne pathogens by growth charac-

Table 29.6	Annual foodborne disease estimates for the United States[a]			
Organism	**Disease[b]**	**Number per year**	**Foods**	
Bacteria				
Bacillus cereus	FP and FI	27,000	Rice and starchy foods, high-sugar foods, meats, gravies, pudding, dry milk	
Campylobacter jejuni	FI	1,963,000	Poultry, dairy	
Clostridium perfringens	FP and FI	248,000	Cooked and reheated meats and meat products	
Escherichia coli O157:H7	FI	63,000	Meat, especially ground meat	
Other enteropathogenic *Escherichia coli*	FI	110,000	Meat, especially ground meat	
Listeria monocytogenes	FI	2,500	Meat and dairy	
Salmonella spp.	FI	1,340,000	Poultry, meat, dairy, eggs	
Staphylococcus aureus	FP	185,000	Meat, desserts	
Streptococcus spp.	FI	50,000	Dairy, meat	
Yersinia enterocolitica	FI	87,000	Pork, milk	
All other bacteria	FP and FI	102,000		
Total bacteria		**4,177,500**		
Protozoa				
Cryptosporidium parvum	FI	30,000	Raw and undercooked meat	
Cyclospora cayetanensis	FI	16,000	Fresh produce	
Giardia lamblia	FI	200,000	Contaminated or infected meat	
Toxoplasma gondii	FI	113,000	Raw and undercooked meat	
Total protozoa		**359,000**		
Viruses				
Norwalk-like viruses	FI	9,200,000	Shellfish, many other foods	
All other viruses	FI	82,000		
Total viruses		**9,282,000**		
Total Annual Foodborne Diseases		**13,818,500**		

[a] Estimates are based on data provided by the Centers for Disease Control and Prevention, Atlanta, GA, USA and are typical of recent years.
[b] FP, food poisoning: FI, food infection.

teristics as well as the use of genetic fingerprinting techniques such as PCR, nucleic acid probes, nucleic acid sequencing, and other molecular diagnostic methods (◻Section 24.12) may be necessary to link specific organisms to foodborne diseases and individual illnesses.

 29.4 Concept Check

Foodborne diseases include food poisoning and food infection. Food poisoning results from the action of microbial toxins, and food infections are due to the growth of microorganisms in the body. Specialized techniques are used to sample microorganisms in food.

◆ Distinguish between food infection and food poisoning.

◆ Describe the sampling of a solid food such as meat for the presence of microorganisms.

◆ What methods are used for the identification of foodborne pathogens?

29.5 Staphylococcal Food Poisoning

A very common food *poisoning* is caused by *Staphylococcus aureus*. This organism and other members of the genus are small, gram-positive cocci (Figure 29.4●) (◻

Section 12.19). As we discussed in Section 26.9, staphylococci are found on the skin and in the respiratory tract of nearly all humans and are often opportunistic pathogens. *S. aureus* is frequently associated with food poisoning because it can grow in many common foods and some strains produce several heat-stable protein superantigen enterotoxins (◻ Section 22.16). If food that contains toxin is ingested, gastroenteritis characterized by nausea, vomiting, and diarrhea, occurs within 1–6 hours.

Epidemiology

Each year, an estimated 185,000 cases of staphylococcal food poisoning occur in the United States (Table 29.6). The foods most commonly involved are custard- and cream-filled baked goods, poultry, meat and meat products, gravies, egg and meat salads, puddings, and creamy salad dressings. If such foods are kept refrigerated after preparation, they usually remain safe because *S. aureus* growth is significantly reduced at low temperatures. Foods of this type, however, are often kept in kitchens at room temperature or outdoors at picnics. The food, if inoculated with *S. aureus* from an infected food handler, supports rapid bacterial growth and enterotoxin production. Even if the toxin-containing foods are reheated again before eating, the toxin is relatively

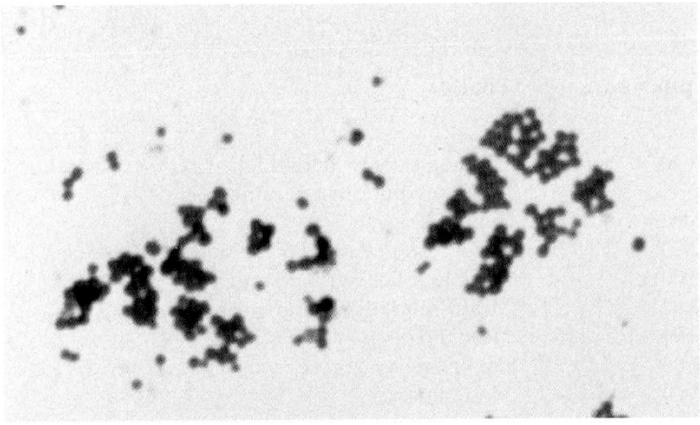

● **Figure 29.4** *Staphylococcus aureus.* In this gram stain, the individual gram-positive cocci are about 0.8 μm in diameter. Staphylococci can divide on several different axes, giving rise to the typical "grape-like" clusters of cells.

heat-stable and may remain active. Live *S. aureus* need not be present in foods causing illness: The illness is solely due to the preformed toxin.

Staphylococcus aureus strains produce at least seven different but related enterotoxins. Most strains of *S. aureus* produce only one or two of these toxins, and some produce none. However, any one of these toxins can cause staphylococcal food poisoning. These enterotoxins are further classified as *superantigens*. Superantigens work by stimulating large numbers of T cells, which in turn release intercellular mediators called cytokines, activating a general inflammatory response in the intestine that results in gastroenteritis, including massive loss of fluids from the intestine (Section 22.16).

The most common enterotoxin produced by *S. aureus* is enterotoxin A, a peptide encoded by a chromosomal gene. Cloning and sequencing of this gene, the *entA* gene, and of several other *S. aureus* enterotoxin genes, show that the toxins are genetically related. Although the *entA* gene is on the bacterial chromosome, type B and C *S. aureus* enterotoxins may be encoded on plasmids, transposons, or lysogenic bacteriophages. We discussed the importance of accessory genetic elements such as plasmids and bacteriophages as vectors for toxin production in Sections 21.10 and 21.11.

Diagnosis, Treatment, and Prevention

Assays based on the detection of either enterotoxin (ELISA detection of enterotoxin; Section 24.10) or *S. aureus* exonuclease (an enzyme that degrades DNA) are available to detect *S. aureus* in food. However, these rapid tests are qualitative, only confirming the presence or absence of *S. aureus* products above the detection limits of the assay. In addition, the exonuclease test does not indicate the presence of exotoxin, and thus is not predictive for food poisoning. To obtain quantitative data and determine the extent of bacterial contamination, bacterial plate counts are required. For staphylococcal counts, a high-salt medium (either sodium chloride or lithium

chloride at a final concentration of 7.5%) is used. Compared to most bacteria present in foods, staphylococci can thrive in habitats with high-salt and low-water activity (Section 29.1).

S. aureus food poisoning can be quite severe, but is self-limiting, usually resolving within 48 hours after onset. Severe cases may require treatment for dehydration. Treatment with antibiotics is not useful because the disease is caused by a preformed toxin, not an active infection.

Staphylococcal food poisoning can be prevented by careful sanitation and hygiene measures both in food production and food preparation steps and by storing foods at low temperatures to inhibit bacterial growth. Foods susceptible to colonization by *S. aureus* and kept for several hours above 4°C (refrigerator temperature) should be discarded rather than eaten.

29.5 Concept Check

Staphylococcal food poisoning results from the ingestion of preformed enterotoxin, a superantigen produced by *Staphylococcus aureus* when growing in foods. In many cases, *S. aureus* cannot be cultured from the contaminated food.

◆ Identify the symptoms of staphylococcal food poisoning and explain the activity of staphylococcal enterotoxins.

◆ Will antibiotics affect the outcome or the severity of staphylococcal food poisoning? Explain.

29.6 Clostridial Food Poisoning

Both *Clostridium perfringens* and *Clostridium botulinum* cause serious food poisoning. Members of the *Clostridium* genus are anaerobic endospore formers (Section 12.20). Canning and cooking procedures kill living organisms but do not kill endospores. Under appropriate anaerobic conditions, the endospores germinate and toxin is produced.

Clostridium perfringens Food Poisoning

Clostridium perfringens is an anaerobic, gram-positive spore-forming rod commonly found in soil (Figure 29.5●). It also lives in small numbers in the intestinal tract of many animals and humans and is therefore found in sewage (Section 12.20). *C. perfringens* is the most prevalent reported cause of *food poisoning* in the United States, with an estimated 248,000 annual cases (Table 29.6).

The disease results from the ingestion of a large dose of *Clostridium perfringens* ($>10^8$ cells) in contaminated cooked and uncooked foods, especially meat, poultry, and fish. Large numbers of *C. perfringens* can grow in meat dishes cooked in bulk (heat penetration in these situations is often slow and insufficient) and then left at 20–40°C for short time periods. Endospores of *C. perfringens* germinate under anoxic conditions, such as in a sealed container, and grow quickly in the meat. However, the toxin is not yet present.

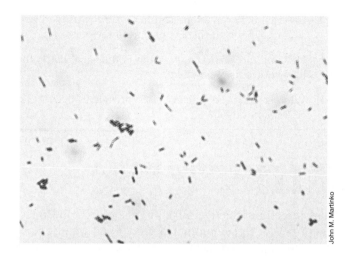

● **Figure 29.5 Gram stain of *Clostridium perfringens*.** The individual gram-positive rods are about 1 μm in diameter.

After consumption of the contaminated food, the living *C. perfringens* begins to sporulate in the intestine, triggering production of the perfringens enterotoxin (∞ Table 21.4). The enterotoxin alters the permeability of the intestinal epithelium, leading to diarrhea and intestinal cramps, usually with no fever or vomiting. The onset of perfringens food poisoning begins about 7–15 hours after consumption of the contaminated food, but usually resolves within 24 hours, and fatalities are rare.

Diagnosis, Treatment, and Prevention

Diagnosis of perfringens food poisoning is made by isolation of *C. perfringens* from the gut or, more reliably, by a direct enzyme-linked immunosorbent assay (ELISA) to detect *C. perfringens* enterotoxin in feces (∞ Section 24.10). Because *C. perfringens* food poisoning is self-limiting, treatment is usually not necessary, although antitoxins are available (∞ Section 22.13). Prevention of perfringens food poisoning requires measures to prevent contamination of raw and cooked foods and control of cooking and canning procedures to ensure proper heat treatment of all foods. Cooked foods should be refrigerated as soon as possible to rapidly lower temperatures and inhibit *C. perfringens* growth.

Botulism

Botulism is a severe food poisoning; it is often fatal and occurs following the consumption of food containing the exotoxin produced by the anaerobic, gram-positive rod *Clostridium botulinum*. This bacterium normally inhabits soil or water, but its endospores may contaminate raw foods before harvest or slaughter. If the foods are properly processed so that the *C. botulinum* endospores are removed or killed, no problem arises; but if viable endospores are present, they may initiate growth and toxin production. Even a small amount of the resultant neurotoxin can be poisonous.

We discussed the nature and activity of botulinum toxin in Section 21.10 (∞ Figure 21.20). Botulinum toxin

is a neurotoxin that causes flaccid paralysis, usually affecting the autonomic nerves that control body functions such as respiration and heartbeat. At least seven distinct types of botulinum toxin are known. The toxins are destroyed by high heat (80°C for 10 minutes), and so thoroughly cooked food, even if contaminated with toxin, *may* be harmless.

Most cases of botulism occur as a result of eating foods that are not cooked after processing (Figure 29.6a●). For example, nonacid, home-canned vegetables (e.g., home-canned corn and beans) are often used without cooking when making cold salads. Smoked and fresh fish, vacuum packed in plastic, are also often eaten without cooking. Under such conditions, *C. botulinum* endospores germinate, and the resulting cells produce toxin. If these foods are consumed, then ingestion of even

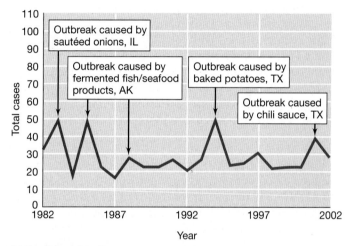

(a) Foodborne botulism

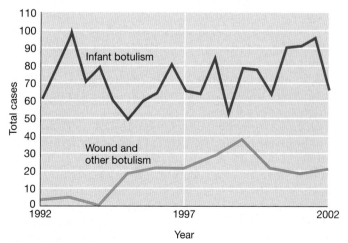

(b) Infant, wound, and other botulism

● **Figure 29.6 The incidence of botulism in the United States.** (a) Foodborne botulism. In years with high numbers of cases, major outbreaks that account for the increase are indicated. (b) Other botulism, including infant botulism. More than half of the cases of infant botulism in the United States occur in California. Wound botulism and other botulism includes all other sources. Data are from the Centers for Disease Control and Prevention, Atlanta, GA, USA.

a small amount will result in this severe and highly dangerous type of food poisoning.

Infant botulism occurs when endospores of *Clostridium botulinum* are ingested, sometimes from raw honey, but more often from no identifiable source (Figure 29.6b). If the infant's normal flora is not well developed or if the infant is undergoing antibiotic therapy, endospores may germinate in the infant's intestine and *C. botulinum* cells may grow and release toxin. Most cases of infant botulism occur between the first week of life and 2 months of age; infant botulism is rare in children older than 6 months when the normal intestinal flora is more developed (∞ Section 21.4). About 70% of all botulism cases in the United States are infant botulism. Wound botulism also occasionally occurs, presumably due to endospores introduced by introduction of a foreign substance by a parenteral route. Wound botulism is commonly associated with illicit injectable drug use (Figure 29.6b).

All forms of botulism are quite rare, with at most six cases occurring per 10 million individuals per year in the United States in recent years. Over the last decade there have been fewer than 155 cases or less each year, but up to 25% of all cases are fatal. Death occurs from respiratory paralysis or cardiac arrest due to the paralyzing action of the botulinum neurotoxin (∞ Section 21.10).

Diagnosis, Treatment, and Prevention

Diagnosis of botulism is by demonstrating botulinum toxin in patient serum or by finding toxin or live *Clostridium botulinum* in suspected food products. Laboratory findings are coupled with clinical observations including neurological signs of localized paralysis (impaired vision and speech) beginning 18–24 hours after ingestion of contaminated food. Treatment involves administration of antitoxin if the diagnosis is early (∞ Section 22.13) and mechanical ventilation for symptoms of flaccid respiratory paralysis (∞ Section 21.10). Prevention requires maintaining careful controls over canning and preservation methods. Heating susceptible foods to destroy endospores, or boiling for 20 minutes destroys the toxin. Home-prepared foods are the most common source of individual foodborne botulism outbreaks.

In infant botulism, *C. botulinum* and toxin are often found in bowel contents. Infant botulism is usually self-limiting, and most infants recover with only supportive therapy, such as assisted ventilation. Occasional deaths may occur due to respiratory failure. Honey is an occasional source of *C. botulinum* endospores. Therefore, feeding honey to children under 2 years of age is not recommended.

 29.6 Concept Check

Clostridium food poisoning results from ingestion of toxins produced by microbial growth in foods or due to microbial growth and toxin production in the body. Perfringens food poisoning is quite common and is usually a self-limiting gastrointestinal disease. Botulism is a rare but very serious disease, with significant mortality.

◆ Describe the events that lead to *C. perfringens* food poisoning. What is the likely outcome of the poisoning?

◆ Describe the events that lead to botulism. What is the likely outcome of botulism?

◆ Why is infant botulism more prevalent than other types of botulism?

III FOOD INFECTION

Food infection is active infection resulting from ingestion of pathogen-contaminated food. Food may contain sufficient numbers of viable pathogens to cause infection and disease in the host. Food infection is a very common type of foodborne illness (Table 29.6), and we begin with *Salmonella* food infection, a typical and widespread example. Many food infection agents also cause waterborne diseases (∞ Chapter 28).

29.7 Salmonellosis

Although sometimes called food poisoning, **salmonellosis** is a gastrointestinal disease due to foodborne *Salmonella* infection. Symptoms begin only after the pathogen colonizes the intestinal epithelium.

Salmonella are gram-negative facultatively aerobic rods related to *Escherichia coli*, *Shigella*, and other enteric *Bacteria* (∞ Section 12.11). *Salmonella* normally inhabit the gut of animals and are thus found in sewage. Virtually all *Salmonella* are pathogenic for humans. One, *S. typhi*, causes the serious human disease typhoid fever, but it is very rare in the United States, with most of the 500 foodborne cases imported from other countries. However, a number of *Salmonella* species cause foodborne gastroenteritis. In all, over 2000 *serovars*, or individual variants, of various *Salmonella* species are known to be pathogenic for humans. *S. typhimurium* is the most common agent of salmonellosis.

Epidemiology

The incidence and prevalence of *reported* salmonellosis has been steady over the last decade, with about 40,000–45,000 documented cases each year (Figure 29.7●). However, less than 4% of the total cases of salmonellosis are probably reported, and the estimated number of actual cases of salmonellosis is over 1.3 million every year (Table 29.6).

The ultimate sources of the foodborne salmonellas are the intestinal tracts of humans and warm-blooded animals, and several mechanisms may introduce these organisms into the food supply. The organism may reach food by fecal contamination from food handlers. Food production animals such as chickens and cattle may also harbor *Salmonella* strains that are pathogenic to humans and may pass the bacteria to finished fresh foods such as

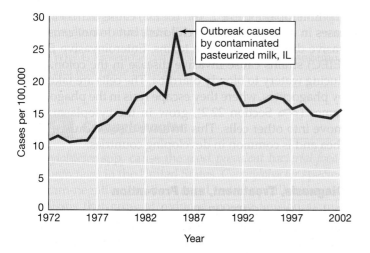

● **Figure 29.7 Prevalence of salmonellosis in the United States, 1972–2002.** The data are presented as the number of cases per 100,000 individuals. The total number of *reported* cases in 2002 was 44,264. Epidemiologic investigations suggest that only about 3% of all cases of salmonellosis are properly identified and reported, indicating that there are about 1.3 million cases of salmonellosis each year. The high incidence in 1984 resulted from several thousand cases caused by contamination of pasteurized milk that was mixed with raw (unprocessed) milk in a major dairy. About 95% of all cases are foodborne. Data are from the Centers for Disease Control and Prevention, Atlanta, GA, USA.

eggs, meat, and dairy products. *Salmonella* food infections are often traced to products such as custards, cream cakes, meringues, pies, and eggnog made with uncooked eggs. Other foods commonly implicated in salmonellosis outbreaks are meats and meat products such as meat pies, cured but uncooked sausages and meats, poultry, milk, and milk products.

The most common salmonellosis is a *Salmonella*-induced *enterocolitis*. Ingestion of food containing viable *Salmonella* results in colonization of the small and large intestine. Onset of the disease occurs 8–48 hours after ingestion. Symptoms include the sudden onset of headache, chills, vomiting, and diarrhea, followed by a fever that lasts a few days. The disease normally resolves without intervention in 2 to 3 days. However, even after recovery, patients shed *Salmonella* in feces for several weeks. Some patients recover and remain asymptomatic, but shed organisms for months or even years, resulting in a chronic carrier condition (∞ Section 25.3).

Salmonellosis may also cause septicemia (a blood infection) and enteric or *typhoid fever*, a disease characterized by systemic infection and high fever lasting several weeks. Mortality can approach 15% in untreated typhoid fever.

Diagnosis, Treatment, and Prevention

Diagnosis of foodborne salmonellosis is made by observation of clinical symptoms, history of recent food consumption, and by culture of the organism from feces. Several selective media (∞ Section 24.2) and tests for the presence of *Salmonella* are commonly done on animal

food products, such as raw meat, poultry, eggs, and powdered milk, because *Salmonella* from food production animals is the usual source of food contamination.

For enterocolitis, treatment is usually unnecessary, and antibiotic treatment does not shorten the course of the disease or eliminate the carrier state. Antibiotic treatment, however, significantly reduces the length and severity of septicemia and typhoid fever. Mortality due to typhoid fever can be reduced to less than 1% with appropriate antibiotic therapy. Several strains of *Salmonella* are resistant to multiple antimicrobial drugs.

Cooked or canned foods that become contaminated by an infected food handler can support the growth of *Salmonella* if the foods are held for long periods of time without heating or refrigeration. Cooked foods heated to 70°C for at least 10 minutes are considered safe if consumed immediately, or if held at 50°C or stored at 10°C or lower. *Salmonella* infections are more common in summer than in winter, probably because warm environmental conditions favor the growth of microorganisms in foods (Figure 29.4).

Although local laws and enforcement vary, because of the lengthy carrier state, infected individuals are often banned from work as food handlers until their feces are negative for *Salmonella* in three successive cultures.

🛑 29.7 Concept Check

There are more than 1.3 million cases of salmonellosis every year in the United States. The disease results from infection with ingested *Salmonella* introduced into the food chain from food production animals or food handlers.

◆ Describe salmonellosis food infection. How does it differ from food poisoning?

◆ How might *Salmonella* contamination of food production animals be contained?

29.8 Pathogenic *Escherichia coli*

Most strains of *Escherichia coli* are not pathogenic and are common commensals found in the intestines of humans. A few strains, however, are potential foodborne pathogens. All pathogenic strains act on the intestine and several are characterized by their ability to produce potent enterotoxins (∞ Section 21.11). The short, gram-negative rods are classified as enteric *Bacteria* (∞ Section 12.11). There are about 200 known pathogenic *E. coli* that can cause life-threatening diarrheal disease and urinary tract infections. The pathogenic strains are divided into several categories based primarily on the toxins they produce and the diseases they cause.

Enterohemorrhagic *Escherichia coli* (EHEC)

Enterohemorrhagic *Escherichia coli* (EHEC) produce *verotoxin*, an enterotoxin similar to one produced by *Shigella dysenteriae*, the Shiga toxin (∞ Table 21.4). After

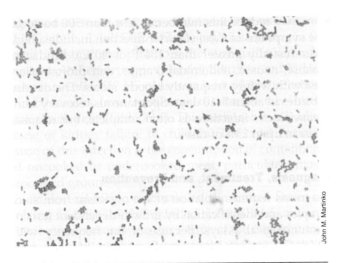

John M. Martinko

● **Figure 29.9 Gram stain of *Listeria monocytogenes*.** The short, gram-positive rods are about 0.5 μm in diameter.

are ready-to-eat processed foods such as meat products and unpasteurized dairy products that are stored for long periods, even at refrigerator temperature (4°C).

Listeria monocytogenes is an intracellular pathogen. It enters the body through the gastrointestinal tract after ingestion of contaminated food. Uptake of the pathogen by phagocytes results in growth and proliferation of the bacterium, lysis of the phagocyte, and spread to surrounding cells. Immunity to *L. monocytogenes* is mainly cell-mediated via T_H1 cells (∞ Section 22.8). Individuals having weakened cellular immunity, including the elderly, neonates, patients undergoing immunosuppressive drug treatment (e.g., steroid treatment), or those who have immunosuppressive diseases such as AIDS, have increased susceptibility to listeriosis (∞ Section 26.14).

Although exposure to *L. monocytogenes* is undoubtedly very common, acute *listeriosis* is quite rare. The acute disease is characterized by septicemia, often leading to meningitis. Acute listeriosis has a mortality rate of about 20%. Although there are only about 2500 cases of acute listeriosis each year, about 500 cases end in death. Nearly all diagnosed cases require hospitalization.

Diagnosis, Treatment, and Prevention

The diagnosis of listeriosis is accomplished by culturing *Listeria monocytogenes* from the blood or spinal fluid. *L. monocytogenes* can be identified in food by direct culture or by a variety of molecular methods such as ribotyping (∞ Section 11.11) and the polymerase chain reaction (PCR) (∞ Section 7.9, Section 24.12, and Table 24.9). Antibiotic treatment with penicillin, ampicillin, or a trimethoprim-sulfamethoxazole combination may be effective.

Prevention measures include recalling contaminated food and taking steps to limit *L. monocytogenes* contamination at the food-processing site. Since *L. monocytogenes* is susceptible to heat and radiation, raw food and food-handling equipment can be readily decontaminated. However, without sterilizing the finished food product, the risk of food contamination cannot be completely eliminated because of the widespread distribution of the pathogen.

Individuals who are immunocompromised are usually advised to avoid nonpasteurized dairy products and ready-to-eat processed meat. Spontaneous abortion is also a frequent outcome of listeriosis. Therefore, to protect the fetus, pregnant women may also be advised to avoid foods that may transmit *L. monocytogenes*.

29.10 Concept Check

Listeria monocytogenes is an environmentally ubiquitous microorganism. In normal individuals, *Listeria* seldom causes infection. However, in immunocompromised individuals, *Listeria* can cause serious disease and even death.

◆ What is the likely outcome of *Listeria* exposure in normal individuals?

◆ What populations are most susceptible to *Listeria* infection? Why?

◆ Describe the pathology of listeriosis.

29.11 Other Foodborne Infectious Diseases

A number of other microorganisms and infectious agents—bacteria, viruses, parasites, and otherwise—contribute to foodborne diseases, and we consider a few of them here.

Bacteria

Table 29.6 lists several other bacteria that cause human foodborne disease. *Yersinia enterocolitica* is commonly found in the intestines of domestic animals and causes foodborne infections due to contaminated meat and dairy products. *Y. enterocolitica* causes enteric fever, a severe life-threatening infection. *Bacillus cereus* produces two enterotoxins that cause diarrhea and vomiting. *B. cereus* grows in high-carbohydrate foods such as rice. Endospores of this gram-positive rod germinate and, as the organism grows in food that is left at room temperature, pathogenic amounts of toxin are produced. Reheating may kill the *B. cereus*, but the toxin may remain active. *B. cereus* may also cause a food infection similar to that caused by *Clostridium perfringens* (Section 29.6). *Shigella* spp. cause nearly 100,000 cases of severe foodborne invasive gastroenteritis called *shigellosis* each year. Several members of the *Vibrio* genus cause food poisoning after consumption of contaminated shellfish.

Viruses

The largest number of annual foodborne infections are thought to be caused by viruses. In general, viral foodborne illness consists of gastroenteritis characterized by diarrhea, often accompanied by nausea and vomiting. Recovery is spontaneous and rapid, usually within 24–48 hours ("24-hour bug"). *Norwalk-like viruses* (⌘ Table 25.8 and Section 28.8) are responsible for most of these mild foodborne infections in the United States (Table 29.6), accounting for over 9 million of the estimated 13 million cases of food infection per year. Rotavirus, astrovirus, and hepatitis A (⌘ Section 26.11) collectively cause 100,000 cases of foodborne disease each year. These viruses inhabit the gut and are often transmitted to food or water with fecal matter. As with many foodborne infections, proper food handling, handwashing, and a source of clean water to prepare fresh foods are essential to prevent infection.

Protozoa

Important foodborne protozoan diseases are listed in Table 29.6. Protozoan parasites including *Giardia lamblia*, *Cryptosporidium parvum*, and *Cyclospora cayetanensis* can be spread via food, presumably contaminated by fecal matter in untreated water used to wash, irrigate, or spray crops. Fresh foods such as fruits are often implicated as the source of these protozoans. We discussed giardiasis and cryptosporidiosis in the previous chapter (⌘ Section 28.6 and Figures 28.11 and 28.12). Cyclosporiasis is an acute gastroenteritis and is an important emerging disease. In the United States, most cases appear to be transmitted by eating fresh produce, often imported from other countries.

Toxoplasma gondii is a protozoan spread through cat feces, but also found in raw or undercooked meat. In most individuals, the *toxoplasmosis* infection causes a mild, self-limiting gastroenteritis. However, prenatal infection can lead to a variety of complications, including blindness and stillbirth. Immunocompromised patients also exhibit signs of acute toxoplasmosis.

Prions, BSE, and nvCJD Disease

Prions are proteins, presumably of host origin, that adopt novel conformations, inhibiting normal protein function and causing disruption in neural tissue (⌘ Section 9.14). Human prion diseases are characterized by a number of neurological symptoms including depression, loss of motor coordination, and dementia.

A foodborne variety of prion disease in humans is known as "new variant Creutzfeldt-Jakob Disease" (nvCJD) and has been linked to consumption of meat products from cattle afflicted with *bovine spongiform encephalopathy (BSE)*, a prion disease commonly called "mad cow disease." The nvCJD is a slow-acting degenerative nervous system disorder with a latent period that extends for years after exposure to the BSE prion (⌘ Section 9.14). Nearly 200 people in the United Kingdom and other European countries have acquired nvCJD. However, nvCJD linked to domestic meat consumption

has not been observed in the United States. Although the mechanism of prion infection is not entirely clear, BSE prions consumed in the meat products from affected cattle trigger structurally and functionally related human proteins to assume an altered conformation, resulting in protein dysfunction and disease (⌘ Figure 9.29). The terminal stages of both BSE and nvCJD are characterized by large vacuoles in brain tissue, giving the brain a "spongy" appearance, from which BSE derives its name (Figure 29.10●).

In the United Kingdom and Europe, about 180,000 cattle have been diagnosed with BSE. Several cattle with BSE have been found in Canada. Small numbers of affected cattle have also been found in the United States during routine testing of brains from slaughtered cattle. In Europe and North America, all cattle known or suspected to have BSE have been destroyed. Bans on feeding cattle with meat and bone meal appear to have stopped the development of new cases of BSE in Europe and have kept the incidence of this disease very low in North America. The infecting prions were probably transferred to food production animals through meat and bone meal feed derived from cattle or other animals not approved for human consumption.

Diagnosis of BSE is done by testing using a prion-susceptible mouse strain or by immunohistochemical or micrographic analysis of biopsied neural tissue (Figure 29.10).

29.11 Concept Check

Over 200 different infectious agents cause foodborne disease. Viruses cause the vast majority of foodborne illnesses. A number of bacteria, protozoans, and prions also cause foodborne illnesses.

◆ Identify the viruses most likely to be involved in foodborne illnesses.

◆ How might *prion* contamination of food production animals be prevented in the United States?

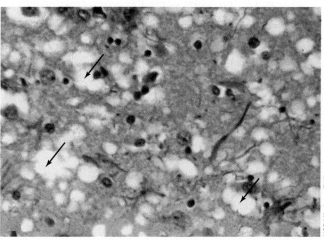

● **Figure 29.10 A brain section from a cow with bovine spongiform encephalopathy (BSE).** Note the vacuoles appearing as holes in the brain, giving it a characteristic spongelike appearance.

REVIEW QUESTIONS

1. Identify and define the three major categories of food with respect to their perishability (∞ Section 29.1).

2. Identify the major methods used to preserve food. Provide an example of a food preserved by each method (∞ Section 29.2).

3. Identify the major categories of fermented foods (∞ Section 29.3).

4. Distinguish between foodborne *infection* and foodborne *poisoning* (∞ Section 29.4).

5. Outline the pathogenesis of staphylococcal food poisoning. Suggest methods for prevention of this disease (∞ Section 29.5).

6. Identify the two major types of clostridial food poisoning. Which is most prevalent? Which is most dangerous? Why (∞ Section 29.6)?

7. What are the possible sources of *Salmonella* spp. that cause food infections (∞ Section 29.7)?

8. What measures are used to control the growth of *Escherichia coli* O157:H7 in ground meat in production plants? By consumers (∞ Section 29.8)?

9. *Campylobacter* causes more foodborne infections than any other bacterium. Identify at least one reason why this is true (∞ Section 29.9).

10. Identify the food sources of *Listeria monocytogenes* infections. Identify the individuals who are at high risk for listeriosis (∞ Section 29.10).

11. Why are viral agents so commonly associated with foodborne disease (∞ Section 29.11)?

APPLICATION QUESTIONS

1. Identify optimum storage conditions for perishable, semiperishable and nonperishable food products. Consider economic factors such as the cost of storage and the value of the food item.

2. For a food of your choice, devise a way to preserve the food by lowering the water activity without drying.

3. Perfringens food poisoning involves ingestion of *Clostridium perfringens* followed by growth and sporulation in the intestine of the host. Sporulation triggers toxin production. Is this disease truly a food poisoning, or might it be classified as a food infection?

4. Improperly handled potato salads are often the source of staphylococcal food poisoning or salmonellosis. Explain several means by which a potato salad could become inoculated with either *Staphylococcus aureus* or *Salmonella* spp.

5. *Clostridium botulinum* requires an anoxic environment for production of botulinum toxin. Identify methods of food preservation that create the anoxic environment necessary for growth of *C. botulinum*. Conversely, identify methods of food preservation that create an oxic environment and discourage the growth of *C. botulinum*. What other factors influence the growth of *C. botulinum*?

6. Why aren't antibiotics generally used to treat salmonellosis? Explain your answer based on the habitat of the organism and access to antibiotics. Also consider the issue of potential antibiotic resistance.

7. Indicate the precautions necessary to prevent infection with pathogenic *Escherichia coli*. Concentrate on *E. coli* O157:H7 and safe food handling, cooking, and consumption.

8. Devise a plan to eliminate *Campylobacter* contamination from a poultry flock or from the finished poultry product. Explain the benefits of *Campylobacter*-free poultry and explain the problems that your plans might encounter.

9. Listeriosis normally occurs only when there is a breakdown in T_H1 cell-mediated immunity. Indicate why this is so. Devise a vaccine to protect against listeriosis. Would your vaccine be of use in the listeriosis-prone population?

10. Indicate potential reasons for the high incidence of viral foodborne disease, especially with Norwalk-like viruses. Devise a plan to eliminate Norwalk-like viruses from the food supply.

11. Indicate the problems inherent in tracking a latent infectious agent like the BSE prion. Can prion diseases be eliminated, and, if so, how?

30

INDUSTRIAL MICROBIOLOGY

In industrial microbiology, economical production is only possible if it is done on a very large scale. In order to make a competitively priced product by microbial fermentation, the growth medium used must be inexpensive and the growth vessels extremely large. Shown here is the inside of a large industrial fermentor complete with fittings for cooling and mixing.

Table 30.1	Fermentor sizes for various industrial processes
Size of fermentor (liters)	**Product**
1–20,000	Diagnostic enzymes, substances for molecular biology
40–80,000	Some enzymes, antibiotics
100–150,000	Penicillin, aminoglycoside antibiotics, proteases, amylases, steroid transformations, amino acids, wine, beer
200,000–500,000	Amino acids (glutamic acid), wine, beer

of heat are vital for successful operation, the fermentor is fitted with an external *cooling jacket* through which steam (for sterilization) or cooling water (for cooling) can be run. For very large fermentors, insufficient heat transfer occurs through the jacket, and so *internal coils* must be provided through which either steam or cooling water can be piped.

A critical part of the fermentor is the *aeration system*. With large-scale equipment, transfer of oxygen throughout the growth medium is critical, and elaborate precautions must be taken to ensure proper aeration. Oxygen is poorly soluble in water, and in a fermentor with a high microbial population density, there is a tremendous oxygen demand by the culture.

Two separate installations are used to ensure adequate aeration: an aeration device, called a *sparger*, and a stirring device, called an *impeller* (Figure 30.4b). The sparger is typically just a series of holes in a metal ring or a nozzle through which filter-sterilized air (or oxygen-enriched air) can be passed into the fermentor under high pressure. The air enters the fermentor as a series of tiny bubbles from which the oxygen passes by diffusion into the liquid. In small fermentors use of a sparger alone may be sufficient to ensure adequate aeration. But in industrial-size fermentors, *stirring* of the fermentor with an impeller is essential (Figure 30.4c). Stirring accomplishes two things: It *mixes the gas bubbles* through the liquid and it *mixes the organism* through the liquid, thus ensuring uniform access of microbial cells to the nutrients.

Fermentation Control and Monitoring

Any microbial fermentation must be monitored to ensure that it is proceeding properly, but it is especially important that industrial fermentors be monitored carefully because there is such a major expense involved. In most cases, it is necessary not only to measure growth and product formation but also to *control* the process by altering environmental parameters as the process proceeds. Environmental factors that are frequently controlled include temperature, oxygen concentration, pH, cell mass, levels of key nutrients, and product concentration.

During growth and product formation in a large-scale fermentation, it is essential to obtain data in real time (see Figure 30.5b). For instance, it may be desirable to

alter one or more of the environmental parameters as the fermentation progresses or to feed a nutrient at a rate that exactly balances growth. Computers process these types of data on-line and then respond accordingly by adding nutrients at the right time to maintain high product yield.

Computers are also used to *model* fermentation processes. Mathematical models can be used to test the effect of various parameters on growth and product yield quickly and interactively. With the models, industrial microbiologists can modify the parameters to see how each affects the process. In this way, many variations in the fermentation can be studied inexpensively on screen, rather than expensively at the pilot plant stage or in the industrial plant.

 30.3 Concept Check

Large-scale industrial fermentations present several engineering problems. Aerobic processes require mechanisms for stirring and aeration. The microbial process must be continuously monitored to ensure satisfactory yields of the desired product.

◆ What types of devices are used to ensure proper aeration in a large-scale fermentation?

◆ What parameters in an industrial fermentation need to be monitored and what adjustments need to be made by computerized control?

30.4 Fermentation Scale-Up

An important aspect of industrial microbiology is the transfer of a process from small-scale laboratory equipment to large-scale commercial equipment, a procedure called **scale-up**. An understanding of the problems in scale-up is extremely important because biocatalytic processes rarely behave the same way in large-scale fermentors (Figure 30.5●) as in small-scale laboratory equipment (Figure 30.6●). Scale-up of an industrial process is the task of the *biochemical engineer*, one who is familiar with gas transfer, fluid dynamics, mixing, and thermodynamics. Scale-up requires knowledge not only of the biology of the producing organism, but also of the physics of fermentor design and operation.

Much of the problem in scale-up involves proper aeration and mixing. These essential ingredients in the industrial process are much easier to accomplish in the small laboratory flask than in the large industrial fermentor. Oxygen transfer especially is much more difficult to obtain in a large fermentor, and because most industrial fermentations are aerobic, effective oxygen transfer is essential. With the rich culture media used in industrial processes, a high biomass is obtained, leading to a high oxygen demand. If aeration is reduced, even for a short period, the culture may experience temporary anoxic conditions, with serious metabolic consequences and reduction in product yield.

(a)

(b)

● **Figure 30.5 Industrial scale fermentations.** (a) A large industrial fermentation plant. Only the tops of the fermentors, which can be several stories high, are visible. (b) Computer control room for a large fermentation plant.

The Scale-Up Process

Transferring an industrial process from the laboratory to the commercial fermentor involves several stages. Things begin in the *laboratory flask*, a very small-scale operation but typically the first indication that a process of commercial interest is possible. From here things are transferred to the *laboratory fermentor*, a small-scale fermentor, generally of glass and of 1 to 10 l in size, in which the first efforts at scale-up are made (Figures 30.4a and see 30.6a). In the laboratory fermentor, it is possible to test variations in medium, temperature, pH, and so on, inexpensively because little cost is involved for either equipment or culture medium.

When tests in the laboratory fermentor are successful, the process moves into the *pilot plant stage*, usually carried out in equipment 300–3000 l in size. Here, the conditions more closely approach the commercial scale; however, cost is not yet a major factor. Finally, the process is moved to the *commercial fermentor* itself, typically 10,000–500,000 l in volume (Figures 30.5a and 30.6b). In all stages, aeration is very closely monitored. As scale-up proceeds from flask to production fermentor, oxygen dynamics are carefully measured at each

step to determine how volume increases affect oxygen demand in the fermentation.

■ **30.4 Concept Check**

Scale-up is the process of gradually converting a useful industrial fermentation from laboratory scale to production scale. Aeration is a particularly critical aspect to monitor during scale-up studies.

◆ What are the differences in size among a typical laboratory fermentor, a pilot plant fermentor, and a commercial fermentor?

II MAJOR INDUSTRIAL PRODUCTS FOR THE HEALTH INDUSTRY

We now consider the products of industrial microbiology, beginning with the antibiotics. Of the microbial products manufactured commercially, the most important for the health industry are the antibiotics. Antibiotic production is a huge industry worldwide and one where many important principles for growing large-scale microbial cultures were first developed.

30.5 Antibiotics: Isolation and Characterization

As discussed in Chapter 20, antibiotics are chemicals produced by microorganisms that kill or inhibit the growth of other microorganisms. The development of antibiotics as agents for treatment of infectious disease has probably had more impact on the practice of medicine than any other single development. Antibiotics are typical secondary metabolites (see Section 30.2). Commercially useful antibiotics are produced primarily by filamentous fungi and by *Bacteria* of the actinomycete group (∞ Section 12.24). Table 30.2 lists the most important antibiotics produced by large-scale industrial fermentation.

Search for New Antibiotics

Although pharmaceutical companies currently do much of their drug discovery using computer modeling (∞ Section 20.13), traditionally, antibiotics were discovered by *screening*. In this approach, a large number of isolates of possible antibiotic-producing microorganisms are obtained from nature in pure culture (Figure 30.7a●), and these isolates are then tested for antibiotic production by seeing whether they produce any diffusible materials that are inhibitory to the growth of test bacteria. The test bacteria used are selected from a variety of bacterial types but are chosen to be representative of, or related to, bacterial pathogens.

The classical procedure for testing new microbial isolates for antibiotic production is the *cross-streak* method. Those isolates that show evidence of antibiotic

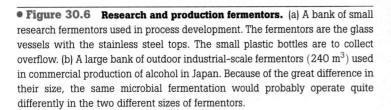

(a)

(b)

● **Figure 30.6   Research and production fermentors.** (a) A bank of small research fermentors used in process development. The fermentors are the glass vessels with the stainless steel tops. The small plastic bottles are to collect overflow. (b) A large bank of outdoor industrial-scale fermentors (240 m³) used in commercial production of alcohol in Japan. Because of the great difference in their size, the same microbial fermentation would probably operate quite differently in the two different sizes of fermentors.

production are then studied further to determine if the antibiotics they produce are new. Most of the isolates obtained produce known antibiotics, so the industrial microbiologist must quickly identify such organisms so that time and resources are not wasted in studying them. Once an organism producing a new antibiotic is discovered, the antibiotic is produced in sufficient amounts for structural analyses and then tested for toxicity and ther-

apeutic activity in infected animals. Unfortunately, most new antibiotics fail these animal tests, but a few prove to be medically useful and are produced commercially. However, with estimates of the number of different antibiotics produced by species of the genus *Streptomyces* alone at over 100,000, research to discover new antibiotics occurs on a continuous basis.

Purification

An antibiotic that is to be produced commercially must first be produced successfully in large-scale industrial fermentors (scale-up, see Section 30.4). The next challenge is to *purify* the product efficiently. Because of the relatively small amounts of antibiotic present in the fermentation liquid, elaborate methods for extraction and purification of the antibiotic are necessary (Figure 30.8●). If the antibiotic is soluble in an organic solvent, it may be relatively simple to purify it by extraction into a small volume of the solvent. If the antibiotic is not solvent soluble, then it must be removed from the fermentation liquid by adsorption, ion exchange, or chemical precipitation (Figure 30.8). In all cases, the goal is to eventually obtain a crystalline product of high purity. Depending on the process, further purification steps may be necessary to remove traces of microbial cell material if it co-purifies with the antibiotic.

Increasing the Yield

Rarely do antibiotic-producing strains just isolated from nature produce the desired antibiotic at sufficiently high concentration that commercial production can begin

Table 30.2	Some antibiotics produced commercially[a]
Antibiotic	**Producing microorganism[b]**
Bacitracin	*Bacillus licheniformis* (EFB)
Cephalosporin	*Cephalosporium* spp. (F)
Chloramphenicol	Chemical synthesis (formerly produced microbially by *Streptomyces venezuelae*) (A)
Cycloheximide	*Streptomyces griseus* (A)
Cycloserine	*Streptomyces orchidaceus* (A)
Erythromycin	*Streptomyces erythreus* (A)
Griseofulvin	*Penicillium griseofulvin* (F)
Kanamycin	*Streptomyces kanamyceticus* (A)
Lincomycin	*Streptomyces lincolnensis* (A)
Neomycin	*Streptomyces fradiae* (A)
Nystatin	*Streptomyces noursei* (A)
Penicillin	*Penicillium chrysogenum* (F)
Polymyxin B	*Bacillus polymyxa* (EFB)
Streptomycin	*Streptomyces griseus* (A)
Tetracycline	*Streptomyces rimosus* (A)

[a] See Chapter 12 and Figure 20.12 for information on the structures of these antibiotics.

[b] EFB, endospore-forming bacterium; F, fungus; A, actinomycete.

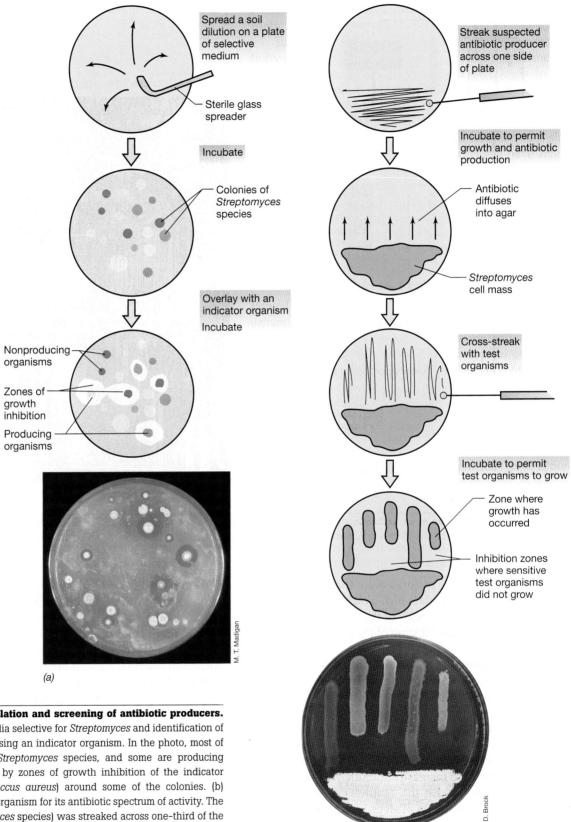

● **Figure 30.7 Isolation and screening of antibiotic producers.**
(a) Isolation using media selective for *Streptomyces* and identification of
antibiotic producers using an indicator organism. In the photo, most of
the colonies are of *Streptomyces* species, and some are producing
antibiotics as shown by zones of growth inhibition of the indicator
organism (*Staphylococcus aureus*) around some of the colonies. (b)
Method of testing an organism for its antibiotic spectrum of activity. The
producer (a *Streptomyces* species) was streaked across one-third of the
plate and the plate incubated. After good growth was obtained, the test
bacteria were streaked perpendicular to the *Streptomyces*, and the
plate was further incubated. The failure of several organisms to grow
near the mass growth of *Streptomyces* indicates that the *Streptomyces*
produced an antibiotic active against these bacteria. Test organisms
(left to right): *Escherichia coli, Bacillus subtilis, Staphylococcus aureus,
Klebsiella pneumoniae, Mycobacterium smegmatis.*

Proteases, Amylases, and High-Fructose Syrup

Enzymes are produced commercially from both fungi and bacteria. The microbial enzymes produced in the largest amounts on an industrial basis are the bacterial **proteases,** used as additives in laundry detergents. Most laundry detergents today contain enzymes, chiefly proteases, but also *amylases, lipases, reductases,* and others. Many of these enzymes are isolated from alkaliphilic bacteria (Section 6.13), mainly species of *Bacillus* such as *Bacillus licheniformis* (Table 30.4). These enzymes, which have pH optima between 9 and 10, remain active at the alkaline pH of laundry detergent solutions.

Other important enzymes manufactured commercially are amylases and glucoamylases, which are used in the production of glucose from starch. The glucose so produced can then be converted by the enzyme *glucose isomerase* to produce fructose, which is about twice as sweet as glucose. The final result is the production of a *high-fructose syrup* from corn, wheat, or potato starch. High-fructose syrups are a big business, since they have a major market in the production of soft drinks. Worldwide production of high-fructose syrups is nearly 10 billion kilograms per year.

Extremozymes: Enzymes from Prokaryotes That Inhabit Extreme Environments

In Chapters 2 and 6 we considered aspects of microbial growth at high temperature and discovered that some prokaryotes, called *hyperthermophiles*, grow optimally at very high temperatures. Hyperthermophiles can grow at such high temperatures because they produce heat-stable macromolecules (Sections 6.10 and 6.12) including enzymes (see Figure 30.15b). The term **extremozyme** has been coined to describe enzymes that function at some environmental extreme, such as high temperatures or low pH (Figure 30.15●). The organisms that produce extremozymes are called *extremophiles* (Table 2.1).

Because many industrial processes operate best at high temperatures, extremozymes from hyperthermophiles are widely used in both industry and research. Besides the *Taq* and *Pfu* DNA polymerases used in the polymerase chain reaction (PCR) described in Section 7.9, extremely thermostable proteases, amylases, cellulases, pullulanases (Figure 30.15b), and xylanases have been isolated and characterized from various hyperthermophiles. However, it is not only thermal stable enzymes that have found a market. Extremozymes that are cold-active (from psychrophiles), active in the presence of high salt (from halophiles), or active at high or low pH (from alkaliphiles and acidophiles, respectively) (Figure 30.15a) are known, and several have been commercialized. The great specificity of enzymes and their ability to distinguish between chiral isomers make those that also function at environmental extremes particularly useful in various industrial applications.

Table 30.4	**Microbial enzymes and their applications**		
Enzyme	**Source**	**Application**	**Industry**
Amylase (starch-digesting)	Fungi	Bread	Baking
	Bacteria	Starch coatings	Paper
	Fungi	Syrup and glucose manufacture	Food
	Bacteria	Cold-swelling laundry starch	Starch
	Fungi	Digestive aid	Pharmaceutical
	Bacteria	Removal of coatings (desizing)	Textile
	Bacteria	Removal of stains; detergents	Laundry
Protease (protein-digesting)	Fungi	Bread	Baking
	Bacteria	Spot removal	Dry cleaning
	Bacteria	Meat tenderizing	Meat
	Bacteria	Wound cleansing	Medicine
	Bacteria	Desizing	Textile
	Bacteria	Household detergent	Laundry
Invertase (sucrose-digesting)	Yeast	Soft-center candies	Candy
Glucose oxidase	Fungi	Glucose removal, oxygen removal	Food
		Test paper for diabetes	Pharmaceutical
Glucose isomerase	Bacteria	High-fructose corn syrup	Soft drink
Pectinase	Fungi	Pressing, clarification	Wine, fruit juice
Rennin	Fungi	Coagulation of milk	Cheese
Cellulase	Bacteria	Fabric softening, brightening; detergent	Laundry
Lipase	Fungi	Breaks down fat	Dairy, laundry
Lactase	Fungi	Breaks down lactose to glucose and galactose	Dairy, health foods
DNA polymerase	Bacteria	DNA replication in polymerase chain	Biological research;
	Archaea	reaction (PCR) technique (Section 7.9)	forensics

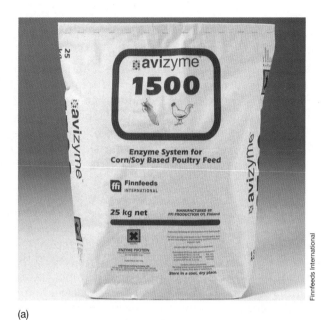

(a)

(b)

● **Figure 30.15 Examples of extremozymes, enzymes that function under environmentally extreme conditions.** (a) Acid-tolerant enzymes. An enzyme mixture used as a feed supplement for poultry. The enzymes function in the stomach of the bird to digest fibrous materials in the feed, thereby improving the nutritional value of the feed and promoting more rapid growth of the bird. (b) Thermostable enzymes. Thermostability of the enzyme pullulanase from *Pyrococcus woesei*, a hyperthermophile whose growth temperature optimum is 100°C (⬡⬡ Section 13.6). Calcium improves the heat stability of this enzyme; however, at 110°C, the enzyme denatures (⬡⬡Section 3.8).

Immobilized Enzymes

For some biocatalytic processes it is desirable to fix soluble enzymes onto a solid surface. These are called **immobilized enzymes**. Immobilization not only makes it easier to carry out the enzymatic reaction under large-scale continuous conditions, but also helps stabilize the en-

zyme to denaturation. A number of different immobilized enzymes have been employed in industrial processes, and a good example is in the starch-processing industry. As previously described, starch can be converted to high-fructose corn syrup by sequential treatment with amylase and glucose isomerase. The practical conversion of glucose to fructose is about 50%, but this can be increased somewhat by separating out the fructose and recycling the remaining glucose.

There are three basic approaches to enzyme immobilization (Figure 30.16●):

1. *Bonding of the enzyme to a carrier.* The bonding can be through adsorption, ionic bonding, or covalent bonding. Carriers used include modified celluloses, activated carbon, clay minerals, aluminum oxide, and glass beads (Figure 30.16).

2. *Cross-linking (polymerization) of enzyme molecules.* Linking of enzyme molecules with each other is usually done by chemical reaction with a cross-linking agent such as glutaraldehyde. Cross-linking of enzymes involves the chemical reaction of amino groups of the enzyme protein with glutaraldehyde. Aldehydes are also denaturing agents, but if the reaction is carried out properly, the enzyme molecules can be linked in such a way that most enzymatic activity is maintained.

3. *Enzyme inclusion, the incorporation of the enzyme into a semipermeable membrane.* Enzymes can be enclosed in microcapsules, gels, semipermeable polymer membranes, or fibrous polymers such as cellulose acetate (Figure 30.16).

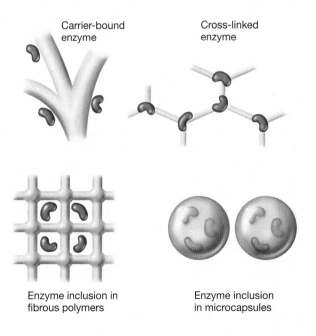

Carrier-bound enzyme

Cross-linked enzyme

Enzyme inclusion in fibrous polymers

Enzyme inclusion in microcapsules

● **Figure 30.16 Procedures for the immobilization of enzymes.** In all cases, enzyme molecules are shown in red. The actual procedure used varies with the enzyme, the reaction, and the scale.

Each of these methods for the use of immobilized enzymes has advantages and disadvantages, and the procedure used depends on the enzyme, the particular industrial application, and the scale of the operation.

30.9 *Concept Check*

Microorganisms are ideal for the large-scale production of enzymes. Many enzymes are used in the laundry industry to remove stains from clothing, and thermostable and alkali-stable enzymes have many advantages in these markets. Enzymes from extremophiles are desirable for biocatalyses under extreme conditions. When an enzyme is used in a large-scale process, it may be desirable to immobilize it by bonding it to an inert substrate.

◆ How are enzymes of use in the laundry industry?

◆ What key enzyme is needed to produce high-fructose syrups?

◆ What is an *extremozyme*?

III MAJOR INDUSTRIAL PRODUCTS FOR THE FOOD AND BEVERAGE INDUSTRIES

Although industrial microbiology as a discipline can trace its roots to the antibiotic era, antibiotics are only one class of products made on an industrial scale. Many food products and supplements are made by industrial processes, as we have already seen. We consider some others of these now, beginning with the production of alcohol for the beverage industry and for use as an industrial solvent.

30.10 Alcohol and Alcoholic Beverages

Most fruit juices and grains will undergo a natural fermentation catalyzed by wild yeasts present on these materials. From these natural fermentations, strains of yeasts have been selected for more controlled production, and today alcoholic beverage production is a large industry worldwide. The most important alcoholic beverages are *wine*, produced by the fermentation of fruit juice; *beer*, or *ale*, produced by the fermentation of malted grains; and *distilled beverages*, produced by concentrating alcohol from a fermentation by distillation. The biochemistry of alcohol fermentation by yeast was discussed in Section 5.10.

Wine Varieties and Production

Most wine is made from grapes, and thus wine manufacture occurs in parts of the world where high-quality grapes can be grown most economically. These include many countries of the European Union, the United States (Figure 30.17●), New Zealand and Australia, and South America.

There are a great number of different wines, and their quality and character vary considerably. *Dry wines* are wines in which the sugars of the juice are practically all fermented, whereas in *sweet wines* some of the sugar is left or additional sugar is added after the fermentation. A *fortified wine* is one to which brandy or some other alcoholic spirit is added after the fermentation; sherry and port are the best-known fortified wines. A *sparkling wine*, such as champagne, is one in which considerable carbon dioxide is present, arising from a final fermentation by the yeast directly in the sealed bottle.

The production of wine begins in the early fall with the harvesting of grapes. The grapes are crushed and the juice, called *must*, is squeezed out. Depending on the grapes used and on how the must is prepared, either white or red wine may be produced (Figure 30.18●). A white wine is made either from white grapes or from the juice of red grapes from which the skins, containing the red coloring matter, have been removed. In the making of red wine, the *pomace* (skins, seeds, and pieces of stem) is left in during the fermentation. In addition to the color difference, red wine has a stronger flavor than white because of larger amounts of *tannins*, chemicals that are extracted into the juice from the grape skins during the fermentation.

The yeasts involved in wine fermentation are of two types: *wild yeasts*, which are present on the grapes as they are taken from the field and are transferred to the juice, and the *cultivated wine yeast, Saccharomyces ellipsoideus*, which is added to the juice to begin the fermentation. Wild yeasts are less alcohol-tolerant than commercial wine yeasts and can also produce undesirable compounds affecting quality of the final product. Thus, it is the practice in most wineries to kill the wild yeasts present in the must by adding sulfur dioxide (listed on the bottle as "sulfites") at a level of about 100 parts per million (ppm). *Saccharomyces ellipsoideus* is resistant to this concentration of sulfur dioxide and is added as a starter culture from a pure culture grown on sterilized grape juice. The fermentation is carried out in vats of various sizes, from 200-l casks to 200,000-l tanks made of oak, cement, stone, or glass-lined metal (see Figure 30.17b). The fermentor must be constructed so that the large amount of carbon dioxide produced during the fermentation can escape but air cannot enter, and this is accomplished by fitting the vessel with a special one-way valve.

With a red wine, after 3–5 days of fermentation, sufficient tannin and color have been extracted from the pomace and the wine is drawn off for further fermentation in a new tank, usually for another week or two. The next step is called *racking*; the wine is separated from the sediment, which contains yeast cells and precipitate, and then stored at lower temperature for aging, flavor development, and further clarification. The final clarification may be hastened by the addition of materials called *fining agents*, such as casein, tannin, or bentonite clay. Alternatively, the wine may be filtered through diatomaceous earth, asbestos, or membrane filters. The wine is then bottled and either stored for further aging or sold.

(a)

(b)

(c)

● **Figure 30.17 Commercial wine making in California (USA).**
(a) Equipment for transporting grapes to the winery for crushing. (b) Large tanks where the main wine fermentation takes place. (c) Barrels where the aging process takes place. Wooden cask storage can last for several years in the case of certain red wines.

Red wine is usually aged for several years or more (Figure 30.17c), but white wine is typically sold without much aging. During aging, complex chemical changes occur, including reduction of bitter components, resulting in improvement in flavor and odor, or *bouquet*. The final alcohol content of wine varies from 6 to 15% depending on the sugar content of the grapes, length of the fermentation, and strain of wine yeast used.

Brewing: Making the Wort

The manufacture of alcoholic beverages made from malted grains is called **brewing** (Figure 30.19●). Typical malt beverages include beer, ale, porter, and stout. *Malt* is prepared from germinated barley seeds, and it contains natural enzymes that digest the starch of grains and convert it to sugar. Since brewing yeasts are unable to digest starch (∞ Figure 3.6b), the malting process is essential for the generation of fermentable substrates from cereal grains.

The fermentable liquid from which beer and ale are made is prepared by a process called *mashing*. The grain of the mash may consist only of malt, or other grains such as corn, rice, or wheat may be added. The mixture of ingredients in the mash is cooked and allowed to steep in a large mash tub at warm temperatures. During the heating period, enzymes from the malt cause digestion of the starches and liberate sugars, which will be fermented by the yeast. Proteins and amino acids are also liberated into the liquid, as are other nutrient ingredients necessary for the growth of yeast.

After cooking, the aqueous extract, called *wort*, is separated by filtration from the husks and other grain residues of the mash. *Hops*, an herb derived from the female flowers of the hops plant, are added to the wort at this stage. Hops is a flavoring ingredient, but it also has antimicrobial properties, which probably help to prevent contamination in the subsequent fermentation. The wort is then boiled for several hours, usually in large copper kettles (Figure 30.19a,b), during which time desired ingredients are extracted from the hops, proteins present in the wort that are undesirable from the point of view of beer stability are coagulated and removed, and the wort is sterilized. The wort is filtered again, cooled, and transferred to the fermentation vessel.

Brewing: The Fermentation Process

Brewery yeast strains are of two major types: *top fermenting* and *bottom fermenting*. The main distinction between the two is that top-fermenting yeasts remain uniformly distributed in the fermenting wort and are carried to the top by the CO_2 gas generated during the fermentation, whereas bottom-fermenting yeasts settle to the bottom. Top yeasts are used in the brewing of *ales*, and bottom yeasts are used to make *lager beers*. Bottom yeasts are usually given the species designation *Saccharomyces carlsbergensis*, and top yeasts are called *Saccharomyces cerevisiae*. Fermentation by top yeasts usually occurs at higher temperatures (14–23°C) than that by bottom yeasts (6–12°C) and is accomplished in a shorter period of time (5–7 days for top fermentation versus 8–14 days for bottom fermentation).

After completion of lager beer fermentation by bottom yeasts, the beer is pumped off into large tanks where it is stored at a cold temperature (about −1°C) for several weeks (Figure 30.19c). Following this, the beer is filtered and placed in storage tanks (Figure 30.19d) from which packaging occurs. Top-fermented ale is stored for only short periods at a higher temperature (4–8°C), which assists in development of the characteristic ale flavor. Some

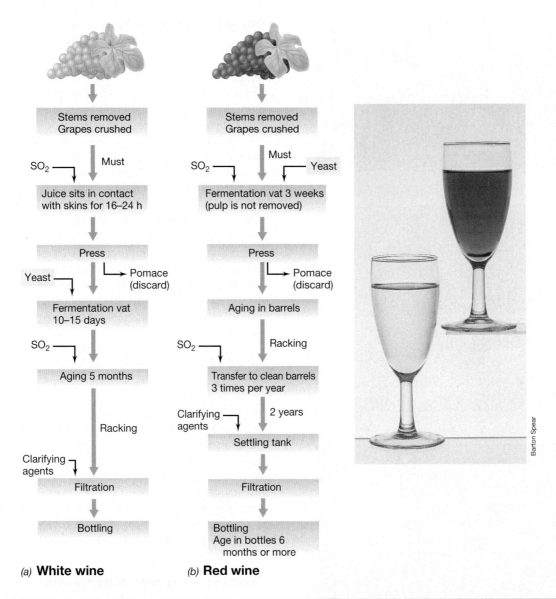

(a) **White wine**

(b) **Red wine**

● **Figure 30.18 Wine production.** (a) White wine. White wines can vary from nearly colorless to straw-colored depending on the grapes used. (b) Red wine. Red wines can vary in color from a faint red to a deep, rich burgundy. The photograph shows a glass of chenin blanc, a typical white wine (left), and a glass of a light red wine (rosé, right). Other typical varieties of white wine include chablis, Rhine wine, sauterne, and chardonnay, while other typical red wines include burgundy, chianti, claret, zinfandel, cabernet, and merlot.

additional details on the brewing process are discussed in the Microbial Sidebar, Home Brew.

Distilled Alcoholic Beverages

Distilled alcoholic beverages are made by heating a fermented liquid at a high temperature that volatilizes most of the alcohol. The alcohol is then condensed and collected, a process called *distilling*. A product much higher in alcohol content can be obtained by this process than is possible by direct fermentation. Virtually any alcoholic liquid can be distilled, and each yields a characteristic distilled beverage. The distillation of malt brews yields *whiskey*, distilled wine yields *brandy*, distillation of fermented molasses yields *rum*, distillation of fermented grain or potatoes yields *vodka*, and distillation of grain and juniper berries yields *gin* (Figure 30.20●).

The distillate contains not only alcohol but also other volatile products arising either from the yeast fermentation or from the ingredients themselves. Some of these other products are desirable flavor ingredients, whereas others are undesirable. To eliminate the latter, the distilled product is almost always aged, usually in wooden barrels. During the aging process, undesirable products are removed and desirable new flavor ingredients develop. The fresh distillate is usually colorless, whereas the aged product is often brown or yellow (Figure 30.20). The character of the final product is partly determined by the manner and length of aging (aging times of 10 years or more are not uncommon for some distilled spirits), and the whole process of manufacturing distilled alcoholic beverages is highly complex. To a great extent, the process is carried out by traditional methods that have been found

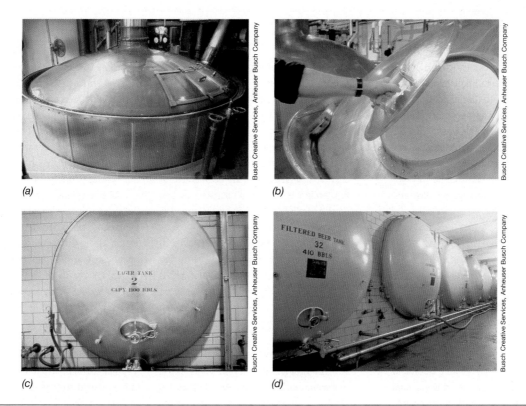

(a)   (b)

(c)   (d)

● **Figure 30.19** **Brewing beer in a commercial brewery.** (a, b) The copper brew kettle is where the wort is mixed with hops and then boiled. From the brew kettle the liquid is passed to large fermentation tanks where yeast ferments glucose to ethanol plus CO_2. (c) If a lager, the beer is then stored for several weeks at low temperature in tanks where settling of particulate matter including yeast cells occurs. (d) The beer is then filtered and placed in storage tanks from which it is packaged into kegs, bottles, or cans.

to yield a particular product, rather than by scientifically proven methods.

Commodity Ethanol

Production of alcohol (ethanol) as a *commodity chemical* is a major biocatalytic process, and today over 30 billion liters of alcohol are produced yearly worldwide from the fermentation of a variety of materials. In the United States most ethanol is obtained by yeast fermentation from corn starch. The starch is treated with enzymes (see Section 30.9) to yield glucose, which is the fermentable substrate. Ethanol is stripped from the fermentation broth by distillation (Figure 30.21●). In other countries, for example in

● **Figure 30.20** **Color of distilled spirits.** Gin or vodka (not shown) are colorless. However, aging in wooden casks yields a distinctive amber or yellow color to certain distilled spirits. Left to right, dark rum, brandy, whiskey.

Brazil, a major ethanol producer, corn and a variety of other fermentable materials are used. These include sugar cane, whey, sugar beets, and even wood chips and waste paper. The cellulose in wood is treated to release glucose, which is then fermented to alcohol.

Ethanol is used as an industrial solvent and also for the production of a gasoline supplement. In the United States *gasohol* is produced by adding 10% ethanol to lead-free gasoline. The combustion of gasohol produces lower amounts of carbon monoxide and nitrogen oxides than pure gasoline. Hence gasohol is marketed as a cleaner burning fuel and gasoline extender and its use is encouraged in major cities where automobile pollution is extensive. Various yeasts have been used in commodity ethanol production, including species of *Saccharomyces*, *Kluyveromyces*, and *Candida*, but most ethanol in the United States is produced by *Saccharomyces*.

30.10 *Concept Check*

Alcoholic beverages are produced by yeast from the fermentation of sugar to ethyl alcohol and CO_2. Wine is produced from grape juice, beer from malted grain, and distilled spirits from the distillation of fermented solutions. Commodity alcohol is used as a gasoline additive and industrial solvent.

◆ How do wines differ from beer in terms of the amount of alcohol they contain?

◆ What are the major differences between a *beer* and an *ale*?

Microbial Sidebar ◆ Home Brew[a]

The amateur brewer can make many kinds of beer, from English bitters and India pale ale to German bock and Russian Imperial stout. The necessary equipment and supplies, including yeast, can be purchased from a local beer and winemakers shop. The Home Wine and Beer Trade Association, P.O. Box 1373, Valrico, FL 33595, (http://www.hwbta.org) can supply the address of a nearby shop.

The brewing process can be divided into three basic stages: (1) *making the wort*, (2) *carrying out the fermentation*, and (3) *bottling and aging*. The character of a brew depends on many factors: the proportion of malt, sugar, hops, and grain; the kind of yeast; the temperature and duration of the fermentation; and how the aging process is carried out. The fermentor itself consists of a 20-l (5-gal) glass jar or carboy that can be fitted with a tightly fitting closure. In order to have a good quality beer, it is essential that *everything* be sterilized that comes into contact with the wort. This includes the fermentor, tubing, stirring spoon, and bottles. An easy procedure is to use a sterilizing rinse consisting of 50–60 ml of liquid bleach in 20 l of water. Soak the items for 15 minutes and then rinse lightly with hot water or air dry.

1. **Making the wort** (Figure 1). In commercial brewing, the wort is made by extracting fermentable sugars and yeast nutrients from malt, sugar, and hops. Many home brewers make their own wort from malt, but a reasonably satisfactory beer can be made with hop-flavored malt extract purchased ready-made. Malt extracts come in a variety of flavors and colors, and the kind of beer depends on the type of malt extract used. A simple recipe for making the wort uses 5–6 lb (2.25–2.75 kg) of hop-flavored malt extract and 20 l of water. The malt ex-

tract and 5–6 l of water are brought to a boil for 15 minutes in an enamel or stainless steel container (aluminum heating kettles must be avoided because of the inhibiting action of metals leached from aluminum containers) (Figure 1). The hot wort is then poured into 14–15 l of clean, cold water that has already been added to the fermentor. After the temperature has dropped below 30°C, the yeast is added to initiate the fermentation.

2. **Carrying out the fermentation** (Figure 2). Add two packs of fresh beer yeast to the cooled wort and cover the fermentor with a rubber stopper into which a plastic hose has been inserted. The hose is directed into a bucket containing water. During the initial 2–3 days of the fermentation, large amounts of CO_2 will be given off, which will exit through the hose. The water trap is to prevent wild yeasts or bacteria from the air from getting back into the fermentor. After about 3 days, the activity will diminish as the fermentable sugars are used up. At this time, the rubber stopper and hose are replaced with an inexpensive fermentation lock. The fermentation lock, which can be purchased at the home brew store, prevents contamination while permitting the small amount of gas still being produced to escape. Allow the beer to ferment for 7–10 days at 10–15°C or higher.

3. **Bottling and aging** (Figure 3). The fermentation should be allowed to proceed for the full 7–10 days, even if the vigorous fermentation action ceases earlier. By this time, most of the yeast should have settled to the bottom of the fermentor. Carefully siphon the beer off the yeast layer, allowing the beer but not the sediment to run into sanitized glass beer bottles. The bottles used should accept standard crown caps, and new, clean caps should be used. Before capping, add three-quarters of a teaspoon

(but no more) of corn syrup to each 12-to 16-ounce (350-to 465-ml) bottle. Once the bottles are capped, turn each one upside down once to mix the syrup and then allow the beer to age upright at room temperature for at least 7–10 days. After this aging period, the beer may be stored at a cooler temperature.

All homemade beer will improve if it is allowed to age for several weeks. Aging tends to make beer smoother and remove bitter components. Using the same basic production equipment, several different types of beer can be made, each with its own distinctive taste and character (the reference listed in the footnote contains a number of beer recipes). Dark beers, which typically contain more alcohol than lighter beers, require more malt for their production, and are usually brewed from a combination of different malts such as those obtained from darker varieties of grain or those that have been roasted to caramelize the sugars and yield a darker color. A typical American style light lager (Figure 4, left) contains about 3.5% alcohol (by volume), whereas a Munich style dark (Figure 4, right) contains 4.25% alcohol, and bock beers contain about 5% alcohol.

The trend toward "individuality" in beer is evident not only by the growing number of home brewers, but also by the fact that major brewers in the United States are feeling more and more competition from new, usually very small, breweries called *microbreweries*. Although total production by a microbrewery may pale by comparison to that of a major brewer, the products themselves often have their own distinctive character and local appeal. Part of these differences probably has to do with the smaller scale on which the brewing takes place but also undoubtedly has to do with the use of different sources of ingredients, water, yeast strains, and brewing procedures. ■

Figure 1

Figure 2

Figure 3

Figure 4

[a]Source: Burch, B. 1992. *Brewing Quality Beers—The Home Brewer's Essential Guidebook*, 2nd edition. Joby Books, Fulton, CA

● **Figure 30.21** **Ethanol production plant, Nebraska, USA.** Here, glucose obtained from corn is fermented by the yeast, *Saccharomyces cerevisiae*, to ethanol plus CO_2. The large tank in the left foreground is the ethanol storage tank and the tanks and pipes in the background are for distilling ethanol from the fermentation liquor.

30.11 Vinegar Production

Vinegar is the product resulting from the conversion of ethyl alcohol to the key ingredient, *acetic acid*, by the **acetic acid bacteria**. Key genera of acetic acid bacteria include *Acetobacter* and *Gluconobacter*. Vinegar can be produced from any substance that contains ethanol, although the usual starting material is wine, beer, or alcoholic apple juice (hard cider). Vinegar can also be produced from a mixture of pure alcohol in water, in which case it is called *distilled vinegar*, the term *distilled* referring to the alcohol from which the product is made rather than the vinegar itself. Vinegar is used as a flavoring ingredient in salads and other foods, and because of its acidity, it is also used in pickling. Meats and vegetables properly pickled in vinegar can be stored unrefrigerated for years.

The acetic acid bacteria are an interesting group of prokaryotes (∞ Section 12.8). These are strictly aerobic bacteria that differ from most other aerobes in that some of them, such as species of *Gluconobacter*, do not oxidize their organic electron donors completely to CO_2 and water (Figure 30.22●). Thus, when provided with ethyl alcohol as electron donor, they oxidize it via quinones only to acetic acid, which accumulates in the medium. Acetic acid bacteria are quite acid tolerant and are not killed by the acidity that they produce. There is a high oxygen demand during growth, and the main problem in the production of vinegar is to ensure sufficient aeration of the medium.

Vinegar Production

There are three different processes for the production of vinegar. The *open-vat (Orleans) method*, which is the original process, is still used in France where it was first devel-

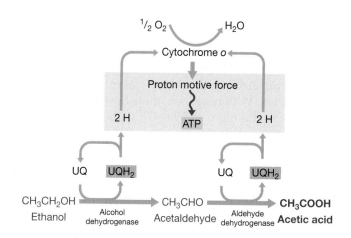

● **Figure 30.22** **The chemistry of vinegar production.** Oxidation of ethanol to acetic acid, the key process in the production of vinegar. UQ, ubiquinone.

oped. Wine is placed in shallow vats with considerable exposure to the air, and the acetic acid bacteria develop as a slimy layer on the top of the liquid. This process is not very efficient because the bacteria come in contact with both the air and the substrate only at the surface.

The second process is the *trickle (Quick vinegar) method*, in which the contact between the bacteria, air, and substrate is increased by trickling the alcoholic liquid over beechwood twigs or wood shavings packed loosely in a vat or column while a stream of air enters at the bottom and passes upward. The bacteria grow on the surface of the wood shavings and thus are maximally exposed both to air and liquid. The vat is called a *vinegar generator* (Figure 30.23●), and the whole process is operated in a

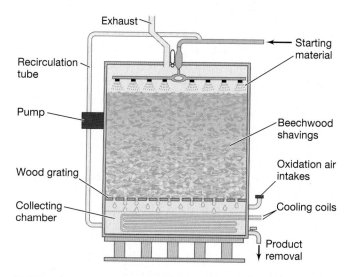

● **Figure 30.23** **Diagram of a vinegar generator.** The alcoholic juice is allowed to trickle through the wood shavings, and air is passed up through the shavings from the bottom. Acetic acid bacteria develop on the wood shavings and convert alcohol to acetic acid. The acetic acid solution accumulates in the collecting chamber and is recycled through the generator until the acetic acid content reaches at least 4%, the minimum for a product to be labeled as "vinegar."

continuous fashion. The life of the wood shavings in a vinegar generator is long, from 5 to 30 years, depending on the kind of alcoholic liquid used in the process.

Finally, the *bubble method* is basically a submerged fermentation such as was already described for antibiotic production. With proper aeration, the efficiency of the bubble method is high, and 90–98% of the alcohol is converted to acetic acid.

Although acetic acid can be easily made chemically from alcohol, the microbial product, *vinegar*, is a distinctive material, the flavor being due in part to other substances present in the starting material and produced in the fermentation. For this reason, the microbial process, especially using the vinegar generator, has not been supplanted by a chemical process.

 30.11 Concept Check

The active ingredient in vinegar is acetic acid, which is produced by acetic acid bacteria oxidizing an alcohol-containing fruit juice. Adequate aeration is the most important consideration in ensuring a successful vinegar process.

◆ Why is O_2 necessary in vinegar production?

◆ Why does vinegar produced by the trickle method have a more distinctive taste than vinegar produced by the bubble method?

30.12 Citric Acid and Other Organic Compounds

Many organic chemicals are produced by microorganisms in sufficient yields that they can be manufactured commercially by fermentation. For example, *citric acid*, is widely used in the food industry as a supplement in beverages, confectionaries, and other foods, and in the leavening of bread, and is also used industrially in the treatment of certain metals, as a detergent additive in place of phosphates, and in various pharmaceutical applications. Other acids include *itaconic acid*, used in the manufacture of acrylic resins, and *gluconic acid*, used in the form of calcium gluconate to treat calcium deficiencies in humans and industrially as a washing and water-softening agent. All of these acids are produced by fungi.

Important bacterially produced organic chemicals include *sorbose*, which is produced when *Acetobacter* oxidizes sorbitol and is used in the manufacture of *ascorbic acid* (vitamin C), and *lactic acid*, used in the food industry to acidify foods and beverages. Lactic acid is produced by various species of lactic acid bacteria (∞ Section 12.19).

We focus here on the citric acid fermentation. Historically, the citric acid fermentation was of great importance because it was the first submerged-culture *aerobic* industrial fermentation. As the citric acid fermentation was perfected, the lessons learned were applied to the

penicillin and other important antibiotic fermentations. Thus to some degree, we owe our current success with the large-scale production of antibiotics to the pioneering work done on the citric acid fermentation. Overall, citric acid production is a big business, as over 550,000 tons of citric acid are produced per year worldwide with a total value approaching $1 billion U.S.

Citric Acid Production: The Link to Iron

Citric acid is produced microbiologically by a fermentation using the mold *Aspergillus niger*. Although citric acid is normally considered in connection with the citric acid cycle (∞ Section 5.13), in certain organisms such as *A. niger*, excretion of large amounts of citric acid can be obtained. The fermentation is carried out aerobically in large fermentors, and a key requirement for high citric acid yield is that the medium be *iron deficient*. The iron deficiency makes cells of *A. niger* overproduce citric acid as a chelator to scavenge iron (Figure 30.24a●; see also Figure 5.1). Therefore, the medium used for citric acid production is treated to remove most of the iron, and the fermentors themselves are made of stainless steel or are glass lined to prevent leaching of iron from the fermentor walls at the low pH values generated by citric acid accumulation.

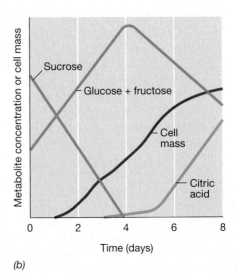

● **Figure 30.24 Citric acid fermentation.** (a) Structure of citric acid. Note how the ionized form, citrate, contains three carboxylic acid groups, which can chelate ferric iron (Fe^{3+}). (b) Kinetics of the citric acid fermentation. Sucrose is degraded by the enzyme *sucrase* to yield glucose plus fructose. See text for further details.

Citric Acid Production: Growth Media, Conditions, and Purification

Growth media used for citric acid production contain any of a variety of starting materials including starch from potatoes, starch hydrolysates, glucose syrup from saccharified starch, sucrose (Figure 30.24b), sugarcane syrup, sugarcane molasses, and sugar beet molasses. If starch is used, amylase (Table 30.4), formed by the producing fungus or added to the fermentation broth, hydrolyzes the starch to sugars. The sugars are catabolized through the glycolytic pathway (∞ Section 5.10) and enter the citric acid cycle where citrate production occurs.

Most citric acid today is produced by submerged processes in large fermentors. However, because *A. niger* is a strict aerobe, it is crucial to this fermentation to make sure that the culture stays properly aerated. Citric acid is produced in this way as a typical secondary metabolite (see Figure 30.2). During the growth phase, sucrose is broken down into glucose and fructose, and by the time the stationary phase is reached, large amounts of these hexoses remain and are converted to citric acid to counter iron starvation (Figure 30.24).

Citric acid is purified following removal of fungal biomass by adding lime (CaO); this precipitates citric acid as calcium citrate. The latter is concentrated by filtration and treated with sulfuric acid to form a solution of citric acid and calcium sulfate ($CaSO_4$, a solid). Following a second filtration to remove crystals of $CaSO_4$, crystals of citric acid are formed by evaporation.

30.12 Concept Check

A number of organic chemicals are produced commercially by use of microorganisms, of which the most important economically is citric acid, produced by *Aspergillus niger*.

◆ Why is citric acid considered a *secondary* metabolite (see Figure 30.2b)?

◆ What is the relationship between iron and citric acid production by *A. niger*?

30.13 Yeast as a Food and Food Supplement

Yeasts are among the most extensively used microorganisms in industry. Besides their importance to the baking and beverage industries, yeasts are grown for the cells themselves or for compounds extracted from the cells (Table 30.5). Unlike in the brewing and baking industries, where *fermentation* is important (∞ Microbial Sidebar: The Products of Yeast Fermentation, Chapter 5), the production of yeast cells requires *respiration* for maximum production of cell material. For food and food supplement yeast, derivatives of *Saccharomyces cerevisiae* are used almost exclusively, and we consider their large scale production here.

Table 30.5	Industrial uses of yeast and yeast products[a]
Production of yeast cells	
Baker's yeast, for bread making	
Dried food yeast, for food supplements	
Dried feed yeast, for animal feeds	
Yeast products	
Yeast extract, for microbial culture media	
B vitamins, vitamin D	
Enzymes for food industry: invertase (sucrase), galactosidase	
Biochemicals for research: ATP, NAD^+, RNA	
Fermentation products from yeast	
Ethanol, for industrial alcohol and as a gasoline extender	
Glycerol	
Beverage alcohol	
Beer, wine	
Distilled beverages	
Whiskey, brandy, vodka, rum	

[a] ∞ Microbial Sidebar, The Products of Yeast Fermentation, Chapter 5 and Sections 5.10 and 5.13.

Yeast Cell Production

Yeast for baking or nutritional purposes is cultured in large aerated fermentors in a medium containing molasses as a major ingredient. Molasses contains large amounts of sugar as the source of carbon and energy, and also contains minerals, vitamins, and amino acids used by the yeast. To make a complete medium for yeast growth, phosphoric acid (a phosphorus source) and ammonium sulfate (a source of nitrogen and sulfur) are added.

Fermentation vessels for yeast production range from 40,000 to 200,000 liters. Beginning with the pure stock culture, several intermediate stages are needed to scale up the inoculum to a size sufficient to inoculate the final stage (Figure 30.25a●). It is undesirable to add all the molasses to the fermentor at once because this results in a sugar excess. The yeast then ferments much of the sugar to alcohol plus CO_2 rather than turning it into yeast cells. Therefore, only a small amount of the molasses is added initially, and then as the yeast culture grows and consumes this sugar, more molasses is added in controlled "feedings."

At the end of the growth period, the yeast cells are recovered from the broth by centrifugation. The cells are then washed by dilution with water and recentrifuged until they are light in color. Baker's yeast is marketed in two ways, either as compressed cakes or as a dry powder. *Compressed yeast cakes* (Figure 30.25b) are made by mixing the centrifuged yeast with emulsifying agents, starch, and other additives that give it a suitable consistency and reasonable shelf life, and the product is then formed into cubes or blocks of various sizes for domestic or commercial use. A compressed yeast cake contains about 70% moisture and thus must be stored in the refrigerator so its activity is maintained.

biotransformations are preferably carried out microbially rather than chemically (⚬⚬ Section 30.8).

11. List three different kinds of enzymes that are produced commercially. For each enzyme, list the organism used in commercial production, the activity of the enzyme, and how the enzyme is used commercially (⚬⚬ Section 30.9).

12. What is *high-fructose syrup*, how is it produced, and what is it used for in the food industry (⚬⚬ Section 30.9)?

13. What are extremozymes? What industrial uses do they have (⚬⚬ Section 30.9)?

14. In what way is the manufacture of *beer* similar to the manufacture of *wine*? In what ways do these two processes differ? How does the production of *distilled alcoholic beverages* differ from that of beer and wine (⚬⚬ Section 30.10)?

15. What is the active ingredient in vinegar? From what substance is it made? Why is *Gluconobacter* suitable for the production of vinegar, whereas *Escherichia coli* is not (*Hint*: think of metabolic differences here) (⚬⚬ Section 30.11)?

16. Give two reasons why stainless steel fermentors are used in the industrial production of citric acid (⚬⚬ Section 30.12).

17. Why are yeasts of such great industrial importance (⚬⚬ Section 30.13)?

18. What part of the mushroom is actually consumed as food? What is contained within this structure (⚬⚬ Sections 30.14 and 14.12)?

APPLICATION QUESTIONS

1. As a researcher in a pharmaceutical company you are assigned the task of finding and developing an antibiotic effective against a new bacterial pathogen. Outline a plan for this process, starting from isolation of the low-yield producing organism to high-yield industrial production of the new antibiotic.

2. A partially consumed bottle of an "organic" (containing no preservatives) red wine is recapped and stored under refrigeration for 2 months. On tasting the wine again, you notice a distinct bitter taste, making the wine undrinkable. Using information presented in this chapter and in Section 12.8, describe (a) what microbially-mediated process occurred in the wine, and (b) a very easy way in which this process could have been prevented.

3. You wish to produce high yields of the amino acid phenylalanine for use in production of the sweetener *aspartame*. The overproducing organism you wish to use is not subject to feedback inhibition by phenylalanine but is subject to typical repression of phenylalanine biosynthesis enzymes by excess phenylalanine. Applying the principles of enzyme regulation studied in Chapter 8 and microbial genetics in Chapter 10, describe *two classes* of mutants you could isolate that would overcome this problem and detail the genetic lesions each would have.

31

GENETIC ENGINEERING AND BIOTECHNOLOGY

From a scientific standpoint, the ability to manipulate DNA in a test tube and then place the altered DNA back into an organism and have it expressed, is an extremely useful and powerful technique. Here, through genetic engineering techniques, soybean plants have been made resistant to the commonly used herbicide, glyphosate.

◆ What is a *cloning vector*?

◆ How can DNA sequencing tell us about protein structure?

31.2 Hosts for Cloning Vectors

When *in vitro* genetic techniques were introduced in Chapter 10, it was in the context of the genetic analysis of *Bacteria*. The primary goal there was to understand the organism. By contrast, in biotechnology genetic engineering is used for commercial ends. For example, we may wish to produce large amounts of a valuable protein, produce a new vaccine, or introduce a specific gene into a plant or an animal. In these cases it is essential to consider the *host* that we will use, which will also guide our choice of a vector.

To obtain large amounts of cloned DNA, an ideal host should grow rapidly in an inexpensive culture medium. The host should also be nonpathogenic, be capable of taking up DNA, be genetically stable in culture, and have the appropriate enzymes to allow replication of the vector. It is also helpful if there is considerable genetic information on the host already available and a wealth of tools for its genetic manipulation. The most useful hosts for cloning are typically microorganisms that grow well and for which we have much information. These include the bacteria *Escherichia coli* and *Bacillus subtilis,* and the yeast *Saccharomyces cerevisiae* (Figure 31.2●). Complete genome sequences are available for all of these organisms (∞Chapter 15).

Prokaryotic Hosts

Although most molecular cloning has been done in *Escherichia coli*, this host has potential disadvantages (Figure 31.2). *E. coli* is an excellent choice for initial cloning work, but this bacterium is problematic as a producing organism because it is found in the human intestinal tract,

and wild-type strains are potentially pathogenic. Also, even nonpathogenic strains produce endotoxins (∞ Sections 4.9 and 21.12) that can contaminate products, an especially bad situation with pharmaceutical injectables. In addition, *E. coli* retains extracellular proteins in its periplasmic space, making isolation and purification of recombinant proteins potentially difficult. However, modified *E. coli* strains have been developed for which most of these problems have been eliminated. Thus, because of the extensive knowledge of its genetics and biochemistry (∞Section 10.19), *E. coli* remains the organism of choice for most cloning studies.

The gram-positive organism *Bacillus subtilis* is also used as a cloning host (Figure 31.2). *B. subtilis* is not a potential pathogen, does not produce endotoxin, does produce spores (which is helpful in industrial microbial processes ∞Section 30.1), lacks a periplasm, and naturally secretes proteins into the medium. Although the technology for cloning in *B. subtilis* is not nearly as well developed as that for *E. coli,* several plasmids and phages suitable for cloning have been developed, and transformation (∞Section 10.7) is a well-developed procedure in *B. subtilis*. Disadvantages of using *B. subtilis* as a cloning host, however, include plasmid instability: It can be difficult to maintain plasmid replication over many subcultures of the organism. Also, foreign DNA is not well maintained in *B. subtilis* cells and so the cloned DNA is often unexpectedly lost.

Often organisms used as hosts for cloning must have specific genotypes to be effective. For instance, if the vector carries the gene for β-galactosidase as a selectable marker (∞Sections 10.16 and 15.1), then the host must have a mutation disabling this gene. Because the bacteriophage M13 infects only bacteria with F pili (∞Sections 10.8 and 16.3), hosts used with M13-derived vectors must contain and express genes on the F plasmid. These types of considerations and others, such as the ability to select for transformants, must be taken into account when selecting a host.

Eukaryotic Hosts

Cloning in eukaryotic microorganisms has some important applications, especially in understanding the details of gene regulation in eukaryotic systems. The yeast *Saccharomyces cerevisiae* has been well studied and is used extensively as a cloning host (Figure 31.2). Plasmid vectors, as well as yeast artificial chromosomes (YACs, ∞Section 15.1), have been developed for yeast.

One important advantage of eukaryotic cells as hosts for cloning vectors is that they already possess the complex RNA and posttranslational processing systems required for the production of gene products in higher organisms. Thus, unlike when eukaryotic DNA is cloned into prokaryotes, these systems do not have to be engineered into the vector. Posttranslational processing, in particular, can create some molecular cloning problems (see Section 31.5).

For many applications, gene cloning in *mammalian cells* is desirable. Mammalian cell culture systems can be

Cloning Hosts

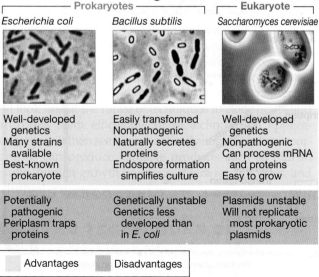

Prokaryotes		Eukaryote
Escherichia coli	*Bacillus subtilis*	*Saccharomyces cerevisiae*
Well-developed genetics	Easily transformed	Well-developed genetics
Many strains available	Nonpathogenic	Nonpathogenic
Best-known prokaryote	Naturally secretes proteins	Can process mRNA and proteins
	Endospore formation simplifies culture	Easy to grow
Potentially pathogenic	Genetically unstable	Plasmids unstable
Periplasm traps proteins	Genetics less developed than in *E. coli*	Will not replicate most prokaryotic plasmids

☐ Advantages ☐ Disadvantages

● **Figure 31.2 Hosts for molecular cloning.** A summary of advantages and disadvantages of common cloning hosts.

handled in some ways like microbial cultures and find wide use in research on human genetics, cancer, infectious disease, and physiology. A disadvantage of mammalian cells as hosts is that they are expensive and difficult to produce under large-scale conditions, and expression levels of cloned genes are often low. Insect cell lines are simpler to grow, and vectors have been developed from an insect DNA virus, the baculovirus (see Section 31.4).

For some applications it is important that the eukaryotic host be a plant cell tissue culture, or even an entire plant. Indeed, genetic engineering has a large number of applications in plant agriculture (see Section 31.10). Regardless of host type, it is necessary that the host be able to take up DNA. We discussed this problem in prokaryotic hosts earlier (⚭Section 10.6), and here we turn our attention to eukaryotic hosts.

Transfection of Eukaryotic Cells

Eukaryotic microorganisms and animal and plant cells can take up DNA in a process that resembles bacterial transformation (⚭Section 10.7). Because the word *transformation* in mammalian cells is used to describe the conversion of cells to the malignant (tumorous, cancerous) state (⚭Section 9.12), the introduction of DNA into mammalian cells has been called *transfection*.

Transfection of cultured animal cells was originally accomplished by precipitating DNA in such a way that the cells would take it up by phagocytosis (⚭Sections 14.10 and 22.2) because they do not have cell walls. In yeast, which is an important organism for genetic engineering, transfection at low efficiencies can be mediated by various methods. However, a technique called *electroporation* is widely applied to all types of eukaryotic cells and can be used whether or not the cell wall of the organism is removed. Electroporation involves exposing host cells to pulsed electrical fields in the presence of cloned DNA. The electrical treatment opens small pores in their membranes through which the DNA can enter. The pores generated are insufficient to cause cell damage or lysis, but sufficient to allow cloned DNA to enter.

In addition to electroporation, a high velocity microprojectile "gun" can be used to incorporate DNA into cells. The *particle gun* operates somewhat like a shotgun. A small steel cylinder containing a gunpowder charge is used to fire nucleic acid–coated particles at the target cells (Figure 31.3●). The particles bombard the cell, piercing cell walls and membranes without actually killing the cells. The nucleic acid entering the cells can then recombine with host DNA. The particle gun has been used successfully to transfect yeast, algae, a variety of plant cells, and even mitochondria and chloroplasts. Moreover, unlike electroporation, the particle gun can be used to get DNA into intact tissue, such as plant seeds. Finally, in animal cells, DNA can be injected into the nucleus using micropipets, a technique called *microinjection*.

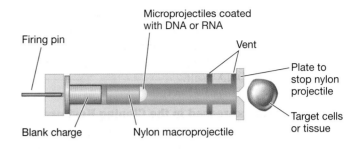

● **Figure 31.3 Nucleic acid gun for transfection of certain eukaryotic cells.** The inner workings of the gun show how nucleic acids coating metal pellets are projected at target cells.

🛑 *31.2 Concept Check*

The choice of a cloning host depends on the final application. In many cases the host can be a prokaryote, but in others it is essential that the host be a eukaryote. Any host must be able to take up DNA, and there are a variety of techniques by which this can be accomplished, both natural and artificial.

◆ Why does molecular cloning require a host?

◆ Describe three mechanisms by which cells can take up DNA.

31.3 Finding the Right Clone

Genetic engineering begins with the isolation of a clone containing a gene of interest. A variety of manipulations can then be performed on this gene, depending on the application, before the final result is achieved. But first it is necessary to identify the host containing the original clone.

Methods to clone DNA include making *gene libraries* from total genomic DNA (⚭Section 10.15), or cloning a DNA fragment made by PCR (⚭Section 7.9). One usually clones from PCR products when the gene of interest has already been identified and the goal is simply to obtain a single clone. However, in a gene library there are thousands or tens of thousands of clones, and typically, only one or a few may be the genes of interest.

One can select for hosts containing a plasmid vector by selecting for a vector marker, such as antibiotic resistance, so that only these cells form colonies (⚭Section 10.15). For host cells containing a viral vector, one simply looks for plaques (⚭Section 10.16). These colonies or plaques can also be screened for vectors that contain foreign DNA inserts by looking for the inactivation of a vector gene (⚭Figures 10.37 and 15.1). In cloning a single DNA fragment generated by PCR or purified by some other means, these simple selections or screenings are usually sufficient. However, with a heterogeneous collection of DNA fragments, as in a gene library, identifying cells carrying cloned DNA is only the first step. The biggest challenge remains *finding the clone* that has the gene of interest. One must examine colonies of bacteria or plaques from viral-infected cells growing on agar

(∞Section 7.8). However, knowledge of the amino acid sequence of a protein can be useful, too, as it can be used to literally *construct* a gene *de novo*.

One can use the amino acid sequence of a protein to design and synthesize an oligonucleotide probe that encodes it. This process is in effect **reverse translation** and is illustrated in Figure 31.9●. Unfortunately, degeneracy of the genetic code (∞Section 7.14) complicates this picture somewhat. Most amino acids are encoded by more than one codon (∞Table 7.5), and codon usage varies from organism to organism. Thus, the best region of a gene to synthesize as a probe is one that encodes part of the protein rich in amino acids specified by only a *single codon* (for example, methionine, AUG; tryptophan, UGG) or at most two codons (for example, phenylalanine, UUU, UUC; tyrosine, UAU, UAC; histidine, CAU, CAC). This strategy increases the chances that the probe will be exactly complementary or nearly complementary to the mRNA or gene of interest. If the complete amino acid sequence of the protein is not known, then one typically uses the sequence at the amino terminus of the protein, where sequencing of the protein begins.

For certain small proteins there may be good reason to synthesize the entire gene. Many mammalian proteins (including high-value therapeutic peptide hormones) are the products of posttranslational processing (∞Section 8.3) and might, therefore, be quite small. Thus, to produce a peptide hormone, it would be more efficient to construct a gene that encoded just the final hormone and not the larger precursor protein from which it was derived. In addition,

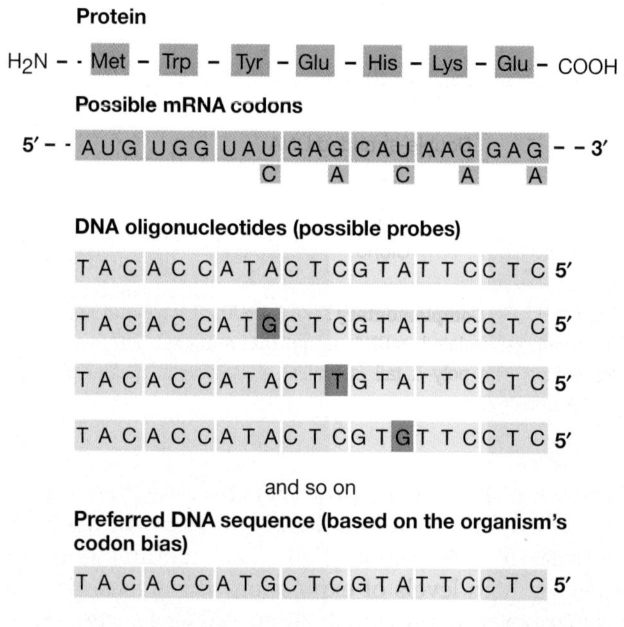

Protein

H₂N – · Met – Trp – Tyr – Glu – His – Lys – Glu · COOH

Possible mRNA codons

5′ – · AUG UGG UAU GAG CAU AAG GAG · – 3′
              C   A   C   A   A

DNA oligonucleotides (possible probes)

T A C A C C A T A C T C G T A T T C C T C 5′

T A C A C C A T G C T C G T A T T C C T C 5′

T A C A C C A T A C T T G T A T T C C T C 5′

T A C A C C A T A C T C G T G T T C C T C 5′

and so on

Preferred DNA sequence (based on the organism's codon bias)

T A C A C C A T G C T C G T A T T C C T C 5′

● **Figure 31.9 Deducing the best sequence of an oligonucleotide probe from the amino acid sequence of the protein: Reverse translation.** Because of degeneracy of the genetic code (∞Section 7.14), many probes are possible. If codon usage by the same organism is known, then a preferred sequence can be selected. It is not essential that complete accuracy be achieved because a small amount of mismatch can be tolerated, especially with long probes.

chemical synthesis not only permits the acquisition of genes that cannot be obtained otherwise but also allows synthesis of *modified genes* that may make useful new proteins. Techniques for the synthesis of DNA molecules are now well developed, and it is possible to synthesize genes encoding proteins over 200 amino acid residues in length (600 nucleotides). The synthetic approach was first used in a major way for production of the human hormone insulin in bacteria, as will be discussed in Section 31.6. It should also be noted that a constructed gene is free of introns, and thus the mRNA made from it does not need to be processed. Also, promoters and other regulatory sequences can easily be built into the gene upstream of the coding sequences and codon bias (∞Section 7.14) accounted for.

With the use of all these techniques, a large number of different human and viral proteins have been expressed at high yield under the control of bacterial regulatory systems. These include, among others, *insulin, virus antigens, interferon*, and *somatostatin* (see Sections 31.6 and 31.7).

Protein Folding and Stability

The synthesis of a protein in a new host is sometimes accompanied by problems other than those whose solutions we have already discussed. For example, some proteins are susceptible to degradation by intracellular proteases and may be destroyed before they can be isolated. Moreover, some eukaryotic proteins are toxic to the prokaryotic host, and therefore the host for the cloning vector may be killed before a sufficient amount of the product is synthesized. Further engineering of either the host or the vector may be necessary to eliminate these problems.

Sometimes when foreign proteins are massively overproduced they form *inclusion bodies* inside the host. Although inclusion bodies are relatively easy to purify because of their size, the protein found in these bodies can be very difficult to solubilize. In many cases these bodies form because the protein is incorrectly folded. One potential solution to this problem is to use a host that overproduces molecular chaperones that aid in folding (∞Section 7.17). However, folding problems can often be solved if the protein from the cloned gene is made as a **fusion protein** along with a protein encoded by the vector. This involves fusing the two genes to yield, following transcription and translation, a single, but cleavable protein. Fusion proteins often simplify purification of the protein of interest if the other part of the fusion protein is one for which rapid and inexpensive purification techniques are available.

Several special *fusion vectors* are now available to encode fusion proteins. The "cloned protein" is then released from the fusion protein after purification by special proteases. Figure 31.10● shows an example of a fusion vector that is also an expression vector. In some cases the cloned protein can also be removed from the fusion protein by specific chemical treatment. Fusion systems can also be used for purposes other than achieving increased protein stability. One advantage of making a fusion pro-

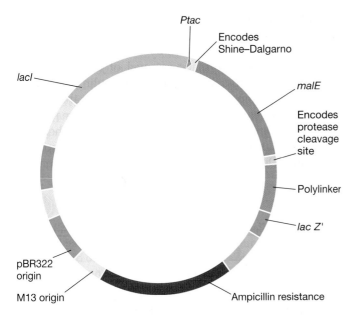

● **Figure 31.10 An expression vector for fusions.** This vector was developed by the New England Biolabs Company. The gene to be cloned is inserted at the polylinker site (◯◯Figure 15.1), so it is in frame with the *malE* gene, which encodes the maltose-binding protein. This insertion inactivates the *lacZ'* gene (◯◯Figure 15.1). The fused gene is under control of the hybrid *tac* promoter (*Ptac*). The plasmid also contains the *lacI* gene, which encodes the *lac* repressor. Therefore, an inducer must be added to the cells in order to turn on the *tac* promoter. The fusion protein is easily purified by methods involving the affinity of the protein for maltose. Once purified, the two portions of the fusion protein can be separated by a very specific protease (factor Xa). The plasmid contains a gene conferring ampicillin resistance on its host. In addition to the plasmid origin of replication, there is a bacteriophage M13 origin. Therefore, this is a *phagemid* and can be propagated either as a plasmid or as a phage.

tein is that the host portion can be engineered to contain the bacterial sequence encoding the *signal peptide* that enables transport of the protein across the cytoplasmic membrane (◯◯Section 7.17). This makes possible a bacterial expression system that not only synthesizes the mammalian protein, but also actually *excretes* it. By employing the right strains and vectors, the desired protein can exceed 200,000 molecules per cell and make up as much as 40% of the protein molecules in a cell.

 31.5 Concept Check

It is possible to achieve very high levels of expression of mammalian genes in prokaryotes. However, the expressed gene must be free of introns. This can be accomplished by using reverse transcriptase to synthesize cDNA from the mature mRNA encoding the protein of interest. It can also be accomplished by synthesizing an entirely synthetic gene from knowledge of the amino acid sequence of the protein of interest. Fusion proteins are often used to stabilize or solubilize the cloned protein.

◆ What major advantage does cloning mammalian genes from mRNA or synthetic genes have over PCR amplification and cloning of the native gene?

◆ How is a fusion protein made?

▐ ▌ PRACTICAL APPLICATIONS OF GENETIC ENGINEERING

We present here only a few of the many applications of genetic engineering in biotechnology. These will focus on existing applications that are of importance to medicine and agriculture. In addition to these, many other exciting applications are emerging or in development. These include, in particular, the enhancement of industrial fermentations (◯◯Chapter 30) and the use of genetically engineered prokaryotes in bioremediation (◯◯Section 19.18) and other areas of environmental biotechnology.

31.6 Production of Insulin: The Beginnings of Commercial Biotechnology

The most economically robust area of biotechnology is the production of *human proteins*. Many proteins from mammalian cells have high pharmaceutical value. However, these proteins are typically present in very small amounts in normal tissue, and it is therefore extremely costly to purify them. Moreover, even if the protein can be produced in cell cultures (◯◯Section 9.3), this is a much more expensive and difficult method of obtaining it than growing microbial cultures that produce it in high yield. Therefore, the biotechnology industry has genetically engineered microorganisms to produce these proteins. A classic example is *insulin* and is discussed in detail here.

The Biology of Insulin

An early and dramatic commercial success in biotechnology was the production of the hormone **insulin**. Many hormones are small proteins. These molecules are critical for the proper control of mammalian metabolism and have important therapeutic uses. Insulin is a protein produced in the pancreas that is vital for the regulation of glucose metabolism in the body. *Juvenile diabetes*, a disease characterized by insulin deficiency, afflicts millions of people (◯◯Section 22.15). The standard treatment for juvenile diabetes is periodic injections or oral administration of insulin.

Because insulins of most mammals are similar in structure, it has been possible in the past to treat human diabetes by use of insulin isolated commercially from beef or pork pancreas. However, nonhuman insulin is not as effective as human insulin, and the isolation process is expensive and complex. Cloning of a human insulin gene in bacteria solved this problem. Cloning and production began using *Escherichia coli*, but today, most genetically engineered insulin is produced in yeast. The final product, biosynthetic human insulin, is identical in every respect to insulin purified from the human pancreas.

Despite being a small protein, insulin was a particular challenge to the biotechnology industry. Insulin in its with a host chromosome, the foreign DNA can be expressed and can confer new properties on the plant.

With the use of *Agrobacterium tumefaciens*, a number of transgenic plants have been produced. Most successes have come with broadleaf crop plants (dicots) such as tomato, potato, tobacco, soybean, alfalfa, and cotton. *A. tumefaciens* has also been used to produce transgenic trees, such as walnut and apple. Transgenic crop plants from the grass family (monocots) have been more difficult to generate using *A. tumefaciens*, but other methods of introducing DNA, such as the particle gun (see Section 31.2), have been used successfully here.

Herbicide and Insect Resistance

Major areas targeted for genetic improvement in plants include herbicide, insect, and microbial disease resistance, as well as improved product quality. More than a thousand different field trials on more than 30 different plant species have been carried out in recent years. The first genetically modified (GM) crop to be grown commercially was tobacco grown in China in 1992. By the year 2000, over 100 million acres (44 million hectares) of GM crops were planted worldwide. Of these, 58% were in soybeans, 23% were in corn, 12% were in cotton, and 6% were in canola. Almost all the GM soybeans and canola planted were herbicide resistant, whereas corn and cotton were herbicide resistant or insect resistant, or both.

Herbicide resistance is genetically engineering into a crop plant so that it will not be killed by the toxic chemical. Many herbicides inhibit a key plant enzyme or protein necessary for growth. For example, the herbicide *glyphosate* kills plants by inhibiting the activity of an enzyme necessary for making aromatic amino acids. Such a herbicide kills both weeds and crop plants and thus must be used as a "preemergence herbicide," that is, before the crop plants emerge from the ground. However, some bacteria contain a form of the enzyme that is naturally resistant to glyphosate. A gene encoding a resistant enzyme from *Agrobacterium* has been cloned, modified for expression in plants, and transferred into important crop plants, such as soybeans. When sprayed with glyphosate, plants containing the bacterial gene grow as well as unsprayed control plants (Figure 31.14●). Herbicide resistant soybeans are now widely planted in the United States.

Genetic engineering has also been used to protect plants from virus infection. For example, it has been discovered that transgenic plants that express the coat protein gene of a virus become resistant to infection by that virus. Although the mechanism of resistance is unknown, the presence of viral coat protein in plant cells apparently interferes with the uncoating of viral particles containing that coat protein, and this interrupts the virus replication cycle.

● **Figure 31.14 Transgenic plants: Herbicide resistance.** The photograph shows a portion of a field of soybeans that has been treated with *Roundup*, a glyphosate–based herbicide manufactured by Monsanto. The plants on the right are normal soybeans; those on the left have been genetically engineered to express glyphosate resistance.

Insect Resistance: Bt Toxin

Insect resistance has also been genetically introduced into plants. One example already in wide use involves the genes encoding the toxic protein of *Bacillus thuringiensis*. This bacterium produces a crystalline protein called *Bt-toxin* (◌◌◌ Section 12.20) that is toxic to moth and butterfly larvae. Certain strains of *B. thuringiensis* produce additional proteins toxic to beetle and fly larvae and mosquitoes.

Biotechnologists are using several different approaches to enhance the use of Bt-toxin for pest control in plants. One approach is to develop a single Bt-toxin that is effective against many different insects. This can be done because the protein has separate domains for specificity and toxicity. The toxic domain is highly conserved in all the various Bt-toxins. Genetic engineers can thus make a gene that encodes the toxic domain and one of several different specificity domains, to yield a series of toxins, each best suited for a particular plant or pest situation.

An effective approach to achieve transgene expression and stability is to transfer the gene directly into the plant genome. For example, a natural Bt-toxin gene has been cloned into a plasmid vector under control of a chloroplast rRNA promoter and transferred into tobacco plant chloroplasts by particle bombardment (see Section 31.2). With this methodology, transgenic plants were obtained that expressed this protein at levels that were extremely toxic to larvae from a number of insect species (Figure 31.15●).

Although transgenic Bt-toxin looked at first to be a great success, there have been some reports of insect resistance. Resistance to insecticides and herbicides is a

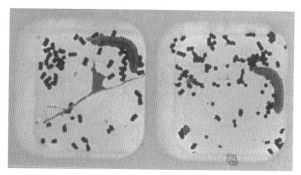

(a)

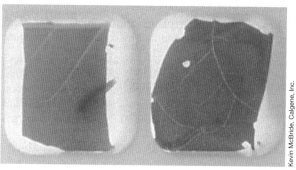

(b)

Kevin McBride, Calgene, Inc.

● **Figure 31.15 Transgenic plants: Insect resistance** (a) The results of two different assays to determine the effect of beet armyworm larvae on tobacco leaves from normal plants. (b) The results of similar assays, but using tobacco leaves taken from transgenic plants that express Bt-toxin in their chloroplasts.

common problem in agriculture, and the fact that a product has been produced by genetic engineering does not protect it from this problem. This emphasizes that many approaches must be used for pest control in agriculture, and Bt-toxin is just one of many tools being developed by agricultural biotechnologists. Nevertheless, Bt-toxin-resistant crops are still widely planted in the United States.

Other Uses of Plant Biotechnology

Not all genetic engineering is directed toward making plants disease resistant. Genetic engineering can be used in a variety of ways for developing GM plants with desirable consumer-oriented characteristics. For instance, the first GM food grown for sale in the U.S. market was a tomato that underwent delayed spoilage. This benefit increased the shelf life of the vegetable. In addition, transgenic plants can be genetically engineered to produce commercial or pharmaceutical products, as has been done with microorganisms (see Section 31.6) and animals (see Section 31.9). For example, crop plants such as tobacco and tomatoes have been engineered to produce a number of different products, such as the human protein interferon. Transgenic crop plants can also be used

to produce human antibodies efficiently and inexpensively. These antibodies, coined "plantibodies," have potential as anticancer or antivirus drugs, and some are undergoing clinical trials. Plants are useful in producing these types of products because they typically modify proteins correctly and because crop plants can be efficiently grown and harvested in large amounts.

Crop plants are also being developed for the production of vaccines. For instance, a recombinant tobacco mosaic virus (∞ Section 16.7) has been engineered whose coat contains antigens of *Plasmodium vivax*, the organism that causes malaria (∞ Section 27.5). This recombinant virus could be used to produce a malaria vaccine in large amounts and at low cost by simply harvesting infected tobacco or tomato plants grown in fields. Another interesting approach is to produce a vaccine in an edible plant product. Such *edible vaccines* now under development could immunize humans against diseases caused by enteric bacteria, including cholera and diarrhea (∞ Section 28.5).

Although public acceptance of GM crops remains high in the United States, there have been some concerns over the contamination of human food with GM corn, so far approved only for animal food. In Europe there has been considerable public concern over GM organisms, and hence several EU countries have imposed bans on the growth or importation of GM crops. Some countries that still accept GM foods have introduced legislation to require all GM foods to be labeled as such. What these practices show is that even good science and potential commercial success does not always ensure public acceptance of an agricultural product.

Because of increased public concern, the acreage planted in many GM crops in the United States has leveled off in recent years. Most concerns center around either the perception of adverse effects of foreign genes on humans or domesticated animals, or the potential "escape" of transgenes from transgenic plants into native plants. At present, supporting evidence for either of these scenarios is scant. However, concerns remain and have served to control the speed at which transgenic plants enter the marketplace.

 31.10 Concept Check

Genetic engineering is being used to make plants resistant to disease, to improve product quality, and to use crop plants as a source of recombinant proteins and even vaccines. One commonly used cloning vector for plants is the Ti plasmid of the bacterium *Agrobacterium tumefaciens*. This plasmid can transfer DNA into plant cells. Commercial plants whose genomes have been modified using *in vitro* genetic techniques are called *genetically modified (GM)* organisms.

◆ Transfer of DNA to plants by the Ti plasmid most resembles what form of bacterial gene transfer (∞ Chapter 10)?

◆ What is a *transgenic plant*?

Table A1.2 — Microbiologically important reduction potentials[a]

Redox pair	E_0' (V)
SO_4^{2-}/HSO_3^-	−0.52
CO_2/formate	−0.43
$2 H^+/H_2$	−0.41
$S_2O_3^{2-}/HS^- + HSO_3^-$	−0.40
Ferredoxin ox/red	−0.39
Flavodoxin ox/red[b]	−0.37
$NAD^+/NADH$	−0.32
Cytochrome c_3 ox/red	−0.29
CO_2/acetate$^-$	−0.29
S^0/HS^-	−0.27
CO_2/CH_4	−0.24
FAD/FADH	−0.22
SO_4^{2-}/HS^-	−0.217
Acetaldehyde/ethanol	−0.197
Pyruvate$^-$/lactate$^-$	−0.19
FMN/FMNH	−0.19
Dihydroxyacetone phosphate/glycerolphosphate	−0.19
$HSO_3^-/S_3O_6^{2-}$	−0.17
Flavodoxin ox/red[b]	−0.12
HSO_3^-/HS^-	−0.116
Menaquinone ox/red	−0.075
APS/AMP + HSO_3^-	−0.060
Rubredoxin ox/red	−0.057
Acrylyl-CoA/propionyl-CoA	−0.015
Glycine/acetate$^-$ + NH_4^+	−0.010
$S_4O_6^{2-}/S_2O_3^{2-}$	+0.024
Fumarate^{2-}/succinate^{2-}	+0.033
Cytochrome b ox/red	+0.035
Ubiquinone ox/red	+0.113
AsO_4^{3-}/AsO_3^{3-}	+0.139
Dimethyl sulfoxide (DMSO)/dimethylsulfide (DMS)	+0.16
$Fe(OH)_3 + HCO_3^-/FeCO_3$	+0.20
$S_3O_6^{2-}/S_2O_3^{2-} + HSO_3^-$	+0.225
Cytochrome c_1 ox/red	+0.23
NO_2^-/NO	+0.36
Cytochrome a_3 ox/red	+0.385
Chlorobenzoate$^-$/benzoate$^-$ + HCl	+0.297
NO_3^-/NO_2^-	+0.43
SeO_4^{2-}/SeO_3^{2-}	+0.475
Fe^{3+}/Fe^{2+}	+0.77
Mn^{4+}/Mn^{2+}	+0.798
O_2/H_2O	+0.82
ClO_3^-/Cl^-	+1.03
NO/N_2O	+1.18
N_2O/N_2	+1.36

[a] Data from Thauer, R. K., K. Jungermann, and K. Decker, 1977. Energy conservation in anaerobic chemotrophic bacteria. *Bacteriol. Rev.* 41:100–180.
[b] Separate potentials are given for each electron transfer in this potentially two-electron transfer.

LIST OF GENERA AND HIGHER ORDER TAXA[a]

Domain *Archaea*
Phylum AI. *Crenarchaeota*
  Class I. *Thermoprotei*
    Order I. *Thermoproteales*
      Family I. *Thermoproteaceae*
        Genus I. *Thermoproteus*
        Genus II. *Caldivirga*
        Genus III. *Pyrobaculum*
        Genus IV. *Thermocladium*
        Genus V. *Vulcanisaeta*
      Family II. *Thermofilaceae*
        Genus I. *Thermofilum*
    Order II. *Caldisphaerales*
      Family I. *Caldisphaeraceae*
        Genus I. *Caldisphaera*
    Order III. *Desulfurococcales*
      Family I. *Desulfurococcaceae*
        Genus I. *Desulfurococcus*
        Genus II. *Acidilobus*
        Genus III. *Aeropyrum*
        Genus IV. *Ignicoccus*
        Genus V. *Staphylothermus*
        Genus VI. *Stetteria*
        Genus VII. *Sulfophobococcus*
        Genus VIII. *Thermodiscus*
        Genus IX. *Thermosphaera*
      Family II. *Pyrodictiaceae*
        Genus I. *Pyrodictium*
        Genus II. *Hyperthermus*
        Genus III. *Pyrolobus*
    Order IV. *Sulfolobales*
      Family I. *Sulfolobaceae*
        Genus I. *Sulfolobus*
        Genus II. *Acidianus*
        Genus III. *Metallosphaera*
        Genus IV. *Stygiolobus*
        Genus V. *Sulfurisphaera*
        Genus VI. *Sulfurococcus*
Phylum AII. *Euryarchaeota*
  Class I. *Methanobacteria*
    Order I. *Methanobacteriales*
      Family I. *Methanobacteriaceae*
        Genus I. *Methanobacterium*
        Genus II. *Methanobrevibacter*
        Genus III. *Methanosphaera*
        Genus IV. *Methanothermobacter*
      Family II. *Methanothermaceae*
        Genus I. *Methanothermus*
  Class II. *Methanococci*
    Order I. *Methanococcales*
      Family I. *Methanococcaceae*
        Genus I. *Methanococcus*
        Genus II. *Methanothermococcus*
      Family II. *Methanocaldococcaceae*
        Genus I. *Methanocaldococcus*
        Genus II. *Methanotorris*
  Class III. *Methanomicrobia*
    Order I. *Methanomicrobiales*
      Family I. *Methanomicrobiaceae*
        Genus I. *Methanomicrobium*
        Genus II. *Methanoculleus*
        Genus III. *Methanofollis*
        Genus IV. *Methanogenium*
        Genus V. *Methanolacinia*
        Genus VI. *Methanoplanus*
      Family II. *Methanocorpusculaceae*
        Genus I. *Methanocorpusculum*

      Family III. *Methanospirillaceae*
        Genus I. *Methanospirillum*
      Genera incertae sedis[b]
        Genus I. *Methanocalculus*
    Order II. *Methanosarcinales*
      Family I. *Methanosarcinaceae*
        Genus I. *Methanosarcina*
        Genus II. *Methanococcoides*
        Genus III. *Methanohalobium*
        Genus IV. *Methanohalophilus*
        Genus V. *Methanolobus*
        Genus VI. *Methanomethylovorans*
        Genus VII. *Methanimicrococcus*
        Genus VIII. *Methanosalsum*
      Family II. *Methanosaetaceae*
        Genus I. *Methanosaeta*
  Class IV. *Halobacteria*
    Order I. *Halobacteriales*
      Family I. *Halobacteriaceae*
        Genus I. *Halobacterium*
        Genus II. *Haloarcula*
        Genus III. *Halobaculum*
        Genus IV. *Halobiforma*
        Genus V. *Halococcus*
        Genus VI. *Haloferax*
        Genus VII. *Halogeometricum*
        Genus VIII. *Halomicrobium*
        Genus IX. *Halorhabdus*
        Genus X. *Halorubrum*
        Genus XI. *Halosimplex*
        Genus XII. *Haloterrigena*
        Genus XIII. *Natrialba*
        Genus XIV. *Natrinema*
        Genus XV. *Natronobacterium*
        Genus XVI. *Natronococcus*
        Genus XVII. *Natronomonas*
        Genus XVIII. *Natronorubrum*
  Class V. *Thermoplasmata*
    Order I. *Thermoplasmatales*
      Family I. *Thermoplasmataceae*
        Genus I. *Thermoplasma*
      Family II. *Picrophilaceae*
        Genus I. *Picrophilus*
      Family III. *Ferroplasmaceae*
        Genus I. *Ferroplasma*
  Class VI. *Thermococci*
    Order I. *Thermococcales*
      Family I. *Thermococcaceae*
        Genus I. *Thermococcus*
        Genus II. *Palaeococcus*
        Genus III. *Pyrococcus*
  Class VII. *Archaeoglobi*
    Order I. *Archaeoglobales*
      Family I. *Archaeoglobaceae*
        Genus I. *Archaeoglobus*
        Genus II. *Ferroglobus*
        Genus III. *Geoglobus*
  Class VIII. *Methanopyri*
    Order I. *Methanopyrales*
      Family I. *Methanopyraceae*
        Genus I. *Methanopyrus*
Domain *Bacteria*
Phylum BI. *Aquificae*
  Class I. *Aquificae*
    Order I. *Aquificales*
      Family I. *Aquificaceae*
        Genus I. *Aquifex*
        Genus II. *Calderobacterium*
        Genus III. *Hydrogenobaculum*

[a]The list of genera and higher order taxa shown here are the organisms recognized as of 2005. Genera or higher-order taxa in quotation marks are recognized taxa whose names have not yet been validated. Because bacterial taxonomy is a work in progress, updates to the list shown here occur as new genera and species are described and as new data support new taxonomic arrangements. For further discussion on bacterial taxonomy, see Sections 11.10–11.13.

[b]Taxa of uncertain affiliation.

Genus IV. *Hydrogenobacter*
Genus V. *Hydrogenothermus*
Genus VI. *Persephonella*
Genus VII. *Sulfurihydrogenibium*
Genus VIII. *Thermocrinis*
Genera incertae sedis[b]
Genus I. *Balnearium*
Genus II. *Desulfurobacterium*
Genus III. *Thermovibrio*
Phylum BII. *Thermotogae*
Class I. *Thermotogae*
Order I. *Thermotogales*
Family I. *Thermotogaceae*
Genus I. *Thermotoga*
Genus II. *Fervidobacterium*
Genus III. *Geotoga*
Genus IV. *Marinitoga*
Genus V. *Petrotoga*
Genus VI. *Thermosipho*
Phylum BIII. *Thermodesulfobacteria*
Class I. *Thermodesulfobacteria*
Order I. *Thermodesulfobacteriales*
Family I. *Thermodesulfobacteriaceae*
Genus I. *Thermodesulfobacterium*
Genus II. *Thermodesulfatator*
Phylum BIV. *Deinococcus-Thermus*
Class I. *Deinococci*
Order I. *Deinococcales*
Family I. *Deinococcaceae*
Genus I. *Deinococcus*
Order II. *Thermales*
Family I. *Thermaceae*
Genus I. *Thermus*
Genus II. *Marinithermus*
Genus III. *Meiothermus*
Genus IV. *Oceanithermus*
Genus V. *Vulcanithermus*
Phylum BV. *Chrysiogenetes*
Class I. *Chrysiogenetes*
Order I. *Chrysiogenales*
Family I. *Chrysiogenaceae*
Genus I. *Chrysiogenes*
Phylum BVI. *Chloroflexi*
Class I. *Chloroflexi*
Order I. *Chloroflexales*
Family I. *Chloroflexaceae*
Genus I. *Chloroflexus*
Genus II. *Chloronema*
Genus III. *Heliothrix*
Genus IV. *Roseiflexus*
Family II. *Oscillochloridaceae*
Genus I. *Oscillochloris*
Order II. *Herpetosiphonales*
Family I. *Herpetosiphonaceae*
Genus I. *Herpetosiphon*
Class II. *Anaerolieae*
Order I. *Anaerolinaeles*
Family I. *Anaerolinaceae*
Genus I. *Anaerolinea*
Genus II. *Caldilinea*
Phylum BVII. *Thermomicrobia*
Class I. *Thermomicrobia*
Order I. *Thermomicrobiales*
Family I. *Thermomicrobiaceae*
Genus I. *Thermomicrobium*
Phylum BVIII. *Nitrospirae*
Class I. *Nitrospira*
Order I. *Nitrospirales*
Family I. *Nitrospiraceae*
Genus I. *Nitrospira*
Genus II. *Leptospirillum*
Genus III. *Magnetobacterium*
Genus IV. *Thermodesulfovibrio*
Phylum BIX. *Deferribacteres*

Class I. *Deferribacteres*
Order I. *Deferribacterales*
Family I. *Deferribacteraceae*
Genus I. *Deferribacter*
Genus II. *Denitrovibrio*
Genus III. *Flexistipes*
Genus IV. *Geovibrio*
Genera incertae sedis[b]
Genus I. *Synergistes*
Genus II. *Caldithrix*
Phylum BX. *Cyanobacteria*
Class I. *Cyanobacteria*
Subsection I. *Subsection 1*
Family I. Family 1.1
Form genus I. *Chamaesiphon*[c]
Form genus II. *Chroococcus*
Form genus III. *Cyanobacterium*
Form genus IV. *Cyanobium*
Form genus V. *Cyanothece*
Form genus VI. *Dactylococcopsis*
Form genus VII. *Gloeobacter*
Form genus VIII. *Gloeocapsa*
Form genus IX. *Gloeothece*
Form genus X. *Microcystis*
Form genus XI. *Prochlorococcus*
Form genus XII. *Prochloron*
Form genus XIII. *Synechococcus*
Form genus XIV. *Synechocystis*
Subsection II. *Subsection 2*
Family I. Family 2.1
Form genus I. *Cyanocystis*
Form genus II. *Dermocarpella*
Form genus III. *Stanieria*
Form genus IV. *Xenococcus*
Family II. Family 2.2
Form genus I. *Chroococcidiopsis*
Form genus II. *Myxosarcina*
Form genus III. *Pleurocapsa*
Subsection III. *Subsection 3*
Family I. Family 3.1
Form genus I. *Arthrospira*
Form genus II. *Borzia*
Form genus III. *Crinalium*
Form genus IV. *Geitlerinema*
Genus V. *Halospirulina*
Form genus VI. *Leptolyngbya*
Form genus VII. *Limnothrix*
Form genus VIII. *Lyngbya*
Form genus IX. *Microcoleus*
Form genus X. *Oscillatoria*
Form genus XI. *Planktothrix*
Form genus XII. *Prochlorothrix*
Form genus XIII. *Pseudanabaena*
Form genus XIV. *Spirulina*
Form genus XV. *Starria*
Form genus XVI. *Symploca*
Genus XVII. *Trichodesmium*
Form genus XVIII. *Tychonema*
Subsection IV. *Subsection 4*
Family I. Family 4.1
Form genus I. *Anabaena*
Form genus II. *Anabaenopsis*
Form genus III. *Aphanizomenon*
Form genus IV. *Cyanospira*
Form genus V. *Cylindrospermopsis*
Form genus VI. *Cylindrospermum*
Form genus VII. *Nodularia*
Form genus VIII. *Nostoc*
Form genus IX. *Scytonema*
Family II. Family 4.2
Form genus I. *Calothrix*
Form genus II. *Rivularia*
Form genus III. *Tolypothrix*
Subsection V. *Subsection 5*

---

[c]The taxonomic position of the cyanobacteria in *Bergey's Manual* is left open. The term "form genus" refers to a group of cyanobacteria with very characteristic morphology found worldwide. However, not all isolates of such a type may actually fit into the same genus. In some cases, pure cultures of form genera are not available.

Family I. Subsection 5.1
  Form genus I. *Chlorogloeopsis*
  Form genus II. *Fischerella*
  Form genus III. *Geitleria*
  Form genus IV. *Iyengariella*
  Form genus V. *Nostochopsis*
  Form genus VI. *Stigonema*
Phylum BXI. *Chlorobi*
  Class I. *Chlorobia*
    Order I. *Chlorobiales*
      Family I. *Chlorobiaceae*
        Genus I. *Chlorobium*
        Genus II. *Ancalochloris*
        Genus III. *Chlorobaculum*
        Genus IV. *Chloroherpeton*
        Genus V. *Pelodictyon*
        Genus VI. *Prosthecochloris*
Phylum BXII. *Proteobacteria*
  Class I. *Alphaproteobacteria*
    Order I. *Rhodospirillales*
      Family I. *Rhodospirillaceae*
        Genus I. *Rhodospirillum*
        Genus II. *Azospirillum*
        Genus III. *Inquilinus*
        Genus IV. *Magnetospirillum*
        Genus V. *Phaeospirillum*
        Genus VI. *Rhodocista*
        Genus VII. *Rhodospira*
        Genus VIII. *Rhodovibrio*
        Genus IX. *Roseospira*
        Genus X. *Skermanella*
        Genus XI. *Thalassospira*
        Genus XII. *Tistrella*
      Family II. *Acetobacteraceae*
        Genus I. *Acetobacter*
        Genus II. *Acidiphilium*
        Genus III. *Acidisphaera*
        Genus IV. *Acidocella*
        Genus V. *Acidomonas*
        Genus VI. *Asaia*
        Genus VII. *Craurococcus*
        Genus VIII. *Gluconacetobacter*
        Genus IX. *Gluconobacter*
        Genus X. *Kozakia*
        Genus XI. *Muricoccus*
        Genus XII. *Paracraurococcus*
        Genus XIII. *Rhodopila*
        Genus XIV. *Roseococcus*
        Genus XV. *Rubritepids*
        Genus XVI. *Stella*
        Genus XVII. *Teichococcus*
        Genus XVIII. *Zavarzinia*
    Order II. *Rickettsiales*
      Family I. *Rickettsiaceae*
        Genus I. *Rickettsia*
        Genus II. *Orientia*
      Family II. *Anaplasmataceae*
        Genus I. *Anaplasma*
        Genus II. *Aegyptianella*
        Genus III. *Cowdria*
        Genus IV. *Ehrlichia*
        Genus V. *Neorickettsia*
        Genus VI. *Wolbachia*
        Genus VII. *Xenohaliotis*
      Family III. *Holosporaceae*
        Genus I. *Holospora*
      Genera incertae sedis[b]
        Genus I. *Caedibacter*
        Genus II. *Lyticum*
        Genus III. *Odyssella*
        Genus IV. *Pseudocaedibacter*
        Genus V. *Symbiotes*
        Genus VI. *Tectibacter*
    Order III. *Rhodobacterales*
      Family I. *Rhodobacteraceae*
        Genus I. *Rhodobacter*

Genus II. *Ahrensia*
Genus III. *Albidovulum*
Genus IV. *Amaricoccus*
Genus V. *Antarctobacter*
Genus VI. *Gemmobacter*
Genus VII. *Hirschia*
Genus VIII. *Hyphomonas*
Genus IX. *Jannaschia*
Genus X. *Ketogulonicigenium*
Genus XI. *Leisingera*
Genus XII. *Maricaulis*
Genus XIII. *Methylarcula*
Genus XIV. *Oceanicaulis*
Genus XV. *Octadecabacter*
Genus XVI. *Pannonibacter*
Genus XVII. *Paracoccus*
Genus XVIII. *Pseudorhodobacter*
Genus XIX. *Rhodobaca*
Genus XX. *Rhodothalassium*
Genus XXI. *Rhodovulum*
Genus XXII. *Roseibium*
Genus XXIII. *Roseinatronobacter*
Genus XXIV. *Roseivivax*
Genus XXV. *Roseobacter*
Genus XXVI. *Roseovarius*
Genus XXVII. *Rubrimonas*
Genus XXVIII. *Ruegeria*
Genus XXIX. *Sagittula*
Genus XXX. *Silicibacter*
Genus XXXI. *Staleya*
Genus XXXII. *Stappia*
Genus XXXIII. *Sulfitobacter*
Order IV. *Sphingomonadales*
  Family I. *Sphingomonadaceae*
    Genus I. *Sphingomonas*
    Genus II. *Blastomonas*
    Genus III. *Erythrobacter*
    Genus IV. *Erythromicrobium*
    Genus V. *Erythromonas*
    Genus VI. *Novosphingobium*
    Genus VII. *Porphyrobacter*
    Genus VIII. *Rhizomonas*
    Genus IX. *Sandaracinobacter*
    Genus X. *Sphingobium*
    Genus XI. *Sphingopyxis*
    Genus XII. *Zymomonas*
Order V. *Caulobacterales*
  Family I. *Caulobacteraceae*
    Genus I. *Caulobacter*
    Genus II. *Asticcacaulis*
    Genus III. *Brevundimonas*
    Genus IV. *Phenylobacterium*
Order VI. *Rhizobiales*
  Family I. *Rhizobiaceae*
    Genus I. *Rhizobium*
    Genus II. *Agrobacterium*
    Genus III. *Allorhizobium*
    Genus IV. *Carbophilus*
    Genus V. *Chelatobacter*
    Genus VI. *Ensifer*
    Genus VII. *Sinorhizobium*
  Family II. *Aurantimonadaceae*
    Genus I. *Aurantimonas*
    Genus II. *Fulvimarina*
  Family III. *Bartonellaceae*
    Genus I. *Bartonella*
  Family IV. *Brucellaceae*
    Genus I. *Brucella*
    Genus II. *Mycoplana*
    Genus III. *Ochrobactrum*
  Family V. *Phyllobacteriaceae*
    Genus I. *Phyllobacterium*
    Genus II. *Aminobacter*
    Genus III. *Aquamicrobium*
    Genus IV. *Fluvibacter*
    Genus V. *Candidatus Liberibacter*[d]

[d]In bacterial taxonomy, candidatus status is for organisms known to exist by 16S rRNA gene sequencing and other key properties, but which are not yet in pure culture.

Genus VI. *Mesorhizobium*
Genus VII. *Nitratireductor*
Genus VIII. *Pseudaminobacter*
Family VI. *Methylocystaceae*
Genus I. *Methylocystis*
Genus II. *Albibacter*
Genus III. *Methylopila*
Genus IV. *Methylosinus*
Genus V. *Terasakiella*
Family VII. *Beijerinckiaceae*
Genus I. *Beijerinckia*
Genus II. *Chelatococcus*
Genus III. *Methylocapsa*
Genus IV. *Methylocella*
Family VIII. *Bradyrhizobiaceae*
Genus I. *Bradyrhizobium*
Genus II. *Afipia*
Genus III. *Agromonas*
Genus IV. *Blastobacter*
Genus V. *Bosea*
Genus VI. *Nitrobacter*
Genus VII. *Oligotropha*
Genus VIII. *Rhodoblastus*
Genus IX. *Rhodopseudomonas*
Family IX. *Hyphomicrobiaceae*
Genus I. *Hyphomicrobium*
Genus II. *Ancalomicrobium*
Genus III. *Ancylobacter*
Genus IV. *Angulomicrobium*
Genus V. *Aquabacter*
Genus VI. *Azorhizobium*
Genus VII. *Blastochloris*
Genus VIII. *Devosia*
Genus IX. *Dichotomicrobium*
Genus X. *Filomicrobium*
Genus XI. *Gemmiger*
Genus XII. *Labrys*
Genus XIII. *Methylorhabdus*
Genus XIV. *Pedomicrobium*
Genus XV. *Prosthecomicrobium*
Genus XVI. *Rhodomicrobium*
Genus XVII. *Rhodoplanes*
Genus XVIII. *Seliberia*
Genus XIX. *Starkeya*
Genus XX. *Xanthobacter*
Family X. *Methylobacteriaceae*
Genus I. *Methylobacterium*
Genus II. *Microvirga*
Genus III. *Protomonas*
Genus IV. *Roseomonas*
Family XI. *Rhodobiaceae*
Genus I. *Rhodobium*
Genus II. *Roseospirillum*
Order VII. *Parvularculales*
Family I. *Parvularculaceae*
Genus I. *Parvularcula*

Class II. *Betaproteobacteria*
Order I. *Burkholderiales*
Family I. *Burkholderiaceae*
Genus I. *Burkholderia*
Genus II. *Cupriavidus*
Genus III. *Lautropia*
Genus IV. *Limnobacter*
Genus V. *Pandoraea*
Genus VI. *Paucimonas*
Genus VII. *Polynucleobacter*
Genus VIII. *Ralstonia*
Genus IX. *Thermothrix*
Genus X. *Wautersia*
Family II. *Oxalobacteraceae*
Genus I. *Oxalobacter*
Genus II. *Duganella*
Genus III. *Herbaspirillum*
Genus IV. *Janthinobacterium*
Genus V. *Massilia*
Genus VI. *Oxalicibacterium*
Genus VII. *Telluria*

Family III. *Alcaligenaceae*
Genus I. *Alcaligenes*
Genus II. *Achromobacter*
Genus III. *Bordetella*
Genus IV. *Brackiella*
Genus V. *Derxia*
Genus VI. *Kerstersia*
Genus VII. *Oligella*
Genus VIII. *Pelistega*
Genus IX. *Pigmentiphaga*
Genus X. *Sutterella*
Genus XI. *Taylorella*
Family IV. *Comamonadaceae*
Genus I. *Comamonas*
Genus II. *Acidovorax*
Genus III. *Alicycliphilus*
Genus IV. *Brachymonas*
Genus V. *Caldimonas*
Genus VI. *Delftia*
Genus VII. *Diaphorobacter*
Genus VIII. *Hydrogenophaga*
Genus IX. *Hylemonella*
Genus X. *Lampropedia*
Genus XI. *Macromonas*
Genus XII. *Ottowia*
Genus XIII. *Polaromonas*
Genus XIV. *Ramlibacter*
Genus XV. *Rhodoferax*
Genus XVI. *Variovorax*
Genus XVII. *Xenophilus*
Genera incertae sedis[b]
Genus I. *Aquabacterium*
Genus II. *Ideonella*
Genus III. *Leptothrix*
Genus IV. *Roseateles*
Genus V. *Rubrivivax*
Genus VI. *Schlegelella*
Genus VII. *Sphaerotilus*
Genus VIII. *Tepidimonas*
Genus IX. *Thiomonas*
Genus X. *Xylophilus*
Order II. *Hydrogenophilales*
Family I. *"Hydrogenophilaceae"*
Genus I. *Hydrogenophilus*
Genus II. *Thiobacillus*
Order III. *Methylophilales*
Family I. *Methylophilaceae*
Genus I. *Methylophilus*
Genus II. *Methylobacillus*
Genus III. *Methylovorus*
Order IV. *Neisseriales*
Family I. *Neisseriaceae*
Genus I. *Neisseria*
Genus II. *Alysiella*
Genus III. *Aquaspirillum*
Genus IV. *Chromobacterium*
Genus V. *Eikenella*
Genus VI. *Formivibrio*
Genus VII. *Iodobacter*
Genus VIII. *Kingella*
Genus IX. *Laribacter*
Genus X. *Microvirgula*
Genus XI. *Morococcus*
Genus XII. *Prolinoborus*
Genus XIII. *Simonsiella*
Genus XIV. *Vitreoscilla*
Genus XV. *Vogesella*
Order V. *Nitrosomonadales*
Family I. *Nitrosomonadaceae*
Genus I. *Nitrosomonas*
Genus II. *Nitrosolobus*
Genus III. *Nitrosospira*
Family II. *Spirillaceae*
Genus I. *Spirillum*
Family III. *Gallionellaceae*
Genus I. *Gallionella*
Order VI. *Rhodocyclales*

Genus VII. *Mesophilobacter*
Genus VIII. *Rhizobacter*
Genus IX. *Rugamonas*
Genus X. *Serpens*
Family II. *Moraxellaceae*
 Genus I. *Moraxella*
 Genus II. *Acinetobacter*
 Genus III. *Psychrobacter*
Family III. Incertae sedis[b]
 Genus I. *Enhydrobacter*
Order X. *Alteromonadales*
 Family I. *Alteromonadaceae*
 Genus I. *Alteromonas*
 Genus II. *Aestuariibacter*
 Genus III. *Alishewanella*
 Genus IV. *Colwellia*
 Genus V. *Ferrimonas*
 Genus VI. *Glaciecola*
 Genus VII. *Idiomarina*
 Genus VIII. *Marinobacter*
 Genus IX. *Marinobacterium*
 Genus X. *Microbulbifer*
 Genus XI. *Moritella*
 Genus XII. *Pseudoalteromonas*
 Genus XIII. *Psychromonas*
 Genus XIV. *Shewanella*
 Genus XV. *Thalassomonas*
 Family II. Incertae sedis[b]
 Genus I. *Teredinibacter*
Order XI. *Vibrionales*
 Family I. *Vibrionaceae*
 Genus I. *Vibrio*
 Genus II. *Allomonas*
 Genus III. *Catenococcus*
 Genus IV. *Enterovibrio*
 Genus V. *Grimontia*
 Genus VI. *Listonella*
 Genus VII. *Photobacterium*
 Genus VIII. *Salinivibrio*
Order XII. *Aeromonadales*
 Family I. *Aeromonadaceae*
 Genus I. *Aeromonas*
 Genus II. *Oceanimonas*
 Genus III. *Oceanisphaera*
 Genus IV. *Tolumonas*
 Family II. Incertae sedis: *Succinivibrionaceae*[b]
 Genus I. *Succinivibrio*
 Genus II. *Anaerobiospirillum*
 Genus III. *Ruminobacter*
 Genus IV. *Succinimonas*
Order XIII. *Enterobacteriales*
 Family I. *Enterobacteriaceae*
 Genus I. *Escherichia*
 Genus II. *Alterococcus*
 Genus III. *Arsenophonus*
 Genus IV. *Brenneria*
 Genus V. *Buchnera*
 Genus VI. *Budvicia*
 Genus VII. *Buttiauxella*
 Genus VIII. *Calymmatobacterium*
 Genus IX. *Cedecea*
 Genus X. *Citrobacter*
 Genus XI. *Edwardsiella*
 Genus XII. *Enterobacter*
 Genus XIII. *Erwinia*
 Genus XIV. *Ewingella*
 Genus XV. *Hafnia*
 Genus XVI. *Klebsiella*
 Genus XVII. *Kluyvera*
 Genus XVIII. *Leclercia*
 Genus XIX. *Leminorella*
 Genus XX. *Moellerella*
 Genus XXI. *Morganella*
 Genus XXII. *Obesumbacterium*
 Genus XXIII. *Pantoea*
 Genus XXIV. *Pectobacterium*
 Genus XXV. *Phlomobacter*

Genus XXVI. *Photorhabdus*
Genus XXVII. *Plesiomonas*
Genus XXVIII. *Pragia*
Genus XXIX. *Proteus*
Genus XXX. *Providencia*
Genus XXXI. *Rahnella*
Genus XXXII. *Raoultella*
Genus XXXIII. *Saccharobacter*
Genus XXXIV. *Salmonella*
Genus XXXV. *Samsonia*
Genus XXXVI. *Serratia*
Genus XXXVII. *Shigella*
Genus XXXVIII. *Sodalis*
Genus XXXIX. *Tatumella*
Genus XL. *Trabulsiella*
Genus XLI. *Wigglesworthia*
Genus XLII. *Xenorhabdus*
Genus XLIII. *Yersinia*
Genus XLIV. *Yokenella*
Order XIV. *Pasteurellales*
 Family I. *Pasteurellaceae*
 Genus I. *Pasteurella*
 Genus II. *Actinobacillus*
 Genus III. *Gallibacterium*
 Genus IV. *Haemophilus*
 Genus V. *Lonepinella*
 Genus VI. *Mannheimia*
 Genus VII. *Phocoenobacter*
Class IV. *Deltaproteobacteria*
 Order I. *Desulfurellales*
 Family I. *Desulfurellaceae*
 Genus I. *Desulfurella*
 Genus II. *Hippea*
 Order II. *Desulfovibrionales*
 Family I. *Desulfovibrionaceae*
 Genus I. *Desulfovibrio*
 Genus II. *Bilophila*
 Genus III. *Lawsonia*
 Family II. *Desulfomicrobiaceae*
 Genus I. *Desulfomicrobium*
 Family III. *Desulfohalobiaceae*
 Genus I. *Desulfohalobium*
 Genus II. *Desulfomonas*
 Genus III. *Desulfonatronovibrio*
 Genus IV. *Desulfothermus*
 Family IV. *Desulfonatronumaceae*
 Genus I. *Desulfonatronum*
 Order III. *Desulfobacterales*
 Family I. *Desulfobacteraceae*
 Genus I. *Desulfobacter*
 Genus II. *Desulfatibacillum*
 Genus III. *Desulfobacterium*
 Genus IV. *Desulfobacula*
 Genus V. *Desulfobotulus*
 Genus VI. *Desulfocella*
 Genus VII. *Desulfococcus*
 Genus VIII. *Desulfofaba*
 Genus IX. *Desulfofrigus*
 Genus X. *Desulfomusa*
 Genus XI. *Desulfonema*
 Genus XII. *Desulforegula*
 Genus XIII. *Desulfosarcina*
 Genus XIV. *Desulfospira*
 Genus XV. *Desulfotignum*
 Family II. *Desulfobulbaceae*
 Genus I. *Desulfobulbus*
 Genus II. *Desulfocapsa*
 Genus III. *Desulfofustis*
 Genus IV. *Desulforhopalus*
 Genus V. *Desulfotalea*
 Family III. *Nitrospinaceae*
 Genus I. *Nitrospina*
 Order IV. *Desulfarcales*
 Family I. *Desulfarculaceae*
 Genus I. *Desulfarculus*
 Order V. *Desulfuromonales*
 Family I. *Desulfuromonaceae*

Genus I. *Desulfuromonas*
Genus II. *Desulfuromusa*
Genus III. *Malonomonas*
Genus IV. *Pelobacter*
Family II. *Geobacteraceae*
Genus I. *Geobacter*
Genus II. *Trichlorobacter*
Order VI. *Syntrophobacterales*
Family I. *Syntrophobacteraceae*
Genus I. *Syntrophobacter*
Genus II. *Desulfacinum*
Genus III. *Desulforhabdus*
Genus IV. *Desulfovirga*
Genus V. *Thermodesulforhabdus*
Family II. *Syntrophaceae*
Genus I. *Syntrophus*
Genus II. *Desulfobacca*
Genus III. *Desulfomonile*
Genus IV. *Smithella*
Order VII. *Bdellovibrionales*
Family I. *Bdellovibrionaceae*
Genus I. *Bdellovibrio*
Genus II. *Bacteriovorax*
Genus III. *Micavibrio*
Genus IV. *Vampirovibrio*
Order VIII. *Myxococcales*
Suborder I. *Cystobacterineae*
Family I. *Cystobacteraceae*
Genus I. *Cystobacter*
Genus II. *Anaeromyxobacter*
Genus III. *Archangium*
Genus IV. *Hyalangium*
Genus V. *Melittangium*
Genus VI. *Stigmatella*
Family II. *Myxococcaceae*
Genus I. *Myxococcus*
Genus II. *Corallococcus*
Genus III. *Pyxicoccus*
Suborder II. *Sorangineae*
Family I. *Polyangiaceae*
Genus I. *Polyangium*
Genus II. *Byssophaga*
Genus III. *Chondromyces*
Genus IV. *Haploangium*
Genus V. *Jahnia*
Genus VI. *Sorangium*
Suborder III. *Nannocystineae*
Family I. *Nannocystaceae*
Genus I. *Nannocystis*
Genus II. *Plesiocystis*
Family II. *Haliangiaceae*
Genus I. *Haliangium*
Family III. *Kocueriaceae*
Genus I. *Kocueria*
Class V. *Epsilonproteobacteria*
Order I. *Campylobacterales*
Family I. *Campylobacteraceae*
Genus I. *Campylobacter*
Genus II. *Arcobacter*
Genus III. *Dehalospirillum*
Genus IV. *Sulfurospirillum*
Family II. *Helicobacteraceae*
Genus I. *Helicobacter*
Genus II. *Sulfurimonas*
Genus III. *Thiovulum*
Genus IV. *Wolinella*
Family III. *Nautiliaceae*
Genus I. *Nautilia*
Genus II. *Caminibacter*
Family IV. *Hydrogenimonaceae*
Genus I. *Hydrogenimonas*
Phylum BXIII. *Firmicutes*
Class I. *Clostridia*
Order I. *Clostridiales*
Family I. *Clostridiaceae*
Genus I. *Clostridium*
Genus II. *Acetivibrio*

Genus III. *Acidaminobacter*
Genus IV. *Alkaliphilus*
Genus V. *Anaerobacter*
Genus VI. *Anaerotruncus*
Genus VII. *Bryantella*
Genus VIII. *Caminicella*
Genus IX. *Caloramator*
Genus X. *Caloranaerobacter*
Genus XI. *Coprobacillus*
Genus XII. *Dorea*
Genus XIII. *Faecalibacterium*
Genus XIV. *Hespellia*
Genus XV. *Natronincola*
Genus XVI. *Oxobacter*
Genus XVII. *Parasporobacterium*
Genus XVIII. *Sarcina*
Genus XIX. *Soehngenia*
Genus XX. *Tepidibacter*
Genus XXI. *Thermobrachium*
Genus XXII. *Thermohalobacter*
Genus XXIII. *Tindallia*
Family II. *Lachnospiraceae*
Genus I. *Lachnospira*
Genus II. *Acetitomaculum*
Genus III. *Anaerofilum*
Genus IV. *Anaerostipes*
Genus V. *Butyrivibrio*
Genus VI. *Catenibacterium*
Genus VII. *Catonella*
Genus VIII. *Coprococcus*
Genus IX. *Johnsonella*
Genus X. *Lachnobacterium*
Genus XI. *Pseudobutyrivibrio*
Genus XII. *Roseburia*
Genus XIII. *Ruminococcus*
Genus XIV. *Shuttleworthia*
Genus XV. *Sporobacterium*
Family III. *Peptostreptococcaceae*
Genus I. *Peptostreptococcus*
Genus II. *Anaerococcus*
Genus III. *Filifactor*
Genus IV. *Finegoldia*
Genus V. *Fusibacter*
Genus VI. *Gallicola*
Genus VII. *Helcococcus*
Genus VIII. *Micromonas*
Genus IX. *Peptoniphilus*
Genus X. *Sedimentibacter*
Genus XI. *Sporanaerobacter*
Genus XII. *Tissierella*
Family IV. *Eubacteriaceae*
Genus I. *Eubacterium*
Genus II. *Acetobacterium*
Genus III. *Anaerovorax*
Genus IV. *Mogibacterium*
Genus V. *Pseudoramibacter*
Family V. *Peptococcaceae*
Genus I. *Peptococcus*
Genus II. *Carboxydothermus*
Genus III. *Dehalobacter*
Genus IV. *Desulfotobacterium*
Genus V. *Desulfonispora*
Genus VI. *Desulfosporosinus*
Genus VII. *Desulfotomaculum*
Genus VIII. *Pelotomaculum*
Genus IX. *Syntrophobotulus*
Genus X. *Thermoterrabacterium*
Family VI. *Heliobacteriaceae*
Genus I. *Heliobacterium*
Genus II. *Heliobacillus*
Genus III. *Heliophilum*
Genus IV. *Heliorestis*
Family VII. *Acidaminococcaceae*
Genus I. *Acidaminococcus*
Genus II. *Acetonema*
Genus III. *Allisonella*
Genus IV. *Anaeroarcus*

Genus V. *Anaeroglobus*
Genus VI. *Anaeromusa*
Genus VII. *Anaerosinus*
Genus VIII. *Anaerovibrio*
Genus IX. *Centipeda*
Genus X. *Dendrosporobacter*
Genus XI. *Dialister*
Genus XII. *Megasphaera*
Genus XIII. *Mitsuokella*
Genus XIV. *Papillibacter*
Genus XV. *Pectinatus*
Genus XVI. *Phascolarctobacterium*
Genus XVII. *Propionispira*
Genus XVIII. *Propionispora*
Genus XIX. *Quinella*
Genus XX. *Schwartzia*
Genus XXI. *Selenomonas*
Genus XXII. *Sporomusa*
Genus XXIII. *Succiniclasticum*
Genus XXIV. *Succinispira*
Genus XXV. *Veillonella*
Genus XXVI. *Zymophilus*
Family VIII. *Syntrophomonadaceae*
Genus I. *Syntrophomonas*
Genus II. *Acetogenium*
Genus III. *Aminobacterium*
Genus IV. *Aminomonas*
Genus V. *Anaerobaculum*
Genus VI. *Anaerobranca*
Genus VII. *Caldicellulosiruptor*
Genus VIII. *Carboxydocella*
Genus IX. *Dethiosulfovibrio*
Genus X. *Pelospora*
Genus XI. *Syntrophospora*
Genus XII. *Syntrophothermus*
Genus XIII. *Thermaerobacter*
Genus XIV. *Thermanaerovibrio*
Genus XV. *Thermohydrogenium*
Genus XVI. *Thermosyntropha*
Order II. *Thermoanaerobacteriales*
Family I. *Thermoanaerobacteriaceae*
Genus I. *Themoanaerobacterium*
Genus II. *Ammonifex*
Genus III. *Caldanaerobacter*
Genus IV. *Carboxydibrachium*
Genus V. *Coprothermobacter*
Genus VI. *Gelria*
Genus VII. *Moorella*
Genus VIII. *Sporotomaculum*
Genus IX. *Thermacetogenium*
Genus X. *Thermanaeromonas*
Genus XI. *Thermoanaerobacter*
Genus XII. *Thermoanaerobium*
Genus XIII. *Thermovenabulum*
Family II. *Thermodesulfobiaceae*
Genus I. *Thermodesulfobium*
Order III. *Halanaerobiales*
Family I. *Halanaerobiaceae*
Genus I. *Halanaerobium*
Genus II. *Halocella*
Genus III. *Halothermothrix*
Family II. *Halobacteroidaceae*
Genus I. *Halobacteroides*
Genus II. *Acetohalobium*
Genus III. *Halanaerobacter*
Genus IV. *Halonatronum*
Genus V. *Natroniella*
Genus VI. *Orenia*
Genus VII. *Selenihalanaerobacter*
Genus VIII. *Sporohalobacter*

Class II. *Mollicutes*
Order I. *Mycoplasmatales*
Family I. *Mycoplasmataceae*
Genus I. *Mycoplasma*
Genus II. *Eperythrozoon*
Genus III. *Haemobartonella*
Genus IV. *Ureaplasma*

Order II. *Entomoplasmatales*
Family I. *Entomoplasmataceae*
Genus I. *Entomoplasma*
Genus II. *Mesoplasma*
Family II. *Spiroplasmataceae*
Genus I. *Spiroplasma*
Order III. *Acholeplasmatales*
Family I. *Acholeplasmataceae*
Genus I. *Acholeplasma*
Genus II. *Phytoplasma*
Order IV. *Anaeroplasmatales*
Family I. *Anaeroplasmataceae*
Genus I. *Anaeroplasma*
Genus II. *Asteroleplasma*
Order V. Incertae sedis[b]
Family I. *Erysipelotrichaceae*
Genus I. *Erysipelothrix*
Genus II. *Bulleidia*
Genus III. *Holdemania*
Genus IV. *Solobacterium*

Class III. *Bacilli*
Order I. *Bacillales*
Family I. *Bacillaceae*
Genus I. *Bacillus*
Genus II. *Amphibacillus*
Genus III. *Anoxybacillus*
Genus IV. *Exiguobacterium*
Genus V. *Filobacillus*
Genus VI. *Geobacillus*
Genus VII. *Gracilibacillus*
Genus VIII. *Halobacillus*
Genus IX. *Jeotgalibacillus*
Genus X. *Lentibacillus*
Genus XI. *Marinibacillus*
Genus XII. *Oceanobacillus*
Genus XIII. *Paraliobacillus*
Genus XIV. *Saccharococcus*
Genus XV. *Salibacillus*
Genus XVI. *Ureibacillus*
Genus XVII. *Virgibacillus*
Family II. *Alicyclobacillaceae*
Genus I. *Alicyclobacillus*
Genus II. *Pasteuria*
Genus III. *Sulfobacillus*
Family III. *Caryophanaceae*
Genus I. *Caryophanon*
Family IV. *Listeriaceae*
Genus I. *Listeria*
Genus II. *Brochothrix*
Family V. *Paenibacillaceae*
Genus I. *Paenibacillus*
Genus II. *Ammoniphilus*
Genus III. *Aneurinibacillus*
Genus IV. *Brevibacillus*
Genus V. *Oxalophagus*
Genus VI. *Thermicanus*
Genus VII. *Thermobacillus*
Family VI. *Planococcaceae*
Genus I. *Planococcus*
Genus II. *Filibacter*
Genus III. *Kurthia*
Genus IV. *Planomicrobium*
Genus V. *Sporosarcina*
Family VII. *Sporolactobacillaceae*
Genus I. *Sporolactobacillus*
Genus II. *Marinococcus*
Family VIII. *Staphylococcaceae*
Genus I. *Staphylococcus*
Genus II. *Gemella*
Genus III. *Jeotgalicoccus*
Genus IV. *Macrococcus*
Genus V. *Salinicoccus*
Family IX. *Thermoactinomycetaceae*
Genus I. *Thermoactinomyces*
Family X. *Turicibacteraceae*
Genus I. *Turicibacter*
Order II. *Lactobacillales*

Family I. *Lactobacillaceae*
  Genus I. *Lactobacillus*
  Genus II. *Paralactobacillus*
  Genus III. *Pediococcus*
Family II. *Aerococcaceae*
  Genus I. *Aerococcus*
  Genus II. *Abiotrophia*
  Genus III. *Dolosicoccus*
  Genus IV. *Eremococcus*
  Genus V. *Facklamia*
  Genus VI. *Globicatella*
  Genus VII. *Ignavigranum*
Family III. *Carnobacteriaceae*
  Genus I. *Carnobacterium*
  Genus II. *Agitococcus*
  Genus III. *Alkalibacterium*
  Genus IV. *Allofustis*
  Genus V. *Alloiococcus*
  Genus VI. *Desemzia*
  Genus VII. *Dolosigranulum*
  Genus VIII. *Granulicatella*
  Genus IX. *Isobaculum*
  Genus X. *Lactosphaera*
  Genus XI. *Marinilactibacillus*
  Genus XII. *Trichococcus*
Family IV. *Enterococcaceae*
  Genus I. *Enterococcus*
  Genus II. *Atopobacter*
  Genus III. *Melissococcus*
  Genus IV. *Tetragenococcus*
  Genus V. *Vagococcus*
Family V. *Leuconostocaceae*
  Genus I. *Leuconostoc*
  Genus II. *Oenococcus*
  Genus III. *Weissella*
Family VI. *Streptococcaceae*
  Genus I. *Streptococcus*
  Genus II. *Lactococcus*
Family VII. Incertae sedis[b]
  Genus I. *Acetoanaerobium*
  Genus II. *Oscillospira*
  Genus III. *Syntrophococcus*

Phylum BXIV. *Actinobacteria*
Class I. *Actinobacteria*
Subclass I. *Acidimicrobidae*
  Order I. *Acidimicrobiales*
    Suborder IV. *Acidimicrobineae*
    Family I. *Acidimicrobiaceae*
      Genus I. *Acidimicrobium*
Subclass II. *Rubrobacteridae*
  Order I. *Rubrobacterales*
    Suborder V. *Rubrobacterineae*
    Family I. *Rubrobacteraceae*
      Genus I. *Rubrobacter*
      Genus II. *Conexibacter*
      Genus III. *Solirubrobacter*
      Genus IV. *Thermoleophilum*
Subclass III. *Coriobacteridae*
  Order I. *Coriobacteriales*
    Suborder VI. *Coriobacterineae*
    Family I. *Coriobacteriaceae*
      Genus I. *Coriobacterium*
      Genus II. *Atopobium*
      Genus III. *Collinsella*
      Genus IV. *Cryptobacterium*
      Genus V. *Denitrobacterium*
      Genus VI. *Eggerthella*
      Genus VII. *Olsenella*
      Genus VIII. *Slackia*
Subclass IV. *Sphaerobacteridae*
  Order I. *Sphaerobacterales*
    Suborder VII. *Sphaerobacterineae*
    Family I. *Sphaerobacteraceae*
      Genus I. *Sphaerobacter*
Subclass V. *Actinobacteridae*
  Order I. *Actinomycetales*
    Suborder VIII. *Actinomycineae*

Family I. *Actinomycetaceae*
  Genus I. *Actinomyces*
  Genus II. *Actinobaculum*
  Genus III. *Arcanobacterium*
  Genus IV. *Mobiluncus*
  Genus V. *Varibaculum*
Suborder IX. *Micrococcineae*
Family I. *Micrococcaceae*
  Genus I. *Micrococcus*
  Genus II. *Arthrobacter*
  Genus III. *Citricoccus*
  Genus IV. *Kocuria*
  Genus V. *Nesterenkonia*
  Genus VI. *Renibacterium*
  Genus VII. *Rothia*
  Genus VIII. *Stomatococcus*
  Genus IX. *Yania*
Family II. *Bogoriellaceae*
  Genus I. *Bogoriella*
Family III. *Rarobacteraceae*
  Genus I. *Rarobacter*
Family IV. *Sanguibacteraceae*
  Genus I. *Sanguibacter*
Family V. *Brevibacteriaceae*
  Genus I. *Brevibacterium*
Family VI. *Cellulomonadaceae*
  Genus I. *Cellulomonas*
  Genus II.*Oerskovia*
  Genus III. *Tropheryma*
Family VII. *Dermabacteraceae*
  Genus I. *Dermabacter*
  Genus II. *Brachybacterium*
Family VIII. *Dermatophilaceae*
  Genus I. *Dermatophilus*
  Genus II. *Kineosphaera*
Family IX. *Dermacoccaceae*
  Genus I. *Dermacoccus*
  Genus II. *Demetria*
  Genus III. *Kytococcus*
Family X. *Intrasporangiaceae*
  Genus I. *Intrasporangium*
  Genus II. *Arsenicicoccus*
  Genus III. *Janibacter*
  Genus IV. *Knoellia*
  Genus V. *Ornithinicoccus*
  Genus VI. *Ornithinimicrobium*
  Genus VII. *Nostocoidia*
  Genus VIII. *Terrabacter*
  Genus IX. *Terracoccus*
  Genus X. *Tetrasphaera*
Family XI. *Jonesiaceae*
  Genus I. *Jonesia*
Family XII. *Microbacteriaceae*
  Genus I. *Microbacterium*
  Genus II. *Agreia*
  Genus III. *Agrococcus*
  Genus IV. *Agromyces*
  Genus V. *Aureobacterium*
  Genus VI. *Clavibacter*
  Genus VII. *Cryobacterium*
  Genus VIII. *Curtobacterium*
  Genus IX. *Frigoribacterium*
  Genus X. *Leifsonia*
  Genus XI. *Leucobacter*
  Genus XII. *Mycetocola*
  Genus XIII. *Okibacterium*
  Genus XIV. *Plantibacter*
  Genus XV. *Rathayibacter*
  Genus XVI. *Rhodoglobus*
  Genus XVII. *Salinibacterium*
  Genus XVIII. *Subtercola*
Family XIII. *Beutenbergiaceae*
  Genus I. *Beutenbergia*
  Genus II. *Georgenia*
  Genus III. *Salana*
Family XIV. *Promicromonosporaceae*
  Genus I. *Promicromonospora*

Genus II. *Cellulosimicrobium*
Genus III. *Xylanibacterium*
Genus IV. *Xylanimonas*
Suborder X. *Corynebacterineae*
Family I. *Corynebacteriaceae*
Genus I. *Corynebacterium*
Family II. *Dietziaceae*
Genus I. *Dietzia*
Family III. *Gordoniaceae*
Genus I. *Gordonia*
Genus II. *Skermania*
Family IV. *Mycobacteriaceae*
Genus I. *Mycobacterium*
Family V. *Nocardiaceae*
Genus I. *Nocardia*
Genus II. *Rhodococcus*
Family VI. *Tsukamurellaceae*
Genus I. *Tsukamurella*
Family VII. *Williamsiaceae*
Genus I. *Williamsia*
Suborder XI. *Micromonosporineae*
Family I. *Micromonosporaceae*
Genus I. *Micromonospora*
Genus II. *Actinoplanes*
Genus III. *Asanoa*
Genus IV. *Catellatospora*
Genus V. *Catenuloplanes*
Genus VI. *Couchioplanes*
Genus VII. *Dactylosporangium*
Genus VIII. *Pilimelia*
Genus IX. *Spirilliplanes*
Genus X. *Verrucosispora*
Genus XI. *Virgisporangium*
Suborder XII. *Propionibacterineae*
Family I. *Propionibacteriaceae*
Genus I. *Propionibacterium*
Genus II. *Luteococcus*
Genus III. *Microlunatus*
Genus IV. *Propioniferax*
Genus V. *Propionimicrobium*
Genus VI. *Tessaracoccus*
Family II. *Nocardioidaceae*
Genus I. *Nocardioides*
Genus II. *Aeromicrobium*
Genus III. *Actinopolymorpha*
Genus IV. *Friedmanniella*
Genus V. *Hongia*
Genus VI. *Kribbella*
Genus VII. *Micropruina*
Genus VIII. *Marmoricola*
Genus IX. *Propionicimonas*
Suborder XIII. *Pseudonocardineae*
Family I. *Pseudonocardiaceae*
Genus I. *Pseudonocardia*
Genus II. *Actinoalloteichus*
Genus III. *Actinopolyspora*
Genus IV. *Amycolatopsis*
Genus V. *Crossiella*
Genus VI. *Kibdelosporangium*
Genus VII. *Kutzneria*
Genus VIII. *Prauserella*
Genus IX. *Saccharomonospora*
Genus X. *Saccharopolyspora*
Genus XI. *Streptoalloteichus*
Genus XII. *Thermobispora*
Genus XIII. *Thermocrispum*
Family II. *Actinosynnemataceae*
Genus I. *Actinosynnema*
Genus II. *Actinokineospora*
Genus III. *Lechevalieria*
Genus IV. *Lentzea*
Genus V. *Saccharothrix*
Suborder XIV. *Streptomycineae*
Family I. *Streptomycetaceae*
Genus I. *Streptomyces*
Genus II. *Kitasatospora*
Genus III. *Streptoverticillium*

Suborder XV. *Streptosporangineae*
Family I. *Streptosporangiaceae*
Genus I. *Streptosporangium*
Genus II. *Acrocarpospora*
Genus III. *Herbidospora*
Genus IV. *Microbispora*
Genus V. *Microtetraspora*
Genus VI. *Nonomuraea*
Genus VII. *Planobispora*
Genus VIII. *Planomonospora*
Genus IX. *Planopolyspora*
Genus X. *Planotetraspora*
Family II. *Nocardiopsaceae*
Genus I. *Nocardiopsis*
Genus II. *Streptomonospora*
Genus III. *Thermobifida*
Family III. *Thermomonosporaceae*
Genus I. *Thermomonospora*
Genus II. *Actinomadura*
Genus III. *Spirillospora*
Suborder XVI. *Frankineae*
Family I. *Frankiaceae*
Genus I. *Frankia*
Family II. *Geodermatophilaceae*
Genus I. *Geodermatophilus*
Genus II. *Blastococcus*
Genus III. *Modestobacter*
Family III. *Microsphaeraceae*
Genus I. *Microsphaera*
Family IV. *Sporichthyaceae*
Genus I. *Sporichthya*
Family V. *Acidothermaceae*
Genus I. *Acidothermus*
Family VI. *Kineosporiaceae*
Genus I. *Kineosporia*
Genus II. *Cryptosporangium*
Genus III. *Kineococcus*
Suborder XVII. *Glycomycineae*
Family I. *Glycomycetaceae*
Genus I. *Glycomyces*
Order II. *Bifidobacteriales*
Family I. *Bifidobacteriaceae*
Genus I. *Bifidobacterium*
Genus II. *Aeriscardovia*
Genus III. *Falcivibrio*
Genus IV. *Gardnerella*
Genus V. *Parascardovia*
Genus VI. *Scardovia*
Family II. Incertae sedis[b]
Genus I. *Actinobispora*
Genus II. *Actinocorallia*
Genus III. *Excellospora*
Genus IV. *Pelczaria*
Genus V. *Turicella*
Phylum BXV. *Planctomycetes*
Class I. *Planctomycetacia*
Order I. *Planctomycetales*
Family I. *Planctomycetaceae*
Genus I. *Planctomyces*
Genus II. *Gemmata*
Genus III. *Isosphaera*
Genus IV. *Pirellula*
Phylum BXVI. *Chlamydiae*
Class I. *Chlamydiae*
Order I. *Chlamydiales*
Family I. *Chlamydiaceae*
Genus I. *Chlamydia*
Genus II. *Chlamydophila*
Family II. *Parachlamydiaceae*
Genus I. *Parachlamydia*
Genus II. *Neochlamydia*
Family III. *Simkaniaceae*
Genus I. *Simkania*
Genus II. *Rhabdochlamydia*
Family IV. *Waddliaceae*
Genus I. *Waddlia*
Phylum BXVII. *Spirochaetes*

Class I. *Spirochaetes*
    Order I. *Spirochaetales*
        Family I. *Spirochaetaceae*
           Genus I. *Spirochaeta*
           Genus II. *Borrelia*
           Genus III. *Brevinema*
           Genus IV. *Clevelandina*
           Genus V. *Cristispira*
           Genus VI. *Diplocalyx*
           Genus VII. *Hollandina*
           Genus VIII. *Pillotina*
           Genus IX. *Treponema*
        Family II. *Serpulinaceae*
           Genus I. *Serpulina*
           Genus II. *Brachyspira*
        Family III. *Leptospiraceae*
           Genus I. *Leptospira*
           Genus II. *Leptonema*
Phylum BXVIII. *Fibrobacteres*
Class I. *Fibrobacteres*
    Order I. *Fibrobacterales*
        Family I. *Fibrobacteraceae*
           Genus I. *Fibrobacter*
Phylum BXIX. *Acidobacteria*
Class I. *Acidobacteria*
    Order I. *Acidobacteriales*
        Family I. *Acidobacteriaceae*
           Genus I. *Acidobacterium*
           Genus II. *Geothrix*
           Genus III. *Holophaga*
Phylum BXX. *Bacteroidetes*
Class I. *Bacteroidetes*
    Order I. *Bacteroidales*
        Family I. *Bacteroidaceae*
           Genus I. *Bacteroides*
           Genus II. *Acetofilamentum*
           Genus III. *Acetomicrobium*
           Genus IV. *Acetothermus*
           Genus V. *Anaerophaga*
           Genus VI. *Anaerorhabdus*
           Genus VII. *Megamonas*
        Family II. *Rikenellaceae*
           Genus I. *Rikenella*
           Genus II. *Alistipes*
           Genus III. *Marinilabilia*
        Family III. *Porphyromonadaceae*
           Genus I. *Porphyromonas*
           Genus II. *Dysgonomonas*
           Genus III. *Tannerella*
        Family IV. *Prevotellaceae*
           Genus I. *Prevotella*
Class II. *Flavobacteria*
    Order I. *Flavobacteriales*
        Family I. *Flavobacteriaceae*
           Genus I. *Flavobacterium*
           Genus II. *Aequorivita*
           Genus III. *Arenibacter*
           Genus IV. *Bergeyella*
           Genus V. *Capnocytophaga*
           Genus VI. *Cellulophaga*
           Genus VII. *Chryseobacterium*
           Genus VIII. *Coenonia*
           Genus IX. *Croceibacter*
           Genus X. *Empedobacter*
           Genus XI. *Gelidibacter*
           Genus XII. *Gillisia*
           Genus XIII. *Mesonia*
           Genus XIV. *Muricauda*
           Genus XV. *Myroides*
           Genus XVI. *Ornithobacterium*
           Genus XVII. *Polaribacter*

GLOSSARY

Only the major terms and concepts are included. If a term is not here, consult the index.

ABC transporter A membrane transport system consisting of three proteins, one of which hydrolyzes ATP, one of which binds the substrate, and one of which functions as the transport channel through the membrane.

Abscess A localized infection characterized by production of pus.

Acetotrophic Splitting of acetate into CH_4 plus CO_2 by certain methanogens.

Acetyl-CoA pathway A pathway of autotrophic CO_2 fixation widespread in obligate anaerobes including methanogens, homoacetogens, and sulfate-reducing bacteria.

Acetylene reduction assay Method of measuring activity of nitrogenase by substituting acetylene for the natural substrate of the enzyme, N_2. Acetylene is reduced to ethylene or ethane, depending on the nitrogenase system involved.

Acid fastness A staining property of *Mycobacterium* species where cells stained with hot carbol fuchsin do not decolorize with acid-alcohol.

Acid mine drainage Acidic water containing H_2SO_4 derived from the microbial oxidation of iron sulfide minerals.

Acidophile An organism that grows best at acidic pH values.

Activation energy Energy needed to make substrate molecules more reactive; enzymes function by lowering activation energy.

Activator protein A regulatory protein that binds to specific sites on DNA and stimulates transcription; involved in positive control.

Active immunity An immune state achieved by self-production of antibodies. Compare with *Passive immunity*.

Active site The portion of an enzyme that is directly involved in binding substrate(s).

Active transport The energy-dependent process of transporting substances into or out of the cell in which the transported substances are chemically unchanged.

Acute In reference to infections, short-term, usually characterized by dramatic onset and rapid recovery.

Adaptive immunity (antigen-specific immunity) The acquired ability to recognize and destroy an individual pathogen or its products relying on previous exposure to the pathogen or its products.

Adherence A property of bacteria that allows them to stick to host surfaces.

Aerobic secondary wastewater treatment Digestive reactions carried out by microorganisms under aerobic conditions to treat wastewater containing low levels of organic materials.

Aerobe An organism that grows in the presence of O_2; may be facultative, obligate, or microaerophilic.

Aerosol Suspension of particles in airborne water droplets.

Aerotolerant Of an anaerobe, not being inhibited by O_2.

Agglutination Reaction between antibody and particle-bound antigen resulting in visible clumping of the particles.

Agretope The portion of a processed antigen that is recognized by MHC protein.

Algae Phototrophic eukaryotic microorganisms.

Alkaliphile An organism that grows best at high pH.

Allergy A harmful immune reaction, usually caused by a foreign antigen in food, pollen, or chemicals, which results in immediate-type or delayed-type hypersensitivity.

Allosteric enzyme An enzyme that contains two combining sites, the active site (where the substrate binds) and the allosteric site (where an effector molecule binds).

Ameboid movement A type of motility in which cytoplasmic streaming moves the organism forward.

Aminoacyl-tRNA synthetase An enzyme that catalyzes the attachment of the correct amino acid to the correct tRNA.

Aminoglycoside An antibiotic such as streptomycin, containing amino sugars linked by glycosidic bonds.

Anabolism The biochemical processes involved in the synthesis of cell constituents from simpler molecules, usually requiring energy.

Anaerobe An organism that grows in the absence of O_2; some may even be killed by O_2 (obligate or strict anaerobes).

Anaerobic respiration Use of an electron acceptor other than O_2 in an electron transport-based oxidation leading to a proton motive force.

Anammox The term that describes anoxic ammonia oxidation.

Anaphylatoxins The C3a and C5a fractions of complement that act to mimic some of the reactions of anaphylaxis.

Anaphylaxis (anaphylactic shock) A violent allergic reaction caused by an antigen–antibody reaction.

Anergy The inability to produce an immune response to specific antigens due to neutralization of effector cells.

Anoxic Absence of oxygen. Usually used in reference to a microbial habitat.

Anoxic secondary wastewater treatment Digestive and fermentative reactions carried out by microorganisms under anoxic conditions to treat wastewater containing high levels of insoluble organic materials.

Anoxygenic photosynthesis Use of light energy to synthesize ATP by cyclic photophosphorylation without O_2 production.

Antibiogram A report indicating the sensitivity of clinically isolated microorganisms to the antibiotics in current use.

Antibiotic A chemical substance produced by a microorganism that kills or inhibits the growth of another microorganism.

Antibiotic resistance The acquired ability of a microorganism to grow in the presence of an antibiotic to which the microorganism is usually sensitive.

Antibody A protein, produced by B lymphocytes, present in serum or other body fluid that combines specifically with antigen. An immunoglobulin.

Antibody-mediated immunity Immunity resulting from direct interaction with antibodies; also called humoral immunity.

Anticodon A sequence of three bases in transfer RNA that base-pairs with a codon in messenger RNA during protein synthesis.

Antigen A molecule capable of interacting with specific components of the immune system.

Antigen-presenting cell (APC) A macrophage, dendritic cell, or B cell that presents processed antigen peptides to a T cell.

Antigenic determinant The portion of an antigen that interacts with an immunoglobulin or T cell receptor. Also called an *epitope*.

Antigenic drift In influenza virus, minor changes in viral proteins (antigens) due to gene mutation.

Antigenic shift In influenza virus, major changes in viral proteins (antigens) due to gene reassortment.

Antimicrobial Harmful to microorganisms by either killing or inhibiting growth.

Antimicrobial agent A chemical that kills or inhibits the growth of microorganisms.

Antimicrobial drug resistance The acquired ability of a microorganism to grow in the presence of an antimicrobial drug to which the microorganism is usually susceptible.

Antiparallel In reference to nucleic acids; one strand runs $5' \rightarrow 3'$, the other $3' \rightarrow 5'$.

Antiseptic (germicide) A chemical agent that kills or inhibits growth of microorganisms and is sufficiently nontoxic to be applied to living tissues.

Antiserum A serum containing antibodies.

Antitoxin An antibody that specifically interacts with and neutralizes a toxin.

Apoptosis Programmed cell death.

Archaea A phylogenetic domain of prokaryotes consisting of the methanogens, most extreme halophiles and hyperthermophiles, and extreme acidophiles such as *Thermoplasma*.

Artificial chromosomes Cloning vectors which can carry very large inserts of foreign DNA and exist in the cell very much like a cellular chromosome. The most widely used are bacterial artificial chromosomes (BACs) and yeast artificial chromosomes (YACs).

Aseptic technique Manipulation of sterile instruments or culture media in such a way as to maintain sterility.

ATP Adenosine triphosphate, the principal energy carrier of the cell.

Attenuation In a pathogen, a decrease or loss of virulence. Also, a mechanism for controlling gene expression. Typically transcription is terminated after initiation but before a full-length mRNA is produced.

Autoantibody An antibody that reacts to self antigens.

Autoclave A sterilizer that destroys microorganisms with temperature and steam under pressure.

Autoimmunity Immune reactions of a host against its own self antigens.

Autolysis The lysis of a cell brought about by the activity of the cell itself.

Autoradiography Detection of radioactivity in a sample, for example, a cell or gel, by placing it in contact with a photographic film.

Autotroph An organism able to use CO_2 as a sole source of carbon.

Auxotroph An organism that has developed a nutritional requirement through mutation. Contrast with a *Prototroph*.

B lymphocyte (B cell) A lymphocyte that has immunoglobulin surface receptors, produces immunoglobulin, and may present antigens to T cells.

Bacteremia The transient appearance of bacteria in the blood.

Bacteria All prokaryotes that are not members of the domain *Archaea*.

Bacteriocidal Capable of killing bacteria.

Bacteriocins Agents produced by certain bacteria that inhibit or kill closely related species.

Bacteroid A swollen, deformed *Rhizobium* cell found in the root nodule; capable of nitrogen fixation.

Bacteriophage A virus that infects prokaryotic cells.

Bacteriorhodopsin A protein containing retinal that is found in the membranes of certain extremely halophilic *Archaea* and that is involved in light-mediated ATP synthesis.

Bacteriostatic Capable of inhibiting bacterial growth without killing.

Barophile An organism that lives optimally at high hydrostatic pressure.

Barotolerant An organism able to tolerate high hydrostatic pressure, although growing better at 1 atm.

Base composition In reference to nucleic acids, the proportion of the total bases consisting of guanine plus cytosine or thymine plus adenine base pairs. Usually expressed as a guanine + cytosine (GC) value, for example, 60% GC.

Batch culture A closed-system microbial culture of fixed volume.

Beta-lactam antibiotic An antibiotic such as penicillin that contains the four-membered heterocyclic β-lactam ring.

Binomial nomenclature system The system for naming organisms in which an organism is given a genus name and a species epithet.

Biocatalysis The use of microorganisms to synthesize a product or carry out a specific chemical transformation.

Biochemical oxygen demand (BOD) The amount of dissolved oxygen consumed by microorganisms for complete oxidation of organic and inorganic material in a water sample.

Biofilm Microbial colonies encased in an adhesive, usually polysaccharide material and attached to a surface.

Biogeochemistry Study of microbially mediated chemical transformations of geochemical interest, for example, nitrogen or sulfur cycling.

Bioinformatics The use of computer programs to analyze, store, and access DNA and protein sequences.

Biological warfare The use of biological agents to kill or incapacitate a population.

Bioremediation Use of microorganisms to remove or detoxify toxic or unwanted chemicals in an environment.

Biosynthesis The production of needed cellular constituents from other (usually simpler) molecules.

Biosynthetic penicillin Production of a particular form of penicillin by supplying the producing organism with specific side chain precursors.

Biotechnology The use of living organisms to carry out defined chemical processes for industrial application.

Biotransformation In industrial microbiology, use of microorganisms to convert a substance to a chemically modified form.

Black smoker A deep-sea hydrothermal vent emitting superheated 250–400°C water and minerals.

Botulism Food poisoning due to ingestion of food containing botulism toxin produced by *Clostridium botulinum*.

Brewing The manufacture of alcoholic beverages such as beer from the fermentation of malted grains.

Broad-spectrum antibiotic An antibiotic that acts on both gram-positive and gram-negative *Bacteria*.

Calvin cycle The biochemical route of CO_2 fixation in many autotrophic organisms.

Canning The process of sealing food in a closed container and heating to destroy living organisms.

Capsid The protein coat of a virus.

Capsomere An individual protein subunit of the virus capsid.

Capsule A dense, well-defined polysaccharide or protein layer closely surrounding a cell.

Carboxysomes Polyhedral cellular inclusions of crystalline ribulose bisphosphate carboxylase (RubisCO), the key enzyme of the Calvin cycle.

Carcinogen A substance that causes the initiation of tumor formation. Frequently a mutagen.

Carrier An individual that harbors infectious organisms but does not show symptoms of disease.

Catabolism The biochemical processes involved in the breakdown of organic or inorganic compounds, usually leading to the production of energy.

Catabolite repression Repression of a variety of unrelated enzymes when cells are grown in a medium containing glucose.

Catalysis Increase in rate of a chemical reaction.

Catalyst A substance that promotes a chemical reaction without itself being changed in the end.

CD4 cells T helper cells. They are targets for HIV infection.

Cell The fundamental unit of life.

Cell-mediated immunity An immune response generated by interactions with antigen-specific T cells. Compare with *Humoral immunity*.

Centers for Disease Control and Prevention (CDC) An agency of the United States Public Health Service that tracks disease trends, provides disease information to the public and to health-care professionals, and forms public policy regarding disease prevention and intervention.

Chemiosmosis The use of ion gradients, especially proton gradients, across membranes to generate ATP. See *Proton motive force*.

Chemokine A small soluble protein produced by a variety of cells that modulates inflammatory reactions and immunity in target cells.

Chemolithotroph An organism obtaining its energy from the oxidation of inorganic compounds.

Chemoorganotroph An organism obtaining its energy from the oxidation of organic compounds.

Chemostat A continuous culture device controlled by the concentration of limiting nutrient and dilution rate.

Chemotaxis Movement toward or away from a chemical.

Chemotherapeutic agent An antimicrobial agent that can be used internally.

Chemotherapy Treatment of infectious disease with chemicals or antibiotics.

Chloramine A water purification chemical made by combining chlorine and ammonia at precise ratios.

Chlorination A highly effective disinfectant procedure for drinking water using chlorine gas or other chlorine-containing compounds as disinfectant.

Chlorine A chemical fed in its gaseous state to disinfect water. A residual level is maintained throughout the distribution system.

Chlorophyll and bacteriochlorophyll Pigments of phototrophic organisms consisting of light-sensitive magnesium tetrapyrroles.

Chloroplast The chlorophyll-containing organelle of phototrophic eukaryotes.

Chlorosomes Cigar-shaped structures enclosed by a nonunit membrane and containing the light-harvesting bacteriochlorophyll (c, c_s, d, or e) in green sulfur bacteria and in *Chloroflexus*.

Chromogenic Producing color; a chromogenic colony is a pigmented colony.

Chromosome A genetic element carrying genes essential to cellular function. Prokaryotes typically have a single chromosome consisting of a circular DNA molecule. Eukaryotes typically have several chromosomes, each containing a linear DNA molecule.

Chronic In reference to infections, long-term, usually characterized by gradual onset, mild or subclinical disease, and lack of permanent recovery.

Cidal Lethal or killing.

Cilium Short, filamentous structure that beats with many others to make a cell move.

Cirrhosis Breakdown of the normal liver architecture resulting in fibrosis.

Citric acid cycle A cyclical series of reactions resulting in the conversion of acetate to CO_2 and NADH. Also called the *Tricarboxylic acid cycle* or the *Kreb's cycle*.

Clarifier (coagulation basin) A reservoir in which the suspended solids of raw water are coagulated and removed.

Class I MHC protein Antigen-presenting molecule found on all nucleated vertebrate cells.

Class II MHC protein Antigen-presenting molecule found on macrophages, B lymphocytes, and dendritic cells in vertebrates.

Clonal deletion For T-cell selection in the thymus, the killing of useless or self-reactive clones.

Clonal selection A theory that each B or T lymphocyte, when stimulated by antigen, divides to form a clone of itself.

Clone A population of cells all descended from a single cell. Also, a number of copies of a DNA fragment obtained by allowing an inserted DNA fragment to be replicated by a phage or plasmid.

Cloning vectors Genetic elements into which genes can be recombined and replicated.

Coagulation The formation of large insoluble particles from much smaller, colloidal particles by the addition of aluminum sulfate and anionic polymers.

Coccoid Sphere-shaped.

Coccus A spherical bacterium.

Codon A sequence of three bases in messenger RNA that encodes a specific amino acid.

Coenzyme A low-molecular-weight molecule that participates in an enzymatic reaction by accepting and donating electrons or functional groups. Examples: NAD^+, FAD.

Coliform Gram-negative, nonsporing, facultative rod that ferments lactose with gas formation within 48 hours at 35°C.

Colonization Multiplication of a microorganism after it has attached to host tissues or other surfaces.

Colony A macroscopically visible population of cells growing on solid medium, arising from a single cell.

Cometabolism The metabolic transformation of a substance while a second substance serves as primary energy or carbon source.

Commodity chemicals Chemicals such as ethanol that have low monetary value and thus are sold primarily in bulk.

Common-source epidemic An epidemic resulting from infection of a large number of people from a single contaminated source.

Compatible solutes Organic compounds that serve as cytoplasmic solutes to balance water relations for cells growing in environments of high salt or sugar.

Competence Ability to take up DNA and become genetically transformed.

Complement A series of proteins that react in a sequential manner with antibody–antigen complexes to amplify or potentiate antibody activity.

Complement fixation The consumption of complement by an antibody–antigen reaction.

Complementary Nucleic acid sequences that can base-pair with each other.

Complementarity-determining regions (CDRs) Variations in amino acid sequence that occur within the variable domains of Igs or TCRs that provide most of the molecular contacts with antigen (also known as *hypervariable regions*).

Complex media Culture media whose precise chemical composition is unknown. Also called *undefined media*.

Concatemer A DNA molecule consisting of two or more separate molecules linked end to end to form a long, linear structure.

Congenital syphilis Syphilis contracted by an infant from its mother during birth.

Conjugation Transfer of genes from one prokaryotic cell to another by a mechanism involving cell-to-cell contact.

Consensus sequence A nucleic acid sequence in which the base present in a given position is that base most commonly found when many experimentally determined sequences are compared.

Consortium A two-membered (or more) bacterial culture (or natural assemblage) in which each organism benefits from the others.

Contagious Transmissible.

Cortex The region inside the spore coat of an endospore, around the core.

Covalent bond A nonionic chemical bond formed by a sharing of electrons between two atoms.

Crista Inner membrane in a mitochondrion; site of respiration.

Culture A particular strain or kind of organism growing in a laboratory medium.

Culture medium An aqueous solution of various nutrients suitable for the growth of microorganisms.

Cutaneous Relating to the skin.

Cyanobacteria Prokaryotic oxygenic phototrophs containing chlorophyll *a* and phycobilins.

Cyst A resting stage formed by some bacteria and protozoa in which the whole cell is surrounded by a thick-walled chemically and physically resistant coating; not the same as a spore or endospore.

Cytochrome Iron-containing porphyrin complexed with proteins, which functions as an electron carrier in the electron transport system.

Cytokine A small, soluble protein produced by a leukocyte that modulates inflammatory reactions and immunity in target cells.

Cytoplasm Cellular contents inside the cytoplasmic membrane, excluding the nucleus.

Cytoplasmic membrane The permeability barrier of the cell, separating the cytoplasm from the environment.

Cytoskeleton Cellular scaffolding in which microfilaments define the shape of eukaryotic cells.

Decontamination Treatment that renders an object or inanimate surface safe to handle.

Defined media Culture media whose exact chemical composition is known. Compare with *Complex media*.

Degeneracy In relation to the genetic code, the fact that more than one codon can code for the same amino acid.

Deletion Removal of a portion of a gene.

Denaturation Irreversible destruction of a macromolecule, as for example, the destruction of a protein by heat.

Denaturing gradient gel electrophoresis (DGGE) An electrophoretic technique that allows resolving nucleic acid fragments of the same size but that differ in sequence.

Dendritic cell A type of leukocyte having phagocytic and antigen-presenting properties, found in lymph nodes and spleen.

Denitrification Microbial conversion of nitrate into nitrogen gases under anoxic conditions.

Dental caries Tooth decay resulting from bacterial infection.

Dental plaque Bacterial cells encased in a matrix of extracellular polymers and salivary products, found on the teeth.

Deoxyribonucleic acid (DNA) A polymer of nucleotides connected via a phosphate–deoxyribose sugar backbone; the genetic material of cells and some viruses.

Desiccation Drying.

Dideoxynucleotide A nucleotide lacking the 3'-hydroxyl group on the deoxyribose sugar. Used in the Sanger method of DNA sequencing.

Differential media A growth medium that allows identification of microorganisms based on phenotypic properties.

Differentiation The modification of a cell in terms of structure and/or function occurring during the course of development.

Diploid In eukaryotes, an organism or cell with two chromosome complements, one derived from each haploid gamete.

Disease Injury to the host that impairs host function.

Disinfectant An antimicrobial agent used only on inanimate objects.

Disinfection The elimination of microorganisms from inanimate objects or surfaces.

Disproportionation The splitting of a chemical compound into two new compounds, one more oxidized and one more reduced than the original compound.

Distribution system Water pipes, storage reservoirs, tanks, and other means used to deliver drinking water to consumers or store it before delivery.

DNA fingerprinting Use of genetic engineering to determine the origin of DNA in a sample of tissue.

DNA gyrase An enzyme found in most prokaryotes that introduces negative supercoils in DNA.

DNA library See *gene library*.

DNA polymerase An enzyme that synthesizes a new strand of DNA in the 5' → 3' direction using an antiparallel DNA strand as a template.

Domain The highest level of biological classification. The three domains of biological organisms are the *Bacteria*, the *Archaea*, and the *Eukarya*. Also used to describe a region of a protein having a defined structure and function.

Doubling time The time needed for a population to double. See also *Generation time*.

Downstream position Refers to nucleic acid sequences on the 3' side of a given site on the DNA or RNA molecule. Compare with *Upstream position*.

Early protein Proteins synthesized soon after virus infection.

Ecology Study of the interrelationships between organisms and their environments.

Ecosystem A community of organisms and their natural environment.

Ecotype A population of genetically identical cells sharing a particular resource within an ecological niche.

Effluent water   Treated wastewater discharged from a wastewater treatment facility.

Ehrlichiosis   One of a group of emerging tick-transmitted disease caused by rickettsias of the *Ehrlichia* genus.

Electron acceptor   A substance that accepts electrons during an oxidation–reduction reaction.

Electron donor   A compound that donates electrons in an oxidation–reduction reaction.

Electron transport phosphorylation   Synthesis of ATP involving a membrane-associated electron transport chain and the creation of a proton motive force. Also called *Oxidative phosphorylation*. See also *Chemiosmosis*.

Electrophoresis   Separation of charged molecules in an electric field.

Electroporation   The use of an electric pulse to enable cells to take up DNA.

ELISA   Enzyme-linked immunosorbent assay. An immunoassay that uses specific antibodies to detect antigens or antibodies in body fluids. The antibody-containing complexes are visualized through enzyme coupled to the antibody. Addition of substrate to the enzyme–antibody–antigen complex results in a colored product.

Emerging infection   An infectious disease that has increased in incidence in the last 20 years or threatens to increase in incidence in the future.

Enantiomer   One form of a molecule that is the mirror image of another form of the same molecule.

Endemic   A disease that is constantly present in low numbers in a population, usually in low numbers. Compare with *Epidemic*.

Endergonic reaction   A chemical reaction requiring an input of energy to proceed.

Endocytosis   A process in which a particle such as a virus is taken intact into an animal cell. Phagocytosis and pinocytosis are two kinds of endocytosis.

Endoplasmic reticulum   An extensive array of internal membranes in eukaryotes.

Endospore   A differentiated cell formed within the cells of certain gram-positive bacteria that is extremely resistant to heat as well as to other harmful agents.

Endosymbiosis   The process by which mitochondria, chloroplasts, and hydrogenosomes originated from the descendants of ancient prokaryotic organisms from the domain *Bacteria*.

Endotoxin   The lipopolysaccharide portion of the cell envelope of certain gram-negative *Bacteria*, which is a toxin when solubilized. Compare with *Exotoxin*.

Enriched media   Media that allow metabolically fastidious organisms to grow because of the addition of specific growth factors.

Enrichment bias   The skewing of enrichment culture results toward "weed" species.

Enrichment culture   Use of selective culture media and incubation conditions to isolate specific microorganisms from natural samples.

Enteric   Intestinal.

Enterotoxin   A protein released extracellularly by a microorganism as it grows that produces immediate damage to the small intestine of the host.

Entropy   A measure of the degree of disorder in a system; entropy always increases in a closed system.

Enzyme   A catalyst, usually composed of protein, that promotes specific reactions or groups of reactions.

Epidemic   A disease occurring in an unusually high number of individuals in a population at the same time. Compare with *Endemic*.

Epidemiology   The study of the occurrence, distribution, and control of diseases.

Epitope   The portion of an antigen that is recognized by an immunoglobulin or a T-cell receptor.

Escherichia coli O157:H7   An enterotoxigenic strain of *E. coli* spread by fecal contamination of animal or human origin to food and water.

Eukarya   The phylogenetic domain containing all eukaryotic organisms.

Eukaryote   A cell or organism having a unit membrane-enclosed (true) nucleus and usually other organelles.

Evolution   Change in a line of descent over time leading to the production of new species or varieties within a species.

Evolutionary distance   In phylogenetic trees, the sum of the physical distance on a tree separating organisms; this distance is inversely proportional to evolutionary relatedness.

Exergonic reaction   A chemical reaction that proceeds with the liberation of energy.

Exons   The coding sequences in a split gene. Contrast with *Introns*, the intervening noncoding regions.

Exotoxin   A protein released extracellularly by a microorganism as it grows that produces immediate host cell damage. Compare with *Endotoxin*.

Exponential growth   Growth of a microorganism where the cell number doubles within a fixed time period.

Exponential phase   A period during the growth cycle of a population in which growth increases at an exponential rate.

Expression   The ability of a gene to function within a cell in such a way that the gene product is formed.

Expression vector   A cloning vector that contains the necessary regulatory sequences allowing transcription and translation of a cloned gene or genes.

Extreme halophile   An organism whose growth is dependent on large amounts (generally >10%) of NaCl.

Extremophile   An organism that grows optimally under one or more chemical or physical extremes, such as high or low temperature or pH.

Facultative   A qualifying adjective indicating that an organism is able to grow in either the presence or absence of an environmental factor (for example, "facultative aerobe").

FAME   Fatty acid methyl ester.

Feedback inhibition   A decrease in the activity of the first enzyme of a biochemical pathway caused by buildup of the final product of the pathway.

Fermentation   Catabolic reactions producing ATP in which organic compounds serve as both primary electron donor and ultimate electron acceptor and ATP is produced by substrate-level phosphorylation.

Fermentation (industrial)   A large-scale microbial process.

Fermenter   An organism that carries out the process of fermentation.

Fermentor   A growth vessel, usually quite large, used to culture microorganisms for the production of some commercially valuable product.

Ferredoxin   An electron carrier of low reduction potential; small protein containing iron-sulfur clusters.

Fever   An abnormal increase in body temperature.

Filamentous   In the form of very long rods, many times longer than wide.

Insertion A genetic phenomenon in which a piece of DNA is inserted into the middle of a gene.

Insertion sequence (IS elements) The simplest type of transposable element. Has only genes involved in transposition.

Integrating vector A cloning vector that becomes integrated into a host chromosome.

Integration The process by which a DNA molecule becomes incorporated into another genome.

Interferons Cytokine proteins produced by virus-infected cells that induce signal transduction in nearby cells, resulting in transcription of antiviral genes and expression of antiviral proteins.

Interleukin (IL) Soluble cytokine or chemokine mediator secreted by leukocytes.

Interspecies hydrogen transfer The process by which organic matter is degraded by the interaction of several groups of microorganisms in which H_2 production and H_2 consumption are closely coupled.

Introns The intervening noncoding sequences in a split gene. Contrasted with *Exons*, the coding sequences.

Invasiveness Pathogenicity caused by the ability of a pathogen to enter the body and spread.

Ionophore A compound that can cause the leakage of ions across membranes.

Irradiation In food microbiology, the exposure of food to ionizing radiation to inhibit microorganisms and insect pests, or to retard growth or ripening.

Isomers Two molecules with the same molecular formula but which differ structurally.

Isotopes Different forms of the same element containing the same number of protons and electrons but differing in the number of neutrons.

Jaundice Production and release of excess bilirubin in the liver due to destruction of liver cells, resulting in yellowing of the skin and whites of the eye.

Joule (J) A unit of energy equal to 10^7 ergs; 1000 Joules equal 1 kilojoule (kJ).

Kilobase (kb) A 1000-base fragment of nucleic acid. A *kilobase pair* (kbp) is a fragment containing 1000 base pairs.

Kinase An enzyme that adds a phosphoryl group, usually from ATP, to a compound.

Koch's Postulates A set of criteria for proving that a given microorganism causes a given disease.

Lag phase The period after inoculation of a culture before growth begins.

Laser tweezers A device used to obtain pure cultures in which a single cell is optically trapped with a laser and moved away from contaminating organisms into sterile growth medium.

Late protein Proteins synthesized toward the end of virus infection.

Latent virus A virus present in a cell, yet not causing any detectable effect.

Leaching Removal of valuable metals from ores by microbial action.

Leukocidin A substance able to destroy phagocytes.

Leukocyte A nucleated cell found in the blood. A white blood cell.

Lichen A fungus and an alga (or a cyanobacterium) living in symbiotic association.

Lipid Water-insoluble organic molecules important in structure of the cytoplasmic membrane and (in some organisms) the cell wall. See also *Phospholipid*.

Lipopolysaccharide (LPS) Complex lipid structure containing unusual sugars and fatty acids found in most gram-negative *Bacteria* and constituting the chemical structure of the outer membrane.

Listeriosis Gastrointestinal food infection caused by *Listeria monocytogenes* that may lead to bacteremia and meningitis.

Lophotrichous Having a tuft of polar flagella.

Lower respiratory tract Trachea, bronchi, and lungs.

Luminescence Production of light.

Lyme disease An emerging tick-transmitted disease caused by the spirochete *Borrelia burgdorferi*.

Lymph A fluid similar to blood that lacks red blood cells and travels through a separate circulatory system (the lymphatic system) containing lymph nodes.

Lymphocyte A white blood cell involved in antibody formation or cellular immune responses.

Lyophilization (freeze-drying) The process of removing all water from frozen food under vacuum.

Lysin An antibody that induces lysis.

Lysis Loss of cellular integrity with release of cytoplasmic contents.

Lysogen A prokaryote containing a prophage. See also *Temperate virus*.

Lysogenic pathway A series of steps that after virus infection, leads to a genetic state (lysogeny) where the viral genome is replicated as a prophage along with that of the host.

Lysosome A cell organelle containing digestive enzymes.

Lytic pathway A series of steps after virus infection that lead to virus replication and the destruction (lysis) of the host cell.

Macromolecule A large molecule (polymer) formed by the connection of a number of small molecules (monomers).

Macrophage Large, noncirculating phagocytic cells involved in both phagocytosis and the antibody production process.

Magnetosomes Small particles of Fe_3O_4 present in cells that exhibit magnetotaxis (magnetic bacteria).

Magnetotaxis Directed movement of bacterial cells by a magnetic field.

Major histocompatibility complex (MHC) A gene cluster, or complex, encoding cell surface proteins important for antigen presentation to T cells in the immune response.

Malaria An insect-transmitted disease characterized by recurrent episodes of fever and anemia caused by the protozoan *Plasmodium* spp., usually transmitted between mammals through the bite of the *Anopheles* mosquito.

Malignant In reference to a tumor, an infiltrating metastasizing growth no longer under normal growth control.

Mast cells Tissue cells adjoining blood vessels throughout the body that contain granules with inflammatory mediators.

Medium (plural media) In microbiology, the nutrient solution(s) used to grow microorganisms.

Megabase (Mb) One million nucleotide bases (or base pairs, abbreviated Mbp).

Meiosis In eukaryotes, reduction division, the process by which the change from diploid to haploid occurs.

Membrane Any thin sheet or layer. See especially *Cytoplasmic membrane*.

Memory cell A differentiated B lymphocyte capable of rapid conversion to an antibody-producing plasma cell on subsequent stimulation with antigen.

Meningitis Inflammation of the meninges (brain tissue), sometimes caused by *Neisseria meningitidis* and characterized by sudden onset of headache, vomiting, and stiff neck, often progressing to coma within hours.

Meningococcemia Fulminant disease caused by *Neisseria meningitidis* and characterized by septicemia, intravascular coagulation, and shock.

Meningoencephalitis Invasion, inflammation, and destruction of brain tissue by the ameba *Naegleria fowlerii* or a variety of other pathogens.

Mesophile Organism living in the temperature range near that of warm-blooded animals and usually showing a growth temperature optimum between 25 and 40°C.

Messenger RNA (mRNA) An RNA molecule transcribed from DNA that contains the genetic information necessary to encode a particular protein.

Metabolism All biochemical reactions in a cell, both anabolic and catabolic.

Metazoa Multicellular organisms.

Methanogen A methane-producing prokaryote; member of the *Archaea*.

Methanogenesis The biological production of methane (CH_4).

Methanotroph An organism capable of oxidizing methane.

Methylotroph An organism capable of oxidizing organic compounds that do not contain carbon–carbon bonds; if able to oxidize CH_4, also a methanotroph.

MIC Minimum inhibitory concentration—the minimum concentration of a substance necessary to prevent microbial growth.

Microaerophilic An organism requiring O_2 but at a level lower than that in air.

Microenvironment The immediate physical and chemical surroundings of a microorganism.

Micrometer One-millionth of a meter, or 10^{-6} m (abbreviated μm), the unit used for measuring microorganisms.

Microorganism A microscopic organism consisting of a single cell or cell cluster, also including the viruses, which are not cellular.

Microtubules Tubes that are the structural entity for eukaryotic flagella, play a role in maintaining cell shape and function as mitotic spindle fibers.

Minus (negative)-strand nucleic acid An RNA or DNA strand that has the opposite sense of (would be complementary to) the mRNA of a virus.

Mitochondrion Eukaryotic organelle responsible for the processes of respiration and electron transport phosphorylation.

Mitosis A highly ordered process by which the nucleus divides in eukaryotes.

Mixotroph An organism able to assimilate organic compounds as carbon sources while using inorganic compounds as electron donors for energy metabolism.

Molds Filamentous fungi.

Molecular chaperone A protein that helps other proteins fold or refold properly.

Molecular cloning Isolation and incorporation of a fragment of DNA into a vector where it can be replicated.

Molecule Two or more atoms chemically bonded to one another.

Monoclonal antibody An antibody produced from a single clone of B cells. This antibody has uniform structure and specificity.

Monocytes Circulating white blood cells that contain many lysosomes and can differentiate into macrophages.

Monomer A building block of a polymer.

Monotrichous Having a single polar flagellum.

Morbidity Incidence and prevalence of disease in a population, including both fatal and nonfatal cases.

Morphology The shape of an organism.

Mortality Incidence and prevalence of death in a population.

Most probable number (MPN) Serial dilution of a natural sample to determine the highest dilution yielding growth.

Motif A conserved amino acid sequence found in all peptide antigens that bind to a given MHC protein.

Motility The property of movement of a cell under its own power.

Mucous membrane Layers of epithelial cells that interact with the external environment.

Mucus Soluble glycoproteins secreted by epithelial cells that coat the mucous membrane.

Mushrooms Filamentous fungi that produce large, often edible structures called fruiting bodies.

Mutagen An agent that induces mutation, such as radiation or certain chemicals.

Mutant A strain differing from its parent because of mutation.

Mutation An inheritable change in the base sequence of the genome of an organism.

Mycorrhiza A symbiotic association between a fungus and the roots of a plant.

Mycosis Infection caused by fungi.

Myeloma A malignant tumor of a plasma cell (antibody-producing cell).

Natural killer (NK) cell A specialized lymphocyte that recognizes and destroys foreign cells or infected host cells in a nonspecific manner.

Negative selection In T-cell selection, T cells that interact with self antigens in the thymus are deleted. See *Clonal deletion*.

Neutralization Interaction of antibody with antigen that reduces or blocks the biological activity of the antigen.

Neutrophil (polymorphonuclear leukocyte) (PMN) A type of leukocyte exhibiting phagocytic properties, a granular cytoplasm (granulocyte), and a multilobed nucleus.

Neutrophile An organism that grows best around pH 7.

Nitrification The microbiological conversion of ammonia to nitrate.

Nitrogen fixation Reduction of nitrogen gas to ammonia ($N_2 + 8\,H \longrightarrow 2\,NH_3 + H_2$) by the enzyme nitrogenase.

Nodule A tumorlike structure produced by the roots of symbiotic nitrogen-fixing plants. Contains the nitrogen-fixing microbial component of the symbiosis.

Nonnucleoside reverse transcriptase inhibitor (NNRTI) A nonnucleoside compound that inhibits the action of retroviral reverse transcriptase by binding directly to the catalytic site.

Nonperishable (stable) foods Foods of low water activity that have an extended shelf life and are resistant to spoilage by microorganisms.

Nonpolar Possessing hydrophobic (water-repelling) characteristics and not easily dissolved in water.

Nonsense mutation A mutation that changes a sense codon into one that does not code for an amino acid.

Normal microbial flora Microorganisms that are usually found associated with healthy body tissue.

Northern blot Hybridization of a single strand of nucleic acid (DNA or RNA) to RNA fragments immobilized on a filter. Compare with *Southern blot* and *Western blot*.

Nosocomial infection Hospital-acquired infection.

Nucleic acid   A polymer of nucleotides. See *Deoxyribonucleic acid* and *Ribonucleic acid*.

Nucleic acid probe   A strand of nucleic acid that can be labeled and used to hybridize to a complementary molecule from a mixture of other nucleic acids. In clinical microbiology or microbial ecology, a short oligonucleotide of unique sequence used as a hybridization probe for identifying specific genes.

Nucleocapsid   The complex of nucleic acid and proteins of a virus, not including, if present, an envelope.

Nucleoid   The aggregated mass of DNA that makes up the chromosome of prokaryotic cells.

Nucleoside   A nucleotide minus phosphate.

Nucleoside reverse transcriptase inhibitor (NRTI)   A nucleoside analog compound that inhibits the action of viral reverse transcriptase by competing with nucleosides.

Nucleotide   A monomeric unit of nucleic acid, consisting of a sugar, a phosphate, and a nitrogenous base.

Nucleus   A membrane-enclosed structure in eukaryotes containing the genetic material (DNA) organized in chromosomes.

Nutrient   A substance taken by a cell from its environment and used in catabolic or anabolic reactions.

Obligate   A qualifying adjective referring to an environmental condition always required for growth (for example, "obligate anaerobe").

Oligonucleotide   A short nucleic acid molecule, either obtained from an organism or synthesized chemically.

Oligotrophic   Describing a habitat in which nutrients are in low supply.

Oncogene   A gene whose expression causes formation of a tumor.

Open reading frame (ORF)   A sequence of DNA which, if transcribed, could be translated to yield a protein of known length and composition. A *functional* ORF is one that actually encodes a protein in the cell.

Operator   A specific region of the DNA at the initial end of a gene, where the repressor protein binds and blocks mRNA synthesis.

Operon   A cluster of genes whose expression is controlled by a single operator. Typical of prokaryotic cells.

Opportunistic infection   An infection usually observed only in an individual with a dysfunctional immune system.

Opportunistic pathogen   An organism that causes disease in the absence of normal host resistance.

Opsonization   Promotion of phagocytosis by a specific antibody in combination with complement.

Organelle   A membrane-enclosed structure found in eukaryotic cells.

Orthologs   Genes found in one organism that are similar to those in another organism but differ because of speciation. See also *Paralogs*.

Osmosis   Diffusion of water through a membrane from a region of low solute concentration to one of higher concentration.

Outbreak   The occurrence of a large number of cases of a disease in a short period of time.

Oxic   Containing oxygen; aerobic. Usually used in reference to a microbial habitat.

Oxidation   A process by which a compound gives up electrons (or H atoms) and becomes oxidized.

Oxidation-reduction (redox) reaction   A pair of reactions in which one compound becomes oxidized while another becomes reduced and takes up the electrons released in the oxidation reaction.

Oxidative (electron transport) phosphorylation   The nonphototrophic production of ATP at the expense of a proton motive force formed by electron transport.

Oxygenic photosynthesis   Use of light energy to synthesize ATP and NADPH by noncyclic photophosphorylation with the production of O_2 from water.

Palindrome   A nucleotide sequence on a DNA molecule in which the same sequence is found on each strand but in the opposite direction.

Pandemic   A worldwide epidemic.

Paralogs   Genes within an organism whose similarity is the result of gene duplication at some time during the evolution of the organism. See also *Orthologs*.

Parasite   An organism able to live in or on a host and cause disease.

Passive immunity   Immunity resulting from transfer of antibodies or immune cells from an immune to a nonimmune individual.

Pasteurization   Reduction of the microbial load in heat-sensitive liquids to kill disease-producing microorganisms and reduce the number of spoilage microorganisms.

Pathogen   An organism, usually a microorganism, that causes disease.

Pathogen-associated molecular pattern (PAMP)   A unique structural components of a microbe or virus recognized by a pattern recognition molecule.

Pathogenicity   The ability of a pathogen to cause disease.

Pattern recognition molecule (PRM)   A membrane-bound protein that recognizes a pathogen-associated molecular pattern, such as a unique component of a microbial cell surface structure.

Penicillin   A class of antibiotics that inhibit bacterial cell wall synthesis; characterized by a β-lactam ring.

Peptide bond   A type of covalent bond joining amino acids in a polypeptide.

Peptidoglycan   The rigid layer of the cell walls of *Bacteria*, a thin sheet composed of *N*-acetylglucosamine, *N*-acetylmuramic acid, and a few amino acids.

Perishable food   Fresh food generally of high water activity that has a very short shelf life due to potential for spoilage by growth of microorganisms.

Periplasmic space   The area between the cytoplasmic membrane and the outer membrane in gram-negative *Bacteria*.

Peritrichous flagellation   In flagellar arrangements, having flagella attached to many places on the cell surface.

Phage   See *Bacteriophage*.

Phagemid   A cloning vector that can replicate either as a plasmid or as a bacteriophage.

Phagocyte   One of a group of cells that recognizes, ingests, and degrades pathogens and pathogen products.

Phagocytosis   Ingestion of particulate material such as bacteria by protozoa and phagocytic cells of higher organisms.

Phenotype   The observable characteristics of an organism. Compare with *Genotype*.

Phosphodiester bond   A type of covalent bond linking nucleotides together in a polynucleotide.

Phospholipid   Lipids containing a substituted phosphate group and two fatty acid chains on a glycerol backbone.

Photoautotroph   An organism able to use light as its sole source of energy and CO_2 as its sole carbon source.

Photoheterotroph An organism using light as a source of energy and organic materials as carbon source.

Photophosphorylation Synthesis of energy-rich phosphate bonds in ATP using light energy.

Photosynthesis The use of light energy to drive the incorporation of CO_2 into cell material. See also *Anoxygenic photosynthesis* and *Oxygenic photosynthesis*.

Phototaxis Movement of a cell toward light.

Phototroph An organism that obtains energy from light.

Phylogeny The ordering of species into higher taxa and the construction of evolutionary trees based on evolutionary (natural) relationships.

Phytanyl A branched-chain hydrocarbon containing 20 carbon atoms, commonly found in the lipids of *Archaea*.

Pickling The process of acidifying food, typically with acetic acid, to prevent microbial growth and spoilage.

Pilus A fimbria-like structure that is present on fertile cells, both Hfr and F$^+$, and is involved in DNA transfer during conjugation. Sometimes called a *sex pilus*. See also *Fimbria*.

Pinocytosis In eukaryotes, phagocytosis of soluble molecules.

Plague An endemic disease in rodents caused by *Yersinia pestis* that is occasionally transferred to humans through the bite of a flea.

Plaque A zone of lysis or cell inhibition caused by virus infection on a lawn of cells.

Plasma The liquid portion of the blood with cells removed and clotting proteins deactivated.

Plasma cell A large, differentiated, short-lived B lymphocyte specializing in abundant (but short-term) antibody production.

Plasmid An extrachromosomal genetic element that is not essential for growth and has no extracellular form.

Platelet A noncellular disc-shaped structure containing protoplasm found in large numbers in blood and functioning in the blood-clotting process.

Plus-strand nucleic acid An RNA or DNA strand that has the same sense as the mRNA of a virus.

Point mutation A mutation that involves one or only a very few base pairs.

Polar Possessing hydrophilic characteristics and generally water-soluble.

Polar flagellation In flagellar arrangements, having flagella attached at one end or both ends of the cell.

Poly-β-hydroxybutyrate (PHB) A common storage material of prokaryotic cells consisting of a polymer of β-hydroxybutyrate (PHB) or other β-alkanoic acids (PHA).

Polyclonal antibodies A mixture of antibodies made by many different B cell clones.

Polyclonal antiserum A mixture of antibodies to a variety of antigens or to a variety of determinants on a single antigen.

Polymer A large molecule formed by polymerization of monomeric units. In water purification, a chemical in liquid form used as a coagulant to produce flocculation in the clarification process.

Polymerase chain reaction (PCR) A method used to amplify a specific DNA sequence *in vitro* by repeated cycles of synthesis using specific primers and DNA polymerase.

Polymorphism The occurrence of multiple alleles at a locus in frequencies that cannot be explained by the occurrence of recent random mutations.

Polymorphonuclear leukocyte (PMN) Motile white blood cells containing many lysosomes and specializing in phagocytosis. Characterized by a distinct segmented nucleus. Also, a neutrophil.

Polynucleotide A polymer of nucleotides bonded to one another by phosphodiester bonds.

Polypeptide Several amino acids linked together by peptide bonds.

Polysaccharide A long chain of monosaccharides (sugars) linked by glycosidic bonds.

Porins Protein channels in the outer membrane of gram-negative *Bacteria* through which small to medium-sized molecules can flow.

Porters Membrane proteins that function to transport substances into and out of the cell.

Positive selection In T-cell selection, T cells that interact with self MHC protein in the thymus are stimulated to grow and develop.

Potable In water purification, drinkable; safe for human consumption.

Precipitation A reaction between antibody and soluble antigen resulting in visible antibody–antigen complexes.

Prevalence The *proportion* of individuals in a population having a disease.

Pribnow box The consensus sequence TATAAT located approximately 10 base pairs upstream from the transcriptional start site. A binding site for RNA polymerase.

Primary antibody response Antibodies made on first exposure to antigen; mostly of the class IgM.

Primary metabolite A metabolite excreted during the growth phase.

Primary producer An organism that uses light or an inorganic compound to synthesize new organic material from CO_2.

Primary structure In an informational macromolecule, such as a polypeptide or a nucleic acid, the *precise sequence* of monomeric units.

Primary transcript An unprocessed RNA molecule that is the direct product of transcription.

Primary wastewater treatment Physical separation of wastewater contaminants, usually by separation and settling.

Primer A molecule (usually a polynucleotide) to which DNA polymerase can attach the first deoxyribonucleotide during DNA replication.

Prion An infectious agent whose extracellular form may contain no nucleic acid.

Probe See *Nucleic acid probe*.

Prochlorophyte A prokaryotic oxygenic phototroph that contains chlorophylls *a* and *b* but lacks phycobilins.

Prokaryote A cell or organism lacking a nucleus and other membrane-enclosed organelles and usually having its DNA in a single circular molecule. Members of the *Bacteria* and the *Archaea*.

Promoter The site on DNA where the RNA polymerase binds and begins transcription.

Prophage The state of the genome of a temperate virus when it is replicating in synchrony with that of the host, typically integrated into the host genome.

Prophylactic Treatment, usually immunological or chemotherapeutic, designed to protect an individual from a future attack by a pathogen.

Prostheca A cytoplasmic extrusion from a cell such as a bud, hypha, or stalk.

Prosthetic group The tightly bound, nonprotein portion of an enzyme; not the same as a *Coenzyme*.

Protease inhibitor A compound that inhibits the action of viral protease by binding directly to the catalytic site, preventing viral protein processing.

Protein A polymeric molecule consisting of one or more polypeptides.

Proteobacteria A large phylum of *Bacteria* that includes many of the common gram-negative bacteria, including *Escherichia coli*.

Proteome The total complement of proteins present in a cell, tissue, or organism at any one time.

Proteomics The large scale or genomewide study of the structure, function, and regulation of the proteins of an organism.

Proteorhodopsin A light-sensitive retinal-containing protein found in some marine *Bacteria* that catalyzes ATP formation.

Proton motive force An energized state of a membrane created by a protein gradient and usually formed through action of an electron transport chain. See also *Chemiosmosis*.

Protoplasm The complete cellular contents, cytoplasmic membrane, cytoplasm, and nucleus/nucleoid.

Protoplast A cell from which the wall has been removed.

Prototroph The parent from which an auxotrophic mutant has been derived. Contrast with *Auxotroph*.

Protozoa Unicellular eukaryotic microorganisms that lack cell walls.

Provirus See *Prophage*.

Psychrophile An organism able to grow at low temperatures and showing a growth temperature optimum of <15°C.

Psychrotolerant Able to grow at low temperature but having a growth temperature optimum of >15°C.

Public health The health of the population as a whole.

Pure culture A culture containing a single kind of microorganism.

Pyogenic Pus-forming; causing abscesses.

Pyrite A common iron ore, FeS_2.

Pyrogenic Fever-inducing.

Quarantine The practice of restricting the movement of individuals with highly contagious serious infections to prevent spread of the disease.

Quaternary structure In proteins, the number and arrangement of individual polypeptides in the final protein molecule.

Quinolones Synthetic antibacterial compounds that interact with DNA gyrase and prevent supercoiling of bacterial DNA.

Quorum sensing Regulatory networks in prokaryotes that respond to population density.

Rabies A usually fatal neurological disease caused by the rabies virus that is usually transmitted by the bite or saliva of an infected carnivore.

Raw water Surface water or groundwater that has not been treated in any way (also called untreated water).

Radioimmunoassay (RIA) An immunological assay employing radioactive antibody or antigen for the detection of antigen or antibody binding.

Radioisotope An isotope of an element that undergoes spontaneous decay with the release of radioactive particles.

Reaction center A photosynthetic complex containing chlorophyll (or bacteriochlorophyll) and other components, within which occurs the initial electron transfer reactions of photophosphorylation.

Reading-frame shift See *Frameshift*.

Recalcitrant Resistant to microbial attack.

Recombinant DNA A DNA molecule containing DNA originating from two or more sources.

Recombination Process by which genetic elements in two separate genomes are brought together in one unit.

Redox See *Oxidation–reduction reaction*.

Reduction A process by which a compound accepts electrons to become reduced.

Reduction potential (E_0') The inherent tendency, measured in volts, of the oxidized compound of a redox pair to become reduced.

Reductive dechlorination Removal of Cl as Cl^- from an organic compound by reducing the carbon atom from C—Cl to C—H.

Reemerging infection Infectious disease, thought to be under control, that produces a new epidemic.

Regulation Processes that control the rates of synthesis of proteins, such as induction and repression.

Regulon A set of operons that are all controlled by the same regulatory protein (repressor or activator).

Replacement vector A cloning vector, such as a bacteriophage, in which some of the DNA of the vector can be replaced with foreign DNA.

Replication Synthesis of DNA using DNA as a template.

Replicative form A double-stranded DNA molecule that is an intermediate in the replication of single-stranded DNA viruses.

Reporter gene A gene incorporated into a vector because the product it encodes is easy to detect.

Repression The process by which the synthesis of an enzyme is inhibited by the presence of an external substance, the repressor.

Repressor protein A regulatory protein that binds to specific sites on DNA and blocks transcription; involved in negative control.

Reservoir In epidemiology, sites in which viable infectious agents remain and from which infection of individuals may occur.

Respiration Catabolic reactions producing ATP in which either organic or inorganic compounds are primary electron donors and organic or inorganic compounds are ultimate electron acceptors.

Response regulator protein One of the members of a two-component system; a regulatory protein that is phosphorylated by a sensor protein (see *Sensor protein*).

Restriction endonucleases (restriction enzymes) Enzymes that recognize and cleave specific DNA sequences, generating either blunt or single-stranded (sticky) ends.

Retrovirus A virus containing single-stranded RNA as its genetic material, which produces a complementary DNA by activity of the enzyme reverse transcriptase.

Reverse electron transport The energy-dependent movement of electrons against the thermodynamic gradient to form a strong electron donor from a weaker electron donor.

Reverse gyrase A topoisomerase present in hyperthermophilic prokaryotes that introduces positive supercoils in DNA.

Reverse transcription The process of copying information found in RNA into DNA.

Reverse translation The mental process of using a codon table and the amino acid sequence of a protein to obtain a possible sequence of the mRNA or the gene that encoded the protein.

Rheumatic fever An inflammatory autoimmune disease triggered by an immune response to infection by *Streptococcus pyogenes*.

Rhizosphere The region immediately adjacent to plant roots.

Ribonucleic acid (RNA) A polymer of nucleotides connected via a phosphate–ribose backbone; involved in protein synthesis or as genetic material of some viruses.

Ribosomal RNA (rRNA) Type of RNA found in the ribosome; some rRNAs participate actively in the process of protein synthesis.

Ribosome A cytoplasmic particle composed of ribosomal RNA and protein, which is part of the protein-synthesizing machinery of the cell.

Riboswitch A messenger RNA that can bind a specific small molecule near its 5′ end that alters its secondary structure and makes it unavailable for translation.

Ribozyme An RNA molecule that can catalyze a chemical reaction.

Rickettsia Obligate intracellular parasites that cause disease, including typhus, Rocky Mountain spotted fever, and ehrlichiosis.

RNA editing Modification of the RNA transcript of a protein-encoding gene by a process other than splicing, yielding a molecule with the required coding properties.

RNA life A hypothetical ancient life form lacking DNA and protein, in which RNA had both a genetic coding and a catalytic function.

RNA polymerase An enzyme that synthesizes RNA in the 5′ → 3′ direction using an antiparallel 3′ → 5′ DNA strand as a template.

RNA processing The conversion of a precursor RNA to its mature form.

Rocky Mountain spotted fever A tick-transmitted disease caused by *Rickettsia rickettsii*, causing fever, headache, rash, and gastrointestinal symptoms.

Root nodule A tumorlike growth on certain plant roots that contains symbiotic nitrogen-fixing bacteria.

Rumen The forestomach of ruminant animals in which cellulose digestion occurs.

S-layer A paracrystalline outer wall layer composed of protein or glycoprotein and found in many prokaryotes.

Salmonellosis Enterocolitis caused by any of more than 2000 variants of *Salmonella* species.

Sanitizers Agents that reduce, but may not eliminate, microbial numbers to a safe level.

Scale-up Conversion of an industrial process from a small laboratory setup to a large commercial fermentation.

Scarlet fever Disease characterized by high fever and a reddish skin rash resulting from an exotoxin produced by cells of *Streptococcus pyogenes*.

Screening Any of a number of procedures that permits the sorting of organisms by phenotype or genotype by allowing growth of some types but not others.

Secondary antibody response Antibody made on second (subsequent) exposure to antigen; mostly of the class IgG.

Secondary metabolite A product excreted by a microorganism near the end of the growth phase or during the stationary phase.

Secondary structure The initial pattern of folding of a polypeptide or a polynucleotide, usually the result of hydrogen bonding.

Secondary treatment In sewage treatment, either the aerobic or anaerobic decomposition of sewage following the removal of nondegradable objects by primary treatment.

Secretion vector A DNA vector in which the protein product is both expressed in and secreted (excreted) from the cell.

Sediment In water purification, soil, sand, minerals, and other large particles found in raw water. In large bodies of water (lakes, the oceans), the materials (mud, rock, and the like) that form the bottom surface of the water body.

Selection Placing organisms under conditions where the growth of those with a particular genotype will be favored.

Selective medium A growth medium that enhances the growth of certain organisms while inhibiting the growth of others due to an added media component.

Semiconservative replication DNA synthesis yielding new double helices, each consisting of one parental and one progeny strand.

Semiperishable food Food of intermediate water activity that has a limited shelf life due to potential for spoilage by growth of microorganisms.

Semisynthetic penicillin A natural penicillin that has been chemically altered.

Sensitivity In immunodiagnostics, the lowest amount of antigen that can be detected in an immunological assay.

Sensor protein One of the members of a two-component system; a kinase found in the cell membrane that phosphorylates itself in response to an external signal and then passes the phosphoryl group to a response regulator protein (see *Response regulator protein*).

Septicemia Infection of the bloodstream by microorganisms.

Serology The study of antigen–antibody reactions *in vitro*.

Serum Fluid portion of blood remaining after the blood cells and materials responsible for clotting are removed.

Sewage Liquid effluents contaminated with human or animal fecal material.

Sexually transmitted infection (STI) An infection that is usually transmitted by sexual contact.

Shine–Dalgarno sequence A short stretch of nucleotides on a prokaryotic mRNA molecule upstream of the translational start site that binds to ribosomal RNA and thereby brings the ribosome to the initiation codon on the mRNA.

Shotgun sequencing Sequencing of DNA from previously cloned small fragments of a genome in a random fashion followed by computational methods to reconstruct the entire genome sequence.

Shuttle vector A cloning vector that can replicate in two different organisms; used for moving DNA between unrelated organisms.

Sickle-cell anemia A genetic trait that confers resistance to malaria but causes a reduction in the number and efficiency of red blood cells by reducing the oxygen-binding affinity of hemoglobin.

Siderophore An iron chelator that can bind iron present at very low concentrations.

Signal sequence A short stretch of amino acids found at the beginning of proteins destined to be excreted from the cell. The signal sequence is usually rich in hydrophobic amino acids, which helps transport the entire polypeptide through the membrane.

Signature sequence Short oligonucleotides of unique sequence found in 16S ribosomal RNA of a particular group of prokaryotes.

Single-cell protein Protein derived from microbial cells for use as food or a food supplement.

Site-directed mutagenesis A technique whereby a gene with a specific mutation can be constructed *in vitro*.

16S rRNA A large polynucleotide (~1500 bases) that functions as a part of the small subunit of the ribosome of

PHYLOGENY OF THE LIVING WORLD—*ARCHAEA*

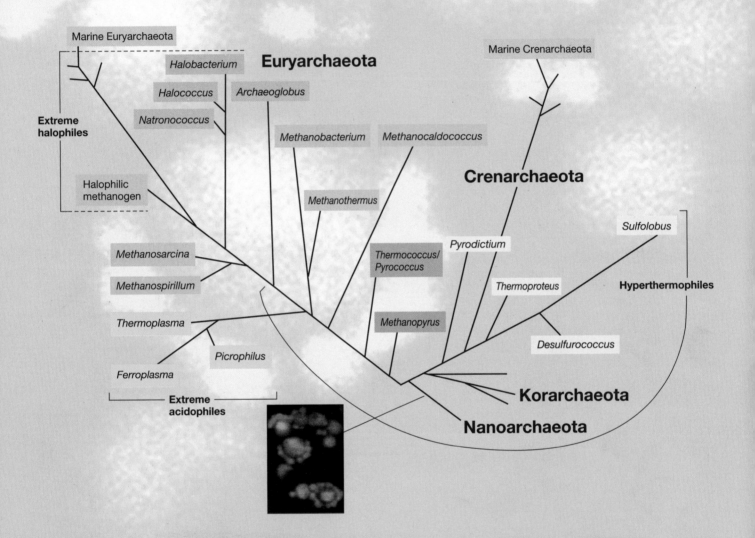

Marine Euryarchaeota

Halobacterium **Euryarchaeota**

Halococcus *Archaeoglobus*

Marine Crenarchaeota

**Extreme
halophiles**

Natronococcus

Methanobacterium *Methanocaldococcus*

Crenarchaeota

Halophilic
methanogen

Methanothermus

Sulfolobus

Methanosarcina

Pyrodictium

*Thermococcus/
Pyrococcus*

Methanospirillum

Thermoproteus **Hyperthermophiles**

Thermoplasma

Methanopyrus

Picrophilus

Desulfurococcus

Ferroplasma

Korarchaeota

**Extreme
acidophiles**

Nanoarchaeota

PHYLOGENETIC TREE OF *ARCHAEA*. This tree is derived from 16S ribosomal RNA sequences. Four major phyla of *Archaea* can be defined: the Crenarchaeota, which consists of both hyperthermophiles and cold-dwelling species; the Euryarchaeota, which contains methanogenic and extremely halophilic prokaryotes; the Korarchaeota, which are, as far as is known, hyperthermophiles; and the Nanoarchaeota, tiny, parasitic hyperthermophiles. See Sections 11.5–11.9 for further information on ribosomal RNA-based phylogenies. *Data for the tree obtained from the Ribosomal Database project* **http://rdp.cme.msu.edu**